ICHEP 2000
Volume II

Proceedings of the 30th International Conference
on

High Energy Physics

ICHEP 2000

Volume II

Proceedings of the 30th International Conference on High Energy Physics

Osaka, Japan

27 July – 2 August 2000

Editors

C. S. Lim

Kobe University, Japan

Taku Yamanaka

Osaka University, Japan

Published by

World Scientific Publishing Co. Pte. Ltd.

P O Box 128, Farrer Road, Singapore 912805

USA office: Suite 1B, 1060 Main Street, River Edge, NJ 07661

UK office: 57 Shelton Street, Covent Garden, London WC2H 9HE

Library of Congress Cataloging-in-Publication Data
Proceedings of the 30th International Conference on High Energy Physics, Osaka, Japan,
27 July–2 August 2000 / editors C.S. Lim, Taku Yamanaka.
p. cm.
ISBN: 9810249756 (Vol. 1) : alk. paper
9810249764 (Vol. 2) : alk. paper
9810245335 (set) : alk. paper
1. Particles (Nuclear physics)--Congresses. I. Lim, C. S. II. Yamanaka, Taku,
1957– . III. International Conference on High Energy Physics (30th : 2000 : Osaka,
Japan)

QC793.3.H5 I56 2000
539.7'2--dc21 2001017781

British Library Cataloguing-in-Publication Data
A catalogue record for this book is available from the British Library.

Printed in Singapore by World Scientific Printers

CONTENTS

VOLUME I

PLENARY SESSIONS

PA-03 Hard Interactions (high Q^2: DIS, jet, perturbative QCD)

PA-06 Light Flavor Physics (Kaon, Muon)

VOLUME II

PA-07 Heavy Flavor Physics (charm, bottom, top and τ)

PA-08 Neutrino Physics

PA-10 Beyond the Standard Model (theory)

PA-11 Search for New Particles and New Phenomena

PA-12 New Detectors and Techniques

PA-15 Field Theory

Editors would like to thank Dr. Kiyotomo Kawagoe for helping us prepare the proceedings.

Parallel Session 7

Heavy Flavor Physics
(charm, bottom, top, and τ)

Conveners: Patricia R. Burchat (Stanford),
Nobuhiko Katayama (KEK) and
Yasuhiro Okada (KEK)

B_S MIXING AT SLD

T. USHER
REPRESENTING THE SLD COLLABORATION
Stanford Linear Accelerator Center, Stanford, Ca. 94309, USA
E-mail: usher@slac.stanford.edu

A preliminary 95% C.L. exclusion on the oscillation frequency of $B_s^0 - \overline{B_s^0}$ mixing is presented by combining three analyses of a sample of 400,000 hadronic Z^0 decays collected by the SLD experiment at the SLC between 1996 and 1998. All three analyses exploit the large forward-backward asymmetry of polarized $Z^0 \to b\bar{b}$ decays, as well as information from the hemisphere opposite that of the reconstructed B decay, to determine the b-hadron flavor at production. The three analyses differ in their reconstruction of the proper time and flavor of the b-hadron at decay. The first analysis performs a full reconstruction of a cascade D_s meson and a partial reconstruction of the b-hadron. In the second analysis, semileptonic decays are selected and the B decay point is reconstructed by vertexing a lepton with a partially reconstructed cascade D meson. The third analysis reconstructs B decay vertices inclusively using a topological technique, with separation between B_s^0 and $\overline{B_s^0}$ decays obtained by exploiting the $B_s^0 \to D_s^-$ cascade charge structure. The results of these analyses are combined to exclude the following values of the $B_s^0 - \overline{B_s^0}$ mixing oscillation frequency: $\Delta m_s < 7.6$ ps^{-1} and $11.8 < \Delta m_s < 14.8$ ps^{-1} at the 95% confidence level.

1 Introduction

Transitions between B^0 and $\overline{B^0}$ mesons take place via second order weak interactions. In the Standard Model, a measurement of the oscillation frequency Δm_d for B_d^0–$\overline{B_d^0}$ mixing determines, in principle, the value of the CKM matrix element $|V_{td}|$, which is parameterized in terms of the Wolfenstein parameters ρ and η, both of which are currently poorly constrained. However, theoretical uncertainties in calculating hadronic matrix elements are large and thus limit the current usefulness of precise Δm_d measurements [1]. Some of these uncertainties cancel when one considers the ratio between Δm_d and Δm_s. Thus, combining measurements of both Δm_d and Δm_s translates into a measurement of the ratio $|V_{td}|/|V_{ts}|$ and provides a stronger constraint on the parameters ρ and η.

2 Experimental Method

Experimentally, a measurement of the time dependence of B^0–$\overline{B^0}$ mixing requires three ingredients: (i) the B decay proper time has to be reconstructed, (ii) the B flavor at production (initial state $t = 0$) needs to be determined, as well as (iii) the B flavor at decay (final state $t = t_{\text{decay}}$). At SLD, the time dependence of B_s^0–$\overline{B_s^0}$ mixing has been studied using three different methods, known as "D_s+Tracks," "Lepton+D," and "Charge Dipole." These analysis were performed over a sample of some 400,000 hadronic Z^0 decays collected between 1996 and 1998, with an average electron polarization of $P_e = 73\%$. The complete details of these three analyses can be found in the references [2,3].

The technique for determining the flavor of the b-hadron at production is the same for all three analyses. Primarily this consists of combining opposite hemisphere jet charge information with that gained from exploiting the large longitudinal polarization of the electron beam. This can be further augmented, when available, by other flavor sensitive quantities from the hemisphere opposite the selected vertex.

The three analyses differ in their determination of the proper time and flavor of the b-hadron at decay and trade off resolution and B_s purity against statistics. The "D_s+Tracks" analysis enjoys the best resolu-

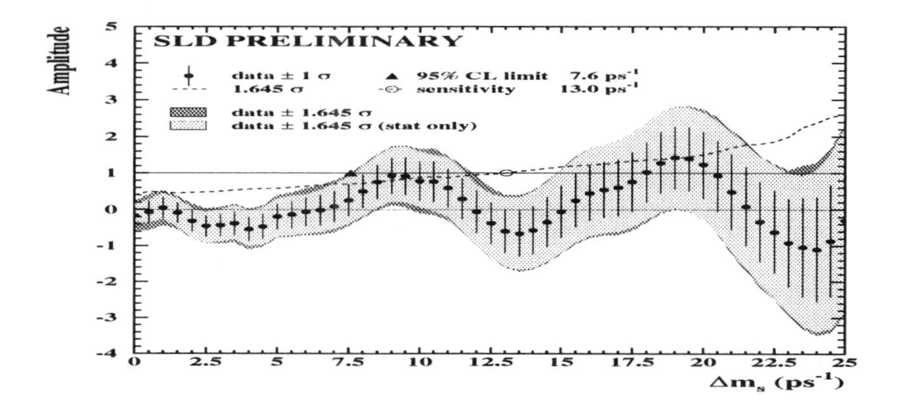

Figure 1. Combined amplitude fit for the SLD data

tion (core decay length resolution $\sigma_l < 50\mu$m) and B_s purity ($f_{B_s} = 38\%$) but suffers the worst statistics with only 361 candidate decays. At the opposite end is the "Charge Dipole" analysis which, with 8556 reconstructed decays, enjoys the highest statistics, but suffers the worst resolution (core $\sigma_l < 80\mu$m) and B_s purity ($f_{B_s} = 15\%$).

3 Results

The study of the time dependence of $B_s^0 - \overline{B_s^0}$ mixing is carried out using the amplitude method described in the reference [4]. In this method, instead of fitting for Δm_s directly, the fit is to the amplitude A of oscillation at fixed values of Δm_s. The result should give $A = 1$ near the true value of Δm_s and $A = 0$ elsewhere. The result of combining the amplitude plots for the three SLD analyses is displayed in Fig. 1. The measured values are consistent with $A = 0$ for the whole range of Δm_s up to 25 ps^{-1} and no evidence is found for a preferred value of the mixing frequency. The following ranges of B_s^0-$\overline{B_s^0}$ oscillation frequencies are excluded at 95% C.L.: $\Delta m_s < 7.6$ ps^{-1} and $11.8 < \Delta m_s < 14.8$ ps^{-1}. The combined sensitivity to set a 95% C.L. lower limit is found to be at a Δm_s value of 13.0 ps^{-1}. These results are preliminary.

Acknowledgments

We thank the personnel of the SLAC accelerator department and the technical staffs of our collaborating institutions for their outstanding efforts. This work was supported by the Department of Energy, the National Science Foundation, the Instituto Nazionale di Fisica of Italy, the Japan-US Cooperative Research Project on High Energy Physics, and the Science and Engineering Research Council of the United Kingdom.

References

1. P. Paganini, F. Parodi, P. Roudeau, and A. Stocchi, *Measurements of the ρ and η Parameters of the V_{CKM} Matrix and Perspectives*, Phys. Scripta **58**, 556 (1998)
2. K. Abe *et al.*, *Time Dependent $B_s^0 - \overline{B_s^0}$ Oscillations Using Exclusively Reconstructed D_s^+ Decays at SLD*, SLAC-PUB-8598, August 2000.
3. K. Abe *et al.*, *Time Dependent $B_s^0 - \overline{B_s^0}$ Mixing Using Inclusive and Semileptonic B Decays at SLD*, SLAC-PUB-8568, August 2000.
4. H.-G. Moser and A. Roussarie, Nucl. Inst. and Meth. **A384**, 491 (1997).

REVIEW ON B^0- $\overline{B^0}$ MIXING AND B-LIFETIMES MEASUREMENTS AT CDF/LEP/SLD

A. STOCCHI

Laboratoire de l'Accélérateur Linéaire IN2P3-CNRS et Université de Paris-Sud, BP 34, F-91898 Orsay Cedex, (stocchi@lal.in2p3.fr)

A review of the results on B^0- $\overline{B^0}$ mixing and b-lifetimes obtained by CDF, LEP and SLD collaborations is presented with special emphasis on B^0_s- $\overline{B^0_s}$ mixing.

1 Introduction

In the last decade, new weakly decaying B-hadrons have been observed ($B^0_s, B_c, \Lambda^0_b, \Xi_b$) and their production and decay properties have been intensively studied. In this context CDF (operating at TeVatron), ALEPH, DELPHI, L3, OPAL (operating at LEP) and SLD (operating at SLC) experiments have played a central role. This has been made possible owing to the excellent performance both of the machines and of the detectors. Above all, these measurements would have not been possible without the development of Silicon detectors.

2 B hadron lifetimes

The measurement of the lifetimes of the different B hadrons is an important test of the B decay dynamics[1].

The main improvements, since last year, have been obtained in the determination of B^0_d and B^+ lifetimes. The results[2] are given in Table 1. In future no improvements are really expected before the start of the new phase at TeVatron. Few conclusions can be drawn. The charged B-mesons live longer than the neutrals. This effect is now established at 3.5 σ. The lifetimes of the two neutral B mesons (B^0_d, B^0_s) are equal within 1σ. To observe any possible (and unexpected difference) new data are needed. The b-baryon lifetime problem still exists (since 3-4 years). b-baryons live shorter than b-mesons, as expected, but the magnitude of the effect is now more than 3 σ away from the lower edge of the predictions[1]. This should push for a better understanding of the theory independently of a possible improvement of the experimental accuracy.

3 Lifetime Difference: $\Delta\Gamma_s$

The ratio between the differences of the widths and of the masses of the B^0_s-$\overline{B}^0_s$ system mass eigenstates is (naively): $\Delta\Gamma_s/\Delta m_s \sim 3/2\pi(m_b/m_t)^2$. If Δm_s is too large (and so, difficult to measure), $\Delta\Gamma_s$ can eventually gives access to it. Unfortunately the theoretical error attached to the evaluation of $\Delta\Gamma_s$ is still quite large, of the order of 50 %. Recent theoretical calculations predict $\Delta\Gamma_s/\Gamma_s$ in the range (5-10) %[3].

From the experimental point of view, the combination of LEP and CDF results gives[4] (assuming $\tau(B^0_d) = \tau(B^0_s)$):

$$\Delta\Gamma_s/\Gamma_s = 0.16^{+0.16}_{-0.13}; < 0.31 \text{ at } 95\% \text{ C.L.}$$

4 $B^0 - \overline{B^0}$ mixing : $\Delta m_d, \Delta m_s$

In the Standard Model, $B^0_q - \overline{B}^0_q$ ($q = d, s$) mixing can be expressed by the following formula:

$$\Delta m_q = G_F^2/6\pi^2 m_W^2 \eta_c S(m_t^2/m_W^2)|V_{tq}|^2 \\ m_{Bq} f_{Bq}^2 B_{Bq} \qquad (1)$$

where $S(m_t^2/m_W^2)$ is the Inami-Lim function, m_t is the $\overline{MS}$ top mass and η_c is a QCD correction factor obtained at NLO order in perturbative QCD. The measurement of Δm_d, (Δm_s) gives access to V_{td}, (V_{ts}) CKM matrix elements and thus to the $\overline{\rho}$ and $\overline{\eta}$

Table 1. Lifetime ratios results

Lifetime ratios	Osaka 2000	Tampere 1999	Theory
$\tau(B^-)/\tau(B^0)$	1.070 ± 0.020	1.065 ± 0.023	1.0 - 1.1
$\tau(B^0_s)/\tau(B^0)$	0.945 ± 0.039	0.937 ± 0.040	0.99 - 1.01
$\tau(b-bary)/\tau(B^0)$	0.780 ± 0.035	0.773 ± 0.036	0.9 - 1.0

parameters[a] of the Wolfenstein parametrization. We can write:

$$\begin{aligned}
&\Delta m_d \sim A^2\lambda^6[(1-\bar{\rho})^2+\bar{\eta}^2)]f^2_{Bd}B_{Bd}\\
&\Delta m_s \sim A^2\lambda^4 f^2_{B_s}B_{Bs}\\
&\rightarrow \Delta m_s \sim 1/\lambda^2\Delta m_d \sim 20\Delta m_d \qquad (2)\\
&\Delta m_d/\Delta m_s \sim \lambda^2/\xi^2[(1-\bar{\rho})^2+\bar{\eta}^2]\\
&\text{with } \xi = \frac{f_{Bs}\sqrt{B_{Bs}}}{f_{Bd}\sqrt{B_{Bd}}}
\end{aligned}$$

The interest of measuring both Δm_d and Δm_s comes from the fact that the ratio ξ is better determined from theory than the individual quantities entering into its expression.

The analyses presented here measure Δm_q by looking at the time dependence behaviour of the oscillations :

$$P(B^0_q \rightarrow \overset{(-)}{B^0_q}) = \frac{1}{2\tau}e^{-t/\tau}(1(\pm)cos\Delta m_q t) \quad (3)$$

4.1 Δm_d results

The time variation of the B_d oscillation has been observed for the first time at LEP. In the last years the precision has impressively improved down to 3 % giving[5] :

$$\Delta m_d = 0.486 \pm 0.015\,\text{ps}^{-1} \quad LEP/SLD/CDF$$

which is an average of 26 measurements !

The recent CLEO χ_d measurement is in agreement with this value giving[5] the final result of : $\Delta m_d = 0.487 \pm 0.014\text{ps}^{-1}$. Improvements on this result are expected in the coming years from B-factories.

4.2 Δm_s results

Since B_s mesons are expected to oscillate 20 times faster than B_d, it is fundamental to have the best possible resolution on the proper time reconstruction. For details on the analyses see the contributions of P. Coyle[6] and T. Usher[6] in these proceedings. No experiment has observed an oscillation signal. A procedure has been set to combine the different analyses and to get a limit or eventually to quantify the evidence for a "combined" signal. This is done in the framework of the amplitude method[7] which consists in modifying the last part of eq.(3) using: $1 \pm A \cos \Delta m_s t$. At any given value of Δm_s, A and σ_A are measured. A = 1 and not compatible with A = 0 indicates an oscillation signal at the corresponding value of Δm_s. The values of Δm_s excluded at 95 % C. L. are those satisfying $A(\Delta m_s) + 1.645\sigma_A(\Delta m_s) < 1$. It is also possible to define the sensitivity as the value of Δm_s corresponding to 1.645 $\sigma_A(\Delta m_s) = 1$. The main actors in this work are LEP and SLD collaborations.

Figure 1 shows the evolution of the combined sensitivity which has dramatically improved during the years. Figure 2 gives the combined plot of the amplitude values as a function of Δm_s[5]. The results are:

$\Delta m_s > 14.9$ ps^{-1}at 95% C.L

sensitivity at 17.9 ps^{-1}.

A "signal" bump is visible at around $\Delta m_s = 17.7$ ps^{-1} with a significance at 2.5 σ level. The probability of a background fluctuation greater of equal to the one observed, and at any Δm_s value, has been evaluated to be about 2.5 %. This result is still expected to improve during next months by continuing the progress in LEP/SLD analyses.

The impact of this result on the determination of the unitarity triangle parameters

[a] $\bar{\rho}$ and $\bar{\eta}$ are related to the original ρ and η parameters $\bar{\rho} = \rho(1-\lambda^2/2), \bar{\eta} = \eta(1-\lambda^2/2)$.

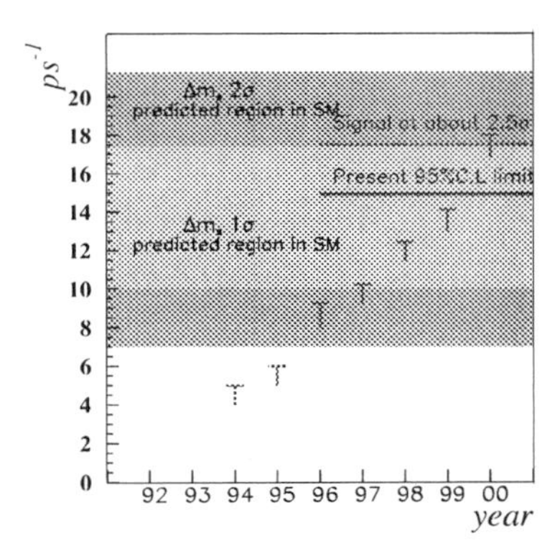

Figure 1. The evolution of the combined Δm_s sensitivity over the years.

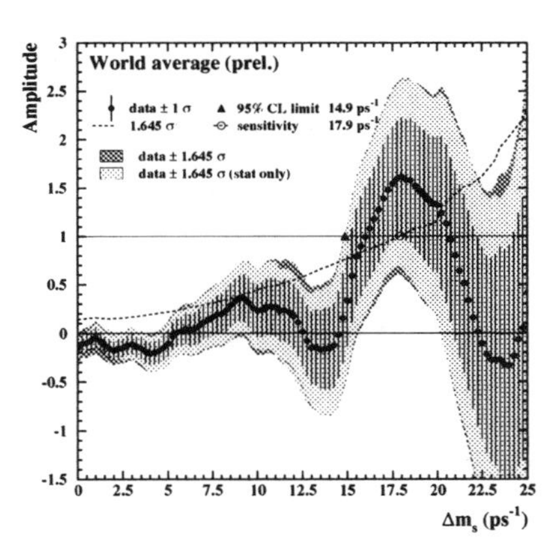

Figure 2. Variation of the combined amplitude versus Δm_s. The points with error bars are the data ; the lines show the 95 % C. L. curves (in dark when systematics are included). The dotted curve indicates the sensitivity.

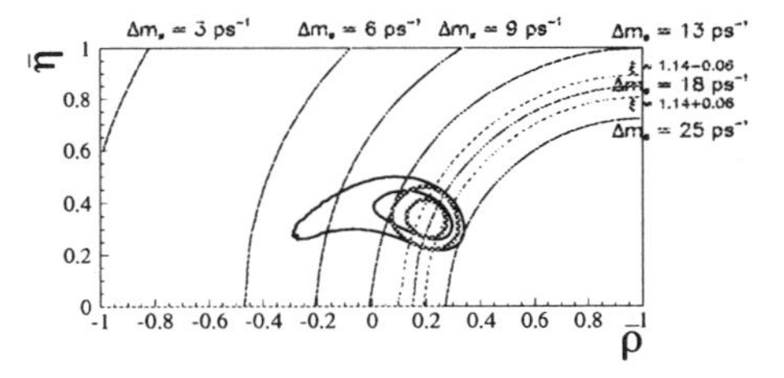

Figure 3. The allowed regions for $\bar{\rho}$ and $\bar{\eta}$ using the constraints given by the measurements of $|\epsilon_k|, |V_{ub}|/|V_{cb}|$ and Δm_d at 68 % and 95 % probability are shown by the thin contour lines. The constraint due to Δm_s is not included. Selected regions for $\bar{\rho}$ and $\bar{\eta}$ when the constraint due to Δm_s is included have been superimposed using thick lines.

is shown in Figure 3. Using the constraint coming from the measurements of V_{ub}, Δm_d, $|\epsilon_K|$ and Δm_s we obtain[8]:

$$\begin{aligned}\bar{\rho} &= 0.206 \pm 0.043; \bar{\eta} = 0.339 \pm 0.044 \\ sin2\beta &= 0.723 \pm 0.069; sin2\alpha = -0.28 \pm 0.27 \\ &\text{and } \gamma = (58.5 \pm 6.9)^{\circ}\end{aligned}$$

Conclusions

The different B-lifetimes have been measured at the few percent level ($\sim$ 1.5 % for B^+ and B_d^0 and $\sim$ 4% for B_s^0 and Λ_b^0). A clear experimental hierarchy has been established. The frequency of the B_d^0 meson oscillation (Δm_d) has been measured with a 3 % precision. As far as the B_s oscillation is concerned the combined sensitivity is now at 17.9 ps $^{-1}$ and a possible signal with a 2.5 σ significance at about Δm_s = 17.7 ps^{-1} has been observed. This result is still expected to be improved during the coming months. Let's wait a bit for claiming the observation of $B_s^0 - \overline{B_s^0}$ oscillations !

Acknowledgement

Congratulations to all the collegues of CDF/LEP/SLD which have made all of it possible ! Thanks to P. Roudeau for the careful reading of the manuscript.

References

1. G. Bellini, I. Bigi, P.J. Dornan, *Phys. Rep.* 289 (1997) 1 and ref. therein; M. Neubert, C.T. Sachrajda, *Nucl. Phys.* **B483** (1997) 339.
2. B lifetime WG http://claires.home.cern.ch/claires/lepblife.html
3. D. Becirevic et al., hep-ph/0006135.; M. Beneke et al. *Phys. Lett.* **B459**(1999) 631.
4. $\Delta\Gamma_s$ WG http://lepbosc.web.cern.ch /LEPBOSC/deltagamma_s/
5. B-oscillations WG http://www.cern.ch /LEPBOSC/combined_results/osaka_2000/
6. T. Usher and P. Coyle, these proceedings
7. H.G. Moser, A. Roussarie, NIM A384 (1997) 491.
8. M. Ciuchini et al., paper submitted at this conference (num. 908)

DETERMINATION OF B^0-$\bar{B}^0$ MIXING FROM THE TIME EVOLUTION OF DILEPTON AND $B^0 \to D^{(*)-}\ell^+\nu$ EVENTS AT BELLE

J. SUZUKI

KEK, High energy Accelerator Research Organization, Oho 1-1 Tsukuba, Ibaraki, Japan
E-mail: jsuzuki@bmail.kek.jp

We report a determination of the B^0-$\bar{B}^0$ mixing parameter Δm_d in 5.1 fb^{-1} of data collected at Belle. We obtain $\Delta m_d = 0.456 \pm 0.008$ (stat.) ± 0.030 (syst.) ps^{-1} (preliminary) for the dilepton events and $\Delta m_d = 0.488 \pm 0.026$ (stat.) ps^{-1} (preliminary) for the $B^0 \to D^{(*)-}\ell^+\nu$ mode. These are the first determinations of Δm_d from time evolution using Υ(4S) decays.

1 Introduction

The phenomenon of B^0-$\bar{B}^0$ mixing occurs via second order box diagrams in the standard model, and its rate is regarded as a fundamental parameter of the B meson system.

At the Belle experiment, the Υ(4S) moves along the electron beam direction (z axis) with a Lorentz boost of $\beta\gamma = 0.425$ by the asymmetric e^+e^- collision. Consequently, the decay vertex separation of the two B mesons is typically 200 μm, which is large enough to be measured by a silicon vertex detector (SVD).

The time evolution of same-flavor (B^0B^0, $\bar{B}^0\bar{B}^0$) and opposite-flavor ($B^0\bar{B}^0$) decays given a $B^0\bar{B}^0$ initial state are given by

$$N_{same} \propto e^{-|\Delta t|/\tau_{B^0}} \left[1 - \cos(\Delta m_d \Delta t)\right], \quad (1)$$

$$N_{opp} \propto e^{-|\Delta t|/\tau_{B^0}} \left[1 + \cos(\Delta m_d \Delta t)\right], \quad (2)$$

where Δm_d is the mass difference between the two mass eigenstates of the B^0 meson, τ_{B^0} is the average B^0 meson life time, and Δt is the proper time difference between the two B-meson decays. In this paper, we determine Δm_d from the time evolution of dilepton and $B^0 \to D^{(*)-}\ell^+\nu$ [1] events in 5.1 fb^{-1}.

2 Dilepton events

The dilepton analysis uses high-momentum leptons for the B flavor tagging and vertexing. Leptons are required to satisfy 1.1 GeV/c $< p^* <$ 2.3 GeV/c and $30^\circ < \theta < 135^\circ$, where p^* is momentum at the Υ(4S) rest frame and θ is polar angle. To measure a precise vertex position, the lepton tracks must have SVD hits. For the background suppression, the opening angle between leptons at the Υ(4S) rest frame is required to be $-0.8 < \cos\theta^*_{\ell\ell} < 0.95$, and $J/\psi \to \ell^+\ell^-$ candidates whose invariant mass is consistent with the world average J/ψ mass are removed. After applying the above criteria, we found 7,418 same-sign (SS) and 35,633 opposite-sign (OS) dilepton events. Since the Υ(4S) is boosted along the z axis, the proper time difference Δt is given by $\frac{1}{\beta\gamma}\Delta z = \frac{1}{\beta\gamma}(z_1 - z_2)$, where z_i is z-vertex position of a B meson.

The mixing parameter Δm_d is extracted by fitting the time evolution of the SS and OS dilepton samples to both signal and background probability density functions (p.d.f). The signal is defined as one where both leptons are primary leptons from semileptonic decay of B^0 or B^+. The signal p.d.f for B^0 are parameterized with a convolution of eqs. (1) and (2) with the detector response function obtained using $J/\psi \to \ell^+\ell^-$ data. The p.d.f. of B^+ is given by a exponential function smeared with the response function. The background for dilepton events arise from various sources: at least one secondary or fake lepton, or continuum. The background p.d.f's are estimated from Monte Carlo simulation, where we correct the detector resolution, the fake lepton rate, and

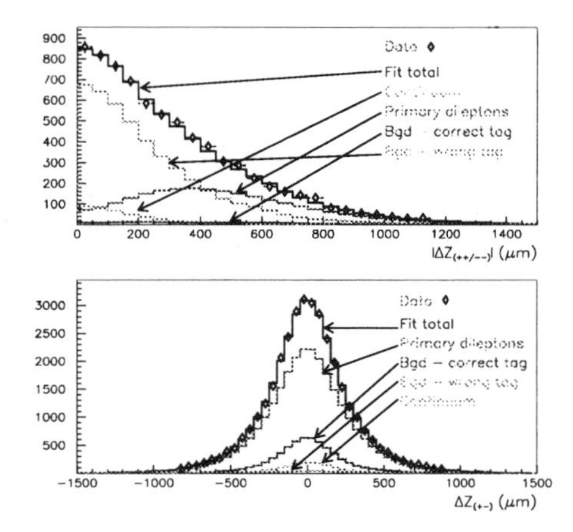

Figure 1. Δz distribution of dileptons for data together with fit result. The upper plot shows the distributions for SS, and the lower plots for OS dileptons. Signal and background dileptons obtained from the fit are also shown.

the branching fraction of D^0 and D^+ [2] to account for the difference between the data and Monte Carlo events. They are classified according to their origin: unmixed-B^0 pairs, mixed-B^0 pairs, B^+B^-, and continuum. Unmixed- and mixed-B^0 pairs are separated because their relative amount and the shapes depend on the mixing parameter.

From the fit, we find $\Delta m_d = 0.456 \pm 0.008$ ps^{-1}. Figure 1 shows the Δz distributions for the data together with the fitted results. Figure 2 shows the OS and SS asymmetry, $(N_{\rm OS}-N_{\rm SS})/(N_{\rm OS}+N_{\rm SS})$, for the data with the fitted curve, where the negative Δz region is folded into the positive region for display purposes.

We list systematic uncertainties on this measurement in Table 1. The sources are constants in the fit such as $f_\pm/f_0$ (branching ratio to B^0 or B^+ of the Υ(4S)) [3], τ_{B^0} and τ_{B^+}/τ_{B^0}, the response function, and the background Δz distribution. The total systematic uncertainties are ± 0.030 ps^{-1}.

The dilepton analysis is described in detail in Ref.5.

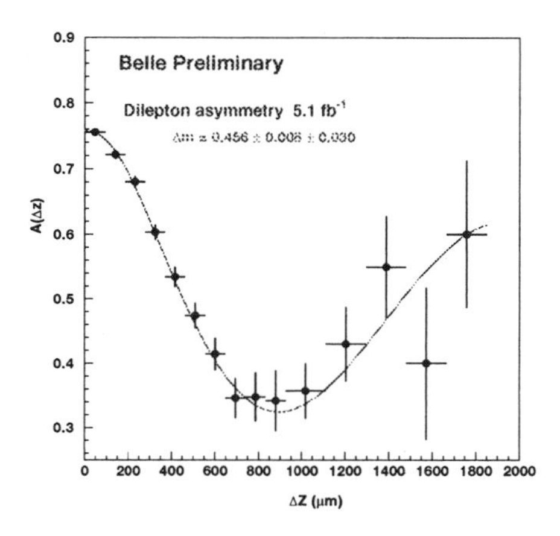

Figure 2. OS and SS dilepton asymmetry vs Δz. The asymmetry is defined as $A(\Delta z) = (N_{\rm OS} - N_{\rm SS})/(N_{\rm OS} + N_{\rm SS})$. The points are the data. The smooth curve is obtained from the fit result.

Table 1. Summary of the systematic uncertainties.

Source (range)	
$f_\pm/f_0$ (1.07 ± 0.09) [3]	± 0.012
B_d^0 life time (1.56 ± 0.04 ps) [4]	± 0.007
$\tau_{B^\pm}/\tau_{B_d^0}$ (1.04 ± 0.04) [4]	± 0.022
response function	± 0.011
background fake rate (±35%)	± 0.007
$\mathcal{B}(B \to D^0 X)$ (±4.6%) [2]	$< \pm 0.001$
$\mathcal{B}(B \to D^\pm X)$ (±14.3%) [2]	$< \pm 0.001$
continuum components (±10%)	± 0.001
background resolution (±18 μm)	$^{+0.000}_{-0.007}$
total	± 0.030

3 $B^0 \to D^{(*)-}\ell^+\nu$ decays

We reconstruct two semileptonic decay chains: $B^0 \to D^{*-}\ell^+\nu$, $D^{*-} \to \bar{D}^0\pi^-$, $\bar{D}^0 \to K^+\pi^-$, $K^+\pi^-\pi 0$ or $K^+\pi^+\pi^-\pi^-$, and $B^0 \to D^-\ell^+\nu$, $D^- \to K^+\pi^-\pi^-$. The source of the largest background is a combinatorial background of D^* or D^-.

The Δm_d measurement requires the flavor and vertex reconstruction of the other B (tagging side) from tracks unused in the $D^{(*)}\ell$ reconstruction. The flavor of the tagging side is identified using two methods. The

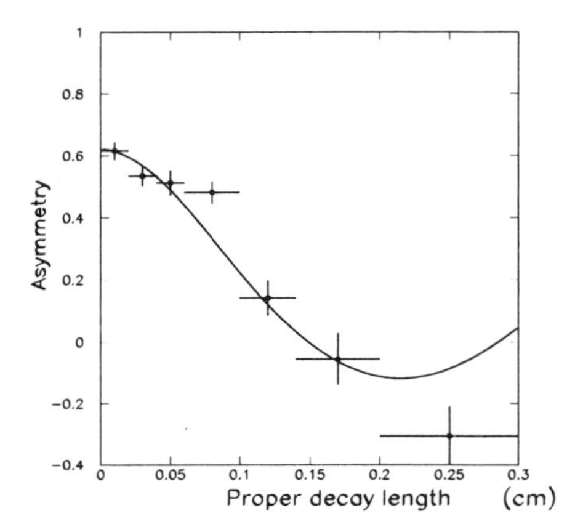

Figure 3. Asymmetry as a function of proper decay length for $B^0 \to D^{*}\ell^{+}\nu$. The points are the data. The solid curve is the fit result.

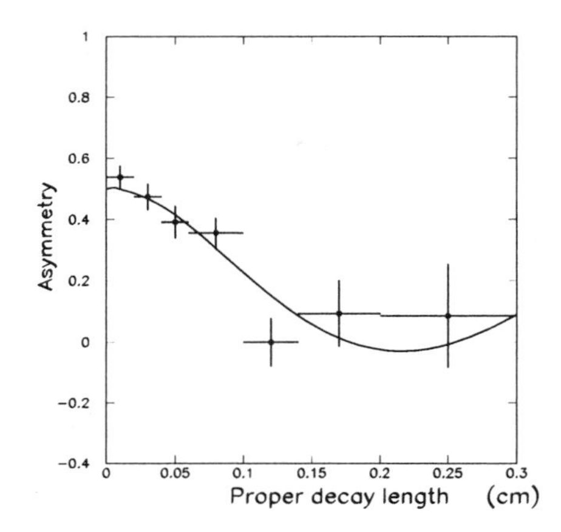

Figure 4. Asymmetry as a function of proper decay length for $B^0 \to D^{+}\ell^{+}\nu$ The points are the data. The solid curve is the fit result.

first method looks for leptons with $p^* > 1.1$ GeV/c (lepton tag). If it fails, we search for charged kaons (kaon tag). The B flavor can be tagged using the lepton charge or the sum of the kaon charges. Events are classified to opposite-flavor (OF) or same-flavor (SF) state, according to the flavor for the $D^{(*)}\ell$ side and that for the tagging side.

The mixing parameter Δm_d is obtained with fitting OF and SF proper decay length $(\frac{1}{\beta\gamma}(z_{D\ell}-z_{tag}))$ distribution. We find $\Delta m_d = 0.488 \pm 0.026$ (stat.) from the fit. The asymmetry which is given by $(N_{\rm OF}-N_{\rm SF})/(N_{\rm OF}+N_{\rm SF})$ as a function of absolute proper decay length is shown in Figures 3 for the $D^*\ell\nu$ mode and 4 for the $D\ell\nu$ mode. The systematic uncertainty is under study.

The wrong flavor tag fraction w is also obtained from the fit. The oscillation is diluted by the wrong tag, and the signal p.d.f's in eqs. (1) and (2) include a dilution factor $(1-2w)$ in front of $\cos(\Delta m_d \Delta t)$. We find $w = 0.07 \pm 0.02$ (stat.) for lepton tag and $w = 0.20 \pm 0.02$ (stat.) for kaon tag. The $\sin 2\phi_1$ measurement at Belle employs these flavor tag techniques and uses these w results.

4 Summary

We have measured the B^0-$\bar{B}^0$ mixing parameter Δm_d from the time evolution of dilepton and $B^0 \to D^{(*)-}\ell^+\nu$ yields in $\Upsilon(4S)$ decays using a sample comprising 5.1 fb^{-1} of data. The preliminary result for dilepton events is $\Delta m_d = 0.456 \pm 0.008$ (stat.) $\pm$ 0.030 (syst.) ps^{-1}, and that for $B^0 \to D^{(*)-}\ell\nu$ is $\Delta m_d = 0.488 \pm 0.026$ (stat.) ps^{-1}.

These results are consistent with the world average Δm_d. This is the first determination of Δm_d using the time evolution of B^0-mesons produced in $\Upsilon(4S)$ decays.

References

1. Throughout this paper a reference to a particular charge state also implies its charge conjugate.
2. L. Gibbons *et al.*, Phys. Rev. **D56** 3783 (1997).
3. J. P. Alexander *et al.*, hep-ex/00006002.
4. C. Caso *et al.*, Eur. Phys. J. **C3** 1 (1998).
5. A. Abashian *et al.*, Belle-CONF-001, Contributed paper #284.

MEASUREMENTS OF HEAVY MESON LIFETIMES WITH BELLE

H. TAJIMA
Department of Physics, University of Tokyo, 7-3-1 Hongo, Bunkyo-ku Tokyo 113-033 Japan
E-mail: tajima@phys.s.u-tokyo.ac.jp

Charmed and beauty meson lifetimes have been measured using 2.75 fb^{-1} (D mesons) and 5.1 fb^{-1} (B mesons) of data collected with the Belle detector at KEKB. The results are $\tau(\overline{B}^0) = (1.50 \pm 0.05 \pm 0.07)$ ps, $\tau(B^-) = (1.70 \pm 0.06^{+0.11}_{-0.10})$ ps, $\tau(D^0) = (414.8 \pm 3.8 \pm 3.4)$ fs, $\tau(D^+) = (1040^{+23}_{-22} \pm 18)$ fs and $\tau(D_s^+) = (479^{+17}_{-16}{}^{+6}_{-8})$ fs, where the first error is statistical and the second error is systematic. The lifetime ratios are measured to be $\tau(B^-)/\tau(\overline{B}^0) = 1.14 \pm 0.06^{+0.06}_{-0.05}$, $\tau(D^+)/\tau(D^0) = 2.51 \pm 0.06 \pm 0.04$ and $\tau(D_s^+)/\tau(D^0) = 1.15 \pm 0.04^{+0.01}_{-0.02}$. The mixing parameter y_{CP} is also measured to be $y_{CP} = 0.03^{+0.15}_{-0.18}{}^{+0.05}_{-0.08}$ for $\overline{B}^0$ and $y_{CP} = (1.0^{+3.8}_{-3.5}{}^{+1.1}_{-2.1})\%$ for D^0, corresponding to 95% confidence intervals, $-0.36 < y_{CP} < 0.35$ and $-7.0\% < y_{CP} < 8.7\%$, respectively. All results are preliminary.

Measurements of individual heavy meson lifetimes provide useful information for the theoretical understanding of heavy meson decay mechanisms. In particular, experimental results[1] yield $\tau(D_s^+)/\tau(D^0) = 1.191 \pm 0.024$, which is inconsistent with the theoretically expected range[2] of 1.00–1.07. Moreover, measurements of the differences of lifetimes for neutral mesons decaying into CP-mixed states and CP-eigenstates can be used to study the $y \equiv \Delta\Gamma/2\Gamma$ and $x \equiv \Delta M/\Gamma$ particle-antiparticle mixing parameters.

The parameter y_{CP}, defined as

$$y_{CP} \equiv \frac{\Gamma(\text{CP even}) - \Gamma(\text{CP odd})}{\Gamma(\text{CP even}) + \Gamma(\text{CP odd})}$$

is related to y and x by the expression

$$\begin{aligned} y_{CP} &= \frac{\tau(D^0 \to K^-\pi^+)}{\tau(D^0 \to K^-K^+)} - 1 \\ &\approx y\cos\phi - \frac{A_{mix}}{2} x \sin\phi, \\ y_{CP} &= 1 - \frac{\tau(\overline{B}^0 \to D^{*+}\ell^-\overline{\nu}, \overline{B}^0 \to D\pi)}{\tau(\overline{B}^0 \to J/\psi K_S)} \\ &\approx y\cos 2\phi_1, \end{aligned}$$

where $\phi(\phi_1)$ is a CP-violating weak phase due to the interference of decays with and without mixing, and A_{mix} is a state-mixing CP-violating parameter ($A_{mix} \approx 4\mathcal{R}e(\epsilon)$). The FOCUS experiment reports $y_{CP} = (3.42 \pm 1.39 \pm 0.74)\%$[3], while CLEO gives $y' \cos\phi = (-2.5^{+1.4}_{-1.6})\%$[4], $x' = (0.0 \pm 1.5 \pm 0.2)\%$ and $A_{mix} = 0.23^{+0.63}_{-0.80}$ using $D^0 \to K^+\pi^-$, where $y' = y\cos\delta - x\sin\delta$ and $x' = x\cos\delta + y\sin\delta$; δ is a strong phase between $D^0 \to K^+\pi^-$ and $\overline{D}^0 \to K^+\pi^-$ decays. These results may be an indication of a large $SU(3)$-breaking effect in $D^0 \to K^\pm\pi^\mp$ decays[5].

This report mainly describes the $\overline{B} \to D^*\ell^-\overline{\nu}$ analysis. The D lifetime analyses are described in Ref. 6.

Candidate $\overline{B} \to D^*\ell^-\overline{\nu}$ decays are selected by applying kinematic constraints on events with a lepton and a $D^* \to D^0\pi$ decay chain, where $D^0 \to K^-\pi^+$, $K^-\pi^+\pi^0$ and $K^-\pi^+\pi^+\pi^-$ decays are used. First, the D^0 decay vertex is determined and then the decay vertex of the $\overline{B} \to D^*\ell^-\overline{\nu}$ candidate is calculated using the lepton and the inferred D^0 track. The vertex point of the accompanying B meson is determined from the remaining tracks, after the rejection of K_S daughters and badly measured tracks. When the reduced χ^2 of the vertex fit is worse than 20, the track that gives the largest contribution to the χ^2 is removed and the vertex fit is repeated. This procedure is iterated until the χ^2 requirement is satisfied. Since the method does not properly treat displaced charm vertices and their daughter tracks, a degradation of the vertex resolution and a bias on the vertex position is introduced. An interaction point constraint is applied to the vertex fit for both B mesons in order to improve the vertex resolution. The typical Δz resolution is 100 μm. The proper-time difference is approximated as $\Delta t \approx \Delta z / c(\beta\gamma)_\Upsilon$ where $(\beta\gamma)_\Upsilon$ is $\beta\gamma$ of the $\Upsilon(4S)$ in the laboratory frame.

The likelihood function for $\overline{B} \to D^*\ell^-\overline{\nu}$ lifetime fit is defined as

$$L(\tau_0, \tau_-, S_t, f_t, S_{BG}, \mu_{BG}, \lambda_{BG}, f_{\lambda BG})$$
$$= \prod_i \int_{-\infty}^{\infty} d(\Delta t')[p^i_{SIG}(\Delta t') + p^i_{BG}(\Delta t')],$$

$$p^i_{SIG}(\Delta t') = (f^i_0 \frac{e^{-\frac{|\Delta t'|}{\tau_0}}}{2 \cdot \tau_0} + f^i_- \frac{e^{-\frac{|\Delta t'|}{\tau_-}}}{2 \cdot \tau_-})$$
$$[(1 - f_t) \frac{e^{-\frac{(\Delta t_i - \Delta t' - \mu)^2}{2\sigma_i^2}}}{\sqrt{2\pi}\sigma_i} + f_t \frac{e^{-\frac{(\Delta t_i - \Delta t' - \mu_t)^2}{2(\sigma_t^i)^2}}}{\sqrt{2\pi}\sigma^i_t}],$$

$$p^i_{BG}(\Delta t') = \sum_k f^i_k \frac{e^{-\frac{(\Delta t_i - \Delta t' - \mu^k_{BG})^2}{2(S^k_{BG}\sigma_i)^2}}}{\sqrt{2\pi}S_{BG}\sigma_i}$$
$$[(1 - f^k_{\lambda BG}) \cdot \delta(\Delta t') + f^k_{\lambda BG} \frac{\lambda^k_{BG}}{2} e^{-\lambda^k_{BG}|\Delta t'|}],$$

where: τ_0 and τ_- are the $\overline{B}^0$ and B^- lifetimes; σ_i and σ^i_t are the main and tail parts of the Δt resolution calculated event-by-event from the track error matrix as described below; f_t denotes the fraction of the tail part of the signal resolution function and is determined from the fit. μ and μ_t are the biases due to the charm meson daughter tracks, determined from the MC simulation; S_{BG}, μ^k_{BG}, λ^k_{BG}, $f^k_{\lambda BG}$ are background-shape parameters, determined from the fit (fake D^*), data (fake lepton) or MC (random $D^*\ell$); f^i_0, f^i_- and f^i_k are fractions of the $\overline{B}^0$ and B^- signals and background contributions that are calculated event-by-event using the measured ΔM_{D^*} value. The $\overline{B} \to D^*X\ell^-\overline{\nu}$ background fractions are estimated from the known branching fractions and included in f^i_0 and f^i_-, since the effect of the missing X is found to be negligible.

The Δt resolution is a convolution of the Δz resolution and the error due to the kinematic approximation ($\Delta t \approx \Delta z/c(\beta\gamma)_\Upsilon$) σ_K:

$$\sigma_i^2 = [\sigma_{\Delta z}/c(\beta\gamma)_\Upsilon]^2 + \sigma_K^2.$$

The Δz resolution $\sigma_{\Delta z}$ is calculated from the vertex resolutions of the reconstructed (σ_z^{rec}) and associated (σ_z^{asc}) B mesons:

$$\sigma^2_{\Delta z} = (S_{det}\sigma_z^{rec})^2 + (S^2_{det} + S^2_{charm})(\sigma_z^{asc})^2,$$

where S_{det} is a global scaling factor that accounts for any systematic bias in the resolution calculation from the track-helix errors, and S_{charm} is a scaling factor to account for the degradation of the vertex resolution of the associated B meson due to contamination of charm daughters. If the reduced χ^2 (χ^2/n) of the vertex fit is worse than 3, the corresponding vertex error (σ_z^{rec} or σ_z^{asc}) is scaled by $[1 + \alpha(\chi^2/n - 3)]$. This χ^2/n-dependent scaling is essential to account for events with large errors. We use the value of $S_{det} = 0.99 \pm 0.04$ determined from the D^0 lifetime fit in the z direction. The values for σ_K, S_{charm} and α are determined from the MC. σ^i_t is calculated in a similar manner. The associated parameter S_t is determined in the fit along with f_t. Figure 1 shows the $\Delta t_{rec} - \Delta t_{gen}$ distribution and resolution function for MC signal events.

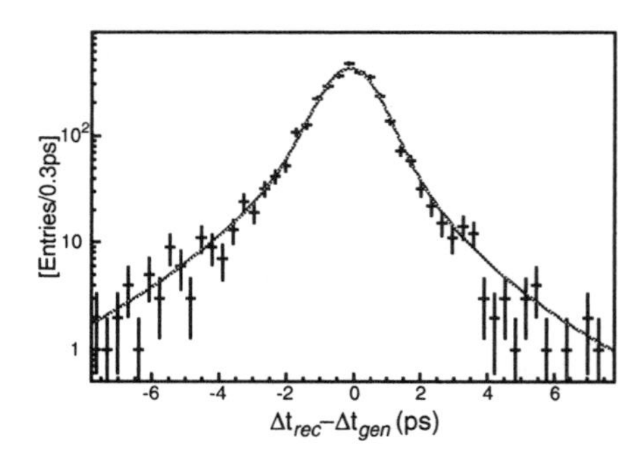

Figure 1. The $\Delta t_{rec} - \Delta t_{gen}$ distribution and resolution function for MC signal events.

The likelihood function for the hadronic modes is defined as

$$L(\tau_{sig}, S_{BG}, S_t^{BG}, \mu_{BG}, \mu_t^{BG}, f_t^{BG}, \lambda_{BG}, f_{\lambda BG})$$
$$= \prod_i \int_{-\infty}^{\infty} d(\Delta t')[p^i_{SIG}(\Delta t') + p^i_{BG}(\Delta t')],$$

$$p^i_{SIG}(\Delta t') = (1 - f^i_{BG})\frac{e^{-\frac{|\Delta t'|}{\tau_{sig}}}}{2 \cdot \tau_{sig}}$$

$$[(1-f_t)\frac{e^{-\frac{(\Delta t_i - \Delta t' - \mu)^2}{2\sigma_i^2}}}{\sqrt{2\pi}\sigma_i} + f_t \frac{e^{-\frac{(\Delta t_i - \Delta t' - \mu_t)^2}{2(\sigma_t^i)^2}}}{\sqrt{2\pi}\sigma_t^i}],$$

$$p^i_{BG}(\Delta t') = f^i_{BG}[(1 - f_t^{BG})\frac{e^{-\frac{(\Delta t_i - \Delta t' - \mu_{BG})^2}{2(S_{BG}\sigma_i)^2}}}{\sqrt{2\pi}S_{BG}\sigma_i}$$

$$+ f_t^{BG}\frac{e^{-\frac{(\Delta t_i - \Delta t' - \mu_t^{BG})^2}{2(S_t^{BG}\sigma_i)^2}}}{\sqrt{2\pi}S_t^{BG}\sigma_i}]$$

$$[(1 - f_{\lambda BG}) \cdot \delta(\Delta t') + f_{\lambda BG}\frac{\lambda_{BG}}{2}e^{-\lambda_{BG}|\Delta t'|}].$$

The fraction of background f^i_{BG} is calculated from the ΔE and M_b values for each event. The background shape parameters S_{BG}, S_t^{BG}, μ_{BG}, μ_t^{BG}, f_t^{BG}, λ_{BG} and $f_{\lambda BG}$ are determined from the fit. We use $S_{det} = 0.94 \pm 0.04$ in the $\overline{B} \to J/\psi \overline{K}$ analysis to account for slightly different kinematic properties from $\overline{B} \to D^*\ell^-\overline{\nu}$, $D\pi$ decays.

Figure 2 shows the Δt distributions and fit results for $\overline{B}^0 \to D^{*+}\ell^-\overline{\nu}$ and $B^- \to J/\psi K^-$ events. Table 1 summarizes the measurement results. The main sources of systematic errors are uncertainties in the resolution function and the Δt dependence of the reconstruction efficiency. All results are preliminary.

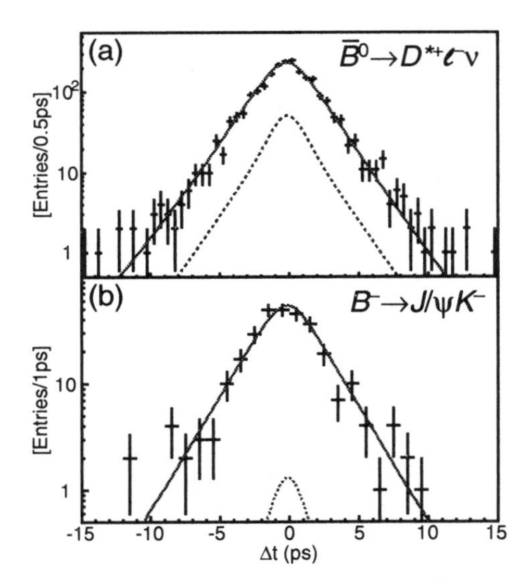

Figure 2. The Δt distributions and fit results for (a) $\overline{B}^0 \to D^{*+}\ell^-\overline{\nu}$ and (b) $B^- \to J/\psi K^-$ events. The dotted curve represents the background.

References

1. H.W.K. Cheung, hep-ex/9912021.
2. I.I. Bigi and N.G. Uraltsev, Z. Phys. C **62** (1994) 623.
3. J.M. Link *et al.* hep-ex/0004034.
4. R. Godang *et al.* hep-ex/0001060.
5. S. Bergmann *et al.* hep-ph/0005181.
6. A. Abashian *et al.* BELLE-CONF-0002.

Table 1. Summary of lifetime measurements.

(a) B lifetime measurements.

Mode	Lifetime
$\overline{B}^0 \to D^{*+}\ell^-\overline{\nu}$	$(1.50 \pm 0.06^{+0.06}_{-0.04})$ ps
$\overline{B}^0 \to D^{*+}\pi^-$	$(1.55^{+0.18+0.10}_{-0.17-0.07})$ ps
$\overline{B}^0 \to D^+\pi^-$	$(1.41^{+0.13}_{-0.12} \pm 0.07)$ ps
$\overline{B}^0 \to J/\psi\overline{K}^{*0}$	$(1.56^{+0.22+0.09}_{-0.19-0.15})$ ps
$\overline{B}^0$ combined	$(1.50 \pm 0.05 \pm 0.07)$ ps
$\overline{B}^0 \to J/\psi K_S$	$(1.54^{+0.28+0.11}_{-0.24-0.19})$ ps
$B^- \to D^{*0}\ell^-\overline{\nu}$	$(1.54 \pm 0.10^{+0.14}_{-0.07})$ ps
$B^- \to D^0\pi^-$	$(1.73 \pm 0.10 \pm 0.09)$ ps
$B^- \to J/\psi K^-$	$(1.87^{+0.13+0.07}_{-0.12-0.14})$ ps
B^- combined	$(1.70 \pm 0.06^{+0.11}_{-0.10})$ ps
$\tau(B^-)/\tau(\overline{B}^0)$	$1.14 \pm 0.06^{+0.06}_{-0.05}$
y_{CP}	$0.03^{+0.15+0.05}_{-0.18-0.08}$

(b) D lifetime measurements.

Mode	Lifetime
$D^0 \to K^-\pi^+$	$(414.8 \pm 3.8 \pm 3.4)$ fs
$D^0 \to K^-K^+$	$(410.5 \pm 14.3^{+9.7}_{-5.9})$ fs
$D^+ \to K^-\pi^+\pi^+$	(1049^{+25+16}_{-24-19}) fs
$D^+ \to \phi\pi^+$	(974^{+68+26}_{-62-18}) fs
D^+ combined	$(1040^{+23}_{-22} \pm 18)$ fs
$D_s^+ \to \phi\pi^+$	$(470 \pm 19^{+5}_{-7})$ fs
$D_s^+ \to \overline{K}^{*0}K^+$	(505^{+34+8}_{-33-12}) fs
D_s^+ combined	(479^{+17+6}_{-16-8}) fs
$\tau(D^+)/\tau(D^0)$	$2.51 \pm 0.06 \pm 0.04$
$\tau(D_s^+)/\tau(D^0)$	$1.15 \pm 0.04^{+0.01}_{-0.02}$
y_{CP}	$(1.0^{+3.8+1.1}_{-3.5-2.1})$ %

PRELIMINARY *BABAR* RESULTS ON B^0 MIXING WITH DILEPTONS AND ON LIFETIME WITH PARTIALLY RECONSTRUCTED B^0 DECAYS

BABAR COLLABORATION,
CHRISTOPHE YÈCHE
CEA Saclay, DAPNIA/SPP, Bat 141, 91191 Gif-Sur-Yvette cedex, France
E-mail: yeche@slac.stanford.edu

With an integrated luminosity of 7.7 fb^{-1} collected on resonance by *BABAR* at the PEP-II asymmetric B Factory, we measure the difference in mass between the neutral B eigenstates, Δm_{B^0}, to be $(0.507 \pm 0.015 \pm 0.022) \times 10^{12}\ \hbar\, s^{-1}$ with dileptons events and presents the preliminary results for the B^0 lifetime, $\tau_{B^0} = 1.55 \pm 0.05 \pm 0.07$ ps and $\tau_{B^0} = 1.62 \pm 0.02 \pm 0.09$ ps obtained from partial reconstruction of the two B^0 decay processes respectively $B^0 \to D^{*-}\pi^+$ and $B^0 \to D^{*-}\ell^+\nu_\ell$.

1 Introduction

This paper presents three analyses designed to select large samples of B^0 mesons using inclusive reconstruction techniques. The first method proposes a precise measurement of the mixing parameter Δm_{B^0} using direct dilepton events which represent 4% of the $\Upsilon(4S) \to B\bar{B}$ decays. The two other methods provide a measurement of the B^0 lifetime by selecting $B^0 \to D^{*-}\pi^+$ and $B^0 \to D^{*-}\ell^+\nu_\ell$ decays; while the two techniques are different in detail, they both share the common feature of making no attempt to reconstruct the D^0 produced in the $D^{*-} \to D^0\pi^-$ decay, thereby achieving high efficiency compared to the exclusive reconstruction.

2 Measurement of B^0 mixing with dileptons

2.1 *Selection of dilepton events and determination of Δt*

In this study[1] of the oscillation frequency Δm_{B^0}, the flavor of the B meson at decay is determined by the sign of leptons produced in semileptonic B decays. For this analysis, electron and muon candidates are required to pass the *very tight* selection criteria fully described in reference[2]; electrons are essentially selected by specific requirements on energy deposited in the Electromagnetic Calorimeter and muons are identified by the use of information provided by the the Instrumented Flux Return.

Non $B\bar{B}$ events (radiative Bhabhas, two-photon and continuum events) are suppressed by applying cuts on the Fox-Wolfram ratio of second to zeroth order moments, on the event squared invariant mass, the event aplanarity and the number of charged tracks. Finally, events with a lepton coming from J/ψ decays are rejected.

The discrimination between direct and cascade leptons is based on a neural network which combines five discriminating variables, all calculated in the $\Upsilon(4S)$ center of mass system: the momenta of the two leptons with highest momenta, the total visible energy, the missing momentum of the event and the opening angle between the two leptons.

The combined effect of the above cuts gives, from Monte Carlo events, a signal purity of 78%. The main source of background consists of $B\bar{B}$ events (12% direct-cascade events). The total number of selected on-resonance events is 36631 (10742 electron pairs, 7836 muon pairs, and 18053 electron-muon pairs). The z coordinate of the B decay vertex is determined by taking the z position of the point of closest approach of the track to an estimate of the position for $\Upsilon(4S)$ decay, obtained by minimizing a χ^2 based on the relative position of the tracks and the beam spot

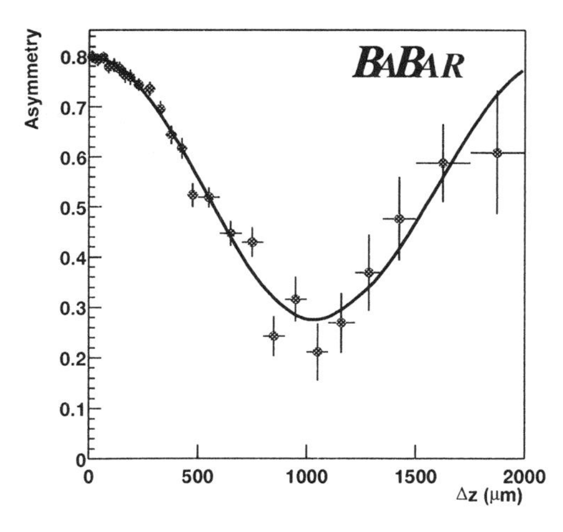

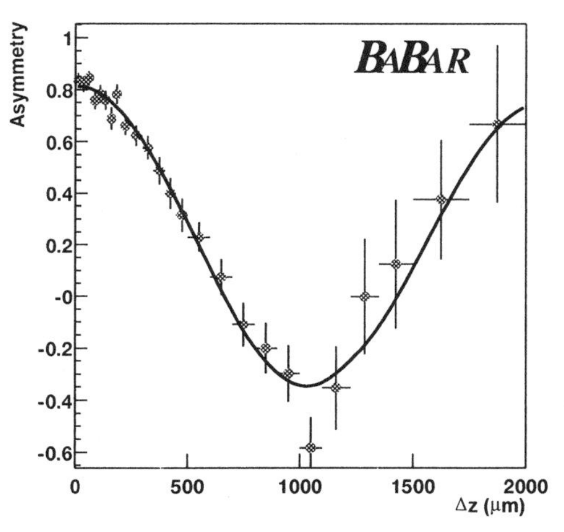

Figure 1. Time-dependent asymmetry $A_{\ell\ell}(|\Delta t|)$ for (a) the inclusive dilepton sample and (b) the dilepton sample enriched with soft pions with a method similar of section 3. The curve represents the result of the fit.

in tranverse plane. A two-Gaussian fit to the resulting Δz resolution function from simulated dilepton events gives $\sigma_n = 87\,\mu$m and $\sigma_w = 195\,\mu$m for the narrow and wide Gaussian, respectively, with 76% of the events in the narrow Gaussian. Then, the time difference between the two B decay times is defined as $\Delta t = \Delta z/(<\beta\gamma> c)$, with $<\beta\gamma>= 0.554$.

2.2 *Measurement of Δm_{B^0}*

The value of Δm_{B^0} is extracted with a χ^2 minimization fit to the dilepton asymmetry:

$$A_{\ell\ell}(|\Delta t|) = \frac{N(\ell^+,\ell^-) - N(\ell^\pm,\ell^\pm)}{N(\ell^+,\ell^-) + N(\ell^\pm,\ell^\pm)}. \quad (1)$$

The fit function takes into account the various time distributions of the dilepton signal and the cascade lepton and the non-$B\overline{B}$ background. The time dependence of this last and its absolute normalization is obtained from off-resonance data. In the fit, three additional parameters are left free: the fraction of charged B, the mistag fraction and the time-dependence of the mistagged events. From a data sample equivalent to $7.73\,fb^{-1}$, we obtain $\Delta m_{B^0} = (0.507 \pm 0.015 \pm 0.022) \times 10^{12}\,\hbar\,s^{-1}$ (see Figure 1) , the main sources of systematic uncertainties are related to the time dependence of the cascade and mis-identified lepton and to the uncertainty on the resolution function (see Table 1).

Table 1. Systematic uncertainty on Δm_{B^0}

Source of systematic uncertainty	$\sigma(\Delta m_{B^0})$ ($10^{12}\,\hbar\,s^{-1}$)
Non-$B\overline{B}$ background	0.005
Mis-Identification	0.011
Cascade events	0.009
Boost approximation	0.001
Beam spot motion ($\leq 20\,\mu$m)	0.001
Δz resolution function	0.009
Tails of the resolution function	0.004
Time-dependence of the resolution function	0.006
Sensitivity to Γ^+ and Γ^0 (PDG 98 $\pm1\sigma$)	0.010
Total	0.022

3 Measurement of B^0 lifetime with partially reconstructed B^0

3.1 *Event selection and determination of Δz*

In studies[3] of the decays $B^0 \to D^{*-}\pi^+$ and $B^0 \to D^{*-}\ell^+\nu_\ell$ reported here, no attempt is made to reconstruct the D^0 decays. Thus, in the hadronic channel, a search is made for a pair of oppositely-charged pions (π_f, π_s) and, assuming that their origin is a B^0 meson and using the beam energy as a constraint, calculates the missing mass M_{miss}. This should be the D^0 mass if the hypothesis was correct. The signal region is taken to be the interval $M_{miss} > 1.854$ GeV/c^2. In the case of the semileptonic decay, due to the limited phase space available in the decay $D^{*-} \to D^0\pi^-$, the D^{*-} four-momentum can be computed by approximating its polar and azimuth angles with those of the slow pion π_s, and parametrizing its momentum as a linear function of the π_s momentum. Then a cut is applied on the invariant mass of the neutrino M_ν^2 estimated from the B^0, D^{*-} and ℓ^+ four-momenta (${M_\nu}^2 > -2\,(\mathrm{GeV}/c^2)^2$). The methods for rejection of the non $B\bar{B}$ background and the identification of the lepton are very similar to those described in section 2.1. For the $B^0 \to D^{*-}\pi^+$, the combinatorial background is reduced by using a Fisher discriminant method combining topological variables relating the position of the tracks and the pseudo-direction of the D^0. The z of the first B^0 is obtained by fitting a vertex between the slow pion and the fast pion or the direct lepton with the beam spot constrained. A fit with the other tracks outside an exclusion cone around the D^0 is performed to determine the z of the second B^0.

3.2 τ_{B^0} measurement

The B^0 lifetime is determined by means of an unbinned maximum likelihood fit, accounting for the event-by-event error determined by the vertex reconstruction algorithm. The fit function is the sum of the probability density function (pdf) for B^0 $B^\pm$, and combinatorial background. The different features of this background (time-dependence, resolution function, etc) are deduced both from Monte Carlo and data with wrong charge association. The results are:

$$\tau_{B^0} = 1.55 \pm 0.05 \pm 0.07\,\mathrm{ps} \quad (D^*\pi),$$
$$\tau_{B^0} = 1.62 \pm 0.02 \pm 0.09\,\mathrm{ps} \quad (D^*\ell\nu).$$

The dominant systematics come from the uncertainty on the fraction and the time-dependence of the backgrounds, the resolution function and the bias due to tracks related to the unconstructed D^0.

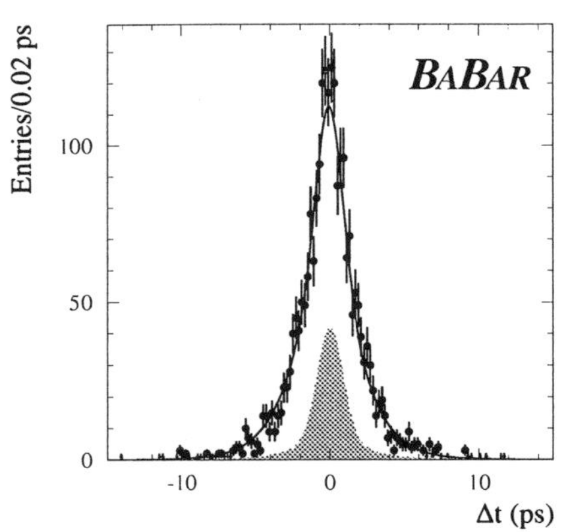

Figure 2. Distribution of Δt in ps for $(D^*\pi)$ events. The continuous line shows the result of the fit and the shaded area shows the contribution of the combinatorial background (27% of the sample).

4 Conclusions

We present a preliminary study of the $B^0\overline{B}^0$ oscillation frequency with an inclusive sample of dilepton events where an accuracy already comparable with the current world average is obtained. The partial reconstuction of $D^{*-}\pi^+$ and $D^{*-}\ell^+\nu_\ell$ gives measurements of the B^0 lifetime and will allow another determination of Δm_{B^0} in the future.

References

1. *BABAR* "Measurement of the time dependence of $B^0\overline{B}^0$ oscillations using inclusive dilepton events", SLAC-PUB-8054.
2. *BABAR* "The first year of the *BABAR* experiment at PEP-II", SLAC-PUB-8539.
3. *BABAR* "Measurement of the $\boldsymbol{B^0}$ meson properties using partially reconstructed $\boldsymbol{B^0}$ to $\boldsymbol{D^{*-}\ \pi^+}$ and $\boldsymbol{B^0}$ to $\boldsymbol{D^{*-}\ \ell^+\ \nu_\ell}$ decays", SLAC-PUB-8053.

MEASUREMENTS OF B^0 AND $B^{\pm}$ LIFETIMES AND B^0-$\bar{B}^0$ MIXING WITH FULLY RECONSTRUCTED B DECAYS IN *BABAR*

FERNANDO MARTíNEZ-VIDAL
LPHNE, IN2P3-CNRS/Universités Paris 6 & 7, France
E-mail: martinef@SLAC.Stanford.EDU

Representing the *BABAR* Collaboration

Time-dependent $B^0\overline{B}^0$ flavor oscillations and B^0 and B^+ lifetimes are studied in a sample of fully reconstructed B mesons collected with the *BABAR* detector, running at the PEP-II asymmetric e^+e^- B Factory with center-of-mass energies near the $\Upsilon(4S)$ resonance. This is the first time that time-dependent mixing and lifetime measurements have been performed at $\Upsilon(4S)$ energies.

1 Introduction

Preliminary measurements of time-dependent $B^0\overline{B}^0$ flavor oscillations and B^0 and B^+ lifetimes have been performed with the *BABAR* detector. These analyses exploit the copious production of B meson pairs in $\Upsilon(4S)$ decays, produced by asymmetric e^+e^- collisions at the PEP-II B Factory at SLAC. These measurements can be used to test theoretical models of heavy quark decay and to constrain the Unitarity Triangle (via the sensitivity to the value of the Cabibbo-Kobayashi-Maskawa matrix [1] element V_{td}). The data set, collected from January to June, 2000, has an integrated luminosity of 8.9 fb^{-1} on the $\Upsilon(4S)$ resonance and 0.8 fb^{-1} collected 40 MeV below the $B\overline{B}$ threshold. This corresponds to about $(10.1 \pm 0.4) \times 10^6$ produced $B\overline{B}$ pairs. The resolution function and mistag rates determined from data in the analyses described here are also used in *CP* asymmetry measurements [2].

2 Experimental Method

The *BABAR* detector is described in detail elsewhere [3]. The analyses described here use all the detector capabilities, including high resolution tracking and calorimetry, particle identification and vertexing.

At PEP-II the B meson pairs produced in the decay of the $\Upsilon(4S)$ resonance are moving in the lab frame along the beam axis (z direction) with a Lorentz boost of $\beta_z\gamma = 0.56$. One B (B_{REC}) is fully reconstructed in an all-hadronic ($B^0 \to D^{(*)-}\pi^+$, $D^{(*)-}\rho^+$, $D^{(*)-}a_1^+$, $J/\psi K^{*0}$ and $B^- \to D^{(*)0}\pi^-$, $J/\psi K^-$, $\psi(2S)K^-$) or semileptonic decay mode ($B^0 \to D^{*-}\ell^+\nu$) [a]. A total of about 2600 neutral, and a similar number of charged, B candidates is reconstructed in hadronic decay modes, with an average purity close to 90%. The background is mainly combinatorial. About 7500 B^0's are reconstructed in semileptonic modes, with an average purity of $\sim 84\%$. Backgrounds to the semileptonic mode are due to combinatorics, fake leptons, $c\bar{c}$ events, and charged B decays from $B^- \to D^{*+}(n\pi)l^-\nu$. Fig. 1 shows the beam-energy substituted B mass (m_{ES}) distributions for the hadronic sample (left) and the $D^* - D^0$ mass distribution for the semileptonic sample (right) [4,5].

The separation between the two B vertices along the boost direction, $\Delta z = z_{\text{REC}} - z_{\text{TAG}}$, is measured and used to estimate the decay time difference, $\Delta t \approx \Delta z / \beta_z\gamma c$. The B_{TAG} vertex is determined via an inclusive procedure applied to all tracks not associated with the B_{REC} meson [4]. The typical

[a] Throughout this paper, conjugate modes are implied.

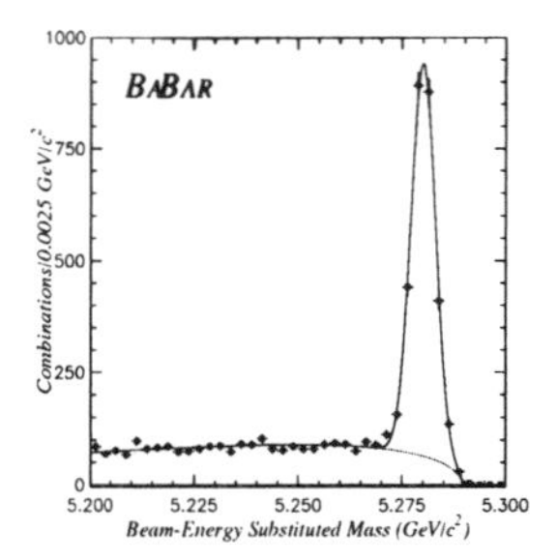

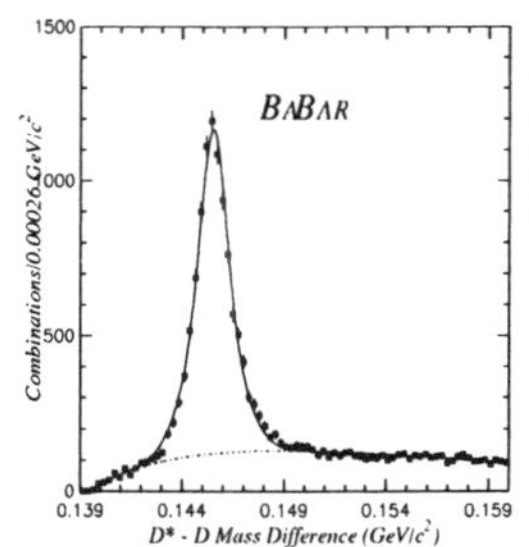

Figure 1. Left: Beam-energy substituted mass distribution ($m_{\rm ES}$) for all the hadronic B^0 modes. Right: $D^* - D^0$ mass difference distribution for the $B^0 \to D^{*-}\ell^+\nu$ sample.

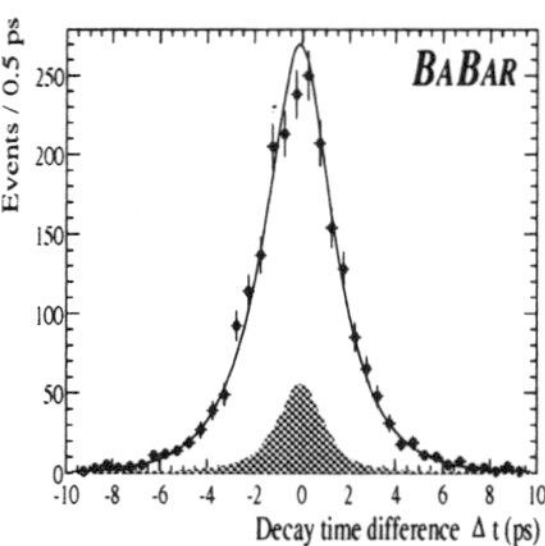

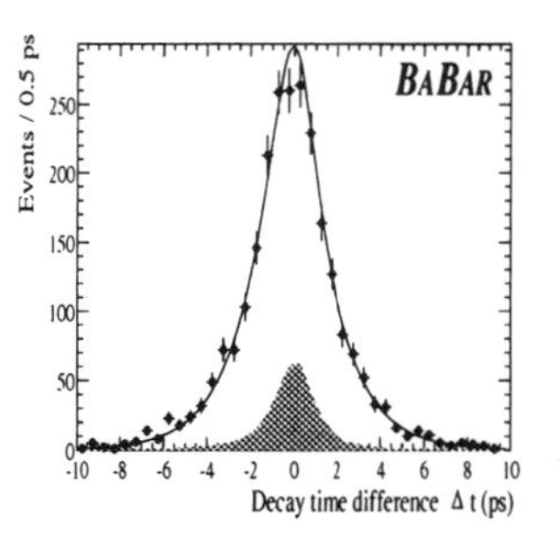

Figure 2. Δt distributions for $B^0/\overline{B}^0$ (left) and B^+/B^- (right) candidates in the signal region ($m_{\rm ES}>$ 5.27 GeV/c^2). The result of the lifetime fit is superimposed. The background is shown by the hatched area.

separation between the two vertices is $\Delta z = \beta_z\gamma c\tau_B \approx 260\,\mu$m, to be compared to the experimental resolution $\sim 110\,\mu$m. The Δt resolution is dominated by the precision on the $B_{\rm TAG}$ vertex, and has little dependence on the decay mode of the $B_{\rm REC}$. An event-by-event Δz resolution is computed and modified to fit the data by convolution with three Gaussians, core, tail and outlier. Most of the events, $\sim 70\%$, are in the core Gaussian, with $\sigma \sim 0.6$ ps.

3 Lifetime Measurements

The B^0 and B^+ lifetimes are extracted from a simultaneous unbinned maximum likelihood fit to the Δt distributions of the signal candidates, assuming a common resolution function. Only hadronic decays from a subsample of 7.4 fb^{-1} integrated luminosity (on-resonance) have been used. An empirical description of the Δt background shape is assumed, using $m_{\rm ES}$ sidebands with independent parameters for neutral and charged mesons. Fig. 2 shows the Δt distributions with the fit result superimposed. Table 1 summarizes the contributions to the total error (see [4] for details).

Table 1. Summary of uncertainties for the B Lifetime measurements.

Source	$\sigma(\tau_B^0)$	$\sigma(\tau_B^+)$	$\sigma(\tau_B^+/\tau_B^0)$
Data statistics	0.052	0.049	0.044
MC statistics	0.016	0.014	0.014
Δt resolution	0.018	0.021	0.009
z scale	0.015	0.016	-
other sources	0.008	0.018	0.012
Total systematics	0.029	0.035	0.021

4 Time-dependent $B^0\overline{B}^0$ mixing

A time-dependent $B^0\overline{B}^0$ mixing measurement requires the determination of the flavor of both B's. The $B_{\rm REC}$ flavor is known if it has been correctly reconstructed, and the flavor of the $B_{\rm TAG}$ is determined by exploiting the correlation between the flavor of the $B_{\rm TAG}$ meson and the charge of its decay products [5]. If there is an identified lepton its charge is used; otherwise the summed charge of identified kaons provides the tag. An event with no tagging leptons or kaons can still be tagged by use of a neural net that exploits the flavor information carried by other decay products, such as soft leptons from charm semileptonic decays and soft pions from D^* decays.

The effective flavor tagging efficiency is given by $Q = \sum_i \epsilon_i(1-2w_i)^2$ where the sum is over tagging categories, each characterized by a tagging efficiency ϵ_i and a probability to mis-identify the B flavor, w_i. Q is related to the statistical significance of the measurement ($1/\sigma^2_{stat} \sim N_{B_{\rm TAG}}Q$).

From the time-dependent rate of mixed (N_{mix}) and unmixed (N_{unmix}) events, the mixing asymmetry $a(\Delta t) = (N_{unmix} - N_{mix})/(N_{unmix} + N_{mix})$ is calculated as a function of Δt and fit to the expected cosine distribution,

$$a(\Delta t) \propto (1 - 2w) \cos \Delta m_d \Delta t \otimes \mathcal{R}(\Delta t|\hat{a}),$$

where $\hat{a}$ are the parameters of the resolution function [5]. A simultaneous unbinned likelihood fit to all the tagging categories, assuming a common resolution function, allows the determination of both Δm_d and the mistag rates, w_i. An empirical description of the backgrounds is determined by fitting to background control samples taken from data, allowing for the following components: i) zero lifetime, ii) non-zero lifetime with no mixing, iii) non-zero lifetime with mixing (only for semileptonic decays). Fig. 3 shows the $a(\Delta t)$ distributions with the fit result superimposed. Table 2 summarizes all the contributions to the total error (see [5] for details).

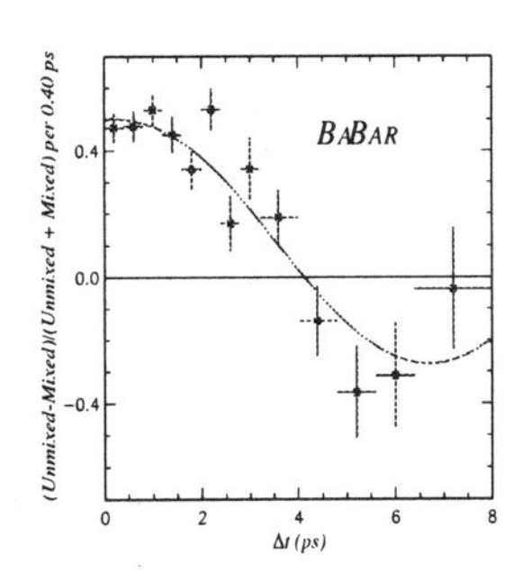

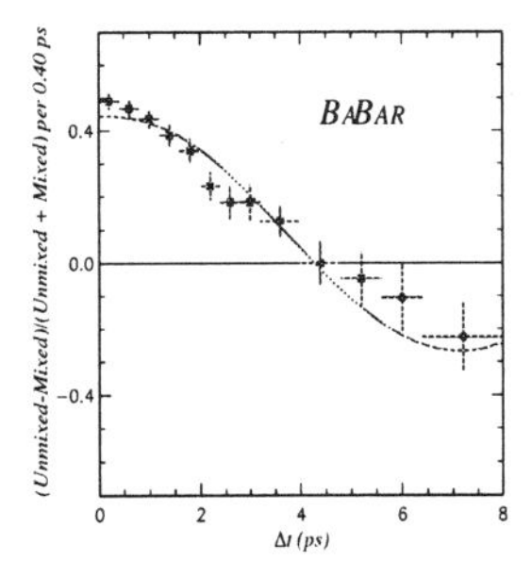

Figure 3. Time-dependent asymmetry $a(\Delta t)$ between unmixed and mixed events for (left) hadronic B candidates with $m_{\rm ES} > 5.27\,{\rm GeV}/c^2$and (right) for $B \to D^* \ell \nu$ candidates.

Table 2. Break-down of uncertainties for the B Mixing measurements.

Source	$\sigma(\Delta m_d)$ hadronic	$\sigma(\Delta m_d)$ semileptonic
Data statistics	0.031	0.020
MC statistics	0.011	0.009
Δt resolution	0.011	0.012
background	0.005	0.013
other sources	0.008	0.008
Total systematics	0.018	0.022

5 Results

The preliminary results for the B meson lifetimes are

$$\tau_{B^0} = 1.506 \pm 0.052 \text{ (stat)} \pm 0.029 \text{ (syst) ps},$$
$$\tau_{B^+} = 1.602 \pm 0.049 \text{ (stat)} \pm 0.035 \text{ (syst) ps}$$

and for their ratio is

$$\tau_{B^+}/\tau_{B^0} = 1.065 \pm 0.044 \text{ (stat)} \pm 0.021 \text{ (syst)}.$$

From the hadronic B^0 sample we measure the $B^0\overline{B}^0$ oscillation frequency:

$$\Delta m_d = 0.516 \pm 0.031 \text{ (stat)} \pm 0.018\text{(syst)}\ \hbar\text{ps}^{-1}$$

and from the $D^{*-}\ell^+\nu$ sample the result is

$$\Delta m_d = 0.508 \pm 0.020 \text{ (stat)} \pm 0.022\text{(syst)}\ \hbar\text{ps}^{-1}.$$

Combining the two Δm_d results, we obtain the preliminary result:

$$\Delta m_d = 0.512 \pm 0.017\text{(stat)} \pm 0.022\text{(syst)}\ \hbar\text{ps}^{-1}.$$

The mistag rates and Δt resolution function extracted from these fits are used in the *BABAR CP* violation asymmetry analysis [2]. The effective flavor tagging efficiency is found to be $Q \approx 28\%$.

The above results are consistent with previous measurements [6] and are of similar precision. They are also compatible with other *BABAR* measurements [7]. Significant improvements are expected in the near future with the accumulation of more data and further systematic studies.

References

1. M. Kobayashi and T. Maskawa, Prog. Theor. Phys. **49** (1973) 652.
2. *BABAR*-CONF-00/01[b].
3. *BABAR*-CONF-00/17[b].
4. *BABAR*-CONF-00/07[b].
5. *BABAR*-CONF-00/08[b].
6. D.E. Groom, *et al.*, Eur. Phys. Jour. C **15**(2000) 1.
7. Ch. Yèche, these proceedings.

[b] *BABAR* Collaboration, B. Aubert *et al.*, contributed paper to this Conference.

PRODUCTION, SPECTROSCOPY AND DECAY OF ORBITALLY EXCITED B MESONS AT LEP

PAULINE GAGNON

Indiana University - CERN EP Division, Geneva 23, CH-1211, Switzerland
E-mail: pauline.gagnon@cern.ch

This paper contains a review of the most recent results from LEP on production rates, masses and widths of orbitally excited B meson states denoted B_J^*. It also provides possible explanations for the various discrepancies observed in these experimental results.

1 Introduction

Heavy Quark Symmetry predicts four B_J^* states with total spin and parity $J^P = 0^+, 1^+, 1^+, 2^+$. They form two doublets characterised by the total angular momentum of the light quark q, namely $j_q = 1/2$ or $3/2$. Members of the same doublets have similar properties. To conserve parity and angular momentum, the B_J^* states can only decay to $B^*\pi$ or $B\pi$ as shown in Fig. 1. S-wave transitions proceed rapidly and correspond to broad states whereas P-wave transitions require more time, hence narrow states. All B_J^* states are expected to overlap in mass even in the hypothetical case of perfect detector resolution. Although all eight $B_J^* \to B^{(*)}\pi\pi$ transitions are theoretically allowed, they are expected to be phase-space suppressed. If dipion transitions exist, they would add more transitions to the ones shown in Fig. 1.

The first observations were reported by OPAL[1] and DELPHI[2] in 1994, and ALEPH[3] in 1995. These initial measurements could not resolve the four underlying states but provided an average B_J^* mass of about 5715 ± 11 MeV and production rate around $(27 \pm 6)\%$. Clearly, larger data samples and improved methods were needed to disentangle the four B_J^* states. B_J^* decays are of particular interest for B oscillations and CP-violation studies[1,4] since the pion from the B_J^* decay provides a clear tag of b quark production flavour. CDF[5] has also investigated this for future oscillation studies.

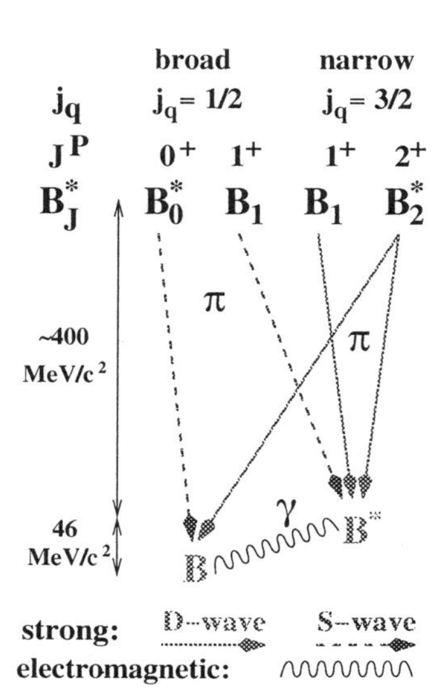

Figure 1. The five allowed transitions for B_J^* states.

All data presented here were obtained in $e^+e^- \to Z^0$ collisions at LEP I. The B_J^* reconstruction technique proceeds as follows: Each event is separated in two using the thrust axis. A b-tag is applied to enhance the $Z^0 \to b\bar{b}$ content in the sample. B mesons are reconstructed either from exclusive decay modes or inclusively, by combining all tracks belonging to a secondary displaced vertex. The B meson is finally associated with a pion coming from the primary vertex to reconstruct the B_J^* meson. From the shape of the mass distribution of the $B\pi$ combinations after background subtraction, one tries to extract the contributions of the four different B_J^* states in the final B_J^* sample.

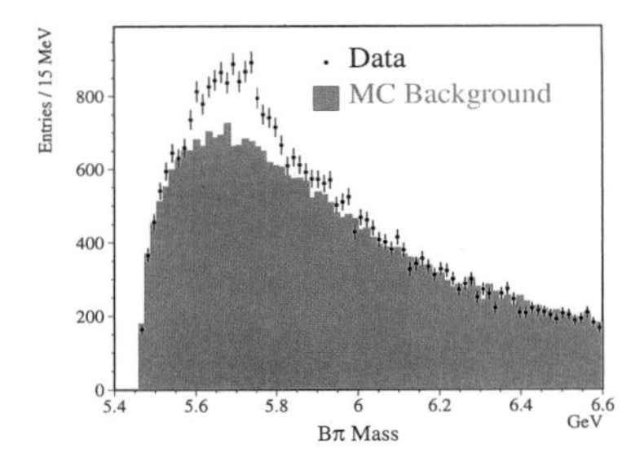

Figure 2. $B\pi$ mass distribution for the inclusive B meson sample from L3.

Figure 3. (top) $B\pi$ mass distribution for inclusive B meson samples from OPAL (centre) after background subtraction and (bottom) efficiency correction for the $B^*\pi$-depleted (left) and -enriched (right) samples.

2 B_J^* reconstruction

ALEPH[4] uses 404 fully reconstructed B mesons from $B \to D^*X$, $X = \pi^\pm$, $\rho^\pm$, $a_1^\pm$, and $B \to J/\psi(\psi')X$, $X = K^\pm, K^*$. These B mesons are associated with a pion identified as originating from the primary vertex. Using charge correlations, one forms right- and wrong-sign samples. The background shape and normalisation are extracted from data using the wrong-sign sample.

Alternatively, L3[6] and OPAL[7] use an inclusive sample of partially reconstructed B mesons associated with a pion to form a much larger B_J^* sample at the cost of degraded resolution. The B meson reconstruction quality is checked with $B\gamma$ combinations using the known difference of 46 MeV between B^* and B meson masses. The background normalisation is obtained from the $B\pi$ mass distribution in the side-band region as shown for L3 in Fig. 2. The background shape for the L3 analysis is taken from Monte Carlo.

OPAL[7] uses two mutually-exclusive samples, one enriched, the other depleted in B^* content to separate $B_J^* \to B^*\pi$ from $B\pi$. This is achieved by looking for a photon likely to originate from a $B^* \to B\gamma$ and assigning a probability weight to each B meson to come from a B^* decay. No attempt is made to reconstruct $B_J^* \to B^*\pi \to B\pi\gamma$ from $B\pi\gamma$ due to the overwhelming combinatorial background. The $B\pi$ mass distributions are shown in Fig. 3 before and after background subtraction and correction due to the B_J^* efficiency dependence on the $B\pi$ mass. The background shape is taken from dedicated background samples from data. The model-independent result is $BR(B_J^* \to B^*\pi(X)) = 0.85^{+0.26}_{-0.27} \pm 0.12\ (syst)$.

3 Mass fit results

A fit to the shape of the $B\pi$ mass distribution after background subtraction yields the masses, widths and total production rate of the four B_J^* states. The number of B_J^* events available to ALEPH, L3 and OPAL are 45 ± 13, 2784 ± 274 and 20833 ± 388 with mass resolutions of 2-5, 45 and 40 MeV, respectively. ALEPH[4] measures $\frac{BR(b \to B_J^*)}{BR(b \to B_{u,d})} = (31 \pm 9\ ^{+6}_{-5}\ (syst))\%$ while L3[6] obtains $(32 \pm 3\ \pm 6\ (syst))\%$

Many assumptions are made to reduce the large number of free parameters for the mass splitting and production rates of each doublet. These are taken from various HQET models and based on D^{**} studies. The interested reader should refer to the original papers for details or slides for a summary. The fit results are shown in Fig. 4 for Aleph, and Fig. 5 for OPAL. L3 only provides Fig. 2.

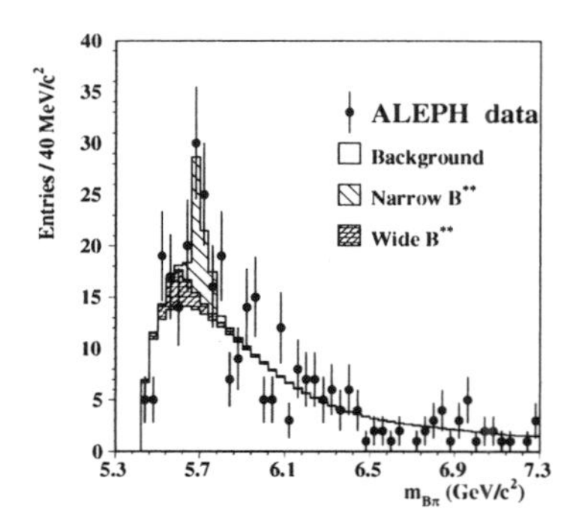

Figure 4. Bπ mass distribution after background subtraction with fit results for ALEPH.

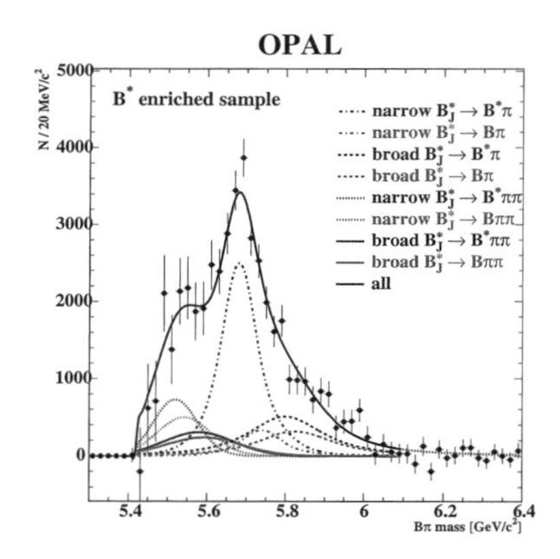

Figure 5. Bπ mass distribution after background subtraction and efficiency correction for OPAL with fit results.

4 Discussion and conclusion

L3 and OPAL results strongly disagree on the mass of the broad states as shown in Table 1. L3 finds them roughly 100 MeV below the narrow states whereas OPAL sees them 100 MeV *above* the narrow peak. L3 also sees a small excess at 5937 ± 21 MeV, with a width of 50 ± 22 MeV and production rate of $(3.4 \pm 1.4)\%$ attributed to B', a radially excited B meson. No such state is seen by OPAL.

ALEPH only fits for the narrow states found at 5727^{+10}_{-12} MeV. By construction, they assume the broad states would be below the narrow states whereas OPAL left this as a free fit parameter. The OPAL fit places the broad states above the narrow peak. This might not be irreconcilable if one considers that both OPAL and ALEPH see a broad peak about 100 MeV below the same narrow peak. The lower broad peak in OPAL data is attributed to transitions containing two pions where one pion is not reconstructed, causing a shift downward to satellite peaks.

The core issue is the fit robustness. Given the large number of free fit parameters, the fit stability requires a clear understanding of the background shape and B_J^* reconstruction efficiencies. Both OPAL and ALEPH use data to check the background shape whereas L3 takes it from the simulation. While all

	L3	OPAL
Γ_{narrow} (MeV)	24 ± 31	18 ± 30
Γ_{broad} (MeV)	70 ± 33	129 ± 68
M_{narrow} (MeV)	5756 ± 8	5738 ± 9
M_{broad} (MeV)	5658 ± 16	5839 ± 40
$b \to B_J^*$	0.32 ± 0.06	-
$B_J^* \to B^*\pi$	-	0.74 ± 0.21
$B_J^* \to B^{(*)}\pi\pi$	-	0.33 ± 0.15

Table 1. *Fit results given for* $B_1(3/2)$ *and* $B_0(1/2)$ *for the narrow and broad doublet, respectively.*

experiments use the same Monte Carlo, only OPAL corrects the shape of the $M_{B\pi}$ distribution to account for the observed efficiency dependence on $M(B\pi)$. More studies will be required to resolve these issues and confirm or refute the existence of B' and di-pion contributions seen only by L3 and OPAL, respectively. The final word will probably come from the Tevatron experiments.

References

1. OPAL, Z. Phys. C66 (1995) 19.
2. DELPHI, Phys. Lett. B345 (1995) 598.
3. ALEPH, Z. Phys. C69 (1996) 393.
4. ALEPH, PLB 425, 1-2 (1998) 215
5. CDF, FERMILAB-Pub-99/330-E.
6. L3, Phys. Lett. B 465, 1-4 (1999) 323
7. http://opal.web.cern.ch/Opal/pubs/pr/

B PHYSICS AT CDF

CHRISTOPH PAUS FOR THE CDF COLLABORATION
Massachusetts Institute of Technology
77 Massachusetts Avenue, Cambridge, MA 02139-4307, USA
E-mail: paus@mit.edu

From 1992 to 1995 the CDF experiment has taken 110 pb^{-1} of $\text{p}\overline{\text{p}}$ collisions at a center-of-mass energy of 1.8 TeV. These data gave rise to a variety of important B physics measurements. Most importantly B mass, lifetime and mixing measurements, the observation of the last missing meson, the B_c, and a measurement of the CP violation parameter $\sin 2\beta$. The highlights of those results are described and a perspective for the upcoming Run II period is given.

1 Introduction

From 1992-1995 CDF has collected 110 pb^{-1} of $\text{p}\overline{\text{p}}$ collisions at a center-of-mass energy of 1.8 TeV. A large number of B physics measurements have been performed and the highlights will be summarized in Section 2. Measurements of B hadron masses, lifetimes and mixing, competitive with the LEP experiments have established the CDF experiment in the realm of B physics. The measurement of the CP violation parameter $\sin 2\beta$, which is still competitive with the first preliminary results of BaBar and Belle[1], demonstrates that CDF will contribute in a significant way when Run II will start in March 2001.

In the first two years the Tevatron will deliver 2 fb^{-1} to each, CDF and DØ, and an anticipated total of about 15 fb^{-1} each will be accumulated until the end of the program. Apart from the increased statistics, CDF will further enhance the B physics potential due to the upgrade of the triggering, tracking and particle identification systems. Apart from measurements of B_u and B_d mesons, CDF and DØ are in a unique position[a] to study heavier B hadrons like B_s, Λ_B and B_c. These measurements complement the program at the B factories and apart from just improving the understanding of CP violation they have a potential to observe effects of new physics beyond the Standard Model. In Section 3 the detector upgrades and the prospects for Run II are discussed.

[a] HERA-B will probably also observe B_s mesons but with lower statistics.

2 Run I Highlights

Due to the large number of B physics results from Run I this report restricts itself mostly to measurements of B hadron masses, lifetimes, mixing and CP violation. B hadron data sample from Run I are obtained by triggering on leptons. Semileptonic B decays have the advantage that the trigger rate is significantly reduced through the clean charged lepton signal. On the other hand the neutrino production decreases the quality of the proper decay length reconstruction.

B hadron mass and lifetime measurements are straightforward and CDF has measured the masses and lifetimes of all B mesons and the Λ_B baryon. CDF governs the world average of the heavy B hadrons since they are not produced at the e^+e^- machines at the $\Upsilon(4\text{S})$ resonance. The masses of those heavy hadrons are

$$\begin{aligned} m(\text{B}_\text{s}) &= 5.3699 \pm 0.0023 \pm 0.0013 \text{ GeV}, \\ m(\Lambda_\text{B}) &= 5.621 \pm 0.004 \pm 0.003 \text{ GeV}, \\ m(\text{B}_\text{c}) &= 6.40 \pm 0.39 \pm 0.13 \text{ GeV}, \end{aligned}$$

together with their lifetimes as measured and published by CDF

$$\tau(\text{B}_\text{s}) = 1.36 \pm 0.10 \text{ ps},$$

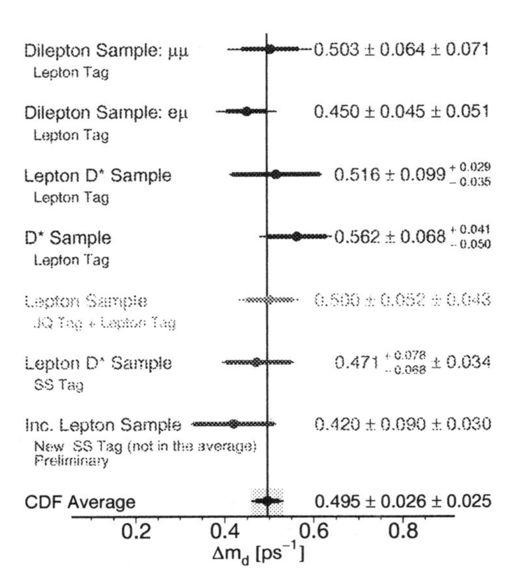

Figure 1. Summary of B_d mixing measurements.

$$\tau(\Lambda_B) = 1.32 \pm 0.17 \text{ ps},$$
$$\tau(B_c) = 0.46 \pm 0.17 \text{ ps}.$$

B mixing and CP violation are both due to the weak transitions between quarks and depend on the complex elements of the well known Cabibbo-Kobayashi-Maskawa matrix. Those measurements are particularly challenging since they require the identification of the flavor of the *B* meson at production. In Run I identification of the flavor at production or *B* flavor tagging has been mostly based on opposite side lepton and jet charge tags and on same side pion tags. A summary of the B_d mixing results is listed in Figure 1. A limit on B_s mixing has been determined to be $\Delta m_s > 5.8$ ps^{-1} at 95% confidence level. To determine the CP violation parameter $\sin 2\beta$, the asymmetry

$$A(t) = \frac{N_{B_d \to J/\psi K_S} - N_{\overline{B}_d \to J/\psi K_S}}{N_{B_d \to J/\psi K_S} + N_{\overline{B}_d \to J/\psi K_S}}$$

in the gold plated decay $B_d \to J/\psi K_S$ is measured. The asymmetry in the Standard Model is given by $-\sin 2\beta \sin \Delta m_d t$ and is

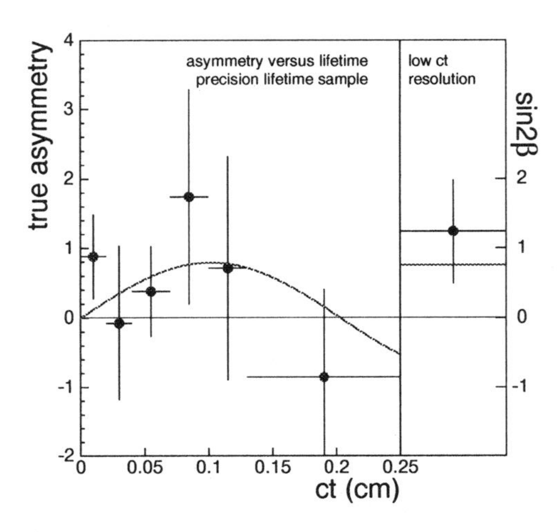

Figure 2. CP asymmetry measurement in $B_d \to J/\psi K_S$.

measured by CDF to be[3]

$$\sin 2\beta = 0.79 \pm 0.39 \pm 0.16$$

which translates to $\sin 2\beta$ being positive at the 93% confidence level. The corresponding time dependent asymmetry is shown in Figure 2. This result has the same statistical precision as, and is consistent with, the results from BaBar and Belle. All measurements so far agree with the Standard Model expectation of about 0.7.

3 Prospects for Run II

The Run II program is going to start in March 2001 providing 2 fb^{-1} to each of the Tevatron experiments in the first two years. CDF is presently preparing for the commissioning run which will allow to see first collisions from September to November 2000.

3.1 Detector Upgrades

To improve the overall detector performance CDF has undergone a major upgrade[2] in the last 5 years. Most important for *B* physics

- The silicon detector with five layers of two sided readout extending from 2.4 cm to

10.7 cm radius with an expected point resolution of 12 μm. Two additional similar silicon layers at 20 cm and 28 cm and a single sided layer at 1.6 cm radius bring the proper time resolutions for B hadrons decaying hadronically to 45 fs. The silicon detector covers a range of $|\eta| < 2$.

- A new time-of-flight detector consisting of Bicron plastic scintillators readout at both ends with 19-staged Hamamatsu PMT's. They are installed inside the solenoid magnet on top of the central outer tracker. The TOF system provides a 2σ separation between pions and kaons for momenta smaller 1.6 GeV being complementary with the dE/dx measurement in the tracker.
- The new trigger system including a track trigger at Level-1 with transverse momenta as low as 1.5 GeV and a secondary vertex trigger at Level-2. The combination of both will allow to trigger on purely hadronic B decays.
- The muon detector coverage has been doubled to cover a range of $|\eta| < 2$.

3.2 B-Physics Prospects

CP violation and B mixing analyses are in the spotlight for Run II. Both critically depend on the flavor tag. Apart from the increased statistics and the improved impact parameter resolutions, the improvement of the B flavor tagging significantly enhances the CDF B physics sensitivity. Depending on the analysis the effective event statistics improves by a factor of up to two just due to the particle identification capability of the new time-of-flight detector. Expectations for some interesting measurements in Run II based on 2 fb^{-1} follow:

- CDF will collect $\approx$ 10k $B_d \to J/\psi K_S$ events resulting in a measurement of $\sin 2\beta$ with an error of ± 0.07.
- The CP violation parameter γ will be measured in a combination of $B_d \to \pi^- \pi^+$ and $B_s \to K^- K^+$ modes with roughly 5k and 10k events, respectively. They allow a measurement of γ with an error better than $\pm 10^\circ$. This result has been obtained assuming the signal to background ratio is as good as 1:2. This number has a large uncertainty which might increase the error of this measurement.
- B_s mixing is measured from $B_s \to D_s^- \pi^+$ decays and with 23k events CDF will be sensitive up to $x_s \approx 65$ which is well above the Standard Model expectation. Probably B_s mixing (x_s= 20-30) will be observed in the first few month of data taking as shown in Figure 3.
- $\Delta\Gamma_s/\Gamma_s$ has been studied in $B_s \to J/\psi\, \phi$ decays. An event sample of 4k events will allow to obtain a precision of ± 0.05.

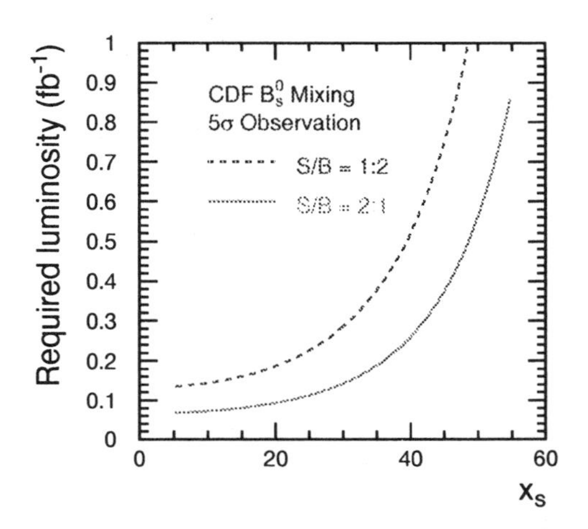

Figure 3. Required luminosity for a five sigma observation of B_s mixing.

References

1. H. Aihara and D. Hitlin these proceedings.
2. CDF Run II Technical Design Report. On the WEB under: http://www-cdf.fnal.gov/upgrades/tdr/tdr.html.
3. T. Affolder *et al.*, CDF Collaboration, Phys. Rev. **D61**, 072005 (2000) 35.

RADIATIVE B DECAYS AT CLEO

T. E. COAN
Physics Department, Southern Methodist University,
Dallas, TX 75275, USA
E-mail: coan@mail.physics.smu.edu

We report on the status of a variety of radiative B decays studied by the CLEO detector with 9.7×10^6 $B\bar{B}$ pairs.

1 Introduction

Flavor changing neutral currents (FCNC) are well known to be forbidden at tree level in the Standard Model (SM). At higher order, however, loop diagrams (box and penguin diagrams) can generate effective flavor changing neutral currents, i.e., $b \to s$ and $b \to d$ transitions. The rates for such transitions in the SM are functions of the top quark mass as well as the masses of the W and and Z gauge bosons. Additionally, these rates are also sensitive to the exchange of heavy non-SM particles such as charged Higgs. Deviations from SM rates for $b \to s$ and $b \to d$ transitions are then a signature of physics beyond the SM. Hence, measurements of $b \to s$ and $b \to d$ transitions are an effective low energy probe of physics at a much higher energy scale. This report emphasizes studies of reactions mediated by electromagnetic penguin diagrams.

2 General Experimental Strategy

All data is taken at a symmetric e^+e^- collider in the vicinity of the $\Upsilon(4S)$ resonance, just above threshold for $B\bar{B}$ production so that B mesons are produced nearly at rest. Approximately two-thirds of the data is taken at the resonance and one-third is taken slightly below the resonance to study continuum and to perform background subtraction. The total luminosity, summed over on and off resonance data, is $14.0\,\text{fb}^{-1}$, corresponding to 9.7×10^6 $B\bar{B}$ pairs.

Two observables are particularly useful for reconstructing B candidates. The first is the difference in energy between a B candidate and the beam energy, $\Delta E = E_B - E_{beam}$. The second is the beam-constrained B-mass, $M_B = \sqrt{E_{beam}^2 - p_B^2}$, where p_B is the momentum of the B candidate. Additionally, the shape of spherical $B\bar{B}$ events is used to distinguish them from jet-like continuum events.

3 Exclusive Modes

CLEO first observed the exclusive mode $B \to K^*\gamma$ and continues to collect statistics. The K^* always decays to $K\pi$ and CLEO reconstructs four modes $(K^\pm\pi^\pm, K_S^0, K^\pm\pi^0, K_S\pi^0)$ to be consistent with the $K^*(890)$ resonance. No other K^* resonances lie in this mass range. The $K\pi$ system and a hard photon are required to be consistent with the B mass using the variables ΔE and M_B. The major background is from continuum ($e^+e^- \to q\bar{q}$ with $q = u, d, c, s$) events accompanied by a hard photon from initial state radiation or events of the type $e^+e^- \to (\pi^0, \eta)X$ with $\pi^0, \eta \to \gamma\gamma$. Both backgrounds are suppressed with appropriate events shape and π^0, η vetoes. Summed over all K^* modes, the yield versus M_B distribution is shown in figure 1.

Separated into neutral and charged B modes, the measured branching fractions are: $\mathcal{B}(B^+ \to K^{*+}\gamma) = (4.5 \pm 0.7 \pm 0.3) \times 10^{-5}$ and $\mathcal{B}(B^0 \to K^{*0}\gamma) = (3.8 \pm 0.9 \pm 0.3) \times 10^{-5}$.

CLEO's large sample of $B \to K^*\gamma$ de-

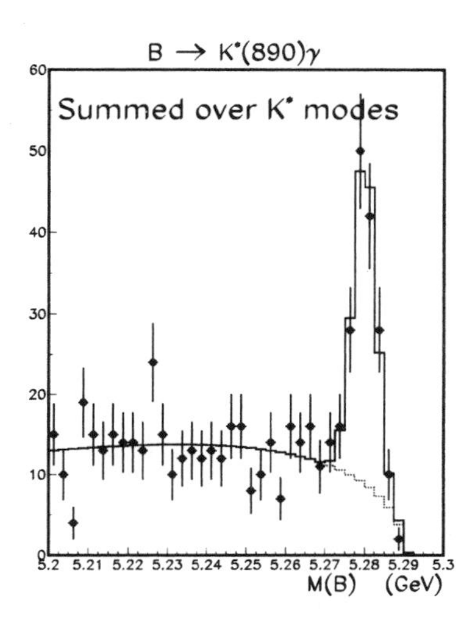

Figure 1. The beam-constrained B mass distribution for $B \to K^*\gamma$ summed over the four K^* modes discussed in the text.

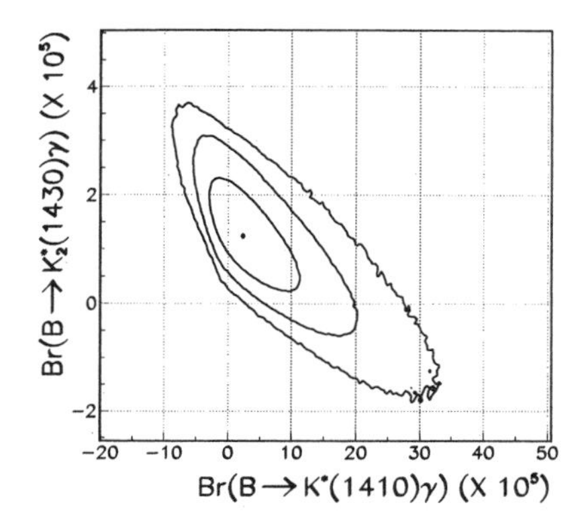

Figure 2. Likelihood contours for the simultaneous fit of $B \to K_2^*(1430)\gamma$ and $B \to K^*(1410)\gamma$ rates. The central point shows the location of the maximum likelihood. The 1, 2, and 3 standard deviations from this maximum are indicated by the contours.

cays allows a search for direct CP violation by measuring the fractional CP asymmetry parameter $A_{CP} = (\mathcal{B}(b) - \mathcal{B}(\bar{b})/(\mathcal{B}(b) + \mathcal{B}(\bar{b}),$ where $\mathcal{B}(b)$ is the branching fraction of B^- and $\bar{B}^0$ into $K^*\gamma$ and $\mathcal{B}(\bar{b})$ is the branching fraction of B^+ and B^0 into $K^*\gamma$. Here, self-tagging K^* decay modes are used to produce the result, summed over charged and neutral B decay modes, $A_{CP} = +0.08 \pm 0.13 \pm 0.03$.

CLEO has searched for resonances heavier than $K^*(890)$, the $K_2^*(1430)$ and the $K^*(1410)$, each of which has an appreciable branching fraction to the final state $K\pi$: $\mathcal{B}(K_2^*(1430) \to K\pi) = 50 \pm 1\%$ and $\mathcal{B}(K^*(1410) \to K\pi) = 7 \pm 1\%$. These modes can be distinguished because of their different helicity distributions and decay widths. Fitting the M_B distribution for both decay modes yields the branching fraction results: $\mathcal{B}(B \to K_2^*(1430)\gamma) = (1.7 \pm 0.6 \pm 0.1) \times 10^{-5}$ and $\mathcal{B}(B \to K^*(1410)\gamma) < 12.7 \times 10^{-5}$ at 90% CL.

The likelihood contours for the fit are shown in figure 2. These results imply that $R \equiv \mathcal{B}(B \to K^*(1430)\gamma)/\mathcal{B}(B \to K^*(890)\gamma) = 0.4 \pm 0.1$, favoring models that use relativistic form factors[2] to predict these rates and disfavoring those with non-relativistic form factors[3].

CLEO has also searched for $b \to d\gamma$ transitions of the form $B \to \rho\gamma$ and $B \to \omega\gamma$ which are a means to limit the CKM ratio $|V_{td}/V_{ts}|$ through the ratio of branching fractions $R \equiv \mathcal{B}(B \to \rho(\omega)\gamma)/\mathcal{B}(B \to K^*\gamma) = \xi|V_{td}/V_{ts}|^2$, where ξ is the ratio of $B \to \rho$ and $B \to K^*\gamma$ form factors and lies in the range 0.6–0.9. The ΔE vs. $M(\pi\pi)$ distributions for $B^0 \to \rho^0\gamma$ and $B^+ \to \rho^+\gamma$ candidates are shown in figure 3. The branching fraction limits are $\mathcal{B}(B^0 \to \rho^0\gamma) < 1.7 \times 10^{-5}$ and $\mathcal{B}(B^+ \to \rho^+\gamma) < 1.3 \times 10^{-5}$ at the 90% CL. This corresponds to $R < 0.32$ at the 90%CL. For the choice $\xi = 0.6$, this implies $|V_{td}/V_{ts}| < 0.72$ at 90% CL. The search for $B \to \omega\gamma$ yields the limit $\mathcal{B}(B^0 \to \omega\gamma) < 0.92 \times 10^{-5}$ at the 90% CL.

4 Inclusive Modes

CLEO has measured the inclusive branching fraction $\mathcal{B}(b \to s\gamma)$ by determining the hard photons energy spectrum and then performing an ON/OFF resonance subtraction. Backgrounds for this analysis are from continuum with initial state radiation and from continuum with a high energy π^0, η, or ω,

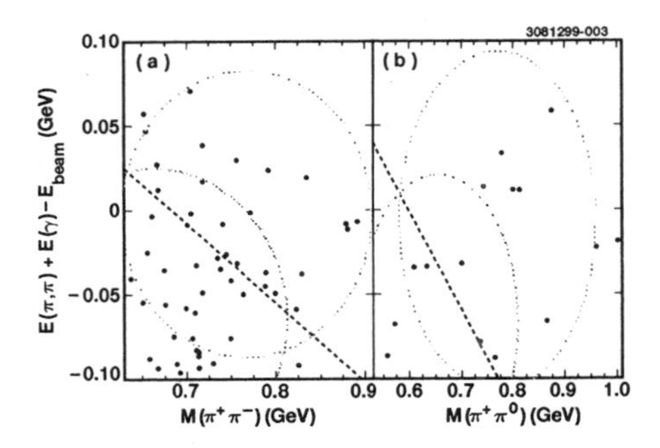

Figure 3. The ΔE vs. $M(\pi\pi)$ distribution for a) $B^0 \to \rho^0\gamma$ and for b) $B^+ \to \rho^+\gamma$ candidates. Dots above the slanted line have passed all cuts. The central dotted ovals are the limits that contain 90% of the $B \to \rho\gamma$ candidates. The partial ovals are the limits that contain 90% of the background $B \to K^*\gamma$ events.

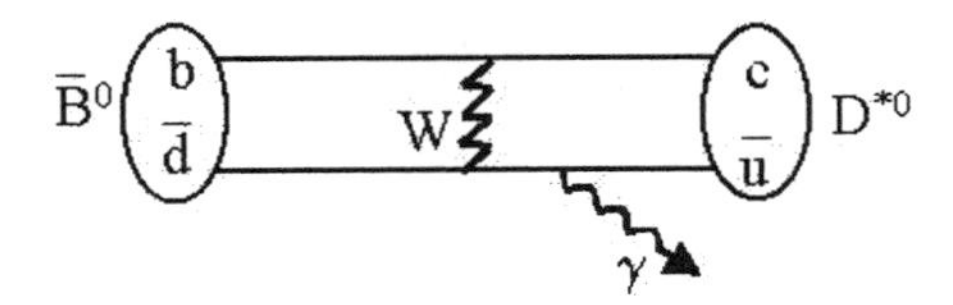

Figure 4. Feynman diagram for the decay $\bar{B}^0 \to D^{*0}\gamma$.

where one of the daughter photons escapes detection. The ON-resonance background is suppressed by two methods. The first method uses a neural net (NN) technique based on event shape variables to separate $B\bar{B}$ events from non-$B\bar{B}$ events. The second technique uses a pseudo-reconstruction method and a second NN. The end result is that events are weighted by a NN output. Afterwards, OFF-resonance data is subtracted. Photon energies E_γ between $2.1\,\text{GeV} < E_\gamma < 2.7\,\text{GeV}$ are used for the final result. Using only $3.1\,\text{fb}^{-1}$ of ON-resonance data, CLEO measures $\mathcal{B}(b \to s\gamma) = (3.15 \pm 0.35 \pm 0.32 \pm 0.26) \times 10^{-4}$, consistent with SM calculations.

CLEO has searched for direct CP violation in $b \to s\gamma$ decays for the full data sample of 9.7×10^6 $B\bar{B}$ events by constraining the fractional CP violation parameter A_{CP}. Although in the SM, A_{CP} is expected to be less than 1%, some non-SM physics scenarios[4,5] permit $A_{CP} < 10 - 40\%$. Events are flavor tagged using either the charge of a high momentum lepton from the "other" B or by using a pseudo-reconstruction technique similar to the one used in the inclusive $b \to s\gamma$ analysis. The typical mis-tag rate is 10%. CLEO finds $-0.22 < A_{CP} < +0.09$ at the 90% CL.

5 Non-penguin Radiative Decays

Finally, CLEO has searched for radiative B decays not mediated by electromagnetic diagrams, as shown in figure 4. In the SM such decays are expected to be small but their observation would permit the importance of W exchange diagrams in B decays to be gauged. Here, the D^{*0} decays by π^0 or γ emission and the D^0 is reconstructed in the $K^-\pi^+, K^-\pi^+\pi^0, K^-\pi^+\pi^-\pi^+$ final states. No events are found in $M_B - \Delta E$ space which leads to the branching fraction limit[6] $\mathcal{B}(\bar{B}^0 \to D^{*0}\gamma) < 5 \times 10^{-5}$ at the 90% CL.

Acknowledgments

The kind assistance of T. Skwarnicki is noted.

References

1. T.E. Coan *et al.*, Phys. Rev. Lett. **84**, 5283 (2000).
2. S. Veseli and M.G. Olsson, Phys. Lett. B **367**, 309 (1996).
3. A. Ali *et al.*, Phys. Lett. B **298**, 195 (1993).
4. A. Kagan and M. Neubert, Phys. Rev. D, **58**, 094012 (1998).
5. M. Aoki *et al.*, Phys. Rev. D **60**, 035004 (1999).
6. M. Artuso *et al.*, Phys. Rev. Lett. **84**, 4292 (2000).

FIRST RESULTS IN EXCLUSIVE RADIATIVE PENGUIN DECAYS AT BABAR

COLIN JESSOP FOR THE BABAR COLLABORATION
P.O. Box 4349, Stanford, CA94304, USA
E-mail: jessop@slac.stanford.edu

We present a preliminary measurement of the branching fraction of the exclusive penguin decay $B^0 \to K^{*0}\gamma$ using $(8.6 \pm 0.3) \times 10^6$ $B\overline{B}$ decays

$$\mathcal{B}(B^0 \to K^{*0}\gamma) = (5.42 \pm 0.82(stat.) \pm 0.47(sys.)) \times 10^{-5} .$$

In addition we search for the related penguin decays with a lepton pair in the final state, $B^+ \to K^+l^+l^-$, $B^0 \to K^{*0}\ l^+l^-$. We find no evidence for these decays in $3.7 \pm\ 0.1 \times\ 10^6$ $B\overline{B}$ decays and set preliminary 90 % C.L upper limits of

$$\begin{aligned}
\mathcal{B}(B^+ \to K^+e^+e^-) &< 12.5 \times 10^{-6},\\
\mathcal{B}(B^+ \to K^+\mu^+\mu^-) &< \ 8.3 \times 10^{-6},\\
\mathcal{B}(B^0 \to K^{*0}e^+e^-) &< 24.1 \times 10^{-6},\\
\mathcal{B}(B^0 \to K^{*0}\mu^+\mu^-) &< 24.5 \times 10^{-6}.
\end{aligned}$$

1 Introduction

In the Standard Model the exclusive decay $B^0 \to K^{*0}\gamma$ proceeds by the $b \to s\gamma$ loop "penguin" diagram. Precise measurements of decay modes involving these transitions and modes with the related $b \to d\gamma$ transition such as $B^0 \to \rho\gamma$ will allow measurements of the top quark couplings V_{ts} and V_{td}. The strength of these transitions may also be enhanced by the presence of non-Standard Model contributions [1]. In the first year of running the *BABAR* experiment has accumulated a dataset comparable to the world's largest to date, and this will increase by an order of magnitude over the next few years. A comprehensive program to study these decays is now underway. The first step in this program is the preliminary measurement of the branching fraction of the exclusive decay mode $B^0 \to K^{*0}\gamma$ using the leading decay mode, $K^{*0} \to K^+\pi^-$. Here K^{*0} refers to the $K^{*0}(896)$ resonance, and charge conjugate channels are assumed throughout. We also present a search for the rarer and as yet unobserved exclusive penguin decays $B \to K\ell^+\ell^-$ and $B \to K^*\ell^+\ell^-$, where ℓ is either an electron or muon.

The data were collected with the *BABAR* detector at the PEP-II asymmetric e^+e^- storage ring. The results presented in this paper are based upon an integrated luminosity of 7.5 fb^{-1} of data corresponding to $(8.6 \pm 0.3) \times 10^6$ $B\overline{B}$ meson pairs recorded at the $\Upsilon(4S)$ energy ("on-resonance") and 1.1 fb^{-1} below the $\Upsilon(4S)$ energy ("off-resonance"). We compute quantities in both the laboratory frame and the rest frame of the $\Upsilon(4S)$. Quantities computed in the rest frame are denoted by an asterisk; eg. E_b^* is the energy of the e^+ and e^- beams which are symmetric in the $\Upsilon(4S)$ rest frame.

2 Measurement of $\mathcal{B}(B^0 \to K^{*0}\gamma)$

We begin the selection by requiring a good photon candidate in the calorimeter with an energy $2.20\,\text{GeV} < E_\gamma^* < 2.85\,\text{GeV}$. We veto photons from π^0's. We next reconstruct the K^{*0} from K^+ and π^- candidates. We consider all pairs of tracks in the event. A track is identified as a kaon by the ring imaging Cherenkov detector (DIRC) and we require $806\,\text{MeV} < M_{K^+\pi^-} < 986\,\text{MeV}$. The B^0 candidates are reconstructed from the K^{*0} and γ candidates. There are backgrounds from continuum $q\overline{q}$ production with the high energy photon originating from initial state radiation or from a π^0 or η. We exploit event topology differences between signal and background to reduce the continuum contribution [5]. In the rest frame of the $\Upsilon(4S)$ the $B\overline{B}$ pairs are produced approximately at rest and therefore decay isotropically while the $q\overline{q}$ pair recoil against each other in a jet-like topology.

Since the B^0 mesons are produced via $e^+e^- \to \Upsilon(4S) \to B\overline{B}$ the energy of the B^0 is given precisely by the beam energy, E_b^* . We reconstruct the B^0 candidate substituting E_b^* for the measured energy of the candidate daughters. We define the difference of the beam energy and energy of the B^0 daughters, $\Delta E^* = E_b^* - E_{K^*}^* - E_\gamma^*$ and require $-200\,\text{MeV} < \Delta E^* < 100\,\text{MeV}$. The B^0 mass is given by, $M_{ES} = \sqrt{E_{beam}^{*2} + |p_B^{*2}|}$, where $|p_B^*|$ is the momentum of the B^0 candidate calculated using the measured momenta of the charged daughters and the energy of the photon. Figure 1 shows the M_{ES} of the candidates. The background is determined empiracally by fitting the ARGUS function [4] to off-resonance data. We find a signal of 48.4 ± 7.3 events with the error coming from the statistical error of the fit.

The efficiency for the selection of $B^0 \to K^{*0}\gamma$ candidates is $(15.6 \pm 0.3)\%$. The branching fraction is measured to be $\mathcal{B}(B^0 \to K^{*0}\gamma) = (5.42 \pm 0.82(stat.) \pm 0.47(sys.)) \times 10^{-5}$ consistent both with previous measurements [2] and with the standard model expectations [3]. The total systematic error of 8.6% is a quadratic sum of several uncorrelated components given in Table 1 [5].

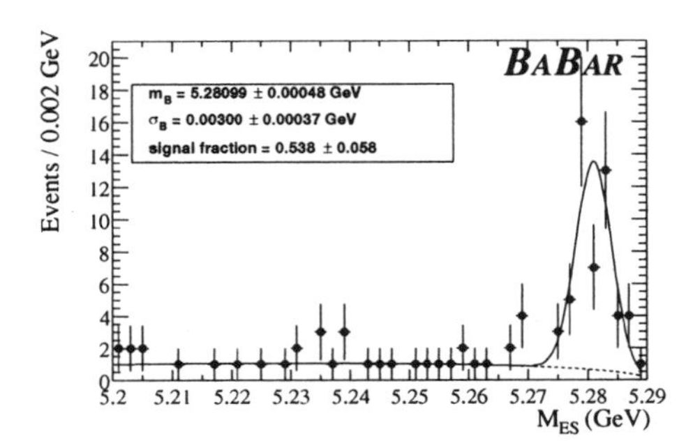

Figure 1. The M_{ES} projection for $B^0 \to K^{*0}\gamma$ $K^{*0} \to K^+\pi^-$ candidates from $(8.6 \pm 0.3) \times 10^6$ $B\overline{B}$ decays.

Table 1. The fractional systematic uncertainties in the measurement of $\mathcal{B}(B^0 \to K^{*0}\gamma)$.

Uncertainty	$\Delta B/B$ %
Tracking efficiency	5.0
Luminosity	3.6
Kaon-id efficiency	3.0
Track resolution	3.0
Energy Resolution	2.5
Background shape	2.3
Monte Carlo Statistics	1.9
Calorimeter energy scale	1.0
Calorimeter efficiency	1.0
$\pi^0\eta$ veto	1.0
Merged π^0 modeling	1.0
Total	8.6

3 Search for $B^+ \to K^+l^+l^-$, $B^0 \to K^{*0}\, l^+l^-$

We search in both the electron and muon channels using a subset of the data corresponding to $3.7 \pm\ 0.1 \times\ 10^6$ $B\overline{B}$ decays. The main goal of our study is to test the performance of a "blind" analysis in which the event selection is optimized without use of the signal or sideband regions in the data. The dominant backgrounds come from random leptons and kaons in $B\overline{B}$ and continuum processes, and from $B \to J/\psi K^{(*)}$ or $B \to \psi(2S)K^{(*)}$ with J/ψ or $\psi(2S) \to \ell^+\ell^-$. The B candidates are reconstructed from

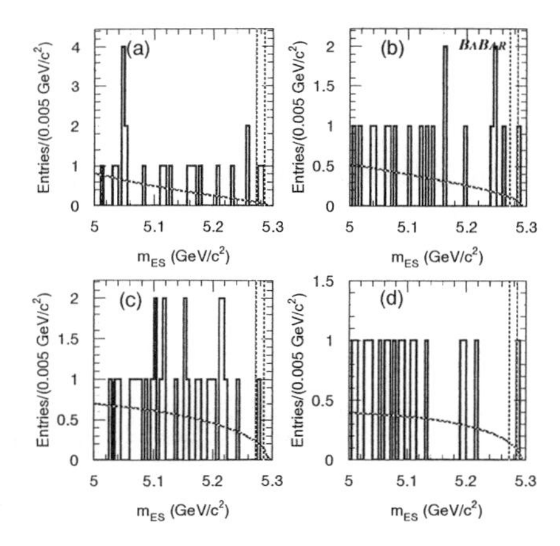

Figure 2. m_{ES} for data after all other event selection criteria are applied: (a) $B^+ \to K^+e^+e^-$, (b) $B^+ \to K^+\mu^+\mu^-$, (c) $B^0 \to K^{*0}e^+e^-$, and (d) $B^0 \to K^{*0}\mu^+\mu^-$. The shape of the fit (the ARGUS function) is obtained from the large statistics sample of fast parameterized Monte Carlo events. The lines indicate the signal region.

K^{*0}, electron and muon candidates. The K^{*0} is reconstructed in the $K^+\pi^-$ final state as above. Electron candidates are identified using the ratio of the deposited calorimeter energy to the associated charged track momentum. Muons are identified by their depth of penetration into the muon detector. Continuum backgrounds are suppressed using event shape variables [6]. The backgrounds from $B \to J/\psi K^{(*)}$ and $B \to \psi(2S)K^{(*)}$ are eliminated by cutting in the ΔE^* vs. $M_{\ell^+\ell^-}$ plane [6]. Figure 2 shows the $m_{\rm ES}$ distributions for the four modes. No evidence of a signal is observed and table 2 gives the derived limits on these processes. The derivation of the limits takes into account the systematic uncertainties given in table 3 [6].

Table 2. Signal efficiencies, the number of observed events, the number of estimated background events, and upper limits on the branching fractions.

Mode	Eff. (%)	Obs. evts	Bkg. est.	$\mathcal{B}/10^{-6}$ (90% C.L.)
$K^+e^+e^-$	13.1	2	0.20	< 12.5
$K^+\mu^+\mu^-$	8.6	0	0.25	< 8.3
$K^{*0}e^+e^-$	7.7	1	0.50	< 24.1
$K^{*0}\mu^+\mu^-$	4.5	0	0.33	< 24.5

Table 3. Summary of the systematic uncertainties given as a percentage error on the branching fraction

	$(\Delta\mathcal{B}/\mathcal{B})$ (%)			
	Kee	$K\mu\mu$	K^*ee	$K^*\mu\mu$
Track eff.	7.5	7.5	10.0	10.0
Lepton-id	4.0	5.0	4.0	5.0
Kaon-id	3.0	3.0	3.0	3.0
ΔE eff.	2.0	2.5	3.3	1.5
m_{ES} eff.	3.0	3.0	3.0	3.0
Vertex eff.	3.0	3.0	4.0	4.0
Fisher eff.	3.0	3.0	3.0	3.0
Luminosity	3.6	3.6	3.6	3.6
MC stat.	3.4	4.0	4.3	6.1
Total	11.7	12.3	14.2	14.8

References

1. J. Hewett and J. Wells, Phys. Rev. D **55**, 5549 (1997).
2. CLEO Collaboration Phys. Rev. Lett. **84**, 5283 (2000).
3. Ali *et al.*Z. Phys. C **63**, 437 (1994) S. Narison, Phys. Lett. B **327**, 354 (1994).
4. H. Albrecht *et al.*(ARGUS), Z. Phys. C **48**, 543 (1990).
5. "A measurement of $\mathcal{B}(B^0 \to K^{*0}\gamma)$" BABAR Collaboration. SLAC-PUB-8534,hep-ex/0008055.
6. "Search for $B^+ \to K^+l^+l^-$, $B^0 \to K^{*0}\, l^+l^-$" BABAR Collaboration. SLAC-PUB-8538.hep-ex/0008059.

STUDIES OF RADIATIVE B MESON DECAYS WITH BELLE

MIKIHIKO NAKAO

KEK, High Energy Accelerator Research Organization, Oho 1-1, Tsukuba, Ibaraki, Japan
E-mail: mikihiko.nakao@kek.jp
for the Belle Collaboration

We have studied radiative B meson decays using a 5.1 fb^{-1} data sample collected at the $\Upsilon(4S)$ resonance with the Belle detector at the KEKB e^+e^- collider. The inclusive branching fraction $\mathcal{B}(b\to s\gamma) = (3.34 \pm 0.50^{+0.34}_{-0.37}{}^{+0.26}_{-0.28}) \times 10^{-4}$ is measured using a technique to subtract the background contribution that requires a relatively small amount of off-resonance data. We measure the exclusive branching fractions to the $K^*\gamma$ final states to be $\mathcal{B}(B^0\to K^*(892)^0\gamma) = (4.94 \pm 0.93^{+0.55}_{-0.52}) \times 10^{-5}$ and $\mathcal{B}(B^+\to K^*(892)^+\gamma) = (2.87 \pm 1.20^{+0.55}_{-0.40}) \times 10^{-5}$. We searched for $B \to \rho\gamma$ decays and obtained an upper limit of $\mathcal{B}(B \to \rho\gamma)/\mathcal{B}(B\to K^*\gamma) < 0.28$ (90% C.L.), where Belle's good high momentum kaon identification is used to reduce the contribution from $B\to K^*\gamma$ to a negligible level.

1 Introduction

Radiative B meson decays are known to be sensitive to the non Standard Model (SM) predictions. Especially $\mathcal{B}(b\to s\gamma)$ is calculated up to the next-to-leading order correction and the existing experimental results already provide stringent limits. The $b\to d\gamma$ process may provide another sensitivity to non-SM predictions, a precise measurement of $|V_{td}/V_{ts}|$ and the direct CP violation within the SM framework. The exclusive channel $B \to \rho\gamma$ can be distinguished from the $B\to K^*\gamma$ signal using particle identification devices.

We have analyzed a data sample of 5.3×10^6 $B\bar{B}$ events corresponding to an integrated luminosity of 5.1 fb^{-1} collected at the $\Upsilon(4S)$ resonance with the Belle detector at the KEKB e^+e^- collider. For background estimations we analyzed a 0.6 fb^{-1} data sample taken 60 MeV below the resonance.

2 Analysis

We have performed measurements of inclusive $\mathcal{B}(b\to s\gamma)$ and exclusive $\mathcal{B}(B\to K^*\gamma)$, and then a search for $B\to\rho\gamma$. The analysis description is given elsewhere[1] and here we describe the essential part only.

The inclusive reconstruction is performed by summing up multiple exclusive final states. We reconstruct strange final state X_s candidates that include one charged kaon or K^0_S and up to 4 pions of which no more than one is a π^0. We tested this method using a Monte Carlo simulation that consists of a $K^*(892)\gamma$ exclusive decay sample and an inclusive $b\to s\gamma$ decay sample for $M_{X_s} > 1.1$ GeV/c^2. The inclusive sample is generated to follow the mass spectrum of a spectator model[2]. We found around 60% of events fall into one of the reconstructed combinations.

From the X_s and photon candidates, we form two independent kinematic variables, the beam constrained mass M_B and the energy difference ΔE in the $\Upsilon(4S)$ rest frame. First, we select candidates in a loose window of $|\Delta E|$ and M_B and we require the X_s and photon directions are back-to-back. In the case of multiple candidates, we select the best candidate that has a minimum absolute value of ΔE, and then apply a cut on M_B.

The dominant hard photon backgrounds come from the $e^+e^-\to q\bar{q}\gamma$ initial state radiation process and photons from the continuum $e^+e^-\to q\bar{q}$ process followed by neutral hadron decays. We optimize the cuts to maximize the signal significance $S/\sqrt{S+B}$ where S and B are the expected number of signal and background events, respectively. Contin-

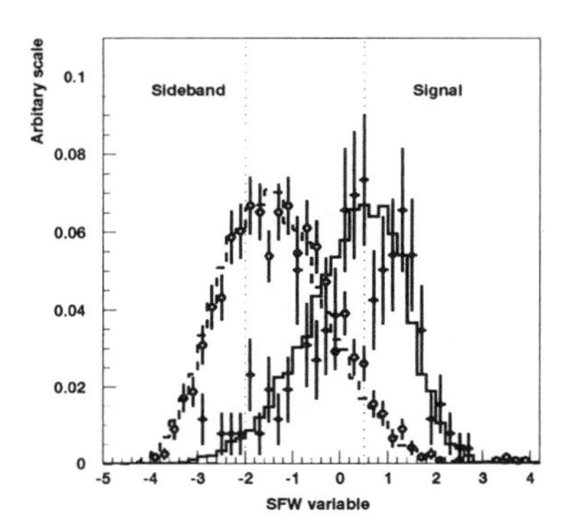

Figure 1. The SFW variable distribution. The background distribution of off-resonance data (open circles) and the Monte Carlo expectation (dashed histogram) are shown. The signal distribution of $B \to D\pi$ data (solid circles) and signal Monte Carlo (solid histogram) are compared.

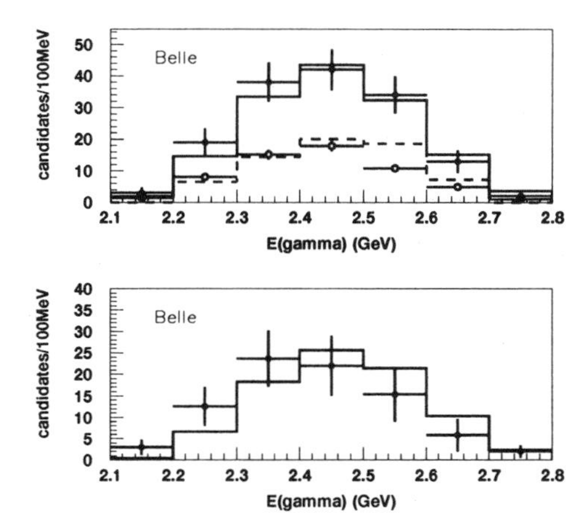

Figure 2. The energy spectrum of photon candidates in the inclusive $b \to s\gamma$ analysis before (top) and after (bottom) background subtraction. The data points (solid circles) are compared with signal Monte Carlo (solid histogram).

uum backgrounds are further rejected using a new empirical event shape variable, which we call the *Super Fox-Wolfram* (SFW) variable (see Figure 1).

The background contribution is subtracted using events from a SFW sideband region (< -2), where only 2% of signal remains. Very little correlation is found between the photon candidate energy and the SFW variable. The absolute scale is determined from the off-resonance data ratio between the signal and sideband region, with relaxed cuts.

Figure 2 shows the photon energy spectrum. We obtain 152 candidate events where 60.0 ± 6.5 background events are expected.

The exclusive $B \to K^*\gamma$ analysis is performed by forming a B meson candidate with a photon candidate and a $K^*(892)$ candidate in the $K^+\pi^-$, $K^0_S\pi^0$, $K^0_S\pi^+$ or $K^+\pi^0$ channels (charge conjugated modes are implied).

The $B \to \rho\gamma$ search is performed in a similar way for $\rho^0(770)$ and $\rho^+(770)$. We apply a tighter set of cuts to improve the expected $S/\sqrt{B}$ ratio, on kaon rejection, continuum background rejection and the signal box. In addition, we explicitly reject K^{*0} candidates from the ρ^0 sample.

Continuum background is rejected with a likelihood ratio constructed from the SFW variable, the B meson flight direction and the K^*/ρ decay helicity angle. We apply cuts on the likelihood ratio to accept 65% and 40% signals for $K^*\gamma$ and $\rho\gamma$ analyses, respectively.

Then we extract the signal yield of $B \to K^*\gamma$ decays by fitting the M_B distribution with a Gaussian signal function and a threshold function for background. The fit is shown in Figure 3. For the $B \to \rho\gamma$ search, we count the number of candidates in the signal windows. We find 0 $\rho^0\gamma$ and 3 $\rho^+\gamma$ candidates.

3 Results

The reconstruction efficiencies are tested with data. For every kind of final state particle, we select a different sample with large statistics. We compare the yield ratio between data and Monte Carlo to estimate how well our efficiencies are reproduced. We use radiative Bhabha events, inclusive η, ϕ, D^0, D^+, D_s decays and $B^- \to D^0\pi^-$ decays to test the photon detection efficiency, tracking

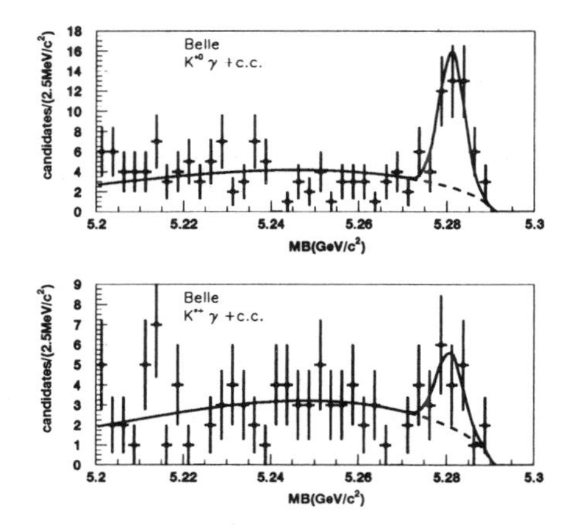

Figure 3. The beam constrained mass (M_B) distributions for the $B^0 \to K^*(892)^0\gamma$ (top) and $B^+ \to K^*(892)^+\gamma$ (bottom) channels.

and particle identification efficiencies and the SFW variable and likelihood ratio efficiencies, respectively.

Finally we obtain the $b \to s\gamma$ branching fraction as

$$\mathcal{B}(b \to s\gamma) = (3.34 \pm 0.50^{+0.34+0.26}_{-0.37-0.28}) \times 10^{-4}$$

where the first error is the statistical error, the second error is the systematic error of our measurement. The third error is a theoretical error, obtained by varying the b-quark Fermi momentum parameter in the MC sample.

In the exclusive analysis, we obtain $38.3 \pm 7.2^{+2.0}_{-1.4}$ and $11.0 \pm 4.6^{+1.6}_{-0.8}$ event yield respectively where the second errors are the systematic errors in the fitting procedure. We obtain the exclusive branching fractions:

$$\mathcal{B}(B^0 \to K^{*0}\gamma) = (4.94 \pm 0.93^{+0.55}_{-0.52}) \times 10^{-5}$$
$$\mathcal{B}(B^+ \to K^{*+}\gamma) = (2.87 \pm 1.20^{+0.55}_{-0.40}) \times 10^{-5}.$$

For the $B \to \rho\gamma$ search, we estimate 1.1 ± 0.4 and 1.8 ± 0.5 background events. The expected $K^*\gamma$ background is 0.08 events for $B^0 \to \rho^0\gamma$ and 0.25 events for $B^+ \to \rho^+\gamma$ searches. We determine upper limits:

$$\mathcal{B}(B^0 \to \rho^0\gamma) < 0.56 \times 10^{-5} \quad (90\% \text{ C.L.})$$
$$\mathcal{B}(B^+ \to \rho^+\gamma) < 2.27 \times 10^{-5} \quad (90\% \text{ C.L.}).$$

We also performed the $K^*\gamma$ reconstruction using the same tight likelihood ratio cut to cancel out major systematic uncertainties. Using the prescription[3] $\mathcal{B}(B \to \rho\gamma) \equiv 2\mathcal{B}(B \to \rho(770)^0\gamma)$, we obtain

$$\mathcal{B}(B \to \rho\gamma)/\mathcal{B}(B \to K^*\gamma) < 0.28 \quad (90\% \text{ C.L.}).$$

4 Discussions

The inclusive $b \to s\gamma$ branching fraction is consistent with the Standard Model prediction[4]. It is also consistent with the recent unpublished CLEO result[5] and the ALEPH result[6]. The exclusive $B \to K^*\gamma$ branching fractions are both in agreement with the only existing result from CLEO[3]. The error of the exclusive analysis is dominated by statistics which we expect to quickly improve in coming runs.

The $B \to \rho\gamma$ search does not show a signal. We demonstrate that the Belle's good kaon identification provides essential separation from $K^*\gamma$, and extend the existing limit[3] with a smaller data sample. We may require a data set 10 times larger to observe a significant signal of $B \to \rho\gamma$.

References

1. A. Abashian *et al.*, Belle Collaboration, Contributed paper #281 to this conference.
2. A. Ali, C. Greub, Phys. Lett. **B259**, 182 (1991).
3. T. Coan *et al.*, CLEO Collaboration, Phys. Rev. Lett. **84**, 5283 (2000).
4. K. Chetyrkin, M. Misiak, M. Münz, Phys. Lett. **B400**, 206 (1997); Erratum ibid. **B425**, 414 (1998).
5. S. Glenn *et al.*, CLEO Collaboration, contribution paper to the 29th ICHEP Conference, CLEO-CONF-98-17, 1998.
6. R. Barate *et al.*, ALEPH Collaboration, Phys. Lett. **B429**, 169 (1998).

STATUS OF THEORETICAL $\bar{B} \to X_S\gamma$ AND $\bar{B} \to X_S L^+L^-$ ANALYSES

MIKOŁAJ MISIAK
Theory Division, CERN, CH-1211 Geneva 23, Switzerland

Status of the theoretical $\bar{B} \to X_s\gamma$ and $\bar{B} \to X_s l^+l^-$ analyses is reviewed. Recently completed perturbative calculations are mentioned. The level at which non-perturbative effects are controlled is discussed.

The present talk will be devoted to discussion of the SM predictions only. Let us begin with $\bar{B} \to X_s\gamma$. Since the completion of NLO QCD calculations[1] 4 years ago, many new analyses have been performed. They include evaluation of non-perturbative Λ^2/m_c^2 corrections[2] and the leading electroweak corrections.[3,4] None of these results exceeds half of the overall $\sim 10\%$ uncertainty, and there are cancellations among them. In consequence, the prediction for $BR[\bar{B} \to X_s\gamma]$ remains almost unchanged: $(3.29 \pm 0.33) \times 10^{-4}$. This prediction agrees very well with the measurements of CLEO[5], ALEPH[6] and BELLE[7], whose combined result is $(3.21 \pm 0.40) \times 10^{-4}$.

The dominant contribution to the perturbative $b \to s\gamma$ amplitude originates from charm-quark loops. After including QCD corrections, the top-quark contribution is less than half of the charm-quark one, and it comes with an opposite sign. This fact should be remembered when one attempts to extract $|V_{ts}|$ from $b \to s\gamma$. The u-quark contribution is suppressed with respect to the charm one by $|V_{ub}V_{us}|/|V_{cb}V_{cs}| \simeq 2\%$.

The results of CLEO, ALEPH and BELLE have to be understood as the ones with subtracted intermediate ψ background, i.e. the background from $\bar{B} \to \psi X_s$ followed by $\psi \to X'\gamma$. This background gives more than 4×10^{-4} in the "total" BR, but gets suppressed when only high-energy photons are counted. A rough estimate[8] of the effect of the photon energy cutoff on this background can be made when X_s in $\bar{B} \to \psi X_s$ is assumed to be massless, and the non-zero spin of ψ is ignored. Then,[a] the intermediate ψ background is less than 5%, for the present experimental cutoff $E_\gamma > 2.1$ GeV in the $\bar{B}$-meson rest frame.[b] However, the background grows fast when the cutoff goes down.

The photon energy cutoff *will* have to go down by at least 200 or 300 MeV in the future. With the present one, non-perturbative effects related to the unknown $\bar{B}$-meson shape function[4] considerably weaken the power of $b \to s\gamma$ for testing new physics. For the same reason, future measurements of $\bar{B} \to X_s\gamma$ should rely as little as possible on theoretical predictions for the precise shape of the photon spectrum above $E_\gamma \sim 2$ GeV.

A systematic analysis of non--perturbative effects in $\bar{B} \to X_s\gamma$ at order $\mathcal{O}(\alpha_s(m_b))$ is missing. There is no straightforward method to perform such an analysis, because there is no obvious operator product expansion for the matrix elements of the 4-quark operators, in the presence of one or more hard gluons (i.e. the gluons with momenta of order m_b). At present, we have only intuitive arguments to convince ourselves that such non-perturbative effects are probably significantly smaller than the overall $\sim 10\%$ theoretical uncertainty in $BR[\bar{B} \to X_s\gamma]$, when the energy cutoff is between 1 and 2 GeV, and when the intermediate $\psi^{(\prime)}$ contribution(s) are subtracted.

---

[a] The $\psi \to X\gamma$ spectrum is available from the ancient MARK II data[9]. New results are expected soon from the BES experiment in Beijing.

[b] A further suppression (to less than 1%) is found when X_s is not treated as massless but the measured[10] mass spectrum is used.

As far as the decay $\bar{B} \to X_s l^+ l^-$ is concerned (for $l = e$ or μ), the best control over non-perturbative effects can be achieved in the region of low dilepton invariant mass ($\hat{s} \equiv m^2_{l^+l^-}/m_b^2 \in [0.05,\ 0.25]$). The present prediction[11] for the branching ratio integrated over this domain is $(1.46 \pm 0.19) \times 10^{-6}$. The quoted uncertainty is only the perturbative one. The non-perturbative Λ^2/m_c^2 and Λ^2/m_b^2 contributions[12] have been included in the central value. They are around 2% and 5%, respectively.

A calculation of $\mathcal{O}(\alpha_s)$ terms in all the relevant Wilson coefficients $C_i(m_b)$ has been recently completed[11], up to small effects originating from 3-loop RGE evolution of C_9. However, the perturbative uncertainty in the above-mentioned prediction remains close to $\sim$13%, because 2-loop matrix elements of the 4-quark operators are unknown.

The low-$\hat{s}$ branching ratio is as sensitive to new physics as the forward-backward or energy asymmetries, i.e. $\sim$100% effects are observed when $C_7(m_b)$ changes sign.

The background from $\bar{B} \to \psi X_s$ followed by $\psi \to l^+ l^-$ is removed by the cutoff $\hat{s} < 0.25$. Analogous contributions from virtual $c\bar{c}$ states are, in principle, included in the calculated Λ^2/m_c^2 correction. An independent verification of this fact can be performed with help of dispersion relations and the factorization approximation.[13] Indeed, for $\hat{s} < 0.25$, the difference between results obtained with help of the two methods is quite small, and can be attributed to higher-order perturbative effects.

On the other hand, the background from $\bar{B} \to \psi X_s$ followed by $\psi \to X' l^+ l^-$ has never been studied. Most probably, for $\hat{s} < 0.25$, it is less important than the analogous background in the case of $\bar{B} \to X_s \gamma$. Experiment-based calculations of these backgrounds are awaited, because they are essential for performing theoretical estimates of similar non-perturbative contributions from other $c\bar{c}$ states.

References

1. K. Chetyrkin, M. Misiak and M. Münz, *Phys. Lett.* B **400**, 206 (1997); C. Greub, T. Hurth and D. Wyler, *Phys. Rev.* D **54**, 3350 (1996); K. Adel and Y.P. Yao, *Phys. Rev.* D **49**, 4945 (1994); A. Ali and C. Greub, *Phys. Lett.* B **361**, 146 (1995).
2. G. Buchalla, G. Isidori and S.J. Rey, *Nucl. Phys.* B **511**, 594 (1998); Z. Ligeti, L. Randall and M.B. Wise, *Phys. Lett.* B **402**, 178 (1997); A.K. Grant et al., *Phys. Rev.* D **56**, 3151 (1997); A. Khodjamirian et al., *Phys. Lett.* B **402**, 167 (1997); M.B. Voloshin, *Phys. Lett.* B **397**, 275 (1997).
3. P. Gambino and U. Haisch, hep-ph/0007259; K. Baranowski and M. Misiak, *Phys. Lett.* B **483**, 410 (2000); A. Strumia, *Nucl. Phys.* B **532**, 28 (1998); A. Czarnecki and W. Marciano, *Phys. Rev. Lett.* **81**, 277 (1998).
4. A.L. Kagan and M. Neubert, *Eur. Phys. J.* C **7**, 5 (1999).
5. S. Ahmed et al., hep-ex/9908022.
6. R. Barate et al., *Phys. Lett.* B **429**, 169 (1998).
7. M. Nakao, this conference; A. Abashian et al., preprint BELLE-CONF-0003.
8. A. Khodjamirian and M. Misiak, in preparation.
9. D.L. Scharre et al., *Phys. Rev.* D **23**, 43 (1981).
10. R. Balest et al., *Phys. Rev.* D **52**, 2661 (1995).
11. C. Bobeth, M. Misiak and J. Urban, *Nucl. Phys.* B **574**, 291 (2000).
12. G. Buchalla, G. Isidori and S.-J. Rey, *Nucl. Phys.* B **511**, 594 (1998); G. Buchalla and G. Isidori, *Nucl. Phys.* B **525**, 333 (1998); A.F. Falk, M. Luke and M.J. Savage, *Phys. Rev.* D **49**, 3367 (1994).
13. F. Krüger and L.M. Sehgal, *Phys. Lett.* B **380**, 199 (1996).

Z PENGUINS AND RARE B DECAYS

GINO ISIDORI
INFN, Laboratori Nazionali di Frascati, I-00044 Frascati, Italy
E-mail: isidori@lnf.infn.it

Rare B decays of the type $b \to s\ \ell^+\ell^-(\nu\bar{\nu})$ are analyzed in a generic scenario where New Physics effects enter predominantly via Z penguin contributions. We show that this possibility is both phenomenologically allowed and well motivated on theoretical grounds. The important role played in this context by the lepton forward-backward asymmetry in $B \to K^*\ell^+\ell^-$ is emphasized.

Flavour-changing neutral-current (FCNC) processes provide a powerful tool in searching for clues about non-standard flavour dynamics. In the present talk we focus on a specific class of non-standard $\Delta B = 1$ FCNC transitions: those mediated by the Z-boson exchange and contributing to rare B decays of the type $b \to s\ \ell^+\ell^-(\nu\bar{\nu})$. As we shall show, these are particularly interesting for two main reasons: i) there are no stringent experimental bounds on these transitions yet; ii) it is quite natural to conceive extensions of the SM where the Z-mediated FCNC amplitudes are substantially modified, even taking into account the present constraints on $\Delta B = 2$ and $b \to s\gamma$ processes.

In a generic extension of the Standard Model where new particles appear only above some high scale $M_X > M_Z$, we can integrate out the new degrees of freedom and generate a series of local FCNC operators already at the electroweak scale. Those relevant for $b \to s\ \ell^+\ell^-(\nu\bar{\nu})$ transitions can be divided into three wide classes: generic dimension-six operators, magnetic penguins and FCNC couplings of the Z boson.[1] The latter are dimension-four operators of the type $\bar{b}_{L(R)}\gamma^\mu s_{L(R)}Z_\mu$, that we are allowed to consider due to the spontaneous breaking of $SU(2)_L \times U(1)_Y$. Their coefficients must be proportional to some symmetry-breaking term but do not need to contain any explicit $1/M_X$ suppression for dimensional reasons, contrary to the case of dimension-six operators and magnetic penguins. This naive argument seems to suggest that FCNC couplings of the Z boson are particularly interesting and worth to be studied independently of the other effects. Actually the requirement of naturalness in the size of the $SU(2)_L \times U(1)_Y$ breaking terms implies that also the adimensional couplings of the non-standard Z-mediated FCNC amplitudes must vanish in the limit $M_X \to \infty$. Nonetheless, as we will illustrate below with an explicit example, the above naive dimensional argument remains a strong indication of an independent behaviour of these couplings with respect to the other FCNC amplitudes.

1 Z penguins within SUSY models

An explicit example where the largest deviations from the SM, in the sector of FCNC, are generated by the Z boson exchange can be realized within supersymmetric models with generic flavour couplings. Within this context, assuming R parity conservation and minimal particle content, FCNC amplitudes involving external quark fields turn out to be generated only at the quantum level. Moreover, assuming the natural link between trilinear soft-breaking terms and Yukawa couplings, sizable $SU(2)_L$- and flavour-breaking effects can be expected in the up sector due to the large Yukawa coupling of the third generation. Thus the potentially dominant non-SM effects in the effective $Z\bar{b}s$ vertex turn out to be generated by chargino-up-squarks loops and have a pure left-handed structure, like in the SM.[2]

Similarly to the $Z\bar{s}d$ case,[3] the first non-vanishing contribution appears to the second order in a simultaneous expansion of chargino and squark mass matrices in the basis of electroweak eigenstates. The potentially largest effect arises when the necessary $SU(2)_L$ breaking ($\Delta I_W = 1$) is equally shared by the $\tilde{t}_R - \tilde{u}^s_L$ mixing and by the chargino-higgsino mixing, carrying both $\Delta I_W = 1/2$. For a numerical evaluation, normalizing the SUSY result to the SM one (evaluated in the 't Hooft-Feynman gauge) and varying the parameters in the allowed ranges, leads to:[1,2]

$$\left| \frac{Z_{sb}^{\rm SUSY}}{Z_{sb}^{\rm SM}} \right| \lesssim 2.5 \left|(\delta^U_{RL})_{32}\right| \left(\frac{M_W}{M_2} \right) . \quad (1)$$

The coupling $(\delta^U_{RL})_{32} = (M^2_{\tilde{U}})_{t_R s_L}/M^2_{\tilde{u}_L}$, which represents the analog of the CKM factor V_{ts} in the SM case, is not very constrained at present and can be of $\mathcal{O}(1)$ with an arbitrary CP-violating phase. Note, however, that vacuum stability bounds imply $|(\delta^U_{RL})_{32}| \lesssim \sqrt{3}m_t/M_S$, where M_S denotes the generic scale of sparticle masses. Therefore the SUSY contribution to the Z penguin decouples as $(M_Z/M_S)^2$ in the limit $M_S/M_Z \to \infty$.

As it can be checked by the detailed analysis of Lunghi *et al.*,[2] in the interesting scenario where the left-right mixing of up-type squarks is the only non-standard source of flavour mixing, Z penguins are largely dominant with respect to other supersymmetric contributions to $b \to s\ \ell^+\ell^-$. The reason for that can be traced back to the dimensional argument discussed before.[1,3]

2 Experimental bounds

The dimension-four effective FCNC couplings of the Z boson relevant for $b \to s$ transitions can be described by means of the following effective Lagrangian

$$\mathcal{L}^Z_{FC} = \frac{G_F}{\sqrt{2}} \frac{e}{\pi^2} M_Z^2 \frac{\cos\Theta_W}{\sin\Theta_W} Z^\mu$$
$$\times \left(Z^L_{sb}\ \bar{b}_L\gamma_\mu s_L + Z^R_{sb}\ \bar{b}_R\gamma_\mu s_R \right) + \text{h.c.}, \quad (2)$$

where $Z^{L,R}_{sb}$ are complex couplings. Evaluated in the 't Hooft-Feynman gauge, the SM contribution to these coupling is $Z^R_{sb}|_{\rm SM} = 0$, $Z^L_{sb}|_{\rm SM} = V^*_{tb}V_{ts}C_0(x_t)$, where $x_t = m_t^2/m_W^2$ and $C_0(x)$ is a loop function[4] of $O(1)$.[a]

Constraints on $|Z^{L,R}_{sb}|$ can be obtained from the experimental upper bounds on exclusive and inclusive $b \to s\ \ell^+\ell^-(\nu\bar{\nu})$ transitions. The latter are certainly more clean form the theoretical point of view although their experimental determination is quite difficult. At present the most significant information from exclusive decays is given by[5] $\mathcal{B}(B \to X_s\ell^+\ell^-) < 4.2 \times 10^{-5}$, leading to[1] $(|Z^L_{sb}|^2 + |Z^R_{sb}|^2)^{1/2} \lesssim 0.15$. Within exclusive channels the most stringent information can be extracted from $B \to K^*\mu^+\mu^-$, where the experimental upper bound[6] on the non-resonant branching ratio ($\mathcal{B}^{\rm n.r.} < 4.0 \times 10^{-6}$) lies only about a factor two above the SM expectation.[7] Taking into account the uncertainties on the hadronic form factors, this implies[1] $|Z^{L,R}_{bs}| \lesssim 0.13$.

Interestingly the above bounds leave open the possibility of large deviations from the SM expectations. In the optimistic case where Z^L_{bs} or Z^R_{bs} were close to saturate these bound, we would be able to detect the presence of non-standard dynamics already by observing sizable rate enhancements in the exclusive modes. In processes like $B \to K^*\ell^+\ell^-$ and $B \to K\ell^+\ell^-$, where the standard photon-penguin diagrams provide a large contribution, the enhancement could be at most of a factor 2-3. On the other hand, in processes like $B \to K^*\nu\bar{\nu}$, $B \to K\nu\bar{\nu}$ and $B_s \to \ell^+\ell^-$, where the photon-exchange amplitude is forbidden, the maximal enhancement could reach a factor 10.

[a] Although $Z^L_{sb}|_{\rm SM}$ is not gauge invariant, we recall that the leading contribution to both $b \to s\ \ell^+\ell^-$ and $b \to s\ \nu\bar{\nu}$ amplitudes in the limit $x_t \to \infty$ is gauge independent and is generated by the large x_t limit of $Z^L_{sb}|_{\rm SM}$ ($C_0(x_t) \to x_t/8$ for $x_t \to \infty$).[4]

3 FB asymmetry in $B \to K^* \mu^+ \mu^-$

If the new physics effects do not produce sizable deviations in the magnitude of the $b \to Z^* s$ transition, it will be hard to detect them from rate measurements, especially in exclusive channels. A much more interesting observable in this respect is provided by the forward-backward (FB) asymmetry of the emitted leptons, also within exclusive modes. In the $\bar{B} \to \bar{K}^* \mu^+ \mu^-$ case this is defined as

$$\mathcal{A}_{FB}^{(\bar{B})}(s) = \frac{1}{d\Gamma(\bar{B} \to \bar{K}^* \mu^+ \mu^-)/ds} \int_{-1}^{1} d\cos\theta \frac{d^2\Gamma(\bar{B} \to \bar{K}^* \mu^+ \mu^-)}{ds\, d\cos\theta} \mathrm{sgn}(\cos\theta) \,, \quad (3)$$

where $s = m^2_{\mu^+\mu^-}/m^2_B$ and θ is the angle between μ^+ and $\bar{B}$ momenta in the dilepton center-of-mass frame. Assuming that the leptonic current has only a vector (V) or axial-vector (A) structure, then $\mathcal{A}_{FB}$ provides a direct measure of the A-V interference. Since the vector current is largely dominated by the γ-penguin amplitude and the axial one is very sensitive to the Z exchange, $\mathcal{A}_{FB}$ is an excellent tool to probe the $Z\bar{b}s$ vertex.

Employing the usual notations for the Wilson coefficients of the SM effective Hamiltonian relevant to $b \to s\ \ell^+\ell^-$ transitions, $\mathcal{A}_{FB}^{(\bar{B})}(s)$ turns out to be proportional to

$$\mathrm{Re}\left\{ C_{10}^* \left[s\ C_9^{\mathrm{eff}}(s) + \alpha_+(s) \frac{m_b C_7}{m_B} \right] \right\}, \quad (4)$$

where $\alpha_+(s)$ is an appropriate ratio of hadronic form factors.[1,8] The overall factor ruling the magnitude of $\mathcal{A}_{FB}^{(\bar{B})}(s)$ is affected by sizable uncertainties. Nonetheless there are at least 3 features of this observable that provide a clear short-distance information:

i) Within the SM $\mathcal{A}_{FB}^{(\bar{B})}(s)$ has a zero in the low s region ($s_0|_{\mathrm{SM}} \sim 0.1$).[8] The exact position of s_0 is not free from hadronic uncertainties at the 10% level, nonetheless the existence of the zero itself is a clear test of the relative sign between C_7 and C_9. The position of s_0 is essentially unaffected by possible new physics effects in the $Z\bar{b}s$ vertex.

ii) The sign of $\mathcal{A}_{FB}^{(\bar{B})}(s)$ around the zero is fixed unambiguously in terms of the relative sign of C_{10} and C_9:[1] within the SM one expects $\mathcal{A}_{FB}^{(\bar{B})}(s) > 0$ for $s > s_0$. This prediction is based on a model-independent relation among the form factorsthat has been overlooked in most of the recent literature. Interestingly, the sign of C_{10} could change in presence of a non-standard $Z\bar{b}s$ vertex leading to a striking signal of new physics in $\mathcal{A}_{FB}^{(\bar{B})}(s)$, even if the rate of $\bar{B} \to \bar{K}^* \ell^+ \ell^-$ was close to its SM value.

iii) In the limit of CP conservation one expects $\mathcal{A}_{FB}^{(\bar{B})}(s) = -\mathcal{A}_{FB}^{(B)}(s)$. This holds at the per-mille level within the SM, where C_{10} has a negligible CP-violating phase, but again it could be different in presence of new physics in the $Z\bar{b}s$ vertex. In this case the ratio $[\mathcal{A}_{FB}^{(\bar{B})}(s) + \mathcal{A}_{FB}^{(B)}(s)]/[\mathcal{A}_{FB}^{(\bar{B})}(s) - \mathcal{A}_{FB}^{(B)}(s)]$ could be different from zero, for s above the charm threshold, reaching the 10% level in realistic models.[1]

Acknowledgements

I am grateful to G. Buchalla and G. Hiller for the enjoyable collaboration on this subject.

References

1. G. Buchalla, G. Hiller, G. Isidori, hep-ph/0006136.
2. E. Lunghi *et al.*, *Nucl. Phys.* **B 568**, 120 (2000).
3. G. Colangelo and G. Isidori, *JHEP* **09**, 009 (1998).
4. T. Inami and C.S. Lim, *Prog. Theor. Phys.* **65**, 297 (81).
5. S. Glenn *et al.* (CLEO Collaboration), *Phys. Rev. Lett.* **80**, 2289 (1998).
6. T. Affolder *et al.* (CDF Collaboration), *Phys. Rev. Lett.* **83**, 3378 (1999).
7. A. Ali, P. Ball, L.T. Handoko and G. Hiller, *Phys. Rev.* **D 61**, 074024 (2000).
8. G. Burdman, *Phys. Rev.* **D 57**, 4254 (1998).

STATUS OF THE HERA-B EXPERIMENT

B. SCHMIDT FOR THE HERA-B COLLABORATION
DESY / HERA-B, Notkestrasse 85, D22607 Hamburg, Germany
E-mail: bernhard.schmidt@desy.de

HERA-B is a fixed target experiment using the halo of the 920 GeV proton beam of HERA on an internal wire target. The aim of the experiment is to trigger on rare B to J/ψ decays, using a highly selective trigger system and to measure the CP violating parameter $sin(2\beta)$. The specific problems of the experiment arise from the extreme background conditions which put unprecedented requirements on the detector components and the triggering and read out system. After 6 years of intense R&D, HERA-B has been finally completed and is now in its commissioning phase.

1 Introduction

HERA-B was proposed in 1994[1] as a fixed target hadronic B-factory with the special goal to measure the CP violating decay asymmetry of neutral B's decaying to the CP eigenstate $J/\psi K_s$. The experiment was designed to cope with 4×10^7 interactions per second, a highly selective trigger scheme should allow to record 200 000 direct J/ψ and about $100 J/\psi$ from B-decays per hour. Running at these design rates for 10^7 seconds, HERA-B would be able to reconstruct 1500 CP violating 'golden decays' and to measure $sin(2\beta)$ with meaningful accuracy. It was clear from the beginning, that the experiment was entering a new regime of particle flux, radiation load and event rates, fully equivalent to conditions expected for the LHC experiments but to be faced about 7 years earlier. The main components of the detector and the triggering and read out electronics are therefore designed at the forefront of technology. Meanwhile the experiment is fully set up and passed an extensive commissioning phase. The present status of the main components and the results obtained so far will be briefly reported in the following sections.

2 Hardware status

HERA-B is of the type of a forward spectrometer with extended particle identification capabilities. The forward angle of 10 - 150 mrad (vertical) and 220 mrad (bending plane) covers about 80 % of the full phase space. The hardware of the HERA-B detector as it was foreseen for the first running period has been completely installed and partially commissioned during 1999 and 2000. The **Wire Target**[2] has now been operated routinely since several years and works to the design specifications in single and multiple wire mode. Cohabitation problems at high rates with the e-p-experiments at HERA could be successfully solved. The **Vertex Detector System (VDS)**[3] consists of 8 superlayers with in total 64 doubly sided silicon detectors. Seven of them are installed in Roman pots and approach the beam as close as 12 mm. The VDS has been built in accordance with the Design Report, and the installation is essentially completed since March 2000. With incomplete instrumentation, the VDS - including all its critical auxiliary systems - is running since several years without any serious problem. The commissioning of the VDS hardware and software is not yet finished. So far, the following performance has been achieved: The hit efficiency is above 97% for 97 out of 116 detector planes, the stand-alone tracking efficiency is better than 95% for tracks with momenta of more than 1 GeV and the primary vertex resolution is 0.070 mm and 0.50 mm in transverse resp. longitudinal direction in good agreement with MC predictions.

Charged particle tracks are followed in the section between the VDS and the Muon system by means of the Inner- and Outer Tracker chambers. Both detectors are key elements for the HERA-B spectrometer and have to provide the hit information for the first level track trigger. The **Inner Tracker (ITR)**[4] covers distances from 5 to 28 cm from the beam axis, corresponding to roughly the forward hemisphere in the CM system. It has to deal with particle flux up to $2.5 \times 10^4 mm^{-2} sec^{-1}$ and radiation doses of 1 Mrad per year. It is build from Microstrip Gas Chambers (MSGC) with Gas Electron Multipliers (GEM) as pre-amplification elements. With more than 10^5 read out channels, it is the worlds largest Micro Pattern detector used in high flux hadronic environment. It has been completed in spring 2000 after six years of continuous struggle and intense R&D. Commissioning has not been finished yet, the overall performance has to be improved further. The chambers are running stably with the HV settings close to the limits. The spatial resolution of 80 μm is in good agreement with the design value. The hit efficiency for minimum ionizing particles is still on the low side ($80 - 90\%$) and needs further improvements. For this years' running, the ITR did not contribute to the trigger due to problems in the trigger part of the front-end electronics. The electronics will be exchanged and improved during the upcoming long shut down. The **Outer Tracker (OTR)**[5] is made from about 1000 honeycomb drift chamber modules with more than $1000 m^2$ detector area and about 115000 read out channels. It covers the backward hemisphere in the CM system, corresponding to distances from 25-200 cm from the beam axis. The OTR was completed this year after solving several fatal problems. All chamber modules were build and installed in less than 9 months. It was operated routinely for both, tracking and triggering. The average hit efficiency of 90 is tolerable, further improvements are hoped for. The spatial resolution is of the order $350 - 400 \mu m$ and still suffering from imperfect alignment. As for the ITR, the operation conditions are close to the HV limit and the detector was continually loosing channels at a rate of 3 - 4 per day. One of the chambers, which got additional efforts during installation, was running perfectly stable for several months. From this, it can be expected that the detector stability can be considerably improved by applying identical measures to all chambers in the upcoming shut down.

Particle identification in HERA-B is based on the RICH, the ECAL and the Muon Detector. The **RICH** uses a $2.5m$ long C_4F_{10} radiator and multi-anode photomultipiers to detect the Cerenkov photons. The RICH was operated routinely throughout the full running period and worked close to its design specifications. The following energy ranges for particle separation could be achieved : $e - \pi(3.4 - 15 GeV/c)$, $\pi - K(12 - 54 GeV/c)$ and $K - p(23 - 85 GeV/c)$. On top of the particle identification, the location of the Cerenkov ring provided useful information about the track angle and helped considerably to commission the tracking detectors. The **ECAL**[6,7] is of the lead(tungsten)/scintillator sandwich type, read out by about 6000 PM's. The calorimeter has been largely completed this year and was successfully calibrated using a very clean π^0 signal with a width of about $9 MeV/c^2$. The commissioning phase was burdened by minor but obstinate problems with stability and hot channels. Up to now the ECAL was used as the main source for lepton pair pre-triggers for the trigger system. Electrons could be positively identified by requiring an associated Bremsstrahlungs cluster. The **Muon System** consists of several layers of tube, pad and gas pixel chambers, embedded in iron loaded concrete absorbers. All components of the Muon System were commissioned during this years' running. Whereas the efficiency of the tube chambers turned out to be sufficiently high, the pad read out suffered from a much too low hit efficiency of only 70%.

The **Trigger System and DAQ** of HERA-B consists of several consecutive steps[8]. The First Level Trigger (FLT)[9,10] is a hardware track trigger searching for lepton pair candidates with an invariant mass above $2GeV/c^2$ within a maximum latency of $12\mu sec$. The FLT consists of about 60 custom made processors, interconnected and linked to the detector hardware by more than 1200 high speed optical links. Pre-triggers (or track seeds) are derived from either the ECAL clusters or pad coincidences in the Muon System. In future, a pre-trigger on high-pt hadrons will be additionally implemented[11,12,13]. The full FLT and Pre-trigger system (except the high-pt part) has been set up in 2000 and underwent an intense commissioning and debugging phase. It could be demonstrated, that the track trigger is technically working, correlations between the FLT messages and offline tracks could be clearly demonstrated. Unfortunately, the concept of the track trigger is highly susceptible to imperfections of both, the detectors and the data transmission network. The limited hit efficiency of the track detectors, unresolved problems with alignment and geometry mapping as well as technical problems with the optical links did not allow to operate the FLT in pair trigger mode with significant efficiency.

The second level trigger (SLT) is a software trigger running on a farm of about 240 Pentium PC's. The SLT is able to digest input rates up to 50 kHz and will be used to refine the FLT decision and to apply cuts on detached vertices. For this years' running, the SLT was used in a variety of different modes for commissioning detector parts as well as for physics data taking. The system runs perfectly stable and fulfills the design specifications. After passing the SLT, the events are build and online reconstructed at a second 190 PC farm[14] , the final logging rate to tape is about 50 Hz. The complete DAQ chain was fully set up and successfully operated this year.

3 Running 2000 and outlook

The running 2000 was mainly dedicated to commissioning the detector components and the trigger and read out system. As part of this and in parallel, several 'physics runs' collecting both, minimum bias and J/ψ triggered data, have been performed. Except for short dedicated periods, the interaction rate was kept low ($5-10$ MHz). For J/ψ triggering, mainly ECAL seeds were used and a sample of about 10000 $J/\psi \rightarrow ee$ could be collected. During the last months, the Muon pre-trigger system came into operation and allowed for the collection of some very clean $J/\psi \rightarrow \mu\mu$ data. The poor statistics collected so far did not allow to see evidence for detached lepton pairs and for an estimation of the B cross section. Beginning in September 2000, HERA-B has to face a long shut down of more than 9 months caused by the machine upgrade to produce higher luminosity for the e-p experiments at HERA. This period will be used for a variety of repair and overhaul work on main components of the experiment. Considerable efforts will have to be spent to bring the detected $B \rightarrow J/\psi X$ rate to the order of magnitude of the design value by the end of 2001. The implementation of the high-pt hardware trigger is of essence[15], since this and other additional trigger schemes will play a fundamental role for the future prospects of the experiment.

References

1. E. Hartouni *et al.*, An Experiment to Study CP-Violation in the B-System Using an Internal Target at the HERA Proton Ring, Design Report DESY-PRC

95/01 (1995)

2. K. Ehret, Nucl. Instrum. Meth. **A446** (2000) 190
3. C.Bauer et al., Nucl. Instrum. Meth. **A447** (2000) 61
4. B. Schmidt *et al.*,MSGC Development for HERA-B,(physics/9804035), *Proc. 36th Workshop of the Eloisatron Project of New Detectors* , Erice, Italy, (1997) pp 270-287
5. M. Capeáns, Nucl. Instrum. Meth. **A446** (2000) 317
6. B.Bobchenko et al., *HERA-B electromagnetic calorimeter,* Proc. of VIII Int. Conf. on Calorimetry in High Energy Physics, June 13-19 1999,LIP,Lisbon,Portugal, World Scientifics, pp 511-517
7. A. Zoccoli et al., Nucl. Instrum. Meth. **A446** (2000) 246
8. E. Gerndt et al., Nucl. Instrum. Meth. **A446** (2000) 264 and references therein
9. E. Gerndt, *The HERA-B Hardware Trigger in 1999*, to be published in Nucl. Instrum. Meth. (2000)
10. D. Ressing, Nucl. Instrum. Meth. **A384** (1996) 131
11. V. Balagura et al., Nucl. Instrum. Meth. **A368** (1995) 252
12. C. Leonidopoulos et al., Nucl. Instrum. Meth. **A427** (1999) 465
13. V.Eiges et al., *High-Pt trigger at the HERA-B experiment*, Proc. of 7th Conf. on Instrumentation for Colliding Beam Physics, November 15-19 1999, Hamamatsu, Japan (2000)
14. A. Gellrich et al., *Full Online Event Reconstruction at HERA-B,* Proc. of CHEP2000, Feb 7-11 2000, Padova, Italy (2000)
15. R.Chistov et al.,*Physics Program with a High-Pt Trigger of HERA-B*, Proc. of 3d Int. Conf. on B Physics and CP Violation, December 3-7 1999, Taipei, Taiwan (1999)

THE PHYSICS REACH OF THE BTEV EXPERIMENT

J.N. BUTLER

Fermi National Accelerator Laboratory, Box 500, Batavia, IL 60510-0500, USA
E-mail: butler@fnal.gov

The BTeV program at the Tevatron will do a complete study of CP violation, mixing, and rare decays in the beauty and charm systems. The physics reach for CP violation and mixing has been determined by detailed simulations of key signals. Efficiencies and backgrounds in modes such as $B^o \to \pi^+\pi^-$, $B^o \to \rho\pi$, $B_s \to D_s K$, and $B_s \to D_s\pi$ are presented. The projected uncertainties on the CKM angles α, β, and γ and the reach in the B_s mixing parameter x_s will be given.

1 Goals of BTeV

BTeV[1] will begin to take data at the Fermilab Tevatron Collider well after the first measurements of CP violation in B decays have been made. The fundamental question – whether the Standard Model's mechanism for CP violation explains the complete pattern of CP violation – will not have been answered. BTeV is designed to

- check the consistency of the Standard Model by making the very difficult measurements that will still be incomplete circa 2006; and
- search for rare and Standard Model violating decays of beauty and charm particles to look for deviations from the Standard Model that could be new physics.

Some of the key states which must be studied to accomplish these goals are shown in Table 1. The last four columns show the features of the detector which are especially crucial to each measurement. The heading "decay time" is meant to emphasize the superb resolution required to follow the fast oscillation of the B_s.

2 BTeV Detector Layout and Key Design Features

A schematic of the BTEV spectrometer is shown elsewhere[2]. The two Tevatron beams collide in the center of the detector. Some of the key design features of the detector are:

- A dipole located on the IR, which gives BTeV TWO spectrometer arms – one covering the forward proton rapidity region and the other covering the forward anti-proton rapidity region;
- A precision vertex detector based on planar arrays of silicon pixels;
- A secondary vertex trigger at Level 1 which makes BTeV especially efficient for states having only hadrons;
- Strong particle identification based on a Ring Imaging Cerenkov Counter and a toroidal muon spectrometer.
- A lead tungstate calorimeter for photon and π^o reconstruction; and
- A very high capacity data acquisition system which frees BTeV from needing to make excessively restrictive choices at the trigger level.

This detector, its trigger, and its data acquisition system, will be capable of addressing whatever problems are likely to arise in beauty and charm quark physics.

The b-quark cross section at 2 TeV is at least 100μb. BTeV is designed to run at a luminosity of $2\times10^{32}/(\text{cm}^2\text{-s})$. This results in 2×10^{11} $b-\bar{b}$ pairs per 10^7 s of running ("1 Snowmass year"). The bunch spacing is 132 ns. The length of the luminous region will be 20-30 cm and its transverse dimension will have a σ of about 50μm. The average number of interactions/beam-crossing is 2.

Table 1. Summary of Key CKM Tests

Quantity	Decay Mode	Vertex Trigger	K/π sep	γ det	Decay time σ
$\sin 2\alpha$	$B^o \to \rho\pi \to \pi^+\pi^-\pi^o$	√	√	√	
$\sin 2\alpha$	$B^o \to \pi^+\pi^-$, $B_s \to K^+K^-$	√	√		√
$\cos 2\alpha$	$B^o \to \rho\pi \to \pi^+\pi^-\pi^o$	√	√	√	
$\text{sign}(\sin 2\alpha)$	$B^o \to \rho\pi$, $B^o \to \pi^+\pi^-$	√	√	√	
$\sin\gamma$	$B_s \to D_sK^-$	√	√		√
$\sin\gamma$	$B^- \to \bar{D^o}K^-$	√	√		
$\sin\gamma$	$B \to K^+\pi^-$, $K_s\pi^-$	√	√		
$\sin 2\chi$	$B_s \to J/\psi\eta'$, $J/\psi\eta$		√	√	√
$\sin 2\beta$	$B^o \to J/\psi K_s$				
$\cos 2\beta$	$B^o \to J/\psi K^*$, $B_s \to J/\psi\phi$				
x_s	$B_s \to D_s\pi^-$	√	√		√
$\Delta\Gamma$ for B_s	$B_s \to J/\psi\eta'$, K^+K^-, $D_s\pi^-$	√	√	√	√

3 The Simulation and Typical Sensitivities

The physics reach of BTeV has been studied using a simulation based on GEANT3[3]. We have used a very realistic description of the detector which includes vacuum pipes, cables, supports, etc. We generated and tracked all primary particles and secondaries from the interactions and decays. Events were generated with Pythia with a Poisson distribution with a mean of two interactions per beam crossing. The simulation used 60 LINUX CPUs and ran for about 4 months. Most of the cycles were devoted to detailed simulation of backgrounds due to B decays.

The BTeV trigger is based on using the pixel detector and a very large farm of advanced processors at the lowest level (Level 1) to inspect every crossing – 7.6 million/s – to look for evidence of particles detached from the primary interaction vertex. The trigger decision is based on a minimum of "N" tracks, typically 2, having an normalized impact parameter of at least "M" standard deviations, typically 6, with respect to the most likely primary vertex. The Level 1 trigger must fire less than 1% of the time. The Level 1 trigger efficiency for states with at least two particles in the decay vertex ranges from 50% to 75%. For $B^o \to \rho^o\pi^o$ it is 56%. Even for the single prong decay, $B^- \to K_s\pi^-$, the trigger is 27% efficient.

A variety of states and their dominant backgrounds have been studied using realistic analysis cuts to reduce the backgrounds to acceptable levels. The sensitivities for several important physics signals are given in Table 2. More details and results for additional states are given in reference 1.

Table 2. Physics Reach of BTeV in 10^7 Seconds of Running

Reaction	Br(B)($\times 10^{-6}$)	# of events	S/B	Parameter	Error or (Value)
$B^o \to \pi^+\pi^-$	4.3	24,000	3	Asymmetry	0.024
$B_s \to D_s K^-$	300	13,100	7	γ	7°
$B^o \to J/\psi\,(\mu\mu)K_s$	445	80,500	10	$\sin 2\beta$	0.025
$B_s \to D_s\pi^-$	3000	103,000	3	x_s	up to (75)
$B^- \to D^o(K^+\pi^-)K^-$	0.17	300	1		
$B^- \to D^o(K^+K^-)K^-$	1.1	1,800	>10	γ	10°
$B^- \to K_s\pi^-$	12.1	8,000	1		
$B^o \to K^+\pi^-$	18.8	108,000	20	γ	$<5^o$
$B^o \to \rho^+\pi^-$	28	9,400	4.1		
$B^o \to \rho^o\pi^o$	5	1,350	0.3	α	$\approx 10^o$
$B_s \to J/\psi(\mu\mu)\eta$	330	1,920	15		
$B_s \to J/\psi(\mu\mu)\eta'$	670	7,280	30	χ	0.033

Acknowledgments

This work received partial support from Fermilab which is operated by University Research Association Inc. under Contract No. DE-AC02-76CH03000 with the United States Department of Energy.

References

1. The full BTeV proposal may be found at http://www-btev.fnal.gov.
2. S. Stone, in these proceedings.
3. GEANT: CERN Program Library Long Writeup W5013.

PRODUCTION OF CHARMONIA AND UPSILON MESONS AT HERA

A. BERTOLIN

on behalf of the ZEUS and H1 Collaborations.
Università and Sezione INFN di Padova,
Dipartimento di Fisica G. Galilei, Via Marzolo 8, 35131 Padova, Italy
E-mail: bertolin@pd.infn.it

Updated measurements of inelastic J/ψ differential cross–sections are compared to the latest theoretical predictions. For the first time, measurements in the low z region are presented. The ψ' to J/ψ inelastic cross section ratio is presented and compared to naive QCD expectations. Elastic Υ measurements are also presented.

1 Inelastic J/ψ

Inelastic J/ψ photoproduction was studied at HERA by the ZEUS and H1 collaborations using the reaction $e^+p \to e^+J/\psi(\to \mu^+\mu^-)X$. Fig. 1 shows the differential cross–sections $d\sigma/dz$ measured by ZEUS and H1 for three different p_T selections. All data sets show cross–sections increasing with z and having a similar shape. The lower pair of full lines show a prediction based on the colour–singlet (CS) and colour–octet (CO) models [1] for the $p_T > 1$ GeV selection. The difference between the lines reflects the uncertainty arising from the determination of the CO matrix elements. The CO calculation falls well below the data. The shape, however, is well described, as shown by the upper pair of full lines, where the prediction has been arbitrarily scaled by a factor of 3 to agree with the measurements. The CO matrix elements used in this calculation come from a fit to the CDF data on prompt J/ψ production.
The CS model has been calculated at NLO [2] for the direct photon process only. Results for $p_T > 1$ GeV are shown by the dotted line in Fig. 1. The predicted cross–section agrees reasonably well with the data, but the result is sensitive to the assumed value of the mass of the charm quark. A value of 1.4 GeV was used, raising it to 1.55 GeV would reduce the predicted cross–section by 35 %.
The CO matrix elements have also been recently determined from an analysis [3] of B meson decays to J/ψ in the context of a non–relativistic QCD Lagrangian. The data come from the CLEO collaboration. The resulting prediction for the $p_T > 2$ GeV selection is shown by the dashed line in Fig. 1. There is good agreement at high z, but the prediction falls below the data for the lower z values.
At low z $d\sigma/dz$ is particularly sensitive to the resolved photon process enhanced by the color octet contribution, as can be seen in Fig. 2. Good agreement in shape is found between the H1 measurements and the theoretical prediction [1] but the computation has been rescaled by a factor of 3 to agree with the measurements.
Fig. 3 compares the measured p_T^2 distribution with the aforementioned CS NLO calculation [2]. In the validity range of the calculation, given by $p_T > 1$ GeV, the NLO result successfully describes the shape and normalization of the p_T^2 dependence.

2 Inelastic charmonium

The inelastic ψ' to J/ψ cross–section ratio has been measured for the first time by the HERA experiments [4] for central z values. The result, Fig. 4, is in good agreement with the naive LO CS model expectation shown by the horizontal line. Fig. 4 also shows the elastic ψ' to J/ψ cross–section ratios as measured by ZEUS, H1 and low energy experiments [5]. At the present level of accuracy no process dependence (inelastic vs elastic) nor energy dependence is seen.

3 Elastic Υ

ZEUS and H1 measurements of the elastic Υ photoproduction cross–section [6], $\gamma p \to \Upsilon(1S)p$, are shown in Fig. 5. This is again a measurement performed at HERA for the first time. Modeling the process by the emission of a colorless gluon pair by the elastically scattered proton and due to the high scale given by the Υ mass itself the measured cross–section can be explained by perturbative QCD calculations given by the continuous lines in Fig. 5.

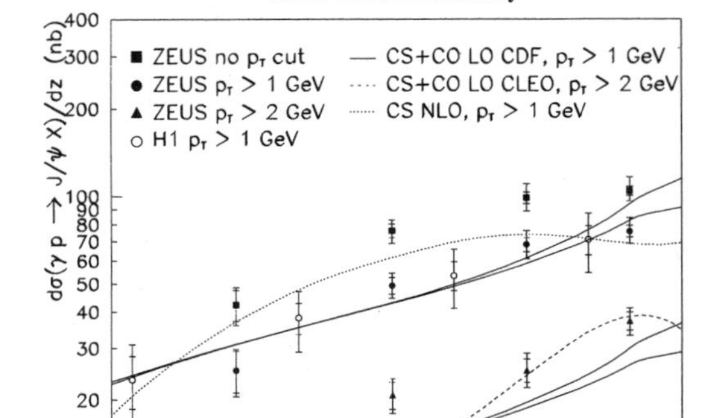

Figure 1. Differential cross–sections $d\sigma/dz$ measured by ZEUS and H1 for three different p_T selections, the theoretical predictions are explained in the text.

4 Conclusions

Accurate measurements of the inelastic J/ψ differential cross–sections have been presented, the HERA data are in good agreement with the direct photon process as computed in the color singlet model framework. First measurements in the low z region, sensitive to the resolved photon process, have also been presented. We also report on measurements of the ψ' to J/ψ inelastic cross–section ratio and on elastic Υ photoproduction. Most of these measurements, still statistically limited, will be improved on the near future by adding more recent data. Much more will come on a longer time scale due to the forthcoming HERA luminosity upgrade, completed by the fall 2001.

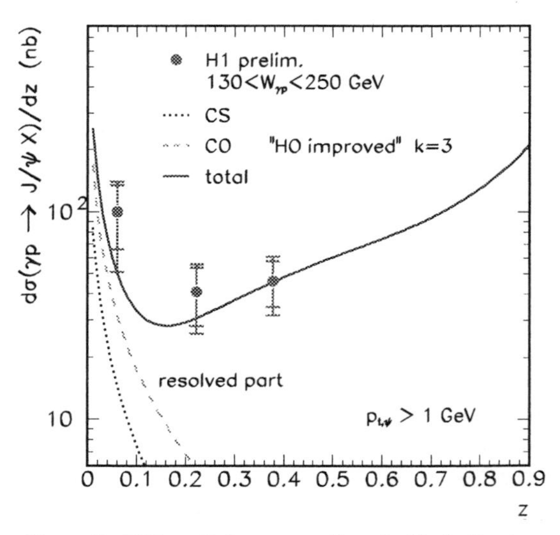

Figure 2. Differential cross–section $d\sigma/dz$ in the low z region measured by H1, the theoretical predictions are explained in the text.

References

1. M. Cacciari and M. Krämer, *Phys. Rev. Lett.* **76** (1996) 4128;
P. Ko J. Lee and H. S. Song, *Phys. Rev.* D **54** (1996) 4312;
B. A. Kniehl et al., *Eur. Phys. J.* C **6** (1999) 493.

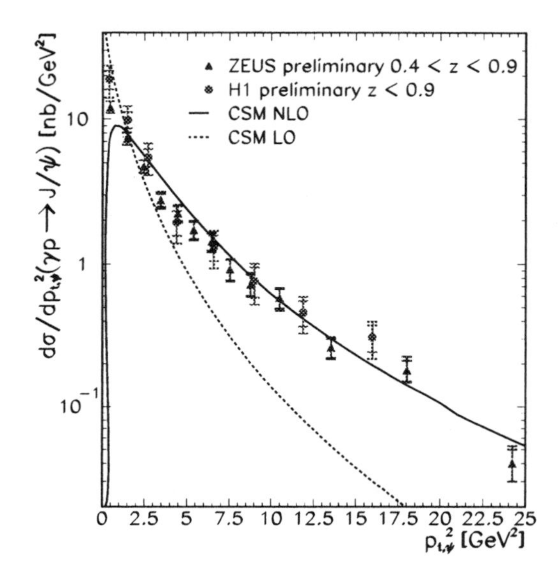

Figure 3. Differential cross-sections $d\sigma/dp_T^2$ measured by ZEUS and H1, the theoretical predictions are explained in the text.

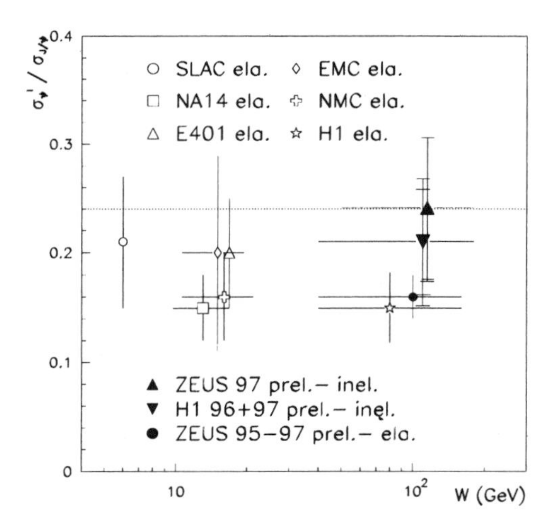

Figure 4. Inelastic ψ' to J/ψ cross-section ratio, measured for the first time by the HERA experiments, together with measurements of the elastic cross-section ratio by ZEUS, H1 and low energy experiments. The horizontal line is explained in the text.

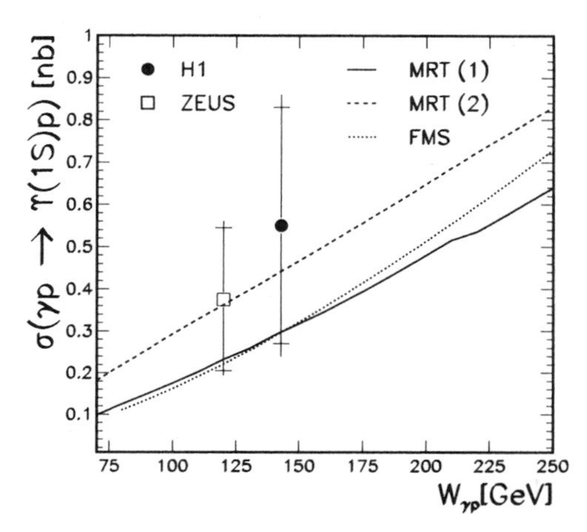

Figure 5. Measurements of the elastic Υ photoproduction cross-section by ZEUS and H1.

2. M. Krämer et al., *Phys. Lett.* B **348** (1995) 657;
M. Krämer, *Nucl. Phys.* B **459** (1996) 3.
3. M. Beneke et al., hep–ph 0001062, to appear in *Phys. Rev.* D.
4. ZEUS Collab., *High Energy Physics 99*, contributed paper 504, Tampere, Finland, July 1999;
H1 Collab., *High Energy Physics 99*, contributed paper 157aj, Tampere, Finland, July 1999;
5. A. Bruni for the ZEUS Collab., *Proceedings of the 6th International Workshop on DIS and QCD*, Brussels, Belgium, April 98, 336;
H1 Collab., C. Adloff et al., *Phys. Lett.* B **421** (1998) 385.
6. ZEUS Collab., J. Breitweg et al., *Phys. Lett.* B **437** (1998) 432; H1 Collab., C. Adloff et al., *Phys. Lett.* B **483** (2000) 23.
7. L. Frankfurt et al., *JHEP* **02** (1999) 2;
A. D. Martin et al., *Phys. Lett.* B **454** (1999) 339.

OPEN CHARM AND BEAUTY PRODUCTION AT HERA

FELIX SEFKOW
(on behalf of the H1 and ZEUS Collaborations)
Physik-Institut der Universität Zürich, Winterthurerstr. 190, CH-8057 Zürich, Switzerland
E-mail: felix.sefkow@desy.de

Selected new results from the H1 and ZEUS collaborations on ep interactions at 300 - 318 GeV centre-of-mass energy are presented. The full pre-upgrade integrated luminosity of HERA of 110 pb^{-1} is used. Charm cross sections are measured up to high values of x_B and Q^2 and are found to be well described by NLO QCD in the 3 flavour scheme. Orbitally excited D mesons are observed; radial excitations are searched for, but are not seen. The first b cross section measurement is confirmed with a lifetime based method, establishing the excess over NLO QCD.

1 Charm

Thanks to the excellent HERA performance the available statistics has strongly increased. ZEUS now has a signal of 27,000 D^* decays in the "golden" mode $D^* \to D^0\pi^+ \to K^-\pi^+\pi^+$. This wealth of data (similarly at hand for H1) allows perturbative QCD to be tested with charm production data in an extended kinematic range and opens the possibility for charm spectroscopy at HERA.

In QCD, heavy quark production in ep interactions predominantly proceeds via boson gluon fusion (BGF), where a quark antiquark pair is created in the interaction of a photon with a gluon in the proton (3 Flavour scheme). At four-momentum transfers much higher than the charm quark mass, $Q^2 \gg m_c^2$, such a description becomes inaccurate, and a treatment in terms of charm densities in the proton may be more adequate.

The single-differential D^* cross sections measured in deep inelastic scattering (DIS) by ZEUS [1] as a function of Bjorken-x and Q^2 now cover a range up to $x_B \simeq 0.1$ and $Q^2 \simeq 1000\,\text{GeV}^2$. (Fig. 1) They are compared with NLO QCD calculations in the 3 Flavour scheme [2], which use as input gluon densities from global fits [3] or a parameterization extracted from scaling violations of the proton structure function F_2, measured at HERA. Good agreement is seen, showing that the BGF picture provides an overall consistent description of charm production and inclusive DIS up to high x_B and Q^2.

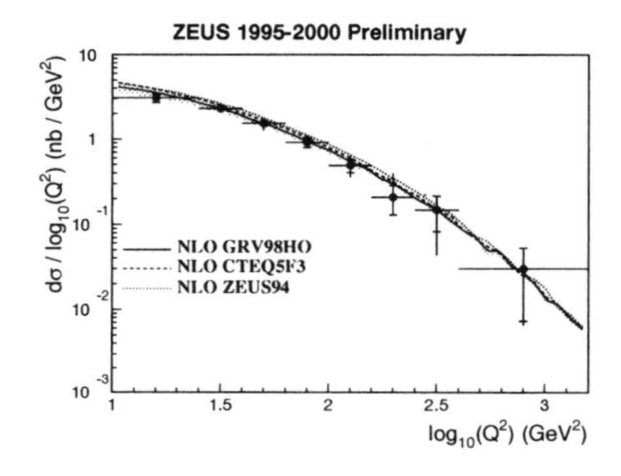

Figure 1. D^* cross section in DIS vs. NLO QCD in the 3 Flavour scheme (shaded: $m_c = 1.3 - 1.6\,\text{GeV}$).

The spectrum of non-strange D mesons is only partially established experimentally. Apart from the lowest mass D and D^* states, the narrow excited P-wave mesons $D_1(2420)$ and $D_2^*(2460)$ have been firmly identified, with spin-parity $J^P = 1^+$ and 2^+. A narrow state interpreted as radially excited $D^{*\prime\pm}$ has been observed by DELPHI [4], but was not confirmed by OPAL and CLEO searches [5].

ZEUS report [6] the observation of orbitally excited D_1^0 and D_2^{*0} mesons in the decay channel $D_J^{(*)0} \to D^{*+}\pi^-$ + c.c. From a fit to the invariant mass (Fig. 2) and π^- helicity angle distributions they extract relative production rates of

$$\frac{D_1^0 \to D^{*+}\pi^-}{D^{*+}} = 3.40 \pm 0.42\,^{+0.78}_{-0.63}\,\%$$

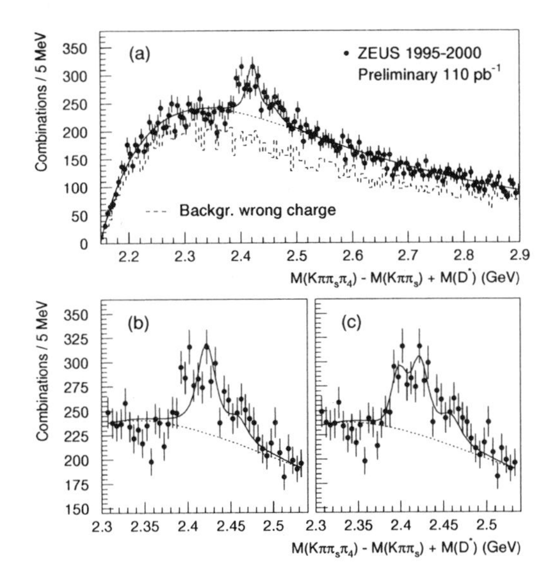

Figure 2. Mass difference distribution for D^{**0} candidates. The curves show fits using mass and helicity angle information.

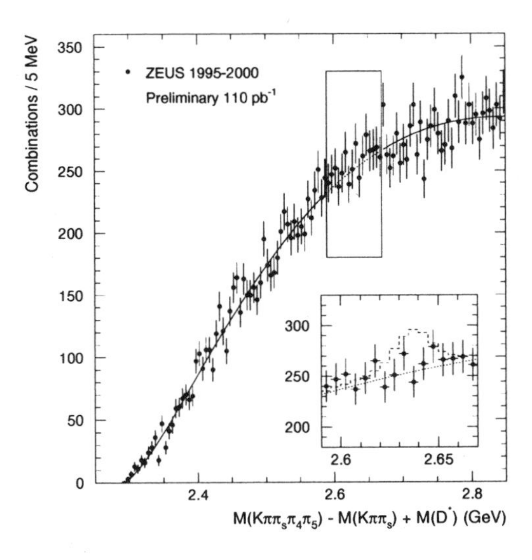

Figure 3. Mass difference distribution for $D^{*\prime\pm}$ candidates. The search region contains 91 ± 75 candidates over the fitted background. The insert shows a Monte Carlo signal normalized to the quoted limit.

$$\frac{D_2^{*0} \to D^{*+}\pi^-}{D^{*+}} = 1.37 \pm 0.40\,^{+0.96}_{-0.33}\,\%$$

A narrow enhancement ($\sim 4\sigma$) is seen at $m_{D^*\pi} = 2398$ MeV and included in the fit (Fig. 2c), but no definite interpretation of this signal is given yet.

Radially excited states are searched for in the channel $D^{*\prime\pm} \to D^{*+}\pi^+\pi^-$ + c.c. No signal is seen (Fig. 3), so that an upper limit is quoted:

$$\frac{D^{*\prime+} \to D^{*+}\pi^+\pi^-}{D^{*+}} < 2.3\,\% \text{ (at 95\% C.L.)}$$

which indicates that the search has a sensitivity corresponding to about the size of the claimed DELPHI signal. Since at HERA almost all charmed mesons originate from prompt charm production, and feed-down from beauty can be neglected, a rather tight limit on $D^{*\prime\pm}$ production in charm fragmentation can be set (at 95% C.L.):

$$f(c \to D^{*\prime+}) \cdot \mathrm{BR}(D^{*\prime+} \to D^{*+}\pi^+\pi^-) < 0.7\%$$

2 Beauty

Beauty production at HERA is suppressed with respect to charm by two orders of magnitude. The measurements so far rely on inclusive semi-leptonic decays, using as signature the high mass of the b quark by observing the transverse momentum p_T^{rel} of the lepton relative to a jet, and also its long lifetime by observing tracks from secondary vertices. The first measurement by H1 [7], using the p_T^{rel} method, revealed a b photoproduction cross section almost a factor of 2 above theoretical prediction [8] (Fig. 4).

The new H1 measurement [9] also uses photoproduction dijet events, where now at least one muon is measured in the two-layer silicon vertex detector. The signed impact parameter δ is determined in the plane transverse to the beam, axis, and the distribution is decomposed by a maximum likelihood fit which adjusts the relative contributions from beauty, charm and fake muons to the sample (Fig. 5). The fit describes the data well and translates into a b cross section that, using

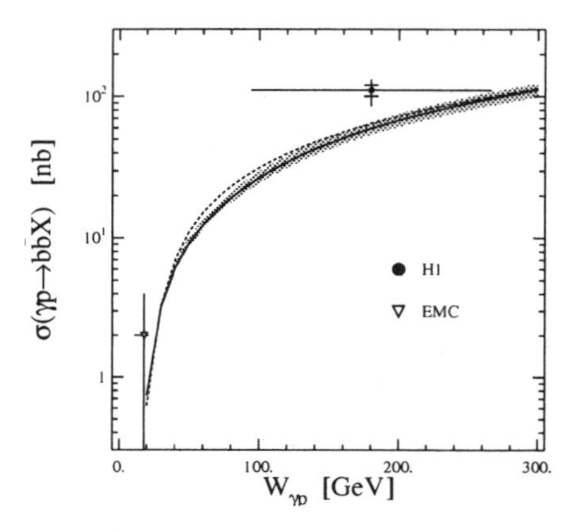

Figure 4. b photoproduction cross section vs. NLO QCD, using different proton structure functions (shaded: scale uncertainty).

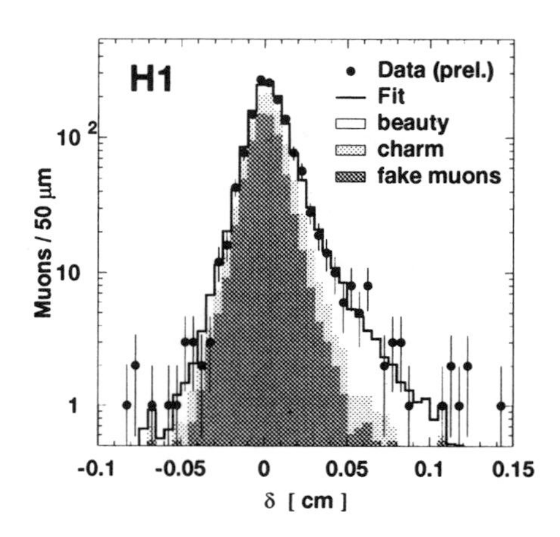

Figure 5. Signed impact parameter distribution for muons. (The sign depends on whether the track intersects the jet axis upstream or downstream (+) of the primary event vertex.)

an independent signature and new data, confirms the published result, based on 1996 data and a different set of cuts. The δ spectrum for a sample with higher b purity, obtained by a cut $p_T^{rel} > 2\,\text{GeV}$ (Fig. 6) agrees well with the prediction from the δ fit to the full sample. Since the two observables are consistent, they can be combined in a likelihood fit of the two-dimensional (δ, p_T^{rel}) distribution. The result, averaged with the published number, is

$$\sigma(ep \to b\bar{b}X \to \mu X) = (170 \pm 25)\,\text{pb}$$

in the range $Q^2 < 1$ GeV^2, $0.1 < y < 0.8$, $p_T(\mu) > 2$GeV, $35^\circ < \theta(\mu) < 130^\circ$. This is higher than the NLO QCD prediction of (104 $\pm$ 17) pb based on [8]. Such a discrepancy between experiment and NLO QCD is now established in both ep and $\bar{p}p$ interactions.

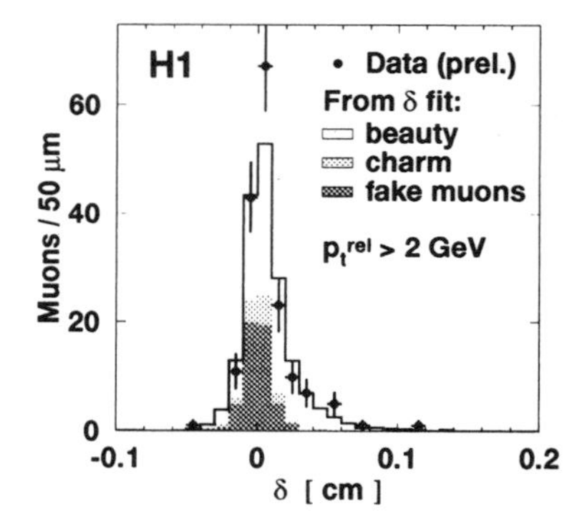

Figure 6. Impact parameter distribution for muons with $p_T^{rel} > 2\,\text{GeV}$. with the absolute prediction from the fit to the full sample.

References

1. ZEUS Coll., *$D^{*\pm}$ Production in Deep Inelastic Scattering*, contrib. paper no. 449.
2. B.W. Harris and J. Smith, *Phys. Rev.* D**57**, 2806 (1998).
3. M. Glueck, E. Reya and A Vogt, hep-ph/9806404;
   H. Lai *et al.*, hep-ph/9903282.
4. DELPHI Coll., *Phys. Lett.* B**426**, 231 (1998).
5. OPAL Coll., contrib. paper to ICHEP 98, OPAL PN 352;
   CLEO Coll., hep-ex/9901008.
6. ZEUS Coll., *Production of P-Wave Charm Mesons at HERA*, contrib. paper no. 448.
7. H1 Coll., *Phys. Lett.* B**467**, 156 (1999).
8. S. Frixione, M.L. Mangano, P. Nason and G. Ridolfi, *Phys. Lett.* B**348**, 633 (1995).
9. H1 Coll., *Measurement of the Beauty Production Cross Section at HERA Using Lifetime Information*, contrib. paper no. 311.

TOP PHYSICS FROM RUN 1 AND RUN 2 PROSPECTS AT CDF

STEVEN R. BLUSK, FOR THE CDF COLLABORATION
University of Rochester, Rochester, New York 14628

We present a summary of top quark physics results from Run 1 at CDF using the Run 1 data sample of 106 pb^{-1}. In addition to the precursory measurements of the top quark mass and $t\bar{t}$ cross section, we have performed a number of other analyses which test the consistency of the $t\bar{t}$ data sample with the standard model (SM). Deviations from SM expectations could provide hints for new physics. We find that the data are consistent with the SM. While the Run 1 data are statistically limited, we have shown that the systematic uncertainties are under control and thus have layed the groundwork for higher precision tests of the SM in Run 2. This report describes the Run 1 top quark analyses and expectations and prospects for top quark measurements in Run 2.

1 Introduction

In $p\bar{p}$ collisions at the Tevatron ($\sqrt{s}$=1.8 TeV), top quark pairs are produced through the strong interaction with an expected cross section (at NLO) of 5.1 pb [1]. Single top quarks are also expected to be produced through a t-W-b electroweak vertex with an expected total cross section of $\approx$1/2 that of $t\bar{t}$ [2]. Within the SM, the top quark is expected to decay with a lifetime of $\approx 10^{-24}$ seconds into a W boson and a b quark. $t\bar{t}$ final states are classified according to the decays modes of the two W bosons. Dilepton final states consist of events where both W bosons decay to an e or μ (BR=5%). Lepton + jets final states include events where one of the W bosons decays leptonically (e or μ) and the other hadronically (BR=30%). The All-Jets mode includes events in which both W-bosons decay hadronically (BR=44%).

2 $t\bar{t}$ Cross Section

Cross section measurements have been made in all three decay channels. In the dilepton channel [3], we observe 9 events with an expected background of 2.5±0.5 events, which leads to a $t\bar{t}$ cross section of $8.2^{+4.4}_{-3.4}$ pb. In the lepton+$\geq$3 jets channel, there are 29 (25) events which are SVX (SLT) tagged with expected backgrounds of 8.1 (13.2) events, leading to a measurement of $5.7^{+1.9}_{-1.5}$ pb [4]. For the All-Jets mode, we measure $7.6^{+3.5}_{-2.7}$ pb [5]. Results from all three channels are combined to obtain a $t\bar{t}$ cross section of $6.5^{+1.7}_{-1.4}$ pb [6] which is within one standard deviation from the theoretical prediction.

3 Top Quark Mass

The most precise measurements in the top quark sector thus far have been in the mass. In the dilepton channel, we use a weighting technique which compares the observed $\not{E}_T$ in each event to the expected value as a function of the assumed top mass. Using a likelihood technique we extract a top mass of M_{top} =167.4±11.4 GeV/c^2 [7]. In the lepton+$\geq$4 jets events, we perform a 2C fit of the final state particles to the decay chain, which results in a measured top mass of 176.1 ± 7.4 GeV/c^2 [8]. Full reconstruction of events in the All-Jets mode is also performed from which we measure M_{top} =186.0±11.5 GeV/c^2 [5]. The result from combining all three measurements is 176.1±6.6 GeV/c^2, roughly 35 times the mass of the next heaviest quark!

4 The $t\bar{t}$ Invariant Mass ($M_{t\bar{t}}$)

The $M_{t\bar{t}}$ analysis [9] proceeds in a similar way to the top mass analysis. To improve the resolution on the four momenta of the final state particles (and thus $M_{t\bar{t}}$), we constrain the top quark mass to 175 GeV/c^2 in the fit. We also require when we remove this constraint that the fitted top quark mass lie in

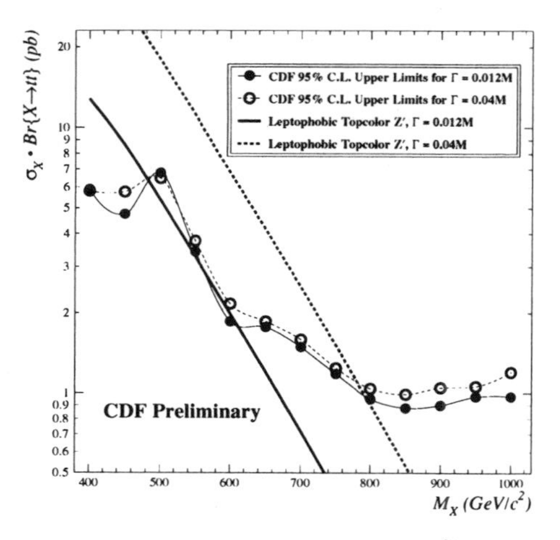

Figure 1. The 95% CL limits on $\sigma_X \dot{B}R(X \to t\bar{t})$ as a function of the mass of X for two different values of the full width of the heavy object. The data are compared to the prediction for a leptophobic topcolor Z' for full widths of 0.012 $M_{Z'}$ and 0.04 $M_{Z'}$.

the range from 150-200 GeV/c^2. The data do not show an excess above the SM prediction, and we therefore present limits on the cross section times branching ratio (see Fig. 1). At the 95% confidence level (CL), the data rule out a topcolor Z' with mass less than 480 (780) GeV/c^2 and natural width equal to 0.012 (0.04) $M_{Z'}$.

5 Top $\mathrm{P_T}$

Like the $M_{t\bar{t}}$ analysis, we use l+4 jet data and constrain the top quarks mass to 175 GeV/c^2. Because of the strong correlation between the top and antitop quarks' $\mathrm{P_T}$, we use only the hadronically decaying top quark. We measure the fraction of top quarks produced in four bins of true $\mathrm{P_T}$: 0-75 GeV/c^2, 75-150 GeV/c^2, 150-225 GeV/c^2, and 225-300 GeV/c^2. First, we determine initial response functions which give the distribution of reconstructed $\mathrm{P_T}$ in each of the four true $\mathrm{P_T}$ bins. The data are then fit to a combination of the four Monte Carlo (MC) reconstructed $\mathrm{P_T}$ distributions using an iterative procedure to minimize the sensitivity of the final result to the initial assumptions of the true top $\mathrm{P_T}$ distribution. Within the limited statistics, the data are consistent with SM expectations. We measure the 95% CL limit for the fraction of top quarks with true $\mathrm{P_T}$ larger than 225 GeV/c to be 0.114.

6 W Helicity in Top Decays

The V-A structure of the t-W-b vertex results in a specific prediction for the W polarization in top decays. At tree level, we expect the fraction of longitudinal W bosons, F_0, to be 70.1±1.6%. The $\mathrm{P_T}$ spectrum of the leading lepton is sensitive to the W polarization. Using MC distributions of longitudinal and left-handed W's, we fit the data to extract the fraction F_0. Using both the lepton+jets and dilepton data samples, we measure $F_0 = 91 \pm 37(stat) \pm 13\%$ [11].

7 Rare Decays

The FCNC decays $t \to Zq$ and $t \to \gamma q$ are strongly suppressed in the S.M. at the level of $\sim 10^{-12}$, and therefore an observation of such events is a signature of new physics. We have performed searches for these decays [12] and find one event in each channel, consistent with background expectations. We therefore set 95% CL limits of 33% and 3.2% respectively for these two FCNC decays.

8 Single Top Production

We have searched for single top in the lepton+jets data. One analysis searches for events in both the W-gluon fusion and the s-channel W^* processes. We select W+1,2,3 jet events which have a SVX b-tag and a top invariant mass, $M_{l\nu b}$ in the range 140 to 210 GeV/c^2. We observe 65 events with an expected background of 62.5±11.5 events. We expect to 4.3 signal events. Fitting the $H_T = \sum \mathrm{E_T}(lepton, \not{E}_T, jets)$ distribution in data to MC signal and background distributions, we extract a cross section limit of 13.5 pb at 95% CL. A second analysis which looks just for the W-gluon fusion process selects W+2 jet events with an SVX tag and the same cut on $M_{l\nu b}$. An interesting and exploitable feature of these events is that, unlike the backgrounds, the product of the leading lepton's charge (Q) and the pseudo-rapidity of the untagged jet (η) peaks at positive $Q \times \eta$. We observe 15 events with an expected background of 12.9±2.1 events (we expect 1.2±0.3 signal events). From a fit of the $Q \times \eta$ distribution in data to MC signal and background distributions, we extract a 95% CL limit of 15.4 pb.

9 Run 2 Expectations

Run 2 will provide ≈40-50 times more $t\bar{t}$ events than Run 1. In addition to a large reduction in statistical uncertainties, systematic uncertainties such as the jet energy scale and MC modelling will also be reduced. For example, the large sample of $Z \rightarrow b\bar{b}$ events can be used to check the $b-jet$ energy scale. The invariant mass of the two untagged jets in double SVX tagged W+4 jet events can be used to check the light quark jet energy scale. A comparison of extra jets in a high purity top sample can be used to put constraints on gluon radiation in the MC simulation. Moreover, we expect to undertake new physics analyses in Run 2, such as studying the spin correlations in $t\bar{t}$ events. Given the size of the Run 2 data sample, we have made projections for the precision we can expect for a variety of measurements. Some of these projections are given in Table 1. Run 2 and a future Run 3 will clearly provide very rich top samples with which to probe the SM and beyond.

Acknowledgments

We thank the Fermilab staff and our CDF collaborators for their vital contributions to these physics analyses.

Table 1. Projections for the expected precision for measurements with an integrated luminosity of 2 fb^{-1}.

Measurement	Precision
M_{top}	1.5%
$t\bar{t}$ cross section	9%
Single top cross section	24%
V_{tb} (from Single top)	13%
F_0	5.5%
$\sigma * BR(X \rightarrow t\bar{t})$	0.1 pb at 1 TeV
$BR(t \rightarrow \gamma c)$	$<2.8\text{x}10^{-3}$
$BR(t \rightarrow Zc)$	$<1.3\text{x}10^{-2}$
$BR(t \rightarrow Hb)$	< 12%

References

1. E. Berger and H. Contopanagos, *Phys. Rev.* D **57**, 253 (1998), *and references therein.*
2. T. Stelzer, Z. Sullivan and S. Willenbrock, *Phys. Rev.* D **58**, 094021 (1998), *and references therein.*
3. F. Abe *et al.*, Phys. Rev. Lett. **80**, 2779 (1998).
4. F. Abe *et al.*, Phys. Rev. Lett. **80**, 2773 (1998).
5. F. Abe *et al.*, Phys. Rev. Lett. **79**, 1992 (1997).
6. F. Ptohos (for the CDF Collaboration), Proceedings of the International Europhysics Conference on High Energy Physics 99, Tampere, Finland, July 17, 1999.
7. F. Abe *et al.*, Phys. Rev. Lett. **82**, 271 (1999).
8. T. Affolder *et al.*, preprint hep-ex/0006028 (submitted to *Phys. Rev.* D); F. Abe *et al.*, Phys. Rev. Lett. **80**, 2767 (1998).
9. T. Affolder *et al.*, preprint hep-ex/0003005 (accepted in Phys. Rev. Lett.)
10. T. Affolder *et al.*, to be submitted to *Phys. Rev. Lett.*
11. T. Affolder *et al.*, *Phys. Rev. Lett.* **84** 216 (2000).
12. T. Affolder *et al.*, *Phys. Rev. Lett.* **80** 2525 (1998).

RECENT RESULTS ON THE TOP QUARK FROM DØ

DHIMAN CHAKRABORTY
(FOR THE DØ COLLABORATION)
Department of Physics and Astronomy, State University of New York at Stony Brook, USA
E-mail: dhiman@fnal.gov

We present results of two studies of the top quark recently completed by the DØ collaboration. They are based on data obtained during the "Run 1" of the Tevatron (1992-96) which delivered about 125 pb^{-1} of $p\bar{p}$ collisions at $\sqrt{s}$ = 1.8 TeV. A search for electroweak production of (single) top quarks puts upper limits of 39 pb on the s−channel process and 58 pb on the t-channel process. A search for charged Higgs bosons in decays of pair-produced top quarks, based on $H^+ \rightarrow \bar{\tau}\nu_\tau$ decays, finds no evidence of signal, and results in the exclusion of a part of the $[m_{H^+}, \tan\beta]$ parameter space.

1 Introduction

At the Tevatron $p\bar{p}$ collider, most top quarks are pair-produced via strong interaction through an intermediate gluon. This was the mode used in its observation[1] and subsequent studies of its properties, including measurements of the $t\bar{t}$ production cross section of 5.9 ± 1.7 pb by the DØ collaboration[2] and $6.5^{+1.7}_{-1.4}$ pb by the CDF collaboration,[3] and of the top quark mass of 174.3 ± 5.1 GeV jointly by the two collaborations.[4] A second production mode is predicted to exist, where top quarks are produced singly through an electroweak Wtb vertex.[5] Measurement of the electroweak production of single top quarks could provide the magnitude of the CKM matrix element V_{tb}, since the cross section is proportional to $|V_{tb}|^2$.

The Higgs sector of the Standard Model (SM) consists of a single complex doublet scalar field which, after spontaneous symmetry breaking, yields a single neutral Higgs boson whose mass is a free parameter of the model. The simplest extension of the SM Higgs sector involves addition of a second doublet and is an integral part of many theories beyond the SM, including supersymmetry. Such two-Higgs-doublet models lead to a total of 5 physical particles, h^0, H^0, A^0, H^+, H^-. According to the SM, a top quark should decay almost exclusively to a W boson and a b quark, i.e., $B(t \rightarrow W^+b) \approx 1$. However, if a charged Higgs with mass $M_{H^+} < m_t - m_b$ exists, then the decay $t \rightarrow H^+b$ could compete with $t \rightarrow W^+b$, depending on M_{H^+} and $\tan\beta$, where $\tan\beta$ is the ratio of the vacuum expectation values of the two doublet fields. Observation of a charged Higgs boson would immediately open doors to physics beyond the SM.

2 Search for single top production

Figure 1 shows the two main modes predicted by the SM for the production of single top quarks at the Tevatron. We denote the s-channel process $q'\bar{q} \rightarrow tb$ shown in Fig. 1(a) and its charge conjugate by "tb", and the t-channel process $q'\bar{q} \rightarrow tqb$, shown in Fig. 1(b) and its charge conjugate by "tqb". For m_t = 175 GeV, the next-to-leading-order (NLO) cross sections for these two processes are 0.73 ± 0.04 pb and 1.70 ± 0.19 pb respectively. The final state consists of two b jets, either zero or one light quark jet, and the decay products of the W boson. We have searched for both modes, with the W boson decaying into $e\nu$ or $\mu\nu$. The data available were 91.9 ± 4.1 pb^{-1} for the e and 88.0 ± 3.9 pb^{-1} for the μ channels.

The analysis[6] relies on identification of a high-p_T e or μ, and of jets, as well as evidence of an escaping neutrino through an imbalance in the transverse component of the

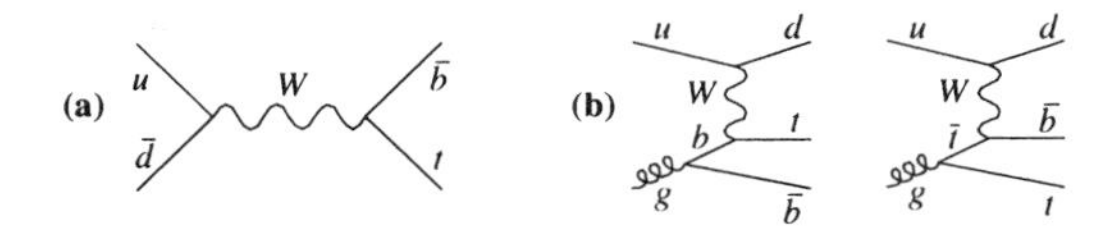

Figure 1. Leading order Feynman diagrams for single top production at the Tevatron: (a) the s-channel, (b) the t-channel.

	e channel	μ channel
	Acceptances (%)	
tb	0.255 ± 0.022	0.122 ± 0.011
tqb	0.168 ± 0.015	0.083 ± 0.008
	Numbers of Events	
tb	0.18 ± 0.03	0.08 ± 0.01
tqb	0.28 ± 0.05	0.13 ± 0.03
W+jets	5.59 ± 0.64	1.12 ± 0.17
QCD	5.92 ± 0.58	0.40 ± 0.09
$t\bar{t}$	1.14 ± 0.35	0.45 ± 0.14
Total Bkgd	12.65 ± 0.93	1.97 ± 0.24
Data	12	5

Table 1. Signal acceptances and numbers of events expected to pass all selection criteria.

total momentum for online and offline selection of signal candidates. At least one of the jets is required to be tagged as a b jet candidate by an associated muon (from semileptonic decay of a b hadron). Cuts are imposed on various linear combinations of the lepton p_T, jet E_T's, and $\not{E}_T$, to further enhance signal relative to the various sources of background. The resultant numbers for expected signal, background (W+jets, QCD, $t\bar{t}$), and the data, are shown in Table 1.

To calculate upper limits on cross sections for the tb and tqb processes, we use a Bayesian approach, with a uniform prior for the signal cross section and a multivariate Gaussian prior for the other quantities. We calculate the likelihood function in each decay channel and combine them to obtain the 95% CL upper limits:

- $\sigma(p\bar{p} \to tb + X) < 39$ pb,
- $\sigma(p\bar{p} \to tqb + X) < 58$ pb.

3 Search for charged Higgs bosons in decays of top quarks

This search[7] is based on the model where the charged Higgs coupling to up- (down-) type quarks and neutral (charged) leptons is directly proportional to the fermion mass, inversely (directly) proportional to $\tan\beta$, and decreases as M_{H^+} increases.[8] The decay $t \to H^+b$ is thus favored for smaller m_{H^+}, and can dominate $t \to W^+b$ if $\tan\beta$ is either larger or smaller than $\sqrt{\frac{m_t}{m_b}}$ by a factor of $\sim$10 (for $M_{H^+} = 100$ GeV). Our search window in the $[m_{H^+}, \tan\beta]$ parameter space, shown in Fig. 2, is bounded in M_{H^+} by the direct lower limit from LEP[9] below and by m_t above, and on either side in $\tan\beta$ by the requirement that the leading order calculations forming the basis of our analysis lend themselves to perturbative treatment. We assume that $B(t \to W^+b) + B(t \to H^+b) = 1$ and that H^+ can decay only to fermion pairs. Then, within the range of M_{H^+} under consideration, $B(H^+ \to c\bar{s}) + B(H^+ \to t^*\bar{b} \to W^+b\bar{b}) \approx 1$ if $\tan\beta < 0.5$, and $B(H^+ \to \bar{\tau}\nu_\tau) \approx 0.96$ if $\tan\beta > 3$. In a previous indirect search,[10] we covered all these three dominant decay modes of H^+ via disappearance of the SM $t\bar{t}$ signal. The present search aims to detect the appearance of $t \to H^+b$ through an excess in the $\tau + \not{E}_T+$ jets final state compared to the SM prediction. Its sensitivity is therefore limited to the region of $\tan\beta > 1$.

The data sample is 62.2 ± 3.1 pb^{-1}. Selection of events proceeds through multijet $+\not{E}_T$ triggers and three steps of offline filtering: preselection after corrections to jets and $\not{E}_T$, an artificial neural network trained to separate signal from background using $\not{E}_T$ in conjunction with angular correlations among the jets, and identification of at least one of

Source	Number of events
SM $t\bar{t}$	1.1 ± 0.3
W+jets (non-$t\bar{t}$)	0.9 ± 0.3
QCD multijet	3.2 ± 1.5
SM total	5.2 ± 1.6
Data	3

Table 2. The numbers of events expected from significant SM processes and that observed in data.

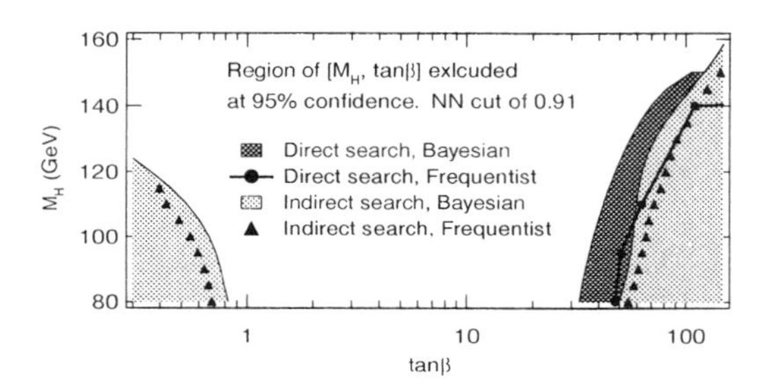

Figure 2. The 95% CL exclusion boundary in the $[m_{H^+}, \tan\beta]$ plane for $m_t = 175$ GeV, and $\sigma(t\bar{t}) = 5.5$ pb, represented by the dark shaded region in addition to the light shaded region on the right.

the jets as a τ decay candidate. The results of the final selection are shown in Table 2.

For the largest values of $B(t \to H^+ b)$ and $B(H^+ \to \bar{\tau}\nu_\tau)$, we expect $\sim 16 \pm 2$ events for $M_{H^+} = 100$ GeV. Since we observe only 3 events (consistent with the SM), we can rule out regions of the parameter space where both $B(t \to H^+ b)$ and $B(H^+ \to \bar{\tau}\nu_\tau)$ are large. A Bayesian interpretation of the results assuming the prior probability density to be uniform in $[m_{H^+}, \log_{10}(\tan\beta)]$ over our search window, and other quantities to have Gaussian distributions, excludes a significant portion of the plane, as shown in Fig. 2. Results of a frequentist interpretation, as well as the results from our previous indirect search are also shown.

4 Summary

Neither our search for electroweak production of single top quarks, nor that for a charged Higgs boson in decays of pair-produced top quarks shows any conclusive evidence of signal, thus leading to upper limits on the rates of the respective processes. These and several other studies that are severely limited by the quantity and the quality of data currently available will benefit greatly from the increases in integrated luminosity, center-of-mass energy, and detector efficiency expected of the upcoming run of the Tevatron collider scheduled to commence in the Spring of 2001.

References

1. F. Abe ***et al.***, (CDF Collaboration) ***Phys. Rev. Lett.*** **74**, 2626 (1995); S. Abachi ***et al.***, (DØ Collaboration) ***Phys. Rev. Lett.*** **74**, 2632 (1995).
2. B. Abbott ***et al.***, (DØ Collaboration) ***Phys. Rev.*** D **60**, 2626 (1999).
3. M. Gallinaro, for the CDF Collaboration, ***The 14th Int. Workshop on High Energy Physics and Quantum Field Theory***, Moscow, Russia (1999).
4. L. Demortier ***et al.***, CDF and DØ Collaborations, ***Fermilab-TM-2084*** (1999).
5. S.S.D. Willenbrock and D.A. Dicus, ***Phys. Rev.*** D **34**, 155 (1986).
6. B. Abbott ***et al.***, (DØ Collaboration) Fermilab-Pub-00/188-E submitted to ***Phys. Rev. Lett.***, and references therein.
7. B. Abbott ***et al.***, (DØ Collaboration) Fermilab-Pub-00/215-E to be submitted to ***Phys. Rev. Lett.***
8. J.F. Gunion, H.E. Haber, G.Kane, and S. Dawson, "The Higgs Hunter's Guide", Addison-Wesley (1990).
9. ALEPH, DELPHI, L3 and OPAL Collaborations, ***Recontres de Moriond***, Les Arcs, France (2000).
10. B. Abbott ***et al.***, (DØ Collaboration) ***Phys. Rev. Lett.*** **82**, 4975 (1999).

$D^0 - \bar{D}^0$ MIXING AND CP VIOLATION FROM FOCUS EXPERIMENT

HWANBAE PARK
Department of Physics, Korea University, Seoul, Korea
E-mail: hbpark@hep.korea.ac.kr

Measurement results on D^0 - $\bar{D}^0$ mixing and CP violation are presented. FOCUS is the fixed target experiment at Fermilab and results are based on a high statistics photo-produced charm sample collected during 96-97 run. We reconstructed more than 1 million charmed particles and compared the lifetimes of two D^0 meson decays to $K^-\pi^+$ and K^-K^+. We obtained a mixing parameter, $y_{\rm cp}$, is $(3.42 \pm 1.39 \pm 0.74)$%. We also searched CP asymmetries in $D^+ \to K^-K^+\pi^+$, $D^0 \to K^-K^+$ and $D^0 \to \pi^-\pi^+$ decay modes. We did not see any evidence of CP violation by comparing the decay rates for particle and antiparticle.

1 Introduction

The FOCUS (Photoproduction of Charm with an Upgraded Spectrometer) is the successor of the E687 experiment[1] with a significantly upgraded spectrometer and designed to study charm physics. Charm particles are produced by the interaction of roughly 180 GeV high energy photons with a segmented beryllium oxide target.

During the 96–97 fixed target run at Fermilab we collected more than 6.3 billion events and reconstructed more than 1 million charm particles in $D \to K\pi, K2\pi$ and $K3\pi$ decay modes. From this sample, we measured the lifetime differences in the D^0 meson system and searched for CP violation in singly Cabibbo suppressed D meson decays. Since the Standard Model predictions for these processes are extremely small, any observation of a signal could be a clear indication of physics beyond the Standard Model.

2 $D^0 - \bar{D}^0$ mixing

In hadronic decays of the neutral charm meson there is an interference term between the mixing and the doubly Cabibbo suppressed paths. By direct comparison of a lifetime difference between weak eigenstates we can search for charm mixing. With assumption of CP is conserved, then

$$y_{\rm cp} = \frac{\tau(D \to K\pi)}{\tau(D \to KK)} - 1 \qquad (1)$$

where $D^0 \to K^-K^+$ is a CP even eigenstate while $D^0 \to K^-\pi^+$ is a mixture state of equal CP even and odd.

For analysis the sample was selected by requiring D^* tagging (*tagged sample*), which satisfy the mass difference between D^* and D is less than 3 MeV around 145.4 MeV or requiring more stringent cuts on particle identification for kaons and pions, momentum asymmetry $(\frac{|P_1-P_2|}{P_1+P_2})$ between two daughter tracks, the resolution in decay proper time and requirement of primary vertex inside the target material (*inclusive sample*). The cuts are chosen not to bias the reduced proper time distribution. The tagged sample has clean signals while the inclusive sample gives larger sample events. After all of cuts applied, we have 119738 signal events in $K\pi$ and 10331 in KK from the combination of two samples. The mass plots for the $K\pi$ and KK candidates used in this analysis are shown in Figure 1. The reflection background coming from $K\pi$ decays in the KK mass distribution are clearly seen and the amount of the reflection is obtained by a mass fit to the signal sample and the reflection mass shape is obtained from a high statistics Monte Carlo sample. We assume that time evolution of the reflection is described by the lifetime of

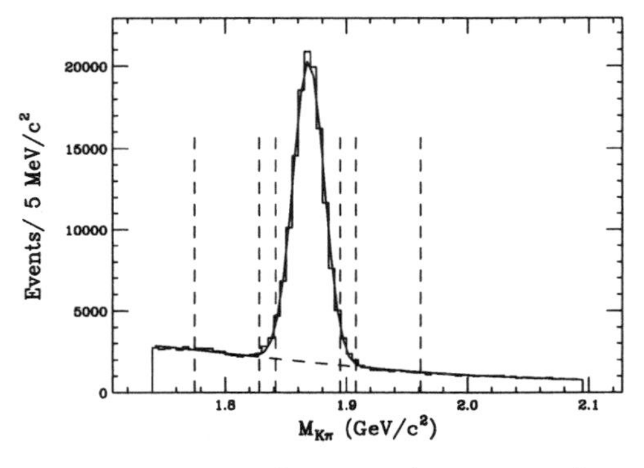

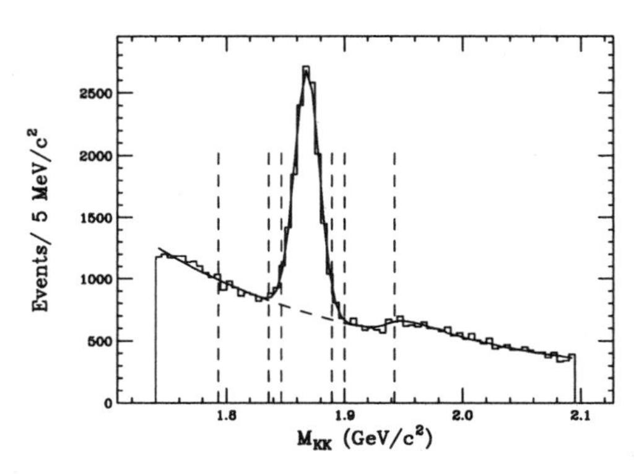

Figure 1. Signal for $D^0 \to K^-\pi^+$ and K^-K^+ with a detachment cut of $\ell/\sigma > 5$. The reflection in the background at higher masses in KK is due to contamination from misidentified $D^0 \to K^-\pi^+$. The yields are 119738 and 10331 for $K^-\pi^+$ and K^-K^+ signal events, respectively. The vertical dashed lines indicate the signal and sideband regions used for the lifetime and $y_{\rm cp}$ fits.

$K\pi$ and fit the reduced proper time distribution of the $K\pi$ and KK samples at the same time. There are four fit parameters such as $K\pi$ lifetime, the lifetime difference between $K\pi$ and KK and two background levels for the $K\pi$ and KK. The signal contributions for the $K^-\pi^+$, K^-K^+ and the reflection from misidentified $D^0 \to K^-\pi^+$ in the reduced proper time histograms are described by $f(t')\exp(-t'/\tau)$ in the fit likelihood. $f(t')$ is a function for any deviation between a pure exponential signal due to acceptance and absorption variation. The background number parameters are either floated or fixed to the number of events in mass sidebands using a Possion term, which ties the background level to that observed in the sidebands, in the fit likelihood. The bin width of the reduced proper time is 200 fs. The fit results to the observed proper time distribution for $K\pi$ and KK are shown in Figure 2. By changing the selection cuts and trying different fitting methods the systematic errors are estimated. We tested the particle identification hypothesis for kaon candidates and the minimum detachment required between primary and secondary vertices. The former affects the level of reflection backgrounds and the latter affects the amount of non-charm backgrounds. Since the results could be affected by various charm reflections which produce curved mass distributions and are therefore not properly subtracted from symmetrically placed sidebands, we check this effect by reducing sideband width by half. We also tried two different options of background handling as stated in the previous paragraph. The differences in fitted $y_{\rm cp}$ is added quadratically. We tried other variations of selection and fitting and found that results are nearly identical to standard fits. We obtained

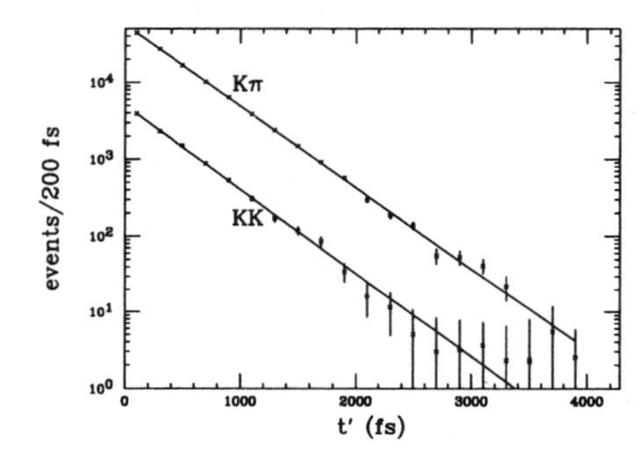

Figure 2. Signal versus reduced proper time for $D^0 \to K^-\pi^+$ and K^-K^+. The fit is over 20 bins of 200 fs bin width. The data is background subtracted and includes the (very small) Monte Carlo correction.

$$y_{\rm cp} = (3.42 \pm 1.39 \pm 0.74)\% \qquad (2)$$

Our result[2] is consistent with that of E791 measurement[3] but sign of our measure-

Table 1. CP asymmetry in D decays.

	$D^+ \to K^-K^+\pi^+$	$D^0 \to K^-K^+$	$D^0 \to \pi^-\pi^+$
FOCUS	$+0.006 \pm 0.011 \pm 0.005$	$-0.001 \pm 0.022 \pm 0.015$	$+0.048 \pm 0.039 \pm 0.025$
E791	-0.014 ± 0.029	$-0.010 \pm 0.049 \pm 0.012$	$-0.049 \pm 0.078 \pm 0.025$

ment is opposite to that of CLEO.[4] The theoretical expectation for the strong phase involved in this process is varied[5] and caution is needed in combining the $y_{\rm cp}$ and the y'(which is a rotational transformation of mixing parameters x and y that depends on a strong phase shift) into one mixing parameter, y.

3 CP violation

It is well known that CP violating effects occur in a decay process only if the decay amplitude is the sum of two different parts, whose phases are made of a weak and a strong contribution. The expected asymmetries are around 10^{-3}. We look at the Cabibbo suppressed decay modes which have the largest branching fractions and select decay modes, $D^+ \to K^-K^+\pi^+$, $D^0 \to K^-K^+$ and $D^0 \to \pi^-\pi^+$. We use the sign of the bachelor pion in the $D^{*\pm}$ decay to tag the neutral D as either a D^0 or a $\bar{D}^0$. In photoproduction fixed target experiment we must account for the different production rates of charm particles and antiparticles. This is done using Cabibbo favored modes $D^0 \to K^-\pi^+$ and $D^+ \to K^-\pi^+\pi^+$. This way also gives an advantage, which most of the corrections due to inefficiencies cancel out by scaling decay rate of singly Cabibbo suppressed decay by Cabibbo favored decay modes. We assume that there is no measurable CP violation in the Cabibbo favored decays. The CP asymmetry can be written as

$$A_{\rm cp} = \frac{\eta(D) - \eta(\bar{D})}{\eta(D) + \eta(\bar{D})} \tag{3}$$

where η is (considering for example the decay mode $D^0 \to K^-K^+$)

$$\eta(D) = \frac{N(D^0 \to K^-K^+)}{N(D^0 \to K^-\pi^+)} \tag{4}$$

and $N(D^0 \to K^-K^+)$ is the efficiency corrected number of candidate decays. To tag the favor of the neutral D meson D^* tagging is required and it gives the statistical error in the neutral decay modes are larger than those of charged decay mode. Our measurements[6] are summarized in the Table 1 and results from E791 experiment[7] are also presented. FOCUS measurement is 2–3 times better than the the previous measurements by the E791. We found no evidence for CP violation from the singly Cabibbo suppressed decay modes.

References

1. P.L. Frabetti *et al.*, Nucl. Instr. Meth. A320, 519 (1992).
2. J.M. Link *et al.*, Phys. Lett. B**485**, 62(2000).
3. E.M. Aitala *et al.*, Phys. Rev. Lett. **83**, 32 (1999).
4. R. Godang *et al.*, Phys. Rev. Lett. **84**, 5038 (2000).
5. S. Bergmann, Y. Grossman, Z. Ligeti, Y. Nir and A.A. Retrov, [hep-ph/0005181].
6. J.M. Link *et al.*, [hep-ex/0005073].
7. E.M. Aitala *et al.*, Phys. Lett. B**403**, 377 (1997); Phys. Lett. B**83**, 32 (1999).

RECENT RESULTS FROM SELEX

J. S. RUSS

Carnegie Mellon University, Pittsburgh, PA 15213 USA
on behalf of the SELEX Collaboration
E-mail: russ@cmphys.phys..cmu.edu

The SELEX experiment (E781) is 3-stage magnetic spectrometer for the study of charm hadroproduction at large x_F using 600 Gev Σ^-, π^- and p beams. New precise measurements of the Λ_c, D^0, and D_s lifetimes are presented. We also report results on Λ_c and D_s production by Σ^-, π^- and p beams at $x_F > 0.2$. The data agree with expectations from color-drag models to explain charm particle/antiparticle production asymmetries.

1 Introduction

Charm physics explores QCD phenomenology in both perturbative and nonperturbative regimes. Charm lifetime measurements test models based on $1/M_Q$ QCD expansions and evaluate corrections from non-spectator W-annihilation and Pauli interference to perturbative QCD matrix elements. Production studies test leading order (LO) and next to leading order (NLO) perturbative QCD. The parton-level processes are symmetric between c and $\bar{c}$, but fixed-target data on hadrons show significant asymmetries. Experimental data from different incident hadrons (π, p and Σ^-) may help to illuminate hadron-scale physics. [1]

2 Features of the Selex spectrometer

The SELEX experiment at Fermilab is a 3-stage magnetic spectrometer [2]. The negative Fermilab Hyperon Beam at 600 GeV had about equal fluxes of π^- and Σ^-. The positive beam was 92% protons. For charm momenta in a range of 100-500 Gev/c mass resolution is constant and primary (secondary) vertex resolution is typically 270 (560) μm. A RICH detector labelled all particles above 25 GeV/c, greatly reducing background in charm analyses. [3] Our charm analysis requires: (i) primary/secondary vertex separation $L \geq 8\sigma$ (σ is the combined error); (ii) the decay tracks extrapolated to the primary vertex z position to miss such that the second-largest transverse miss distance $\geq 20\mu$m; (iii) the secondary vertex to lie outside any charm target by at least 0.5 mm; (iv) decays to occur within a given fiducial region; and (v) proton and kaon tracks to be identified by the RICH.

3 Measurements of the Λ_c, D^0 and D_s lifetimes

The lifetime hierarchy for the charm system presents a challenge to HQET and pQCD methodologies due to the low charm quark mass. We report here lifetimes for the following charm hadrons in the decay modes listed: (i) $\Lambda_c^+ \to pK^-\pi^+$; (ii) $D^0 \to K^-\pi^+$ and $K^-\pi^+\pi^-\pi^+$ + c.c.; and (iii) $D_s \to K^*(892)K$ and $\phi\pi$.

SELEX charm signals are extracted by the sideband subtraction method with a fixed signal region ($\pm 2.5\sigma_M$, i.e., 20 MeV/c^2) centered on the charm mass. Sidebands of equal width are defined above and below the charm mass region. The background under the charm peak is the average of the two sideband regions.

π/K misidentification causes mixing of charm signals. In SELEX this is significant only for the D_s peak. For both D_s modes kaon momenta are $\leq$160 Gev/c to reduce misidentification. Any KKπ event having a

Table 1. PRELIMINARY SELEX lifetimes (fs)

Charm Particle	τ	σ_{stat}	σ_{syst}
Λ_c^+	198.1	7.0	5.5
D^0	407.0	6.0	4.3
D_s	475.6	17.5	4.4

pseudo-$D^{\pm}$ mass in the interval 1867 $\pm$ 20 Mev/c^2 is removed to eliminate an artificial lengthening of the D_s lifetime, even though some of these are real D_s events.

We make a binned maximum likelihood fit simultaneously to signal and sideband regions in reduced proper time $t^* = M(L - 8\sigma)/pc$. SELEX results are given in Table 1. The D_s/D^0 lifetime ratio R, sensitive to the W-annihilation amplitude in the weak decay, is 1.17 $\pm$ 0.057, consistent with other recent published results.

4 Charm and Anticharm Hadroproduction

For Σ^- and proton beams (valence q, not $\overline{\text{q}}$) ${\Lambda_c}^+$ production is favored over $\overline{\Lambda_c}^-$ at all x_F, and the difference increases dramatically at large x_F. The efficiency difference for charm and anticharm hadrons in SELEX is small, at most 3%. π^- production (valence q and $\overline{\text{q}}$) shows a small asymmetry that becomes consistent with zero at large x_F. The data are shown in Fig. 1. For D_s the Σ^- beam produces far more D_s^- (s-quark) than D_s^+ ($\overline{s}$ quark), with the difference increasing at large x_F. For pions, the yield is smaller and the integral yield difference between D_s^- and D_s^+ for $x_F \geq 0.2$ is consistent with zero. These asymmetry patterns suggest a connection between a charm or anticharm quark and remnants of the beam fragmentation jet, like in the quark-gluon string model. There is no evidence in the large-x_F events for a diffractive partner to the leading charm hadron, again consistent with a color-drag picture.

The p_T distributions for Λ_c^+ production from the three different beam hadrons have

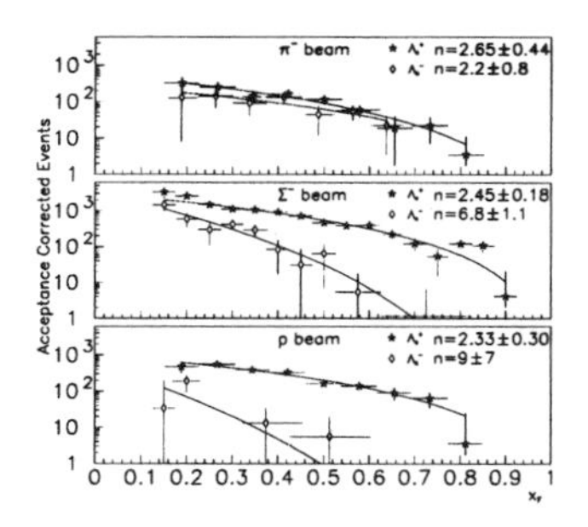

Figure 1. x_F dependence of Λ_c^+ and $\overline{\Lambda_c}^-$ production by different beams

identical gaussian distributions out to $p_T^2 \leq 2$ (GeV/c)^2, after which the good-statistics Σ^- data show a change to power-law behavior, as is expected from QCD and has been observed in D-meson production by π^-.

5 Summary

SELEX has explored charm hadroproduction in the large x_F region using different beams. Results favor the color-drag explanation of production asymmetry. We used the data to measure preliminary charm lifetimes: $\tau(\Lambda_c) = 198.1 \pm 7.0 \pm 5.5$ fs, $\tau(D^0) = 407.0 \pm 6.0 \pm 4.3$ fs, and $\tau(D_s)$ of $475.6 \pm 17.5 \pm 4.4$ fs. The SELEX result for the ratio τ_{D_s}/τ_{D^0} is 1.17 $\pm$ 0.057, 3.3σ from unity.

References

1. S.Frixione,M. Mangano, P. Nason and G. Ridolfi "Heavy quark production" in Heavy Flavour II", A. Buras and M. Lindner eds. World Scientific Publishing Singapore 1997; I.Bigi and N. Uraltsev, Z. Physik C 62 (1994) 623;
2. J. Russ, *et al.* , *in Proceedings of the 29th International Conference on High Energy Physics,* edited by A. Astbury *et al.* (World Scientific, Singapore, 1999) VolII, p. 1259;hep-ex/9812031.
3. J. Engelfried, *et al.*, Nucl.Instr.and Methods **A431**, 53, 1999

FIRST RESULTS OF E835 2000 DATA RUN

KEITH E GOLLWITZER
For the E835 Collaboration
Fermi National Accelerator Laboratory, PO Box 500, Batavia IL 60510,USA
E-mail: gollwitz@fnal.gov

During 2000, E835 has been taking data which include scans of the triplet P states of charmonium, χ_{c2}, χ_{c1} and χ_{c0}. Preliminary analyses of the reactions $\bar{p}p \to \chi_{c2}, \chi_{c1}, \chi_{c0} \to \gamma J/\psi \to \gamma(e^+e^-)$ have been performed. The preliminary results for mass, Γ and $\Gamma_{\bar{p}p}/\Gamma$ for χ_{c2}, χ_{c1} and χ_{c0} are presented and compared with previous measurements done by E760 in 1990 and E835 in 1997.

1 Introduction

Fermilab experiment E835 has taken data in 2000 continuing the study of charmonium produced in antiproton-proton annihilations. As in the earlier Fermilab charmonium experiments, E760 (1990-1991) and E835 (1996-1997), precision spectroscopic measurements have been made using the stochastically cooled circulating antiproton beam ($\Delta p/p \approx 10^{-4}$) intersecting a molecular hydrogen gas jet target for luminosities of 1-4 $\times 10^{-31}$ cm^{-2} s^{-1}. The E760/E835 method of stepping the beam energy through a resonance and using the detector to identify electromagnetic decays of the charmonium states allows us to effectively map out the excitation curve (the convolution of the beam energy distribution and the Breit-Wigner resonance).

The electromagnetic decays are used to identify charmonium candidate events in the presence of the large hadronic background ($\approx$ 60 mb). One of the best signatures is when there is a transition from a charmonium resonance to J/ψ, which in turn decays to e^+e^-. The e^+e^- kinematics from a J/ψ decay are easy to trigger upon and can be quickly analyzed. We report preliminary results of the triplet P states through the study of the reactions

$$\bar{p}p \to \chi_{c2}, \chi_{c1}, \chi_{c0} \to \gamma J/\psi \to \gamma(e^+e^-), \quad (1)$$

where the resonances have been scanned separately.

2 Detector

The E835 detector[1] is non-magnetic and has large acceptance, covering the complete azimuth (ϕ) and polar angles (θ) from 3° to 70°. The cylindrical central part of the detector consists of three scintillator hodoscopes (ϕ segmented into 8, 24 and 32 elements), two 2-layer sets of straw tubes (ϕ measurement), two 2-layer scintillating fiber detectors (θ measurement), a 16-cell threshold Čerenkov counter and a 1280 lead-glass array (20θ by 64ϕ) forming the central electromagnetic calorimeter (CCAL). In the forward direction there are an 8 element scintillator hodoscope and a planar electromagnetic calorimeter (FCAL) consisting of 144 lead-glass counters. The luminosity is measured by detecting recoil protons from the jet target which have been elastically scattered by the antiproton beam at 86.5° to the beam direction.

3 Event Selection

We require that the final state e^+e^- of reactions 1 to be within the electron/pion discriminating Čherenkov counter: $15° < \theta < 65°$. In the trigger, a *candidate track* is defined by coincidence between the azimuthally corresponding elements of two of the central hodoscopes and a cell of the Čherenkov. In addition, summed CCAL counters signals are formed and sums above θ-dependent

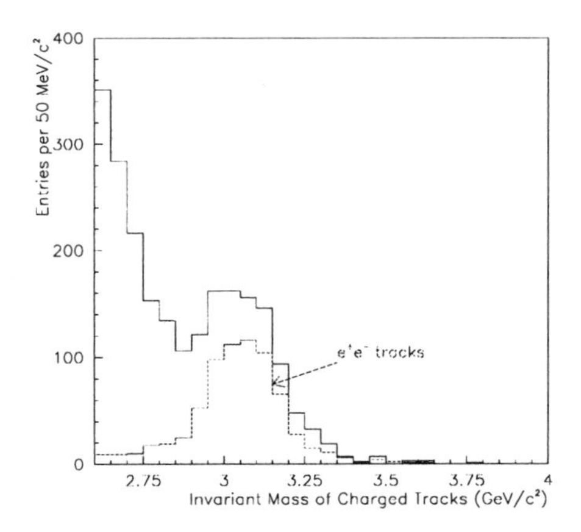

Figure 1. Invariant mass for all pairs of charged tracks (open) and identified e^+e^- tracks (shaded). Data are from the χ_{c1} scan.

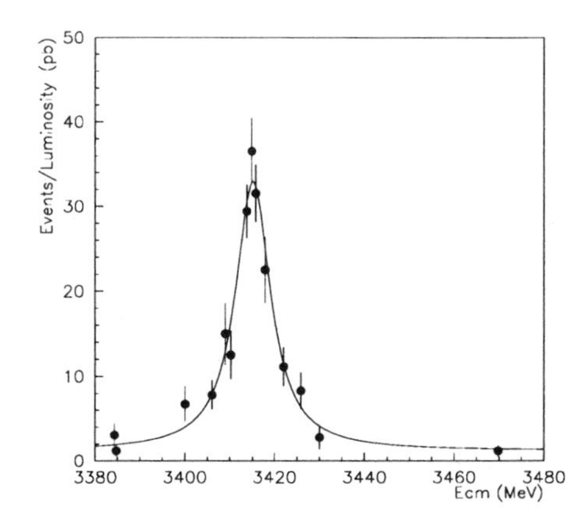

Figure 2. Candidate events per pb^{-1} for the χ_{c0} scan.

thresholds are defined as *energetic particles.* The overall trigger for reactions 1 requires that there are at least two candidate tracks and, independently, at least two well separated energetic particles; the latter effectively requires an invariant mass greater than 1.8 GeV/c^2 in the CCAL.

The first step of the analysis is to associate each candidate track with an energetic particle in the CCAL; the open histogram in Figure 1 shows the invariant mass of pairs of associated charged tracks. For each associated charged track, further cuts on the pulse heights from each scintillator counter and Čherenkov cell are performed to remove charged hadrons and photon conversion pairs. In addition, the characteristics of the CCAL signals (shape analysis) are used to ensure that only one electromagnetic shower contributed to each energy deposit. The invariant masses of events with two associated tracks that satisfy these criteria are shown as the shaded plot in Figure 1.

A third, single electromagnetic shower energy deposit, in either the CCAL or FCAL, is assigned to be the the transition photon of reactions 1. A $5C$ kinematical fit is performed to select the final candidate events.

4 Preliminary Results

The number of candidate events per pb^{-1} at the peaks of the resonances ranges from 35 to 600 while the background is $\approx$ 2 events per pb^{-1}. In Figure 2, the χ_{c0} scan data and excitation curve are shown. The overall acceptance-efficiency is estimated to be 0.3. The data are then fit to excitation curves, Breit-Wigner resonances convoluted with the beam energy distributions, and a flat background.

The preliminary results for the mass, Γ and $\Gamma_{\bar{p}p}/\Gamma$ for the charmonium triplet P states are given in Table 1; all errors are statistical. The previous results of E760[2] and E835[3] are included for comparisons. In addition, the integrated luminosities used for performing each scan are given.

5 Discussion

All of the preliminary results are in agreement with the previous measurements. The χ_{c2} and χ_{c1} scans are statistically the same as the previous measurements and possibly later in the running in 2000 further scan statistics may be collected. The increased amount of integrated luminosity expended to the χ_{c0}

Table 1. Comparison of preliminary results with previous E760 and E835 results.

	Experiment	$\int L\ dt$ (pb^{-1})	Mass (MeV/c^2)	Γ (MeV)	$\Gamma_{\bar{p}p}/\Gamma$ (10^{-4})
χ_{c2}	E835(2000 prelim.)	1.115	3555.9 $\pm$ 0.1	2.3 $\pm$ 0.2	0.89 $\pm$ 0.08
	E760(1990)	1.160	3556.15 $\pm$ 0.07	1.98 $\pm$ 0.17	1.00 $\pm$ 0.08
χ_{c1}	E835(2000 prelim.)	1.330	3510.64 $\pm$ 0.03	0.94 $\pm$ 0.08	0.87 $\pm$ 0.08
	E760(1990)	1.030	3510.53 $\pm$ 0.04	0.88 $\pm$ 0.11	0.86 $\pm$ 0.10
χ_{c0}	E835(2000 prelim.)	29.289	3415.2 $\pm$ 0.4	9.4 $\pm$ 1.1	5.5 $\pm$ 0.4
	E835(1997)	3.522	3417.4 $\pm$ 1.9	16.6 $\pm$ 5.2	4.8 $\pm$ 0.9

scan has decreased the statistical errors significantly.

In 1997, the χ_{c0} resonance was near an unstable operating beam energy of the Accumulator. For the current data taking, the Accumulator lattice has been changed such that the unstable operation point has been moved away from the χ_{c0}.

The $\Gamma_{\bar{p}p}/\Gamma$ for χ_{c2} and χ_{c1} are nearly the same and it is surprising that the same quantity for χ_{c0} is larger. All E760/E835 measurements for the $\chi_c \to \bar{p}p$ branching fractions are dependent upon the accepted values for the branching fractions[4] $\chi_c \to \gamma J/\psi$. These branching fractions for χ_{c2} and χ_{c1} are known to better than 8% while the χ_{c0} radiative transition is known to 27%. It should be noted that the BES collaboration[5] has reported the χ_{c0} branching fraction to $\bar{p}p$ to be larger than the χ_{c2} and χ_{c1}. However, BES measurements for all of the triplet P state branching fractions to $\bar{p}p$ are smaller by nearly a factor of two than the E760 and E835 measurements.

E835 will be analyzing other final states from the decays of the triplet P states. In addition, E835 has taken, and will be taking, data at the ψ' and the h_c (singlet P wave state of charmonium, ^{1}P$_1$).

Acknowledgments

I would like to thank my colleagues from Fermilab, INFN & University of Ferrara, INFN & University of Genova, University of California at Irvine, University of Minnesota, Northwestern University and INFN & University of Torino for their collaborative effort in the experiment's preparation, data taking, analyses and help in preparing this talk and paper. We would like to thank the support staffs of the participating institutions and in particular to the effort of the operation of the Accumulator by the Fermilab Antiproton Source Department and the Beams Division. This work is supported by the U. S. Department of Energy and the Italian Istituto Nazionale di Fisica Nucleare.

References

1. M. Ambrogiani *et al.* [E835 Collaboration], Phys. Rev. **D60**, 032002 (1999).
2. T. A. Armstrong *et al.* [E760 Collaboration], Nucl. Phys. **B373**, 35 (1992).
3. M. Ambrogiani *et al.* [E835 Collaboration], Phys. Rev. Lett. **83**, 2902 (1999).
4. D. E. Groomi *et al.* Eur. Phys. J. **C15**, 1 (2000).
5. J. Z. Bai *et al.* [BES Collaboration], Phys. Rev. Lett. **81**, 3091 (1998).

CLEO RESULTS ON HEAVY MESON MIXING

HARRY N. NELSON

Physics Department, University of California, Broida Hall (Bldg. 572), Santa Barbara, CA 93106-9530, USA

E-mail: hnn@hep.ucsb.edu

We discuss recent CLEO results on $D^0 - \overline{D}^0$ and $B_d^0 - \overline{B}_d^0$ mixing. The principal results are that for the D^0 system, allowing for CP violations, the mixing amplitude $x' < 2.9\%$ (95% C.L.), and for the B_d^0 system, $\chi = 0.198 \pm 0.013 \pm 0.014$. We make projections for future sensistivity to $D^0 - \overline{D}^0$ mixing, and to $\sin(2\beta + \gamma)$.

1 Introduction

The $D^0 - \overline{D}^0$ system is unlike other systems that mix, such as $K^0 - \overline{K}^0$, $B_d^0 - \overline{B}_d^0$, and $B_s^0 - \overline{B}_s^0$ in at least two respects: first, the Standard Model contributions are thought to be extremely small, so non-Standard contributions might be obvious; second, $D^0 - \overline{D}^0$ is the only system that consists of up type quarks. New physics that differentiates between up and down type quarks could be revealed by study of D^0–$\overline{D}^0$. Indeed, there are numerous models, with relevant particles as massive as 100 TeV, that predict large D^0–$\overline{D}^0$ mixing.[1]

We describe recent results from CLEO II.V on $D^0 - \overline{D}^0$ mixing, where we use the sequence $D^{*+} \to D^0 \pi_s^+$, where the charged pion is 'slow' in momentum, followed by the appearence of a $K^+\pi^-$ final state, and the sequence formed by application of charge conjugations. We also describe results on B_d^0–$\overline{B}_d^0$ mixing at the $\Upsilon(4s)$. Following the decay of the $\Upsilon(4s)$ to $B_d^0 \overline{B}_d^0$, we tag one B_d^0 with a semileptonic decay, and the other by a partial reconstruction of the exclusive final state $D^{*+}\pi^-$.

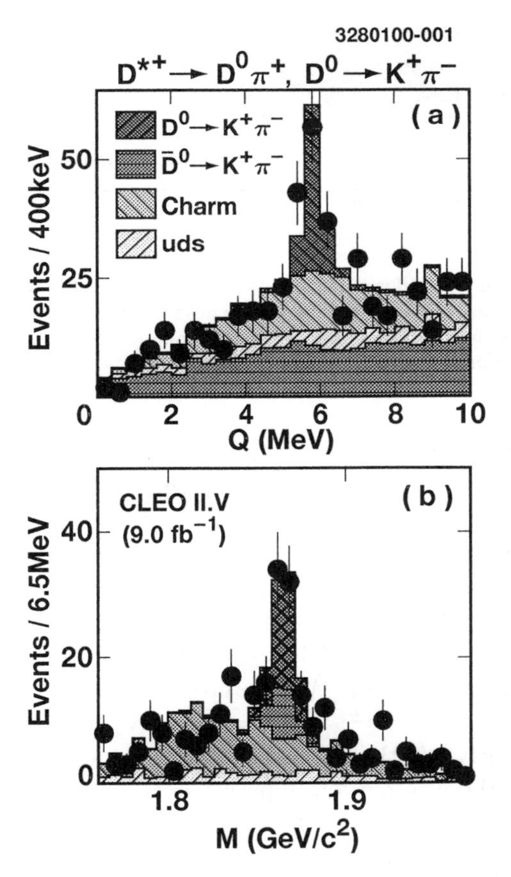

Figure 1. Signal for the 'wrong sign' decay $D^0 \to K^+\pi^-$, projected onto Q (a) and M (b). The fit for the signal and various backgrounds are given by the hatched and colored histograms.

2 D^0-$\overline{D}^0$ Mixing

The final configuration of the CLEO II detector, known as CLEO II.V, took 9.0 fb^{-1} of e^+e^- collisions between 1996 and 1999.

CLEO II.V featured a vertex detector

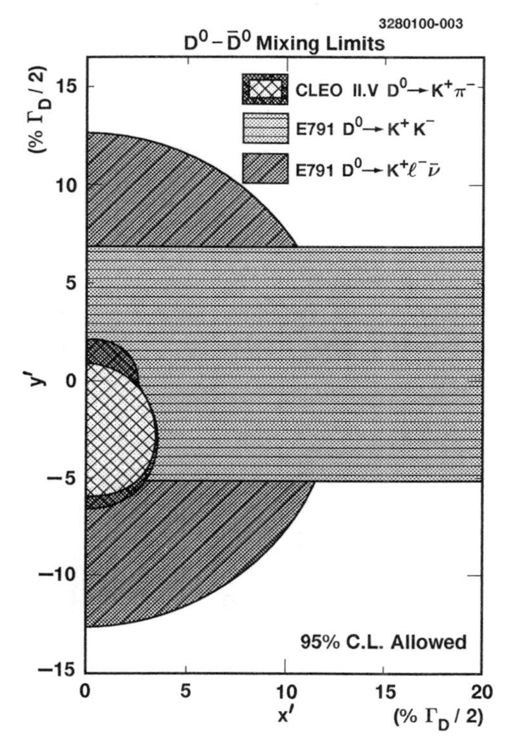

Figure 2. Allowed regions for the mixing amplitudes, x' and y', at the 95% C.L. Those nearest the origin are from this work, where the inner(outer) region requires(does not require) CP conservation.

(SVX) consisting of three layers of double-sided silicon, and helium as the drift chamber gas. The measurement of z by the SVX narrowed the resolution in Q, the energy released in the $D^{*+} \to D^0\pi^+$ decay to $\sigma_Q = 190\,\text{KeV}$, or about 1/4 that obtained in earlier CLEO work. The use of helium narrowed the resolution in M, the mass reconstructed in $D^0 \to K^{\mp}\pi^{\pm}$, to $\sigma_M = 6.4\,\text{MeV}$, which is nearly 1/2 that of earlier CLEO work. Both of these resolution improvements are important for detecting the signal of $D^0 \to K^+\pi^-$, which is shown in Fig. 1.

The fit in Fig. 1 indicates $44.8^{+9.7}_{-8.7}$ signal events.[2] The rate of 'wrong-sign' decay, relative to 'right-sign decay' is $0.332^{+0.063}_{-0.065} \pm 0.040\,\%$, which is close to $\tan^4\theta_C$.

We analyze the decay times of the events in the signal region to deduce results on the $D^0 \to \overline{D}^0$ mixing amplitudes. The normalized amplitude $x(y)$ describes transitions through off(on)-shell intermediate states. The existence of a direct decay amplitude for $D^0 \to K^+\pi^-$ complicates the analysis. The direct decay contributes a purely exponential distribution of decay times. The direct decay amplitude might have a strong phase shift δ relative to the favored decay amplitude, $\overline{D}^0 \to K^+\pi^-$, and so the interference of mixing and direct decay contributes decay times according to the distribution $y'te^{-t}$, where $y' = y\cos\delta - x\sin\delta$. Pure mixing then contributes decay times according to the distribution $(1/4)(x^2+y^2)t^2e^{-t}$.

Fits to our data result in the allowed regions in Fig. 2. In our principal results we allow CP violation simultaneously in all three terms of the time evolution: in direct decay, interference, and in mixing.

New physics would most probably appear in the amplitude x. When $\delta = 0$, $x' = x$, and our limit is $x < 2.9\%$ at 95% C.L. At roughly $x \sim 1.0\%$, D^0–$\overline{D}^0$ mixing would surpass K^0–$\overline{K}^0$ mixing as the most tight constraint on flavor changing neutral currents.

Our technique is now limited by wrong-sign 'background' from the direct decay. We then predict that for future work at the B-factories, this technique will give new sensitivity only as the one-quarter power of the integrated luminosity, and will reach about $x' = 0.7\%$ at 1000 fb^{-1}. In contrast, there are techniques that might be background-free at the ψ'', and so the scaling with luminosity would go as the one-half power, and hit about $x = 0.1\%$ at 1000 fb^{-1}.

3 B^0_d-$\overline{B}^0_d$ Mixing

At CESR, the B's do not move sufficiently to allow the measurement of decay times. Thus, we are sensitive only to the effect of mixing after integration over the decay time variables, in particular, $\chi = \Gamma(B^0_d \to \overline{B}^0_d)/[\Gamma(B^0_d \to B^0_d) + \Gamma(B^0_d \to \overline{B}^0_d)] \approx (x^2 +$

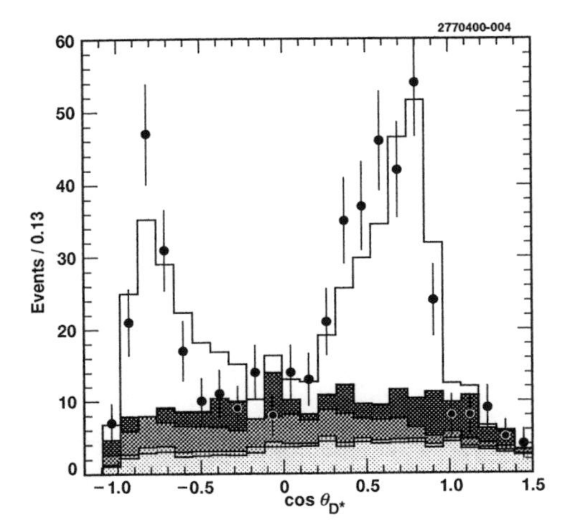

Figure 3. Like sign events, which contain the B mixing signal, as a function of the decay angle of the D^{*+}, $\cos\theta^*$. The data are the points, and the fit results are full histograms, with backgrounds colored.

$y^2)/2/(1+x^2)$, where x and y were described earlier, but in this case are for the $B_d^0-\overline{B}_d^0$ system.

We look for the process $\Upsilon(4s)\to B_d^0\overline{B}_d^0\to \overline{B}_d^0\overline{B}_d^0$, followed by one $\overline{B}_d^0\to X\ell^-\overline{\nu}_\ell$, and the other $\overline{B}_d^0\to D^{*+}h_W^-$, and the processes obtained by charge conjugations. The hadron from the W^-, h_W^-, can be either a π^- or a ρ^-, and is fully reconstructed. We reconstruct only the slow pion from $D^{*+}\to D^0\pi_s^+$; there is sufficient information to reconstruct all of the decay kinematics with zero constraints.[3]

The complete CLEO-II data set of 9.1 fb^{-1} taken on the $\Upsilon(4s)$ resonance is used to measure the mixing signal, and 4.4 fb^{-1} taken off-resonance is used to estimate various backgrounds. We observe 458 mixed or 'like sign' events ($\ell^\pm h_W^\pm$), shown in Fig. 3, and 1524 unmixed or 'unlike sign' events ($\ell^\pm h_W^\mp$).[4]

After correction, we find $\chi = 0.198 \pm 0.013\pm 0.014$. The principal contributions to the systematic error come from cases where the $\overline{B}_d^0\to D^{*+}h_W^-$ decay is a mis-tagged B_d^0 decay (0.009), charged B background (0.007), and uncertainty over two body background (0.006).

If we assume that $|y|\ll x$, as is theoretically expected, we can conclude that $\Delta m = 0.523\pm 0.029\pm 0.031\text{ps}^{-1}$. Comparison of the charge states of the like sign events allows us to restrict CP violation in B_d^0 state mixing by $|\text{Re}(\epsilon_B)|<3.4\%$, 95% C.L.

At an accelerator where the B decay times can be measured, the events used here to measure mixing can be used to measure $\sin(2\beta+\gamma)$.[5] The work in Ref. 5 omitted a form factor suppression in the path through Cabibbo-suppressed decay, and so resulted in an optimistic projection. Using the results here, and with improvements expected at the B-factories from better tagging and use of decay time dependence, we can project that an error on $\sin(2\beta+\gamma)$ of about 1/3 can be reached with 200 fb^{-1}.

Acknowledgments

This work is based on the Ph. D. dissertations of David Asner and Andrew Foland. I would like to thank the conference organizers for their delightful conference, and particularly Taku Yamanaka for his extra attention, and the help desk for guiding me to a bicycle shop.

References

1. H.N. Nelson, hep-ex/9908021 (unpublished).
2. R. Godang *et al.*, *Phys. Rev. Lett.* **84**, 5038 (2000).
3. G. Brandenburg *et al.*, *Phys. Rev. Lett.* **80**, 2762 (1998).
4. B. H. Behrens *et al.*, hep-ex/0005013.
5. *The Babar Physics Book*, SLAC-R-0504, P.F. Harrison and H.R. Quinn, ed., Stanford, California (1998), pp. 425-434.

Λ_C PRODUCTION AND DECAY

ROY A. BRIERE
Carnegie Mellon University, 5000 Forbes Ave., Pittsburgh, PA 15217
E-mail: rbriere@andrew.cmu.edu

We present CLEO results on the Λ_c lifetime, $\mathcal{B}(\Lambda_c \to pK\pi)$ using $\Lambda_c - D^{(*)}$ correlations, and correlated production of the Λ_c baryon in $e^+e^- \to c\bar{c}$ fragmentation.

The Λ_c is more than just a c quark in a baryon. Naively, one might expect that $\Gamma_{tot}(\Lambda_c) = \Gamma_{tot}(D^0)$, but W exchange and Pauli interference alter the picture. Ideally, experiments would provide one precise absolute Λ_c branching ratio, but a cleaner determination of $\mathcal{B}(\Lambda_c \to pK\pi)$ is still needed. Finally, independent fragmentation of quarks in $e^+e^- \to c\bar{c}$ is usually assumed, but contrary evidence from charm fragmentation will be presented below.

CLEO data sets include 4.7 fb^{-1} of data on and just below the $\Upsilon(4S)$, used for the $\mathcal{B}(\Lambda_c \to pK\pi)$ analysis, and taken with a straw-tube chamber as the innermost tracker. An additional 9.0 fb^{-1}, taken after installation of a silicon vertex detector, is used for the Λ_c lifetime. The statistics-limited correlated fragmentation measurement uses both data sets. The CLEO II detector is described in detail elsewhere.[1,2]

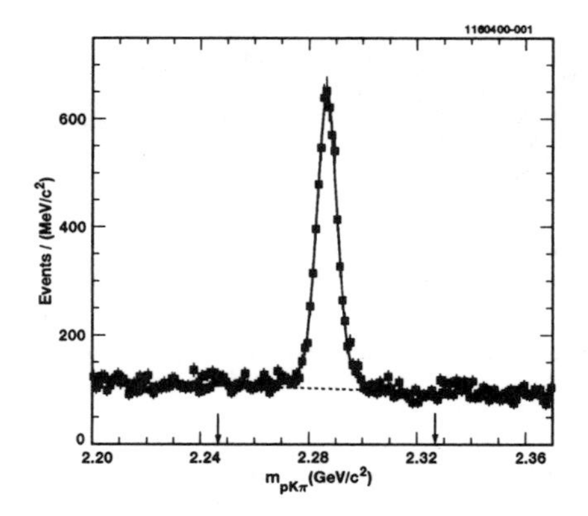

Figure 1. $\Lambda_c \to pK\pi$ mass peak with fit to a double-gaussian signal and linear background.

1 Λ_c Lifetime

CLEO recently published precise D meson lifetimes;[3] the method used there is very similar to that presented here.

We use $\Lambda_c \to pK\pi$ candidates with momenta > 2.6 GeV/c. The mean Λ_c momentum in the sample is 3.3 GeV/c, corresponding to $\gamma\beta c\tau \simeq 75\mu$m. Vertex resolution along flight direction is about 80 μm. We use only the vertical components of decay length, l_y, and momentum, p_y, to extract the proper decay time as $(m_{\Lambda_c}/cp_y)l_y$. Here, l_y is measured from the beam spot centroid which is well-determined for each machine fill. The approximate xyz beam spot size is (in std. deviations) $350\mu\text{m} \times 7\mu\text{m} \times 10^4\,\mu\text{m}$, with z along the beam direction. Note that the y beam size is much less than the vertex resolution. Finally, we veto $D^+ \to K^-\pi^+\pi^+$ background which may enter due to $\pi \leftrightarrow p$ mis-ID and could easily bias our results, since the D^+ lifetime is much larger than that of the Λ_c. Figure 1 shows the final Λ_c signal in data, yielding 5513 ± 120 events.

We perform an unbinned maximum likelihood fit to all $pK\pi$ combinations within ± 40 MeV of the nominal Λ_c mass. The probability an event is signal or background is assigned based on the mass-dependent ratio of these components in the fit shown in Figure 1. The lifetime fit has seven free parameters, including the Λ_c lifetime, a global scale factor for the predicted per-candidate decay length resolution, and the fraction of signal with a wide resolution function (width fixed

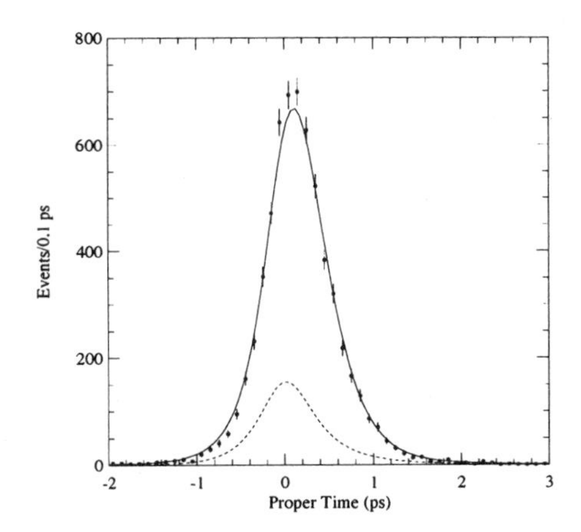

Figure 2. Λ_c lifetime data (points) shown with full fit (solid) and background (dashed) curves.

at 8 ps). Others parameters describe the fraction and lifetime of mis-reconstructed events, the fraction of prompt (zero-lifetime) background, and the mean lifetime of non-prompt background. Figure 2 shows the results of our fit.

We find $\tau(\Lambda_c) = 177.6 \pm 5.9 \pm 4.0$ fs. The largest systematic errors include 2.2 fs from a check where we fit to a zero-lifetime sample of $\gamma\gamma \to 4\pi$ (no bias was detected and the statistical accuracy is used), 2.7 fs from statistical accuracy of Monte-Carlo fit tests, and 1.8 fs from dependence on Λ_c mass. In all, nine detailed systematics were considered.

This is lower and more precise than the current world average of 206±12 fs.[4] The best previous single measurement was from E687: $215 \pm 16 \pm 8$ fs. Both SELEX and FOCUS also presented new results at this conference.

2 $\Lambda_c \to pK\pi$ Branching Ratio

It is difficult to measure an absolute Λ_c branching ratio. The central problem involves determining how many Λ_c were produced in the data sample. Since a nice summary of past techniques and their acknowledged limitations has been given by Burchat,[4] we will not repeat it here. Our technique is based on $c\bar{c}$ charm-tagging.[5]

Most of the time (corrections will be made for exceptions), one has a Λ_c and an $\bar{N}$ in one hemisphere, and a $\bar{c}$ hadronizing in the other hemisphere. An opposite hemisphere topology for particles A, B is denoted as $(A|B)$ below, where the B serves to tag the $\bar{c}$ quark and a $|$ separates the particles into hemispheres, for example $e^+e^- \to (c|\bar{c}) \to (\Lambda_c\bar{p}|\bar{D})$. We can measure the branching ratio as $\mathcal{B}(\Lambda_c \to pK\pi) = (\bar{p}\Lambda_c(\to pK\pi)|\bar{D})/(\bar{p}|\bar{D})$ with the $\bar{D}$ meson tagging the $\bar{c}$. In this expression, we must correct the numerator for events with no $\bar{D}$ tag, $(\bar{p}\Lambda_c|N\bar{\Lambda}_c)$, and fake protons. We must also correct the denominator for events with baryons in opposite hemispheres, $(\bar{p}D|\bar{D}N)$, as well as fake protons.

Three types of $\bar{c}$ tags are used. First, we can fully reconstruct one of $\bar{D}^0 \to K^+\pi^-$, $D^- \to K^+\pi^-\pi^-$, or $D_s^- \to \phi\pi^-$. We can also look for the soft transition pion from $D^{*-} \to \bar{D}\pi^-_{soft}$ or the e^- from $\bar{D} \to Xe^-\nu_e$ (both via the angular correlation with the jet thrust axis). The results for $\mathcal{B}(\Lambda_c \to pK\pi)$ are $(4.9 \pm 0.5)\%$ from $\bar{D}$ tags, $(5.2 \pm 1.3)\%$ from π_{soft} tags, and $(5.6 \pm 2.5)\%$ from e^- tags. Combining all three tag methods, we find: $\mathcal{B}(\Lambda_c \to pK\pi) = (5.0 \pm 0.5 \pm 1.2)\%$.

Systematic errors are dominated by knowledge of the tag proton efficiency and spectrum (15%) and event selection and MC modeling $(12 - 5\%)$; 11 other sources of error were also carefully evaluated.

3 Λ_c Production Correlations

Many analyses are simplified if one can assume that the c and $\bar{c}$ in $e^+e^- \to c\bar{c}$ fragment independently. To what extent this assumption is valid must be tested empirically.

In this analysis, we measure how often we see $(\Lambda_c^+|\Lambda_c^-)$ in opposite hemispheres of an event as compared to $(\Lambda_c^+|\bar{D})$. In each case, we normalize to the total number of reconstructed tags to obtain

Table 1. Fragmentation double ratios from data.

	$\Lambda_c \to pK\pi$	$\Lambda_c \to p\bar{K}^0_S$
$\frac{(\Lambda_c^+\|\Lambda_c^-)/\Lambda_c^-}{(\Lambda_c^+\|\bar{D}^0)/\bar{D}^0}$	3.11 ± 0.91	3.40 ± 1.10
$\frac{(\Lambda_c^+\|\Lambda_c^-)/\Lambda_c^-}{(\Lambda_c^+\|D^-)/D^-}$	2.71 ± 0.87	2.93 ± 1.08
$\frac{(\Lambda\|\Lambda_c^-)/\Lambda}{(\Lambda\|\bar{D}^0)/\bar{D}^0}$	3.58 ± 0.23	2.97 ± 0.34
$\frac{(\Lambda\|\Lambda_c^-)/\Lambda}{(\Lambda\|D^-)/D^-}$	3.17 ± 0.24	2.63 ± 0.32

Table 2. Fragmentation test with Monte Carlo.

	$\Lambda_c \to pK\pi$	$\Lambda_c \to p\bar{K}^0_S$
$\frac{(\Lambda_c^+\|\Lambda_c^-)/\Lambda_c^-}{(\Lambda_c^+\|\bar{D}^0)/\bar{D}^0}$	0.85 ± 0.85	1.40 ± 0.99
$\frac{(\Lambda_c^+\|\Lambda_c^-)/\Lambda_c^-}{(\Lambda_c^+\|D^-)/D^-}$	0.62 ± 0.63	1.61 ± 1.22
$\frac{(\Lambda\|\Lambda_c^-)/\Lambda}{(\Lambda\|\bar{D}^0)/\bar{D}^0}$	1.55 ± 0.18	1.08 ± 0.27
$\frac{(\Lambda\|\Lambda_c^-)/\Lambda}{(\Lambda\|D^-)/D^-}$	1.26 ± 0.15	0.88 ± 0.22

the ratio $(\Lambda_c^+|TAG)/TAG$, where $TAG = \Lambda_c^-, \bar{D}^0, \text{or} D^-$.

If fragmentation is independent, then one expects that $(\Lambda_c^+|\Lambda_c^-)/\Lambda_c^- = (\Lambda_c^+|\bar{D})/\bar{D}$. Thus, we may divide these expressions to form a double ratio which would equal one in the absence of any correlations.

Whether or not such a relation holds is related to whether or not baryon number conservation occurs via the production of an anti-baryon in the Λ_c hemisphere as is often supposed. If not, that is if there is a tendency for the anti-baryon to be opposite, then we expect that some fraction of the time it will include the $\bar{c}$ quark, leading to correlations in fragmentation. Thus, although usually ignored until now, some degree of correlation should not be too surprising. Finally, since full reconstruction of two charm states per event in golden modes has limited efficiency, one can also look at $(\Lambda|TAG)$ as an indicator of Λ_c correlation productions. Using a Λ to indicate the likely presence of a Λ_c gives us larger statistics at the cost of some additional systematics in interpreting the data.

We summarize the results as a list of various double ratios from data (separately for two Λ_c modes) in Table 1. We remind the reader that a deviation from unity indicates possible correlations. In Table 2, we list the same quantities for Monte Carlo, where correlations are absent in the event generator. This provides a cross-check of the analysis, as well as an indication that the detector does not introduce large spurious deviations from unity.

The data provide consistent evidence for production correlations; with additional data is should be possible to quantify the effect further and update the standard Monte Carlo generators to include this effect.

4 Conclusions

CLEO has a long history of charm baryon physics, including the discovery of more states than all other experiments combined. The contributions included here help to address several important topics including the lifetime hierarchy of charmed particles, a new method for an absolute determination of $\mathcal{B}(\Lambda_c \to pK\pi)$, and a demonstration of correlated fragmentation of $c\bar{c}$ jets.

Acknowledgments

We thank our colleagues at CLEO and CESR for their hard work and thank the U.S. Department of Energy for support.

References

1. Y. Kubota *et al.*, *Nucl. Instrum. Methods Phys. Res.* A **320**, 66 (1992).
2. T. Hill, *Nucl. Instrum. Methods Phys. Res.* A **418**, 32 (1998).
3. G. Bonvicini *et al.*, *Phys. Rev. Lett.* **82**, 4586 (1999).
4. Particle Data Group, D.E. Groom *et al.*, *Eur. Phys. J.* C **15**, 1 (2000).
5. D.E. Jaffe *et al.*, CLNS 00/1664 (submitted to *Phys. Rev.* D).

LEPTONIC DECAYS OF THE $\mathrm{D_s}$ MESON AT LEP

N.MARINELLI

Imperial College, Blackett Laboratory, Prince Consort Road, London SW7 2BW
E-mail: n.marinelli@ic.ac.uk

Results on BR($\mathrm{D_s} \to \tau\nu$) from the LEP experiments are presented here. With a recent measurement the ALEPH collaboration has also explored the channel $\mathrm{D_s} \to \mu\nu$ showing the first evidence at LEP for such a signal. The $\mathrm{D_s}$ decay constant is extracted from the measurements of the two branching ratios showing consistency between all the results.

1 Introduction

The leptonic decay $\mathrm{D_s} \to l\nu$ is the second–generation analog of the charged pion leptonic decay and it proceeds via the Cabibbo allowed annihilation of the c and $\bar{\mathrm{s}}$ quarks in the $\mathrm{D_s}$ meson. Standard Model predictions for the BR($\mathrm{D_s} \to l\nu$) branching fraction are based on the following expression:

$$\frac{G_F^2}{8\pi}\tau_{\mathrm{D_s}} f_{\mathrm{D_s}}^2 \mid \mathrm{V_{cs}} \mid^2 m_{\mathrm{D_s}} m_l^2 \left(1 - \frac{m_l^2}{m_{\mathrm{D_s}}^2}\right)^2, \quad (1)$$

where $\tau_{\mathrm{D_s}}$ is the mean lifetime of the $\mathrm{D_s}$, $\mathrm{V_{cs}}$ is the relevant CKM matrix element, $m_{\mathrm{D_s}}$ is the $\mathrm{D_s}$ mass, m_l is the mass of the lepton and $f_{\mathrm{D_s}}$ is the decay constant. Extracting $f_{\mathrm{D_s}}$ from the measurement of the branching fraction is the ultimate purpose of such an analysis because it allows verification of various theoretical calculations [1,2,3]. The experimentally accessible channels are $\mathrm{D_s} \to \tau\nu$ and $\mathrm{D_s} \to \mu\nu$, $\mathrm{D_s} \to e\nu$ being strongly helicity suppressed.

2 Analysis Methods

The general strategy for selecting events in the $\mathrm{D_s} \to \tau\nu$ channel ($\tau \to e\nu\bar{\nu}$ or $\tau \to \mu\nu\bar{\nu}$) is searching for hemispheres containing a high momentum lepton associated with large missing energy. ALEPH [4] looks for the decay chain $\mathrm{e^+e^-} \to \mathrm{c\bar{c}}, \mathrm{c} \to \mathrm{D_s}, \mathrm{D_s} \to \tau\nu$ distinguishing between muons and electrons, while DELPHI and L3 [5,6] reconstruct the decay chain $\mathrm{e^+e^-} \to \mathrm{c\bar{c}}, \mathrm{c} \to \mathrm{D_s^*} \to \mathrm{D_s}\gamma, \mathrm{D_s} \to \tau\nu$, by requiring a high energy photon in the hemisphere containing the lepton.

The selection is optimized for $\mathrm{Z} \to \mathrm{c\bar{c}}$ events while rejecting $\mathrm{Z} \to \mathrm{b\bar{b}}$ where the $\mathrm{D_s}$'s have a softer energy spectrum and are harder to detect. The rejection of the background, mostly due to semileptonic decays in charm and bottom events, is achieved mainly by cutting on the reconstructed $\mathrm{D_s}$ energy or on the mass difference M($\mathrm{D_s}\gamma$)-M($\mathrm{D_s}$). Lifetime based variables are used to discriminate the signal from the b background, exploiting the fact that in charm hemispheres containing a $\mathrm{D_s} \to l\nu$ decay all particles but the lepton originate from the primary vertex. In the ALEPH analysis the physical background from $\mathrm{D^+} \to l\nu$ is indistinguishable from the $\mathrm{D_s} \to l\nu$ so it is treated as signal and related to $f_{\mathrm{D_s}}$ using the prediction $f_{\mathrm{D_s}}/f_{\mathrm{D^+}} = 1.11^{+0.06}_{-0.05}$ [3].

The background due to $\mathrm{b\bar{b}}$, $\mathrm{c\bar{c}}$ and light flavour events still present in the selected sample has to be subtracted in order to extract the $\mathrm{D_s} \to \tau\nu$ component. This is estimated from the M($\mathrm{D_s}\gamma$) Monte Carlo distribution for DELPHI and L3, while ALEPH use a linear combination of hemisphere variables suitably chosen with the help of the Monte Carlo simulation, according to their discriminant power (Linear Discriminant Variables technique). Two variables, $\mathrm{U_b}$ and $\mathrm{U_c}$, are optimized in order to discriminate the signal from background in $\mathrm{b\bar{b}}$ and $\mathrm{c\bar{c}}$ events respectively, as shown in Fig. 1.

A fit is then performed to the data with

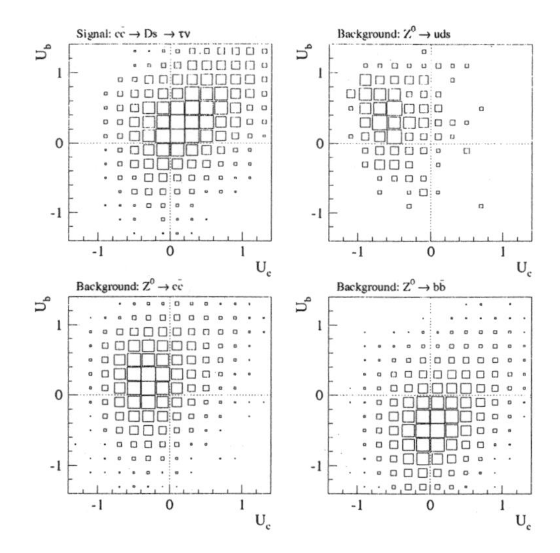

Figure 1. Monte Carlo U_b vs U_c distributions in the electron channel for the ALEPH $D_s \to \tau\nu$ analysis.

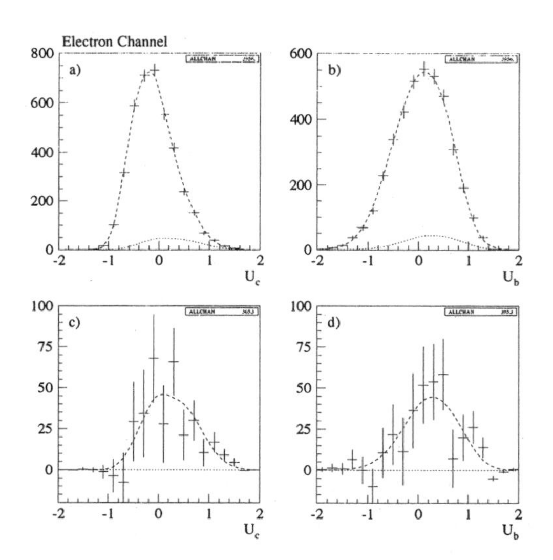

Figure 2. Projections of the fit to the data in the electron channel for the ALEPH $D_s \to \tau\nu$ analysis. (a) U_c distribution; (b) U_b distribution. (c) and (b) are the same distributions after subtracting the fitted background.

a function consisting of a signal plus three background (uds,$c\bar{c}$, $b\bar{b}$) components. The relative normalizations of the various contributions are fixed according to Eq. 1, the measured charm production rates [7,8], the predicted f_{D_s}/f_{D^+} and other quantities [9]. In Fig. 2 the result of the fit is shown, projected along the two variables U_b and U_c.

The first evidence of the $D_s \to \mu\nu$ signal has been observed by ALEPH [4]. The analysis technique follows what is done for the $D_s \to \tau\nu$ channel except that also the $M(\mu\nu)$ invariant mass is calculated and used as a third discriminant variable in the fit together with another pair of variables U_b, U_c, suitably optimized to distinguish $c\bar{c} \to D_s \to \mu\nu$ from $b\bar{b}$ and $c\bar{c}$ background. This analysis is also sensitive to $D_s \to \tau\nu$ decays so the extracted branching fraction refers to the combined signal. In Fig. 3 the three-dimensional fit is projected along the $M(\mu\nu)$ axis.

The main source of systematic error, common to all the analyses, is the uncertainty on the charm hadron production rates, (mostly due to $P(c\to D_s)$ for ALEPH and $P(c \to D_s^*)$ for DELPHI and L3, since they directly govern the number of D_s (D_s^*) mesons present in the sample. Other important systematic uncertainties on the ALEPH measurements are due to the fragmentation of the charm quark, which directly affects the charmed hadron reconstructed energy, and the detector resolution.

3 Results

The measured branching fractions are summarized in Table 1.

The extracted values for f_{D_s} are shown in Fig. 4. The quoted ALEPH result is obtained by the combination of $f_{D_s} = (273 \pm 18 \pm 42)$ MeV from the $D_s \to \tau\nu$ analysis and $f_{D_s} = (291 \pm 25 \pm 38)$ MeV from the $D_s \to \mu\nu$ analysis. The LEP measurements are compared with two more results from fixed target experiments and with the Lattice QCD prediction showing overall good agreement.

Table 1. Experimental results on BR($D_s \rightarrow \tau\nu$) and BR($D_s \rightarrow \mu\nu$).

	BR($D_s \rightarrow \tau\nu$)
ALEPH $\tau \rightarrow e\nu\bar{\nu}$	$(5.86 \pm 1.18 \pm 2.09)\%$
ALEPH $\tau \rightarrow \mu\nu\bar{\nu}$	$(5.78 \pm 0.85 \pm 1.76)\%$
ALEPH Combined	$(5.79 \pm 0.76 \pm 1.78)\%$
DELPHI	$(8.5 \pm 4.2 \pm 2.6)\%$
L3	$(7.4 \pm 2.8 \pm 2.4)\%$
	BR($D_s \rightarrow \mu\nu$)
ALEPH	$(0.68 \pm 0.11 \pm 0.18)\%$

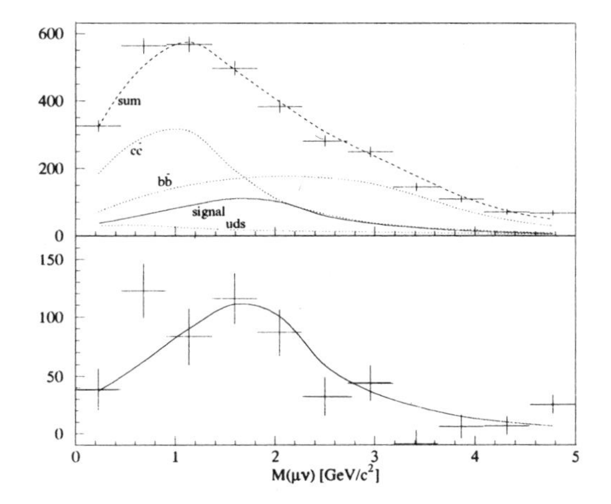

Figure 3. M($\mu\nu$) projection of fit to data in the ALEPH $D_s \rightarrow \mu\nu$ channel, before and after the background subtraction.

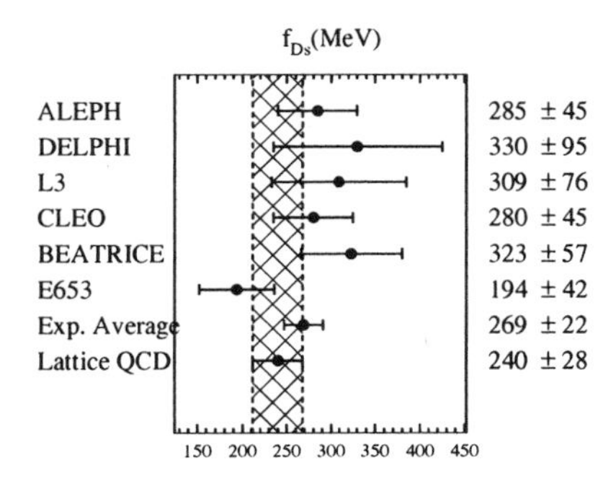

Figure 4. Experimental results for f_{D_s} from the LEP experiments compared with fixed target experiments and theoretical predictions.

References

1. J.Richman and P.Burchat, *Rev. Mod. Phys.* **67** (1995) 893, hep–ph/9508250.
2. S. Narison, *Nucl. Phys. Proc. Suppl.* **74** (1999) 304, hep-ph/9811208
3. T. Draper, *Nucl. Phys. Proc. Suppl.* **73** (1999) 43, hep-lat/9810065
4. ALEPH Collaboration, Contributed Paper to ICHEP 2000, Osaka, Japan.
5. DELPHI Collaboration, Contributed Paper to ICHEP 1997, Jerusalem, Israel.
6. L3 Collaboration, *Phys. Lett.* B**396** (1997) 327.
7. The LEP and SLD Collaborations, "A combination of Preliminary Electroweak Measurements and Constraints on the Standard Model", CERN-EP/2000-016.
8. ALEPH Collaboration, *Phys. Lett.* B**388** (1996) 648.
9. C. Caso *et al.* Particle Data Group, *Eur. Phys. J.* B **478** (2000) 31.

LATEST B_S MIXING RESULTS FROM ALEPH

P. COYLE

Centre de Physique des Particules de Marseille, France
E-mail: coyle@cppm.in2p3.fr

B_s mixing results from the ALEPH experiment, based on analysis of about 4M Z^0 decays at LEP are presented. Three recently updated analyses; fully reconstructed B_s, D_s-lepton and inclusive-lepton are discussed. The combination of all three analyses yields $\Delta m_s > 10.6$ ps^{-1} at 95% CL, with a sensitivity of 12.8 ps^{-1}.

1 Introduction

Particle anti-particle mixing in neutral mesons is a consequence of flavour non-conservation in charged weak-current interactions. A measurement of the oscillation frequency for the case of the B_s meson (Δm_s) when combined with the already well measured B_d oscillation frequency (Δm_d) would provide an important constraint on the ratio of CKM matrix elements V_{ts}/V_{td} and thus on the length of the right side of the unitarity triangle. Standard Model expectations for the value of Δm_s range between $10 - 20$ ps^{-1}, close to the limit of the current experimental reach.

2 Selections

Since the statistical significance of an oscillation signal depends exponentially on the proper time resolution, optimisation of the decay length resolution and correct treatment of the event-by-event estimate of the uncertainty on the decay length resolution is crucial. ALEPH have pursued three different selections which trade resolution with sample size in order to optimise the sensitivity at large Δm_s. These analyses are improved versions of previously published analyses. They have been performed on the recently reprocessed LEP1 data, and therefore benefit from improved track reconstruction and particle identification.

2.1 Fully Reconstructed B_s

The best proper time resolution ($\sigma_t \approx 0.08$ ps) is obtained with this selection[1]. About 20 B_s candidates are reconstructed in the channels $B_s \to D_s^{(*)}\pi$ and $B_s \to D_s^{(*)}a_1$, with $D_s \to \phi\pi$ and K^*K. Events proceeding via the D_s^* have a slightly lower reconstructed B_s mass due to the missing photon. The expected B_s purity is about 40%.

2.2 D_s-lepton

A larger data sample is isolated using events containing a fully reconstructed D_s correlated with an opposite sign lepton in the same hemisphere[1]. A total of 297 D_s-lepton candidates are reconstructed in eight different D_s decay channels. The average fraction of B_s in the sample is also about 40% with a significant contamination from $B \to D_sDX$ events in which the D decays semi-leptonically. For sub-samples of the data the B_s fraction is significantly larger, for example when the p_t of the lepton is greater than 2 GeV/c^2 the B_s fraction is above 85%. The presence of the neutrino in the decay smears the decay length resolution and introduces a momentum uncertainty of $\approx 10\%$ ($\sigma_t \approx 0.2$ ps).

2.3 Inclusive lepton

Compared to the previously published inclusive lepton analysis [2], in this new analysis [3] the data sample is increased by a factor 2.2 ($\approx$ 74k) while preserving a similar or slightly better decay length resolution. This is achieved by no longer cutting directly on the lepton p_t but rather by using an event by event weight based on a discriminating variable constructed from various properties of the lepton (p,p_t, ν energy, impact parameter etc). In addition a new vertexing algorithm is applied in which an estimate of the B_s decay direction is incorporated into the vertex fit as a 'pseudo' track. This algorithm yields an improved decay length resolution and therefore allows the sample size to be enlarged.

Another important improvement is that the estimate of the uncertainty on the decay length, derived from the track error matrices, is parametrized as a function of various quantities such as the mass at the charm vertex and the B flight distance.

3 Results

For all three selections an improved initial state tag ($\approx$ 24%) is achieved using a neural net[4] based on quantities from both hemispheres of the event such as primary and secondary vertex charges, jet charges, charge of an opposite hemi. lepton, charge of kaons at the decay vertex (same hemi.) or the primary vertex (opposite hemi.) etc.

The results are presented within the framework of the "amplitude fit", in which the B_s oscillation amplitude $\mathcal{A}$ is measured at each fixed value of Δm_s, using a maximum likelihood fit. Assuming $f_s = 9.7 \pm 1.2\%$ [5] values of $\Delta m_s > 0/6.4/11.1$ ps^{-1} are excluded at 95% CL (systematic uncertainties included) for the fully reconstructed/D_s-lepton/ inclusive lepton analyses respectively. Although the fully reconstructed analysis does not provide any exclusion, its uncertainty on the amplitude at large Δm_s is competitive despite the very small number of events selected. Compared to analyses from other experiments the ALEPH inclusive-lepton analysis provides the best limit and sensitivity currently available.

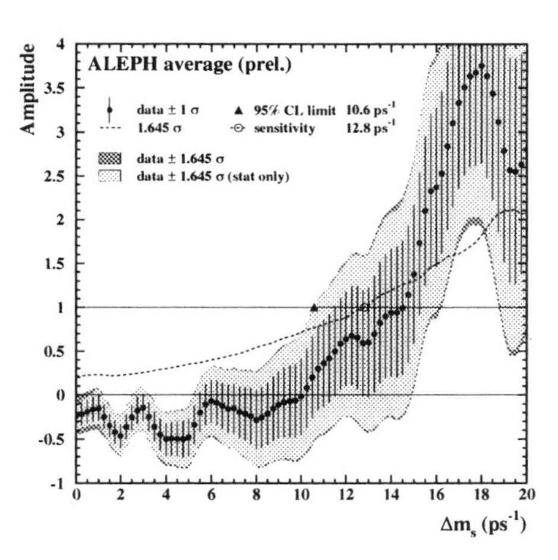

Figure 1. Amplitude spectrum for the combined ALEPH B_s mixing analyses.

The amplitude spectrum for the combination of all three analyses is shown in Fig.1 and yields $\Delta m_s > 10.6$ ps^{-1} at 95% CL. For this combination, events common to the D_s-lepton and inclusive-lepton analyses are removed from the inclusive-lepton. The sensitivity of all analyses combined, defined by $1.645\sigma_{\mathcal{A}}(\Delta m_s^{sens}) = 1$ is 12.8 ps^{-1}. If the amplitude spectrum is converted to a negative delta loglikelihood curve, a minimum at $\Delta m_s = 17.2$ ps^{-1} is obtained with a depth relative to infinity of 3 units; corresponding to about a 2.4σ effect.

References

1. Aleph Collab., ALEPH/2000-029, contributed paper PL-03/128.
2. Aleph Collab.,, *Eur. Phys. J.* C**7**, 553 (1999).
3. Aleph Collab., ALEPH/2000-059, contributed paper PL-03/234.
4. Aleph Collab., ALEPH/2000-045, contributed paper PL-03/173.
5. B Oscillation LEP working group, `http://www/cern.ch/LEPBOSC/combined_results/osaka_2000`

MEASUREMENTS OF INCLUSIVE AND EXCLUSIVE B DECAYS TO CHARMONIUM WITH *BABAR*

GERHARD RAVEN
ON BEHALF OF THE *BABAR* COLLABORATION
University of California at San Diego, La Jolla, CA92093, USA

Using 8.5M $B\overline{B}$ events recorded by the *BABAR* detector the yield of inclusive J/ψ, the branching ratios of $\psi(2S)$ and χ_c are presented. Combining the charmonium state with either a $K^{\pm}$,K^0_S, $K^{*\pm}$ or K^{*0} exclusive B decays are reconstructed and their branching ratios determined. Using the fully reconstructed decays, both the B^0 and $B^{\pm}$ masses and their difference is measured. Finally the contributions of *CP* even and odd amplitudes in the decay $B \to J/\psi K^*$ are determined from an angular analysis.

1 Introduction

Observation of charmonium mesons in B decays is a crucial component of the measurement of time-dependent *CP*-violating asymmetries[1].

For the analyses described here, a sample of 7.7 fb^{-1} collected at the $\Upsilon(4S)$ resonance with an additional 1.2 fb^{-1} collected below the $B\overline{B}$ threshold are used. The number of $B\overline{B}$ events is determined by counting the number of hadronic events selected both on and off resonance. The continuum contribution to the on-resonance sample is estimated by rescaling the number of off-resonance hadronic events by the ratio of the number of observed $\mu^+\mu^-$ events in the two samples. This procedure yields a total of $8.46 \pm 0.14 \cdot 10^6$ selected $B\overline{B}$ events.

2 Inclusive decays of B to Charmonium

Events containing a J/ψ are selected by requiring two identified leptons of opposite charge. Electrons are selected by requiring the observed energy in the calorimeter to match the measured momentum, the shape of the calorimeter cluster and the observed ionization in the tracking detectors. Muons are identified by requiring a minimum ionizing signal in the calorimeter and by their penetration into and observed cluster shape in the instrumented flux return. The number of J/ψ events is determined by fitting the invariant mass distribution to a pdf obtained from a simulation which includes both final state radiation and bremsstrahlung. The fits yield $4920 \pm 100 \pm 180$ $J/\psi \to e^+e^-$ and $5490 \pm 90 \pm 90$ $J/\psi \to \mu^+\mu^-$ signal events.

Events containing $\psi(2S)$ decays are reconstructed in both the leptonic decays of the $\psi(2S)$ and its decays to $J/\psi\pi^+\pi^-$. In case of the former, the number of signal events is extracted in similar fashion to the J/ψ; for the latter, a fit to the mass difference between the $\psi(2S)$ and the J/ψ candidates is performed. We find $131 \pm 29 \pm 2$ decays to e^+e^-, 125 ± 19 to $\mu^+\mu^-$, 126 ± 44 to $J/\psi(\mu^+\mu^-)\pi^+\pi^-$ and 162 ± 23 to $J/\psi(e^+e^-)\pi^+\pi^-$.

The χ_{c1} and χ_{c2} are reconstructed by combining a J/ψ candidate with a photon. The signal yield is determined by fitting the mass difference between the χ_c and J/ψ candidates. We fit simultaneously for a χ_{c1} and possible χ_{c2} component. The shape of the signal is taken from the simulation, and the mass difference between the χ_{c1} and χ_{c2} is fixed to the PDG value[4]. We find $129 \pm 26 \pm 13$ χ_{c1} and 3 ± 21 χ_{c2} candidates in which $J/\psi \to e^+e^-$ and $204 \pm 47 \pm 12$ χ_{c1} and 47 ± 21 χ_{c2} candidates in which $J/\psi \to \mu^+\mu^-$.

The branching ratios of $B \to \psi(2S)X$ and $B \to \chi_{c1}X$ are determined[2] by measuring their rates relative to the measured J/ψ yield, and

Table 1. Measured Inclusive Branching Ratios

Mode	Br ($\times 10^{-2}$)
$\psi(2S)$	$0.25 \pm 0.02 \pm 0.02$
χ_{c1}	$0.39 \pm 0.04 \pm 0.04$
χ_{c2}	< 0.24 (90%CL)

a limit is set on the decay to $\chi_{c2}X$. The results are summarized in Table 1.

3 Exclusive decays of B to Charmonium

As the exclusive decays in general have very little background, lepton identification is required for only one of the two J/ψ decay products. After including photons compatible with bremsstrahlung from one of the leptons, the charmonium states are selected in a window around their expected mass[4], and the observed momenta are refined by a kinematic fit constraining the charmonium masses.

The charmonium candidates are then combined with either a K^+, a K^0_S (either $\pi^+\pi^-$ or $\pi^0\pi^0$), K^{*0} (either $K^+\pi^-$ or $K^0_S\pi^0$) or K^{*+} (either $K^0_S\pi^+$ or $K^+\pi^0$) to form a B candidate. The two most significant observables used to identify the signal are ΔE, the difference in energy between the reconstructed B decay and $\sqrt{s}/2$, and the energy-substituted B mass, $m_{\rm ES} = \sqrt{(\sqrt{s}/2)^2 - {P^*_B}^2}$ where P^*_B is the center of mass momentum of the B candidate. An example of these distributions is shown in figure 1. In the case of multiple candidates per event, only the candidate with the smallest $|\Delta E|$ is selected.

The signal yields are determined by fitting the $m_{\rm ES}$ distribution with the sum of a Gaussian and an ARGUS function[5]; for the K^* modes a likelihood fit is performed to all modes simultaneously, taking into account the cross-feed between the decays.

Systematic uncertainties considered include the number of produced B events (3.6%), the signal fit (0.9–8.6%), uncertainties on the measured tracking (2.5% per track), neutral (0.6–11%) and particle ID (2.5–8.8%) efficiencies, the tracking resolution (0.6–2.6%), the branching ratios of secondary decays (2.2–13.1%) and MC statistics (0.5–5.8%). The observed yields and branching ratios[3] are summarized in Table 2.

Table 2. Measured Exclusive Branching Ratios

Mode	Yield	Br ($\times 10^{-4}$)
$B^\pm \to J/\psi K^{*\pm}$	126 ± 12	$13.2 \pm 1.4 \pm 2.1$
$B^0 \to J/\psi K^{*0}$	188 ± 14	$13.8 \pm 1.1 \pm 1.8$
$B^\pm \to J/\psi K^\pm$	445 ± 21	$11.2 \pm 0.5 \pm 1.1$
$B^0 \to J/\psi K^0$		
$K^0_S \to \pi^+\pi^-$	93 ± 10	$10.2 \pm 1.1 \pm 1.3$
$K^0_S \to \pi^0\pi^0$	14 ± 4	$7.5 \pm 2.0 \pm 1.2$
$B^\pm \to \psi(2S) K^\pm$	73 ± 8	$6.3 \pm 0.7 \pm 1.2$
$B^0 \to \psi(2S) K^0$		
$K^0_S \to \pi^+\pi^-$	23 ± 5	$8.8 \pm 1.9 \pm 1.8$
$B^\pm \to \chi_{c1} K^\pm$	44 ± 9	$7.7 \pm 1.6 \pm 0.9$

4 Measurement of B meson masses

The masses of the B mesons are measured using fully reconstructed decays of $B^0 \to J/\psi K^0_S(\pi^+\pi^-)$, $B^0 \to J/\psi K^{*0}(K^+\pi^-)$ and $B^\pm \to J/\psi K^\pm$. These modes are chosen for their small backgrounds and good knowledge of the masses of their decays products.

The invariant mass of the B candidates is derived by fitting the decay products to a common vertex, constraining the J/ψ and K^0_S masses to their nominal values. Uncertainties in the magnetic field and the alignment of the tracking detectors could introduce a bias in the momentum measurement. Their effect is quantified by comparing the reconstructed J/ψ and K^0_S masses with the PDG values[4]. The effect of background on the measurement has been estimated by removing separately the N events with the smallest and the largest mass, where N is the number of background events determined from the sidebands.

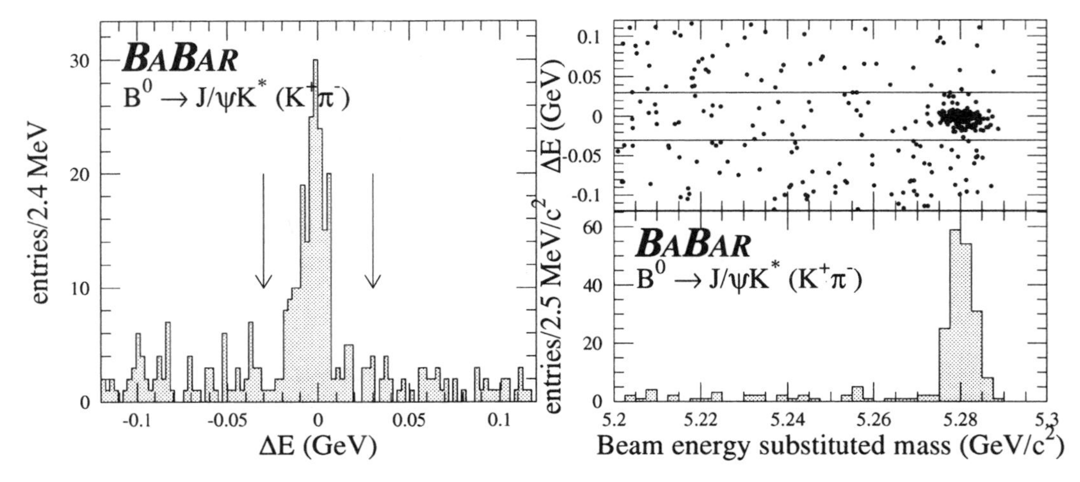

Figure 1. Example of the ΔE (left) and $m_{\rm ES}$ (right) distributions for the decay $B \to J/\psi K^*(K^+\pi^-)$.

The resulting masses[3] are:

$$m(B^0) = 5279.0 \pm 0.8 \pm 0.8\,\mathrm{MeV/c^2}$$
$$m(B^\pm) = 5278.8 \pm 0.6 \pm 0.4\,\mathrm{MeV/c^2}$$

where the first error is the quadratic sum of the statistical and uncorrelated systematic errors and the second error is the correlated systematic error.

The mass difference between B^0 and $B^\pm$ mesons is evaluated by fitting the $m_{\rm ES}$ distributions of the three above-mentioned channels. The use of $m_{\rm ES}$ has the advantage that it reduces the uncertainty in the momentum scale, whilst the uncertainty due to the beam energy cancels in the difference. The mass difference is determined[3] to be:

$$m(B^0) - m(B^\pm) = 0.28 \pm 0.21 \pm 0.04\,\mathrm{MeV/c^2}.$$

5 Angular analysis of $B \to J/\psi K^*$

The decay of $B \to J/\psi K^*$ proceeds through two *CP* even amplitudes $(A_0, A_\parallel)$ and one *CP* odd amplitude $(A_\perp)$. The contribution of $A_\perp$ must be known before a value of sin2β can be determined from this decay channel.

The relative contributions of the three amplitudes are determined using an unbinned extended likelihood fit to the decay angles, imposing the constraint $|A_0|^2 + |A_\parallel|^2 + |A_\perp|^2 = 1$. The *CP* odd fraction is found to be $|A_\perp|^2 = 0.13 \pm 0.06 \pm 0.02$ whereas the longitudinal polarization Γ_L/Γ is given by $|A_0|^2 = 0.60 \pm 0.06 \pm 0.04$. Sources of systematic uncertainties include the knowledge of the background, the acceptance corrections, the cross-feed amongst $B \to J/\psi K^*$ modes and the contribution from heavier K^* mesons.

References

1. *BABAR* collaboration, "A study of time-dependent *CP*-violating asymmetries in $B^0 \to J/\psi K^0_S$ and $B^0 \to \psi(2S) K^0_S$ decays", *BABAR*-CONF-00/01, SLAC-PUB-8540; D. Hitlin, these proceedings
2. *BABAR* collaboration, "Measurement of inclusive production of charmonium states in B meson decays", *BABAR*-CONF-00/04, SLAC-PUB-8526.
3. *BABAR* collaboration, "Exclusive B decays to charmonium final states", *BABAR*-CONF-00/05, SLAC-PUB-8527.
4. Particle Data Group, D.E. Groom *et al.*, Eur. Phys. Jour. C **15** (2000) 1.
5. ARGUS Collaboration, H. Albrecht *et al.*, Phys. Lett. **B254** (1991) 288.

STUDIES OF B MESON DECAYS TO FINAL STATES CONTAINING CHARMONIUM WITH BELLE

S. SCHRENK

Physics Department - Virginia Tech, Blacksburg, Virginia 24061-0435 USA
and
KEK (Belle), 1-1 Oho, Tsukuba, Ibaraki 305 Japan
E-mail: schrenk@bmail.kek.jp

We present preliminary results from Belle on B meson decays to final states containing charmonium. Results include the yield from several inclusive decays, the first observation of $B \to J/\psi K_1(1270)$, and measurement of the polarization for $B \to J/\psi K^{*0}$.[1]

1 Introduction

Current interest in decays of the B meson to states with charmonium focuses on their use for the measurement of CP violation. Since these measurements are limited by statistics, it is desirable to utilize as many decay modes as possible. This means finding new exclusive decays and understanding the CP content of modes such as $J/\psi K^*$.

The results presented here are based on a subset of the 6.2 fb^{-1} of data delivered in the past year by the asymmetric e^+e^- collider KEK-B operating at the $\Upsilon(4S)$ resonance.

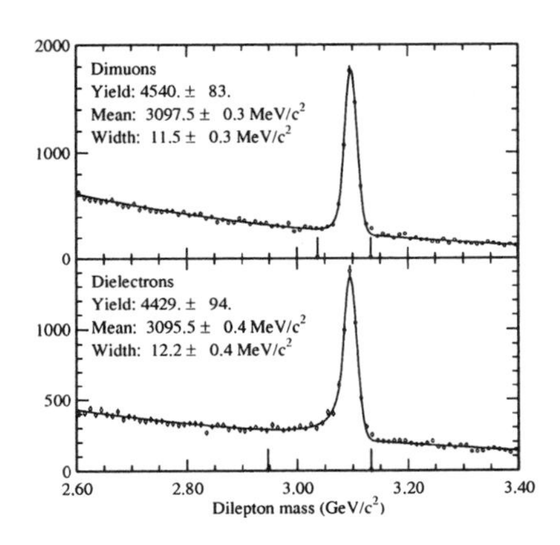

Figure 1. Dilepton invariant mass plots for $J/\psi \to l^+l^-$ with 'tight' lepton identification (6.2 fb^{-1}).

2 Inclusive $B \to$ *Charmonium*

The first step is to find charmonia inclusively. We reconstruct J/ψ (ψ') through the dilepton channel. We can then build on the J/ψ to identify ψ' and χ_c states.

2.1 J/ψ

We reconstruct J/ψ via the decay $J/\psi \to l^+l^-$, where l is either an electron or muon. This decay mode has a relatively high branching fraction and is experimentally clean. For most modes we require both leptons to be positively identified. For $B \to J/\psi K$ we require one lepton to be positively identified and the other to pass a hadron veto test. To account for final state radiation and Bremsstrahlung the four-momentum of any photon within 0.05 radians of the initial direction of an electron is added to the electron four-momentum. This significantly reduces the radiative tail in the line shape. We also require the J/ψ candidate's center-of-mass (CM) momentum to be less than 2.0 GeV/c, the maximum for a J/ψ from a B meson. The inclusive spectra for $B \to J/\psi, J/\psi \to e^+e^-$ and $J/\psi \to \mu^+\mu^-$ are shown in Fig. 1.

2.2 ψ'

We reconstruct ψ' two different ways. The first is identical to that used for J/ψ, except that the CM momentum must be less than 1.7 GeV/c. Fig. 2 shows the combined dimuon and dielectron line shape.

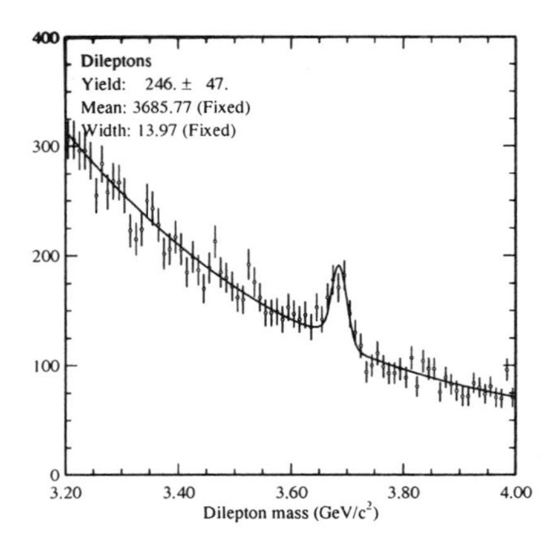

Figure 2. Dilepton invariant mass plot for $\psi' \to l^+l^-$ with tight lepton identification.

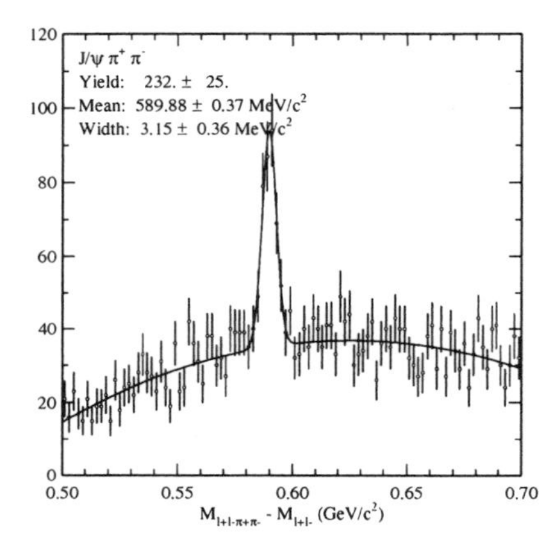

Figure 3. Mass difference between ψ' candidate and J/ψ candidate for the decay $\psi' \to J/\psi\pi^+\pi^-$. J/ψ candidate formed with 'tight' lepton identification.

We also reconstruct ψ' through the decay to $J/\psi\pi^+\pi^-$. It has been observed that the dipion mass $M_{\pi\pi}$ favors a high value.[2] We thus reduce background with little loss in efficiency by requiring $M_{\pi\pi} > 0.4$ GeV/c^2. The mass difference between the ψ' candidate and J/ψ candidate is shown in Fig. 3.

2.3 χ_c

We reconstruct the χ_c through its decay to $J/\psi\gamma$. To reduce the background we require that the energy of the γ be greater than 60 MeV and that when the γ is combined with other photons it not form a π^0 candidate. We

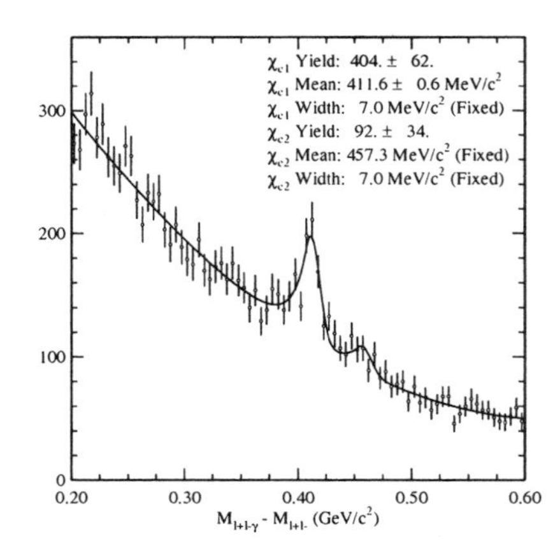

Figure 4. Mass difference between χ_c candidate and J/ψ candidate for the decay $\chi_c \to J/\psi\gamma$. J/ψ candidate formed with 'tight' lepton identification.

fit the resulting distribution with a line shape taken from Monte Carlo. We allow for the presence of both χ_{c1} and χ_{c2}. Fig. 4 shows the mass difference distribution.

3 Exclusive $B \to$ *Charmonium*

Once we find a charmonium candidate we then try to reconstruct a B meson exclusively. Details of reconstructing charmonia + $\{K, \pi\}$ can be found in Aihara's contribution to these proceedings.

Two kinematic variables are important for exclusive B reconstruction at the $\Upsilon(4S)$ resonance. Both rely on the fact that there is only enough energy at the $\Upsilon(4S)$ to create a $B\bar{B}$ pair. The first is ΔE, the difference between the reconstructed B candidate energy and one half of the center of mass energy. The second is the "Beam Constrained Mass," $\mathrm{M}_B \equiv \sqrt{\mathrm{E}_B^2 - \mathrm{P}_{cand}^2}$ where E_B is the CM beam energy and P is the candidate momentum.

4 Observation of $B \to J/\psi K_1(1270)$

We look for the $K_1(1270)$ resonance in the modes: $B^+ \to K^+\pi^+\pi^-$, $B^0 \to K^+\pi^-\pi^0$, and $B^0 \to K^0\pi^+\pi^-(K^0 \to \pi^+\pi^-)$.

A scatter plot of $M_{\pi\pi}$ *vs.* $M_{K\pi\pi}$ shows

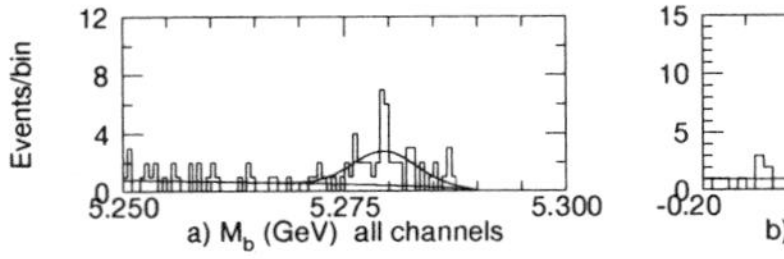

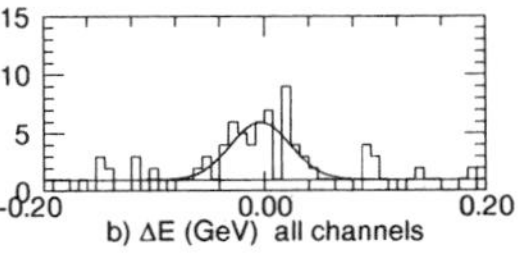

Figure 5. $K_1(1270)$ center of mass.

a clustering in the region near $M_{\pi\pi} \approx 0.75$ GeV and $M_{K\pi\pi} \approx 1.3$ GeV/c that is characteristic of the decay of $K_1(1270)$ to $K\rho$. Threshold effects distort both the $K_1(1270)$ and ρ line shapes. Therefore, we make an asymmetric selection around the ρ mass peak ($0.62 \leq M_{\pi\pi} \leq 0.84$ GeV).

Figure 5 shows the projection of the Beam Constrained Mass and ΔE for events with $M_{\pi\pi}$ in the ρ and $M_{K\pi\pi}$ in the $K_1(1270)$ signal regions. We normalize to the yield for $B^+ \to J/\psi K^+$ in the same data set. Many of the systematic errors then cancel. The largest systematic errors are the $K_1(1270)$ branching fraction (14%), other $K\pi\pi$ resonances (13%), non-resonant $K\pi\pi$ (10%) and reconstruction of the pions (5% for $\pi^\pm$ and 10% for π^0).

We find the branching fractions to be: $\mathcal{B}(B^0 \to K_1^0(1270)) = (1.4 \pm 0.4 \pm 0.4) \times 10^{-3}$ and $\mathcal{B}(B^+ \to K_1^+(1270)) = (1.5 \pm 0.4 \pm 0.4) \times 10^{-3}$.

5 Polarization in $B \to J/\psi K^*$

The decay $B \to K^{*0}$, $K^{*0} \to K_S\pi^0$ is useful for CP studies if the decay is dominated by even or odd CP. We measure the polarization using the modes $K^* \to K^+\pi^-$, $K_S\pi^+$, and $K^+\pi^0$. The $K\pi$ invariant mass distribution is shown in Fig. 6. First we select $J/\psi K^*$ events with a B candidate energy and Beam Constrained Mass consistent with expected values. We then do unbinned likelihood fits for the polarization in both the Helicity and Transversity bases (the later basis gives directly the CP-odd component of the decay[3]). The largest errors come from the

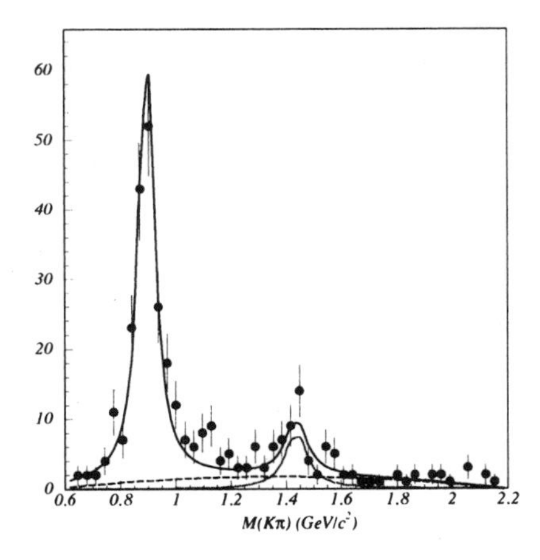

Figure 6. $K\pi$ invariant mass histogram. The $K_2^*(1430)$ can be seen above the $K^*(892)$ peak.

efficiency function used in the fit, the polarization from "feed across" and the polarization of non-resonant $K\pi$. With the Helicity basis we find the fraction of longitudinal polarization to be $\Gamma_L/\Gamma = 0.51 \pm 0.06 \pm 0.04$, in good agreement with previous CLEO and CDF measurements. With the transversity basis we find the P odd component to be $|A_\perp|^2 = 0.27 \pm 0.11 \pm 0.05$. This number is marginally larger than previously reported values.

6 Conclusion

We have presented yields for several inclusive decays of B mesons to states with charmonium, the first observation of $B \to J/\psi K_1(1270)$ and measurements of the polarization of $B \to J/\psi K^*$ in both the Helicity and Transversity bases.

References

1. This talk covers abstracts #1018 and #1019.
2. D. Coffman *et. al.*, Phys. Rev. Lett. **68**, 282 (1992).
3. I.Dunietz, H.Quinn, A.Snyder, W.Toki and H.J.Lipkin, Phys. Rev. **D43**, 2193 (1991).

HADRONIC B DECAYS TO CHARM FROM CLEO

SHELDON STONE

Physics Department, Syracuse University, Syracuse, N. Y. 13244-1130, USA
E-mail: stone@phy.syr.edu

Recent results include the measurement of $B \to D^{(*)}\pi^+\pi^-\pi^-\pi^o$ decays, with rates of ~1.8%, observation of a significant $\omega\pi^-$ substructure that appears to result from the decay of the ρ' resonance, whose mass and width we determine as 1418 MeV and 382 MeV, respectively. Using B to charmonium decays we find that the B^o fraction in $\Upsilon(4S)$ decays is nearly half, that the CP asymmetries in $J/\psi K^{\mp}$ and $\psi' K^{\mp}$ are small, and we also report on the first observation of $B \to \eta_c K$.

1 Introduction

Understanding hadronic decays of the B is crucial to insuring that decay modes used for measurement of CP violation truly reflect the underlying quark decay mechanisms expected theoretically. Most of the decay rate is hadronic. The semileptonic branching ratio for e^-, μ^- and τ^- totals ~25%.[1] Currently, measured exclusive branching ratios for hadronic B decays total only a small fraction of the hadronic width. The measured modes for the $\overline{B}^o$, including $D^+(n\pi)^-$, $D^{*+}(n\pi)^-$, where $3 \geq n \geq 1$, $D^{+(*)}D_s^{-(*)}$ and J/ψ exclusive, totals only about 10%.[1] Thus our understanding of hadronic B decay modes is not yet well based in data. It is also interesting to note that the average charged multiplicity in hadronic B^o decays is 5.8±0.1.[2] Since this multiplicity contains contributions from the D^+ or D^{*+} normally present in $\overline{B}^o$ decay, we expect a sizeable, approximately several percent, decay rate into final states with four pions.[3]

2 First Observation of the ρ' in B Decays

We have made the first statistically significant observations of six hadronic B decays shown in Table 1 that result from studying the reaction $B \to D^{(*)}\pi^+\pi^-\pi^-\pi^o$.[4]

A significant fraction of the final state is $D^{(*)}\omega\pi^-$, and there is a low-mass resonant substructure in the $\omega\pi^-$ mass. (See Fig. 1) A simple Breit-Wigner fit assuming a single resonance and no background gives a mass of 1418±26±19 MeV with an intrinsic width of 382±41±32 MeV.

Table 1. Measured Branching Ratios

Mode	$\mathcal{B}$ (%)
$\overline{B}^o \to D^{*+}\pi^+\pi^-\pi^-\pi^o$	1.72±0.14±0.24
$\overline{B}^o \to D^{*+}\omega\pi^-$	0.29±0.03±0.04
$\overline{B}^o \to D^{+}\omega\pi^-$	0.28±0.05±0.03
$B^- \to D^{*o}\pi^+\pi^-\pi^-\pi^o$	1.80±0.24±0.25
$B^- \to D^{*o}\omega\pi^-$	0.45±0.10±0.07
$B^- \to D^{o}\omega\pi^-$	0.41±0.07±0.04

We determine the spin and parity of the $\omega\pi^-$ resonance (denoted A temporarily) by considering the decay sequence $B \to A\ D$; $A \to \omega\pi$ and $\omega \to \pi^+\pi^-\pi^o$.

The angular distributions are shown in Fig. 2. Here θ_A is the angle between the ω direction in the A rest frame and the A direction in the B rest frame; θ_ω is the orientation of the ω decay plane in the ω rest frame, and χ is the angle between the A and ω decay planes.

The data are fit to the expectations for the various J^P assignments. The ω polarization is very clearly transverse ($\sin^2\theta_\omega$) and that infers a 1^- or 2^+ assignment. The $\cos\theta_A$ distribution prefers 1^-, as does the fit to all

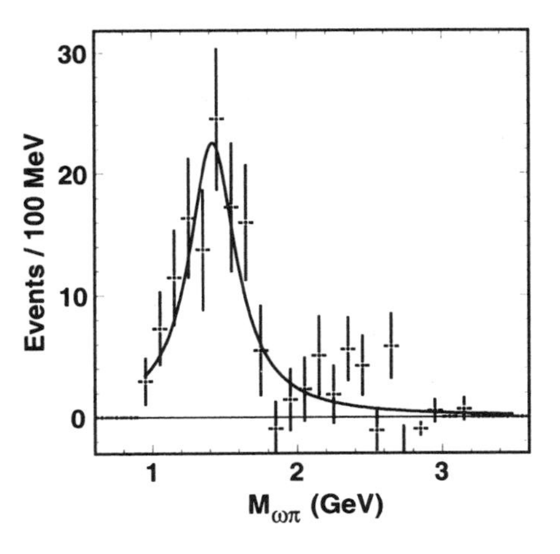

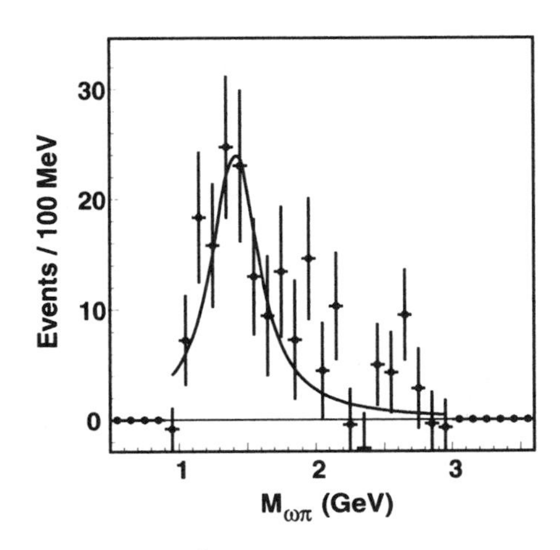

Figure 1. (left) The background subtracted $\omega\pi^-$ mass spectrum from $\overline{B}^o \to D^{*+}\omega\pi^-$ decays fit to a simple Breit-Wigner shape. (right) Same for $B \to D\omega\pi^-$ decays (D^o and D^+ are summed).

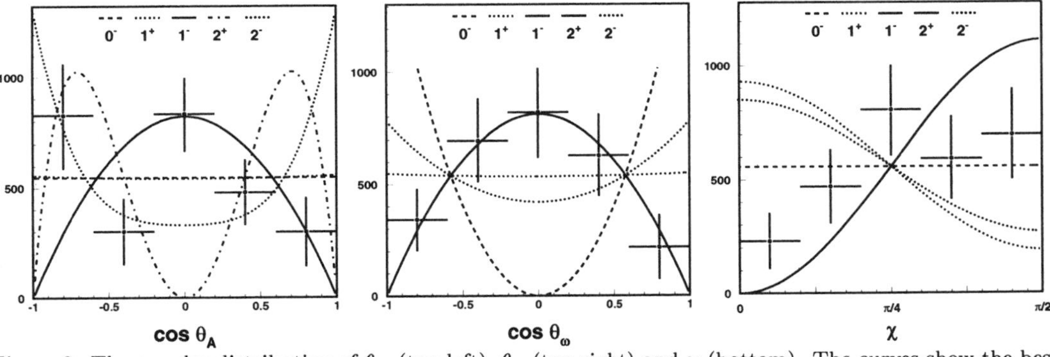

Figure 2. The angular distribution of θ_A (top-left), θ_ω (top-right) and χ (bottom). The curves show the best fits to the data for for different J^P assignments. The 0^- and 1^+ are almost indistinguishable in $\cos\theta_A$, while the 1^- and 2^+ are indistinguishable in $\cos\theta_\omega$ and χ. The vertical axis gives efficiency corrected events, 104 events are used.

three projections. Thus this state is likely to be the elusive ρ' resonance.[5] These are by far the most accurate and least model dependent measurements of the ρ' parameters. The ρ' dominates the final state. (Thus the branching ratios for the $D^{(*)}\omega\pi^-$ apply also for $D^{(*)}\rho'^-$.)

Heavy quark symmetry predicts equal partial widths for $D^*\rho'$ and $D\rho'$. We measure the relative rates to be $1.06\pm0.17\pm0.04$; consistent within our relatively large errors.

Factorization predicts that the fraction of longitudinal polarization of the D^{*+} is the same as in the related semileptonic decay $B \to D^*\ell^-\bar{\nu}$ at four-momentum transfer q^2 equal to the mass-squared of the ρ'

$$\frac{\Gamma_L\,(B \to D^{*+}\rho'^-)}{\Gamma\,(B \to D^{*+}\rho'^-)} = \frac{\Gamma_L\,(B \to D^*\ell^-\bar{\nu})}{\Gamma\,(B \to D^*\ell^-\bar{\nu})}\Big|_{q^2=m^2_{\rho'}} \qquad (1)$$

Our measurement of the D^{*+} polarization is $(63\pm9)\%$. The model predictions in semileptonic decays for a q^2 of 2 GeV^2, are between 66.9 and 72.6%.[6] Fig. 3 shows the measured polarizations for the $D^{*+}\rho'^-$, the $D^{*+}\rho^-$,[7] and the $D^{*+}D_s^{*-}$ final states. The latter is based on a new measurement using partial reconstruction of the D^{*+}.[8] Thus this prediction of factorization is satisfied.

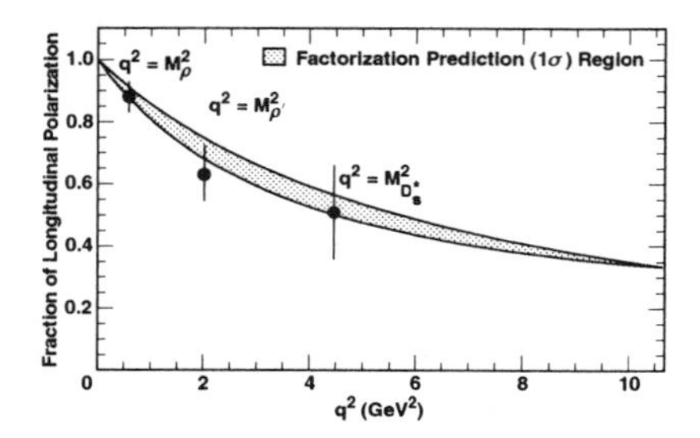

Figure 3. Measured D^{*+} polarization versus model predictions.

3 New Results from $B \to$ *Charmonium* Decays

3.1 $\frac{f_{oo}}{f_{+-}} = \frac{\mathcal{B}(\Upsilon(4S) \to B^o \overline{B}^o)}{\mathcal{B}(\Upsilon(4S) \to B^+ B^-)}$

We measured an important although often overlooked, parameter that effects every B^o or B^+ branching ratio measured on the $\Upsilon(4S)$ resonance. We determined the ratio of the number of $B^o \overline{B}^o$ pairs versus the number of $B^- B^+$ pairs by measuring the ratio of widths of two-body final states of the type $\psi^{(\prime)} K^{(*)}$. Since the ψ is I=0 and the B and K are I=1/2, the decay is pure ΔI=1/2 and therefore the individual channels, such as ψK, must have the same decay widths for B^o and B^+. We find $\frac{f_{oo}}{f_{+-}} = 1.04 \pm 0.07 \pm 0.04$, which yields nearly equal rates for charged and neutral B's, i.e. $f_{oo} = 0.49 \pm 0.02 \pm 0.01$.[9]

3.2 *Limits on CP Violation in* $B^{\mp} \to J/\psi$ *(or* ψ'*)* $K^{\mp}$

In the Standard Model only one decay amplitude dominates for decays of the type $B^{\mp} \to$ *charmonium* $K^{\mp}$. CLEO measured asymmetries consistent with zero for the J/ψ and ψ' cases: $(1.8 \pm 4.3 \pm 0.4)\%$ and $(2.0 \pm 9.1 \pm 1.0)\%$, respectively.[10]

3.3 *First Measurement of* $\mathcal{B}(B \to \eta_c K)$

We observe exclusive final states with the $c\bar{c}$ being the η_c.[11] The rates are $\mathcal{B}(B^- \to \eta_c K^-) = (6.9^{+2.6}_{-2.1} \pm 0.8 \pm 2.0) \times 10^{-3}$ and $\mathcal{B}(B^o \to \eta_c K^o) = (10.9^{+5.5}_{-4.2} \pm 1.2 \pm 3.1) \times 10^{-3}$. These rates are large, similar to those for the $J/\psi K$ final state.

4 Conclusions

Recent CLEO data has vastly increased our knowledge about hadronic B decays. We have seen the elusive ρ' and have the best measurement of its mass and width. Factorization when applied to polarizations appears to work in D^{*+} plus ρ, ρ' and D_s^* decays. The rates of the $\Upsilon(4S)$ into charged and neutral B's has been accurately determined and the first observation of $B \to \eta_c K$ decays has been made. Much however remains to be learned.

Acknowledgments

We thank N. Isgur, J. Rosner and J. Schechter for useful discussions. We thank the National Science Foundation for support.

References

1. C. Caso *et al.*, *The European Physical Journal* **C3** 1(1998).
2. G. Brandenburg *et al.* (CLEO), *Phys. Rev.* **D61**, 072002(2000).
3. ARGUS previously reported a signal of 28±10 events in $D^{*+}\pi^+\pi^-\pi^-\pi^o$, giving a branching ratio of (3.4±1.8)%. H. Albrecht *et al.*, *Z. Phys.* **C 48**, 543 (1990).
4. M. Artuso *et al.* [hep-ex/0006018].
5. A. B. Clegg and A. Donnachie, *Z. Phys.* **C 62**, 455 (1994).
6. N. Isgur and M. B. Wise, *Phys. Lett.* **B237**, 527 (1990); M. Wirbel *et al.*, *Z. Phys.* **C29**, 627 (1985); N. Neubert, *Phys. Lett.* **B264**, 455 (1991).
7. M. Artuso *et al.* (CLEO), *Phys. Rev. Lett.* **82**, 3020 (1999).
8. S. Ahmed *et al.*, [hep-ex/0008015].
9. J.P. Alexander *et al.*, [hep-ex/0006002].
10. G. Bonvicini *et al.*, [hep-ex/0003004].
11. K.W. Edwards *et al.*, [hep-ex/0007012].

N_C IN B-HADRON DECAYS

G. BARKER

IEKP, Universität Karlsruhe, Kaiserstr. 12, Postfach 6980, D-76128 Karlsruhe

E-mail: Gary.Barker@cern.ch

The current experimental status of measuring n_c, the mean number of charm(and anti-charm) quarks in b-hadron decay, is reviewed. A preliminary new analysis of the 'wrong-sign' charm rate from the DELPHI Collaboration is presented together with a new, preliminary world average value of $n_c = 1.154 \pm 0.036$.

1 Introduction

The quantity n_c, the mean number of charm (and anti-charm) quarks per b-quark decay, is of interest in order to test the compatibility of measurements with the theory of b-quark decay in the n_c against $\mathrm{BR}(b \to X\ell\nu)$ plane.

2 Measuring n_c

Current measurements of n_c use one of three main analysis techniques, the first of which is the **open-charm counting** method. Here, experiments count the number of D-hadron states reconstructed in the standard channels listed in Table 1. In general therefore, experiments measure the inclusive production rate for specific D-hadron states scaled by the standard decay channel branching ratio. n_c is then extracted from the data using the following relationship, [a]

$$n_c = BR(b \to \bar{D}^0, D^-, D_s^-, \bar{\Lambda}_c^- X) + 2 \cdot BR(b \to (c\bar{c})X), \quad (1)$$

where the second term is the branching ratio into charmonium. Published results[2] from this technique come from CLEO, ALEPH, DELPHI and OPAL.

The remaining two methods study b-hadron decay into **double-open charm** i.e. the process $b \to c\bar{c}s$ containing a 'wrong-sign'(W.S.) or 'upper-vertex' D-hadron from the decay of the W together with a right-sign(R.S.) D from the 'lower-vertex'. In an

[a]In this note charge conjugate reactions are implied.

Table 1. Exclusive D-channels reconstructed in open-charm counting analyses. Branching ratio values are from the PDG98[1].

Channel	Branching Ratio
$D^0 \to K^-\pi^+$	0.0385 ± 0.0009
$D^+ \to K^-\pi^+\pi^+$	0.091 ± 0.006
$D_s^+ \to \phi\pi^+$	0.036 ± 0.009
$\Lambda_c^+ \to pK^-\pi^+$	0.050 ± 0.013

inclusive approach, DELPHI[3] fits simultaneously for $BR(b \to no\ charm + X)$ and $BR(b \to double\ charm + X)$ using a charged track impact parameter based b-tag distribution. n_c can then be extracted using the following relationship,

$$n_c = 1 - BR(b \to no\ charm + X) + BR(b \to double\ charm + X) + 2 \cdot BR(b \to (c\bar{c})X). \quad (2)$$

$BR(b \to no\ charm + X)$ refers to the measured quantity and so includes contributions from $b \to u$, and $b \to sg$ transitions plus decays of the b-quark to charmonium. In **exclusive** approaches, the W.S. D-hadron is reconstructed so that in Equation 2 $BR(b \to double\ charm + X) = \sum_i BR(W^- \to D_i X)$ where $D_i = (\bar{D}^0, D^-, D_s^-, \bar{\Lambda}_c^-)$. $BR(b \to no\ charm + X)$ can be taken from the inclusive analysis. From this method there are published results from CLEO[4] and ALEPH[5] and a new preliminary result from DELPHI which is outlined in the next section.

Table 2. Inclusive b-branching fractions to exclusive D final states from open charm counting analyses. The second quoted error is due to D-branching ratio uncertainties listed in Table 1.

Decay channel	LEP	CLEO
BR($b \to D^0, \bar{D}^0 X$)	$59.3 \pm 2.3 \pm 1.4\%$	$64.9 \pm 2.4 \pm 1.5\%$
BR($b \to D^+, D^- X$)	$22.5 \pm 1.0 \pm 1.5\%$	$24.0 \pm 1.3 \pm 1.6\%$
BR($b \to D_s^+, D_s^- X$)	$17.3 \pm 1.1 \pm 4.3\%$	$11.8 \pm 0.9 \pm 2.9\%$
BR($b \to \Lambda_c^+, \bar{\Lambda}_c^- X$)	$10.2 \pm 1.0 \pm 2.7\%$	$5.5 \pm 1.3 \pm 1.4\%$

3 The DELPHI W.S. Charm Analysis

The new DELPHI analysis[6], exclusively reconstructs both R.S. and W.S. D^0 and D^+ mesons in b-hadron decays from the 1994-95 data set. From these candidates, the R.S. and W.S. components are separated into two tagged samples by use of a novel b-hadron decay flavour tag from the DELPHI inclusive b-physics package, BSAURUS[7]. Finally, the number of W.S. D-candidates in the W.S.-tagged sample ($N_{W.S.}$) is extracted by a fit to the data, exploiting the fact that the D-momentum spectrum in the B-rest frame is softer for W.S. D's than for R.S.. The number of R.S. D-candidates in the R.S.-tagged sample ($N_{R.S.}$), can follow simply from counting since the W.S. contamination is low ($\sim 1\%$).

The preliminary result is,

$$\frac{BR(B \to D^0 X)}{BR(B \to \bar{D}^0 X}) = \Delta(D^0) \cdot \frac{N_{W.S.}(D^0)}{N_{R.S.}(D^0)} = 12.9 \pm 2.8\% \quad (3)$$

$$\frac{BR(B \to D^+ X)}{BR(B \to D^- X)} = \Delta(D^+) \cdot \frac{N_{W.S.}(D^+)}{N_{R.S.}(D^+)} = 12.3 \pm 6.7\%, \quad (4)$$

where the errors are statistical only. Using the inclusive branching ratios for D^0 and D^+ from Table 2, the following W.S. branching ratios are extracted

$$BR(B \to D^0 X) = 6.8 \pm 1.3 \pm 1.1\% \quad (5)$$

$$BR(B \to D^+ X) = 2.5 \pm 1.3 \pm 1.1\%, \quad (6)$$

where the second error is systematic and includes contributions from the efficiency correction factors $\Delta(D^0), \Delta(D^+)$ in Equations 3 and 4 , D-momentum shapes, B-decay model dependence and background corrections.

4 n_c Averaging[8]

4.1 *Open-Charm Counting*

The average inclusive D-branching fraction results from open-charm counting experiments at LEP and CLEO are summarised in Table 2. The corresponding value for n_c follows by applying Equation 1 after accounting for the unmeasured charm-strange baryon contribution according to, $BR(b \to \Xi_c^{0,+}, \Xi_c^{0,-} X) = (0.4 \pm 0.3) \cdot BR(b \to \Lambda_c^+, \bar{\Lambda}_c^-$ $X)$. The resulting global average from this method is $n_c = 1.144 \pm 0.059$.

4.2 *Double-Open Charm*

Using Equation 2 the **inclusive** analysis of DELPHI [3] corresponds to $n_c = 1.151 \pm 0.048$. The results and averages from the **exclusive** analyses are collected in Table 3. By applying Equation 2, in the way described in Section 2, the corresponding average value of n_c is 1.196 ± 0.050.

5 Conclusion

The current averages of n_c obtained from open-charm counting (Section 4.1) and the double open-charm methods (Section 4.2), are seen to be in good agreement. A global correlated average across all cur-

Table 3. Double open-charm measurements from the exclusive analyses of CLEO[4], ALEPH[5] and DELPHI(see Section 3). Results are presented as W.S. charm inclusive branching ratios for each D-hadron type. The second quoted error is due to D-branching ratio uncertainties listed in Table 1.

	$BR(W^- \to \bar{D}^0, D^- X)$	$BR(W^- \to D_s^- X)$	$BR(W^- \to \bar{\Lambda}_c^- X)$
CLEO	$0.072 \pm 0.020 \pm 0.002$	$0.078 \pm 0.031 \pm 0.029$	0.009 ± 0.006
ALEPH	$0.110 \pm 0.038 \pm 0.007$	$0.168 \pm 0.044 \pm 0.051$	-
DELPHI	$0.092 \pm 0.022 \pm 0.002$	-	-
Average	0.085 ± 0.014	0.087 ± 0.043	0.009 ± 0.006

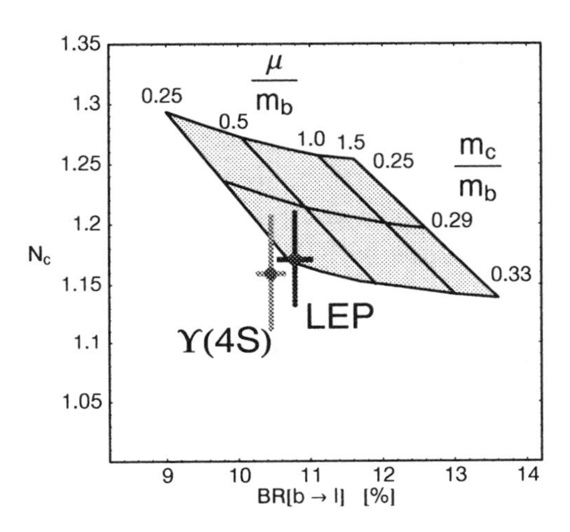

Figure 1. The current n_c against BR($b \to X\ell\nu$) plot. Here, the $\Upsilon(4s)$ branching ratio is taken from the PDG98[1] and the LEP value($10.79 \pm 0.25\%$) is the current average reported by Blyth[10], scaled to the $\Upsilon(4s)$ assuming the semi-leptonic width is the same for all B-species. The theory bound is from Neubert and Sachrajda[9].

rently available measurements gives the value $n_c = 1.154 \pm 0.036$. The present status of agreement between data and theory in the n_c against BR($b \to X\ell\nu$) plane is shown in Figure 1 where the n_c points have been calculated by applying the averaging procedure of Section 4 to LEP and CLEO data independently. The figure illustrates that, within the current errors, there is compatibility between results from LEP and the $\Upsilon(4S)$ but that the precision on n_c needs to improve before conclusions can be drawn regarding agreement with theory.

References

1. C. Caso et al., *Eur. Phys.* J **C3**, 1 (1998).
2. L. Gibbons et al., *Phys. Rev.* D **56**, 3783, (1997)
   D. Gibaut et al., *Phys. Rev.* D **53**, 4734, (1996)
   G. Crawford et al., *Phys. Rev.* D **45**, 752, (1992)
   D. Buskulic et al.,*Phys. Lett.* B **388**, 648 (1998)
   P. Abreu et al., CERN-EP/99-66, accepted by Eur. Phys. J. C.
   G. Alexander et al., *Z. Phys.* C **72**, 1., (1996).
3. DELPHI Collaboration, *Phys. Lett.* B **426**, 193 (1998).
4. T. E. Coan et al., *Phys. Rev. Lett.* **80**, 1150, (1998).
5. R. Barate et al., *Eur. Phys. J.*C **4**,557, (1998).
6. C. Schwanda, DELPHI 2000-105 CONF 404, Contrib. to ICHEP2000.
7. T. Allmendinger, G. Barker, M. Feindt, C. Haag, M. Moch, DELPHI 2000-069 PHYS 868.
8. P. Roudeau, detailed in CERN report, *Combined results on b-hadron production rates and decay properties.* In preparation.
9. M. Neubert and C. T. Sachrajda, *Nucl.Phys.* B **483**, 339, (1997).
10. S. Blyth, in Proc. of *ICHEP 2000*, Osaka, Japan, 2000.

STUDIES OF B MESON DECAYS TO CHARMED FINAL STATES WITH BELLE

K. HANAGAKI

Department of Physics, Princeton University, Princeton, NJ, 08544, USA
E-mail: kazu@bmail.kek.jp

We report on studies of B meson decays to charmed final states at Belle. The observation of the Cabibbo suppressed decays $B \to D^{(*)}K$, the search for $\bar{B^0} \to D_S^- \pi^+$, and the measurement of branching ratios $B(\bar{B^0} \to D^{(*)+}l^-\bar{\nu})$ are presented.

1 Introduction

Measurements of the angle ϕ_3 (or γ) and the length of the side (V_{ub}/V_{cb}) of the CKM unitarity triangle [1] are crucial, in addition to measurements of ϕ_1 (or β) and ϕ_2 (or α), in a test of Standard Model describing CP violation.

The decay $B \to D^{(*)}K$ provides a theoretically clean method to access the ϕ_3. Assuming the factorization, the Cabibbo suppressed decay $B \to D^{(*)}K^-$ is related at tree-level to the Cabibbo allowed $B \to D^{(*)}\pi^-$ by

$$R \equiv \frac{B(B \to D^{(*)}K^-)}{B(B \to D^{(*)}\pi^-)} \simeq \tan^2(\theta_c)(f_K/f_\pi)^2 \ ,$$

where θ_c is Cabibbo mixing angle, and f_K and f_π are the kaon and pion decay constants, respectively. Correcting for the difference in the available phase space by the corresponding τ lepton decays, we estimate the R to be about 0.074.

Since the decay $\bar{B^0} \to D_S^- \pi^+$ (charge conjugate modes are implied in this report) occurs via the transition $b \to uW^-(\to \bar{c}s)$, measuring $B(\bar{B^0} \to D_S^- \pi^+)$ enables us to extract $|V_{ub}|$. Using the measured value of $B(\bar{B^0} \to D_S^- D^+) = (8.0 \pm 3.0) \times 10^{-3}$ [2] and $|V_{cb}/V_{ub}|^2 = 0.0064$, one can naively estimate the $B(\bar{B^0} \to D_S^- \pi^+)$ to be $(5.1 \pm 2.3) \times 10^{-5}$.

The exclusive semileptonic decay $\bar{B^0} \to D^{(*)+}l^-\bar{\nu}$ is among the cleanest modes to measure $|V_{cb}|$ when interpreted the framework of heavy quark effective theory [3].

The data used in the analysis of $B \to D^{(*)}K$ and $\bar{B^0} \to D_S^- \pi^+$ ($\bar{B^0} \to D^{(*)+}l^-\bar{\nu}$) corresponds to an integrated luminosity of 5.3 fb^{-1} (2.7 fb^{-1}) accumulated at the $\Upsilon(4S)$ resonance and recorded in the Belle detector at the KEKB e^+e^- asymmetric collider.

2 $B \to D^{(*)}K$ decays

We reconstruct D^0 candidates using the $K^-\pi^+$, $K^-\pi^+\pi^0$, and $K^-\pi^+\pi^+\pi^-$ decay modes, and D^+ candidates in the $K^-\pi^+\pi^+$, $K_S\pi^+$, $K_S\pi^+\pi^0$, and $K_S\pi^+\pi^+\pi^-$ modes with the help of K/π identification for charged kaons. Candidate π^0's are formed from pairs of γ's with the energy greater than 30 MeV each. Two sigma cuts are imposed on the reconstructed D^0 mass. $B^- \to D^0K^-/D^0\pi^-$ events are selected by combining a D^0 and a K^- or π^- surviving the K/π identification cut. We calculate the beam constrained mass, $M_B \equiv \sqrt{E^{*2}_{beam} - p^{*2}_B}$, where p_B is the B^- momentum and E_{beam} is the beam energy in the $\Upsilon(4S)$ rest frame. The $\Delta E \equiv E^*_{D^0} + E^*_{h^-} - E^*_{beam}$, where $E^*_{D^0}$ is the D^0 energy and $E^*_{h^-}$ is the energy of K^-/π^- *assuming the pion mass*, is also used to select B^- decays. Event shape cuts are imposed to reduce continuum backgrounds.

Figure 1 shows the ΔE after requiring $5.27 < M_B < 5.29$ GeV$/c^2$. In the D^0K^- enriched sample by the K^- identification, two peaks exist at $\Delta E = -49$ MeV corresponding to D^0K^- and $\Delta E = 0$ by $D^0\pi^-$ feed down. To obtain the signal yield, we fit the ΔE

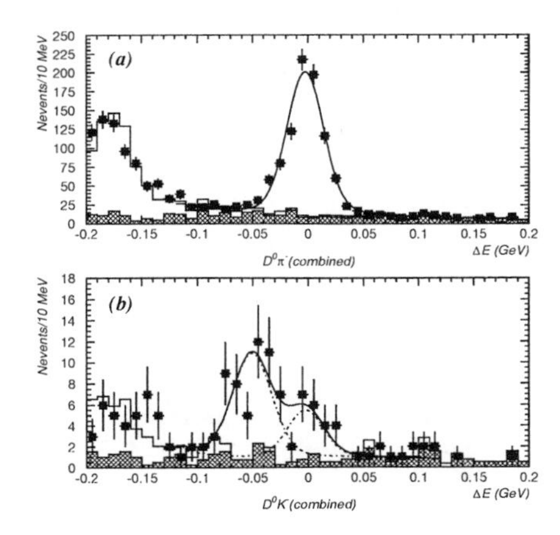

Figure 1. The ΔE distributions of the (a) $B^- \to D^0\pi^-$ and the (b) $B^- \to D^0K^-$ enriched sample. The dots with the error bars represent the data, the curves are the results of the fit, the opened histograms show the background level expected by MC, and the hatched histograms indicate the continuum component.

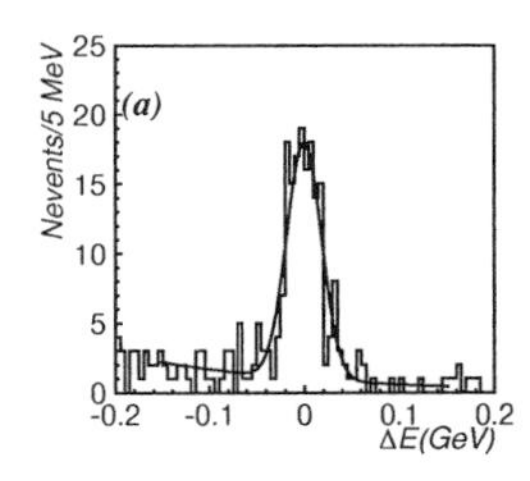

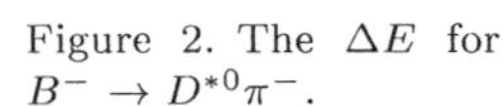

Figure 2. The ΔE for $B^- \to D^{*0}\pi^-$.

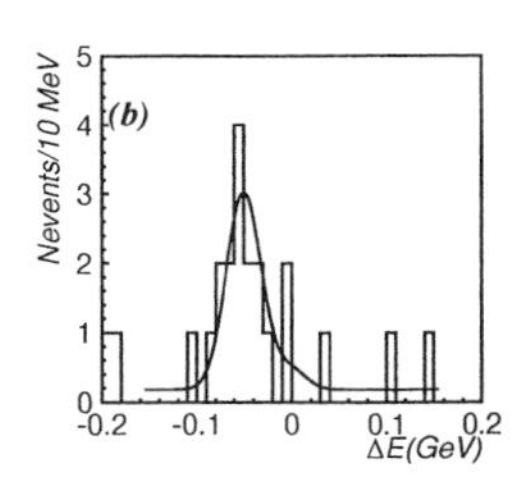

Figure 3. The ΔE for $B^- \to D^{*0}K^-$.

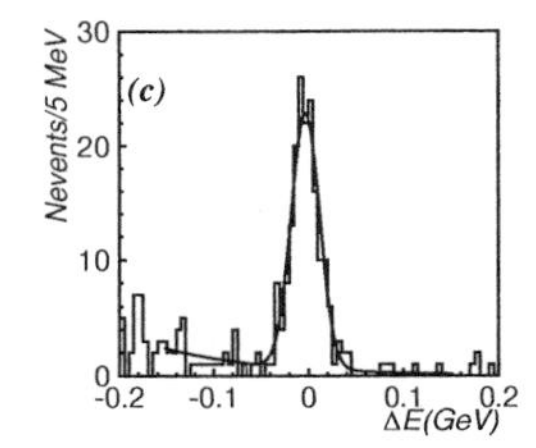

Figure 4. The ΔE for $\bar{B^0} \to D^{*+}\pi^-$.

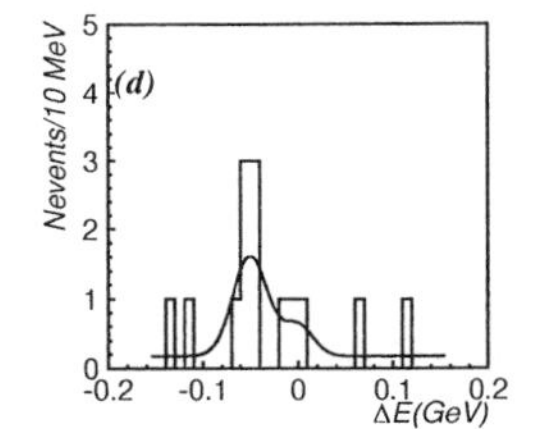

Figure 5. The ΔE for $\bar{B^0} \to D^{*+}K^-$.

distributions with a double Gaussian for signal and a MC determined background shape. The fit yields 48.7 ± 8.4 D^0K^- signal events and 832 ± 34 $D^0\pi^-$ events. Using the detection efficiency estimated by MC, 11 (13) to 26 (32)% depending on the D^0's decay mode for $B^- \to D^0K^-(\pi^-)$, we find the preliminary value of the R to be $0.081 \pm 0.014 \pm 0.011$, where the first (second) error is statistical (systematic). The systematic error is dominated by the uncertainty in estimating the number of signal events by the fit.

D^{*0} and D^{*+} candidates are reconstructed in the $D^{*0} \to D^0\pi^0$, $D^{*+} \to D^0\pi^+$, and $D^{*+} \to D^+\pi^0$ decay modes. A 4 sigma cut on the D meson mass and a 3 sigma cut on the $D\pi - D$ mass difference are imposed. M_B is calculated again with the pion mass assumption for a prompt K^-/π^-. Figures 2 through 5 show the ΔE distributions for $B^- \to D^{*0}h^-$ and $\bar{B^0} \to D^{*+}h^-$ processes. To extract the signal yields, the ΔE shapes are fitted with a single Gaussian for signal and linear background function. The statistical significance of $B^- \to D^{*0}K^-$ ($\bar{B^0} \to D^{*+}K^-$) signal is 4.5 (3) standard deviations. We find the preliminary ratios of $B(B^- \to D^{*0}K^-)/B(B^- \to D^{*0}\pi^-) = 0.134^{+0.045}_{-0.038} \pm 0.015$ and $B(\bar{B^0} \to D^{*+}K^-)/B(\bar{B^0} \to D^{*+}\pi^-) = 0.062^{+0.030}_{-0.024} \pm 0.013$.

3 A Search for $\bar{B^0} \to D_S^-\pi^+$

D_S's are reconstructed in the $D_S^- \to \phi(\to K^+K^-)\pi^-$ decay mode with the help of K/π identification. The mass cut of ± 10 MeV/c^2 which corresponds to 2 (2.5) sigma is applied in the D_S^- (ϕ) selection. We require $|\cos\theta_{helicity}| > 0.3$, where $\theta_{helicity}$ is the angle between D_S^- and K^+ in the ϕ's rest frame. Event shape cuts are used to suppress continuum backgrounds. As similar to $B \to DK$ analysis, the M_B and ΔE are calculated from the D_S^- and a prompt π^+. The signal region is defined as $5.274 < M_B < 5.2844$ GeV/c^2 and $|\Delta E| < 0.03$ GeV, which are 2 sigma cuts.

The ΔE vs M_B distribution of the remaining events is shown in Fig. 6. No events are observed in the signal region while 0.6

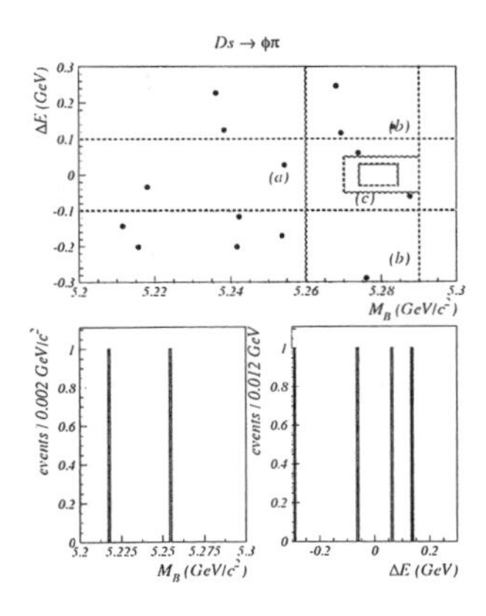

Figure 6. The ΔE vs M_B distribution in $\bar{B^0} \to D_S^- \pi^+$ search (top). The smallest box indicates the signal region. The lower plots show the M_B with $\Delta E < 0.05$ GeV, and ΔE with $5.27 < M_B < 5.29$ GeV$/c^2$.

background events (dominated by continuum events) are predicted by the sideband of M_B and ΔE. Using the efficiency of 17% by MC, we preliminarily set the 90% confidence level upper limit of $B(\bar{B^0} \to D_S^- \pi^+) < 2.1 \times 10^{-4}$.

4 $\bar{B^0} \to D^{(*)+} l^- \bar{\nu}$ Decays

We form $\bar{B^0} \to D^{*+} l^- \bar{\nu}$ decays from pairs of leptons and D^* meson candidates. Leptons are identified e^- (or μ^-) with the momentum between 1.0 (1.4) GeV$/c$ and 2.45 GeV$/c$ at the $\Upsilon(4S)$ rest frame. D^0 candidates are reconstructed in the $D^0 \to K^-\pi^+$ and $D^0 \to K^-\pi^+\pi^0$ decay modes. D^{*+}'s are selected by combining the D^0 candidate and a slow pion with the 2 MeV$/c^2$ cut on the mass difference. The missing mass squared defined as $M^2_{miss} \equiv (P_B - P_{D^*l})^2$, where P_B and P_{D^*l} are 4D momentum vector for B and D^*l system, enables to suppress background events. Using the efficiency estimated by MC, $\sim 1\%$ for $D^0 \to K^-\pi^+\pi^0$, and 5.6 (4.8)% for $D^0 \to K^-\pi^+$ plus e^- (μ^-), the preliminary branching ratio is measured to be $(4.74 \pm 0.25 \pm 0.51)\%$ where the first (second) error is statistical (systematic).

$\bar{B^0} \to D^+ l^- \bar{\nu}$ decays are reconstructed from pairs of positively identified leptons with the laboratory momentum greater than 0.8 GeV$/c$ and D^+ candidates. Using the particle identification devices, D^+ candidates are reconstructed in the $D^+ \to K^-\pi^+\pi^+$ decay channel, and required to be $1.0 < P_{D^+} < 2.5$ GeV$/c$ to reduce feed down and continuum background events. Using the hermeticity of the Belle detector, we extract neutrino's momentum from the missing momentum and energy in each event. This enables to fully reconstruct $\bar{B^0} \to D^+ l^- \bar{\nu}$ events as shown in Fig. 7. Using the efficiency of 2.21%, we obtain the preliminary result of $B(\bar{B^0} \to D^+ l^- \bar{\nu}) = (2.07 \pm 0.21 \pm 0.31)\%$.

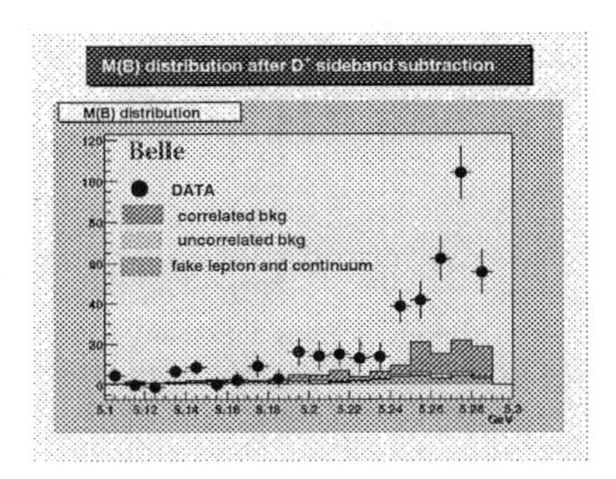

Figure 7. The M_B distribution for $\bar{B^0} \to D^+ l^- \bar{\nu}$. The dots show the data, and the histograms represent some expected background contributions.

Acknowledgments

I gratefully acknowledge the efforts of collaborators in Belle and KEKB group.

References

1. N. Cabibbo, *Phys. Rev. Lett.* **10**, 531 (1963); M. Kobayashi and T. Maskawa, *Prog. Theo. Phys.* **49**, 652 (1973).
2. Particle Data Group (C. Caso *el al.*), *Euro. Phys. J.* C **3**, 1 (1998).
3. N. Isgur and M.B. Wise, *Phys. Lett.* B **232**, 113 (1989); M. Neubert, *Phys. Rep* **245**, 259 (1994).

B DECAYS TO $D_S^{(*)}$ AND D^*

GLORIA VUAGNIN
representing the BABAR Collaboration

Sezione INFN di Trieste, Area di Ricerca, Padriciano 99, 34012 Trieste, Italy
E-mail: gloria.vuagnin@trieste.infn.it

The e^+e^- annihilation data recorded with the *BABAR* detector has been used to study B^0 decays to $D_s^{(*)+}$ and D^{*-} mesons. The production fraction of inclusive $D_s^{(*)+}$ and the corresponding momentum spectra have been determined. Exclusive decays $B^0 \to D^{*-}D_s^{(*)+}$ have been identified with a partial reconstruction technique and their branching ratios have been measured. Fully reconstructed B^0 decays in the hadronic modes $B^0 \to D^{*-}\pi^+$ and $B^0 \to D^{*-}\rho^+$ have been also studied and the measurement of their absolute branching ratios is reported.

1 Introduction

The study of $D_s^{(*)+}$ production in B^0 decays allows us to understand the mechanisms leading to the creation of $c\bar{s}$ quark pairs. The precise measurement of the momentum spectrum determines the fraction of two body and multibody decay modes, and consequently helps to understand the $b \to c\bar{c}s$ transitions. In this study we report a new measurement of the inclusive $D_s^{(*)+}$ production rate in B^0 decays and the branching fraction of two specific two-body B^0 decay modes involving a $D_s^{(*)+}$ meson. We also have performed a study, with full reconstruction, of the decay modes $B^0 \to D^{*-}\pi^+$ and $B^0 \to D^{*-}\rho^+$ and measured the corresponding branching fractions. These measurements are interesting for testing factorization models of B decays to open charm. Throughout this paper, conjugate modes are implied.

2 The dataset

The data were collected with the *BABAR* detector while operating in the PEP-II storage ring at the Stanford Linear Accelerator Center. For the inclusive $D_s^{(*)+}$ production in B^0 decays and the $B^0 \to D^{*-}D_s^{(*)+}$ branching fraction measurements we used a data sample equivalent to 7.7 fb^{-1} of integrated luminosity collected while running on the $\Upsilon(4S)$ resonance and a sample of 1.2 fb^{-1} collected 40 MeV below the $B\bar{B}$ threshold. The measurements of the branching fractions $B^0 \to D^{*-}\pi^+$ and $B^0 \to D^{*-}\rho^+$ use a subset of the same data sample corresponding to an integrated luminosity of 5.2 fb^{-1}.

3 Inclusive $D_s^{(*)+}$ production in B^0 decay

The D_s^+ mesons are reconstructed in the decay mode $D_s^+ \to \phi\pi^+$ where $\phi \to K^+K^-$. Particle identification is crucial to obtain a clean sample. Three charged tracks combining to from a common vertex are considered to be a D_s^+ candidate. Two of this tracks, with opposite charge, are required to be identified as kaons and their invariant mass must be within 8 MeV of the nominal ϕ mass. In this decay channel, the ϕ meson is polarized longitudinally which means the helicity angle of the decay, θ_H has a $\cos^2\theta_H$ dependence[1]. The requirement $|cos\theta_H| > 0.3$ keeps 97.5% of the signal while rejecting 30% of the background. The D_s^{*+} are reconstructed in the decay channel $D_s^{*+} \to D_s^+\gamma$ where $D_s^+ \to \phi\pi^+$. $\phi\pi^+$ combinations within 2.5σ of the nominal D_s^+ mass are taken as D_s^+ candidate. Photons must have a minimum energy of at least 50 MeV. The number of D_s^+

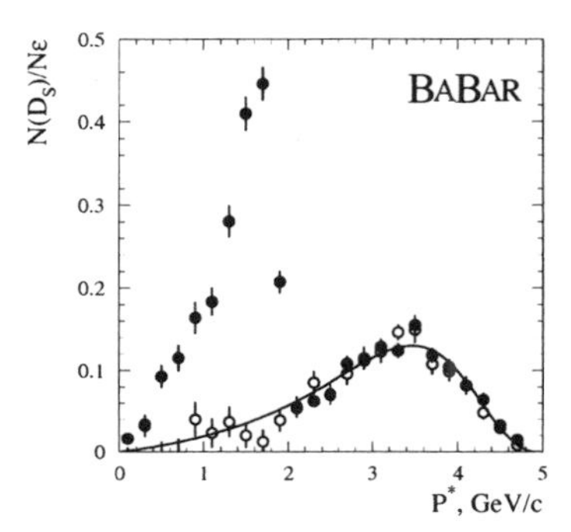

Figure 1. Momentum spectrum efficiency-corrected for D_s^+. Solid circles indicate the on resonance data point, while open circles are for data collected off resonance scaled according to the luminosity. The solid line shows the result of the fit with a Peterson fragmentation function.

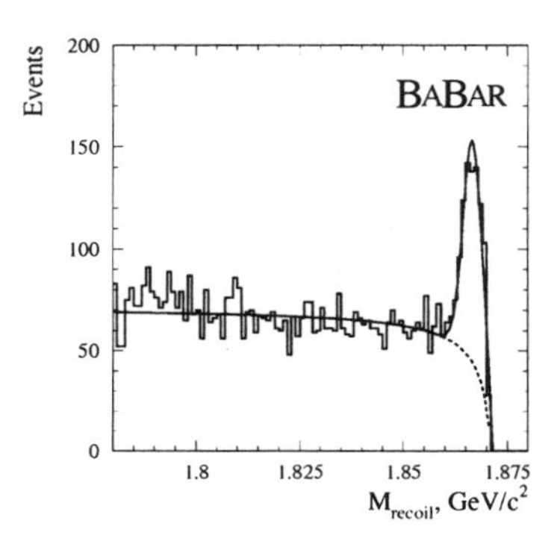

Figure 2. Missing mass distributions for the $D_s^{\pm} - \pi$ systems before background subtraction.

mesons is extracted by a Gaussian fit of the $\phi\pi^+$ invariant mass distribution for different momentum ranges in the $\Upsilon(4S)$ rest frame. Similarly, the number of D_s^{*+} is extracted by fitting the mass difference $m_{D_s^{*+}} - m_{D_s^+}$ distribution. The efficiency-corrected number of reconstructed D_s^+ as a function of their momentum is shown in Fig. 1.

In order to determine the $D_s^{(*)+}$ momentum spectrum for the continuum, on resonance data with momentum higher than 2.45 GeV/c and off resonance data, scaled according to the luminosity ratio, have been fitted after efficiency correction using the Peterson fragmentation function. The momentum spectrum of $D_s^{(*)+}$ produced in B^0 decays is obtained by subtracting the value of the fit function from the on resonance data after correcting for efficiency. The measured branching fractions are $(11.9 \pm 0.3 \pm 1.1) \times 10^{-2}$ and $(6.8 \pm 0.7 \pm 0.8) \times 10^{-2}$ for $B^0 \to D_s^+ X$ and $B^0 \to D_s^{*+} X$ respectively, assuming a $D_s^+ \to \phi\pi^+$ branching fractions of $3.6 \pm 0.9\%$.

4 Measurement of $B^0 \to D^{*-} D_s^{(*)+}$ branching fractions

The measurement of the branching fractions for the decays $B^0 \to D^{*-} D_s^+$ and $B^0 \to D^{*-} D_s^{*+}$ uses a partial reconstruction technique. The $D_s^{(*)+}$ are fully reconstructed, but no attempt is made to identify the $\bar{D}^0$ coming from the D^{*-} decay. Instead, we combine a $D_s^{(*)+}$ candidate with a π^- and assume their origin is a B^0 meson. We then calculate the missing invariant mass which should be the $\bar{D}^0$ mass if our hypothesis is correct. The yield of $B^0 \to D^{*-} D_s^{(*)+}$ is evaluated by fitting the missing mass distribution (Fig. 2) with a Gaussian and a background function[1]. The measured branching fractions are $(7.1 \pm 2.4 \pm 2.5 \pm 1.8) \times 10^{-3}$ for the cannel $B^0 \to D^{*-} D_s^+$ and for $B^0 \to D^{*-} D_s^{*+}$ $(2.5 \pm 0.4 \pm 0.5 \pm 0.6) \times 10^{-2}$ assuming a $D_s^+ \to \phi\pi^+$ branching fractions of $3.6 \pm 0.9\%$[2].

5 Measurement of $B^0 \to D^{*-}\pi+$ and $B^0 \to D^{*-}\rho^+$ branching fractions

B^0 candidates in the channel $D^{*-}\pi^+$ and $D^{*-}\rho^+$ are fully reconstructed using the decay chain $D^{*-} \to \bar{D}^0\pi^-$, followed by $\bar{D}^0 \to K^+\pi^-$. The ρ^+ is seen in its decay to $\pi^+\pi^0$. The selection of events is based on a few simple criteria. Tracks are required to originate from the beam spot and no particle identification is used. Photons with energy greater than 30 MeV are combined to form π^0 candidates. Kaons and pions with opposite charge and coming from the same vertex must have an invariant mass within $\pm 2.5\sigma$

of the nominal D^0 mass to form a D^0 candidate. The D^0 candidates are required to have a momentum greater than 1.3 GeV/c in the $\Upsilon(4S)$ frame and are combined with a pion to form a charged D^* candidate. We require $\Delta m = m(\bar{D}^0\pi^-) - m(\bar{D}^0)$ to be within 2.5σ of the nominal mass difference $D^{*-} - \bar{D}^0$. The D^{*-} is combine with a π^+ candidate, with a momentum greater than 500 MeV/c or a ρ^+ to form B^0 candidates. In the decay $B^0 \to D^{*-}\pi^+$ the longitudinal polarization of the D^{*-} is used to reduce background[3]. For the $B^0 \to D^{*-}\rho^+$ mode, ρ^+ candidate are selected requiring the $\pi^+\pi^0$ invariant mass within 150 MeV/c^2 of the ρ^+ nominal mass. Event shape variables are also used to remove continuum background.

For correctly reconstructed B^0 mesons, the energy of the B^0 candidate, $E^*_{B^0}$ must be equal to the beam energy E^*_b were both are evaluated at the $\Upsilon(4S)$ frame. We define $\Delta E = E^*_{B^0} - E^*_b$. The beam energy substituted mass, m_{ES} is defined as $m^2_{ES} = (E^*_b)^2 - (\sum_i \boldsymbol{p}_i)^2$, where $\boldsymbol{p}_i$ is the momentum of the ith daughter of the B candidate. The variables ΔE and m_{ES} are used to define the signal and sideband regions. For both modes, the region between 5.2 and 5.3 GeV/c^2 in m_{ES} and between ± 300 MeV in ΔE is used to study signal and background properties. By staying below $|\Delta E| = m_\pi$, we avoid correlated background from B decays where a real final state pion is either not included in the reconstruction or a random one is added to the observed state.

The measurement of branching fractions requires an estimate of the number of signal events. A Gaussian distribution and a background function[4], which parametrize how the phase space approach zero as the energy approaches E^*_b, are used to fit the m_{ES} distribution obtained requiring $|\Delta E| < 2.5\sigma_{\Delta E}$ as shown in Fig. 3. Based on the fitted yield of signal events the preliminary results for the branching fractions for $B^0 \to D^{*-}\pi^+$ and $B^0 \to D^{*-}\rho^+$ are $(2.9 \pm 0.3 \pm 0.3) \times 10^{-3}$ and $(11.2 \pm 1.1 \pm 2.5) \times 10^{-3}$ respectively. The branching fraction for $B^0 \to D^{*-}\rho^+$ includes all non-resonant and quasi-two-body contributions that lead to a $\pi^+\pi^0$ invariant mass in the ρ band. However, the acceptance for non-resonant $D^{*-}\pi^+\pi^0$ decays is about 15% of $D^{*+}\rho^+$ so that, combined with the known branching fraction for this mode, the non-resonant contribution to our result is expected to be quite small. Both branching fraction results compare well with previous measurements and with the world average[2].

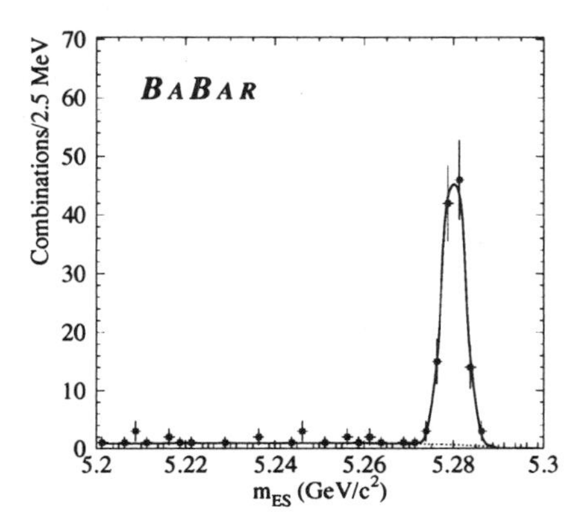

Figure 3. Distribution of m_{ES} for $|\Delta E| < 2.5\sigma_{\Delta E}$ for the cannel $B^0 \to D^{*-}\pi^+$.

References

1. *BABAR* collaboration, "Study of inclusive $D_s^{(*)\pm}$ production in B decays and measurement of $B^0 \to D^{*-}D_s^{(*)+}$ decays using a partial reconstruction technique", *BABAR*-CONF-00/13, SLAC-PUB-8535.
2. C. Caso *et al.*, Eur. Phys. Jour. C **3** (1998) 1.
3. *BABAR* collaboration, "Measurement of the branching fractions of $B^0 \to D^{*-}\pi^+$ and $B^0 \to D^{*-}\rho^+$", *BABAR*-CONF-00/06, SLAC-PUB-8528.
4. ARGUS Collaboration, H. Albrecht *et al.*, Z. Phys. **C48** (1990) 543; superseded results in *op cit.*, Phys. Lett. **B185** (1987) 218; Phys. Lett. **B182** (1986) 95.

MEASUREMENT OF $|V_{cb}|$ WITH $\bar{B}^0 \to D^{*+}\ell^-\bar{\nu}$ AT CLEO

K. M. ECKLUND

Lab of Nuclear Studies, Cornell University, Ithaca, New York, 14850, USA
E-mail: kme@mail.lns.cornell.edu

We determine the CKM matrix element $|V_{cb}|$ using a sample of 3.33 million $B\bar{B}$ events in the CLEO detector at CESR. We determine the yield of reconstructed $B \to D^{*+}\ell\nu$ decays as a function of $w = v_B \cdot v_{D^*}$, and from this we obtain the differential decay rate $d\Gamma/dw$. By extrapolating $d\Gamma/dw$ to $w = 1$, we extract the product $|V_{cb}|\mathcal{F}(1)$, where $\mathcal{F}(1)$ is the form factor at $w = 1$ and is predicted accurately by theory. We find $\mathcal{F}(1)|V_{cb}| = 0.0424 \pm 0.0018(\text{stat.}) \pm 0.0019(\text{syst.})$. We also integrate the differential decay rate over w to obtain $\mathcal{B}(B \to D^{*+}\ell\nu) = (5.66 \pm 0.29 \pm 0.33)\%$. All results are preliminary.

1 Introduction

The CKM matrix element $|V_{cb}|$ sets the length of the base of the familiar unitarity triangle. One strategy for determining $|V_{cb}|$ uses the decay $B \to D^*\ell\nu$. The rate for this decay, however, depends not only on $|V_{cb}|$ and well-known weak decay physics, but also on strong interaction effects, which are parametrized by form factors. These effects are difficult to quantify, but Heavy Quark Effective Theory (HQET) offers a method for calculating them at the kinematic point at which the final state D^* is at rest with respect to the initial B meson ($w \equiv v_{D^*} \cdot v_B = 1$; w is the relativistic boost γ of the D^* in the B rest frame). In this analysis, [1] we take advantage of this information: we measure $d\Gamma/dw$ for these decays, and extrapolate to obtain the rate at $w = 1$. The rate at this point is proportional to $[|V_{cb}|\mathcal{F}(1)]^2$ where $\mathcal{F}(w)$ is the form factor. Combined with the theoretical results, this gives $|V_{cb}|$.

The analysis uses $\bar{B}^0 \to D^{*+}\ell\nu$ decays. We divide the reconstructed events into bins of w. In each bin we extract the yield of $D^*\ell\nu$ decays using a fit to the distribution $\cos\theta_{B-D^*\ell}$, where

$$\cos\theta_{B-D^*\ell} = \frac{2E_B E_{D^*\ell} - m_B^2 - m_{D^*\ell}^2}{2|\mathbf{p}_B||\mathbf{p}_{D^*\ell}|}. \quad (1)$$

The angle $\cos\theta_{B-D^*\ell}$ is thus the reconstructed angle between the D^*-lepton combination and the B meson, computed with the assumption that the only missing mass is that of the neutrino. This distribution distinguishes $B \to D^*\ell\nu$ decays from decays such as $B \to D^{**}\ell\nu$, since $D^*\ell\nu$ decays are concentrated in the physical region, $-1 \le \cos\theta_{B-D^*\ell} < 1$, while the larger missing mass of the $D^{**}\ell\nu$ decays allows them to populate $\cos\theta_{B-D^*\ell} < -1$. Given the $D^*\ell\nu$ yields as a function of w, we fit for a parameter describing the form factor and the normalization at $w = 1$. This normalization is proportional to the product $[|V_{cb}|\mathcal{F}(1)]^2$.

2 Event Reconstruction

We do our analysis with 3.33 million $B\bar{B}$ events (3.1 fb^{-1}) produced on the $\Upsilon(4S)$ resonance at the Cornell Electron Storage Ring and detected in the CLEO II detector. [2]

We reconstruct D^{*+} candidates in the decay chain $D^{*+} \to D^0\pi^+$ followed by $D^0 \to K^-\pi^+$. D^0 candidates must have $|m_{K\pi} - 1.865| \le 0.020$ GeV. For a D^{*+}, we require $\Delta m = m_{K\pi\pi} - m_{K\pi}$ to be within 2 MeV of the D^{*+}-D^0 mass difference.

Lepton candidates are either electron candidates ($0.8 < p_e \le 2.4$ GeV) identified using the CsI calorimeter, or muon candidates ($1.4 < p_\mu \le 2.4$ GeV) which penetrate the muon system beyond 5 interaction lengths.

Exact reconstruction of w requires knowledge of the flight direction of the B meson. While this is unknown, our knowledge of $\cos\theta_{B-D^*\ell}$ limits it relative to that of the $D^*-\ell$ combination. We therefore compute w using the directions at each end of the range and we then take the average, giving a w resolution of 0.03. We divide our sample into 10 equal bins from 1.0 to 1.51.

3 Extracting the $D^*\ell\nu$ Yields

At this stage, our sample of candidates contains not only $D^*\ell\nu$ events, but also $B \to D^*X\ell\nu$ decays and various backgrounds. (In the following, we refer to $B \to D^{**}\ell\nu$ and non-resonant $B \to D^*\pi\ell\nu$ collectively as $D^*X\ell\nu$ decays.)

In order to disentangle the $D^*\ell\nu$ from the $D^*X\ell\nu$ decays, we use a binned maximum likelihood fit to the $\cos\theta_{B-D^*\ell}$ distribution. In this fit, the normalizations of the various background distributions are fixed and we allow the normalizations of the $D^*\ell\nu$ and the $D^*X\ell\nu$ events to float.

The distributions of the $D^*\ell\nu$ and $D^*X\ell\nu$ decays come from our signal Monte Carlo. The background shapes and normalizations are determined from data and/or Monte Carlo. We divide these backgrounds into five classes: continuum, combinatoric, uncorrelated, correlated and fake lepton.

Continuum background from $e^+e^- \to q\bar{q}$ events comprises about 4% of the events within the range $-1 < \cos\theta_{B-D^*\ell} \leq 1$ (the "signal region"). We estimate this background using data taken below the $\Upsilon(4S)$ resonance.

Combinatoric background events are those in which one or more of the particles in the D^* candidate does not come from a true D^* decay. This background contributes 6% of the events in the signal region. We take the $\cos\theta_{B-D^*\ell}$ distribution of combinatoric background events from the Δm sideband ($0.155 < \Delta m \leq 0.165$ GeV), normalized by a fit to the Δm distribution.

Uncorrelated background, which accounts for approximately 4% of the events in the signal region, arises when the D^* and lepton come from the decays of different B mesons. We estimate this background using Monte Carlo normalized to the inclusive yields of D^{*+} mesons and leptons observed in data.

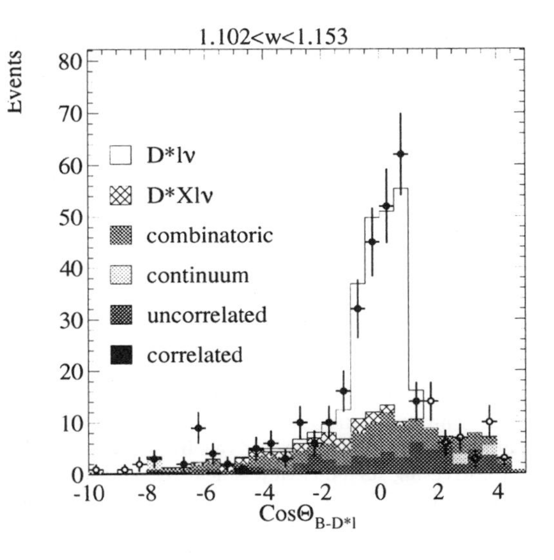

Figure 1. The event yields (circles) in the third w bin with the results of the fit superimposed. The fit range is $-8 \leq \cos\theta_{B-D^*\ell} < 1.5$, and is indicated with the solid circles.

Correlated background events are those in which the D^* and lepton are daughters of the same B, but the decay was not $B \to D^*\ell\nu$ or $B \to D^*X\ell\nu$. (*e.g.* $B \to D^*\tau\nu$ followed by leptonic τ decay.) This background accounts for less than 0.5% of the events in the signal region and is provided by Monte Carlo simulation.

Candidates with a fake lepton are negligible, due to excellent lepton identification.

Having obtained the distributions in $\cos\theta_{B-D^*\ell}$ of the signal and background components, we fit for the yield of $D^{*+}\ell\nu$ events in each w bin. A representative fit is shown in Figure 1. The quality of the fits is good, as is agreement between the data and fit distributions outside the fitting region.

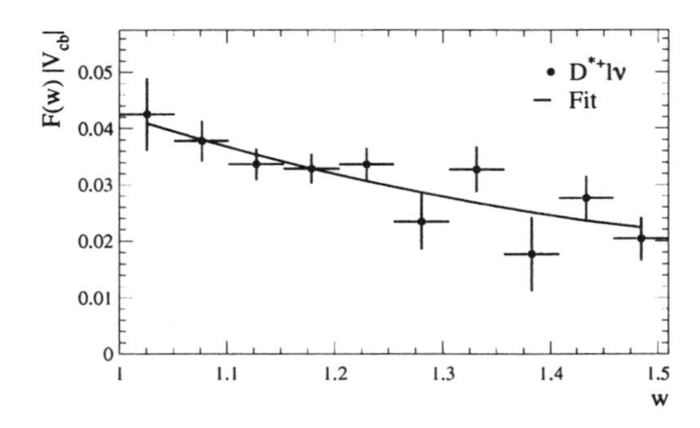

Figure 2. The results of the fit to the w distribution. The figure displays $|V_{cb}|\mathcal{F}(w)$, where the data points (solid circles) are derived from the yields after correcting for efficiency, smearing, and all terms in the differential decay rate apart from $|V_{cb}|\mathcal{F}(w)$. The curve shows the result of the fit.

4 The $|V_{cb}|$ Fit

The partial width for $B \to D^*\ell\nu$ decays is given by [3]

$$\frac{d\Gamma}{dw} = \frac{G_F^2}{48\pi^3}\left[|V_{cb}|\mathcal{F}(w)\right]^2 \mathcal{G}(w) \qquad (2)$$

where $\mathcal{G}(w)$ is a known kinematic function, and the form factor $\mathcal{F}(w)$ describes the hadronic $B \to D^*$ transition.

We use the result of recent work, [4,5] which uses dispersion relations to constrain the shape of the form factor. The parameterization has a single shape parameter ρ^2, the slope of the form factor at $w = 1$. It also depends on the form factor ratios R_1 and R_2, which depend only weakly on w; we use the values measured by CLEO, [6] consistent with theoretical expectations. [7]

In our analysis, we assume that the form factor has the functional form derived from dispersion relations and fit our yields as a function of w for $\mathcal{F}(1)|V_{cb}|$ and ρ^2. Our simple χ^2 fit adjusts $|V_{cb}|$ and ρ^2 to match the observed number of events in each w-bin, accounting for efficiency and w smearing.

The result of the fit is shown in Figure 2. We find

$$|V_{cb}|\mathcal{F}(1) = 0.0424 \pm 0.0018 \pm 0.0019 \qquad (3)$$

Table 1. The fractional systematic uncertainties (given in %).

Source	$\lvert V_{cb}\rvert\mathcal{F}(1)$	ρ^2	Γ
Comb. Bkgd	1.4	1.8	1.2
Uncorr. Bkgd	0.7	0.9	0.7
Corr. Bkgd	0.4	0.3	0.5
Slow π eff	3.1	3.7	2.9
K, π & ℓ eff	1.0	0.0	1.9
Lepton ID	1.1	0.0	2.1
$< p_B >$ & m_B	0.3	0.5	0.4
$D^*X\ell\nu$ model	0.2	1.9	1.9
Number of $B\bar{B}$	0.9	0.0	1.8
Subtotal	3.8	4.7	5.0
$\mathcal{B}(D^* \to D\pi)$	0.4	0.0	0.7
$\mathcal{B}(D \to K\pi)$	1.2	0.0	2.3
τ_B	1.0	0.0	2.1
$R_1(1)$ and $R_2(1)$	1.4	12.0	1.8
Subtotal	2.1	12.0	3.7
Total	4.4	13	6.2

$$\rho^2 = 1.67 \pm 0.11 \pm 0.22 \text{ and} \qquad (4)$$

$$\chi^2 = 3.1/8 \text{ dof.} \qquad (5)$$

with a correlation coefficient between $|V_{cb}|\mathcal{F}(1)$ and ρ^2 of 0.90. These parameters give $\Gamma = 0.0366 \pm 0.0018 \pm 0.0023$ ps^{-1}, and $\mathcal{B}(B \to D^{*+}\ell\nu) = (5.66 \pm 0.29 \pm 0.33)\%$. The quality of the fit is excellent. We note that the slope is higher than that found in the previous CLEO analysis [8] because of the curvature introduced into our form factor. If we use a linear form factor and the same subset of the data, we obtain results compatible with the earlier analysis.

5 Systematic Uncertainties

The systematic uncertainties are summarized in Table 1. The dominant systematic uncertainties arise from our background estimations and from our knowledge of the slow pion reconstruction efficiency.

6 Conclusions

We have fit the w distribution of $B \to D^*\ell\nu$ decays for the slope of the form factor and $|V_{cb}|\mathcal{F}(1)$. Our result in Equation 3 implies

$$|V_{cb}| = 0.0464 \pm 0.0020 \pm 0.0021 \pm 0.0021, \quad (6)$$

where we have used $\mathcal{F}(1) = 0.913 \pm 0.042$. [9] These results are consistent with LEP measurements [10] of $\mathcal{F}(1)|V_{cb}|$ and ρ^2, but somewhat higher. Our analysis benefits from small backgrounds and good resolution in w. These results are preliminary.

References

1. J. P. Alexander *et al.* (CLEO Collaboration), ICHEP 00-770, hep-ex/0007052.
2. Y. Kubota *et al.* (CLEO Collaboration), Nucl. Instrum. Methods Phys. Res., Sect. A **320**, 66 (1992).
3. J. D. Richman and P. R. Burchat, Rev. Mod. Phys., **67**, 893 (1995).
4. C. G. Boyd, B. Grinstein, R.F. Lebed, Phys. Rev. D **56**, 6895 (1997) (hep-ph/9705252).
5. I. Caprini,L. Lellouch and M. Neubert, Nucl. Phys. B **530**, 153 (1998) (hep-ph/9712417).
6. J. Duboscq *et al.*(CLEO Collaboration) PRL **76**, 3898 (1996).
7. M. Neubert, Physics Reports, **245**, 259 (1994).
8. B. Barish *et al.*(CLEO Collaboration), Phys. Rev. **D 51**, 1014 (1995).
9. BaBar Physics Book, P. F. Harrison and H. R. Quinn, editors, SLAC-R-504 (1998).
10. E. Barberio, these proceedings.

INCLUSIVE B-HADRON SEMILEPTONIC DECAYS AT LEP AND EXTRACTION OF $|V_{\rm cb}|$

SIMON C. BLYTH

Department of Physics, Carnegie Mellon University, Pittsburgh, USA
E-mail: simon.blyth@cern.ch

Recent results on inclusive semileptonic branching fractions of B hadrons produced in Z decays at LEP are presented. New measurement techniques and approaches to b-decay modelling uncertainties are reported. The method of combination of the LEP results is described and a comparison with results obtained at the $\Upsilon(4S)$ is made. An extraction of the CKM matrix element $|V_{cb}|$ is presented.

1 Introduction

Measurements of the b-hadron semileptonic branching fraction $BR(b \to \ell^- \bar{\nu}_\ell X)$ provide important tests of the modelling of heavy hadron dynamics and lead to the most direct determination of $|V_{cb}|$. The cascade branching fraction $BR(b \to c \to \ell^+ \nu_\ell X)$ is important as cascade decays are the principal background to the direct decays and this ratio is a vital input to other measurements.

2 LEP measurements

Common features of the analyses of ALEPH[1], DELPHI[2], L3[3] and OPAL[4] are to divide the selected hadronic events into two hemispheres by a plane perpendicular to the thrust axis and to use lifetime information to construct hemisphere b-tags. These tags provide unbiased, high-purity samples of b-hadron hemispheres opposite to which prompt leptons can be selected. The tagging purities are determined from the data by double-tag methods. Direct and cascade branching ratios can then be obtained from fits to the measured lepton (p,p_t) spectra, using models for the expected fragmentation of b-quarks and for the expected spectra of the assumed components, as illustrated in Fig. 1. The poorly known lepton spectra of the components lead to the largest uncertainties in this method.

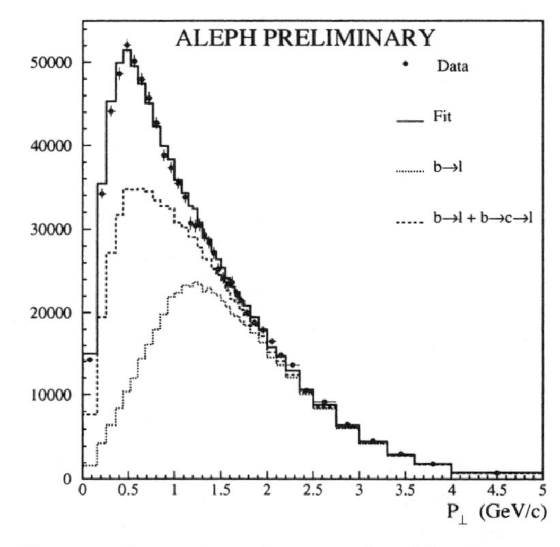

Figure 1. Comparison of measured and fitted p_t spectra together with corresponding expectations of contributions from $b \to \ell^- \bar{\nu} X$ and $b \to c \to \ell^+ \nu X$.

The principal approach adopted to reduce model dependencies is the use of charge correlation information in both di-lepton[1,2] and also single lepton events where weighted hemisphere charges[1,2,4] provide the second charge.

The preliminary ALEPH[1] charge correlation analysis simultaneously uses information from three independent samples. In two of them charge correlations between selected leptons and charge estimators from the opposite hemispheres are used and in the third only the number of leptons observed opposite to tightly $b-$tagged hemispheres is used, avoiding introduction of large spectra shape dependencies. The charge estimators are the lepton charge of a high p_t lepton and a hemisphere charge, constructed from momentum and impact parameter weighted charges.

DELPHI[2] have used four different analyses. One of them uses the charge correlation between the b quark and the lepton produced in its decay, where the quark charge is determined by reconstruction of the b-hadron decay vertex. Neural networks trained to separate b-hadron decay and fragmentation particles and to distinguish b and $\bar{b}$ jets were used.

L3[3] uses a double tag method with high p_t lepton and lifetime b-tags to determine tagging efficiencies, and hence BR($b \to \ell\bar{\nu}X$), together with R_b.

OPAL[4] combines the discriminating power of 8 input variables with two neural networks trained to distinguish direct and cascade decays from backgrounds, and performs a 2D fit to the outputs.

3 Lepton spectra modelling

Uncertainties arising from modelling the $b \to \ell$ spectra are estimated using a well defined prescription[5] of reweighting the lepton momentum in the b-hadron rest frame, obtained from the ACCMM model at which results are quoted, in order to use the other benchmark models ISGW(+1σ) and ISGW**(-1σ).

ALEPH[1] also uses the spectra from exclusive predictions and branching ratio measurements of B mesons into D,D^* and D^{**}, to give the expected $b \to \ell$ lepton spectra and variations in it. This method causes a slight reduction in uncertainties compared to the above prescription for their p_t fit analysis.

4 Results and comparisons

The direct and cascade branching ratio results are combined in a LEPEWWG fit including the additional parameters R_b, BR($c \to \ell^+$) and the average mixing parameter $\bar{\chi}$; the BR($b \to \ell^-$) measurements and the combined results are shown in Fig. 2 and

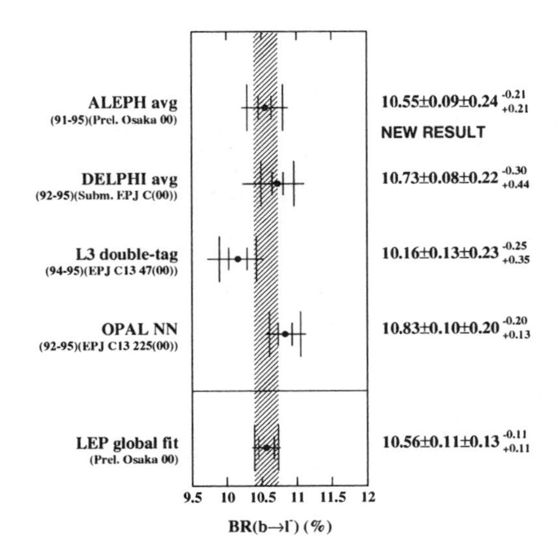

Figure 2. BR($b \to \ell^-$) results from the LEP experiments, together with the global fit result of the LEPHF EW working group. Uncertainties from statistical, systematic and modelling sources are given.

Table 1.

Table 1. Average BR results (%) with statistical, systematic and modelling uncertainties.

$b \to \ell^-$	$10.56 \pm 0.11 \pm 0.13 \pm 0.11$
$b \to c \to \ell^+$	$7.98 \pm 0.13 \pm 0.14 \pm 0.10$

Common input parameter values and systematic definitions are used for all measurements and the interdependencies and correlations between them are accounted for in the combination.

A comparison of the LEP results with results obtained at the $\Upsilon(4S)$ resonance[6] is shown in Fig. 3 together with HQET expectations[7]. The LEP BR($b \to \ell^-$) result has been corrected by the lifetime ratio $(\tau_{B^0} + \tau_{B^-})/2\tau_b = 1.021 \pm 0.013$[8], where τ_b is the average b-hadron lifetime measured at LEP. This correction accounts for the presence of B_s mesons and b-baryons in the LEP samples. Both the semileptonic branching ratio and charm multiplicity[9] results are in good agreement, but towards the lower end of expectations for m_c/m_b and μ/m_b.

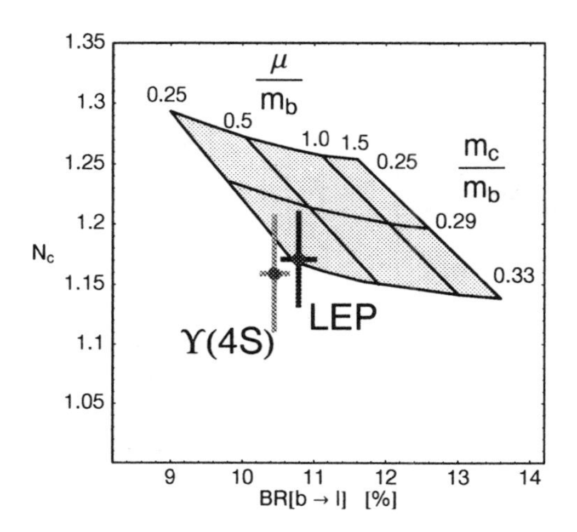

Figure 3. Charm multiplicity per b hadron decay N_c vs BR($b \to \ell$) from Z^0 and $\Upsilon(4S)$ results compared with HQET expectations for ranges of renormalisation scale μ/m_b and quark masses m_c/m_b.

5 Extraction of $|V_{cb}|$

HQET implemented through the operator product expansion[10] provides a relation between $\Gamma(b \to \ell^- \bar{\nu} X_c) = \mathrm{BR}(b \to \ell^- \bar{\nu} X_c)/\tau_b$ and $|V_{cb}|$, where X_c represents all states containing a charmed quark. Allowing $|V_{cb}|$ to be extracted from the inclusive semileptonic branching ratio, the charmless branching ratio $\mathrm{BR}(b \to \ell^- \bar{\nu} X_u) = (0.174 \pm 0.057)\%$[11] and the average b-hadron lifetime $\tau_b = 1.564 \pm 0.014$ ps[12].

$$|V_{cb}^{incl.}| = (40.70 \pm 0.45 \pm 2.03(th)) \times 10^{-3}$$

The 5% theory error[13] on this extraction[14] has been inflated by a factor of 2 compared to the original estimate[10].

6 Conclusion

Previously reported discrepancies between $\mathrm{BR}(b \to \ell^- \bar{\nu} X)$ measurements at the Z^0 and the $\Upsilon(4S)$ are now reduced to the 1.5σ level. The inclusive branching ratio provides our most precise determination of $|V_{cb}|$.

Acknowledgments

I would like to thank the representatives of the LEP experiments and of the various LEP working groups for providing their results and assistance.

References

1. ALEPH Collab., "Inclusive semileptonic branching ratios of b hadrons produced in Z decays", ICHEP2000-182.
2. DELPHI Collab., "Measurement of the semileptonic b branching fractions and average b mixing parameter in Z decays", Submitted to Euro. Phys. J.
3. L3 Collab., "Measurement of R_b and $\mathrm{Br}(b \to \ell\nu X)$ at LEP Using Double-Tag Methods.", M. Acciarri *et al.*, Eur. Phys. J. **C13** (2000) 47-61.
4. OPAL Collab., "Measurements of inclusive semileptonic branching fractions of b hadrons in Z^0 decays.", G. Abbiendi *et al.*, Eur. Phys. J. **C13** (2000) 225-240.
5. The LEP Experiments: ALEPH, DELPHI, L3 and OPAL, Nucl. Instrum. Meth., **A378** (1996) 101.
6. Review of Particle Physics, C. Caso *et al.*,(Particle Data Group), Euro. Phys. J. **C3** (1998) 1.
7. M. Neubert, C.T. Sachrajda, Nucl. Phys. **B483** (1997) 339.
8. Private uncorrelated ratio using LEP B lifetime WG updates for Osaka.
9. P. Roudeau private averages for Osaka.
10. I.I. Bigi, M. Shifman, N. Uraltsev, Annu. Rev. Nucl. Part. Sci. 47(1997) 591.
11. LEP $|V_{ub}|$ WG update for Osaka.
12. LEP B lifetime WG update for Osaka.
13. ALEPH, CDF, DELPHI, L3, OPAL, SLD, "Combined results on b-hadron production rates, lifetimes, oscillations and semileptonic decays", 19 March 2000, CERN-EP-2000-096/SLAC-PUB-8492.
14. LEP $|V_{cb}|$ WG update for Osaka.

V_{cb} MEASUREMENT AT LEP

E. BARBERIO

CERN/EP Geneva CH

This note summarises the LEP $|V_{cb}|$ extraction from $\overline{B}_d{}^0 \rightarrow D^{*+}\ell^-\overline{\nu}_\ell$ decays, and presents new measurements of narrow and broad excited charmed states in B meson semileptonic decays.

1 Introduction

Within the framework of the Standard Model of electroweak interactions, the elements of the Cabibbo-Kobayashi-Maskawa mixing matrix are free parameters, constrained only by the requirement that the matrix be unitary. Heavy Quark Effective Theory (HQET) provides a means to determine $|V_{cb}|$ with relatively small theoretical uncertainties, by studying the exclusive $\overline{B}_d{}^0 \rightarrow D^{*+}\ell^-\overline{\nu}_\ell$ decay process as function of the recoil kinematics of the D^{*+} meson.

The decay rate is parameterised as a function of the variable w, defined as the product of the four-velocities of the D^{*+} and the $\overline{B}^0$ mesons. This is related to the square of the four-momentum transfer from the $\overline{B}^0$ to the $\ell^-\overline{\nu}_\ell$ system, q^2, by

$$w = \frac{m^2_{D^{*+}} + m^2_{B^0} - q^2}{2m_{B^0}m_{D^{*+}}},$$

and its value ranges from 1.0, when the D^{*+} is produced at rest in the $\overline{B}^0$ rest frame, to about 1.50. Using HQET, the differential partial width for this decay is given by

$$\frac{d\Gamma}{dw} = \mathcal{K}(w)\mathcal{F}^2(w)|V_{cb}|^2$$

where $\mathcal{K}(w)$ is a known phase space term and $\mathcal{F}(w)$ is the hadronic form factor for the decay. Although the shape of this form factor is not known, its magnitude at zero recoil, $w = 1$, can be estimated using HQET. In the heavy quark limit ($m_b \rightarrow \infty$), $\mathcal{F}(w)$ coincides with the Isgur-Wise function [1,2] which is normalised to unity at the point of zero recoil. Corrections to $\mathcal{F}(1)$ have been calculated to take into account the effects of finite quark masses and QCD corrections [3] yielding $\mathcal{F}(1) = 0.88 \pm 0.05$ [4]. The unknown function $\mathcal{F}(w)$ is approximated with an expansion around $w = 1$ [5] and parameterized in terms of the variable ρ, which is the slope parameter at zero recoil. Theoretical predictions restrict ρ^2 in the range $0.14 < \rho^2 < 1.54$.

2 Exclusive semileptonic B decays at LEP

Since the phase-space factor $\mathcal{K}(w)$ tends to zero as $w \rightarrow 1$, the decay rate vanishes at $w = 1$ and the accuracy of the extrapolation relies on achieving a reasonably constant reconstruction efficiency in the region close to $w = 1$. At the Υ(4S), experiments have the advantage that w is approximately the boost of the D^* in the lab frame, which implies a good resolution for recontructing w. However, they suffer from low reconstruction efficiency near $w = 1$ because of a low efficiency to recontruct the slow pion from the low energy $D^{*+} \rightarrow \pi^+D^0$ decay. At LEP, B^0 are produced with a large variable boost (about 30 GeV in average). This makes the reconstruction of w tricky, requiring reconstruction of neutrino four-momentum, giving a relatively poor resolution. By contrast, LEP experiments benefit from an efficiency which varies only mildly with w, right down to $w = 1$.

The $|V_{cb}|$ determination from the LEP experiments is described in [6 7 8]. Due to the need to reconstruct the neutrino 4-momentum, at LEP, it is more difficult to distinguish signal events from B semileptonic decays to charm excited states which then decay to a D^{*+} (hereafter referred to as D^{**}). These decays are the major source of physics background and systematics for LEP analyses.

The D^{**} can be either a resonant narrow state, D_1 or D_2^*, or broad and/or non-resonant state, D_1(broad) or D_0^*(broad). The existence of resonant narrow states is well established [9] and a signal of a broad resonance has been seen by CLEO [10]. DELPHI, using data collected at the Z^0 from 1992 and 1995, performed a search for broad and non-resonant D^{**} states in B decays [11]. The DELPHI measurement of the B semileptonic branching fraction into D_1 and D_2^* is in good agreement with previous results [12 13]. The average of the DELPHI, ALEPH and CLEO narrow states

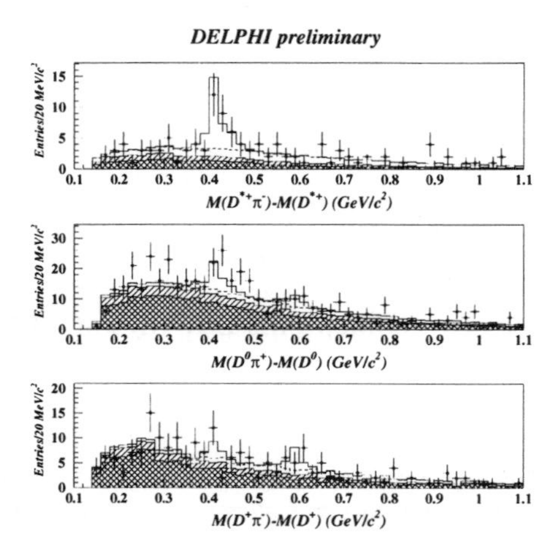

Figure 1. Invariant mass difference distributions for $\bar{B} \to D^{*+}\pi^-\ell^-X, D^0\pi^+\ell^-X, D^+\pi^-\ell^-X$ decay channels. The solid line is a fit to the data. The double-hatched histogram is the fake D background, the hatched histogram the contribution from fragmentation particles and the dashed lines are the contribution from non-narrow states.

branching fractions shows that the ratio [11]

$$R_{**} = \frac{Br(\bar{B}\to D_2^*\ell\bar{\nu})}{Br(\bar{B}\to D_1\ell\bar{\nu})}$$

is smaller than one, in disagreement with HQET models where infinite quark mass is assumed [14], but in agreement with models which take into account finite quark mass corrections [15]. In these models the B semileptonic branching fraction for D_0^*(broad) is predicted to be smaller that for the D_1, contrary to observation. This may indicate that the exceses of non-narrow D^{**} states, Figure 1, is due to a non-resonant D^{**} production, as predicted in [16].

Since HQET models predict the variation of the D^{**} form factors as a function of w only for resonant states (whose existence is established), in our evaluation of the systematic uncertainty from B to D^{**} decays we do not consider non-resonant D^{**} states.

The branching ratios of the B semileptonic D^{**} decays are taken from [4]. For the variation of the D^{**} form factors as a function of w, HQET predicts that in the infinite charm mass limit, the rate near $w = 1$ is suppressed by a further factor $(w^2 - 1)$ when compared with the signal [17]. In this case, the rate uncertainty would have a large effect on the slope with only a small influence on V_{cb} [7]. However, as discussed

Parameter	Value	Reference
R_b	$(21.68 \pm 0.07)\%$	19
f_d	$(40.0 \pm 1.0)\%$	18
$\tau(\bar{B}_d^0)$	(1.55 ± 0.02) ps	9
x_E	0.702 ± 0.008	19
$Br(D^{*+}\to D^0\pi^+)$	(68.3 ± 1.4) %	9
$Br(\bar{B}\to\tau\bar{\nu}_\tau D^{*+})$	$(1.27 \pm 0.21)\%$	see text
$Br(B^-\to D^{*+}\pi^-\ell\bar{\nu})$	(1.25 ± 0.16) %	4
$Br(\bar{B}_d^0 \to D^{*+}\pi^0\ell\bar{\nu})$	$(0.58 \pm 0.08)\%$	4
$Br(B_s\to D^{*+}K\ell\bar{\nu})$	$(0.61 \pm 0.22)\%$	4

Table 1. Values of the most relevant parameters affecting the measurement of V_{cb} .

before, models in this extreme case fail to predict the ratio R_{**} which is known from experiment to satisfy $R_{**} < 0.6$ [4]. A more precise treatment which accounts for $\mathcal{O}(1/m_c)$ corrections is proposed in [15]. Several possible approximations of the form factors are provided, depending on five different expansion schemes and on three different input parameters. To be conservative, each scheme proposed was tested. For each scheme, the input parameters were varied over the range consistent with the measured value of R_{**}. The systematic error due to the D^{**} background was computed as half the difference between the two extreme results.

3 Conclusion

The three LEP analyses have been performed using different inputs. In order to combine them, the central value of each analysis is adjusted according to the difference between the used and desired parameter values and the associated systematic error. The systematic error itself is then scaled to reflect the desired uncertainty on the input parameter. The common set of inputs are listed in Table 1. After this corrections, the LEP average gives:

$$\mathcal{F}(1)|V_{cb}| = (35.0 \pm 0.7_{stat} \pm 1.4_{sys}) \times 10^{-3}$$
$$\rho = 1.05 \pm 0.08 \pm 0.15$$

The confidence level of the fit is 65%. The error ellipses for the corrected measurements and for the LEP average are shown in Figure 2. The main uncertainty comes from the $b \to D^{**}\ell^-\bar{\nu}_\ell$ contribution (mainly the shape), which is fully correlated between experiments. The theoretical estimate $\mathcal{F}(1)$=0.88 ± 0.05 is used to

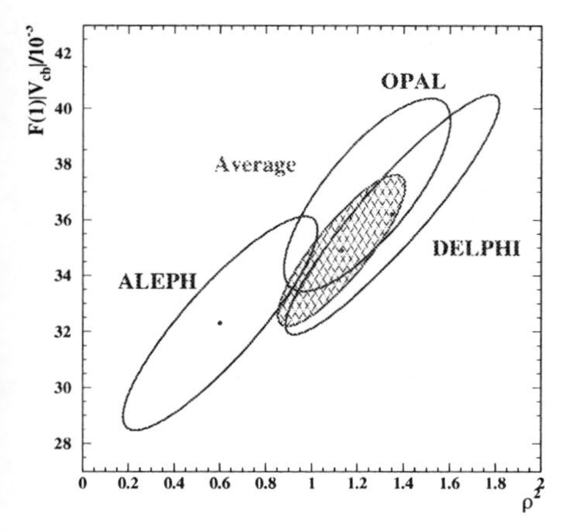

ure 2. Corrected $\mathcal{F}(1)|V_{cb}|$ values from the LEP experiments d LEP average. The values shown in the plot have been adted by the LEP working group and are those used to perform average. The original values can be found in [6 7 8] .

tain:

$$|V_{cb}| = (39.8 \pm 1.8_{exp} \pm 2.2_{theo}) \times 10^{-3}.$$

At LEP, $|V_{cb}|$ is determined also from the inclusive meson semileptonic decay [4]. The B meson semileptnic branching fraction provided by [19] for this conence is used to get:

$$|V_{cb}| = (40.7 \pm 0.5_{exp} \pm 2.0_{theo}) \times 10^{-3}.$$

is measurement is combined with the LEP average $|V_{cb}|$ obtained from $\overline{B}_d^0 \rightarrow D^{*+}\ell^-\overline{\nu}_\ell$ decays. The st important source of correlations between these o methods, are theoretical uncertainties in the evaltion of the average momentum of the b-quark ine the b-hadron. Theoretical uncertainties in the delling of $b \rightarrow \ell$ decays and the exact amount of $\rightarrow D^{**}$ decays are also taken as fully correlated. Untainties from lepton identification and background ntribute to the correlated error but to a much lesser tent. The combined value is (Figure 3):

$$|V_{cb}| = (40.4 \pm 1.8) \times 10^{-3}$$

ere within the total error of 1.8, 1.0 comes from rrelated sources.

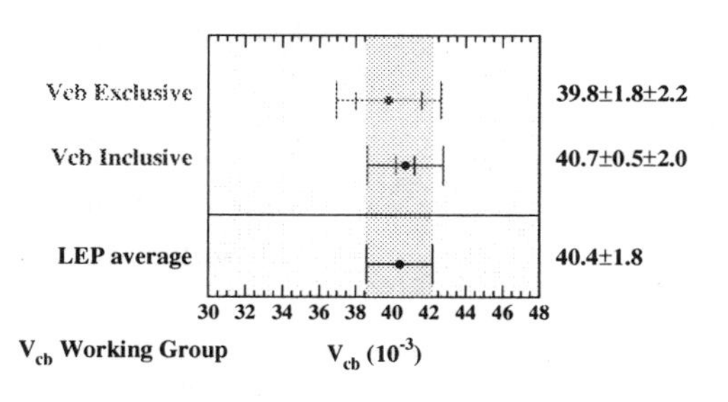

Figure 3. $|V_{cb}|$ LEP average

References

1. N.Isgur and M.Wise, Phys.Lett. **B232** (1989); N.Isgur and M.Wise, Phys.Lett. **B237** (1990).
2. A.F.Falk, H.Georgi, B.Grinstein and M.B.Wise, Nucl.Phys. **B343** (1990).
3. M.Luke, Phys.Lett. **B252** (1990).
4. ALEPH, CDF, DELPHI, L3, OPAL, SLD, CERN-EP-2000-096.
5. Caprini, Lellouch and Neubert, Nucl.Phys. **B530** (1998).
6. ALEPH coll, Phys.Lett. **B359** (1995).
7. DELPHI Coll, contr. 4-518 to HEP 99.
8. OPAL coll., Phys.Lett. **B482** (2000).
9. Particle Data Group, C.Caso *et al.*, Eur.Phys.J. **C3** (1998).
10. CLEO coll, CLEO CONF 99-6, hep-ex/9908009 (1999).
11. DELPHI coll., DELPHI CONF 405, contributed paper for ICHEP2000.
12. ALEPH coll, Zeit.Phys. **C73** (1997).
13. CLEO coll, Phys.Rev.Lett. 80 (1998).
14. V.Morenas et all., Phys.Rev. **D56** (1997); M.Q.Huang and Y.B.Dai, Phys.Rev. **D59** (1999); M.Oda et all., hep-ph/0005102 (2000).
15. A.K.Leiboich, Z.Ligeti, I.W.Steward, M.B.Wise, Phys.Rev. **D57** (1998); Phys.Rev. Lett. **78** (1997).
16. N.Isgur, Phys.Rev. **D60** (1999).
17. The BaBar Physics Book, SLAC-R-504, Chap. 8.
18. LEP B Oscillations Working Group.
19. LEP/SLD Electroweak Heavy Flavour Results Summer 2000 Conferences, http://lepewwg.web.cern.ch/LEPEWWG/heavy/.

RARE B MESON DECAYS IN CLEO

R. STROYNOWSKI

Physics Department, Southern Methodist University, Dallas, Texas 75275-0175, USA
E-mail: ryszard@mail.physics.smu.edu

We report the measurements of rare, 2-body, charmless B mesons decays made with the full CLEO data sample. All four $K\pi$ modes have been now observed as well as the $\pi^+\pi^-$ and $\pi^+\pi^0$ final states. In addition to the decays into π or K with η, η' new final states consisting of pseudoscalar and a vector particle have been observed. Branching fractions for rare B decays reach a $0.5 - 2.0 \times 10^{-5}$ level. New limits have been set on fully leptonic B decays.

1 Introduction

One of the main goals of the B physics studies today is to construct a coherent description of the weak quark couplings and their phases and to establish CP violation outside the kaon sector. Charmless hadronic B decays are expected to proceed via a combination of loop ("penguin") and tree diagrams. In principle, the ratios of the branching fractions for such decays can allow to extract the information on angles of the unitarity triangle. In practice, there are both theoretical and experimental problems: 1) there is usually more than one contributing diagram and the resulting interference necessitates measurements of many related modes and 2) both types of diagrams are heavily suppressed and the corresponding branching fractions are very small. B decays to fully leptonic final states are either strongly suppressed or, as in the case of the decay $B \to e\mu$, forbidden. Studies of such states provide strong limits on the contribution of new physics.

All results presented here are based on a complete CLEO data sample of about 13 fb^{-1} collected on $\Upsilon(4S)$ resonance and in the continuum below the $b\bar{b}$ threshold. The analyses follow the general scheme based on maximum likelihood fits described for past searches of rare 2-body B decays [1].

2 B Decays to $K\pi$ and $\pi\pi$ Final States

We have made the first significant observation of all B decays to $K\pi$ and to the $\pi^+\pi^-$ final states [2]. The measured branching fractions are listed in Table 1. Upper limits for the $\pi^\pm\pi^0$, $\pi^0\pi^0$ and for all KK channels are small. The rate for B decays to $\pi^+\pi^-$ is smaller than that for $\pi^\pm\pi^\mp$. This is inconsistent with the expectations for the strong phase enhancement of this process. The hierarchy of rates for the decays into KK, $K\pi$ and $\pi\pi$ indicates large contribution of the gluonic penguin processes.

Table 1. Measured Branching Ratios

Mode	$\mathcal{B}(10^{-6})$
$B \to K^\pm\pi^\mp$	$17.2^{+2.5}_{-2.4}\pm1.2$
$B \to K^0\pi^\pm$	$18.2^{+4.6}_{-4.0}\pm1.6$
$B \to K^\pm\pi^0$	$11.6^{+3.0+1.4}_{-2.7-1.3}$
$B \to K^0\pi^0$	$14.6^{+5.9+2.4}_{-5.1-3.3}$
$B \to \pi^\pm\pi^\mp$	$4.3^{+1.6}_{-1.4}\pm0.5$
$B \to \pi^\pm\pi^0$	< 12.7
$B \to \pi^0\pi^0$	< 5.7
$B \to K^+K^-$	< 1.9
$B \to K^\pm K^0$	< 5.1
$B \to K^0\bar{K^0}$	< 17

3 Decays into η, η'

The updated results [3] for the B decay branching fractions into *eta* and *eta'* confirm past CLEOobservation that the rate of decays into η' is larger than that into η. The intrinsic charm content of the η' has been proposed, but no enhancement of the corresponding η_c production has been observed [4]. The pattern of the branching fractions is compatible with the that resulting from constructive interference of gluonic penguin diagrams for $\eta' K$ and ηK^* and destructive for ηK and $\eta' K^*$.

4 B Decays to Pseudoscalar - Vector Final States

The exclusive hadronic decays to pseudoscalar-vector final states are of special interest. They allow for the determination of the angle α of the unitarity triangle. Such determination will require a full Dalitz analysis of the final states with different charge combinations. CLEO data are not yet sensitive enough to perform such analysis. We report here on the search for the B decays to pion or kaon and a vector meson ρ, K^* or ω. All charge combination were searched for in a combined fits allowing for separation of pions and kaons. The likelihood contours for such fits are illustrated in Fig. 1. There are three channels for which we obtain measurements of the branching fractions with significance greater than 5σ: $B^- \to \pi^-\rho^0$, $B^0 \to \pi^\pm\rho^\mp$ and $B^- \to \pi^-\omega$. The results are listed in Table 2. Final states with other charge combinations or with a kaon and a vector meson have branching fractions below CLEO sensitivity. The measurements and the estimated upper limits indicate a generally good agreement with theoretical factorization based models[6] which predict low rates for the $\Delta S = 1$ transitions. Less understood is the ratio of the decay rates of the charged and neutral B decays into $\pi\rho$, which is smaller than expected.

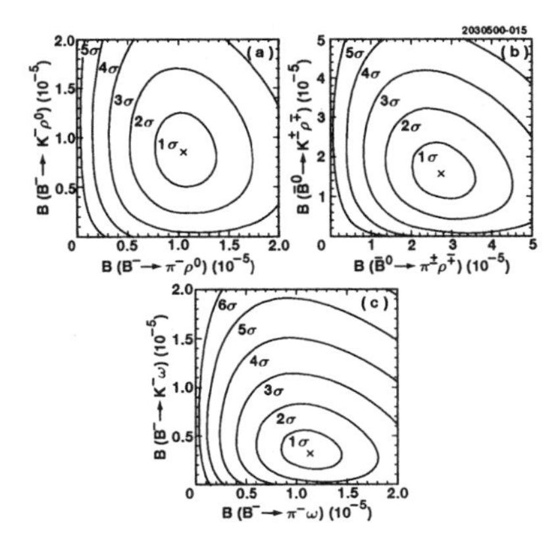

Figure 1. Likelihood contours and n standard deviations.

Table 2. Pseudoscalar-Vector Branching Fractions

Mode	$\mathcal{B}(10^{-6})$
$B^- \to \pi^-\rho^0$	$10.4^{+3.3}_{-3.4}\pm 2.1$
$B^0 \to \pi^\pm\rho^\mp$	$27.6^{+8.4}_{-7.4}\pm 4.2$
$B^- \to \pi^-\omega$	$11.3^{+3.3}_{-2.9}\pm 1.4$

5 $B \to \phi K$ Decay

The B decay to ϕK can only proceed via a gluonic penguin process in which a gluon splits into two strange quarks $g \to s\bar{s}$. The final state is simple. Since the ϕ meson is narrow, the experimental signature of such decay is clean. Inclusive $B \to \phi X$ rate was calculated [7] to be $(0.6 - 2.0) \times 10^{-4}$. The fraction of the inclusive rate going into kaon is expected to be 6 - 10 %. The CLEO fit, illustrated in Fig. 2, gives the a branching fraction $BR(B \to \phi K) = 6.2^{+2.0+0.7}_{-1.8-1.7} \times 10^{-6}$, which is in good agreement with the expectations.

6 Leptonic B Decays

In the Standard Model B decays into lepton pairs are highly suppressed. The expected branching fractions of $\mathcal{B}(B^0 \to e^+e^-) = 2.6 \times 10^{-15}$ and $\mathcal{B}(B^0 \to \mu^+\mu^-) = 1.1 \times 10^{-10}$ are too small to be observed at CLEO. Lep-

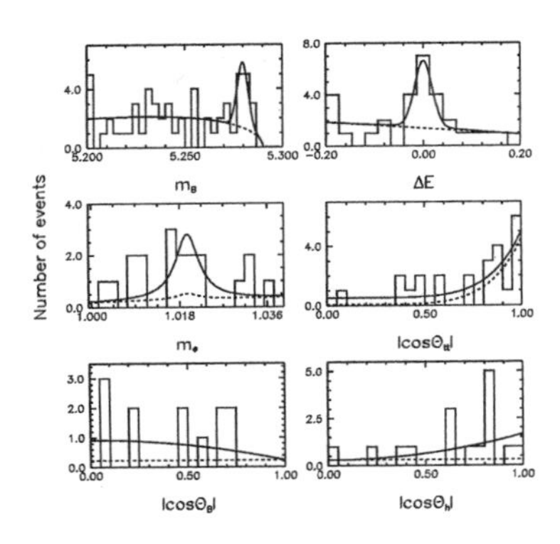

Figure 2. Fit projections onto six kinematical variables used to select candidate events. The dashed line represents the background.

ton number violating decay $B \to e^{\pm}\mu^{\mp}$ is forbidden. The observation of fully leptonic decays at CLEO would provide an evidence for new physics. The decay $B \to \tau\nu_\tau$ is particularly interesting because of its simple dependence on the decay constant f_B and the CKM matrix element V_{ub}. The experimental signature for e^+e^-, $\mu^+\mu^-$ and $e\mu$ decays is a pair of charged leptons with invariant mass end total energy compatible with that of the parent B meson. The search for the $\tau\nu$ decays is complicated by the missing neutrinos. We used a sample of about 8200 fully reconstructed B mesons to search for a one prong τ decays from the other B. In all cases we find no evidence for fully leptonic decays [8] and the corresponding upper limits on the branching fractions are listed in Table 3.

Table 3. Upper limits for leptonic B decays

Mode	$\mathcal{B}$
$B^0 \to e^+e^-$	8.3×10^{-7}
$B^0 \to \mu^+\mu^-$	6.1×10^{-7}
$B^0 \to e^{\pm}\mu^{\mp}$	15×10^{-7}
$B^+ \to \tau\nu_\tau$	8.4×10^{-4}

7 Conclusions

Over past two years CLEO measured or set upper limits on over 60 exclusive charmless hadronic and fully leptonic B decays. All measured branching fractions are small and their pattern indicates existence of many contributing and interfering diagrams. The general agreement with theoretical expectations is probably due to the insufficient experimental precision. The experimenter's view of the theoretical situation is best illustrated in Fig. 7.

References

1. T.E. Coan *et al.*, *Phys. Rev.* D **59**:111101 (1999).
2. D. Cronin-Hennessy *et al.*, *Phys. Rev. Lett.***85**,515 (2000).
3. S.J. Richichi *et al.*, *Phys. Rev. Lett.***85**,520 (2000).
4. S. Stone, this proceedings.
5. C.P. Jessop *et al.*, hep-ex/0006008.
6. A. Ali *et al.*,*Phys. Rev.* D **59**:014005 (1999).
7. N. G. Deshpande and X. He, *Phys. Lett.***B336**, 471 (1994).
8. T. Bergfeld *et al.*, hep-ex/0007042.

STUDIES OF CHARMLESS HADRONIC DECAYS OF B MESONS WITH BELLE

PAOTI CHANG

Department of Physics, National Taiwan University, No. 1, Sec. 4, Roosevelt Rd., Taipei, Taiwan, ROC

E-mail: pchang@phys.ntu.edu.tw

We report preliminary results on studies of various charmless hadronic B decays using a 5.3 fb^{-1} data sample collected at the $\Upsilon(4S)$ resonance with the Belle detector at the KEKB e^+e^- storage ring. Final states with $B \to K^+h$ and $B \to \pi^+h$ can be clearly separated using the high momentum particle identification system of the Belle detector. The obtained branching ratios of all $B \to K\pi$ as well as the 90% C.L. upper limits of $B \to \pi\pi$ and $B^0 \to K^+K^-$ will be given. We also report the first observation of $B^+ \to \phi K^+$ decay and its corresponding branching ratio is $(1.72^{+0.67}_{-0.54} \pm 0.18) \times 10^{-5}$.

1 Introduction

B meason decays to charmless hadronic final states provide a rich sample to test Standard Model and to probe for new physics[1]. It has been suggested that an isospin analysis can be performed to extract the second CP violation angle $\phi_2(\alpha)$ using the decay rates of $B^+ \to K^+\pi^0, B^+ \to K^0\pi^+$ and $B^0 \to K^+\pi^-$, along with the rates and the CP asymmetry in $B^0 \to K^0_S\pi^0$. The third CP violation angle $\phi_3(\gamma)$ may be also determinable via the $\pi\pi$ and $K\pi$ branching ratios. The rare decay, $B^+ \to \phi K^+$, is also interesting because there is no tree level contribution to the decay amplitude within the frame work of standard model. Any larger than expected branching fraction may suggest new physics.

2 Data Analysis

In this paper we report the results of $B \to hh$ (h stands for K^+, π^+, or K_S) and $B^+ \to \phi K^+$ searches using 5.3 fb^{-1} of data (5.1 fb^{-1} for h^+h^- mode). This data sample was collected by the Belle detector[2] on the $\Upsilon(4S)$ resonance at the KEKB asymmetric e^+e^- storage ring[3]. The beam energy is 3.5 GeV for positrons and 8.0 GeV for electrons. The detailed of the Belle detector is described in Ref. 2.

B candidates are identified using the beam constrained mass, m_B, and energy difference, ΔE. Hadron identification is provided by the ACC and CDC. Since the charged hadrons directly from B two-body decays have momenta above 1.5 GeV/c, TOF information is not used. We combine the K/π probability from the ACC and dE/dx to form a K/π likelihood $L(K)/L(\pi)$. K-π separation is then achieved by cutting on the likelihood ratio, $L_K/(L_\pi + L_K)$. Charged tracks with likelihood ratio greater than 0.6 are identified as kaons and less than 0.4 as pions (see Fig. 1). As for kaons from ϕ decays, we apply a rather loose cut, likelihood ratio (TOF information is used) greater than 0.1, to reduce combinatoric background. The efficiencies and fake rates of kaon/pion identification are determined using continuum D^* decays: $D^{*+} \to D^0\pi_s^+, D^0 \to K^-\pi^+$. The identification efficiency and the fake rate for high momentum kaons are 78% and 18% (true K fakes π), respectively; the corresponding values for pions are 90% and 7% (true π fakes K).

Candidate $\pi^0 \to \gamma\gamma$ decays are identified by combining two photon clusters and requiring their combined mass be within 16.2 MeV/c^2 of the nominal π^0 mass. We then perform a π^0 mass constraint using the error matrix for each photon cluster. Candidate $K_s \to \pi^+\pi^-$ decays are reconstructed by as-

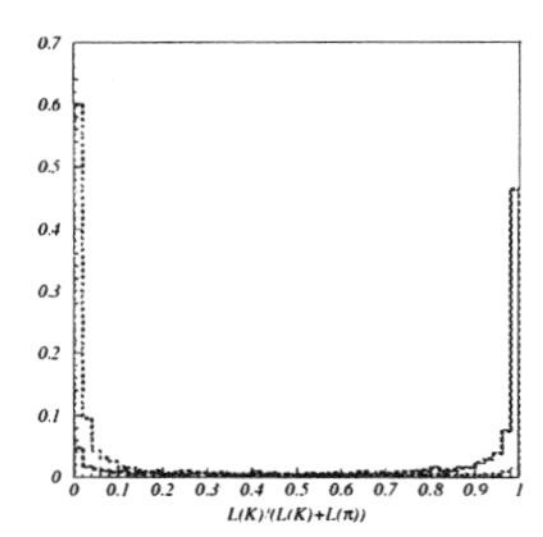

Figure 1. Kaon likelihood ratio for high momentum kaons/pions from D^0 decays in the D^* sample. The solid line is for kaons and the dashed one is for pions.

sociating two oppositely charged tracks and requring the two track mass to be within 30 MeV/c^2 of the K^0 mass and the two track converging point to be distinct transeversly from the run average beam position.

To further distinguish signals from background, we rely on shape variables to discriminate $B\bar{B}$ events and jet-like events from the $e^+e^- \to q\bar{q}$ continuum. Besides convential shape variables (see Ref. 4), Belle has developed a set of variables (SFW) extended from the Fox-Wolfram moments. The detailed description of all event shape variables and the SFW usage can be found in Ref. 4.

In the h^+h^- (including $K_s^0 h^+$) analyses, we multiply the probability density of each variable ($\cos\theta_B$, $\cos\theta_{hh}$, and SFW) to form a signal likelihood and a background likelihood. The signal and background probability density functions (PDFs) are determined using simulated signal events and data in the B sideband region, respectively. The signal and background likelihoods are then combined to form a likelihood ratio:

$$LR(B\bar{B}) = \frac{L(B\bar{B})}{L(B\bar{B}) + L(q\bar{q})}.$$

A cut at 0.8 in this likelihood ratio is applied to suppress the continuum background. Futhermore, the data sample $B^+ \to D^0\pi^+$ is used to check the event shape distributions and the reconstruction efficiency. We then fit to $M_b/\Delta E$ distributions after $\Delta E/M_b$ cuts are applied. Since we estimate quiet a sizeable $K^+\pi^-/\pi^+\pi^-$ feed down in the $\pi^+\pi^-/K^+\pi^-$ sample, we fit both $K^+\pi^-$ and $\pi^+\pi^-$ components in the ΔE distribution to obtain signal yields. Fig. 2 shows the ΔE spectrum with the fitted curves for the $K^+\pi^-$ sample. And all the h^+h^- results are given in the summary table.

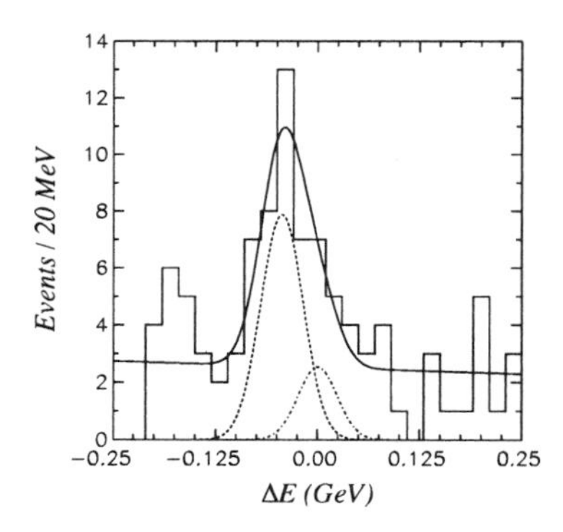

Figure 2. ΔE distribution for the $K^+\pi^-$ sample. Superimposed are the fitted curves for $K^+\pi^-$ part (dashed line centered around -45 MeV/c^2), $\pi^+\pi^-$ part (dashed line centered at zero), and both components plus background (solid line).

As for $h\pi^0$ case, a maximum likelihood (ML) fit is performed to obtain signal yields. Four variables used in the ML fit are: M_b, ΔE, $\cos\theta_B$, and a fisher discriminant which contains SFW variables, *Sphericity*, and $\cos\theta_{\text{thrust}}$. The $K^+\pi^0$ feed down in the $\pi^+\pi^0$ sample is estimated by adding an extra $K\pi^0$ component in the fit. This component has a signal shape but is shifted in ΔE by -40 MeV and its normalization is determined from the size of the observed $B^+ \to K^+\pi^0$ signal, the kaon identification efficiency, and the K to π fake rate. Figure 3 displays the projections of the fit; the branching ratio results and their significances are listed in Table 1.

Since the continuum background in $B \to \phi K$ decays is small after the kaon identification is performed, we apply rather loose cuts to reduce the backgrounds: both $|\cos\theta_{\text{thrust}}|$ and $|\cos\theta_B|$ less than 0.8, ϕ $|\cos\theta_{\text{heli}}| > 0.5$.

Table 1. Summary table for various measurements of charmless B decays. The first errors in signal yields, branching ratios, and CLEO results are statistical and the second systematic. The systematic uncertainty includes the systematic errors of the fit and the reconstruction efficiency.

Mode	Yield	Sign.	Eff.(%)	Br($\times 10^{-5}$)	UL.($\times 10^{-5}$)	CLEO($\times 10^{-5}$)
$K^+\pi^-$	$25.6^{+7.5}_{-6.8} \pm 3.8$	4.4	28	$1.74^{+0.51}_{-0.46} \pm 0.34$	-	$1.72^{+0.25}_{-0.24} \pm 0.12$
$\pi^+\pi^-$	$9.3^{+5.3}_{-5.1} \pm 2.0$	1.9	28	$0.63^{+0.39}_{-0.35} \pm 0.16$	1.65	$0.43^{+0.16}_{-0.14} \pm 0.05$
K^+K^-	$0.8^{+3.1}_{-0.8}$	-	20	-	0.6	< 0.19
$K^0\pi^+$	$5.7^{+3.4}_{-2.7} \pm 0.6$	2.4	13	$1.66^{+0.98+0.22}_{-0.78-0.24}$	3.4	$1.82^{+0.46}_{-0.40} \pm 0.16$
K^0K^+	$0.0^{+0.5}_{-0.0}$	-	11	-	0.8	< 0.51
$K^+\pi^0$	$32.3^{+9.4+2.4}_{-8.4-2.2}$	5.0	31	$1.88^{+0.55}_{-0.49} \pm 0.23$	-	$1.12^{+0.30+0.14}_{-0.27-0.13}$
$\pi^+\pi^0$	$5.4^{+5.7+1.0}_{-4.4-1.1}$	1.3	30	$0.33^{+0.35}_{-0.27} \pm 0.07$	1.01	< 1.27
$K^0\pi^0$	$10.8^{+4.8+0.7}_{-4.0-0.5}$	3.9	19	$2.10^{+0.93+0.25}_{-0.78-0.23}$	-	$1.46^{+0.59+0.24}_{-0.51-0.33}$
ϕK^+	$9.2^{+3.6}_{-2.9} \pm 0.8$	5.4	10	$1.72^{+0.67}_{-0.54} \pm 0.18$	-	$0.64^{+2.5+0.5}_{-2.1-2.0}$

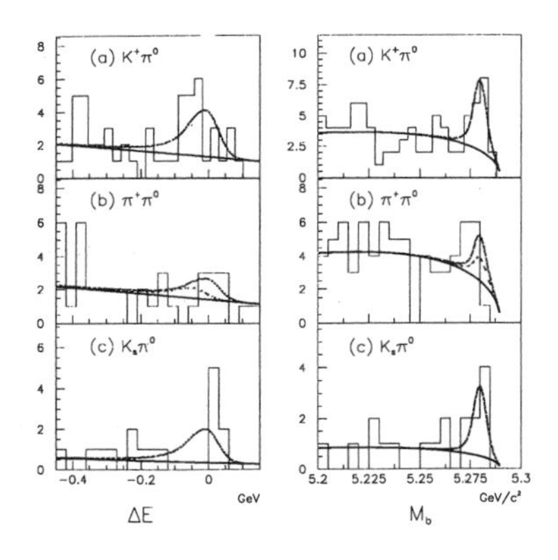

Figure 3. The projections of the likelihood fits on ΔE and M_b distributions for (a) $K^+\pi^0$, (b) $\pi^+\pi^0$, and (c) $K_S\pi^0$ cases. Dashed lines in (b) indicate the $K^+\pi^0$ feed down expectation.

A clear $B^+ \to \phi K^+$ signal is seen (Fig. 4) and the branching ratio is measured to be $1.72^{+3.6}_{-2.9} \pm 0.8$, indicating the first evidence of pure $b \to s s \bar{s}$ transition.

3 Conclusion

Table 1 summarizes our measurements of various charmless B decays and also shows the corresponding CLEO numbers. In general, Belle and CLEO results are consistent with each other. We confirm that the branching ratios of $B \to K\pi$ are larger than that of

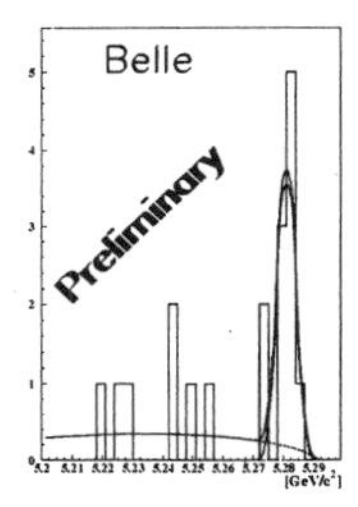

Figure 4. The M_b distribution after shape variable cuts for $B^+ \to \phi K^+$ search.

$B \to \pi\pi$. A large than theoretically predicted $B^0 \to K^0\pi^0$ signal has been observed in both experiments. However, Belle comes up with the first $B^+ \to \phi K^+$ measurement, whose lower limit is larger than the CLEO central value. Other rare B decay searches are on going and we expect to have more fruitful results with more data to come.

References

1. M. Neubert, hep-ph/0001334 (2000).
2. Belle Collaboration, Technical Design Report, KEK Report 95-1, 1995.
3. KEKB accelerator group, KEKB B Factor Design Report, KEK Report 95-7, 1995.
4. ICHEP 2000 Contributed papers 287, 289, 292.

STUDIES OF CHARMLESS TWO-BODY, QUASI-TWO-BODY AND THREE-BODY B DECAYS

THERESA J. CHAMPION
ON BEHALF OF THE BABAR COLLABORATION
University of Birmingham, Edgbaston,
Birmingham, England.
E-mail: tjc@SLAC.Stanford.edu

Preliminary results are presented on a search for several exclusive charmless hadronic B decays, from data collected by the *BABAR* detector near the $\Upsilon(4S)$ resonance. These include two-body decay modes $h^{\pm}h^{\mp}$, three-body decay modes with final states $h^{\pm}h^{\mp}h^{\pm}$ and $h^{\pm}h^{\mp}\pi^0$, and quasi-two-body decay modes with final states X^0h and $X^0K^0_S$, where $h = \pi$ or K and $X^0 = \eta'$ or ω. The measurement of branching fractions for four decay modes, and upper limits for nine modes are presented.

1 Introduction

Charmless hadronic decays of B mesons provide rich opportunities for exploring a number of CP violation phenomena. In particular, several of these decay modes offer the future possibility of measuring directly the CKM angle α of the Standard Model [1].

Preliminary searches have been carried out for a number of charmless hadronic B decays using the initial *BABAR* data sample. The data consist of $7.7\,\text{fb}^{-1}$ taken at the $\Upsilon(4S)$ resonance, and $1.2\,\text{fb}^{-1}$ taken below the $B\overline{B}$ threshold. The number of $B\overline{B}$ events has been determined from hadronic event selection[2] to be $8.46 \pm 0.14 \times 10^6$.

For all decay modes a simple cut-based analysis has been performed, and additionally for the two-body modes a global likelihood fit has been carried out. A "blind" analysis methodology has been adopted throughout, so that the signal region for each mode remained hidden until all decisions concerning event selection had been taken. The signal region has been represented using the two variables $m_{\text{ES}} = \sqrt{(\sqrt{s}/2)^2 - P_B^{*2}}$ and $E_B^* - \sqrt{s}/2$, where E_B^* and P_B^* are the energy and 3-momentum of the B in the cms. Fig.1 shows the distribution of events in the $m_{\text{ES}} - \Delta E$ plane for the mode $B^+ \to \eta' K^+$.

In order to facilitate understanding of the data, use was made of suitable calibration modes $B^+ \to \overline{D}^0\pi^+, \overline{D}^0 \to K^+\pi^-$ and $\overline{D}^0 \to K^+\pi^-\pi^0$. These have similar final state kinematics to the modes of interest, but with $\sim \times 10$ higher branching ratios.

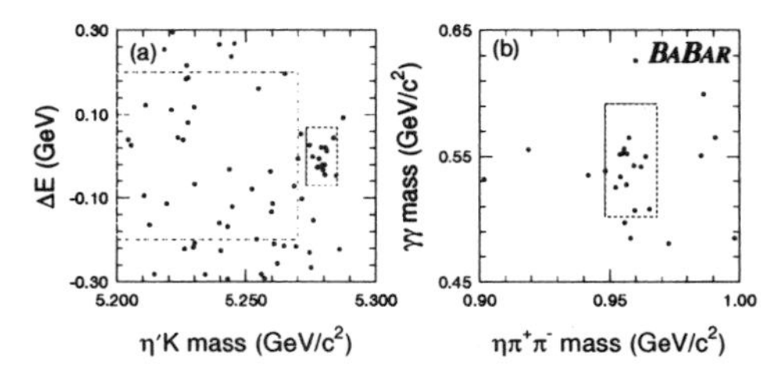

Figure 1. Kinematics of $B^+ \to \eta' K^+$.

2 Event Selection

The most significant issues for event selection, common to all the rare charmless decay modes, are effective kaon identification and the suppression of continuum background. Additional features of individual modes, such as the reconstructed mass and helicity angle distributions of intermediate resonances, have been used for selection where appropriate (for details see [2],[3]).

2.1 Kaon Identification

Excellent kaon identification is essential for all the decay modes of interest. The primary

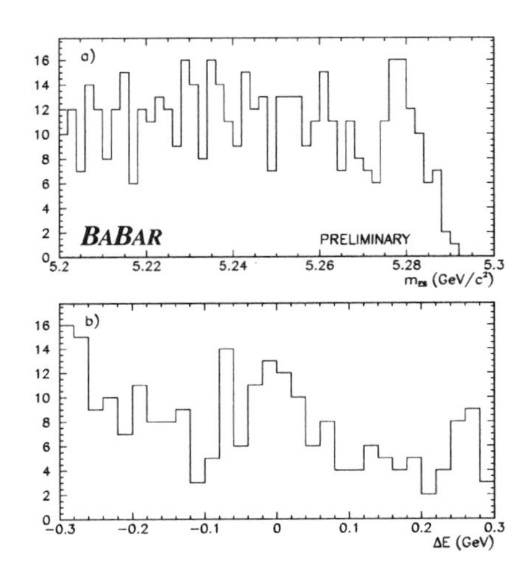

Figure 2. (a) m_{ES} and (b) ΔE distributions for $B^0 \to \rho^{\mp}\pi^{\pm}$.

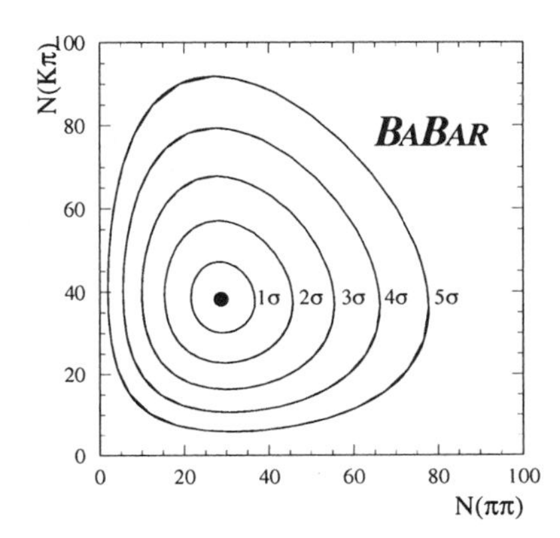

Figure 3. Likelihood contours for $\pi^{\pm}\pi^{\mp}$ and $K^{\pm}\pi^{\mp}$.

system in *BABAR* for K/π discrimination is the DIRC (Detector of Internally Reflected Cherenkov light), which achieves separation at $> 2\sigma$ for momenta up to 4 GeV[4]. Further information for particle identification is provided by $\mathrm{d}E/\mathrm{d}x$ from the Drift Chamber and the Silicon Vertex Tracker.

2.2 Background Characterisation and Suppression

All rare charmless B decays suffer from high levels of background from continuum events. This background can be substantially reduced by exploiting the differences in topology between $B\overline{B}$ events and continuum events. In the cms, a $B\overline{B}$ event is approximately isotropic, and there is no correlation between the decay topologies of the two Bs. A continuum event, however, exhibits a two-jet structure, so that the directions of the decay products are highly correlated. A number of event shape variables have been used in the analyses to discriminate signal events from continuum background. Following background discrimination cuts, a significant amount of background remained. This has been estimated using both on- and off-resonance data. In the on-resonance case, the background has been characterised by fitting the m_{ES} distribution to an ARGUS function[5] using the sidebands, and extrapolating to the signal region. In the off-resonance case, the number of events in the sidebands and signal region were counted directly.

3 Analysis

For the simple counting analyses, cut optimisation was performed with respect to the predicted sensitivity of a measurement, using the expected signal yield and the estimated efficiency. Fig.2 shows the distributions of m_{ES} and ΔE for the mode $B^0 \to \rho^{\mp}\pi^{\pm}$.

For the two-body modes, a global likelihood fit was carried out. A likelihood function was constructed to determine the signal and background yields, using the parameters m_{ES}, ΔE, the Fisher Discriminant formed from a set of event shape variables, and the Cherenkov angles of the two decay products. The probability density functions used in the fit were obtained from studies of the data where possible. The likelihood contours for $\pi^{\pm}\pi^{\mp}$ and $K^{\pm}\pi^{\mp}$ are shown in Fig.3.

The results of searches for 13 decay modes are presented in Table 1.

Table 1. Summary of Results.

Decay mode	yield	*BABAR* BR/10^{-6}	CLEO[6] BR/10^{-6}
$B^0 \to \pi^\pm\pi^\mp$	25 ± 8	$9.3^{+2.6+1.2}_{-2.3-1.4}$	$4.3^{+1.6}_{-1.4} \pm 0.5$
$B^0 \to K^\pm\pi^\mp$	26 ± 8	$12.5^{+3.0+1.3}_{-2.6-1.7}$	$17.2^{+2.5}_{-2.4} \pm 1.2$
$B^0 \to K^\pm K^\mp$	1 ± 4	< 6.6	< 1.9
$B^+ \to \omega h^+$	6 ± 4	< 24	$< 14.3 \pm 3.6$
$B^0 \to \omega K^0_S$	0	< 14	< 21
$B^+ \to \eta' K^+$	12 ± 4	$62 \pm 18 \pm 8$	80 ± 10
$B^0 \to \eta' K^0_S$	1 ± 1	< 112	$< 89 \pm 18$
$B^+ \to K^{*0}\pi^+$	10 ± 5	< 28	< 16
$B^+ \to \rho^0 K^+$	11 ± 5	< 29	< 17
$B^+ \to K^+\pi^-\pi^+$	19 ± 6	< 66	< 28
$B^+ \to \rho^0\pi^+$	25 ± 8	< 39	$< 10.4 \pm 3.4$
$B^+ \to \pi^+\pi^-\pi^+$	5 ± 6	< 22	< 41
$B^0 \to \rho^\mp\pi^\pm$	36 ± 10	$< 48.5 \pm 13.4^{+5.8}_{-5.2}$	$< 27.6 \pm 8.4$

References

1. P. F. Harrison and H. R. Quinn, eds., "The *BABAR* Physics Book", SLAC-R-405 (1998).
2. *BABAR* Collaboration, B. Aubert *et al.*, "Measurements of charmless three-body and quasi-two-body B decays", *BABAR*-CONF-00/15, SLAC-PUB-8537.
3. *BABAR* Collaboration, B. Aubert *et al.*, "Measurement of branching fractions for two-body charmless B decays to charged pions and kaons at *BABAR*", *BABAR*-CONF-00/14, SLAC-PUB-8536.
4. *BABAR* Collaboration, B. Aubert *et al.*, "The first year of the BABAR experiment at PEP-II", BABAR-CONF-00/17, SLAC-PUB-8539.
5. ARGUS Collaboration, H.Albrecht *et al.*, Phys. Lett. B **254** (1991) 288.
6. Particle Data Group, D.E. Groom *et al.*, Eur.Phys.Jour. **C15**, 1 (2000).

CONSTRAINTS ON γ AND STRONG PHASES FROM $B \to \pi K$ DECAYS

ANDRZEJ J. BURAS
Technische Universität München, Physik Department, D-85748 Garching, Germany
E-mail: aburas@ally.t30.physik.tu-muenchen.de

ROBERT FLEISCHER
Deutsches Elektronen-Synchrotron DESY, Notkestr. 85, D-22607 Hamburg, Germany
E-mail: Robert.Fleischer@desy.de

As we pointed out recently, the neutral decays $B_d \to \pi^{\mp} K^{\pm}$ and $B_d \to \pi^0 K$ may provide non-trivial bounds on the CKM angle γ. Here we reconsider this approach in the light of recent CLEO data, which look very interesting. In particular, the results for the corresponding CP-averaged branching ratios are in favour of strong constraints on γ, where the second quadrant is preferred. Such a situation would be in conflict with the standard analysis of the unitarity triangle. Moreover, constraints on a CP-conserving strong phase $\delta_{\rm n}$ are in favour of a negative value of $\cos\delta_{\rm n}$, which would be in conflict with the factorization expectation. In addition, there seems to be an interesting discrepancy with the bounds that are implied by the charged $B \to \pi K$ system: whereas these decays favour a range for γ that is similar to that of the neutral modes, they point towards a positive value of $\cos\delta_{\rm c}$, which would be in conflict with the expectation of equal signs for $\cos\delta_{\rm n}$ and $\cos\delta_{\rm c}$.

1 Introduction

In order to obtain direct information on the angle γ of the unitarity triangle of the CKM matrix, $B \to \pi K$ decays are very promising. In the following, we focus on our analysis Ref. 1, making use of the most recent CLEO data[2]. Because of the small ratio $|V_{us}V_{ub}^*/(V_{ts}V_{tb}^*)| \approx 0.02$, $B \to \pi K$ modes are dominated by QCD penguin topologies. Due to the large top-quark mass, we have also to care about electroweak (EW) penguins. In the case of $B_d^0 \to \pi^- K^+$ and $B^+ \to \pi^+ K^0$, these topologies contribute in colour-suppressed form and are hence expected to play a minor role, whereas they contribute in colour-allowed form to $B^+ \to \pi^0 K^+$ and $B_d^0 \to \pi^0 K^0$ and may here even compete with tree-diagram-like topologies.

So far, strategies to probe γ through $B \to \pi K$ decays have focused on the following two systems: $B_d \to \pi^{\mp} K^{\pm}$, $B^{\pm} \to \pi^{\pm} K$ ("mixed")[3,4], and $B^{\pm} \to \pi^0 K^{\pm}$, $B^{\pm} \to \pi^{\pm} K$ ("charged")[5]. Recently, we pointed out that also the neutral combination $B_d \to \pi^{\mp} K^{\pm}$, $B_d \to \pi^0 K$ is very promising[6].

2 Constraints on γ

Interestingly, already CP-averaged branching ratios may lead to highly non-trivial constraints on γ. Here the key quantities are

$$R \equiv \frac{\text{BR}(B_d \to \pi^{\mp} K^{\pm})}{\text{BR}(B^{\pm} \to \pi^{\pm} K)} = 0.95 \pm 0.28 \quad (1)$$

$$R_{\rm c} \equiv \frac{2\text{BR}(B^{\pm} \to \pi^0 K^{\pm})}{\text{BR}(B^{\pm} \to \pi^{\pm} K)} = 1.27 \pm 0.47 \quad (2)$$

$$R_{\rm n} \equiv \frac{\text{BR}(B_d \to \pi^{\mp} K^{\pm})}{2\text{BR}(B_d \to \pi^0 K)} = 0.59 \pm 0.27, \quad (3)$$

where we have also taken into account the CLEO results reported in Ref. 2. If we employ the $SU(2)$ flavour symmetry and certain dynamical assumptions, concerning mainly the smallness of FSI effects, we may derive a general parametrization[6] for (1)–(3),

$$R_{(\rm c,n)} = R_{(\rm c,n)}(\gamma, q_{(\rm c,n)}, r_{(\rm c,n)}, \delta_{(\rm c,n)}), \quad (4)$$

where $q_{(\rm c,n)}$ denotes the ratio of EW penguins to "trees", $r_{(\rm c,n)}$ is the ratio of "trees" to QCD penguins, and $\delta_{(\rm c,n)}$ is the CP-conserving strong phase between "tree" and QCD penguin amplitudes. The parameters $q_{(\rm c,n)}$ can be fixed through theoretical arguments: in the "mixed" system, we have

$q \approx 0$, as EW penguins contribute only in colour-suppressed form; in the charged[5] and neutral[6] $B \to \pi K$ systems, q_c and q_n can be fixed through the $SU(3)$ flavour symmetry without dynamical assumptions. The $r_{(c,n)}$ can be determined with the help of additional experimental information: in the "mixed" system, r can be fixed through arguments based on "factorization", whereas r_c and r_n can be determined from $B^+ \to \pi^+\pi^0$ by using only the $SU(3)$ flavour symmetry.

At this point, a comment on FSI effects is in order. Whereas the determination of q and r as sketched above may be affected by FSI effects, this is *not* the case for $q_{c,n}$ and $r_{c,n}$, since here $SU(3)$ suffices. Nevertheless, we have to assume that $B^+ \to \pi^+ K^0$ or $B_d^0 \to \pi^0 K^0$ do *not* involve a CP-violating weak phase:

$$\begin{aligned} A(B^+ \to \pi^+ K^0) &= -\,|\tilde{P}|e^{i\delta_{\tilde{P}}} \\ &= A(B^- \to \pi^- \overline{K^0}). \end{aligned} \tag{5}$$

This relation may be affected by rescattering processes such as $B^+ \to \{\pi^0 K^+\} \to \pi^+ K^0$:

$$A(B^+ \to \pi^+ K^0) = -\,|\tilde{P}|e^{i\delta_{\tilde{P}}}\left[1 + \rho_c\, e^{i\theta} e^{i\gamma}\right],$$

where ρ_c is doubly Cabibbo-suppressed and is naively expected to be negligibly small. In the "QCD factorization" approach[7], there is no significant enhancement of ρ_c through rescattering processes. However, there is still no theoretical consensus on the importance of FSI effects. In the charged $B \to \pi K$ strategy to probe γ, they can be taken into account through $SU(3)$ flavour-symmetry arguments and additional data on $B^\pm \to K^\pm K$ decays. The present experimental upper bounds on these modes are not in favour of dramatic effects. In the case of the neutral strategy, FSI effects can be included in an *exact manner* with the help of the mixing-induced CP asymmetry $\mathcal{A}_{\text{CP}}^{\text{mix}}(B_d \to \pi^0 K_S)$[6].

In contrast to $q_{(c,n)}$ and $r_{(c,n)}$, the strong phase $\delta_{(c,n)}$ suffers from large hadronic uncertainties and is essentially unknown. However, we can get rid of $\delta_{(c,n)}$ by keeping it as

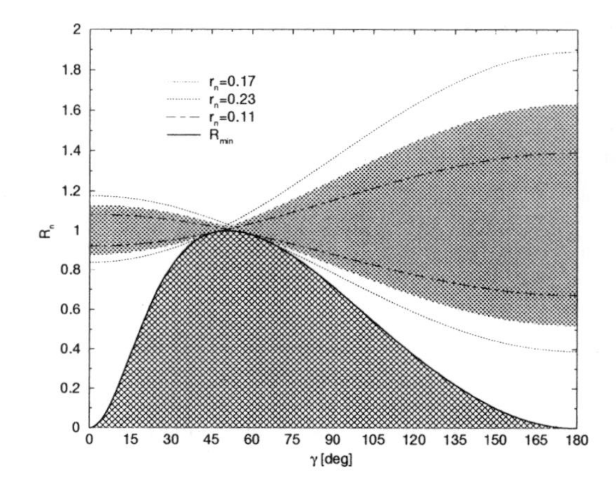

Figure 1. The dependence of the extremal values of R_n (neutral $B \to \pi K$ system) on γ for $q_n = 0.63$.

a "free" variable, yielding minimal and maximal values for $R_{(c,n)}$:

$$R_{(c,n)}^{\text{ext}}\Big|_{\delta_{(c,n)}} = \text{function}(\gamma, q_{(c,n)}, r_{(c,n)}). \tag{6}$$

Keeping in addition $r_{(c,n)}$ as a free variable, we obtain another – less restrictive – minimal value for $R_{(c,n)}$:

$$R_{(c,n)}^{\text{min}}\Big|_{r_{(c,n)},\delta_{(c,n)}} = \kappa(\gamma, q_{(c,n)}) \sin^2\gamma. \tag{7}$$

In Fig. 1, we show the dependence of (6) and (7) on γ for the neutral $B \to \pi K$ system[a]. Here the crossed region below the R_{min} curve, which is described by (7), is excluded. On the other hand, the shaded region is the allowed range (6) for R_n, arising in the case of $r_n = 0.17$. Fig. 1 allows us to read off immediately the allowed region for γ for a given value of R_n. Using the central value of the present CLEO result (3), $R_n = 0.6$, the R_{min} curve implies $0° \leq \gamma \leq 21° \vee 100° \leq \gamma \leq 180°$. The corresponding situation in the $\overline{\varrho}$–$\overline{\eta}$ plane is shown in Fig. 2, where the crossed region is excluded and the circles correspond to $R_b = 0.41 \pm 0.07$. As the theoretical expression for q_n is proportional to $1/R_b$, the constraints in the $\overline{\varrho}$–$\overline{\eta}$ plane are actually more appropriate than the constraints on γ.

[a]The charged $B \to \pi K$ curves look very similar.

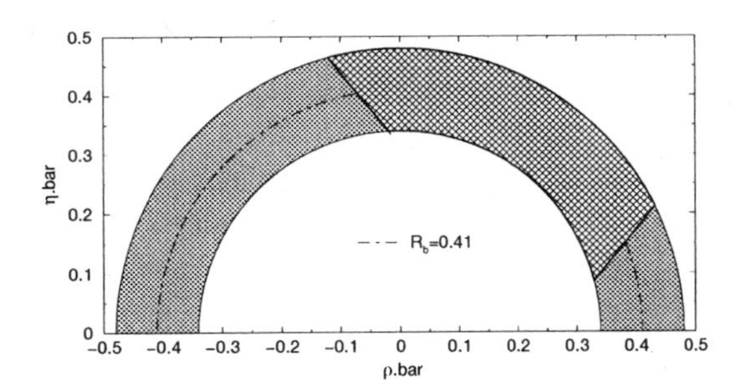

Figure 2. The constraints in the $\overline{\varrho}$–$\overline{\eta}$ plane implied by (7) for $R_{\rm n} = 0.6$ and $q_{\rm n} = 0.63 \times [0.41/R_b]$.

If we use additional information on the parameter $r_{\rm n}$, we may put even stronger constraints on γ. For $r_{\rm n} = 0.17$, we obtain, for instance, the allowed range $138° \leq \gamma \leq 180°$. It is interesting to note that the $R_{\rm min}$ curve is only effective for $R_{\rm n} < 1$, which is favoured by the most recent CLEO data[2]. A similar pattern is also exhibited by the first BELLE results[8] presented at this conference, yielding $R_{\rm n} = 0.4 \pm 0.2$.

For the central value $R_{\rm c} = 1.3$ of (2), (7) is not effective and $r_{\rm c}$ has to be fixed in order to constrain γ. Using $r_{\rm c} = 0.21$, we obtain $87° \leq \gamma \leq 180°$. Although it is too early to draw definite conclusions, it is important to emphasize that the most recent CLEO results on $R_{\rm (c,n)}$ prefer the second quadrant for γ, i.e. $\gamma \geq 90°$. Similar conclusions were also obtained using other $B \to \pi K$, $\pi\pi$ strategies[9]. Interestingly, such a situation would be in conflict with the standard analysis of the unitarity triangle[10], yielding $38° \leq \gamma \leq 81°$.

3 Constraints on Strong Phases

The $R_{\rm (c,n)}$ allow us to determine $\cos\delta_{\rm (c,n)}$ as functions of γ, thereby providing also constraints on the strong phases $\delta_{\rm (c,n)}$[1]. Interestingly, the present CLEO data are in favour of $\cos\delta_{\rm n} < 0$, which would be in conflict with "factorization". Moreover, they point towards a positive value of $\cos\delta_{\rm c}$, which would be in conflict with the theoretical expectation of equal signs for $\cos\delta_{\rm c}$ and $\cos\delta_{\rm n}$.

4 Conclusions and Outlook

If future data should confirm the "puzzling" situation for γ and $\cos\delta_{\rm c,n}$ favoured by the present $B \to \pi K$ CLEO data, it may be an indication for new-physics contributions to the EW penguin sector, or a manifestation of flavour-symmetry-breaking effects. In order to distinguish between these possibilities, further studies are needed. As soon as CP asymmetries in $B_d \to \pi^\mp K^\pm$ or $B^\pm \to \pi^0 K^\pm$ are observed, we may even *determine* γ and $\delta_{\rm (c,n)}$. Here we may also arrive at a situation, where the $B \to \pi K$ observables do not provide any solution for these quatities[11], which would be an immediate indication for new physics. We look forward to new data from the B-factories.

References

1. A.J. Buras and R. Fleischer, *Eur. Phys. J.* C **16**, 97 (2000).
2. CLEO Collaboration, hep-ex/0001010; R. Stroynowski, these proceedings.
3. R. Fleischer, *Phys. Lett.* B **365**, 399 (1996); M. Gronau and J.L. Rosner, *Phys. Rev.* D **57**, 6843 (1998).
4. R. Fleischer and T. Mannel, *Phys. Rev.* D **57**, 2752 (1998).
5. M. Neubert and J.L. Rosner, *Phys. Lett.* B **441**, 403 (1998); *Phys. Rev. Lett.* **81**, 5076 (1998).
6. A.J. Buras and R. Fleischer, *Eur. Phys. J.* C **11**, 93 (1999).
7. M. Beneke *et al.*, *Phys. Rev. Lett.* **83**, 1914 (1999); M. Beneke, these proceedings.
8. BELLE Collabor., BELLE-CONF-0005 and 0006; P. Chang, these proceedings.
9. W.-S. Hou and K.-C. Yang, *Phys. Rev.* D **61**, 073014 (2000); W.-S. Hou, these proceedings.
10. A. Ali and D. London, DESY 00-026 [hep-ph/0002167].
11. R. Fleischer and J. Matias, *Phys. Rev.* D **61**, 074004 (2000).

POSSIBILITY OF LARGE FSI PHASES IN LIGHT OF $B \to K\pi$ & $\pi\pi$ DATA

GEORGE W.S. HOU

Department of Physics, National Taiwan University, Taipei, Taiwan, R.O.C.

After briefly reviewing how data indicated that factorization seems to work in observed two body charmless modes, *if one takes* $\gamma \gtrsim 90^\circ$, we point out that the $K^0\pi^0$ mode seems too large. This and other hints suggest that perhaps not only γ is large, but rescattering phase δ could be sizable as well.

1 Path to $\gamma > 90^\circ$ and Factorization

CLEO data has driven B phenomenology in the classic way in the last 3 years.

<u>**1997**</u>: $\bar{K}^0\pi^- > K^-\pi^+ \simeq 1.5 \times 10^{-5}$
This lead to the Fleischer–Mannel bound and a boom in theory activity, leading eventually to model-indep. methods for extracting γ.

<u>**1998**</u>: $\bar{K}^0\pi^- \simeq K^-\pi^+ \lesssim K^-\pi^0 \simeq 1.5 \times 10^{-5}$
First equality precipitated suggestion for large γ; strength of $K\pi^0$ indicated EWP.[1]

<u>**1999**</u>: Multiple modes emerge, *e.g.*
⋆ $\rho^0\pi$, $\rho\pi$, $\omega\pi$: $\exists\ b \to u$ tree (T).
⋆ ωK disappear: As it should.
⋆ $K\pi^0 \simeq \frac{2}{3}\,(\bar{K}^0\pi \simeq K\pi)$: EWP at work!
⋆ $\pi\pi \sim \frac{1}{4}\,K\pi \Longrightarrow$ Large γ!
⋆ $K^0\pi^0 \sim K\pi,\ K^0\pi \Longrightarrow$ <u>Problem</u>.

The host of emerging modes in 1999 lead to the observation[2] that "*Factorization works in observed two body charmless rare B decays,* **if** $\cos\gamma \lesssim 0$." To stay low key, only the sign change in $\cos\gamma$ was initially advocated, but emboldened by this observation, quantification was sought. Using known factorization formulas *etc.*,[3] a "global fit" of more than 10 modes gave[4] $\gamma \simeq 105^\circ$, which is in some conflict with the "CKM Fit" value[5] of $\gamma \simeq 58.5^\circ \pm 7.1^\circ$. However, by end of 1999, all B practioners have switched to $\gamma \gtrsim 80^\circ$–90°, as reflected in the 5 rare B theory talks here.

We have gained from *hadronic* rare B modes new knowledge on CKM. That hadronization does not mask this (factorization works!) is quite astonishing. At this Conference, first physics results have been reported from (asymmetric) B Factories. Both BaBar and Belle have collected data comparable to CLEO II+ II.V. Belle[6] confirms CLEO results on $K\pi$ modes and $\pi^-\pi^+$, while BaBar[7] is at some variance, finding $\pi^-\pi^+ \simeq K^-\pi^+$. One surprise is the ϕK mode, where Belle reports a sizable rate with *lower bound* above the new CLEO central value,[8] which is just above its own previous limit. The age of competition has obviously arrived, and we look forward to healthy and at same time explosive developments in near future.

It is said[9] that $\rho^0\pi \lesssim \omega\pi$ at present no longer supports $\gamma > 90^\circ$ as it was[2] in early 1999 when $\rho^0\pi > \omega\pi$ was reported by CLEO. Likewise, $\eta' K^0 > \eta' K^-$ is also at odds with $\gamma > 90^\circ$. We caution that the number of $\omega\pi$ events reported by BaBar[7] indicates a rate lower than CLEO's, while the $\rho\pi$ rates seem considerably larger. As for the $\eta' K$ modes, they are not yet fully understood. As stressed by Golutvin,[10] clearly we "NEED MORE DATA!", which will arrive in due course.

2 Problems?

Besides the strength of $\eta' K$ modes, another problem was apparent by summer 1999: $K^0\pi^0$ seems too large![11] Playing games [a] with "central values" from CLEO, we might also note that $\pi^-\pi^+ < \pi^-\pi^0$ seems a bit small, while the direct CP asymmetries $a_{\rm CP}$ in $K^-\pi^+$, $K^-\pi^0$ and $\bar{K}^0\pi^-$ modes give a "pattern" that is different from SM expectations with only S.D. rescattering phases.

[a] We are well aware of the "Central Value Syndrome" sufferred by theorists, but advocate that these games still stimulate the field by exploring possibilities.

As mentioned, $K\pi^0/K\pi \simeq 0.65$ confirms *constructive* EWP-P interference for $K\pi^0$ in SM. From the operators and the π^0 w.f. (sign traced to $d\bar{d} \to \pi^0$) one expects *destructive* EWP-P interference in $K^0\pi^0$,[11]

$$\frac{\overline{K}^0\pi^0}{\overline{K}^0\pi^-} \approx \frac{1}{2}\left|1 - r_0\,\frac{1.5a_9}{a_4 + a_6 R}\right|^2 \approx \frac{1}{3}, \quad (1)$$

where $r_0 = f_\pi F_0^{BK}/f_K F_0^{B\pi} \simeq 0.9$ and $R = 2m_K^2/(m_b - m_d)(m_s + m_d)$. Hence $K^0\pi^0 > K\pi^0$ is very hard to reconcile.

Could this be due to Final State Interactions (FSI) alone? Taking $\gamma = 60°$, $\pi^-\pi^+$ can be accounted for via $\pi^-\pi^+ \leftrightarrow \pi^0\pi^0$ rescattering, but the $K\pi$ modes fit data poorly. Thus, we still need to call on the service of γ.

3 Large γ and δ?

Let us set up a simple formalism for what we mean by FSI, or $\delta \neq 0$, which goes beyond factorization. Our *Ansatz* simply extends naive factorization amplitudes A_I by adding δ_I to model hadronic phases in final state. [b] The $B \to \pi\pi$ amplitudes become,

$$\begin{aligned}
\mathcal{A}(B \to \pi\pi) &= A_0 e^{i\delta_0} + A_2 e^{i\delta_2}, \\
\mathcal{A}(B \to \pi^0\pi^0) &= \tfrac{1}{\sqrt{2}} A_0 e^{i\delta_0} - \sqrt{2} A_2 e^{i\delta_2}, \\
\mathcal{A}(B \to \pi\pi^0) &= \tfrac{3}{\sqrt{2}} A_2 e^{i\delta_2}, \qquad (2)
\end{aligned}$$

where $I = 0,\ 2$ stand for final state isospin. For $K\pi$ modes, we have the amplitudes

$$\begin{aligned}
\mathcal{A}(B \to K\pi) &= A_{\frac{3}{2}} e^{i\delta_{\frac{3}{2}}} - A^-_{\frac{1}{2}} e^{i\delta_{\frac{1}{2}}}, \\
\mathcal{A}(B \to K^0\pi^0) &= \sqrt{2} A_{\frac{3}{2}} e^{i\delta_{\frac{3}{2}}} + \tfrac{1}{\sqrt{2}} A^-_{\frac{1}{2}} e^{i\delta_{\frac{1}{2}}}, \\
\mathcal{A}(B \to K\pi^0) &= \sqrt{2} A_{\frac{3}{2}} e^{i\delta_{\frac{3}{2}}} + \tfrac{1}{\sqrt{2}} A^+_{\frac{1}{2}} e^{i\delta_{\frac{1}{2}}}, \\
\mathcal{A}(B \to K^0\pi) &= -A_{\frac{3}{2}} e^{i\delta_{\frac{3}{2}}} + A^+_{\frac{1}{2}} e^{i\delta_{\frac{1}{2}}}, \qquad (3)
\end{aligned}$$

where $A^\mp_{\frac{1}{2}} \equiv A_{\frac{1}{2}} \mp B_{\frac{1}{2}}$ and A_I (B_I) are $\Delta I = 1$ (0) amplitudes for final state isospin I. It is tempting to use SU(3), since $\delta_2 \cong \delta_{\frac{3}{2}}$ holds. However, $(\pi\pi)_{I=0}$ has **1** in addition to **8** and **27** contributions, so in principle $\delta_0 \neq \delta_{\frac{1}{2}}$. Furthermore, $K\bar{K}$ has yet to be seen.

[b] S.D. quark-level rescatterings are included in A_I.

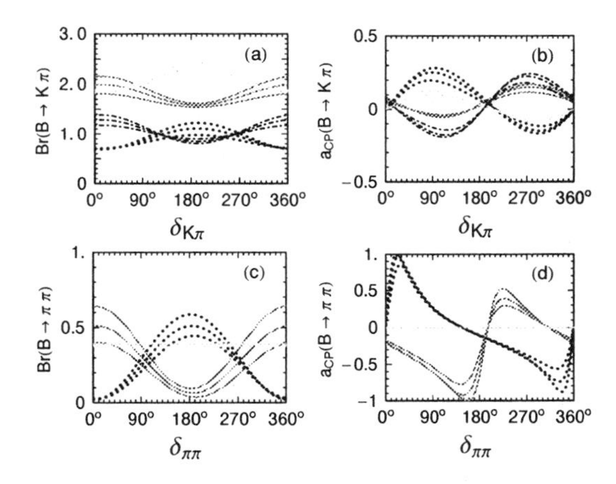

Figure 1. Brs and $a_{\rm CP}$s for $K\pi$ and $\pi\pi$ vs. δ. For curves from: (a) up (down) to down (up) for $K^-\pi^+, \overline{K}^0\pi^{-,0}$ ($K^-\pi^0$) at $\delta = 180°$; and (b) up (down) to down (up) for $K^-\pi^{+,0}$ ($\overline{K}^0\pi^{-,0}$) at $\delta = 90°$; and (c) down to up at $\delta = 180°$; and (d) down to up for $\pi^-\pi^+$ ($\pi^0\pi^0$) at $\delta = 160°$ (20°), are for $\gamma = 130°, 110°$, and $90°$, respectively.

We plot in Fig. 1 the Brs and $a_{\rm CP}$s vs. the phase differences $\delta_{K\pi} = \delta_{\frac{3}{2}} - \delta_{\frac{1}{2}}$ and $\delta_{\pi\pi} = \delta_2 - \delta_0$, respectively, for several large γ values. A numerical exercise illustrates the point. With CLEO data only (Belle/BaBar results are still too preliminary), one has $K\pi : K^0\pi : K\pi^0 : K^0\pi^0 = 1 : 1.06 : 0.67 : \mathbf{0.85}$. Taking, *e.g.* $\gamma = 110°$, we read off from Fig. 1(a) and find the ratio $1 : 0.94 : 0.65 : 0.35$, and $K^0\pi^0$ is clearly a problem. Allowing $\delta_{K\pi} \sim 90°$, the ratio becomes $1 : 1.12 : 0.61 : 0.47$. This is far from resolving the problem, but the *trend* is good. What's more, the central values for $a_{\rm CP}$ in $K\pi$, $K^0\pi$, $K\pi^0$ modes become -0.04, 0.13 and -0.16, which fits the pattern of experimental central values -0.04, 0.17 and -0.29 very well. We see from Fig. 1(c) that taking $\delta_{\pi\pi} \sim \delta_{K\pi}$ gives $\pi\pi < \pi\pi^0$. There are some further dramatic consequences:

- $a_{\rm CP}^{\bar{K}^0\pi^0} \sim -a_{\rm CP}^{K^-\pi^0}$ large.
- $\pi^0\pi^0 \sim \pi\pi \sim 3\text{–}5 \times 10^{-6} \lesssim \pi\pi^0$.
  (still satisfy CLEO bound)
- $a_{\rm CP}^{\pi\pi}$, $a_{\rm CP}^{\pi^0\pi^0} \sim -60\%, -30\%$ possible.

measurable in a couple of years.

4 Remarks and Comments

Our approach of elastic $2 \to 2$ rescattering may be too simplistic, since $B \to i \to K\pi$ involves many intermediate staes i and can be highly inelastic. [c] Furthermore, our elastic rescattering is of form $K^-\pi^+ \to \bar{K}^0\pi^0$. Such "charge exchange" for $p_{K,\pi} \sim 2.5$ GeV is rather counter-intuitive. We stress, however, that our approach is *phenomenological*: data indicates large γ plus simple factorization works; we then make a minimal extension with FSI phases, without pretending to know their origin. They could be effective parameters arising from e.g. annihilation diagrams.[12] But if they genuinely arise from L.D. physics, they would then pose a real problem for PQCD, which argues that long distance effects are $1/m_b$ suppressed.[13]

We offer some brief comments:

• Some people define $\delta = \delta_{\rm P} - \delta_{\rm T}$. For us this mixes elastic and inelastic phases.

• Many works *force* $K^0\pi$ to be pure P, which is a strong assumption *vs.* Eqs. (2) and (3).

• To account for large $K^0\pi^0$, some recent work[14] invoke a "$a^c_{\frac{3}{2}}$" amplitude that is 8 to 10 times larger than factorization result (which arises from the isospin violating EWP), where c indicates it arises from charm intermediate states. We point out, however, that $b \to c\bar{c}s$ is purely $\Delta I = 0$ and cannot generate $I = \frac{3}{2}$ final state.

• $D\bar{D} \to \pi\pi$ rescattering has been used[15] to illustrate the importance of inelastic rescattering. While it is fine as an illustration, a single channel is not quite meaningful because of existence of many channels.

We believe that inelastic phases are *impossible* to understand precisely because there are too many channels. A statistical approach[16] of averaging over large number of random inelastic phases again gives $\delta \sim 10°$–$20°$. In contrast, $2 \to 2$ elastic rescattering is unique for two-body final states. To the least it effectively models hadronic interactions beyond (naive) factorization, which seems to work in two body charmless B decays so far.

[c] "Pomeron" exchange suggest $\pi/2$ phase shift for all channels hence $\delta \simeq 0$, while "Regge" exchange is subleading, typically giving $\delta \sim 10°$–$20°$.

5 Conclusion

Data indicates that factorization seems to work for the first 10-20 or so two body rare B modes, *if* we take $\gamma \gtrsim 90°$. An exception is the strength of $K^0\pi^0$ mode. We find a coherent picture where γ is large, but so is some effective FSI phase δ. The picture can account for the current central values of $\pi^-\pi^+$ vs. $\pi^-\pi^0$ and the pattern seen in $a_{\rm CP}$'s for $K\pi$ modes. It also gives consequences that are testable in next couple of years: Large $a_{\rm CP}$ in $\bar{K}^0\pi^0$ and $K^-\pi^0$; $\pi^0\pi^0 \sim \pi^-\pi^+$ with $a_{\rm CP}$ as large as -30% and -60%, respectively.

References

1. N.G. Deshpande *et al.*, *Phys. Rev. Lett.* **82**, 2240 (1999).
2. X.G. He, W.S. Hou, K.C. Yang, *Phys. Rev. Lett.* **83**, 1999 (1100).
3. A. Ali, G. Kramer, C.D. Lü, *Phys. Rev.* **D 58**, 094009 (1998); Y.H. Chen et al., *Phys. Rev.* **D 60**, 094014 (1999).
4. W.S. Hou, J.G. Smith, F. Würthwein, hep-ex/9910014.
5. A. Stocchi, these proceedings.
6. P. Chang, these proceedings.
7. T. Champion, these proceedings.
8. R. Stroynowski, these proceedings.
9. H.Y. Cheng, these proceedings.
10. A. Golutvin, Plenary talk, these proceedings.
11. W.S. Hou, K.C. Yang, *Phys. Rev. Lett.* **84**, 4806 (2000).
12. H.n. Li, these proceedings.
13. M. Beneke, these proceedings.
14. Y.F. Zhou *et al.*, hep-ph/0006225.
15. Z.Z. Xing, hep-ph/0007136
16. M. Suzuki, L. Wolfenstein, *Phys. Rev.* **D 60**, 074019 (1999).

IMPLICATIONS OF RECENT MEASUREMENTS OF NONLEPTONIC CHARMLESS B DECAYS

HAI-YANG CHENG
Institute of Physics, Academia Sinica, Taipei, Taiwan 115, R.O.C.
E-mail: phcheng@ccvax.sinica.edu.tw

Implications of recent measurements of hadronic charmless B decays are discussed.

1 $B \to \pi\pi$, πK Decays

Both B factories BELLE [1] and BABAR [2] have reported at this conference the preliminary results of $B \to K\pi$ and $B \to \pi^+\pi^-$ (see Table I). For the unitary angle $\gamma \sim 60°$, the ratio $R = \mathcal{B}(B \to K^\pm\pi^\mp)/\mathcal{B}(B \to \pi^+\pi^-)$ is conventionally predicted to lie in the region $1.3-2.0$, to be compared with the experimental values of 4.0 ± 1.6, 2.8 ± 2.0 and 1.3 ± 0.5 obtained by CLEO [3], BELLE and BABAR, respectively. Hence, the expected ratio R is smaller than the CLEO and BELLE results and in agreement with BABAR. Of course, more data are needed to clarify the issue.

If the CLEO or BELLE data are taken seriously, we may ask the question of how to accommodate the data of $K\pi$ and $\pi\pi$ simultaneously. A fit to $\pi^+\pi^-$ yields $F_0^{B\pi}(0) < 0.25$ for $|V_{ub}/V_{cb}| = 0.09$ and $\gamma = 60°$. The $K\pi$ rates will then become too small compared to the data. By contrast, a fit to $K\pi$ modes usually implies a too large $\pi^+\pi^-$ rate. There are several possibilities that the CLEO or BELLE data of $K^\pm\pi^\mp$ and $\pi^+\pi^-$ can be accommodated: (1) $\gamma \sim 60°$ and $F_0^{B\pi}(0) < 0.25$ with a smaller strange quark mass, say $m_s(m_b) = 60$ MeV. The idea is that the $K\pi$ mode will receive a sizable $(S-P)(S+P)$ penguin contributions, while the $\pi\pi$ decay is not much affected. However, a rather smaller m_s is not consistent with recent lattice calculations. (2) $\gamma \sim 60°$ and $F_0^{B\pi}(0) = 0.30$ with the following cases: (i) a smaller V_{ub}, say $|V_{ub}/V_{cb}| \approx 0.06$, so that the $\pi^+\pi^-$ rate is suppressed. However, this small V_{ub} is not favored by data. (ii) a large nonzero isospin $\pi\pi$ phase shift difference of order, say $70°$, can yield a substantial suppression of the $\pi^+\pi^-$ mode [4]. However, $\pi^0\pi^0$ will be substantially enhanced by the same strong phase. The CLEO new measurement [3] $\mathcal{B}(B \to \pi^0\pi^0) < 5.7 \times 10^{-6}$ indicates that the strong phase cannot be too large. (iii) a large inelasticity for $\pi^+\pi^-$ and D^+D^- modes so that the former is suppressed whereas the latter is enhanced. (3) a large γ, say $\gamma \sim (110-130)°$, and $F_0^{B\pi}(0) = 0.30$. Several calculations [5] based on generalized or QCD improved factorization imply $\gamma > 100°$. This scenario is interesting and popular, but it encounters two problems: (i) It is in conflict with the unitary angle $\gamma = (58.5 \pm 7.1)°$ extracted from the global CKM fit[6]. (ii) The CLEO data other than $K\pi$ and $\pi\pi$ do not strongly support a large γ (see below). (4) $\gamma \sim 90°$ and $F_0^{B\pi}(0) = 0.25$ as assumed in a recent PQCD analysis [7]. In this work, the $\pi^+\pi^-$ rate is small because of the small form factor $F_0^{B\pi}(0)$, and $K\pi$ rates are enhanced by large penguin effects owing to steep μ dependence of the leading-order penguin coefficients $c_4(\mu)$ and $c_6(\mu)$ at the hard scale $t < m_b/2$.

As shown in [8], the nonfactorized term is dominated by hard gluon exchange in the heavy quark limit as soft gluon contributions to χ_i are suppressed by orders of $\Lambda_{\rm QCD}/m_b$. However, there is an additional chirally enhanced corrections to the spectator-interaction diagram, which are logarithmically divergent [9]. For example, an additional $(V-A)(V-A)$ spectator interac-

Table 1. Experimental values of the branching ratios (in units of 10^{-6}) for $B \to K\pi$ and $\pi\pi$.

Decay	CLEO [10,3]	BELLE [1]	BABAR [2]
$K^\pm\pi^\mp$	$17.2^{+2.5}_{-2.4} \pm 1.2$	$17.4^{+5.1}_{-4.6} \pm 3.4$	$12.5^{+3.0+1.3}_{-2.6-1.7}$
$K^0\pi^\pm$	$18.2^{+4.6}_{-4.0} \pm 1.6$	$16.6^{+9.8+2.2}_{-7.8-2.4}$	
$K^\pm\pi^0$	$11.6^{+3.0+1.4}_{-2.7-1.3}$	$18.8^{+5.5}_{-4.9} \pm 2.3$	
$K^0\pi^0$	$14.6^{+5.9+2.4}_{-5.1-3.3}$	$21.0^{+9.3+2.5}_{-7.8-2.3}$	
$\pi^\pm\pi^\mp$	$4.3^{+1.6}_{-1.4} \pm 0.5$	$6.3^{+3.9}_{-3.5} \pm 1.6$	$9.3^{+2.6+1.2}_{-2.3-1.4}$

tion proportional to the twist-3 wave function ϕ_σ^π will contribute to $B \to \pi\pi$, πK. Consequently, the nonfactorized contribution to the coefficient $a_2(\pi\pi)$, for example, can be large [11]; its real part lies in the range 0.17–0.25. We find that even in the leading-twist approximation, the same logarithmically divergent integral also appears in the charm quark mass corrections to the spectator-interaction diagram in $B \to J/\psi K(K^*)$ decays [11]. As a result, $\mathrm{Re}\, a_2(J/\psi K)$ is in the vicinity of 0.22.

2 $B \to \rho\pi$, $\omega\pi$ Decays

The class-III decays $B^\pm \to \rho^0\pi^\pm$, $\omega\pi^\pm$ are tree-dominated and sensitive to $(N_c^{\rm eff})_2$ appearing in a_2; their branching ratios decrease with $(N_c^{\rm eff})_2$. The present data[12]

$$\mathcal{B}(B^\pm \to \rho^0\pi^\pm) = (10.4^{+3.3}_{-3.4} \pm 2.1) \times 10^{-6},$$
$$\mathcal{B}(B^\pm \to \omega\pi^\pm) = (11.3^{+3.3}_{-2.9} \pm 1.5) \times 10^{-6}, \quad (1)$$

imply $(N_c^{\rm eff})_2 < 3$ as in $B \to D\pi$ decays.

The decay rate of $\rho^0\pi^\pm$ is sensitive to γ, while $\omega\pi^\pm$ is not. For example, $\mathcal{B}(B^\pm \to \rho^0\pi^\pm)/\mathcal{B}(B^\pm \to \omega\pi^\pm) \sim 1$ for $\gamma \sim 60°$, and $\mathcal{B}(B^\pm \to \rho^0\pi^\pm)/\mathcal{B}(B^\pm \to \omega\pi^\pm) > 1$ for $\gamma > 90°$ if $A_0^{B\omega}(0) = A_0^{B\rho}(0)$. Therefore, a large γ preferred by the previous measurement [13] $\mathcal{B}(B^\pm \to \rho^0\pi^\pm) = (15 \pm 5 \pm 4) \times 10^{-6}$, is no longer favored by the new measurement of $\rho^0\pi^\pm$.

3 $B \to \phi K$ Decays

The previous limit[13] for the branching ratio of $B^\pm \to \phi K^\pm$ is 0.59×10^{-5} at 90% C.L. However, CLEO has also seen a 3σ evidence for the decay $B \to \phi K^*$. Its branching ratio, the average of ϕK^{*-} and ϕK^{*0} modes, is reported to be [14] $\mathcal{B}(B \to \phi K^*) = (1.1^{+0.6}_{-0.5} \pm 0.2) \times 10^{-5}$. Theoretical calculations based on factorization indicate that the branching ratio of ϕK is similar to that of ϕK^*. Therefore, it is difficult to understand the non-observation of ϕK.

An observation of the ϕK signal was reported at this conference to be $(6.4^{+2.5+0.5}_{-2.1-2.0}) \times 10^{-6}$ by CLEO [3] and $(17.2^{+6.7}_{-5.4} \pm 1.8) \times 10^{-6}$ by BELLE [1]. The decay amplitude of the penguin-dominated mode $B \to K\phi$ is governed by $[a_3 + a_4 + a_5 - \frac{1}{2}(a_7 + a_9 + a_{10})]$, where a_3 and a_5 are sensitive to nonfactorized contributions. In the absence of nonfactorized effects, we find[15] $\mathcal{B}(B^\pm \to \phi K^\pm) = (6.3-7.3) \times 10^{-6}$, which is in good agreement with the CLEO result, but smaller than the BELLE measurement.

4 $B \to K\eta'$, $K^*\eta$ Decays

The decays $B \to K^{(*)}\eta(\eta')$ involve interference between the penguin amplitudes arising from $(\bar{u}u + \bar{d}d)$ and $\bar{s}s$ components of the η or η'. The branching ratios of $K\eta'$ $(K^*\eta)$ are anticipated to be much greater than $K\eta$ $(K^*\eta')$ modes owing to the presence of constructive interference between two comparable penguin amplitudes arising from non-strange and strange quarks of the $\eta'(\eta)$.

The measured branching ratios of the decays $B \to \eta' K$ are

$$\mathcal{B}(B^\pm \to \eta' K^\pm) = (80^{+10}_{-\ 9} \pm 7) \times 10^{-6},$$
$$\mathcal{B}(B^0 \to \eta' K^0) = (89^{+18}_{-16} \pm 9) \times 10^{-6}, \quad (2)$$

by CLEO [16] and $(62 \pm 18 \pm 8) \times 10^{-6}$, $< 1.12 \times 10^{-4}$, respectively by BABAR [2]. The earlier theoretical predictions in the range of $(1-2) \times 10^{-5}$ are too small compared to experiment. It was realized later (for a review, see e.g. [17]) that $\eta' K$ gets enhanced because of (i) the small running strange quark mass at the scale m_b, (ii) the sizable $SU(3)$ breaking in the decay constants f_8 and f_0, (iii) an enhancement of the form factor $F_0^{B\eta'}(0)$ due to the smaller mixing $\eta - \eta'$ mixing angle $-15.4°$ rather than $\approx -20°$, (iv) contribution from the η' charm content, and (v) constructive interference in tree amplitudes. It was also realized not long ago that [18] the above-mentioned enhancement is partially washed out by the anomaly effect in the matrix element of pseudoscalar densities, an effect overlooked before. As a consequence, the net enhancement is not very large; we find[15] $\mathcal{B}(B^\pm \to K^\pm \eta') = (40-50) \times 10^{-6}$, which is still smaller than the CLEO result but consistent with the BELLE measurement. This implies that we probably need an additional (but not dominant !) SU(3)-singlet contribution to explain the $B \to K\eta'$ puzzle.

Finally, it is worth remarking that if $\gamma > 90°$, the charged mode $\eta' K^-$ will get enhanced, while the neutral mode $\eta' K^0$ remains stable [4]. The present data of $K\eta'$ cannot differentiate between $\cos\gamma > 0$ and $\cos\gamma < 0$.

5 $B \to \omega K$ and ρK Decays

The published CLEO result [14] of a large branching ratio $(15^{+7}_{-6} \pm 2) \times 10^{-6}$ for $B^\pm \to \omega K^\pm$ imposes a serious problem to the generalized factorization approach: The observed rate is enormously large compared to naive expectation [4]. The destructive interference between a_4 and a_6 terms renders the penguin contribution small. It is thus difficult to understand the large rate of ωK. Theoretically, it is expected that[4] $\mathcal{B}(B^- \to \omega K^-) \gtrsim 2\mathcal{B}(B^- \to \rho^0 K^-) \sim 2 \times 10^{-6}$, which now agrees with the new measurement [3] of $\mathcal{B}(B^- \to \omega K^-) < 7.9 \times 10^{-6}$.

References

1. BELLE Collaboration, P. Chang in these proceedings.
2. BABAR Collaboration, T.J. Champion in these proceedings.
3. CLEO Collaboration, R. Stroynowski in these proceedings.
4. Y.H. Chen *et al., Phys. Rev.* **D60**, 094014 (1999).
5. D.S. Du *et al.,* hep-ph/0005006; T. Muta *et al.,* hep-ph/0006022.
6. A. Stocchi in these proceedings.
7. Y.Y. Keum, H.n. Li, and A.I. Sanda, hep-ph/0004173.
8. M. Beneke *et al., Phys. Rev. Lett.* **83**, 1914 (1999).
9. M. Beneke in these proceedings.
10. D. Cronin-Hennessy *et al.,* hep-ex/0001010.
11. H.Y. Cheng and K.C. Yang, in preparaion.
12. C.P. Jessop *et al.,* hep-ex/0006008.
13. CLEO Collaboration, M. Bishai *et al.,* CLEO CONF 99-13, hep-ex/9908018.
14. T. Bergfeld *et al., Phys. Rev. Lett.* **81**, 272 (1998).
15. H.Y. Cheng and K.C. Yang, *Phys. Rev.* **D62**, 054029 (2000).
16. S.J. Richichi *et al., Phys. Rev. Lett.* **85**, 520 (2000).
17. H.Y. Cheng and B. Tseng, *Phys. Rev.* **D58**, 094005 (1998).
18. A.L. Kagan and A.A. Petrov, hep-ph/9707354; A. Ali and C. Greub, *Phys. Rev.* **D57**, 2996 (1998).

QCD FACTORIZATION FOR $B \to \pi K$ DECAYS

M. BENEKE

Institut für Theoretische Physik E, RWTH Aachen, D - 52056 Aachen, Germany
E-mail: mbeneke@physik.rwth-aachen.de

G. BUCHALLA

Theory Division, CERN, CH-1211 Geneva 23, Switzerland

M. NEUBERT

Newman Laboratory of Nuclear Studies, Cornell University, Ithaca, NY 14853, USA

C.T. SACHRAJDA

Dept. of Physics and Astronomy, University of Southampton, Southampton SO17 1BJ, UK

We examine some consequences of the QCD factorization approach to non-leptonic B decays into πK and $\pi\pi$ final states, including a set of enhanced power corrections. Among the robust predictions of the approach we find small strong-interaction phases (with one notable exception) and a pattern of CP-averaged branching fractions, which in some cases differ significantly from the current central values reported by the CLEO Collaboration.

1 Introduction

The observation of B decays into πK and $\pi\pi$ final states has resulted in a large amount of theoretical and phenomenological work that attempts to interpret these observations in terms of the factorization approximation (FA), or in terms of general parameterizations of the decay amplitudes. A detailed understanding of these amplitudes would help us to pin down the value of the CKM angle γ using only data on CP-averaged branching fractions. Theoretical work on the heavy-quark limit has justified the FA as a useful starting point[1,2], but predicts important and computable corrections. Here we discuss the most important consequences of this approach for the πK and $\pi\pi$ final states.

To leading order in an expansion in powers of Λ_{QCD}/m_b, the $B \to \pi K$ matrix elements obey the factorization formula

$$\begin{aligned}\langle \pi K|Q_i|B\rangle &= f_+^{B\to\pi}(0)\, f_K\, T^{\text{I}}_{K,i} * \Phi_K \\ &+ f_+^{B\to K}(0)\, f_\pi\, T^{\text{I}}_{\pi,i} * \Phi_\pi \qquad (1)\\ &+ f_B f_K f_\pi\, T^{\text{II}}_i * \Phi_B * \Phi_K * \Phi_\pi\,,\end{aligned}$$

where Q_i is an operator in the weak effective Hamiltonian, $f_+^{B\to M}(0)$ are semi-leptonic form factors of a vector current evaluated at $q^2 = 0$, Φ_M are leading-twist light-cone distribution amplitudes, and the $*$-products imply an integration over the light-cone momentum fractions of the constituent quarks inside the mesons. When the hard-scattering functions T are evaluated to order α_s^0, Eq. (1) reduces to the conventional FA. The subsequent results are based on kernels including all corrections of order α_s. A detailed justification of (1) is given in Ref. [2]. Compared to our previous discussion of $\pi\pi$ final states[1] the present analysis incorporates three new ingredients:

i) the matrix elements of electroweak (EW) penguin operators (for πK modes);

ii) hard-scattering kernels for general, asymmetric light-cone distributions;

iii) the complete set of "chirally enhanced" $1/m_b$ corrections.[1]

The second and third items have not been considered in other[3] generalizations of Ref. [1] to the πK final states. The third one, in particular, is essential for estimating some of the theoretical uncertainties of the approach.

We now briefly present the input to our

calculations. Following Ref. [1], we obtained the coefficients $a_i(\pi K)$ of the effective factorized transition operator defined analogously to the case of $\pi\pi$ final states, but augmented by coefficients $a_{7-10}(\pi K)$ related to EW penguin operators and electro-magnetic penguin contractions of current–current and QCD penguin operators. A sensible implementation of QCD corrections to EW penguin matrix elements implies that one departs from the usual renormalization-group counting, in which the initial condition for EW penguin coefficients is treated as a next-to-leading order (NLO) effect. Our NLO initial condition hence includes the α_s corrections computed in Ref. [4].

Chirally enhanced corrections arise from twist-3 two-particle light-cone distribution amplitudes, whose normalization involves the quark condensate. The relevant parameter, $2\mu_\pi/m_b = -4\langle\bar{q}q\rangle/(f_\pi^2 m_b)$, is formally of order $\Lambda_{\rm QCD}/m_b$, but large numerically. The coefficients a_6 and a_8 are multiplied by this parameter. There are also additional chirally enhanced corrections to the spectator-interaction term in (1), which turn out to be the more important effect. In both cases, these corrections involve logarithmically divergent integrals, which violate factorization. For instance, for matrix elements of $V-A$ operators the hard spectator interaction is now proportional to ($\bar{u} \equiv 1-u$)

$$\int_0^1 \frac{du}{\bar{u}}\frac{dv}{\bar{v}}\,\Phi_K(u)\left(\Phi_\pi(v) + \frac{2\mu_\pi}{m_b}\frac{\bar{u}}{u}\right)$$

when the spectator quark goes to the pion. (Here we used that the twist-3 distribution amplitudes can be taken to be the asymptotic ones when one neglects twist-3 corrections without the chiral enhancement.) The divergence of the v-integral in the second term as $\bar{v} \to 0$ implies that it is dominated by soft gluon exchange between the spectator quark and the quarks that form the kaon. We therefore treat the divergent integral $X = \int_0^1 (dv/\bar{v})$ as an unknown parameter (different for the penguin and hard scattering contributions), which may in principle be complex owing to soft rescattering in higher orders. In our numerical analysis we set $X = \ln(m_B/0.35\,\mathrm{GeV}) + r$, where r is chosen randomly inside a circle in the complex plane of radius 3 ("realistic") or 6 ("conservative"). Our results depend on the B-meson parameter[1] λ_B, which we vary between 0.2 and 0.5 GeV. Finally, there is in some cases a non-negligible dependence of the coefficients $a_i(\pi K)$ on the renormalization scale, which we vary between $m_b/2$ and $2m_b$.

2 Results

We take $|V_{ub}/V_{cb}| = 0.085$ and $m_s(2\,\mathrm{GeV}) = 110\,\mathrm{MeV}$ as fixed input, noting that ultimately the ratio $|V_{ub}/V_{cb}|$, along with the CP-violating phase $\gamma = \arg(V_{ub}^*)$, might be extracted from a simultaneous fit to the $B \to \pi K$ and $B \to \pi\pi$ decay rates.

2.1 SU(3) breaking

Bounds[5,6] on the CKM angle γ derived from ratios of πK branching fractions rely on an estimate of $SU(3)$ flavour-symmetry violations. We find that "non-factorizable" $SU(3)$-breaking effects (i.e., effects not accounted for by the different decay constants and form factors of pions and kaons in the conventional FA) do not exceed a few percent at leading power.

2.2 Amplitude parameters

The approach discussed here allows us to obtain the decay amplitudes for the $\pi\pi$ and πK final states in terms of the form factors and the light-cone distribution amplitudes. The $\pi^0\pi^0$ final state is very poorly predicted and will not be discussed here. We write

$$\mathcal{A}(B^0 \to \pi^+\pi^-) = T\,[e^{i\gamma} + (P/T)_{\pi\pi}]$$

and parameterize the πK amplitudes by[6]

$$\begin{aligned} \mathcal{A}(B^+ \to \pi^+ K^0) &= P\left(1 - \varepsilon_a\, e^{i\eta} e^{i\gamma}\right), \\ -\sqrt{2}\,\mathcal{A}(B^+ \to \pi^0 K^+) &= P\Big[1 - \varepsilon_a\, e^{i\eta} e^{i\gamma} \\ &\quad - \varepsilon_{3/2}\, e^{i\phi}(e^{i\gamma} - q\, e^{i\omega})\Big], \end{aligned} \qquad (2)$$

Table 1. Parameters for the $B \to \pi K$ amplitudes as defined in (2), for conservative variation of all input parameters (see text).

	Range, NLO	LO
$-\varepsilon_a\, e^{i\eta}$	$(0.017\text{–}0.020)\, e^{i\,[13,21]^\circ}$	0.02
$\varepsilon_{3/2}\, e^{i\phi}$	$(0.20\text{–}0.38)\, e^{i\,[-30,7]^\circ}$	0.36
$q\, e^{i\omega}$	$(0.53\text{–}0.63)\, e^{i\,[-7,3]^\circ}$	0.64
$\varepsilon_T\, e^{i\phi_T}$	$(0.20\text{–}0.29)\, e^{i\,[-19,3]^\circ}$	0.33
$q_C\, e^{i\omega_C}$	$(0.00\text{–}0.22)\, e^{i\,[-180,180]^\circ}$	0.06
$(P/T)_{\pi\pi}$	$(0.19\text{–}0.29)\, e^{i\,[-1,23]^\circ}$	0.16

$$-\mathcal{A}(B^0 \to \pi^- K^+) = P\Big[1 - \varepsilon_a\, e^{i\eta} e^{i\gamma} - \varepsilon_T\, e^{i\phi_T}(e^{i\gamma} - q_C\, e^{i\omega_C})\Big],$$

and $\sqrt{2}\,\mathcal{A}(B^0 \to \pi^0 K^0) = \mathcal{A}(B^+ \to \pi^+ K^0) + \sqrt{2}\,\mathcal{A}(B^+ \to \pi^0 K^+) - \mathcal{A}(B^0 \to \pi^- K^+)$. Table 1 summarizes the numerical values for the amplitude parameters for the conservative variation of X, and variation of the other parameters as explained above. The LO results correspond to the conventional FA at the fixed scale $\mu = m_b$. They are strongly scale dependent. In comparison, the scale-dependence of the NLO result is small, with the exception of $q_C\, e^{i\omega_C}$. One must keep in mind that the ranges may overestimate the true uncertainty, since the parameter X may ultimately be constrained from a subset of branching fractions. This is true in particular for the quantity $\varepsilon_{3/2}$ in Table 1, which can be extracted from data.[6]

2.3 Ratios of CP-averaged rates

Since the form factor $f_+(0)$ is not well known, we consider here only ratios of CP-averaged branching ratios . We display these as functions of the CKM angle γ in Fig. 1.

Table 1 shows that the corrections with respect to the conventional FA are significant (and important to reduce the renormalization-scale dependence). Despite this fact, the *qualitative* pattern that emerges

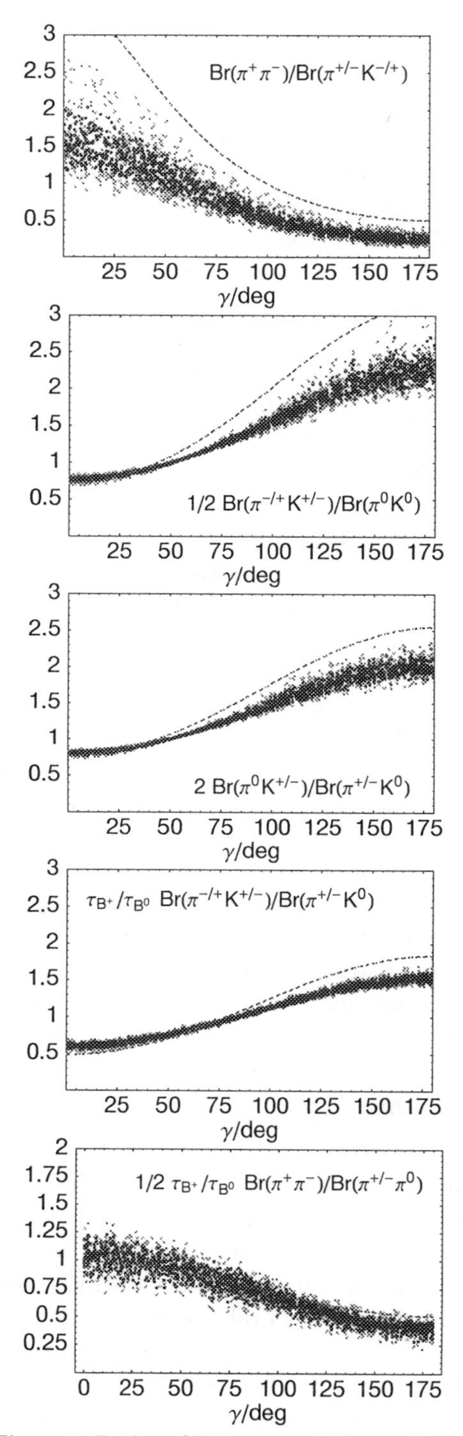

Figure 1. Ratios of CP-averaged $B \to \pi K$ and $\pi\pi$ decay rates. The scattered points cover a realistic (dark) and conservative (light) variation of input parameters. The dashed curve is the LO result, corresponding to conventional factorization.

for the set of πK and $\pi\pi$ decay modes is similar to that in conventional factorization. In particular, the penguin–tree interference is constructive (destructive) in $B \to \pi^+\pi^-$ ($B \to \pi^- K^+$) decays if $\gamma < 90°$. Taking the currently favoured range $\gamma = (60 \pm 20)°$, we find the following robust predictions:

$$\frac{\mathrm{Br}(\pi^+\pi^-)}{\mathrm{Br}(\pi^\mp K^\pm)} = 0.5\text{–}1.9 \quad [0.25 \pm 0.10]$$

$$\frac{\mathrm{Br}(\pi^\mp K^\pm)}{2\mathrm{Br}(\pi^0 K^0)} = 0.9\text{–}1.4 \quad [0.59 \pm 0.27]$$

$$\frac{2\mathrm{Br}(\pi^0 K^\pm)}{\mathrm{Br}(\pi^\pm K^0)} = 0.9\text{–}1.3 \quad [1.27 \pm 0.47]$$

$$\frac{\tau_{B^+}}{\tau_{B^0}}\frac{\mathrm{Br}(\pi^\mp K^\pm)}{\mathrm{Br}(\pi^\pm K^0)} = 0.6\text{–}1.0 \quad [1.00 \pm 0.30]$$

The first ratio is in striking disagreement with current CLEO data[7] (square brackets). The near equality of the second and third ratios is a result of isospin symmetry.[6] We find $\mathrm{Br}(B \to \pi^0 K^0) = (4.5 \pm 2.5) \times 10^{-6}\,(V_{cb}/0.039)^2 (f_+^{B\to\pi}(0)/0.3)^2$ almost independently of γ. This is three time smaller than the central value reported by CLEO.

2.4 *CP asymmetry in $B \to \pi^+\pi^-$ decay*

The stability of the prediction for the $\pi^+\pi^-$ amplitude suggests that the CKM angle α can be extracted from the time-dependent mixing-induced CP asymmetry in this decay mode, without using isospin analysis. Fig. 2 displays the coefficient S of $-\sin(\Delta M_{B_d} t)$ as a function of $\sin(2\alpha)$ for $\sin(2\beta) = 0.75$. For some values of S there is a two-fold ambiguity (assuming all angles are between 0° and 180°). A consistency check of the approach could be obtained, in principle, from the coefficient of the $\cos(\Delta m_{B_d} t)$ term.

3 Conclusions

We have examined some of the consequences of the QCD factorization approach to B decays into πK and $\pi\pi$ final states, leaving a detailed discussion to a subsequent publication. Here we have focused on robust predictions for ratios of CP-averaged decay rates. Our result for the ratio of the $B \to \pi^+\pi^-$ and $B \to \pi^\mp K^\pm$ decay rates is in disagreement with the current experimental value, unless the weak phase γ were significantly larger than 90°.

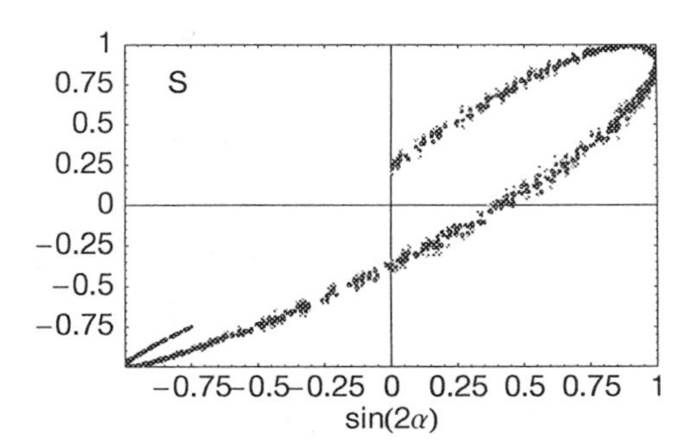

Figure 2. Mixing-induced CP asymmetry in $B \to \pi^+\pi^-$ decays. The lower band refers to values $45° < \alpha < 135°$, the upper one to $\alpha < 45°$ (right) or $\alpha > 135°$ (left). We assume $\alpha, \beta, \gamma \in [0, \pi]$.

Note added. At this conference results on the CP averaged branching ratios have been presented by the BABAR and BELLE collaborations for the first time (T.J. Champion; P. Chang, these proceedings). The results for $\mathrm{Br}(\pi^+\pi^-)/\mathrm{Br}(\pi^\mp K^\pm)$ are 0.74 ± 0.29 (BABAR) and 0.36 ± 0.26 (BELLE). Final states with neutral particles have been measured only by BELLE, but continue to have large errors.

References

1. M. Beneke, G. Buchalla, M. Neubert, C.T. Sachrajda, *Phys. Rev. Lett.* **83**, 1914 (1999).
2. M. Beneke, G. Buchalla, M. Neubert, C.T. Sachrajda, hep-ph/0006124.
3. D. Du, D. Yang, G. Zhu, hep-ph/0005006; T. Muta, A. Sugamoto, M. Yang, Y. Yang, hep-ph/0006022.
4. A.J. Buras, P. Gambino, U.A. Haisch, *Nucl. Phys.* **B570**, 117 (2000).
5. R. Fleischer, T. Mannel, *Phys. Rev.* **D57**, 2752 (1998).
6. M. Neubert, J.L. Rosner, *Phys. Lett.* **B441**, 403 (1998); M. Neubert, *JHEP* **02**, 014 (1999).
7. D. Cronin-Hennessy *et al.* (CLEO Collaboration), hep-ex/0001010.

PENGUIN ENHANCEMENT AND $B \to K(\pi)\pi$ DECAYS IN PERTURBATIVE QCD

HSIANG-NAN LI

Department of Physics, National Cheng-Kung University, Tainan, Taiwan 701, ROC
E-mail: hnli@mail.ncku.edu.tw

We present perturbative QCD (PQCD) factorization theorem for exclusive B meson decays and predictions for the branching ratios and CP asymmetries of the $B \to K\pi$ and $\pi\pi$ modes. It is found that penguin contributions are enhanced, such that the $B \to K\pi$ and $\pi\pi$ data can be explained simultaneously using the unitarity angle $\phi_3 \sim 90^o$.

Perturbative QCD (PQCD) factorization theorem for exclusive heavy-meson decays has been developed, and applied to various decay modes successfully [1,2,3,4]. The predictions have been found to be well consistent with experimental data. It has been proposed that the unitarity angle ϕ_3 can be determined from the decays $B \to K\pi$ and $\pi\pi$ [5,6,7]. The contributions to these modes involve inteference between penguin and tree amplitudes. We shall show that the above amplitudes and relevant strong phases can be calculated in the PQCD framework.

Exclusive B meson decays involve three scales: the W boson mass M_W, at which the matching conditions of the effective weak Hamiltonian to the full Hamiltonian are defined, the typical scale t of a hard amplitude, which reflects the specific dynamics of a decay mode, and the factorization scale $1/b$, with b being the conjugate variable of parton transverse momenta. Above the factorization scale, perturbative corrections produce two types of large logarithms: $\ln(M_W/t)$ and $\ln(tb)$. The former are summed by renormalization-group equations to give the evolution from M_W down to t described by the Wilson coefficients. The latter are summed to give the evolution from t to $1/b$.

Dynamics below $1/b$ is regarded as being nonperturbative, and parametrized into meson wave functions. The wave functions, though not calculable, are universal. This universality is attributed to the cancellation of nonfactorizable soft gluon contributions at leading power of $1/M_B$, M_B being the B meson mass. We extract wave functions from experimental data, and then employ these wave funcitons to make model-independent predictions for other processes.

Parton transverse momenta should be considered in order to smear infrared divergences contained in heavy-to-light transition form factors. Because of the inclusion of parton transverse momenta, double logarithms are generated from the overlap of collinear and soft enhancements in radiative corrections to meson wave functions. The resummation [8] of these double logarithms leads to a Sudakov form factor, which suppresses the long-distance contributions in the large b region. This suppression guarantees the applicability of PQCD to exclusive decays around M_B [1].

With all the large logarithms organized, the remaining finite radiative corrections are absorbed into the hard b quark decay amplitude. This hard amplitude contains all possible Feynman diagrams, such as factorizable and nonfactorizable diagrams. The annihilation topology is also included. Therefore, the conventional factorization asummption (FA) [9] for two-body B meson decays, in which nonfactorizable and annihilation contributions are neglected, is not necessary. It has been shown that factorizable annihilation contributions are in fact important, and give large strong phases in the PQCD approach [3].

An essential difference between the FA and PQCD approaches is that the hard scale at which Wilson coefficients are evaluated is chosen arbitrarily as m_b or $m_b/2$ in the former, m_b being the b quark mass, but as hard gluon momenta in the latter. This choice leads to an enhancement of penguin contributions by nearly 50% compared to those in the FA approach, which is crucial for the explanation of the data of all $B \to K\pi$ and $\pi\pi$ modes using a smaller angle $\phi_3 \sim 90^o$. Note that an angle ϕ_3 larger than 110° must be adopted in order to explain the above data in the FA approach [10].

Recently, Beneke *et al.* proposed an alternative formalism for two-body charmless B meson decays [11]. They claimed that factorizable contributions are not calculable in PQCD, but nonfactorzable contributions are in the heavy quark limit. Hence, the former are treated in the same way as FA: they are expressed as products of Wilson coefficients and model form factors. The latter, calculated as in the PQCD approach, are written as the convolutions of hard amplitudes with three (B, π, π) meson wave functions. Annihilation diagrams are neglected as in FA. Hence, this formalism can be regarded as a mixture of the FA and PQCD approaches.

Our predictions for the branching ratio of each $B \to K\pi$ mode corresponding to $\phi_3 = 90^o$ are,

$$\begin{aligned} B(B^+ \to K^0\pi^+) &= 20.22 \times 10^{-6} , \\ B(B^- \to \bar{K}^0\pi^-) &= 19.79 \times 10^{-6} , \\ B(B_d^0 \to K^+\pi^-) &= 22.74 \times 10^{-6} , \\ B(\bar{B}_d^0 \to K^-\pi^+) &= 15.50 \times 10^{-6} , \\ B(B^+ \to K^+\pi^0) &= 11.40 \times 10^{-6} , \\ B(B^- \to K^-\pi^0) &= 7.89 \times 10^{-6} , \\ B(B_d^0 \to K^0\pi^0) &= 8.81 \times 10^{-6} , \\ B(\bar{B}_d^0 \to \bar{K}^0\pi^0) &= 9.25 \times 10^{-6} , \\ B(B_d^0 \to \pi^\pm\pi^\mp) &= 4.6 \times 10^{-6} , \end{aligned} \tag{1}$$

consistent with the CLEO data [12],

$$\begin{aligned} B(B^\pm \to K^0\pi^\pm) &= (18.2^{+4.6}_{-4.0} \pm 1.6) \times 10^{-6} , \\ B(B_d^0 \to K^\pm\pi^\mp) &= (17.2^{+2.5}_{-2.4} \pm 1.2) \times 10^{-6} , \\ B(B^\pm \to K^\pm\pi^0) &= (11.6^{+3.0+1.4}_{-2.7-1.3}) \times 10^{-6} , \\ B(B_d^0 \to K^0\pi^0) &= (14.6^{+5.9+2.4}_{-5.1-3.3}) \times 10^{-6} , \\ B(B_d^0 \to \pi^\pm\pi^\mp) &= (4.3^{+1.6+0.5}_{-1.5-0.5}) \times 10^{-6} . \end{aligned} \tag{2}$$

Acknowledgements

The work was supported in part by the National Science Council of R.O.C. under the Grant No. NSC-89-2112-M-006-004 and in part by Grant-in Aid for Special Project Research (Physics of CP Violation) from Ministry of Education, Science and Culture of Japan.

References

1. H-n. Li and H.L. Yu, Phys. Rev. Lett. **74**, 4388 (1995); Phys. Rev. D **53**, 2480 (1996).
2. T.W. Yeh and H-n. Li, Phys. Rev. D **56**, 1615 (1997).
3. Y.Y. Keum, H-n. Li, and A.I. Sanda, hep-ph/0004004; hep-ph/0004173.
4. C. D. Lü, K. Ukai, and M. Z. Yang, hep-ph/0004213.
5. M. Gronau, J.L. Rosner, and D. London, Phys. Rev. Lett. **73**, 21 (1994).
6. R. Fleischer and T. Mannel, Phys. Rev. D **57**, 2752 (1998).
7. M. Neubert and J. Rosner, Phys. Lett. B **441**, 403 (1998).
8. J.C. Collins and D.E. Soper, Nucl. Phys. B **193**, 381 (1981).
9. M. Bauer, B. Stech, M. Wirbel, Z. Phys. C **34**, 103 (1987); Z. Phys. C **29**, 637 (1985).
10. W.S. Hou, J.G. Smith, and F. Würthwein, hep-ex/9910014.
11. M. Beneke, G. Buchalla, M. Neubert, and C.T. Sachrajda, Phys. Rev. Lett. **83**, 1914 (1999); hep-ph/0006124.
12. CLEO Coll., Y. Kwon *et al.*, hep-ex/9908039.

LIFETIME PATTERN OF HEAVY HADRONS

B. GUBERINA, B. MELIĆ AND H. ŠTEFANČIĆ

Theoretical Physics Division, Rudjer Bošković Institute,
P.O.Box 180, HR-10002 Zagreb, Croatia
E-mails: guberina@thphys.irb.hr, melic@thphys.irb.hr, shrvoje@thphys.irb.hr

We discuss the lifetime pattern of weakly decaying heavy hadrons.

1 Introduction

A lot of physical observables in heavy-quark decays are described using the inverse heavy-quark mass expansion in terms of a few basic quantities, i.e., quark masses and hadronic expectation values of several local operators [1]. The remarkable fact that the expansion is applicable even to the the charmed case enables one to connect charmed and beauty sectors. In this presentation [a] we relate lifetimes of heavy hadrons and obtain some interesting predictions.

Large lifetime differences between charmed hadrons, shown in Fig. 1, cannot be explained by taking just $\mathcal{O}(1/m_c^2)$ corrections into account. The diversity among charmed hadron lifetimes is attributed to the effects of four-quark operator ($\mathcal{O}^{4q}$) contributions of the order of $1/m_c^3$. For charmed mesons [2] and baryons [3], the theoretical findings have been also confirmed by experiment. Analogously, a significant spread of doubly charmed baryon lifetimes is predicted [4], Fig. 1.

For beauty hadron decays there is a rapid convergence of the $1/m_b$ expansion and the leading $\mathcal{O}(1/m_b^2)$ corrections introduce a difference of just $2 - 3\%$ between lifetimes of beauty hadrons. But, owing to the peculiarity of the $1/m_Q$ expansion, exhibited also in charmed decays, there is still some place for $\mathcal{O}(1/m_Q^3)$ operators to play a significant role in beauty decays. In a recent work [5] we have found an enhancement of the $\mathcal{O}^{4q}$ contributions in beauty baryon decays. Such an enhancement brings the $\tau(\Lambda_b)/\tau(B)$ ratio much closer to the experimental value compared with the standard nonrelativistic-model estimation and predicts a much larger spread among beauty-baryon lifetimes, Fig. 2.

2 Connecting charm and beauty

As in the calculation of the $\mathcal{O}^{4q}$ contributions one relies on the questionable nonrelativistic models, it is a challenge to search for some model-independent determination of such contributions. Having in mind the same formalism applied in the treatment of both charmed and beauty decays, and large effects which the $\mathcal{O}^{4q}$ contributions exhibit in the lifetimes of charmed as well as beauty hadrons, discussed in Sec. 1, we try to perform a model-independent analysis searching for an explicit connection between charmed and beauty sectors.

Similar ideas have already been applied, first by Voloshin [6], to obtain the lifetime differences between beauty hyperons, and then by us [5], where we have obtained, in a moderate model-dependent analysis, an enhancement of four-quark contributions for beauty baryons with predictions shown in Fig. 2.

We group heavy hadrons exhibiting the same type of $\mathcal{O}^{4q}$ contributions in pairs and consider their decay-rate differences applying the SU(2) isospin symmetry and the heavy-quark symmetry (HQS) [7]. Cabibbo suppressed modes and mass corrections in final states are neglected. We start our investiga-

[a]Talk given by B. Melić at ICHEP2000, July 27 - August 2, Osaka, Japan

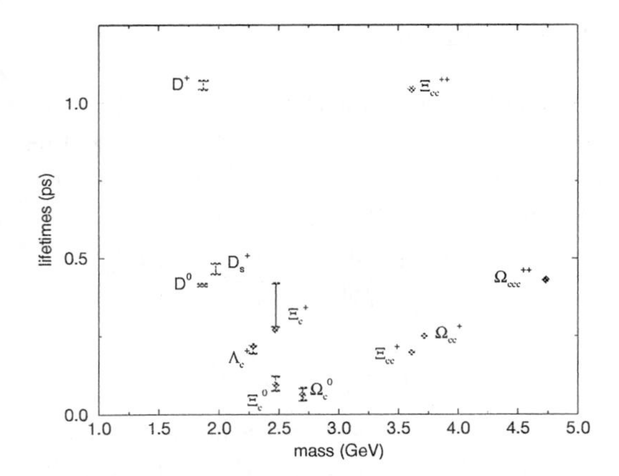

Figure 1. Lifetimes of weakly decaying charmed hadrons. Diamonds denote theoretical predictions.

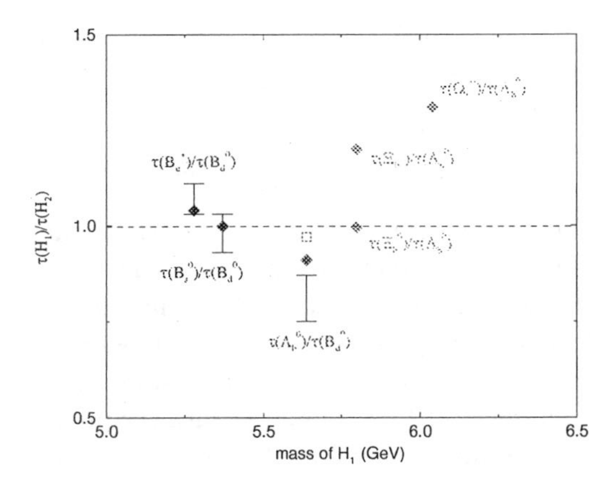

Figure 2. Lifetimes of weakly decaying beauty hadrons. Diamonds denote theoretical predictions. The square stands for the standard nonrelativistic model prediction of the $\tau(\Lambda_b)/\tau(B)$ ratio.

tion by treating different heavy-hadron sectors separately.

2.1 *Heavy mesons*

The first pair of considered mesons form D^+ and B^-. In decays of both of these mesons the negative Pauli interference occurs. The second pair of particles having the effects of a weak exchange form D^0 and B^0. The idea is to combine decay rates in such a manner that the effects of the $\mathcal{O}^{4q}$ contributions are isolated:

$$\Gamma(D^+) - \Gamma(D^0) = \frac{G_F^2 m_c^2}{4\pi}|V_{cs}|^2|V_{ud}|^2 \times \left[\langle D^+|\mathcal{L}_{\rm PI}^{cd}|D^+\rangle - \langle D^0|\mathcal{L}_{\rm exc}^{cu}|D^0\rangle\right]$$

$$\Gamma(B^-) - \Gamma(B^0) = \frac{G_F^2 m_b^2}{4\pi}|V_{cb}|^2|V_{ud}|^2 \times \left[\langle B^-|\mathcal{L}_{\rm PI}^{bu}|B^-\rangle - \langle B^0|\mathcal{L}_{\rm exc}^{bd}|B^0\rangle\right] .$$

$\mathcal{L}$'s are parts of the weak Lagrangian involving Cabibbo-leading nonleptonic as well as semileptonic parts.

Assuming the isospin symmetry in the heavy-quark limit, the $\mathcal{O}^{4q}$ contributions get reduced and we obtain the following relation:

$$R^{BD} = \frac{\Gamma(B^-) - \Gamma(B^0)}{\Gamma(D^+) - \Gamma(D^0)} = \frac{m_b^2}{m_c^2}\frac{|V_{cb}|^2}{|V_{cs}|^2}\left[1 + \mathcal{O}(1/m_{b,c})\right] . \quad (1)$$

Starting from this expression, we may check the standard formalism of inclusive decays, expressed by Eq. (1), against experimental data on heavy-meson lifetimes [8]. The results are

$$R^{BD} = 0.020 \pm 0.007 ,$$
$$R^{BD}_{\rm exp} = 0.030 \pm 0.011 ,$$

and they are consistent within errors.

2.2 *Heavy baryons*

In baryon decays we find Ξ_c^+ and Ξ_b^- experiencing the negative interference, and Ξ_c^0 and Ξ_b^0 exhibiting the weak exchange. There are also different nonleptonic and semileptonic contributions from the operators involving s-quark, but they cancel in the decay-rate differences:

$$\Gamma(\Xi_c^+) - \Gamma(\Xi_c^0) = \frac{G_F^2 m_c^2}{4\pi}|V_{cs}|^2|V_{ud}|^2 \times \left[\langle \Xi_c^+|\mathcal{L}_{\rm PI}^{cu}|\Xi_c^+\rangle - \langle \Xi_c^0|\mathcal{L}_{\rm exc}^{cd}|\Xi_c^0\rangle\right]$$

$$\Gamma(\Xi_b^-) - \Gamma(\Xi_b^0) = \frac{G_F^2 m_b^2}{4\pi}|V_{cb}|^2|V_{ud}|^2 \times \left[\langle \Xi_b^-|\mathcal{L}_{\rm PI}^{bd}|\Xi_b^-\rangle - \langle \Xi_b^0|\mathcal{L}_{\rm exc}^{bu}|\Xi_b^0\rangle\right] .$$

Applying SU(2) symmetry and HQS as before, we obtain to order $1/m_{c,b}$

$$R^{bc} = \frac{\Gamma(\Xi_b^-) - \Gamma(\Xi_b^0)}{\Gamma(\Xi_c^+) - \Gamma(\Xi_c^0)} = \frac{m_b^2}{m_c^2} \frac{|V_{cb}|^2}{|V_{cs}|^2} \quad (2)$$

This relation can serve as a test of the model-dependent predictions, presented in Figs.1 and 2. If we calculate R^{bc} using model-dependent approach [3,5] consistently with approximations made in this analysis, we obtain a difference of 12% compared with the prediction obtained from Eq. (2). By performing the complete calculation with the mass corrections and Cabibbo-suppressed modes included, we can judge the order of neglected corrections to be less than 10%.

The relation (2) enables us to obtain a prediction for the lifetime difference between beauty hyperons, using measured lifetimes of singly-charmed baryons Ξ_c^+ and Ξ_c^0 [3]:

$$\Gamma(\Xi_b^-) - \Gamma(\Xi_b^0) = -(0.14 \pm 0.06)\ \mathrm{ps}^{-1}.$$

The prediction can be compared with the values from [6] and [5], where $SU(3)_f$ symmetry and HQS were used. All results appear to be consistent with each other.

2.3 Doubly-heavy baryons

Finally, a similar procedure applies to doubly-heavy baryons. The obtained expression now relates doubly beauty baryons with doubly-charmed baryons to order $1/m_{c,b}$:

$$R^{bbcc} = \frac{\Gamma(\Xi_{bb}^-) - \Gamma(\Xi_{bb}^0)}{\Gamma(\Xi_{cc}^{++}) - \Gamma(\Xi_{cc}^+)} = \frac{m_b^2}{m_c^2} \frac{|V_{cb}|^2}{|V_{cs}|^2} \quad (3)$$

Unfortunately, there is still no experimental evidence for doubly-heavy baryon lifetimes to check the above relation. However, we can use it to calculate the splitting among doubly-beauty hyperons Ξ_{bb}^- and Ξ_{bb}^0, by taking existing theoretical predictions for doubly-charmed lifetimes [4]. The prediction for the doubly-beauty lifetime spread is

$$\Gamma(\Xi_{bb}^-) - \Gamma(\Xi_{bb}^0) = -0.073\,\mathrm{ps}^{-1}.$$

3 Conclusions

By inspection of all three relations (1), (2) and (3) we can see that there exists a universal behavior in decays of heavy hadrons summarized by the expression:

$$\frac{\Gamma(B^-) - \Gamma(B^0)}{\Gamma(D^+) - \Gamma(D^0)} = \frac{\Gamma(\Xi_b^-) - \Gamma(\Xi_b^0)}{\Gamma(\Xi_c^+) - \Gamma(\Xi_c^0)} = $$
$$= \frac{\Gamma(\Xi_{bb}^-) - \Gamma(\Xi_{bb}^0)}{\Gamma(\Xi_{cc}^{++}) - \Gamma(\Xi_{cc}^+)} = \frac{m_b^2}{m_c^2} \frac{|V_{cb}|^2}{|V_{cs}|^2}.$$

This relation connects all sectors of weakly decaying heavy hadrons that are usually treated separately: mesons and baryons, charmed and beauty particles, and brings some order in the otherwise rather intricate pattern of heavy-hadron lifetimes. The predictions we have obtained, although burdened with some approximations, if experimentally confirmed would indicate that four-quark operators can account for the greatest part of the decay rate differences among heavy hadrons.

This work was supported by the Ministry of Science and Technology of the Republic of Croatia under the contract No. 00980102.

References

1. I. Bigi, hep-ph/0001003
2. B. Guberina, S. Nussinov, R.D. Peccei, R. Rückl, *Phys. Lett.* B **89**, 111 (1979).
3. B. Guberina, B. Melić, *Eur. Phys. J.* C **2**, 697 (1998).
4. B. Guberina, B. Melić, H. Štefančić, *Eur. Phys. J.* C **9**, 213 (1999); Erratum, *ibidem* C **13** 551 (2000).
5. B. Guberina, B. Melić, H. Štefančić, *Phys. Lett.* B **469**, 253 (1999).
6. M.B. Voloshin, *Phys. Rept.* **320**, 275 (1999).
7. B. Guberina, B. Melić, H. Štefančić, *Phys. Lett.* B **484**, 43 (2000).
8. Partical Data Group, C. Caso et al, *Eur. Phys. J.* 1 C **3**, 1 (1998).

Parallel Session 8

Neutrino Physics

Conveners: Kenzo Nakamura (KEK),
Sandip Pakvasa (Hawaii) and
Paolo Strolin (CERN & Naples)

NEUTRINO MASS: THE PRESENT AND THE FUTURE

BORIS KAYSER
National Science Foundation, 4201 Wilson Blvd., Arlington VA 22230, USA
E-mail: bkayser@nsf.gov

We argue that the evidence for neutrino mass is quite compelling. This mass raises a number of questions, which we enumerate, about neutrinos. Then we focus on one of these questions—the issue of the possible neutrino mass spectra. In particular, we explain that one can have a four-neutrino spectrum which does not require significant sterile-neutrino involvement in either the atmospheric or solar neutrino oscillations.

Before we discuss the physics of neutrinos with mass, let us step back and ask whether the evidence that neutrinos *do* have mass is really convincing. We believe that it is. The most compelling single piece of evidence is the observed violation of the equality

$$\phi(\nu_\mu \text{ Up}) = \phi(\nu_\mu \text{ Down}) \ . \quad (1)$$

Here, $\phi(\nu_\mu \text{ Up})$ is the total flux of atmospheric muon neutrinos observed by an underground detector to be coming *upward* from all directions below the horizontal at the location of the detector, while $\phi(\nu_\mu \text{ Down})$ is the corresponding total flux observed to be coming *dowmward* from all directions above the horizontal. The atmospheric neutrinos are produced by cosmic rays in the earth's atmosphere all around the world, and so are incident on the detector from all directions. In considering our expectations for the relationship between $\phi(\nu_\mu \text{ Up})$ and $\phi(\nu_\mu \text{ Down})$, let us suppose that nothing—neither neutrino oscillation nor anything else—decreases or increases the atmospheric ν_μ flux as the neutrinos travel from their points of origin to the detector. Then, as illustrated by the "Sample ν_μ path" in Fig. 1, any ν_μ that enters the sphere S defined in the figure caption will eventually exit this sphere. Thus, since we are dealing with a steady-state situation, the total ν_μ fluxes entering and exiting S per unit time must be equal. Now, for neutrino energies $E >$ a few GeV, the flux of cosmic rays that create the atmospheric neutrinos is known to be isotropic. Thus, at these energies, the atmospheric muon neutrinos are being produced at the same rate everywhere around the earth. Thanks to this spherical symmetry, the equality betwen the ν_μ fluxes entering and exiting S must hold, not only for S as a whole, but at each point of S. In particular, it must hold at the location of the detector. But, as is clear from Fig. 1, a ν_μ entering S through the detector must be part of the downward flux $\phi(\nu_\mu \text{ Down})$. One exiting S through the detector must be part of $\phi(\nu_\mu \text{ Up})$. Thus, the equality of the ν_μ fluxes entering and exiting S at the detector implies that $\phi(\nu_\mu \text{ Down}) = \phi(\nu_\mu \text{ Up})$.[1] With a bit more effort, but no additional assumptions, one can show that this equality must hold not only for the integrated downward and upward fluxes, but angle by angle. That is, the flux coming down from zenith angle θ_z must equal that coming up from angle $\pi-\theta_z$.[2]

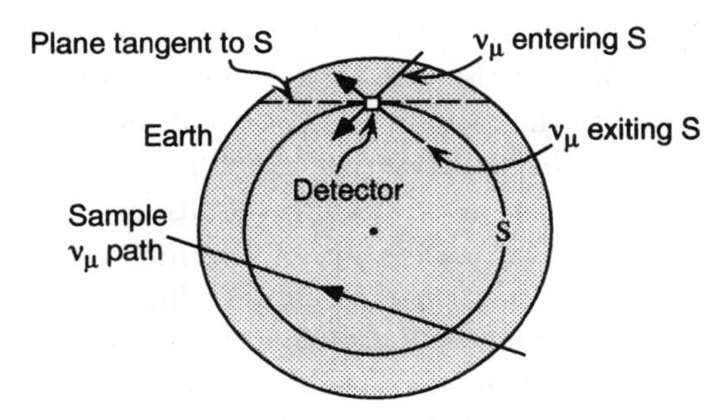

Figure 1. Atmospheric muon neutrino fluxes at an underground detector. S is a sphere centered at the center of the earth and passing through the detector.

The Super-Kamiokande detector (Super-

K) finds that for multi-GeV atmospheric muon neutrinos,[3]

$$\frac{\phi(\nu_\mu \text{ Up}; -1.0 < \cos\theta_z < -0.2)}{\phi(\nu_\mu \text{ Down}; +0.2 < \cos\theta_z < +1.0)} = 0.54 \pm 0.04\ , \tag{2}$$

in strong disagreement with the expected equality of upward and downward fluxes. Thus, some mechanism must be changing the atmospheric ν_μ flux while the neutrinos travel to the detector. As we see, this conclusion follows merely from the isotropy of the cosmic rays, the fact that the earth is round, and the fact that the ratio in Eq. (2) is not unity.

The most attractive candidate for the mechanism that is altering the atmospheric ν_μ flux is the oscillation of the muon neutrinos into neutrinos of another flavor. Indeed, neutrino oscillation fits the detailed atmospheric neutrino data very well.[3] Barring the exotic (albeit intriguing) possibility of extra spatial dimensions, neutrino oscillation implies neutrino mass.

Amusingly, an alternative candidate, neutrino decay, also fits the detailed atmospheric neutrino data well.[4] To be sure, decay within the time that a neutrino takes to traverse the earth is theoretically less likely than oscillation. Nevertheless, it is interesting that the decay model[4] survives all the comparisons with data that have so far been made. Future long-baseline neutrino experiments capable of distinguishing between the sinusoidal dependence on (distance/energy) that is characteristic of oscillation and the exponential dependence that is characteristic of decay would discriminate between the two possibilities. The decay hypothesis would also be tested by more accurate information on the rate of neutral current (NC) events induced by atmospheric neutrinos in an underground detector. If ν_μ oscillates to ν_τ, the NC event rate will be the same as if there were no oscillation or decay. But if neutrino decay is playing a prominent role, then the electroweak-active neutrino flux is reduced by the decay process, and so the NC event rate will be lower than when there is no oscillation or decay.[5]

Both the oscillation and decay explanations of the behavior of atmospheric neutrinos imply neutrino mass and mixing. Strong further evidence for mass and mixing comes from the behavior of the solar neutrinos, which can be successfully explained in terms of matter-enhanced or perhaps vacuum oscillation.[3] Finally, there is unconfirmed evidence for $\overset{(-)}{\nu}_\mu \to \overset{(-)}{\nu}_e$ oscillation in the LSND experiment.[6] As we have seen, the evidence for mass and mixing from the atmospheric neutrinos is very strong indeed.

That neutrinos have mass means that there is some spectrum of three or more neutrino mass eigenstates, $\nu_1, \nu_2, \nu_3, \ldots$, which are the neutrino analogues of the charged-lepton mass eigenstates, e, μ, and τ. That neutrinos mix means that the neutrino state $|\nu_\ell\rangle$ coupled by the weak interaction to the particular charged-lepton mass eigenstate ℓ (e, μ, or τ) is not one of the neutrino mass eigenstates $|\nu_m\rangle$, but some linear combination of the neutrino mass eigenstates. That is,

$$|\nu_\ell\rangle = \sum_m U^*_{\ell m} |\nu_m\rangle\ , \tag{3}$$

where U is the unitary leptonic mixing matrix, often called the Maki-Nakagawa-Sakata matrix.[7]. The neutrino $|\nu_\ell\rangle$ is called the neutrino of "flavor" ℓ. The decays $Z \to \nu_\ell \overline{\nu_\ell}$ are known to produce only three distinct neutrinos of definite flavor: ν_e, ν_μ, and ν_τ. However, there may be more than three neutrinos ν_m of definite *mass*. If, for example, there are four neutrino mass eigenstates, then one linear combination of them,

$$|\nu_S\rangle = \sum_m U^*_{sm} |\nu_m\rangle\ , \tag{4}$$

must not couple to the Z, and hence must not enjoy normal weak interactions. Consequently, this linear combination is referred to as a "sterile" neutrino.

Having learned that neutrinos almost certainly have mass and mix, we would like to learn the answers to the following questions:

- How many neutrino flavors, active and sterile, are there? Equivalently, how many neutrino mass eigenstates are there?
- What are the masses, M_m, of the mass eigenstates ν_m?
- Is each neutrino of definite mass a Majorana particle ($\overline{\nu_m} = \nu_m$), or a Dirac particle ($\overline{\nu_m} \neq \nu_m$)?
- What are the elements $U_{\ell m}$ of the leptonic mixing matrix?
- Does the behavior of neutrinos, in oscillation and other contexts, violate CP invariance?
- What are the electromagnetic properties of neutrinos? In particular, what are their dipole moments?
- What are the lifetimes of the neutrinos?

What we already know about these questions, and how we might learn more, are discussed in a previous paper.[8] Here, we would only like to add to that discussion some comments on the possible neutrino mass spectra and mixings suggested by the data on oscillation.

It is generally believed that if the atmospheric, solar, and LSND neutrinos all genuinely oscillate, then nature must contain at least four nondegenerate neutrino mass eigenstates.[9] As explained previously, the four corresponding neutrino flavor eigenstates must then be ν_e, ν_μ, ν_τ, and a neutrino which is sterile, ν_S. Thus, if the atmospheric, solar, and LSND oscillations are all genuine, then nature contains a fourth neutrino quite different from the three neutrinos already familiar to us.

If the so-far unconfirmed oscillation seen in the LSND experiment is set aside, then the oscillations of the atmospheric and solar neutrinos can be explained in terms of just three neutrinos. The $(\text{Mass})^2$ spectrum of these neutrinos can, for example, be as shown in Fig. 2. The height of this entire spectrum

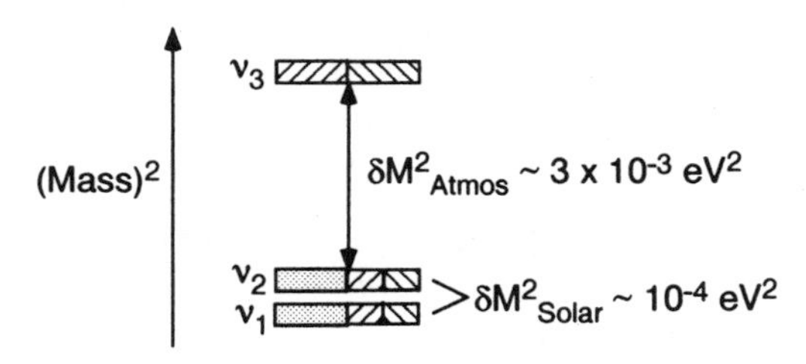

Figure 2. A three-neutrino $(\text{Mass})^2$ spectrum that accounts for the atmospheric and solar neutrino oscillations. The neutrinos ν_1, ν_2, and ν_3 are mass eigenstates. The rough flavor content of each is indicated as follows: The ν_e fraction of a mass eigenstate is dotted, the ν_μ fraction is shown by right-leaning hatching, and the ν_τ fraction by left-leaning hatching.

above $(\text{Mass})^2 = 0$ is completely undetermined, because neutrino oscillation probabilities depend only on $(\text{Mass})^2$ *splittings* and not on the individual underlying masses.[8] The splitting $\delta M^2_{\text{Atmos}} \sim 3 \times 10^{-3}$ eV^2 between mass eigenstates ν_3 and ν_2 is chosen to yield the observed atmospheric neutrino oscillation. The smaller splitting $\delta M^2_{\text{Solar}} \sim 10^{-4}$ eV^2 between ν_2 and ν_1 is chosen, in this example, to be consistent with large-mixing-angle MSW neutrino flavor conversion in the sun. The flavor content of the mass eigenstates is chosen in the same way.[10] An alternative spectrum in which the two closely-spaced mass eigenstates are at the top of the picture, rather than at the bottom, is also possible.

If we try to explain all reported oscillations, including the one seen by LSND, then, as already stated, the neutrino spectrum must contain at least four states. Until recently, it has been argued that, to be consistent with all oscillation data, both positive and negative, any such four-neutrino spectrum must be of the "2+2" variety.[11] That is, as illustrated in Fig. 3, it must consist of two pairs of neutrinos, with the members of each pair closely spaced, and with an "LSND gap" of order 1 eV^2 between the two pairs. As

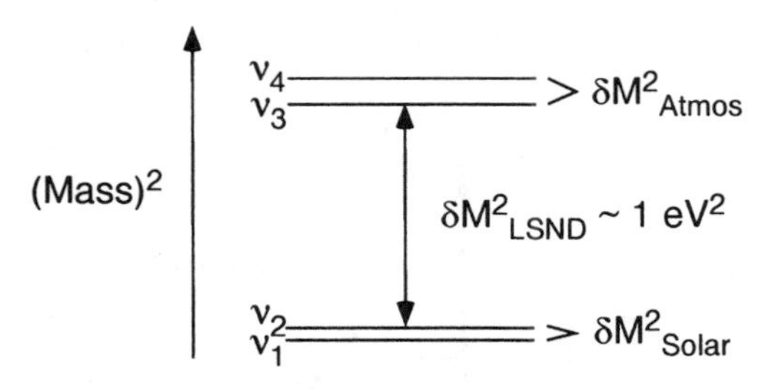

Figure 3. A four-neutrino spectrum of the "2+2" variety. The neutrinos ν_1, ν_2, ν_3, and ν_4 are mass eigenstates. The splitting $\delta M^2_{\rm LSND}$ is the one called for by the LSND oscillation. An alternate spectrum with $\delta M^2_{\rm Solar}$ at the top and $\delta M^2_{\rm Atmos}$ at the bottom is also possible.

previously explained, whenever there are four neutrino mass eigenstates, one linear combination of them must be a sterile neutrino, ν_S. An interesting feature of the "2+2" four-neutrino schemes is that they predict that ν_S plays a significant role either in the atmospheric neutrino oscillation or in the solar one. However, analyses of the Super-K atmospheric neutrino data disfavor atmospheric neutrino oscillation into a sterile neutrino at the 99% confidence level,[12] and are fully compatible with oscillation into an active neutrino. Furthermore, recent Super-K analyses of all the solar neutrino data disfavor solar neutrino oscillation into a sterile neutrino (either by the MSW effect or in vacuum).[3] Thus, at least to some degree, the data disfavor a major involvement of ν_S in either the atmospheric or solar oscillation.[13] This raises an interesting question: Suppose that, indeed, neither the neutrino state into which the atmospheric neutrinos oscillate, nor the one into which the solar ones do, is to any significant extent sterile. Would that rule out *all* four-neutrino explanations of the neutrino oscillation data? The answer to this question is "no".[14] The LSND experiment is now reporting[6] a somewhat lower oscillation probability than it did earlier. Thanks to this lower value, it is now possible to account for all the oscillation data with the "3+1" four-neutrino spectum shown in Fig. 4.[14] This spectrum contains three

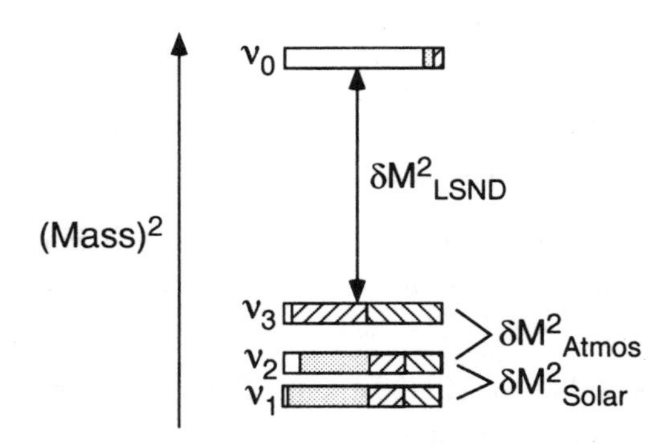

Figure 4. A "3+1" spectrum consistent with the neutrino oscillation data. The neutrinos ν_1, ν_2, ν_3, and ν_0 are mass eigenstates. Their active flavor content is indicated as in Fig. 2, and their sterile flavor content by white regions. The very small active content of ν_0, and the very small sterile content of $\nu_1 - \nu_3$, are exaggerated.

light, relatively closely-spaced mass eigenstates, ν_1, ν_2, and ν_3. These mass eigenstates are essentially fully active, and explain the atmospheric and solar oscillations in the same way as the three neutrinos in Fig. 2 do. Thus, no sterile neutrino plays a significant role in either of these oscillations. However, the spectrum of Fig. 4 also contains a fourth mass eigenstate, ν_0, which is almost totally sterile, and which has a (Mass)^2 separated from those of ν_1, ν_2, and ν_3 by an LSND gap of order 1 eV^2. In the past, a 3+1 spectrum of this kind was excluded by an incompatibility between LSND and the negative searches for ν_e and ν_μ disappearance. To understand this incompatibility, we note that if the spectrum is like that of Fig. 4, then the mass splittings between ν_1, ν_2, and ν_3 are invisible in any oscillation experiment with a distance to energy ratio, L/E, like that of LSND, for which $\delta M^2_{\rm Atmos,\, Solar} \times (L/E) << 1$. Thus, in any such experiment, there seem to be only two neutrinos: ν_0, and the ν_1-ν_2-ν_3 complex, whch appears to be only one neutrino. Hence, for any such experiment, the probabilities $P(\nu_\ell \to \nu_{\ell'})$ of the oscillations $\nu_\ell \to \nu_{\ell'}$ are described by the two-neutrino formulae[15]

$$P(\nu_\ell \to \nu_{\ell' \neq \ell}) = 4P_\ell P_{\ell'} \times \sin^2 \left[1.27 \delta M^2 (\mathrm{eV}^2) \frac{L\,(\mathrm{km})}{E\,(\mathrm{GeV})}\right] \quad (5)$$

and

$$P(\nu_\ell \to \nu_\ell) = 1 - 4P_\ell(1 - P_\ell) \times \sin^2 \left[1.27 \delta M^2 (\mathrm{eV}^2) \frac{L\,(\mathrm{km})}{E\,(\mathrm{GeV})}\right] . \quad (6)$$

Here, $P_\ell \equiv |U_{\ell H}|^2$, where ν_H is the heavier of the two neutrino mass eigenstates, and δM^2 is the $(\mathrm{Mass})^2$ splitting between these eigenstates. For the "quasi-two-neutrino"spectrum of Fig. 4, $\delta M^2 = \delta M^2_{\mathrm{LSND}}$ and $P_\ell = |U_{\ell 0}|^2$. In particular, P_e is the ν_e (dotted) fraction of ν_0, and P_μ is the ν_μ (right-leaning hatched) fraction. From Eq. (5), we see that for any assumed value of $\delta M^2_{\mathrm{LSND}}$, the LSND $\overset{(-)}{\nu_\mu} \to \overset{(-)}{\nu_e}$ oscillation determines an allowed range for $P_e P_\mu$. But, from Eq. (6), we see that for the same assumed $(\mathrm{Mass})^2$ splitting, the negative searches at reactors for ν_e disappearance through oscillation place an upper limit on P_e. Similarly, the negative searches at accelerators for ν_μ disappearance place an upper limit on P_μ. With the $\overset{(-)}{\nu_\mu} \to \overset{(-)}{\nu_e}$ oscillation probability reported earlier by LSND, there was no value of $\delta M^2_{\mathrm{LSND}}$ for which the LSND-allowed range for $P_e P_\mu$ was not incompatible with the upper limit on $P_e P_\mu$ coming from the negative searches for ν_e and ν_μ disappearance. However, with the new, smaller oscillation probability being reported now by LSND, there are several values of $\delta M^2_{\mathrm{LSND}}$ for which the LSND-allowed range for $P_e P_\mu$ and the upper limit on this quantity from the negative searches for disappearance are not incompatible. Thus, the 3+1 spectrum of Fig. 4 is a possible explanation of all the present neutrino oscillation data, even though it does not imply substantial sterile-neutrino involvement in either the atmospheric or solar neutrino oscillations. It will be interesting to see whether this spectrum can withstand future tests.

In conclusion, the evidence for neutrino mass has become quite convincing. However, we are just beginning to learn how many neutrinos there are, whether there are any sterile neutrinos, and what the neutrino masses and mixings are. While oscillation data already constrain the neutrino mass spectrum and neutrino mixing, a fair number of possibilities remain. In neutrino physics, interesting years lie ahead.

Acknowledgments

It is a pleasure to thank the Aspen Center for Physics for its hospitality while some of the work reported here was done.

References

1. This argument is taken from B. Kayser in *Proc. 17th Int. Workshop on Weak Interact. and Neutrinos (WIN '99)*, eds. C.A. Dominguez and R. D. Viollier (World Scientific, Singapore, 2000) p. 339.
2. See, for example, P. Fisher, B. Kayser, and K.S. McFarland, in *Ann. Rev. Nucl. Part. Sci.* **49**, eds. C. Quigg, V. Luth, and P. Paul (Annual Reviews, Palo Alto, CA, 1999) p. 481.
3. E. Kearns, in these Proceedings.
4. V. Barger, J. Learned, P. Lipari, M. Lusignoli, S. Pakvasa, and T. Weiler, *Phys. Lett.* **B 462,** 109 (1999).
5. We thank C. Baltay for pointing out that if the data show the NC event rate to be the same as in the absence of oscillation or decay, but decay is actually occurring, then the decay products must be able to produce (or at least mimic) NC events. We thank S. Pakvasa for pointing out the constraints that limit the degree to

which the decay products can actually do this. Finally, we thank S. Pakvasa for a general discussion of decay models, and both him and J. Learned for comments on the relevant present data, which are reported in S. Fukuda *et al.,* hep-ex/0009001.

6. G. Mills, to appear in the Proceedings of the XIX Int. Conf. on Neutrino Physics and Astrophysics, held in Sudbury, Canada, June 2000.
7. Z. Maki, M. Nakagawa, and S. Sakata *Prog. Theor. Phys.* **28**, 870 (1962).
8. B. Kayser, hep-ph/0010065, to appear in the Proceedings of the XIX Int. Conf. on Neutrino Physics and Astrophysics, held in Sudbury, Canada, June 2000.
9. C. Giunti, *Nucl. Instrum. Meth.* **A 451**, 51 (2000).
10. V. Barger *et al., Phys. Lett.* **B 437,** 107 (1998).
11. S.M. Bilenky, C. Giunti, and W. Grimus, *Eur. Phys. J.* **C 1**, 247 (1998); S.M. Bilenky *et al., Phys. Rev.* **D 60**, 073007 (1999); V. Barger, S. Pakvasa, T.J. Weiler, and K. Whisnant, *Phys. Rev.* **D 58**, 093016 (1998); V. Barger, T.J. Weiler, and K. Whisnant, *Phys. Lett.* **B 427,** 97 (1998).
12. S. Fukuda *et al.,* Ref. 5.
13. See, however, M.C. Gonzalez-Garcia and C. Peña-Garay, hep-ph/0009041.
14. V. Barger, B. Kayser, J. Learned, T. Weiler, and K. Whisnant, *Phys. Lett.* **B 489**, 345 (2000).
15. See, for example, B. Kayser and R. Mohapatra, to appear in *Current Aspects of Neutrino Physics,* ed. D. Caldwell (Springer-Verlag, 2000).

SOLAR AND ATMOSPHERIC NEUTRINO OSCILLATIONS

M. C. GONZALEZ-GARCIA
Inst. de Física Corpuscular
C.S.I.C. - Univ. de València, Spain

I review the status of neutrino masses and mixings in the light of the solar and atmospheric neutrino data in the framework of two–, three– and four neutrino mixing.

1 Indications for Neutrino Mass: Two–Neutrino Analysis

I first review the present experimental status for solar and atmospheric neutrinos and the results of the different analysis in the framework of two–neutrino oscillations.

1.1 Solar Neutrinos

The sun is a source of $\nu_e's$ which are produced in the different nuclear reactions taking place in its interior. Along this talk I will use the ν_e fluxes from Bahcall–Pinsonneault calculations [1] which I refer to as the solar standard model (SSM). These neutrinos have been detected at the Earth by seven experiments which use different detection techniques [2,3]: The chlorine experiment at Homestake, the water cerencov experiments Kamiokande and Super–Kamiokande (SK) and the radiochemical Gallex, GNO and Sage experiments. Due to the different energy threshold for the detection reactions, these experiments are sensitive to different parts of the solar neutrino spectrum. They all observe a deficit between 30 and 60 % which seems to be energy dependent mainly due to the lower Chlorine rate. To the measurements of these six ex-

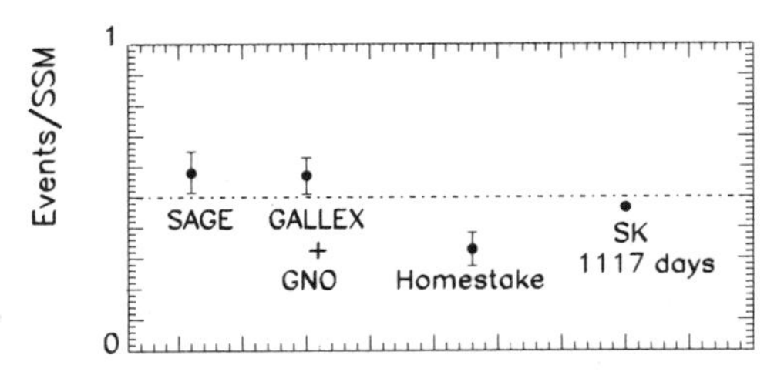

periments we have to add also the new results from SNO. They are however still not in the form of definite measured rates which could be included in this analysis.

SK has also presented their results after 1117 days of data taking on [3]:

– The recoil electron energy spectrum: SK has measured the dependence of the even rates on the recoil electron energy spectrum divided in 18 bins starting at 5.5 MeV. They have also reported the results of a lower energy bin 5 MeV $< E_e <$5.5 MeV, but its systematic errors are still under study and it is not included in their nor our analysis. The spectrum shows no clear distortion with $\chi^2_{flat} = 13/(17dof)$.

–The Zenith Angle Distribution (Day/Night Effect) which measures the effect of the Earth Matter in the neutrino propagation. SK finds few more events at night than during the day but the corresponding Day–Night asymmetry $A_{D/N} = -0.034 \pm 0.022 \pm 0.013$ is only 1.3σ away from zero.

In order to combine both the Day–Night information and the spectral data SK has also presented separately the measured recoil energy spectrum during the day and during the night. This will be referred in the following as the day–night spectra data which contains 2×18 data bins.

The most generic and popular explanation of the solar neutrino anomaly is in terms of neutrino masses and mixing leading to oscillations of ν_e into an active (ν_μ and/or ν_τ) or sterile neutrino, ν_s. In Fig. 1 I show the allowed two–neutrino oscillation regions obtained in our updated global analysis of the solar neutrino data [4]. We show the possible

Table 1. Best fit points and GOF for the allowed solutions for combinations of observables.

		Active				Sterile
		SMA	LMA	LOW	VAC-QVO	SMA
Rates	$\Delta m^2/\text{eV}^2$	5.5×10^{-6}	1.9×10^{-5}	$9. \times 10^{-8}$	9.7×10^{-11}	4.1×10^{-6}
	$\tan^2\theta$	0.0015	0.29	0.65	0.51 (1.94)	0.0015
	Prob (%)	50 %	8 %	0.5 %	2 %	19 %
Rates	$\Delta m^2/\text{eV}^2$	5.0×10^{-6}	3.2×10^{-5}	$1. \times 10^{-7}$	8.6×10^{-10}	3.9×10^{-6}
+Spec$_D$	$\tan^2\theta$	0.00058	0.33	0.67	1.5 (QVO)	0.0006
+Spec$_N$	Prob (%)	34 %	59 %	40 %	29 %	30%

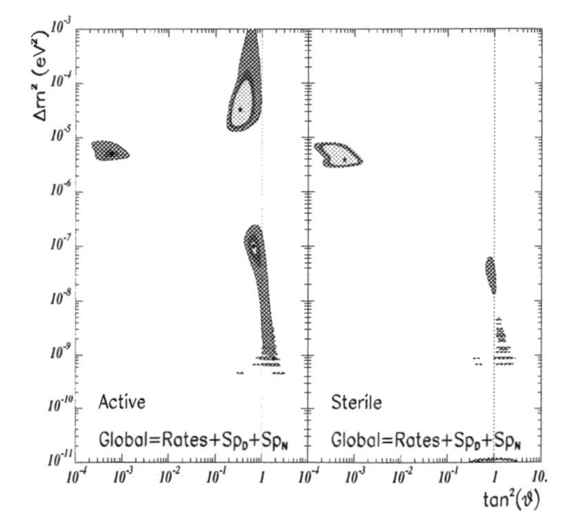

Figure 1. Presently allowed solar neutrino parameters for two-neutrino oscillations by the global analysis from Ref. [4]. The plotted regions are 90%, 95 and 99% CL.

solutions in the full parameter space for oscillations including both MSW and vacuum, as well as quasi-vacuum oscillations (QVO) and matter effects for mixing angles in the second octant (the so called dark side). These results have been obtained using the general expression for the survival probability found by numerically solving the evolution equation in the Sun and the Earth matter valid in the full oscillation plane. In the case of active–active neutrino oscillations we find three allowed regions for the global fit: the SMA solution, the LMA and LOW-QVO solution. For sterile neutrinos only the SMA solution is allowed. For oscillations into an sterile neutrino there are differences partly due to the fact that now the survival probability depends both on the electron and neutron density in the Sun but mainly due to the lack of neutral current contribution to the water cerencov experiments. In Table 1 I give the values of the parameters in these minima as well as the GOF corresponding to each solution. There are some points concerning these results that I would like to stress:

(a) Despite giving a worse fit to the observed total rates, once the day–night spectra data is included the LMA gives the best fit. This is mainly driven by the flatness of the spectrum and it was already the case with the last year data.

(b) The GOF of the LOW solution has increased considerably as it describes the spectrum data very well despite it gives a very bad fit to the global rates. Notice also that LOW and QVO regions are connected at the 99 %CL and they extend into the second octant so maximal mixing is allowed at 99 % CL for Δm^2 in the LOW-QVO region.

(c) As for the SMA the result from the correct statistically combined analysis shown in Fig. 1 and in Table 1 indicates that the SMA can describe the full data set with a probability of 34% but it is now shifted to smaller mixing angles to account for the flatter spectrum.

(d) Similar statement holds for the SMA solution for sterile neutrinos.

Thus the conclusion is that from the statistical point of view all solutions are acceptable since they all provide a reasonable GOF to the full data set. LMA and LOW-QVO solutions for oscillations into active neutrino seem slightly favoured over SMA solutions for oscillations into active or sterile neutrinos but these last two are not ruled out.

1.2 Atmospheric Neutrinos

Atmospheric showers are initiated when primary cosmic rays hit the Earth's atmosphere. Secondary mesons produced in this collision, mostly pions and kaons, decay and give rise to electron and muon neutrino and antineutrinos fluxes. Atmospheric neutrinos can be detected in underground detectors by direct observation of their charged current interaction inside the detector. These are the so called contained events. SK has divided their contained data sample into sub-GeV events with visible energy below 1.2 GeV and multi-GeV above such cutoff. On average, sub-GeV events arise from neutrinos of several hundreds of MeV while multi-GeV events are originated by neutrinos with energies of the order of several GeV. Higher energy muon neutrinos and antineutrinos can also be detected indirectly by observing the muons produced in their charged current interactions in the vicinity of the detector. These are the so called upgoing muons. Should the muon stop inside the detector, it will be classified as a "stopping" muon, (which arises from neutrinos of energies around ten GeV) while if the muon track crosses the full detector the event is classified as a "through-going" muon which is originated by neutrinos with energies of the order of hundred GeV.

At present the atmospheric neutrino anomaly (ANA) can be summarized in three observations:

– There has been a long-standing deficit of about 60 % between the predicted and observed ν_μ/ν_e ratio of the contained events [5] now strengthened by the high statistics sample collected at the SK experiment [6].

– The most important feature of the atmospheric neutrino data at SK is that it exhibits a *zenith-angle-dependent* deficit of muon neutrinos which indicates that the deficit is larger for muon neutrinos coming from below the horizon which have traveled longer distances before reaching the detector.

– The deficit for thrugoing muons is smaller that for stopping muons, *i.e.* the deficit decreases as the neutrino energy grows.

The most likely solution of the ANA involves neutrino oscillations. In principle we can invoke various neutrino oscillation channels, involving the conversion of ν_μ into either ν_e or ν_τ (active-active transitions) or the oscillation of ν_μ into a sterile neutrino ν_s (active-sterile transitions) [7]. Oscillations into electron neutrinos are nowadays ruled out since they cannot describe the measured angular dependence of muon-like contained events [7]. Moreover the most favoured range of masses and mixings for this channel have been excluded by the negative results from the CHOOZ reactor experiment [8].

In Fig. 2 I show the allowed neutrino oscillation parameters obtained in our global fit [7] of the full data set of atmospheric neutrino data on vertex contained events at IMB, Nusex, Frejus, Soudan, Kamiokande [5] and SK experiments [6] as well as upward going muon data from SK, Macro and Baksan experiments in the different oscillation channels.

The two panels corresponding to oscillations into sterile neutrinos in Fig. 2 differ in the sign of the Δm^2 which was assumed in the analysis of the matter effects in the Earth for the $\nu_\mu \to \nu_s$ oscillations. Concerning the quality of the fits our results show that the best fit to the full sample is obtained for the $\nu_\mu \to \nu_\tau$ channel although from the global analysis oscillations into sterile neutrinos cannot be ruled out at any reasonable CL. Due to matter effects the distribution for upgoing muons in the case of $\nu_\mu \to \nu_s$ are flatter than for $\nu_\mu \to \nu_\tau$. Data show a

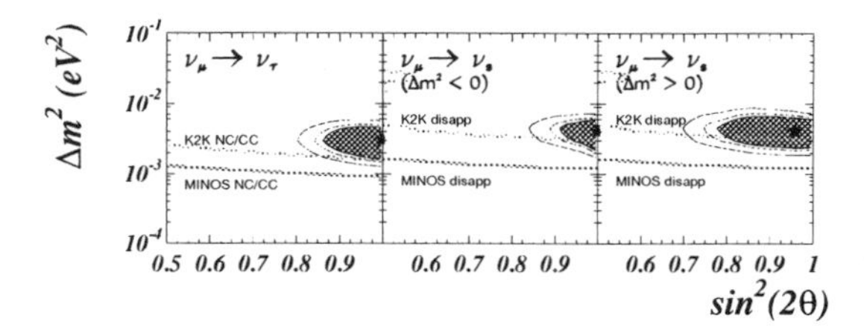

Figure 2. Allowed atmospheric oscillation parameters all for experiments, combined at 90 (shadowed area), 95 % and 99 % CL (thin solid line) for all possible oscillation channels, from Ref. [7]. The expected sensitivity for upcoming long-baseline experiments is also displayed.

somehow steeper angular dependence which can be better described by $\nu_\mu \to \nu_\tau$. This leads to the better quality of the global fit in this channel. Pushing further this feature SK collaboration has presented an analysis of the angular dependence of the through-going muon data in combination with the up-down asymmetry of partially contained events and the neutral current enriched events which seems to disfavour the possibility $\nu_\mu \to \nu_s$ at the 3–σ level [6].

2 Three–Neutrino Oscillations

In the previous section I have discussed the evidences for neutrino masses and mixings as usually formulated in the two–neutrino oscillation scenario. Let us now fit all the different evidences in a common three–neutrino framework and see what is our present knowledge of the neutrino mixing and masses. Here I present a brief summary of such analysis performed in Ref. [9] and I refer to that publication for further details and references.

The evolution equation for the three neutrino flavours can be written as:

$$-i\frac{d\nu}{dt} = \left[U\frac{M_\nu}{2E}U^\dagger + H_{int}\right] , \qquad (1)$$

where M_ν is the diagonal mass matrix for the three neutrinos and U is the unitary matrix relating the flavour and the mass basis. H_{int} is the Hamiltonian describing the neutrino interactions. In general U contains 3 mixing angles and 1 or 3 CP violating phases depending on whether the neutrinos are Dirac or Majorana. I will neglect the CP violating phases as they are not accessible by the existing experiments. In this case the mixing matrix can be conveniently chosen in the form

$$U = R_{23}(\theta_{23}) \times R_{13}(\theta_{13}) \times R_{12}(\theta_{12}) , \quad (2)$$

where R_{ij} is a rotation matrix in the plane ij. With this the parameter set relevant for the joint study of solar and atmospheric conversions becomes five-dimensional:

$$\begin{aligned} \Delta m^2_\odot \equiv \Delta m^2_{21} , \qquad & \Delta m^2_{atm} \equiv \Delta m^2_{32}, \\ \theta_\odot \equiv \theta_{12} , \qquad & \theta_{atm} \equiv \theta_{23} , \ \theta_{reac} \equiv \theta_{13} , \end{aligned} \quad (3)$$

where all mixing angles are assumed to lie in the full range from $[0, \pi/2]$.

In general the transition probabilities will present an oscillatory behaviour with two oscillation lengths. However from the required hierarchy in the splittings $\Delta m^2_{atm} \gg \Delta m^2_\odot$ indicated by the solutions to the solar and atmospheric neutrino anomalies it follows that:

– For solar neutrinos the oscillations with the atmospheric oscillation length are averaged out and the survival probability takes the form:

$$P^{3\nu}_{ee,MSW} = \sin^4\theta_{13} + \cos^4\theta_{13} P^{2\nu}_{ee,MSW} \quad (4)$$

where $P^{2\nu}_{ee,MSW}$ is obtained with the modified sun density $N_e \to \cos^2\theta_{13} N_e$. So the analyses of solar data constrain three of the five independent oscillation parameters: $\Delta m^2_{21}, \theta_{12}$ and θ_{13}.

– Conversely for atmospheric neutrinos, the solar wavelength is too long and the corresponding oscillating phase is negligible. As a consequence the atmospheric data analysis restricts Δm^2_{32}, θ_{23} and θ_{13}, the latter being the only parameter common to both solar and atmospheric neutrino oscillations and which may potentially allow for some mutual influence.

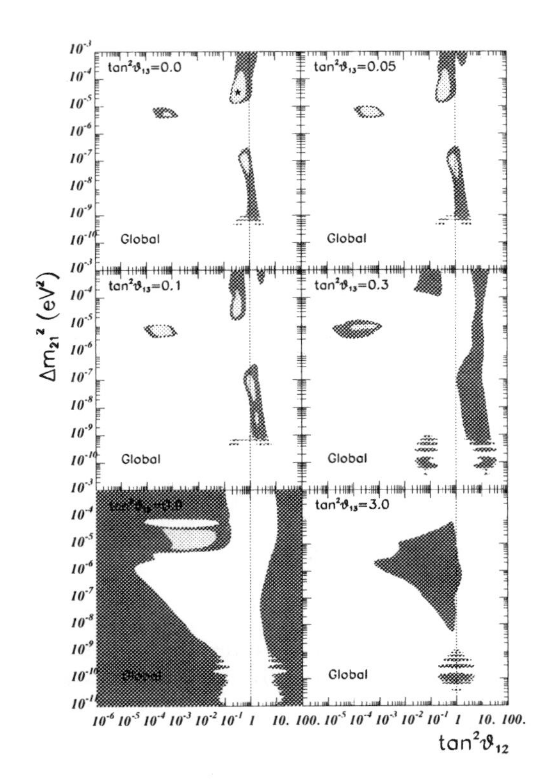

Figure 3. Allowed regions at 90% and 99% CL for the oscillation parameters Δm_{21} and $\tan^2(\theta_{12})$ from the global analysis of the solar neutrino data in the framework of three–neutrino oscillations for different values of the angle θ_{13} from Ref. [9].

Therefore solar and atmospheric neutrino oscillations decouple in the limit $\theta_{13} = 0$. In this case the values of allowed parameters can be obtained directly from the results of the analysis in terms of two–neutrino oscillations presented in the first section. Deviations from the two–neutrino scenario are then determined by the size of the mixing θ_{13}. This angle is constrained by the CHOOZ reactor experiment which imposes an strong lower limit on the probability

$$P_{ee}^{CHOOZ} = 1 - \sin^2(2\theta_{13})\sin^2\left(\frac{\Delta m_{32}^2 L}{4E_\nu}\right) \quad (5)$$

> 0.91 at 90 % CL for $\Delta m_{32}^2 > 10^{-3}$ eV^2.

The first question to answer is how the presence of this new angle affects the analysis of the solar and atmospheric neutrino data

Data Set	$\tan^2\theta_{13}$	
	min	limit 99%
Solar	0.0	3.5 (62°)
Atmos	0.026	0.57 (37°)
Atm+CHOOZ	0.005	0.08 (16°)
Atm+Solar	0.015	0.52 (36°)
Atm+Solar+Chooz	0.005	0.085 (16°)

[9]. In Fig. 3 I show the allowed regions for the oscillation parameters Δm_{21} and $\tan^2(\theta_{12})$ from our global analysis of the solar neutrino data in the framework of three–neutrino oscillations for different values of the angle θ_{13}. The allowed regions for a given CL are defined as the set of points satisfying the condition $\chi^2(\Delta m_{12}^2, \tan^2\theta_{12}, \tan^2\theta_{13}) - \chi^2_{min} \leq \Delta\chi^2(\text{CL, 3 dof})$ where, for instance, $\Delta\chi^2(\text{CL, 3 dof})$=6.25, 7.83, and 11.36 for CL=90, 95, and 99 % respectively. The global minimum used in the construction of the regions lays in the LMA region and corresponds to $\tan^2\theta_{13} = 0$, this is, for the "decoupled" scenario. Notice that the only difference between the first panel in Fig. 3 and the active oscillations solution in Fig. 3 is due to the different numbers of dof used in the definition of the regions. The behaviour of the regions illustrate the "tension" between the data on the total event rates which favour smaller θ_{13} values and the day–night spectra which allow larger values. It can also be understood as the "tension" between the energy dependent and constant pieces of the electron survival probability in Eq. (4).

As seen in the figure the effect is small unless large values of θ_{13} are involved. From Fig. 3 we find that as $\tan^2\theta_{13}$ increases all the allowed regions disappear, leading to an upper bound on $\tan^2\theta_{13}$ for any value of Δm_{21}^2, independently of the values taken by the other parameters in the three–neutrino mixing matrix. The corresponding 90 and 99 % CL bounds are tabulated in Table 2.

As for the atmospheric neutrino data in Fig. 4 I show the $(\tan^2\theta_{23},\ \Delta m_{32}^2)$ allowed regions, for different values of $\tan^2\theta_{13}$ from the global analysis of the atmospheric neutrino data. The upper-left panel, $\tan^2\theta_{13} = 0$, corresponds to pure $\nu_\mu \to \nu_\tau$ oscillations, and one can note the exact symmetry of the contour regions under the transformation $\theta_{23} \to \pi/4 - \theta_{23}$. This symmetry follows from the fact that in the pure $\nu_\mu \to \nu_\tau$ channel matter effects cancel out and the oscillation probability depends on θ_{23} only through the double-valued function $\sin^2(2\theta_{23})$. For non-vanishing values of θ_{13} this symmetry breaks due to the three-neutrino mixing structure even if matter effects are neglected We see that the analysis of the full atmospheric neutrino data in the framework of three-neutrino oscillations clearly favours the $\nu_\mu \to \nu_\tau$ oscillation hypothesis. As a matter of fact the best fit corresponds to a small value of $\theta_{13} = 9^\circ$. With our sign assignment we find that for non-zero values of θ_{13} the allowed regions become larger in the second octant of θ_{23}. No region of parameter space is allowed (even at 99% C.L.) for $\tan^2\theta_{13} > 0.6$. Larger values of $\tan^2\theta_{13}$ would imply a too large contribution of $\nu_\mu \to \nu_e$ and would spoil the description of the angular distribution of contained events. The mass difference relevant for the atmospheric analysis is restricted to lay in the interval: $1.25 \times 10^{-3} < \Delta m^2_{32}/\text{eV}^2 < 8 \times 10^{-3}$ at 99 % CL. Thus it is within the range of sensitivity of the CHOOZ experiment and as a consequence the angle θ_{13} is further constrained when we include in the analysis the results from this reactor experiment. This is illustrated in Table2 where one sees that the limit on $\tan^2\theta_{13}$ is strengthen when the CHOOZ data is combined with the atmospheric neutrino results.

One can finally perform a global analysis in the five dimensional parameter space combining the full set of solar, atmospheric and reactor data. As an illustration of such analysis I present in Table2 the resulting bounds on θ_{13}. The final results from the joint solar, atmospheric, and reactor neutrino data

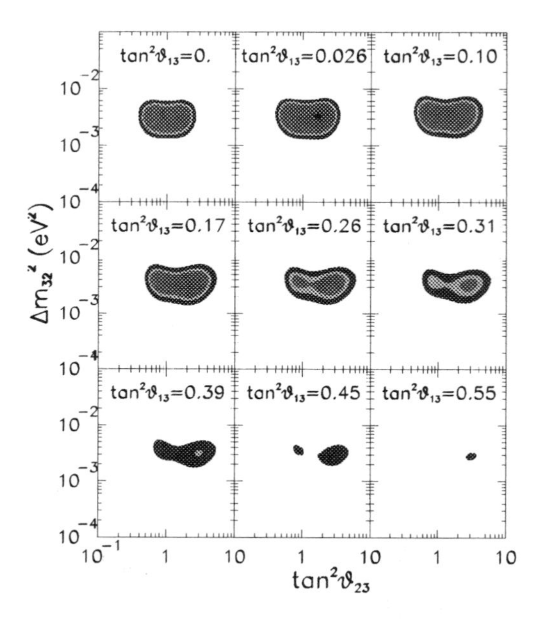

Figure 4. 90, 95 and 99% CL three-neutrino allowed regions in $(\tan^2\theta_{23},\ \Delta m^2_{32})$ for different $\tan^2\theta_{13}$ values, for the combination of global analysis of atmospheric neutrino data from Ref. [9]. The best-fit point is denoted as a star.

analysis lead to the following allowed ranges of parameters at 99% CL

$$\begin{aligned} 1.1 \times 10^{-3} < \Delta m^2_{32}/\text{eV}^2 &< 7.3 \times 10^{-3} \\ 0.33 < \quad \tan^2\theta_{23} &< 3.8 \\ \tan^2\theta_{13} &< 0.085 \quad \text{(if solar LMA)} \\ \tan^2\theta_{13} &< 0.135 \quad \text{(if solar SMA)} \end{aligned} \tag{6}$$

In conclusion we see that from our statistical analysis of the solar data it emerges that the status of the large mixing-type solutions has been further improved with respect to the previous SK data sample, due mainly to the substantially flatter recoil electron energy spectrum. In contrast, there has been no fundamental change, other than further improvement due to statistics, on the status of the atmospheric data. For the latter the oscillation picture clearly favours large mixing, while for the solar case the preference is still not overwhelming. Both solar and atmospheric data favour small values of the additional θ_{13} mixing and this behaviour is strengthened by the inclusion of the reactor limit.

3 Four–Neutrino Oscillations

Together with the results from the solar and atmospheric neutrino experiments we have one more evidences pointing out towards the existence of neutrino masses and mixing: the LSND results. These three evidences can be accommodated in a single neutrino oscillation framework only if there are at least three different scales of neutrino mass-squared differences which requires the existence of a light sterile neutrino. Here I present a brief update of the analysis performed in Ref. [10] of solar neutrino data in such framework of four–neutrino mixing and I refer to that publication for further as well as for the relevant references.

In four-neutrino schemes the rotation U relating the flavor neutrino fields to the mass eigenstates fields is a 4×4 unitary mixing matrix, which contains, in general 6 mixing angles (I neglect here the CP phases). Existing bounds from negative searches for neutrino oscillations performed at collider as well as reactor experiments impose severe constrains on the possible mass hierarchies as well as mixing structures for the four–neutrino scenario. In particular they imply:

(a) Four-neutrino schemes with two pairs of close masses separated by a gap of about 1 eV which gives the mass-squared difference responsible for the oscillations observed in the LSND experiment, can accommodate better the results of all neutrino oscillation experiments.

(b) In the study of solar and atmospheric neutrino oscillations only four mixing angles are relevant and the U matrix can be written as $U = R_{34}\,R_{24}\,R_{23}\,R_{12}$ We choose solar neutrino oscillations to be generated by the mass-square difference between ν_2 and ν_1. With this choice the survival of solar ν_e's mainly depends on the mixing angle ϑ_{12} and it is independent of ϑ_{34}. The mixing ϑ_{23} and ϑ_{24} determine the relative amount of transitions into sterile ν_s or active ν_μ and ν_τ only through the combination $\cos\vartheta_{23}\cos\vartheta_{24}$ ($c^2_{23}c^2_{24}$). We distinguish the following limiting cases:

- $c^2_{23}c^2_{24} = 0$ corresponding to the limit of pure two-generation $\nu_e \to \nu_a$ transitions.
- $c^2_{23}c^2_{24} = 1$ for which we have the limit of pure two-generation $\nu_e \to \nu_s$ transitions.
- If $c^2_{23}c^2_{24} \neq 1$, solar ν_e's can transform in the linear combination ν_a of active ν_μ and ν_τ.

In the general case of simultaneous $\nu_e \to \nu_s$ and $\nu_e \to \nu_a$ oscillations the corresponding probabilities are given by

$$P_{\nu_e\to\nu_s} = c^2_{23}c^2_{24}\,(1 - P_{\nu_e\to\nu_e})\,, \tag{7}$$
$$P_{\nu_e\to\nu_a} = \left(1 - c^2_{23}c^2_{24}\right)(1 - P_{\nu_e\to\nu_e})\,. \tag{8}$$

where $P_{\nu_e\to\nu_e}$ takes the standard two–neutrino oscillation for Δm^2_{21} and θ_{12} but computed for the modified matter potential $A \equiv A_{CC} + c^2_{23}c^2_{24}A_{NC}$ Thus the analysis of the solar neutrino data in the four–neutrino mixing schemes is equivalent to the two–neutrino analysis but taking into account that the parameter space is now three–dimensional $(\Delta m^2_{12}, \tan^2\vartheta_{12}, c^2_{23}c^2_{24})$.

I first present the results of the allowed regions in the three–parameter space for the global combination of observables. In Fig. 5 I show the sections of the three–dimensional allowed volume in the plane $(\Delta m^2_{21}, \tan^2(\vartheta_{12}))$ for different values of $c^2_{23}c^2_{24}$. The global minimum used in the construction of the regions lies in the LMA region and for pure active oscillations value of $c^2_{23}c^2_{24} = 0$. As seen in Fig. 5 the SMA region is always a valid solution for any value of $c^2_{23}c^2_{24}$. This is expected as in the two–neutrino oscillation picture this solution holds both for pure active–active and pure active–sterile oscillations. Notice, however, that the statistical analysis is different: in the two–neutrino picture the pure active–active and active–sterile cases are analyzed separately, whereas in the four–neutrino picture they are taken into account simultaneously in a consistent scheme. We see that in this "unified" framework, since the GOF of the SMA solution for pure sterile oscillations is worse than for SMA pure active oscillations (as discussed in the first section), the corresponding allowed region is smaller as they are now defined with respect to a common minimum.

On the other hand, the LMA, LOW and QVO solutions disappear for increasing values of the mixing $c_{23}^2 c_{24}^2$. I list in Table 2 the maximum allowed values of $c_{23}^2 c_{24}^2$ for which each of the solutions is allowed at a given CL. We see that at 95 % CL the LMA solution is allowed for maximal active-sterile mixing $c_{23}^2 c_{24}^2 = 0.5$ while at 99% CL all solutions are possible for this maximal mixing case.

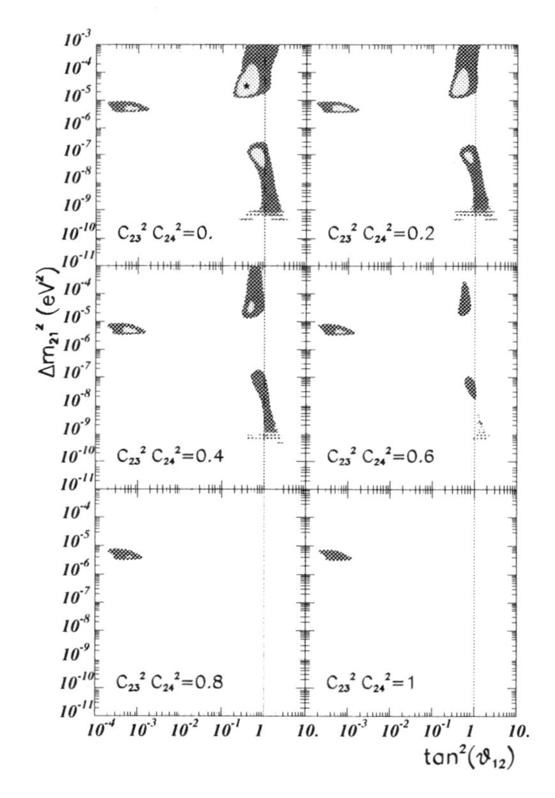

Figure 5. Results of the global analysis for the allowed regions in Δm_{21}^2 and $\sin^2\vartheta_{12}$ for the four–neutrino oscillations. The different panels represent the allowed regions at 99% (darker) and 90% CL (lighter). The best–fit point in the three parameter space is plotted as a star.

Table 2. Maximum allowed value of $c_{23}^2 c_{24}^2$ for the different solutions to the solar neutrino problem

CL	SMA	LMA	LOW	QVO
90	0.9	0.44	0.3	forbidden
95	all	0.53	0.44	0.28
99	all	0.72	0.77	0.88

This work was supported by grants DGICYT-PB98-0693 and PB97-1261, GV99-3-1-01, and ERBFMRXCT960090 and HPRN-CT-2000-00148 of the EU.

References

1. J. N. Bahcall, S. Basu and M. H. Pinsonneault, Phys. Lett. **B433**, 1 (1998).
2. B. T. Cleveland *et al.*, Astrophys. J. **496**, 505 (1998); SAGE Collab, talk by V. Gavrin at Neutrino 2000; GALLEX and GNO Collab., talk by E. Belloti at Neutrino 2000; SNO Collab, talk by A. B. McDonald at Neutrino 2000, Sudbury, Canada, June 2000.
3. See T. Takeuchi at these proceedings.
4. M.C. Gonzalez-Garcia, C. Peña-Garay, hep-ph/0009041.
5. NUSEX Collab., M. Aglietta *et al.*, Europhys. Lett. **8**, 611 (1989); Fréjus Collab., Ch. Berger *et al.*, Phys. Lett. **B227**, 489 (1989); IMB Collab., R. Becker-Szendy *et al.*, Phys. Rev. **D46**, 3720 (1992); Kamiokande Collab., H. S. Hirata *et al.*, Phys. Lett. **B280**, 146 (1992) and Phys. Lett. **B335**, 237 (1994); Soudan Collab., W. W. M. Allison *et al.*, Phys. Lett. **B449**, 137 (1999); MACRO Collab., talk by F. Ronga at these proceedings.
6. See T. Toshito at these proceedings.
7. N. Fornengo *et al.*, Nucl. Phys. **B580** (2000) 58; M.C. Gonzalez-Garcia *et al.*, Nucl. Phys. **B543**, 3 (1999) and Phys. Rev. **D58** (1998) 033004.
8. CHOOZ Collaboration, M. Apollonio *et al.*. Phys. Lett. **B420**, 397 (1998).
9. M.C. Gonzalez-Garcia *et al.* hep-ph/0009350.
10. C. Giunti,*et al.* Phys. Rev. **D62**, 013005 (2000).

ATMOSPHERIC NEUTRINO RESULTS FROM SOUDAN 2

EARL PETERSON
(FOR GEOFF PEARCE AND THE SOUDAN 2 COLLABORATION)
School of Physics and Astronomy, University of Minnesota,
116 Church St. SE, Minneapolis MN, 55455 USA
E-mail: eap@mnhep.hep.umn.edu

We present preliminary results from a 5.1 kTy exposure of Soudan 2; the atmospheric neutrino flavor ratio, R, using a our entire sample of quasi-elastic neutrino interactions and an L/E analysis for oscillations based on a high-resolution subsample. The flavor ratio (tracks/showers) is $0.68 \pm 0.11(stat.) \pm 0.06(syst.)$. The (L/E) analysis produces a best-fit value, based on a Feldman-Cousins analysis, of $sin^2(2\Theta)$ of 0.90 and Δm^2 of 7.910^{-3}.

1 The Soudan 2 Detector

The Soudan 2 detector is a 963 ton fine-grained gas tracking calorimeter located in the Soudan Underground Mine State Park, Soudan, Minnesota. It consists of 224 1 meter x 1 meter x 2.7 meter iron modules weighing 4.3 tons each. The detector is surrounded by a 1700 m^2 active shield mounted on the cavern walls. More details of the module construction and performance can be found in References [1,2]. Reference [3] contains more information about the shield.

2 Data Analysis

The data described in this report come from a 5.1 kiloton-year exposure taken between April 1989 and January 2000. The data analysis consists of isolating a sample of contained events (in which all tracks and showers are located within the fiducial volume, 20 cm inside the detector). Events are then scanned by physicists to finalize the containment. Monte Carlo events are interspersed with data.

The Monte Carlo sample used in this analysis is 5.45 times the size of the expected neutrino sample (based on the Bartol 96 flux) and reproduces the actual performance of the Soudan 2 detector to a high degree of accuracy.

During scanning, events are classified into one of three categories: single track, single shower, and multiprong. Short single highly-ionizing tracks are classified as protons and discarded. Events with two or more particles (other then recoil nucleons) emerging from the primary vertex, or single track events which are charged pions having visible scatters, are classified as multiprong. Some multiprong events whose flavor is clear are used (along with events with recoils) for the L/E analysis, discussed below.

3 The Flavor Ratio

Contained events are a mixture of neutrino interactions and background processes. Neutral particles which originate with the interaction of cosmic ray muons in the rock surrounding the detector cavern are the principal source of background. Such events are usually accompanied by large numbers of charged particles which strike the active shield. Shield activity therefore provides a tag for background events. An event with zero shield hits is referred to as 'gold'; such an event is a neutrino candidate. Events with two or more shield hits are referred to as 'rock' events; they comprise a shield-tagged background sample.

Some muon interactions in the rock produce contained events *unaccompanied* by shield hits, due either to shield inefficiency or because the interaction did not produce any

Table 1. Data used in the R calculation. The Monte Carlo numbers in parentheses are normalized to the detector exposure. The error on R is statistical only.

Number of Gold Tracks	133
Number of Gold Showers	193
Number of MC Tracks	1097 (193.1)
Number of MC Showers	1017 (179.0)
Corrected Number of ν Tracks	105.1 ± 12.7
Corrected Number of ν Showers	142.3 ± 13.9
Value of R	0.68 ± 0.11

charged particles which entered the shield. The number of such interactions is determined by examining the distributions of event depths in the detector. We fit the depth distributions to determine the amount of background present in the gold sample. Table 1 shows the flavor ratios with and without the background subtraction.

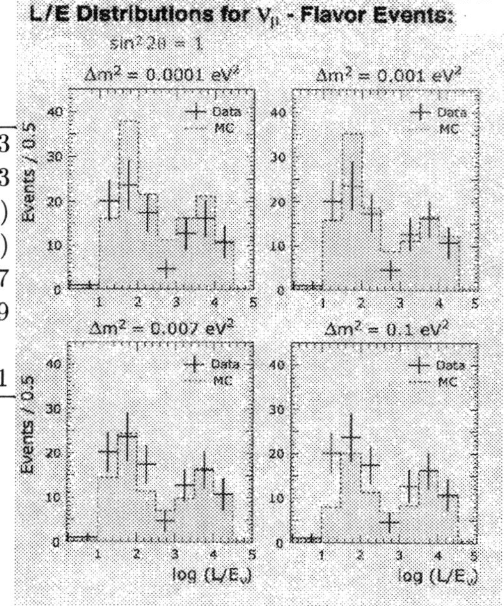

Figure 1. Preliminary L/E distributions for ν_μ CC and ν_e CC background subtracted data (crosses) and the Monte Carlo expectation (dashed histogram). The (unoscillated) MC is normalized to 5.1 kiloton-years data via the ν_e sample.

4 Neutrino Oscillation Analysis

The ability to identify an oscillation signature in an L/E distribution is mainly limited by the measurement of the incident neutrino direction. The neutrino directional measurement is smeared by detector resolution, target Fermi motion, and the failure to image all final state particles. We have found that by using events with recoils or placing energy cuts on the data we can obtain a subsample of events which have the potential for good directional measurement, and hence better L/E determination [4]. The preliminary cuts that isolate this sample are:

- **Tracks and Showers**

  $P_{lept} > 150$ MeV/c if a recoil is present

  $P_{lept} > 600$ MeV/c if no recoil is present

- **Multiprongs**

  $E_{vis} > 700$ MeV

  $P_{vis} > 450$ MeV/c

  $P_{lept} > 250$ MeV/c .

The high-resolution cuts are very effective at eliminating background, which is predominantly low energy. The L/E distributions for the background subtracted preliminary data are compared to the Monte Carlo expectation in Figure 1. The most obvious feature is the overall deficit of ν_μ CC events compared with expectation.

If we compare our data to Monte Carlo samples that include neutrino oscillations (of $\nu_\mu \rightarrow \nu_\tau$, where ν_τ CC interactions are ignored) preliminary indications are that $\sin^2(2\theta) = 1$ and $\Delta m^2 = 7x10^{-3}$ eV^2 provide a satisfactory fit.

A Feldman-Cousins fit to the data sample produces the exclusion plot in Figure 2. The 90%-confidence limit and best-fit points are indicated.

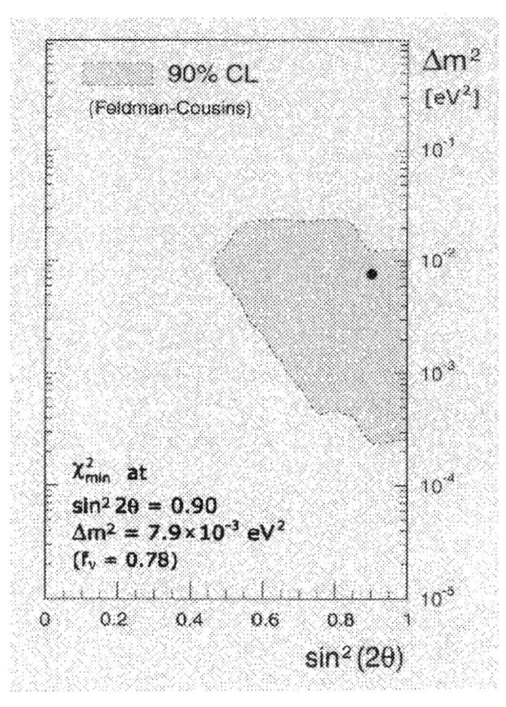

Figure 2. Preliminary results from a Feldman-Cousins fit to the Soudan 2 high-resolution 5.1 kTy data sample.

Acknowledgements

We acknowledge the support of the U.S. Department of Energy, the State and University of Minnesota and the U.K. Particle Physics and Astronomy Research Council. We would also like to thank the Minnesota Department of Natural Resources for allowing us to use the Soudan Underground Mine State Park.

References

1. Soudan 2 Collaboration: W.W.M Allison *et al.*, Nucl. Instrum. Methods A 376 (1996) 36.
2. Soudan 2 Collaboration: W.W.M Allison *et al.*, Nucl. Instrum. Methods A 381 (1996) 385.
3. W.P. Oliver *et al.*, Nucl. Instrum. Methods A 276 (1989) 371.
4. H. Tom *et al.* 'Search for Neutrino Oscillation Effects Using Neutrino Zenith Angle and L/E Distributions in Soudan 2', Internal Memo PDK-699, March 1998.

ATMOSPHERIC NEUTRINO RESULTS FROM MACRO

F.RONGA

Macro Collaboration

INFN Laboratori Nazionali di Frascati PO Box 13 Frascati Italy

Email :Ronga@LNF.INFN.IT

We present an update of the measurement of the the flux and angular distribution of atmospheric muon neutrinos using the MACRO detector at the Gran Sasso Laboratory. The different data sets are in agreement with $\nu_\mu \to \nu_\tau$ oscillations with maximum m ixing and $\Delta m^2 \sim 10^{-3} \div 10^{-2} eV^2$. The two flavor sterile neutrino oscillations are ruled out at 98%.

1 Introduction

This paper is about the update to March 2000 of the atmospheric neutrino results already published by the MACRO collaboration [1,2].

The active detectors in MACRO are streamer tube chambers, which are used for tracking, and liquid scintillator counters used for timing. The requirement of a reconstructed track selects muon events.

Three different topologies of neutrino events are analyzed up to now: *Up Through* events (high energy events), *Internal Up* events and *Internal Down* events together with *Up Stop* events (low energy events).

The *Up Through* tracks come from ν_μ interactions in the rock below MACRO. The muon crosses the whole detector ($E_\mu > 1$ GeV). The time information provided by scintillator counters permits to know the flight direction (time-of-flight method). The median neutrino energy for this kind of events is of the order of 50 GeV. The data have been collected with different detector configurations starting in 1989 with a small part of the apparatus.

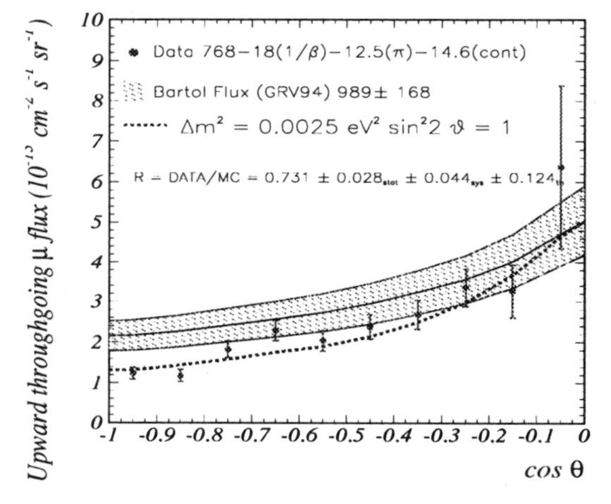

Figure 1. Zenith distribution of flux of upgoing muons with energy greater than 1 GeV for data and Monte Carlo for the combined MACRO data. The shaded region shows the expectation for no oscillations with the 17% normalization uncertainty. The solid line shows the prediction for an oscillated flux with $\sin^2 2\theta = 1$ and $\Delta m^2 = 0.0025$ eV^2.

The *Internal Up* events come from ν interactions inside the apparatus. Since two scintillator layers are intercepted, the time-of-flight method is applied to identify the upward going events. The median neutrino energy for this kind of events is around 3.5 GeV.

The *Up Stop* and the *Internal Down* events are due to external interactions with upward-going tracks stopping in the detector (*Up Stop*) and to neutrino induced downgoing tracks with the vertex in the lower part of MACRO (*Internal Down*). These events are identified by means of topological criteria. The lack of time information prevents to distinguish the two sub-samples. The median neutrino energy is around 4.2 GeV. An almost equal number of *Up Stop* and *Internal*

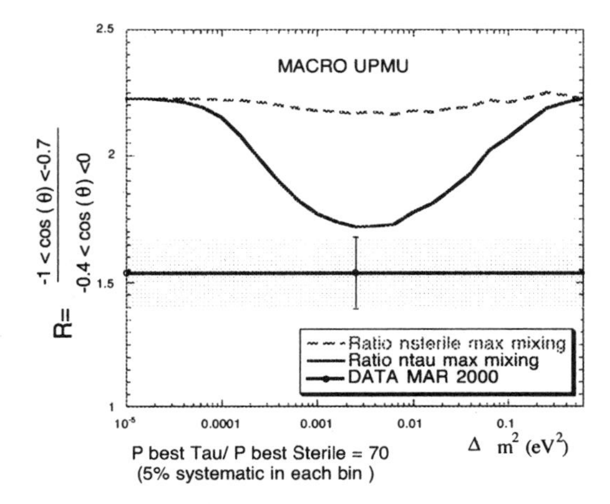

Figure 2. The ratio between the data in two bins and the comparison with the ν_s and ν_τ oscillations with maximum mixing

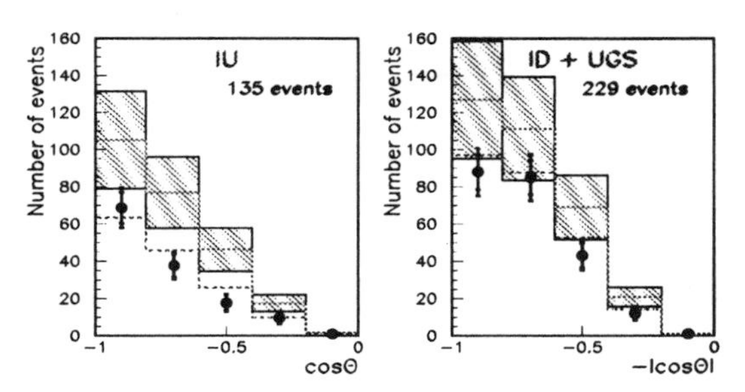

Figure 3. Zenith angle (θ) distribution for IU and $UGS + ID$ events. The background-corrected data points (black points with error bars) are compared with the Monte Carlo expectation (25% uncertainty) assuming no oscillation (full line) and two-flavor oscillation (dashed line) using maximum mixing and $\Delta m^2 = 2.5 \times 10^{-3}\ eV^2$.

Down is expected if neutrinos do not oscillate. In case of oscillations it is not expected a reduction in the flux of the *Internal Down* events while it is expected a reduction in the number of the *Up Stop* events similar to the one expected for the *Internal Up*.

Only the data collected with the full MACRO (live-time around 5.1 years) have been used in the low energy event analysis.

2 Upward through-going muons

The measured muon velocity is calculated with the convention that muons going down through the detector are expected to have $1/\beta$ near +1 while muons going up through the detector are expected to have $1/\beta$ near -1. Upward going muons are selected with the requirement $-1.25 \leq 1/\beta \leq -0.75$.

Removing the backgrounds, the observed number of upward through-going muons integrated over all zenith angles is 723.

The total systematic uncertainty on the expected flux of muons, adding the errors from the Bartol neutrino flux, neutrino cross-section and muon propagation in quadrature is $\pm 17\%$. This theoretical error in the prediction is mainly a scale error that doesn't change the shape of the angular distribution. The number of expected events integrated over all zenith angles is 989, giving a ratio of the observed number of events to the expectation of 0.73 $\pm$0.028(stat) $\pm$0.044(systematic) $\pm$0.12(theoretical).

Figure 1 shows the zenith angle distribution of the measured flux of upgoing muons with energy greater than 1 GeV for all MACRO data compared to the Monte Carlo expectation for no oscillations and with a $\nu_\mu \to \nu_\tau$ oscillated flux with $\sin^2 2\theta = 1$ and $\Delta m^2 = 0.0025$ eV^2.

The shape of the angular distribution has been tested with the hypothesis of no oscillations normalizing data and predictions. The χ^2 is 24.3, for 9 degrees of freedom. Also $\nu_\mu \to \nu_\tau$ oscillations are considered. The best χ^2 in the physical region of the oscillation parameters is 11.2 for Δm^2 around $0.0025 eV^2$ and maximum mixing (the minimum is outside the physical region).

Using the matter effect it is possible to discriminate between the $\nu_\mu - \nu_s$ oscillation and the $\nu_\mu - \nu_\tau$ computing the ratio between the number of events in the two angular regions shown in Figure 2. The angular regions are chosen according to a Montecarlo study which provides the bins which should be used

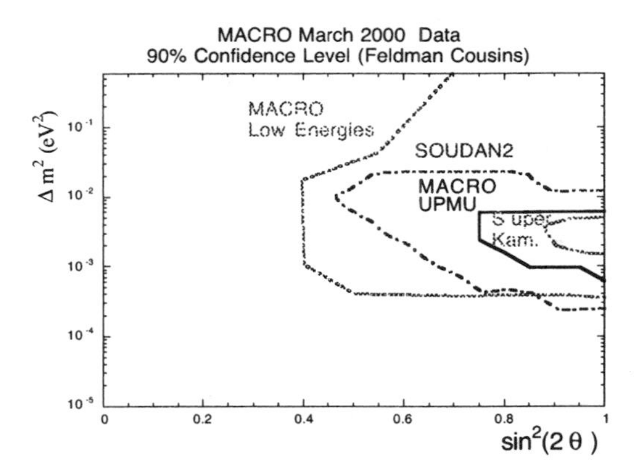

Figure 4. The 90% confidence level regions of the experiments with positive indication of oscillation for atmospheric neutrinos. The MACRO limits are computed using the Feldman-Cousins procedure

to have the best discrimination between the two kind of oscillations. Using this ratio the statistical significance is higher than in the case of a χ^2 test with data binned in 10 bins, but some features of the angular distribution could be lost. The ratio is insensitive to most of the errors on the theoretical prediction of the ν flux and cross section[3]. From the plot in Figure 2 using the statistical error and a systematic error of 5% for each angular region (mainly due to the acceptance) the ratio between the best probability for tau neutrino and the best for sterile neutrino is 70.

3 Loe energy events

The analysis of the *Internal Up* events is similar to the analysis of the *Up Through* events. The main difference is due to the requirement that the interaction vertex should be inside the apparatus. After the background subtraction (6 events) 135 events are classified as *Internal Up* events

The *Internal Down* and the *Up Stop* events are identified by topological constraints. The main requirement is the presence of a reconstructed track crossing the bottom scintillator layer. After background subtraction (9 events), 229 events are classified as *Internal Down* and *Up Stop* events.

The angular distributions of data and predictions are compared in Figure 3. The low energy samples show an uniform deficit of the measured number of events over the whole angular distribution with respect to the predictions, while there is good agreement with the predictions based on neutrino oscillations.

The theoretical errors coming from the neutrino flux and cross section uncertainties almost cancel if the ratio between the measured number of events $\frac{IU}{(ID+UGS)}$ is compared with the expected one. The partial error cancellation arises from the nearly equal energy distributions of parent neutrinos for the IU and the ID+UGS events. The measured ratio is $\frac{IU}{ID+UGS} = 0.59 \pm 0.07_{stat}$, while the one expected without oscillations is $0.75 \pm 0.04_{sys} \pm 0.04_{theo}$. The probability (one-sided) to obtain a ratio so far from the expected one is 3%, nearly independent of the neutrino flux and neutrino cross sections.

4 Conclusions

The three different neutrino event topologies support the two flavor $\nu_\mu \to \nu_\tau$ oscillations. Figure 4 shows the confidence level regions for the high energy events and the low energy events. The preferred regions are in agreement with Superkamiokande and Soudan2. The two flavor $\nu_\mu \to \nu_s$ oscillations (with any mixing) is disfavored at level of 98% respect the $\nu_\mu \to \nu_\tau$ with maximum mixing.

References

1. M. Ambrosio *et al.* (MACRO Collab.), Phys. Lett. B**434**, 451 (1998).
2. M. Ambrosio *et al.* [MACRO Collaboration], Phys. Lett. B**478** (2000) 5.
3. P.Lipari, talk at the Neutrino 2000 conference

ATMOSPHERIC NEUTRINO RESULTS FROM SUPER-KAMIOKANDE

TOSHIYUKI TOSHITO
FOR THE SUPER-KAMIOKANDE COLLABORATION
Kamioka Observatory, Institute for Cosmic Ray Research, University of Tokyo, Japan.
E-mail: toshi@suketto.icrr.u-tokyo.ac.jp

We present atmospheric neutrino results from a 71 kiloton year (1140 days) exposure of the Super-Kamiokande detector. Our data are well explained by $\nu_\mu \to \nu_\tau$ 2-flavor oscillations. We update the 3 flavor analysis of contained events. Also we have been attempting to discriminate between the possible oscillating partners of the muon neutrino as being either the tau neutrino or the sterile neutrino. These tests use expected differences due to neutral currents and matter effects to discriminate the possibilities. We find no evidence favoring sterile neutrinos, and reject the hypothesis at 99% confidence level.

1 Introduction

Atmospheric neutrino are produced as decay products in hadronic showers resulting from collisions of cosmic-rays with nuclei in the upper atmosphere. Production of electron and muon neutrino is dominated by the processes $\pi^\pm \to \mu^\pm + \nu_\mu(\bar{\nu_\mu})$ followed by $\mu^\pm \to e^\pm + \bar{\nu_\mu}(\nu_\mu) + \nu_e(\bar{\nu_e})$. That gives an expected ratio of the flux of ν_μ to the flux of ν_e of about 2. Vertically downward-going neutrinos travel about 15km while vertically upward-going neutrinos travel about 13,000km before interacting in the detector. Thanks to good geometrical symmetry of the earth, we can expect up-down symmetry of neutrino flux. Details in prediction of atmospheric neutrino flux is discussed on Ref. 1. Neutrino oscillation occurs if a finite mass difference and mixing angle exists. The oscillation probability between two neutrino flavors is given by $P = \sin^2 2\theta \cdot \sin^2(1.27\frac{L(\mathrm{km})}{E_\nu(\mathrm{GeV})}\Delta m^2(\mathrm{eV}^2))$, where θ is the mixing angle, L is the flight length of the neutrino, E_ν is the neutrino energy, Δm^2 is the mass squared difference. The range of energy of observable atmospheric neutrino is from a few hundred MeV to the order of 100GeV. The broad energy spectrum and flight distances makes measurement of atmospheric neutrino sensitive to neutrino oscillation with Δm^2 down to the order of $10^{-4}\mathrm{eV}^2$. Several recent underground experiments report atmospheric neutrino results in terms of neutrino oscillation[2,3,4]. This paper reports on recent results of the Super-Kamiokande.

2 Event sample

Super-Kamiokande is a 50 kiloton water Cherenkov detector constructed under Mt. Ikenoyama located at the central part of Japan, giving it a rock over-burden of 2,700 m water-equivalent. The fiducial mass of the detector for atmospheric neutrino analysis is 22.5 kiloton. Neutrino events interacting with the water are observed as fully-contained (FC) or partially-contained (PC) events according to the amount of anti-counter activity. FC events with only one reconstructed ring are subdivided into e-like and μ-like based on likelihood analysis of the reconstructed Cherenkov ring. Our 1140 live-days data are summarized in Table 1 and their zenith angle distributions are shown in Figure 1. The flavor ratio is evaluated by taking the double ratio using MC expectation without oscillation as: $R \equiv \frac{(\mu\mathrm{-like}/e\mathrm{-like})_{\mathrm{DATA}}}{(\mu\mathrm{-like}/e\mathrm{-like})_{\mathrm{MC}}}$. The observed values are $0.652^{+0.019}_{-0.018} \pm 0.051$ for sub-GeV and $0.668^{+0.035}_{-0.033} \pm 0.079$ for multi-GeV samples. These small double ratio is consistent with our previous result[2] and indecates a deficit of muon neutrino explained by neutrino oscillation. Neutrino events interacting with rock surrounding the detector can be ob-

Table 1. Event summary for 1140days contained sample

sub-GeV ($E_{vis} < 1.33$GeV)

	Data	MC(Honda flux)
1ring e-like	2531	2402.6
1ring μ-like	2486	3620.9
multi ring	1885	2321.5
Total	6902	8345.0

multi-GeV FC($E_{vis} > 1.33$GeV)

	Data	MC(Honda flux)
1ring e-like	576	555.4
1ring μ-like	502	738.7
multi ring	1198	1470.1
Total	2276	2764.2

Partially Contained

	Data	MC(Honda flux)
Total	665	945.1

served as upward through going muons or stopping muons. Details in the analysis can be found on Ref. 2.

3 $\nu_\mu \to \nu_\tau$ oscillation analysis

Allowed oscillation parameter region using contained and upward going muon samples is shown in Fig. 2. The minimum χ^2 including unphysical region (in physical region) is found to be 135.3(135.4) with 152 degrees of freedom (d.o.f.) at $\Delta m^2 = 3.2 \times 10^{-3}(3.2 \times 10^{-3})\text{eV}^2$, $\sin^2 2\theta = 1.01(1.00)$. The deficit of upward going μ-like data is well explained by assuming $\nu_\mu \to \nu_\tau$ oscillation (Figure 1). χ^2 for no oscillation was found to be 316.2 for 154 d.o.f.

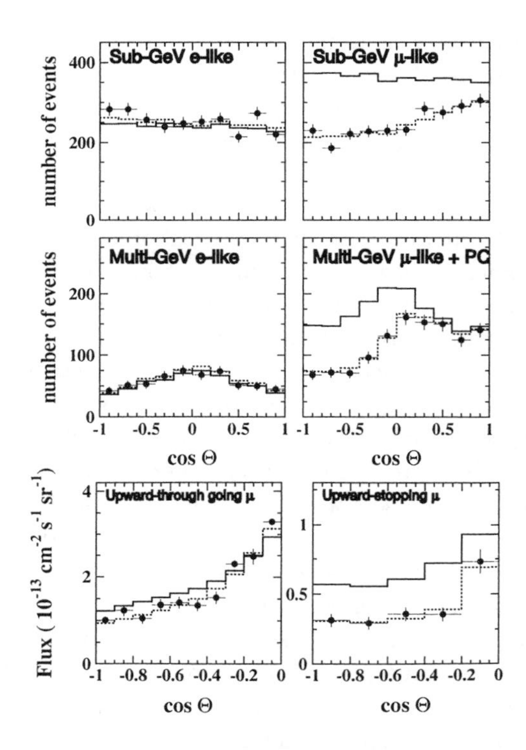

Figure 1. Zenith angle distribution of Super-Kamiokande 1140 days FC, PC and UPMU samples. Dots, solid line and dashed line correspond to data, MC with no oscillation and MC with best fit oscillation parameters, respectively.

4 Three flavor oscillation analysis

The results described in the last section agree well with the 2-flavor $\nu_\mu \to \nu_\tau$ oscillation hypothesis. The data, however, does not preclude the possibility of 3-flavor oscillation. If the mass-degeneracy condition, $\Delta m^2_{23} \gg \Delta m^2_{12}$ is assumed (not unreasonable for $\Delta m^2_{23} = \Delta m^2_{atm} > 10^{-3}\text{eV}^2$ and $\Delta m^2_{12} = \Delta m^2_{sol} < 10^{-4}\text{eV}^2$), the 3-flavor neutrino oscillation probability is given to a good approximation as: $P(\nu_\mu \leftrightarrow \nu_e) = \sin^2 2\theta_{13} \cdot \sin^2\theta_{23} \cdot \Psi$, $P(\nu_\mu \leftrightarrow \nu_\tau) = \cos^4\theta_{13} \cdot \sin^2 2\theta_{23} \cdot \Psi$, $P(\nu_e \leftrightarrow \nu_\tau) = \sin^2 2\theta_{13} \cdot \cos^2\theta_{23} \cdot \Psi$, $\Psi = \sin^2(1.27 \cdot \Delta m^2 \cdot L/E)$. In these

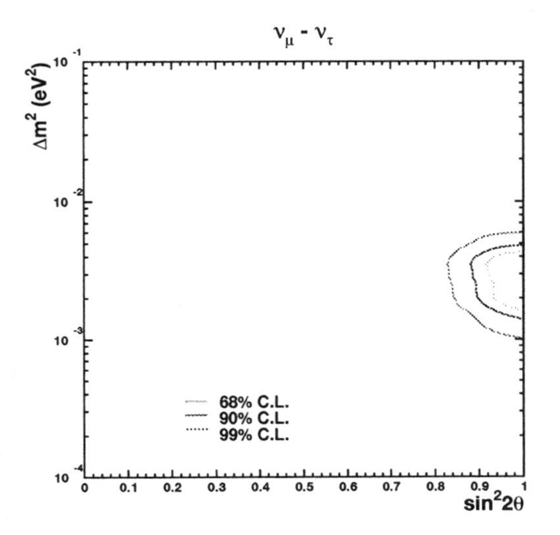

Figure 2. 68,90 and 99% confidence level allowed regions for $\nu_\mu \rightarrow \nu_\tau$ oscillation obtained by Super-Kamiokande 1140 days result.

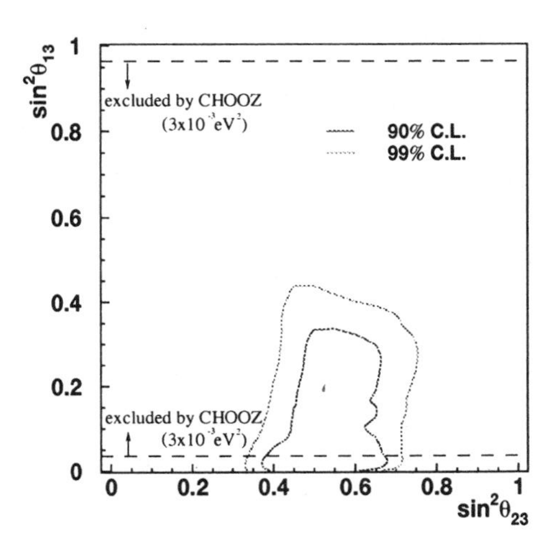

Figure 3. 90% and 99% confidence level allowed region for 3-flavor oscillation obtained by Super-Kamiokande 1140 days FC and PC sample. The limit on θ_{13} from the CHOOZ is shown as well.

equations, $\Delta m^2 = \Delta m^2_{23}$, and Δm^2_{12} is set to zero. Note that there are only three oscillation parameters ($\sin^2\theta_{13}$,$\sin^2\theta_{23}$,Δm^2). The allowed parameter region for $\sin^2\theta_{13}$ and $\sin^2\theta_{23}$ is shown in Fig. 3. The best-fit parameter values are $(\sin^2\theta_{13}, \sin^2\theta_{23}, \Delta m^2) = (0.02, 0.55, 2.5 \times 10^{-3}\text{eV}^2)$; this is consistent with the CHOOZ[5] result. Our result is also consistent with pure 2-flavor $\nu_\mu \rightarrow \nu_\tau$ oscillation hypothesis with maximal mixing $((\sin^2\theta_{13}, \sin^2\theta_{23}) = (0, 0.5))$.

5 Study on $\nu_\mu \rightarrow \nu_\tau$ and $\nu_\mu \rightarrow \nu_{sterile}$

Some models predict that ν_μ oscillates into "sterile" neutrino (ν_s) that does not interact even via neutral current (NC). If the observed deficit of ν_μ is due to $\nu_\mu \rightarrow \nu_s$ oscillation, then the number of events produced via NC interaction for up-going neutrino should also be reduced. Moreover, in the case of $\nu_\mu \rightarrow \nu_s$ oscillation, matter effect will suppress oscillation in the high energy ($E_\nu > 15$GeV) region[6]. We used the following data sample to observe these effects: (a) NC enriched sample; (b) the high-energy ($E_{vis} > 5$GeV) PC sample; and (c) up-through-going muons. Zenith angle distributions for each sample is shown in Fig. 4. The hypothesis test is performed using the up($\cos\Theta < -0.4$)/down($\cos\Theta > 0.4$) ratio in samples (a) and (b) and the vertical($\cos\Theta < -0.4$)/horizontal($\cos\Theta > -0.4$) ratio in sample (c). Fig. 5 shows excluded regions obtained by combined ((a),(b)and(c)) analysis along with the allowed regions from 1ring-FC sample analysis assuming $\nu_\mu \rightarrow \nu_\tau$ and $\nu_\mu \rightarrow \nu_s$. The results show that $\nu_\mu \rightarrow \nu_s$ oscillation is disfavored at 99%.

6 Summary

Super-Kamiokande results from 1140 days of contained events and up-going muons events give 90% C.L. allowed parameter regions of $\sin^2 2\theta > 0.88$ and $1.5 \times 10^{-3} < \Delta m^2 < 5 \times 10^{-3}\text{eV}^2$. The results of 3-flavor oscillation analysis is consistent with the CHOOZ and 2-flavor $\nu_\mu \rightarrow \nu_\tau$ oscillation. Finally, $\nu_\mu \rightarrow \nu_s$ oscillation is disfavored at 99% C.L..

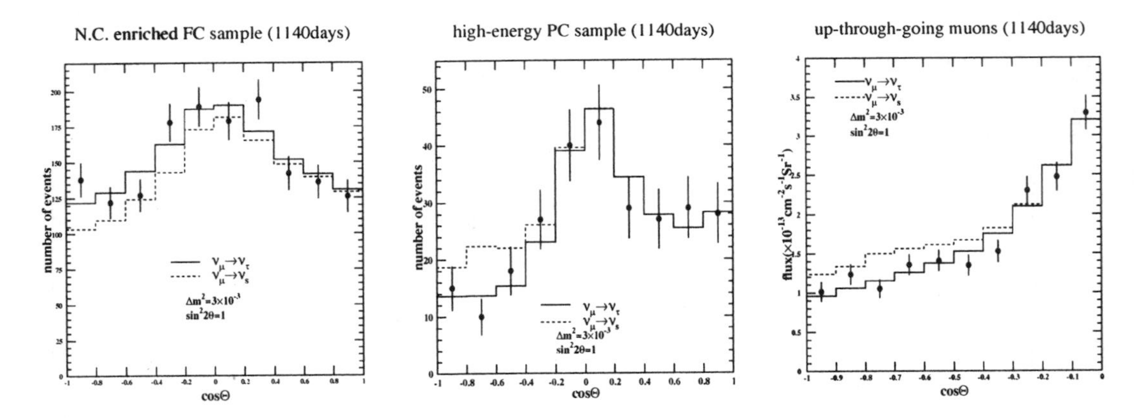

Figure 4. Zenith angle distributions of left: NC enriched sample, center: high-energy PC sample, right: up-through-going muon sample.

References

1. M. Honda *et al.*, *Phys. Rev.* D**52**, 4985 (1995); V. Agrawal *et al.*, *Phys. Rev.* D**53**, 1313 (1996).
2. Y. Fukuda *et al.*, *Phys. Lett.* B**433**, 9 (1998); *Phys. Lett.* B**436**, 33 (1998); *Phys. Rev. Lett.* **81**, 1562 (1998); *Phys. Rev. Lett.* **82**, 2644 (1999); *Phys. Lett.* B**467**, 185 (1999).
3. M. Ambrosio *et al.*, *Phys. Lett.* B**434**, 451 (1998); F. Ronga, for the MACRO Collaboration, in these Proceedings.
4. W.W.M.Allison *et al.*, *Phys. Lett.* B**391**, 491 (1997); G. Pearce, for the Soudan 2 Collaboration, in these Proceedings.
5. M. Apollonio *et al.*, *Phys. Lett.* B**466**, 415 (1999).
6. E. Akhmedov *et al.*, *Phys. Lett.* B**300**, 128 (1993); P. Lipari and M. Lusignoli, *Phys. Rev.* D**58**, 73005 (1998); Q. Y. Liu and A. Yu. Smirnov, *Nucl. Phys.* B**524**, 505 (1998); Q. Y. Liu *et al.*, *Phys. Lett.* B**440**, 319 (1998).

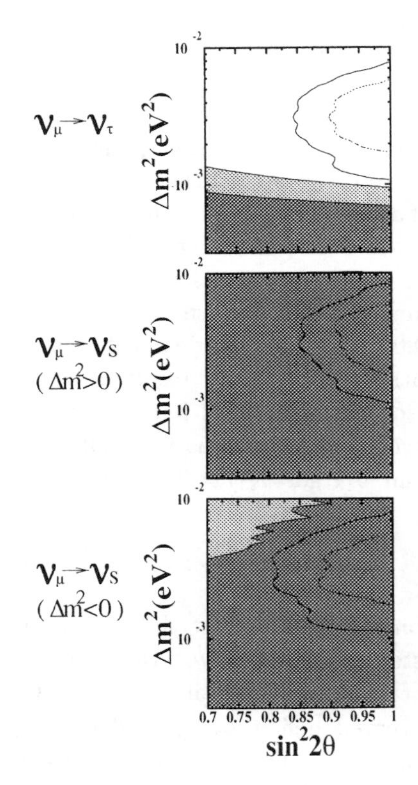

Figure 5. Excluded regions for three oscillation modes. The light(dark) gray region is excluded at 90(99)% C.L. by NC enriched sample and high-energy sample analysis. Thin dotted(solid) line indicates the 90(99)% C.L. allowed regions from 1ring-FC sample analysis.

SOLAR NEUTRINO RESULTS FROM SUPER-KAMIOKANDE

Y. TAKEUCHI
for the Super-Kamiokande Collaboration
Kamioka Observatory, ICRR, Univ. of Tokyo, Higashi-Mozumi, Kamioka-cho, Yoshiki-gun, Gifu 506-1205, Japan
E-mail: takeuchi@icrr.u-tokyo.ac.jp

Latest Super-Kamiokande results of the solar neutrino flux, day/night results, energy spectrum measurements, and oscillation analyses are reported. The observation period spans May 31, 1996 to April 24, 2000, which corresponds to a detector live time of 1117 days. Our preliminary results indicate 1.3σ difference between day and night flux, and the energy spectrum expressed as data/(BP98 SSM) is consistent with a flat spectrum with $\chi^2/\text{D.O.F.} = 13.7/17$. Comparing global-flux oscillation analysis and SK day and night spectra, MSW SMA region, Just-So region and 2-flavor sterile solutions are disfavored at 95% C.L.

1 Introduction

Super-Kamiokande[a] is a large cylindrical imaging water Čerenkov detector, which is constructed from 50000-ton of purified water, 11146 of 20-inch PMTs, 1867 of 8-inch PMTs, and so on. It is located about 1000m underground (2700m water equivalent) in the Mozumi mine in Kamioka town.

The detection method for Solar neutrinos in Super-Kamiokande is:

$$\nu + \mathrm{e}^- \rightarrow \mathrm{e}^- + \nu.$$

The detector observes the Čerenkov photons emitted from the recoil electron in the water. Using this interaction, direction and energy of an event can be observed in real-time. Therefore, Super-Kamiokande has provided unique solar neutrino measurements[1], that is, directional measurement of solar neutrinos, energy spectrum, difference of the fluxes in daytime and in nighttime (day/night analysis), etc. These information are necessary to obtain flux-independent evidence for the solar neutrino oscillation. In this report, latest Super-Kamiokande results of the solar neutrino flux, day/night results, energy spectrum measurements are reported.

[a]SUPER KAMIOKA Nucleon Decay Experiment

2 Data Set and Analysis

The observation period spans May 31, 1996 to April 24, 2000, which corresponds to a detector live time of 1117 days. During this period, we obtained about 2×10^9 events as raw data, and reduced to 192204 events ($5.0 \leq$ Energy $<$ 20MeV, 22.5kton fid. vol.) by applying fiducial volume cut, energy cut, first reduction, spallation event cut, second reduction, and gamma cut.

The fiducial volume cut at 22.5kt and energy cut at 5.0 MeV reduce number of events by 2 order. In the first reduction step, simple noise events (flash PMT events, electrical noise, etc.) are removed by factor 2. The reduction factor for solar neutrino M.C. events is about a few percent. Possible systematic error of first reduction step is estimated $\pm 1\%$ for flux measurements.

In the second reduction step, more powerful noise reductions are applied. They reduce number of events, especially in low-energy region, by factor about 10. For solar neutrino M.C. events, reduction factor is 30%–15%, depending on energy. Possible systematic error of second reduction is $+1.9\%-1.3\%$.

Spallation cut and gamma cut remove physical (not noise) events which become background of solar neutrino events. Spallation events are caused by energetic cosmic-ray muons in the Super-Kamiokande detector, and they are removed by the maximum likelihood method using muon energy, time, and distance information. The removal factors for data and solar neutrino M.C. are 10 and 20%, respectively. Systematic error is estimated at 0.1%. Gamma cut removes events coming form outside fiducial volume, like gamma-rays from surrounding 20-inch PMTs, etc. It reduces data by factor 10 at the low-energy side. Reduction power is 15%–8% for solar neutrino M.C., depending on energy. Systematic error of gamma cut is estimated $\pm 0.5\%$ for flux measurements.

Total systematic error for flux measurements is +3.3% −2.8% for 1117day data set.

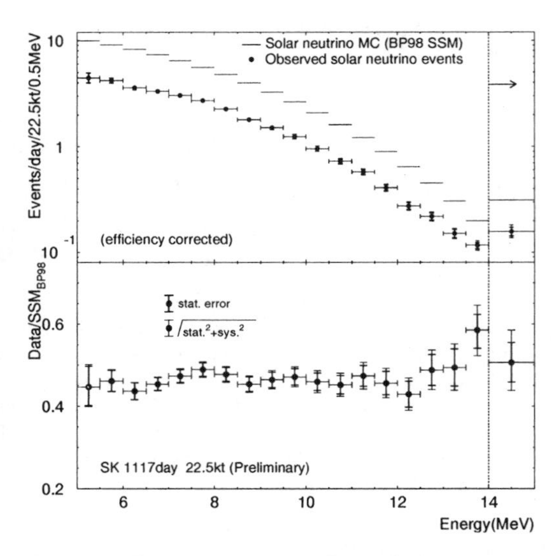

Figure 1. Energy spectrum of recoil electrons produced by ^{8}B and hep solar neutrinos. In upper plot, the data are corrected for trigger and analysis efficiencies. The filled circles and lines represent observed and expected spectrum, respectively. The lower plot shows the spectrum normalized bin-by-bin to the BP98 SSM prediction. The 14–20MeV region is combined into a single bin.

3 Results

The solar neutrino flux for the energy range 5.5–20 MeV has been measured to be $(2.40 \pm 0.03(\text{stat.}) \ {}^{+0.08}_{-0.07}\ (\text{sys.})) \times 10^{6}\ \text{cm}^{-2} \cdot \text{s}^{-1}$. The break-down of the flux in various bins of zenith angle, θ_z, is shown in Table 1. The asymmetry between the day and night flux (defined as $(\text{day} - \text{night})/\,[0.5 \cdot (\text{night} + \text{day})]$) is $-0.034 \pm 0.022\,(\text{stat})^{+0.013}_{-0.012}\,(\text{sys})$. This is $1.3\,\sigma$ from no asymmetry, including systematic errors.

The recoil electron energy spectrum from the elastic scattering of solar neutrinos is shown in Figure 1. The lower plot in this figure shows the spectrum normalized bin-by-bin to the BP98 SSM prediction. The χ^2 between the observed spectrum and a flat spectrum is 13.7 for 17 degrees of freedom, considering correlated systematic errors.

Figure 2 shows significance of possible day/night asymmetry and energy spectrum distortion as a function of the detector live time.

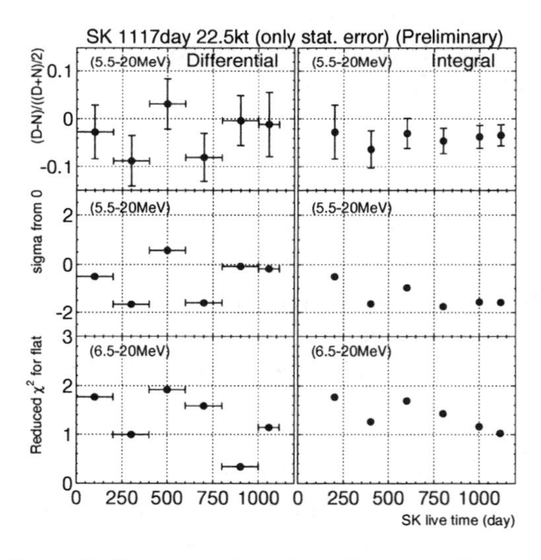

Figure 2. Time variation of significances. The top, middle and bottom plots show asymmetry between the day and night flux, sigma of the day/night ratio from no asymmetry, and reduced χ^2 between the observed spectrum and a flat spectrum, respectively. The left side and right side figures show the differential plots in each 200day period and the integral plots from the beginning of the experiment, respectively.

Table 1. Observed solar neutrino flux ratios in each zenith angle region: the ratio of observed and expected number of events (3rd column), and 1σ error of systematic error (4th column). Systematic errors are relative to DAY-flux. The zenith angle is the angle between the reconstructed recoil electron direction and the upward direction.

	Zenith angle	Data/SSM$_{BP98}$	δ_i
ALL	$-1 \leq \cos\theta_z \leq 1$	$0.465^{+0.005}_{-0.005}$	——
DAY	$-1 \leq \cos\theta_z \leq 0$	$0.456^{+0.007}_{-0.007}$	——
NIGHT	$0 < \cos\theta_z \leq 1$	$0.472^{+0.007}_{-0.007}$	——
N1	$0 < \cos\theta_z \leq 0.2$	$0.491^{+0.018}_{-0.018}$	$^{+1.3}_{-1.3}\%$
N2	$0.2 < \cos\theta_z \leq 0.4$	$0.468^{+0.017}_{-0.016}$	$^{+1.3}_{-1.3}\%$
N3	$0.4 < \cos\theta_z \leq 0.6$	$0.476^{+0.015}_{-0.014}$	$^{+1.3}_{-1.3}\%$
N4	$0.6 < \cos\theta_z \leq 0.8$	$0.467^{+0.016}_{-0.015}$	$^{+1.3}_{-1.3}\%$
N5	$0.8 < \cos\theta_z \leq 1$	$0.454^{+0.016}_{-0.016}$	$^{+1.3}_{-1.3}\%$
N4'	$0.6 < \cos\theta_z \leq 0.838$	$0.477^{+0.014}_{-0.014}$	$^{+1.3}_{-1.3}\%$
CORE	$0.838 < \cos\theta_z \leq 1$	$0.430^{+0.016}_{-0.016}$	$^{+1.3}_{-1.3}\%$

4 Oscillation analysis

In this report, the following definitions of χ^2 are used:

$$\chi_a^2 = \sum_{\mathrm{D,N}}\sum_{i=1}^{18}\left\{\frac{R_{ob,i} - R_{ex,i}\cdot\alpha\cdot F_i(\epsilon)}{\sigma_i}\right\}^2 + \left\{\frac{\epsilon}{\sigma_{cor}}\right\}^2$$

$$\chi_b^2 = \sum_{\mathrm{D,N}}\sum_{i=1}^{18}\left\{\frac{R_{ob,i} - R_{ex,i}\cdot\alpha\cdot F_i(\epsilon)}{\sigma_i}\right\}^2 + \left\{\frac{\epsilon}{\sigma_{cor}}\right\}^2 + \left\{\frac{1-\alpha}{\sigma_\alpha}\right\}^2$$

$$\chi_c^2 = \sum_{\mathrm{Ga,Cl,SK}}\frac{(R_{ob,I} - R_{ex,I})(R_{ob,J} - R_{ex,J})}{\sigma_{I,J}^2}$$

$$\sigma_i = \sqrt{\sigma_{stat,i}^2 + \sigma_{uncor,i}^2}$$

$$\sigma_{I,J}^2 = \sigma_{th,I,J}^2 + \delta_{I,J}\cdot\sigma_{ex,I}\cdot\sigma_{ex,J}$$

Here, R_{ob} is observed ratio of $(\frac{Data}{SSM})$ in each bin. R_{ex} is expected ratio of $(\frac{with\ oscillation}{no\ oscillation})$. D, N and "i=1–18" means day spectrum, night spectrum, and the energy binning above 5.5MeV in Figure 1, respectively. Total data points for χ_a^2 and χ_b^2 is 36. σ_{stat}, σ_{uncor}, and σ_{cor} correspond statistical, systematical uncorrelated, and systematical correlated errors, respectively. $F_i(\epsilon)$ is the response function. ϵ is shift factor of the correlated error. α and σ_α represent flux normalization factor and theoretical error on flux. $\sigma_{th,I,J}$ and σ_{ex} are theoretical correlation matrix and experimental error, respectively[2]. χ_a^2 is used for the flux independent test via day spectrum and night spectrum. A flux constraint is applied to the day spectrum and night spectrum test in χ_b^2. χ_c^2 is used for flux-global analysis.

Figure 3 and Figure 4 show results from our neutrino oscillation analysis based on the SK 1117days' data set, using above χ^2 definitions. In these figure, 3 regions are just overlaid. From Fig. 3, MSW small angle solutions and Just-So solutions are disfavored at 95% C.L. by comparing global-flux fit and SK day and night spectrum analysis. From Fig. 4, 2-flavor sterile solutions are disfavored at 95% C.L. by comparing global-flux fit and SK day and night spectrum. From only Super-Kamiokande, that is, SK day and night spectrum with flux constraint analysis, large mixing regions are favored.

5 Conclusion

The latest solar neutrino flux and energy spectrum measurements are reported. Our oscillation analyses based on the latest mea-

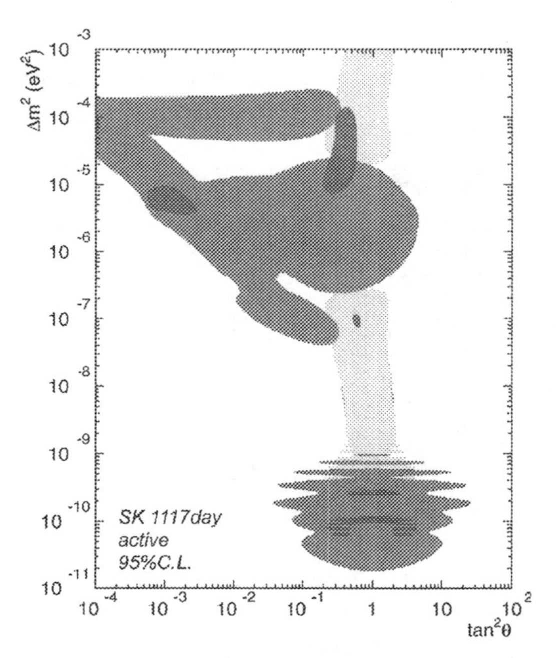

Figure 3. Parameter regions in 2-flavor active case. The dark(green), middle(red), and light(blue) color areas correspond to allowed by flux-global(χ_c^2), excluded by SK day/night spectrum(χ_a^2), and allowed by SK day/night spectrum with flux constraint(χ_b^2) regions, respectively. Confidence levels are 95%.

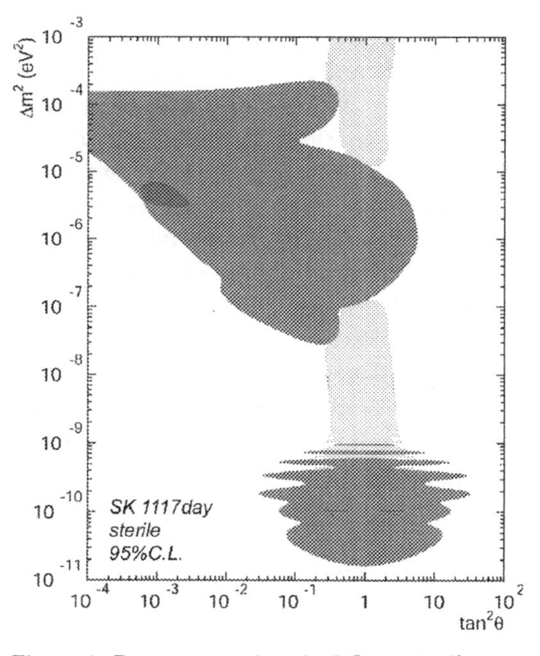

Figure 4. Parameter regions in 2-flavor sterile case. The color definitions are same as Figure 3.

surements are reported also. All measurements are consistent with those published previously. In particular, we have found that the day *vs* night flux difference is at the 1.3 σ level, and that the recoil electron energy spectrum (normalized bin-by-bin to BP98 SSM) is consistent with a flat spectrum with $\chi^2 = 13.7/17$ D.O.F.. Comparing global-flux fit and SK day and night spectra oscillation analysis, we have found that MSW SMA region, Just-So region and 2-flavor sterile solutions are disfavored at 95% C.L.

Acknowledgments

The author acknowledges the cooperation of the Kamioka Mining and Smelting Company. The Super-Kamiokande has been built and operated from funding by the Japanese Ministry of Education, Science, Sports and Culture, and the United States Department of Energy.

This work was supported in part by Grant-in-Aid for Scientific Research (B) of the Japanese Ministry of Education, Science and Culture.

References

1. Super-Kamiokande Collaboration, Phys. Rev. Lett. **81** (1998) 1158; Phys. Rev. Lett. **82** (1999) 1810; Phys. Rev. Lett. **82** (1999) 2430.
2. G. L. Fogli and E. Lisi, Astroparticle Phys. **3** (1995) 185.

FIRST SOLAR NEUTRINO OBSERVATIONS AT THE SUDBURY NEUTRINO OBSERVATORY

JOSHUA R. KLEIN, FOR THE SNO COLLABORATION

Department of Physics and Astronomy, University of Pennsylvania, Philadelphia, Pennsylvania, 19104-6396

We present here the first observations of solar neutrinos from the Sudbury Neutrino Observatory. SNO is a water Cerenkov detector which uses heavy water (D_2O) as both the interaction and detection medium, providing sensitivity both to charged-current (exclusive ν_e) and neutral current (inclusive for all active flavors) reactions. Based upon the radial distribution of events in the detector, the angular distribution of the events with respect to the Sun, and the energy distribution, we conclude that SNO is seeing ^{8}B solar neutrinos via the charged current reaction on deuterium.

1 Introduction

The Sudbury Neutrino Observatory (SNO) has been designed primarily to resolve the outstanding questions regarding the flux of neutrinos produced by the Sun. SNO approaches these questions directly by comparing the solar ν_e flux with the flux of all weakly interacting neutrino flavors. The comparison is done using heavy water (D_2O) as both the interaction and the detection medium. Neutrinos interact in three ways in D_2O:

- $\nu_e + d \rightarrow p + p + e^-$ (charged current)
- $\nu_x + d \rightarrow p + n + \nu_x$ (neutral current)
- $\nu_{e,(\mu,\tau)} + e \rightarrow \nu_{e,(\mu,\tau)} + e^-$ (elastic scattering)

The charged current (CC) reaction is sensitive only to ν_e's while the neutral current (NC) reaction occurs for all weakly interacting neutrino flavors. Therefore comparison of the neutrino flux measured via the NC reaction and that measured via the CC reaction tells us whether the solar neutrino flux contains a significant fraction of non-electron neutrinos, and thus whether neutrinos from the Sun are oscillating into other flavors.

There are further advantages to this method. The CC reaction is a sensitive measure of the incident neutrino energy—far more so than the elastic scattering (ES) reaction which produces a recoil electron energy spectrum that is nearly flat. SNO will therefore be able to produce a high sensitivity measurement of the solar ν_e spectrum above ~ 5 MeV. Optimistically, a sensitive measurement of this spectrum can tell us unambiguously whether neutrino oscillations are occurring in an energy-dependent way, thereby distinguishing different oscillation scenarios.

Lastly, the ES reaction, which like the NC reaction is sensitive to all neutrino flavors (though predominantly ν_e) provides both a second measurement of the ν_e/ν_x flux ratio as well as a way of comparing SNO's flux measurements directly to other experiments.

2 The SNO Detector

The SNO detector is located 2 km underground in INCO Ltd.'s Creighton Mine. The D_2O volume is contained in a 12 m diameter acrylic vessel (AV), surrounded by a 2.5 m H_2O shield. These target volumes are viewed by ~ 9500 photomultiplier tubes. Further details of the detector design can be found elsewhere [1].

SNO detects neutrinos in two ways. For both the CC and ES reactions, a relativistic electron is produced directly, and the resultant Cerenkov light is detected by the surrounding PMTs. In pure D_2O, events from the NC reaction are detected when the liberated neutron is captured on deuterium, pro-

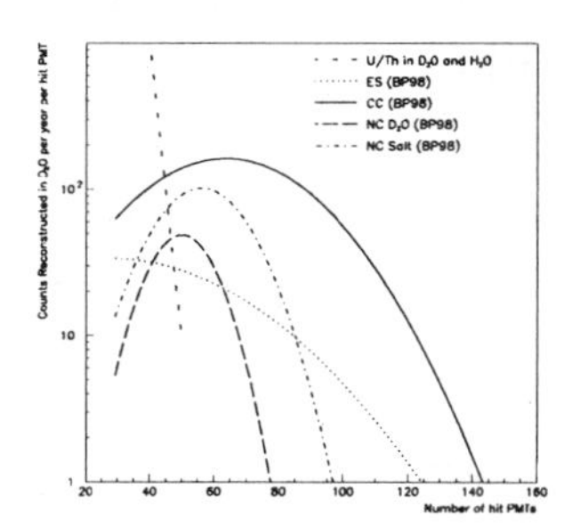

Figure 1. Monte Carlo simulation of signals in SNO, using 9 hits/MeV scale, and assuming Standard Solar Model fluxes.

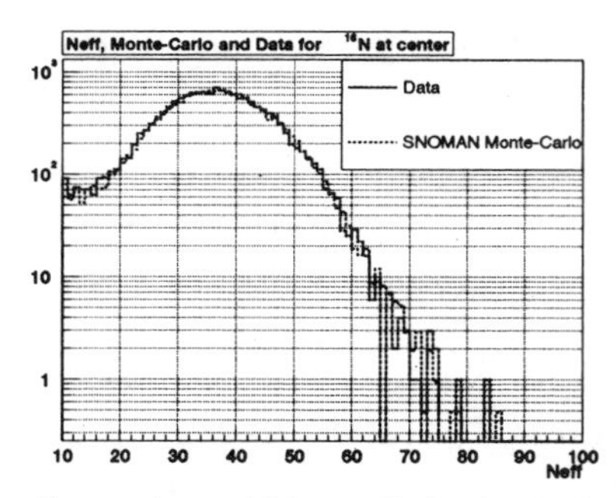

Figure 2. Comparison of Monte Carlo simulation of ^{16}N source using preliminary measurements of detector optical parameters to measurements made with the actual source at the center of the detector.

ducing a 6.25 MeV γ. In future phases of the experiment, this neutral current signal will be enhanced, first by the addition of Cl to the heavy water, increasing both the capture efficiency for the neutrons and the energy released, and then by the installation of discrete proportional counters optimized for direct observation of neutrons.

Figure 1 summarizes the simulated response of the detector in the pure D_2O configuration to both the neutrino signals and a sample of the backgrounds due to radioactivity in the water, using Standard Solar Model neutrino fluxes. For comparison, the expected response to neutrons with the Cl added is also indicated. The distributions plotted here are in terms 'NHIT'—the number of tubes hit in each event, which is correlated with the event energies.

3 Detector Running

Water fill of the SNO detector was completed in May, 1999, initiating a commissioning period during which several improvements were made to detector conditions. A lowering of channel thresholds to roughly 1/4 of a photoelectron, an increase in phototube gain by $\sim 40\%$, as well as the commissioning of magnetic coils to partially cancel the Earth's field inside the SNO cavity, all combined to increase the detector response by $\sim 25\%$, up to a total of roughly 9 PMT hits per MeV of deposited energy. This energy response is somewhat better than our expectations based upon Monte Carlo simulations.

In addition to increasing the detector response, several other changes were made. The hardware threshold for the main analysis trigger was lowered by ~ 1 MeV to 2 MeV, the lowest threshold ever achieved by a large scale water Cerenkov detector. Additional phototubes were installed to allow later analyses to remove events caused by light emission in the neck of the AV. Finally, the radon sealing for the D_2O volume was improved, bringing radon levels to below design goals.

At the beginning of November, 1999, the detector conditions were frozen and the acquisition of SNO's first physics data set began.

4 Detector Calibrations

SNO's calibration program includes regularly run electronics calibrations which provide constants for the measurement of charge and time on each channel, and calibrations of phototube response, which provide charge-dependent corrections to the timing and measurements of phototube gains.

Among the most critical of the calibrations is the measurement of the detector's optical properties. The optical parameters are used to reconstruct the energy of events as a function of event position and direction, and are included in the Monte Carlo simulation to better model detector behavior. Using preliminary measurements of the optical parameters, Figure 2 compares the Monte Carlo simulation of SNO's ^{16}N source (which provides a triggered 6.1 MeV γ) to the measurements made with the source at the center of the detector. Only the response due to prompt (direct) light is shown. As a function of position and direction, we have found that these optical parameters yield agreement between the simulation and the data to $\sim$ 2% within the restricted volume in which measurements have so far been made.

5 Backgrounds

At low energies (below about 4 MeV), radioactive impurities (in particular, the decay chains of ^{238}U and ^{232}Th) produce Cerenkov signals in both the H_2O and D_2O. The contribution of these events to the solar neutrino energy region will be measured both by assay and in-situ counting with the detector itself. Preliminary assay measurements show that the level of U in both the light and heavy water is an order of magnitude or so below the design goal, while the level of Th is near or slightly above the design goal. These levels indicate that the background to the NC reaction will be less than 10% of the expected solar neutrino rate.

The cavity walls and the phototube support structure can be the source of high energy γ's which can contaminate the neutrino signal above 4 MeV. A small fraction of these may make it through the H_2O shield, and in addition mis-reconstruction of interactions in the light water can alias events into the heavy water. Preliminary analyses aimed at measuring this source of high energy background show that their contribution to the event sample within the heavy water is less than a few percent of the signal. High energy backgrounds associated with spallation from cosmic rays entering the detector are easily removed through their time correlation with the initiating event. The low rate of muons in SNO ($\sim$ 3/hour) means that these cuts add a negligible dead time.

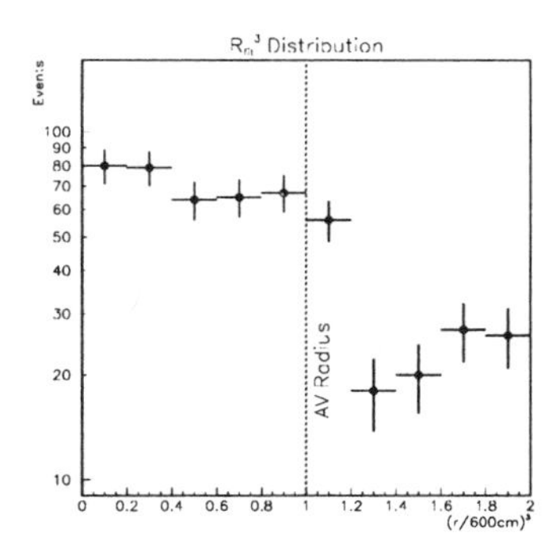

Figure 3. Distribution of reconstructed event positions in the detector.

Finally, with the goal of background levels less than $\sim$ 1% at high energies, SNO must ensure that instrumental backgrounds such as flasher PMT's and light emission in the neck of the acrylic vessel are rejected with very high efficiency. Cuts designed to eliminate these backgrounds have high rejection factors, but the application of the same cuts to calibration data taken with the ^{16}N source shows that, over the NHIT window spanned by the source data, fewer than $\sim$ 0.5% of Cerenkov events are removed.

6 Solar Neutrino Signal

The solar neutrino signal in SNO has several unique signatures. The restricted volume of the heavy water should show an excess of events as compared to the outer light water volume due to the charged current interactions on deuterium. Figure 3 plots the distribution of reconstructed event positions at high energy as a function of R^3 in the detector, clearly showing the excess of within the D_2O volume.

The distribution of event directions also shows a clear solar neutrino signal, in this

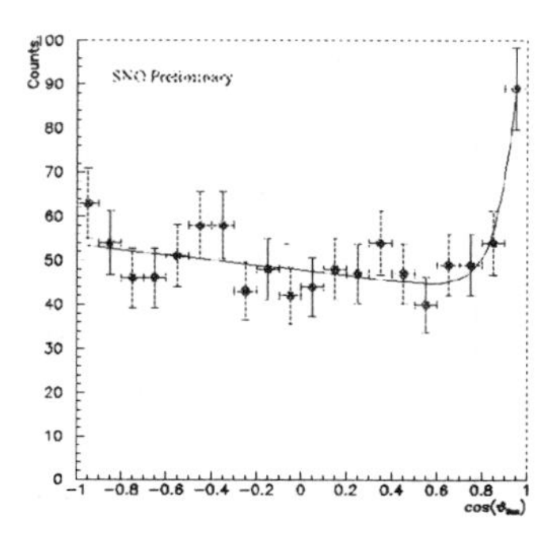

Figure 4. Distribution of reconstructed event directions with respect to the solar position.

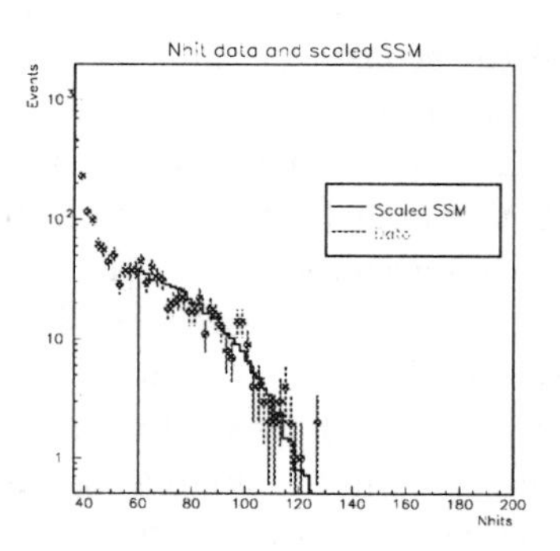

Figure 5. Comparison of event NHIT to scaled prediction of Monte Carlo using a ^{8}B initial ν spectrum.

case from the elastic scattering reaction. Figure 4 plots the correlation of event directions with the solar position, showing a clear peak toward the solar direction. While the charged current reaction also has a correlation with the sun of opposite sign (it falls away as $(1 - \frac{1}{3}\cos\theta_{\odot})$), the statistics in Figure 4, which includes low energy background events, is not sufficient to clearly see this effect.

Lastly, as described in Section 1, a plot of the spectrum of the number of PMT hits in each event should look very much like the incident ^{8}B spectrum. Figure 5 compares the NHIT spectrum of events in the detector within a restricted fiducial volume to a scaled Monte Carlo simulation of the expected spectrum based on an initial ^{8}B neutrino spectrum. As the figure shows, the events in the detector look very much like what we expect for a ^{8}B solar signal.

7 Conclusions

The SNO detector looks to be in excellent shape—the levels of high energy and instrumental backgrounds are at or near our goals, and the level of low energy backgrounds as measured by water assays look low enough to permit accurate measurements of the total flux of active neutrinos via the NC reaction in future phases of the experiment. Based upon the radial, directional, and energy distributions of the events in the detector, the data in the region of interest appear to be dominated by ^{8}B solar neutrinos detected via the charged current and elastic scattering reactions, with very little background. After full studies of detector systematics are completed, the pure D_2O run will provide an accurate measurement of the solar ν_e flux at the Earth.

Acknowledgments

This research has been financially supported in Canada by the Natural Sciences and Engineering Research Council, Industry Canada, National Research Council of Canada, Northern Ontario Heritage Fund Corporation and the Province of Ontario, in the United States by the Department of Energy, and in the United Kingdom by the Science and Engineering Research Council and the Particle Physics and Astronomy Research Council. The heavy water has been loaned by AECL with the cooperation of Ontario Hydro. The provision of INCO of an underground site is greatly appreciated.

References

1. Boger *et al*, *Nucl. Inst. Meth.* **A449**, 172, (2000).

RECENT RESULTS FROM THE K2K (KEK-TO-KAMIOKA) NEUTRINO OSCILLATION EXPERIMENT

MAKOTO SAKUDA
Institute of Particle and Nuclear Studies, High Energy Accelerator Research Organization (KEK), Japan
E-mail: makoto.sakuda@kek.jp

We report the latest results of the oscillation search in the ν_μ disappearance mode from data taken from June, 1999, to June, 2000, with the K2K experiment. We observed 27 fully-contained events in the 22.5 kton fiducial volume of Super-Kamiokande (SK), while the corresponding expected number of events is estimated to be $40.3^{+4.7}_{-4.6}$ in the case of no oscillations.

1 Introduction

The evidence for $\nu_\mu \to \nu_\tau$ oscillations has become stronger with improved statistics on atmospheric neutrinos from SK [2]. The K2K experiment [1] will search for both $\nu_\mu \to \nu_\tau$ oscillations (disapearance mode) and $\nu_\mu \to \nu_e$ oscillations (appearance mode) in the parameter region $\Delta m^2 > 3 \times 10^{-3}\ eV^2$, by sending a well-defined ν_μ beam from KEK to SK over the distance of 250km.

2 K2K experiment

2.1 Beamline

The neutrino beam for the K2K neutrino oscillation experiment was successfully commissioned in March, 1999 [3]. After an engineering run in March, K2K started data-taking in June, 1999.

A proton beam of 12 GeV is accumulated in 9 bunches in the main ring of the KEK proton synchrotron, and is extracted in a single turn of 1.1 μsec from the main ring[4]. The repetition rate is 1 spill per 2.2 seconds. The proton beam is transported to a 66cm-long aluminum target imbedded in the first magnetic horn. The beam size at the target was measured to be about 2 cm in diameter (full width). The diameter of the target was 2.0cm in diameter in June and 3.0cm in diameter after October, 1999. The main ring intensity of about about 7.2×10^{12} protons per spill is near the designed maximum [4] and about 5.4×10^{12} protons are delivered to the target.

Positively charged particles (π^+ and K^+) produced in the target are focused to a nearly parallel beam in an axial magnetic field generated by a pair of horns, and are guided to a decay region of 200m in length. This results in a wide-band neutrino beam with a mean energy of 1.4 GeV and an angular divergence about 20 mradians. The flux and spectrum are unchanged within 1 km (4mrad) from the center of SK. The beam Monte Carlo program considers the proton beam profile before the target and generates $\pi^{\pm,0}$, $K^{\pm,0}$ and secondary protons according to the existing data [5]. GEANT3.15 is used to trace the secondary particles in the target, through the magnetic horns, and the decay tunnel.

Two segmented parallel-plate ionization chambers (SPIC) monitor the profiles and two current transformers (CT) measure the number of protons before the target. There is also a SPIC and silicon pad detectors after the beam dump, 200m downstream of the target. During the experiment, we monitor the intensity and the profile before the target and those after the beam dump every spill in order to assure stable operation of the beamline. A gas-Cherenkov detector (Pion monitor) was placed after the second horn once in June and once in November, 1999. It measures the Cherenkov light of charged pions at various gas pressures in order to ex-

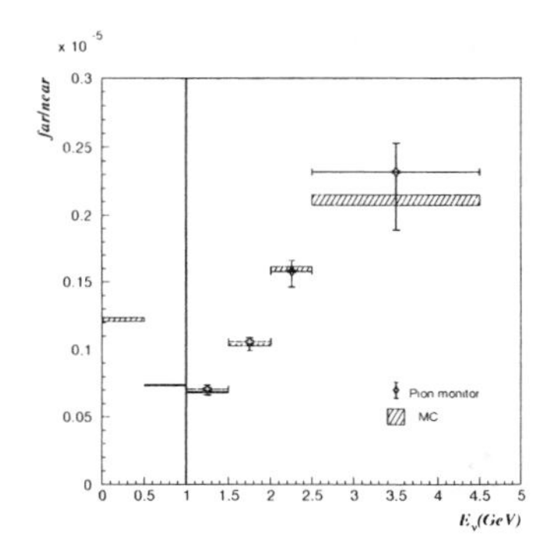

Figure 1. Ratio of the neutrino spectrum, $\Phi_{SK}(E_\nu)/\Phi_{KEK}(E_\nu)$, estimated by pion monitor (diamond with error bars) and by beam Monte Carlo (hatched box).

tract the momentum and angle distribution of charged pions for $p > 2$ GeV/c. From the measurements of the pion monitor, the ratio of the neutrino spectrum at SK and KEK, $\Phi_{SK}(E_\nu)/\Phi_{KEK}(E_\nu)$, was estimated for $E_\nu > 1$ GeV. The MC prediction agrees with pion monitor data as shown in Fig.1. This assures the validity of the beam MC calculation.

2.2 Neutrino Detectors

The near neutrino detector at KEK is located 300m downstream of the target. It is comprised of a fine-grained detector (FGD) and a 1kton water Cherenkov detector. The far detector is SK, 250km away. The near detector is designed to measure the ν_μ flux, the energy spectrum, the x-y profile and ν_e contamination in the ν_μ beam. The 1kton detector measures neutrino interactions with the neutrino beam using the same water Cherenkov-detector technique as SK and aims at reducing the systematic errors in the oscillation analysis. The fine-grained detector consists of the water target, a scintillating-fiber (SciFi) tracking detector, scintillating counters surrounding the SciFi detector, lead-glass counters, and a muon range detector (MUC). The FGD has good tracking capability, thus complementing the 1kton detector, and allows discrimination between different types of interactions. The SciFi tracking detector [6] consists of twenty $2.4m \times 2.4m$ modules. Each module is placed after a 6cm-thick water target of $2.4m \times 2.4m$ area, contained in 1.8mm-thick aluminum tubes. The position resolution of each module is about 1mm. MUC consists of 12 iron plates and drift chambers, covering $7.6m \times 7.6m$ area. The energy of the muon (E_μ) can be measured by MUC with an accuracy of $\Delta E_\mu = 150\ MeV$. The electrons are identified by lead-glass counters with $\Delta E/E = 10\%/\sqrt{E(GeV)}$.

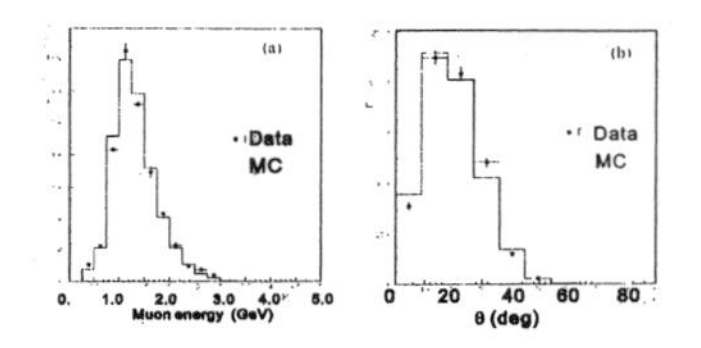

Figure 2. (a) Energy and (b) angular distribution of muons in SciFi events.

2.3 Events at Near Detector and SK

We call an event "SciFi" type ("MUC" type) when the vertex of a muon track produced in charged-current process is found in the SciFi (MUC) detector. We show in Fig.2 the energy and angular distribution of muons in SciFi events. The MC simulation (histogram) reproduces data well. The normalization in this fugure is arbitrary. The MC simulation considers the neutrino spectrum predicted by the beam MC, neutrino interactions and the secondary interactions in the detector. Since the weight of the iron plates of MUC amounts to 1000tons, many charged-current interactions occur in the iron. We show in Figs.3(a)(b) the neutrino profile measured with MUC. From the fit to the profile, the center in x and y coordinates was estimated for every 2 days and plotted in Figs.3(c)(d). This shows that the neutrino beam correctly points to SK within 1mrad throughout the experiment.

We estimate the number of SK events (N_{SK}^{pred}) for the case of no oscillations using the observed number of events at the near detector at KEK (N_{KEK}):

$$N_{SK}^{pred} = N_{KEK} \cdot R \cdot \frac{\epsilon_{SK}}{\epsilon_{KEK}} \cdot \frac{POT_{SK}}{POT_{KEK}}, \quad (1)$$

where R is the event ratio SK/KEK, calculated by MC, ϵ_{KEK} and ϵ_{SK} are the detection efficiency at KEK and SK detectors, POT_{KEK} and POT_{SK} are the number of protons on target on which the data analysis is based at KEK and at SK. The event ratio (R) is defined as

$$R = \frac{\int \Phi_{SK}(E_\nu) \cdot N_{target}^{SK} \cdot \sigma(E_\nu) dE_\nu}{\int \Phi_{KEK}(E_\nu) \cdot N_{target}^{KEK} \cdot \sigma(E_\nu) dE_\nu}. \quad (2)$$

The expected numbers (N_{SK}^{pred}) as estimated by near detectors are:
$N_{SK}^{pred} = 40.3^{+4.7}_{-4.6}$ ($1kton$), $41.5^{+6.2}_{-6.4}$ (MUC), and $40.1^{+4.9}_{-5.4}$ ($SciFi$). The statistical errors are negligible in the given errors. Since the target material (water) is common and the detector type is the same for 1kton detector and SK, some of the systematic errors cancel in the ratio and thus the total error is the smallest of the three numbers. The expected numbers estimated by different types of near detectors agree well. The total systematic error in N_{SK}^{pred} for FC events is about 12%, due mainly to the uncertainty in $\Delta R = {}^{+6\%}_{-7\%}$, $\Delta\epsilon_{KEK} \sim 8\%$ and $\Delta\epsilon_{SK} \sim 3\%$.

We analyze the data at SK from a total of 2.29×10^{19} protons incident on target. Two types of beam-induced events are selected: the first are called "fully-contained (FC)" events, where the significant light is detected in the inner detector only; the other is called "outer detector (OD)" events, where light is detected in the outer detector. FC-in (FC-out) stands for FC events with the interaction vertex inside (outside) the fiducial volume (22.5kton) of SK. FC events will give the cleanest sample of beam-induced events and will be used in the full spectral analysis.

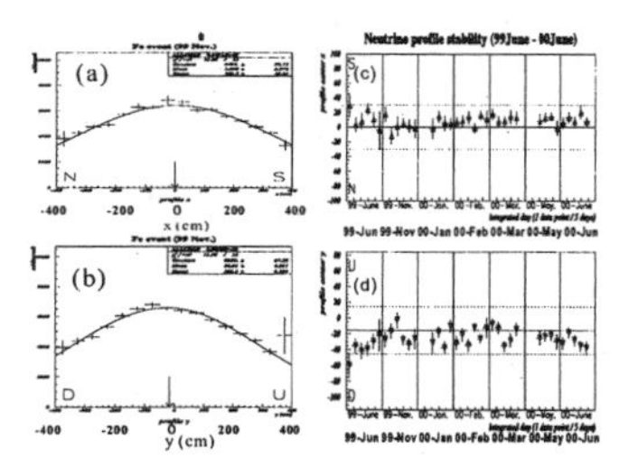

Figure 3. (a) Horizontal and (b) vertical beam profile measured with MUC. The peak positions in x and y profiles are plotted for every 2 days from June '99 to June '00. The solid holizontal lines show the direction to SK and the dashed lines correspond to the range of ±1mrad.

OD events are useful to check the consistency of the event rate as compared to FC events.

We search for those beam-induced events at SK for $\pm 500\mu sec$ from the time when the neutrino beam should reach SK, using the universal time supplied by the Global-Positioning System (GPS) [7]. The GPS system provides the timestamps at both near detector and SK. Beam spill times (T_{KEK}) are recorded at the near detector, and the time of each event trigger (T_{SK}) is recorded at SK. The time difference ($T_{SK} - T_{KEK}$) was checked to be less than 0.3 μsec by an atomic clock.

For each SK event, we calculate the time difference $\Delta T = T_{SK} - T_{KEK} - T_{TOF}$, where $T_{TOF} = 833.3$ μsec is the flight time of neutrinos from KEK to SK. The time difference ΔT of FC events over $\pm 500\mu sec$ and $|\Delta T| < 5\mu sec$ is plotted in Fig.4. FC events from KEK neutrinos are found at the expected arrival time. Only two background events are detected outside the expected beam gate, consistent with the rate of the atmosphric neutrino events. The background contamination inside the beam gate is less than 10^{-3} events. The number of beam-induced events at SK is summarized in Table 3. FC events are decomposed into events with a single Cherenkov ring (1-ring) and multi-rings. 1-ring events are further clasified into 14 muon-like events and 1 electron-like event. The

Table 1. Summary of SK events from June 1999 to June 2000. The mixing parameter $sin^2 2\theta = 1$ is assumed.

	Data	No oscillation	$\Delta m^2 = 3 \times 10^{-3}$	5×10^{-3}	7×10^{-3} $(eV)^2$
FC-in	27	$40.3^{+4.7}_{-4.6}$	$26.6^{+3.4}_{-3.3}$	$17.8^{+2.3}_{-2.2}$	14.9±1.9
1-ring	15	24.3±3.6	14.4±2.3	9.4±1.5	8.6±1.4
μ-like	14	21.9±3.5	12.4±2.1	7.5±1.3	6.8±1.2
e-like	1	2.4±0.5	2.1±0.4	1.9±0.4	1.8±0.4
multi-ring	12	16.0±2.7	12.2±2.1	8.4±1.5	6.3±1.1
FC-out	16	17.2	11.2	7.6	6.7
OD events	23	44.5	31.0	20.3	15.7

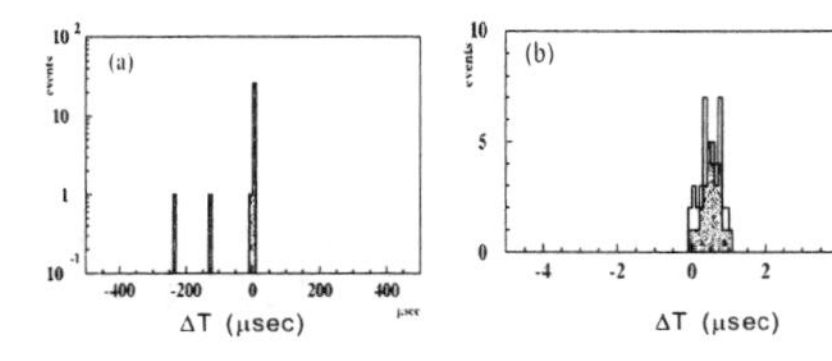

Figure 4. Time difference $\Delta T = T_{SK} - T_{KEK} - T_{TOF}$ of FC events within (a) $|\Delta T| < 500\mu sec$ and $|\Delta T| < 5\mu sec$.

observed and expected numbers for FC-out events and OD events are given in the table as a reference. We include in the table the expected number of SK events for the case of no oscillations and three typical oscillation parameters. The observed number of FC-in events (1-ring and multi-ring) is slightly lower than that expected for no oscillations.

3 Summary

The K2K experiment has been running smoothly since June, 1999, and collected 2.6×10^{19} protons on target including a running period which ends in June, 2000. The interaction rates were measured by near detectors (SciFi, MUC and 1kton) and they are consistent with one another. We found from the neutrino beam profile measured at the near detector that the neutrino beam correctly points to the SK direction within 1mrad throughout the experiment. We have established the procedure to estimate the number of SK events. We analysed data at SK from 2.3×10^{19} protons on the target. The 27 fully-contained events are observed in the 22.5 kton fiducial volume of SK, while the expected number is $40.3^{+4.7}_{-4.6}$ in the case of null oscillations. Our data disfavor null oscillation at the 2σ level. As we collect more data, we will perform the oscillation analysis with the energy spectrum. We will continue data-taking for the next few years untill we accumulate 10^{20} protons on target.

References

1. K.Nishikawa, Nucl.Phys.B (Proc.Suppl.)77(1999)198.
2. T.Toshito, this proceedings.
3. H.Noumi et al., Nucl.Instrum.Meth.A398(1997)399.
4. K.Takayama, KEK-preprint 99-57, August, 1999.
5. Our π production model is nearly the same as the one, except for an overall normalization factor, which is described in Y. Cho et al., Phys.Rev.D4 (1971)1967.
6. A.Suzuki et al., hep-ex/0004024, April, 2000.
7. H.G.Berns and R.J.Wilkes, IEEE NS 47, 340(2000).

RESULTS FROM DONUT

NAKAMURA M. FOR THE DONUT COLLABORATION
NAGOYA University, Furo-cho, Chikusa-ku, NAGOYA, 464-8602, Japan.
E-mail: nakamura@flab.phys.nagoya-u.ac.jp

Among 203 located prompt neutrino interactions, five ν_τ^{CC} events were detected. The background from charged charm decays and hadronic secondary interactions is estimated to be 0.5. We conclude that these are the evidence of the first observation of $\nu_\tau + N \to \tau + X$.

1 Introduction

DONUT is an hybrid emulsion experiment which intend to detect ν_τ charge current interactions [1]. A prompt neutrino beam, created by dumping 800GeV/c protons from Tevatron , was exposed to an emulsion-counter hybrid experimental set up of DONUT.

So called an Emulsion Cloud Chamber (ECC) type target, which is a stack of passive materials (stainless steel plates of 1 mm thick for DONUT) and emulsion plates ($100\mu m$ thick emulsion layer coated on both sides of a plastic base, $200\mu m$ or $800\mu m$ thick), was utilized as the neutrino target. Adding to the intrinsic high position resolution of emulsion, the automatic emulsion read-out systems developed by Nagoya group makes the ECC not only to be utilized as very precise vertex detector but also as a tool to measure the momentum of charged particles (by multiple Coulomb scattering measurement) and to identify the particle kind, especially in the case of electrons(by detecting its rapid energy loss and/or its starting stage of electromagnetic cascade shower) [2].

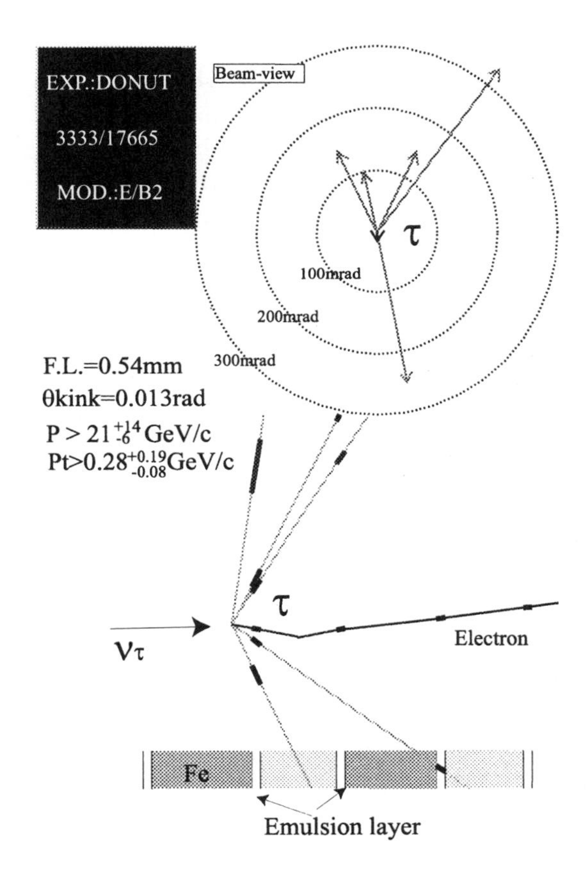

Figure 1. One DONUT ν_τ^{CC} event. The decay is identified as $\tau \to e\nu_\tau\overline{\nu_e}$. One charged particle emitted to the most forward direction decayed into an electron. The daughter was accompanied by 2 electron pairs after passing through a material of $\sim 0.4Xo$. The developed electromagnetic showers created a vast hits in the SFT. The measured momentum by multiple scattering is considered to be the lower limit because of the energy loss by bremstralung.

2 Emulsion analysis

The analysis flow starts from the vertex location in the emulsion target. We have counter-reconstructed 699 events in our fiducial volume of the ECC target. Suspending events of which vertex is at the regions with quite high muon background density, vertex location of 451 events were tried until spring 2000. Among them, we have succeeded to locate 262 events. After suspending bad alignment events (59 events), we applied systematic decay search sequence to the remained 203 events.

The decay topologies of the τ is categorized into "long" and "short" decays. In long decays, the parent τ track can be recognized in at least one emulsion layer. A decay can be detected as an angular reflection of a track (kink topology). In short decays, only the decay daughter can be recognized as a track which has non-zero impact parameter with respect to the neutrino interaction vertex. In both cases practical emulsion resolution, $\sigma_{position} = 0.3\mu m$, is especially suited for their detection.

For long decay search, after selecting kink candidates ($\theta_{kink} \geq 5mrad$ with good χ^2 daughter track), the decay Pt was calculated by using the measured kink angle and the momentum measured by multiple Coulomb scattering method. For short decay search, instead of the real impact parameter, the nearest distance of any two tracks was measured. The combinations, which have nearest, distance greater than four times measurement error were selected to measure the momentum of the corresponding tracks. Then the minimum Pt (Pt_{min}) was calculated. Pt_{min} is defined as measured momentum times minimum kink angle by assuming that the decay point is at the most downstream of the vertex material plate.

In order to reject hadronic interaction background, $Pt \geq 0.25\ GeV/c$ and $Pt_{min} \geq 0.1\ GeV/c$ was required for long and short decay candidates, respectively. Five for long and two for short events were remained. For these events, the existence of leptons from the neutrino interaction was checked in order to reject charm events. At the same time, also it was checked whether the daughter is a lepton or not. One long event had an electron from the interaction vertex and was identified as a charged charm production in ν_e^{CC}. Also one short event was identified as a charged charm production in $\overline{\nu_\mu^{CC}}$, because of the μ^+ from the interaction vertex. Also in this event one of the three daughters was identified as μ^-.

As a result, four long and one short events were remained as τ samples. The observed number is consistent with the expectation of 4.1 long and 0.8 short decays. The background from charm decays and hadronic secondary interactions are estimated to be 0.2 and 0.2 events for long and 0.07 and 0.03 events for short, respectively. Charged charm kink decay becomes a background if the lepton from the ν^{CC} interaction is missed. In DONUT, muon is identified by the most downstream muon identifier and almost electron is identified in the ECC.

In two events, the daughter was identified as an electron, therefore the decay was identified as $\tau \rightarrow e\nu_\tau\overline{\nu_e}$. These two events are free from the hadronic secondary interaction background. In Fig.1, one of these two events is shown.

3 Conclusion

The evidence of the ν_τ charge current interactions has been successfully detected by DONUT hybrid emulsion detector.

Acknowledgments

I wish to express special thanks to the organizers of ICHEP2000.

References

1. Nakamura M., *Nucl. Phys.*B **77** (1999) 259.
2. T. Nakano, The Ultra Track Selecter, Nagoya University note, in preparation. S.Aoki et al., *Nucl. Instr. and Meth.* **B51**(1990)466.

RECENT RESULTS FROM CHORUS

M. KOMATSU
Physics Dept., Nagoya University, 464-8602, Japan

THE CHORUS COLLABORATION

We have compleated the "Phase I" oscillation analysis for the entire CHORUS data set, taken in the years 1994-1997, the search for ν_τ charged current events has been performed for both muonic and hadronic decays of the τ lepton. No τ candidate has been found in "Phase I" analysis. The $\nu_\mu \to \nu_\tau$ mixing is excluded down to $sin^2 2\theta_{\mu\tau} = 6.8 \times 10^{-4}$ for large $\Delta m^2_{\mu\tau}$ using the statistical treatment given by ref. [3]. The analysis continues ("Phase II") with a new techniques, developed in DONUT experiment to improve the sensitivity.

1 Experiment

CHORUS is an emulsion-electronic detector "hybrid" experiment designed to search for $\nu_\mu \to \nu_\tau$ oscillation through the observation of charged current interactions $\nu_\tau N \to \tau^- X$, followed by the decay of the τ lepton, the emulsion technique isuued to detect the decay topology, kinematical informations from the electronic detectors. The hybrid setup is described in detail elsewhere [1].

The experiment was performed in the CERN Wide Band Neutrino Beam, which contains mainly ν_μ, essentially without prompt ν_τ contamination. The data taking has been performed for four years of running from 1994 to 1997.

2 Phase I analysis

During the four years of operation the emulsion target has been exposed to the neutrino beam for an integrated intensity which corresponds to 5.06×10^{19} protons on target(PoT).

The search for ν_τ interactions has been performed for the muonic and the hadronic one prong decay. The information of the electronic detectors has been used to define two distinct data sets, the so called 1μ and 0μ samples. For each sample different kinematical selections are applied to reduce scanning load, while keeping a high sensitivity in both samples. The event must contain at least one negative muon or a hadron track, as possible decay product of a τ.

For these samples, the analysis in the emulsion target has been performed with guidance from scintillating fiber tracker system. The tracks are followed in target emulsion stacks, where the search for the τ decay is performed. The detailed event selection is described elsewhere [2].

The Phase I analysis results are summarised in Table 1. We observed no τ candidate event in the Phase I analysis with 0.1 and 1.1 expected background events, for 1μ and 0μ samples respectively.

Limit on $\nu_\mu \to \nu_\tau$ oscillation have been computed on the basis of zero candidates observed both in the 1μ and 0μ samples.

For the determination of N_τ, the upper limit on the number of ν_τ candidates given the null observation, we have used the method proposed by Junk [3] which allows the combination of different channels, taking into account the errors on the background and on the signal. The limit obtained on the $\nu_\mu \to \nu_\tau$ oscillation probability is

$$P_{\mu\tau} \leq 3.4 \cdot 10^{-4} \tag{1}$$

The 90% C.L. excluded region in the $(sin^2 2\theta_{\mu\tau}, \Delta m^2_{\mu\tau})$ parameter space is represented in Figure 1. Full mixing between ν_μ and ν_τ is excluded at 90% C.L. if $\Delta m^2_{\mu\tau} > 0.6 eV^2$; large $\Delta m^2_{\mu\tau}$ values are excluded at 90% C.L. for $sin^2 2\theta_{\mu\tau} > 6.8 \cdot 10^{-4}$.

Table 1. Data flow chart

Protons on target	$5.06\ 10^{19}$	
1μ: events with 1 negative muon and vertex predicted in emulsion	713,000	
1μ: $p_\mu < 30GeV/c$ and angular cuts	477,600	
1μ: events scanned	355,395	
1μ: vertex located	143,742	
1μ: events selected for eye-scan	11,398	
1μ: kink candidates after eye-scan	0	
0μ with vertex predicted in emulsion(CC contamination)	335,000	(140,000)
0μ with 1 negative track (P=1-20GeV and angular cuts)	122,400	
0μ: events scanned	85211	
0μ: vertex located(corrected number after reprocessing)	23,206	(20,081)
0μ: events selected for eye-scan	2,282	
0μ: kink candidates after eye-scan	0	

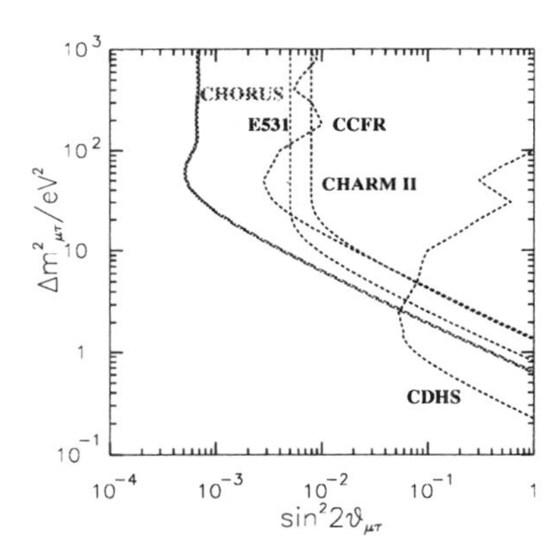

Figure 1. Present result of excluded region

NOMAD [4] result can not be directly compared to ours, since the statistical treatment of the data is different [5]. To compare with NOMAD result, using the same statistical treatment as the NOMAD experiment, we obtain $P_{\mu\tau} \leq 2.2 \cdot 10^{-4}$,$\Delta m^2_{\nu\tau} > 0.5eV^2$ and $sin^2 2\theta_{\mu\tau} > 4.4 \cdot 10^{-4}$.

3 Phase II analysis

We continue the analysis with a new scanning method, so called "Netscan", developed in the DONUT experiment, to reach our design sensitivity. We expect to achieve higher kink finding efficiency and electron identification to search electronic decay mode of the τ. The data taking speed in the emulsion scanning for CHORUS is about 6,000 events/month for $1.5 \times 1.5 \times 6.3mm^3$ perpendicularly and along the beam. We have achieved $0.25\mu m$ in position and $0.58mrad$ in angule. Efficiencies are being studues on charm decays detected using the same techniques. In a pilot analysis, we have treated 8,974 ν_μ charged current interactions, and observed 196 charm candidates.

The phase II analysis has started with the aim of reaching the design sensitivity of $P_{\mu\tau} = 10^{-4}$.

References

1. E. Eskut *et al.*, CHORUS Collaboration, Nucl. Instr. and Meth. **A401** 7 (1997)
2. E. Eskut *et al.*, CHORUS Collaboration, Phys. Lett. **B434** (1998) 205.
3. T. Junk, Nucl. Instr. and Meth. **A434** 435 (1999).
4. J. Astier *et al.*, NOMAD Collaboration, preprint CERN/EP/2000/049.
5. G.J. Feldman, R.D. Cousins, Phys. Rev. **D57** 3873 (1998)

LATEST RESULTS ON NEUTRINO OSCILLATION FROM THE NOMAD EXPERIMENT

C. RODA FOR THE NOMAD COLLABORATION
Universitá di Pisa and I.N.F.N. Pisa, Italy, E-mail chiara.roda@cern.ch

The latest results on ν_τ appearance search in $\nu_\mu \rightarrow \nu_\tau$ and $\nu_e \rightarrow \nu_\tau$ oscillation in NOMAD are reported. The analysis, carried out on the full data sample, is based on kinematic criteria. No evidence of oscillations is found. In the two-family oscillation scenario this sets a limit on $\sin^2 2\theta_{\mu\tau} < 4.0 \times 10^{-4}$ at large Δm^2 and $\Delta m^2 < 0.7\ eV^2/c^4$ for $\sin^2 2\theta_{\mu\tau} = 1$ at 90% C.L.. If the results are interpreted in the $\nu_e \rightarrow \nu_\tau$ oscillation hypothesis the corresponding limits are $\sin^2 2\theta_{e\tau} < 2.0 \times 10^{-2}$ at large Δm^2 and $\Delta m^2 < 6\ eV^2/c^4$ for $\sin^2 2\theta_{e\tau} = 1$ at 90% C.L..

NOMAD is a ν_τ appearance experiment : oscillations of neutrinos in the beam are detected looking for ν_τ charged current (CC) interactions in the NOMAD active target. The estimated relative beam composition is $\nu_\mu : \bar{\nu}_\mu : \nu_e : \bar{\nu}_e = 1.00 : 0.061 : 0.0094 : 0.0024$, with average neutrino energies of 23.5, 19.2, 37.1, and 31.3 GeV, respectively[1]. The prompt ν_τ component is negligible[2].

The core of the NOMAD detector[3] consists of a series of drift chambers placed in a 0.4 Tesla magnetic field, acting both as target and as spectrometer. Downstream of the drift chambers are placed a transition radiation detector, a preshower detector, an electromagnetic calorimeter, a hadronic calorimeter and a muon detector. The full data correspond to about 1.35×10^6 ν_μ charged current interactions.

Identification of ν_τCC interactions is performed searching for τ decays to $e^-\bar{\nu}_e\nu_\tau$, $h^-(n\pi^0)\nu_\tau$ and $h^-h^+h^-(n\pi^0)\nu_\tau$, amounting to 82.5% of the total τ decay branching ratio. The oscillation analysis starts by selecting well reconstructed neutrino interactions in the fiducial volume of the active target. A first filter against ν_μCC interactions is then applied rejecting events containing a well identified muon.

The two sources contributing to the background of the oscillation signal, CC and neutral current (NC) interactions, are characterized by different kinematics. In CC events the lepton is well isolated, its momentum in the plane transverse to the beam direction balances that of the hadron jet and the missing transverse momentum is small and randomly directed. NC events are instead characterized by high missing transverse momentum and by τ decay candidates embedded in the hadron jet. Signal events have intermediate characteristics between these extremes : the missing transverse momentum is generated by the neutrino(s) produced in the τ decay, and isolation of the charged decay product(s) is smeared since their momentum has a component transverse to the τ.

In the analysis the kinematic characteristics of signal and background are described using likelihood functions built from multidimensional probability density functions. Events are then classified using the logarithm of the ratio ($\ln \lambda$) between the likelihood functions for signal and background hypotheses. The high end of this function is subdivided into various bins characterized by different signal to background ratios, which are then regarded as independent analyses. This method allows the use of the information on the shape of the distribution even with a limited statistics. The various bins and the different decay channels, are combined within the Unified Approach[5].

A signal in NOMAD appears as an excess of events incompatible with the expected background. It is thus crucial to have reliable

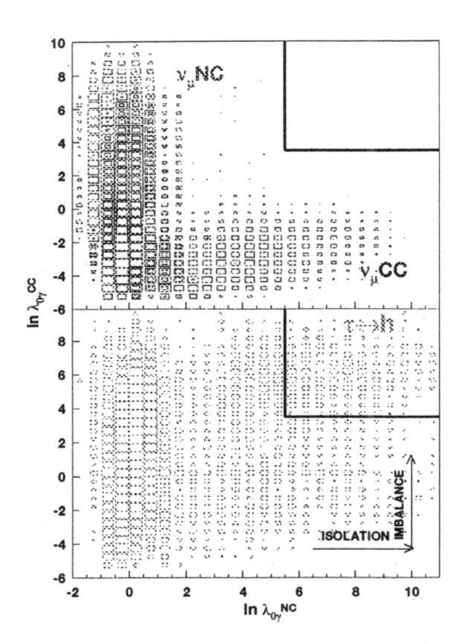

Figure 1. Scatter plot of $\ln \lambda_{CC}$ vs $\ln \lambda_{NC}$ for the (0γ) topology, for MC ν_μ NC and CC interactions (upper plot) and for MC $\tau^- \to h^-(n\pi^0)\nu_\tau$ (lower plot). The lines on the upper right corner delimit the "blind region" (BOX) and the four signal bins.

and unbiased background predictions. Reliable prediction of suppression factors as large as 10^5 are obtained by correcting the MC efficiencies using a method based on data themselves. Samples of "simulated" events are generated from ν_μ CC interactions where the identified muon is replaced by a MC neutrino, electron or τ decay of appropriate momentum. These samples, obtained both from data and MC events, are referred to as Data Simulator events (DS) and Monte Carlo Simulator events (MCS) respectively. Final efficiencies, of each background source and τ decay mode, are then calculated as $\epsilon = \epsilon_{MC} \times \epsilon_{DS}/\epsilon_{MCS}$.
In order to have an unbiased background prediction a "blind analysis" is performed. In this analysis method the kinematic region most sensitive to the signal ("blind region") in each decay channel is analyzed only once all analysis procedures have been defined and background predictions are validated on data control samples. Two data samples are used as the main control samples : the search for τ^+, where, given the beam composition no signal is expected, and the search for τ^- outside the "blind region". Among all validated analyses in a give decay channel, the most sensitive one is chosen.

The oscillation analysis[4] is based on the

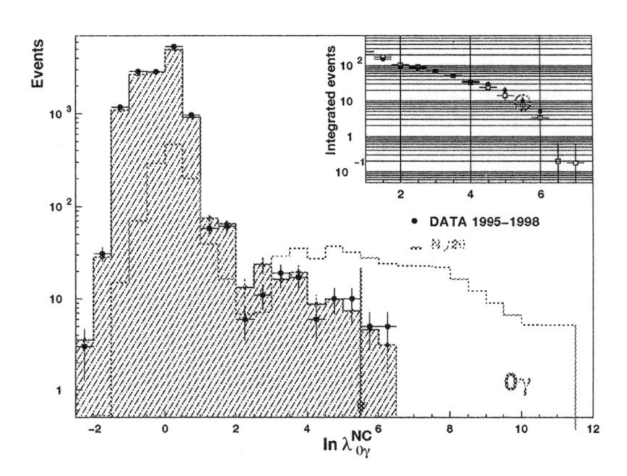

Figure 2. Distribution of $\ln \lambda_{NC}$ for the final 0γ candidates for Data (dots), total expected background (hatched histogram) and expected τ signal, for $P_{osc} = 1$, scaled down by a factor 20 (histogram). The boundary of the "blind region" is indicated by an arrow. The inset gives, the running integral of $\ln \lambda_{NC}$ for data (dots) and expected background (open squares); the encircled point is at the edge of the "blind region".

$e^-\bar{\nu}_e\nu_\tau$, $h^-(n\pi^0)\nu_\tau$ and $h^-h^+h^-(n\pi^0)\nu_\tau$ τ decay modes. Two independent analyses are performed for each channel, one for low multiplicity (LM) and one for deep inelastic (DIS) samples. The analysis on the DIS samples for the decay channels $\tau^- \to e^-\bar{\nu}_e\nu_\tau$ and $\tau^- \to h^-(n\pi^0)\nu_\tau$ are the most sensitive ones. In particular a new $\tau^- \to h^-(n\pi^0)\nu_\tau$ analysis has recently updated the results of the oscillation search.
The $\tau^- \to h^-(n\pi^0)\nu_\tau$ signal is looked for within three possible topologies : single π (0γ) and $\pi\pi^0$ where the π^0 is detected as one (1γ) or two (2γ) electromagnetic clusters. The candidate selection is performed using three likelihood functions, each one optimized for a particular topology. In each event the final candidate is defined as the

one which maximizes the logarithm of the ratio of the likelihood functions for correct and random combinations of tau decay products. Background rejection is achieved with

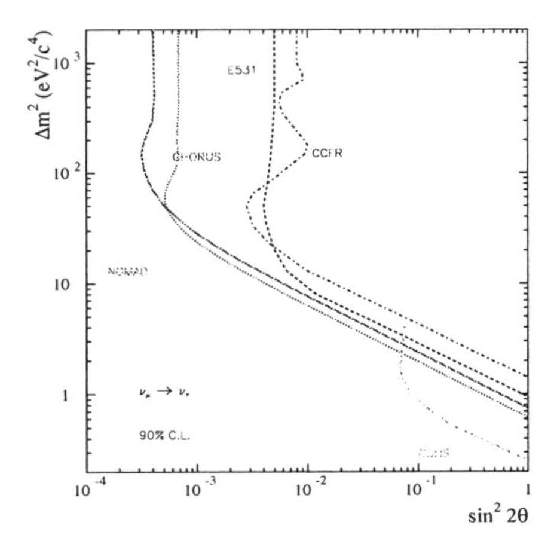

Figure 3. Excluded regions at 90% C.L. in the ($\sin^2 2\theta$,Δm^2) plane in the two family oscillation scenario for the $\nu_\mu \rightarrow \nu_\tau$ hypothesis. The NOMAD result is shown together with results obtained from other experiments[6].

two likelihood functions, one mostly based on unbalance to reject CC ($\mathcal{L}_{CC}$), and a second one mostly based on isolation against NC ($\mathcal{L}_{NC}$). The "blind region" defined in the two dimensional plane $[\ln \lambda_{CC}, \ln \lambda_{NC}]$, is shown in Fig.1 for signal and background events. The final sensitivity is obtained subdividing this region into four bins for each decay topology. Data in the "blind region" are found to be consistent with background prediction (Fig. 2).

Final results are calculated as frequentist confidence intervals[5] combining all decay channels and signal bins. The overall analysis reaches a sensitivity on the oscillation probability of 2.6×10^{-4}, compatible with the design sensitivity.

Data in all analysis bins are consistent with background expectations thus no evidence of oscillation is found. The limit obtained on the oscillation probability is $P_{osc}(\nu_\mu \rightarrow \nu_\tau) < 2.0 \times 10^{-4}$ at 90% C.L.. This result corresponds to $\sin^2 2\theta_{\mu\tau} < 4.0 \times 10^{-4}$ at large Δm^2 and $\Delta m^2 < 0.7\ eV^2/c^4$ for $\sin^2 2\theta_{\mu\tau} = 1$. In the absence of signal events the probability to obtain an upper limit smaller or equal to 2.0×10^{-4} is 46%. The excluded region at 90% C.L. in the ($\sin^2 2\theta_{\mu\tau}$,Δm^2) plane is shown in figure 3.

In the two-family approximation this result can be reinterpreted in terms of $\nu_e \rightarrow \nu_\tau$ oscillation under the assumption that any ν_τ originates from oscillation of the small ν_e component in the beam. In this hypothesis we obtain at 90% C.L. $P_{osc}(\nu_e \rightarrow \nu_\tau) < 1.0 \times 10^{-2}$, which gives $\sin^2 2\theta_{e\tau} < 2.0 \times 10^{-2}$ at large Δm^2 and $\Delta m^2 < 6\ eV^2/c^4$ for $\sin^2 2\theta_{e\tau} = 1$. In absence of signal events the probability to obtain an upper limit smaller or equal to 1.0×10^{-2} is 48%.

References

1. G. Collazuol et al., presented at NOW98 Workshop, Amsterdam, 7-9 September 1998, CERN Preprint OPEN-98-032.
2. M.C. Gonzales-Garcia, J.J. Gomez-Cadenas, Phys. Rev. D 55 (1997) 1297; B. Van de Vyver, Nucl. Instr. and Meth. A 385 (1997) 91.
3. J. Altegoer et al., Nucl. Instr. and Meth. A 404 (1998) 96.
4. P. Astier et al., Phys. Lett. B 483 (2000) 387.
5. G.J. Feldman, R.D. Cousins, Phys. Rev. D 57 (1998) 3873.
6. L. Ludovici, CHORUS Collaboration, Proceedings of Neutrino 2000, 6-21 June 2000, Sudbury, Ontario, Canada. E531 Collaboration, N. Ushida et al., Phys. Rev. Lett. 57 (1986) 2897; CCFR Collaboration, K.S. McFarland et al., Phys. Rev. Lett. 75 (1995) 3993; CDHS Collaboration, F. Dydak et al., Phys. Lett. B 134 (1984) 281.

NEUTRINO OSCILLATION EXPERIMENTS AT FERMILAB

ADAM PARA
Fermilab, Pine St., Batavia IL 60510, USA
E-mail: para@fnal.gov

Neutrino oscillations provide an unique opportunity to probe physics beyond the Standard Model. Fermilab is constructing two new neutrino beams and detectors to provide decisive test of two of the recent positive indications for neutrino oscillations: the MiniBOONE experiment will settle the LSND controversy, MINOS will provide detailed information in the region indicated by the SuperK results.

1 Introduction

There are several experimental indications that neutrinos may undergo oscillations:

1. The solar neutrino deficit.

   The flux of solar ν_e measured in several experiments is about 50% of the flux expected in the Standard Solar Model. This large discrepancy is unlikely to be caused by our ignorance of the physics of the Sun; it can be interpreted as a result of $\nu_e \rightarrow \nu_x$ oscillations.

2. Atmospheric neutrinos.

   The SuperKamiokande detector has been used to detect interactions of atmospheric neutrinos. The results show a depletion of the ν_μ interaction rate as a function of the zenith angle, while the ν_e interaction rate is consistent with expectations[1]. The observed depletion is consistent with the hypothesis of neutrino oscillations and strongly suggests $\nu_\mu \rightarrow \nu_\tau$ oscillations with nearly maximal mixing angle and a Δm^2 in the range $0.003 - 0.01 eV^2$ [1]

3. LSND Experiment at LAMPF

   LAMPF neutrino beam comes from stopped pions produced by 800 MeV proton beam A liquid scintillator detector recorded an excess of 82.8 ± 23.7 $\overline{\nu}_e$ interactions above the expected background of 17.3 ± 4 events. This result is consistent with the hypothesis of $\overline{\nu}_\mu \rightarrow \overline{\nu}_e$ oscillations, if the Δm^2 is in the range $0.3 - 2\ eV^2$.

Given the potential importance of the neutrino mass sector, the following questions pose a challenge to experiments in the near future:

- Are all three indications really examples of neutrino oscillations?
  Observation of oscillatory behavior as a function of distance and/or neutrino energy, would be a particularly convincing proof.

- What are the oscillation modes?

- What are the oscillation parameters?
  What are the patterns of the Δm^2_{ij}? The mixing angles? What are the elements of the lepton mixing matrix? Are there dominant and sub-dominant oscillation modes corresponding to each oscillation frequency (Δm^2)?

Studies of oscillations in the solar neutrino region require extra-terrestrial distances and/or very low energy neutrino sources, such as nuclear reactors. Oscillations in the atmospheric neutrinos region or the region indicated by the LSND experiment lend themselves to studies with neutrino beams produced in the laboratory. Such investigations of the neutrino oscillations are an important part of the scientific program at Fermilab.

2 Fermilab Accelerators and Neutrino Beams

The flagship of the Fermilab high energy physics program is the Tevatron Collider. Recent upgrades of the accelerator infrastructure were specifically designed to boost the luminosity of the Collider and at the same time to enable a fixed-target program, like neutrino experiments, to ru at the same time as the Collider experiments. Two parts the accelerator complex are being used to produce neutrino beams: the 8 GeV Booster and the newly constructed Main Injector.

2.1 8 GeV Booster Neutrino Beam (MiniBOONE)

The Booster is the oldest part of the accelerator complex at Fermilab. Upgraded to 400 MeV injection in 1993, it is capable of delivering up to 5×10^{12} protons per pulse with the repetition rate up to 7.5Hz.

A Booster[3] neutrino beam is under construction. An 8 GeV proton beam is extracted onto a beryllium target. A magnetic horn focuses secondary pions and kaons, its focusing power being optimized for 3 GeV secondaries. The secondary beam has a relatively short decay path which can be varied from 25 to 50 m. A variable decay path will provide additional information on the ν_e component of the beam.

The high proton flux and an efficient horn focusing will provide a high flux neutrino beam, yielding over 2,000,000 ν_μ interactions per kton-year at a distance 500 m from the source. The neutrino flux will have a maximum around $E_\nu = 1$ GeV and an exponential high energy tail falling to the 10% level at $E_\nu = 3$ GeV.

2.2 NuMI: Neutrino Beams from the Main Injector (MINOS)

The Main Injector accelerator is a 150 GeV proton synchrotron constructed to replace the original Fermilab Main Ring. It is expected to serve as a high intensity, fast-cycling accelerator for antiproton production, as an injector into the Tevatron and simultaneously to support fixed target experiments using 120 GeV protons.

It is expected that the Main Injector will be able to deliver 3.6×10^{20} protons per year onto the NuMI target, while also supporting antiproton production for the Tevatron Collider experiments.

The high intensity and high repetition rate of the Main Injector allows for neutrino beams of unprecedented intensity, thus creating an opportunity for long baseline neutrino oscillation experiments. The Main Injector will accelerate 6 batches of 8×10^{12} protons each with a repetition rate of 1.9 secs. One of these batches will be used for antiproton production, while the remaining five batches will be extracted onto the segmented carbon target.

Secondary pions and kaons are collected and focussed by a system of two parabolic magnetic horns, and subsequently produce a neutrino beam by decaying inside a 675 m long decay pipe. The beam optics are designed to allow selection of the neutrino beam energy by moving the focusing elements (horns) in a manner similar to a zoom lens. The energy spectra of three possible beam configurations are shown in Fig.1, together with a spectrum of a perfect beam, where all of the secondary particles were collected and allowed to decay. The NuMI beam design provides an overall efficiency about 50%, in all three beam configurations.

3 MiniBOONE Experiment: Checking the LSND Result

The primary goals of this experiment are:

- Unambiguously confirm or disprove the existence of the neutrino oscillation signal suggested by the LSND experiment

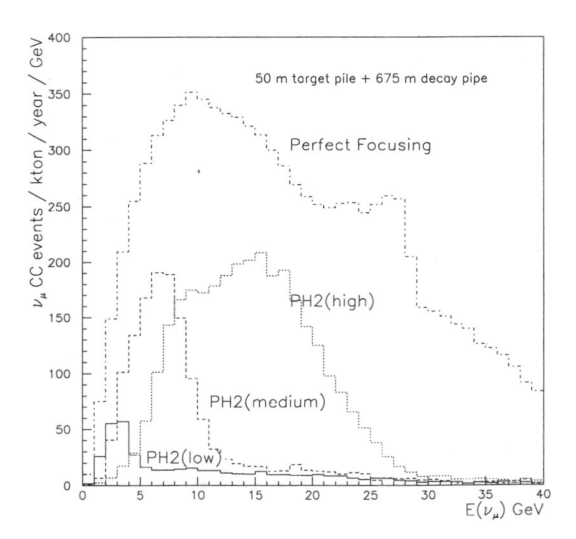

Figure 1. Neutrino beam spectra for different NuMI beam configurations

- Provide precise measurements of the oscillation parameters, should the existence of the effect be established, or improve the existing limits if the effect is not confirmed

The detector will consist of a 12 m diameter spherical tank filled with 769 tons of mineral oil, located at the distance of 500 m from the neutrino source. Cerenkov light emitted by particles produced in the neutrino interactions will be detected by 1220 eight-inch photomultiplier tubes. The pattern of the Cerenkov rings will be used as a primary tool to identify electrons and muons. The outer 50 cm of the volume is optically isolated from from the main detector volume; it is viewed by 292 outward-pointing photomultipliers and will serve as a veto against cosmic rays.

The Booster neutrino beam is expected to yield 500,000 ν_μ quasi-elastic charged current (CC) interactions in the MiniBOONE fiducial volume per year of operation.. The expected signal of the neutrino oscillations,

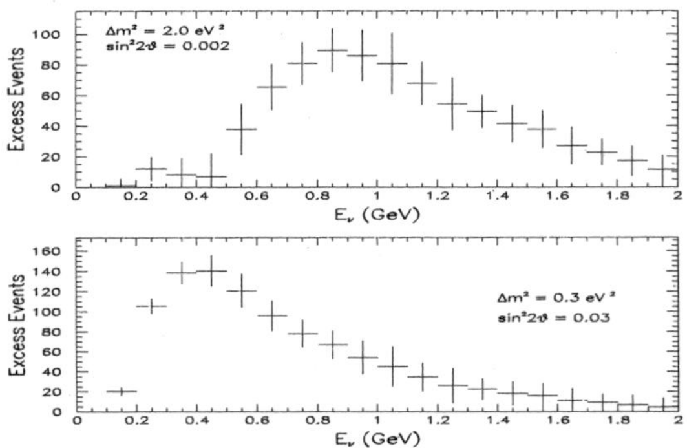

Figure 2. Expected spectra of the excess ν_e interactions for two possible oscillation scenarios

predicted from the LSND results, will consist of a sample of 1000 identified ν_e CC interactions. The main background will be due to the intrinsic ν_e component of the beam: it is expected that there will be 1275 events from muon decays and 425 events from kaons decays. The background calculations can be experimentally verified by changing the length of the decay region. An additional handle will be provided by the fact, that the sample of ν_e interactions due to $\nu_\mu \to \nu_e$ oscillations will have an energy distribution different from the ν_e component of the beam (see Fig.2).

The large size of the oscillation signal predicted by LSND will enable a precise measurement of the underlying oscillation parameters whereas an absence any signal will lead to greatly improved limits on possible $\nu_\mu \to \nu_e$ oscillations. This is shown in Fig.3.

4 MINOS Experiment: Measuring Oscillation Parameters in the SuperK region

The MINOS[4] experiment is designed to investigate neutrino oscillations in the region indicated by the atmospheric neutrino experiments. Two detectors, functionally identical, will be placed in the NuMI neutrino beam: one at Fermilab and the second one in the

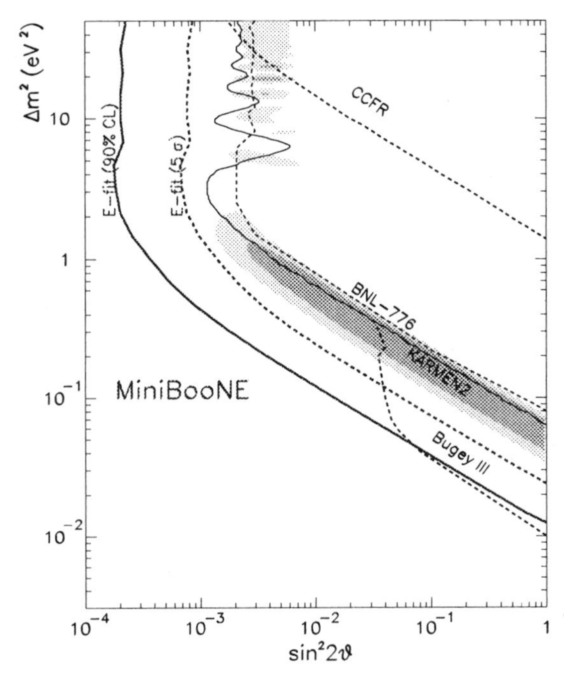

Figure 3. Sensitivity of the MiniBOONE experiment

Soudan iron mine, 732 km away.

4.1 Two Detectors Neutrino Oscillation Experiment

Two identical detectors placed in the same neutrino beam make the oscillation experiment relatively easy. The observed interactions of ν_μ can be divided into two classes: "CC"-like, with an identified μ track, and "NC"-like, muonless. In the absence of oscillations the ratio of the observed numbers of the CC and NC-like events in the two detectors must be the same, provided that the same classification algorithm is used. This remains quite true, even if the beam spectra at the two detector locations differ slightly, as the ratio $\frac{\sigma^{NC}}{\sigma^{CC}}$ is energy independent.

If ν_μ undergoes oscillations, some fraction, ξ, of the original ν_μ beam will arrive at the far detector in an 'oscillated' form. The CC interaction of this oscillated neutrino will not, in general, produce a μ in a the final state and they will be classified as "NC"-like interactions. The interaction cross section of this neutrino may be potentially reduced by a factor η with respect to the $\sigma^{CC}_{\nu_\mu}$.

A particularly sensitive measure of the oscillations is the double ratio $R \equiv \left(\frac{NC}{CC}\right)_{near} / \left(\frac{NC}{CC}\right)_{far}$:

$$R = \frac{1}{1-\xi}\left(1 + \frac{\eta\xi\frac{\sigma^{CC}}{\sigma_{NC}}}{1+\varepsilon\frac{\sigma^{CC}}{\sigma_{NC}}}\right) \qquad (1)$$

where ε is the fraction of the ν_μ CC interactions misclassified os NC events.

R combines the sensitivities of the disappearance experiment, the $\frac{1}{1-\xi}$ term, and the appearance experiment, the $\eta\frac{\sigma^{CC}}{\sigma_{NC}}$ term. In addition R has very small systematic uncertainty, as most of the neutrino flux uncertainties cancel. The value of ξ will be determined from the disappearance experiment, hence the value of R will provide additional information about the oscillation mode through the value of η:

$$\eta = \begin{cases} 1 & \nu_\mu \to \nu_e \\ 0.2-0.3 & \nu_\mu \to \nu_\tau, \text{ spectrum dependent} \\ 0 & \nu_\mu \to \nu_{sterile} \end{cases} \qquad (2)$$

4.2 MINOS detectors

The MINOS experiment will consist of two nearly identical detectors. One is located at the Fermilab site, some 500 meters behind the decay pipe. The second detector is located in northern Minnesota, 732 km from Fermilab in a new cavern, which is under construction in the Soudan mine, close to the existing Soudan II detector.

The far MINOS detector will consist of two supermodules, 2.7 kton each. They will be constructed from steel octagons, 8 m in diameter, with a toroidal magnetic field of about 1.5 T. Steel plates, 2.5 cm thick, will be interspersed with planes of scintillator strips to provide calorimetric measurement of the deposited energy with energy $\Delta E/E \sim 0.6/\sqrt{E}$ (E in GeV). The active detector elements will consist of strips of extruded scintillator, 1 cm thick, 4 cm wide and up to 8 m long. A wavelengthshifting fiber runs the length of each strip to collect the light and transmits it to a Hamamatsu M16 photomultiplier. The fine granularity of the scintillator strips will allow them to be used as a tracking detector to measure muon trajectories and determine the muon momentum from the curvature in the magnetic field.

The near detector, on the Fermilab site, will be as similar as practical to the far detector, except for its overall size.

The neutrino beam line and the MINOS detectors are under construction and data taking is expected to commence in 2003. The low energy beam is currently planned for the initial run of the experiment.

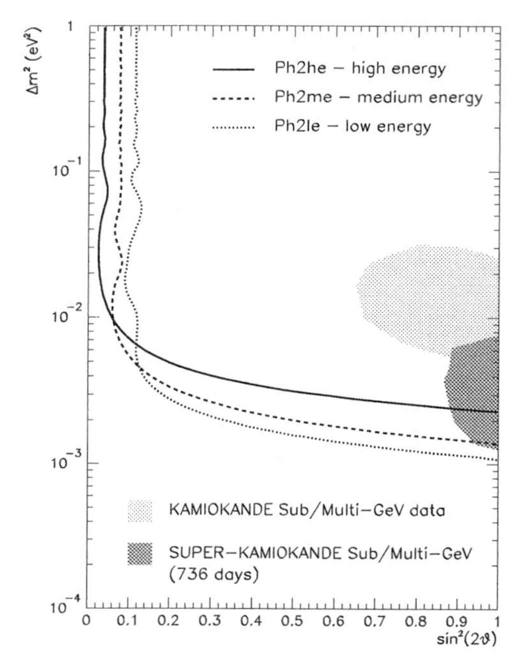

Figure 4. 90% C.L. limits on the $\nu_\mu \to \nu_\tau$ oscillations parameters for 2 years exposure

4.3 MINOS Physics Measurements

Two massive detectors and an intense neutrino beam constitute a powerful tool to investigate neutrino oscillations, especially when the beam energy can be chosen to maximize the oscillation signal. MINOS will perform several independent measurements, which will provide a clear and complete picture of the neutrino oscillations in the SuperK region. These measurements fall into three different categories:

- Firm evidence for ν_μ oscillations.

  In the presence of the SuperK-indicated effect we expect at least two evidences for the oscillations:

  - A value of double ratio R (Eq.1) different from one. The sensitivity of this measurement for different possible Δm^2 depends on the selected beam energy, as shown in Fig.4
  - A change of the ν_μ charged curent interaction rate and a characteristic **oscillatory** modification of the total energy spectrum observed at the far detector, as shown in Fig.5

  Near/far detector comparison will reduce the systematic uncertainties. The neutrino beam spectrum measurement with the near detector will constrain the predicted neutrino flux at the far detector.

- Measurement of the oscillation parameters: Δm^2 and $\sin^2 2\theta$.

  Fits of the observed depletion of the CC energy spectrum in Fig.5 will provide a precise estimate of the oscillation parameters. The expected precision of this determination depends somewhat on the oscillation scenario and on the choice of the beam. Two years exposure of the MINOS detectors will yield measurements with the precision illustrated in Fig.6.

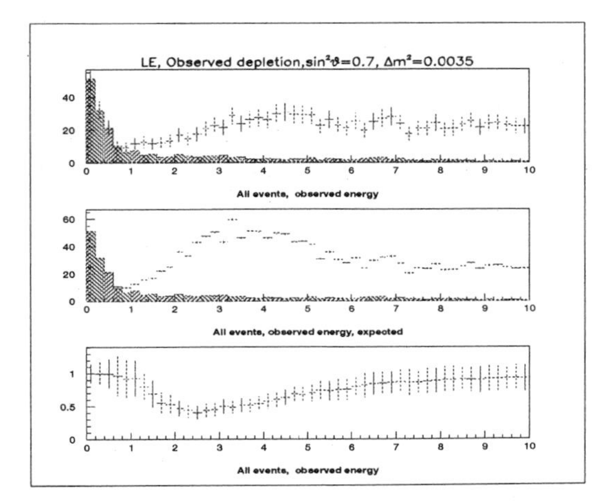

Figure 5. Observed spectrum of the identified ν_μ CC events in the presence (top) or absence (middle) of oscillations. Shaded histogram represents contribution of mis-identified NC events. Ratio of the observed and the expected distributions is shown at the bottom for 2 years exposure. It is a direct measurement of the $\nu_\mu$survival probability.

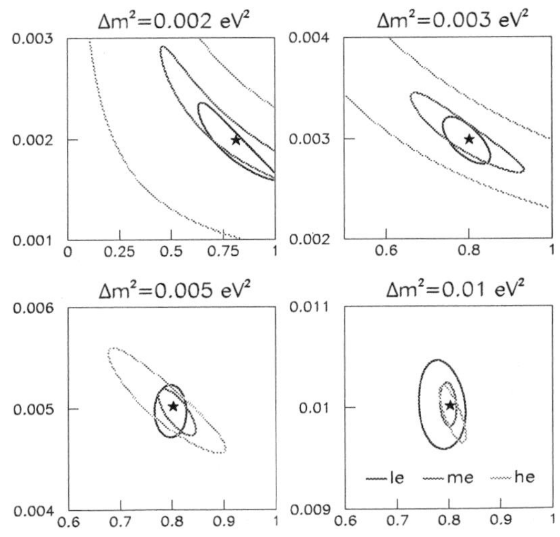

Figure 6. The 68% C.L. error contours in the $sin^2 2\theta = 0.8$ and Δm^2 plane for the expected oscillation signal with $sin^2 2\theta = 0.8$ and different Δm^2. Different contours represent measurements with different beams: low (le), medium (me) and high (he) energy and two years exposure.

- Determination of the oscillation mode(s)
  - $\nu_\mu \rightarrow \nu_{sterile}$?
    The large mixing angle indicated by the SuperK results leads to a significant contribution of the appearance term to R in Eq.1. The measurement of R will provide a decisive demonstration for or against sterile neutrinos as a dominant oscillation mode over the entire region of SuperK.
  - $\nu_\mu \rightarrow \nu_e$?
    The fine granularity of the MINOS detector will allow for identification of electromagnetic showers with an efficiency of $15 - 20\%$. The background from mis-identified NC interactions and the intrinsic ν_e component of the beam (expected to be of the order of 0.6% of the beam) will be measured with high accuracy by the near detector. Fig.7 shows the sensitivity of the MINOS detector to this oscillation mode in comparison with the limits from CHOOZ experiment.
  - $\nu_\mu \rightarrow \nu_\tau$?
    Indirect evidence for this oscillation mode will be provided by the measurement of R (Eq.1). For relatively high Δm^2, above 5×10^{-3} eV^2 a significant sample of CC ν_τ interactions can be identified by exclusive decay modes, like $\tau \rightarrow \pi$. The near detector will again be crucial in reliable determination of the unavoidable background from NC interactions.

5 Conclusions and Outlook

New experiments under construction at Fermilab will help to clarify the situation with

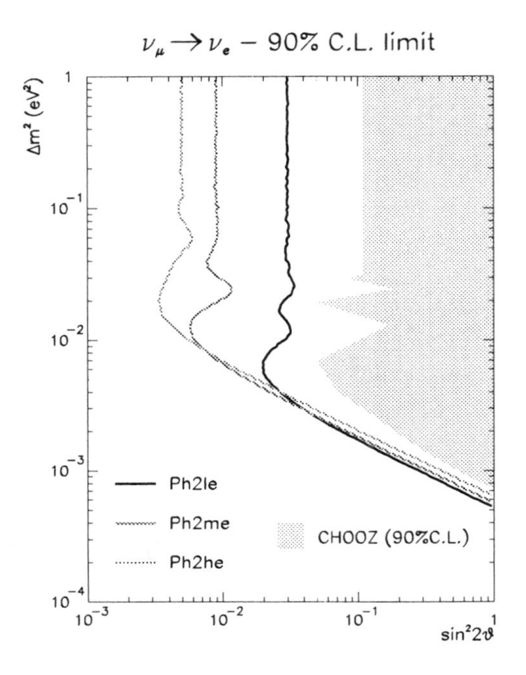

Figure 7. The 90% C.L. limits for $\nu_\mu \rightarrow \nu_e$ oscillation parameters for two years exposure with different beams: low (Ph2le), medium (Ph2me) and high (Ph2he) energy.

neutrino oscillations in the $\Delta m^2 > 0.001\text{eV}^2$ region. The MiniBOONE experiment will settle, within the coming 2-3 years, the issue of the LSND results either by precise determination of the underlying oscillation parameters or by setting limits far outside the LSND-allowed region. Within the ext 5-6 years the MINOS experiment will decisively establish the phenomenon of neutrino oscillations and measure precisely the corresponding mixing angles and Δm^2 values in the region indicated by the SuperK experiment. The question of the dominant oscillation mode: $\nu_\mu \rightarrow \nu_\tau$ or $\nu_\mu \rightarrow \nu_{sterile}$ will be settled. The sub-dominant mode $\nu_\mu \rightarrow \nu_e$ will be established or the existing CHOOZ limit will be significantly improved.

6 Acknowledgements

It is a pleasure to thank the organizers and to congratulate them for the flawless organization of such a pleasant and stimulating conference. I am indebted to my collegues in MINOS and MiniBOONE collaborations for valuable discussions.

References

1. Y. Fukuda *et al*, *Phys. Rev. Lett.* **81**, 1562 (1998), see also T.Toshito, these proceedings
2. C. Athanassopoulos *at al*, *Phys. Rev.* C **55**, 2079 (1997), for the latest results see also: R. Imlay, these proceedings
3. http://www-boone.fnal.gov/
4. http://www.hep.anl.gov/ndk/hypertext/minos_tdr.html

THE CERN TO GRAN SASSO NEUTRINO OSCILLATIONS EXPERIMENTS

A BETTINI

Gran Sasso National Laboratory and Padova University and INFN

The CERN to Gran Sasso neutrino oscillations program will be briefly described; I'll also give some information on the complementary approach using atmospheric muon neutrinos to improve our knowledge of the oscillation phenomenon.

Experiments on electron-neutrinos from the Sun and on atmospheric electron- and muon-neutrinos have shown that, very likely, neutrinos oscillate. This implies that, contrary to the assumptions of the Standard Model: 1. electron-, muon-, and tau-neutrinos are not stationary states; 2. at least two of the three stationary states (ν_1,ν_2, and ν_3) have non zero mass; 3. Flavour lepton numbers are not conserved. From the CHOOZ experiment and from SuperK we know that the atmospheric neutrino oscillation are likely mainly $\nu_\mu \leftrightarrow \nu_\tau$ with, possibly, a small component of $\nu_\mu \leftrightarrow \nu_e$. The square mass difference, as measured by Superk is 1.5×10^{-3} eV^2 $< \Delta m^2 < 5 \times 10^{-3}$ eV^2.

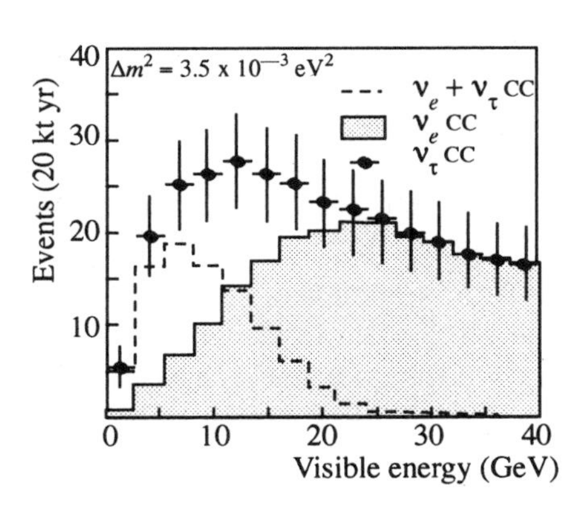

Figure 1. ICARUS visible energy distribution in the electron channel

We must now aim to definitively proof the oscillation phenomenon, to observe the new flavour (τ ?) appearance, to measure the mixing matrix elements and the square mass differences and to establish the very nature of the neutrinos whether they are Majorana or Dirac particles. In the program we are developing at the INFN Gran Sasso National Laboratory (LNGS) we are including new experiments on solar neutrino, double beta decay, neutrinos from supernovae, atmospheric muon-neutrinos and on the CNGS[1] (CERN Neutrinos to Gran Sasso) beam. Let me recall that LNGS consists in three large underground halls, about 100 m long, 20 m $\times$ 20 m across, easily accessible to large pieces of apparatus through the adjacent freeway tunnel. They have been oriented toward CERN to be able to host experiments on a neutrino beam. Services and shops are hosted in external buildings and halls. A review of the current program have led to the decision to decommission part of the running experiments, making almost half of the laboratory space free for new ones by mid 2001.

The CNGS program will allow a major forward step with the construction at CERN of an artificial, well controlled neutrino source and at LNGS of experiments, both optimised for tau-neutrino appearance. The neutrino beam[1], has been designed by a CERN-INFN group. The pions and kaons produced by the 400 GeV SPS proton beam will be focussed by a horn plus reflector system that is followed by a 1 km long decay tunnel, a hadron stop and muon detectors for beam characteristics determination. A close detector station is not needed for appearance experiments. Running in the "shared" mode,

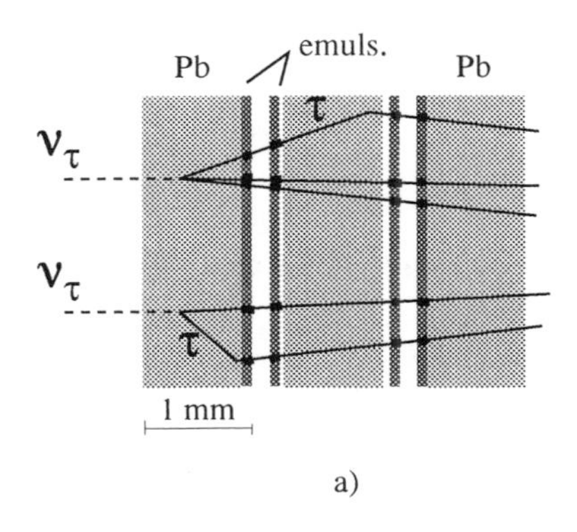

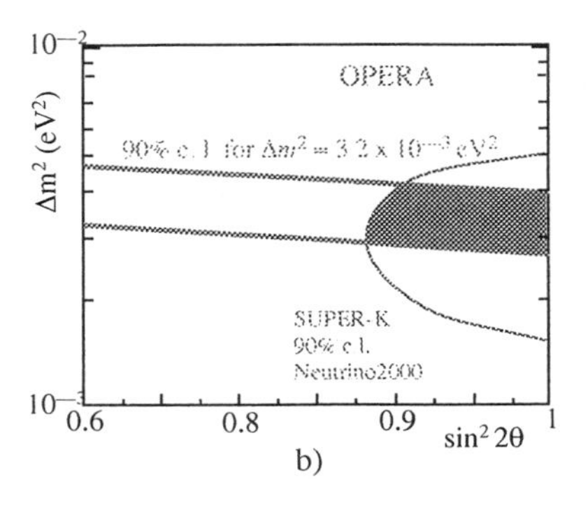

Figure 2. a) OPERA cell structure, and b) parameters determination

the beam will give 3200 CC ν_μ interactions per year in a kt fiducial mass detector at LNGS corresponding to 25 ν_τ interactions for $\Delta m^2 = 3.5 \times 10^{-3}$ eV^2 and maximum mixing (yields will be 1.7 times larger in a dedicated mode operation).

The charged daughters of τ's will be detected, in one or more decay channels: $\tau^- \to \mu^- \bar{\nu_\mu} \nu_\tau (18\%)$; $e^- \bar{\nu_e} \nu_\tau (18\%)$; $h^- \nu_\tau n \pi^0$ (50%); $2\pi^- \pi^+ \nu_\tau n \pi^0$ (14%). Two main background rejection tools are available: 1. the direct observation of τ decays requiring micrometer scale granularity and sub-micron resolution, provided only by the emulsion technique (as used by CHORUS and DONUT and proposed by OPERA); 2. the use of kinematic selection requiring good particle identification and good resolution in momentum unbalance (as used by NOMAD and proposed by ICARUS).

ICARUS[2] is a liquid argon time projection chamber providing bubble chamber quality 3D images of the events, continuous sensitivity, self-triggering capability, high granularity calorimetry and dE/dx measurement. The R&D program performed between 1991 and '95 on a 3 t detector solved all the major technical problems with the detector continuously running for several years, showing the reliability of the technique. The present phase aims to show the feasibility of a large scale detector. A 600 t unit is under construction and is expected to be operational by the end of this year. This will be the main milestone to proceed in the program. The safety issues connected with the installation of a large cryogenic volume underground are also being studied. The project foresees the construction of a large modular detector followed by spectrometer/calorimeter units to cover a broad physics program, of which I'll discuss only the CNGS ν_τ appearance part.

The electron channel will be the main one (but not the only) for the ICARUS search for τ appearance, that will be made by searching for an excess at low electron energies. The background due the ν_e component of the beam ($< 1\%$) will be small and the superior e/π^0 separation capability will be exploited to cut out ν_μ neutral current events. Fig. 1 shows the visible energy distribution as expected in a 20 kt yr exposure for $\Delta m^2 = 3.5 \times 10^{-3}$ eV^2 and maximum mixing. After kinematical cuts, 35 τ events will be detected with a residual background of 5 events. If the systematic uncertainty in background calculation, presently being studied, will, as expected, not be large, a 20 kt yr exposure will have more than 4σ discovery potential for $\Delta m^2 > 2 \times 10^{-3}$ eV^2.

OPERA[3] design is based on the ECC concept, which combines in one cell the high precision tracking capability of nuclear emulsion and the large target mass given by lead

plates. Fig. 2a shows the cell structure (1 mm thick Pb plate followed by a film made by two emulsion layers 50 μm thick on either side of a 200 μm plastic base). Two topologies of τ events are shown: "long", where the τ production and decay are separated by at least one film (detected via the decay angle), and "short", where they are not (detected via impact parameter).
Several cells are sandwiched in a structure called a *brick*, the basic element used to build vertical planar structures called *walls*. A wall followed by a set of tracker planes makes up a *module*. A *supermodule* is made of a sequence of modules, and a downstream *muon spectrometer*. Overall three supermodules are foreseen, providing a sensitive mass of around 2 kt.
The target trackers, made of scintillator strips, are used to identify the fired brick, for general tracking and for shower energy. A muon spectrometer is made of a dipole magnet (1.5 T) preceded and followed by precision trackers (drift tubes) and low resolution tracking planes (RPC) embedded in the magnet. It provides measurement of muon momentum and charge (to reduce charm background) and beam characteristics. Fired bricks will be removed and processed on day by day basis.
The ECC technique, that has recently lead DONUT to discover the ν_τ, will be further developed. The 176 000 m^2 of emulsion sheets will be industrially produced with an automatic chain and a new generation of automatic scanning/measuring devices will be produced by the collaboration. All the one-prong decay channels will, almost evenly, contribute to the final OPERA statistics. In a five year run, 18 identified τ decays with 0.6 background events are expected for $\Delta m^2 = 3.2\times10^{-3}$ eV^2 (the best fit of SuperK), 44 for 5×10^{-3} eV^2 and 4 for 1.5 $\times10^{-3}$ eV^2. Fig. 2b) shows the expected accuracy in the parameter space.

Atmospheric neutrino experiments are com-

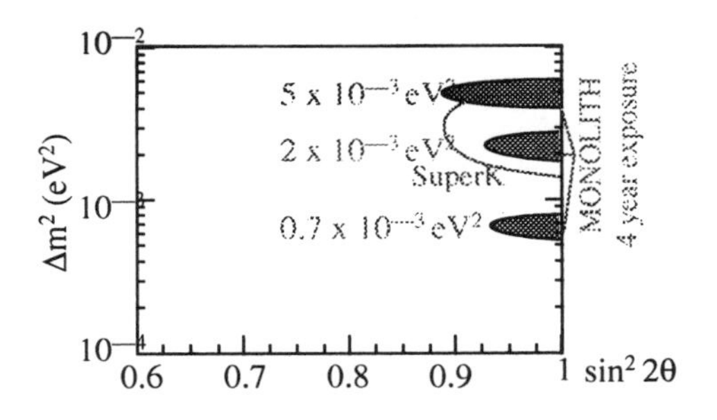

Figure 3. MONOLITH parameters determination

plementary to CNGS, as they become easier for lower values of Δm^2, especially below 1-2 $\times10^{-3}$ eV^2. To improve on SuperK one must measure as accurately as possible muon-neutrino energy E and flight length L, to have L/E. L is obtained from the neutrino direction, but in practice the μ direction is measured. To have a good correlation, one must use only μ's above a GeV or so, were the cosmic ray flux is low. As a consequence several kt mass detectors are needed. MONOLITH proposal is a 35 kt spectrometer made of 8 cm thick horizontal Fe magnetised (1.3 T) plates. The interleaved tracking planes have 1 cm spatial resolution and good (1 ns) timing, for up/down discrimination.
For a given direction, down-going ν_μ do not oscillate, while upward going do. The ratio between the two fluxes is almost MonteCarlo independent. Its measurement as a function of the angle with the superior MONOLITH L/E resolution, will allow the detection of the first oscillation period. It will also result in a substantial improvement in the knowledge of Δm^2, as shown in Fig. 3.

References.

[1] CERN 98-02, INFN/AE-98/05 and CERN-SL/99-034(DI),INFN/AE-99/05
[2] LNGS-P21/99, CERN/SPSC 99-25 and LNGS-P21/99.Add.1 and 2, CERN/SPSC 99-40
[3] CERN/SPSC 2000-028; SPSC/P318; LNGS P25/2000. July 10, 2000

REACTOR NEUTRINO OSCILLATION RESULTS AND THE STATUS OF KAMLAND

JUNPEI SHIRAI
Research Center for Neutrino Science, Tohoku University
Aoba-ku, Sendai-shi, 980-8578, JAPAN
E-mail: shirai@awa.tohoku.ac.jp

Recent results from reactor neutrino experiments and the present status of KamLAND experiment, now under construction in the Kamioka underground laboratory in Japan are presented.

1 Introduction

Non-zero mass of the neutrino discovered by Superkamiokande[1] in the studies of atmospheric neutrino, has suggested $\nu_\mu \to \nu_\tau$ or $\nu_\mu \to \nu_s$ oscillation but not $\nu_\mu \to \nu_e$ oscillation. As for the $\nu_e(\overline{\nu}_e)$ oscillation, there are two observations ; the solar neutrino deficits[2] and LSND results[3]. The former is explained by four candidate solutions which should be identified by further studies, and the latter has not been confirmed by other experiments. It is thus very important to study $\nu_e(\overline{\nu}_e)$ oscillation to make clear the situation.

Reactor neutrino experiments have played a unique role of studying $\overline{\nu}_e \to \overline{\nu}_x$ oscillation. This paper makes a brief report on the latest reactor neutrino oscillation experiments followed by an overview of a highly sensitive search which is going to be performed by KamLAND (Kamioka Liquid scintillator Anti-Neutrino Detector) with the present status.

2 Reactor neutrino experiment

Power reactors provide a unique experimental method to study $\overline{\nu}_e$ oscillation, because (i) the reactor is a pure $\overline{\nu}_e$ source, (ii) the $\overline{\nu}_e$ energy is much lower than that utilized in accelerator based experiment, and is preferable to search for small Δm^2, and (iii) $\overline{\nu}_e$ flux from a reactor is well understood with an uncertainty of around 3%.

The last item should be emphasized because it means that a front detector is not required to measure the $\overline{\nu}_e$ flux. The $\overline{\nu}_e$ is detected through the inverse-beta decay,

$$\overline{\nu}_e p \to e^+ n. \tag{1}$$

Thus, the reactor neutrino experiment, which is a disappearance experiment, provides a very reliable way to study $\overline{\nu}_e \to \overline{\nu}_x$ oscillation.

Expectation of the $\overline{\nu}_e$ flux and its spectrum requires the following information which is provided by power companies; (i) isotopic abundance of the fuel elements (the initial condition and the variation), and (ii) continuous values of the thermal power to know instantaneous fission rates

Based on the above informations, the $\overline{\nu}_e$ flux and the spectrum are calculated by computer codes. Although contribution to the uncertainties of the $\overline{\nu}_e$ flux from the fission rates of the fuel elements amounts to be less than 1%, the calculated flux is quite different between different codes which has caused dominant uncertainties. Therefore, experimental data about the beta spectrum emitted from the fission elements have been used to estimate the $\overline{\nu}_e$ flux.

There are two long-baseline (around 1km) reactor experiments recently performed by Chooz[4] and Palo Verde[5], using a large volume (more than 10 tons) liquid scintillator. The former experiment has estimated the total normalization error of 2.7% including the cross section of (1) and the detector efficiencies.

Comparing the observed and estimated positron spectra they found no evidence for $\overline{\nu}_e \rightarrow \overline{\nu}_x$ oscillation. The excluded regions[5] at 90%CL in Δm^2 vs $\sin^2(2\theta)$ plane have extended Δm^2 to around $10^{-3}eV^2$ at large $\sin^2 2\theta$ and consistent with the atmospheric neutrino oscillation observed by Superkamiokande not including $\nu_\mu(\overline{\nu}_\mu) \rightarrow \nu_e(\overline{\nu}_e)$.

The experiments have shown that the experimental techniques have been established for using a large volume liquid scintillator in long-baseline reactor neutrino experiments suggesting a good feasibility of a larger size liquid scintillator detector to perform $\nu_e(\overline{\nu}_e)$ experiment with much higher sensitivities.

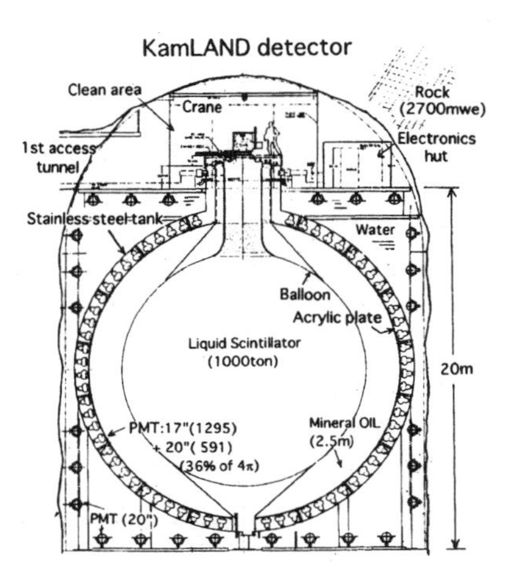

Figure 1. The KamLAND detector.

3 KamLAND experiment

KamLAND is a multi-purpose detector with a 1000 ton liquid scintillator aiming at measuring low energy neutrinos and anti-neutrinos with energies ranging from several MeV to sub-MeV. It is under construction in an underground laboratory located 1000m beneath the top of the Mt. Ikenoyama in the Kamioka mine in Japan where the Kamiokande detector used to be.

The location of the detector is nicely suited for studying very long-baseline reactor $\overline{\nu}_e$ oscillation experiment. The expected $\overline{\nu}_e$ flux at KamLAND is $1.3 \times 10^6/cm^2/s$, of which 80% comes from reactors locating at 175±35 km.

Figure 1 shows the detector. The central part is an 18m-diameter stainless steel tank containing the liquid scinntillator filled in a 13m-diameter plastic balloon. The balloon is suspended in mineral oil and viewed by 1886 17" and 20" PMTs at the inner surface of the tank covering 36% of 4π. The outside of the tank is a pure-water Čerenkov counter with 260 20" PMTs installed on the surrounding wall to identify cosmic ray muons.

The $\overline{\nu}_e$ are detected through the reaction (1) taking place in the liquid scintillator as a correlated pair of prompt e^+ signal and a delayed γ signal emitted from neutron capture process, np$\rightarrow \gamma$(2.2MeV)d, which takes place 180μs in average after the prompt one.

The newly developed 17"PMTs in the central detector have excellent characteristics of a one-photon signal with a typical time resolution of 1.5ns(σ) and a clear one-photon peak in puse height spectrum with a peak-to-valley ratio of 3.4. It has also very good linearity up to 500 photo-electron (p.e.) and does not show saturation above 10000 p.e. The liquid scintillator is a mineral oil-based pseudocumene with PPO (1.5g/l) as a fluor. The light output amounts to 190 p.e. for 1 MeV energy deposit in the center of the detector.

The experiment is carried out in very low environmental radioactivities. The heavy shields of the rock overburden (2700mwe) reduces the flux of cosmic ray muons by a factor 10^5 to around 0.3Hz penetrating the whole detector. The water vessel, stainless steel tank and the mineral oil (2.5m thick) further reduce the environmental radioactivities of surrounding rocks. The liquid scintillator and the mineral oil have very low radio-impurities, 10^{-13} gU/g or less in delivery.

The remaining radio-impurities are removed by a purification system using water extraction and nitrogen purging.

Figure 2 shows the expected e^+ energy spectrum in the KamLAND detector for the reactor $\overline{\nu}_e$ as well as backgorunds from the material inside the detector and from surrounding rocks. We can expect 450 reactor $\overline{\nu}_e$ events per year which might be overlapped in a region below 2.5 MeV by the earth $\overline{\nu}_e$ spectrum[6] (not shown here).

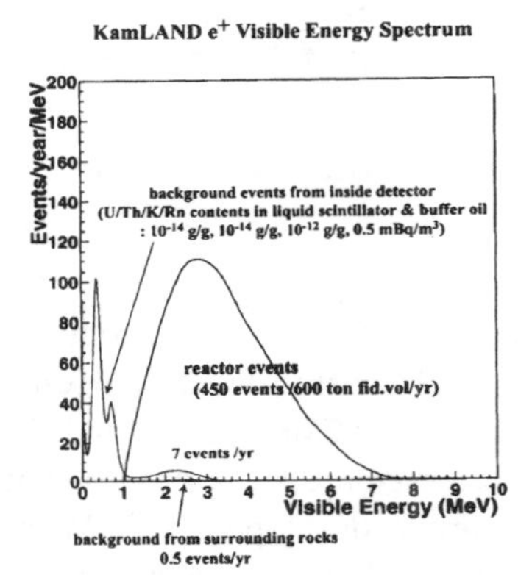

Figure 2. The expected e^+ spectrum of the reactor $\overline{\nu}_e$ events and backgrounds in the KamLAND detector.

Figure 3 shows the sensitivity of the $\overline{\nu}_e \rightarrow \overline{\nu}_x$ oscillation by 3 year operation of KamLAND with different background conditions. The search region extends to below $6 \times 10^{-6} eV^2$ at $\sin^2 2\theta$ around 1 and covers completely the LMA solution to the solar neutrino problem.

4 Status of KamLAND experiment

The KamLAND experiment was approved in 1997 by Monbusho (the ministry of education, science and culture) as one of the Center-of-Excellence research projects. Construction started in 1998 from removal of the old Kamiokande detector, enlarging the cavern and the main access road, as well as R&D works for 17"PMT, the liquid scintillaor, the balloon and the purification system followed

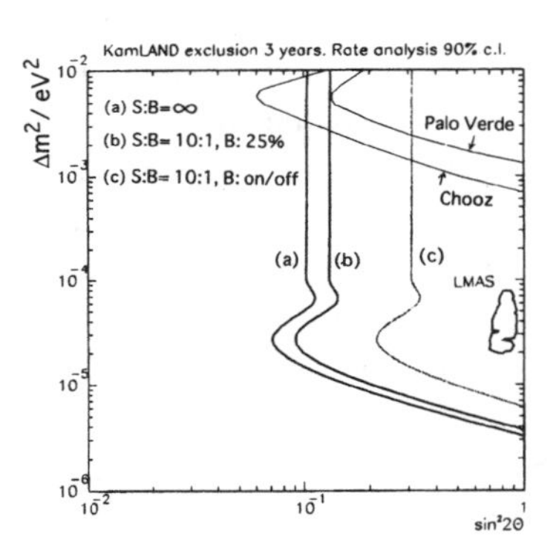

Figure 3. Sensitivity (90%CL) of the reactor $\overline{\nu}_e$ oscillation experiment by KamLAND and excluded regions by recent long-baseline reactor experiments

by mass production of PMTs and construction jobs of facilities lasting about 3 years. Now construction of the spherical tank (Figure 4) and the purification system (Figure 5) has been completed. A full-scale test balloon was installed in the tank (Figure 6) and filling test was successfully performed. At present PMTs are being installed in the tank (Figure 7). So far all the jobs are on schedule and they are now entering the final stage. After the balloon installation and oil filling, data taking will be started in the early stage of 2001.

Figure 4. The 3000m^3 spherical tank of the KamLAND detector.

Acknowledgments

The author would like to show special thanks to the organizing committee for providing

Figure 5. Water extraction and nitrogen purging towers of the purification system.

with an opportunity to attend the conference. The KamLAND experiment is supported by the Japanese Ministry of Education, Science and Culture and the U.S. Department of Energies.

References

1. Y.Fukuda et al., *Phys. Rev. Lett.* **81**, 1562 (1998).
2. Y.Fukuda et al., Phys. Rev. Lett. **82**, 2430 (1999).
3. C.Athanassopoulos et al., *Phys. Rev.* **C54**, 2685 (1996) and *Phys. Rev. Lett.* **77**, 3082 (1996).
4. M.Apollonio et al., *Phys. Lett.* **B 338**, 383 (1998) and *Phys. Lett.* **B 466**, 415 (1999).
5. F.Boehm et al., STANFORD-HEP-00-03, Mar 2000. 19pp., hep-ex/0003022.
6. R.S.Raghavan et al., *Phys. Rev. Lett* **80**, 635 (1998).

Figure 6. A full-scale test balloon in the spherical tank being inflated for a trial test of the liquid filling by using water.

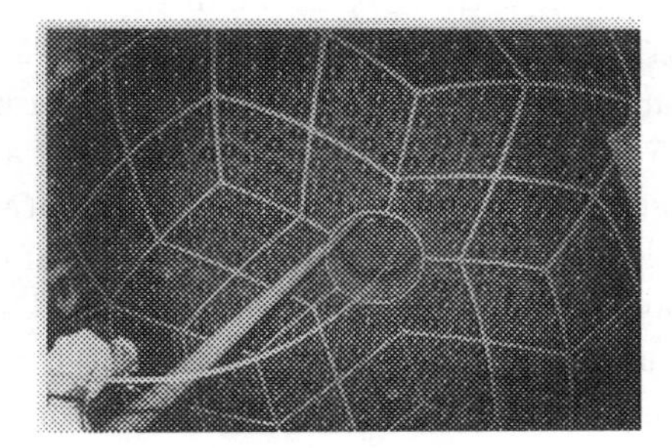

Figure 7. PMT installation in the spherical tank.

FINAL NEUTRINO OSCILLATION RESULTS FROM LSND

R.L. IMLAY
REPRESENTING THE LSND COLLABORATION
Lousiana State University Baton Rouge LA 70803 USA
E-mail: imlay@phzeus.phys.lsu.edu

The LSND experiment at Los Alamos has conducted searches for $\bar{\nu}_\mu \to \bar{\nu}_e$ oscillations using $\bar{\nu}_\mu$ from μ^+ decay at rest and for $\nu_\mu \to \nu_e$ oscillations using ν_μ from π^+ decay in flight. For the $\bar{\nu}_\mu \to \bar{\nu}_e$ search, a total excess of $83.3 \pm 21.2 \pm 12.0$ events is observed with e^+ energy between 20 and 60MeV. If attributed to neutrino oscillations, the most favored allowed region is a band from 0.2 to 2.0eV^2.

The LSND experiment has published evidence for neutrino oscillations for both $\bar{\nu}_\mu \to \bar{\nu}_e$ [1] and $\nu_\mu \to \nu_e$ [2] oscillations. In this report we present final oscillation results for the entire 1993-1998 data sample that combine the two oscillation searches in a global analysis and that makes use of a new event reconstruction that has greatly improved the spatial resolution. An excess of events consistent with neutrino oscillations is observed.

The primary oscillation search in LSND is for $\bar{\nu}_\mu \to \bar{\nu}_e$ oscillations, where the $\bar{\nu}_\mu$ arise from μ^+ decay at rest in the beam stop and the $\bar{\nu}_e$ are identified through the reaction $\bar{\nu}_e p \to e^+ n$. We use a two-fold signature of a positron and a correlated 2.2MeV γ from neutron capture on a free proton. The positron selection criteria for this primary oscillation search follow. In order to eliminate muon decay events, it is required that there be no event within 8μs in the future or within 12μs in the past. The particle identification parameter, χ_p, is required to lie in the range $-1.5 < \chi_p < 0.5$, where the precise range is determined by maximizing the acceptance divided by the square root of the beam-off background. There must be fewer than 4 veto hits associated with the event and the time of the nearest veto hit must be more than 30ns from the event time. The positron energy is required to satisfy $20 < E_e < 60$MeV. The reconstructed position must be more than 35cm from the nearest phototube surface. Finally, it is required that there be no more than

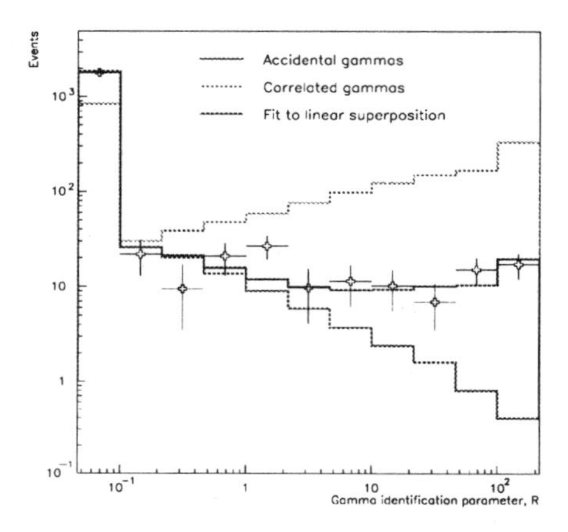

Figure 1. The R_γ distribution for events that satisfy the positron selection criteria.

one γ with $R_\gamma > 10$ (see below) in order to reduce the background from cosmic-ray neutrons, which typically knock-out additional neutrons.

The 2.2MeV γ identification is based on the likelihood ratio, R_γ, which is defined to be the likelihood that the γ is correlated divided by the likelihood that the γ is accidental. R_γ depends on three quantities: the number of hit phototubes associated with the γ, the distance between the reconstructed γ and positron positions, and the time between the γ and positron. Note that with the new reconstruction, the correlated γ efficiency has increased while the accidental γ efficiency has decreased. For $R_\gamma > 10$, the correlated and accidental efficiencies are 0.393 and 0.003, respectively, compared to 0.230 and 0.006 for

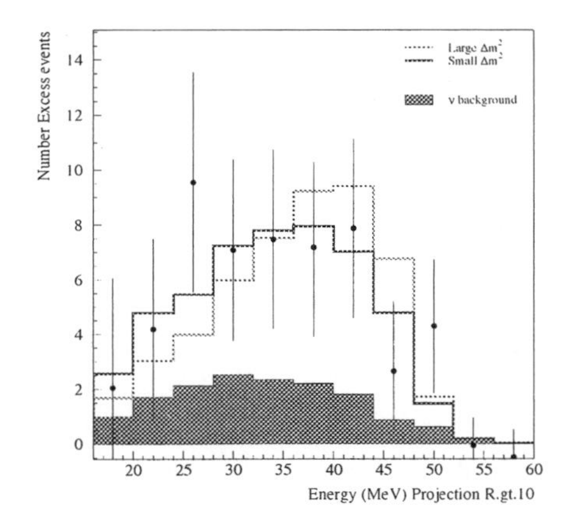

Figure 2. The energy distribution of events with $R_\gamma > 10$. The shaded region shows the estimated neutrino background. The curves show the expected distributions from a combination of neutrino background plus neutrino oscillations at high or low Δm^2.

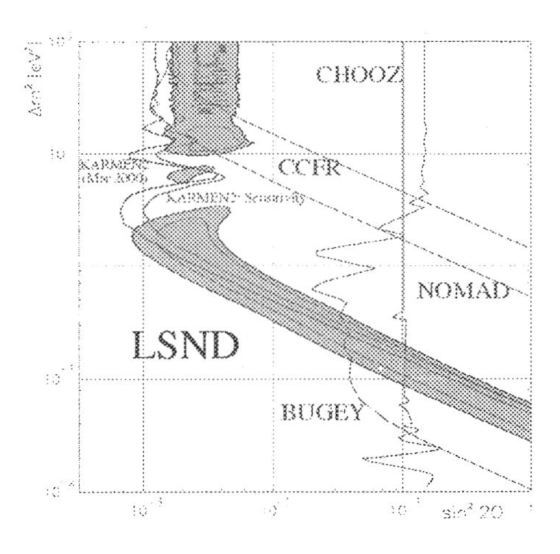

Figure 3. A Δm^2 vs. $\sin^2 2\theta$ oscillation parameter fit. Allowed 90% and 99% regions are shown.

the old reconstruction.

A fit to the R_γ distribution, shown in Fig. 1, gives a beam on-off excess of 113.3±21.2 events with a correlated neutron. Subtracting the neutrino background from μ^- decay at rest followed by $\bar{\nu}_e p \to e^+ n$ scattering (21.6 events) and π^- decay in flight followed by $\bar{\nu}_\mu p \to \mu^+ n$ scattering (8.4) events leads to a total excess of 83.3±21.2 events or an oscillation probability of $(0.25 \pm 0.06 \pm 0.04)\%$.

A fairly clean sample of oscillation candidates can be obtained by requiring $R_\gamma > 10$. Fig. 2 displays the energy distribution for $R_\gamma > 10$. The shaded region shows the estimated neutrino background while the curves show the expected distributions from a combination of neutrino background plus neutrino oscillations at high or low Δm^2. The data agree well with the oscillation hypothesis.

The secondary oscillation search in LSND is for $\nu_\mu \to \nu_e$ oscillations, where the ν_μ arise from π^+ decay in flight in the beam stop and the ν_e are identified through the reaction $\nu_e C \to e^- X$. The electron selection criteria for this primary oscillation search is almost the same as for the primary search, except that the electron energy is required to be in the range $60 < E_e < 200$MeV and there must be no associated 2.2MeV γ.

A Δm^2 vs. $\sin^2 2\theta$ oscillation parameter fit for the entire data sample, $20 < E_e < 200$MeV, is shown in Fig. 3. The fit includes both $\bar{\nu}_\mu \to \bar{\nu}_e$ and $\nu_\mu \to \nu_e$ oscillations, as well as all known neutrino backgrounds. The allowed regions correspond to 90% and 99% CL, while the curves are 90% CL limits from the Bugey reactor experiment, CCFR, NOMAD, and KARMEN. The most favored allowed region is the band from 0.2 to 2.0eV^2.

The LSND experiment provides evidence for neutrino oscillations from both the primary $\bar{\nu}_\mu \to \bar{\nu}_e$ oscillation search and the secondary $\nu_\mu \to \nu_e$ oscillation search. The MiniBooNE experiment at Fermilab, which is presently under construction, will provide a definitive test of the LSND results.

References

1. C. Athanassopoulos *et al.*, *Phys. Rev.* *C***54**, 2685 (1996).
2. C. Athanassopoulos *et al.*, *Phys. Rev.* *C***58**, 2489 (1998).

FOUR NEUTRINO OSCILLATION ANALYSIS OF ATMOSPHERIC NEUTRINO DATA AND APPLICATION TO LONG BASELINE EXPERIMENTS

OSAMU YASUDA

Department of Physics, Tokyo Metropolitan University
1-1 Minami-Osawa Hachioji, Tokyo 192-0397, Japan
E-mail: yasuda@phys.metro-u.ac.jp

Analyis of the Superkamiokande atmospheric neutrino data is presented in the framework of four neutrinos without imposing constraints of Big Bang Nucleosynthesis. Implications to long baseline experiments are briefly discussed.

1 Four neutrino analysis of atmospheric neutrinos

Four neutrino mixing schemes have caught much interest, since they are the simplest scenario which accounts for the solar and atmospheric neutrino problems and the LSND data[1] in the framework of neutrino oscillations. To reconcile the data at 90%CL of LSND, Bugey and CDHSW one has to have two degenerate massive states[2,3] ($m_1^2 \simeq m_2^2 \ll m_3^2 \simeq m_4^2$). For simplicity I assume $\Delta m_{21}^2 = \Delta m_{\odot}^2$, $\Delta m_{43}^2 = \Delta m_{\rm atm}^2$ and I adopt the notation in [2] for the 4×4 MNS matrix:

$$\begin{pmatrix} \nu_e \\ \nu_\mu \\ \nu_\tau \\ \nu_s \end{pmatrix} = \begin{pmatrix} U_{e1} & U_{e2} & U_{e3} & U_{e4} \\ U_{\mu 1} & U_{\mu 2} & U_{\mu 3} & U_{\mu 4} \\ U_{\tau 1} & U_{\tau 2} & U_{\tau 3} & U_{\tau 4} \\ U_{s1} & U_{s2} & U_{s3} & U_{s4} \end{pmatrix} \begin{pmatrix} \nu_1 \\ \nu_2 \\ \nu_3 \\ \nu_4 \end{pmatrix}.$$

For the range of the Δm^2 suggested by the LSND data at 90%CL, which is given by 0.2 eV^2 $\lesssim \Delta m_{\rm LSND}^2 \lesssim$ 2 eV^2 when combined with the data of Bugey E776 and KARMEN2, the constraint by the Bugey data gives $|U_{e3}|^2 + |U_{e4}|^2 \ll 1$. Also $|U_{s3}|^2 + |U_{s4}|^2$ has to be very small[2,4] if one demands that the number N_ν of effective neutrinos in Big Bang Nucleosynthesis (BBN) be less than four. In this case the solar neutrino deficit is explained by $\nu_e \leftrightarrow \nu_s$ oscillations with the Small Mixing Angle (SMA) MSW solution and the atmospheric neutrino anomaly is accounted for by $\nu_\mu \leftrightarrow \nu_\tau$. However, some people have given conservative estimate for N_ν and if their estimate is correct then $|U_{s3}|^2 + |U_{s4}|^2 \ll 1$ may no longer hold. Recently the Superkamiokande group has reported their result[5] on the solar neutrino experiment which indicates that the SMA and the Vacuum Oscillation solutions are disfavored. Meanwhile Giunti, Gonzalez-Garcia and Peña-Garay[6] have analyzed the solar neutrino data in the four neutrino scheme without BBN constraints. Their updated results show that the SMA solution exists for $0 \le c_s \lesssim 0.8$ ($c_s \equiv |U_{s1}|^2 + |U_{s2}|^2$), while the Large Mixing Angle (LMA) and LOW solutions survive only for $0 \le c_s \lesssim 0.4$ and $0 \le c_s \lesssim 0.2$, respectively. In this talk I will present some of the updated results of my work[7] on the four neutrino oscillation analysis of the Superkamiokande atmospheric neutrino data and I will briefly give some implications to long baseline experiments, assuming that the solar neutrino deficit is solved by the LMA solution. For details of the analysis and the references see [7]. In the analysis of atmospheric neutrinos, the effect of $\Delta m_{\odot}^2$ is negligible, so I assume $\Delta m_{21}^2 = 0$ and $U_{e3} = U_{e4} = 0$ for simplicity. To avoid contradiction with the CDHSW data, I will take Δm_{32}^2=0.3eV^2 as a reference value. The best fit to the atmospheric neutrino data[8] for 1144 days is obtained for $\Delta m_{43}^2 = 2.0 \times 10^{-3}$eV^2, $(\theta_{24}, \theta_{34}, \theta_{23}) = (45°, -30°, 20°)$, $\delta_1 = 45°$ and the allowed region at 90%CL is obtained

(The allowed region for 1144 day data is almost the same as for 990 day data). As a sample let me show the result for the case of $\theta_{24} = 45°$, $\delta_1 = 90°$. The shadowed area in Fig.1 (a) is the 90%CL allowed region projected on the $(\theta_{34}, \theta_{23})$ plane for various values of Δm^2_{43}. If I demand that the solar neutrino deficit be solved by the LMA solution, then $c_s \lesssim 0.4$ which is depicted in Fig.1 (b). Combining the results on the atmospheric neutrinos and the solar neutrinos, the allowed region becomes the shadowed area in Fig.1 (c). It turns out that if I require $c_s \lesssim 0.4$ then the solution prefers relatively large θ_{23} for any value of δ_1.

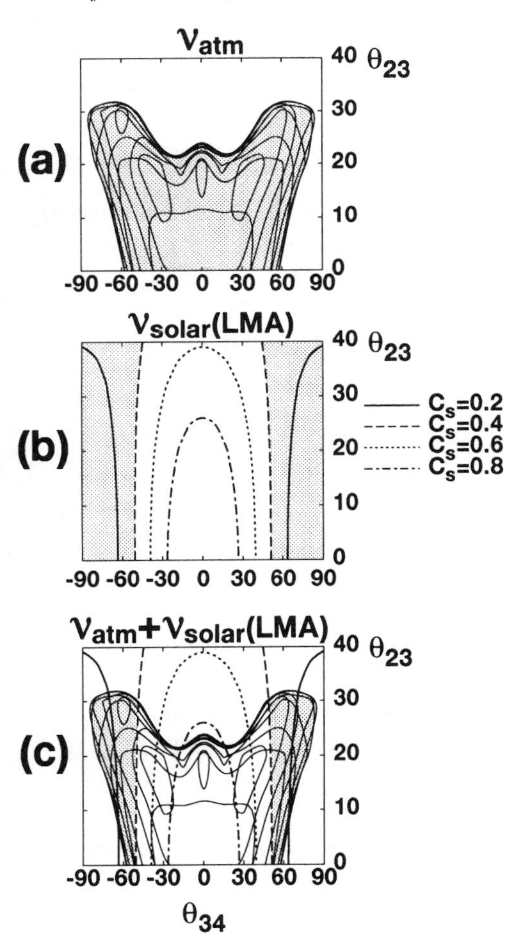

Figure 1. Allowed region for $\delta_1 = \pi/2$, $\theta_{24} = \pi/4$

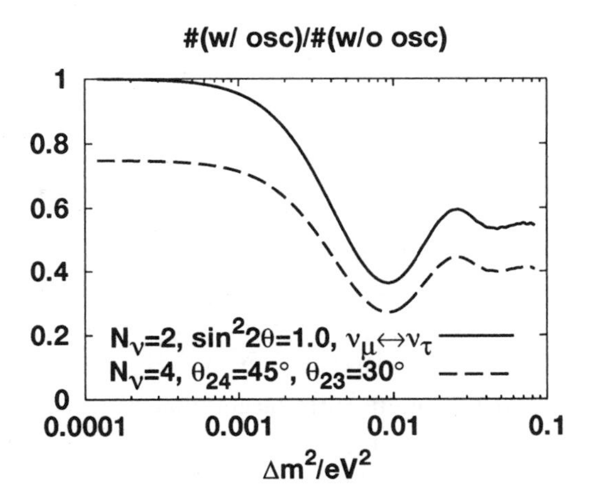

Figure 2. The ratio r at K2K

2 Application to long baseline experiments

First implication of the present scheme is the deficit which is expected to be seen at the K2K experiments. Ignoring the matter effect, I have for the neutrino energy $E \sim 1$GeV and the path length L=250km

$$P(\nu_\mu \to \nu_\mu) = 1 - 2|U_{\mu 2}|^2(1 - |U_{\mu 2}|^2) \\ -4|U_{\mu 3}|^2|U_{\mu 4}|^2 \sin^2\left(\frac{\Delta m^2_{\rm atm}L}{4E}\right), \quad (1)$$

where $|U_{\mu 2}| = |c_{24}s_{23}|$ is not necessarily small. The behavior of the ratio $r \equiv$#(CC + NC events with oscillations)/ #(CC + NC events without oscillations) as a function of $\Delta m^2_{\rm atm}$ is given in Fig.2. Because of the contribution of the first term on the RHS of (1) the ratio is lower for the present four neutrino scheme than for the standard two flavor case with the same $\Delta m^2_{\rm atm}$.

Second implication is the possible measurement of CP violation at the JHF experiment[9] which is expected to start from 2006. Since there is no strong constraint on the mixing angles θ_{24}, θ_{34}, θ_{23}, CP violation in the channel $\nu_\mu \to \nu_s$ could be large. If we measure the neutral current π^0 production for ν_μ and $\bar{\nu}_\mu$ beams and compare them[10],

Table 1. Yields of NC π^0 at JHF with 10^{21} POT

	no osc	$\delta_1 = \frac{\pi}{2}$	$\delta_1 = 0$	$\delta_1 = -\frac{\pi}{2}$
N_ν	393	357	311	282
$N_{\bar\nu}$	199	146	158	184

then the absolute value of the ratio

$$R \equiv \frac{\left.\frac{N_\nu(\delta_1)}{N_{\bar\nu}(\delta_1)}\right|_{\rm dat} - \left.\frac{N_\nu(\delta_1=0)}{N_{\bar\nu}(\delta_1=0)}\right|_{\rm MC}}{\left.\frac{N_\nu(\delta_1)}{N_{\bar\nu}(\delta_1)}\right|_{\rm dat} + \left.\frac{N_\nu(\delta_1=0)}{N_{\bar\nu}(\delta_1=0)}\right|_{\rm MC}}$$

could be significantly larger than the statistical fluctuation (~ 0.05 for 10^{21} POT) (The yields is given in Table 1 for an optimistic set of the oscillation parameters ($\delta_1 = 90°$, $\theta_{24} = 40°$, $\theta_{23} = 30°$, $\theta_{34} = 60°$, $\Delta m^2_{\rm atm} = 1.6 \times 10^{-3}{\rm eV}^2$). The ratio R is depicted in Fig.3 where the same set of the oscillation parameters as in Table 1 and the Wide Band Beam[9] is assumed in calculations).

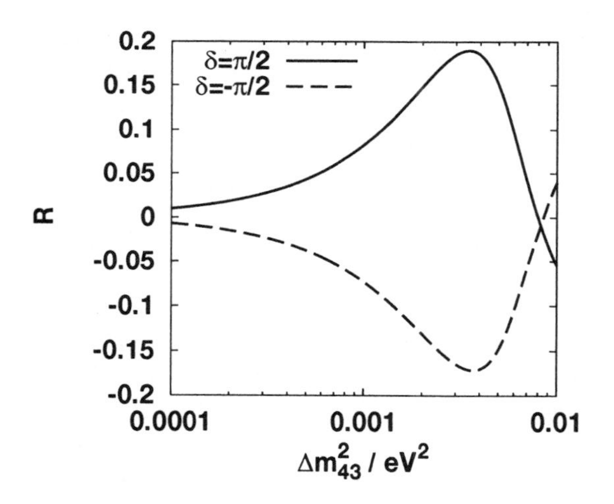

Figure 3. The ratio R at JHF

3 Conclusions

I have presented my result on the four neutrino oscillations of the Superkamiokande atmospheric neutrino data without assuming the BBN constraints. By combining the analysis on the solar neutrino data by Giunti et al. and assuming the LMA solar solution, I found that there is relatively large contribution of $\Delta m^2_{\rm LSND}$ in the atmospheric neutrino oscillations. It suggests that the number of events at K2K is less than the one for the standard two flavor scenario for the same $\Delta m^2_{\rm atm}$ and that CP violation could be measured at JHF after running for several years with 10^{21}POT/yr.

Acknowledgments

I would like to thank T. Nakaya for pointing out the NC π^0 channel to me to measure CP violation. This research was supported in part by a Grant-in-Aid for Scientific Research of the Ministry of Education, Science and Culture, #12047222, #10640280.

References

1. G. Mills, talk at Neutrino2000, http://nu2000.sno.laurentian.ca/G.Mills/.
2. N. Okada and O. Yasuda, Int. J. Mod. Phys. **A 12**, 3669 (1997).
3. S.M. Bilenky , C. Giunti and W. Grimus, Eur. Phys. J. **C1**, 247 (1998).
4. S.M. Bilenky et al., Astropart. Phys. **11**, 413 (1999).
5. Y. Suzuki, talk at Neutrino2000, http://nu2000.sno.laurentian.ca/Y.Suzuki/.
6. C. Giunti et al., Phys. Rev. **D62**, 013005 (2000); M. C. Gonzalez-Garcia, talk in these proceedings, http://ichep2000.hep.sci.osaka-u.ac.jp/scan/0728/pa08/gonzalez_garcia/.
7. O. Yasuda, hep-ph/0006319.
8. H. Sobel, talk at Neutrino2000, http://nu2000.sno.laurentian.ca/H.Sobel/.
9. Y. Itow et al., Letter of Intent, http://www jhf.kek.jp/JHF_WWW/LOI/jhfnu_loi.ps.
10. T. Nakaya, private communication.

THE EFFECT OF A SMALL MIXING ANGLE IN THE ATMOSPHERIC NEUTRINOS

S. MIDORIKAWA

Faculty of Engineering, Aomori University, 2-3-1 Kobata, Aomori 030-0943, Japan
E-mail: midori@aomori-u.ac.jp

The effect of matter enhanced neutrino oscillations on atmospheric neutrinos is investegated systematically in the framework of the one mass dominant model of three flavors. The resonance conditions of neutrinos crossing the earth are determined by the three parameters, namely, the zenith angle, $\Delta m^2/E$, and the mixing angle θ_3 of the electron neutrinos with tau neutrinos. The values of the triplet parameters are given numerically and the feasibility to determine the mixing angle is discussed.

It was more than ten years ago that the atmospheric neutrino anomaly [1] could be explained by the $\nu_\mu \leftrightarrow \nu_\tau$ oscillations[2].

This interpretation has been confirmed by the observation of the zenith angle dependence of the neutrino fluxes by the SuperKamiokande[3]. While the zenith angle dependence of the electron-like events is consistent with the theoretical predictions[4], that of the muon-like events disagree with the theory, especially for the up-going events.

The neutrino state with flavor basis and that of mass basis is related to each other by the mixing matrics U. The results of the Super-Kamiokande shows the almost $\nu_\mu \to \nu_\tau$ oscillations, meaning that $U_{e3} \equiv \sin\theta_3 \simeq 0$, which is consistent with the CHOOS results[5].

Although θ_3 is small, it will reaveal a sizable effect when neutrinos go through resonance oscillations[6] in the earth. Matter enhanced oscillations has also been applied to the atmospheric neutrinos[7]. However, it seems that no systmatic study has not been made yet, especially for the three flavor model of neutrinos.

The purpose of this paper is to reveal the conditions that amplify the effect of θ_3 through matter oscillations so that we can determine this small mixing angle.

I assume that the masses of neutrinos are hierarchical. Furthermore I assume that the atmospheric neutrino oscillations are derived by $\delta m_{32}^2 \equiv m_3^2 - m_2^2$. The solar neutrino oscillations are attributed to $\delta m_{21}^2 \equiv m_2^2 - m_1^2$, and is much smaller than δm_{32}^2. Thus, we can safely put $\delta m_{21}^2 = 0$. The only relevant mass parameter is $\Delta m^2 \equiv \delta m_{31}^2 = \delta m_{32}^2$. The parameters contained in the model are θ_2, θ_3, and Δm^2. In calculating the matter effect of the earth, I use the density profile referred to as the PREM model [8].

The prominent feature of our model is that the survival probability of electron neutrinos $P(\nu_e \to \nu_e)$ is completely determined by the three parameters, *i.e.*, θ_3, $\Delta m^2/E$ where E is the neutrino energy, and the zenith angle z. It is natural to define the resonance condition as

$$P(\nu_e \to \nu_e) = 0 \qquad (1)$$

The resonance states form curves in the the three parameter space (θ_3, $\Delta m^2/E$, $\cos z$) corresponding to discretization of the resonance wavelength. In a previous work[9] I solve eq.(1) numerically, and show the values of the three parameters at the resonance.

To proceed further, asuume that $\Delta m^2 = 3 \times 10^{-3}\ eV^2$ and $\theta_2 = 45°$, and use the fluxes of Honda et al. of ref.4. As an examle, choose $\cos z = -0.9$. The resonance occurs[9] at $\theta_3 = 6.4°$ and $\Delta m^2/E = 6.1 \times 10^{-4}$, corresponding to $E = 5\ GeV$. Fig.1 represents how the $(\nu_e + \bar{\nu_e})$ flux destorts as θ_3 varies. The matter effects appear as an excess of the upward-going ν_e, which decrease as θ_3 devi-

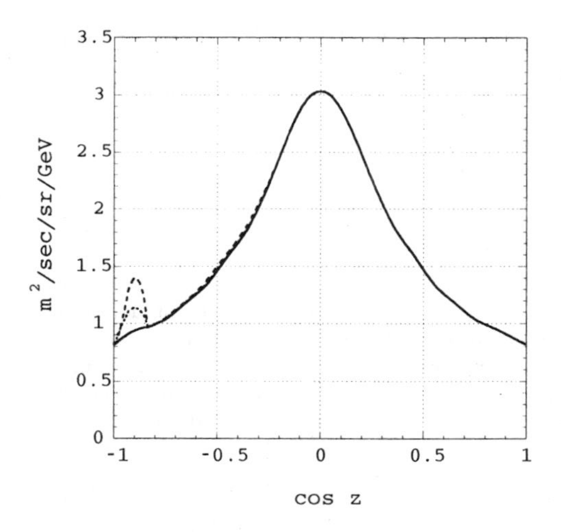

Figure 1. Zenith angle dependence of the $\nu_e + \bar{\nu_e}$ fluxes at E = 5 GeV. solid line: no oscillations. dashed line: $\theta_3 = 6.4°$. dotted line: $\theta_3 = 3°$.

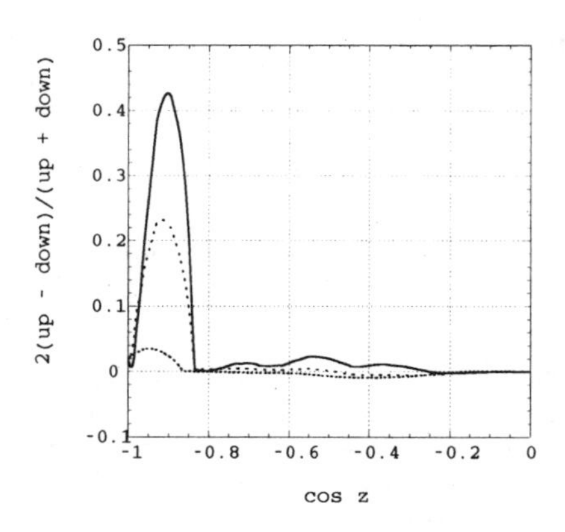

Figure 2. Zenith angle dependence fo the up-down asymmetry of $\nu_e + \bar{\nu_e}$ flux. The notation is the same as in Fig.1.

ates from the resonance.

It is more approapriate to use the up-down asymmetry of the flux [10], since the ambiguity of the flux ratio is much less than the absolute fluxes[11]. The result is shown in Fig.2. The up-down ratio amounts to 0.43 at $\theta_3 = 6.4°$, which could be accessible in the feature experimets. If we can measure θ_3 by the above method, this in turn becomes the experimental test of the MSW effect.

This research was supported in part by the Grant-in-Aid for Scientific Research of the Ministry of Education, Scinece and Culture of Japan No. 10640281.

References

1. K.S. Hirata et al., Phys. Lett. B**205**, 416 (1988).
2. J. Learned, S. Pakvasa, and T. Weiler, Phys. Lett. B**207**, 79 (1988); V. Berger and K. Whisnant, Phys. Lett. B**209**, 365 (1988); K. Hidaka, M. Honda, and S. Midorikawa, Phys. Rev. Lett. **61**, 1537 (1988).
3. Y. Fukuda et al., Phys. Rev. Lett. **81**, 1562 (1998).
4. G. Barr et al., Phys. Rev. D**39**, 3532 (1989); V. Agrawal et al., Phys. Rev. D**53**, 1314 (1996). M. Honda et al., Phys. Lett. B**248**, 193 (1990); M. Honda et al., Phys. Rev. D**52**, 4985 (1995).
5. M. Apollonio et al., Phys. Lett. B**420**, 397 (1998).
6. S.P. Mikheyev and A.Yu. Smirnov, Yad. Fiz. **42**, 1441 (1985) [Sov. J. Nucl. Phys. **42**, 913 (1985)]; L. Wolfenstein, Phys. Rev. D**17**, 2369 (1978).
7. E.D. Carlson, Phys. Rev. D**34**, 1454 (1986). J. Pantaleone, Phys. Rev. Lett. **81**, 5060 (1998); M.V. Chizhov and S.T. Petcov, Phys. Rev. Lett. **83**, 1096 (1999); E.Kh. Akhmedov et al., Nucl. Phys. B**3542**, 1454 (1986); T. Ohlsson and H. Snellman, Phys. Lett. B**474**, 153 (2000);
8. A.D. Dziewonski and D.L. Anderson, Physics of the Earth and Planetary Interiors **25**, 297 (1981).
9. S. Midorikawa, hep-ph/005008, to be published in the 1st Workshop on *Neutrino Oscillations and their Origin*, February 11-13, 2000.
10. J.W. Flanagan et al., Phys. Rev. D**57**, 2649 (1998).
11. T.K. Gaisser et al., Phys. Rev. D**54**, 5578 (1996).

PHYSICS AT NEUTRINO FACTORIES

K. J. PEACH
CLRC - Rutherford Appleton Laboratory, Chilton near Didcot, Oxon. OX11 0QX, UK
E-mail: Ken.Peach@rl.ac.uk

There is increasing interest in using intense neutrino beams from a high-energy muon storage ring - the Neutrino Factory - to make precise measurements of the lepton mixing matrix, including the T-violating phase, as well as a diverse programme of other physics.

1 Introduction

The interest in neutrino beams from decays of stored muons has been stimulated by two recent developments. Firstly, the evidence for neutrino oscillations with large mixing angles has become stronger, and it is possible that CP-violation (or T-violation) in the lepton sector might be experimentally accessible with terrestrial experiments. Secondly, technical developments originally in the context of the design of a high-energy muon collider indicate that a 'Neutrino Factory' based on a high-intensity muon storage ring is feasible. Over the past two years, there have been several studies[?] and workshops[?] devoted to the neutrino factory. A three-stage scenario has emerged for facilities based on high intensity muon storage rings[a]. The first stage is a neutrino factory, requiring less cooling than a collider. The second stage would be a 'Higgs factory' in the range 100-500GeV, and the final stage could be a multi-TeV muon collider.

2 The Neutrino Factory

The neutrino factory consists of a series of technically challenging components.
(a) A high intensity proton source (to a good approximation, the muon flux is proportional to the power incident on the target);
(b) a high power target ($\simeq$ 4MW);
(c) a pion and muon production, capture and decay channel;
(d) a muon cooling channel for subsequent acceleration and storage;
(e) the acceleration of the muons to the stored energy, likely to be 20-50Gev;
(f) a muon storage ring, yielding $\simeq 10^{21}$ muon decays.
There are fierce technical challenges in many of these areas, particularly in the targetry and cooling, and there is a significant R&D activity underway to address these issues - see the talk by Tigner[?].

The stored muon energy is critical, determining both physics potential and cost, since a large part of the cost comes from RF requirements of the acceleration stage. For physics reasons, the muon energy should be well above the threshold for τ appearance experiments from electron and muon flavours, and the neutrino cross-section, and thus the statistics, grows linearly with energy. (In the absence of background, the reduction in the oscillation amplitude with increasing energy is offset by the increase in flux through the decreased beam divergence.) A 'figure of merit' entering the cost and scientific potential optimisation for oscillation studies is $E_\mu \times I_\mu \times M_D$, where E_μ and I_μ are the energy and intensity of the stored muons, and M_D is the target mass of the neutrino detector.

3 Neutrino oscillations

The current status of the lepton mixing matrix has been extensively reviewed at this conference[?]. Briefly, for three generations of neutrinos, there are seven parame-

[a]Muon storage rings have, of course, been used for muon g-2 experiments - see the talk by Carey[?]

ters which describe the Maki-Nagawa-Sakata (MNS) neutrino mixing matrix U connecting the mass eigenstates $|\nu_i>$ and flavour eignestates $|\nu_\alpha>$ through $|\nu_\alpha>= U_{\alpha i}|\nu_i>$, and which govern the neutrino masses and oscillations. These may be expressed as two independent mass differences (Δm^2_{21} and Δm^2_{32}), three mixing angles (θ_{12}, θ_{23} and θ_{13}), one T-violating phase (δ) and an absolute mass scale. In addition, if neutrinos are Majorana particles, there are two additional phases (because two fewer phases may be absorbed into the lepton field definition); however, these do not contribute to neutrino oscillations but may be measurable through, for example, neutrinoless double β decay.

A neutrino factory in principle can make 12 independent measurements, namely

$$\mathrm{e} \to \mathrm{e}\ ;\ \mathrm{e} \to \mu\ ;\ \mathrm{e} \to \tau$$

$$\mu \to \mathrm{e}\ ;\ \mu \to \mu\ ;\ \mu \to \tau$$

for both neutrinos and antineutrinos. Thus, if a detector could provide both lepton flavour and lepton charge identification, six of the parameters could be over determined, leaving only the absolute mass-scale (and any Majorana phases) unconstrained. Note that (see for example Lindner[?]) the phase can only be determined from appearance experiments.

Neutrinos from stored muons have four advantages over conventional neutrino sources. Firstly, the muon flux and polarisation in the storage ring, which is well measured by the muon decay electron spectrum, determines the intensity and spectrum of the neutrinos. Secondly, there is a large controlled flux of high-energy ν_e (well above the τ threshold) for $\nu_e \to \nu_\tau$ appearance experiments. Thirdly, neutrino and antineutrino interactions can be studied with comparable statistics and spectra by switching between μ^+ and μ^- in the storage ring. Finally, the contamination from other neutrino flavours is negligible, if the lepton charge is measured at the detector. These are likely to be important in making precision measurements of the neutrino parameters.

The probability that a neutrino of energy E_ν created initially as a flavour eigenstate $|\nu_\alpha>$ is detected as flavour eigenstate $|\nu_\beta>$ after a distance L has two components - a CP-preserving part $P_+(\nu_\alpha \to \nu_\beta)$ which is the same for neutrinos and antineutrinos, and a CP-changing part $P_-(\nu_\alpha \to \nu_\beta)$ which changes sign, where

$$P_+(\nu_\alpha \to \nu_\beta) = \delta_{\alpha\beta} - 4\Sigma_{i>j} Re\left(U_{\alpha i}U^*_{\beta i}U^*_{\alpha j}U_{\beta j}\right)\sin^2\left(\frac{\Delta m^2_{ij}L}{4E_\nu}\right)$$

$$P_-(\nu_\alpha \to \nu_\beta) = 2\Sigma_{i>j} Im\left(U_{\alpha i}U^*_{\beta i}U^*_{\alpha j}U_{\beta j}\right)\sin\left(\frac{\Delta m^2_{ij}L}{2E_\nu}\right).$$

For estimating, it is useful to replace $\frac{\Delta m^2_{ij}L}{4E}$ by $1.27\frac{\Delta m^2_{ij}L}{E}$ with Δm^2_{ij} measured in eV^2, L measured in km, and the neutrino energy E_ν measured in GeV. It is clear from this that, for squared mass differences of $\sim 0.003\mathrm{eV}^2$, the maximum effect occurs at distances between source and detector of several thousand km for energies well above the τ threshold.

As is well-known, a three-generation mixing model cannot survive if all of the present indications of neutrino oscillations are confirmed, and new physics will necessarily be part of the explanation. Under these circumstances, the unique features of the neutrino factory may well be critical. The present situation[?] is that the analysis of the atmospheric neutrino data favours $0.0013\mathrm{eV}^2 < \Delta m^2_{32} < 0.008\mathrm{eV}^2$ with a mixing angle $\sin^2 2\theta_{23}$ close to maximal, and consistent with $\nu_\mu \to \nu_\tau$ oscillations. The solar neutrino data are consistent with a squared mass difference Δm^2_{21} at least an order of magnitude smaller than Δm^2_{32}, with some evidence that $\sin^2 2\theta_{12}$ may be large and θ_{13} small. There is still part of the parameter space consistent with the LSND data that is not excluded by other experiments, with relatively small mixing angle and larger $\Delta m^2 \geq 0.01\mathrm{eV}^2$. Further atmospheric and solar neutrino data, and the results of conventional accelerator experiments (MiniBOONE, MINOS, CERN to Gran Sasso, ...) will refine these measurements and resolve any apparent contradictions. Nevertheless, θ_{13}, the sign of Δm^2_{32} and the phase angle δ are likely to remain poorly constrained.

The magnitude of the interference term which gives rise to CP- or T-Violation is proportional to the product of two terms. The first term is a product of the MNS angles $\cos\theta_{12}\cos^2\theta_{13}\cos\theta_{23}\sin\theta_{12}\sin\theta_{13}\sin\theta_{23}\sin\delta$ which requires that neither θ_{13} nor δ should be too small. The second term $\sin\frac{\Delta m^2_{21}L}{2E}\sin^2\frac{\Delta m^2_{32}L}{4E}+\sin\frac{\Delta m^2_{32}L}{2E}\sin^2\frac{\Delta m^2_{21}L}{4E}$ connects the two independent mass scales; clearly both the sin and $\sin^2$ terms must be significant at the same $\frac{L}{E}$, which means that $\Delta m^2_{21}/\Delta m^2_{32}$ should not be too small. A number of studies[?,?] show that the optimal sensitivity for the measurement of the phase δ is ~3000km, ignoring the LSND mass scale.

There are in principle two different methods which could be used to measure CP- or T-violation. The oscillation probability of charge conjugate pairs could be compared, for example $P(\nu_\mu\to\nu_e)$ and $P(\overline{\nu}_\mu\to\overline{\nu}_e)$, or time-reversed pairs could be compared, for example $P(\nu_\mu\to\nu_e)$ and $P(\nu_e\to\nu_\mu)$ or $P(\overline{\nu}_\mu\to\overline{\nu}_e)$ and $P(\overline{\nu}_e\to\overline{\nu}_\mu)$. In practice, all of these may be needed in order to disentangle asymmetries which arise from the MNS phase from matter effects which arise because of the difference between the neutrino and antineutrino cross-sections in the earth. For this reason, it may be essential to have detectors at two different baselines, so that these effects can be disentangled. Taking one of the baselines as around 3000km, the other could be either much less (a few hundred km) or much more (>6000km).

4 Other neutrino physics

The intense neutrino beams close to the storage ring may also be used for conventional neutrino physics (DIS, charm physics, CKM elements, $\sin^2\theta_W$...) with high statistics[?,?]. Events rates of $\sim 10^7$/kg/year would permit precise studies of structure functions and parton distribution functions with hydrogen, deuterium and polarised targets. In particular, the unpolarised and polarised parton distribution functions (except perhaps the heavy flavour pdf's) could be determined with high precision. This physics programme would benefit from higher neutrino energies.

Acknowledgements

This brief report builds on the work of many people too numerous to acknowledge here. I have relied heavily on the work in the feasibility studies[?] and the workshop proceedings[?], and the references therein. I acknowledge their input.

References

1. *Prospective Study of Muon Storage Rings at CERN*, ed. B. Autin, A. Blondel and J. Ellis, CERN 99-02, ISBN 92-9083-143-0; *Technical Feasibility of a Neutrino Factory*, N. Altenkamp and D. Findley, report to the Fermilab Directorate; *Physics at a Neutrino factory*, ed. S. Geer and H. Schellmann, FERMILAB-FN-692.
2. NUFACT-99, Lyon, July 1999; NUFACT-00, Monterey, May 2000.
3. R. Carey, *New results from the g-2 experiment*, this conference.
4. M. Tigner, *Future Accelerators*, this conference.
5. C. Gonzales Garcia, *Current status of the solar and atmospheric neutrino data*, this conference;
6. M. Lindner, *CP violations in lepton number violation processes and neutrino oscillation*, this conference.

MATTER EFFECTS AND CP-VIOLATION IN NEUTRINO OSCILLATION

M. LINDNER

Theoretische Physik, Physik Department, Technische Universität München, James-Franck-Str., D-85748 Garching b. München, Germany

E-mail: lindner@ph.tum.de

The potential of very long baseline neutrino oscillation experiments to measure precisely leptonic mixing angles, CP-violation and matter effects is discussed.

Almost all neutrino oscillation results are so far described by vacuum oscillations in a two flavour picture. Increased precision in the future will however require oscillation formulae with all flavours and the mixing matrix U:

$$\begin{aligned} P(\nu_{e_l} \to \nu_{e_m}) &= \left|\langle \nu_m | U e^{-iHt} U^\dagger | \nu_l \rangle\right|^2 \\ &= \delta_{lm} - 4\sum_{i>j} \quad \mathrm{Re} J_{ij}^{e_l e_m} \sin^2 \Delta_{ij} \\ &\quad -2\sum_{i>j} \quad \mathrm{Im} J_{ij}^{e_l e_m} \sin 2\Delta_{ij} \qquad (1) \end{aligned}$$

where $J_{ij}^{e_l e_m} := U_{li} U_{lj}^* U_{mi}^* U_{mj}$ and $\Delta_{ij} := \frac{\Delta m_{ij}^2 L}{4E}$, with $\Delta m_{ij}^2 = m_i^2 - m_j^2$. It can be easily seen[1] that the oscillation probabilities in eq. (1) depend for three or more non-degenerate neutrinos on CP-violating phases in the mixing matrix U. Additionally matter effects[2,3] must be taken into account in eq. (1) where appropriate, e.g. for beams crossing the earth. The presence of matter implies a mapping of the vacuum masses and mixings to effective parameters in matter. This leads in the limit where the small solar mass splitting Δm_{21}^2 is ignored to the mappings $\theta_{13} \to \theta_{13,m}$, $m_1^2 \to m_{1,m}^2$, $m_3^2 \to m_{3,m}^2$ such that all Δm_{ij}^2 are affected, while θ_{12}, θ_{23} and the CP-phase δ are unchanged[4]. Note that there are different mappings for neutrinos and antineutrinos due to the opposite sign of the interaction. Matter effects make the general expressions for the oscillation of three or more neutrinos rather lengthy, but it is important to note that this allows tests of matter effects, which are so far experimentally unprobed. Optimal sensitivity is obtained[2] when the resonance condition is fulfilled, i.e. for $2EV \simeq \Delta m_{31}^2 \cos 2\theta_{13}$. For negligible small θ_{13}, $\Delta m_{31}^2 = 3.5 \times 10^{-3}$ eV^2 and typical densities in the earth crust one finds $E_{opt} \simeq 15$ GeV and $E_{opt} \simeq 8$ GeV in the core. Another important point of matter effects is the extraction of the sign of Δm_{31}^2, which is undetermined in vacuum oscillations resulting in mass ordering ambiguities[5,2] as shown in Fig. 1. The extraction of the sign

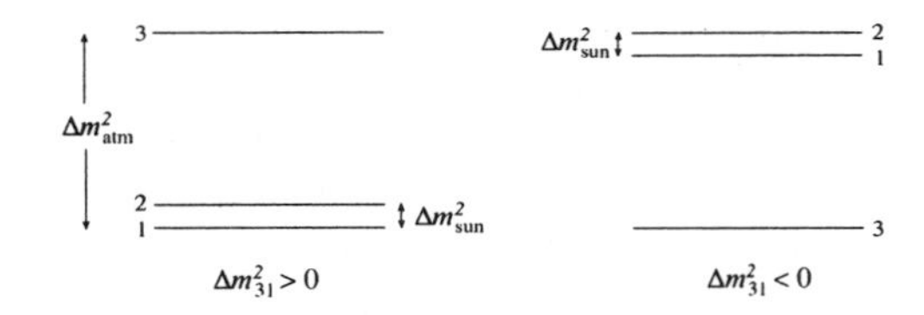

Figure 1. Ordering ambiguities in vacuum oscillation.

of Δm^2 via matter effects is again connected to the resonance condition.

Neutrino factories[6] would be an ideal tool to probe the above effects, but it will take some time to be built and it is interesting what could be achieved meanwhile with conventional neutrino beams. The potential of several experimental setups is therefore compared[7] by showing the precision of the leading oscillation parameters (Δm_{31}^2 and θ_{23}) as well as the sensitivity to the sub-leading parameters (θ_{13} and the sign of Δm_{31}^2). The energy thresholds, resolutions and the dependence on these parameters is also shown. The considered experimental setups include two beams, namely (I) conventional wide band beams and (II) neutrino factory beams which are assumed to point at three different types of detectors, specifically (a) magnetized iron detectors, (b) large water or ice Cherenkov detectors ("neutrino-telescopes") and (c) ring imaging water Cherenkov detectors. Due to their high threshold neutrino-telescopes are usually not considered for neutrino oscillation, but it has been shown recently that the effective threshold can be considerably lower in a high rate oscillation experiment[3,7]. Figs. 2 and 3 show

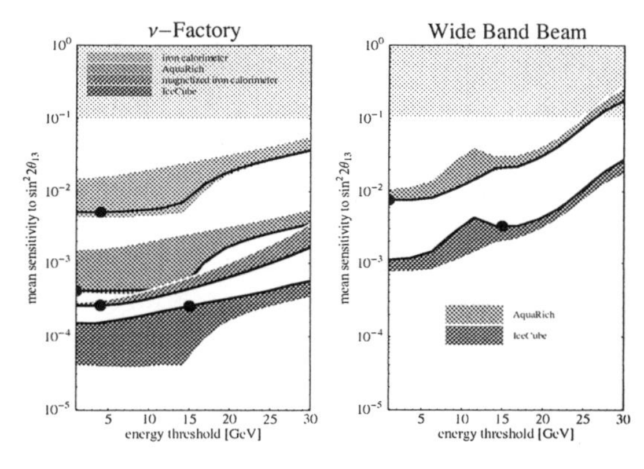

Figure 2. Sensitivity to θ_{13}.

in comparison the sensitivity to θ_{13} and the θ_{13} range where the sign of Δm^2 can be extracted[7]. The figures show that conventional beams offer quite some improvements until a neutrino factory is built. The possibility to extract the sign of Δm^2 implies precise tests of matter effects [2]. CP-violating effects can not be measured in disappearance channels and are most likely measured with neutrino factories. Examples of CP-violating effects are shown[1] in Fig. 4. In summary there is a very promising future for "precision neutrino oscillation physics", resulting in very precise masses, mixings and maybe CP-violation in the lepton sector. Neutrino factories, which will take some time to be realized, would the best option. Meanwhile the remarkable potential of conventional setups should be exploited.

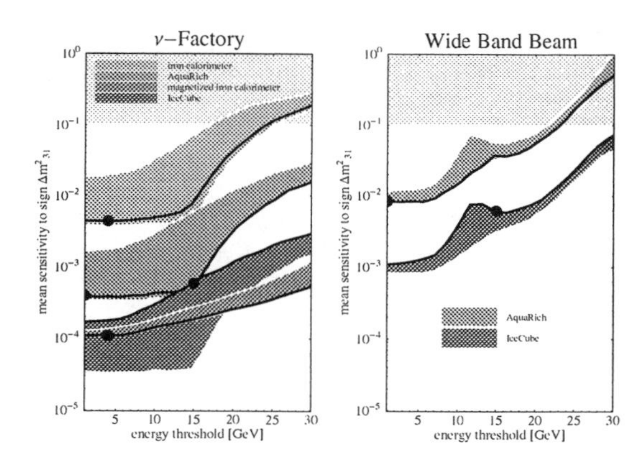

Figure 3. θ_{13} sensitivity range for the sign of Δm^2.

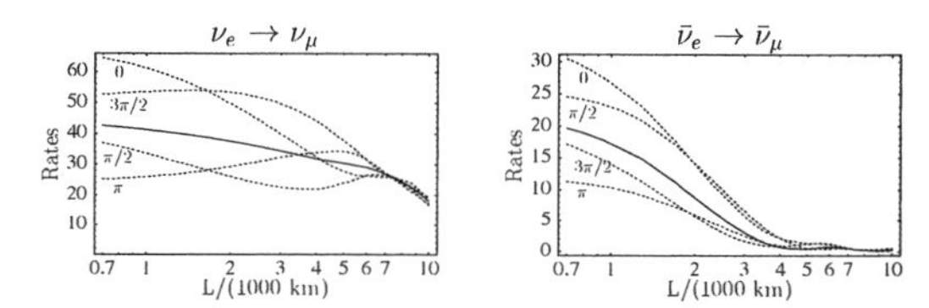

Figure 4. Example of CP violating effects at a neutrino factory.

References

1. K. Dick, M. Freund, M. Lindner and A. Romanino, Nucl. Phys. **B 562** (1999) 29; A. Romanino, Nucl. Phys. **B 574** (2000) 675; A. Donini, M.B. Gavela, P. Hernandez, S. Rigolin, Nucl. Phys. **B 574** (2000) 23; A.M. Gago, V. Pleitez and R. Zukanovich Funchal, Phys. Rev. **D 61** (2000) 016004; V. Barger, S. Geer, R. Raja, K. Whisnant, hep-ph/0007181.
2. M. Freund, M. Lindner, S.T. Petcov and A.Romanino, Nucl.Phys.**B578**(2000)27.
3. K. Dick, M. Freund, P. Huber and M. Lindner, hep-ph/0006090.
4. M. Freund, P. Huber and M. Lindner, hep-ph/0004085.
5. V. Barger, S. Geer, R. Raja and K. Whisnant, Phys.Rev.D62:013004,2000.
6. See e.g. S. Geer, hep–ph/0008155 and references therein.
7. K. Dick, M. Freund, P. Huber and M. Lindner, hep-ph/0008016.

CP VIOLATIONS IN LEPTON NUMBER VIOLATION PROCESSES AND NEUTRINO OSCILLATIONS

K. MATSUDA, N. TAKEDA, T. FUKUYAMA AND H. NISHIURA[†]

Department of Physics, Ritsumeikan Univ., Kusatsu, Shiga 525-8577, Japan

[†]Department of General Education, Junior College of Osaka Institute of Technology, Asahi-ku, Osaka 535-8585, Japan

There appear so many parameters and their mutual correlations in neutrino oscillation, beta (β), and neutrinoless double beta ($(\beta\beta)_{O\nu}$) decays experiments. To get the constraints on these parameters we propose a graphical method which enables us to grasp the constraints effectively and intuitively. To keep illustrative clearance, though this method is valid for more general case, we examine explicitly the case for the CP violating phase factors having ± 1 or $\pm i$. Some constraints derived from oscillation, β and $(\beta\beta)_{0\nu}$ decays experiments are also discussed.

From the recent neutrino oscillation experiments[1] it becomes affirmative that neutrinos have masses. On the other hand, the experiments intending to determine directly neutrino mass are also on going. The upper limit for the neutrino mass (m_ν) from the tritium β decay defined by

$$(m_\nu)^2 \equiv \sum_{j=1}^{3} |U_{ej}|^2 m_j^2 = \quad (1)$$
$$m_1^2 + |U_{e2}|^2(m_2^2 - m_1^2) + |U_{e3}|^2(m_3^2 - m_1^2).$$

is 2.9 eV [2]. The most recent experimental upper bound for the averaged mass $\langle m_\nu \rangle$ for Majorana neutrinos from $(\beta\beta)_{0\nu}$ is given by $\langle m_\nu \rangle < 0.2 eV$ [3]. The next generation experiment GENIUS[4] is anticipated to reach a considerably more stringent limit $\langle m_\nu \rangle < 0.01 - 0.001 eV$. In these situations we should study the many parameters appearing in neutrino physics incorporating all these direct and indirect experiments. To disentangle the complicated correlations among the data from many different kinds of experiments we proposed a graphical method [5]. This method enables us to grasp the geometrical relations among the parameters of masses, mixing angles, CP phases etc. and obtains the constraints on them more easily than the analytical calculations [6]. In this talk we apply this method to obtain the constraints on $\langle m_\nu \rangle$ for the given CP phases. For illustrative purpose we consider the specific values of CP phases which are partly supported by theoretical models [7]. We also discuss some constraints derived from the above three different kinds of experiments.

The averaged mass $\langle m_\nu \rangle$ obtained from $(\beta\beta)_{0\nu}$ is given [8] by the absolute values of averaged complex masses for Majorana neutrinos as

$$\langle m_\nu \rangle = |M_{ee}|. \quad (2)$$

Here the averaged complex mass M_{ee} is, after suitable phase convention, defined by

$$\begin{aligned} M_{ee} &\equiv \sum_{j=1}^{3} U_{ej}^2 m_j \equiv |U_{e1}|^2 m_1 \\ &\quad + |U_{e2}|^2 e^{2i\beta} m_2 + |U_{e3}|^2 e^{2i(\rho-\phi)} m_3, \\ &= |U_{e1}|^2 \widetilde{m_1} + |U_{e2}|^2 \widetilde{m_2} + |U_{e3}|^2 \widetilde{m_3} \quad (3) \end{aligned}$$

Here we have defined the complex masses $\widetilde{m_i} (i = 1, 2, 3)$ by

$$\begin{aligned} &\widetilde{m_1} \equiv m_1, \qquad \widetilde{m_2} \equiv e^{2i\beta} m_2 \equiv \eta_2 m_2, \\ &\widetilde{m_3} \equiv e^{2i(\rho-\phi)} m_3 \equiv e^{2i\rho'} m_3 \equiv \eta_3 m_3. \quad (4) \end{aligned}$$

The CP violating effects are invoked in β , ρ and ϕ appearing in U . U_{aj} is the Maki-Nakagawa-Sakata (MNS) left-handed lepton mixing matrix which combines the weak eigenstate neutrino ($a = e, \mu$ and τ) to the mass eigenstate neutrino with mass m_j (j=1,2 and 3).

The M_{ee} is the "averaged" complex mass of the masses $\widetilde{m_i}(i = 1,2,3)$ weighted by three mixing elements $|U_{ej}|^2(j = 1,2,3)$ with the unitarity constraint $\sum_{j=1}^{3} |U_{ej}|^2 = 1$. Therefore, the position of M_{ee} in a complex mass plane is within the triangle formed by the three vertices $\widetilde{m_i}(i = 1,2,3)$ if the magnitudes of $|U_{ej}|^2(j = 1,2,3)$ are unknown (Fig.1). This triangle is referred to as the complex-mass triangle[5]. The three mixing elements $|U_{ej}|^2(j = 1,2,3)$ indicate the division ratios for the three portions of each side of the triangle which are divided by the parallel lines to the side lines of the triangle passing through the M_{ee}. (Fig.1). The CP violating phases 2β and $2\rho'$ represent the rotation angles of $\widetilde{m_2}$ and $\widetilde{m_3}$ around the origin, respectively. Since $\langle m_\nu \rangle = |M_{ee}|$, the present experimental upper bound on $\langle m_\nu \rangle$ (we denote it $\langle m_\nu \rangle_{max}$.) indicates the maximum distance of the point M_{ee} from the origin and forms the circle in the complex plane.

We proceed to discuss the main theme to demonstrate how our formulation works. First we give the constraints on $\langle m_\nu \rangle$, given $\eta_i = \pm 1$ or $\pm i$. We list up the constraints on $\langle m_\nu \rangle$ for the various combinations of given CP violating factors η_i.(Table 1) Here we demonstrate the graphical method explicitly for $\eta_2 = i$ and $\eta_3 = 1$ case. In this case Fig.1 becomes Fig.2. Here we have imposed the constraint on U_{e3} from the oscillation experiments of CHOOZ and SuperKamiokande [9], $|U_{e3}|^2 < 0.026$. The shaded region is allowed. $\langle m_\nu \rangle$ is the distance from the origin to the shaded region. So the minimum and maximum of $\langle m_\nu \rangle$ is easily estimated from Fig.2. Obviously, the minimum is $\overline{OA}$. The maximum value depends on whether the line PQ crosses the horizontal axis at a point larger than m_2 or not. If $\overline{OP} > m_2$, the maximum is $\overline{OP}$. If $\overline{OP} < m_2$, the maximum becomes m_2. Next we consider the constraints among MNS parameters, (m_ν), and $\langle m_\nu \rangle$. (Fig.3) Taking Eq.(1) and the division rate in Fig.1 into account, we can restrict the allowed re-

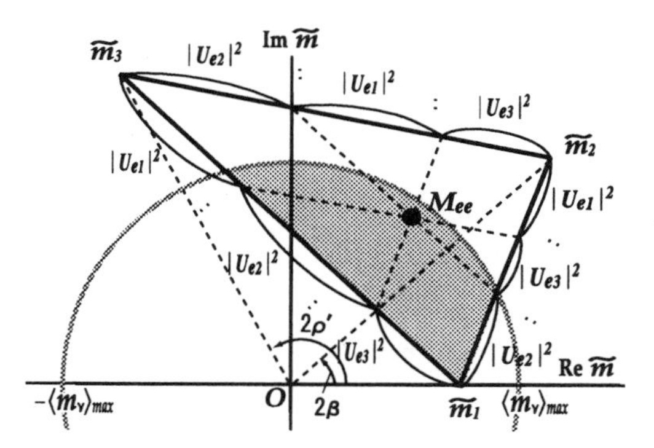

Figure 1. Graphical representations of the complex mass M_{ee} and CP violating phases. The position of M_{ee} is within the triangle formed by the three points $\widetilde{m_i}(i = 1,2,3)$ which are defined in Eq. (4). The allowed position of M_{ee} is in the intersection (shaded area) of the inside of this triangle and the inside of the circle of radius $\langle m_\nu \rangle_{max}$ around the origin.

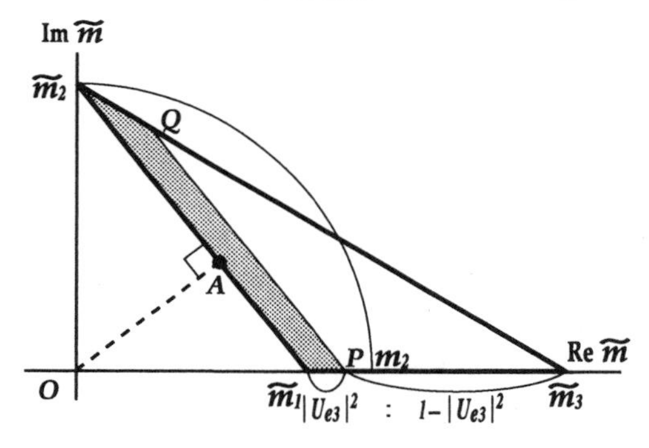

Figure 2. Complex-mass triangle supplemented with $|U_{e3}|^2 < 0.026$ for $\eta_2 = i$ and $\eta_3 = 1$. The shaded region is allowed. The minimum of $|M_{ee}|$ is $\overline{OA}$. The maximum of $|M_{ee}|$ changes depending on whether $|U_{e3}|^2_{max}$ is larger than $(m_2 - m_1)/(m_3 - m_1)$ or not.

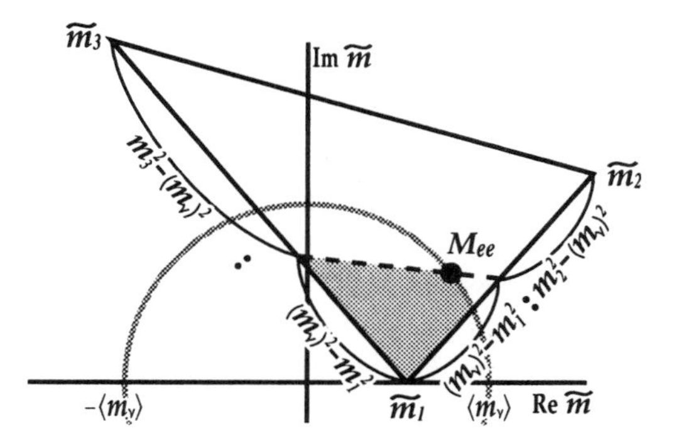

Figure 3. The relation between $\langle m_\nu \rangle$ and (m_ν). The position of M_{ee} must be at the intersection point between the gray circle and the dotted line.

$\eta_3 \backslash \eta_2$	1	-1	$\pm i$
1	$m_1 \le \langle m_\nu \rangle \le A_2$	$0 \le \langle m_\nu \rangle \le \max\begin{pmatrix} m_2 \\ A_1 \end{pmatrix}$	$B_{12} \le \langle m_\nu \rangle \le \max\begin{pmatrix} m_2 \\ A_1 \end{pmatrix}$
-1	$\max\begin{pmatrix} 0 \\ C \end{pmatrix} \le \langle m_\nu \rangle \le \max\begin{pmatrix} m_2 \\ -C \end{pmatrix}$	$0 \le \langle m_\nu \rangle \le A_2$	$\max\begin{pmatrix} 0 \\ D_{21} \end{pmatrix} \le \langle m_\nu \rangle \le \max\begin{pmatrix} m_2 \\ E_{23} \end{pmatrix}$
$\pm i$	(i) $\lvert U_{e3}\rvert^2_{max} < m_1^2/(m_1^2+m_3^2)$, $E_{13} \le \langle m_\nu \rangle \le \max\begin{pmatrix} m_2 \\ E_{23} \end{pmatrix}$ (ii) $\lvert U_{e3}\rvert^2_{max} \ge m_1^2/(m_1^2+m_3^2)$, $B_{13} \le \langle m_\nu \rangle \le \max\begin{pmatrix} m_2 \\ E_{23} \end{pmatrix}$	$0 \le \langle m_\nu \rangle \le \max\begin{pmatrix} m_2 \\ E_{23} \end{pmatrix}$	$B_{12} \le \langle m_\nu \rangle \le A_2$
$\mp i$	————	————	$\max\begin{pmatrix} 0 \\ D_{12} \end{pmatrix} \le \langle m_\nu \rangle \le \max\begin{pmatrix} m_2 \\ E_{23} \end{pmatrix}$

Table 1. Given the $\eta_i = \pm 1$ or $\pm i$, the constraints on the averaged mass $\langle m_\nu \rangle$ are given. Notations are as follows. $|U_{e3}|^2_{max} = 0.026$, $A_i \equiv m_i + |U_{e3}|^2_{max}(m_3 - m_i)$, $B_{ij} \equiv m_i m_j / \sqrt{m_i^2 + m_j^2}$, $C \equiv m_1 - |U_{e3}|^2_{max}(m_1 + m_3)$ $D_{ij} \equiv m_i(m_j - |U_{e3}|^2_{max}(m_j + m_3))/\sqrt{m_i^2 + m_j^2}$ and $E_{ij} \equiv \sqrt{(1 - |U_{e3}|^2_{max})^2 m_i^2 + |U_{e3}|^4_{max} m_j^2}$. The max $(a, b)^T$ indicates the larger value between a and b. The double signs of η_2 and η_3 are in the same order.

gion of M_{ee} corresponding to whether (m_ν) and $\langle m_\nu \rangle$ have the definite values or the upper limits: If both have the definite values M_{ee} is on the black spot. If neither has the definite value, it is in the shaded region. Here the dotted line is the direct line connecting the two points which divides each side by the given ratio. If either one has the definite value, M_{ee} is on the dotted line or circumference of the boarder. For example, assuming $|U_{e3}| = 0$ and that the definite values of $\langle m_\nu \rangle$ and (m_ν) are given, we can restrict β:

$$\cos 2\beta = \frac{\langle m_\nu \rangle^2 - |U_{e1}|^4 m_1^2 - |U_{e2}|^4 m_2^2}{2 m_1 m_2 |U_{e1}|^2 |U_{e2}|^2} \quad (5)$$

Here, m_1 and m_2 can be expressed by only experimental values as $m_1^2 = (m_\nu)^2 - |U_{e2}|^2 \Delta m_{12}^2$ and $m_2^2 = (m_\nu)^2 + |U_{e1}|^2 \Delta m_{12}^2$. In this way, we can obtain the relation between β and ρ' by using only experimental values without the masses m_i which are not directly accessible to experiments.

The work of K.M. is supported by the JSPS Research Fellowship, No.10421.

References

1. T.Kajita, talk presented at Neutrino '98 (Takayama, Japan, June 1998).
2. C. Weinheimer et al., Phys. Lett. **B460** 219 (1999); V.M. Lobashev et al., ibid. 227.
3. L. Baudis et al., Phys. Rev Lett. **83** 41 (1999).
4. J. Hellmig, H.V. Klapdor-Kleingrothaus, Z.Phys. **A 359** 361 (1997); H.V. Klapdor-Kleingrothaus, M. Hirsch, Z.Phys. **A 359** 382 (1997); L. Baudis et al., Phys. Rep. **307** 301 (1998).
5. K. Matsuda, N. Takeda, T. Fukuyama, and H. Nishiura, to appear in Phys. Rev. **D** (hep-ph/0003055).
6. T. Fukuyama, K. Matsuda, and H. Nishiura, Phys. Rev. **D57** 5844 (1998); Mod. Phys. Lett. **A13** 2279 (1998); Mod. Phys. Lett. **A14** 433 (1999).
7. K. Fukuura, T. Miura, E. Takasugi and M. Yoshimura, Phys. Rev. **D61** 073002 (2000).
8. M. Doi, T. Kotani, H. Nishiura, K. Okuda and E. Takasugi, Phys. Lett. **102B** 323 (1981); W.C. Haxton and G.J. Stophenson Jr., Prog. Part. Nucl. Phys. **12** 409 (1984).
9. M. Apollonio et al., Phys. Lett. **B466** 415 (1999) and Phys. Lett. **B338**, 383 (1998).

SELECTED RECENT RESULTS FROM AMANDA

PRESENTED BY D.F. COWEN FOR THE AMANDA COLLABORATION

E. Andrés[11], P. Askebjer[13], X. Bai[1], G. Barouch[11], S.W. Barwick[8], R.C. Bay[7], K.-H. Becker[2], L. Bergström[13], D. Bertrand[3], D. Bierenbaum[8], A. Biron[4], J. Booth[8], O. Botner[12], A. Bouchta[4], M.M. Boyce[11], S. Carius[5], A. Chen[11], D. Chirkin[7,2], J. Conrad[12], J. Cooley[11], C.G.S. Costa[3], D.F. Cowen[10], J. Dailing[8], E. Dalberg[13], T. DeYoung[11], P. Desiati[4], J.-P. Dewulf[3], P. Doksus[11], J. Edsjö[13], P. Ekström[13], B. Erlandsson[13], T. Feser[9], M. Gaug[4], A. Goldschmidt[6], A. Goobar[13], L. Gray[11], H. Haase[4], A. Hallgren[12], F. Halzen[11], K. Hanson[10], R. Hardtke[11], Y.D. He[7], M. Hellwig[9], H. Heukenkamp[4], G.C. Hill[11], P.O. Hulth[13], S. Hundertmark[8], J. Jacobsen[6], V. Kandhadai[11], A. Karle[11], J. Kim[8], B. Koci[11], L. Köpke[9], M. Kowalski[4], H. Leich[4], M. Leuthold[4], P. Lindahl[5], I. Liubarsky[11], P. Loaiza[12], D.M. Lowder[7], J. Ludvig[6], J. Madsen[11], P. Marciniewski[12], H.S. Matis[6], A. Mihalyi[10], T. Mikolajski[4], T.C. Miller[1], Y. Minaeva[13], P. Miočinović[7], P.C. Mock[8], R. Morse[11], T. Neunhöffer[9], F.M. Newcomer[10], P. Niessen[4], D.R. Nygren[6], H. Ogelman[11], C. Pérez de los Heros[12], R. Porrata[8], P.B. Price[7], K. Rawlins[11], C. Reed[8], W. Rhode[2], A. Richards[7], S. Richter[4], J. Rodríguez Martino[13], P. Romenesko[11], D. Ross[8], H. Rubinstein[13], H.-G. Sander[9], T. Scheider[9], T. Schmidt[4], D. Schneider[11], E. Schneider[8], R. Schwarz[11], A. Silvestri[2,4], M. Solarz[7], G.M. Spiczak[1], C. Spiering[4], N. Starinsky[11], D. Steele[11], P. Steffen[4], R.G. Stokstad[6], O. Streicher[4], Q. Sun[13], I. Taboada[10], L. Thollander[13], T. Thon[4], S. Tilav[11], N. Usechak[8], M. Vander Donckt[3], C. Walck[13], C. Weinheimer[9], C.H. Wiebusch[4], R. Wischnewski[4], H. Wissing[4], K. Woschnagg[7], W. Wu[8], G. Yodh[8], S. Young[8]

(1) Bartol Research Institute, University of Delaware, Newark, DE 19716, USA; (2) Fachbereich 8 Physik, BUGH Wuppertal, D-42097 Wuppertal, Germany; (3) Brussels Free University, Science Faculty CP230, B-1050 Brussels, Belgium; (4) DESY-Zeuthen, D-15735 Zeuthen, Germany; (5) Dept. of Technology, Kalmar University, S-39129 Kalmar, Sweden; (6) Lawrence Berkeley National Lab., Berkeley, CA 94720, USA; (7) Dept. of Physics, University of California, Berkeley, CA 94720, USA; (8) Dept. of Physics and Astronomy, University of California, Irvine, CA 92697, USA; (9) Institute of Physics, University of Mainz, D-55099 Mainz, Germany; (10) Dept. of Physics and Astronomy, University of Pennsylvania, Phila., PA 19104, USA; (11) Dept. of Physics, University of Wisconsin, Madison, WI 53706, USA; (12) Dept. of Radiation Sciences, Uppsala University, S-75121 Uppsala, Sweden; (13) Fysikum, Stockholm University, S-11385 Stockholm, Sweden

We present a selection of results based on data taken in 1997 with the 302-PMT Antarctic Muon and Neutrino Detector Array-B10 ("AMANDA-B10") array. Atmospheric neutrinos created in the northern hemisphere are observed indirectly through their charged current interactions which produce relativistic, Cherenkov-light-emitting upgoing muons in the South Pole ice cap. The reconstructed angular distribution of these events is in good agreement with expectation and demonstrates the viability of this ice-based device as a neutrino telescope. Studies of nearly vertical upgoing muons limit the available parameter space for WIMP dark matter under the assumption that WIMPs are trapped in the earth's gravitational potential well and annihilate with one another near the earth's center.

1 Introduction

The AMANDA-B10 high energy neutrino detector consists of 302 optical modules (OMs) on ten strings. Each OM is a PMT with passive electronics housed in a glass pressure vessel. The OMs are deployed in a cylindrical volume about 120 m in diameter and 500 m in height at depths between roughly 1500 and 2000 m below the surface of the South Pole ice cap. In this region the optical properties of the ice are well suited for detecting the Cherenkov light emitted by relativistic charged particles.[1] This light is used to reconstruct individual events.[2] Figure 1 shows a clear candidate upgoing muon observed in 1997 together with diagrams of the entire array and an individual OM. An electrical cable provides high voltage to the PMTs and transmits their signal pulses to the surface electronics. A light diffuser ball connected via fiber optic cable to a laser on the surface is used for calibration purposes.

In January 2000 the AMANDA-B10 detector was upgraded to a total of 19 strings with 667 OMs. This new detector, dubbed "AMANDA-II," is 200 m in diameter and the same height and depth as AMANDA-B10. A proposal exists to construct "IceCube," a kilometer-scale device with 4800 OMs on 80 strings.

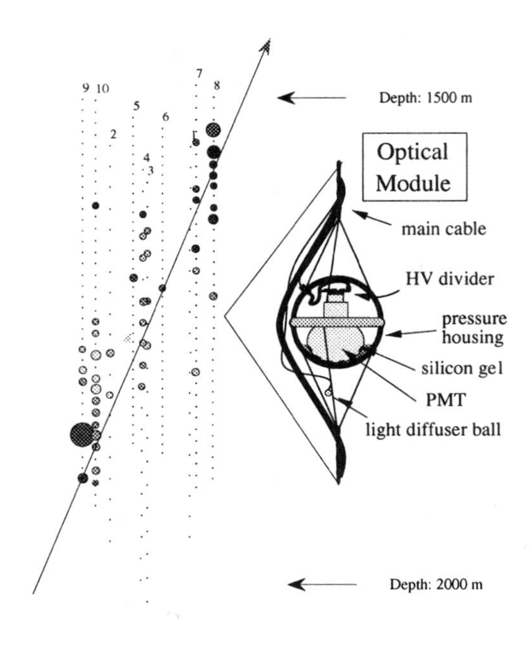

Figure 1. The AMANDA-B10 array with a candidate upgoing muon. The dots and circles represent individual optical modules, one of which is shown on the right of the figure. The circles represent hit channels, with the size of each circle proportional to the pulse amplitude, and the relative photon arrival times proportional to the shading (lighter shading corresponds to earlier arrival times).

2 Atmospheric Neutrinos

The upgoing muon signal from atmospheric neutrino interactions is extracted from a large downgoing muon background due to cosmic ray interactions in the atmosphere above the South Pole. Simulations predict about 20 detected upgoing muons from atmospheric neutrino interactions with $E \gtrsim 40$ GeV and $6 \cdot 10^6$ detected downgoing muons from cosmic rays per day. These events are reconstructed by fitting a track to the measured Cherenkov arrival times via a maximum likelihood method, taking into account the scattering and absorption of Cherenkov light in the ice.

Roughly $5 \cdot 10^{-5}$ of the downgoing muons are misreconstructed as upgoing. This background is greatly reduced by application of six quality criteria which exploit the differences between signal and background in the temporal and spatial profiles of the measured Cherenkov photons. In conjunction with the reconstruction criterion, application of these criteria results in a total background rejection factor of roughly 10^8 while retaining about 5% of the atmospheric neutrino signal, resulting in 188 events in the final sample. The livetime is $1.16 \cdot 10^7$ s for this data. This analysis, along with background studies currently underway, indicate that the final sample con-

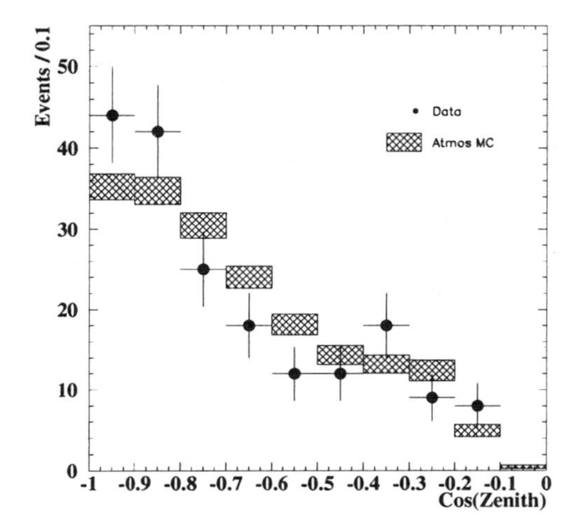

Figure 2. The reconstructed zenith angle distribution for the data (filled circles) and signal Monte Carlo (shaded boxes). The Monte Carlo is normalized to the data. The vertical widths of the boxes indicate the uncertainty computed using binomial statistics.

tains roughly 15% background contamination from cosmic ray muons.

The zenith angle distribution for the 188 events is shown in fig. 2. The Monte Carlo prediction of 235 events is consistent within systematic uncertainties due to the atmospheric neutrino flux and absolute detector sensitivity, and in the figure the Monte Carlo has been normalized to the data. The shape of the data distribution agrees well with expectation, with both distributions showing the reduced acceptance at high zenith angles due to the long and narrow shape of the detector. Along with measurements of AMANDA's angular resolution and offset, performed using downgoing muons which simultaneously trigger AMANDA and the SPASE surface air shower array,[3] the AMANDA detector is demonstrably a functioning neutrino telescope.

3 WIMP Search

The universe is believed to contain substantial amounts of non-baryonic dark matter. A candidate weakly interacting massive particle (WIMP) is the lightest supersymmetric particle of the Minimal Supersymmetric Standard Model, the neutralino. When neutralinos interact with nuclei in the earth, they lose energy and become trapped in the earth's gravitational field. They can then annihilate with one another in the earth's core, generating high energy neutrinos which would produce nearly vertical ($\cos\theta < -0.97$) upgoing muons in AMANDA.

A search for upgoing muons was performed in this narrow angular region and the number detected is consistent with the number expected from atmospheric neutrinos. This measurement can therefore be used to place limits on the flux of neutrinos from neutralino annihilation as a function of various supersymmetric models. Figure 3 compares the AMANDA limit with theoretical predictions[4] and curves derived[5] from limits published by Baksan,[6] MACRO[7] and SuperKamiokande.[8] Although studies of detector sensitivity and the impact of neutrino oscillations are expected to weaken the AMANDA limit, additional data and improved analysis techniques will eventually allow AMANDA to derive improved and very competitive limits on neutralino dark matter in the region depicted in the figure.

4 Conclusions

Using upgoing muons from atmospheric neutrino interactions as a test beam, the AMANDA-B10 detector has been shown to perform well as a neutrino telescope. The detector has already been used to produce a competitive limit on neutralino WIMP annihilation. Initial limits on cosmological diffuse ultrahigh energy (UHE, $E \gtrsim 10$ TeV) neutrino flux, UHE neutrino point sources, UHE neutrinos from gamma ray bursters, relativistic monopoles, and supernova explosions have been presented elsewhere.[1,9] These limits are indicative of the great discovery potential of AMANDA-B10 and especially that

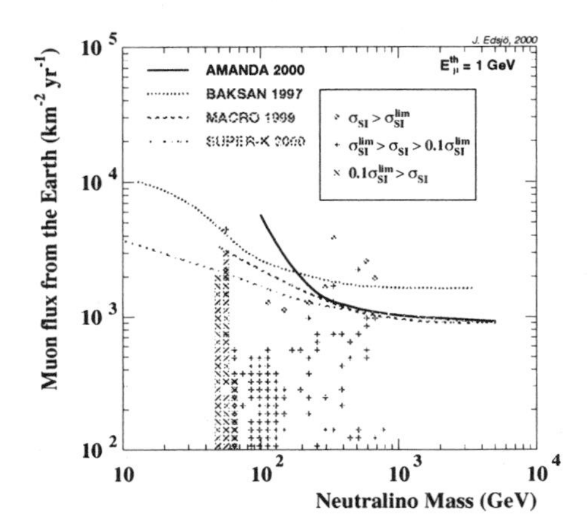

Figure 3. Theoretical predictions of high energy muon fluxes due to neutralino annihilation in the earth's core with limits from several experiments. The different symbols correspond to different sets of theoretical input parameters.[4] The AMANDA 90% CL limit is corrected to correspond to a threshold of 1 GeV.

of its larger successors, AMANDA-II and ultimately IceCube.

Acknowledgments

This research was supported by the U.S. NSF office of Polar Programs and Physics Division, the U. of Wisconsin Alumni Research Foundation, the U.S. DoE, the Swedish Natural Science Research Council, the Swedish Polar Research Secretariat, the Knut and Alice Wallenberg Foundation, Sweden, the German Ministry for Education and Research, the US National Energy Research Scientific Computing Center (supported by the U.S. DoE), U.C.-Irvine AENEAS Supercomputer Facility, and Deutsche Forschungsgemeinschaft (DFG). C.P.H. received support from the EU 4th framework of Training and Mobility of Researchers, contract ERBFMBICT91551 and D.F.C. acknowledges the support of the NSF CAREER program. P. Loaiza was supported by a grant from the Swedish STINT program. P. Desiati was supported by the Koerber Foundation (Germany).

References

1. See HE3.1.06, 3.2.11, 4.1.14, 4.1.15, 4.2.05, 4.2.06, 4.2.07, 5.3.05, 5.3.06, 6.3.01, 6.3.02 and 6.3.07 in Proc. 26th Intl. Cosmic Ray Conf. (ICRC99), Salt Lake City, Aug. 1999 and at krusty.physics.utah.edu/icrc1999/.
2. E. Andres *et al.*, Astropart. Phys. **13** (2000) 120.
3. J.E. Dickinson *et al.*, Nucl. Inst. Meth. **A440** (2000) 95.
4. L. Bergström, J. Edsjö, and P. Gondolo, Phys. Rev. **D58** (1998) 103519.
5. J. Edsjö, private communication.
6. M.M. Boliev *et al.*, Nucl. Phys. **B48** (Proc. Suppl.) (1998) 83.
7. T. Montaruli *et al.*, HE 4.2.03 Proc. 26th Intl. Cosmic Ray Conf. (ICRC99) Salt Lake City, Aug. 1999 and M. Ambrosio *et al.*, astro-ph/0002492 (submitted to Ap. J.).
8. A. Okada *et al.*, astro-ph/0007003 and these proceedings.
9. S.W. Barwick *et al.*, to be published in Proc. of the XIX Intl. Conf. on Neutrino Physics and Astrophysics (Neutrino 2000), Sudbury, Canada, Jun. 2000 and at alumni.laurentian.ca/www/physics/nu2000/.

RECENT PROGRESS OF THE ANTARES PROJECT

JUAN JOSÉ HERNÁNDEZ
(on behalf of the ANTARES Collaboration)
IFIC – Instituto de Física Corpuscular
CSIC – Universitat de València, apdo. 22085, E-46071 Valencia, Spain

The ANTARES collaboration aims to build, deploy and operate a high energy cosmic neutrino detector of large surface under the Mediterranean sea. The ANTARES design for a 0.1 km^2 high energy cosmic neutrino detector is briefly explained and some of the results recently obtained with a demonstrator string immersed at 1100 m are shortly reviewed.

1 Introduction

The interest of the detection and study of high energy neutrinos as cosmic messengers has been extensively discussed in the literature [1]. Neutrino telescopes use the Earth as a shield against atmospheric muons and take advantage of polar ice or sea and lake water as active media for the detection of the Cherenkov light produced by the neutrino-induced muons [2]. At present, two first-generation cosmic neutrino detectors are actually operating [3].

The ANTARES collaboration aims to build a high energy cosmic neutrino detector under the Mediterranean sea. To this end a series of deployments of instrumented strings have been carried out which have led to the measurement of the relevant environmental properties [4]. Subsequently, a suitable site (30 km off the coast of La Seyne sur Mer in Southern France) for the installation of a 0.1 km^2 detector has been selected. Furthermore, a detailed proposal for the construction and deployment of such a detector has been written [5]. In this note, we briefly describe the main features of the ANTARES proposal for a 0.1 km^2 detector and report on the latest progress made by our collaboration.

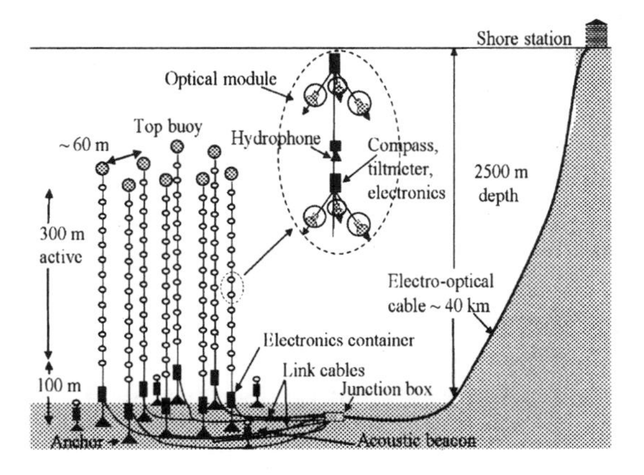

Figure 1. The ANTARES 0.1 km^2 detector concept.

2 The ANTARES detector design

A schematic drawing of the future ANTARES 0.1 km^2 detector is shown in figure 1. The detector consists of a total of 13 strings anchored to the sea bed and with the necessary buoyancy to stay taut. The horizontal separation between strings is around 60 m and their position on the sea bed follows approximately a spiral curve in order to avoid reconstruction ambiguities due to symmetries. The strings are in turn composed of storeys, whose main component are 3 optical modules. There are 30 storeys in each string, one every 12 m, starting 100 m above the sea bed. The optical modules in the storeys are pressure-resistant spheres housing one large-photocathode photomultiplier (PMT) and its associated electronics. Silicon gel is used to hold the PMT inside the optical module and to ensure a good optical coupling between the PMT and the sphere. The PMTs, which point downwards, have their axes making an angle of 45° with respect to the horizontal plane with their projections onto this plane being separated by 120°.

The string also contains several acoustic devices (rangemeters) which together with receivers on the sea bed (transponders) are able to determine accurately the profile of the string and the position of its components. A series of tiltmeters along the string completes the positioning system. A number of light sources along the string and in the optical modules act as well-known optical beacons in order to perform the time calibration of the detector.

3 Demonstrator string

In November 1999 a demonstrator string was immersed at a depth of around 1100 m in a site situated 37 km off the coast of Marseilles. This site (42° 59 N, 5° 17 E), different from the final ANTARES site, was chosen by virtue of its accesibility to a suitable electro-optical cable. The string (see figure 2) has a total length of 354 m and consists of 16 storeys, separated 14.6 m vertically from each other. Each storey has two 17" Benthos spheres, 1.5 m horizontally apart. Some of the spheres were instrumented: there were seven photomultipliers (six 8" and one 10") and six tiltmeters. The string also contained another 11 tiltmeters (satellites in the figure), a system for acoustic positioning and some other measuring devices. Although it does not have the final design of the 0.1 km^2-detector string, the demonstrator string had many components in common with it and allowed a wide range of studies, such as the test of the deployment techniques of a full-scale string, the checking of the data retrieval through electro-optical cable, working experience on the positioning systems, recording of down-going muon events, etc.

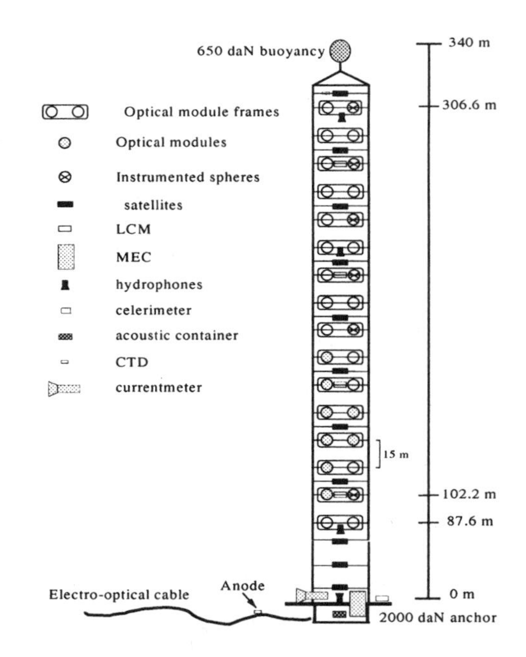

Figure 2. Schematic representation of the ANTARES demonstrator string.

3.1 Positioning systems

The string contained a total of 17 active tiltmeters (11 in the string spacers and 6 in Benthos spheres) which allowed to deduce the string shape and its movement in space. The string could be reconstructed as a straight line inclined at 2.5° to the vertical and with a negligible twist, headings being stable to within 2° over one week. The tilt of the string was stable to 0.2° over this same period. The reconstruction accuracy of this system alone was better than 10 cm in the horizontal coordinates and $\sim$1 cm in the vertical position.

In addition, an acoustic system of 3 rangemeters along the string and 4 acoustic transponders around its base was also available. The transponders were located on the sea floor at around 200 m from the string anchor. All these devices were able to emit and receive soundwaves (40–60 kHz) and measure their arrival time. A sound velocimeter with an accuracy better than 5 cm/s completed the set-up. Measurements were performed every 5 or 10 seconds. The distance between rangemeters or between transponders could be measured with residuals of the order of 1 cm. The 12 different transponder-rangemeter distances were simultaneously fit, giving residuals of the order of 5 cm, well within specifications.

3.2 Reconstructed events

More than 50000 coincidences in the seven photomultipliers were recorded. These coincidences are due to atmospheric (down-going) single and multiple muon events. Although with the information delivered by just one string the muon tracks cannot be reconstructed in space, a hyperbolic fit to the altitude versus time pattern of the hit PMTs allows the determination of their azimuthal angle.

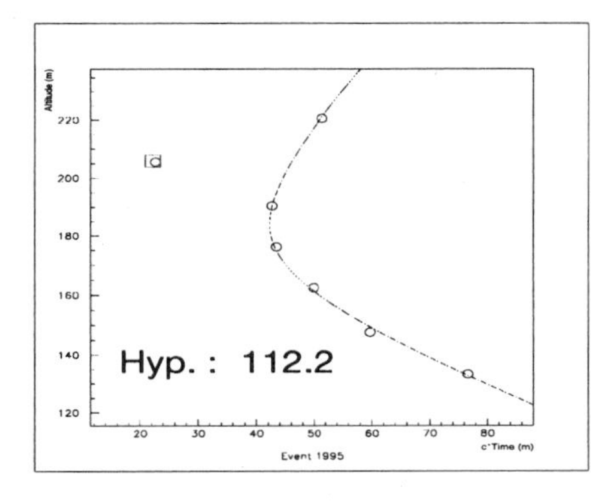

Figure 3. Example of reconstructed down-going muon with its hyperbolic fit superimposed. The boxed hit is due to ^{40}K.

An example of reconstructed track can be seen in figure 3. The value extracted for the azimuthal angle is given in the figure and the corresponding hyperbolic fit is shown superimposed. The hit in a box is most likely due to ^{40}K, identified as such by the reconstruction software and excluded from the fit. The distributions of timing residuals of all reconstructed events ($\sigma \sim 6$ ns) and of the azimuthal angle of the tracks are in agreement with the Monte Carlo expectations.

Conclusions

Important progress has been made by the ANTARES collaboration towards the installation of a cosmic neutrino telescope in the Mediterranean sea. A site with the appropiate characteristics has been selected, a variety of checks of the different components of such a detector have been performed and a proposal for a 0.1 km^2 detector has been put forward. Recently, a full-scale demonstrator string with some of the instrumentation to be employed in the final design has been deployed, operated and recovered. Extensive checks were made with this demonstrator string and very useful experience has been gained. Several tens of thousands atmospheric muon events were recorded and analysed.

The ANTARES collaboration aims to have 13 strings deployed by the end of 2003. This 0.1 km^2 detector will be a major step towards a 1 km^3 detector in the Mediterranean sea.

References

1. See for instance: F.Halzen, Phys. Rep. **333-334** (1-6) (2000) 349; T.K.Gaisser, F.Halzen and T.Stanev, Phys. Rep. **271** (5-6) (1996) 355, Phys. Rep. **258** (1995) 173
2. The idea dates back to the 60's, see: M.A. Markov in proceedings. *Proc.* ICHEP 60, Rochester (1960).
3. See for instance: BAIKAL: V.A. Balkanov, ICRC 99, Salt Lake City, USA, (1999); AMANDA: D.F. Cowen's, these
4. See for instance: P. Amram et al., Astr. Phys. **13** (2000) 127; N. Pallanque-Delabrouille, ICRC 99 *Proc.*, Salt Lake City, USA, (1999).
5. ANTARES proposal, astro-ph/9907432.

STATUS OF BOREXINO

M. PALLAVICINI

Istituto Nazionale di Fisica Nucleare, via Dodecaneso, 33, 16146 Genova, Italy
E-mail: pallas@ge.infn.it

Experiment Borexino, that is now under construction at the Gran Sasso laboratory in Italy, is devoted to provide the first direct measurement of the solar ^7Be neutrinos. It is expected to start data taking at the end of 2001. In this paper a brief description of its status and construction phases is given.

1 Introduction

Borexino is a solar neutrino detector whose aim is to provide the first, direct, real time measurement of the ^7Be ν flux.

This measurement could give a clue in order to solve the long standing solar neutrino problem, i.e. the well known disagreement between the theoretical calculations of the solar neutrino flux and the experimental measurements performed at different energies and with different techniques.

Indeed, the copious analysis of the data of the existing solar neutrino experiments[1] have pointed out the paradox of the missing ^7Be neutrinos, whose flux is not compatible with the ^8B flux measured by SuperKamiokande[5].

This scenario suggests either that non-standard physics is at work or that significant modification to the Standard Solar Model are required. This latter possibility, though it cannot be completely excluded, is now considered very unlikely, expecially because of the impressive results of Helio-Sismology[6].

The exciting scenario that this situation depicts is that non-standard effects like vacuum neutrino oscillations or resonant oscillations driven by the MSW effect[7] in the solar matter could give a reasonable solution to this problem. The different theoretical explanations predict very different ^7Be flux, so it is crucial to get a direct measurement of this quantity.

The aim of Borexino is to achieve this extremely difficult task. As it is shown in Table 1, the expected number of neutrino events in Borexino is very different for the pure vacuum, Large Mixing Angle and Small Mixing Angle MSW solutions. A good measurement with 10 % error could disentangle between these solutions.

Table 1. Expected number of ^7Be events in Borexino for different possible scenarios.

Model	Rate (ev/day)
MSS (BP 98)	50
MSW s.m.a.	12
MSW l.m.a.	32
MSW low	26
Vacuum	38

2 The experimental apparatus

Borexino is a large calorimeter, based on 300 t of ultra-pure organic scintillator that is both the neutrino target and the detector medium.

The scintillator (PPO dissolved in Pseudocumene) is contained in a transparent Nylon vessel (diameter 8.5 m) which is itself contained in large Stainless Steal Sphere (SSS, diameter 13.7 m) filled with pure Pseudocumene (PC) buffer. The whole detector, shown in Fig. 1, is installed inside a large Water Tank (WT, 18 m diameter and 17 m maximal height) which is filled with 3000 t of ultrapure water.

The ^7Be neutrinos are detected by observing the scintillator light emitted by elastically scattered electrons. This scintillator light is detected by 2214 PMTs mounted on the SSS. Being the ^{7}Be neutrinos monochromatic, the energy spectrum of the scattered electrons is almost flat up to about 600 KeV and then drops quickly. The number of photons collected, taking into account the scintillator yield (about 10.000 photons per MeV), the geometrical coverage and the PMT efficiency, is about 400 ph./MeV which gives a good energy resolution in the whole "neutrino window". For each event, the total charge and the absolute time of each photon hit is recorded. The energy deposit position is determined by triangulation techniques and the time distribution of the photon allows pulse shape discrimination to reject 90% of α decays.

To reduce the effect of the external backgrounds (expecially γ rays emitted by the PMT glass and by the SSS and radioactive background coming from the Nylon vessel), only the very inner 100 t of scintillator will define a "fiducial volume" for neutrino detection.

The main experimental problems come from the extreme radio-purity that is required in order to achieve the measurement. Indeed, Borexino is the first experiment that tries to identify a very low neutrino flux (about 50 events/day according to the Standard Solar Model) in an energy range that lies completely within the radioactive background range. In order to be able to perform the measurement, ^{14}C, the Uranium and Thorium chains contaminants, the cosmogenic ^{40}K must all be below the limits given in Table 2. Besides, all materials surrounding the detector must be carefully selected for radio-purity, because γ rays from external sources are also dangerous. For this reason all PMTs have been built from special glass and all materials used for their incapsulation and installation have been carefully checked. Special care have been devoted to the selection of the Nylon for the internal

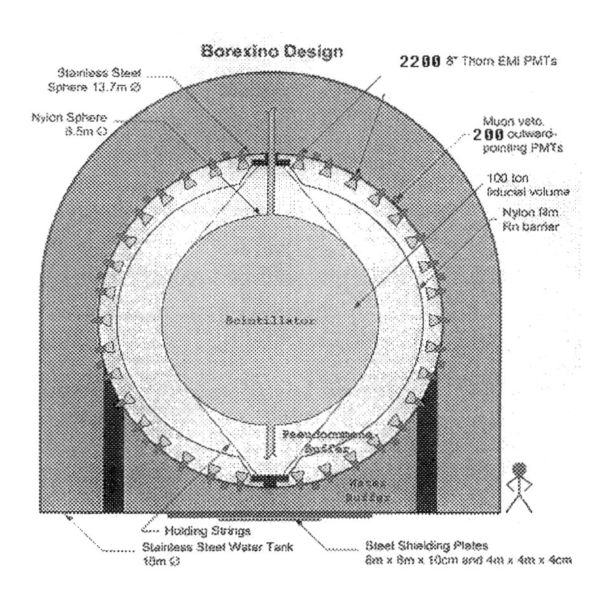

Figure 1. Conceptual drawing of the Borexino detector.

vessel, which must be very transparent, robust enough to withstand the buoyancy due to the density difference between pure PC and PPO/PC solution, and must have a very low permeability to Radon. Another Nylon shroud is foreseen to reduce the Radon diffusion toward the internal "fiducial volume".

The possibility to achieve this extremely high radio purity has been also checked, with very encouraging results, at the Counting Test Facility (CTF, see [2]-[4] for details).

Borexino is installed in the Laboratori Nazionali del Gran Sasso in Italy under 3500 w.e.m., where the cosmic rays flux is naturally reduced by a factor 10^6 with respect to sea level. Though, this flux is still about 1 event per m^2 per hour, which is definitely too much for Borexino. To reduce this background, the Cerenkov light produced by muons in the WT water will be detected by 200 additional PMTs which will provide an efficient muon veto. The muon reduction factor using the external muon veto and the internal detector selection criteria is expected to be 10^4.

Another crucial part of the experimental system is constituted by the fluid handling and purification systems. Pseudocumene purity is extremely important both from the radioactive point of view (U and Th concentrations, ^{40}K, dissolved Radon) and from the optical point of view because light mean free path must be of the order of several meters to allow an effective photon detection (dissolved oxygen is particularly dangerous and must be removed by means of nitrogen bubbling).

Borexino is equipped with several purification systems that can effectively remove these contaminants. In particular, there is an ultra-pure nitrogen plant, a water extraction system, a silica gel column, and distillation column. All these devices are already installed in Hall C at LNGS and will be operational at the beginning of 2001.

Table 2. Expected number of background events in Borexino for different radioactive contaninants, assuming concentration $10^{-16}g/g$ of U and Th chains $10^{-14}g/g$ of ^{40}K and $10^{-18}g/g$ of ^{14}C

Decay	Element	0.25-0.8 MeV
$\beta-\gamma$	^{238}U,^{234}Th chains	11.4
$\beta-\gamma$	^{40}K	1.5
α	^{238}U,^{234}Th chains	6.4
	Total	19.3

3 Conclusion

Borexino is expected to start data taking at the end of 2001. Right now (sep. 2000) the WT is finished, the SSS is finished and its precision cleaning is on-going. PMT sealing and packaging has started. Cable and PMT installation are scheduled to be completed by spring 2001 and then pseudocumene filling will continue for 6 months.

References

1. R. Davis Jr. *Prog. Part. Nucl. Phys.* **32**, 13 (1994)
   B. T. Cleveland et. al., *Astrophys. J.* **495** (1998)
   P. Anselmann et. al., *Phys Lett.* B **342**, 440 (1995)
   J. N. Abdurashitov et. al., *Phys Lett.* B **328**, 234 (1994)

2. G. Alimonti et. al., *Nucl. Instr. & Meth.* A **406**, 411 (1998)
3. G. Alimonti et. al., *Phys. Lett.* B **422**, 349 (1998)
4. G. Bellini et. al., *Nucl. Phys. Proc. Suppl.* **48**, 363 (1996)
5. Y. Fukuda et. al., *Phys. Rev. Lett.* **77**, 1683 (1996)
6. S. Basu et. al., *Astrophys. J.* **460**, 1064 (1996)
   S. Basu et. al., *Bull. Astron. Soc. India* **24**, 147 (1996)
7. L. Wolfenstein, *Phys. Rev.* D **17**, 2369 (1978)
   S.P. Mikheyev and A. Yu. Smirnov, *Yad. Fiz.* D **42**, 1441 (1985)
   S.P. Mikheyev and A. Yu. Smirnov, *Nuovo Cimento* C **9**, 17 (1986)

LENS: SPECTROSCOPY OF LOW ENERGY SOLAR NEUTRINOS

S. SCHÖNERT

on behalf of the LENS Collaboration

Max-Planck-Institut für Kernphysik, Saupfercheckweg 1, D–69117 Heidelberg and Research Center for Cosmic Neutrinos, Institute for Cosmic Ray Research, University of Tokyo, 5-1-5 Kashiwa-no-ha, Kashiwa, Chiba, 277-8582, Japan

E-mail: stefan.schoenert@mpi-hd.mpg.de

The LENS experiments will measure energy resolved sub-MeV solar electron-neutrinos (ν_e) in real time via inverse $\beta-$transition populating an isomeric state in the daughter nuclei. The subsequent de-excitation provides a delayed coincidence tag which discriminates against background. A liquid scintillation detector loaded with 20 t of Yb would yield an event rate of 190 pp- and 175 ^{7}Be neutrinos per year. Essential information on neutrino mixing and masses can be derived.

A flux deficit of solar neutrinos has been established unambiguously by the Homestake, (Super-)Kamiokande, Sage and Gallex experiments providing redundantly evidence for non-standard neutrino properties [1 2]. In particular, the gallium experiments, the only experiments which assay the primary pp-ν_e flux are the cornerstones for this conclusion.

Combined analysis of the solar neutrino data restrict neutrino oscillation parameters to several limited regions (c.f. Ref.[3] in this proceedings). Each solution makes a particular prediction about the energy and time dependence of the spectrum. They are best distinguishable at sub-MeV energies thus making the spectroscopy of pp-, ^{7}Be- and CNO-ν_e's the essential task for future experiments. Full information separately for charged and neutral current contributions are needed at low energies.

The scientific goal of LENS (**L**ow **E**nergy **N**eutrino **S**pectroscopy) is the energy-resolved measurement of the solar electron-neutrino (ν_e) spectrum at low energies via charged current interaction[4]. The ν_e flux of the primary p-p, ^{7}Be and CNO reactions in the sun will be detected by inverse β transition to an excited state of the daughter nuclei. The favoured nuclear system under investigation is ^{176}Yb-^{176}Lu (Fig. 1. A neutrino event is tagged by a coincidence measurement of a prompt electron yielding the neutrino energy and a delayed 72 keV γ-ray from isomeric de-excitation with a life time of 50 ns.

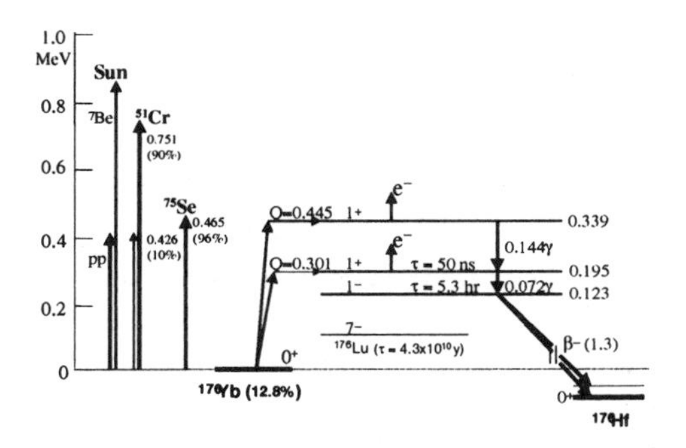

Figure 1. Level structure and tag for Yb-LENS

The present detector design aims at a modular structure of Yb-loaded liquid scintillation elements, This gives a high segmentation in order to discriminate against background caused by random coincidences of ^{14}C and other sources. According to current design studies the detector will contain 20 tons of natural Yb dissolved at 5-10% weight into a liquid scintillator.

The cross section for neutrino interaction has been derived from measurements of the Gamov-Teller matrix elements (B(GT)) via charge exchange reactions (p,n) and (^{3}He,t) on ^{176}Yb. The measurements were cali-

Table 1. Gamow-Teller resonance energies and strengths from $^{176}Yb(^3He,t)^{176}Lu$. The threshold energy for the various levels is given by $Q_\nu = E(level) + 106.2$ keV.

	Yb Level	B(GT)
1	194.5 keV	0.20(4)
2	338.9 keV	0.11(2)
3	3070 keV	0.62(8)

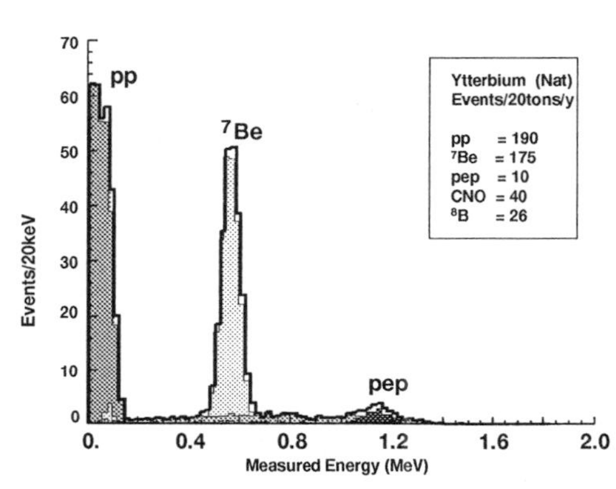

Figure 2. Solar neutrino spectrum assuming a standard solar neutrino flux.

brated with the known weak matrix element of ^{164}Dy-^{164}Ho and with the known Fermi matrix element of the isobaric analog states[5]. According to this values (Tab. 1) 190 pp- and 175 ^{7}Be-ν_e per year and 20 t Yb (nat) would be detected for a (non-existing) full standard solar model neutrino flux.

Though B(GT) measurements give reliable estimates, we plan to measure the cross section to the *two* excited levels with *two* different ν_e-sources: ^{51}Cr (3-8 MCi, τ = 40 d, E_ν = 751 keV (90%), 426 keV (10%)) and ^{75}Se (1 MCi, τ = 172 d, E_ν = 463 keV (96%)). Fig. 3 shows the detector response to ν_e-sources.

The LENS project is now in a pilot phase with the main focus on prototype detector development to be operated at the LNGS underground laboratories, Italy. Since February 2000 a formal collaboration has been established with the following contributing institutions: College de France, Dapnia CEA/Saclay (France), MPI für Kernphysik Heidelberg (Germany), LNGS INFN (Italy), RCNP Univ. Osaka (Japan), INR Moscow, Inst. of Phys. Chem. Moscow, Inst. of Chem. Technology Moscow (Russia), Bell Labs, Brookhaven Nat. Lab., Los Alamos Nat. Lab., Virginia Tech. (US).

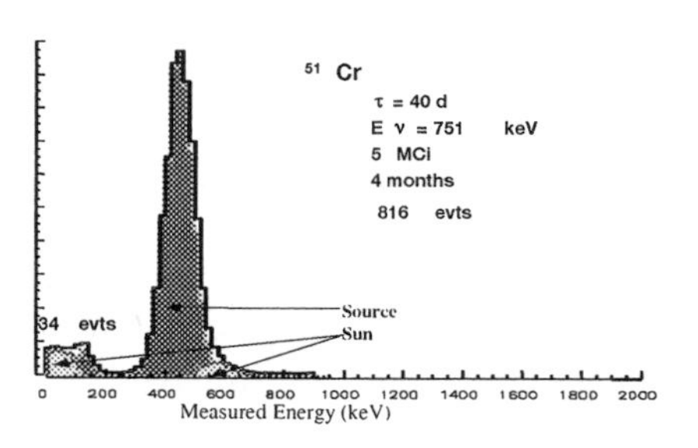

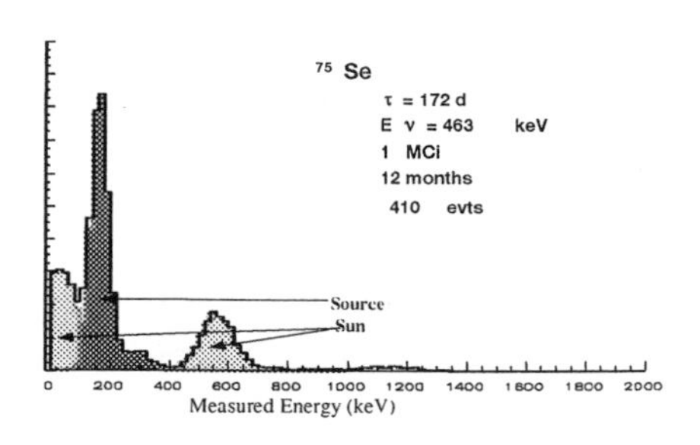

Figure 3. Response of Yb-LENS to MCi ν_e-calibration sources.

References

1. T. Kirsten, *Rev. of Mod. Phys.* Vol. 71, No. 4, 1213-1232 (1999).
2. for example: V. Berezinsky *et al*, *Phys. Lett.* B **185**, 365 (1996).
3. M.C. Gonzalez-Garcia, ICHEP2000
4. R. S. Raghavan, *Phys. Rev. Lett.* 78, 3618 (1997).
5. M. Fujiwara *et al.*, subm. to *Phys. Rev. Lett.*, Bhattacharya *et al.*, subm. to *Phys. Rev. Lett.*.

SIGNATURE NEUTRINOS FROM ULTRAHIGH-ENERGY PHOTONS

ALEXANDER KUSENKO
Department of Physics and Astronomy, UCLA, Los Angeles, CA 90095-1547
and
RIKEN BNL Research Center, Brookhaven National Laboratory, Upton, NY 11973
E-mail: kusenko@ucla.edu

At high red shift, the temperature of cosmic microwave background is sufficiently high for the ultrahigh-energy photons to pair-produce muons and pions through interactions with the background photons. At the same time, the radio background and magnetic fields are too weak to drain energy out of the electromagnetic cascade before the muons and pions are produced. Decays of the energetic muons and pions yield neutrinos with some distinctive spectral properties that can be detected and can indicate the presence of ultrahigh-energy photons at high red shift. The neutrino signature can help identify the origin of cosmic rays beyond the Greisen-Zatsepin-Kuzmin cutoff.

The origin of ultrahigh-energy cosmic rays [1], with energies beyond the Greisen-Zatsepin-Kuzmin (GZK) cutoff [2], remains an outstanding puzzle [3]. Many proposed explanations invoke new sources, such as superheavy relic particles [4,5] or topological defects [6,7], which can generate photons at both low and high red shifts. To understand the origin of the ultrahigh-energy cosmic rays (UHECR), it is crucial to distinguish such sources from more conventional astrophysical ones [8]. The latter tend to produce more protons than photons. In addition, the "astrophysical" candidate sources like, *e. g.*, active galactic nuclei, have formed at relatively low red shift. In contrast, topological defects could operate at much higher red shifts.

Sources of ultrahigh-energy photons that were active at red shift $z > 3$ can be identified by observation of neutrinos produced in interactions of energetic and background photons [9]. This may help understand the origin of UHECR.

At red shift z the cosmic microwave background radiation (CMBR) has temperature $T_{CMB}(z) = 2.7(1+z)$K. Because of this, at high red shift the photon-photon interactions can produce pairs of muons and charged pions, whose decays generate neutrinos. This is in sharp contrast with the $z \lesssim 1$ case, where the photons do not produce neutrinos[a] as they lose energy mainly by scattering off the radio background through electron-positron pair production and subsequent electromagnetic cascade [11]. The ratio of the CMBR density to that of universal radio background (RB) increases at higher z, and the process $\gamma\gamma_{CMB} \rightarrow \mu^+\mu^- \rightarrow e^+e^-\bar{\nu}_\mu\nu_\mu\bar{\nu}_e\nu_e$ can produce neutrinos. The threshold for this lowest-energy neutrino-generating interaction is $\sqrt{s} > 2m_\mu = 0.21$GeV, or

$$E_\gamma > E_{\rm th}(z) = \frac{10^{20}\text{eV}}{1+z} \qquad (1)$$

At $z < 1$ the main source of energy loss for photons is electromagnetic cascade that involves e^+e^- pair production (PP) on the radio background photons [11,3]. The radio background is generated by normal and radio galaxies. Its present density [12] is higher than that of CMB photons in the same energy range. The radio background determines the mean interaction length for the e^+e^- pair production. At red shift z, however, the comoving density of CMB photons is the same, while the comoving density of radio background is lower. Models of cosmological evolution of radio sources [13] predict a

[a] Neutrinos can be produced by sources of ultrahigh-energy protons through pion photoproduction [10]. Here we only consider sources of ultrahigh-energy photons.

sharp drop in the density of radio background at red shift $z \gtrsim 2$. Let z_R be the value of red shift at which the scattering of high-energy photons off CMBR dominates over their scattering off RB. Based on the models of RB [13], we take $z_R \sim 3$. Another source of energy losses in the electromagnetic cascade is the synchrotron radiation by the electrons in the intergalactic magnetic field (IGMF) [b]. This is an important effect for red shift $z < z_M$, where $z_M \sim 5$ corresponds to the time when IGMF is weak, and the synchrotron losses are not significant.

Let us now consider the propagation of photons at $z > z_{\rm min} = \max(z_R, z_M)$. In particular, we are interested in the neutrino-generating process $\gamma\gamma_{CMB} \to \mu^+\mu^-$. The threshold for this reaction is given in eq. (1). For $\sqrt{s} > 2m_{\pi^\pm} = 0.28$ GeV the charged pion production and decay can also contribute to the neutrino flux. Although the cross section for the electron pair production is higher than that for the muon pair production, neutrinos are nevertheless produced. This is because the high-energy photons are continuously regenerated in the electromagnetic cascade [3]. Since the energies of the two interacting photons are vastly different, either the electron or the positron produced in the reaction $\gamma\gamma_{CMB} \to e^+e^-$ has energy close to that of the initial photon. This electron undergoes inverse Compton scattering (ICS) and produces a photon with a comparable energy. As a result, the electromagnetic cascade creates a mixed beam of photons and electrons with comparable fluxes. Thanks to the regeneration of high-energy photons, the energy attenuation length $\lambda_{\rm eff}$ is much greater than the pair production interaction length $\lambda(\gamma\gamma_{CMB} \to e^+e^-)$.

For energies in the range of interest, $\lambda_{\rm eff} \gg \lambda(\gamma\gamma_{CMB} \to \mu^+\mu^-)$. Therefore, in the absence of dense radio background all photons with $E > E_{\rm th}$ pair-produce muons and pions before their energy is reduced by the cascade. Due to the kinematics, one of the two muons has a much higher energy than the other, in full analogy with the e^+e^- case. Muons decay before they can interact with the photon background. Each energetic muon produces two neutrinos and an electron. The latter can regenerate a photon via ICS. This process can repeat until the energy of a regenerated photon decreases below the threshold for muon pair production.

The flux of neutrinos relative to the observed flux of UHECR depends on the time evolution of the source. Sources of UHE photons, whether they are topological defects [7] or decaying relic particles with cosmologically long lifetimes [4], produce high-energy photons at some rate $\dot{n}_X$. One can parameterize [7] this rate as $\dot{n}_X = \dot{n}_{X,0}(t/t_0)^{-m}$, with $m = 0$ for decaying relic particles, $m = 3$ for ordinary string and necklaces, and $m \geq 4$ for superconducting strings [6,7].

Sources with $m = 3$ are of particular interest as possible candidates for the origin of UHECR [6,7]. The neutrinos produced by such sources at high red shift have energies [9] up to 10^{18} eV with a sharp cutoff below 10^{19}eV. The flux of 10^{18} eV neutrinos [9] is predicted to be $\sim 10^{-16}{\rm cm}^{-2}{\rm s}^{-1}{\rm sr}^{-1}$. This flux will be accessible to several experiments in the near future [14].

In summary, sources of ultrahigh-energy photons that operate at red shift $z > z_{\rm min} \sim 5$ produce neutrinos with energy $E_\nu \sim 10^{18}$eV. The flux depends on the evolution index m of the source. A distinctive characteristic of this type of neutrino background is a cutoff below 10^{19}eV due to the universal radio background and magnetic fields at $z < z_{\rm min}$. This is in contrast with sources of ultrahigh-energy protons that can produce neutrinos with energies up to the GZK cutoff and beyond.

I thank V. Berezinsky for very helpful discussions. This work was supported in part by the US Department of Energy grant DE-

[b] I thank V. Berezinsky for pointing this out to me.

FG03-91ER40662, Task C, as well as by a grant from UCLA Council on Research.

References

1. M. Takeda *et al.*, Phys. Rev. Lett. **81**, 1163 (1998); M.A. Lawrence, R.J. Reid and A.A. Watson, J. Phys. G **G17**, 733 (1991); D. J. Bird *et al.*, Phys. Rev. Lett. **71**, 3401 (1993); Astrophys. J. **424**, 491 (1994); N. Hayashida *et al.*, Astrophys. J. **522**, 225 (1999) [astro-ph/0008102].
2. K. Greisen, *Phys. Rev. Lett.* **16**, 748 (1966); G. T. Zatsepin and V. A. Kuzmin, *Pisma Zh. Eksp. Teor. Fiz.* **4**, 114 (1966).
3. For review, see, *e.g.*, P. L. Biermann, J. Phys. G **G23**, 1 (1997); P. Bhattacharjee and G. Sigl, Phys. Rept. **327**, 109 (2000).
4. V. Berezinsky, M. Kachelriess, and A. Vilenkin, Phys. Rev. Lett. **79**, 4302 (1997); V. A. Kuzmin and V. A. Rubakov, Phys. Atom. Nucl. **61**, 1028 (1998) [Yad. Fiz. **61**, 1122 (1998)]; V. Kuzmin and I. Tkachev, JETP Lett. **68**, 271 (1998); M. Birkel and S. Sarkar, Astropart. Phys. **9**, 297 (1998); D.J. Chung, E.W. Kolb, and A. Riotto, Phys. Rev. Lett. **81**, 4048 (1998); Phys. Rev. **D59**, 023501 (1999); K. Benakli, J. Ellis, and D. V. Nanopoulos, Phys. Rev. **D59**, 047301 (1999); Phys. Rev. **D59**, 123006 (1999); G. Gelmini and A. Kusenko, Phys. Rev. Lett. **84**, 1378 (2000). J. L. Crooks, J. O. Dunn, and P. H. Frampton, astro-ph/0002089.
5. For review, see, *e.g.*, V. A. Kuzmin and I. I. Tkachev, Phys. Rept. **320**, 199 (1999).
6. A. Vilenkin, Phys. Rept. **121**, 263 (1985); A. Vilenkin and E. P. S. Shellard, *Cosmic strings and other topological defects*, Cambridge University Press, Cambridge, England, 1994; M. B. Hindmarsh and T. W. Kibble, Rept. Prog. Phys. **58**, 477 (1995).
7. C. T. Hill and D. N. Schramm, Phys. Rev. **D31**, 564 (1985); C. T. Hill, D. N. Schramm, and T. P. Walker, Phys. Rev. **D36**, 1007 (1987). P. Bhattacharjee,
C. T. Hill, and D. N. Schramm, Phys. Rev. Lett. **69**, 567 (1992). V. Berezinsky, P. Blasi and A. Vilenkin, Phys. Rev. **D58**, 103515 (1998); V. S. Berezinsky and A. Vilenkin, hep-ph/9908257.
8. E. Waxman, Phys. Rev. Lett. **75**, 386 (1995); G. R. Farrar and P. L. Biermann, Phys. Rev. Lett. **81**, 3579 (1998); A. Dar, A. De Rujula and N. Antoniou, astro-ph/9901004. G. R. Farrar and T. Piran, Phys. Rev. Lett. **84**, 3527 (2000); E. Ahn, G. Medina-Tanco, P. L. Biermann, and T. Stanev, astro-ph/9911123.
9. A. Kusenko and M. Postma, hep-ph/0007246.
10. C. T. Hill and D. N. Schramm, Phys. Lett. **B131**, 247 (1983); S. Yoshida, G. Sigl, and S. Lee, Phys. Rev. Lett. **81**, 5505 (1998); G. Sigl, S. Lee, P. Bhattacharjee, and S. Yoshida, Phys. Rev. **D59**, 043504 (1999)
11. V. Berezinsky, Sov. J. of Nucl. Phys. **11**, 222 (1970).
12. R. J. Protheroe and P. L. Biermann, Astropart. Phys. **6**, 45 (1996).
13. J. J. Condon, Astrophys. J. **284**, 44 (1984).
14. R.M.Baltrusaitis *et al.*, Phys. Rev. **D31**, 2192 (1985); D. B. Cline and F. W. Stecker, astro-ph/0003459; J. Alvarez-Muniz and F. Halzen, astro-ph/0007329.

NEUTRINOS FROM STELLAR COLLAPSE: EFFECTS OF FLAVOUR MIXING

D. INDUMATHI

The Institute of Mathematical Sciences, CIT Campus, Chennai 600 113, India
E-mail: indu@imsc.ernet.in

We discuss neutrino emission from supernovae and the effect of non-vanishing masses and mixings among neutrino flavours on the detection of such neutrinos by a water Cerenkov detector.

1 Neutrinos from stellar collapse

Gravitational collapse of large mass stars (8–10 $M_\odot$) is associated with copious neutrino emission. The supernova SN1987A was the first (and only) one whose neutrino emission was observed.

The effect of neutrino masses and mixing on the detection of neutrinos from stellar collapse by a water Cerenkov detector has been analysed in Ref. [1,2,3,4]. This talk is based on the work of Dutta et al.[1].

We concentrate here on the so-called cooling neutrinos where approximately equal amounts of energy are radiated in all neutrino (and antineutrino) species. The flux distribution is essentially a thermal spectrum. Electron-type neutrinos (and antineutrinos) have both charged and neutral current interactions with matter; hence they are emitted at lower average energies (temperatures) than μ or τ type neutrinos. The average energy of $\nu_\mu = \nu_\tau = \nu_x = \bar{\nu}_x$ is about 25 MeV while that of electron (anti)neutrinos is about (16)11 MeV. The flux distribution is largely model–independent and is therefore sensitive to neutrino properties such as mass and mixing.

2 Neutrino mixing in dense matter

Analyses of the solar neutrino, atmospheric neutrino and lab neutrino experiments such as CHOOZ have all indicated that possibly neutrinos have masses, mix and therefore oscillate. Their mass-squared differences and mixing angles are constrained by an understanding of all this data.

Within a 3-flavour scenario of mixing, data imply that the mass eigenstates are mixtures of the flavour eigenstates,

$$U^v = U_{23}(\psi) \times U_{phase} \times U_{13}(\phi) \times U_{12}(\omega) \ ,$$

with $\delta_{13} \gg \delta_{12}$, where δ_{ij} is the mass squared difference, $m_i^2 - m_j^2$. The CHOOZ experiment [5] limits the angle $\phi < 6°$. Solar neutrino data constrains $\delta_{12} \leq 10^{-5}$ eV^2 but allows for both small and large ω, with and without MSW matter effects [6]:

1. $\sin^2(2\omega) = 6.0 \times 10^{-3}, \delta m_{12}^2 = 5.4 \times 10^{-6}$ eV^2 (SMA). The small angle MSW solution.
2. $\sin^2(2\omega) = 0.76, \delta m_{12}^2 = 1.8 \times 10^{-5}$ eV^2 (LMA). The large angle MSW solution.
3. $\sin^2(2\omega) = 0.96, \delta m_{12}^2 = 7.9 \times 10^{-8}$ eV^2 (LMA-V). The large angle vacuum solution.

In highly dense matter in the supernova core, matter effects [7] drive the mixing angle ϕ for neutrinos to

$$\phi_m \stackrel{A\to\infty}{\to} \frac{\pi}{2} \ .$$

Here A is the usual matter dependent term. On the other hand, for antineutrinos,

$$\phi_m \stackrel{A\to\infty}{\to} 0 \ .$$

Hence ν_e and $\bar{\nu}_e$ are produced as pure mass eigenstates so that the survival probability of $|\nu_e\rangle$ is simply the projection of $|\nu_3\rangle$ onto

Table 1. Events with electron energy greater than 8 MeV. Maximal values of ϕ and ω are assumed for mixing in the 3-flavour case.

	Channel	No Osc.	Osc.
νe	forward	5	7
$\bar{\nu}_e p$	isotropic	272	323
$\nu_e, \bar{\nu}_e O$	backward	5	32

the ν_e state in the detector, $P_{ee} = \sin^2 \phi$ (provided there are no Landau Zener (LZ) jumps). The adiabatic survival probability is small because of ϕ. On the other hand, $\overline{P}_{ee} = \cos^2 \phi \cos^2 \omega$ and depends on the (12) mixing angle as well.

As a consequence, the fluxes of neutrinos detected on earth, F_i, are not the same as those emitted at the supernova F_i^0. For example,

$$F_e = F_e^0 - (1 - P_{ee})(F_e^0 - F_x^0) .$$

Here $F_x = F_\mu = F_\tau$. Because of the mixing, the colder electron spectrum acquires an admixture of the hotter μ, τ spectra and vice versa. The net result is an enhanced ν_e and $\bar{\nu}_e$ event rate and a reduced ν_x and $\bar{\nu}_x$ rate (the latter may or may not be observable).

3 Interaction at the detector

There are three kinds of interaction of the neutrinos: with electrons, protons and oxygen in the detector. The first is forward peaked while the second has the largest cross-section and is approximately isotropic. While the CC interaction with oxygen increases rapidly with energy [8], it has a threshold of (11.4) 15.4 MeV for electron (anti)neutrinos. Also, these events are almost completely backward peaked for a hot spectrum.

On mixing, the otherwise colder electron neutrinos acquire a hot admixture of $\nu_{\mu,\tau}$ neutrinos which can excite the oxygen channel. This leads to a large increase in the number of backward peaked events and is a strong indicator of mixing.

4 Results

We use the supernova model based on the model of Ref. [9]. The number of events obtained for a 1 kTon detector for a supernova 10 kPc away is shown in Table 1 for 3-flavour mixing. The 4-flavour analysis is more complicated as they depend on detailed constraints from solar neutrino analysis.

We have also computed the events as a function of the detected electron energy. Here we include the effects of non-adiabaticity as well as the 4-flavour analysis including a sterile neutrino. In the 3-flavour case, the lower doublet is separated by the mass squared difference suggested by the solar neutrino analysis, while the upper eigenstate is separated from the doublet by the larger atmospheric neutrino scale. In the 4-flavour case, we use a two-doublet model with a large (1 eV^2) gap between the lower two and upper two doublets, motivated by the LSND result. The separation in the lower and upper doublets is determined by the solar and atmpospheric scales respectively.

Fig. 1 shows the event rates for neutrinos. The large increase with (adiabatic) mixing in the oxygen channels is clearly seen. When non-adiabatic effects are included, there is not much change in the 3-flavour case, while there is severe depletion in the 4-flavour case. The extent of depletion is not very sensitive to ω.

Fig. 2 shows the event rates for antineutrinos. The 3-flavour mixing results in an increase in the event rate for $\bar{\nu}_e$ at larger energies. In the 4-flavour case, there is actual depletion of the spectrum due to loss into the sterile channel, proportional to ω.

A combination of detection of both isotropic and forward events at lower energies can help distinguish the 3 and 4 flavour solutions for the non-adiabatic case.

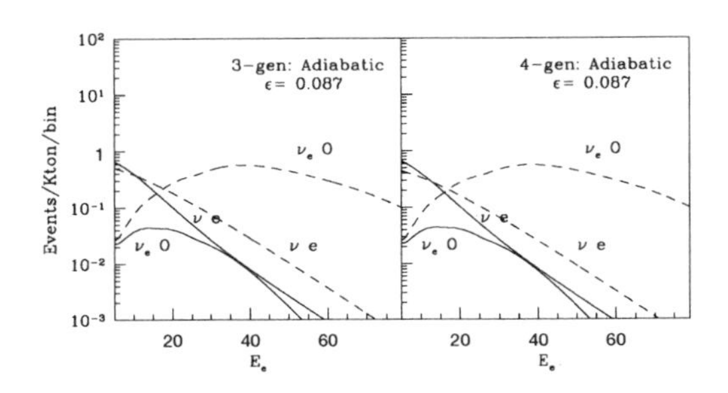

Figure 1. Event rates for neutrinos; $\epsilon = \sin\phi$ as a function of the electron energy E_e. (Dashed) solid lines represent (no) mixing cases.

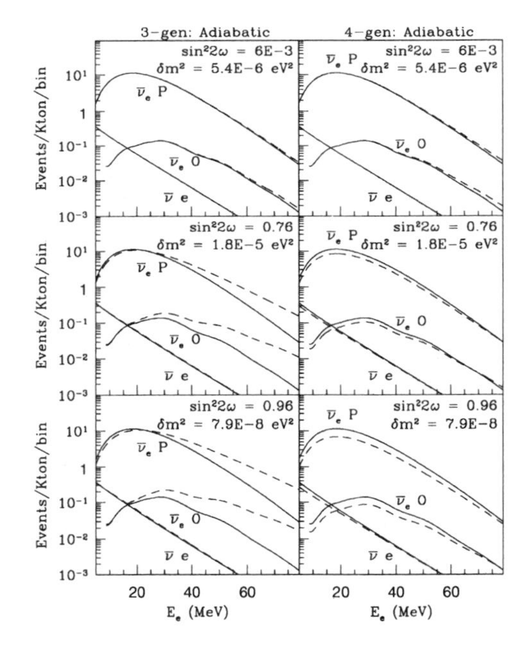

Figure 2. Antineutrino event rates for $\epsilon = 0.087$ as a function of electron energy E_e for different values of ω.

5 Conclusions

- Neutrino mixing and oscillations give rise to characteristic signatures of neutrino emissions from supernovae.
- Neutrinos and antineutrinos of all flavours are emitted; hence different mixing angles can be probed.
- Large ν_e-$\nu_{\mu,\tau}$ or $\bar{\nu}_e$-$\bar{\nu}_{\mu,\tau}$ mixing is signalled by backward peaked events.
- In some parts of the parameter space, as shown in the figures, it is possible to distinguish 3- and 4-flavour mixing.

Finally, how frequent are supernovae? This is not very well understood, although once every 30 years in nearby galaxies is considered not unreasonable. One must therefore be patient.

References

1. Gautam Dutta, D. Indumathi, M.V.N. Murthy, G. Rajasekaran, *Phys. Rev.* D **61**, 013009 (2000); hep-ph/0006171, to appear in *Phys. Rev.* D (2000).
2. J. F. Beacom and P. Vogel, *Phys. Rev.* D **58**, 053010 (1998); *ibid* 093012.
3. Sandhya Choubey and Kamales Kar, preprint hep-ph/9905327.
4. Amol S.Dighe and Alexei Yu. Smirnov, *Phys. Rev.* D **62**, 033007 (2000).
5. M. Apollonio et al., CHOOZ Collab., *Phys. Lett.* B **420**, 397 (1998); *Phys. Lett.* B **466**, 415 (1999).
6. J.N. Bahcall, P.I. Krastev, and A.Yu. Smirnov, *Phys. Rev.* D **58**, 96016 (1998).
7. T. K. Kuo and J. Pantaleone, *Phys. Rev.* D **37**, 298 (1988).
8. W. C. Haxton, *Phys. Rev.* D **36**, 2283 (1987).
9. T. Totani, K. Sato, H.E. Dalhed, J. R. Wilson, *Astrophys. J.* **496**, 216 (1998).

Parallel Session 9

Particle Astrophysics and Cosmology (excluding neutrino)

Conveners: Keith A. Olive (Minnesota) and
Michel Spiro (Saclay)

RECENT RESULTS AND OBSERVATIONAL STATUS OF TEV GAMMA-RAY ASTRONOMY

A. KOHNLE
Max-Planck-Institut für Kernphysik, Postfach 103980, D-69029 Heidelberg, Germany
E-mail: Antje.Kohnle@mpi-hd.mpg.de

TeV gamma-ray astronomy has come of age in the last ten years: there are several well-established and well-studied sources in the TeV sky; presently there are nine imaging atmospheric Cherenkov telescope (IACT) facilities and several next-generation instruments coming online 2001 to 2004 with an order of magnitude improved sensitivity and lower energy threshold. Solar farm experiments have already demonstrated energy thresholds of 50 GeV, and large field-of-view and high duty cycle detectors exist to monitor bright transients and search for gamma-ray burst emission.

1 Introduction

The aims of TeV gamma-ray astronomy are manifold, both in astrophysics and particle physics: to find the sources of cosmic rays, the understanding of the acceleration mechanisms in gamma-ray sources and the physical source parameters, the understanding of flux variability, the search for new source classes with expected TeV emission, "observational cosmology", such as the determination of the infrared extragalactic photon background from gamma-ray attenuation and cascading, and the search for dark matter and for quantum gravity effects.

From the wealth of observations and theoretical models, only a very limited sample can be given here. The reader is referred to [1 2 3] for recent more extensive reviews of the field.

2 The IACT Technique

Cosmic gamma-ray photons interact with the atmosphere and produce electromagnetic cascades with ~1000 electrons and positrons at the shower maximum in ~8 km altitude for a 1 TeV primary photon. The electrons and positrons emit Cherenkov radiation with an angle of ~1 degree, so that the signal on the ground is a short, few ns long flash of light with a few 100 ph/m^2 covering an area of 240 m diameter. Charged particles, a background a factor of >1000 above the photon signal, also produce air showers with electromagnetic subcascades. Cherenkov telescopes image the light onto a closely packed matrix of photomultiplier tubes in the focal plane. On a statistical basis, one can reject hadronic events due to their more irregular shape and on the basis of directional information. The IACT observatories worldwide (Whipple, Crimea, Shalon, Cangaroo, HEGRA, CAT, Durham, Tactic, Seven TA) have energy thresholds of 0.25 to 1 TeV. As one example, the HEGRA telescope system has an energy flux sensitivity of 10^{-11} erg/(cm^2 s) for a 5 σ detection for 1 h of observation.

3 Sources

The present TeV source catalog comprises 13 sources. Four of these are confirmed, i.e., seen with high (> 5 σ) significance by more than one group; the others are typically 5 to 7 σ detections of a single group. There are also interesting non-detections, i.e., upper limits at the current sensitivity of instruments, of diffuse emission from the Galactic plane, pulsed emission from pulsars, and non-blazar type AGNs. Only a few of the sources can be described here.

3.1 Galactic Sources

The Galactic sources comprise three plerions (Crab, Vela, PSR1706-44), three shell-type supernova remnants (SN1006, RXJ1713.7-3946, Cas-A), and one X-ray binary (Cen-X3). The Crab and PSR1706-44 are confirmed sources.

The "standard candle" of TeV astronomy is the Crab Nebula, the remains of a supernova explosion in the year 1054 AD. It was the first detected source (by the Whipple group in 1989), and is the strongest persistent source known. The current most sensitive instruments detect a few gammas per minute from the Crab. The spectrum from radio to TeV energies can be understood in the framework of synchrotron-inverse Compton models, where the electrons emitted from the pulsar magnetosphere are accelerated in a termination shock and then emit synchrotron radiation in the magnetic field of the nebula. TeV emission results from inverse Compton interactions of the synchrotron-emitting electrons with various photon fields (synchrotron, dust emission, cosmic microwave background). There is now good agreement between the TeV spectra of different groups [4], whereby the spectrum can be fit by a single power law. The HEGRA spectrum between 1 and 20 TeV is $dN/dE = 2.79 \pm 0.02 \pm 0.5 \cdot 10^{-11} (E/TeV)^{-2.59 \pm 0.03 \pm 0.05}$ ph/(cm^2 s TeV).

It is widely believed that shell-type supernova remnants (SNRs) are the primary sources of cosmic rays up to $\sim 10^{15}$ eV. Detailed models of diffusive shock acceleration in different types of SNRs make predictions of the expected TeV fluxes [5 6 7]. There are now three TeV detections of shell-type SNRs, and although the results are not inconsistent with a hadronic origin, all results are consistent with electron acceleration. There are also interesting upper limits of SNRs that start to constrain the parameter space of hadronic models [8].

The first detection of TeV emission from a shell-type SNR was SN1006 by the Cangaroo experiment at energies above 1.7 TeV [9]. The ASCA X-ray image of SN1006 shows two bright rims of non-thermal emission implying the presence of electrons with energies up to 100 TeV. The TeV gamma-ray emission is localized along the NE rim, and can be modeled by energetic electrons generating inverse Compton radiation seeded by the cosmic microwave background (see e.g. [10]).

Recently, HEGRA has reported a 5 σ detection of the ~300 year old SNR Cassiopeia A in a very deep exposure of 232 h [11]. The integral flux of 33 milli-Crab makes this the weakest TeV source detected to date. Models of the source emission show that the spectral index might provide discrimination between a leptonic and a hadronic origin of the TeV flux [12].

3.2 Extragalactic Sources

All extragalactic TeV sources belong to the blazar class of active galactic nuclei (AGNs). AGNs are believed to consist of an accretion disc surrounding a central massive black hole and relativistic outflows perpendicular to the disc, called jets. In blazars, the jets are assumed to be closely aligned with the viewing angle. Models for TeV emission from blazar jets divide into two classes depending on the assumed jet content (hadronic models: p e^-, leptonic models: e^+ e^-) [13]. Hadronic models are attractive as they involve the acceleration of cosmic rays up to 10^{19} eV, whereas leptonic models naturally explain the current data.

Of the six TeV blazars detected to date (Mkn 421, Mkn 501, 1ES2344+514, PKS2155-304, PKS1959+650, 3C66A), Mkn 421 and Mkn 501 are confirmed by many groups, and we will concentrate on these here. The first outstanding feature of blazars is their flux variability. In 1997, Mkn 501 underwent a giant outburst lasting many months, with a mean source strength above

1 Crab and flares up to 10 Crab. Variability was seen on timescales down to hours. In Mkn 421, short flares with doubling times of 15 to 30 minutes have been detected [14]. This places severe constraints on the size of the emission volume and implies bulk Lorentz factors of 5 to 10. The Whipple group was able to place limits on quantum gravity effects with a relevant energy scale of $E_{QG}/\xi > 4 \cdot 10^{16}$ GeV by looking for an energy-dependent time dispersion of the TeV photons [15].

HEGRA and Whipple detected no significant change in the spectral index for the different flux states during the 1997 outburst of Mkn 501 [16]. CAT sees a hardening of the spectrum with increasing flux [17], the effect being most marked at low energies. These results are not necessarily in conflict due to the different energy thresholds and different data sets. HEGRA sees a softening of the spectral index by 0.44±0.1 of the Mkn 501 low state spectrum in 1998 and 1999 compared to the 1997 flare spectrum [18]. Modeling the spectral evolution with flux intensity and during a flare interprets the results in terms of particle injection, acceleration, cooling and escape.

Since 1995, multi-wavelength campaigns with instruments in other wavelength bands, in particular with X-ray satellites, have been carried out to simultaneously observe TeV blazars. These show a tight correlation of the X-ray and TeV flux, with larger amplitude variations in the TeV region. Analysis of contemporaneous HEGRA and RXTE Mkn 501 1997 data show the TeV flux rising quadratically with the X-ray flux [19]. These results can be interpreted in terms of leptonic synchrotron-self-Compton models. Simultaneously fitting the X-ray and the TeV spectrum, one obtains values of the bulk Lorentz factor, the size of the emission region, and the magnetic field strength.

The time-averaged spectrum of Mkn 501 shows an intriguing curvature, which may be expected due to gamma-ray attenuation by pair creation with photons of the extragalactic background light ($\gamma_{EBL} + \gamma_{TeV} \rightarrow e^+ + e^-$) [18, 20]. The cross-section is maximal for $E_{TeV} \cdot E_{EBL} \sim 3m^2c^4$. The TeV attenuation can thus be used to derive the infrared background, a probe of the history of star-formation in our Universe. However, one clearly needs a larger sample of sources at different redshifts to make more detailed studies.

3.3 *Dark Matter Search*

Radiation from wimp annihilation or relic particle decay from the Galactic halo or close galaxies with high M/L ratios (M87, dwarf spheroidal galaxies) could lead to a detectable TeV signal for next-generation instruments, as well as sensitivity to some regions of parameter space [21, 22]. In particular, neutralino annihilation gives rise to line radiation via $\chi\chi \rightarrow \gamma\gamma$ and $\chi\chi \rightarrow \gamma Z^\circ$, which would be a "smoking gun" signal.

4 The Future of Very High Energy Astronomy

Ideally, one would like to lower the energy threshold and extend the energy range (with good energy resolution), increase the sensitivity (with good angular resolution and background rejection), and increase the duty cycle and the field-of-view. This can only be achieved with a combination of detection techniques that are tradeoffs between the various objectives.

4.1 *Low energy Threshold Instruments*

These detectors exploit the larger but cruder mirror areas of existing arrays of solar heliostats to achieve a low energy threshold. Secondary optics in the tower image the light from the heliostats onto photomultiplier tubes. This approach is being pursued by the CELESTE, STACEE, Solar-2, and GRAAL collaborations. First results show great promise: CELESTE has detected both the Crab and Mkn 421 at high (7-8 σ) significance in 1999/2000 observations with an energy threshold of 50 GeV [23]. STACEE has detected the Crab at 7 σ significance 1998/1999 with an energy threshold of 190 GeV, and is now starting data-taking with 48 heliostats [24].

4.2 Large Aperture / High Duty Cycle Instruments

These instruments, though having lower sensitivity, are ideal for monitoring of bright transients and for gamma-ray burst searches. The Tibet-III scintillator array at 4300 m a.s.l. achieves an energy threshold of a few TeV. MILAGRO is a water Cherenkov experiment with an energy threshold below 1 TeV. ARGO will start operation in 2001 at the Tibet-III site using resistive plate counters. The potential of these intruments is shown by the detection of the Crab and Mkn 501 by the Tibet experiment [25] [26], and by the possible detection of the gamma-ray burst GRB970417a by the Milagro prototype detector, with a $2 \cdot 10^{-3}$ chance probability [27].

4.3 Next-generation IACT Experiments

Four next-generation IACT experiments, Cangaroo-III, H.E.S.S., MAGIC, and VERITAS, are coming online in the years 2001 to 2004 with an order of magnitude improvement in sensitivity and an order of magnitude lower energy threshold (30 to 100 GeV). With these instruments, GeV - TeV astronomy has an exciting future ahead: high source statistics of known sources, searches for new sources believed to emit high energy radiation, detailed spectroscopy, and mapping of extended emission. This will ultimately lead to a better understanding of nonthermal processes in the Universe.

References

1. M. Catanese and T. Weekes, *PASP* **111, 764**, 1193 (1999).
2. R.A. Ong, *Phys. Rep.* **305**, 93 (1998).
3. F.A. Aharonian and C.W. Akerlof, *Annu. Rev. Nucl. Part. Sci.* **47**, 273 (1997).
4. F.A. Aharonian et al., *ApJ*, in press
5. L.O'C. Drury, F.A. Aharonian and H.J. Völk, *A & A* **287**, 959 (1994).
6. E.G. Berezhko and H.J. Völk, *Astropart. Phy.* **7**, 183 (1997).
7. M.G. Baring, D.C. Ellison, S.P. Reynolds, I.A. Grenier and P. Goret, *ApJ* **513**, 311 (1999).
8. J.H. Buckley et al., *A & A* **329**, 639 (1998).
9. T. Tanimori et al., *ApJ* **497**, L25 (1998).
10. A. Mastichiadis and O.C. de Jager, *A & A* **311**, L5 (1996).
11. G. Pühlhofer et al., Poster presented at the *International Symposium on High Energy Gamma-Ray Astronomy* , Heidelberg, Germany, June 26-30, 2000.
12. A.M. Atoyan, R.J. Tuffs, F.A. Aharonian and H.J. Völk, *A & A* **354**, 915 (2000).
13. M. Sikora, in *Proc. International Symposium on High Energy Gamma-Ray Astronomy* , (AIP, Germany, 2000) in press.
14. J.A. Gaidos et al., *Nature* **383**, 319 (1996).
15. S.D. Biller et al., *Phys.Rev.Lett.* **83**, 2108 (1999).
16. F.A. Aharonian et al., *A & A* **349**, 11 (1999).
17. A. Djannati-Atai et al., *A & A* **350**, 17 (1999).
18. F.A. Aharonian et al., *ApJ* in press.
19. H. Krawczynski, P.S. Coppi, T. Maccarone and F.A. Aharonian, *A & A* **353**, 97 (2000).
20. F. Krennrich et al., *ApJ* **511**, 149 (1999).
21. L. Bergström, P. Ullio and J.H. Buckley, *Astropart. Phys.* **9**, 137 (1998).
22. E.A. Baltz, C. Briot, P. Salati, R. Taillet and J. Silk, *Phys.Rev.* D **61**, 23514, (2000)
23. M. De Naurois et al., in *Proc. International Symposium on High Energy Gamma-Ray Astronomy* , (AIP, Germany, 2000) in press.
24. S. Oser et al., submitted to ApJ, astro-ph/0006304
25. M. Amenomori et al., *ApJ* **525**, L93 (1999).
26. M. Amenomori et al., *ApJ* **532**, 302 (2000).
27. R. Atkins et al., *ApJ* **533**, L119 (2000).

THE PIERRE AUGER EXPERIMENT AT THE SOUTHERN HEMISPHERE

M. KLEIFGES FOR THE AUGER COLLABORATION
Forschungszentrum Karlsruhe, Institut für Kernphysik, Postfach 3640, 76021 Karlsruhe, Germany,
E-mail: matthias.kleifges@hpe.fzk.de

The international Pierre Auger Project will measure the cosmic ray energy spectrum above 10^{19} eV as a function of the mass of the primaries and will determine the source distribution. Existing experiments have proven the existence of particles with such high energy, but their origin could not be identified due to the low event rate and no satisfying explanation for the acceleration mechanism is known. The combination of water Čerenkov and fluorescence detector stations spread over an area of 3000 km^2 in Argentina is the approach of the Auger collaboration to overcome the limitations of the present situation. We present the design of the experiment with emphasis on the fluorescence detector electronics and readout system.

1 Introduction

Recent results from AGASA[1] and other[2] experiments show that cosmic ray particles with energies $\geq 10^{20}$ eV hit the atmosphere roughly once per second. To explain this observation an efficient particle accelerator or particle factory is required as a source of these particles. Cosmic rays of energies below 10^{16} eV can be explained with the Fermi shock acceleration at supernovae remnants, but the accelaration to the extremely high energy region is still a mystery. Indeed, we have good reasons to believe that particles with energies $\geq 10^{19}$ eV have an entirely different origin than those of lower energy.
Greisen, Zatsepin and Kuz'min pointed out that protons, nuclei, and photons with energies above 5×10^{19} interact strongly with the cosmic microwave background[3]. Due to this interaction the propagation of cosmic rays over distances larger than 100 Mpc should reduce their energy below the GZK cutoff energy of 5×10^{19} eV. The sources of the ultra-high energetic particles recently observed are therefore expected to exist within a distance below some 100 Mpc from earth.
So far, 13 events above 10^{20} eV have been recorded[4]. More experimental results are necessary to investigate the nature of these extremely high energy particles, to understand the propagation in the universe and to find astrophysical sources.
In the following sections we present the design of the Pierre Auger southern experiment, which is currently being built near Malargue in the Mendoza province of Argentina. Later, a second experiment is planned in the northern hemisphere to obtain full sky exposure.

2 Hybrid Concept

The Pierre Auger observatory intends to measure the energy spectrum, the arrival direction, and the isotope composition of primary particles above 10^{19} eV[5]. The very low integrated particle flux of only 1 event per km^2 and century for energies above 5×10^{19} eV demands a very large detector area. The observatory is a hybrid system consisting of 1600 surface detector (SD) stations and an independent fluorescence detector (FD) system. The SD measures the lateral distribution of secondary particles using water Čerenkov detectors equally spaced on a grid over an area of more than 3000 km^2[6]. The FD system observes the longitudinal development of the shower profile through 4 optical detectors — called "FD eye" station. During clear dark nights each eye collects the fluorescence light emitted isotropically along the shower track as a time sequence of light . The

design provides a central FD eye station with a 360° field of view and three perimeter eyes with 180° field of view looking towards the center of the SD array.
The advantage of our hybrid detector concept compared to similar experiments is the possibility to study many parameters of individual events simultaneously and thereby to improve the energy and angular accuracy of either subsystem.
In the following section we describe the design of the FD system in more detail: the telescope setup, the readout system and the front end electronics.

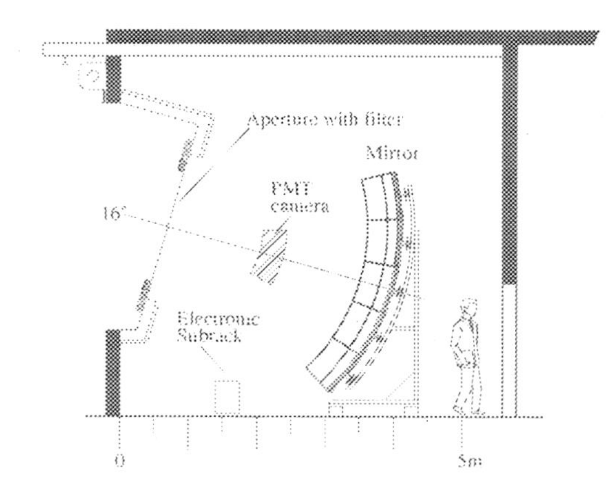

Figure 1. Schematic view of Schmidt telescope with aperture system, mirror, camera and front end electronics.

3 Structure of Fluorescence Detector Stations

3.1 Telescope Setup

Each FD eyes consists of 6 or 12[a] wide-angle Schmidt telescopes shown in figure 1. The light from a $30° \times 30°$ section of the sky enters the telescope through the 2.2 m wide aperture system which contains an only UV-light transparent optical filter and a corrector lens. The incident light is focused onto a camera located in the focal plane of a spherical mirror assembled from 36 mirror elements. The camera detects light with 440 photomultipliers (PMTs) arranged in a matrix of 20x22 pixels.

The telescopes are housed inside a building with a stabilized room temperature of $21° \pm 3°$ centigrade and a humidity level below 70 %.

3.2 Overview of the readout scheme

The layout of the readout chain shown in figure 2 follows the structure of the pixel matrix. Following the signal chain from the PMTs of the camera to the telecommunication link the readout consists of 4 main parts:

- 6 or 12 cameras each with 440 PMTs and the head electronics with an active voltage divider,
- Mirror sub-racks, commercial 9U VME crates containing the analog and digital front end modules,
- DAQ subnet, linking the Mirror-PC with the Eye-PC via a LAN switch, and
- Eye network linking the readout system to local slow control system and via the telecommunication tower to central data acquisition system.

These parts perform the task of shaping the PMT signals, digitize and store them, generate a trigger signal by pattern recognition and manage the readout of the stored data. The Mirror PCs of the DAQ subnet compress the data, improve the trigger decision and collect data of the same event from different telescopes. For these tasks we use low-cost industry PCs, which are operated without harddisk under LINUX. The final trigger decision is made by the Eye PC before the data are transferred to the central data acquisition system (CDAS) in Malargue via a bidirectional radio link.
A GPS clock synchronizes all FD mirror subracks. The overall synchronization with the

[a]6 units for the perimeter eyes and 12 for the central eye

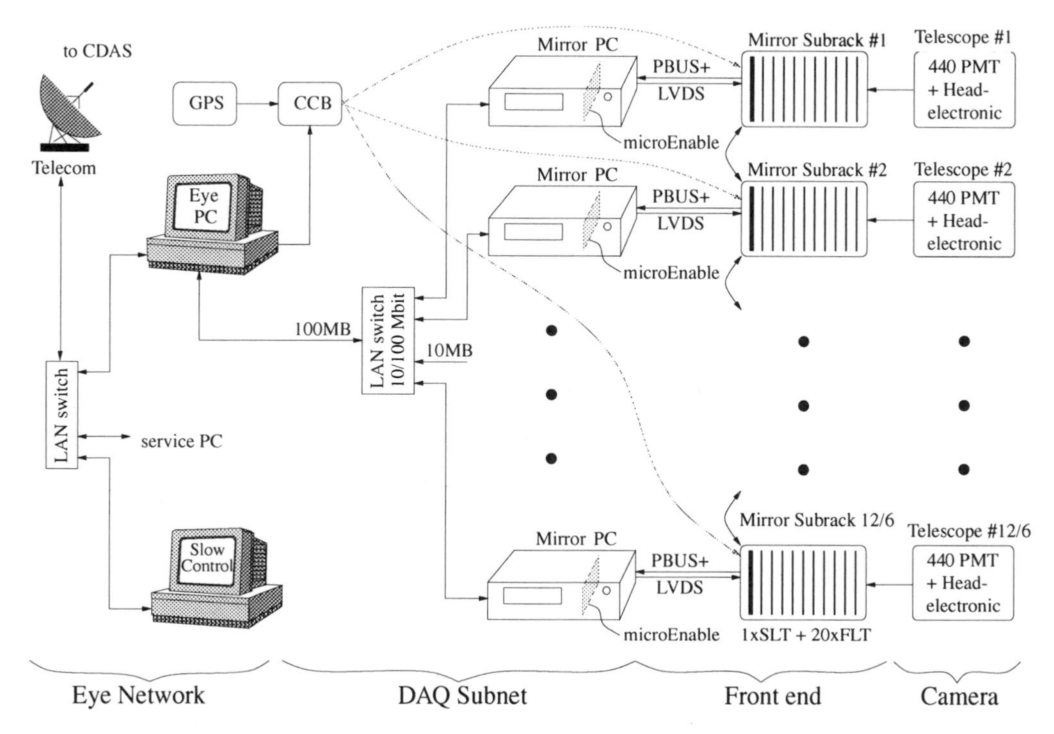

Figure 2. Readout scheme for a FD eye station: Eye network, DAQ subnet, front end electronic and camera.

surface detectors, which also use GPS clocks, has to be better then 120 ns.

4 Multilevel Trigger System

All functions of the digital electronics are implemented in reprogrammable FPGA logic in order to achieve high flexibility, cost-effectiveness and ease of maintenance. The high data rate from the sampling of the PMT signals is reduced through a multilevel trigger system (see table 1).

Table 1. Expected performance of the multilevel trigger system

Trigger Level	rate [Hz]
pure data	10^7 per pixel
First Level Trigger	100...200 per pixel
Second Level Trigger	0.1...1 per mirror
Software trigger	≤ 0.02 per eye

The 20 First Level Trigger (FLT) modules in each sub-rack evaluate the individual pixel status (on/off). This information is used by the Second Level Trigger (SLT) module in the sub-rack to identify track segments and short light tracks. The third trigger level rejects pixels of inconsistent time order and demands a well defined track length and energy by software evaluation.

4.1 First Level Trigger Module

The analog and digital part of the front end electronics consist of separate modules which are adjacent to each other and connected through three multipin connectors. The analog part is developed at INFN Milan, Pavia and Torino. It contains line receivers, filters and amplifier stages[7].

Optimal measurement of the shower profile requires a 10 MHz sampling of each PMT

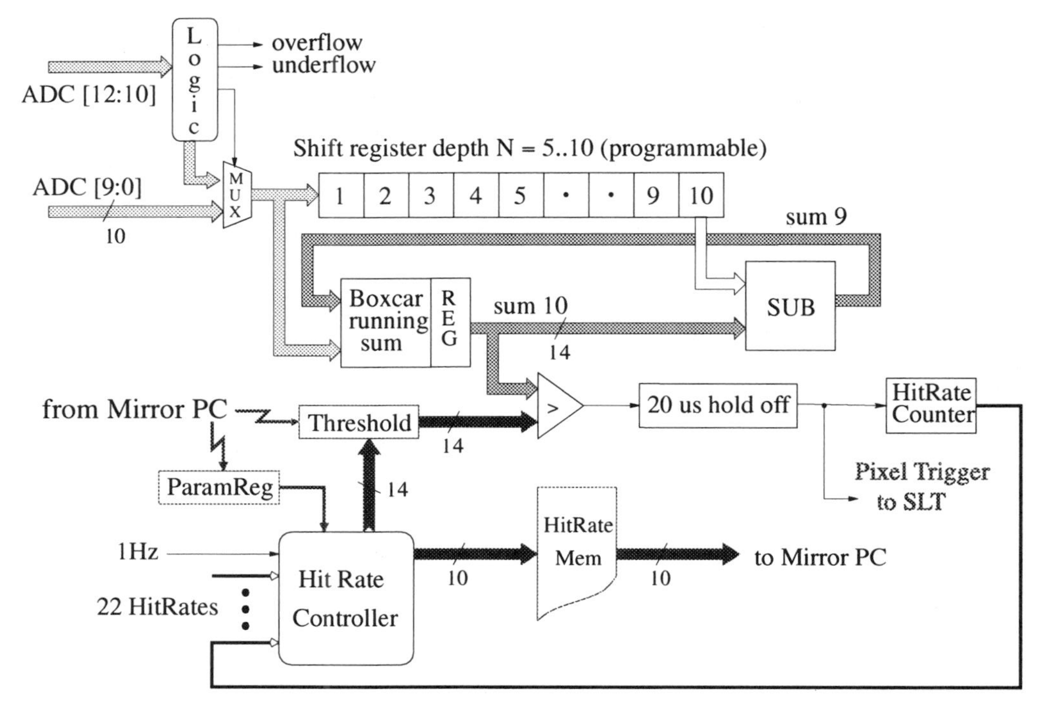

Figure 3. Scheme for generating the pixel trigger and regulation of its rate.

signal. Therefore, each FLT module contains 22 ADCs (12-bit) to continuously digitize the PMT signals of one camera column. The results are stored in 32K $\times$ 16 bit SRAM's. The address space of each RAM is divided into 32 pages, of 1024 words each. As long as there is no trigger from the SLT each page operates like a circular buffer memory to hold the ADC values of the last $100\mu s$. In case of a trigger all modules switch synchronously to the next unused memory page.
Another task of the module is to identify pixels with light levels above the natural night sky brightness. The sliding sum of the last 10 samples works as a digital filter and is compared with a threshold value according to the scheme of figure 3. Each pixel above the threshold generates a (pixel) trigger for a 20 μs long overlap coincidence time.
The trigger rate (or hit rate) is measured separately for each channel. It is used to adjust the individual thresholds in a feedback loop to keep the hit rate constant and compensate varying sky light conditions.

4.2 Second Level Trigger

The pixel triggers generated for each channel are analyzed by the FPGA logic of the SLT. The main task of this logic is to generate an internal trigger if the pattern of triggered pixels follows a straight line. The algorithm regards 5 pixels as a straight track if they are shaped like the fundamental patterns in figure 4 and those created by rotation and mirror reflection.

The algorithm accepts also track segments where one pixel out of 5 is missing. This ac-

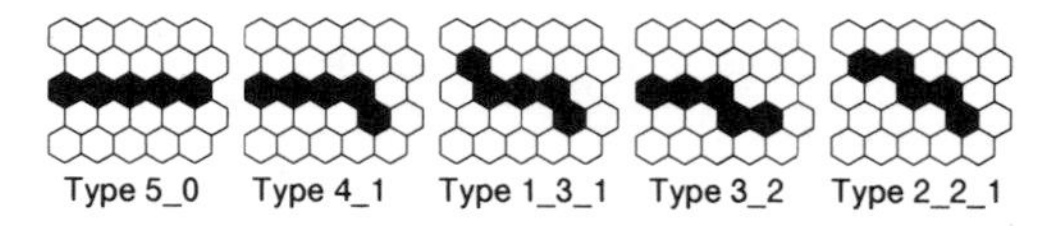

Figure 4. Basic topological types of pattern used for pattern recognition algorithm by the SLT module.

counts for the situations that a PMT is defective or has not received a sufficient amount of light as the track touched only the edge of the pixel.
Every 100 ns the logic reads the pixel status of two adjacent FLT modules and maps it to a pipelined memory structure of 22 rows and 5 columns. Instead of checking the full camera image of 22x20 pixels at once, the pattern are found by a combinatorial logic inside a smaller 22x5 submatrix contained in the memory structure. Every 50 ns 2102 combinations – grouped into 108 pattern classes of identical shape – are checked. If a track segment is found a trigger is generated and the pattern class number 1...108 is saved.

5 Status and Outlook

At present, the collaboration has started to build a prototype experiment. It consists of 40 Čerenkov stations positions at a region east of Malargue and 2 prototype FD telescope systems looking towards it. Several SD stations have already been deployed.
The first eye building for the FD prototype at Los Leones is going to be completed end of October 2000. After installing all components (mirrors, aperture , camera, electronics) the hybrid system will be tested in spring 2000 and step by step enlarged to the full size.

References

1. M. Takeda et al., *Phys. Rev. Lett* **81**, 1163 (1998)
2. James W. Cronin, *Rev. Mod. Phys.* **71**, No.2 (1999) 165
3. K. Greisen, *Phys. Rev. Lett* **16**, 748 (1966); G.T. Zatsepin and V.A. Kuzmin *JETP Lett.* **4**, 178 (1966).
4. M.Nagano and A.A. Watson, *Rev. Mod. Phys.* **72**,(2000) 689
5. The Pierre Auger Observatory Design Report, 2nd edition, March 97, The Auger Collaboration.
6. D.F. Nitz, "Triggering and Data Acquisition Systems for the Auger Observatory", Proceedings of the 10th. IEEE Real Time Conference, Beaune, France, 1997, 61-65.
7. S. Argiro et al., "The analog Signal Processor of the Auger Fluorescence Detector Prototype", presented at the Elba meeting on Advanced Detectors 2000, to be published in NIM
8. H. Gemmeke, A. Grindler, H. Keim, M. Kleifges, N. Kunka, Z. Szadkowski, D. Tcherniakhovski, "Design of the Trigger System for the Auger Fluorescence Detector", contribution to the 11th IEEE Real Time Conference, Santa Fee 1999, USA, IEEE-NSS 47 (2000) 519.

CHAOTIC INFLATION IN SUPERGRAVITY

MASAHIRO KAWASAKI
Research Center for the Early Universe, Graduate School of Science, University of Tokyo, Tokyo 113-0033, JAPAN
E-mail: kawasaki@resceu.s.u-tokyo.ac.jp

It is shown that chaotic inflation naturally takes place in the framework of supergravity if we assume hat the Kähler potential has a shift symmetry of the inflaton chiral multiplet and introduce a small breaking parameter.

1 Introduction

The inflationary universe [1] is a new paradigm in modern cosmology. Inflation not only solves the longstanding problems such as the horizon and flatness problems in cosmology, but also accounts for the origin of density fluctuations [2] as observed by the Comic Background Explorer(COBE) satellite [3]. Among various types of inflation models proposed so far, the chaotic inflation model [4] is the most attractive since it can realize an inflationary expansion even in the presence of large quantum fluctuations at the Planck time. Thus, the chaotic inflation is free from the initial value problem which is serious in other inflation models like new and hybrid inflation models. Furthermore, the chaotic inflation takes place around the gravitational scale so that it does not receive so-called flatness problem (the longevity problem).

On the other hand, supersymmetry (SUSY) [5] is widely discussed as the most interesting candidate for the physics beyond the standard model since it ensures the stability of the large hierarchy between the electroweak and the Planck scales against radiative corrections. This kind of stability is also very important to keep the flatness of the inflaton potential at the quantum level. Therefore, it is quite natural to consider the inflation model in the framework of supergravity.

However, the above two ideas, i.e. chaotic inflation and supergravity, have not been naturally realized simultaneously. In the minimal supergravity the potential exponentially grows beyond the gravitational scale M_G ($\simeq 2.4 \times 10^{18}$ GeV) while the inflaton field φ must take a value much larger than the gravitational scale M_G to cause the chaotic inflation. Thus, the exponential growth of the potential makes it very difficult to incorporate the chaotic inflation in the framework of supergravity. In fact, all of the existing models [6,7] for chaotic inflation use rather specific Kähler potential, and one needs a fine tuning in the Kähler potential. Thus, it is very important to find a natural chaotic inflation model.

Here, we propose a natural chaotic inflation model where the form of Kähler potential is determined by a symmetry. With this Kähler potential the inflaton φ may have a large value $\varphi \gg M_G$ to begin the chaotic inflation.
[This paper is based on the work with M. Yamaguchi and T. Yanagida [8].]

2 Model

The present model is based on the Nambu-Goldstone-like shift symmetry of the inflaton chiral multiplet $\Phi(x, \theta)$. Namely, we assume that the Kähler potential $K(\Phi, \Phi^*)$ is invariant under the shift of Φ,

$$\Phi \to \Phi + i\, CM_G, \tag{1}$$

where C is a dimensionless real parameter. Thus, the Kähler potential is a function of

$\Phi + \Phi^*$, i.e $K(\Phi, \Phi^*) = K(\Phi + \Phi^*)$. It is now clear that the supergravity effect $e^{K(\Phi+\Phi^*)}$ does not prevent the imaginary part of the scalar components of Φ from having a larger value than M_G. We identify it with the inflaton field φ. However, as long as the shift symmetry is exact, the inflaton φ never has a potential and hence it never causes the inflation. Therefore, we have to introduce a small breaking term of the shift symmetry in the theory. We introduce a small mass term in the superpotential,

$$W = mX\Phi, \tag{2}$$

where $X(x, \theta)$ is a new chiral multiplet. Notice that the present model possesses $U(1)_R$ symmetry under which

$$\begin{aligned} X(\theta) &\to e^{-2i\alpha} X(\theta e^{i\alpha}), \\ \Phi(\theta) &\to \Phi(\theta e^{i\alpha}), \end{aligned} \tag{3}$$

and Z_2 symmetry under which

$$\begin{aligned} X(\theta) &\to -X(\theta e^{i\alpha}), \\ \Phi(\theta) &\to -\Phi(\theta e^{i\alpha}). \end{aligned} \tag{4}$$

The above superpotential is not invariant under the shift symmetry of Φ. However, we should stress that the present model is completely natural in 't Hooft's sense [9], since we have an enhanced symmetry (the shift symmetry) in the limit $m \to 0$. That is, we consider that the small parameter m is originated from small breaking of the shift symmetry in a more fundamental theory. We consider that as long as $m \ll \mathcal{O}(1)$, the corrections from the breaking term eq.(2) to the Kähler potential are negligibly small. (The Kähler potential may have the induced breaking terms such as $K \simeq |m\Phi|^2 + \cdots$. However, these breaking terms are negligible in the present analysis as long as $|\varphi| \lesssim m^{-1}M_G^2$.) Then, we assume that the Kähler potential has the shift symmetry eq.(1) and the above $U(1)_R \times Z_2$ symmetry neglecting the breaking effects,

$$\begin{aligned} K(\Phi, \Phi^*, X, X^*) &= K[(\Phi + \Phi^*)^2, XX^*] \\ &= \frac{1}{2}(\Phi + \Phi^*)^2 + XX^* + \cdots. \end{aligned} \tag{5}$$

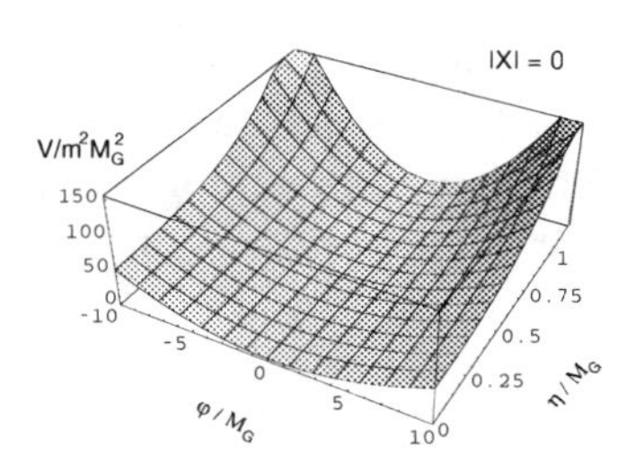

Figure 1. Potential for $\Phi = (\eta + i\varphi)/\sqrt{2}$.

3 Dynamics of inflation

The potential $V(\Phi, X)$ given by

$$\begin{aligned} V(\Phi, X) = m^2 e^K \big[&|\Phi|^2(1 + |X|^4) \\ &+ |X|^2 \left\{1 - |\Phi|^2 + (\Phi + \Phi^*)^2 \right. \\ &\left. \times (1 + |\Phi|^2)\right\} \big], \end{aligned} \tag{6}$$

where we have neglected higher order terms in the Kähler potential eq.(5) and X denotes also the scalar component of the superfield X. Here and hereafter we set the gravitational scale unity. Now, we decompose the complex scalar field Φ into two real scalar fields as,

$$\Phi = \frac{1}{\sqrt{2}}(\eta + i\varphi). \tag{7}$$

The potential shown in Fig. 1 for $X = 0$. Because of the presence of e^K factor in eq.(6) η and $|X|$ should be taken as $|\eta|, |X| \lesssim \mathcal{O}(1)$. On the other hand, φ can take a value much larger than $\mathcal{O}(1)$ since e^K does not contain φ. For $\eta, |X| \ll \mathcal{O}(1)$, we can rewrite the potential as

$$V(\eta, \varphi, X) \simeq \frac{1}{2}m^2\varphi^2(1 + \eta^2) + m^2|X|^2. \tag{8}$$

Since the initial values of the inflaton $\varphi(0)$ is determined so that $V(\varphi(0)) \sim \frac{1}{2}m^2\varphi(0)^2 \sim 1$, $\varphi(0) \sim m^{-1} \gg 1$. (Notice that one has only to demand $\varphi(0) \gtrsim 15.0$ in order to solve the flatness and horizon problems. [1]) For such large φ the effective mass of η becomes much larger than m and hence it quickly settles down to $\eta = 0$. On the other hand, the X field has a relatively light mass $\sqrt{2}m$ and slowly roll down toward the origin ($X = 0$). However, because the initial value of X is much smaller than $\varphi(0)$, the field value of X becomes much smaller than φ during inflation. In addition, higher order terms in Kähler potential may induce the effective mass of order the Hubble parameter for X. Thus we can take $\eta \simeq 0$ and $X \simeq 0$. Then, the potential eq.(8) is written as

$$V(\varphi, X) \simeq \frac{1}{2} m^2 \varphi^2. \tag{9}$$

This is the potential just required for chaotic inflation.

4 Observational Implications

Since the resultant potential (9) is the same as that studied previously, it is easy to estimate the observable quantities such as density fluctuations and its spectral index.

First, the density fluctuations produced by this chaotic inflation is estimated as

$$\frac{\delta\rho}{\rho} \simeq \frac{1}{5\sqrt{3}\pi} \frac{m}{2\sqrt{2}} \varphi^2. \tag{10}$$

The normalization at the COBE scale ($\delta\rho/\rho \simeq 2 \times 10^{-5}$ for $\varphi_{\rm COBE} \simeq 14$ [3]) gives

$$m \simeq 10^{13} \text{ GeV} \simeq 10^{-5}. \tag{11}$$

The spectral index is

$$n_s \simeq 0.96 \tag{12}$$

at the COBE scale. It is also known that the chaotic inflation produces significant amount of gravitational waves [11] that gives a contribution to Cosmic Microwave Background (CMB) anisotropies. The relative contribution to the CMB quadrupole anisotropies of gravitational waves (T) to scalar density fluctuations (S) is given by [12]

$$\frac{T}{S} \simeq \frac{25}{\varphi} \simeq 0.1, \tag{13}$$

which may be detectable in future satellite experiments [13,14].

5 Reheating

After the inflation ends, the inflaton field φ begins to oscillate and its successive decays cause reheating of the universe. In the present model the reheating takes place efficiently if we introduce the following superpotential:

$$W = \lambda X H \bar{H}, \tag{14}$$

where H and $\bar{H}$ are a pair of Higgs doublets whose R-charge are assumed to be zero and λ is a constant. Owing to this superpotential, $D_X W$ is changed to $D_X W = (m\Phi + \lambda H\bar{H})(1 + |X|^2)$. $|H|$ and $|\bar{H}|$ take values $\lesssim \mathcal{O}(1)$ due to the factor of $e^{K(H,\bar{H})}$ as the X field. Therefore, the $m\Phi$ term dominates $D_X W$ unless $\lambda \gtrsim \mathcal{O}(1)$, since $|\Phi(0)| \sim m^{-1}$ at the beginning of the universe, and the chaotic inflation begins. Once the inflation takes place, H and $\bar{H}$ acquire masses of the order the Hubble scale and rapidly go to zero. Thus, the above superpotential eq.(14) does not affect the dynamics of the inflation. Then, we have the coupling of the inflaton φ to the Higgs doublets as

$$L \sim \lambda m \varphi H \bar{H}, \tag{15}$$

which gives the reheating temperature

$$T_R \sim 10^9 \text{ GeV} \left(\frac{\lambda}{10^{-5}}\right) \left(\frac{m}{10^{13}\text{GeV}}\right)^{1/2}. \tag{16}$$

In order to avoid the overproduction of gravitinos, the reheating temperature T_R must be lower than 10^9GeV [15][see Fig. 2], which requires the small coupling $\lambda \lesssim 10^{-5}$. The small coupling λ is naturally understood in 't Hooft's sense [9] provided that $H\bar{H}$ is even under the Z_2 symmetry in eq.(4).

6 Conclusion

We have shown that a chaotic inflation successfully takes place if we assume that the Kähler potential has the shift symmetry of the inflaton chiral multiplet Φ and introduce a small breaking term of the shift symmetry

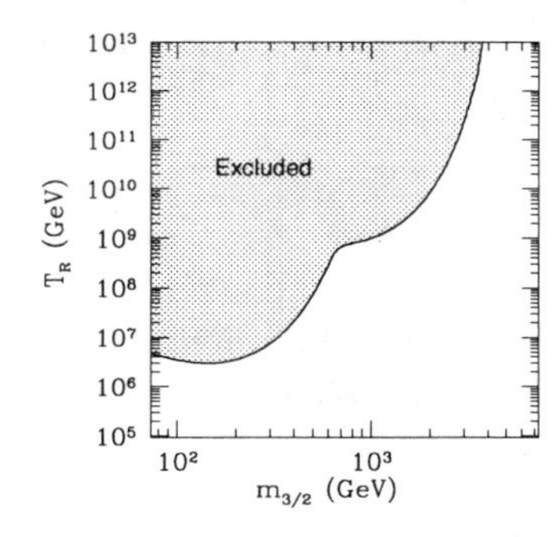

Figure 2. Constraint on the reheating temperature.

in the superpotential eq.(2). Furthermore, the chaotic inflation predicts the spectral index of the density fluctuations $n_s \simeq 0.96$, and it also produces gravitational waves which might be detectable in future astrophysical observations [13,14]. The reheating temperature after the chaotic inflation can be as low as 10^9 GeV which avoids the overproduction of gravitinos.

References

1. See, for example, A.D. Linde, Particle Physics and Inflationary Cosmology, (Harwood, Chur, Switzerland, 1990).
2. A. Guth and So-Y. Pi, Phys. Rev. Lett. **49**, 1110 (1982);
   S. W. Hawking, Phys. Lett. **115B**, 295 (1982);
   A. A. Starobinsky, Phys. Lett. **117B**, 175 (1982).
3. C.L. Bennett *et al.*, Astrophys. J. **464**, 1 (1996).
4. A. D. Linde, Phys. Lett. **129B**, 177 (1983).
5. See, for a review, H. P. Nilles, Phys. Rep. **110**, 1 (1984).
6. A. S. Goncharov and A. D. Linde, Phys. Lett. **139**, 27 (1984); Class. Quantum Grav. **1**, L75 (1984).
7. H. Murayama, H, Suzuki, T. Yanagida, and J. Yokoyama, Phys. Rev. **D50**, R2356 (1994).
8. M. Kawasaki, Masahide Yamaguchi, T. Yanagida, Phys. Rev. Lett. in press, hep-ph/0004243.
9. G. 't Hooft, in *Recent developments in gauge theories*, edited by G. 't Hooft *et al.* (Plenum Press, Cargèse, 1980).
10. See, for a review, D. H. Lyth and A. Riotto, Phys. Rep. **314**, 1 (1999).
11. V.A. Rubakov, M.V. Sazhin, A.V. Veryaskin, Phys. Lett. **B115**,189 (1982);
    A.A. Starobinsky, Quantum Gravity, Proc 2nd Seminar Quantum Theory of Gravitation, Inst. Nucl. Res. USSR Acad. Sci, Moscow, p58 (1982).
12. A.A. Starobinsky, Sov. Astron. Lett. **42**,152 (1985).
13. `http://map.gsfc.nasa.gov`
14. `http://astro.estec.esa.nl /SA-general/Projects/Planck/`
15. M. Yu. Khlopov and A.D. Linde, Phys. Lett. **138B**, 265 (1984);
    J. Ellis, G.B. Gelmini, J.L. Lopez, D.V. Nanopoulos and S. Sarker, Nucl. Phys. **373**, 399 (1992);
    M. Kawasaki and T. Moroi, Prog. Theor. Phys. **93**, 879 (1995).

SUPERSYMMETRIC DARK MATTER AND CONSTRAINTS FROM LEP

KEITH A. OLIVE

TH Division, CERN, Geneva, Switzerland

and

Theoretical Physics Institute, School of Physics and Astronomy,

University of Minnesota, Minneapolis MN, USA

E-mail: olive@umn.edu

Accelerator constraints on the parameter space of the Minimal Supersymmetric extension of the Standard Model are analyzed and contrasted with the parameter space yielding dark matter which is allowed by the cosmological constraints on the relic density. The most important accelerator limits are those from searches for charginos $\chi^{\pm}$, neutralinos χ_i and Higgs bosons at LEP. Constraints derived from $b \to s\gamma$ decay are also incorporated. It is found that $m_\chi > 51$ GeV and $\tan\beta > 2.7$ if all soft supersymmetry-breaking scalar masses are universal, including those of the Higgs bosons, and that these limits weaken to $m_\chi > 47$ GeV and $\tan\beta > 2.1$ if non-universal scalar masses are allowed.

It is will known that supersymmetry with unbroken R-parity offers a natural cold dark matter candidate with a cosmologically significant relic density [1]. Indeed, one of the reasons that supersymmetric dark matter is the main focus of many direct detection searches for dark matter, is that over a wide range of the supersymmetric parameter space (with mass parameters $\lesssim 1$ TeV), the relic density (in units of the critical density) takes values of order $\Omega_\chi \sim$ a few tenths. In fact, until recently, some of the strongest constraints on the susy parameter space came from the cosmological upper limit $\Omega_\chi h^2 \leq 0.3$. In addition, if the neutralino is responsible for the dark matter, then we should have $\Omega_\chi h^2 > 0.1$ The final runs at LEP [2] combined with the current bounds from $b \to s\gamma$ [3], further constrain the available susy parameter space. Here I will briefly summarize the combined cosmological and accelerator constraints [4].

In the analysis below, I will distinguish between two simplifications of the minimal supersymmetric standard model (MSSM). One in which all soft scalar masses (including the Higgs soft masses) are unified at the GUT scale denoted as UHM (for universal Higgs masses) or the constrained MSSM (CMSSM). In the second case, the Higgs soft masses are not unified with the the squark and slepton soft masses which remain universal at the GUT scale. This case will be denoted as nUHM or simply the MSSM. In addition to the parameters common to both models, the ratio of the two Higgs vevs, $\tan\beta$, and the trilinear soft mass terms, A (also assumed to be unified at the GUT scale, there are several other key parameters which determine the susy particle spectrum.

In the UHM model, there is one common soft scalar mass at the GUT scale, m_o. The gaugino masses are also unified at the GUT scale so that $M_1 = M_2 = M_3 = m_{1/2}$. Finally in the UHM model, the sign of the Higgs mixing mass parameter, μ, is arbitrary. The values of the magnitude of μ and the pseudoscalar Higgs mass, m_A, are determined by the conditions of electro-weak symmetry breaking. In contrast, in the nUHM model, since the soft Higgs masses, $m_1, m_2 \neq m_0$, the values of μ and m_A are free parameters. It is also common to use the SU(2) gaugino mass, M_2 at the weak scale as the free parameter rather than $m_{1/2}$. M_2 is easily related to $m_{1/2}$ through $M_2 = (\alpha_2/\alpha_{GUT}) \times m_{1/2}$. In the figures, $m_o - m_{1/2}$ plots will always refer to UHM models, and $M_2 - \mu$ plots will refer to nUHM models.

The accelerator bounds used in the analysis below can be summarized as follows:

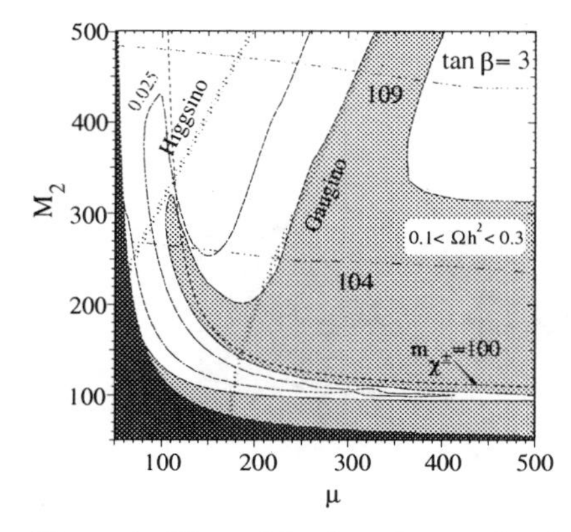

Figure 1. The μ, M_2 plane for $\tan\beta = 3$, $m_0 = 100$ GeV and $m_A = 1$ TeV.

- Chargino mass limit:
  $m_{\chi^\pm} \geq 101 - 102$ GeV
  mainly constrains $m_{1/2}$ (M_2 and μ) in the UHM (nUHM).
- Higgs mass limit:
  $m_h \geq 107 - 112$ GeV; $\tan\beta = 3$
  $m_h \geq 103 - 111$ GeV; $\tan\beta = 5$
  mainly constrains $m_{1/2}$ (m_A, M_2 and A) at low $\tan\beta$.
- $b \to s\gamma$
  mainly constrains $m_{1/2}$ (m_A) particularly at high $\tan\beta$ and $\mu < 0$.

The ranges for the chargino and Higgs mass limits refer to the difference between the established LEP limits and preliminary (but expected limits) [2]. Note that the LEP sensitivity to detecting the Higgs diminishes at high $\tan\beta$. At $\tan\beta \gtrsim 8$, the limit is only 89 GeV. Furthermore, due to theoretical uncertainty in the calculation of the Higgs mass, the constraints imposed are about 3 GeV less than the stated experimental constraint. It is also required that sfermions masses are larger than 98 GeV, and that the neutralino is the LSP.

Constraints from the chargino and Higgs searches are shown in Fig. 1 and Fig. 2, for the nUHM and UHM respectively [4] for selected values of $\tan\beta$ and $sgn(\mu)$. In Fig. 1, contours of $\Omega_\chi h^2 = 0.025, 0.1$ and 0.3 are shown as solid lines, and the preferred region with $0.1 < \Omega_\chi h^2 < 0.3$ is shown light-shaded. The dashed line corresponds to $m_{\chi^\pm} = 100$ GeV. The near-horizontal dot-dashed lines are Higgs mass contours, and the hashed lines are 0.9 Higgsino and gaugino purity contours. The dark shaded region has $m_{\chi^\pm} < m_Z/2$. It is apparent that the bulk of the cosmological region with $0.1 \leq \Omega_\chi h^2 \leq 0.3$ has $\mu \gtrsim M_2$, indicating that LSP dark matter is generically a gaugino: in these regions, it is mainly a Bino. There are, however, small regions at smaller $|\mu|$ (for given M_2), where the LSP is mainly a Higgsino. However, as can be seen in Fig. 1, this Higgsino possibility is under severe pressure from several LEP constraints, including the chargino and Higgs searches. Careful analysis [4] has shown that predominantly Higgsino dark matter is excluded and the gaugino content must be at least 30%.

In Fig 2, the region allowed by the cosmological constraint $0.1 \leq \Omega_\chi h^2 \leq 0.3$, after including coannihilations [5], has medium shading. Dotted lines delineate the announced LEP constraint on the $\tilde{e}$ mass and the disallowed region where $m_{\tilde{\tau}_1} < m_\chi$ has dark shading. The contour $m_{\chi^\pm} = 102$ GeV is shown as a near-vertical dashed line in each panel. Also shown as dot-dashed lines are relevant Higgs mass contours. The long dashed curves in panels (a), (b) represent the anticipated limits from trilepton searches at Run II of the Tevatron [6].

Though not shown in Figs. 1 and 2, contributions $b \to s\gamma$ come from chargino-stop and charged Higgsino exchanges [7]. When $\mu > 0$, these contributions interfere destructively, and the limits are weakened, that is the supersymmetric contributions are with the experimental uncertainty for moderately low $\tan\beta \lesssim 10$ in the UHM. At $\tan\beta = 20$, and $m_0 \lesssim 400$ GeV, even for $\mu > 0$, we obtain $m_{1/2} \gtrsim 200$ GeV from $b \to s\gamma$ [4]. For negative μ, we find $m_{1/2} \gtrsim 230$ GeV at $\tan\beta = 3$ and $m_{1/2} \gtrsim 450 - 500$ GeV at $\tan\beta = 20$.

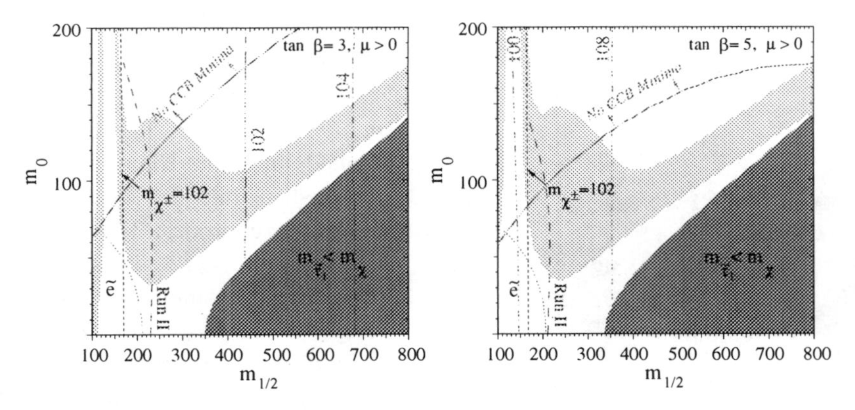

Figure 2. *The $m_{1/2}, m_0$ plane for $\mu > 0$, $A = -m_{1/2}$ and $\tan\beta =$ (a) 3, (b) 5.*

In the nUHM, the constraint from $b \to s\gamma$, can be greatly diminished by choosing large ($\gtrsim 350 - 500$ GeV) m_A.

From the accelerator bounds on the chargino and Higgs masses, one can place lower bounds on the neutralino mass. The limits are summarized in Fig. 3 for $\mu > 0$, under various different assumptions: UHM or nUHM and (in the former case) whether one requires the present vacuum to be stable against transition to a charge- and colour-breaking (CCB) vacuum (best achieved when $A_0 = -m_{1/2}$) or not (UHM$_{\rm min}$, where A_0 is left free). Also, limits are given for both the available 1999 LEP data and for a 'realistic' assessment of the likely sensitivity of data to be taken in 2K [4].

In all cases, for both positive and negative μ, the lower limits on m_χ are relatively insensitive to $\tan\beta$ at large $\tan\beta$. Here, they are determined by the LEP chargino bound, as the LEP Higgs mass bound is weaker than the chargino bound at large $\tan\beta$. In fact, in the two UHM cases shown, the points at which the limiting curves bend upward, as one decreases $\tan\beta$, are precisely the points at which the Higgs mass bound becomes more stringent than the chargino bound. In the UHM cases, the neutralino mass limits are strong at intermediate values of $\tan\beta \simeq 4$–7 because the cosmological bound on the relic density prohibits going to large values of m_0, and ensures that the Higgs bound places a strong constraint. Below this break point, the lower limit on m_χ increases rapidly with decreasing $\tan\beta$. Above this break point, the limit on m_χ is relatively insensitive to the additional theoretical assumptions made, such as UHM vs. UHM$_{\rm min}$ or nUHM. However, in the nUHM cases, because one can increase m_0 sufficiently to weaken the Higgs mass bound, the break point occurs at a lower value of $\tan\beta$. To go to lower values of $\tan\beta$ then requires a substantial increase in m_χ.

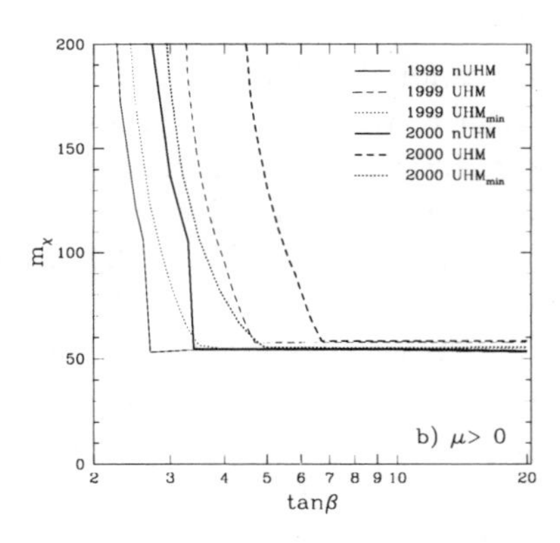

Figure 3. *Lower limits on the neutralino mass m_χ as functions of $\tan\beta$ for $\mu > 0$.*

One can also place an upper limit on the neutralino mass from relic density considerations. In the nUHM, the upper limit on a mostly bino LSP is about 300 GeV [8]. This limit is somewhat soft and can be avoided if the LSP is sitting on a pole (e.g. the pseudo-scalar Higgs pole) or is nearly degenerate with a squark such as the stop [9]. In the UHM, co-annihilations increase the upper limit to about 600 GeV [5].

In the UHM cases with and without the restriction forbidding CCB vacua, the lower limit on the lightest MSSM Higgs mass, in particular, implies lower limits on $\tan\beta$ which are plotted in Fig. 4. Recall that the existing Higgs mass calculations in the MSSM are believed to be accurate to about 3 GeV. Therefore in computing the bounds on $\tan\beta$ one should shift the experimental bound down by 3 GeV before reading the values of $\tan\beta$ off of Fig. 4. We also show in Fig. 4 the lower bound on $\tan\beta$ obtained in the nUHM, which is significantly weaker than in the UHM cases, and essentially independent of the sign of μ.

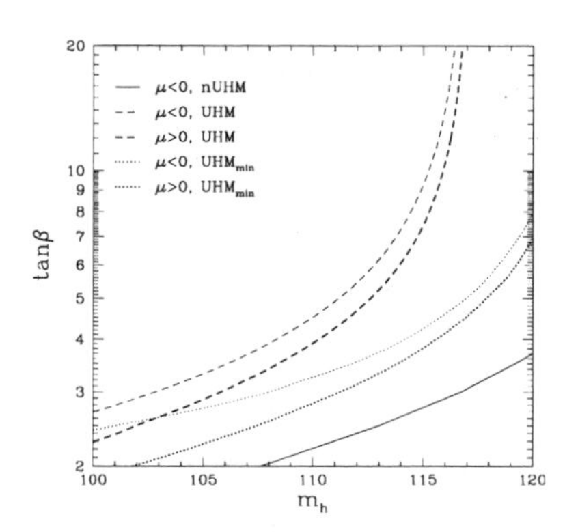

Figure 4. *Lower limit on* $\tan\beta$ *imposed by the experimental and cosmological constraints, as a function of the experimental Higgs mass limit. The UHM,* UHM_{min} *and nUHM labels are the same as in Fig. 3.*

The current 2K limits on $\tan\beta$ are summarized in Table 1.

Acknowledgments

I would like to thank my collaborators J. Ellis, T. Falk, and G. Ganis. This work was supported in part by DOE grant DE–FG02–94ER–40823 at the University of Minnesota.

	UHM	UHM_{min}	nUHM
$\mu < 0$	4.0	3.1	2.1
$\mu > 0$	3.6	2.7	2.1

Table 1. *Limits on* $\tan\beta$, *assuming the 'realistic' 2K energies and luminosities .*

References

1. J. Ellis, J.S. Hagelin, D.V. Nanopoulos, K.A. Olive and M. Srednicki, *Nucl. Phys.* **B238** (1984) 453.
2. See the LEP Joint Supersymmetry Working Group URL: `http://lepsusy.web.cern.ch/lepsusy/Welcome.html`.
3. CLEO Collaboration, M.S. Alam *et al.*, Phys. Rev. Lett. **74** (1995) 2885; S. Ahmed *et al.*, `CLEO CONF 99-10`; ALEPH Collaboration, R. Barate *et al.*, Phys. Lett. **B429** (1998) 169.
4. J. Ellis, T. Falk, G. Ganis, and K. Olive, hep-ph/0004169.
5. J. Ellis, T. Falk and K. A. Olive, Phys. Lett. **B444** (1998) 367; J. Ellis, T. Falk, K. A. Olive and M. Srednicki, Astropart. Phys. **13** (2000) 181.
6. See S. Abel *et al.*, Tevatron SUGRA Working Group Collaboration, hep-ph/0003154, and references therein.
7. M. Ciuchini, G. Degrassi, P. Gambino and G.F. Giudice, Nucl. Phys. **B527** (1998) 21; P. Ciafaloni, A. Romanino and A. Strumia, Nucl. Phys. **B524** (1998) 361; F. Borzumati and C. Greub, Phys. Rev.**D58** (1998) 074004.
8. K.A. Olive and M. Srednicki, *Phys. Lett.* **B230** (1989) 78; *Nucl. Phys.* **B355** (1991) 208; K. Greist, M. Kamionkowski, and M.S. Turner, *Phys. Rev.* **D41** (1990) 3565.
9. C. Boehm, A. Djouadi, and M. Drees, *Phys. Rev.* **D62** (2000) 035012.

MACHOS

J.-F. GLICENSTEIN
DSM/DAPNIA/SPP
CEA Saclay
F-91191 Gif-sur-Yvette, France
E-mail: glicens@hep.saclay.cea.fr

MACHOs have been long standing candidates for Galactic dark matter. In 1986, it was suggested that the microlensing of sources in dense stellar fields could constrain the mass fraction of MACHOs in the dark dalo. After 10 years of experimental search, MACHOs have been ruled out as major contributors to Galactic dark matter over a wide mass range. However, the explanation of observational results towards the Large Magellanic Cloud is still controversial.

1 Introduction

From primordial nucleosynthesis bounds, it is believed that dark matter cannot be fully composed of baryons. However, a large fraction of the Galactic dark matter might be composed of baryons [1]. They have to be hidden either in very cold molecular clouds, or in dark compact objects, the so-called MACHOs (Massive Astrophysical Compact Halo Object). Examples of MACHOs are snowballs, planets, brown dwarfs, red dwarfs, dead stars such as white dwarfs and neutron stars, and black holes. MACHOs are difficult to observe directly, although their direct detection is sometimes possible, e.g section 4.3. In 1986, B.Paczyński [2] showed that MACHOs in the mass range from $[10^{-7} - 10^{2}] M_{\odot}$ could be discovered or strongly constrained by studying the microlensing of resolved stars in the Large Magellanic Cloud (LMC).

2 Microlensing expectations

Gravitational lensing is a consequence of the deflection of light by massive bodies ("lenses"). Compact lenses like MACHOs distort the light beam from background sources and create two images. For sources located in the Magellanic Clouds and lenses in the Galactic halo with masses less than 100 $M_{\odot}$, the typical separation of the images is less than 1 *mas*, too small to be resolved with present ground or space based telescopes. The source is said to be "microlensed". The total flux coming from the source is magnified independently of wavelength (achromaticity). It can be detected if the lens moves in front of the source. If the lens is a single compact object ("point lens") and the effects of the finite size of the source can be neglected ("point source"), the magnification, A, versus time, t, curve is given by

$$A = \frac{u^2+2}{u\sqrt{u^2+4}} \quad (1)$$

$$u^2 = u_o^2 + \left(\frac{t-t_o}{t_E}\right)^2 \quad (2)$$

where u_o, t_o and t_E are parameters. t_E (the "timescale" of the event) is a function of the transverse velocity, of the lens distance and mass m_{MACHO}.

The motion of the Earth ("parallax") has to be taken into account for events with timescale over a few months. The effect of parallax is large when the lens is near the observer or when its mass is small. (Non)-observation of parallax on microlensing candidates constrains lens distances and masses.

More information is provided by binary microlensing events. Binary lenses produce caustics which are sometimes observed. If the radius of the source is known, the time taken by the source to cross the caustic line gives a measurement of the velocity of the projection of the source onto the lens plane (sec. 4.2).

The optical depth τ is the probability of observing a magnification of more than 34% towards a given direction at a given time; it is independent of m_{MACHO}. The contribution of halo lenses to the optical depth towards the LMC is expected to be $\tau^{LMC} \sim 5\ 10^{-7}$ for a standard dark halo fully comprised of MACHOs. The optical depth towards the SMC is in the range $\tau^{SMC} \sim 5 - 7\ 10^{-7}$, depending on the Galactic model. The timescale of events scales as $m_{MACHO}^{1/2}$. For a microlensing event observed towards the LMC, one has:

$$t_E \sim 70 \sqrt{\frac{m_{MACHO}}{M_\odot}} \text{ days} \qquad (3)$$

The measurement of t_E allows to estimate m_{MACHO}. The event rate towards the LMC is $\Gamma_{LMC} \sim 1.6\ 10^{-6}\sqrt{\frac{M_\odot}{m_{MACHO}}}$/star/year, assuming 100% experimental efficiency. Tens of million stars have to be monitored during years to obtain a signal. One has to use crowded fields such as the LMC or the SMC (Small Magellanic Cloud). Resolved stars from these fields are actually bright stars blended with a few fainter stars. Since any star in the blend can be lensed, observed microlensing events are in general chromatic.

3 Observational results

3.1 Early history (before 1999)

By 1992, two experimental groups, the french **EROS** (''**EROS1**'') and the australo-american **MACHO** had started searching for Galactic dark matter with microlensing. **EROS** had a major hardware upgrade (''**EROS2**'') in 1996. Both experiments are monitoring the LMC and the SMC. The first microlensing candidates towards the LMC were reported in 1993 [3 4]. The analysis of the first 2 years of **MACHO** data was published in 1997. A total of 8 candidate events were observed with a typical timescale $t_E = 50$ days (which translates into a typical mass $m_{MACHO} = 0.5 M_\odot$). The measured optical depth was

$$\tau_{MACHO2yr}^{LMC} = 2.9^{+1.4}_{-0.9}\ 10^{-7}. \qquad (4)$$

According to this result, roughly half (and possibly all) of the dark halo mass should be in compact objects. The analysis of **EROS1** gave two microlensing candidates towards the LMC. No microlensing candidate with $t_E < 17$ days was found by either experiment. Since this timescale corresponds to $m_{MACHO} \sim 0.05\ M_\odot$, a strong limit on the contribution of planet-sized objects to Galactic dark matter was set [5].

3.2 Recent results

The analysis of the **EROS** 1996-1998 data taken towards the SMC came in 1999 [6]. Only 1 microlensing candidate was found, while a dark halo made of 0.5 $M_\odot$ objects would contribute 4-6. This translates into an upper limit on the halo mass fraction in 0.5 $M_\odot$ MACHOs $f_{MACHO} < 0.5$(95% CL). This limit is conservative, since it has been realized [10] that the "self-lensing" contribution to the signal towards the SMC may be substantial (see section 4.2). The event found is peculiar: its t_E ($\sim$ 125 days) is longer than the t_E of any event found towards the LMC. The parallax analysis suggests that the lens must be either very close to the SMC or heavy ($m_{lens} > 0.6 M_\odot (95\%\ CL)$). The interpretation as a "self-lensing" event is more natural.

The **MACHO** 1992-1998 LMC data analysis [9] has been presented in 2000. 13 (17) microlensing candidates have been found (depending on the cuts), while the estimated background is 2-4 events (see section 4.2). 55 (70) events were expected for a standard halo full of $\sim 0.5\ M_\odot$ MACHOs. Hence, the **MACHO** collaboration still claims the detection of a $0.15 - 0.9 M_\odot$ MACHO signal, but with a smaller halo mass fraction $f_{MACHO} \sim 0.2$. In terms of optical depth

$$\tau_{MACHO5.7yr}^{LMC} = 1.2^{+0.4}_{-0.3}\ 10^{-7} \qquad (5)$$

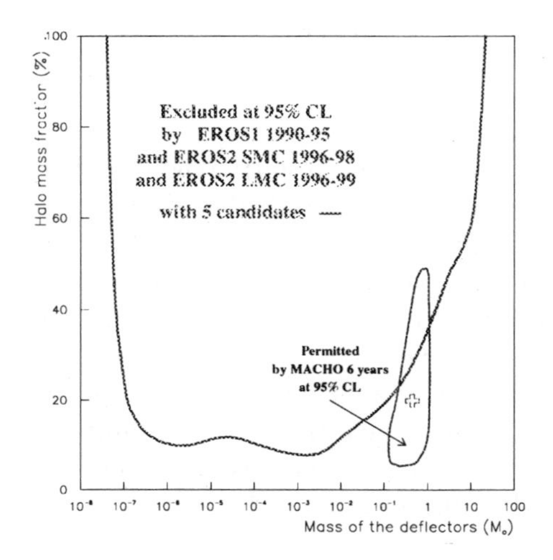

Figure 1. Exclusion/acceptance plot for MACHOs. The red solid curve is the 95% CL exclusion region of the EROS experiment. The blue line is the MACHO 95% CL acceptance contour obtained with their 5.7 year analysis (13 event sample).

Meanwhile, **EROS** [7] has extracted a limit from the **EROS1** LMC (1990-1995), **EROS2** LMC and SMC (1996-1998) combined data. A more stringent limit, taking into account the **EROS2** 1998-1999 data is available from reference [8]. One of the two **EROS1** candidates (LMC-2), which was "magnified" in 1990 was seen to vary again in 1999. Four more microlensing candidates were found in the **EROS2** LMC analysis. So **EROS** has a total of 5 (not especially nice) microlensing candidates (the SMC candidate is considered as self-lensing and not taken into account), while $\sim 30\ 0.5 M_{\odot}$ MACHOs were expected towards the LMC. **EROS** has decided to set an upper limit instead of claiming a Galactic halo signal.

The **EROS** 95% exclusion region is compared with the **MACHO** signal region on figure 1. The results are clearly compatible, but the interpretation is different.

4 Discussion

The excess events seen by the **MACHO** collaboration towards the LMC can be either a signal (sec. 4.3) or a background. In the latter case, it can be no microlensing at all (sec. 4.1) or microlensing by "known" populations (sec. 4.2).

4.1 Variable star backgrounds

Several variable star backgrounds to the microlensing search have been identified. The "blue bumpers" are young, bright, blue stars. Their flux variations are sometimes compatible with microlensing light curves, except for chromaticity. Fortunately, the interpretation of the observed event as the amplification of a faint star blended with the source turns out to be unphysical [8].

Cataclysmic variable bursts (e.g. dwarf novæ) can also be misinterpreted as microlensing events. The **MACHO** group shows evidence that some of its microlensing candidates could be supernovæ exploding in galaxies behind the LMC. These candidates are rejected when their light curves make a better fit to type Ia supernova templates than to microlensing light curves. **EROS** rejects this background by cutting on the asymmetry of the light curve.

Other sources of variable stars backgrounds are likely to exist (e.g **EROS1** LMC-2). However a few "gold plated" microlensing events have been found by **EROS** and **MACHO** towards the SMC and the LMC (e.g **MACHO** alert LMC-99-2, event SMC-98-1). Thus variable stars can explain at most a fraction of the signal.

Known populations of stars contribute to the optical depth towards the LMC and the SMC. For instance, solar mass stars located in the LMC are too faint to be resolved by **EROS** or **MACHO**: they are "dark objects" for the microlensing surveys.

4.2 Self-lensing

The major stellar populations to be considered are the Galactic disk and the various components of the Magellanic Clouds.

The contribution of stars in the Galactic disk to the optical depth is expected to be $\tau^{GD} \sim 10^{-8}$, an order of magnitude less than what is observed by MACHO.

The SMC is known to be elongated along the line of sight. Hence the lensing of a source in the SMC by a lens in the SMC ("self-lensing") is expected to be non negligible. The self-lensing optical depth towards the SMC has been estimated by various authors [10 12] to be $\tau^{SMC} \sim (0.5-2)\ 10^{-7}$. The observation of 1 event corresponds to an optical depth of $\tau^{SMC}_{EROS} \sim 1\ 10^{-7}$ and is clearly compatible with the expectation from self-lensing. Towards the SMC, the self-lensing contribution to the signal is as large as (or larger than) the Galactic halo contribution.

This conclusion is supported by the analysis of binary event SMC-98-1. This event was detected online by the MACHO alert system. The source star is too faint to be on the EROS catalog. A joint effort of the microlensing community led to an intensive photometric follow-up of this event [13]. The measured proper motion (angular velocity) of the source: $\mu \sim 1.4$ km/s/kpc is incompatible with a lens located in the Galactic halo ($\mu \sim 15$ km/s/kpc) and compatible with a lens in the SMC ($\mu \sim 0.5$ km/s/kpc).

The idea that the microlensing signal from the LMC can be explained by self-lensing traces back to Wu [15] and Sahu [16]. The LMC is believed to be a thin disk seen with a tilt angle of ~ 30 deg . The LMC self-lensing models have been analyzed by Gyuk et al. [14]. These authors find a self-lensing optical depth in the range $(0.5-8)\ 10^{-8}$, depending on the parameters of the LMC model with a prefered value of $\tau^{LMC}_{self} \simeq 2.5\ 10^{-8}$. The central value is a factor of 5 smaller than the optical depth measured by the MACHO collaboration. The self-lensing background was estimated by MACHO with the preferred model of Gyuk et al. to be 2-4 events for their 5.7 years analysis [9], giving a Galactic Halo signal of 11-13 events. However several non-standard models of the LMC predict microlensing optical depths compatible with the observations [17 18].

The self-lensing hypothesis can be tested observationally. The spatial distribution of observed candidates should scale like the distribution of sources (roughly flat) if the LMC sources are lensed by Galactic halo lenses. In the self-lensing hypothesis, the spatial distribution of events scales like the distribution of sources times the mass density in the LMC and should be concentrated towards the center of the LMC. The MACHO group [9] has compared the spatial distribution of their events to the predictions of the standard halo model and of the best self-lensing model of Gyuk et al [14]. The data are slightly (at the 2 σ level) in favor of the standard halo hypothesis. However, as seen previously, the self-lensing optical depth predicted by Gyuk is smaller than the MACHO measurement by a factor of 5, so data should be compared to the predictions of other self-lensing models.

4.3 White dwarfs

Assuming a Galactic Halo signal, the mass of the objects detected towards the LMC is $m_{MACHO} \simeq 0.5\ M_{\odot}$, which suggests white dwarfs (WD). These WD are old [11] (≥ 14 Gyr), faint, high proper motion stars. According to cooling models, old hydrogen WD with are still bright enough to be searched for by direct searches. Two white dwarf candidates (when 3.6 were expected for a halo full of WD) were found with two Hubble Deep Fields taken two years apart by Ibata et al. [19]. However Flynn et al [21] combined the results of reference [19] with the results of older photographic surveys and found a much smaller halo mass fraction in WD. A small positive signal (2 candidates found with 20 expected for $f_{WD} = 1$) compatible with MACHO's results was claimed in reference [20]. A signal was also searched for by the EROS group [22]. The EROS data were taken over a

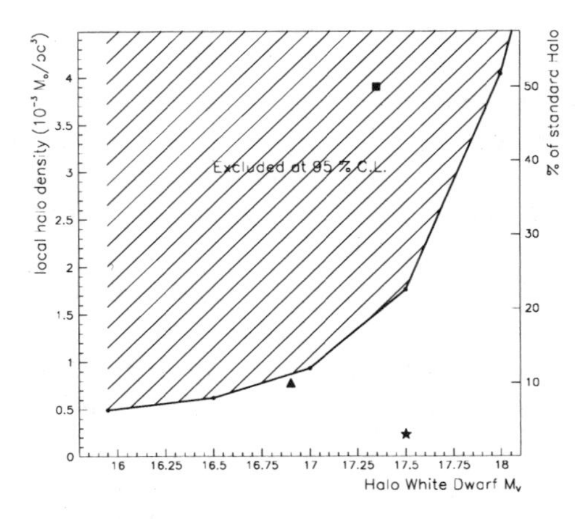

Figure 2. EROS2 exclusion plot for old Galactic white dwarfs. The square, the triangle and the star show respectively the results from Ibata et al. (1999), ibid. (2000) and Flynn et al.(1999)

large area of 440 square degrees, 250 of which have been analyzed. The analysis requires at least one astrometric measurement per year for 3 years. No candidate was found, while 20 were expected, assuming 14 Gyr old WD. As shown on figure 2, this rules out the Ibata et al 1999 result [19] and sets a 95% CL limit on the halo mass fraction in old WD.

5 Conclusion

After ten years of monitoring the Magellanic Clouds, it is now clear that MACHOs of less than a few $M_\odot$ cannot be a major contributor the Galactic mass budget. Strong limits have been set on the contribution to the dark Galactic Halo of low mass ($[10^{-6} - 10^{-2}]$ $M_\odot$) MACHOs ($f_{MACHO} < 0.1$), brown dwarfs ($f_{MACHO} < 0.2$) and 0.5 $M_\odot$ objects ($f_{MACHO} < 0.3$) .

The microlensing candidates seen towards the LMC are still not fully understood. Some of them (though not all of them) may be variable stars. The background from self-lensing should be small unless our understanding of LMC structure is incorrect. The existence of an old protogalactic WD population in the dark halo is still an open question. Direct searches show however that the halo mass fraction in old WD with an hydrogen atmosphere is less than 15 %.

References

1. B.J.Carr in *"The identification of Dark Matter"*, N.J.Spooner and V.Kudryavtsev editors, (World Scientific,Singapore,1997)
2. B.Paczyński, ApJ **304** (1986) 1
3. E.Aubourg et al. (EROS) *Nature* **365**,623 (1993)
4. C.Alcock et al. (MACHO) *Nature* **365**,621 (1993)
5. C.Alcock et al. (EROS & MACHO) *ApJ* **499**,L9 (1998)
6. C.Afonso et al. (EROS) *Astron. Astrophys.* **344**, L63 (1999)
7. T.Lasserre et al. (EROS) *Astron. Astrophys* **335**, L39 (2000)
8. T.Lasserre *PhD thesis*, Saclay report DAPNIA/SPP-00-04-T (2000)
9. C.Alcock et al., (MACHO), to appear in ApJ, astro-ph/ 0001272 (2000)
10. N.Palanque-Delabrouille et al. (EROS) *Astron. Astrophys* **332**,1 (1998)
11. D.Graff et al. *ApJ* **499**, 7 (1998)
12. D.Graff, L.Gardiner *MNRAS* **307**, 507 (1999)
13. C.Afonso et al.,(EROS, MACHO/GMAN, MPS, OGLE and PLANET) *ApJ* **532**, 340 (2000)
14. G.Gyuk,N.Dala and K.Griest *ApJ* **535**, 90 (2000)
15. X.P.Wu *ApJ* **435**, 66 (1994)
16. K.C.Sahu *Nature* **370**, 275 (1994)
17. P.Salati et al. *Astron. Astrophys* **350** L57 (1999)
18. H.Z.Zhao et.al *ApJ* **532**, L37 (2000)
19. R.Ibata et al. *ApJ* **524**, L95 (1999)
20. R.Ibata et al. *ApJ* **532**, L41 (2000)
21. C.Flynn et al., astro-ph/ 9912264, submitted to MNRAS (1999)
22. B.Goldman, astro-ph/ 0008383 (2000)

A LARGE-SCALE SEARCH FOR DARK-MATTER AXIONS

D. KINION AND K. VAN BIBBER
Lawrence Livermore National Laboratory
7000 East Ave., Livermore, CA 94550
E-mail: kinion1@llnl.gov

We review the status of two ongoing large-scale searches for axions which may constitute the dark matter of our Milky Way halo. The experiments are based on the microwave cavity technique proposed by Sikivie, and marks a 'second-generation' to the original experiments performed by the Rochester-Brookhaven-Fermilab collaboration, and the University of Florida group.

1 INTRODUCTION

Axions, a promising cold dark matter candidate, arise from a minimal extension of the Standard Model to enforce Strong-CP conservation. The Peccei-Quinn solution to the Strong-CP problem in QCD [1] involves an approximate $U_{PQ}(1)$ global symmetry which is spontaneously broken at some unkown symmetry-breaking scale f_a. The axion is the associated pseudo-Goldstone Boson.[2]

The properties of the axion depend mainly on the symmetry breaking scale f_a. Its mass is given by

$$m_a[eV] \approx 0.6\ eV \frac{10^7\ GeV}{f_a\ [GeV]} \tag{1}$$

All of the axions couplings are proportional to m_a. The coupling relevant for cavity detectors is the two-photon coupling described by

$$L_{a\gamma\gamma} = g_\gamma \frac{\alpha \phi_a}{4\pi f_a} F_{\mu\nu} \tilde{F}^{\mu\nu} = -g_{a\gamma\gamma} \phi_a \mathbf{E} \cdot \mathbf{B} \tag{2}$$

where α is the fine structure constant, ϕ_a is the axion field, g_γ is a model-dependent constant of order unity, and $g_{a\gamma\gamma} = (\alpha g_\gamma / \pi f_a)$. For the two most important axion models, KSVZ[3] and DFSZ[4], $g_\gamma \sim 0.97$, and $g_\gamma \sim -0.36$ respectively.

Since f_a is unknown and arbitrary, m_a could have any value. Fortunately, astrophysical and cosmological considerations help constrain m_a. The presently allowed mass range, or axion window is $1\ \mu\text{eV} < m_a < 10$ meV.[5,6]

2 THE MICROWAVE CAVITY AXION DETECTOR

To date, the most efficient method of searching for axions is the microwave cavity technique originally proposed by Sikivie.[8] In a static background magnetic field, axions will decay into single photons via the Primakoff effect. The energy of the photons is equal to the rest mass of the axion with a small contribution from its kinetic energy, hence their frequency is given by $hf = m_a c^2(1 + O(10^{-6}))$. A high-Q resonant cavity, tuned to the axion mass serves as the detector for the converted photons. The expected signal power varies with the experimental parameters as[8,9]

$$P_{a\rightarrow\gamma} \propto B^2 V C Q f \rho_a \tag{3}$$

where B is the background magnetic field, V is the cavity volume, C is a mode dependent form factor, Q is the loaded quality factor, f is the resonant frequency, and ρ_a is the local halo axion density. For the parameters of the U.S. experiment, the power from KSVZ axions is typically 5×10^{-22} W.

Since the axion mass is unknown, the frequency of the cavity must be tunable. The scan rate scales as

$$\frac{df}{dt} \propto \frac{f_o^2 Q_u C^2 B^4 V^2}{T_s^2} \tag{4}$$

where T_s is the system noise temperature.

3 THE U.S. DARK-MATTER AXION SEARCH

This section describes the operations and results of a microwave cavity axion search currently operating at Lawrence Livermore National Laboratory. This experiment, a collaboration of LLNL, MIT, Univ. of Florida, LBNL, UCB, Univ. of Chicago, and FNAL, has been operating with greater than 90% live time since February 1996 exploring the region from 0.3 to 3.0 GHz (1.2 to 12.4 μeV) at better than KSVZ sensitivity.[10] The experiment draws heavily on the experience gained in two pilot experiments performed in the late 1980's, one by a collaboration of Rochester-Brookhaven-Fermilab (RBF)[11] and a second at the University of Florida (UF).[12]

Figure 1 is a schematic of the U.S. dark-matter axion detector.

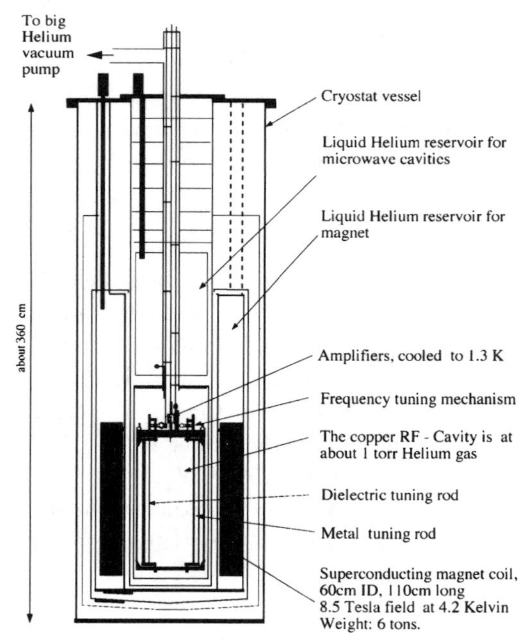

Figure 1. The U.S. Axion search detector.

The magnet employed in this search is a superconducting NbTi solenoid constructed by Wang NMR Inc.[13] The operating field at the center of the coil is usually 7.62 T.

The microwave cavities are right-circular cylinders constructed from stainless steel and copper plated. Two different cavity configurations have been used so far in this experiment. The region from 550 - 810 MHz has been scanned with a single cavity 50 cm in diameter and 1 m long. Recently, a set of four indentical cavities has been operated with the output combined in phase using a Wilkinson power combiner. Each cavity has a 20 cm diameter and is 1 m long. The four-cavity array will be used to search from 810-2000 MHz.

Power-combining multiple cavities allows the entire magnet volume to be utilized as the frequency of the cavities increases. This is possible because the axion signal is coherent on laboratory scales ($\lambda_D \approx$ 10-100 m).

Moving a combination of metal and dielectric rods, running the full length of the cavities, changes the resonant frequency. These rods can move from the center of the cavity to the wall. The single cavity rods were moved using stepper motors followed by a gear reduction of 42000:1. The final step size was approximately 600 nm, corresponding to roughly 500 Hz frequency shifts. This mechanical system was not practical for the four-cavity array, so a new piezoelectric based mechanism was implemented. The stepping resolution with this system was better than 50 nm, corresponding to a frequency resolution better than 100 Hz. All four cavities must have the same frequency for optimal phase-matching.

The cryogenic amplifiers used in this search are double-balanced GaAs HFET amplifiers supplied by NRAO.[14] The *in situ* measured noise temperatures range from 1.7 - 4.5 K.

A double-heterodyne receiver mixes a small bandwidth centered on the cavity frequency down to 35 kHz. This audio fre-

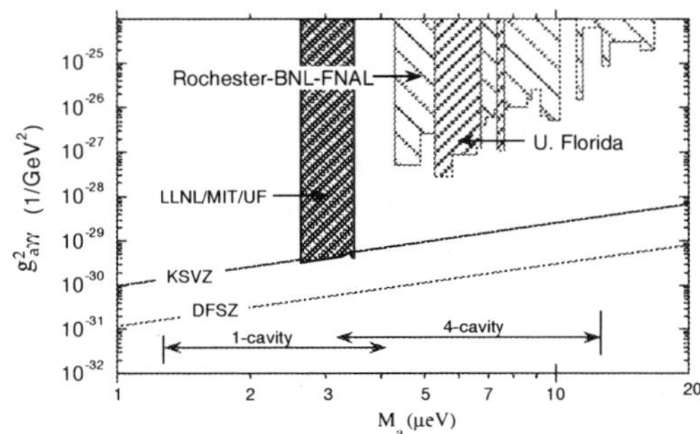

Figure 2. Axion couplings and masses excluded at the 90% confidence level by the U.S. experiments. The solid lines indicate the KSVZ and DFSZ model predictions. The arrows at the bottom indicate the coverage of different cavity configurations. The results from the two pilot experiments are scaled to 90% c.l. and $\rho_a = \rho_{halo}$.

quency signal is then sent to medium and high-resolution spectrum analyzers.

The medium-resolution search channel consists of a Stanford Research Systems[15] FFT spectrum analyzer with a frequency resolution of 125 Hz. These data are coadded and the result searched for Maxwellian peaks a few bins wide (about 700 Hz) characteristic of thermalized axions in the halo.[16]

An independent, high-resolution search channel operates in parallel to explore the possibility of fine-structure in the axion signal.[17,18] The 35 kHz signal passes through a third mixing stage to shift the center frequency to 5 kHz. A PC based DSP takes a single 50 second spectrum and performs an FFT with 20 mHz frequency resolution, about the limit imposed by the Doppler shift due to the earth's rotation. These data are searched for coincidences between different scans, as well as coincidences with peaks in the medium resolution data.

So far, no axion signal has been detected. Based on these results, we exclude at 90% confidence a KSVZ axion of mass between 2.5 and 3.3 μeV, assuming that thermalized axions comprise a major fraction of our galactic halo (ρ_a = 450 MeV/cm^3). This exclusion region and the results from two pilot experiments are shown in Figure 2. For more details see Ref. [][20].

In March 2000, the first data from the four-cavity array was taken. This was a commissioning run in a region with a low form factor. With the arrival of new HFET amplifiers in the 1-2 GHz region, production running with the four-cavity array will commence in Fall 2000.

4 RESEARCH AND DEVELOPMENT

The ultimate goal of this experiment is to scan as much of the axion window as possible with DFSZ sensitivity. Since the expected power from DFSZ axion conversion is an order of magnitude lower than that from KSVZ axions it would take one hundred times longer to reach similar sensitivity. From Equation 4, the scanning rate goes as T_s^{-2}, therefore, an order of magnitude reduction in system noise temperature would allow a scan at DFSZ sensitivity with the same rate as the present scan with KSVZ sensitivity. This is achievable with new dc SQUID based RF amplifiers.

4.1 dc SQUID Amplifiers

In the past two years a group at Berkeley led by John Clarke has developed dc SQUID amplifiers in the 100 - 3000 MHz range specifically for the axion experiment. Noise temperatures as low as 50 mK have been measured at a physical temperature of 30 mK.

The dc SQUID consists of two Josephson junctions connected in parallel on a superconducting loop. The SQUID produces an output voltage in response to a small input flux, and is a very sensitive flux-to-voltage transducer. Detailed computer simulations of the signal and noise properties were made by Tesche and Clarke.[21]

The most common configuration of a dc SQUID amplifier is shown in Figure 3.[22] The superconducting loop is a square washer with a slit on one side. The loop is closed via a superconducting counter-electrode connected to the washer by two resistively-shunted Josephson junctions. Flux is coupled into the SQUID through a microstrip input coil separated from the washer by a thin insulating layer. A microstrip resonator is formed by the open-ended stripline whose impedance is determined by the inductance of the input coil and its ground plane, and the capacitance between them. Near the fundamental frequency of the stripline, the gain of the amplifier is strongly enhanced.

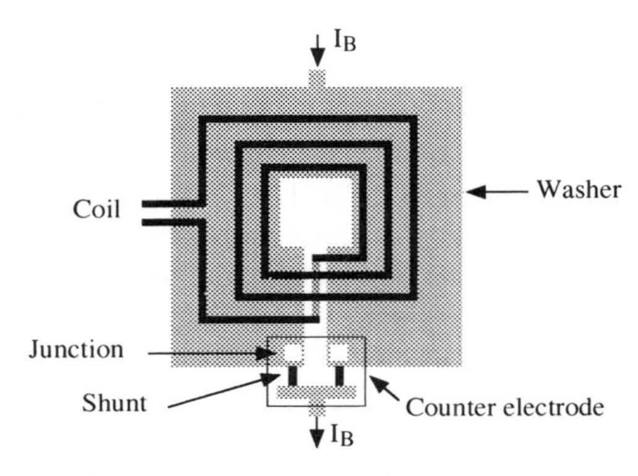

Figure 3. Square-Washer dc SQUID. The signal is coupled to the innermost turn of the coil, the outermost turn is either open-ended, shorted to the washer, or connected to a variable reactance.

The square-washer SQUIDs fabricated at Berkeley had inner and outer dimensions of 0.2 mm x 0.2 mm and 1 mm x 1 mm, and the input coils had a width of 5 μm and lengths ranging from 6 - 71 mm. The resonant frequency of the stripline scales as ℓ^{-1}; the highest frequency amplifier built so had $f > 3$ GHz. Frequencies up to 5-7 GHz should be achievable with the same design.

The bandwidth of these amplifiers has been greatly improved by varying the resonant frequency of the stripline *in situ*. This has been accomplished by connecting a pair of GaAs varactor diodes across the previously open end of the microstrip.[23]

The noise temperature of the SQUID amplifiers was measured using a heated resistor. The dominant source of noise in these devices is the Johnson noise from the resistive shunts across the Josephson junctions. This noise scales linearly with temperature, so the noise temperature of dc SQUIDs is expected to be proportional to their physical temperature until either the quantum limit ($T_a = h\nu/k_B \ln 2$)[24] is reached or hot electron effects in the shunts become dominant.

In an attempt to reduce the noise temperature, SQUIDs were cooled to 0.4 - 0.5 K in a charcoal-pumped, single-shot ^{3}He cryostat.[25] The system noise temperature at 438 MHz was 0.50 ± 0.07 K, of which 0.38 ± 0.07 K was contributed by the postamplifier.

At 500 mK, the noise temperatures are already within a factor of four of the quantum limit, which for a 500 MHz amplifier is approximately 35 mK. Demonstrating a quantum limited amplifier will require a much quieter postamplifier. Toward this end, a second SQUID has been used as a postamplifier to the input SQUID. The maximum power gain at 386 MHz was 33.5 ± 1 dB. Noise temperatures below 100 mK have been measured using cascaded SQUIDs cooled in a dilution refrigerator. Work is continuing to demonstrate quantum-limited noise performance.

4.2 Higher Frequency Cavities

Predictions of m_a from string models are typically $\mathcal{O}(100\ \mu\text{eV})$, requiring cavities with $f_{010} \sim 25$ GHz. The technique of power-combining signals from many small cavities is only practical for frequencies up to ≈ 3 GHz, because the number of cavities required scales as f_{010}^3, where f_{010} is the frequency of the TM_{010} mode. An alternative for reaching higher frequencies is the strategic placement of metal posts inside a single larger cavity.[19] If they are practical, these cavities could extend the mass range of microwave cavity axion searches by another decade.

5 RYDBERG ATOM SINGLE-QUANTUM DETECTOR

Another second-generation axion search is under development at the University of Kyoto. This effort seeks to exploit the extremely low-noise photon counting capability of Rydberg atoms in a Sikivie-type microwave cavity experiment. The initial goal is to sweep out a 10% mass window around $2.4\mu eV$.

An experiment utilizing Rydberg atom single-quantum detection in Kyoto is well along in commissioning ('CARRACK' for Cosmic Axion Research with Rydberg Atoms in a Resonant Cavity in Kyoto).[27] A sketch of the apparatus is shown in Figure 4. The microwave resonator is a single copper cavity (4.5 cm radius, 72.5 cm long) which fits inside a superconducting solenoid (15 cm diameter, 50 cm long, 7 T peak field). Power from the conversion cavity is coupled to a niobium superconducting cavity just above it, where the magnetic field is canceled by a bucking coil. The frequency of both cavities are made to track by means of 6 mm sapphire rods inserted axially into them. The cavities are cooled to < 15 mK by means of a dilution refrigerator.

A beam of rubidium atoms is accelerated, neutralized and directed vertically through the detection cavity. Just before entering the detection cavity, the atoms are excited to a Rydberg state with principal quantum number near 160, by triple optical excitation with three colinear diode laser beams. In the detection cavity, the Rydberg atoms are then Stark-tuned so an E1 $np \rightarrow (n+2)s$ transition is matched to the cavity frequency. After exiting, the Rydberg atoms are selectively ionized by an electric field (around 0.5 V/cm) and the liberated electron is detected and amplified by an electron multiplier ("Channeltron").

Studies have been performed to confirm that the experiment is sensitive to single

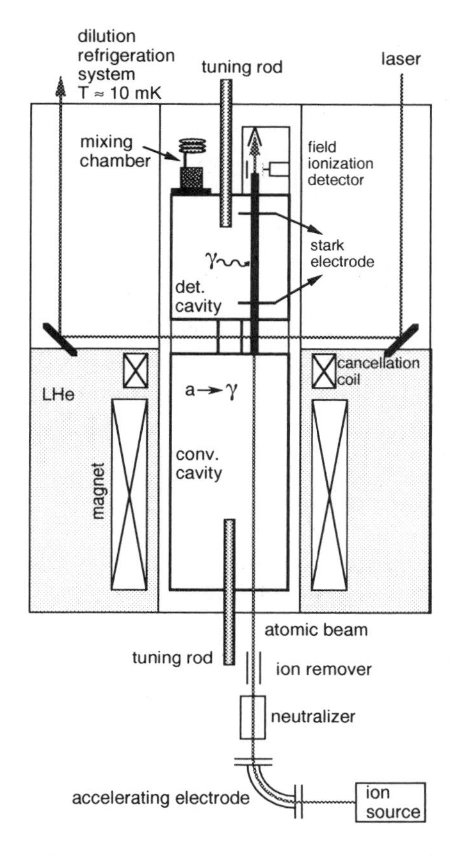

Figure 4. Schematic diagram of the experimental system to search for axions with Rydberg atoms in cooled resonant cavities.

blackbody photons in the < 15 mK range. These include verifying the temperature dependence, and the number and velocity of the Rydberg atoms.[28] Several percent of mass range around 2.4 GHz ($\sim 10\mu eV$) has been swept out, but candidate peaks have not been eliminated yet, nor have potential systematic backgrounds been rejected.

6 CONCLUSIONS

Axions are a well-motivated dark-matter candidate, enjoying both a bounded parameter space as well as experiments capable of searching a large portion of that region definitively. Quantum-limited SQUID amplifiers and Rydberg atom single-quantum detectors will give microwave-cavity axion detectors sensitivity to the most feebly coupled axion models. At the same time, new high frequency cavity designs may extend the mass coverage by at least another decade.

ACKNOWLEDGMENTS

Work performed under the auspices of the U.S. Department of Energy by the University of California Lawrence Livermore National Laboratory under contract W-7405-ENG-48.

References

1. R. Peccei and H. Quinn, *Phys. Rev. Lett.* **38**, 1440 (1977).
2. S. Weinberg, *Phys. Rev. Lett.* **40**, 223 (1978); F. Wilczek, *ibid.* 279 (1978).
3. J.E. Kim, *Phys. Rev. Lett.* **43**, 103 (1979); M.A. Shifman, A.I. Vainshtein, and V.I. Zakharov, *Nucl. Phys.* **166**, 493 (1980).
4. M. Dine, W. Fischler, and M. Srednicki, *Phys. Lett.* **104**, 199 (1981); A.R. Zhitnitsky, *Sov. J. Nucl. Phys.* **31**, 260 (1980).
5. G.G. Raffelt, *Stars as Laboratories for Fundamental Physics*, University of Chicago Press, Chicago (1996).
6. J. Preskill, M. Wise, and F. Wilczek, *Phys. Lett.* **120**, 127 (1983); L. Abbott and P. Sikivie *ibid.*, 133; M. Dine and W. Fischler, *ibid.*, 137; M.S. Turner, *Phys. Rev.* D **33**, 889 (1986).
7. A.D. Linde, *Phys. Lett.* B **201**, 437 (1988).
8. P. Sikivie, *Phys. Rev. Lett.* **51**, 1415 (1983).
9. L. Krauss *et al.*, *Phys. Rev. Lett.* **55**, 1797 (1985).
10. C. Hagmann *et al.*, *Phys. Rev. Lett.* **80**, 2043 (1998).
11. S. DePanfilis *et al.*, *Phys. Rev. Lett.* **59**, 839 (1987), W. Wuensch *et al.*, *Phys. Rev.* D **32**, 2988 (1980).
12. C. Hagmann *et al.*, *Phys. Rev.* D **42**, 1297 (1980).
13. Wang NMR Inc., 550 N. Canyons Parkway, Livermore, CA 94550.
14. E. Daw and R.F. Bradley, *J. Appl. Phys.* **82**, 1925 (1997).
15. Stanford Research Systems,1290-D Reamwood Ave., Sunnyvale, CA 94089.
16. M.S. Turner, *Phys. Rev.* D **42**, 3572 (1990).
17. P. Sikivie and J. Ipser, *Phys. Lett.* B **291**, 288 (1992).
18. P. Sikivie *et al.*, *Phys. Rev. Lett.* **75**, 2911 (1995).
19. C. Hagmann *et al.*, *Rev. Sci. Instr.* **61**, 1076 (1990).
20. H. Peng *et al.*, *Nucl. Instrum. Methods* **A 444**, 569 (2000).
21. C.D. Tesche and J. Clarke, *J. Low Temp. Phys.* **27**, 301 (1977).
22. M. Mück, M.-O. André, J. Clarke, J. Gail, and C. Heiden., *Appl. Phys. Lett.* **72**, 2885 (1998).
23. M. Mück, M.-O. André, J. Clarke, J. Gail, and C. Heiden., Submitted to *App. Phys. Lett.*
24. J. Clarke, *Proc. of the NATO ASI on SQUID Sensors: Fundamentals, Fabrication and Applications*, Kluwer Academic Publishers, H. Weinstock ed. (1996).
25. M. Mück, M.-O. André, J. Clarke, J. Gail, and C. Heiden., *Appl. Phys. Lett.* **75**, 698 (1999).
26. T.F. Gallagher, *Rydberg Atoms*, Cambridge University Press, Cambridge (1994).
27. M. Tada *et al.*, *Proc. of the* 5th *IFT Workshop on Axions* in *Nucl. Phys. B (Proc. Suppl.)* **72**, 164 (1999).
28. S. Matsuki, private communication (1998).

DARK MATTER DETECTION CROSS SECTIONS

R. ARNOWITT AND B. DUTTA

Center For Theoretical Physics, Department of Physics, Texas A&M University, College Station TX 77843-4242

We consider here the spin independent neutralino-proton cross section for a variety of SUGRA and D-brane models with R-parity invariance. The minimum cross section generally is $\gtrsim 1 \times 10^{-(9-10)}$pb (and hence accessible to future detectors) except for special regions of parameter space where it may drop to $\simeq 10^{-12}$pb. In the latter case the gluino and squarks will be heavy ($\gtrsim$1 TeV).

Dark matter detectors have now achieved a sensitivity that they have begun to probe interesting parts of SUSY parameter space. It is thus of interest to see what sensitivity will be needed to explore the entire space. To examine this we consider here models based on gravity mediated supergravity (SUGRA), where the LSP is generally the lightest neutralino ($\tilde{\chi}_1^0$). Neutralinos in the halo of the Milky Way might be directly detected by their scattering by terrestrial nuclear targets. Such scattering has a spin independent and a spin dependent part. For heavy nuclear targets the former dominates, and it is possible to extract then (to a good approximation) the neutralino -proton cross section, $\sigma_{\tilde{\chi}_1^0-p}$. Current experiments (DAMA, CDMS, UKMDC) have sensitivity to halo $\tilde{\chi}_1^0$ for

$$\sigma_{\tilde{\chi}_1^0-p} \gtrsim 1 \times 10^{-6}\,\text{pb} \tag{1}$$

and future detectors (GENIUS, Cryoarray) plan to achieve a sensitivity of

$$\sigma_{\tilde{\chi}_1^0-p} \gtrsim (10^{-9} - 10^{-10})\,\text{pb} \tag{2}$$

We consider here two questions: (1) what part of the parameter space is being tested by current detectors, and (2) what is the smallest value of $\sigma_{\tilde{\chi}_1^0-p}$ the theory is predicting (i.e. how sensitive must detectors be to cover the full SUSY parameter space). The answer to these questions depends in part on the SUGRA model one is considering and also on what range of theoretical and input parameter one assumes. In the following, we examine three models that have been considered in the literature based on grand unification of the gauge coupling constants at $M_G \cong 2 \times 10^{16}$ GeV: (1) Minimal super gravity GUT (mSUGRA) with universal soft breaking at M_G [1]; (2) Nonuniversal soft breaking models for Higgs and third generation scalar masses at M_G, and D-brane models (based on type IIB orientifolds[2]) which allow for nonuniversal gaugino masses and nonuniversal scalar masses at M_G [3].

While each of the above models contain a number of unknown parameters, theories of this type can still make relevant predictions for two reasons: (i) they allow for radiative breaking of $SU(2) \times U(1)$ at the electroweak scale (giving a natural explanation of the Higgs mechanism), and (ii) along with calculating $\sigma_{\tilde{\chi}_1^0-p}$, the theory can calculate the relic density of $\tilde{\chi}_1^0$, i.e $\Omega_{\tilde{\chi}_1^0} = \rho_{\tilde{\chi}_1^0}/\rho_c$ where $\rho_{\tilde{\chi}_1^0}$ is the relic mass density of $\tilde{\chi}_1^0$ and $\rho_c = 3H_0^2/8\pi G_N$ (H_0 is the Hubble constant and G_N is the Newton constant). Both of these greatly restrict the parameter space. In general one has $\Omega_{\tilde{\chi}_1^0} h^2 \sim (\int_0^{x_f} dx \langle \sigma_{\text{ann}} v \rangle)^{-1}$ (where σ_{ann} is the neutralino annihilation cross section in the early universe, v is the relative velocity, $x_f = kT_f/m_{\tilde{\chi}_1^0}$, T_f is the freeze out temperature, $\langle ... \rangle$ means thermal average and $h = H_0/100$ km s^{-1}Mpc^{-1}). The fact that these conditions can be satisfied for reasonable parts of the SUSY parameter space represents a significant success of the SUGRA models.

In the following we will assume $H_0 = (70 \pm 10)$km s^{-1}Mpc^{-1} and matter (m) and baryonic (b) relic densities of $\Omega_m = 0.3 \pm 0.1$ and $\Omega_b = 0.05$. Thus $\Omega_{\tilde{\chi}_1^0} h^2 = 0.12 \pm 0.05$. The calculations given below allow for a 2σ spread, i.e. we take $0.02 \leq \Omega_{\tilde{\chi}_1^0} h^2 \leq 0.25$. It is clear that when the MAP and Planck satelites determine the cosmological parameters accurately, the SUGRA dark matter predictions will be greatly sharpened.

1 Calculational Details

In order to get reasonably accurate results, it is necessary to include a number of theoretical corrections in the analysis. We list here the main ones used in the calculations below: (i) In relating the theory at M_G to phenomena at the electroweak scale, the two loop gauge and one loop Yukawa renormalization group equations (RGE) are used, iterating to get a consistent SUSY spectrum. (ii) QCD RGE corrections are further included below the SUSY breaking scale for contributions involving light quarks. (iii) A careful analysis of the light Higgs mass m_h is necessary (including two loop and pole mass corrections) as the current LEP limits impact sensitively on the relic density analysis for $\tan\beta \leq 5$. (iv) L-R mixing terms are included in the sfermion $(\text{mass})^2$ matrices since they produce important effects for large $\tan\beta$ in the third generation. (v) One loop corrections are included to m_b and m_τ which are again important for large $\tan\beta$. (vi) The experimental bounds on the $b \to s\gamma$ decay put significant constraints on the SUSY parameter space and theoretical calculations here include the leading order (LO) and approximate NLO corrections. We have not in the following imposed $b-\tau$ (or $t-b-\tau$) Yukawa unification or proton decay constraints as these depend sensitively on unknown post-GUT physics. For example, such constraints do not naturally occur in the string models where $SU(5)$ (or $SO(10)$) gauge symmetry is broken by Wilson lines at M_G (even though grand unification of the gauge coupling constants at M_G for such string models is still required).

All the above corrections are under theoretical control except for the $b \to s\gamma$ analysis where a full NLO calculations has not been done. (We expect that while the full analysis might modify the regions of parameter space excluded by the $b \to s\gamma$ experimental constraint, the minimum and maximum values of $\sigma_{\tilde{\chi}_1^0-p}$ would probably not be significantly changed.) The analysis of $\sigma_{\tilde{\chi}_1^0-p}$, taking into account the above theoretical corrections has now been carried out by several groups obtaing results in general agreement [4,5,6,7,8,9,10]. These results are presented below.

Accelerator bounds significantly limit the SUSY parameter space. In the following we assume the LEP bounds [8] $m_h > 104(100)$ GeV for $\tan\beta=3(5)$ and $m_{\chi_1^\pm} > 104(100)$ GeV. (For $\tan\beta > 5$, the m_h bounds do not produce a significant constraint[11].) For $b \to s\gamma$ we assume an allowed range of 2σ from the CLEO data [12]. The Tevatron gives a bound of $m_{\tilde{g}} \geq 270$ GeV(for $m_{\tilde{q}} \cong m_{\tilde{g}}$)[13].

Theory allows one to calculate the $\tilde{\chi}_1^0$-quark cross section and we follow the analysis of [14] to convert this to $\tilde{\chi}_1^0 - p$ scattering. For this one needs the $\pi - N$ σ term,$\sigma_{\pi N}$ and $\sigma_0 = \sigma_{\pi N} - (m_u + m_d)\langle p|\bar{s}s|p\rangle$ and the quark mass ratio $r = m_s/(1/2)(m_u + m_d)$. We use here $\sigma_{\pi N} = 65$ MeV, from recent analyses [15,16] based on new $\pi - N$ scattering data, $\sigma_0 = 30$ MeV[17] and r= 24.4 ± 1.5[18].

2 mSUGRA model

We consider first the mSUGRA model where the most complete analysis has been done. mSUGRA depends on four parameters and one sign: m_0 (universal scalar mass at M_G), $m_{1/2}$ (universal gaugino mass at M_G), A_0 (universal cubic soft breaking mass) , $\tan\beta = \langle H_2\rangle/\langle H_1\rangle$ (where $\langle H_{2,1}\rangle$ gives rise to (up, down) quark masses) and $\mu/|\mu|$(where μ is the Higgs mixing parameter in the superpotential, $W_\mu = \mu H_1 H_2$). One conventionally restricts the range of these parameters by "naturalness" conditions and in the following we assume $m_0 \leq 1$ TeV, $m_{1/2} \leq 600$ GeV (corresponding to $m_{\tilde{g}} \leq 1.5$ TeV, $m_{\tilde{\chi}_1^0} \leq 240$ GeV), $|A_0/m_0| \leq 5$, and $2 \leq tan\beta \leq 50$. Large tanβ is of interest since SO(10) models imply tan$\beta \geq 40$ and also $\sigma_{\tilde{\chi}_1^0-p}$ increases with tanβ. $\sigma_{\tilde{\chi}_1^0-p}$ decreases with $m_{1/2}$ for large $m_{1/2}$, and thus if one were to increase the bound on $m_{1/2}$ to 1 TeV ($m_{\tilde{g}} \leq 2.5$ TeV), the cross section would drop by a factor of 2-3.

The maximum $\sigma_{\tilde{\chi}_1^0-p}$ arise then for large tanβ and small $m_{1/2}$. This can be seen in Fig.1 where $(\sigma_{\tilde{\chi}_1^0-p})_{\max}$ is plotted vs. $m_{\tilde{\chi}_1^0}$ for tanβ=20, 30, 40 and 50. Current detectors obeying Eq (1) are then sampling the parameter space for large tan β, small $m_{\tilde{\chi}_1^0}$ (and also small $\Omega_{\tilde{\chi}_1^0}h^2$) i.e

$$tan\beta \gtrsim 25; \; m_{\tilde{\chi}_1^0} \lesssim 90\,\text{GeV}; \; \Omega_{\tilde{\chi}_1^0}h^2 \lesssim 0.1. \tag{3}$$

To discuss the minimum cross section, it is convenient to consider first $m_{\tilde{\chi}_1^0} \lesssim 150$ GeV ($m_{1/2} \leq 350$) where no coannihilation occurs. The minimum cross section occurs for small tanβ. One finds

$$\sigma_{\tilde{\chi}_1^0-p} \gtrsim 4 \times 10^{-9}\text{pb}; \; m_{\tilde{\chi}_1^0} \lesssim 140\text{GeV} \tag{4}$$

which would be accessible to detectors that are currently being planned (e.g. GENIUS).

For larger $m_{\tilde{\chi}_1^0}$, i.e. $m_{1/2} \gtrsim 150$ the phenomena of coannihilation can occur in the relic density analysis since the light stau, $\tilde{\tau}_1$, (and also $\tilde{e}_R$, $\tilde{\mu}_R$) can become degenerate with the $\tilde{\chi}_1^0$. The relic density constraint can then be satisfied in narrow corridor of m_0 of width $\Delta m_0 \lesssim 25$ GeV, the value of m_0 increasing as $m_{1/2}$ increases [6]. Since m_0 and $m_{1/2}$ increase as one progresses up the corridor, $\sigma_{\tilde{\chi}_1^0-p}$ will generally decrease.

We consider first the case $\mu > 0$[19]. One finds in general that $\sigma_{\tilde{\chi}_1^0-p}$ also decreases as A_0 increases. Fig.2 shows $\sigma_{\tilde{\chi}_1^0-p}$ in the domain of large A_0 and for two values of tanβ. One sees that the smaller tanβ still gives the lower cross section, though the difference is mostly neutralized at larger $m_{1/2}$. (For large tanβ, m_0 also becomes large to satisfy the relic density constraint i.e $m_0 \cong 700$ GeV for tanβ=40, $m_{1/2} = 600$ GeV.) We have in general for this regime

$$\sigma_{\tilde{\chi}_1^0-p} \gtrsim 1 \times 10^{-9}\text{pb; for}\, m_{1/2} \leq 600\text{GeV}, \quad \mu > 0\,,\, A_0 \leq 4m_{1/2}. \tag{5}$$

This is still within the sensitivity range of proposed detectors.

When μ is negative an "accidental" cancellation can occur in part of the parameter space in the coannihilation region which can greatly reduce $\sigma_{\tilde{\chi}_1^0-p}$ [7]. This can be seen in Fig.3, where starting with small tanβ the cross section decreases, leading to a minimum at about tanβ=10, and then increases again for larger tanβ. At the minimum one has $\sigma_{\tilde{\chi}_1^0-p} \cong 1 \times 10^{-12}$ when tanβ=10 and $m_{1/2} = 600$ GeV. More generally one has

$$\sigma_{\tilde{\chi}_1^0-p} < 1 \times 10^{-10}\text{pb} \tag{6}$$

for the parameter domain when $4 \lesssim \tan\beta \lesssim 20$, $m_{1/2} \gtrsim 450$GeV($m_{\tilde{g}} \gtrsim 1.1$TeV), $\mu < 0$. In this domain, $\sigma_{\tilde{\chi}_1^0-p}$ would not be accessible to any of the currently planned detectors. However, mSUGRA also then predicts that this could happen only when the gluino and squarks have masses greater than 1 TeV (and for only a restricted region of tanβ) a result that could be verified at the LHC.

3 Nonuniversal SUGRA Models

In the discussion of SUGRA models with nonuniversal soft breaking, universality for the first two generations of squark and slepton masses at M_G is usually maintained to suppress flavor changing neutral currents. One allows, however, the Higgs and third generation squark and slepton masses to become nonuniversal. We maintain gauge and gaugino mass unification at M_G.

While these models contain a large number of new parameters, their effects on $\sigma_{\tilde{\chi}_1^0-p}$ can be charcterized approximately by the signs of the deviations from universality[9]. One choice can greatly increase $\sigma_{\tilde{\chi}_1^0-p}$, by a factor of 10-100 compared to the universal case, and the reverse choice can reduce $\sigma_{\tilde{\chi}_1^0-p}$ (though by a much lesser amount). Thus it is possible for detectors to probe regions of smaller tanβ with nonuniversal breaking, and detectors obeying Eq. (1) can probe part of the parameter space for tanβ as low as $tan\beta \simeq 4$.

The minimum cross section occurs (as in mSUGRA) at the lowest tanβ and at the largest $m_{1/2}$ i.e. in the coannihilation region. We limit ourselves here to the case where only the Higgs masses are nonuniversal. One finds then results similar to mSUGRA i.e. $\sigma_{\tilde{\chi}_1^0-p} \gtrsim 10^{-9}$ pb for $\mu > 0$, $m_{1/2} \leq 600$ GeV. For $\mu < 0$, there can again be a cancellation of matrix elements reducing the cross section to 10^{-12} pb when $m_{1/2} = 600$ GeV in a restricted part of the parameter space when tan$\beta \simeq 10$.

4 Summary

We have examined here the neutralino-proton cross section for a number of SUGRA type models. In all the models considered, there are regions of parameter space with $\tilde{\chi}_1^0 - p$ cross sections of the size that could be observed with current detectors. Thus with the sensitivity of Eq. (1), detectors would be sampling regions of the parameter space for mSUGRA where tan$\beta \gtrsim 25$, $m_{\tilde{\chi}_1^0} \lesssim 90$GeV and $\Omega_{\tilde{\chi}_1^0}h^2 \lesssim 0.1$. Nonuniversal models can have larger cross sections and so detectors could sample down to tan$\beta \gtrsim 4$, while for the D-brane models considered, detectors could sample down to tan$\beta \gtrsim 15$.

The minimum cross sections these models predict are considerably below current sensitivity. Thus for mSUGRA one finds for $\mu > 0$ that $\sigma_{\tilde{\chi}_1^0-p} \gtrsim 1 \times 10^{-9}$pb for $m_{1/2} \leq$ 600GeV, $\mu > 0$, where $m_{1/2} = 600$ GeV corresponds to $m_{\tilde{g}} \cong 1.5$ TeV, $m_{\tilde{\chi}_1^0} \cong 240$ TeV. This is still in the range that would be accessible to detectors being planned (such as GENIUS or Cryoarray). For $\mu < 0$, a cancellation can occur in certain regions of parameter space allowing the cross sections to fall below this. Thus

$$\sigma_{\tilde{\chi}_1^0-p} < 1 \times 10^{-10}\text{pb for } 4 \lesssim \tan\beta \lesssim 20, \quad \mu < 0,\ m_{1/2} \gtrsim 450\,\text{GeV} \qquad (7)$$

and reaching a minimum of $\sigma_{\tilde{\chi}_1^0-p} \cong 1\times10^{-12}$ pb for tanβ=10, $m_{1/2}$=600 GeV, $\mu < 0$. This domain would appear not to be accessible to future planned detectors. Since $m_{1/2}$=450 GeV corresponds to $m_{\tilde{g}} \cong 1.1$ TeV, this region of parameter space would imply a gluino squark spectrum at the LHC above 1 TeV.

The above results holds for the mSUGRA model. While a full analysis of coannihilation has not been carried out for the nonuniversal and D-brane models, results similar to the above hold for these over large regions of parameter space. Thus for nonuniversal Higgs masses and for the D-brane model one finds $\sigma_{\tilde{\chi}_1^0-p} \gtrsim 10^{-9}$ pb for $\mu > 0$, while a cancellelation allows $\sigma_{\tilde{\chi}_1^0-p}$ to fall to 10^{-12} pb for $\mu < 0$ at tan$\beta \simeq 10$ (with again a gluino/squark mass spectrum in the TeV domain).

5 Acknowledgement

This work was supported in part by NSF grant no. PHY-9722090.

References

1. A.H. Chamseddine, R. Arnowitt and P. Nath, *Phys. Rev. Lett.* **49**, 970 (1982); R. Barbieri, S. Ferrara and C.A. Savoy, *Phys. Lett.* B **119**, 343 (1982); L. Hall, J. Lykken and S. Weinberg, *Phys. Rev.* D **27**, 2359 (1983); P. Nath, R. Arnowitt and A.H. Chamseddine, *Nucl. Phys.* B **227**, 121 (1983).
2. L. Ibanez, C. Munoz and S. Rigolin, *Nucl. Phys.* B **536**, 29. (1998).

3. M. Brhlik, L. Everett, G. Kane and J. Lykken; *Phys. Rev.* D **62**, 035005 (2000).
4. A. Bottino, F. Donato, N. Fornengo and S. Scopel, Phys. Rev. **D 59**, 095004 (1999).
5. A. Bottino, F. Donato, N. Fornengo and S. Scopel, *Astropart. Phys.* **13**, 215 (2000).
6. J. Ellis, T. Falk, K.A. Olive and M. Srednicki, *Astropart. Phys.* **13**, 181 (2000).
7. J. Ellis, A. Ferstl and K.A. Olive; hep-ph/0001005.
8. J. Ellis, T. Falk, G. Ganis and K.A. Olive, hep-ph/0004109.
9. E. Accomando, R. Arnowitt, B. Dutta and Y. Santoso, hep-ph/0001019 (to appear in *Nucl. Phys.*B).
10. R. Arnowitt, B. Dutta and Y. Santoso, hep-ph/0005154.
11. Further analysis of 2000 LEP data may raise the lower bound on m_h, which would exclude part of the parameter space for low tanβ and low $m_{\tilde{\chi}_1^0}$.
12. M. Alam et al., *Phys. Rev. Lett.* **74**, 2885 (1995).
13. D0 Collaboration, *Phys. Rev. Lett.* **83**, 4937 (1999).
14. J. Ellis and R. Flores,*Phys. Lett.* B **263**, 259 (1991); **B 300**, 175 (1993).
15. M. Ollson, hep-ph/0001203.
16. M. Pavan, R. Arndt, I. Stravkovsky, and R. Workman, nucl-th/9912034, *Proc. of 8th International Symposium on Meson-Nucleon Physics and Structure of Nucleon*, Zuoz, Switzerland, Aug., (1999).
17. A. Bottino, F. Donato, N. Fornengo, and S. Scopel, *Astropart. Phys.* **13**, 215 (2000).
18. H. Leutwyler, *Phys. Lett.* B **374**, 163 (1996).
19. We use the ISAJET sign convention for μ and A.

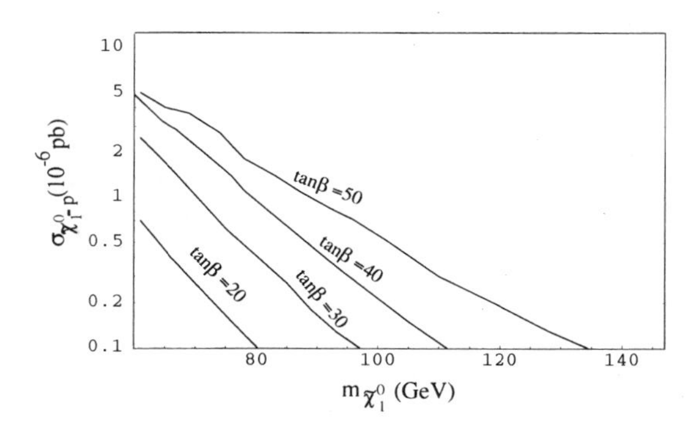

Figure 1. $(\sigma_{\tilde{\chi}_1^0-p})_{\max}$ for mSUGRA obtained by varying A_0 over the parameter space for tanβ = 20, 30, 40, and 50[9]. The relic density constraint has been imposed.

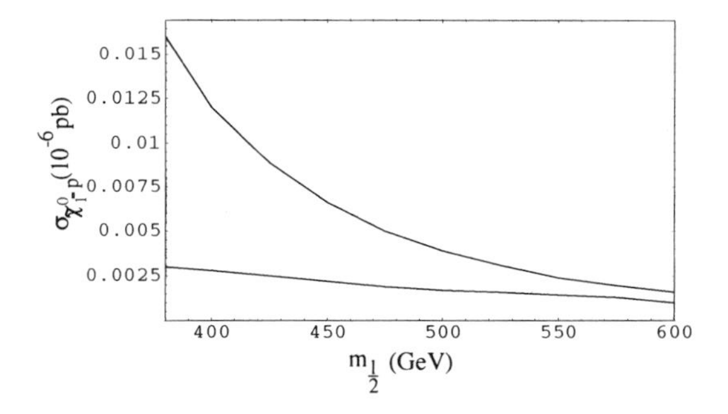

Figure 2. $(\sigma_{\tilde{\chi}_1^0-p})$ for mSUGRA in the coannihilation region for tanβ = 40 (upper curve) and tanβ = 3 (lower curve), $A_0 = 4m_{1/2}$, $\mu > 0$.

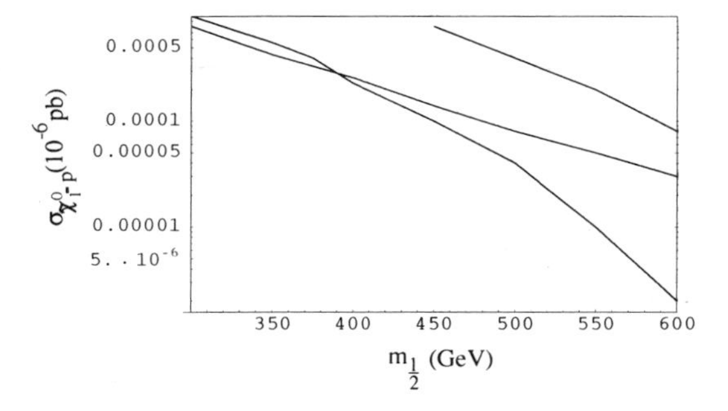

Figure 3. $(\sigma_{\tilde{\chi}_1^0-p})$ for mSUGRA and $\mu < 0$ for (from top to bottom on right) tanβ=20, 5 and 10. Note that for tan$\beta \geq 10$, the curves terminate at the left due to the $b \to s\gamma$ constraint.

DARK MATTER SEARCHES

N.J.C. SPOONER
Department of Physics and Astronomy, University of Sheffield, Hicks Building, Hounsfield Road, Sheffield S3 7RH
E-mail: n.spooner@sheffield.ac.uk

A brief review is presented of the present status of experiments searching for dark matter in the form of Weakly Interacting Massive Particles by direct means. After an outline of the various detector strategies adopted, emphasis is switched to experiments that have recently produced new search results, followed by an outline of prospects for future more sensitive experiments.

1 Detector Strategies

Since realisation in the 1980s that WIMP dark matter particles may be detectable by direct searches in the laboratory several different detector technologies have been developed for the task by a wide variety of groups. Table 1 lists some of these experiments (as known to the author). Of these arguably the most successful have been based on detection of ionisation, scintillation or phonons (a full overview of these and others can be found in several recent international workshops devoted to the subject[1,2,3]). However, there has developed a general commonality amongst the techniques to progress beyond early counting experiments, such as exemplified by low background Ge detectors used in double beta decay experiments, towards techniques that can combine: (a) a means of positively identifying the nuclear recoil events expected from WIMP interactions, and (b) some means of achieving this with sufficient sensitivity eventually to allow the lowest predicted neutralino cross-sections ($\sigma_{WIMP-p} < 10^{-8}$ pb) to be probed - a combination of high (>10s kg) mass and efficient background rejection.

Experiment	Site	Targets
NAIAD (UKDMC)	Boulby	NaI
DAMA (Rome)	Gran Sasso	NaI, CaF_2, Xe
Saclay	Modane	NaI
USC-PNL-Zaragoza	Canfranc, Soudan	NaI, Ge
ELEGANTS-V	Kamioka	NaI
Osaka-Tokushima	Oto-Cosmo	CaF_2
Heidelberg/Moscow	Gran Sasso	Ge
USC-PNL-Zaragoza- TANDAR	Canfranc-Sierra Grande	Ge
Neuchatel-Caltech-PSI	St.Gottard	Ge
ZEPLIN (UKDMC- UCLA-Torino-ITEP)	Boulby	Xe
SIMPLE (CERN-Lisbon-Paris)	Paris	F,Cl,C
Montreal-Chalk River	Montreal	F,Cl,C

CDMS	Stanford	Ge, Si
Edelweiss	Modane	Ge
CRESST	Gran Sasso	sapphire, $CaWO_4$
Milan	Gran Sasso	TeO_2
SALOPARD	Canfranc	Sn
ORPHEUS	Bern	Sn
Tokyo-Osaka	Tokyo	LiF
DRIFT (UKDMC-Temple-Oxy-Surrey)	Boulby	CS_2, Ar, Xe

Table 1. Selected world dark matter experiments

In addition to recoil identification, and hence rejection of electron background, strategies to achieve signal sensitivity also include: (i) measurement of the expected small annual modulation in the WIMP-induced recoil spectrum (of order 5%) due to modulation of the Earth's orbital velocity component parallel to the galactic plane, and, by extension, (ii) measurement of the direction of nuclear recoils within the detector. Detectors with the latter capability could ultimately provide a definitive dark matter signal by correlating events with the diurnal change in orientation of the experiment, due to the Earth's rotation relative to the galactic dark matter halo. Direction sensitivity provides a final goal for experiments that would yield the maximum information on events. However, although examples, such as the DRIFT gas-based experiment, are under study (see Sec. 3) none have so far been run as experiments, so that current examples still rely on recoil identification and/or annual modulation alone.

Various recoil identification techniques are possible but most success has been achieved so far using NaI scintillation detectors (for example DAMA[4] and UKDMC[5]) and Ge or Si thermal-ionisation detectors (for example CDMS[6] and Edelweiss[7]). In NaI nuclear recoils release scintillation photons typically 30% faster than for equivalent energy electron recoils. This allows identification by statistical means following accumulation of sufficient events processed by pulse shape analysis[8]. The latter technique relies on measurement of the ratio of energy released in the phonon and ionisation channels at mK temperatures - the phonon yield being higher for nuclear recoils[9,10].

In the case of an annual modulation search this would best be performed following recoil discrimination, on that component of events known to be dominated by nuclear recoils. This would exclude as much electron background as possible, assumed to be not modulating, and hence improve signal to noise. This has not so far been achieved. However, several annual modulation experiments have been performed without recoil discrimination including in NaI, for instance by DAMA[11] and ELEGANTS[12], and in Ge, for instance the USC-PNL-Zaragoza-Tandar collaboration[13].

2 DAMA, UKDM, Saclay and CDMS

Significant results have recently been announced by several collaborations. The DAMA (Rome) collaboration has been searching for annual modulation using ~100 kg of NaI at Gran Sasso. More than four years of running has now been achieved yielding ~60,000 kg.days of data[14]. By performing a basic noise cut based on pulse time discrimination the collaboration are able to reject photomultiplier events and hence achieve a low energy threshold of ~2 keV. Annual modulation analysis is then performed in the 2-

6 keV region, sufficient to probe for spin independent neutralino interactions at σ_{WIMP-p} = 10^{-5} - 10^{-6} pb, given the measured background level of 1-2 dru (events/kg/d/keV). Fig. 1 reproduces a plot of the count rate residuals versus time for analysis of the full 4-year data set[14].

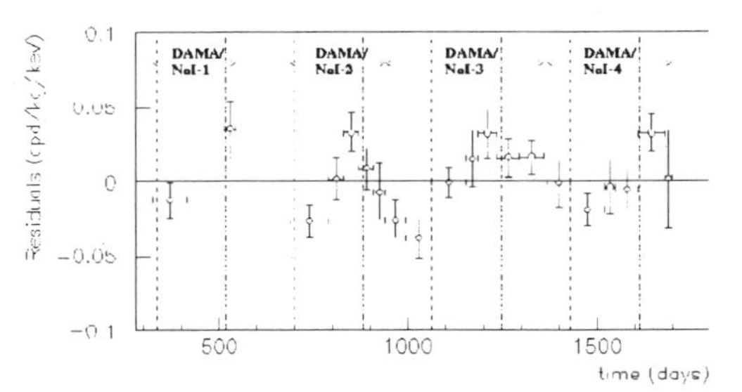

Figure 1. 2-6 keV count residuals from the DAMA 4-year data set[14].

Combining all 4 years, the DAMA group interpret the data as evidence (at 4σ c.l.) for a positive spin independent neutralino signal due to a flux of particles of mass 52 (+10, - 8) GeV of cross section $\zeta\sigma = 7.2$ (+0.4, -0.9) x 10^{-6} pb (where ζ is the halo parameterisation factor[14]). This interpretation would, for instance, require that ~50% of the raw event rate in the 2-3 keV energy bin are due to WIMP interactions alone, with ~3% responsible for the modulation.

The DAMA results have been subject to great comment in the community (see for instance Gerbier[15] and Smith[16]) since despite considerable efforts by DAMA to eliminate systematic effects the possibility remains that the modulation results from an as-yet unidentified background source or artefact of the detector. Anomalous effects in NaI, though not necessarily related to the DAMA effect, have been observed in NaI operated by the UKDMC group and by Saclay. The UKDMC have been operating NaI dark matter detectors since 1990 in which pulse shape discrimination is used to search for a population of fast nuclear recoil events[5]. Following a sensitivity upgrade of the group's main 5 kg detector (DM46) in 1997 by increasing the light collection efficiency, a population of anomalous fast time constant events was observed (ratio of $\tau_{anom}:\tau_{\gamma} = 0.65$). These events have since been observed in many NaI detectors at very similar rates and with similar exponentially falling energy spectra[17]. More importantly analysis has now been completed of data from a 9.7 kg detector of the Saclay group (DM70) which also shows these events. This crystal was originally part of the DAMA array, fabricated by the same manufacturer and with the same construction as the DAMA 9.7 kg crystals. Fig. 2 shows comparison of the UKDMC and Saclay anomalous event spectra.

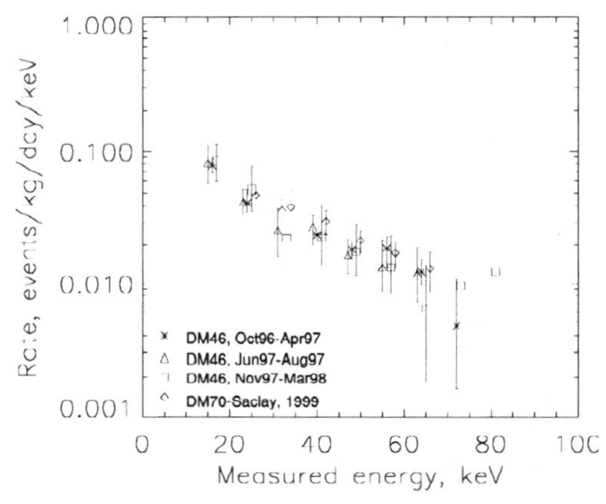

Figure 2. Fast time constant event spectra from UKDMC NaI (5 kg DM46) and Saclay NaI (9.7 kg DM70).

Extensive studies by the UKDMC[18,19] suggest that the fast time constant events may result from diffusion of radon gas to the detector surfaces. This could result in implantation of alpha emitters into the sub-micron surface layer. Simulations show that outward going alphas would deposit energy with a spectral form similar to that observed (unlike inward going alphas), though reasons for the quite high count rate (~0.1 dru) relative to the levels of radon known to be present, and the similarity of rate between quite different NaI detectors, are still being investigated.

It is not possible yet to confirm whether or not the fast time constant events seen in the UKDMC and Saclay NaI are related to the modulating signal in the DAMA array, in particular because pulse shape discrimination is not possible below ~4 keV. However, recent long-term analysis of the UKDMC DM46 detector for the period 1997-1999 has revealed that the rate of the fast events does fluctuate with time with a characteristic quite different from the gamma background rate which remains steady (see Fig. 3). This effect is not inconsistent with possible fluctuating levels of radon. Therefore, it may not be unreasonable to assume such effects are possible in any NaI detector, with important implications for annual modulation searches.

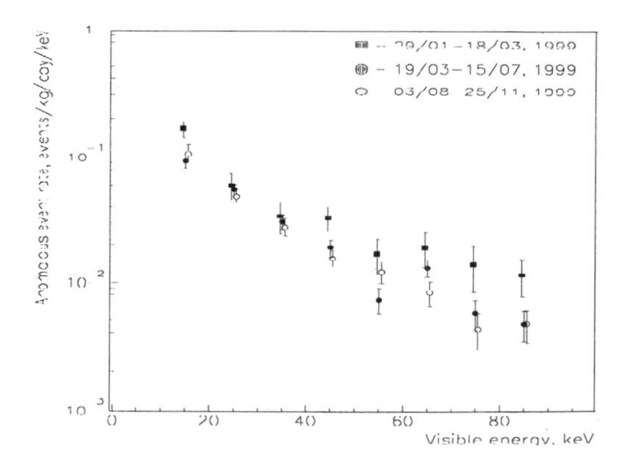

Figure 3. UKDMC fast time constant event spectra from 1999 runs of NaI crystal DM46.

By using the ionisation+thermal technique of recoil discrimination (see Sec. 1) the CDMS consortium have recently obtained first data from their Ge and Si experiment constructed at a shallow (~10m deep) site at Stanford University. Results have now been published for 1.6 kg.days Si and 10.6 kg.days of Ge[6]. Analysis of the Ge data has revealed 17 events lying in a region of low ionisation yield expected for nuclear recoils. The discrimination power is sufficient that in the presence of no WIMP or neutron flux zero counts would be expected. However, the shallow site, despite the presence of a highly efficient muon veto surrounding the detector, ensures that there is a significant neutron background from cosmic ray muon interactions. This is confirmed by the observation that 4 of the events occur in coincidence with recoil-like events in neighbouring detector segments.

The remaining 13 single hit recoils, if interpreted as WIMP events, are compatible with the lower part of the allowed DAMA WIMP region (equivalent to ~2.3 event/kgGe/d - 2 x 10^{-42} cm^2 WIMP-nucleon cross section). However, clearly some or all of these events may be neutrons. By making use of the observed single and double hit ratio CDMS are able to reduce systematic errors in Monte Carlo simulations of muon propagation and neutron production through their complex detector and shielding, sufficient to predict the likely absolute rate of neutrons. Fig. 4 shows the likelihood upper limit on WIMP events that results from subtraction of the neutron events based on this simulation. The result indicates that the DAMA and CDMS results are incompatible at 99% c.l.

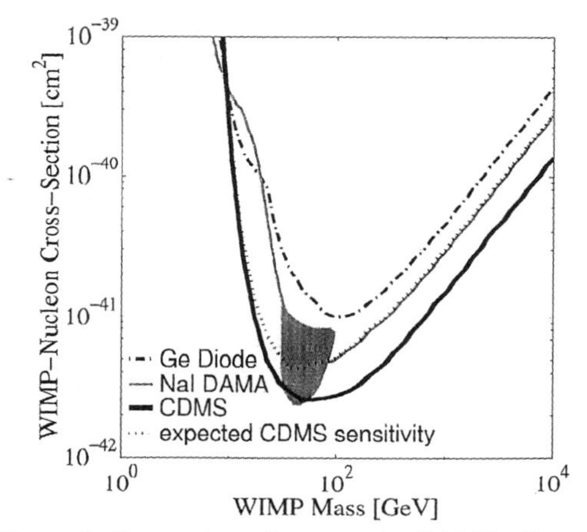

Figure 4. Comparison between the DAMA allowed region and CDMS limit[6].

3 Future plans

The neutron flux at Stanford remains a limiting factor in the CDMS experiment but plans to start operation at Soudan mine from 2001 should resolve this issue. The UKDMC is now operating NaI at Boulby without encapsulation (the NAIAD experiment) having demonstrated that careful preparation of crystal surfaces can eliminate the anomalous events. The collaboration anticipates 50 kg running by early 2001[20]. Meanwhile, the Ge experiment of Heidelberg-Moscow continues to run[21] and several new experiments are expected to become operational in the next few years (see Table 1). Of note are the bolometric experiments Edelweiss[7] and CRESST[22]. The former uses the ionisation+thermal technique in Ge with NTD thermistors and is already running at Modane. Sensitivity estimates for 500 kg.d Ge are comparable with CDMS predictions. CRESST, based at Gran Sasso, has recently developed a recoil discrimination technique based on simultaneous measurement of phonons and light in $CaWO_4$. This supplements their existing technology based on sapphire. They plan upgrades towards 10 kg during 2001.

Several non-bolometric techniques other than NaI are also being developed. These include Superheated Droplet Detectors (for instance, CERN-Paris-Lisbon[23] and Montreal-Chalk River[24] collaborations), liquid Xenon (for instance, DAMA[25] and UKDMC[26]) and low-pressure (10-40 Torr) gas TPC (the UKDMC-Temple-Occidental direction sensitive DRIFT experiment[27]). Liquid Xe scintillator is a particularly powerful technique because interactions produce several processes that can provide high discrimination depending on the dE/dx[28]. These are: i) excitation - resulting in Xe_2^* molecules which decay emitting 175 nm photons with a mixture of 3 ns and 27 ns time constants depending on the dE/dx, and ii) ionisation - resulting in Xe_2^+ ions which, after a delay of ~40 ns for gammas and <3 ns for nuclear recoils, can recombine to give Xe_2^*. The latter can decay as in i) or, if an electric field is applied the recombination can be stopped and the charge drifted and accelerated to produce a second (proportional) scintillation pulse. Thus there are two means of discrimination possible, either conventional pulse shape analysis when recombination is allowed or by scintillation-ionization in which the primary scintillation pulse S1 is followed a few ms later by a secondary pulse S2.

The UKDMC has just started operation at Boulby of a 4 kg detector (ZEPLIN I) based on pulse shape discrimination in Xe. Two-phase (gas and liquid) Xe detectors ZEPLIN II and ZEPLIN III with full scintillation-ionisation are on schedule for operation by mid 2001 in collaboration with UCLA, Torino, ITEP, CERN and Columbia. The latter is 6 kg and incorporates a high electric field in the liquid to enhance the recoil ionisation signal, the former is a 30 kg detector based on an original UCLA/CERN concept[29].

The technique of measuring the direction of WIMP induced recoiling nuclei is potentially very powerful because the motion of the Solar System through the Galactic halo (at ~230 kms^{-1}) ensures a forward-back asymmetry in recoil directions that increases rapidly with recoil energy[30] (>1: 100 above 100 keV). This is a unique feature of WIMP events and allows the prospect of discrimination from all normal isotropic backgrounds by correlating event direction with motion through the halo. The DRIFT collaboration is now building a 1 m^3 experiment using CS_2 for operation from mid 2001 at Boulby[31]. CS_2 produces -ve ions that are drifted to a MWPC readout, the advantage over conventional gases being that track diffusion is much reduced (<1 mm over 1 m) obviating the need for a magnetic field. Background predictions and tests indicate that zero background is a real possibility with DRIFT. The directional signal is so powerful that only a few 10s of events are required to definitively identify a signal as being due to WIMPs.

4 Conclusion

After some years of intense r&d effort significant progress has been made by several groups notably DAMA, CDMS and UKDMC. However, for these examples each has been hit by phenomena with characteristics close to that expected for WIMPs. They have dealt with the situation in different ways. DAMA observe an annual modulation - they interpret this as a WIMP signal and plan to continue running. CDMS suffers from neutron background that naturally produces WIMP-like events - they are subtracting them, then plan to move underground. The UKDMC have observed recoil-like events that probably arise from surface alpha contamination, they are attempting to remove them. The next 2 years should see these issues resolved.

References

1. see authors in *IDM98*, ed. N.J.C. Spooner, V. Kudryavtsev (World Scientific, 1999).
2. see authors in *Dark2000* 10-16 July 2000, Heidleberg, Germany.
3. see authors in *DM2000* 23-25 Feb 2000 Marina del Rey, CA, USA.
4. R. Bernabei et al., *Phys. Lett.* B **389**, 757 (1996).
5. P. F. Smith et al., *Phys. Lett.* B **379**, 299 (1996).
6. CDMS Collaboration, sub to *Phys. Rev. Lett.*, astro-ph 00002471 (2000).
7. J. Gascon, *Dark2000* 10-16 July 2000, Heidleberg, Germany.
8. D. R. Tovey et al *IDM98*, ed. N.J.C. Spooner, V. Kudryavtsev (World Scientific, 1999) 649.
9. N. J. C. Spooner et al., *LTD-4*, ed. N. E. Booth, G. L. Salmon (Editions Frontières, 1991) 165.
10. B. Sadoulet et al., *LTD-4*, ed. N. E. Booth, G. L. Salmon (Editions Frontières, 1991) 147.
11. R. Bernabei et al., *Phys. Lett.* B **450**, 448 (1999).
12. K. Fushimi et al., Proc. *Dark Matter in the Universe* (UAP, 1997) 115.
13. D. Abriola et al. *IDM98*, ed. N.J.C. Spooner, V. Kudryavtsev (World Scientific, 1999).
14. R. Bernabei et al., *Phys. Lett.* B **480**, 23 (2000).
15. G. Gerbier, *DM2000* 23-25 Feb 2000 Marina del Rey, CA, USA.
16. P.F. Smith, *Phys World* July (2000)
17. N. J. C. Spooner et al., *DM2000* 23-25 Feb 2000 Marina del Rey, CA, USA.
18. V. A. Kudryavtsev et al., *Phys. Lett.* B **452**, 167 (1999).
19. N. J. T. Smith et al., *Phys. Lett.* B **467**, 132 (1999); N. J. T. Smith, J. D. Lewin, P. F. Smith, *Phys. Lett.* B **485**, 9 (2000).
20. N. J. C. Spooner et al., *Phys. Lett.* B **473**, 330 (2000).
21. L. Baudis et al., *Nucl. Phys.* B **70**, 106 (1999).
22. M. Bravin et al., *LTD-8*, 15-20 Aug. 1999 Dalfsen, Netherlands.
23. J.J. Collar et al., *IDM96* ed. N.J.C. Spooner (World Scientific, 1997) 563.
24. L.A.Hamel et al., *Nucl. Inst. & Meth.* A **388**, 91 (1997); R. Gornea et al, *Dark2000* 10-16 July 2000, Heidleberg, Germany.
25. R. Bernabei et al., *Phys. Lett.* B **436**, 379 (1998).
26. T. J. Sumner et al., *26th ICRC* **2**, 516 (1999).
27. C. J. Martoff et al., *Nucl. Inst. & Meth.* A **440**, 355 (2000).
28. G. J. Davies et al., *Phys. Lett.* B **320**, 395 (1994).
29. H. Wang et al., *DM2000* 23-25 Feb 2000 Marina del Rey, CA, USA.
30. P. F. Smith and J. D. Lewin, *Phys. Rep.* **187**, 203 (1990).
31. M.Lehner et al., *Dark2000* 10-16 July 2000, Heidleberg, Germany.

COSMIC RAY MEASUREMENTS WITH THE AMS EXPERIMENT IN SPACE

V. CHOUTKO

MIT-LNS, 77 Massachusetts Av., Cambridge MA 02171-9131, USA

E-mail: v.choutko@cern.ch

The measurements of the cosmic ray spectra with the AMS experiment in near Earth orbit during the June 1998 100 hours engineering flight are presented. The antimatter to matter limit of the order of 10^{-6} is established. The systematic study of the particles fluxes below the geomagnetic cutoff is done.

1 Introduction

The Alpha Magnetic Spectrometer (AMS)[1] is a high energy physics experiment scheduled for three years operation on board of the International Space Station.

The physics goals of the AMS are to search for the antimatter in the Universe on the level less than 10^{-8}, to search for the dark matter and to make high statistics measurements of light isotops spectra. over a broad energy range.

The preliminary version of the AMS detector had 100 hours engineering flight STS–91 on board of the space shuttle Discovery.

1.1 AMS Detector

The major elements of AMS (see Fig. 1) consisted of a permanent magnet, a tracker, time of flight hodoscopes, a Cerenkov counter and anticoincidence counters. The permanent magnet had the shape of a cylindrical shell with inner diameter 1.1 m, length 0.8 m and provided a central dipole field of 0.14 Tesla across the magnet bore. The six layers of double sided silicon tracker were arrayed transverse to the magnet axis. The tracker measured the trajectory of relativistic singly charged particles with an accuracy of 20 μ in the bending coordinate and 33 μ in the non-bending one, as well as provided measurements of the energy loss. The time of flight system had four layers, measured singly charged particle transit times with an accuracy of 120 psec and also yielded energy loss measurements. The Aerogel Cerenkov counter was used to make independent velocity measurements. A layer of anticoincidence scintillation counters lined the inner surface of the magnet.

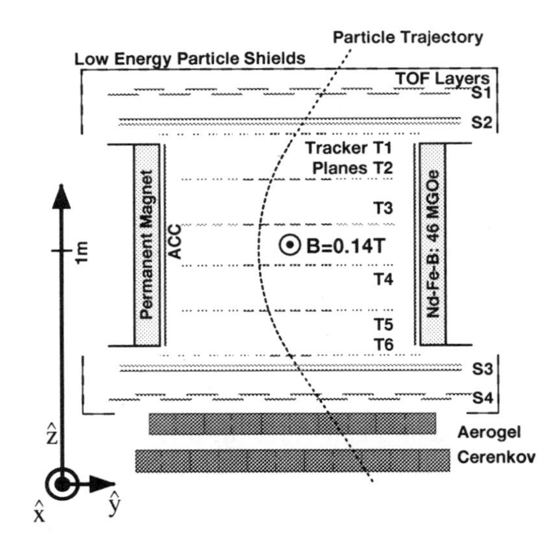

Figure 1. Schematic view of AMS as flown on STS–91

1.2 Test Flight

The flight orbit parameters were 51.7° orbit inclination, 320 to 390 km orbit altitude and 0°, 20°, 45° and 180° shuttle zenith attitude.

2 Measurements of the Cosmic Ray Spectra

Cosmic rays spectra measured by AMS in near Earth orbit were expected to follow rigidity power law, with low energy part modified by solar wind and geomagnetic field. Depending on shuttle coordinates, geomagnetic cutoff values varied from 0.4 to 55 GV.

2.1 H and He Nuclei

Fig. 2 and Fig. 3 show respectively the scaled primary proton and He nuclei spectra as measured by AMS in comparison with some recent balloon-born results[2,3,4,5,6] and with flux used to calculate atmospheric neutrinos[7].

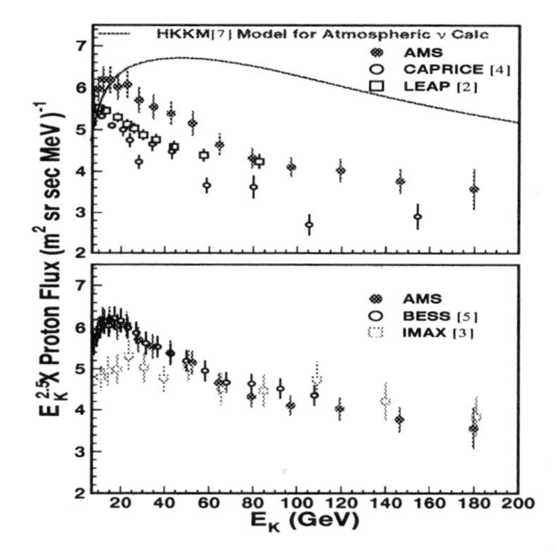

Figure 2. The primary proton spectrum multiplied by $E_K^{2.5}$ in units of $GeV^{2.5}/(m^2\,sec\,sr\,MeV)$ as measured by AMS in comparison with some recent balloon based measurements and with flux used to calculate atmospheric neutrinos.

Fitting AMS data to a power law in rigidity, $\Phi = \Phi_0 R^{-\gamma}$, over the range $10\,GV < R < 200\,GV$ for protons yields $\gamma_p = 2.78 \pm 0.009\,(fit) \pm 0.019\,(sys)$, $\Phi_0^p = 17.1 \pm 0.15\,(fit) \pm 1.3\,(sys) \pm 1.5\,(\gamma)\ GV^{2.78}/(m^2 sec\,sr\,MV)$; and over the range $20\,GV < R < 200\,GV$ for He nuclei yields $\gamma_{He} = 2.740 \pm 0.010\,(stat) \pm 0.016\,(sys)$, $\Phi_0^{He} = 2.52 \pm 0.09\,(fit) \pm 0.13\,(sys) \pm 0.14\,(\gamma)\ GV^{2.74}/(m^2 sec\,sr\,MV)$.

2.2 Electrons and Positrons

Fig. 4 shows the AMS measurement of electron spectrum in comparison with some

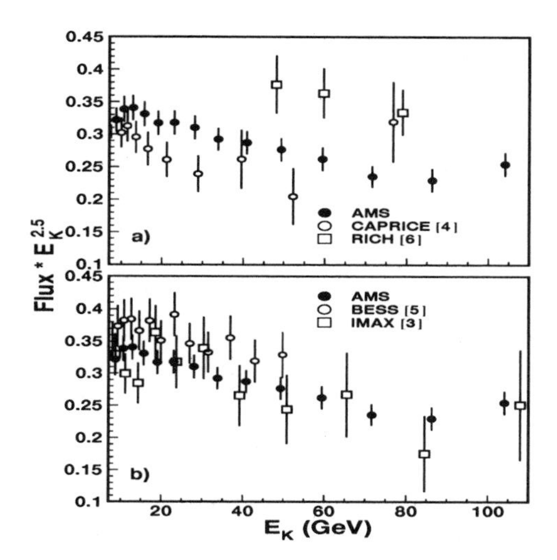

Figure 3. The primary He nuclei spectrum multiplied by $E_K^{2.5}$ in units of $GeV^{2.5}/(m^2\,sec\,sr\,MeV)$ as measured by AMS in comparison with some recent balloon-born measurements.

balloon-born measurements[8,9,10]. Fitting the AMS data to power law, $\Phi = \Phi_0 E^{-\gamma}$ above 10 GeV results in $\gamma = 3.2 \pm 0.1\,(stat)$.

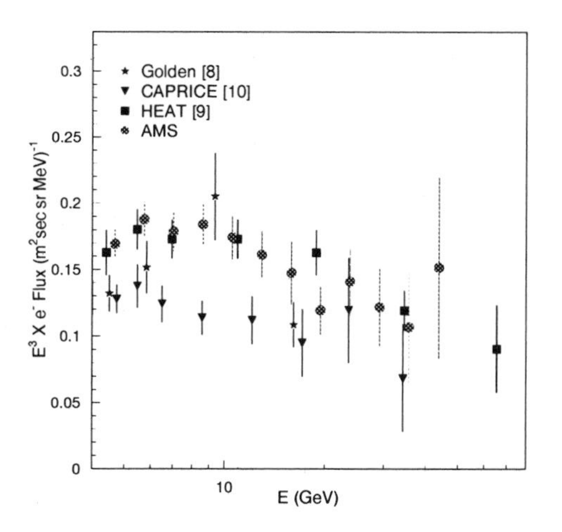

Figure 4. The primary electron spectrum multiplied by E^3 in units of $GeV^3/(m^2\,sec\,sr\,MeV)$ as measured by AMS in comparison with some balloon-born measurements.

Fig. 5 compares the AMS $\frac{e^+}{e^+ + e^-}$ measurement to the previous results[10,9,11].

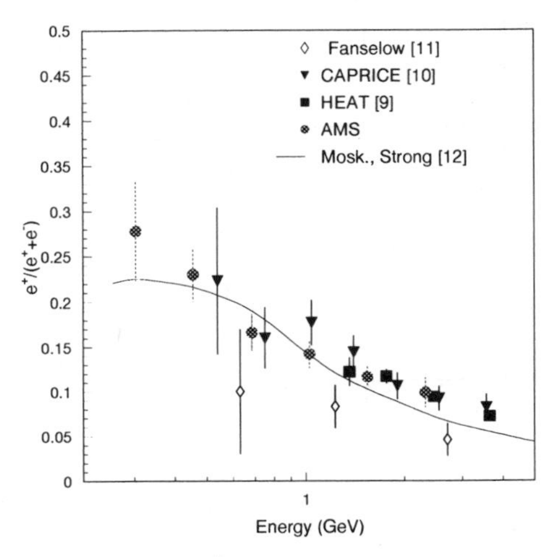

Figure 5. The $\frac{e^+}{e^+ + e^-}$ measurement of cosmic e^+ and e^- in comparison with balloon based results and prediction from[12].

2.3 *Solar Modulation of Cosmic Rays*

Fitting the low energy cosmic rays to solar modulated local interstellar spectra yields solar modulation parameters shown in Table 1.

Table 1. Cosmic ray solar modulation parameters estimated from the AMS data.

Particle Type	SM Parameter (MV)
Protons	600 ± 30
He Nuclei	620 ± 40
Deuterons	650 ± 20
Electrons	400 ± 30

3 Search for Antimatter

Discovery of a single antiHe would be an evidence of existence of primordial antimatter. During the test flight AMS collected about $3{\cdot}10^6$ He nuclei and $2 \cdot 10^5$ ions with charge greater than two. No antimatter candidates was found, which allowed to set the 95% CL spectrum independent[13] antimatter to matter upper limits as shown in Fig. 6.

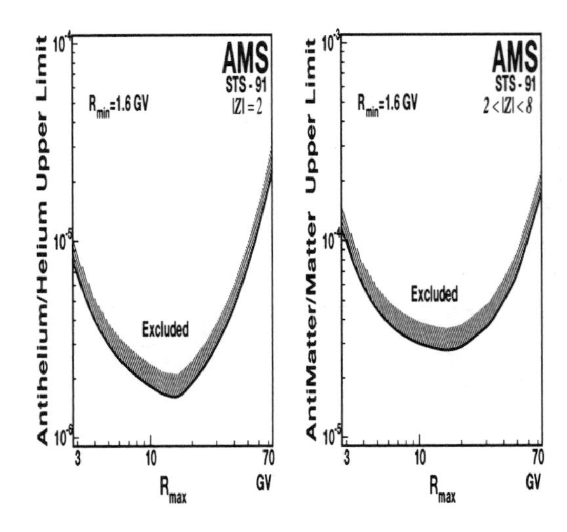

Figure 6. The AMS 95% CL spectrum independent antimatter to matter upper limits for ions with charge less than 8.

4 Second Spectra Measurements

AMS systematically measured the cosmic ray spectra as a function of magnetic latitude[14].

4.1 *Protons*

Fig. 7 shows the protons spectra collected by AMS versus the magnetic latitude. Second proton spectrum was observed, fluxes for downward and upward protons being equal. A substantial increase of the second spectrum flux near magnetic equator was equally seen.

4.2 *Electron and Positrons*

Fig. 8 shows the AMS e^+ and e^- spectra versus the magnetic latitude. Second lepton spectra were observed, $\frac{e^+}{e^-}$ fluxes ratio varying between one at high magnetic latitudes to 4 near magnetic equator.

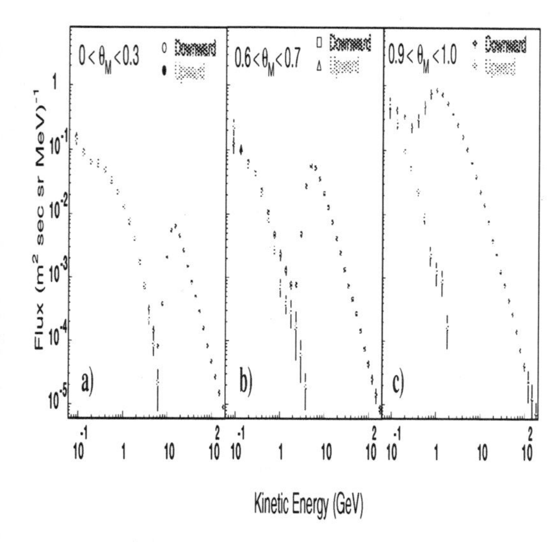

Figure 7. Flux spectra for downward and upward going protons separated according to the geomagnetic latitude at which they were detected.

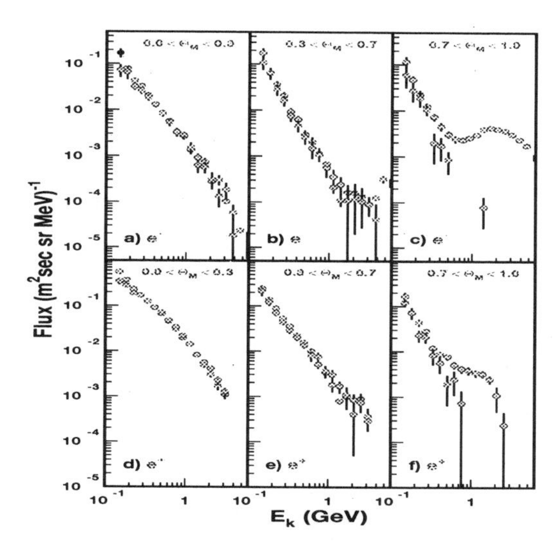

Figure 8. Flux spectra for downward going electrons and positrons, separated according to the geomagnetic latitude at which they were detected.

4.3 Understanding the origin

Tracing particles trajectories back and forth in the geomagnetic field allowed to calculate their life time as well the origin and sink coordinates. Two distinct component were observed: "Short" ($<$ 200 ms life time) and "long" ($>$ 200 ms life time) lived leptons and protons, the majority of the particles being "long" lived ones. "Long" lived particles were originated from two complementary geographic regions as shown in Fig. 9.

References

1. S. Ahlen *et al.* NIM **A350**, 251 (1994).
2. E. S. Seo *et al.* ApJ **378**, 763 (1991).
3. W. Menn *et al.* ApJ **533**, 281 (2000).
4. M. Boezio *et al.* ApJ **518**, 457 (1999).
5. T. Sanuki *et al.* a-ph/0002481 (2000).
6. J. Backley *et al.* ApJ **429**, 736 (1994.)
7. M. Honda *et al.* PhR **D52**, 4985 (1995).
8. R.L. Golden *et al.* ApJ **436**, 769 (1994).
9. S.Q. Barwick *et al.* ApJ **498**, 779 (1998).
10. M. Boezio *et al.* ApJ, **532**, 653 (2000).
11. J.L. Fanselow *et al.* ApJ **158**, 771 (1969).
12. I.V. Moskalenko and A.W. Strong ApJ **493**, 694 (1998).
13. J. Alcaraz *et al.* PhL **B461**, 387 (1999).
14. J. Alcaraz *et al.* PhL **B472**, 215 (2000); J. Alcaraz *et al.* PhL **B484**, 10 (2000).

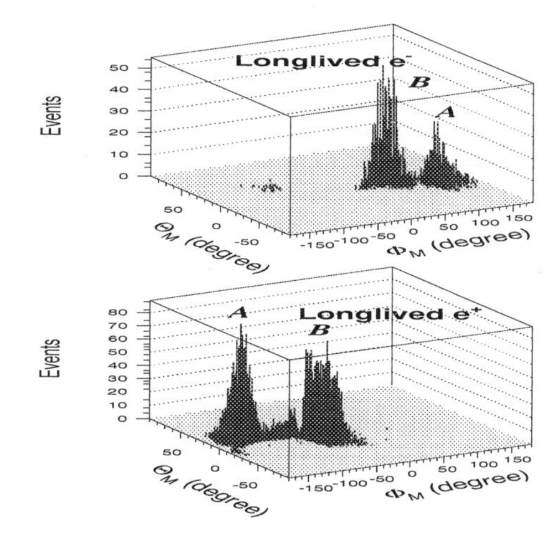

Figure 9. The point of origin of long–lived leptons with energies $<$ 3 GeV and geomagnetic latitude $<$ 0.7.

FIRST RESULTS FROM THE L3+C EXPERIMENT AT CERN

THOMAS HEBBEKER, L3 COLLABORATION
Humboldt University, Invalidenstr. 110, D-10115 Berlin, Germany,
E-mail: hebbeker@physik.hu-berlin.de

The L3+C experiment combines the high precision muon spectrometer of the L3 detector at LEP, CERN, with a small air shower array. Cosmic ray muon events can be measured in the momentum range from 20 to 2000 GeV. Multi-muon events allow - together with a shower energy measurement in the scintillator array - to constrain the chemical composition of the primary nuclei. Until summer 2000 about 8 billion muons and 25 million air shower events have been recorded. Here, we report first results on the momentum spectrum and the muon charge ratio.

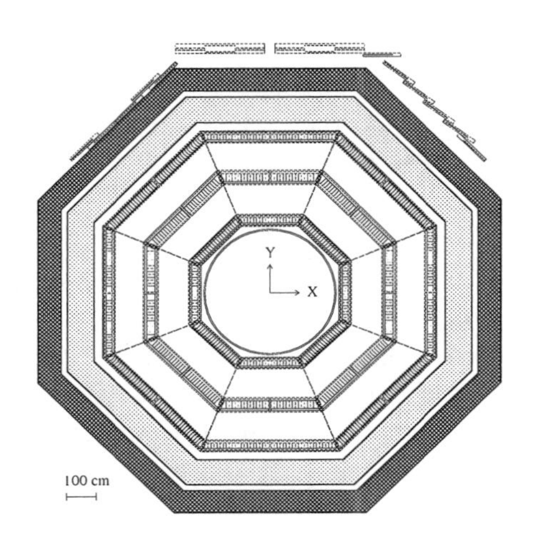

Figure 1. Cross section of the L3 muon spectrometer. The outer octagons correspond to coil and yoke of the magnet. Scintillators are installed on the top.

1 The L3+C Experiment

The L3 detector[1] at CERN's LEP accelerator is located near Geneva (6.02^0 E, 46.25^0 N) at an altitude of 450 m. It is installed underground below a 30 m layer of molasse and contains a huge muon spectrometer with a volume exceeding $1000\,\mathrm{m}^3$ inside a magnetic field of $B = 0.5\,\mathrm{T}$. L3 has been operating very successfully since 1989 to study e^+e^- interactions. The muon spectrometer consists of two groups of 8 'octants' each, with three layers of precise drift chambers, see figure 1.

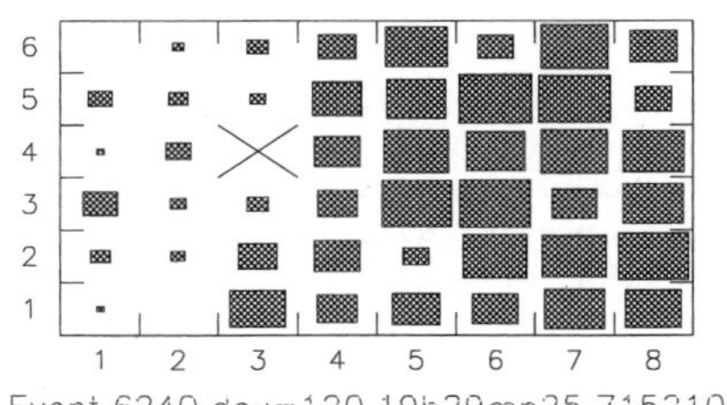

Figure 2. Air shower event in the L3+C detector.

The detector is enclosed in a 12 m diameter magnet (coil and yoke) with a field along the z direction. In order to measure the arrival time of the muons - which is needed to determine the drift time in the chambers - $200\,\mathrm{m}^2$ of scintillators have been placed on top of the L3 magnet.

On a flat roof belonging to a building located at the surface above L3 we have installed 50 scintillators of size $0.5\,\mathrm{m}^2$, distributed over an area of $30 \times 54\,\mathrm{m}^2$. This air shower detector can measure the energy and direction of atmospheric showers. Figure 2 shows an event recorded in the air shower array with an energy of about 1700 TeV.

The L3 muon spectrometer together with the surface array constitutes the L3+C experiment[2].

2 Performance and Data Taking

The molasse layer implies a threshold on the muon momentum (surface value) of 15 GeV

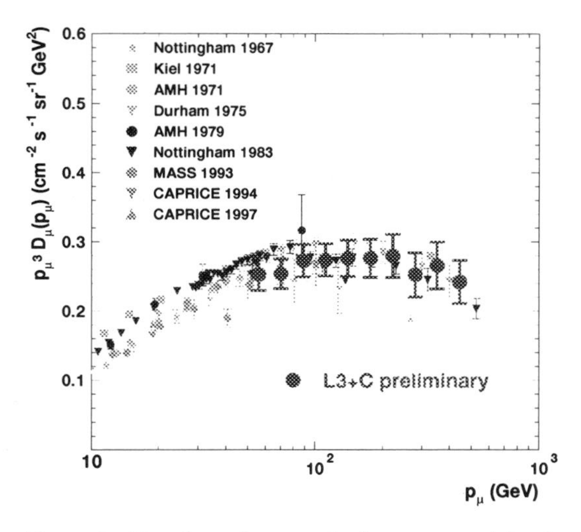

Figure 3. The flux of atmospheric muons weighted with p^3 as a function of momentum p.

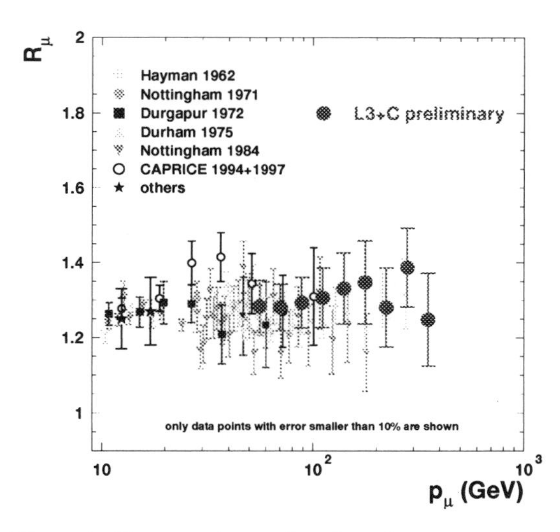

Figure 4. Charge ratio as a function of muon momentum.

and limits the angular resolution to 0.2^0 at 100 GeV. The momentum resolution was found to be 6.6% at 100 GeV. This allows a charge determination up to energies of a few TeV.

The trigger conditions fix the energy threshold of the air shower array to about 10 TeV. The event-by-event energy resolution for showers with its core inside the array is approximately 30%. The zenith angle can be determined via the arrival times of the particles in the different scintillators with a precision of $1^0 - 2^0$.

Cosmic data taking with the muon detector has started in 1999. Between May and November 1999 about 5 billion events were recorded. This number will double in this year's data taking period. The scintillator array was installed this spring and has been fully functional since April 2000. At the time of the conference in Osaka 25 million events had been accumulated. About 30% of these showers are accompanied by muon(s) in the L3 spectrometer.

Data taking of the L3+C experiment will end in autumn 2000.

3 Very First Results

3.1 Muon Momentum Spectrum and Charge Ratio

Data from 1999 with a total livetime of about one month were used to determine a first momentum spectrum in the range $p = 50 - 500$ GeV. Strict quality cuts have been applied. Only muons measured in both an upper and a lower octant were accepted. The zenith angle was restricted to $\theta \leq 10^0$. The measured flux$\times p^3$ is shown in figure 3 together with results from other detectors[3]. The systematic error is the dominant one. So far it amounts to 9%, but we hope to reduce it to 2.5% in the future. The charge ratio was calculated from the same event sample. Figure 4 shows the preliminary results in comparison with other measurements[3]. For the small sample analysed up to now the statistical error dominates at high momenta.

3.2 Multimuon Events

We have started to analyse multi muon events. A beautiful example is displayed in figure 5. We record about 200 events per day

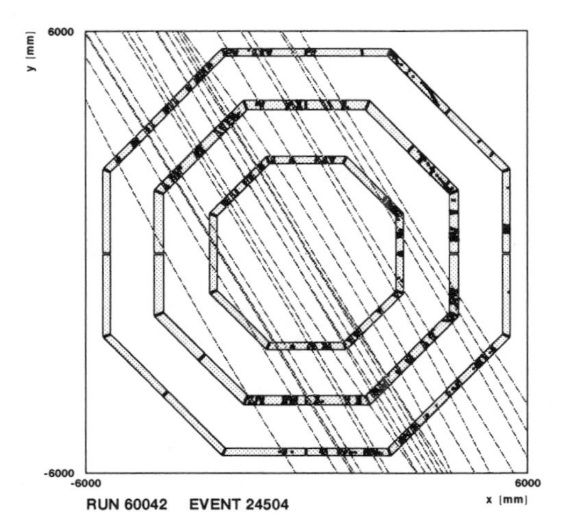

Figure 5. A multimuon event recorded in 1999.

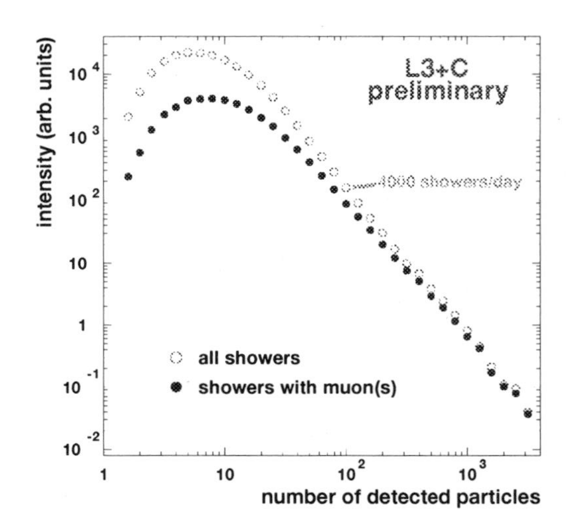

Figure 6. Uncorrected air shower flux as seen by L3+C.

with a muon multiplicity exceeding 10.

3.3 Air Shower Energy Distribution

Based on one week of data taking we have analysed the air shower flux as a function of the total number of particles seen in the scintillator array. The last quantity is proportional to the shower energy (if the core is inside the array), with the following rough conversion factor: N particles $\rightarrow$ N TeV. We observe about 4000 showers with more than 100 particles per day. Figure 6 shows both the flux of all showers and of those which are accompanied by one or more muons in the spectrometer. No correction for trigger efficiency or distance between core and array has been applied. Towards high shower energies the probability to find muons as well reaches 90%. Below approximately 10 TeV the trigger of the array (requiring hits in three adjacent rows) is not fully efficient, therefore the measured flux drops.

4 Summary and Conclusions

A new type of cosmic ray detector, L3+C, combines air shower data with precise muon measurements.

First preliminary results on the muon spectrum and charge ratio up to momenta of 500 GeV have been obtained. A substantial reduction of the statistical and systematic errors is expected in the future. Multi-muon events and shower energy measurements will constrain the chemical composition of the primaries.

Acknowledgments

I thank the German Bundesministerium für Bildung und Forschung for their financial support.

References

1. L3 Collaboration, B. Adeva et al, NIM A **289** (1990) 35.
2. `http://l3www.cern.ch/l3_cosmics/`
3. T. Hebbeker and C. Timmermans, 'A Compilation of High Energy Atmospheric Muon data at Sea Level', to be published.

Parallel Session 10

Beyond the Standard Model (theory)

Convener: Hitoshi Murayama (U.C. Berkeley)

NEUTRINO YUKAWA COUPLINGS AND FCNC PROCESSES IN B DECAYS IN SUSY-GUT

SEUNGWON BAEK, TORU GOTO AND YASUHIRO OKADA
Theory Group, KEK, Tsukuba, Ibaraki, 305-0801, Japan

KEN-ICHI OKUMURA
Institute for Cosmic Ray Research, University of Tokyo, Kashiwa, Chiba, 277-8582, Japan

Flavor changing neutral current and lepton flavor violating processes are studied in the SU(5) SUSY-GUT with right-handed neutrino supermultiplets. Using input parameters motivated by neutrino oscillation, it is shown that the time-dependent CP asymmetry of $b \to s\gamma$ can be as large as 20%. We also show that the B_s–$\bar{B}_s$ mixing can be significantly different from the standard model prediction.

Effects of the physics beyond the standard model (SM) may appear in the flavor physics in quarks and leptons. One of indications is already given by the atmospheric and the solar neutrino anomalies which are interpreted as evidences of neutrino oscillation[1]. A natural way to explain small neutrino masses is the see-saw mechanism with very heavy right-handed neutrinos. This scenario suggests the existence of new sources of flavor mixings in the lepton sector at much higher energy scale than the electroweak scale.

In this work[2] we consider flavor changing neutral current (FCNC) and lepton flavor violation (LFV) processes in the model of a SU(5) supersymmetric (SUSY) grand unified theory (GUT) which incorporates the see-saw mechanism for the neutrino masses. In the SUSY model based on the minimal supergravity the mass matrices of squarks and sleptons are flavor-blind at the Planck scale M_P. However renormalization effects due to Yukawa coupling constants of quarks, leptons and neutrinos induce flavor mixing in the squark/slepton mass matrices. In the context of SUSY-GUT with right-handed neutrinos, the flavor mixing related to the neutrino oscillation can provide the mixing in the squark sector. We show that due to the large mixing of the second and third generations suggested by the atmospheric neutrino data, B_s–$\bar{B}_s$ mixing and the CP asymmetry of the $B \to M_s\gamma$ process, where M_s is a CP eigenstate including the strange quark, have significant deviations from the SM predictions.

The relevant part of the superpotential for the SU(5) SUSY GUT with right-handed neutrino supermultiplets is given by

$$W = \frac{1}{8} f_U^{ij} \Psi_i \Psi_j H_5 + f_D^{ij} \Psi_i \Phi_j H_{\bar{5}} + f_N^{ij} N_i \Phi_j H_5 + \frac{1}{2} M_\nu^{ij} N_i N_j \ , \quad (1)$$

where Ψ_i, Φ_i and N_i are $\mathbf{10}$, $\bar{\mathbf{5}}$ and $\mathbf{1}$ representations of SU(5) gauge group. $i, j = 1, 2, 3$ are the generation indices. $H_{5,\bar{5}}$ are Higgs superfields with $\mathbf{5}$ and $\bar{\mathbf{5}}$ representations. $f_{U,D,N}$ are Yukawa coupling matrices and M_ν is the Majorana mass matrix. Below the GUT scale M_G and the Majorana mass scale M_R the superpotential is written as

$$W = \tilde{f}_U^{ij} Q_i U_j H_2 + \tilde{f}_D^{ij} Q_i D_j H_1 + \tilde{f}_L^{ij} E_i L_j H_1 - \frac{1}{2} \kappa_\nu^{ij} (L_i H_2)(L_j H_2) \ , (2)$$

where κ_ν is obtained by integrating out N_i at M_R. The Yukawa coupling constants $\tilde{f}_U$, $\tilde{f}_D$ and $\tilde{f}_L$ are related to f_U and f_D at M_G. The masses and mixings of the quarks and leptons are determined from the superpotential Eq. (2) at the low energy scale.

As discussed above, the renormalization effects due to the Yukawa coupling constants induce various FCNC/LFV effects from the mismatch between the bases of quark/lepton and squark/slepton masses. In particular the top Yukawa coupling constant is responsible for the running of the $\tilde{q}_L$ and $\tilde{u}_R$ masses. At the same time the $\tilde{e}_R$ mass matrix receives sizable corrections between M_P and M_G scales and LFV processes are induced[3]. In a similar way, if f_N^{ij} is large, the $\tilde{l}_L$ and $\tilde{d}_R$ mass matrices receive sizable flavor changing effects due to the running between M_P and $M_{G,R}$ scales[4]. These are sources of extra contributions to various FCNC/LFV processes.

We calculate the following observables: the CP violation parameter in the K^0–$\bar{K}^0$ mixing ε_K, B_q–$\bar{B}_q$ mass splitting Δm_q ($q = d, s$), the branching ratios of $b \to s\,\gamma$, $\mu \to e\,\gamma$ and $\tau \to \mu\,\gamma$, and the amplitude of the time-dependent CP asymmetry in the $B \to M_s\,\gamma$ process[5], which is written as

$$A_t = \frac{2\text{Im}(\mathrm{e}^{-i\theta_B} c_7 c_7')}{|c_7|^2 + |c_7'|^2}\,, \tag{3}$$

where c_7 and c_7' are the Wilson coefficients in the effective Lagrangian for the $b \to s\,\gamma$ decay $\mathcal{L} = (c_7 \bar{s}\sigma^{\mu\nu} b_R + c_7' \bar{s}\sigma^{\mu\nu} b_L)F_{\mu\nu} + \text{H.c.}$. $\theta_B = \arg M_{12}(B_d)$ where $M_{12}(B_d)$ is the B_d–$\bar{B}_d$ mixing amplitude.

We solved renormalization group equations (RGEs) for Yukawa coupling matrices and the SUSY breaking parameters keeping all the flavor mixings. We specify neutrino parameters as well as the quark/lepton masses and the Cabibbo-Kobayashi-Maskawa (CKM) matrix as follows. The inputs from the neutrino oscillation are two mass differences and the Maki-Nakagawa-Sakata (MNS) matrix. In order to relate these inputs to f_N and M_ν, we work in the basis where $\hat{f}_L$ is diagonal and $f_N = \hat{f}_N V_L$ ($\hat{f}_N$ is diagonal). In this basis $\kappa_\nu = V_L^{\mathrm{T}} \hat{f}_N M_\nu^{-1} \hat{f}_N V_L$ at the matching scale M_R. Once we fix three neutrino masses, V_{MNS}, $\hat{f}_N$ and the unitary matrix V_L we can obtain M_ν. Then using the GUT relation for Yukawa coupling constants, we calculate all squark/slepton mass matrices

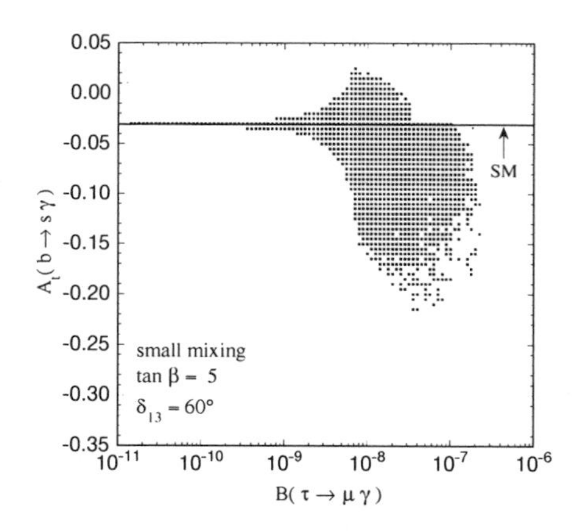

Figure 1. Time-dependent CP asymmetry in $b \to s\gamma$ decay as a function of the branching ratio of $\tau \to \mu\gamma$.

through RGEs. Note that V_L essentially determines the flavor mixing in the $\tilde{d}_R$ and $\tilde{l}_L$.

We consider the following parameter sets, corresponding to the Mikheyev-Smirnov-Wolfenstein solutions for the solar neutrino problem. (i) small mixing angle solution: $\sin^2 2\theta_{12} = 5.5 \times 10^{-3}$, $m_\nu = (2.24, 3.16, 59.2) \times 10^{-3}\,\text{eV}$, (ii) large mixing angle solution: $\sin^2 2\theta_{12} = 1.$ $m_\nu = (4.0, 5.83, 59.5) \times 10^{-3}\,\text{eV}$, In both cases we take $\sin^2 2\theta_{23} = 1$, $\sin^2 2\theta_{13} = 0$ and M_ν to be proportional to a unit matrix with a diagonal element of $M_R = 4 \times 10^{14}\,\text{GeV}$ so that $V_L = V_{\mathrm{MNS}}^\dagger$ at M_R. Free parameters in the minimal supergravity model are the universal scalar mass m_0, the unified gaugino mass M_0, the scalar trilinear parameter A_0, the ratio of two vacuum expectation values $\tan\beta$ and the sign of the Higgsino mass parameter μ. We take $\tan\beta = 5$ and vary other SUSY parameters. We also impose various constraints from SUSY particles search, the measurement of $\text{B}(b \to s\,\gamma)$[6] and the search of $\mu \to e\,\gamma$[7].

Fig. 1 shows A_t as a function of $\text{B}(\tau \to \mu\,\gamma)$ for case (i) with fixed CKM parameters $|V_{ub}/V_{cb}| = 0.08$ and $\delta_{13} = 60°$ where δ_{13} is the phase parameter in the CKM matrix. We see that $|A_t|$ can be as large as 20% when $\text{B}(\tau \to \mu\,\gamma)$ is larger than 10^{-8} level. The large asymmetry arises because the renormalization effect of f_N induces sizable contribution to c_7' through gluino–$\tilde{d}_R$ loop diagrams. Since this asymmetry is suppressed by a factor m_s/m_b in the SM, a sizable asymmetry is a clear signal of new physics beyond the SM.

In Fig. 2 we show allowed regions in the space of $\Delta m_s/\Delta m_d$ and the time-dependent CP asymmetry of $B \to J/\psi\, K_S$ for the case (i) and (ii). Here $|V_{ub}|$ is varied within $0.08 < |V_{ub}/V_{cb}| < 0.1$ and δ_{13} is scanned for the whole range. For the case (i) we see that the deviation of $A_t(B \to J/\psi\, K_S)$ from the SM value is small while $\Delta m_s/\Delta m_d$ can differ from the SM value by 40%. This pattern of deviation is understood as follows. The new contributions to ε_K and $M_{12}(B_d)$ are suppressed due to the small 1-2 and 1-3 mixings in the neutrino sector so that the allowed region of δ_{13} does not change much. The deviation in $\Delta m_s/\Delta m_d$ comes from the SUSY contribution to $M_{12}(B_s)$ induced by the large 2-3 mixing in the neutrino sector. On the other hand we see a correlation between the deviations in the case (ii). Due to the large 1-2 mixing, ε_K can be enhanced even after imposing the $\mathrm{B}(\mu \to e\,\gamma)$ constraint in this case. Consequently the allowed range of δ_{13} by the constraint from ε_K changes. The region with large deviations in both $\Delta m_s/\Delta m_d$ and $A_t(B \to J/\psi\, K_S)$ corresponds to a small δ_{13} region where the constraint from ε_K is satisfied by a large SUSY contribution. This figure means the deviation from the SM may be seen in both cases once $\Delta m_s/\Delta m_d$ and $A_t(B \to J/\psi\, K_S)$ are measured precisely.

In conclusion, we studied the effects of the neutrino Yukawa coupling matrix on FCNC/LFV processes in the SU(5) SUSY-GUT with right-handed neutrino supermultiplets. It is shown that $A_t(B \to M_s\,\gamma)$ can be $\sim 20\%$ when the $\mathrm{B}(\tau \to \mu\,\gamma)$ is about 10^{-7}. We also show that the B_s–$\bar{B}_s$ mixing can be significantly different from the presently allowed range in the SM. Since these signals provide quite different signatures compared to the SM and the minimal supergravity model without GUT and right-handed neutrino interactions, future experiments in B physics and LFV can give us important clues on the interactions at very high energy scale.

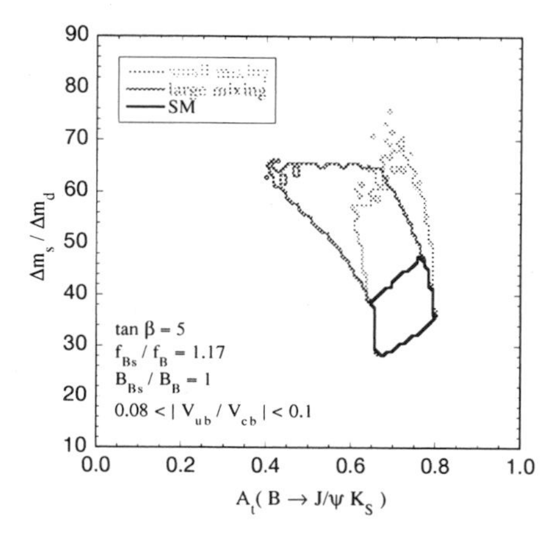

Figure 2. Allowed regions in the space of $\Delta m_s/\Delta m_d$ and the CP asymmetry in $B \to J/\psi K_S$ decay.

References

1. E. Kearns, talk at ICHEP 2000.
2. S. Baek *et al.*, hep-ph/0002141.
3. R. Barbieri and L. J. Hall, Phys. Lett. **B338**, 212 (1994); J. Hisano *et al.*, Phys. Lett. **B391**, 341 (1997); R. Barbieri *et al.*, Nucl. Phys. **B445**, 219 (1995); N. G. Deshpande *et al.*, Phys. Rev. Lett. **77**, 4499 (1996).
4. F. Borzumati and A. Masiero, Phys. Rev. Lett. **57**, 961 (1986); J. Hisano *et al.*, Phys. Lett. **B357**, 579 (1995); Phys. Rev. **D53**, 2442 (1996); J. Hisano *et al.*, Phys. Lett. **B437**, 351 (1998); Phys. Rev. **D59**, 116005 (1999); J. Ellis *et al.*, Eur. Phys. J. **C14**, 319 (2000).
5. D. Atwood *et al.*, Phys. Rev. Lett. **79**, 185 (1997); C. Chua *et al.*, Phys. Rev. **D60**, 014003 (1999).
6. S. Ahmed *et al.*, [CLEO Collaboration], CLEO CONF 99-10 (1999).
7. M. L. Brooks *et al.* [MEGA Collaboration], Phys. Rev. Lett. **83**, 1521 (1999).

SUPERSYMMETRY VERSUS PRECISION EXPERIMENTS REVISITED

GI-CHOL CHO
Department of Physics, Ochanomizu University,
Bunkyo, Tokyo 112-8610, Japan
E-mail: cho@phys.ocha.ac.jp

We study constraints on the Minimal Supersymmetric Standard Model from electroweak experiments. We find that the light sfermions always make the fit worse than the Standard Model, while the light chargino generally make the fit slightly better through the oblique corrections. The best overall fit to the precision measurements are found when the mass of lighter chargino is about 100 GeV and the $SU(2)_L$ doublet sfermions are all much heavier. We find the slight improvement of the fit over the SM, where the total χ^2 of the fit decreases by about one unit.

1 Electroweak observables in the MSSM

Precision measurements of the electroweak observables on the Z-pole at LEP1 and SLC [1], and the W-boson mass at LEP2 and Tevatron [2] are expected to give the stringent constraints on the Minimal Supersymmetric Standard Model (MSSM). The supersymmetric (SUSY) contributions to the electroweak observables are given through the oblique corrections and the process specific vertex/box corrections. It has been shown that the Z-pole observables are conveniently parametrized in terms of two oblique parameters S_Z and T_Z [3], and the $Zf_\alpha f_\alpha$ vertex corrections, where f denotes the fermion species and α is their chirality. The oblique parameters S_Z and T_Z are related to the S and T parameters [4],

$$S_Z \equiv S + R - 0.064x_\alpha, \tag{1}$$

$$T_Z \equiv T + 1.49R - \frac{\Delta\overline{\delta}_G}{\alpha}, \tag{2}$$

where $x_\alpha = \frac{1/\alpha(m_Z^2)-128.90}{0.09}$ and $\Delta\overline{\delta}_G/\alpha$ parametrize the hadronic uncertainty of the QED coupling and the corrections to the μ-decay constant, respectively. The parameter R is introduced as the difference of the Z-boson propagator corrections between $q^2 = m_Z^2$ and $q^2 = 0$. For convenience of later analysis, we introduce ΔS_Z and ΔT_Z as the shifts from S_Z and T_Z at the SM reference point, $m_t = 175$ GeV, $m_{H_{\rm SM}} = 100$ GeV, $\alpha_s(m_Z) = 0.118$ and $1/\alpha(m_Z^2) = 128.90$,

$$\Delta S_Z = \Delta S + \Delta R - 0.064x_\alpha, \tag{3}$$

$$\Delta T_Z = \Delta T + 1.49\Delta R - \frac{\Delta\overline{\delta}_G}{\alpha}. \tag{4}$$

In addition to ΔS_Z and ΔT_Z, we adopt the W-boson mass $m_W(\text{GeV}) = 80.402 + \Delta m_W$ as the third oblique parameter instead of the U-parameter [4]:

$$\Delta m_W = -0.288\Delta S + 0.418\Delta T + 0.337\Delta U + 0.012x_\alpha - 0.126\frac{\Delta\overline{\delta}_G}{\alpha}. \tag{5}$$

So the new physics contributions to the electroweak observables can be summarized by three oblique corrections $\Delta S_Z, \Delta T_Z$ and Δm_W, and the non-oblique corrections Δg_α^f and $\Delta\overline{\delta}_G$.

2 Quantum corrections in the MSSM

First we show the constraints on the oblique parameters from the experiments and study the SUSY contributions to them. Taking account of Δg_L^b which may have the non-trivial m_t-dependence even in the SM, we perform the 5-parameter fit $(\Delta S_Z, \Delta T_Z, \Delta m_W, \Delta g_L^b, \alpha_s(m_Z))$ and find

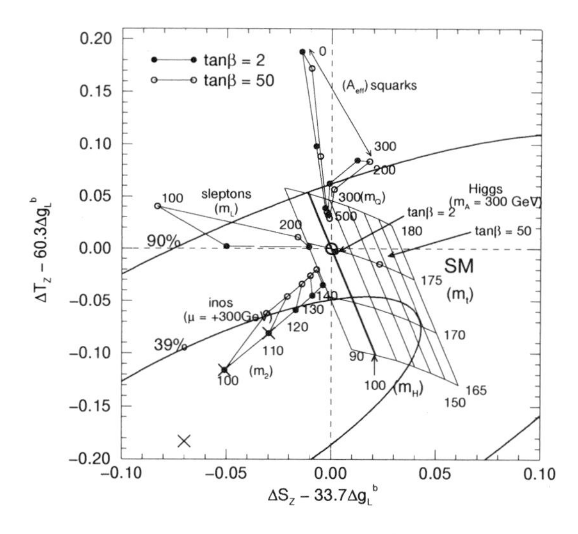

Figure 1. Supersymmetric contributions to $\Delta S_Z - 33.7\Delta g_L^b$ and $\Delta T_Z - 60.3\Delta g_L^b$. The symbol ($\times$) denotes the best fit from the electroweak data. The 39% ($\Delta\chi^2 = 1$) and 90% ($\Delta\chi^2 = 4.61$) contours are shown. The SM predictions are given for $m_t = 165 \sim 180$ GeV and $m_{H_{\rm SM}} = 90 \sim 150$ GeV.

the following constraint:

$$\left.\begin{array}{l}\Delta S_Z - 33.7\Delta g_L^b = -0.070 \pm 0.113 \\ \Delta T_Z - 60.3\Delta g_L^b = -0.183 \pm 0.137\end{array}\right\} \rho = 0.89,$$
$$\Delta m_W = 0.008 \pm 0.046, \qquad (6)$$
$$\chi^2_{\rm min} = 15.4 + \left(\frac{\Delta g_L^b + 0.00086}{0.00077}\right)^2,$$

where d.o.f. = 19 − 5 = 14. We show the supersymmetric contributions to $\Delta S_Z - 33.7\Delta g_L^b$ and $\Delta T_Z - 60.3\Delta g_L^b$ individually for $\tan\beta = 2$ and 50 in Fig. 1. The complete oblique corrections in the MSSM are given by their sum.

The squark and slepton contributions are shown as functions of the soft SUSY breaking mass parameters $m_{\tilde{Q}}$ and $m_{\tilde{L}}$, respectively, by assuming that their universality between the left- and the right-handed components. The effects of the left-right mixing of the sfermions are examined by introducing the effective A-parameter, $A_{\rm eff} \equiv A^t_{\rm eff} = A^b_{\rm eff}$. For example, in the case of the stop, $A_{\rm eff}$ is given by $A_{\rm eff} = A_t - \mu\cot\beta$. It can be seen from the figure that the squark contribution makes the fit worse, because it always makes ΔT_Z larger than the SM. The presence of the left-right mixing (*i.e.* non-zero $A_{\rm eff}$) make the squark contribution to the ΔT_Z-parameter slightly mild. The slepton contribution also makes the fit worse through the negative contribution to ΔS_Z. We find that the contributions from the left-handed sfermions to ΔS_Z and ΔT_Z are much larger than those from the right-handed sfermions.

We show the MSSM Higgs boson contributions to the oblique parameters for the CP-odd Higgs scalar mass $m_A = 300$ GeV and the SUSY breaking mass parameter $m_{\rm SUSY} = 1$ TeV, which appears in the effective Higgs potential. The lightest Higgs boson mass m_h in the figure is 106 GeV for $\tan\beta = 2$, and 129 GeV for $\tan\beta = 50$. We can see that the MSSM Higgs boson contributions to $(\Delta S_Z, \Delta T_Z)$ behave like the SM Higgs boson for $m_A \gtrsim 300$ GeV.

We show the chargino/neutralino contributions to $(\Delta S_Z, \Delta T_Z)$ as a function of the wino mass M_2 and for the higgsino mass $\mu = 300$ GeV. The points with the ($\times$) symbol in the figure are excluded from the direct search limit of the chargino mass (~ 90 GeV) at the LEP2 experiments. The chargino/neutralino contributions show the negative ΔS_Z and ΔT_Z, which may improve the fit over the SM. This is essentially because of ΔR which resides both in ΔS_Z and ΔT_Z. We find that the chargino contribution to ΔR is negative and large as compared to the sfermion contributions. For example, the contribution of the wino-like chargino to ΔR has the singularity $1/\sqrt{4M_2^2/m_Z^2 - 1}$ when M_2 is close to a half of m_Z. On the other hand, when $4M_2^2/m_Z^2 \gg 1$, ΔR is suppressed as m_Z^2/M_2^2, but the coefficient of the wino contribution is found to be about 90 times larger than that of the right-handed slepton contribution. This large negative contribution to ΔR makes both ΔS_Z and ΔT_Z significantly negative when a relatively light chargino exists.

Besides the oblique corrections, we have studied the non-oblique corrections such as the Zff vertex corrections and/or the vertex/box corrections to the μ-decay process in detail [3], and we found no improvement of the fit over the SM through the non-oblique corrections. Then, the best fit of the MSSM may be found when all sfermions and Higgs bosons are heavy enough, while the chargino is relatively light so that the radiative corrections in the MSSM are dominated by the chargino contribution to the oblique parameters. We perform the global fit to all electroweak data in the MSSM by assuming that all sfermions and heavy Higgs bosons masses are 1TeV. Under this assumption, we show the total χ^2 in the MSSM as a function of the lighter chargino mass $m_{\tilde{\chi}_1^-}$ for $\tan\beta = 2$ and $M_2/\mu = 0.1, 1$ and 10 in Fig. 2. The decoupling in the large SUSY mass limit is examined by comparing the MSSM fit with the SM fit at $m_{H_{\rm SM}} = m_h = 106$GeV rather than the SM best fit at $m_{H_{\rm SM}} = 117$GeV. When the lighter chargino mass is around its lower mass bound from the LEP2 experiment, we find the slight improvement of the fit over the SM, where the total χ^2 decreases by about one unit. The case of $\tan\beta = 50$ shows the similar behavior [3].

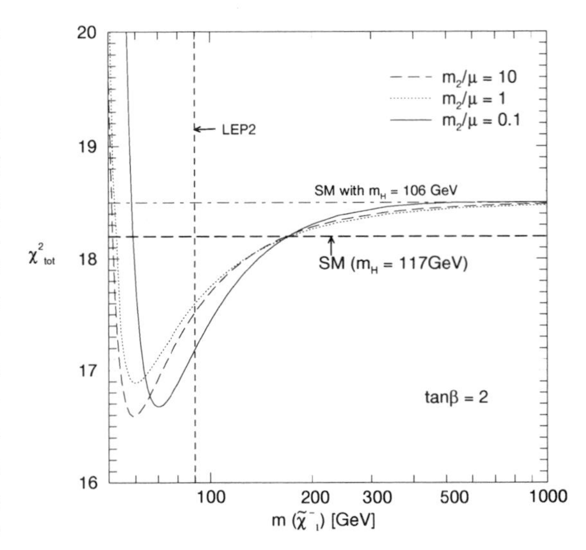

Figure 2. The total χ^2 in the MSSM as a function of the lighter chargino mass $m_{\tilde{\chi}_1^-}$ for $\tan\beta = 2$. The SM best fit ($\chi^2 = 18.2$) is shown by the dashed horizontal line. The dot-dashed horizontal line shows the SM fit using $m_{H_{\rm SM}} = 106$ GeV which is the lightest Higgs boson mass predicted in the MSSM. Three different M_2-μ ratio (10, 1, 0.1) are studied. The bound on $m_{\tilde{\chi}_1^-}$ from the LEP2 experiment is shown by the dashed vertical line.

3 Summary

We have studied constraints on the MSSM from the electroweak precision experiments. Owing to the negative large contribution to ΔT_Z from the light chargino, the improvement of the fit over the SM is expected, if the left-handed sfermions are heavy enough to decouple from the electroweak processes. The global fit of the MSSM show that, if the masses of all sfermions and heavy Higgs bosons are 1 TeV, the total χ^2 in the MSSM decreases by about one unit comparing with the SM when the lighter chargino mass is close to its direct search limit from LEP2.

Acknowledgments

The author would like to thank Kaoru Hagiwara for fruitful collaboration which this report is based upon.

References

1. The LEP Collaborations ALEPH, DELPHI, L3, OPAL, the LEP Electroweak Working Group and the SLD Heavy Flavor Group, CERN-EP/99-15.
2. I. Riu, talk given at the XXXIVth Rencontres de Moriond, March 13-20, 1999.
3. G.C. Cho and K. Hagiwara, Nucl. Phys. **B574** (2000) 623.
4. M.E. Peskin and T. Takeuchi, Phys. Rev. Lett. **65** (1990) 964; Phys. Rev. **D46** (1992) 381.

YUKAWA TEXTURES AND HORAVA-WITTEN M-THEORY

R. ARNOWITT AND B. DUTTA

Center For Theoretical Physics, Department of Physics, Texas A&M University, College Station TX 77843-4242

The general structure of the matter Kahler metric in the $\kappa^{2/3}$ expansion of Horava-Witten M-theory with nonstandard embeddings is examined. It is shown that phenomenological models based on this structure can lead to Yukawa and V_{CKM} hierarchies (consistent with all data) without introducing ad hoc small parameters if the 5-branes lie near the distant orbifold plane and the instanton charges of the physical plane vanish. M-theory thus offers an alternate way of describing these hierarchies, different from the conventional models of Yukawa textures.

1 Introduction

Over the past year considerable progress has been made in understanding Horava-Witten heterotic M-theory [1] with "non-standard" embeddings For a review, see [2]. In this picture, space has an 11 dimensional orbifold structure of the form (to lowest order) $M_4 \times X \times S^1/Z_2$ where M_4 is Minkowski space, X is a 6 dimensional (6D) Calabi-Yau space, and $-\pi\rho \leq x^{11} \leq \pi\rho$. The space thus has two orbifold 10D manifolds $M_4 \times X$ at the Z_2 fixed points at $x^{11} = 0$ and $x^{11} = \pi\rho$ where the first is the visible sector and the second is the hidden sector, each with an a priori E_8 gauge symmetry. In addition there can be a set of 5-branes in the bulk at points $0 < x_n < \pi\rho$, $n = 1...N$ each spanning M_4 (to preserve Lorentz invariance) and wrapped on a holomorphic curve in X (to preserve N=1 supresymmetry).

In general, physical matter lives on the $x^{11} = 0$ orbifold plane and only gravity lives in the bulk. The existence of 5-branes allows one to satisfy the cohomological constraints with E_8 on the $x^{11} = 0$ plane breaking to $G \times H$ where G is the structure group of the Calabi-Yau manifold and H is the physical grand unification group. Thus non standard embeddings allow naturally for physically interesting grand unification groups. We consider here the case $G = SU(5)$, and hence $H = SU(5)$.

Phenomenology has played an important role in guiding the general structure of Horava-Witten theory. Thus the Calabi-Yau manifold is assumed to have a size $r^{-1} \simeq M_G$ to account for the success of grand unification at $O(10^{16}$ GeV) and this then requires the orbifold scale to be $(\pi\rho)^{-1} \simeq 10^{15}$ GeV, to account for the size of the 4D Planck mass. In addition, constructing three generation models has been an important element in string theory from its inception, and recently, three generation models with a Wilson line breaking $SU(5)$ to $SU(3) \times SU(2) \times U(1)$ have been constructed in the M-theory frame work using torus fibered Calabi-Yau manifolds (with two sections)[3]. Also, the general structure (to the first order) of the Kahler metric of the matter field has been constructed[4]. We examine here this structure and show that it can lead to Yukawa textures with all CKM and quark mass data in agreement with experiment, without any undue fine tuning. Thus M-theory leads to a new way of considering the Yukawa sector of the Standard Model. It is also possible to show that a three generation model with a Wilson line (to break SU(5) to the standard model) and possessing some of the basic properties needed for the phenomenology exists for the Calabi-Yau manifold with del-Pezzo base dP_7 [5].

This note is a summary of the above results, and details can be found in [5]

2 Kahler Metric

The bose part of the 11 dimensional gravity multiplet consists of the metric tensor g_{IJ}, the antisymmetric 3-form C_{IJK} and its field strength $G_{IJKL} = 24\partial_{[I}C_{JKL]}$. ($I$, J, J, $K = 1...11$.). The G_{IJKL} obey field equations $D_I G^{IJKL} = 0$ and Bianchi identities

$$(dG)_{11RSTU} = 4\sqrt{2}\pi(\frac{\kappa}{4\pi})^{2/3}[J^0\delta(x^{11}) + J^{N+1}\delta(x^{11}-\pi\rho) + \frac{1}{2}\Sigma_{n=1}^{N} J^n(\delta(x^{11}-x_n) + \delta(x^{11}+x_n))]_{RSTU} \quad (1)$$

Here $(\kappa^{2/9})$ is the 11 dimensional Planck scale, and J^n, $n = 0, 1, ...N+1$ are sources from orbifold planes and the N 5-branes. These equations can be solved perturbatively in powers of $(\kappa^{2/3})$ [4]. The effective 4D theory can then be characterized by a Kahler potential $K = Z_{IJ}\bar{C}^I C^J$, Yukawa couplings Y_{IJK} for the matter fields C^I and gauge functions from the physical orbifold plane x^{11}=0. To first order, Z_{IJ} takes the form [4].

$$Z_{IJ} = e^{-K_T/3}[G_{IJ} - \frac{\epsilon}{2V}\tilde{\Gamma}^i_{IJ}\Sigma_0^{N+1}(1-z_n)^2\beta_i^{(n)}] \quad (2)$$

Here $\epsilon = (\kappa/4\pi)^{2/3}2\pi^2\rho/V^{2/3}$ is the expansion parameter. V is the Calabi-Yau volume, G_{IJ}, $\tilde{\Gamma}^i_{IJ}$ and Y_{IJK} can be expressed in terms of integrals over the Calabi-Yau manifold [4], and K_T is the Kahler potential for the moduli.

3 Phenomenological Yukawa Matrices

If the perturbation analysis is to be a reasonable approximation, the second term of Eq.(2) should be a small correction. A priori one expects G_{IJ}, $\tilde{\Gamma}^i_{IJ}$ and Y_{IJK} to be characteristically of $O(1)$, and the parameter ϵ is not too small. However, the second term will be small if $\beta_i^{(0)}$ were to vanish and if the 5-branes were to be near the distant orbifold plane i.e. $d_n \equiv 1 - z_n$ were small, where $z_n = x_n/\pi\rho$. In the following we will assume then that

$$\beta_i^{(0)} = 0;\ d_n = 1 - z_n \cong 0.1 \quad (3)$$

The condition $\beta_i^{(0)} = 0$ is non trivial, but it is possible to show that a three generation model of a torus fibered Calabi-Yau manifold with Wilson like breaking $SU(5)$ to $SU(3) \times SU(2) \times U(1)$ with del -Pezzo base dP_7 has this property[5].

Eq.(3) then suggests that it is the ϵ term of Eq.(2) that are the third generation contributions to the Kahler metric. A simple phenomenological example for the u and d quark contributions with these properties (and containg the maximum numbers of zeros) is ($f_T \equiv exp(-K_T/3)$):

$$Z^u = f_T \begin{pmatrix} 1 & 0.345 & 0 \\ 0.345 & 0.132 & 0.639d^2 \\ 0 & 0.639d^2 & 0.333d^2 \end{pmatrix};$$

$$Z^d = f_T \begin{pmatrix} 1 & 0.821 & 0 \\ 0.821 & 0.887 & 0 \\ 0 & 0 & 0.276 \end{pmatrix}. \quad (4)$$

with Yukawa matrices diagY^u=(0.0765, 0.536, 0.585 $Exp[\pi i/2]$) and diagY^d=(0.849, 0.11, 1.3).

These expression offer an alternate possibility for generating Yukawa hierarchies. Thus to obtain the physical Yukawa matrices, one must first diagonalize the Kahler metric and then rescale it to unity. Then using the renormalization group equations, one can generate the CKM matrix, and the quark masses. The results are given in the following table:

Quantity	Th. Value	Exp. Value[6]
m_t(pole)	170.5	175± 5
$m_c(m_c)$	1.36	1.1-1.4
m_u(1 GeV)	0.0032	0.002-0.008
$m_b(m_b)$	4.13	4.1-4.5
m_s(1 GeV)	0.110	0.093-0.125[7]
m_d(1 GeV)	0.0055	0.005-0.015
V_{us}	0.22	0.217-0.224
V_{cb}	0.036	0.0381±0.0021[8]
V_{ub}	0.0018	0.0018-0.0045
V_{td}	0.006	0.004-0.013

and sin2β=0.31 and sinγ=0.97. The agreement with experiment is quite good. Note also that m_u/m_d=0.582 and m_s/m_d=20.0 in good agreement with Leutwyler evaluations[9] 0.553± 0.043 and 18.9±0.8.

While the precise choice of entries in $Z^{u,d}$ and $Y^{u,d}$ are chosen to obtain the above results, as shown in [5], the quark mass hierarchies arise naturally from a Kahler metric of the type of Eq. (4). Similarly the smallness of the off diagonal V_{CKM} matrix elements also occur naturally as a consequence of the above model. Thus it is possible for M-theory to generate the Yukawa hierarchies without any undue fine tuning and without introducing ad hoc very small off diagonal entries.

4 Conclusion

Horava-Witten M-theory has now progressed to the point where it offers a fundamental framework for building phenomenological models. Thus it allows for conventional GUT groups (e.g. $SU(5)$, $SO(10)$), accommdates grand unification at $M_G \cong 3\times10^{16}$ GeV, and has three generation models where the GUT group breaks to the Standard Model at M_G by a Wilson line.

M theory with non-standard embeddings also offers new possibility of encoding the Yukawa hierarchies in the Kahler metric. This can happen naturally if the 5-branes cluster near the hidden orbifold plane ($d_n \equiv 1-z_n \simeq 0.1$) and the instanton charges of the physical plane vanish ($\beta_i^{(0)} = 0$). It is possible to construct three generation manifolds possessing these properties[5]. Then if the $\kappa^{2/3}$ term of the Kahler metric is attributed to the third generation of the u-quarks, one can construct models possessing all the experimental hiererchies without any undue fine tuning or ad hoc small parameters. While models of this type are to be viewed as "string inspired" as one can not perform the integrals over the Calabi-Yau manifold, they may give general insights into the nature of the Calabi-Yau manifold.

5 Acknowledgement

This work was supported in part by NSF grant no. PHY-9722090.

References

1. P. Horava and E. Witten, Nucl. Phys. B **460**, 506 (1996); Nucl. Phys. B **475**, 94 (1996); E. Witten, Nucl. Phys. B **471**, 135 (1996).
2. B. Ovrut, hep-th/9905115, Lectures at Asian Pacific Center for Theoretical Physics, Third Winter School, January 21-February 5, 1999, Cheju Island, Korea.
3. R. Donagi, B. Ovrut, T. Pantev and D. Waldram, hep-th/9912208.
4. A. Lukas, B. Ovrut and D. Waldram, JHEP **9904**, 009 (1999).
5. R. Arnowitt and B. Dutta, hep-ph/0006172.
6. Except as otherwise noted, expermental entries are from the Particle Data group, Eupropean Phys. Journ. C **3**, 1 (1998).
7. S. Aoki, hep-ph/9912288.
8. S. Stone, hep-ph/9904350.
9. H. Leutwyler, Phys. Lett. B **374**, 163 (1996).

CONSTRAINTS ON R–PARITY VIOLATION FROM PRECISION ELECTROWEAK MEASUREMENTS

TATSU TAKEUCHI[(1)], OLEG LEBEDEV[(2)], AND WILL LOINAZ[(1,2)]
[(1)] *IPPAP, Physics Department, Virginia Tech, Blacksburg, VA 24061, USA*
[(2)] *Department of Physics, Amherst College, Amherst, MA 01002, USA*

We constrain the size of R–parity violating couplings using precision electroweak data.

1 Introduction

Precision electroweak measurements provide a window to physics beyond the Standard Model by constraining the size of radiative corrections from new particles and interactions. In this contribution, we summarize the constraints from the LEP and SLD data on an R–parity violating extension to the MSSM. Details are presented in Ref. 1.[a]

We extend the MSSM with the addition of the the following terms to the superpotential:

$$\frac{1}{2}\lambda_{ijk}\hat{L}_i\hat{L}_j\hat{E}_k+\lambda'_{ijk}\hat{L}_i\hat{Q}_j\hat{D}_k+\frac{1}{2}\lambda''_{ijk}\hat{U}_i\hat{D}_j\hat{D}_k\ , \tag{1}$$

where $\hat{L}_i$, $\hat{E}_i$, $\hat{Q}_i$, $\hat{U}_i$, and $\hat{D}_i$ are the MSSM superfields defined in the usual fashion and the subscript $i = 1, 2, 3$ is the generation index. We focus our attention on these supersymmetric interactions only and ignore possible R–parity violating soft–breaking terms.[3] This allows us to rotate away any bilinear terms that may be present.[4b]

Since the couplings constants λ_{ijk}, λ'_{ijk}, and λ''_{ijk} are arbitrary and do not have any *a priori* flavor structure, they generically lead to flavor dependent processes and corrections to electroweak observables. In particular, they will give rise to flavor dependent corrections to the $Zf\bar{f}$ vertices which can be well constrained by the Z-pole data from LEP and SLD. Previous works [6] have already placed bounds on λ_{ijk} of $\mathcal{O}(10^{-2})$ from lepton universality in low energy charged current processes, and their effects on Z-pole observables are negligible. λ'_{ijk} and λ''_{ijk} have been less tightly constrained. However, the simulateneous presence of both terms leads to unacceptably fast proton decay so we will assume that only one of these terms is present at a time.

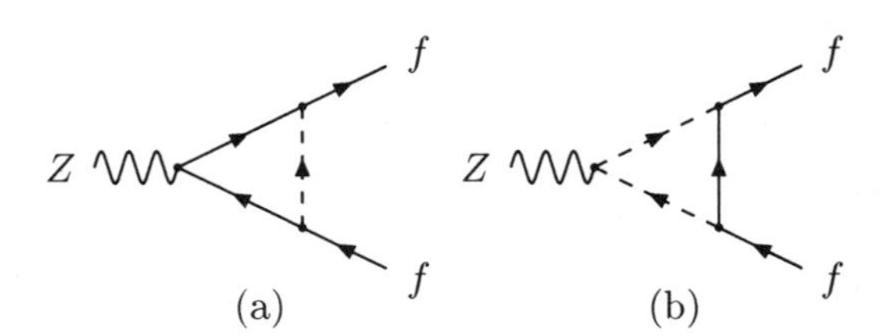

Figure 1. Corrections to the $Zf\bar{f}$ coupling from the R–parity violating interactions in Eq. (1). The sfermions are represented by the dashed lines. Wave-function renormalization corrections are not shown.

2 The Corrections

Corrections to the $Zf\bar{f}$ vertex from the interactions in Eq. (1) are shown schematically in Fig. 1. Of all the possible corrections, it can be shown that only those with the top quark in the internal fermion line are numerically significant. Therefore, we only need to consider the λ'_{i3k} (9 parameters) or the λ''_{3jk} (3 parameters)[c] interactions. In this approximation, the λ'_{i3k} inter-

[a] For a discussion on the constraints on the MSSM itself, see Ref. 2.

[b] An extension of the MSSM with R–parity violation including bilinear and soft–breaking terms is discussed in Ref. 5.

[c] λ''_{ijk} is antisymmetric in the latter two indices.

actions only affect the couplings of the left-handed charged leptons and the right-handed down-type quarks, while the λ''_{3jk} interactions only affect the couplings of the right-handed down-type quarks.

The actual sizes of these corrections depend on the masses of the internal sfermions. As a reference, we choose a common sfermion mass of 100 GeV. In this case, the shifts in the $Zf\bar{f}$ couplings due to the λ'_{i3k} interactions are found to be:[d]

$$\begin{aligned}\delta h^{\not R}_{e_iL} &= 0.0061 \sum_k |\lambda'_{i3k}|^2, \\ \delta h^{\not R}_{d_kR} &= -0.00215 \sum_i |\lambda'_{i3k}|^2.\end{aligned}$$

Similarly, the shifts due to the λ''_{3jk} interactions are:

$$\delta h^{\not R}_{d_jR} = -0.0043 \sum_k |\lambda''_{3jk}|^2.$$

In order to constrain the size of these shifts, all vertex and oblique corrections from within the MSSM must also be included and accounted for consistently. Here, we observe that the majority of the Z–pole observables are parity–violating asymmetries or ratios of partial widths which are all *ratios of coupling constants.*[e] Oblique corrections enter the coupling constants through the ρ-parameter and the effective value of $\sin^2\theta_W$.[8] The dependence on the ρ-parameter cancels in the ratios, isolating the effects of oblique corrections in $\sin^2\theta_W$. For the vertex corrections, we can apply the same approximation as above and neglect those without a heavy internal fermion. Of the corrections that remain, the simplifying assumption that all the sfermion masses are degenerate allows us to either cancel the correction in the ratios of coupling constants, or absorb them into a shift in $\sin^2\theta_W$. The only vertex correction that must be considered independently is the Higgs sector induced correction to the b_L coupling. These considerations allow us to parametrize all the corrections from both within and without the MSSM in terms of just a few parameters which can be fit to the differences of the Z-pole data[9] and ZFITTER[10] predictions.

3 Lepton Universality

The shifts in the left-handed couplings of the charged leptons break lepton universality. Fitting to the leptonic data from LEP and SLD, we find:

$$\begin{aligned}\delta h^{\not R}_{\mu_L} - \delta h^{\not R}_{e_L} &= 0.00038 \pm 0.00056 \\ \delta h^{\not R}_{\tau_L} - \delta h^{\not R}_{e_L} &= -0.00013 \pm 0.00061\end{aligned}$$

The couplings contributing to $\delta h^{\not R}_{e_L}$ are already well constrained from other experiments, so if we neglect them we obtain the following 1σ (2σ) bounds:

$$\begin{aligned}|\lambda'_{23k}| &\le 0.40 \quad (0.50) \\ |\lambda'_{33k}| &\le 0.28 \quad (0.42)\end{aligned}$$

4 Hadronic Observables

The couplings of the right–handed down–type quarks are constrained by the hadronic observables from LEP and SLD. A global fit to all relevant observables yields:

$$\begin{aligned}\delta h^{\not R}_{d_R} &= 0.081 \pm 0.077 \\ \delta h^{\not R}_{s_R} &= 0.055 \pm 0.043 \\ \delta h^{\not R}_{b_R} &= 0.026 \pm 0.010\end{aligned}$$

Note that $\delta h^{\not R}_{d_R}$ and $\delta h^{\not R}_{s_R}$ are positive by more than 1σ, while $\delta h^{\not R}_{b_R}$ is positive by more than 2σ. Since both the λ'_{i3k} and λ''_{3jk} interactions lead to negative shifts, all these couplings are ruled out at the 1σ level. The (2σ) $[3\sigma]$ bounds are:

$$\begin{aligned}|\lambda'_{i31}| &\le (5.8) \quad [8.4] \\ |\lambda'_{i32}| &\le (3.8) \quad [5.9] \\ |\lambda'_{i33}| &\le (\quad\;\;) \quad [1.4]\end{aligned}$$

or

$$|\lambda''_{321}| \le (2.7) \quad [4.1]$$

---

[d]The tree level coupling of fermion f to the Z is normalized to $h_f = I_{3f} - Q_f \sin^2\theta_W$.

[e]The same observation has been used to constrain a variety of models in Ref. 7.

$$|\lambda''_{33i}| \leq (\quad) \; [0.96]$$

Stronger constrains on λ'_{i31} and λ'_{i32} are available from other experiments.

5 Bayesian Limits

If one makes the *a priori* assumption that the MSSM with R–parity violation is the correct underlying theory, one obtains the following 68% (95%) Bayesian confidence limits:

$$\begin{aligned} \delta h^{\not R}_{d_R} &\geq -0.061 \;\; (-0.123) \\ \delta h^{\not R}_{s_R} &\geq -0.031 \;\; (-0.064) \\ \delta h^{\not R}_{b_R} &\geq -0.0046 \;\; (-0.010) \end{aligned}$$

This translates into

$$\begin{aligned} |\lambda'_{i31}| &\leq 5.2 \;\; (7.6) \\ |\lambda'_{i32}| &\leq 3.8 \;\; (5.6) \\ |\lambda'_{i33}| &\leq 1.4 \;\; (2.2) \end{aligned}$$

or

$$\begin{aligned} |\lambda''_{321}| &\leq 2.7 \;\; (3.9) \\ |\lambda''_{33i}| &\leq 1.0 \;\; (1.5) \end{aligned}$$

While these Bayesian bounds are considerablely weaker, they are accompanied by large values of χ^2.

6 The Common Sfermion Mass

To obtain bounds for a common sfermion mass other than the value of $m_{\tilde{f}} = 100$ GeV used in this analysis, the limits should be rescaled by $\sqrt{F(x_0)/F(x)}$, where

$$F(x) \equiv \frac{x}{1-x}\left(1 + \frac{1}{1-x}\ln x\right),$$

and

$$x \equiv \frac{m_t^2}{m_{\tilde{f}}^2}, \qquad x_0 \equiv \frac{m_t^2}{(100\,\mathrm{GeV})^2}.$$

Acknowledgments

Helpful communications with B. Allanach, J. E. Brau, R. Clare, H. Dreiner, D. Muller, Y. Nir, A. Pilaftsis, F. Rimondi, P. Rowson, D. Su, Z. Sullivan, and M. Swartz are gratefully acknowledged. This work was supported in part (O.L. and W.L.) by the U. S. Department of Energy, grant DE-FG05-92-ER40709, Task A.

References

1. O. Lebedev, W. Loinaz, and T. Takeuchi, *Phys. Rev.* D **61**, 115005 (2000); *Phys. Rev.* D **62**, 015003 (2000).
2. G.–C. Cho, in this proceedings (hep-ph/0009022).
3. M. Nowakowski and A. Pilaftsis, *Nucl. Phys.* B **461**, 19 (1996).
4. L. J. Hall and M. Suzuki, *Nucl. Phys.* B **231**, 419 (1984).
5. O. C. W. Kong, in this proceedings (hep-ph/0008251).
6. B. C. Allanach, A. Dedes, and H. K. Dreiner, *Phys. Rev.* D **60**, 075014 (1999).
7. T. Takeuchi, A. K. Grant, and J. L. Rosner, hep-ph/9409211, W. Loinaz and T. Takeuchi, *Phys. Rev.* D **60**, 015005 (1999), O. Lebedev, W. Loinaz, and T. Takeuchi, *Phys. Rev.* D **62**, 055014 (2000); hep-ph/0006031, L. N. Chang, O. Lebedev, W. Loinaz, and T. Takeuchi, hep-ph/0005236.
8. M. E. Peskin and T. Takeuchi, *Phys. Rev. Lett.* **65**, 964 (1990); *Phys. Rev.* D **46**, 381 (1992).
9. D. Abbaneo, et al., CERN-EP-99-015, J. Mnich, CERN-EP-99-143, K. Abe, et al., hep-ex/9908006; hep-ex/9908038. K. Ackerstaff, et al., *Z. Phys.* C **76**, 387 (1997), E. Boudinov, et al., DELPHI 99-98 CONF 285, talks by M. Swartz at *Lepton-Photon '99* and by J. E. Brau, S. Fahey, and G. Quast at *HEP-EPS '99.*
10. The ZFITTER package: D. Bardin, et al., *Z. Phys.* C **44**, 493 (1989); *Nucl. Phys.* B **351**, 1 (1991); *Phys. Lett.* B **255**, 290 (1991); CERN-TH-6443/92; DESY 99-070, hep-ph/9908433.

BILINEAR R-PARITY VIOLATING SUSY: SOLVING THE SOLAR AND ATMOSPHERIC NEUTRINO PROBLEMS

M. HIRSCH, W. POROD AND J. W. F. VALLE

Instituto de Física Corpuscular – C.S.I.C., Departamento de Física Teòrica, Universitat de València, Edificio Institutos de Paterna, Apartado de Correos 22085, 46071 València

M. A. DíAZ

Fac. de Física, Universidad Católica de Chile, Av. Vicuña Mackenna 4860, Santiago, Chile

J. C. ROMÃO

Depto. de Física, Inst. Superior Técnico, Av. Rovisco Pais 1, 1049-001 Lisboa, Portugal

Bilinear R-parity violation is a simple extension of the MSSM allowing for Majorana neutrino masses. One of the three neutrinos picks up mass by mixing with the neutralinos of the MSSM, while the other two neutrinos gain mass from 1-loop corrections. Once 1-loop corrections are carefully taken into account the model is able to explain solar and atmospheric neutrino data for specific though simple choices of the R-parity violating parameters.

1 Introduction

Atmospheric neutrino data so far provides the most compelling evidence for non-zero neutrino masses [1], indicating large mixing among neutrino flavours and a typical Δm^2_{ij} of the order of $\Delta m^2_{ij} \sim$ (few) 10^{-3} eV^2. On the other hand, despite new high quality data, the long standing solar neutrino problem still allows both large and small mixing angle values for neutrinos [2] and $\Delta m^2 \sim \mathcal{O}(10^{-5})$ eV^2 or $\Delta m^2 \sim \mathcal{O}(10^{-7})$ eV^2 [2] (SMA and LMA or LOW solutions of the solar neutrino problem, respectively).

Many attempts to explain neutrino masses can be found in the literature [3]. Here we summarize the work of [4]. It is based on a simple bilinear R-parity violating (BRPV) extension of the MSSM. This model, despite being minimalistic can explain atmospheric and solar neutrino data for specific ranges of model parameters. Its attractiveness lies in the fact that these parameter choices necessary to solve the neutrino problems give at the same time definite predictions for accelerator physics. [5]

2 Bilinear R-parity violating SUSY

In the simplest extension of the MSSM including R-parity violation the superpotential contains just 3 additional bilinear terms [6]

$$W = W_{MSSM} + \epsilon_i \hat{L}_i \hat{H}_u \tag{1}$$

They violate lepton number by one unit and therefore necessarily generate Majorana neutrino masses. Corresponding bilinear R-parity violating terms appear in the soft SUSY breaking terms, but strictly speaking these are not independent parameters because of the tadpole conditions [6].

In this model at tree-level only one neutrino picks up a mass via mixing with the neutralinos. This tree-level mass can be estimated by [7]

$$m_\nu = \frac{M_1 g^2 + M_2 g'^2}{4det(\mathcal{M}_{\chi^0})} |\vec{\Lambda}|^2 \tag{2}$$

Here, M_1 and M_2 are the MSSM gaugino masses, $\mathcal{M}_{\chi^0}$ is the MSSM neutralino mass matrix and $\vec{\Lambda}$ is defined by $\Lambda_i = \epsilon_i v_d + \mu \langle \tilde{\nu}_i \rangle$, with $\langle \tilde{\nu}_i \rangle$ being scalar neutrino vevs.

Since only one neutrino gains mass at tree-level in BRPV, to study solar and atmospheric neutrino problems at the same time, it is necessary to include 1-loop corrections. Details are given in [4].

3 Numerical results

After including 1-loop corrections the BRPV model produces for nearly all choices of parameters a hierarchical mass spectrum. The largest neutrino mass can then usually be estimated by the tree-level value. This is demonstrated in Fig. 1, where we show Δm^2_{atm} as a function of $|\vec{\Lambda}|/(\sqrt{M_2}\mu)$. As the figure shows, correct Δm^2_{atm} can be easily obtained by an appropriate choice of $|\vec{\Lambda}|$.

The solar mass scale, on the other hand, is entirely generated at 1-loop order and therefore depends on the model parameters in a complicated way. Fig. 2 shows one example. The parameter $\epsilon^2|/\vec{\Lambda}|$ is most important for determining the size of the loop corrections, but loops also show a strong dependence on $\tan\beta$.

Turning to the discussion on neutrino angles, we note that as long as the 1-loop corrections are not larger than the tree-level contribution, the flavour composition of the 3rd mass eigenstate is approximately given as

$$U_{\alpha 3} \approx \Lambda_\alpha/|\Lambda|. \tag{3}$$

Since atmospheric and reactor [8] neutrino data tell us that $\nu_\mu \to \nu_\tau$ oscillations are preferred over $\nu_\mu \to \nu_e$ oscillations, we conclude that $\Lambda_e \ll \Lambda_\mu \simeq \Lambda_\tau$ is required for BRPV to fit the data. This is shown in figs. 3 and 4.

For the solar angle the situation is more complex. As explained in [4] there are two cases to distinguish. With the usual minimal supergravity unification assumptions, ratios of ϵ_i/ϵ_j fix the ratios of Λ_i/Λ_j. Since atmospheric (and reactor) neutrino data imply that $\Lambda_e \ll \Lambda_\mu, \Lambda_\tau$ only the small angle solution to the solar neutrino problem can be ob-

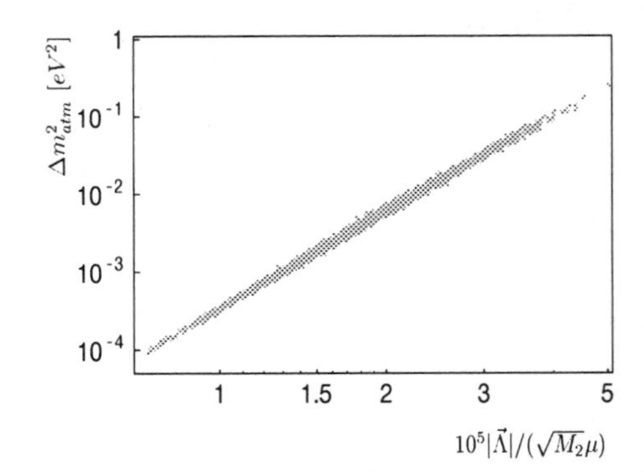

Figure 1. The atmospheric Δm^2 as a function of $|\vec{\Lambda}|/(\sqrt{M_2}\mu)$. The figure shows how the tree-level approximation can be used to fix the largest mass scale in the bilinear model.

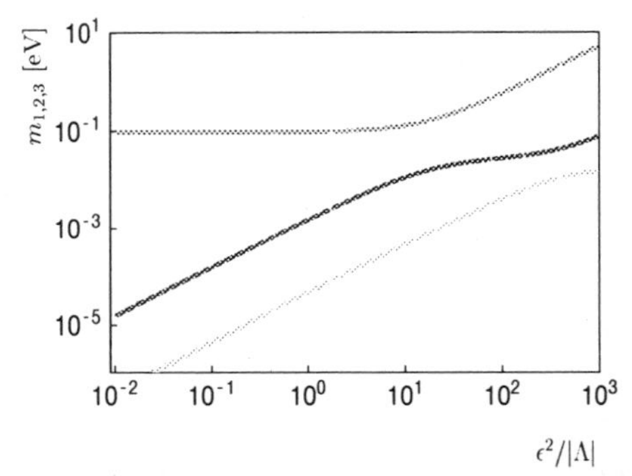

Figure 2. The solar Δm^2 as a function of $\epsilon^2|/\vec{\Lambda}|$ for otherwise fixed parameters of the model. The figure shows how the importance of loop corrections increases with increasing $\epsilon^2|/\vec{\Lambda}|$.

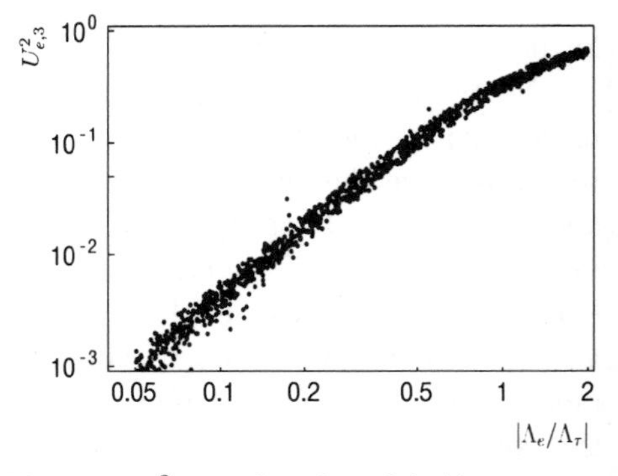

Figure 3. U^2_{e3} as a function of Λ_e/Λ_τ.

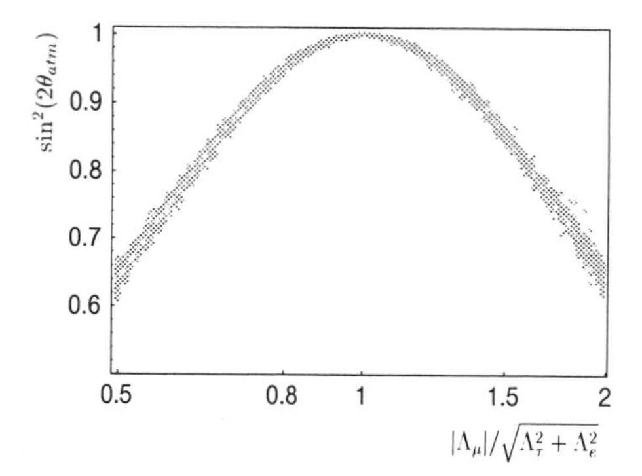

Figure 4. Atmospheric neutrino mixing angle as a function of $\Lambda_\mu/\sqrt{\Lambda_\tau^2+\Lambda_e^2}$. Since $\Lambda_e \ll \Lambda_\mu, \Lambda_\tau$ is required by the reactor neutrino data, $\Lambda_\mu \simeq \Lambda_\tau$ is needed to obtain large atmospheric neutrino mixing.

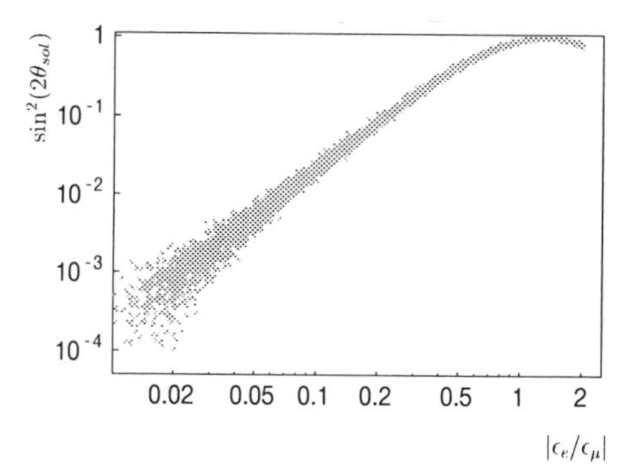

Figure 5. Solar neutrino mixing angle as a function of ϵ_e/ϵ_μ.

tained in this case. However, even a very tiny deviation from universality of soft parameters at the unification scale relaxes this constraint sufficiently, such that also large angle solar solutions can be obtained in our model, see Fig. 5.

4 Conclusions

Bilinear R-parity violating SUSY, despite being a very simple extension of the MSSM can explain atmospheric and solar neutrino data [4], once 1-loop corrections are taken carefully into account. The main attractiveness of the model, however, lies in the fact that it can be tested at future accelerators. In [5] we discuss the definite predictions made for neutralino decays.

Acknowledgments

This work was supported by DGICYT grants PB98-0693 and SB97-BU0475382 (W. P.), by the TMR contracts ERBFMRX-CT96-0090 and ERBFMBICT983000 (M. H.).

References

1. Y. Fukuda *et al.* Phys. Rev. Lett. **81**, 1562 (1998); and hep-ex/9805006
2. N. Fornengo, M. Gonzalez-Garcia & J. Valle, Nucl. Phys. **B580** (2000) 58; M. Gonzalez-Garcia et al, Nucl. Phys. **B573**, 3 (2000); an updated discussion of neutrino data was presented by M. C. Gonzalez-Garcia at this conference, and is avalaible from `http://neutrinos.uv.es//`.
3. A simple search at `http://xxx.lanl.gov/find/hep-ph` reveals hundreds of attempts to explain neutrino masses in the light of atmospheric and solar data
4. M. Hirsch, M.A. Diaz, W. Porod, J.C. Romao and J.W.F. Valle, hep-ph/0004115, to appear in Phys. Rev. **D**; J.C. Romão, M.A. Diaz, M. Hirsch, W. Porod and J.W.F. Valle, Phys. Rev. **D61**, 071703 (2000)
5. M. Hirsch, W. Porod, J.C. Romao and J.W.F. Valle, talk at this conference
6. M.A. Diaz, J.C. Romao and J.W.F. Valle, Nucl. Phys. **B524** (1998) 60
7. M. Hirsch and J. W. F. Valle, Nucl. Phys. **B557**, 60 (1999); M. Hirsch, J. C. Romao and J. W. F. Valle, Phys. Lett. **B486** (2000) 255
8. M. Apollonio et al, Phys.Lett. **B466** (1999) 415; F. Boehm et al, hep-ex/9912050

SOME RECENT RESULTS FROM THE COMPLETE THEORY OF SUSY WITHOUT R-PARITY

OTTO C. W. KONG

Institute of Physics, Academia Sinica, Nankang, Taipei, TAIWAN 11529
E-mail: kongcw@phys.sinica.edu.tw

We review an efficient formulation of the complete theory of supersymmetry without R-parity, where all the admissible R-parity violating terms incorporated. Some interesting recent results will be discussed, including newly identified 1-loop contributions to neutrino masses and electric dipole moments of neutron and electron, resulted from R-parity violating LR squark and slepton mixings.

1 The Generic Supersymmetric Standard Model

Supersymmetry without R-parity is nothing but the generic supersymmetric Standard Model, *i.e.* a theory built with the minimal superfield spectrum incorporating the Standard Model (SM) particles and interactions dictated by the SM (gauge) symmetries and the idea that SUSY is softly broken. The theory is hence generally better motivated than *ad hoc* versions of R-parity violating (RPV) theories. Here, however, RPV parameters come in various forms. The latter includes the more popular trilinear (λ_{ijk}, λ'_{ijk}, and λ''_{ijk}) and bilinear (μ_i) couplings in the superpotential, as well as soft SUSY breaking parameters of the trilinear, bilinear, and soft mass (mixing) types. From a phenomenological point of view, there is the related notion of (RPV) "sneutrino VEV's". In order not to miss any plausible RPV phenomenological features, it is important that all of the RPV parameters be taken into consideration with *a priori* bias. To emphasize the point, we call the model the *complete theory* of SUSY without R-parity.

The most general renormalizable superpotential for the supersymmetric SM (without R-parity) can be written as

$$W = \varepsilon_{ab}\Big[\mu_\alpha \hat{H}_u^a \hat{L}_\alpha^b + h_{ik}^u \hat{Q}_i^a \hat{H}_u^b \hat{U}_k^C + \lambda'_{\alpha jk} \hat{L}_\alpha^a \hat{Q}_j^b \hat{D}_k^C + \frac{1}{2}\lambda_{\alpha\beta k} \hat{L}_\alpha^a \hat{L}_\beta^b \hat{E}_k^C\Big] + \frac{1}{2}\lambda''_{ijk} \hat{U}_i^C \hat{D}_j^C \hat{D}_k^C, \qquad (1)$$

where (a, b) are $SU(2)$ indices, (i, j, k) are the usual family (flavor) indices, and (α, β) are extended flavor index going from 0 to 3. In the limit where $\lambda_{ijk}, \lambda'_{ijk}, \lambda''_{ijk}$ and μ_i all vanish, one recovers the expression for the R-parity preserving case, with $\hat{L}_0$ identified as $\hat{H}_d$. Without R-parity imposed, the latter is not *a priori* distinguishable from the $\hat{L}_i$'s. Note that λ is antisymmetric in the first two indices, as required by the $SU(2)$ product rules, as shown explicitly here with $\varepsilon_{12} = -\varepsilon_{21} = 1$. Similarly, λ'' is antisymmetric in the last two indices, from $SU(3)_C$.

R-parity is exactly an *ad hoc* symmetry put in to make $\hat{L}_0$, stand out from the other $\hat{L}_i$'s as the candidate for $\hat{H}_d$. It is defined in terms of baryon number, lepton number, and spin as, explicitly, $\mathcal{R} = (-1)^{3B+L+2S}$. The consequence is that the accidental symmetries of baryon number and lepton number in the SM are preserved, at the expense of making particles and superparticles having a categorically different quantum number, R-parity. The latter is actually not the most effective discrete symmetry to control superparticle mediated proton decay[1], but is most restrictive in terms of what is admitted in the Lagrangian, or the superpotential alone. On the other hand, R-parity also forbides neutrino masses in the supersymmetric SM. The strong experimental hints for the existence of (Majorana) neutrino masses[2] is an indication of lepton number violation, hence suggestive of R-parity violation.

The soft SUSY breaking part of the Lagrangian is more interesting, if only for the fact that many of its interesting details have been overlooked in the literature. However, we will postpone the discussion till after we address the parametrization issue.

2 Parametrization

Doing phenomenological studies without specifying a choice of flavor bases is ambiguous. It is like doing SM quark physics with 18 complex Yukawa couplings, instead of the 10 real physical parameters. As far as the SM itself is concerned, the extra 26 real parameters are simply redundant, and attempts to related the full 36 parameters to experimental data will be futile. In SUSY without R-parity, the choice of an optimal parametrization mainly concerns the 4 $\hat{L}_\alpha$ flavors. We use here the single-VEV parametrization[3] (SVP), in which flavor bases are chosen such that : 1/ among the $\hat{L}_\alpha$'s, only $\hat{L}_0$, bears a VEV, *i.e.* $\langle \hat{L}_i \rangle \equiv 0$; 2/ $h^e_{jk} (\equiv \lambda_{0jk}) = \frac{\sqrt{2}}{v_0}\,\mathrm{diag}\{m_1, m_2, m_3\}$; 3/ $h^d_{jk} (\equiv \lambda'_{0jk} = -\lambda_{j0k}) = \frac{\sqrt{2}}{v_0}\mathrm{diag}\{m_d, m_s, m_b\}$; 4/ $h^u_{ik} = \frac{v_u}{\sqrt{2}} V^\dagger_{CKM}\,\mathrm{diag}\{m_u, m_c, m_t\}$, where $v_0 \equiv \sqrt{2}\,\langle \hat{L}_0 \rangle$ and $v_u \equiv \sqrt{2}\,\langle \hat{H}_u \rangle$. Thus, the parametrization singles out the $\hat{L}_0$ superfield as the one containing the Higgs. As a result, it gives the complete RPV effects on the tree-level mass matrices of all the states (scalars and fermions) the simplest structure. The latter is a strong technical advantage.

There are, in fact, many subtle issues involved in a consistent formulation of the complete theory. One has to be particularly careful when the perspective of adding the various RPV terms to the MSSM is taken. Such issues will be addressed in a forthcoming review[4], to which the readers are referred.

3 Fermion Sector Phenomenology

The SVP gives quark mass matrices exactly in the SM form. For the masses of the color-singlet fermions, all the RPV effects are parametrized by the μ_i's only. For example, the five charged fermions (3 charged leptons + Higgsino + gaugino), we have

$$\mathcal{M}_C = \begin{pmatrix} M_2 & \frac{g_2 v_0}{\sqrt{2}} & 0 & 0 & 0 \\ \frac{g_2 v_u}{\sqrt{2}} & \mu_0 & \mu_1 & \mu_2 & \mu_3 \\ 0 & 0 & m_1 & 0 & 0 \\ 0 & 0 & 0 & m_2 & 0 \\ 0 & 0 & 0 & 0 & m_3 \end{pmatrix} . \qquad (2)$$

Moreover each μ_i parameter here characterizes directly the RPV effect on the corresponding charged lepton ($\ell_i = e, \mu$, and τ). This, and the corresponding neutrino-neutralino masses and mixings, has been exploited to implement a detailed study of the tree-level RPV phenomenology from the gauge interactions, with interesting results[3].

Neutrino masses and oscillations is no doubt a central aspect of any RPV model. In our opinion, it is particularly important to study the various RPV contributions in a framework that takes no assumption on the other parameters. Our formulation provides such a framework. Interested readers are referred to Refs.[5,6,7,8,9].

4 Interesting Phenomenology from SUSY Breaking Terms

The soft SUSY breaking part of the Lagrangian can be written as

$$\begin{aligned} V_{\rm soft} &= \tilde{Q}^\dagger \tilde{m}_Q^2\, \tilde{Q} + \tilde{U}^\dagger \tilde{m}_U^2\, \tilde{U} + \tilde{D}^\dagger \tilde{m}_D^2\, \tilde{D} + \tilde{L}^\dagger \tilde{m}_L^2 \tilde{L} \\ &+ \tilde{E}^\dagger \tilde{m}_E^2\, \tilde{E} + \tilde{m}_{H_u}^2\, |H_u|^2 + \Big[\frac{M_1}{2} \tilde{B}\tilde{B} + \frac{M_2}{2} \tilde{W}\tilde{W} \\ &+ \frac{M_3}{2} \tilde{g}\tilde{g} + \epsilon_{ab} \Big(B_\alpha\, H_u^a \tilde{L}_\alpha^b + A^U_{ij}\, \tilde{Q}^a_i H^b_u \tilde{U}^C_j \\ &+ A^D_{ij} H^a_d \tilde{Q}^b_i \tilde{D}^C_j + A^E_{ij} H^a_d \tilde{L}^b_i \tilde{E}^C_j + A^{\lambda'}_{ijk} \tilde{L}^a_i \tilde{Q}^b_j \tilde{D}^C_k \\ &+ \frac{1}{2} A^{\lambda}_{ijk} \tilde{L}^a_i \tilde{L}^b_j \tilde{E}^C_k \Big) + \frac{1}{2} A^{\lambda''}_{ijk} \tilde{U}^C_i \tilde{D}^C_j \tilde{D}^C_k + \mathrm{h.c.} \Big] \end{aligned} \qquad (3)$$

where we have separated the R-parity conserving A-terms from the RPV ones (recall $\hat{H}_d \equiv \hat{L}_0$). Note that $\tilde{L}^\dagger \tilde{m}_L^2 \tilde{L}$, unlike the other soft mass terms, is given by a 4×4 matrix. Explicitly, $\tilde{m}^2_{L_{00}}$ corresponds to $\tilde{m}^2_{H_d}$ of the MSSM case while $\tilde{m}^2_{L_{0k}}$'s give RPV mass mixings.

Obtaining the squark and slepton masses is straight forward. The only RPV contribution to the squark masses[7,11] is given by a $-(\mu_i^* \lambda'_{ijk})\frac{v_u}{\sqrt{2}}$ term in the LR mixing part. Note that the term contains flavor-changing ($j \neq k$) parts which, unlike the A-terms ones, cannot be suppressed through a flavor-blind SUSY breaking spectrum. Hence, it has very interesting implications to quark electric dipole moments (EDM's) and related processses such as $b \to s\,\gamma$[10,11,12]. For instance, it contributes to neutron EDM at 1-loop order, through a simple gluino diagram of the d squark. If one naively imposes the constraint for this RPV contribution itself not to exceed the experimental bound on neutron EDM, one gets roughly $\mathrm{Im}(\mu_i^* \lambda'_{i11}) \lesssim 10^{-6}\,\mathrm{GeV}$, a constraint that is interesting even in comparison to the bounds on the corresponding parameters obtainable from asking no neutrino masses to exceed the super-Kamiokande atmospheric oscillation scale[10].

Things in the slepton sector are more complicated. The $1+4+3$ charged scalar masses are given in terms of blocks

$$\begin{aligned}
\widetilde{\mathcal{M}}_{Hu}^2 &= \tilde{m}_{H_u}^2 + \mu_\alpha^* \mu_\alpha + M_Z^2\,\cos 2\beta\left[\frac{1}{2} - \sin^2\theta_W\right] \\
&+ M_Z^2\,\sin^2\beta\,[1-\sin^2\theta_W]\,, \\
\widetilde{\mathcal{M}}_{LL}^2 &= \tilde{m}_L^2 + m_L^\dagger m_L + M_Z^2\,\cos 2\beta\left[-\frac{1}{2} + \sin^2\theta_W\right] \\
&+ \begin{pmatrix} M_Z^2\,\cos^2\beta\,[1-\sin^2\theta_W] & 0_{1\times 3} \\ 0_{3\times 1} & 0_{3\times 3} \end{pmatrix} + (\mu_\alpha^* \mu_\beta)\,, \\
\widetilde{\mathcal{M}}_{RR}^2 &= \tilde{m}_E^2 + m_E m_E^\dagger + M_Z^2\,\cos 2\beta\left[-\sin^2\theta_W\right]\;; \qquad (4)
\end{aligned}$$

and

$$\widetilde{\mathcal{M}}_{LH}^2 = (B_\alpha^*) + \begin{pmatrix} \frac{1}{2}\,M_Z^2\,\sin 2\beta\,[1-\sin^2\theta_W] \\ 0_{3\times 1} \end{pmatrix}\,, \qquad (5)$$

$$\widetilde{\mathcal{M}}_{RH}^2 = -(\mu_i^* \lambda_{i0k})\,\frac{v_0}{\sqrt{2}}\,, \qquad (6)$$

$$(\widetilde{\mathcal{M}}_{RL}^2)^T = \begin{pmatrix} 0 \\ A^E \end{pmatrix}\frac{v_0}{\sqrt{2}} - (\mu_\alpha^* \lambda_{\alpha\beta k})\,\frac{v_u}{\sqrt{2}}\,. \qquad (7)$$

We have to skip the neutral scalar part[7] here. The RPV contributions to the charged, as well as neutral, scalar masses and mixings give rise to new terms in quarks and electron EDM's[11,13], $b \to s\,\gamma$[12], $\mu \to e\,\gamma$[13], and neutrino masses diagrams that have been largely overlooked[6,7]. The last includes diagrams corresponding to a SUSY version of the popular Zee neutrino mass model[14]. Details are to be found in the cited references.

Acknowledgments

The author thanks M. Bisset, K. Cheung, S.K. Kang, Y.-Y. Keum, C. Macesanu, and L.H. Orr, for collaborations on the subject, and colleagues at Academia Sinica for support. He is grateful to the generous hospitality of YITP, Kyoto Univ. and KEK, which helps to make his trip to Osaka possible.

References

1. L.E. Ibáñez and G.G. Ross, Nucl. Phys. **B368**, 3 (1992).
2. M.C. Gonzalez-Garcia, this proceedings.
3. M. Bisset, O.C.W. Kong, C. Macesanu, and L.H. Orr, Phys. Lett. **B430**, 274 (1998); Phys. Rev. **D62**, *035001* (2000).
4. O.C.W. Kong, IPAS-HEP-k008, *manuscript in preparation.*
5. O.C.W. Kong, Mod. Phys. Lett. **A14**, 903 (1999).
6. K. Cheung and O.C.W. Kong, Phys. Rev. **D61**, *113012* (2000).
7. O.C.W. Kong, hep-ph/0004107.
8. S.K. Kang and O.C.W. Kong, IPAS-HEP-k009, *manuscript in preparation.*
9. See also A. Abada and M. Losada, hep-ph/9908352; S. Davidson and M. Losada, JHEP **0005**, 021 (2000).
10. Y.-Y. Keum and O.C.W. Kong, hep-ph/0004110.
11. Y.-Y. Keum and O.C.W. Kong, IPAS-HEP-k006, *manuscript in preparation.*
12. Y.-Y. Keum and O.C.W. Kong, *work in progress.*
13. K. Cheung and O.C.W. Kong, IPAS-HEP-k007, *manuscript in preparation*; K. Cheung, Y.-Y. Keum, and O.C.W. Kong, *work in progress.* See also K. Choi, E.J. Chun, and K. Hwang, hep-ph/0004101; hep-ph/0005262.
14. A. Zee, Phys. Lett. **93B**, 389 (1980).

NEUTRINO MIXING FROM THE CKM MATRIX IN SUSY $SO(10) \times U(2)_F$

MU-CHUN CHEN AND K.T. MAHANTHAPPA

Department of Physics, University of Colorado, Boulder, CO 80309-0390, USA
E-mail: mu-chun.chen@colorado.edu, ktm@verb.colorado.edu

We construct a realistic model based on SUSY $SO(10)$ with $U(2)$ flavor symmetry. A set of symmetric mass textures give rise to very good predictions; 15 masses and 6 mixing angles are predicted by 11 parameters. Both the vacuum oscillation and LOW solutions are favored for the solar neutrino problem.

The flavor problem with hierarchical fermion masses and mixing has attracted a great deal of attention especially since the advent of the atmospheric neutrino oscillation data from Super-Kamiokande indicating non-zero neutrino masses. The non-zero neutrino masses give support to the idea of grand unification based on $SO(10)$ in which all the 16 fermions (including ν_R) can be accommodated in one single spinor representation. Furthermore, it provides a framework in which seesaw mechanism arises naturally. Naively one expects, for symmetric mass textures, six texture zeros in the quark sector. But it has been observed by Ramond, Roberts and Ross[1] that the highest number of texture zeros has to be five, and using phenomenological analyses, they were able to arrive at five sets of up- and down-quark mass matrices with five texture zeros. Our analysis with recent experimental data and using CP conserving real symmetric matrices indicates that only one set remains viable. The aim of this talk, based on Ref.[2], is to present a realistic model based on $SO(10)$ combined with $U(2)$ as the flavor group, utilizing this set of symmetric mass textures for charged fermions. We first discuss the viable phenomenology of mass textures followed by the model which accounts for it, and then the implications of the model for neutrino mixing are presented.

The set of up- and down-quark mass matrix combination is given by, at the GUT scale,

$$M_u = \begin{pmatrix} 0 & 0 & a \\ 0 & b & c \\ a & c & 1 \end{pmatrix} d, \quad M_d = \begin{pmatrix} 0 & e & 0 \\ e & f & 0 \\ 0 & 0 & 1 \end{pmatrix} h \quad (1)$$

with $a \ll b \ll c \ll 1$, and $e \ll f \ll 1$. Symmetric mass textures arise naturally if $SO(10)$ breaks down to the SM group via the left-right symmetric breaking chain $SU(4) \times SU(2)_L \times SU(2)_R$. The $SO(10)$ symmetry relates the up-quark to the Dirac neutrino mass matrices, and the down-quark to the charged lepton mass matrices. To achieve the Georgi-Jarlskog relations, $m_d \simeq 3m_e$, $m_s \simeq \frac{1}{3}m_\mu$, $m_b \simeq m_\tau$, a factor of -3 is needed in the $(2,2)$ entry of the charged lepton mass matrix,

$$M_e = \begin{pmatrix} 0 & e & 0 \\ e & -3f & 0 \\ 0 & 0 & 1 \end{pmatrix} h \quad (2)$$

This factor of -3 can be generated by the $SO(10)$ Clebsch-Gordon coefficients through the couplings to the $\overline{126}$ dimesional representations of Higgses. In order to explain the smallness of the neutrino masses, we will adopt the type I seesaw mechanism which requires both Dirac and right-handed Majorana mass matrices to be present in the Lagrangian. The Dirac neutrino mass matrix is identical to the one of the up-quark in the framework of $SO(10)$

$$M_{\nu_{LR}} = \begin{pmatrix} 0 & 0 & a \\ 0 & b & c \\ a & c & 1 \end{pmatrix} d \quad (3)$$

The right-handed neutrino sector is an unknown sector. We find[2] that if the right-handed neutrino mass matrix has the same texture as that of the Dirac neutrino mass matrix,

$$M_{\nu_{RR}} = \begin{pmatrix} 0 & 0 & \delta_1 \\ 0 & \delta_2 & \delta_3 \\ \delta_1 & \delta_3 & 1 \end{pmatrix} M_R \qquad (4)$$

and if the elements δ_i are of the right orders of magnitudes, determined by $\delta_i = f_i(a,b,c,t)$, then the resulting effective neutrino mass matrix will take the following form

$$M_{\nu_{LL}} = M_{\nu_{LR}}^T M_{\nu_{RR}}^{-1} M_{\nu_{LR}} = \begin{pmatrix} 0 & 0 & t \\ 0 & 1 & 1 \\ t & 1 & 1 \end{pmatrix} \Lambda \qquad (5)$$

Note that $M_{\nu_{LL}}$ has the same texture as that of $M_{\nu LR}$ and $M_{\nu_{RR}}$. That is to say, the seesaw mechanism is form invariant. A generic feature of mass matricies of the type given in Eq.(5) is that they give rise to bi-maximal mixing pattern. After diagonalizing this mass matrix, one can see immediately that the squared mass difference between $m_{\nu_1}^2$ and $m_{\nu_2}^2$ is of the order of $O(t^3)$, while the squared mass difference between $m_{\nu_2}^2$ and $m_{\nu_3}^2$ is of the order of $O(1)$, in units of Λ. For $t \ll 1$, the phenomenologically favored relation $\Delta m_{atm}^2 \gg \Delta m_{\odot}^2$ is thus obtained.

The $U(2)$ flavor symmetry[3] is implemented *á la* the Froggatt-Nielsen mechanism. It simply says that the heaviest matter fields acquire their masses through tree level interactions with the Higgs fields while masses of lighter matter fields are produced by higher dimensional interactions involving, in addition to the regular Higgs fields, exotic vector-like pairs of matter fields and the so-called flavons (flavor Higgs fields). After integrating out superheavy ($\approx M$) vector-like matter fields, the mass terms of the light matter fields get suppressed by a factor of $\frac{<\theta>}{M}$, where $<\theta>$ is the VEVs of the flavons and M is the UV-cutoff of the effective theory above which the flavor symmetry is exact. We assume that the flavor scale is higher than the GUT scale. The heaviness of the top quark and to suppress the SUSY FCNC together suggest that the third family of matter fields transform as a singlet and the lighter two families of matter fields transform as a doublet under $U(2)$. In the flavor symmetric limit, only the third family has non-vanishing Yukawa couplings. $U(2)$ breaks down in two steps: $U(2) \xrightarrow{\epsilon M} U(1) \xrightarrow{\epsilon' M}$ *nothing*, where $\epsilon' \ll \epsilon \ll 1$ and M is the flavor scale. These small parameters ϵ and ϵ' are the ratios of the vacuum expectation values of the flavon fields to the flavor scale. Since $\psi_3\psi_3 \sim 1_S$, $\psi_3\psi_a \sim 2$, $\psi_a\psi_b \sim 2 \otimes 2 = 1_A \oplus 3$, the only relevant flavon fields are in the $A^{ab} \sim 1_A$, $\phi^a \sim 2$, and $S^{ab} \sim 3$ dimensional representations of $U(2)$. Because we are confining ourselves to symmetric mass textures, we use only ϕ^a and S^{ab}. In the chosen basis, the VEVs various flavon fields could acquire are given by

$$\frac{\langle\phi\rangle}{M} \sim O\begin{pmatrix} \epsilon' \\ \epsilon \end{pmatrix}, \quad \frac{\langle S^{ab}\rangle}{M} \sim O\begin{pmatrix} \epsilon' & \epsilon' \\ \epsilon' & \epsilon \end{pmatrix} \qquad (6)$$

Putting everything together, a symmetric mass matrix would have the following built-in hierarchy given by

$$\begin{pmatrix} \epsilon' & \epsilon' & \epsilon' \\ \epsilon' & \epsilon & \epsilon \\ \epsilon' & \epsilon & 1 \end{pmatrix} \qquad (7)$$

Combining $SO(10)$ with $U(2)$, the most general superpotential which repects the symmetry one could write down is given schematically by

$$W = H(\psi_3\psi_3 + \psi_3\frac{\phi^a}{M}\psi_a + \psi_a\frac{S^{ab}}{M}\psi_b) \qquad (8)$$

The superpotential of our model which generates fermion masses is given by

$$W = W_{D(irac)} + W_{M(ajorana)} \qquad (9)$$

$$W_D = \psi_3\psi_3 T_1 + \frac{1}{M}\psi_3\psi_a\left(T_2\phi_{(1)} + T_3\phi_{(2)}\right)$$

$$+\frac{1}{M}\psi_a\psi_b\left(T_4+\overline{C}\right)S_{(2)}+\frac{1}{M}\psi_a\psi_b T_5 S_{(1)}$$

$$W_M=\psi_3\psi_3\overline{C}_1+\frac{1}{M}\psi_3\psi_a\Phi\overline{C}_2+\frac{1}{M}\psi_a\psi_b\Sigma\overline{C}_2$$

where T_i's and $\overline{C}_i$'s are the 10 and $\overline{126}$ dimensional Higgs representations of $SO(10)$ respectively, and Φ and Σ are the doublet and triplet of $U(2)$, respectively. Detailed quantum number assignments and the VEVs acquired by various scalar fields are given in Ref.[2]. This superpotential gives rise to the mass textures given in Eq.(1)-(4). Various entries of these matrices are given in terms of ϵ, ϵ', and ratios of Higgs VEVs. Note that, since we use $\overline{126}$ dimensional representaions of Higgses to generate the heavy Majorana neutrino mass terms, R-parity symmetry is preserved at all energies.

With values of up-quark masses, charged lepton masses and the Cabbibo angle, the input parameters at the GUT scale are determined[2]. The charged fermion mass predictions of our model at M_Z which are summerized in Table[1] including 2-loop RGE effects are in good agreements with the experimental values[2]. The CKM matrix is predicted to be

$$\begin{pmatrix} 0.975 & 0.222 & 0.00354 \\ 0.222 & 0.975 & 0.0367 \\ 0.00474 & 0.0368 & 0.999 \end{pmatrix} \qquad (10)$$

In the neutrino sector, the VO solution to the solar neutrino problem is obtained with $(\delta_1,\delta_2,\delta_3,M_R)=(0.00116, 3.32\times 10^{-5}, 0.0152, 1.32\times 10^{14} GeV)$. The atmospheric and solar squared mass differences are predicted to be $\Delta m^2_{23}=3.11\times 10^{-3}eV^2$ and $\Delta m^2_{12}=2.87\times 10^{-10}eV^2$; the mixing angles are given by $\sin^2 2\theta_{atm}=0.999$, and $\sin^2 2\theta_{\odot}=0.991$. $|U_{e\nu_3}|$ is predicted to be 0.0527 which is below the upper bound 0.16 by the CHOOZ experiment. We can also have the LOW solution with $(\delta_1,\delta_2,\delta_3,M_R)=(0.00115, 2.35\times 10^{-4}, 0.0168, 1.62\times 10^{13} GeV)$. In this case, $\Delta m^2_{23}=3.97\times 10^{-3}eV^2$, and $\Delta m^2_{12}=1.30\times 10^{-7}eV^2$. The mixing angles are

	data at M_z	predictions
m_u	$2.33^{+0.42}_{-0.45} MeV$	$1.917 MeV$
m_c	$677^{+56}_{-61} MeV$	$738.7 MeV$
m_t	$181^{+}_{-}13 GeV$	$184.3 MeV$
$\frac{m_d}{m_s}$	$17 \sim 25$	22.5
m_s	$93.4^{+11.8}_{-13.0} MeV$	$83.15 GeV$
m_b	$3.00^{+}_{-}0.11 GeV$	$3.0141 GeV$
m_e	$0.486847 MeV$	$0.486 MeV$
m_μ	$102.75 MeV$	$102.8 MeV$
m_τ	$1.7467 GeV$	$1.744 GeV$

Table 1. Predictions and values extrapolated from experimental data at M_Z for charged fermion masses.

given by $\sin^2 2\theta_{atm}=0.999$, and $\sin^2 2\theta_{\odot}=0.990$. $|U_{e\nu_3}|$ is predicted to be 0.0743. It is possible to have the LAMSW solution with $(\delta_1,\delta_2,\delta_3,M_R)=(0.00108, 9.87\times 10^{-5}, 0.0224, 2.42\times 10^{12} GeV)$. These parameters predict $\Delta m^2_{23}=9.85\times 10^{-3}eV^2$, and $\Delta m^2_{12}=2.75\times 10^{-5}eV^2$. The mixing angles are $\sin^2 2\theta_{atm}=1.00$, and $\sin^2 2\theta_{\odot}=0.985$. However, $|U_{e\nu_3}|$ is predicted to be 0.158, right at the experimental upper bound. We note that a $|U_{e\nu_3}|$ value of less than 0.158 would lead to $\Delta m^2_{23}>10^{-2}eV^2$ leading to the elimination of the LAMSW solution in our model. This is a characteristic of the LAMSW solution with $\Delta m^2_{12}\geq 10^{-5}eV^2$.

Other aspects of our model including the proton stability and symmetry breaking are under investigation.

Acknowledgments

This work was supported in part by the US DoE Grant No. DE FG03-05ER40894.

References

1. P. Ramond, R. Roberts and G. Ross, *Nucl. Phys.* B **406**, 19 (1993).
2. M.C. Chen and K.T. Mahanthappa, hep-ph/0005292, to appear in *Phys. Rev.* D; this paper contains details and other relevant references.
3. R. Barbieri, L.J. Hall, S. Raby and A. Romanino, *Nucl. Phys.* B **493**, 3 (1997).

WEINBERG MODEL OF *CP* VIOLATION

B. H. J. MCKELLAR
School of Physics, University of Melbourne, Parkville, Vic Australia 3052
E-mail: b.mckellar@physics.unimelb.edu.au

XIAO-GANG HE
Department of Physics, National Taiwan University, Taipei, 10617, Taiwan.
E-mail: hexg@phys.ntu.edu.tw

DARWIN CHANG
Department of Physics, National Tsinhua University, Hsinchu, 300, Taiwan
E-mail: chang@phys.nthu.edu.tw

Using older data, there were many analyses which declared the death of the Weinberg model of spontaneous CP violation. These conclusions were invalid because they were based on optimistic estimates of the accuracy of the calculations. Using more recent data, only now can the model be ruled out even with realistic error estimates, because, in the Weinberg model of CP violation, the recently measured value of $\mathrm{Re}(\epsilon'/\epsilon) = (1.92 \pm 0.25) \times 10^{-3}$ is incompatible with the branching ratio $B(b \to s\gamma) = (3.15 \pm 0.54) \times 10^{-4}$.

1 Introduction

Understanding the origin of CP violation remains a challenge in particle physics. While the Standard Model (SM) of CP violation based on the Kobayashi-Maskawa (KM) mechanism is so far consistent with observations of CP violation in kaon system, there are intriguing hints, such as the baryon asymmetry of the universe, that other sources of CP violation may exist.

An attractive model which has the potential to explain the existing data, and generate the baryon asymmetry, and in addition allows CP symmetry to be broken spontaneously and therefore give an interesting explanation of the origin for CP violation, is the Weinberg model of CP violation with three Higgs doublets[1]. This model has the added virtue of preserving, in a natural way, neutral flavour conservation at tree level.

In this model there are 2 charged and 3 neutral Higgs particles. For our purpose the charged Higgs are more important, and the interaction Lagrangian of the charged Higgs particles with the quarks is

$$L = 2^{3/4} G_F^{1/2} \bar{U} \left[V_{KM} M_D (\alpha_1 H_1^+ + \alpha_2 H^+) R + M_U V_{KM} (\beta_1 H_1^+ + \beta_2 H_2^+) L \right] D + \text{h.c.} , \quad (1)$$

where $R(L) = (1 \pm \gamma_5)/2$, and $M_{U,D}$ are the diagonal up and down quark mass matrices. The parameters α_i and β_i, introduce CP violation through $\mathrm{Im}(\alpha_1 \beta_1^*) = -\mathrm{Im}(\alpha_2 \beta_2^*)$. V_{KM} is a real mixing matrix.

We review the current experimental and theoretical status of each of the observables individually, and then discuss the constraints placed on the Weinberg model by the total ensemble of data. We find[2] that it is not possible to choose the parameters to simultaneously satisfy the constraints imposed by the recently measured value of[3] $\mathrm{Re}(\epsilon'/\epsilon) = (1.92 \pm 0.25) \times 10^{-3}$ and the branching ratio[4] $B(b \to s\gamma) = (3.15 \pm 0.54) \times 10^{-4}$.

2 The Observables

2.1 $\sin 2\beta$ in the Weinberg model

In the Weinberg model of spontaneous CP violation, CP violating contributions to the decay amplitudes and $B^0 - \bar{B}^0$ mixing are proportional to $\mathrm{Im}(\alpha_1\beta_1^*)$ and are suppressed, compared to SM contributions by additional factors of $m_c m_b/m_H^2$ for for the amplitudes, and m_b/m_H and m_b^2/m_H^2 for the mixing. These suppression factors lead to small CP violating phases and result in a very small value[5], $|\sin 2\beta| < 0.05$.

The ALEPH, OPAL and CDF data reported in 1999 gave[6] $\sin 2\beta = 0.91 \pm 0.35$, in conflict with the above limit at the 2σ level. However at this conference, Belle and BaBar reported preliminary results[7], which when averaged with the above, give $\sin 2\beta = 0.49 \pm 0.45$, consistent with the above limit at the 75% level.

2.2 ϵ'/ϵ in the Weinberg model

The dominant contribution to ϵ'/ϵ in the Weinberg model is from the flavor changing gluonic dipole interaction given by

$$H(sdg) = ig_s \tilde{f} m_s \bar{s}\sigma_{\mu\nu}\lambda^a G_a^{\mu\nu}(1-\gamma_5), \quad (2)$$

where the strength of this interaction, $\tilde{f}$, is proportional to $\mathrm{Im}(\alpha_1\beta_1^*)$

The contribution to ϵ'/ϵ is dominated by the lightest charged Higgs, so here we neglect the contribution from the heavier Higgs.

Using result for the matrix element from Ref.[8], and the isospin breaking correction factor of Ref.[9], we obtain

$$\mathrm{Re}\left(\frac{\epsilon'}{\epsilon}\right) = 1.7 \times 10^7 (\mathrm{GeV}^2)\tilde{f}B_0. \quad (3)$$

The parameter B_0 represents the uncertainty in the evaluation of the matrix elements. To produce the recently observed value for ϵ'/ϵ within 3σ, $\tilde{f}B_0$ has to be in the range $(0.69 \sim 1.57) \times 10^{-10}$ (GeV^{-2}).

We conservatively allow B_0 to vary from 0.5 to 2 to allow for possible uncertainties[10]. The most conservative range for $\tilde{f}$ is then $(0.35 \sim 3.1) \times 10^{-10}$ GeV^{-2}.

2.3 ϵ in the Weinberg model

A successful model for CP violation must to be able to produce the experimental value for ϵ. The dominant contribution actually comes from long distance effects, which in turn are generated by CP violation due to the gluonic dipole interaction. Following Ref.[11] we assume the contribution to ϵ is from π, η, η' poles. Many parameters enter the result which are not well known, including the strong coupling constant g_s at the Kaon scale, the strange quark mass m_s, the $\eta - \eta'$ mixing angle, parameters describing the U(3) and SU(3) breaking effects, and the hadronic matrix element of the CP violating interaction.

There are solutions for ϵ with $\tilde{f}$ in the ranges $(0.85 \sim 2.56) \times 10^{-10}$ GeV^{-2} and also with $\tilde{f}$ near -1.7×10^{-11} GeV^{-2}. The allowed range associated with ϵ has a large overlap with that determined from ϵ'/ϵ.

2.4 d_n in the Weinberg model

The experimental bound on the neutron EDM, d_n has been used to provide restrictions on the model, and has been claimed to rule it out[12]. The neutron EDM can be generated by the exchange of neutral and charged Higgs particles[13]. It is not impossible that these contributions may cancel each other and result in a very small neutron EDM. Here we will not entertain this possibility, but instead single out the potentially large valence quark contributions and require that each of them satisfies the experimental constraints.

The charged Higgs boson contribution to the neutron EDM is strongly restricted. The dominant term comes from the down quark EDM, and is proportional to the down quark mass and $\tilde{f}$. Using $\tilde{f} = (0.35 \sim 3.1) \times 10^{-10}$ GeV^{-2} determined from ϵ'/ϵ, we estimate the neutron EDM as $(0.25 \sim 3.5) \times 10^{-24}(m_d/300\mathrm{MeV})e\mathrm{cm}$.

2.5 $b \to s\gamma$ in the Weinberg model

The CP conserving process $b \to s\gamma$ can place constraints on the CP violating parameters of the model[5], because the CP violating amplitudes contribute to the total rate.

There are both SM and Weinberg model contributions, and there is a region in parameter space where the CP conserving contributions of the SM and of charged Higgs mutually cancel. The CP violating amplitudes contribute significantly. The branching ratio increases with Higgs mass for fixed $\tilde{f}$.

The experimental branching ratio[4], $B(b \to s\gamma) = (3.15 \pm 0.54) \times 10^{-4}$, has been confirmed by Belle at this conference. The 95% c.l. upper bound 4.5×10^{-4} for $b \to s\gamma$, requires $|\tilde{f}| \leq 1.7 \times 10^{-11}$ GeV^{-2} for the charged Higgs mass greater than 70 GeV. For larger m_H, or a smaller branching ratio, tighter constraints are placed on $|\tilde{f}|$.

3 Discussion

No constraint is placed on the model by the present results for $\sin 2\beta$ from ALEPH, OPAL, CDF, BaBar and Belle.

If the constituent mass $m_d \sim 300$ MeV is used, the model may be in trouble. If the current quark mass $m_d \sim 10$ MeV is used, the predicted value of d_n satisfies the experimental limit as long as $\tilde{f} < 2.56 \times 10^{-10}$ GeV^{-2}. The values of $\tilde{f}$ from ϵ'/ϵ, ϵ and d_n then have a region of consistency, as do the values constrained by $B(b \to s\gamma)$, ϵ and d_n.

However there is a definite conflict between the limits on $\tilde{f}$ from ϵ'/ϵ and $B(b \to s\gamma)$. The latter requires $|\tilde{f}| \leq 1.7 \times 10^{-11}$ GeV^{-2}, and the former requires 3.5×10^{-11}GeV$^{-2} \leq \tilde{f} \leq 31 \times 10^{-11}$GeV^{-2}. As we have been careful to make very conservative estimates of the allowed range of $\tilde{f}$ (in the hope of finding that there was still a small region of parameter space in which the model is consistent with the data), the gap between these allowed regions for $\tilde{f}$ is unbridgeable. Thus we conclude that the Weinberg model is conservatively but confidently ruled out by the recent data for ϵ'/ϵ and for $B(b \to s\gamma)$.

Acknowledgments

This work was supported in part by grants NSC 89-2112-M-002-016, and NSC 89-2112-M-007-010 of the R.O.C. and by the Australian Research Council.

References

1. S. Weinberg, Phys. Rev. Lett. **37**, 657(1976).
2. D. Chang, X.-G. He and B. McKellar, e-print hep-ph/9908357.
3. P. Debu, in these proceedings.
4. S. Glenn et al., CLEO Collaboration, CLEO CONF 98-17; M. Nakao, Belle Collaboration, these proceedings
5. Y. Grossmann and Y. Nir, Phys. Lett. **B313**, 126(1993).
6. R. Forty, Aleph collaboration, Aleph 99-099/CONF 99-054.
7. H. Aihara (Belle Collaboration); D. Hitlin (BaBar Collaboration), in these proceedings.
8. S. Bertolini, M. Fabbrichesi and E. Gabrielli, Phys. Lett. **B327**,136(1994).
9. J. Donoghue et al., Phys. Lett. **B179**, 361(1986); A. Buras and J. Gerard, Phys. Lett. **B192**, 156(1987).
10. X.-G. He and G. Valencia, Phys. Rev. **D61**, 117501(2000).
11. J. Donoghue and B. Holstein, Phys. Rev. **D32**, 1152(1985); H.-Y. Cheng, Phys. Rev. **D34**, 1397(1986).
12. I.I. Bigi and A. Sanda, Phys. Rev. Lett. **58**, 1604(1987); I. B. Khriplovich, Phys. Lett. **B382**, 145(1996).
13. X.-G. He, B. McKellar and S. Pakvasa, Int. J. Mod. Phys. **A4**, 5011(1989).

SPARTICLE MASSES AND PHENOMENOLOGY BEYOND MSUGRA

XERXES TATA
Dept. of Physics and Astronomy, University of Hawaii, Honolulu, HI 96822, U.S.A.

We rapidly survey proposals for high scale physics that do not yield universality of GUT scale soft-SUSY breaking sparticle masses which is the hallmark of the mSUGRA framework. We then focus on scalar mass non-universality from $SO(10)$ D-terms and on models with an inverted mass hierarchy.

The proliferation of scalars in supersymmetric models leads to new problems not present in the Standard Model. Even if renormalizable, gauge invariant baryon or lepton number violating operators are forbidden by a discrete symmetry (usually R-parity), there are new (generically large) sources for flavour changing neutral currents. Within the popular mSUGRA and gauge-mediated SUSY breaking models, these are suppressed because, neglecting Yukawa couplings, scalars (at least those with identical gauge quantum numbers) are essentially degenerate. Flavour problems are also ameliorated if fermion and sfermion mass matrices are aligned, or simply if sfermions are heavy.

Motivated by the fact that universality of soft SUSY breaking parameters is not a generic property of supergravity models, and further, (unless new symmetries are assumed) radiative corrections destroy any assumed universality at tree level, we surveyed[1] various proposals for high scale physics which lead to non-universality of scalar and/or gaugino masses at the GUT scale. These include:

1. Models with universality at M_{Planck}: Substantial non-universality can be induced at M_{GUT} via renormalization group evolution (RGE) even though there is no large log because the small coupling is compensated by large group theory coefficents.

2. Models with non-universal gaugino masses: This can happen even in a GUT if a field that breaks supersymmetry is not a GUT singlet.

3. Anomaly-mediated SUSY breaking (AMSB): Sparticle Masses are determined by the β functions of the low energy theory. Sparticles with the same gauge quantum numbers have the same mass, except for Yukawa coupling effects.

4. Missing partner models with hypercolour interactions: These provide a solution to the doublet-triplet splitting problem. Since MSSM gauginos do not all come from the same factor of the underlying group, their masses are different.

5. String-based models which yield non-universal scalar masses except when SUSY breaking is dominated by the *vev* of the dilaton field.

6. Models where scalar masses receive additional (non-universal) weak scale contributions from D-terms of a gauge group with rank ≥ 5 even if the additional symmetry is broken at a much higher scale.

7. Inverted Mass Hierarchy (IMH) models, where the third generation of sfermions is at the sub-TeV scale, while the first two generations are much heavier. Since the latter couple only very weakly to the Higgs sector, they contribute to the Higgs self energy only at 2 loops (neglecting their Yukawa couplings), so that they do not destabilize the electroweak scale even if they are heavy.

The ISAJET event generator has recently been upgraded[2], and allows the user to input a variety of non-universal masses via $NUSUGi$ options. To facilitate the simulation of string models, or models with a right-handed neutrino, the $SSBCSC$ option allows one to modify (from M_{GUT}) the scale below which the MSSM is assumed to be the effec-

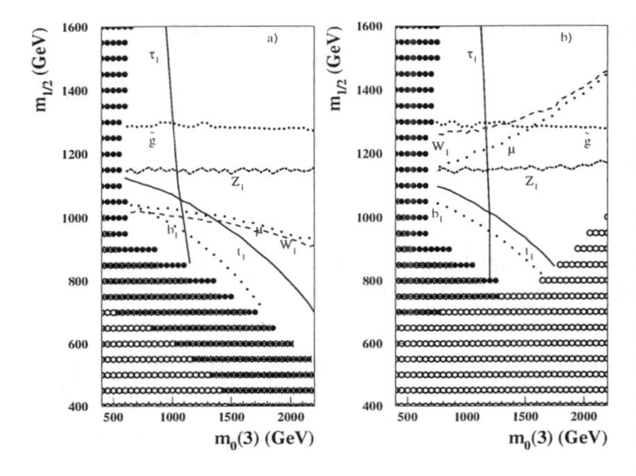

Figure 1. The $m_0(3) - m_{1/2}$ plane of the GUT scale IMH model showing viable regions. Contours are for 1 TeV masses except $\tilde{g}$ (3 TeV) and $\widetilde{Z}_1$ (0.5 TeV). Regions with open circles have no REWSB, those with stars have tachyonic masses, and those with solid circles have a charged or coloured LSP and are excluded. We take $m_0(1,2) = 10$ TeV and illustrate our results for $\tan\beta = 3$ and 35.

tive theory. The minimal AMSB model has also been incorporated.

1 SUSY $SO(10)$ Phenomenology

The interpretation of the super-Kamiokande atmospheric neutrino data as neutrino oscillations provides strong motivation for reconsidering $SO(10)$ models. In the minimal model, there is only one Yukawa coupling (per generation), and the MSSM parameter $\tan\beta$ is large (~ 50). The mSUGRA $SO(10)$ model appears to run into difficulties with radiative electroweak symmetry breaking (REWSB) if m_t is fixed at 175 GeV. If $SO(10)$ D-term contributions are included, $m_{H_u}^2 < m_{H_d}^2$ facilitating REWSB and leading to a calculable model[3] with gauge and Yukawa coupling unification (with $\mu < 0$ yielding better Yukawa unification).

We found[3] the following: (1) the $\tilde{b}_1$ is almost always the lightest of the squarks and may even be accessible at Tevatron upgrades; (2) the neutralino relic density is in the allowed range ($\Omega_{\widetilde{Z}_1} h^2 \sim 0.1 - 0.4$, depending on $\Omega_\Lambda h^2$) over a wide range of model parameters: furthermore, the universe is never too young; (3) direct neutralino detection rates in ^{73}Ge detectors are between 0.01-1 event/kg day over much of the interesting range of parameters; (4) for $\mu < 0$, the rate for $b \to s\gamma$ is generally outside the experimental range, but there is a small region of parameter space (with small $A_t(weak)$ and $m_{\tilde{t}_1} \sim 1$ TeV) in agreement with data; (5) sparticles are generally heavy, and the LHC is required for a definitive exploration of the model. For some ranges of parameters, the $4b$ signal from $b\bar{b}A$ and $b\bar{b}H$ production, the trilepton signal from $\widetilde{W}_1\widetilde{Z}_2$ production or the signal from sbottom pair production might be observable at Tevatron upgrades.

2 Inverted Mass Hierarchy

An IMH can potentially resolve the SUSY flavour problem. If it is present at the GUT scale, negative two loop contributions (involving loops of heavy first and second generation sfermions) to the third generation squared masses can overcome[4] positive one loop gaugino contributions, resulting in tachyonic third generation masses at the weak scale.

Fig. 1 illustrates a snapshot of our examination[5] of such a scenario. Here $m_0(3)$ is the (common) third generation mass parameter at the GUT scale. We see that there is a region of parameter space with sub-TeV third generation sfermions and other sfermions at the 10 TeV scale.[a]

In such scenarios, it is possible[5] that usual strategies for SUSY searches yield no observable signals even at the LHC. If $\tan\beta$ is small, these models possibly also suffer from problems with too large a relic density from the neutralino LSP (except in regions where $m_{\tilde{\tau}_1} \leq 500$ GeV, or where co-annihilation is

[a] Some authors impose additional constraints on the parameter space from fine-tuning considerations.

important). While it appears to be possible to obtain scenarios on the edge of naturalness with sub-TeV third generation masses and the masses of the first two generations of 5-15 TeV this, by itself, is not enough to satisfy constraints from kaon physics.

It is reasonable to ask why there is an IMH in the first place. In a series of pioneering papers, Bagger et al.[6] attempted to generate such a hierarchy dynamically. Within the minimal $SO(10)$ framework with boundary conditions,

$$4m_{16}^2 = 2m_{10}^2 = A_0^2 \sim [\mathcal{O}(10)\ \mathrm{TeV}]^2$$

they showed that RGE (due to third generation Yukawa interactions) causes the corresponding scalar masses to be strongly suppressed provided $m_{1/2}$ is in the sub-TeV range, resulting in a large IMH: quantitatively, values of $S \equiv \langle m(1,2)^2\rangle/\langle m(3)^2\rangle$, where $\langle m(i)^2\rangle$ is the mean squared mass of the ith generation, in the range 100-800 were claimed, depending on M_{ν_R} and the unified Yukawa coupling f_{GUT}. This pretty picture is marred by the fact they did not obtain REWSB, and so were unable to compute sparticle or even quark and lepton masses. We showed[7] that by including D-terms REWSB can be obtained, and all particle masses computed with the introduction of just one additional parameter. However, we were unable to obtain $S \geq 7$, in sharp contrast to earlier studies [6].

Subsequent examination of this framework showed that variation of M_{ν_R} and other parameters still does not allow $S \geq 10$. We traced the difference to the fact that the large values[6] of S were obtained for f_{GUT} = 1-2, which yields m_t = 190-215 GeV, well above the experimental value of 174±5 GeV. In contrast, when we fix m_t = 175 GeV, our value of f_{GUT} is typically ~ 0.5-0.7. By allowing values of $f_{GUT} \sim 1.2$, we were able to obtain S =30-35 but with incorrect t, b and τ masses. It seems that we are unable to obtain $S \geq 10$ largely because of our smaller value of Yukawa coupling,[b] but also due to deviations from the "ideal" boundary conditions caused by D-terms, and the inclusion of sub-TeV terms in the evolution.

We are currently examining the relic density, $b \to s\gamma$ processes and the collider signals within this framework. For a sample slice of parameter space with large S, we find that the LHC should be able to discover sparticles via the usual searches in the multijet plus multi-lepton + $\not{E}_T$ channels. The LHC reach, which is enhanced by requiring tagged b jets in the data sample, corresponds to $m_{\tilde{g}} \sim 1600$ GeV for $m_{\tilde{u}} = 3300$ GeV. Experiments at the Tevatron are unlikely to be able to see sparticle signals, except possibly a light sbottom. For the largest values of S, we found that $m_{1/2} \sim 0.2 m_0(3)$, so that charginos might be too heavy to discover at a 500 GeV e^+e^- collider. It would be interesting to see if gluino loops can contribute significantly to the $b \to s\gamma$ decay rate because squarks are highly non-degenerate.

I am grateful to H. Baer, C. Balázs, M. Brhlik, M. Díaz, J. Ferrandis, P. Mercadante, P. Quintana and Y. Wang for collaborations. This research was supported by the US Dept. of Energy and by NSF.

References

1. H. Baer et al. *JHEP* **04**, 016 (2000).
2. F. Paige et al. hep-ph/0001086.
3. H. Baer et al. *Phys. Rev.* **D61**, 111701 (2000) and hep-ph/0005027.
4. N. Arkani-Hamed and H. Murayama, *Phys. Rev.* **D56**, R6733 (1997); K. Agashe and M. Grässer, *Phys. Rev.* **D59**, 015007 (1999).
5. H. Baer et al. hep-ph/0008061.
6. J. Bagger et al. *Phys. Lett.* **B473**, 264 (2000) and references therein.
7. H. Baer et al. *Phys. Lett.* **B475**, 289 (2000).

[b] A simple analytic approximation[6] suggests that the S would depend *exponentially* on f_{GUT}^2.

ONE-LOOP CONTRIBUTIONS OF THE SUPER-PARTNER PARTICLES TO $E^-E^+ \to W^-W^+$ IN THE MSSM

KAORU HAGIWARA
Theory Group, KEK, Tsukuba, Ibaraki 305-0801, Japan
E-mail: kaoru.hagiwara@kek.jp

SHINYA KANEMURA
Institut für Theoretishce Physik der Universität Karlsruhe, D-76128 Karlsruhe, Germany
E-mail: kanemu@physik.uni-karlsruhe.de

YOSHIAKI UMEDA
II Institut für Theoretishce Physik der Universität Hamburg, D-22761 Hamburg, Germany
E-mail: umeda@mail.desy.de

One-loop contributions of super-partner particles to W-pair production at e^+e^- collision are discussed in the MSSM. To obtain trustworthy results we test our calculation using three methods: (1) sum rules among form factors which result from the BRS invariance, (2) the decoupling theorem, (3) the high-energy stability. We examine the corrections taking into account constraints from the direct search experiments and the precision data. The results for the sfermion contributions are presented.

1 Introduction

We discuss the one-loop super-partner particle contributions to $e^-e^+ \to W^-W^+$ in the MSSM. The SM particles have their partners, such as sfermions and inos. We here concentrate on the sfermion one-loop effects[1]. The sfermions include squarks and sleptons, whose mass matrices are expressed by

$$M_{\tilde{f}}^2 = \begin{bmatrix} m_{Q,L}^2 + m_Z^2 c_{2\beta}(T_{f_L}^3 - \hat{s}^2 Q_f) + m_f^2 & -m_f A_f^{\text{eff}} \\ -m_f A_f^{\text{eff}*} & m_{U,D,E}^2 + m_Z c_{2\beta}\hat{s}^2 Q_f + m_f^2 \end{bmatrix}.$$

The off-diagonal elements $A_{D,E}^{\text{eff}} = A_{D,E} + \mu \tan\beta$ and $A_U^{\text{eff}} = A_U + \mu\cot\beta$ are multiplied by the fermion mass, so that the mixing are important for stops.

2 Calculation

Helicity amplitudes for $e^-(k,\tau)e^+(\overline{k},\overline{\tau}) \to W^-(p,\lambda)W^+(\overline{p},\overline{\lambda})$, where $k,\overline{k},p,\overline{p}$ are momenta, $\tau,\overline{\tau}(=-\tau),\lambda,\overline{\lambda}$ are helicities, may be expressed by using 16 basis tensors as

$$\mathcal{M}_\tau^{\lambda\overline{\lambda}} = \sum_{i=1}^{16} F_{i,\tau}(s,t)\, j_\mu\, T_i^{\mu\alpha\beta} \epsilon_\alpha^*(p,\lambda)\epsilon_\beta^*(\overline{p},\overline{\lambda}). \quad (1)$$

The 16 form factors, $F_{i,\tau}$, include all information of the dynamics, while the other part are determined by the kinematics. For physical W-pair production, 9 basis tensors are enough. The rest are used for processes with unphysical (scalar) W bosons, which are used for the test of $F_{i,\tau}$. For this test, we also need to calculate $e^-(k,\tau)e^+(\overline{k},\overline{\tau}) \to W^-(p,\lambda)w^+(\overline{p})$ (w^+: the Nambu-Goldstone boson), whose amplitudes are decomposed as

$$\mathcal{M}_\tau^{\lambda} = i\sum_{i=1}^{4} H_{i,\tau}(s,t)\, j_\mu\, S_i^{\mu\alpha} \epsilon_\alpha^*(p,\lambda), \quad (2)$$

with four basis tensors and form factors $H_{i,\tau}$.

We employ the $\overline{\text{MS}}$ scheme, and we take $\hat{e}$, $\hat{g}$ and M_W as input SM parameters. The MSSM $\overline{\text{MS}}$ couplings are determined by

$$\frac{1}{\hat{e}^2_{\text{MSSM}}(\mu)} = \frac{1}{\hat{e}^2_{\text{SM}}(\mu)} - \Delta\Pi_{T,\gamma}^{QQ}(0,\mu), \quad (3)$$

$$\frac{1}{\hat{g}^2_{\text{MSSM}}(\mu)} = \frac{1}{\hat{g}^2_{\text{SM}}(\mu)} - \Delta\Pi_{T,\gamma}^{3Q}(0,\mu), \quad (4)$$

where $\Delta\Pi_{T,\gamma}^{QQ}(0,\mu)$ and $\Delta\Pi_{T,\gamma}^{3Q}(0,\mu)$ are non-SM contributions to gauge-boson two-point functions. The SM $\overline{\text{MS}}$ couplings are calculated by using the SM RGE's and experimen-

Test by using the BRS sum rule ($i = 1$, $\tau = -1$)

$\sqrt{s}$	Left-hand-side $\xi_{1j}F_j(s,t)$ / Right-hand-side $C_{\text{mod}}H_i(s,t)$
200 GeV	$-1.385496590672218 \times 10^{-6}$
	$-1.385496590672223 \times 10^{-6}$
1000GeV	$-6.682526871892199 \times 10^{-8}$
	$-6.682526871892053 \times 10^{-8}$

Table 1. The test by the BRS sum rules.

tal values for the effective charges[2]. $M_W = 80.41$GeV is taken from the data.

3 Tests

One difficulty in loop-level calculations is to determine reliability of the results. This is especially so in our process in which a subtle gauge cancellation takes place among diagrams at each level of perturbation. Incomplete treatment for higher order terms can lead artificially large collections. In order to obtain solid results, we test our calculation by the following methods.

3.1 The BRS invariance

Useful sum rules among form factors between the W^-W^+ and the W^-w^+ processes are induced from the BRS invariance;

$$\sum_{j=1}^{16} \xi_{ij} F_{j,\tau}(s,t) = C_{\text{mod}} H_{i,\tau}(s,t), \qquad (5)$$

where ξ_{ij} are determined by the kinematic parameters and C_{mod} differs from 1 at loop levels. We can use them to test $F_{i,\tau}$. In Table 1, values for both sides in the sum rule are shown. They coincide with each other.

3.2 The decoupling theorem

The cross section should be of the SM prediction in the large sfermion-mass limit. We use this fact to test the overall renormalization factor which cannot be tested by the BRS sum rules. In Figure 1, for large-mass limit ($1/M^2 \to 0$: M is a scale of SUSY soft-

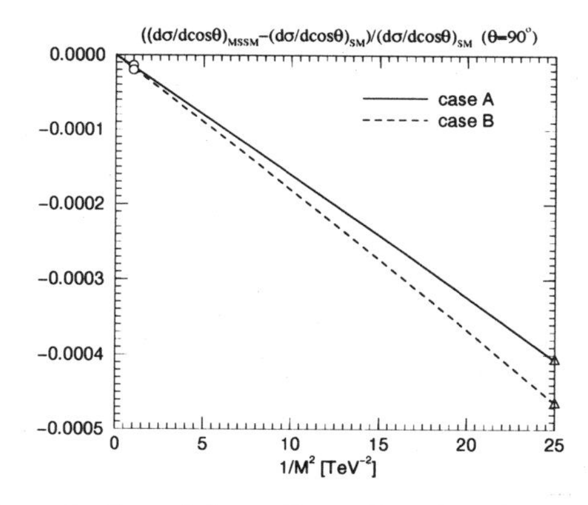

Figure 1. Test of decoupling. Case A and B correspond to non-mixing and stop-mixing cases.

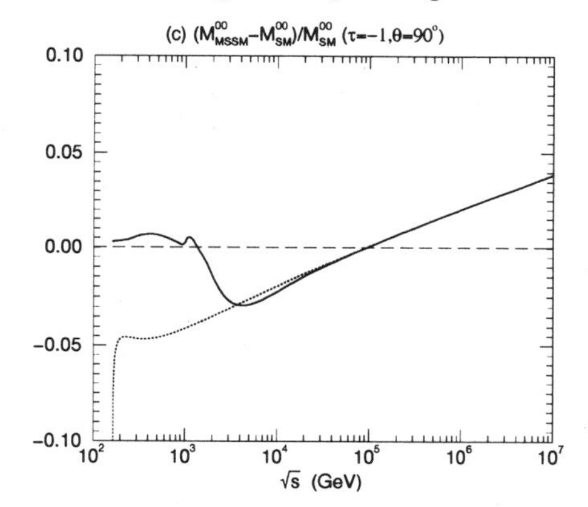

Figure 2. Test for high-energy stability. Real curve and dotted one corresponds to results from the full calculation and the analytic expression, respectively.

breaking masses), the deviation from the SM prediction becomes zero for each case.

3.3 High energy stability

At high energies large gauge cancellation takes place, so it is important to see the high energy stability of the numerical results. We calculate high-energy analytic expression for the amplitude. In Figure 2, the high-energy expression and the full calculation give same results in the high energy limit.

4 Sfermion one-loop effects

In Figure 3, the squark one-loop effects on the 00 helicity amplitude for parameter sets in Table 2. The corrections to the SM predic-

First 2 generations	Case 1	Case 2	Case 3
Input parameters			
$m_{\tilde{Q}} = m_{\tilde{U}} = m_{\tilde{D}}$	300	500	1000
$A^{eff}_{\tilde{f}}$	0	0	0

Table 2. Cases without mixing.

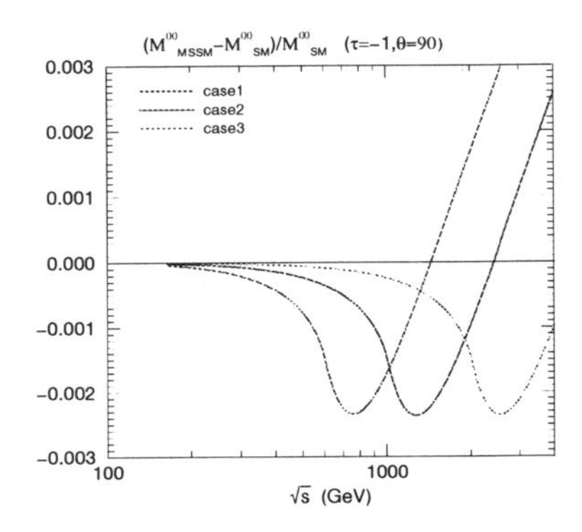

Figure 3. Squark effects of the first two generation. There is no mixing between $\tilde{f}_L$ and $\tilde{f}_R$

tion are negative and the behavior is rather simple. There is a peak slightly after the squark-pair threshold. The corrections to the SM prediction are at most a few times 0.1%.

In Figure 4, effects of the third generation squarks with large stop-mixing are shown. The parameters defined in Table 3 are chosen so as to be the maximal mixing with mixing angle 45°. The corrections are positive. Larger effects appear for larger A^{eff}_{f}. It, however, turns out that such enhancement due to the mixing is strongly constrained by the precision data. In Figure 5, each case in Table 3 is plotted on the S-T parameter plane[1]. The cases for large corrections (case 2, case 3 in Table 3) stay outside the 99% contour and thus they are excluded. After all only smaller corrections than a few times 0.1% are allowed.

5 Conclusion

The sfermion effects on this process is small.

References

1. S. Alam, K. Hagiwara, S. Kanemura, R. Szalapski, and Y. Umeda, To appear in Phys. Rev. **D**, (hep-ph/0002066), and references therein; Nucl. Phys. **B541** (1999) 50.
2. K. Hagiwara, D. Heidt, C.S. Kim and S. Matsumoto, Z. Phys. **C64** (1994) 559.

$\tilde{t}$-$\tilde{b}$ sector:	Case 1	Case 2	Case 3
Input parameters			
$m_{\tilde{Q}} = m_{\tilde{U}} = m_{\tilde{D}}$	300	400	500
$A^{eff}_{\tilde{f}}$	625	1025	1539
Output parameters			
$m_{\tilde{t}_1}$	100	100	100
$m_{\tilde{t}_2}$	478	607	741
$\cos\theta_{\tilde{t}}$	0.708	0.708	0.707

Table 3. Maximal stop-mixing cases.

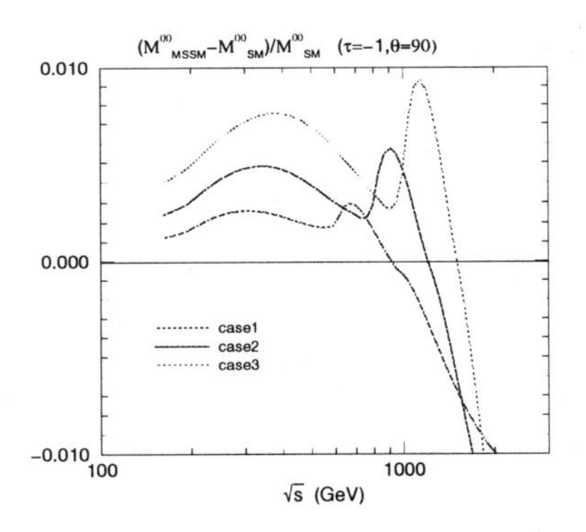

Figure 4. The third-generation squark effects. Maximal stop-mixing cases.

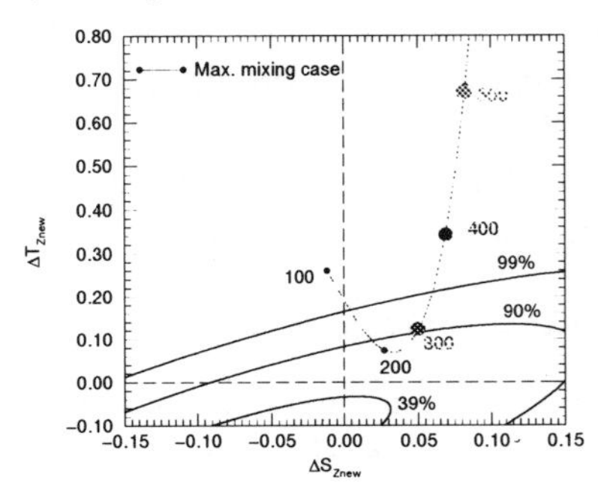

Figure 5. The parameter sets defined in Table 3 on S-T plane. The origin indicates the SM prediction.

Probing Anomalous Top Quark Couplings at the Future Linear Colliders*

T. Huang†

CCAST (world laboratory), P.O. Box 8730, Beijing 100080, China
Institute of High Energy Physics, Chinese Academy of Sciences,
P.O. Box 918, Beijing, 100039, China

Abstract

In terms of an effective Lagrangian we investigate the possibilities of probing anomalous top quark couplings, $t\bar{t}H$, $\gamma t\bar{t}$, $Zt\bar{t}$ and tWb at the future linear colliders. It is found that at a linear collider with a c. m. energy $\sqrt{s} \sim 0.5 - 1.5$ TeV and a high luminosity of $10 - 1000$ fb^{-1}, $e^+e^- \to t\bar{t}H$ is an ideal process in probing anomalous $t\bar{t}H$ couplings. We also study in detail the effects of anomalous couplings on $t\bar{t}$ spin correlations in the top pair production as well as the top quark decay processes with three bases (helicity, beam line and off-diagonal bases). Our results show that with a c. m. energy $\sqrt{s} \sim 0.5 - 1$ TeV and a high luminosity of $1 - 100$ fb^{-1}, the anomalous couplings $\gamma t\bar{t}$, $Zt\bar{t}$ and tWb may be sensitively probed.

One believes that the large top-quark mass, which is close to the order of the weak scale ($m_t \approx v/\sqrt{2}$), makes the third generation to play a significant role in probing the new physics beyond the Standard Model (SM). Thus the linear collider (LC) will have a potential to explore the new physics associated with the Higgs and the top-quark sector.

In order to explore the possibility, we take a model-independent approach by using a linearly realized effective Lagrangian to dimension-6 operators including CP violation. We discuss the process $e^+e^- \to t\bar{t}H$ and the top quark spin correlation to probe the non-standard couplings. Particularly, the process $e^+e^- \to t\bar{t}H$ is an ideal one for probing anomalous coupling $t\bar{t}H$ and hopefully gains some insight for the new physics beyond the SM. The observability of the signal from anomalous couplings $t\bar{t}H$, $\gamma t\bar{t}$, $Zt\bar{t}$ and tWb with c. m. energy $\sqrt{s} \sim 0.5 - 1$ TeV is studied.

In the case of linear realization, the new physics is parameterized by higher dimensional operators which contain the SM fields and are invariant under the SM gauge group. Below the new physics scale Λ, the effective Lagrangian can be written as

$$\mathcal{L}_{eff} = \mathcal{L}_0 + \frac{1}{\Lambda^2}\sum_i C_i O_i + \mathcal{O}(\frac{1}{\Lambda^4}) \qquad (1)$$

where $\mathcal{L}_0$ is the SM Lagrangian. O_i are dimension-6 operators which are $SU_c(3)\times SU_L(2)\times U_Y(1)$ invariant and C_i are coefficients which represent the coupling strengths of O_i [1]. All the operators O_i are hermitian and the coefficients C_i are real and the order of unity. If we assume that the new physics is of the origin associated with the electroweak symmetry breaking, then it is natural to identify the cut-off scale Λ to be the order of $\mathcal{O}(4\pi v)$. Alternatively, based on unitarity argument for massive quark scattering [2], the scale for new physics in the top-quark sector should be below about 3 TeV. There are twelve dimension-six CP even operators. All the operators, which give new contributions to the couplings of $t\bar{t}H$, $\gamma t\bar{t}$, $Zt\bar{t}$ and tWb , are listed in Refs. [3, 4, 5].

Among them, some of the operators are energy independence such as

$$O_{t1} = (\Phi^\dagger\Phi - \frac{v^2}{2})\left[\bar{q}_L t_R \widetilde{\Phi} + \widetilde{\Phi}^\dagger \bar{t}_R q_L\right] \qquad (2)$$

and some are energy-dependent, such as

$$O_{Dt} = (\bar{q}_L D_\mu t_R)D^\mu \widetilde{\Phi} + (D^\mu \widetilde{\Phi})^\dagger (\overline{D_\mu t_R} q_L) \qquad (3)$$

due to the deviative. The energy-dependence of all dimension-6 operators are listed in the table of Refs. [3, 4, 5].

Generally, we can examine the possible constraints on the operators from the measurement $Z \to b\bar{b}$. The observable R_b at the Z pole is calculated to be

$$\begin{aligned} R_b &\equiv \frac{\Gamma(Z \to b\bar{b})}{\Gamma(Z \to \text{hadrons})} \\ &= R_b^{SM}\left[1 + 2\frac{v_b\delta V + a_b\delta A}{v_b^2 + a_b^2}(1 - R_b^{SM})\right], \qquad (4) \end{aligned}$$

where v_b and a_b represent the SM couplings and δV, δA the new physics contributions. If we attribute the difference between R_b^{SM} and R_b^{exp} as the new physics contribution, we obtain the limit at the 1σ (3σ) level

$$-4\times 10^{-3}\ (-8\times 10^{-3}) < \delta V < -5\times 10^{-5}\ (4\times 10^{-3}). \qquad (5)$$

*Invited talk at the XXX International Conference on High Energy Physics, Osaka, July 26-August 2, 2000.

†Written with T. Han, Z.-H. Lin, J.-X. Wang and X. Zhang.

Assuming that there is no accidental cancellation between different operators and noting that $2s_W c_W m_Z/ev \simeq 1$, we obtain the bound for each of them at the 1σ (3σ) level as

$$5\times10^{-5}\ (-4\times10^{-3}) \ < \ \frac{v^2}{\Lambda^2}C^{(1)}_{\Phi q}\ (or\ \frac{v^2}{\Lambda^2}C^{(3)}_{\Phi q}) \ < \ 4\times10^{-3}\ (8\times10^{-3}). \quad (6)$$

On other hand, the constraints from $A^{(b)}_{FB}$ are weaker than R_b.

For O_{t1}, O_{t2}, O_{Dt}, $O_{tW\Phi}$ and $O_{tB\Phi}$ they are not constrained by R_b at tree level and bounds on them can be studied from the argument of partial wave unitarity in Ref. [6]. It is informative to see the ranges of the unitarity bounds for $\Lambda \approx 3-1$ TeV:

$$|C_{t1}|\frac{v^2}{\Lambda^2} \simeq 1.0-3.0, \qquad |C_{t2}|\frac{v^2}{\Lambda^2} \simeq 0.29-2.6,$$
$$C_{Dt}\frac{v^2}{\Lambda^2} \simeq 0.07-0.63 \ \text{or} \ C_{Dt}\frac{v^2}{\Lambda^2} \simeq -(0.04-0.40),$$
$$|C_{tW\Phi}|\frac{v^2}{\Lambda^2}\ or\ |C_{tB\Phi}|\frac{v^2}{\Lambda^2} \simeq 0.02-0.15\,. \quad (7)$$

At present, there is no significant experimental constraint on the CP-odd couplings involving the top-quark sector.

The relevant Feynman diagrams for $e^+e^- \to t\bar{t}H$ production are depicted in Fig. 1, where (a)−(c) are those in the SM and the dots denote the contribution from new interactions. The four-particle vertex (Fig. 1d) should be paid more attention since there is no such vertex in the SM but exists in the effective couplings due to the gauge invariance. We evaluate all the diagrams including interference effects, employing a helicity amplitude package (FDC) developed in [7]. This package has the flexibility to include new interactions beyond the SM. We have not included the QCD corrections to the signal process, which are known to be positive and sizable.

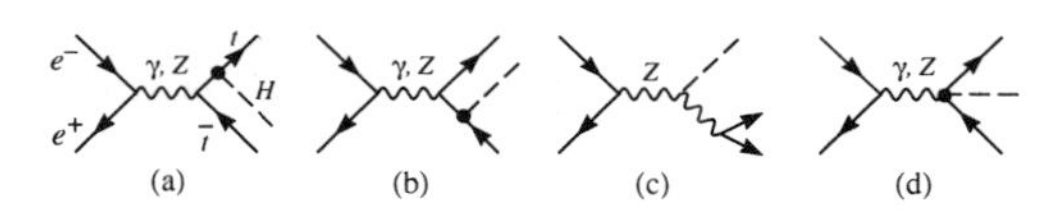

Figure 1. Feynman diagrams for $e^+e^- \to t\bar{t}H$ production. (a)-(c) are those in the SM. The dots denote the contribution from new interactions.

Due to the strong constraints on $O^{(1)}_{\Phi q}$ and $O^{(3)}_{\Phi q}$ from the $Z \to b\bar{b}$ measurement, (see Eq. (6)), the effects of these operators at colliders will be rather small and can be neglected. Following the energy-dependence behavior, we expect that modifications to the SM prediction from different operators would be distinctive at high energies. For the purpose of illustration, we only present results for the operators O_{t1} (energy-independent) and O_{Dt} (most sensitive to energy scale) to represent to others.

The results include the production cross sections versus $\sqrt{s}$, the Higgs mass m_H and the couplings. The effect due to the operator O_{Dt} is insignificant at $\sqrt{s}=0.5$ TeV, which at higher energies the contribution from O_{Dt} is substantial.

To establish the sensitivity limits on the anomalous couplings that may be probed at the future LC experiments, one needs to consider the identification of the final state from $t\bar{t}H$, including the branch ratios and the detection efficiencies. The branching ratio of the leading decay mode $H \to b\bar{b}$ is about 80% ∼ 50% for the mass range of 100 ∼ 130 GeV. We assume 65% efficiency for a single b-tagging to identify four b-jets in the final state. With the desirable consideration we estimate an efficiency factor ϵ_S for detecting $e^+e^- \to t\bar{t}H$ to be $\epsilon_S = 10-30\%$ and a factor ϵ_B for reducing QCD and EW background to be $\epsilon_B = 10\%$. In order to estimate the luminosity (L) needed for probing the effects of the non-standard couplings, we define the significance of a signal rate (S) relative to a background rate (B) in terms of the Gaussian statistics

$$\sigma_S = \frac{S}{\sqrt{B}} \quad (8)$$

for which a signal at 95% (99%) confidence level (C.L.) corresponds to $\sigma_S = 2$ (3). They are calculated as

$$S = L(|\sigma-\sigma_{SM}|)\epsilon_S,$$
$$B = L\left[\sigma_{SM}\epsilon_S + (\sigma_{QCD}+\sigma_{EW})\epsilon_B\right]. \quad (9)$$

Then we obtain the luminosity required for observing the effects of O_{t1} and O_{Dt} at 95% C.L. for 0.5 TeV and 1 TeV for C_{t1} and for 1 TeV and 1.5 TeV for C_{Dt} in Fig. 2 where the two curves are for 10% and 30% of signal detection efficiency, respectively. It can be seen that at a 0.5 TeV collider, one would need rather high integrated luminosity to reach the sensitivity to the anomalous couplings; while at a collider with a higher c.m. energy one can sensitively probe those couplings with a few hundred fb^{-1} luminosity.

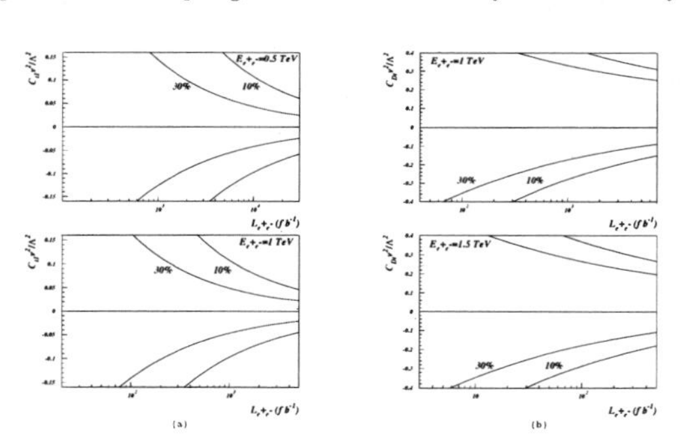

Figure 2. (a) Sensitivity to the anomalous couplings versus the integrated luminosity for a 95% confidence level limits with $m_H = 120$ GeV for (a) O_{t1} at $\sqrt{s}=0.5$ TeV and $\sqrt{s}=1$ TeV and for (b) O_{Dt} at $\sqrt{s}=1$ TeV and $\sqrt{s}=1.5$ TeV.

If there exist effective CP-odd operators beside the SM interaction, then CP will be violated in the Higgs and top-quark sector. By using the similar discussion, one can try to observe the effects of the operators beyond the SM expectation. The CP-violating effect can be parameterized by a cross section asymmetry as

$$A_{CP} \equiv \frac{\sigma((p_1 \times p_3) \cdot p_4 < 0) - \sigma((p_1 \times p_3) \cdot p_4 > 0)}{\sigma((p_1 \times p_3) \cdot p_4 < 0) + \sigma((p_1 \times p_3) \cdot p_4 > 0)} \quad (10)$$

where p_1, p_3 and p_4 are the momenta of the incoming electron, top quark and anti-top quark, respectively. The luminosity required for detecting the effects on the total cross sections and A_{CP} is shown in Fig. 3 versus CP-odd couplings with 95% C.L. for $m_H = 120$ GeV and $\sqrt{s} = 1$ TeV. The solid curves are for the cross sections with efficiency factor $\epsilon_S = 30\%$ and $\epsilon_B = 10\%$ according to Eq. (9). Apparently, the effects on the total cross section due to CP-odd operators are much stronger than that on A_{CP}. In other words, the direct observation of the CP asymmetry would need much lighter luminosity to reach.

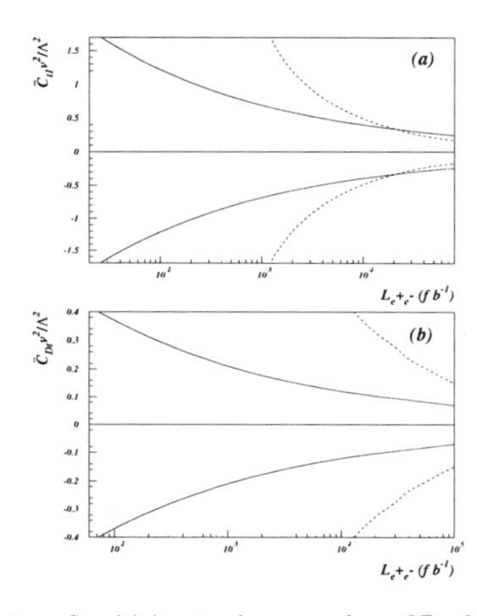

Figure 3. Sensitivity to the anomalous CP-odd couplings versus the integrated luminosity for a 95% confidence level limits and for 30% of detection efficiency at $\sqrt{s} = 1$ TeV , with $m_H = 120$ GeV. The solid line is for the total cross section and the dash line is for the CP asymmetry A_{CP}.

We now discuss the corrections from the anomalous couplings in the top-quark pair production and the top-quark decays. In order to evaluate the possible effects from new physics and study which spin basis is more sensitive to anomalous couplings, we use the generic spin basis suggested by Parke and Shadmi [8]. They are helicity basis, beamline basis and off-diagonal basis. With the above bases and center of mass energy $\sqrt{s}$, we calculate the differential polarized cross section at the tree level and the differential cross section of $e^- e^+ \to t\bar{t} \to (b\bar{l}\nu_{\bar{l}})(\bar{b}l'\nu_{l'})$. In order to see the effects of the anomalous coupling, we define $< S_t >$, $< S_{\bar{t}} >$ and $< S_t S_{\bar{t}} >$ are t, $\bar{t}$ and $t\bar{t}$ correlation functions and calculate the relevant observables. Our results show that with a c.m. energy $\sqrt{s} \sim 0.5 - 1$ TeV and a higher luminosity of $1 - 100\ fb^{-1}$, the anomalous couplings $\gamma t\bar{t}$, $Zt\bar{t}$ and tWb may be sensitively probed [5].

In summary, we have considered a general effective lagrangian to dimension-6 operators including CP violation effects. The constraints on these anomalous couplings has been derived from the $Z \to b\bar{b}$ data and unitarity consideration.

In order to explore the effects of these non-standard couplings, we have studied the process $e^+e^- \to t\bar{t}H$, anomalous couplings ($\gamma t\bar{t}$, $Zt\bar{t}$ and tWb) on $t\bar{t}$ spin correlations in the top pair production as well as the top-quark decay process with three bases (helicity, beamline and off-diagonal basis). We find that the future linear collider experiments should be able to probe those couplings well below their unitarity bounds. To reach a good sensitivity, the higher integrated luminosity needed is about several hundred fb^{-1} for a c.m. $\sqrt{s} \sim 0.5 - 1.5$ TeV.

References

[1] C. J. C. Burgess and H. J. Schnitzer, Nucl. Phys. **B228**, 454 (1983); C. N. Leung, S. T. Love and S. Rao, Z. Phys. **C31**, 433 (1986); W. Buchmuller and D. Wyler, Nucl. Phys. **B268**, 621 (1986); K. Hagiwara, S. Ishihara, R. Szalapski and D. Zeppenfeld, Phys. Rev. **D48**, 2182 (1993).

[2] M. Chanowitz, M. Furman and I. Hinchliffe, Nucl. Phys. **B153**, 402 (1979); T. Appelquist and M. Chanowitz, Phys. Rev. Lett. **59**, 2405 (1987); S. Jager and S. Willenbrock, Phys. Lett. **B435**, 139 (1998); R. S. Chivukula, Phys. Lett. **B439**, 389 (1998).

[3] K. Whisnant, J. M. Yang, B.-L. Young and X. Zhang, Phys. Rev. **D56**, 467 (1997); F. Larios, T. Tait and C.-P. Yuan, Phys. Rev. **D57**, 3106 (1998).

[4] T. Han, T. Huang, Z.-H. Lin, J.-X. Wang and X. Zhang, Phys. Rev. **D61**, 015006 (2000).

[5] T. Huang, Z.-H. Lin, J.-X. Wang and X. Zhang, in preparation.

[6] G. J. Gounaris, D. T. Papadamou, F. M. Renard, Z. Phys. **C76**, 333 (1997).

[7] J.-X Wang, Computer Phys. Commun. 86 (1993) 214-231.

[8] S. Parke and Y, Shadmi, Phys. Lett. **B387**, 199 (1996); G. Mahlon and S, Parke, Phys. Lett. **B411**, 173 (1997).

NEXT-TO-LEADING ORDER SUSY-QCD CALCULATION OF ASSOCIATED PRODUCTION OF GAUGINOS AND GLUINOS

EDMOND L. BERGER AND T. M. P. TAIT

High Energy Physics Division, Argonne National Laboratory, Argonne, IL 60439, USA
E-mail: berger@anl.gov, tait@hep.anl.gov

M. KLASEN

II. Institut für Theoretische Physik, Universität Hamburg, D-22761 Hamburg, Germany
E-mail: michael.klasen@desy.de

Results are presented of a next-to-leading order calculation in perturbative QCD of the production of charginos and neutralinos in association with gluinos at hadron colliders. Predictions for cross sections are shown at the energies of the Fermilab Tevatron and CERN Large Hadron Collider for a typical supergravity (SUGRA) model of the sparticle mass spectrum and for a light gluino model.

1 Motivation

The mass spectrum in typical supergravity and gauge-mediated models of supersymmetry (SUSY) breaking favors much lighter masses for gauginos than for squarks. Because the masses are smaller, there is greater phase space at the Tevatron and greater partonic luminosities for gaugino pair production, and for associated production of gauginos and gluinos, than for squark pair production. In this contribution, we summarize our recent calculations at next-to-leading order (NLO) in perturbative quantum chromodynamics (QCD) of the total and differential cross sections for associated production of gauginos and gluinos at hadron colliders[1,2]. Associated production offers a chance to study the parameters of the soft SUSY-breaking Lagrangian. Rates are controlled by the phases of the $\tilde{\chi}$ and $\tilde{g}$ masses and by mixing in the squark and gaugino sectors. In addition to the potentially large cross section for associated production, the leptonic decay of the gaugino makes this process a good candidate for mass determination of the gluino and for discovery or exclusion of an intermediate-mass gluino.

2 NLO SUSY-QCD Formalism

Associated production of a gluino and a gaugino proceeds in leading order (LO) through a quark-antiquark initial state and the exchange of an intermediate squark in the t-channel or u-channel. At NLO, loop corrections must be included. In addition, there are 2 to 3 parton processes initiated either by quark-antiquark scattering, with a gluon radiated into the final state, or by quark-gluon scattering with a light quark radiated into the final state. For the quark-antiquark initial state, the loop diagrams involve the exchange of intermediate Standard Model or SUSY particles in self-energy, vertex, or box diagrams. Ultraviolet and infrared divergences appear at the upper and lower boundaries of integration over unobserved loop momenta. They are regulated dimensionally and removed through renormalization or cancellation with corresponding divergences in the 2 to 3 parton (real emission) diagrams that have an additional gluon radiated into the final state. In addition to soft divergences, real emission contributions have collinear divergences that are factored into the NLO parton densities. The full treatment is presented in our long paper[2].

3 Tevatron and LHC Cross Sections

To obtain numerical predictions for hadronic cross sections, we choose an illustrative SUGRA model with parameters $m_0 = 100$ GeV, A_0=300 GeV, tan $\beta = 4$, and sign $\mu = +$. Because the gluino, gaugino, and squark masses all increase with parameter $m_{1/2}$ (but are insensitive to m_0), we vary $m_{1/2}$ between 100 and 400 GeV. The resulting masses for $\tilde{\chi}^0_{1...4}$ vary between 31...162, 63...317, 211...665, and 241...679 GeV; $\tilde{\chi}^{\pm}_{1,2}$ are almost degenerate in mass with $\tilde{\chi}^0_{2,4}$. The mass $m_{\tilde{\chi}^0_3} < 0$ inside a polarization sum. Our approach is general, and results can be obtained for any set of gaugino and gluino masses. For our second model, we select one[3] with an intermediate-mass gluino as the lightest SUSY particle (LSP), fixing $m_{\tilde{g}} = 30$ GeV, and $m_{\tilde{q}} = 450$ GeV. We choose a weak sector identical to the SUGRA case. In our paper[2], we also quote results for anomaly mediated, gauge mediated, and gaugino mediated models.

We convolve LO and NLO partonic cross sections with CTEQ5 parton densities in LO and NLO ($\overline{\rm MS}$) along with 1- and 2-loop expressions for α_s, the corresponding values of Λ, and five active quark flavors.

For the SUGRA case, we present total hadronic cross sections in Fig. 1 as functions of the gluino mass. The light gaugino channels should be observable at both colliders. At the Tevatron, for 2 fb^{-1} of integrated luminosity, 10 or more events could be produced in each of the lighter gaugino channels if $m_{\tilde{g}} < 450$ GeV. The heavier Higgsino channels are suppressed by about one order of magnitude and might be observable only at the LHC. As a rough estimate of uncertainty associated with the choice of parton densities, we note that the NLO cross section for $\tilde{\chi}^0_2$ production is lower by 12% at the Tevatron with the CTEQ5 set than for the CTEQ4 set, and 4% lower at the LHC. The impact of the NLO corrections can be seen more readily in

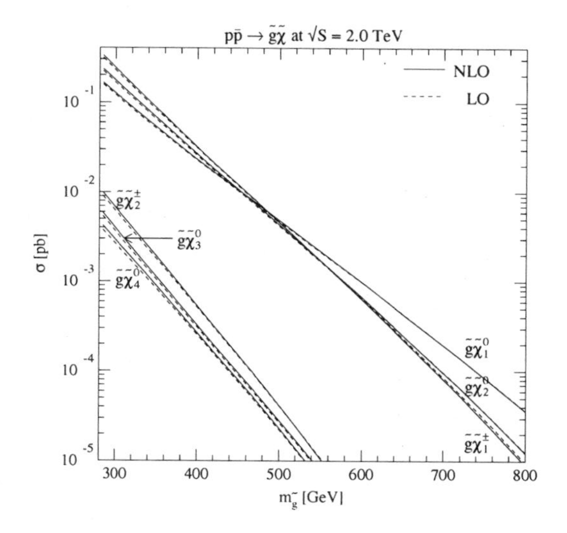

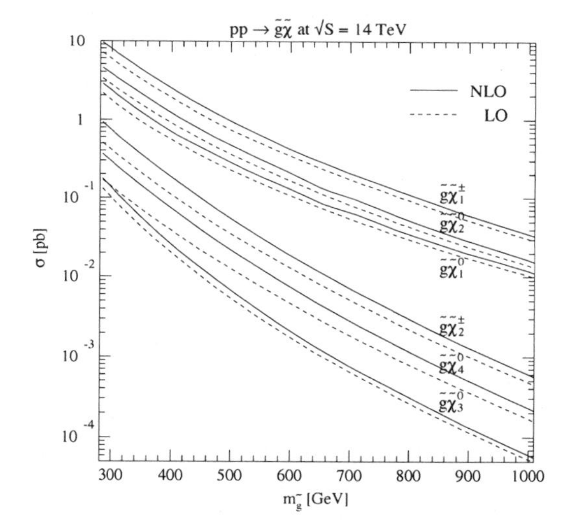

Figure 1. Predicted total hadronic cross sections at Run II of the Tevatron and at the LHC for all six $\tilde{g}\tilde{\chi}$ channels in a typical SUGRA model as functions of the gluino mass.

the ratio of NLO to LO cross sections computed at a renormalizaton scale set equal to the average mass of the final state particles. The NLO effects are moderate (of $\mathcal{O}$ (10%)) at the Tevatron, while at the LHC the NLO contributions can increase the cross sections by as much as a factor of two. The second initial-state channel, initiated by gluon quark scattering, plays a significant role at the energy of the LHC.

For the case of a gluino with mass 30 GeV, the total hadronic cross sections are shown in Fig. 2 as functions of $m_{1/2}$. At the Tevatron, for 2 fb^{-1} of integrated luminosity, 100 or more events could be produced in each of the lighter gaugino channels if $m_{1/2} < 400$ GeV. In this case, NLO enhancement factors lie in the ranges 1.3 to 1.4 at the Tevatron and 2 to 4 at the LHC.

An important measure of the theoretical reliability is the variation of the hadronic cross section with the renormalization and factorization scales. At LO, these scales enter only in the strong coupling constant and the parton densities, while at NLO they appear also explicitly in the hard cross section. The scale dependence is reduced considerably after NLO effects are included. The Tevatron (LHC) cross sections vary by $\pm 23(12)\%$ at LO, but only by $\pm 8(4.5)\%$ in NLO when the scale is varied by a factor of two around the central scale.

For experimental searches, distributions in transverse momentum are important since cuts on p_T help to enhance the signal. In our long paper[2], we show that NLO contributions can have a large impact on p_T spectra at the LHC, owing to contributions from the gq initial state. At the Tevatron the NLO p_T-distribution is shifted moderately to lower p_T with respect to the LO expectation.

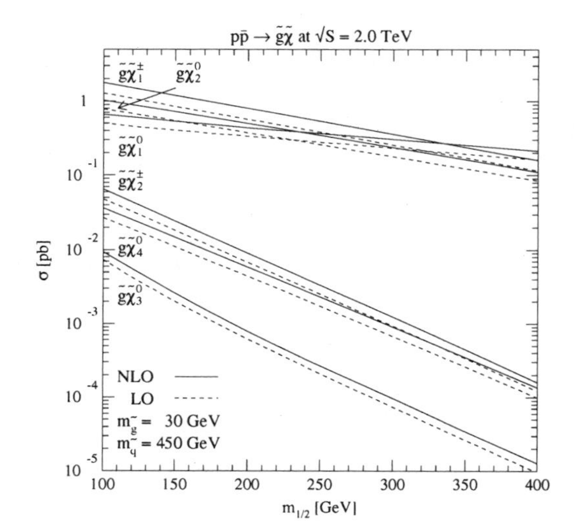

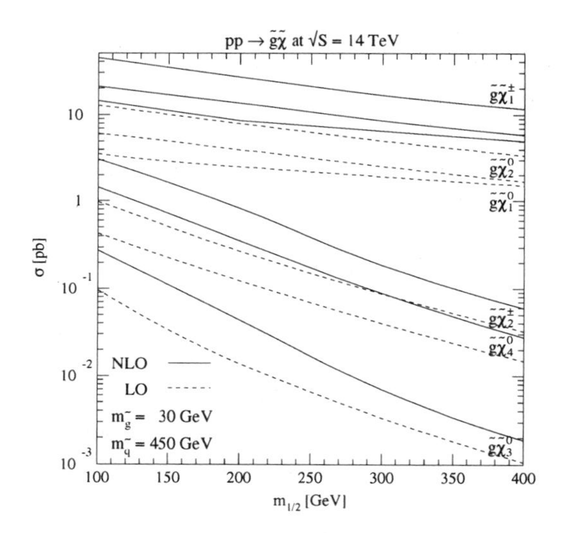

Figure 2. Predicted total hadronic cross sections at Run II of the Tevatron and at the LHC for all six $\tilde{g}\tilde{\chi}$ channels in our model with a gluino of mass 30GeV, as functions of the parameter $m_{1/2}$.

Acknowledgments

Work at Argonne National Laboratory is supported by the U.S. Department of Energy, Division of High Energy Physics, under Contract W-31-109-ENG-38. M.K. is supported by DFG through grant KL 1266/1-1.

References

1. E. L. Berger, M. Klasen, and T. Tait, *Phys. Lett.* B **459**, 165 (1999) and Proceedings of the 35th Rencontres de Moriond on *QCD and High Energy Hadronic Interactions*, Les Arcs, France, 2000, hep-ph/0005199.
2. E. L. Berger, M. Klasen, and T. Tait, hep-ph/0005196, *Phys. Rev.* D, in press.
3. S. Raby, *Phys. Lett.* B **422**, 158 (1998); A. Mafi and S. Raby, hep-ph/9912436.

IMPROVED SUSY QCD CORRECTIONS TO HIGGS BOSON DECAYS INTO QUARKS AND SQUARKS

Y. YAMADA[1], H. EBERL[2], K. HIDAKA[3], S. KRAML[2], W. MAJEROTTO[2]

[1] *Department of Physics, Tohoku University, Sendai 980–8578, Japan*

[2] *Institut für Hochenergiephysik der Österreichischen Akademie der Wissenschaften, A–1050 Vienna, Austria*

[3] *Department of Physics, Tokyo Gakugei University, Koganei, Tokyo 184–8501, Japan*

The $\mathcal{O}(\alpha_s)$ SUSY QCD corrections to the decays of the MSSM Higgs bosons into bottom quarks and squarks can be very large for large $\tan\beta$ in the on–shell renormalization scheme. We improve the calculation by a careful choice of the tree-level Higgs boson couplings in terms of running parameters of quarks and squarks.

1 Introduction

The MSSM has two Higgs doublets (H_1, H_2) which give five physical bosons (h^0, H^0, A^0, $H^{\pm}$). Their couplings to bottom quarks b and squarks $\tilde{b}$ are enhanced for large $\tan\beta$. In this case the decays to b are usually the main modes.[1] Decays to $\tilde{b}$ can be also dominant.[2] Studying these decays is therefore very important.

These decays receive large SUSY QCD corrections.[3,4] When the on–shell scheme is adopted for quarks and squarks, the corrections are often very large and make the perturbation calculation quite unreliable. The large gluon loop correction can be absorbed by using the QCD running quark mass in the coupling. However, the gluino loop correction can also be very large for large $\tan\beta$.

Here we improve[5] the one–loop SUSY QCD corrected widths of the Higgs boson decays into quarks and squarks. The essential point of the improvement is to define appropriate tree–level couplings of the Higgs bosons to b and $\tilde{b}$.

2 Gluino corrections to Higgs–quark couplings

The main part of the large gluino loop corrections to the Higgs decay widths into b originates from the $\bar{b}bH_2$ coupling which is generated by squark–gluino loops.

At tree–level, the $\bar{b}bH_2$ coupling is forbidden by SUSY. However, the interaction $h_b\Delta_b\bar{b}bH_2$ is generated by the loop correction due to the soft SUSY breaking. The squark-gluino loops give $\Delta_b \sim \alpha_s m_{\tilde{g}}\mu/m^2_{\tilde{b}}$. Δ_b can have further contributions from other loop corrections.[6,7]

The effective interactions between Higgs bosons and b, after integrating out the squarks, are properly described by

$$\begin{aligned}\mathcal{L}^{\rm eff}_{\rm int} = & -(h_b/\sqrt{2})\bar{v}[\cos\beta + \Delta_b\sin\beta]\,\bar{b}b \\ & -(h_b/\sqrt{2})[\cos\alpha + \Delta_b\sin\alpha]\,H^0\bar{b}b \\ & +(h_b/\sqrt{2})[\sin\alpha - \Delta_b\cos\alpha]\,h^0\bar{b}b \\ & +(ih_b/\sqrt{2})[\sin\beta - \Delta_b\cos\beta]\,A^0\bar{b}\gamma_5 b \\ & +h_b[\sin\beta - \Delta_b\cos\beta]\,H^-\bar{b}_R t_L + (\text{h.c.}). \end{aligned} \quad (1)$$

The first term of Eq. (1) gives the (non–SUSY) QCD running mass $m_b(Q)_{\rm SM}$. The difference from the SUSY QCD running mass $m_b(Q)_{\rm MSSM} = (h_b/\sqrt{2})\bar{v}\cos\beta$ is enhanced by $\tan\beta$. As a result, the gluino loop correction to m_b can become very large[6] for large $\tan\beta$.

In Eq. (1) the contributions of Δ_b to the Higgs–bottom couplings take forms different from those to m_b. When the tree–

level couplings are given in terms of $m_b(Q)_{\rm SM}$ or the on–shell mass M_b, the corrections by Δ_b can be enhanced very much[7,5,8] for $\tan\beta \gg 1$. This is the main source of the large gluino loop corrections to decay widths to b in the on–shell scheme. Note that Δ_b itself is smaller than one and therefore does not destroy the validity of the perturbation expansion.

We can improve the QCD perturbative expansion by changing the choice of the tree–level Higgs-bottom couplings. For example, when the tree–level $A^0\bar{b}b$ coupling is expressed in terms of $m_b(Q)_{\rm MSSM}$ at $Q = m_A$, the correction from Δ_b becomes very small. We therefore expect that $m_b(Q)_{\rm MSSM}$ is an appropriate parameter for the $A^0 \to \bar{b}b$ decay. This is also the case for the $H^+ \to t\bar{b}$ decay. The H^0 and h^0 decays need a special treatment. For very large m_A, the $H^0\bar{b}b$ and $h^0\bar{b}b$ couplings are properly parametrized by $m_b(Q)_{\rm MSSM}$ and $m_b(Q)_{\rm SM}$, respectively. In general, the appropriate tree–level couplings are given by their linear combinations.

3 Higgs–squark couplings

The large SUSY QCD corrections to the Higgs decays into squarks in the on–shell scheme[4] mainly come from the counterterms for the Higgs–squark couplings, which depend on (m_q, $\theta_{\tilde{q}}$, A_q). As in the decays to quarks, we can improve the perturbation calculation by using SUSY QCD running parameters $m_q(Q)_{\rm MSSM}$ and $A_q(Q)$ in the tree–level couplings. However, the mixing angles $\theta_{\tilde{q}}$ are kept on–shell in order to cancel the $\tilde{q}_1 - \tilde{q}_2$ mixing squark wave function corrections.

4 Numerical results

We calculated[5] the one–loop SUSY QCD corrected widths of the Higgs boson decays to b and $\tilde{b}$, with and without the improvement presented here. In obtaining $m_b(Q)_{\rm MSSM}$ from $m_b(Q)_{\rm SM}$, we express the sbottom parameters in the sbottom–gluino loops in terms of $m_b(Q)_{\rm MSSM}$ and perform an iteration procedure. The large higher-order gluino corrections to m_b are then resummed.[5,8]

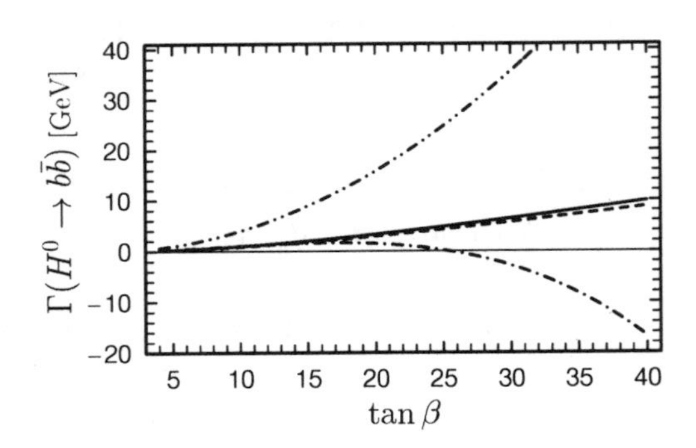

Figure 1. Decay width of $H^0 \to b\bar{b}$ as a function of $\tan\beta$. Dash–dot–dotted, dash–dotted, dashed, and full lines correspond to the on–shell tree–level, on–shell one–loop, improved tree–level, and improved one–loop results, respectively. The SUSY parameters are $(M_{\tilde{Q}}, M_{\tilde{U}}, M_{\tilde{D}}) = (300, 270, 330)$ GeV, $A_t = 150$ GeV, $A_b(Q = m_A) = -700$ GeV, $(m_{\tilde{g}}, \mu, m_A) = (350, 260, 800)$ GeV.

Here we show the tree–level and corrected widths of the decay $H^0 \to b\bar{b}$ in Fig. 1, and those of the decay $H^0 \to \tilde{b}_1\tilde{b}_1^*$ in Fig. 2. One can clearly see that the differences between tree–level and corrected widths decrease dramatically by our method, demonstrating the improvement of the perturbation expansion.

5 Summary

We have improved the SUSY QCD corrections to the Higgs decays into b and $\tilde{b}$. The essential point of the improvement is to define appropriate tree–level couplings of the Higgs bosons to b and $\tilde{b}$, in terms of the running parameters of quarks and squarks. We have also shown the numerical improvement of the SUSY QCD corrected decay widths.

We note that our method will also be useful in studying other processes with Higgs bosons.

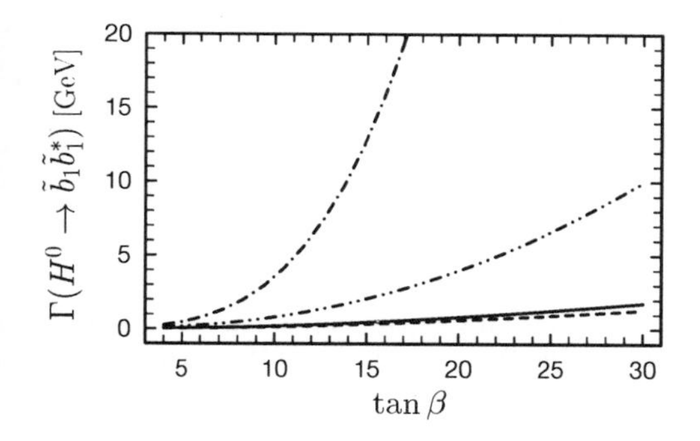

Figure 2. Decay width of $H^0 \to \tilde{b}_1 \tilde{b}_1^*$ as a function of $\tan\beta$. Notations and parameters are the same as in Fig. 1.

Acknowledgments

The work of Y. Y. was supported in part by the Grant–in–aid for Scientific Research from the Ministry of Education, Science, Sports, and Culture of Japan, No. 10740106. H. E., S. K., and W. M. thank the "Fonds zur Förderung der wissenschaftlichen Forschung of Austria", project no. P13139-PHY for financial support.

References

1. A. Djouadi, J. Kalinowski, and P.M. Zerwas, *Z. Phys.* C **70**, 435 (1996).
2. A. Bartl *et al.*, *Phys. Lett.* B **315**, 360 (1993); **389**, 538 (1996).
3. A. Dabelstein, *Nucl. Phys.* B **456**, 25 (1995); J.A. Coarasa, R.A. Jiménez, and J. Solà, *Phys. Lett.* B **389**, 312 (1996); A. Bartl *et al.*, *ibid.* **378**, 167 (1996); R.A. Jiménez and J. Solà, *ibid.* B **389**, 53 (1996).
4. A. Bartl *et al.*, *Phys. Lett.* B **402**, 303 (1997); A. Arhrib *et al.*, *Phys. Rev.* D **57**, 5860 (1998).
5. H. Eberl, K. Hidaka, S. Kraml, W. Majerotto, and Y. Yamada, *Phys. Rev.* D **62**, 055006 (2000).
6. R. Hempfling, *Phys. Rev.* D **49**, 6168 (1994); L.J. Hall, R. Rattazzi, and U. Sarid, *ibid.* D **50**, 7048 (1994); M. Carena, M. Olechowski, S. Pokorski, and C.E.M. Wagner, *Nucl. Phys.* B **426**, 269 (1994).
7. M. Carena, S. Mrenna, and C.E.M. Wagner, *Phys. Rev.* D **60**, 075010 (1999); **62**, 055008 (2000).
8. M. Carena, D. Garcia, U. Nierste, and C.E.M. Wagner, *Nucl. Phys.* B **577**, 88 (2000).

LSP DARK MATTER AND LHC

MIHOKO M. NOJIRI
YITP, Kyoto University, Kyoto 606-8502, Japan
E-mail: nojiri@yukawa.kyoto-u.ac.jp

We show that LHC experiments might well be able to determine all the parameters required for a prediction of the present density of thermal LSP relics from the Big Bang era. If the LSP is an almost pure bino we usually only need to determine its mass and the mass of the $SU(2)$ singlet sleptons. The only requirement to do this is that $m_{\tilde{\ell}_R} < m_{\tilde{\chi}_2^0}$, which is true for most of the cosmologically interesting parameter space. If the LSP has a significant higgsino component, its predicted thermal relic density is smaller than for an equal–mass bino. We show that in this case squark decays also produce significant numbers of $\tilde{\chi}_4^0$ and $\tilde{\chi}_2^{\pm}$. Reconstructing the corresponding decay cascades then allows to determine the higgsino component of the LSP.

1 Introduction

The Minimal Supersymmetric Standard Model (MSSM) is one of the most promising extensions of the Standard Model. If Nature chooses low energy supersymmetry (SUSY), sparticles will be found *for sure*, as they will be copiously produced at future colliders such as the Large Hadron Collider (LHC) at CERN. On the other hand, there are several on-going and future projects searching for LSP Dark Matter. One of them even claims a positive signal,[2] although the current situation is rather contradictory.[3] In any case, it seems very plausible that both SUSY collider signals and LSP Dark Matter in the Universe will be found in future.

Recently interesting possibilities have been pointed out where non–thermal production of Dark Matter is significant.[4,5] Generally the known bound from the thermal LSP density may easily be evaded by assuming a low post–inflationary reheating temperature of the Universe, without endangering the standard successes of Big Bang cosmology.[6] If the reheating temperature is below the neutralino decoupling temperature, the relation between neutralino pair annihilation rates and the mass density of the Universe disappears.

While these non–thermal mechanisms open exciting new possibilities, direct experimental or observational tests of them might be difficult, since they all have to occur before Big Bang nucleosynthesis (BBN). It is therefore interesting to determine the actual LSP relic density, both "locally" (in the solar system) and averaged over the Universe; and the predicted thermal LSP relic density, as precisely as possible. A positive difference between the actual and predicted LSP density would indicate the existence of non–thermal relics, whereas a negative difference would hint at large entropy production below the LSP freeze–out temperature (e.g. due to a low reheating temperature).

The matter density in the Universe divided by the critical density, $\Omega_{\rm matter}$, is claimed to be tightly constrained already; $\Omega_{\rm matter} = 0.35 \pm 0.07$.[7] On the other hand, the thermal relic density of the Universe $\Omega_{\tilde{\chi}_1^0} h^2$ ($h = 0.65 \pm 0.05$) has been calculated through the mass and interaction of the LSP, which is likely to be the lightest neutralino $\tilde{\chi}_1^0$.

The purpose of this paper[a] is to discuss how future LHC experiments can contribute to the determination of the MSSM parameters that are needed to predict the thermal LSP relic density. In Sec. 2, we point out that $\tilde{\chi}_2^0 \to \tilde{e}_R$ is open for most parameters giving a reasonable LSP density in mSUGRA, making the determination of $m_{\tilde{\chi}_2^0}$, $m_{\tilde{\chi}_1^0}$ and $m_{\tilde{e}_R}$

[a] This talk based on the paper[1]

possible at the LHC.[8] The mass density is determined by the LSP and slepton masses, if the LSP is mostly a bino as expected in mSUGRA. In this case $\Omega_{\tilde{\chi}_1^0}$ can be predicted to about 10 to 20% accuracy.

In Secs. 3 we discuss a non–mSUGRA scenario. It is then easy to find cases with comparable higgsino and gaugino masses, $\mu \sim M$, while keeping all squared scalar masses positive at M_X. The LSP then has a significant higgsino component, so that its density cannot be predicted by only studying $\tilde{\chi}_2^0 \to \tilde{e}_R \to \tilde{\chi}_1^0$ decays. We point out that the cascade decay $\tilde{\chi}_2^+ \to \tilde{\nu}_L \to \tilde{\chi}_1^+$ can then often be identified, providing clear evidence that $\mu \sim M$. We also present a detailed case study with $\mu \sim M_2$ to confirm the potential of LHC experiments to analyze $\tilde{\chi}_2^+$ cascade decays; this allows a complete determination of the neutralino mass matrix. Sec. 4 is devoted to discussions.

2 $\Omega_{\tilde{\chi}_1^0}$ in mSUGRA

In the minimal supergravity model one assumes universal soft breaking parameters at the GUT scale: a universal scalar mass m, universal gaugino mass M, etc. The renormalization group evolution of soft breaking squared Higgs masses then leads to consistent breaking of the electroweak symmetry, provided the higgsino mass parameter μ can be tuned independently.

The mass density $\Omega_{\tilde{\chi}_1^0} h^2$ of the LSP is calculated from the pair annihilation cross section by using the expressions [9]

$$\Omega_{\tilde{\chi}_1^0} h^2 = \frac{1.07 \times 10^9/\mathrm{GeV} x_F}{\sqrt{g_*} M_P (a + 3b/x_F)} \tag{1}$$

except near regions of parameter space where special care is needed. Here $M_P = 1.22 \cdot 10^{19}$ GeV is the Planck mass, and a and b are the first two coefficients in the Taylor expansion of the pair annihilation cross section of the LSP with respect to the relative velocity of the LSP pair in its center of mass frame. g^* is the effective number of relativistic degrees of

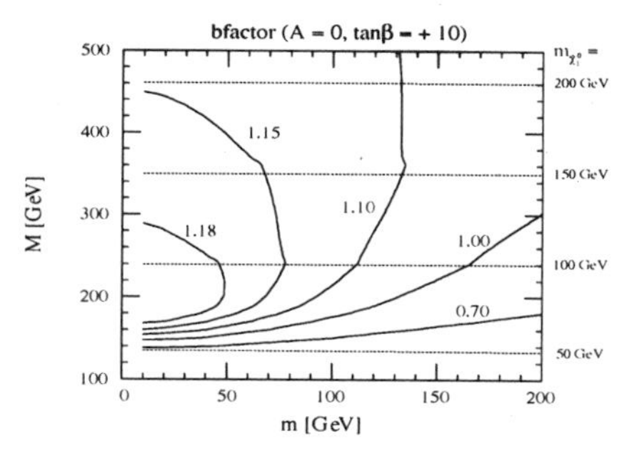

Figure 1. contours of constant b in m and M plane

freedom at LSP freeze–out temperature $T_F = m_{\tilde{\chi}_1^0}/x_F$.

MSUGRA predicts a bino–like LSP $\tilde{\chi}_1^0$ and wino–like $\tilde{\chi}_1^+$ and $\tilde{\chi}_2^0$ for moderate values of m and M (below $\sim$ 500 GeV). This is a rather model independent result.[10] Large positive corrections to squark masses from gaugino loops, together with the large top Yukawa coupling, drive the squared soft breaking Higgs mass m_2^2 negative at the weak scale. One has to make μ large to obtain the correct electroweak symmetry breaking scale, if scalar masses and gaugino masses are of the same order.

If slepton masses are moderate, the LSP is bino–like, and one is sufficiently far away from s–channel poles ($2m_{\tilde{\chi}_1^0} \neq m_Z, m_{\mathrm{Higgs}}$), the mass density is essentially determined by t–channel $\tilde{e}_R$ exchange.[11] This can be seen in Fig. 1 where $b \equiv 10^6$ GeV^2 $\times\, \Omega_{\tilde{\chi}_1^0} h^2 \sigma_{\tilde{B}}$ is plotted. Here $\sigma_{\tilde{B}}$ is the scaled bino pair annihilation cross section in the limit where $m_{\tilde{e}_R} \ll m_{\tilde{e}_L}, m_{\tilde{q}}$ [11],

$$\sigma_{\tilde{B}} \sim \frac{m^2_{\tilde{\chi}_1^0}}{(m^2_{\tilde{e}_R} + m^2_{\tilde{\chi}_1^0})^2} \tag{2}$$

Although corresponding $\Omega_{\tilde{\chi}_1^0} h^2$ changes by more than a factor of 4, the change of b is very small over the wide range of parameter region with $M > 160$ GeV, confirming the bino–like nature of the LSP for the mSUGRA case.

Analyses of sparticle production at the LHC would lead to tight constraints on the predicted thermal relic density $\Omega_{\tilde{\chi}_1^0}h^2$. When the cascade decay $\tilde{q} \to \tilde{\chi}_2^0 \to \tilde{e}_R \to \tilde{\chi}_1^0$ is open, a clean SUSY signal is $ll + jets+$ missing E_T. It was shown [8] that $m_{\tilde{q}}$, $m_{\tilde{\chi}_2^0}$, $m_{\tilde{e}_R}$ and $m_{\tilde{\chi}_1^0}$ can be reconstructed from the upper end points of the m_{jll} and m_{jl} distributions, $m_{jll}^{\max}$ and $m_{jl}^{\max}$; the edge of the m_{ll} distribution, $m_{ll}^{\max}$; and the lower end point of the m_{jll} distribution with $m_{ll} > m_{ll}^{\max}/\sqrt{2}$, $m_{jll}^{\min}$. Here j refers to one of the two hardest jets in the event. In most cases it is chosen such that it has the smaller jll invariant mass; this is meant to select the jet from the primary $\tilde{q} \to \tilde{\chi}_2^0 q$ decay. Those end points are given by the analytical formulae of sparticle masses.

For the previously studied example studied,[8] the so-called "point 5"(m = 100 GeV and M = 300 GeV), the errors on $m_{\tilde{e}_R}$ and $m_{\tilde{\chi}_1^0}$ are strongly correlated and are found to be 12% for $m_{\tilde{\chi}_1^0}$ and 9% for $m_{\tilde{e}_R}$. We find that the corresponding error on $\sigma_{\tilde{B}}$, and hence on the prediction for $\Omega_{\tilde{\chi}_1^0}h^2$, for this parameter point is 20%. If the error (which is dominated by systematics associated with uncertainties of signal distributions) is reduced by a detailed study of various signal distributions, the error on $\sigma_{\tilde{B}}$ may go down below 10%. It is worth noting that $m_{\tilde{e}_R} < mzii$ is prefered when $\Omega_{th} \sim 0.3$ is required. This is because the pair annihilation cross section is large enough to give small mass density of universe when $m_{\tilde{e}_R}$ is close to $m_{\tilde{\chi}_1^0}$, while $m_{\tilde{\chi}_1^0} \sim 0.5 m_{\tilde{\chi}_2^0}$, see Eq.(2). The mass determination would be possible in the wide parameter region.

3 $\Omega_{\tilde{\chi}_1^0}$ in non-mSUGRA scenarios and collider signals

In the previous section, we found that measurements at LHC experiments are sufficient for a prediction of $\Omega_{\tilde{\chi}_1^0}h^2$, if the cascade decay $\tilde{\chi}_2^0 \to \tilde{e}_R \to \tilde{\chi}_1^0$ is open and LSP bino dominance is assumed. Now the question is if LHC experiments can be used to check the assumption that the LSP is mostly a bino. In this and the following Section, we discuss a scenario where the inequality $M_1 < M_2$ is kept, while μ is substantially smaller than the mSUGRA prediction. In such a case Z exchange effects and/or LSP annihilation into W pairs are expected to be more important than in the strict mSUGRA scenario studied in the previous Section, and one needs more information to predict the thermal contribution to $\Omega_{\tilde{\chi}_1^0}h^2$.

The relative size between μ and M is controlled by Higgs sector mass parameters. The MSSM Higgs potential can be written as

$$V = (m_1^2 + \mu^2)H_1^\dagger H_1 + (m_2^2 + \mu^2)H_2^\dagger H_2 ..... \tag{3}$$

Here m_1 and m_2 are soft breaking Higgs masses. In the previous Section we took $m_1 = m_2 = m$ at the GUT scale. However $|\mu| \sim M$ may be achieved by allowing non-universal soft breaking Higgs masses, $m_1, m_2 \neq m$.

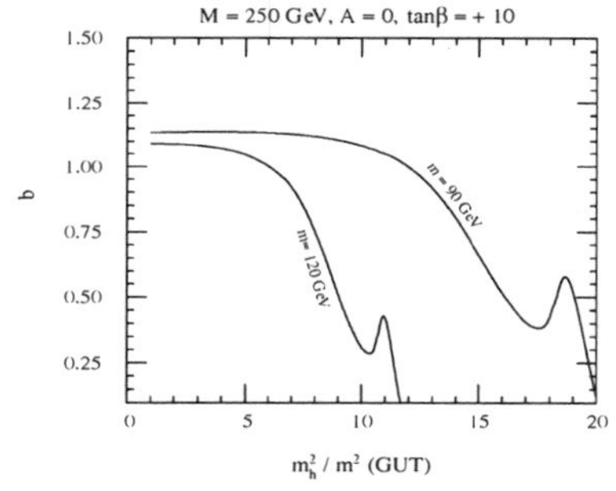

Figure 2. b as function of $(m_h/m)^2$. We fix $M = 250$ GeV, $A = 0$ and $\tan\beta = 10$.

In Fig. 2, we plot b factor vs. $(m_h/m)^2$. By increasing m_h, μ is reduced gradually so that $m_h^2 + \mu^2$ at the GUT scale is roughly constant. Generally $|\mu| \sim M$ can be achieved if $m_h|_{GUT} \gtrsim M$; the precise value is determined by the top Yukawa coupling. These quantities vary substantially once $|\mu|$ falls below M_2 for $m_h/m \sim 4(3)$ for $m = 90(120)$ GeV.

The reduction of $|\mu|$ would alter SUSY signals at colliders significantly. When $|\mu| \lesssim M$, $\tilde{\chi}_4^0$ and $\tilde{\chi}_2^+$ production from the decay of $SU(2)$ doublet squarks becomes important as they have substantial wino component. Namely the channels

$$\tilde{\chi}_2^+ \to \tilde{\nu}_L^{(*)} \to \tilde{\chi}_1^+$$
$$\tilde{\chi}_4^0 \to \tilde{e}_L^{(*)}(\tilde{e}_R^{(*)}) \to \tilde{\chi}_2^0(\tilde{\chi}_1^0) \qquad (4)$$

should be seen in addition to the conventional $\tilde{\chi}_2^0 \to \tilde{\chi}_1^0 ll$ signal. In Fig. 3, we show various $\tilde{q}_L$ decay branching ratios, defined as an average of $\tilde{u}_L$ and $\tilde{d}_L$ branching ratios. As m_h increases we find substantial branching ratios into the heavier neutralino and chargino, once $|\mu|$ becomes comparable to M_2. Discriminating experimentally between scenarios with $|\mu| > M$, where these new signals are small, and $|\mu| < M_2$, where they are expected to be significant, would be important to predict the mass density of the Universe.

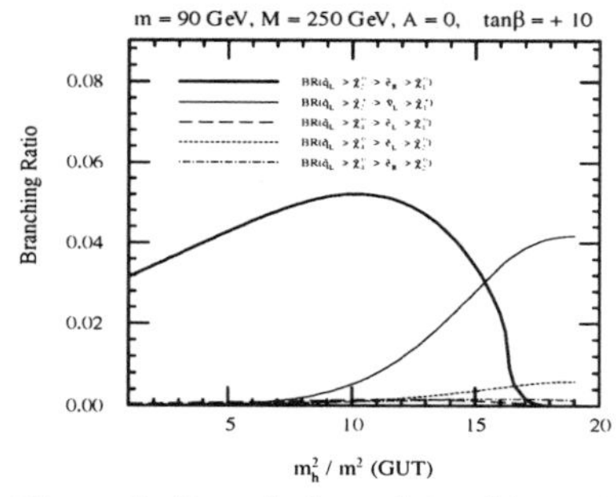

Figure 3. Squark decay branching ratios for $m = 90$ GeV, $M = 250$ GeV, $A = 0$ and $\tan\beta = 10$. We average $\tilde{u}_L$ and $\tilde{d}_L$ decay branching ratios.

We now study leptonic SUSY signals at the LHC for a case where $\tilde{\chi}_2^+$ production from $\tilde{q}_L$ decays is sufficiently common to be detectable. We used ISAJET [12] to generate signal events, while ATLFAST [13] was used to simulate the detector response. For this analysis we choose the MSSM parameter point shown in Table 1. The value of the GUT scale Higgs mass is chosen such that the $\tilde{B}$ component of $\tilde{\chi}_1^0$ $N_{\tilde{B}} = 0.9$, so that effects from its other components on the predicted LSP relic density start to be significant. See Fig. 2.

We now show several SUSY event distributions, after applying the cuts [14] to reduce

m	90.0	$\tan\beta$	10
m_1	360	m_2	360
M	250	μ	199.85
M_1	103.9	M_2	208.75
$m_{\tilde{e}_R}$	139.3	$m_{\tilde{\nu}}$	190.28
$m_{\tilde{\chi}_1^0}$	93.18	$m_{\tilde{\chi}_2^0}$	155.13
$m_{\tilde{\chi}_1^+}$	148.44	$m_{\tilde{\chi}_2^+}$	272.52

Table 1. Mass parameters and relevant sparticle masses in GeV for the point studied in this paper. ISAJET was used to generate this spectrum.

mode	$m_{ll}^{\max}$
D1) $\tilde{q}_L \to \tilde{\chi}_2^0 \to \tilde{e}_R \to \tilde{\chi}_1^0$	50.6
D2) $\tilde{q}_L \to \tilde{\chi}_2^+ \to \tilde{\nu}_L \to \tilde{\chi}_1^+$	121.9
D3) $\tilde{q}_L \to \tilde{\chi}_4^0 \to \tilde{e}_L \to \tilde{\chi}_2^0$	117.9
D4) $\tilde{q}_L \to \tilde{\chi}_4^0 \to \tilde{e}_L \to \tilde{\chi}_1^0$	159.6
D5) $\tilde{q}_L \to \tilde{\chi}_4^0 \to \tilde{e}_R \to \tilde{\chi}_1^0$	174.3

Table 2. End points of m_{ll} distribution (in GeV) for different decay processes.

the SM background to a negligible level. In Fig. 4 we show the di–lepton invariant mass distribution for our representative point. After the subtraction of $e\mu$ events, we see a distribution with at least four edges. They are consistent with those found in Table 2. We then follow the previous analysis,[8] by taking the jets with the first and the second largest P_T and considering their m_{jl} and m_{jll} distributions. Those distributions will contain events from the different decay chains listed in Table 2, but they can easily be separated out by requiring m_{ll} to lie between certain values. Otherwise the analyisis are complitely paralles and we do not repeat this here. The detailed discussion would be found elsewhere.[1]

We now discuss the possibility to identify $\tilde{\chi}_2^+$ decays through the chain D2). In this case most daughter $\tilde{\chi}_1^+$'s would decay further as $\tilde{\chi}_1^+ \to \tilde{\tau}_1 \to \tilde{\chi}_1^0$, producing a τ lepton in the last step of the cascade decay. The τll invariant mass never exceeds $m_{\tilde{\chi}_2^+} - m_{\tilde{\chi}_1^0}$. Hadronic τ decays might be identified by looking for

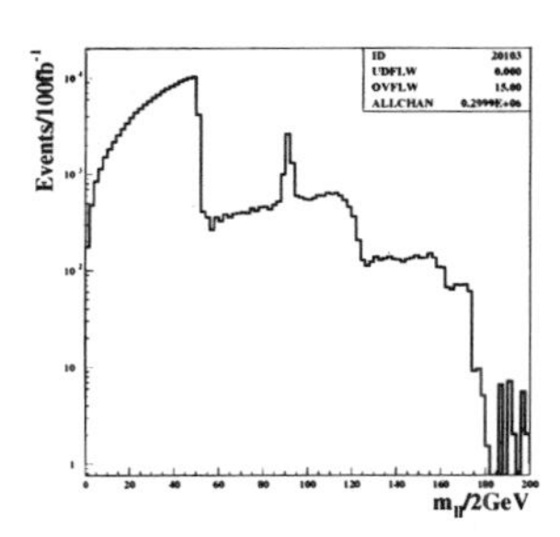

Figure 4. The $m_{e^+e^-} + m_{\mu^+\mu^-} - m_{e^+\mu^-} - m_{e^-\mu^+}$ distribution for the parameter point listed in Table 1.

a narrow jet that is isolated from other jet activity. When we look into the the $m_{j_\tau ll}$ distributions for 55 GeV $< m_{ll} <$ 125 GeV (Fig. 5), we find the events clustered in the region $m_{j_\tau ll} <$ 190 GeV, while no such structure is found for the events with $m_{ll} <$ 55 GeV. The only possible interpretation would be that most ll pairs with 55 GeV$< m_{ll} <$125 GeV stem from the decay of a charged particle, $\tilde{\chi}_2^+$.

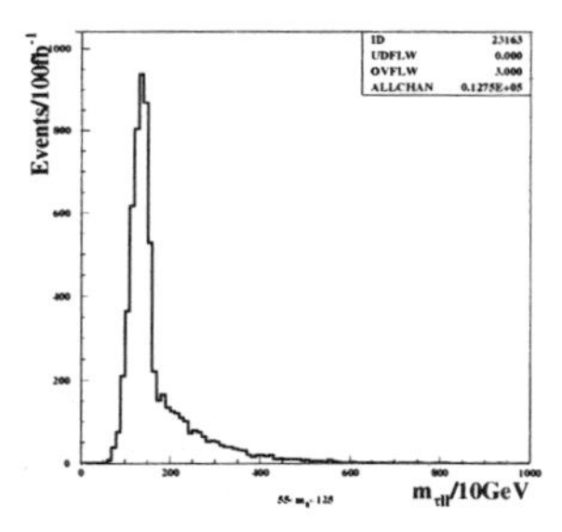

Figure 5. The invariant mass distribution $m_{j_\tau ll}$ for a) $m_{ll} <$ 55 GeV and b) for 55 GeV$< m_{ll} <$ 125 GeV.

Note that there are substantial constraints on –ino masses and slepton masses from $SU(2) \times U(1)$ gauge invariance. The six chargino and neutralino masses are determined by the values of the four parameters M_1, M_2, μ and $\tan\beta$, while $m_{\tilde{e}_L}$ and $m_{\tilde{\nu}}$ are related by $m_{\tilde{\nu}}^2 - m_{\tilde{e}_L}^2 = m_Z^2 \cos^2\theta_W \cos 2\beta$. Therefore the measured edges and end points originating from several decay chains can over–constrain the relevant MSSM parameters. Statistically, it seems possible to measure end points for decay modes D1) and D2). This constrains mass differences among $\tilde{q}_L$, $\tilde{\chi}_2^0$, $\tilde{\chi}_1^0$, $\tilde{e}_R$ (D1) and $\tilde{q}_L$, $\tilde{\chi}_2^+$, $\tilde{\chi}_1^+$, $\tilde{\nu}_L$ (D2). We expect that these masses can be reconstructed with O(10) GeV errors, as was the previous study.[8] However, the corresponding errors on some MSSM parameters are significantly larger.

	M_2	μ	$m_{\tilde{e}_R}$	$m_{\tilde{\nu}}$
μ max	196.8	235.7	147.4	202.6
μ min	218.7	180.1	135.2	185.8

Table 3. Maximal and minimal μ solution satisfying $\Delta\chi^2 \leq 1$

In order to illustrate this point, we list two sets of MSSM parameters which reproduce all kinematic end points within $\Delta\chi^2 =$ 1, We assume error of endpoints is 1% of $M_i^{\rm input}$ for distributions involving a jet, and 0.3 % of $m_{ll}^{\rm max,input}$. In Table 3, we list the solution with maximal and minimal μ (for $\tan\beta \leq 20$) that satisfy $\Delta\chi^2 \leq 1$.

Note that the errors of the dimensionful parameters are strongly correlated, so that solutions with $\Delta\chi^2 < 1$ almost fall onto a one–dimensional line in the seven–dimensional parameter space. For fixed $\tan\beta$ we find two distinct sets of solutions, with $\mu > M$ and $\mu < M$, respectively. $\tan\beta$ cannot be fixed; one can only determine that $\tan\beta \gtrsim 8.65$ where the minimum is achieved when $M_2 \sim \mu$. On the other hand the corresponding chargino and neutralino masses only vary within 15 GeV between the two extreme solutions except for $\tilde{\chi}_3^0$.

Reducing the errors on μ and $\tan\beta$ would be necessary to predict the thermal relic density accurately. The μ (max,min) solutions predict $\Omega_{\tilde{\chi}_1^0} h^2 = 0.160$ and 0.122, respectively, as compared to 0.152 for the input point. Even smaller values of $\Omega_{\tilde{\chi}_1^0} h^2$ are possible if we relax the upper bound on $\tan\beta$, which was imposed "by hand" in this fit.

Here μ determines the size of the higgsino components of the LSP, which begins to be significant in this region of parameter space. The product $\mu \tan\beta$ determines the amount of $\tilde{\tau}_L - \tilde{\tau}_R$ mixing, which reduces the predicted relic density through a reduced $\tilde{\tau}_1$ mass and enhanced $S-$wave annihilation.

Strategies that might be useful for reducing the errors on these two quantities are following. Note that the ratio of the $\tilde{\chi}_2^+$ and $\tilde{\chi}_2^0$ modes increases by more than a factor of three whenswitching from the μ max solution to the μ min solution. This is almost entirely due to the change of μ; the value of $\tan\beta$ is not important here (as long as $\tan^2\beta \gg 1$). On the other hand the branching ratio into e^+e- or $\mu^+\mu-$ are highly $\tan\beta$ dependent because the $\tilde{\chi}_2^0$ decay is sensitive to stau mixing and masses, which depends on $\tan\beta\mu$. Therefore one should first extract information about the $\tilde{\tau}$ sector, e.g. by comparing signals from $\tilde{\chi}_2^0 \to \tau^+\tau^-\tilde{\chi}_1^0$ to those from $\tilde{\chi}_2^0 \to e^+e^-\tilde{\chi}_1^0$. This will give information on the soft breaking masses in the $\tilde{\tau}$ sector as well as on the product $\mu \tan\beta$. This information, together with the result of the kinematic fit described above, will allow one to predict the branching ratios of the decays with reasonable precision. This in turn will allow to translate the measured strengths of the signals from decay chains D1) and D2) into squark branching ratios. Finally, these branching ratios can be used to greatly reduce the error on μ.

Given that the very weak upper bound $\tan\beta \leq 20$ which we imposed in the fit summarized in Table 3 is sufficient to predict $\Omega^{th}_{\tilde{\chi}_1^0} h^2 \simeq 0.135 \pm 0.03$, it seems certain that the strategy outlined above will again allow to predict the thermal relic density to better than 20%. The only loophole occurs if $\tan\beta$ is very large or third generation soft masses are very differnt from that of the first generation. The detailed arguent was given elsewhere.[1]

Whenever LHC experiments find a large sample of SUSY events, it should be possible to either predict the thermal relic density of LSPs with a fairly small error, or else one will be able to conclude that thermal relic LSPs do not contribute significantly to the overall mass density of the Universe. In the latter case one would need physics beyond the MSSM, and/or a non–thermal LSP production mechanism, to explain the Dark Matter in the Universe.

References

1. DAMA Collaboration, Phys. Lett. **B480**, 23 (2000).
2. CDMS Collaboration, Nucl. Instrum. Meth. **A444**, 345 (2000).
3. T. Moroi and L. Randall, Nucl. Phys. **B570**, 455 (2000).
4. K. Enqvist and J. McDonald, Nucl. Phys. **B538**, 321 (1999).
5. G. F. Giudice, E. W. Kolb and A. Riotto, hep–ph/0005123.
6. M. S. Turner, astro–ph/9912211.
7. M. Drees, Y.G. Kim, M.N. Nojiri, D. Toya, K. Hasuko, and T. Kobayashi, hep-ph/0007202.
8. H. Bachacou, I. Hinchliffe and F. E. Paige, Phys. Rev. **D62**, 015009 (2000).
9. E. W. Kolb and M. S. Turner, *The Early Universe*, Addison-Wesley, 1990. K. Griest, M. Kamionkowski and M.S. Turner, Phys. Rev. **D41**, 3565 (1990).
10. T. Falk, Phys. Lett. **B456**, 171 (1999).
11. M. Drees and M. M. Nojiri, Phys. Rev. D**47**, 376 (1993).
12. F. E. Paige, S. D. Protopopescu, H. Baer and X. Tata, hep–ph/9810440.
13. E. Richter-Was et al., ATLFAST2.21, ATLAS Internal Note, PHYS-NO-079.
14. I. Hinchliffe, F. E. Paige, M. D. Shapiro, J. Soderqvist and W. Yao, Phys. Rev. D **55**, 5520 (1997).

LARGE NEUTRINO MIXING IN GRAND UNIFIED THEORIES

F. FERUGLIO
University of Padova and I.N.F.N., Padova, Italy
E-mail: feruglio@pd.infn.it

A non-minimal, semi-realistic version of supersymmetric $SU(5)$ grand unified theory is discussed. The solution of the doublet-triplet splitting problem leads at the same time both to a better agreement between the predicted and observed values of the low-energy strong coupling constant and to a prolongation of the proton lifetime. A $U(1)$ flavor symmetry allows to accommodate a realistic mass spectrum in the charged and in the neutral fermion sectors and protects these results from dangerous radiative corrections or non-renormalizable operators.

Minimal versions of grand unified theories (GUT) are plagued by severe fine-tuning problems.

- Doublet-triplet splitting – By far the most severe obstacle, it requires an unnatural adjustment of one part in 10^{14}. Moreover, the tree-level solution to this problem can be spoiled either by radiative corrections when supersymmetry (SUSY) is broken, or by non renormalizable operators.
- Proton decay – Minimal $SU(5)$ is ruled out by the recent SuperKamiokande data $\tau_p/BR(p \to K^+\bar{\nu}) > 2 \cdot 10^{33}$ ys (90% CL) [1]. Moreover, in the absence of flavour symmetries, non-renormalizable operators contributing to proton decay and originating at the Planck scale M_{Pl} should be depleted by hand by about seven order of magnitudes.
- Wrong mass relations – The equality of down and charged lepton masses at the GUT scale in minimal $SU(5)$ is inaccurate for the first two generations, even though it is correct order-of-magnitude wise. The whole neutrino sector is missing in minimal $SU(5)$.
- Strong coupling constant – Gauge coupling unification in minimal GUT leads to a value of $\alpha_s(M_Z)$ that, although affected by large uncertainties, tends to be too large.

There can also be other issues more specific to the supersymmetric realization of GUT ideas, like for instance the supersymmetric flavour problem. In ref. [2] we discuss a semi-realistic model that addresses and solves the above problems in an extended version of $SU(5)$, supplemented by a $U(1)_Q$ flavour symmetry.

In this model the doublet triplet splitting problem is solved by a variant of the missing partner mechanism employing, beyond 5 and $\bar{5}$, the 50, $\overline{50}$, $75 \equiv Y$ and $1 \equiv X$ $SU(5)$ representations (see table 1) [3]. The multiplet Y, singlet under the flavour symmetry, breaks $SU(5)$ down to $SU(3) \times SU(2) \times U(1)$, while X, characterized by $Q = -1$, is the only field charged under $U(1)_Q$ that acquires a large VEV. In the limit of exact SUSY, the doublets in 5 and $\bar{5}$ remain massless. The mass m_T of the effective triplet suppressing the dimension 5, $|\Delta B| = 1$ operators is proportional to $\langle Y \rangle^2/\langle X \rangle$, where $\langle X \rangle$ is undetermined. Finally, operators like $5\bar{5}X^m Y^n$ $(m, n > 0)$, which potentially could destabilize the doublets, are forbidden by $U(1)_Q$. When SUSY is broken, $\langle X \rangle$ acquires a VEV close to the cut-off Λ of the theory and a μ term can be generated à la Giudice-Masiero from a higher-dimensional term in the Kähler potential.

The spectrum of heavy particles associated to the missing partner mechanism produces two main effects.

- The strong coupling constant $\alpha_s(M_Z)$

Table 1. Chiral Multiplets Quantum Numbers.

Field	$SU(5)$	$U(1)_Q$
H	5	-2
$\overline{H}$	$\bar{5}$	+1
H_{50}	50	2
$H_{\overline{50}}$	$\overline{50}$	-1
Y	75	0
X	1	-1
Ψ_{10}	10	(4,3,1)
$\Psi_{\bar{5}}$	$\bar{5}$	(4,2,2)
Ψ_1	1	(1,-1,0)

receives large threshold corrections from the splitted Y supermultiplet. As a result, $\alpha_s(M_Z)$ is smaller than in minimal $SU(5)$. Indeed values of m_T larger by a factor 20-30 than in minimal $SU(5)$ are required to reconcile the prediction with the data, with a direct advantage for proton decay.

- The model is no longer asymptotically free, due to the large field content. The $SU(5)$ coupling constant blows up at a scale $\overline{\Lambda}$ smaller than the Planck scale. We typically find $\Lambda \leq \overline{\Lambda} \approx 20\ M_{GUT}$.

Fermion masses are obtained from the $U(1)_Q$ charge assignment given in table 1. As well known, abelian charges constrain the spectrum up to unknown coefficients of order one. It is possible to choose these coefficients in order to correctly reproduce quark masses, mixing angles and the CP violating phase. The model predicts $\tan\beta \approx O(1)$, which also moderate the proton decay amplitudes. The neutrino sector of the model is quite similar to the one discussed in ref. [4]. As a consequence of the $U(1)$ assignment and of the see-saw mechanism, a large mixing for atmospheric neutrinos is obtained. Such a mixing is directly related to a large mixing between the right-handed s and b quark fields, via the minimal $SU(5)$ relation $m_e = {m_d}^T$, which is approximately valid also in the present model. This is the reason why a large mixing among leptons is compatible with small quark mixing angles, even in a GUT, where lepton and quarks belong to the same representations of the gauge group. The solar mixing angle is expected to be close to maximal and, numerically, the so called LOW and vacuum oscillation solutions are equally possible. Finally a θ_{13} angle of order 0.05 is predicted. A well known obstacle in minimal $SU(5)$ is the strict equality $m_e = {m_d}^T$, compatible with the third generation, but inexact for the first and the second families. The correction requires order-one adjustments that can be obtained in the present model by allowing, beyond the minimal $\Psi_{10}G_d\Psi_{\bar{5}}\bar{5}$ Yukawa coupling also the non-renormalizable term $1/\Lambda\Psi_{10}F_d\Psi_{\bar{5}}\bar{5}Y$. The Y multiplet differentiates charged leptons from down quarks and we find:

$$m_d \approx \left[G_d + \frac{\langle Y\rangle}{\Lambda}F_d\right] \quad , \tag{1}$$

$$m_e^T \approx \left[G_d - 3\frac{\langle Y\rangle}{\Lambda}F_d\right] \quad , \tag{2}$$

where the 3×3 matrices G_d and F_d are constrained by the flavour symmetry. It is interesting to observe that we can reproduce the relations $m_\tau \approx m_b$, $m_\mu \approx 3m_s$ and $m_e \approx m_d/3$, by taking $\langle Y\rangle/\Lambda$ of order 0.1, in agreement with $\Lambda \leq \overline{\Lambda} \approx 20\ M_{GUT}$. While the predictivity of the model is reduced because non-renormalizable operators are only suppressed by powers of M_{GUT}/Λ, still these corrections could explain the small distortion of the spectrum with respect to the minimal model.

Proton decay dominant amplitudes are derived from the dimension 5 superpotential:

$$w = \frac{1}{m_T}\left[Q\hat{A}QQ\hat{C}L + U^c\hat{B}E^cU^c\hat{D}D^c\right] \quad , \tag{3}$$

which, although formally equal to that of the minimal model, contains four important differences:

- An effective triplet mass m_T, larger by a factor 20-30 than in minimal $SU(5)$ leads to a suppression factor 400-900 in rate.
- An additional Yukawa coupling is allowed by the symmetries of the theory: $\Psi_{10}G_{\overline{50}}\Psi_{10}\overline{50}$. While the couplings of the conventional term $\Psi_{10}G_u\Psi_{10}5$ are restricted by the up quark masses, the couplings of the new term are unconstrained, since $\langle\overline{50}\rangle = 0$. We obtain:

$$\hat{B} = -2\hat{A} = \left[G_u - \frac{c_2\langle Y\rangle}{c_4\langle X\rangle}G_{\overline{50}}\right] , \quad (4)$$

where c_2 and c_4 are dimensionless coefficients. As a consequence, a large region in parameter space exists where a sizeable destructive interference between the G_u and the $G_{\overline{50}}$ contributions can occur.
- Also the $\hat{C}$ and $\hat{D}$ couplings are distorted:

$$\hat{C} = \left[-G_d - \frac{\langle Y\rangle}{\Lambda}F_d\right] , \quad (5)$$

$$\hat{D} = \left[G_d - \frac{\langle Y\rangle}{\Lambda}F_d\right] . \quad (6)$$

This modification, however, has a not-too-large effect on proton decay rates.
- The non-renormalizable operators that could originate at the cut-off scale Λ are controlled by the flavour symmetry and lead to a contribution to the proton decay amplitude that can be comparable to the one coming from the color triplet exchange.

As a result, the predicted range for the proton decay rates considerably extends with respects to that of minimal $SU(5)$, allowing values that are not incompatible with the present limits and are testable in the next generation of experiments. In particular, our numerical estimate gives $8 \cdot 10^{31}$ ys $< \tau_p/BR(p \to K^+\bar{\nu}) < 3 \cdot 10^{34}$ ys and $2 \cdot 10^{32}$ ys $< \tau_p/BR(p \to \pi^+\bar{\nu}) < 8 \cdot 10^{34}$ ys.

In summary, it is a remarkable feature of the model that the presence of the representations 50, $\overline{50}$ and 75, demanded by the missing partner mechanism for the solution of the doublet-triplet splitting problem, directly produces, through threshold corrections at M_{GUT}, a decrease of the value of $\alpha_s(m_Z)$ that corresponds to coupling unification and an increase in the effective mass that mediates proton decay. As a consequence the value of the strong coupling is in better agreement with the experimental value and the proton decay rate is smaller by a factor 400-900 than in the minimal model. The presence of these large representations also has the consequence that the asymptotic freedom of $SU(5)$ is spoiled and the associated gauge coupling becomes non perturbative below M_{Pl}. We argue that this property far from being unacceptable can actually play an important role to obtain better results for fermion masses.

Acknowledgments

I warmly thank Guido Altarelli and Isabella Masina for the enjoyable collaboration on which this talk is based.

References

1. Masato Shiozawa, talk at this conference.
2. G. Altarelli, F. Feruglio and I. Masina, “From minimal to Realistic Supersymmetric $SU(5)$ Grand Unification”, preprint CERN-TH.2000-171, DFPD 00/TH/34, hep-ph/0007254.
3. Z. Berezhiani and Z. Tavartkiladze, Phys. Lett. B396 (1997) 150, hep-ph/9611277.
4. G. Altarelli and F. Feruglio, JHEP 9811 (1998) 021, hep-ph/9809596; Phys. Lett. B451 (1999) 388, hep-ph/9812475.

SOLAR NEUTRINO SOLUTIONS IN NON-ABELIAN FLAVOUR SYMMETRY

MORIMITSU TANIMOTO
Department of Physics, Niigata University, Niigata, JAPAN,
E-mail: tanimoto@muse.hep.sc.niigata-u.ac.jp

We have studied the large mixing angle MSW solution for the solar neutrinos in the non-abelian flavor symmetry. We predict the MNS mixing matrix taking account of the symmetry breakings.

1 LMA-MSW Solution in Non-Abelian Flavor Symmetry

Recent data in S-Kam favor the large mixing angle MSW (LMA-MSW) solution. How does one get the LMA-MSW solution as well as the maximal mixing of the atmospheric neutrinos in theory? It is not easy to reproduce the nearly bi-maximal mixings with LMA-MSW solution in GUT models[1,2,3].

The non-abelian flavor symmetry $S_{3L} \times S_{3R}$ or $O_{3L} \times O_{3R}$ leads to the LMA-MSW solution naturally[4,5]. The mass matrices are

$$M_E \propto \begin{pmatrix} 1 & 1 & 1 \\ 1 & 1 & 1 \\ 1 & 1 & 1 \end{pmatrix}, \quad M_\nu \propto \begin{pmatrix} 1 & 0 & 0 \\ 0 & 1 & 0 \\ 0 & 0 & 1 \end{pmatrix}.$$

The orthogonal matrix diagonalizes M_E is

$$F = \begin{pmatrix} 1/\sqrt{2} & 1/\sqrt{6} & 1/\sqrt{3} \\ -1/\sqrt{2} & 1/\sqrt{6} & 1/\sqrt{3} \\ 0 & -2/\sqrt{6} & 1/\sqrt{3} \end{pmatrix}. \quad (1)$$

The MNS mixing matrix is given as $U_\nu \simeq F^T$, and so we predict $\sin^2 2\theta_\odot = 1$ and $\sin^2 2\theta_{\rm atm} = 8/9$ in the symmetric limit.

In this talk, we discuss masses and flavor mixings of quarks/leptons in the non-abelian flavor symmetry with the SU(5) GUT[4]. We consider $O(3)_{\mathbf{5}^*} \times O(3)_{\mathbf{10}} \times Z_6$ symmetry. Our scenario for fermion masses is

- Neutrinos have degenerate masses.
- Quarks/charged-leptons are massless.
- Symmetry breakings give Δm^2 and other fermion masses.

2 $O(3)_{\mathbf{5}^*} \times O(3)_{\mathbf{10}} \times Z_6$ Symmetry

Quarks and leptons belong to $\mathbf{5}^*$ and $\mathbf{10}$ of the $SU(5)$ GUT and $\mathbf{3}$ of the $O(3)$ symmetry. Higgs H ($\overline{H}$) belong to $\mathbf{5}$ ($\mathbf{5}^*$) of the $SU(5)$ and $\mathbf{1}$ of the $O(3)$. Then, neutrinos have the $O(3)_{\mathbf{5}^*} \times O(3)_{\mathbf{10}}$ invariant mass term

$$\frac{< H >^2}{\Lambda} \nu_L \nu_L \ . \quad (2)$$

The Z_6 symmetry forbids $\psi_{\mathbf{10}}(\mathbf{3})\psi_{\mathbf{10}}(\mathbf{3})H$, which gives degenerate up-quark masses[4].

The flavor symmetry is broken explicitly by $\Sigma^{(i)}_{\mathbf{5}^*}(\mathbf{5},\mathbf{1})$, $\Sigma^{(i)}_{\mathbf{10}}(\mathbf{1},\mathbf{5})$ ($i = 1,2$), which transform as the symmetric traceless tensor $\mathbf{5}$'s of $O(3)$. Dimentionless breaking parameters are given as

$$\sigma^{(1)}_{\mathbf{10},\ \mathbf{5}^*} \equiv \frac{\Sigma^{(1)}_{\mathbf{10},\ \mathbf{5}^*}}{M_f} = \begin{pmatrix} 1 & 0 & 0 \\ 0 & 1 & 0 \\ 0 & 0 & -2 \end{pmatrix} \delta_{\mathbf{10},\ \mathbf{5}^*},$$

$$\sigma^{(2)}_{\mathbf{10},\ \mathbf{5}^*} \equiv \frac{\Sigma^{(2)}_{\mathbf{10},\ \mathbf{5}^*}}{M_f} = \begin{pmatrix} 1 & 0 & 0 \\ 0 & -1 & 0 \\ 0 & 0 & 0 \end{pmatrix} \epsilon_{\mathbf{10},\ \mathbf{5}^*}.$$

Neutrinos get Majorana masses from a superpotential

$$W = \frac{H^2}{\Lambda}\ell(\mathbf{1} + \alpha_i \sigma^{(i)}_{5^*})\ell \ , \quad (3)$$

which yields a diagonal neutrino mass matrix. In order to get the charged lepton masses, we introduce $O(3)_{\mathbf{5}^*}$-triplet $\phi_{\mathbf{5}^*}(\mathbf{3},\mathbf{1})$ and $O(3)_{\mathbf{10}}$-triplet $\phi_{\mathbf{10}}(\mathbf{1},\mathbf{3})$. These VEV's are determined by the superpotential

$$\begin{aligned} W &= Z_{\mathbf{5}^*}(\phi^2_{\mathbf{5}^*} - 3v^2_{\mathbf{5}^*}) + Z_{\mathbf{10}}(\phi^2_{\mathbf{10}} - 3v^2_{\mathbf{10}}) \\ &+ X_{\mathbf{5}^*}(a_{(i)}\phi_{\mathbf{5}^*}\sigma^{(i)}_{\mathbf{5}^*}\phi_{\mathbf{5}^*}) + X_{\mathbf{10}}(a'_{(i)}\phi_{\mathbf{10}}\sigma^{(i)}_{\mathbf{10}}\phi_{\mathbf{10}}) \\ &+ Y_{\mathbf{5}^*}(b_{(i)}\phi_{\mathbf{5}^*}\sigma^{(i)}_{\mathbf{5}^*}\phi_{\mathbf{5}^*}) + Y_{\mathbf{10}}(b'_{(i)}\phi_{\mathbf{10}}\sigma^{(i)}_{\mathbf{10}}\phi_{\mathbf{10}}) \end{aligned}$$

where $Z_{\mathbf{10},\ \mathbf{5}^*}$, $X_{\mathbf{10},\ \mathbf{5}^*}$, $Y_{\mathbf{10},\ \mathbf{5}^*}$ are all singlets of $O(3)_{\mathbf{5}^*} \times O(3)_{\mathbf{10}}$. Minimizing the potential, we get

$$< \phi_{\mathbf{5}^*} > \equiv \begin{pmatrix} 1 \\ 1 \\ 1 \end{pmatrix} v_{\mathbf{5}^*}, \ < \phi_{\mathbf{10}} > \equiv \begin{pmatrix} 1 \\ 1 \\ 1 \end{pmatrix} v_{\mathbf{10}}.$$

Masses of charged leptons arise from a superpotential

$$W = \frac{\kappa_E}{M_f^2} (\bar{e}\phi_{\mathbf{10}})(\phi_{\mathbf{5}^*}\ell)\overline{H}, \tag{4}$$

which is the realization of "Democratic Mass Matrix",

$$M_E \propto \left(\frac{v_{\mathbf{5}^*} v_{\mathbf{10}}}{M_f^2}\right) \begin{pmatrix} 1 & 1 & 1 \end{pmatrix} \begin{pmatrix} 1 \\ 1 \\ 1 \end{pmatrix}. \tag{5}$$

Adding the superpotential containing the flavor symmetry breaking parameters $\sigma^{(i)}_{\mathbf{5}^*,\ \mathbf{10}}$, we get the charged lepton mass matrix:

$$M_E^H \equiv F^T M_E F = \kappa_E \left(\frac{v_{\mathbf{5}^*} v_{\mathbf{10}}}{M_f^2}\right) < \overline{H} > \times \begin{pmatrix} \epsilon_{\mathbf{5}^*}\epsilon_{\mathbf{10}} & \epsilon_{\mathbf{10}}\delta_{\mathbf{5}^*} & \epsilon_{\mathbf{10}} \\ \epsilon_{\mathbf{5}^*}\delta_{\mathbf{10}} & \delta_{\mathbf{5}^*}\delta_{\mathbf{10}} & \delta_{\mathbf{10}} \\ \epsilon_{\mathbf{5}^*} & \delta_{\mathbf{5}^*} & 3 \end{pmatrix}, \tag{6}$$

in which order one coefficients are omitted. The mass ratios are given as

$$\frac{m_\mu}{m_\tau} \simeq \mathcal{O}(\delta_{\mathbf{5}^*}\delta_{\mathbf{10}}), \quad \frac{m_e}{m_\tau} \simeq \mathcal{O}(\epsilon_{\mathbf{5}^*}\epsilon_{\mathbf{10}}).$$

The quark/lepton masses and mixings fix

$$\delta_{\mathbf{10}} \simeq \lambda^2, \ \epsilon_{\mathbf{10}} \simeq \lambda^3 \sim \lambda^4, \ \delta_{\mathbf{5}^*} \simeq \lambda, \ \epsilon_{\mathbf{5}^*} \simeq \lambda^2.$$

3 Neutrino Masses and Mixings

Neutrino masses are given as

$$m_1 \simeq c_\mu(1 + \alpha_1\delta_{\mathbf{5}^*} + \alpha_2\epsilon_{\mathbf{5}^*}),$$
$$m_2 \simeq c_\mu(1 + \alpha_1\delta_{\mathbf{5}^*} - \alpha_2\epsilon_{\mathbf{5}^*}),$$
$$m_3 \simeq c_\mu(1 - 2\alpha_1\delta_{\mathbf{5}^*}), \quad c_\mu = \frac{< H >^2}{\Lambda}$$

which leads to (with $\delta_{\mathbf{5}^*} \simeq \lambda, \ \ \epsilon_{\mathbf{5}^*} \simeq \lambda^2$)

$$\left|\frac{\Delta m^2_{21}}{\Delta m^2_{32}}\right| = \frac{2}{3}\frac{\alpha_2\epsilon_{\mathbf{5}^*}}{\alpha_1\delta_{\mathbf{5}^*}} \frac{1 + \alpha_2\epsilon_{\mathbf{5}^*}}{1 - \frac{1}{2}\alpha_1\delta_{\mathbf{5}^*}} \simeq \lambda^2 \sim \lambda.$$

Putting $\Delta m^2_{32} = 3 \times 10^{-3}\mathrm{eV}^2$, we predict $\Delta m^2_{21} \simeq$ (factor) $\times\ 10^{-4}\mathrm{eV}^2$, which is consistent with the LMA-MSW solution. Flavor mixings come from the charge lepton mass matrix since the neutrino one is diagonal. The charged lepton mass matrix is diagonalized by $V_R^\dagger M_E^H V_L$, in which

$$V_L^\dagger \simeq \begin{pmatrix} 1 & \lambda & \lambda^2 \\ -\lambda & 1 & \lambda \\ -\lambda^2 & -\lambda & 1 \end{pmatrix}. \tag{7}$$

The neutrino mixing matrix is given by $V_L^\dagger F^T$. We predict

$$\sin^2 2\theta_\odot = (1 - \frac{4}{3}\lambda^2)^2 \simeq 0.87$$
$$\sin^2 2\theta_{\mathrm{atm}} = \frac{8}{9}(1 - \lambda^2)(1 + \frac{1}{\sqrt{2}}\lambda - 2\lambda^2)^2 \simeq 0.95$$
$$|U_{e3}| = \frac{2}{\sqrt{6}}\lambda(1 - \frac{1}{\sqrt{2}}\lambda) \simeq 0.14. \tag{8}$$

4 Summary

It is remarked that:

- The solar neutrino mixing $\sin^2 2\theta_\odot$ deviates from the maximal mixing (~ 0.87).
- The atmospheric neutrino mixing $\sin^2 2\theta_{\mathrm{atm}}$ deviates from 8/9 depending phase of λ.
- $\mathbf{U_{e3}}$ is near to the experimental bound of CHOOZ ($\leq \mathbf{0.16}$) .

Neutrino masses are degenerated within a factor 2. For example, we get $m_1 \simeq 0.030\mathrm{eV}$, $m_2 \simeq 0.033\mathrm{eV}$, $m_3 \simeq 0.058\mathrm{eV}$, which is consistent with $\beta\beta_{0\nu}$ decay bound.

References

1. Y. Nomura and T. Yanagida, *Phys. Rev.* D**59**, 017303 (2000).
2. Y. Nomura and T. Sugimoto, *Phys. Rev.* D**61**, 093003 (2000).
3. C.H. Albright and S.M. Barr, *Phys. Lett.* B**461**, 218 (2000).
4. M.Tanimoto, T.Watari and T.Yanagida, *Phys. Lett.* B**461**,345(1999).
5. M. Tanimoto, *Phys. Lett.* B**483**, 417 (2000).

FLAVOR SYMMETRY AND NEUTRINO OSCILLATIONS

YUE-LIANG WU

Institute of Theoretical Physics,
Chinese Academy of Sciences, Beijing 100080, China
E-mail: ylwu@itp.ac.cn

We show how the nearly bi-maximal mixing scenario comes out naturally from gauged $SO(3)_F$ flavor symmetry via spontaneous symmetry breaking. An interesting relation between the neutrino mass-squared differences and the mixing angle, i.e., $\Delta m^2_{\mu e}/\Delta m^2_{\tau\mu} \simeq 2|U_{e3}|^2$ is obtained. The smallness of the ratio (or $|U_{e3}|$) can also naturally be understood from an approximate permutation symmetry. Once the mixing element $|U_{e3}|$ is determined, such a relation will tell us which solution will be favored within this model. The model can also lead to interesting phenomena on lepton-flavor violations.

1 Introduction

The greatest success of the standard model (SM) is the gauge symmetry structure $SU(3)_c \times SU_L(2) \times U_Y(1)$ which has been tested by more and more precise experiments. In the SM, neutrinos are assumed to be massless. The recent evidences for oscillation of atmospheric neutrinos[1] and for the deficit of the measured solar neutrino flux[2] strongly suggest that neutrinos are massive though their masses are small, and new physics beyond the SM is necessary. The scenario most favoured by the current data[3] may comprise just three light neutrinos with nearly bimaximal mixing via MSW solution[4]. It is of interest to note that such a scenario was shown to be naturally obtained from a simple extention of the SM with gauged $SO(3)_F$ flavor symmetry[5]. In this talk I mainly describe the most interesting features resulting from such a simply extended model.

2 The model

For a less model-dependent analysis, we directly start from an $SO(3)_F \times SU(2)_L \times U(1)_Y$ invariant effective lagrangian with three $SO(3)_F$ Higgs triplets

$$\begin{aligned}\mathcal{L} &= \frac{1}{2}g_3' A^k_\mu (\bar{L}_i \gamma^\mu (t^k)_{ij} L_j + \bar{e}_{Ri}\gamma^\mu (t^k)_{ij} e_{Rj}) \\ &+ (Y_{1ij}\bar{L}_i \phi_1 e_{R\,j} + Y_{2ij}\bar{L}_i\phi_2\phi_2^T L^c_j + H.c.) \\ &+ D_\mu\varphi^* D^\mu\varphi + D_\mu\varphi'^* D^\mu\varphi' + D_\mu\varphi''^* D^\mu\varphi'' \\ &- V_\varphi + \mathcal{L}_{SM} \qquad (1)\end{aligned}$$

with effective Yukawa couplings

$$\begin{aligned}Y_{1ij} &= c_1\varphi_i\varphi_j\chi + c_1'\varphi_i'\varphi_j'\chi' + c_1''\varphi_i''\varphi_j''\chi'' \\ Y_{2ij} &= c_0\varphi_i\varphi_j^* + c_0'\varphi_i'\varphi_j'^* + c_0''\varphi_i''\varphi_j''^* + c\delta_{ij}\end{aligned}$$

$\mathcal{L}_{SM}$ denotes the lagrangian of the standard model. $\bar{L}_i(x) = (\bar{\nu}_i, \bar{e}_i)_L$ (i=1,2,3) are the $SU(2)_L$ doublet leptons and $e_{R\,i}$ $(i = 1,2,3)$ are the three right-handed charged leptons. $A^i_\mu(x)t^i$ $(i = 1,2,3)$ are the $SO(3)_F$ gauge bosons with t^i the $SO(3)_F$ generators and g_3' is the corresponding gauge coupling constant. Here $\phi_1(x)$ and $\phi_2(x)$ are two Higgs doublets, $\varphi(x)$, $\varphi'(x)$ and $\varphi''(x)$ are three $SO(3)_F$ Higgs triplets, and $\chi(x)$, $\chi'(x)$ and $\chi''(x)$ are three singlet scalars. The couplings c, c_a, c_a' and c_a'' $(a = 0,1)$ are dimensional constants. The structure of the above effective lagrangian can be obtained by imposing an additional U(1) symmetry [5].

The Higgs potential for the $SO(3)_F$ Higgs triplets has the following general form before symmetry breaking

$$\begin{aligned}V_\varphi &= \frac{1}{2}\mu^2(\varphi^\dagger\varphi) + \frac{1}{2}\mu'^2(\varphi'^\dagger\varphi') + \frac{1}{2}\mu''^2(\varphi''^\dagger\varphi'') \\ &+ \frac{1}{4}\lambda(\varphi^\dagger\varphi)^2 + \frac{1}{4}\lambda'(\varphi'^\dagger\varphi')^2 + \frac{1}{4}\lambda''(\varphi''^\dagger\varphi'')^2 \\ &+ \frac{1}{2}\kappa_1(\varphi^\dagger\varphi)(\varphi'^\dagger\varphi') + \frac{1}{2}\kappa_1'(\varphi^\dagger\varphi)(\varphi''^\dagger\varphi'') \\ &+ \frac{1}{2}\kappa_1''(\varphi'^\dagger\varphi')(\varphi''^\dagger\varphi'') + \frac{1}{2}\kappa_2(\varphi^\dagger\varphi')(\varphi'^\dagger\varphi)\end{aligned}$$

$$+\frac{1}{2}\kappa_2'(\varphi^\dagger\varphi'')(\varphi''^\dagger\varphi) + \frac{1}{2}\kappa_2''(\varphi'^\dagger\varphi'')(\varphi''^\dagger\varphi') \,.$$

As the $SO(3)_F$ flavor symmetry is treated to be a gauge symmetry, one can always express the complex $SO(3)_F$ Higgs triplet field in terms of three rotational fields $\eta_i(x)$ and three amplitude fields $\rho_i(x)$

$$\begin{pmatrix} \varphi_1(x) \\ \varphi_2(x) \\ \varphi_3(x) \end{pmatrix} = e^{i\eta_i(x)t^i}\frac{1}{\sqrt{2}}\begin{pmatrix} \rho_1(x) \\ i\rho_2(x) \\ \rho_3(x) \end{pmatrix} \quad (2)$$

Similar forms are for $\varphi'(x)$ and $\varphi''(x)$. Assuming that only the amplitude fields get VEVs after spontaneous symmetry breaking, namely $< \rho_i(x) >= \sigma_i$, $< \rho_i'(x) >= \sigma_i'$ and $< \rho_i''(x) >= \sigma_i''$, we then obtain the following equations from minimizing the Higgs potential[5]

$$\begin{aligned} &\sigma_1' = \sqrt{\xi}\sigma_1, \ \sigma_2' = \sqrt{\xi}\sigma_2, \ \sigma_3' = -\sqrt{\xi}\sigma_3, \\ &\sigma_1'' = \sqrt{2\xi'}\sigma_1, \ \ \sigma_2'' = -\sqrt{2\xi'}\sigma_2, \ \ \sigma_3'' = 0, \\ &\sigma_3^2 = \sigma_1^2 + \sigma_2^2 = 2\sigma_1^2 = \sigma^2/2 \,. \end{aligned} \quad (3)$$

where we have asummed a global minimum potential energy $V_\varphi|_{min}$ for varying ξ and ξ' at the minimizing point

$$\begin{aligned} V_\varphi|_{min} = &-\sigma^4(\lambda + \lambda'\xi^2 + \lambda''\xi'^2 \\ &+ 2\kappa_1\xi + 2\kappa_1'\xi' + 2\kappa_1''\xi\xi')/4 \end{aligned} \quad (4)$$

with $\sigma^2 = \sigma_1^2 + \sigma_2^2 + \sigma_3^2$, $\xi = \sigma^2/\sigma'^2$ and $\xi' = \sigma^2/\sigma''^2$. It is seen that with these considerations the VEVs are completely determined by the Higgs potential.

It is remarkable that with these relations the mass matrices of the neutrinos and charged leptons are simply given by

$$M_e = \frac{m_\tau}{2}\begin{pmatrix} \frac{1}{2} & \frac{1}{2}i & \frac{1}{\sqrt{2}} \\ \frac{1}{2}i & -\frac{1}{2} & \frac{1}{\sqrt{2}}i \\ \frac{1}{\sqrt{2}} & \frac{1}{\sqrt{2}}i & 1 \end{pmatrix} \quad (5)$$

$$+\frac{m_\mu}{2}\begin{pmatrix} \frac{1}{2} & \frac{1}{2}i & -\frac{1}{\sqrt{2}} \\ \frac{1}{2}i & -\frac{1}{2} & -\frac{1}{\sqrt{2}}i \\ -\frac{1}{\sqrt{2}} & -\frac{1}{\sqrt{2}}i & 1 \end{pmatrix} - \frac{m_e}{2}\begin{pmatrix} 1 & i & 0 \\ i & -1 & 0 \\ 0 & 0 & 0 \end{pmatrix}$$

$$M_\nu = \hat{m}_\nu \begin{pmatrix} 1 & 0 & \frac{1}{\sqrt{2}}\hat{\delta}_- \\ 0 & 1 & 0 \\ \frac{1}{\sqrt{2}}\hat{\delta}_- & 0 & 1+\hat{\Delta}_- \end{pmatrix} \quad (6)$$

3 Nearly Bimaximal Mixing

It is more remarkable that the mass matrix M_e can be diagonalized by a unitary bi-maximal mixing matrix U_e via $D_e = U_e^\dagger M_e U_e^*$ with

$$U_e^\dagger = \begin{pmatrix} \frac{1}{\sqrt{2}}i & -\frac{1}{\sqrt{2}} & 0 \\ \frac{1}{2} & -\frac{1}{2}i & -\frac{1}{\sqrt{2}} \\ \frac{1}{2} & -\frac{1}{2}i & \frac{1}{\sqrt{2}} \end{pmatrix} \quad (7)$$

and $D_e = diag.(m_e, m_\mu, m_\tau)$. The neutrino mass matrix can be easily diagonalized by an orthogonal matrix O_ν via $O_\nu^T M_\nu O_\nu$ with $(O_\nu)_{13} = \sin\theta_\nu \equiv s_\nu$ and $\tan 2\theta_\nu = \sqrt{2}\hat{\delta}_-/\hat{\Delta}_-$. Thus the CKM-type lepton mixing matrix U_{LEP} that appears in the interaction term $\mathcal{L}_W = \bar{e}_L\gamma^\mu U_{LEP}\nu_L W_\mu^- + H.c.$ is given by $U_{LEP} = U_e^\dagger O_\nu$. Explicitly, one has

$$U_{LEP} = \begin{pmatrix} \frac{1}{\sqrt{2}}ic_\nu & -\frac{1}{\sqrt{2}} & \frac{1}{\sqrt{2}}is_\nu \\ \frac{1}{2}c_\nu + \frac{1}{\sqrt{2}}s_\nu & -\frac{1}{2}i & \frac{1}{2}s_\nu - \frac{1}{\sqrt{2}}c_\nu \\ \frac{1}{2}c_\nu - \frac{1}{\sqrt{2}}s_\nu & -\frac{1}{2}i & \frac{1}{2}s_\nu + \frac{1}{\sqrt{2}}c_\nu \end{pmatrix}. \quad (8)$$

The three neutrino masses are found to be

$$\begin{aligned} m_{\nu_e} &= \hat{m}_\nu[1 - (\sqrt{\hat{\Delta}_-^2 + 2\hat{\delta}_-^2} - \hat{\Delta}_-)/2\,] \\ m_{\nu_\mu} &= \hat{m}_\nu \\ m_{\nu_\tau} &= \hat{m}_\nu[1 + \hat{\Delta}_- + (\sqrt{\hat{\Delta}_-^2 + 2\hat{\delta}_-^2} - \hat{\Delta}_-)/2\,]. \end{aligned} \quad (9)$$

The similarity between the Higgs triplets $\varphi(x)$ and $\varphi'(x)$ naturally motivates us to consider an approximate (and softly broken) permutation symmetry between them. This implies that $|\hat{\delta}_-| << 1$. To a good approximation, the mass-squared differences are given by $\Delta m_{\mu e}^2 \equiv m_{\nu_\mu}^2 - m_{\nu_e}^2 \simeq \hat{m}_\nu^2\hat{\Delta}_-(\hat{\delta}_-/\hat{\Delta}_-)^2$ and $\Delta m_{\tau\mu}^2 \equiv m_{\nu_\tau}^2 - m_{\nu_\mu}^2 \simeq \hat{m}_\nu^2\hat{\Delta}_-(2+\hat{\Delta}_-)$, which leads to the approximate relation

$$\begin{aligned} &\frac{\Delta m_{\mu e}^2}{\Delta m_{\tau\mu}^2} \simeq \left(\frac{\hat{\delta}_-}{\sqrt{2}\hat{\Delta}_-}\right)^2 \simeq s_\nu^2 = 2|U_{e3}|^2 << 1 \\ &\quad 0.2 \sim 0.09 \qquad MSW - LMA \\ &\simeq 0.02 \sim 0.002 \quad MSW - LOW \\ &\quad 10^{-7} \qquad\qquad Vacuum\ Oscillation \end{aligned} \quad (10)$$

which implies that once the mixing element $|U_{e3}|$ is determined, such a relation will tell us which solution should be favored.

When going back to the weak gauge and charged-lepton mass basis, the neutrino mass matrix gets the following interesting form

$$M_\nu/\hat{m}_\nu \simeq \begin{pmatrix} 0 & \frac{1}{\sqrt{2}}i & \frac{1}{\sqrt{2}}i \\ \frac{1}{\sqrt{2}}i & \frac{1}{2} & -\frac{1}{2} \\ \frac{1}{\sqrt{2}}i & -\frac{1}{2} & \frac{1}{2} \end{pmatrix} \quad (11)$$

$$+\frac{\hat{\delta}_-}{2}\begin{pmatrix} 0 & -\frac{1}{\sqrt{2}}i & \frac{1}{\sqrt{2}}i \\ -\frac{1}{\sqrt{2}}i & -1 & 0 \\ \frac{1}{\sqrt{2}}i & 0 & 1 \end{pmatrix} + \frac{\hat{\Delta}_-}{2}\begin{pmatrix} 0 & 0 & 0 \\ 0 & 1 & -1 \\ 0 & -1 & 1 \end{pmatrix}.$$

As $(M_\nu)_{ee} = 0$, the neutrinoless double beta decay is forbiden in the model. Thus the neutrino masses can be approximately degenerate and large enough ($\hat{m}_\nu = O(1)$ eV) to play a significant cosmological role. Note that our scenario was shown to remain stable after considering renormalization group effects[5].

4 Lepton Flavor Violations

The mass matrix of gauge fields A_μ^i is

$$M_F^2 = \frac{m_F^2}{3}\begin{pmatrix} 2(\xi_+ + \xi') & 0 & -\sqrt{2}\xi_- \\ 0 & 3\xi_+ + \xi' & 0 \\ -\sqrt{2}\xi_- & 0 & 3\xi_+ + \xi' \end{pmatrix} \quad (12)$$

with $m_F^2 = 3g_3'^2\sigma^2/8$ and $\xi_\pm = (1\pm\xi)/2$. This mass matrix is diagonalized by an orthogonal matrix O_F via $O_F^T M_F^2 O_F$ with $(O_F)_{13} = \sin\theta_F \equiv s_F$ and $\tan 2\theta_F = 2\sqrt{2}\xi_-/(\xi_+ - \xi')$. Denoting the physical gauge fields as F_μ^i, we then have $A_\mu^i = O_F^{ij}F_\mu^j$. In the physical mass basis, we have for gauge interactions

$$\mathcal{L}_F = \frac{g_3'}{2}F_\mu^i \bar{\nu}_L t^j O_F^{ji}\gamma^\mu \nu_L \quad (13)$$
$$+ \frac{g_3'}{2}F_\mu^i\left(\bar{e}_L V_e^i \gamma^\mu e_L - \bar{e}_R V_e^{i*}\gamma^\mu e_R\right)$$

with $V_e^i = U_e^\dagger t^j U_e O_F^{ji}$. Explicitly, we find

$$V_e^1 = \begin{pmatrix} c_F & i\frac{1}{2}s_F & -i\frac{1}{2}s_F \\ -i\frac{1}{2}s_F & \frac{1}{2}c_F + \frac{1}{\sqrt{2}}s_F & \frac{1}{2}c_F \\ i\frac{1}{2}s_F & \frac{1}{2}c_F & \frac{1}{2}c_F - \frac{1}{\sqrt{2}}s_F \end{pmatrix}$$

$$V_e^2 = \begin{pmatrix} 0 & \frac{1}{2} & -\frac{1}{2} \\ \frac{1}{2} & 0 & i\frac{1}{\sqrt{2}} \\ -\frac{1}{2} & -i\frac{1}{\sqrt{2}} & 0 \end{pmatrix} \quad (14)$$

$$V_e^3 = \begin{pmatrix} -s_F & i\frac{1}{2}c_F & -i\frac{1}{2}c_F \\ -i\frac{1}{2}c_F & -\frac{1}{2}s_F + \frac{1}{\sqrt{2}}c_F & -\frac{1}{2}s_F \\ i\frac{1}{2}c_F & -\frac{1}{2}s_F & -\frac{1}{2}s_F - \frac{1}{\sqrt{2}}c_F \end{pmatrix}.$$

Thus the $SO(3)_F$ gauge interactions allow lepton flavor violating process $\mu \to 3e$, its branch ratio is

$$Br(\mu \to 3e) = \left(\frac{v}{\sigma}\right)^4 \frac{2\xi_-^2}{[(3\xi_+ + \xi')(\xi_+ + \xi') - \xi_-^2]^2} \quad (15)$$

with $v = 246$GeV. For $\sigma \sim 10^3 v$, the branch ratio could be very close to the present experimental upper bound $Br(\mu \to 3e) < 1 \times 10^{-12}$. Thus when taking the mixing angle θ_F and the coupling constant g_3' for the $SO(3)_F$ gauge bosons to be at the same order of magnitude as those for the electroweak gauge bosons, we find that masses of the $SO(3)_F$ gauge bosons are at the order of magnitudes $m_{F_i} \sim 10^3 m_W \simeq 80$ TeV. For smaller mising angle θ_F, the $SO(3)_F$ gauge boson masses m_{F_i} could be below 1 TeV.

Acknowledgments

This work was supported in part by Outstanding Young Scientist Research Fund under the grant No. 19625514.

References

1. Y.Fukuda *et al.*, Phys. Rev. Lett. **81**, 1562 (1998); **82**, 2644 (1999).
2. Y.Fukuda *et al.*, Phys. Rev. Lett. **81**, 1158 (1998); **82**, 2430 (1999).
3. Y. Takeuchi, see this Proceedings.
4. L. Wolfenstein, Phys. Rev. **D17**, 2369 (1978); S.P. Mikheyev and A. Yu. Smirnov, Sov. J. Nucl. Phys. 42, 913 (1985).
5. Y.L.Wu, Science in China, Series **A 43** 988 (2000); Phys. Rev. **D60** 073010 (1999); Eur. Phys. J. **C10** 491 (1999); Intern. J. of Mod. Phys. **A14** 4313 (1999); hep-ph/9905222, to be published in J. Phys. G.

THE $b \to X_s\gamma$ DECAY RATE IN NLO, AND HIGGS BOSON LIMITS IN THE CMSSM

W. DE BOER, M. HUBER

Institut für Experimentelle Kernphysik, Universität Karlsruhe, Germany

A.V. GLADYSHEV, D.I. KAZAKOV

Bogoliubov Laboratory of Theoretical Physics, Joint Institute for Nuclear Research, 141 980 Dubna, Moscow Region, Russian Federation

New NLO $b \to X_s\gamma$ calculations have become available. We observe that at large $\tan\beta$ the dominant NLO term of the chargino amplitude, which is proportional to $\mu\tan^2\beta$, changes the sign of this amplitude in a large region of the CMSSM parameter space, so that the preferred sign of the Higgs mixing parameter μ now agrees with the preferred sign of $b-\tau$ unification. We find that the $b \to X_s\gamma$ rate does not constrain the CMSSM anymore, if the higher order contributions and its uncertainties from the incomplete calculations are taken into account.

The Higgs boson mass in the CMSSM is found to be between 110 and 120 GeV for a top mass of 175 GeV. The mean Higgs boson mass value and its dominant errors are: $m_h = 115 \pm 3$ (*stopmass*) $\pm$ 1.5 (*stopmixing*) ± 2 (*theory*) ± 5 (*topmass*) GeV. This Higgs mass range is valid for all $\tan\beta$ values above 20 and decreases for lower $\tan\beta$. The 95% C.L. Higgs mass limit of 113 GeV from LEP implies $\tan\beta > 4.0$ in the CMSSM.

1 Introduction

In a previous paper we showed that the inclusive decay rate $b \to X_s\gamma$ severely constrains the high $\tan\beta$ solution of the Constrained Minimal Supersymmetric Standard Model (CMSSM) [1]. This was mainly caused by the fact that $b-\tau$ Yukawa coupling unification requires a negative sign for the Higgs mixing parameter μ, while the $b \to X_s\gamma$ rate preferred the opposite sign. However, the next-to-leading order (NLO) calculations for the $b \to X_s\gamma$ rate in the MSSM [2] can change the sign of μ. Consequently the allowed parameter space becomes much larger for the high $\tan\beta$ scenario, especially if the uncertainties from the incomplete NLO calculations are taken into account in our global analysis.

2 NLO corrections to $b \to X_s\gamma$

After studying the paper from Ref.[2] in detail[3], we found that the large NLO contributions can change the sign of the chargino amplitude at large $\tan\beta$ for practically the whole parameter space, as shown in Fig. 1[4]. Since the chargino-stop amplitude is of the same order of magnitude as the SM W-t amplitude for most of the parameter space, it interferes strongly: positively for $\mu > 0$ and negatively for $\mu < 0$. In the first case the $b \to X_s\gamma$ rate becomes rapidly too big, while for $\mu < 0$ the value is in the acceptible range, as shown by the top right figure of Fig. 1 for practically the whole parameter space. In addition, for this sign of μ there is no problem with $b-\tau$ unification, electroweak symmetry breaking and gauge coupling unification[4].

In Ref.[2] only the effect of the lightest stop to the NLO contributions was considered and no flavour mixing between the three generations was taken into account. The latter was found to be small in the CMSSM. The effect of the missing contributions of the heavier stop has been studied. The general picture does not change by including heavier stop terms analogous to the light stop terms, although complete calculations including diagrams where both stops contribute simultaneously have not yet been calculated. Given

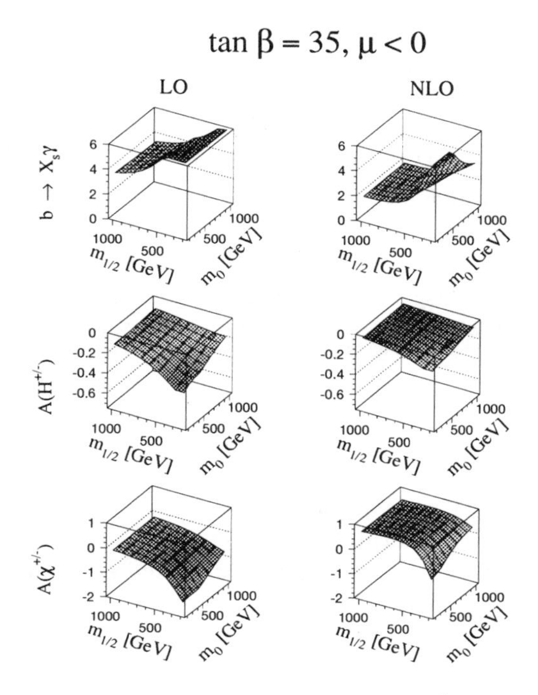

Figure 1. The decay rate (in units of 10^{-4}) and selected amplitudes (in units of 10^{-2}) of the $b \to X_s\gamma$ decay for negative μ and $\tan\beta = 35$. These amplitudes should be compared with the SM amplitude of $-0.56 \cdot 10^{-2}$. Note the sign change in the $\chi^{\pm}$ amplitude.

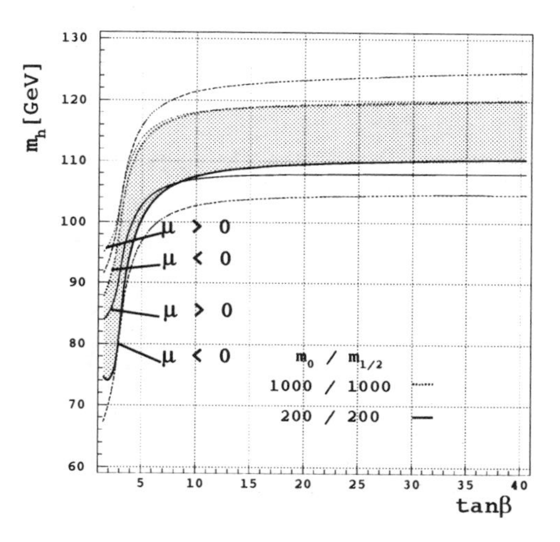

Figure 2. The lightest Higgs boson mass as function of $\tan\beta$, as calculated by the effective potential approach[6]. The shaded band shows the variation of $m_0 = m_{1/2}$ between 200 and 1000 GeV for $\mu < 0$, $m_t = 175$ GeV, and $A_0 = 0$. The maximum (minimum) Higgs boson mass value, shown by the upper (lower) line are obtained for $A_0 = -3m_0$, $m_t = 180$ GeV, $m_0 = m_{1/2} = 1000$ GeV ($A_0 = 3m_0$, $m_t = 170$ GeV, $m_0 = m_{1/2} = 200$ GeV). As can be seen the curves show an asymptotic behaviour for large values of $\tan\beta$.

the uncertainties from the incomplete calculations, we exclude the $b \to X_s\gamma$ constraint and study the Higgs mass prediction in the CMSSM.

The lightest Higgs boson mass m_h is shown as function of $\tan\beta$ in Fig. 2. The shaded band corresponds to the uncertainty from the stop mass and stop mixing for $m_t = 175$ GeV. The upper and lower lines correspond to m_t=170 and 180 GeV, respectively. The Higgs mass varies between 110 and 120 GeV, if the CMSSM parameters are varied in a wide range for $m_t = 175$ GeV. The errors around the central value of 115 GeV are estimated to be: $m_h = 115 \pm 3$ $(stopmass)$ ± 1.5 $(stopmixing)$ ± 2 $(theory)$ ± 5 $(topmass)$ GeV. As can be seen from Fig. 2 this central value is valid for all $\tan\beta > 20$ and decreases for lower $\tan\beta$.

One observes that for a SM Higgs limit of 113 GeV[5] all values of $\tan\beta$ below 4.0 are excluded in the CMSSM. We thank P. Gambino, G.F. Giudice and M. Misiak for helpful discussions on the NLO $b \to X_s\gamma$ rates.

References

1. W. de Boer et al., Phys. Lett. **B 438** (1998) 281.
2. M. Ciuchini et al., Nucl. Phys. **B 534** (1998) 3-20.
3. M. Huber, Preprint IEKP-KA/2000-4, Diplomarbeit, Universität Karlsruhe.
4. W. de Boer et al., hep-ph/0007078
5. The LEP Higgs Working Group, R. Bock et al., CERN-EP-2000-055
   P. Igo-Kemenes, these proceedings
6. M. Carena, M. Quirós and C. Wagner, Nucl. Phys. **B 461** (1996) 407

FULLY SUPERSYMMETRIC CP VIOLATIONS IN THE KAON SYSTEM

S. BAEK, J.-H. JANG, P. KO* AND J. H. PARK
Dep. of Physics, KAIST, Taejon 305-701, KOREA
*E-mail: *pko@charm.kaist.ac.kr*

We show that, on the contrary to the usual claims, fully supersymmetric CP violations in the kaon system are possible through the gluino mediated flavor changing interactions. Both ϵ_K and $\mathrm{Re}(\epsilon'/\epsilon_K)$ can be accommodated for relatively large $\tan\beta$ without any fine tunings or contradictions to the FCNC and EDM constraints.

Recent observation of $\mathrm{Re}(\epsilon'/\epsilon_K)$ by KTeV collaboration, $\mathrm{Re}(\epsilon'/\epsilon_K) = (28 \pm 4) \times 10^{-4}$ [1], nicely confirms the earlier NA31 experiment [2] $\mathrm{Re}(\epsilon'/\epsilon_K) = (23 \pm 7) \times 10^{-4}$. This nonvanishing number indicates unambiguously the existence of CP violation in the decay amplitude ($\Delta S = 1$). Along with another CP violating parameter known for long time, $\epsilon_K = e^{i\pi/4}$ $(2.280 \pm 0.013) \times 10^{-3}$ [3], these two parameters quantifying CP violations in the kaon system can be accommodated by the KM phase in the standard model (SM). The SM prediction for $\mathrm{Re}(\epsilon'/\epsilon_K)$ is about 5×10^{-4} and lies in the lower side of the data, although theoretical uncertainties are rather large [4].

However, it would be interesting to consider a possibility that both ϵ_K and $\mathrm{Re}(\epsilon'/\epsilon_K)$ have their origins entirely different from the KM phase in the SM, in particular in supersymmetric models. In this talk, we argue that all the observed CP violating phenomena in the kaon system in fact can be accommodated in terms of a *single* complex number $(\delta_{12}^d)_{LL}$ that parameterizes the squark mass mixings in the chirality and flavor spaces for relatively large $\tan\beta$ without any fine tuning or any contradictions with experimental data on FCNC, even if we assume $\delta_{\rm KM} = 0$ as in this talk. This talk is based on Ref. [5].

In order to study the gluino mediated flavor changing phenomena in the quark sector, it is convenient to use the so-called mass insertion approximation (MIA) [6]. The parameters $(\delta_{ij}^d)_{AB}$ characterize the size of the gluino-mediated flavor (i, j) and chirality (A, B) changing amplitudes. They may be also CP violating complex numbers.

Now, if one saturates Δm_K and ϵ_K with $(\delta_{12}^d)_{LL}$ alone, the resulting $\mathrm{Re}(\epsilon'/\epsilon_K)$ is too small by more than an order of magnitude, unless one invokes some finetuning [7]. On the other hand, if one saturates $\mathrm{Re}(\epsilon'/\epsilon_K)$ by $|\mathrm{Im}(\delta_{12}^d)_{LR}| \sim 10^{-5}$, the resulting ϵ_K is too small by more than an order of magnitude, unless one invokes some finetuning. Therefore the folklore was that the supersymmetric contributions to $\mathrm{Re}(\epsilon'/\epsilon_K)$ is small. Recently, Masiero and Murayama showed that this conclusion can be evaded in generalized SUSY models [8] with a few reasonable assumptions on the size of the $(\delta_{12}^d)_{LR}$. But in their model, one has to introduce a new CP violating parameter $(\delta_{12}^d)_{LL}$ in order to generate ϵ_K and also predict too large neutron EDM which is very close to the current upper limit.

In the following, we show that there is another generic way to evade this folklore in supersymmetric models if $|\mu \tan\beta|$ is relatively large, say $\sim 10 - 20$ TeV. Moreover, both ϵ_K and $\mathrm{Re}(\epsilon'/\epsilon_K)$ can be generated by a single CP violating complex parameter in the MSSM. In other words, fully supersymmetric CP violations are possible in the kaon system. The argument goes as follows : if $|(\delta_{12}^d)_{LL}| \sim O(10^{-3} - 10^{-2})$ with the phase $\sim O(1)$ saturates ϵ_K, this same parameter can lead to a sizable $\mathrm{Re}(\epsilon'/\epsilon_K)$ through the $(\delta^d_{12})_{LL}$ insertion followed by the FP (LR) mass insertion, which is proportional to

$$(\delta^d_{22})_{LR} \equiv m_s(A^*_s - \mu\tan\beta)/\tilde{m}^2 \sim O(10^{-2}),$$

where $\tilde{m}$ denotes the common squark mass in the MIA. It should be emphasized that the induced $(\delta^d_{12})^{\mathrm{ind}}_{LR} \equiv (\delta^d_{12})_{LL} \times (\delta^d_{22})_{LR}$ is different from the conventional $(\delta^d_{12})_{LR}$ in the literature. The LR mixing $(\delta_{12})^{\mathrm{ind}}_{LR}$ induced by $(\delta^d_{12})_{LL}$ is typically very small in size $\sim O(10^{-5})$, but this is enough to generate the full size of $\mathrm{Re}(\epsilon'/\epsilon_K)$ as shown below. Thus the usual folklore can be simply evaded. Our spirit to generate supersymmetric $\mathrm{Re}(\epsilon'/\epsilon_K)$ is different from Ref. [8], where the LR mass matrix form is assumed to be similar to the Yukawa matrix so that they predict the neutron EDM to be close to the current upper limit. On the other hand, we do not assume any specific flavor structure in trilinear A couplings. Also our model does not suffer from the EDM constraint at all.

Let us first consider the gluino-squark contributions to the $K^0 - \overline{K^0}$ mixing due to two insertions of $(\delta^d_{12})_{LL}$. The corresponding $\Delta S = 2$ effective Hamiltonian is given by

$$\mathcal{H}_{\mathrm{eff}}(\Delta S = 2) = C_1 \overline{d}^\alpha_L \gamma_\mu s^\alpha_L \, \overline{d}^\beta_L \gamma^\mu s^\beta_L$$

with the Wilson coefficient C_1 being

$$C_1 = -\frac{\alpha_s^2}{216\tilde{m}^2}(\delta^d_{12})^2_{LL} \; f_1(x) \qquad (1)$$

Here, $x = m^2_{\tilde{g}}/\tilde{m}^2$ and the loop function $f_1(x)$ is given in [6].

Now we turn to the $\Delta S = 1$ effective Hamiltonian $\mathcal{H}_{\mathrm{eff}}(\Delta S = 1) = \sum_{i=3}^{8} C_i \mathcal{O}_i$. The sdg operator O_8 which is relevant to $\mathrm{Re}(\epsilon'/\epsilon_K)$ is defined as

$$\mathcal{O}_8 = \frac{g_s}{4\pi} m_s \overline{d}^\alpha_L \sigma^{\mu\nu} T^a s^\alpha_R G^a_{\mu\nu}, \qquad (2)$$

and other four quark operators $O_{i=3,\dots,6}$ and the corresponding Wilson coefficients from C_3 to C_8 with a single mass insertion are available in the literature [6]. If we consider the penguin diagram Fig. 1 with the double mass insertion, the Wilson coefficient C_8 is given by

$$C_8^{(2)} = \frac{\alpha_s}{\tilde{m}^2}\frac{m_{\tilde{g}}}{m_s}(\delta^d_{12})^{\mathrm{ind}}_{LR} \; M_8(x) \qquad (3)$$

where the explicit form of the $\sim O(1)$ function $M_8(x)$ can be found in Ref. [5]. Since $C_8^{(2)}$ is proportional to $m_{\tilde{g}}/m_s$, it is very important for generating $\mathrm{Re}(\epsilon'/\epsilon_K)$ even if $(\delta^d_{12})^{\mathrm{ind}}_{LR}$ is fairly small.

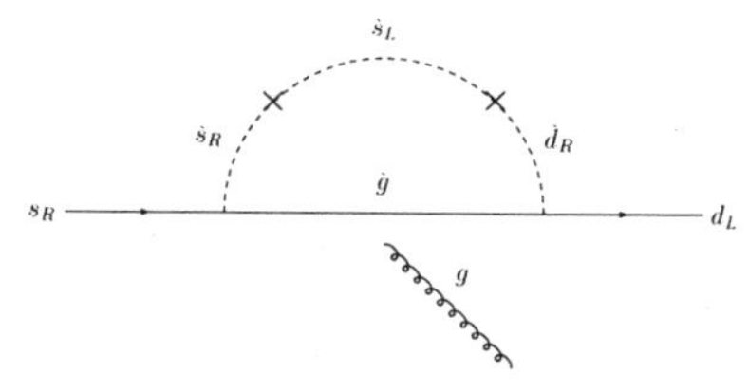

Figure 1. Feynman diagram for $\Delta S = 1$ process. The cross denotes the flavor changing (LL) and the flavor preserving (LR) mixings, respectively.

It is straightforward to calculate ϵ_K and ϵ'/ϵ_K using the same parameters as in Ref. [9] with $m_s(2\text{ GeV}) = 130$ MeV. The corresponding SM prediction for $\mathrm{Re}(\epsilon'/\epsilon_K) = 5.7 \times 10^{-4}$. For those points which satisfy $\Delta m_K(\mathrm{SUSY}) \leq \Delta m_K(\exp)$ and $|\epsilon_K(\mathrm{SUSY}) - \epsilon_K(\exp)| < 1\sigma$, we plot ϵ'/ϵ_K in Fig. 2 as functions of the modulus r and the phase φ of the parameter $(\delta^d_{12})_{LL} \equiv re^{i\varphi}$ for the common squark mass $\tilde{m} = 500$ GeV. The upper (lower) rows correspond to $\widetilde{A_s} \equiv (A_s - \mu^*\tan\beta) = -10(20)$ TeV. It is clear that both ϵ_K and $\mathrm{Re}(\epsilon'/\epsilon_K)$ can be nicely accommodated with a single complex number $(\delta^d_{12})_{LL}$ with $\sim O(1)$ phase in our model without any difficulty, if $|\mu|$ and $\tan\beta$ is relatively large so that $|\widetilde{A_s}|$ becomes a few tens of TeV. It is important to realize that in our model there is no conflict with neutron EDM from $(\delta^d_{12})_{LL}$, since the EDM is generated by another parameter $\left(\delta^d_{11}\right)_{LR}$, which can be taken as real independent of $(\delta^d_{12})_{LL}$. This is in sharp contrast with the case of Ref. [8] where the neutron edm is inevitably close to the current upper limit. Note that we are not assuming any specific flavor structures in

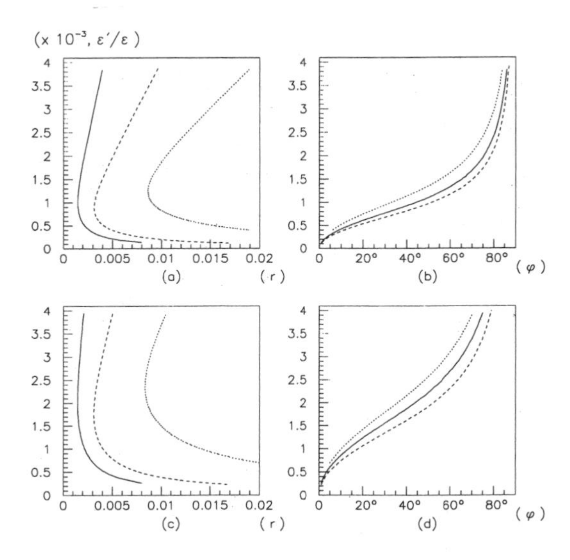

Figure 2. $Re(\epsilon'/\epsilon_K)$ as a function of the modulus r [(a) and (c)] and the phase φ [(b) and (d)] of the parameter $(\delta^d_{LL})_{12}$ with $\widetilde{A_S}$ to be $-10\ TeV$ ((a),(b)) and $-20\ TeV$ ((c),(d)). The common squark mass is chosen to be $\widetilde{m} = 500$ GeV, and the solid, the dashed and the dotted curves correspond to $x = 0.3,\ 1.0,\ 2.0$, respectively.

the A terms at all, unlike many other recent models in the literatures.

In conclusion, we showed that both ϵ_K and $\mathrm{Re}(\epsilon'/\epsilon_K)$ can be accommodated with a single CP violating and flavor changing down-squark mass matrix elements $[(\delta^d_{12})_{LL} \sim 10^{-3}]$ without any fine tuning or any conflict with the data on FCNC processes, if $|\mu \tan\beta| \sim O(10)$ TeV. Our mechanism utilizes this FC LL mass insertion along with the FP LR mass insertion propotional to $(\delta^d_{22})_{LR} \sim 10^{-2}$. The latter is *generically present in any SUSY models* including the MSSM, and thus there is no fine tuning in our model for accommodating both ϵ_K and $\mathrm{Re}(\epsilon'/\epsilon_K)$ in terms of a single $(\delta^d_{12})_{LL}$. Phrasing differently, *the SUSY ϵ_K problem implies SUSY ϵ' problem if $|\mu \tan\beta|(\sim O(10-20)$ TeV) is relatively large.* One can also consider our mechanism in the more minimal SUSY model, where $(\delta_{22})^d_{LR}$ is proportional to $m_b(A_b - \mu\tan\beta)$ so that $A_b - \mu\tan\beta$ may be lowered significantly. The best discriminant between our model and the SM model would be probably the branching ratios for $K \to \pi\nu\nu$ and CP violations in B decays. For example, the time dependent asymmetry in $B^0 \to J/\psi K_S$, $\sin 2\beta$, can be completely different from the SM prediction. Also if the KM phase is nonzero, there will be additional constributions to ϵ_K and $\mathrm{Re}(\epsilon'/\epsilon_K)$ from the SM and other SUSY loop diagrams. All these finer details will be discussed elsewhere in the forthcoming publication [10].

Acknowledgments

This work is supported by BK21 program of Ministry of Education.

References

1. A. Alavi-Harati et al., *Phys. Rev. Lett.* **83**, 22 (1999).
2. G.D.Barr et al., NA31 Collaboration, *Phys. Lett.* B **317**, 233 (1993).
3. Particle Data Group, *Eur. Phys. J.* C **3**, 1 (1998).
4. S. Bosch *et al.*, TUM-HEP-347-99, hep-ph/9904408.
5. S. Baek, J.-H. Jang, P. Ko and J. H. Park, hep-ph/9907572, KAIST-TH 1999/04, to appear in Phys. Rev. D.
6. F. Gabbiani, E.Gabrielli, A. Masiero, L. Silvestrini, *Nucl. Phys.* B **477**, 321 (1996) and references therein.
7. E. Gabrielli *et al.*, *Phys. Lett.* B **374**, 80 (1996).
8. A. Masiero and H. Murayama, *Phys. Rev. Lett.* **83**, 907 (1999).
9. G. Buchalla, A.J.Buras, M.E. Lautenbacher, *Rev. Mod. Phys.* **68**, 1125 (1996).
10. S. Baek, J.-H. Jang, P. Ko and J.H. Park, in preparation.

NEW SUPERSYMMETRIC STANDARD MODEL WITH STABLE PROTON

MAYUMI AOKI

Theory Group, KEK, Tsukuba, Ibaraki 305-0801, Japan
E-mail: mayumi.aoki@kek.jp

NORIYUKI OSHIMO

Department of Physics, Ochanomizu University
Otsuka 2-1-1, bunkyo-ku, Tokyo 112-8610, Japan
E-mail: oshimo@phys.ocha.ac.jp

We discuss a supersymmetric extension of the standard model with an extra U(1) gauge symmetry. In this model, the proton stability is guaranteed by the gauge symmetry without invoking R parity. The gauge symmetry breakdown automatically generates an effective μ term and large Majorana masses for right-handed neutrinos. The supersymmetry-soft-breaking terms for scalar fields could be universal at a very high energy scale, and the electroweak symmetry is broken through radiative corrections.

The supersymmetric standard model (SSM) is considered to be one of the most plausible extensions of the standard model (SM). In particular, the minimal SSM (MSSM) is usually treated as the standard theory around the electroweak scale. However, this MSSM suffers one potentially serious problem. The proton may decay through baryon-number-violating couplings of dimension four, and thus its life time could become unacceptably short. In order to forbid those dangerous couplings, therefore, an *ad hoc* discrete symmetry is imposed on the model through R parity. In the SM, on the other hand, the interactions which induce the proton decay are not allowed by gauge symmetry. The SSM would become more plausible, if the proton can be protected from decay more naturally.

In this report, we present a new SSM [1,2], in which the proton stability is guaranteed by an extra U(1) gauge symmetry, within the framework of a model coupled to $N = 1$ supergravity. This model also provides natural explanations for the μ parameter and neutrino masses which are merely put by hand in the usual SSM.

Keeping the extension of the SM as minimal as possible, the particle contents of the model are taken as shown in Table 1. The extra U(1) gauge symmetry is denoted by U$'$(1), for which the charges of superfields are expressed as Q_Q, Q_{U^c}, etc.. The index i $(= 1, 2, 3)$ of the superfields for quarks and leptons stands for the generation, while the indices j $(= 1 \cdots n_j)$ of H_1 and H_2, k $(= 1 \cdots n_k)$ of S, and l $(= 1 \cdots n_l)$ of K and K^c are attached for possible multiplication.

In addition to the superfields of the MSSM, our model has SM gauge singlets N^c and S for, respectively, right-handed neutrinos and Higgs bosons to break the U$'$(1) symmetry. The superpotential is then required to contain the couplings H_2LN^c and SN^cN^c for giving non-vanishing but tiny masses to the ordinary neutrinos. The μ term in the MSSM is replaced by SH_1H_2. New colored superfields K and K^c are incorporated to cancel a chiral anomaly. Their fermion components can receive masses from SKK^c.

The hypercharges of K and K^c, the U$'$(1) charges of all the superfields, and the numbers n_j, n_k, and n_l are determined by chiral and trace anomalies and necessary couplings of superfields. Barring irrational hypercharges for K and K^c, we obtain $Y_K = \pm\frac{1}{3}$ and $n_j = n_k = n_l = 3$. The superfields with the same quantum numbers are all trip-

Table 1. Particle contents and their quantum numbers. $i = 1, 2, 3$; $j = 1, .., n_j$; $k = 1, .., n_k$; $l = 1, .., n_l$.

	SU(3)	SU(2)	U(1)	U′(1)
Q^i	3	2	$\frac{1}{6}$	Q_Q
U^{ci}	3*	1	$-\frac{2}{3}$	Q_{U^c}
D^{ci}	3*	1	$\frac{1}{3}$	Q_{D^c}
L^i	1	2	$-\frac{1}{2}$	Q_L
N^{ci}	1	1	0	Q_{N^c}
E^{ci}	1	1	1	Q_{E^c}
H_1^j	1	2	$-\frac{1}{2}$	Q_{H_1}
H_2^j	1	2	$\frac{1}{2}$	Q_{H_2}
S^k	1	1	0	Q_S
K^l	3	1	Y_K	Q_K
K^{cl}	3*	1	$-Y_K$	Q_{K^c}

licated. The hypercharge Y_K is either $1/3$ or $-1/3$. However, the proton stability is satisfied only for $Y_K = 1/3$. In this case, allowed couplings of dimension four are given by H_1QD^c, H_2QU^c, H_1LE^c, H_2LN^c, SH_1H_2, SN^cN^c, and SKK^c. The baryon number is conserved. The lowest dimension couplings with baryon-number violation are given by the D terms of $QQU^{c*}E^{c*}$, $QQD^{c*}N^{c*}$, and $QU^{c*}D^{c*}L$, which are of dimension six. The proton decay is adequately suppressed. On the other hand, for $Y_K = -1/3$, the particle contents of one generation can be embedded in the fundamental 27 representation of the E_6 group. As well known, the baryon and lepton numbers are not conserved in couplings of dimension four, such as $U^cD^cK^c$ and LQK^c, inducing a fast proton decay.

Requiring orthogonality between U(1) and U′(1) generators, the U′(1) charges of the superfields are determined up to a normalization factor. In Table 2, we show the U′(1) charges which are normalized to the U(1) charges.

The superpotential is given by

$$\begin{aligned} W = \;& \eta_u^{ijk} H_2^i Q^j U^{ck} + \eta_d^{ijk} H_1^i Q^j D^{ck} \\ & + \eta_\nu^{ijk} H_2^i L^j N^{ck} + \eta_e^{ijk} H_1^i L^j E^{ck} \\ & + \lambda_N^{ijk} S^i N^{cj} N^{ck} + \lambda_H^{ijk} S^i H_1^j H_2^k \\ & + \lambda_K^{ijk} S^i K^j K^{ck}, \end{aligned}$$

Table 2. U′(1) charges of the superfields.

Q_Q	Q_{U^c}	Q_{D^c}	Q_L	Q_{N^c}	Q_{E^c}
$\frac{1}{12}$	$\frac{1}{12}$	$\frac{7}{12}$	$\frac{7}{12}$	$-\frac{5}{12}$	$\frac{1}{12}$
Q_{H_1}	Q_{H_2}	Q_S	Q_K	Q_{K^c}	
$-\frac{2}{3}$	$-\frac{1}{6}$	$\frac{5}{6}$	$-\frac{2}{3}$	$-\frac{1}{6}$	

where all the couplings allowed by gauge symmetry and renormalizability are contained. The couplings are all cubic, and there is no mass parameter.

The model is coupled to $N = 1$ supergravity, which breaks supersymmetry softly in the observable world. The Lagrangian contains, as well as supersymmetric terms, mass terms for scalar bosons and gauge fermions, and trilinear couplings for scalar bosons. In the ordinary scheme, the masses-squared and trilinear coupling constants for the scalar bosons have universal values $m_{3/2}^2$ and A, respectively, at the energy scale a little below the Planck mass. Hereafter, the scalar components of the superfields H_1, H_2, and S are expressed by $\tilde{H}_1$, $\tilde{H}_2$, and $\tilde{S}$.

The parameter values of the model change according to the relevant energy scale. The mass-squared of $\tilde{H}_2^3$ receives large negative contributions through quantum corrections, owing to a large coefficient η_u of $H_2^3Q^3U^{c3}$ related to the top quark mass. As a result, this mass-squared becomes small around the electroweak scale, leading to non-vanishing vacuum expectation values (VEVs) for $\tilde{H}_1^3$ and $\tilde{H}_2^3$. The electroweak symmetry is broken through radiative corrections. On the other hand, for the first two generations, quantum corrections to the masses-squared of $\tilde{H}_1^i$ or $\tilde{H}_2^i$ are not large, so that the VEVs of these scalar bosons vanish. If a coefficient λ_K of $S^3K^3K^{c3}$ is large, the mass-squared of $\tilde{S}^3$ is also driven small. A non-vanishing VEV is induced for $\tilde{S}^3$, and the U′(1) symmetry is broken spontaneously.

The scalar potential is numerically analyzed at the electroweak scale to examine the parameter regions which give a vacuum consistent with experimental results. In particular, the U′(1) symmetry predicts a new neutral gauge boson Z', for which stringent constraints are obtained on the mass and the mixing with the Z boson of the SM. We assume that $\tilde{H}_1^3$, $\tilde{H}_2^3$, and $\tilde{S}^3$ of the third generation have non-vanishing VEVs, which are denoted by v_1, v_2, and v_s. It is shown that sizable regions are allowed for the mass-squared parameters $M_{H_1}^2$, $M_{H_2}^2$, and M_S^2 of the Higgs bosons. The coefficient λ_H of SH_1H_2 should be around $0.1 - 0.4$. Owing to the constraints from the Z' boson, $M_{H_1}^2$ is mostly larger than $(1\ \text{TeV})^2$. The value of $M_{H_2}^2$ is generally smaller than $M_{H_1}^2$ in magnitude, and M_S^2 is always negative. The VEV of $\tilde{S}$ is larger than 1 TeV.

The term $\lambda_H SH_1H_2$ assumes the μ term of the SSM. The effective μ parameter is given by $\lambda_H v_s/\sqrt{2}$, which has an appropriate magnitude for the electroweak symmetry breaking. The terms SN^cN^c induce large Majorana masses for the right-handed neutrinos. The ordinary neutrinos have non-vanishing masses approximately given by $|\eta_\nu|^2 v_2^2/2\sqrt{2}|\lambda_N|v_s$. Taking for $\lambda_N \sim \lambda_H$, the neutrino masses become very small if the coupling constants η_ν for the neutrino Dirac masses are of the order of that for the electron η_e. A typical mass scale of squarks and sleptons is given by the universal value $m_{3/2}$ for the scalar boson masses, which is around M_{H_1} and thus of order 1 TeV. Then, the smallness of the neutron and the electron electric dipole moments, which is another problem in the MSSM, can be explained.

In this model, the lightest Dirac fermion ψ_K in the K and K^c system is stable, having both color and electric charges. In the early universe, after going out of thermal equilibrium, this fermion is bound to become a color-singlet particle. Since its mass is large, the decoupling occurs much earlier than for the up and down quarks. Therefore, the bound state is mainly formed by ψ_K and its anti-particle. This is an electrically neutral meson, and eventually decays into lighter particles. The remnants for ψ_K could exist in the present universe, and may be explored by non-accelerator experiments. However, its relic density depends on various uncertain factors, making a definite prediction difficult.

The energy dependencies of the model parameters are quantitatively described by renormalization group equations. The coupling constant η_u for the top quark mass should be around unity at the electroweak scale, which is obtained for $\eta_u > 0.1$ at the high energy scale. The coupling constant λ_K evolves similarly. If both η_u and λ_K are larger than 0.1 at the high energy scale, the mass-squared parameters $M_{H_2}^2$ and M_S^2 receive large quantum corrections. Taking the universal values as $m_{3/2}^2 \sim (1\ \text{TeV})^2$ and $A \sim 1$ for the mass-squared parameters and the trilinear coupling constants, the values of $M_{H_2}^2$ and M_S^2 become small enough at the electroweak scale to induce SU(2)×U(1)×U′(1) gauge symmetry breaking. The quantum corrections to these parameters also become large, if the gauge fermion is heavy. On the other hand, the energy dependence of $M_{H_1}^2$ is weak and its value is not much different from the universal value. It is shown that there are reasonable parameter regions at the high energy scale, with the masses and trilinear coupling constants for scalar bosons being universal, which give parameter values at the electroweak scale consistent with phenomena. Unfortunately, however, the gauge coupling constants are not unified at an energy scale for possible grand unification, unless the particle contents of the model are modified.

References

1. M. Aoki and N. Oshimo, *Phys. Rev. Lett.* **84**, 5269 (2000).
2. M. Aoki and N. Oshimo, *Phys. Rev.* D **62**, 055013 (2000).

RECONSTRUCTION OF SUPERSYMMETRIC THEORIES AT HIGH ENERGY SCALES

W. POROD

Inst. de Física Corpuscular (IFIC), CSIC, E-46071–València, Spain

E-mail: porod@ific.uv.es

We have studied the reconstruction of supersymmetric theories at high scales by evolving the fundamental parameters from the electroweak scale upwards. Universal minimal supergravity and gauge mediated supersymmetry breaking have been taken as representative alternatives. Pseudo-fixed point structures require the low–energy boundary values to be measured with high precision.

1. When supersymmetry is discovered and its spectrum of particles and their properties are measured, the mechanism of electroweak symmetry breaking must be determined. This will be related to the reconstruction of the supersymmetric theory at high energy scales. This problem will be addressed in this report. In particular, we will explore the bottom-up approach, which is the most unbiased method in this context. First indications might already be given by the particle spectrum and various experimental signatures. We assume that precision measurements at a high luminosity e^+e^- linear collider[1] (LC) are available. More details and an extended list of references are given in Ref.[2].

As representative examples we study minimal supergravity (mSUGRA)[3] and gauge mediated supersymmetry breaking (GMSB)[4]. mSUGRA is characterized by a GUT scale M_U of $O(10^{16}$ GeV) where the gauge couplings unify. M_U is also the scale of supersymmetry breaking which is parameterized by a common gaugino mass parameter $M_{1/2}$, a common scalar mass parameter M_0 and a common trilinear coupling A_0 between sfermions and Higgs bosons. GMSB is characterized by a messenger scale M_m in the range between ~ 10 TeV and $\sim 10^6$ TeV. In this scenario the mass parameters of particles carrying the same gauge quantum numbers squared are universal. The regularity for scalar masses would be observed at the scale M_m while the gaugino mass parameters should unify at 1-loop order at the GUT scale M_U as in the mSUGRA case.

2. The mSUGRA point we have analyzed in detail, is characterized by: $M_{1/2} =$ 190 GeV, $M_0 = 200$ GeV, $A_0 = 550$ GeV, $\tan\beta = 30$, and sign$(\mu) = -$. The modulus of μ is calculated from the requirement of radiative electroweak symmetry breaking. We have checked the compatibility of this point with $b \to s\gamma$ [5] and the ρ-parameter[6]. We have used two-loop renormalization equations[7] to define the parameters at the electroweak scale where we also have calculated the threshold effects[8].

The parameters provide the experimental observables, including the supersymmetric particle spectrum and production cross sections. These observables are endowed with errors as expected from a future LC experiment[9]. The analysis of the entire particle spectrum except the gluino requires LC energies up to 1 TeV and an integrated luminosity of about 1 ab^{-1}. The errors given in Ref.[9] are scaled in proportion to the masses of the spectrum. Moreover, they are inflated conservatively for particles that decay predominantly to τ channels, according to typical reconstruction efficiencies such as given in Ref.[10]. Typically the relative errors expected from measurements at a Linear Collider are $O(10^{-3})$ for weakly interacting particles and $O(10^{-2})$ for strongly interacting particles (for more details see Table 1 in [2]). For the cross sections we use purely statis-

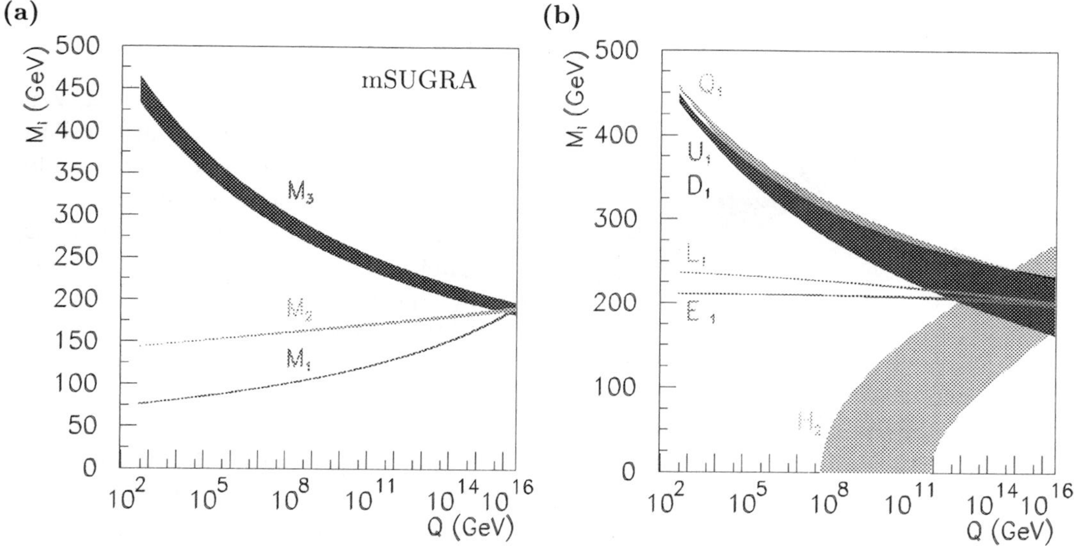

Figure 1. **mSUGRA:** *Evolution of (a) gaugino and (b) sfermion mass parameters in the bottom–up approach. The mSUGRA point probed is characterized by the parameters* $M_0 = 200$ *GeV,* $M_{1/2} = 190$ *GeV,* $A_0 = 550$ *GeV,* $\tan\beta = 30$, *and* $\text{sign}(\mu) = (-)$. *[The widths of the bands indicate the 95% CL.]*

tical errors, assuming a conservative reconstruction efficiency of 20%. In the case of the gluino we assume that LHC can measure its mass within an error of 10 GeV [11].

These observables together with the corresponding errors are interpreted as the experimental data from the experiment and they are used to reconstruct the underlying fundamental parameters. These parameters are evaluated in the bottom-up approach to the grand unification scale within the given errors. The results for the evolution of the mass parameters to the GUT scale M_U are shown in Fig. 1. The left-hand side (a) of the figure presents the evolution of the gaugino parameters M_i which apparently is under excellent control, as is the extrapolation of the slepton mass parameter in Fig. 1(b). The accuracy deteriorates for the squark mass parameters and for the Higgs mass parameter M_{H_2}. This can be understood after inspecting the corresponding RGEs. In case of $M_{\tilde{Q}_1}$ the parameter receives a rather large contribution from M_3 when renormalized from the high scale down to the electroweak scale: The error on M_3 is large compared to $M_{1,2}$ as can be seen from Fig. 1a) and is in turn the source of the error on the squark mass. In case of M_{H_2} large Yukawa couplings lead to a pseudo-fixed point behaviour implying a rather weak dependence of the electroweak parameter on the original parameter at the high scale.

Inspecting Fig. 1(b) leads to the conclusion that the top-down approach eventually may generate an incomplete picture. *Global* fits based on mSUGRA without allowing for deviations from universality, are dominated by $M_{1,2}$ and the slepton mass parameters. Therefore, the structure of the theory in the squark sector is not scrutinized stringently at the unification scale in the top-down approach. By contrast, the bottom-up approach demonstrates very clearly the extent to which the theory can be tested at the high scale.

3. The analysis has been repeated for gauge mediated supersymmetry breaking. Regularity among particles carrying the same gauge quantum numbers squared, should be observed in the evolution of mass parameters at the messenger scale M_m. The evolu-

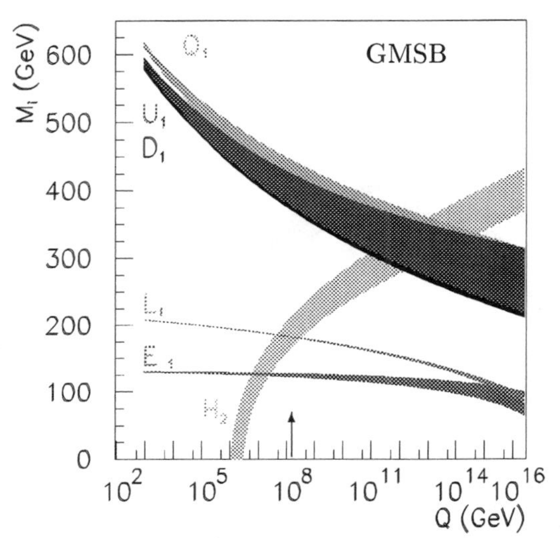

Figure 2. **GMSB:** *Evolution of sfermion mass parameters in the bottom–up approach. The GMSB point has been chosen as* $M_m = 2 \cdot 10^5$ *TeV,* $\Lambda = 28$ *TeV,* $N_5 = 3$, $\tan\beta = 30$, *and* $\text{sign}(\mu) = (-)$. *[The widths of the bands indicate the 95% CL.]*

tion of the sfermion mass parameters of the first/second generation and the Higgs mass parameter M_{H_2} is presented in Fig. 2. It is obvious that M_{H_2} approaches the mass parameter for the left-chiral sleptons M_{L_1} at the GMSB scale which is indicated by the arrow. Moreover, the figure demonstrates clearly that GMSB will not be confused with the mSUGRA scenario as no more regularity can be observed at the GUT scale M_U.

4. The model–independent reconstruction of the fundamental supersymmetric theory at the high scale, the grand unification scale M_U in supergravity or the intermediate scale M_m in gauge mediated supersymmetry breaking, appears feasible. Regular patterns can be observed by evolving the gaugino and scalar mass parameters from the measured values at the electroweak scale to the high scales. The necessary high accuracy requires in addition to the LHC input values high–precision LC values. The future experimental input from LC is particularly important if the universality at the GUT scale is (slightly) broken.

Acknowledgments

I am grateful to G.A. Blair and P.M. Zerwas for a very interesting and fruitful collaboration. This work has been supported by the Spanish 'Ministerio de Educacion y Cultura' under the contract SB97-BU0475382, by Spanish DGICYT grants PB98-0693, and by the EEC under the TMR contract ERBFMRX-CT96-0090.

References

1. H. Murayama and M. E. Peskin, Ann. Rev. Nucl. Part. Sci. **46**, 533 (1996); E. Accomando *et al.*, ECFA/DESY LC Working Group, Phys. Rep. **299**, 1 (1998); P. M. Zerwas, Proceedings, 1999 Cargèse Institute for High-Energy Physics, hep-ph/0003221.
2. G.A. Blair, W. Porod, and P.M. Zerwas, hep-ph/0007107.
3. H.P. Nilles, Phys. Rep. **110**, 1 (1984).
4. G.F. Giudice and R. Rattazzi, Phys. Rep. **322**, 419 (1999).
5. A.L. Kagan and M. Neubert, Eur. Phys. J. **C7**, 5 (1999); D.A. Demir, A. Masiero, and O. Vives, Phys. Rev. **D61**, 075009 (2000).
6. M. Drees and K. Hagiwara, Phys. Rev. **D42**, 1709 (1990).
7. S. Martin and M. Vaughn, Phys. Rev. **D50**, 2282 (1994); Y. Yamada, Phys. Rev. **D50**, 3537 (1994); I. Jack, D.R.T. Jones, Phys. Lett. **B333**, 372 (1994).
8. J. Bagger, K. Matchev, D. Pierce, and R. Zhang, Nucl. Phys. **B491**, 3 (1997).
9. G.A. Blair and U. Martyn, Proceedings, LC Workshop, Sitges 1999, hep-ph/9910416.
10. M.M. Nojiri, K. Fujii, and T. Tsukamoto, Phys. Rev. **D54**, 6756 (1996).
11. I. Hinchliffe et al., Phys. Rev. **D55**, 5520 (1997); Atlas Collaboration, Technical Design Report 1999, Vol. II, CERN/LHCC/99-15, ATLAS TDR 15.

RUNNING COUPLINGS IN EXTRA DIMENSIONS

JISUKE KUBO AND HARUHIKO TERAO

Institute for Theoretical Physics, Kanazawa University, Kanazawa 920-1192, Japan
E-mail: jik@hep.s.kanazawa-u.ac.jp, terao@hep.s.kanazawa-u.ac.jp

GEORGE ZOUPANOS

Physics Department, Nat. Technical University, GR-157 80, Zografou, Athens, Greece
E-mail: George.Zoupanos@cern.ch

The regularization scheme dependence of running couplings in extra compactified dimensions is discussed. We examine several regularization schemes explicitly in order to analyze the scheme dependence of the Kaluza-Klein threshold effects, which cause the power law running, in the case of the scalar theory in five dimensions with one dimension compactified. It is found that in 1-loop order, the net difference in the running of the coupling among the different schemes is reduced to be rather small after finite renormalization. An additional comment concerns the running couplings in the warped extra dimensions which are found to be regularization dependent above TeV scale.

1 RG in large extra dimensions

Recently the extra compactified dimensions have been attracting much attention as possibilities to explain various hierarchy problems; the gauge hierarchy, the Yukawa hierarchy and so on. The effective field theories in extra dimensions contain towers of massive Kaluza-Klein (KK) excitations, whose quantum effects alter the behavior of the running couplings from logarithmic to power. Therefore the traditional picture of gauge coupling unification may be drastically changed [1]. In this talk we examine the running coulings explicitly in several schemes and consider the origin of power law running and their scheme (in)dependence [2].

In practice the notion of running couplings is not well defined in extra dimensions due to nonrenormalizability. We cannot help but defining the field theories by a certain cutoff. Then the Wilson RG is supposed to offer a natural framework to define the RG flows in such cases. There is another reason why the Wilson RG is suitable for the calculation of the β-functions in extra dimensions. The power law behavior of the running coupling is supposed to be generated by successive threshold corrections by tower of massive KK modes. The Wilson RG is faithful to their decoupling effects.

Therefore we apply the Exact RG, which is a continuum formulation of the Wilson RG, to a scalar field theory in $M_4 \times S^1$ space-time. In this formulation a certain cutoff is performed to the internal loop momenta, and the running coupling is defined by variation of the cutoff scale. We derive the β-functions for the four scalar coupling in the 1-loop level, assuming that the coupling is weak enough. We also calculate the β-functions in other schemes: the proper time regularization and the momentum subtraction, and compare these results.

In all schemes we obtain the β-functions as

$$\beta_\lambda = \frac{d\lambda}{dt} = b\epsilon_k(t)\lambda^2, \tag{1}$$

where scale parameter $t = \ln R\Lambda$ is introduced in terms of radius of the compact space R and the cutoff scale Λ. b is the 1-loop coefficient in four dimensions. The function $\epsilon_k(t)$ is dependent on each scheme k, and it's asymptotic form is given by

$$\begin{aligned} \epsilon_k(t) &\to 1 \qquad\quad \text{for } t \ll 0, \\ &\to B(k)e^t \text{ for } t \gg 0, \end{aligned} \tag{2}$$

where B_k is a scheme dependent constant.

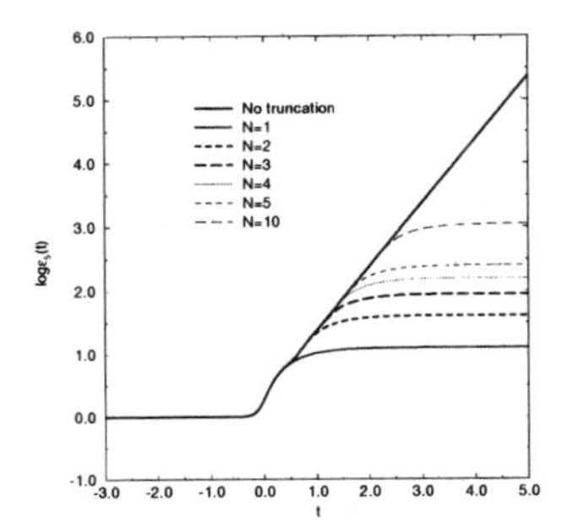

Figure 1. Scale dependence of the β-function coefficient calculated in an ERG scheme.

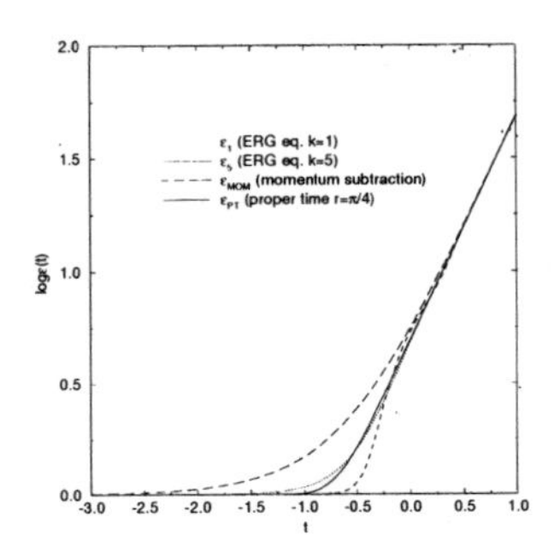

Figure 2. Scheme dependence of the net thershold corrections.

This function is shown in Fig.1 by truncating the KK modes at N-th level. It is seen that the β-function shifts to the five dimensional one smoothly by the KK threshold corrections. We may define the effective dimensions [2] by

$$D_{\text{eff}}(t) = 4 + \frac{d \ln \epsilon_k}{dt}, \tag{3}$$

which shows smooth transition from 4 to 5.

2 Scheme dependence

The β-functions given in Eq. (1) are scheme dependent even in the 1-loop level. However we may redefine the coupling constants in different schemes so that the β-functions match in the asymptotic region. The scheme dependence remained even after this procedure represents the net ambiguity in the RG in extra dimensions. The results obtained by explicit calculations are shown in Fig. 2. It is seen that the unremovable scheme dependence is rather small. In consequence, it is found that the GUT predictions for the low energy gauge couplings [1] are almost scheme independent [2].

3 RG in the warped extra dimensions

The Randall-Sundrum type of compactification [3] is now under the most vigorous investigation. The extra dimension is bounded by two three branes, in one of which the Planck scale is reduced to TeV scale. A peculiarity of this compactification is that the mass scale of the KK tower appears at TeV order, even if the bare mass is Planck scale [4]. If we apply the RG scheme mentioned above to this case, the running couplings are found to show power law behavior too. This should be compared with the results obtained by the PV regularization [5]. However it should be said that the running coupling above TeV scale is totally regularization dependent, since the KK spectrum is influenced by the Planck scale (string) physics.

References

1. K. Dienes, E. Dudas and T. Gherghetta, *Phys. Lett.* **B436**, 3110 (1998); *Nucl. Phys.* **B537**, 47 (1999); D. Ghilencea and G. G. Ross, *Phys. Lett.* **B442**, 165 (1998); T. Kobayashi, et. al., *Nucl. Phys.* **B550**, 99 (1999).
2. J. Kubo, H. Terao and G. Zoupanos, *Nucl. Phys.* **B574**, 459 (2000).
3. L. Randall and R. Sundrum, *Phys. Rev. Lett.* **83**, 3370 (1999).
4. W. D. Goldberger and M. B. Wise, *Phys. Rev.* **D60**, 107505 (1999); S. Chang, et. al., hep-ph/9912498.
5. A. Pomarol, hep-ph/0005293.

UNIFIED FIELD THEORY INDUCED ON THE BRANE WORLD

KEIICHI AKAMA

Department of Physics, Saitama Medical College, Moroyama, Saitama, 350-0496, Japan
E-mail: akama@saitama-med.ac.jp

We show how the gravity, gauge, and matter fields are induced on dynamically localized brane world.

This talk is based on a work done in collaboration with A. Akabane, T. Hattori, K. Katsuura, and H. Mukaida.[1] Here I would like to show how the gravitational, gauge and matter field theories are induced on a brane world. Recently, idea of the brane world attracts much attention with expectation for large extra dimensions, or in connection with the superstring theory.[2-5] The idea that we live in a hyper-surface in a higher dimensional spacetime is very old.[6-14] The present author also presented a model eighteen years ago.[10]

Here, we consider a three dimensional domain wall localized by the double well potential

$$U = -\frac{1}{4}\lambda\left(\Phi^2 - \frac{m^2}{\lambda}\right)^2, \tag{1}$$

for a scalar field Φ in 4+1 dimensional spacetime,[11] where λ and m are constants. The well known kink solution

$$\Phi = \Phi_K(y) \equiv \frac{m}{\sqrt{\lambda}}\tanh\frac{my}{\sqrt{2}}, \tag{2}$$

for this model gives rise to a flat domain wall located at the y=0 hyper-surface.

Such a domain wall is in general curved, and it should be necessarily curved if we want to describe the gravitational dynamics on the brane world. We adopt a curvilinear coordinate system (x^0, x^1, x^2, x^3, y) such that the extra dimension coordinate y vanishes on the brane. Here we assume the whole spacetime is flat, for simplicity. Then, the kink solution $\Phi_K(y)$ becomes an approximate solution for the domain wall. However, the equation of motion

$$\partial_M E G^{MN} \partial_N \Phi = -EU'(\Phi) \tag{3}$$

with $E = \det E_{KM}$ $(K, M = 0, \cdots, 4)$ involves now the vielbein E_{KM} and metric tensor $G_{MN} = E_{KM}E^K{}_N$ for the curvilinear coordinate, and the solution should be distorted from the kink function $\Phi_K(y)$ depending on the metric. Here we assume that the extra dimension coordinate y is taken along the straight normal line perpendicular to the brane at their crossing point. Then the bulk vielbein is given by

$$\begin{pmatrix} E_{k\mu} & E_{k4} \\ E_{4\mu} & E_{44} \end{pmatrix} = \begin{pmatrix} e_{k\mu} - yb_{k\mu} & 0 \\ 0 & -1 \end{pmatrix}, \tag{4}$$

where $e_{k\mu}$ is the induced vielbein on the brane, and the $b_{\mu\nu}$ is the extrinsic curvature of the brane. The vielbein and the extrinsic curvature should obey the Gauss-Codazzi equation which is the embeddability condition of the brane.

We solve the equation of motion (3) by rewriting the field Φ into the form

$$\Phi(x^\mu, y) = \Phi_K(y) + \chi(x^\mu, y)\Phi_k(y)',$$
$$\chi(x^\mu, y) = \chi_0(x^\mu) + \chi_1(x^\mu)y + \cdots. \tag{5}$$

Then the equation of motion reduces to recursion formulae for the coefficients $\chi_i(x^\mu)$. We determine them one by one. The solution is given by

$$\Phi = \Phi_{br} \equiv \Phi_K + \chi\Phi_k',$$
$$\chi = by^2/2 + (b_2 + b^2)y^3/6 + \cdots. \tag{6}$$

where $b = b_{\mu\nu}g^{\nu\mu}$, $b_2 = b_{\mu\nu}b_{\rho\sigma}g^{\nu\rho}g^{\sigma\mu}$ and we put the arbitrary functions χ_0 and χ_1 vanishing for simplicity.

We consider the quantum fluctuations around this classical solution Φ_{br}. For this purpose we expand our dynamical variable Φ

in terms of the complete set $\{\eta_n\}$ of small fluctuation modes of the kink solution Φ_K.

$$\Phi = \Phi_{br} + \varphi_0\eta_0 + \varphi_1\eta_1 + \int \varphi_q\eta_q dq, \quad (7)$$

where $\{\eta_n\}$ consists of

$$\eta_0 = 1/\cosh^2 z, \qquad (z = my/\sqrt{2}) \quad (8)$$

$$\eta_1 = \sinh z/\cosh^2 z, \quad (9)$$

$$\eta_q = e^{iqz}(\tanh^2 z - 1 - q^2 - 3iq\tanh z. \quad (10)$$

Among them, we omit the translation zero mode η_0, because it only translates the position of the brane, and should be considered with other configuration of the brane vielbein $e_{k\mu}$ and the extrinsic curvature $b_{\mu\nu}$. The dynamical degrees of freedom of the zero mode are transferred to those of the brane vielbein and the extrinsic curvature which are constrained by the Gauss-Codazzi equation.

We substitute the expanded form (7) of the field Φ back into the original whole-spacetime action

$$S = \int E\left(G^{MN}\partial_M\Phi\partial_N\Phi/2 - U(\Phi)\right) d^5x, \quad (11)$$

and rearrange the terms with respect to the dynamical variables $e_{k\mu}$, $b_{\mu\nu}$, φ_1, and φ_q. Among the terms, those concerned with $e_{k\mu}$, $b_{\mu\nu}$, and φ_1 are localized around the brane. This means that the fields are trapped on the brane. So we perform the y integration for the trapped sector.

$$S = \int E\Big[\frac{1}{2}G^{MN}(\partial_M\varphi_1\eta_1)(\partial_M\varphi_1\eta_1)$$
$$-\frac{m^4}{4\lambda} - \frac{1}{2}m^2\varphi_1^2\eta_1^2 + \frac{1}{4}\lambda(\Phi_K + \chi\Phi'_K)^4$$
$$-\frac{3}{2}\lambda(\Phi_K + \chi\Phi'_K)^2\varphi_1^2\eta_1^2\Big] dyd^4x \quad (12)$$

They are integrations of the known functions such as

$$\int {\eta_1}^2 dy = \frac{\sqrt{2}}{m}\int\left(\frac{\sinh z}{\cosh^2 z}\right)^2 dz = \frac{2\sqrt{2}}{3m},$$
$$\int y^4\Phi_K'^2\eta_1^2 dy = \frac{2\sqrt{2}}{\lambda m}\int\frac{z^4\sinh^2 z}{\cosh^8 z}dz$$
$$= \frac{2\sqrt{2}}{\lambda m}\left(-\frac{2}{15} - \frac{\pi^2}{45} + \frac{\pi^4}{225}\right), \text{ etc.}$$

Thus we finally obtain the effective action

$$S = \int e\Big[-\frac{2\sqrt{2}m^3}{3\lambda} + \frac{\sqrt{2}m}{3\lambda}\left(1 + \frac{\pi^2}{3}\right)b_2$$
$$+\frac{\sqrt{2}m}{\lambda}\left(-\frac{4}{5} - \frac{\pi^2}{6} + \frac{7\pi^4}{1200}\right)b^2$$
$$+\frac{\sqrt{2}}{3m}\left(g^{\mu\nu}\partial_\mu\varphi_1\partial_\nu\varphi_1 - \frac{3}{2}m^2{\varphi_1}^2\right)$$
$$+\frac{\sqrt{2}}{3m^3}\left(1 + \frac{\pi^2}{12}\right)\partial_\mu\varphi_1\partial_\nu\varphi_1$$
$$\times\{6b^\mu{}_\lambda b^{\lambda\nu} - 4bb^{\mu\nu} - (b_2 - b^2)g^{\mu\nu}\}$$
$$+\frac{\sqrt{2}}{m}\left(-\frac{9}{10} + \frac{\pi^2}{360} - \frac{\pi^4}{300}\right)b^2{\varphi_1}^2$$
$$+\frac{\sqrt{2}}{m}\left(\frac{5}{6} - \frac{\pi^2}{360}\right)b_{\mu\nu}b^{\mu\nu}{\varphi_1}^2\Big]d^4x \quad (13)$$

They consist of the cosmological term, the mass terms of the field $b_{\mu\nu}$, the kinetic and the mass terms of the field $\varphi_1$and the interactions among them. The masses and coupling constants are definitely calculated. The field φ_1 is a prototype of trapped field on the brane. If there exist bosonic or fermionic fields which coupled to our original field Φ in the whole spacetime, their low-lying modes are trapped around the brane, and described by similar effective actions on the brane.

A problem is, however, no kinetic terms are induced for the vielbein $e_{k\mu}$ and the extrinsic curvature $b_{\mu\nu}$. This is because we assumed that the whole space is flat for a technical simplicity. If the whole space is curved, the expression of the whole-space vielbein E_{KM} involves the derivatives of the brane vielbein $e_{k\mu}$ and extrinsic curvature $b_{\mu\nu}$, which give rise to their kinetic terms. Among them the Einstein gravity action is induced for the brane metric $g_{\mu\nu}$. This is nothing but the trapped graviton in its proper treatment. An outstanding feature of the brane world picture is that the extrinsic curvature fields become dynamical within the brane under the constraint of the Gauss-Codazzi equation. Anther effect which contributes to the kinetic terms are the quantum fluctuation of the trapped matters on the brane which is calculated through the quantum loop diagrams with internal lines of these matter fields.[15−17] Interestingly, the kinetic terms due to the quantum effects can be induced even if the whole spacetime is flat.

So far we assumed a specific model of a 3 brane in a 4+1 dimensions. The number of the extra dimension was 1. If the number is grater than 1, the normal connections of the brane naturally take part in the play. They are nothing but the gauge fields of the O(N) rotation group of the extraspace. If we interpret them as the physical gauge fields like photon, gluon, and weak bosons, interesting phenomenological consequences may emerge, since they are severely constrained by the Gauss-Codazzi-Ricci equations.

In conclusion, on the dynamically localized brane world: (i) Low lying small fluctuation modes are trapped with proper field theoretical effective action. (ii) The gravitational field $e_{k\mu}$, the extrinsic curvature $b_{\mu\nu}$, and the gauge fields A_μ^{ab} are induced on the brane. (iii) The masses and coupling constants are definitely calculated. (iv) The kinetic terms of $e_{k\mu}$, $b_{\mu\nu}$, and A_μ^{ab} on the brane are induced through quantum fluctuations of the trapped matters, and they are induced even if the whole spacetime is flat. (v) $e_{k\mu}$, $b_{\mu\nu}$, and A_μ^{ab} should obey the Gauss-Codazzi-Ricci equations inaddition to their individual equations of motion.

References

1. K. Akama, T. Hattori, hep-th/0008133; A. Akabane, K. Akama, T. Hattori, K. Katsuura, and H. Mukaida, in preparation.
2. I. Antoniadis, Phys. Lett. **B246**, 377, (1990); N. Arkani-Hamed, S. Dimopoulos, and G. Dvali, Phys. Lett. **B429**, 263-272,(1998); Phys. Rev. **D59**, 086004,(1999); I. Antoniadis, N. Arkani-Hamed, S. Dimopoulos, and G. Dvali, Phys. Lett. **B436** (1998) 257-263.
3. K. R. Dienes, E. Dudas, T. Gherghetta, Phys. Lett. **B436**, 55-65 (1998); Nucl. Phys. **B537**, 47-108 (1999).
4. C. P. Bachas, JHEP **9811**, 023 (1998).
5. L. Randall, R. Sundrum, Phys. Rev. Lett. **83**, 3370-3373 (1999); **83**, 4690-4693 (1999).
6. C. Fronsdal, Nuovo Cimento **13**, 988, 2560 (1959).
7. D. W. Joseph, Phys. Rev. **126**, 319 (1962).
8. T.Regge and C.Teitelboim, in *Marcel Grossman Meeting on Relativity, 1975*(North Holland, 1977) 77.
9. M. D. Maia, Braz. J. Phys. **8** 429-441 (1978); J. Math. Phys. **25**, 2090 (1984).
10. K. Akama, in *Lecture Notes in Physics, 176, Proceedings, Nara, Japan, 1982*, edited by K. Kikkawa, N. Nakanishi and H. Nariai, (Springer-Verlag, 1983) 267-271 (e-print: hep-th/0001113).
11. V. A. Rubakov and M. E. Shaposhnikov, Phys. Lett. **B125**, 136-138 (1983).
12. M. Visser, Phys. Lett. **B159**, 22 (1985).
13. B. Holdom, Nucl. Phys. **B233**, 413 (1984); M. Pavšič, Class. Quant. Grav. **2**, 869 (1985); G. W. Gibbons and D. L. Wiltshire, Nucl. Phys. **B287**, 717 (1987); Q. Shafi, and C. Wetterich, Nucl. Phys. **B297**, :697 (1988);
14. K. Akama, Prog. Theor. Phys. **78**, 184-188 (1987); **79**, 1299-1304 (1988); **80**, 935-940 (1988). For further references, see Ref. 1.
15. A. D. Sakharov, Dokl. Akad. Nauk SSSR **177**, 70 (1967) [Sov. Phys. Dokl. **12**, 1040 (1968)].
16. K. Akama, Y. Chikashige, T. Matsuki and H. Terazawa, Prog. Theor. Phys. **60**, 868 (1978); K. Akama, Prog. Theor. Phys. **60**, 1900 (1978).
17. A. Zee, Phys. Rev. Lett. **42**, 417 (1979); S. L. Adler, Phys. Rev. Lett. **44**, 1567 (1980).

Parallel Session 11

Search for New Particles and New Phenomena

Conveners: William C. Carithers (Fermilab) and
Luc Pape (CERN)

SEARCH FOR STANDARD MODEL HIGGS BOSON AT LEP2

SHAN JIN

University of Wisconsin-Madison, Madison, WI 53706, USA
Corresponding adress: CERN / EP division, CH-1211 Geneva 23, Switzerland
E-mail: Shan.Jin@cern.ch

In this talk, the search results on Standard Model Higgs boson using LEP data collected at energies up to 202 GeV are reported. No statistically significant excess has been observed when compared to the Standard Model background prediction. The preliminary LEP combined lower bound at 95% confidence level on the mass of the Standard Model Higgs boson is 107.7 GeV/c^2.

1 Introduction

In 1999 the four LEP experiments (ALEPH, DELPHI, L3, OPAL) have collected data at various energies between 192 and 202 GeV, for aprroximately 900 pb^{-1} (about 230 $pb{-}1$ for each experiment).

From combining the earlier data collected by the LEP experiments at center of mass energies up to 189 GeV, a 95% CL lower bound of 95.2 GeV/c^2 on the mass of Standard Model (SM) Higgs boson has been obtained[1]. In this talk, we present an update of the SM Higgs search which includes the new data collected at center-of-mass enrgies uo to 202 GeV[2].

2 Search Topologies

At LEP the SM Higgs boson is expected to be produced mainly via Higgs-strahlung process $e^+e^- \rightarrow HZ$, while contributions from the $WW \rightarrow H$ fusion channel, $e^+e^- \rightarrow H\nu\bar{\nu}$, are typically below 10%. The searches performed by the four LEP collaborations encompass the usual HZ final state topologies:

(1) Four-jet ($b\bar{b}q\bar{q}$) channel

The four-jet final state accounts for about 65% of the Higgs-Strahlung cross section. The events are characterized by two jets from Z decay accompanied by two b jets from the Higgs boson decay. This channel has largest contribution to the search results due to its dominant branch ratio.

(2) Missing energy ($b\bar{b}\nu\bar{\nu}$) channel

The missing energy final state comprises 20% of signal decays. These events are characterized by large missing mass compatible with the Z mass, and two acoplanar b jets.

(3) Leptonic ($b\bar{b}e^+e^-$ and $b\bar{b}\mu^+\mu^-$) channel

The leptonic final state represents 6.7% of the Higgs-Strahlung cross section. The signal events are characterized by two isolated energetic leptons with an invariant mass close to m_Z and a large hadronic recoil mass. Altough this channel has a low braching ratio, the experimental signature is clean and the Higgs boson mass can be reconstructed with a good resolution.

(4) Tau ($b\bar{b}\tau^+\tau^-$ and $\tau^+\tau^-q\bar{q}$) channel

The tau channel has a branching ratio of 9% of signal decays. The signal events are charaterized by two τ's and two jets with a mass of either τ pair or jet pair compatible with the Z mass.

In the four jet and tau channels, the overlap between the SM Higgs and MSSM Higgs analyses needs to be treated carefully in order not to subtract background twice since the same SM Higgs analyses are combined in the MSSM Higgs searches.

The analysis procedures of the four LEP experiments are described in individual documents [3].

Table 1. Numbers of expected events from background and observed events of the four LEP experiments at energies between 192 and 202 GeV.

Bkg. Exp. Evts. Obs.	A	D	L	O
Fourjet:	46.4	175.4	73.5	27.7
	30	161	72	33
Missing	11.0	105.7	145.0	9.3
Energy:	7	108	157	6
Leptonic:	28.5	20.8	29.4	8.1
	26	19	35	10
Tau:	11.9	6.9	22.3	6.4
	11	6	19	4
Total	97.8	308.8	270.7	51.5
	74	294	283	53

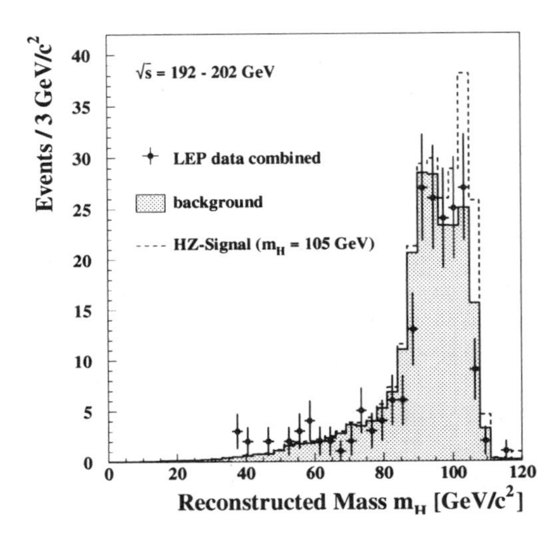

Figure 1. LEP-combined distribution of the reconstructed SM Higgs boson mass in searches conducted at $\sqrt{s}$ between 192 and 202 GeV. The figure displays the data (dots with error bars), the predicted SM background (shaded histogram) and the prediction for a Higgs boson of 105 GeV/c^2 mass (dashed histogram).

3 Results

The numbers of expected events from background and observed events at energies between 192 and 202 GeV are listed in table 1. Fig. 1 shows the distribution of the reconstructed Hihhs masses for a subset of the events in Table 1. The figure has been obtained with the supplementary requirement that the contributions from the four experiments (selecting the most signal-like set of events) be roughly equal. Since all events enter with equal weight, such a distribution does not reflect for example differences in mass resolutions, signal sensitivities and background rates, which charaterize the various search channels and individual experiments. So no quantitative conclusions should be drawn from this figure.

4 Confidence Levels

4.1 Estimator (test-statistic):

An Estimator (or a test-statistic) X quantifies the "signal-ness" of an experiment. $X = -2lnQ$ where $Q = L(s+b)/L(b)$ is the likelihood ratio. Here the likelihood $L(s+b)$ assumes a signal process at Higgs mass m_H in addition to the background process, and the likelihood $L(b)$ assumes no Higgs signal but background process only.

4.2 Confidence Levels CL_{s+b} and CL_b:

Given an observed value of the estimator $X_{observed}$, the confidence level on the *signal+ background* and *background − only* hypotheses are calculated as $CL_{s+b} = P_{s+b}(X \geq X_{observed})$ and $CL_b = P_b(X \geq X_{observed})$. The value P_b is the estimator probability function for experiments with background process only, while P_{s+b} is the estimator probability function for experiments with both background processes and a signal process of a given Higgs mass m_H with the Standard Model cross section.

4.3 Confidence Level CL_s:

The limit on m_H is obtained via the limit on the Higgs cross section (number of signal events). So the confidence level CL_s needs to be defined. Two methods are used in the CL_s calculation in the LEP experiments:

(A) Generalized Bayesian Method[4]:

$CL_s = CL_{s+b}/CL_b$.

This method are used in DELPHI, L3, OPAL and LEP combined results.

(B) Signal Estimator Method[5]:

$CL_s = CL_{s+b}$ - (1 - CL_b) $\cdot e^{-S}$, where S is the number of predicted signal events.

This method is used in the ALEPH results.

Both methods satisfy $CL_s = e^{-S}$ when zero event is observed, which is independent of background prediction. Method (B) gives typically about 0.5 GeV/c^2 better sensitivity for the exclusion on m_H than method (A).

Table 2. 95 % CL median expected limits and observed limits of the LEP experiments

Limits (GeV/c^2)	A	D	L	O
Expected	107.8	105.7	105.0	105.2
Observed	107.7	106.1	107.0	103.0

5 Lower Limit on m_H

The compatibility with background of the result is given by $1 - CL_b$, which is plotted as a function of m_H in Figure 2. This plot shows that there is no evidence for the SM Higgs boson.

A 95 % confidence level lower limit on the SM Higgs mass may be set by identifying the mass region where $CL_s < 0.05$, as shown in Figure 3. The median limit expected in the absence of a signal is 109.1 GeV/c^2 and the limit observed by combining the LEP data is 107.9 GeV/c^2.

The above limits were obtained ignoring existing correlations of the systematic errors between the four experiments, in data sets at different energies and between search channels. A separate study of the effect of correlations results in a downward shift by 150 MeV/c^2 of the ovserved mass limit. So the observed limit is decreased by this amount and 107.7 GeV/c^2 is quoted as the 95 % CL lower bound for the mass of the SM Higgs boson.

The 95% CL expected limits and observed limits for each LEP experiment are listed in Table 2.

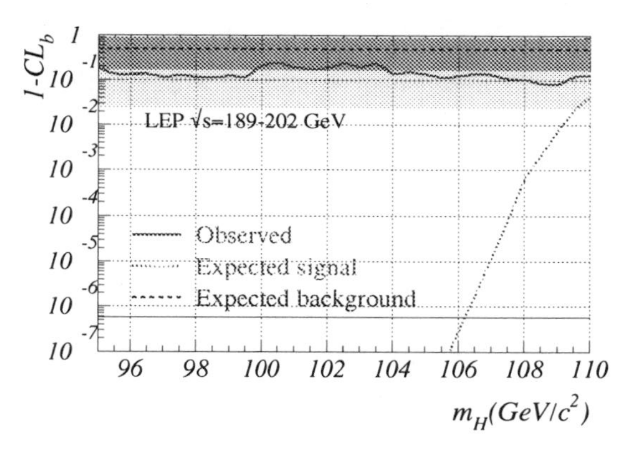

Figure 2. The confidence level $1 - CL_b$ as a function of m_H. The straight dashed line at 50% and shaded bands represent the median result and the $\pm 1\sigma$ and $\pm 2\sigma$ probability bands expected in the absence of a signal. The solid curve is the observed result and dotted curve shows the median result expected for a signal when tested at the "true" mass. The horizontal line at 5.7×10^{-7} indicates the level for a 5 σ discovery.

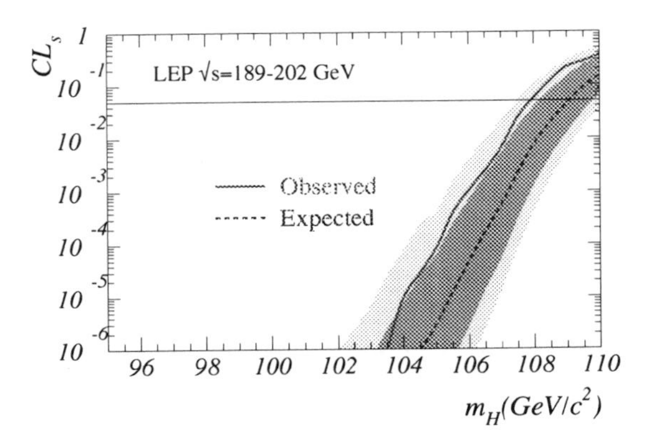

Figure 3. The confidence level CL_s for the signal hypothesis versus m_H. The solid curve is the observed result, the dashed curve the median result expected in the absence of a signal. The shaded areas represent the symmetric 1σ and 2σ probability bands of CL_s in the absence of a signal. The intersections of the curves with the horizontal line at $CL_s = 0.05$ give the mass limits at 95% confidence level.

Conclusion

In the absence of a statistically significant excess in the data collected by LEP experiments in 1999 at energies between 192 and 202 GeV, a preliminary lower bound of 107.7 GeV/c^2 has been obtained at the 95% confidence level.

Acknowledgments

We wish to thank our colleagues from the accelerator divisions for the successful high energy operation of LEP. The results presented in this talk are due to the devoted work of many physicists from four LEP experiments. I am grateful to all of them. This work is supported in part by the United States Department of Energy through grand DE-FG-0295-ER40896.

References

1. ALEP, DELPHI, L3 and OPAL Collab., The LEP Working group for Higgs boson searches, ALEPH 99-081 CONF 99-052, DELPHI 99-142 CONF 327, L3 Note 2442, OPAL Technical Note TN-614.
2. ALEP, DELPHI, L3 and OPAL Collab., The LEP Working group for Higgs boson searches, ALEPH 2000-028 CONF 2000-023, DELPHI 2000-050 CONF 365, L3 Note 2525, OPAL Technical Note TN-646.
3. ALEPH Collab., ALEPH 00-006 CONF 00-003;
   DELPHI Collab., DELPHI 2000-092 CONF 391;
   L3 Collab., L3 Note 2588;
   OPAL Collab., OPAL Physics Note PN 426.
4. A.Read, CERN 2000-005.
5. S.Jin and P. McNamara, CERN 2000-005, physics/9812030.

SUSY HIGGS AT LEP

IAN FISK

Physics Department, University of California, San Diego, La Jolla, CA 92093, USA
E-mail: ifisk@ucsd.edu

The results of the search at LEP for the h and A neutral Higgs bosons and the H^+ and H^- charged Higgs bosons are presented from the 1998 data taken at 189 GeV center-of-mass energy and 1999 data taken at 192 GeV, 196, GeV, 200 GeV, and 202 GeV center-of-mass energies from the Aleph, Delphi, L3, and Opal collaborations as well as the combined results of the four. Together this data set constitutes nearly 900 pb^{-1} above 192 GeV and an additional 700 pb^{-1} at 189 GeV. Results of the four experiments are consistent with the expected Standard Model background and limits on the Higgs boson masses are set. Combining the experiments, limits are set at $m_h > 88.3$ GeV and $m_A > 88.4$ GeV for the neutral Higgs bosons and $m_{H+-} > 78.6$ GeV for the charged Higgs bosons. All results should be considered preliminary.

1 Introduction

1.1 Higgs Production

In the MSSM there are two Higgs doublets which give rise to 5 Higgs bosons: h, H, H^+, H^-, and A. h and A are the most often predicted to be accessible at LEP energies. The main neutral Higgs production mechanisms are Higgs-strahlung production when an h is radiated by a Z boson and pair production when the h and A are produces as a pair from the decay of a Z. The Higgs-strahlung process is analgous to Standard Model Higgs production and dominant in the MSSM for low values of $\tan(\beta)$, the ratio of the Higgs vacuum expectation values. The pair production process is dominant at high values of $\tan(\beta)$. In the intermediate region the search must have good sensitivity to both processes.

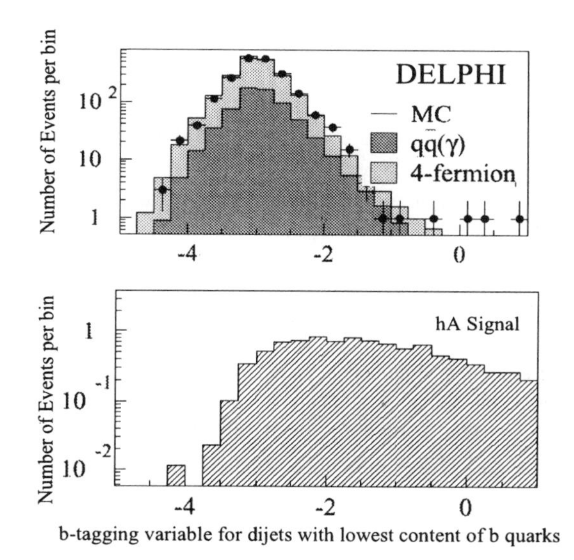

Figure 1. Example of the descriminating power of the b-tagging variable between background and signal.

At Higgs masses visible at LEP the dominant decay of the h and A bosons is to b quarks: over 90% at high $\tan(\beta)$ and the h decays to b's over 80% at low $\tan(\beta)$. The ability to identify b quarks will be the most powerful tool to destinguish signal from background. Figure 1 shows the b-tagging variable used by the Delphi experiment[1]. The region on the right has excellent separation between signal and background. The second highest decay mode of the neutral Higgs is to τ leptons representing approximately 5% to 8% of the total.

The two charged Higgs bosons are usually predicted to be heavy and are most often investigated outside the context of the MSSM. The predicted production mechanism is the decay of a Z or γ into a H^+ H^- pair. At masses visible at LEP the charged Higgs would decay to a τ and ν_τ or a c and an s quark. The percentage into each branching ratio is left as a free parameter of the search.

2 The Higgs Search

2.1 Selections

All experiments perform searches for hZ production, investigating approximately 98% of the available decay modes: $hZ \to q\bar{q}q\bar{q}$, $hZ \to q\bar{q}\tau^+\tau^-$, $hZ \to \tau^+\tau^-q\bar{q}$, $hZ \to q\bar{q}\nu\bar{\nu}$, $hZ \to q\bar{q}e^+e^-$, and $hZ \to q\bar{q}\mu^+\mu^-$. These searches are identical to the Standard Model Higgs search and are described in [1,2,3,4].

Dedicated searches have been performed for h and A pair production investigating approximately 97% of the available decay modes: $hA \to b\bar{b}b\bar{b}$, $hA \to \tau^+\tau^-b\bar{b}$, and $hA \to b\bar{b}\tau^+\tau^-$.

2.2 Final Variable

To reduce easily eliminated background a pre-selection is applied. Once a sample of events is chosen a final variable can be created. The final variable is made of a combination of descriminating information. The b-tagging variable, like that shown in Figure 1 is the most important input to the neutral Higgs searches, but improved distinguishing power can be achieved by adding additional information like production angle. In the charged Higgs search the recontructed Higgs mass is most descriminating input variable.

Information in the final variable is combined using a variety of techniques by the four LEP experiments: likelihood ratio techniques, neural network techniques, combining probabilities, or a series of linear cuts[1,2,3,5,6]. Examples of final variables from Aleph and Opal are given in Figure 2[2,3].

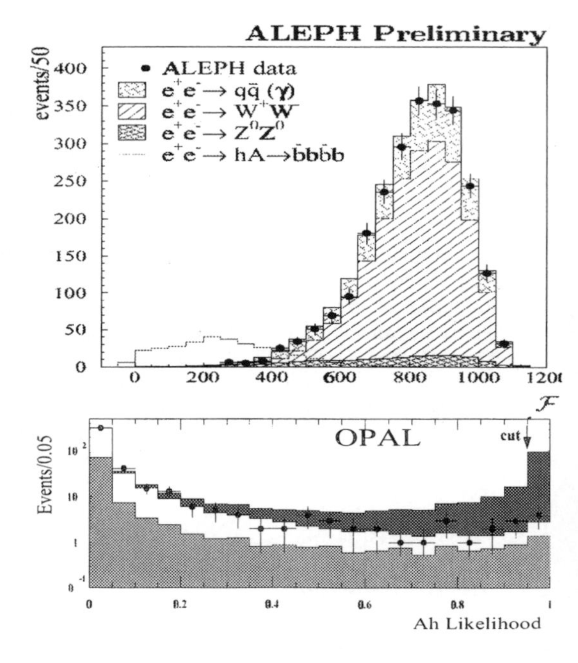

Figure 2. Examples of the hA final variable for the four jet Higgs search from the Opal and Aleph collaborations.

3 Results

Using the final variable and the statistical methods described in [7], the presence or absence of a Higgs signal in the data collected can be quantified. The two quantities calculated are CL_B, the data consistency with a background only hypothesis and the Higgs mass when the Confidence Level with which the Higgs can be excluded crosses 95%.

3.1 Limits

The CL limits on the mass of the h and A independent of $\tan(\beta)$ are given in Table 1 [7] as well as the number of events collected and expected in the 1999 data set.

The sensitivity of the search improves when the experiments are combined together. The observed $\tan(\beta)$ independent limits are $m_h > 88.3$ GeV and $m_h > 88.4$ GeV. With expected limits of 90.8 GeV and 91.1 GeV on the h and A mass respectively. Figure 3 [7] shows the combine limit in the scanned plane for the more conservative maximal h mass choice of MSSM parameters. As can be seen in the figure, the region of $\tan(\beta)$ between 0.7 and 1.8 is excluded at the 95% confidence level [7].

For the Charged Higgs the combined

Table 1. Table of the expected background and observed data events for all four LEP experiments as well as the observed and expected lower mass limits on the h and A.

Exp.	A	D	L	O
$\mathcal{L}(\text{pb}^{-1})$	237	228	233	214-217
$\text{hA} \to \text{b}\bar{\text{b}}\text{b}\bar{\text{b}}$				
Bkg. Exp.	46.4	51.1	13.7	10.5
Data	30	47	10	16
$\text{hA} \to \text{b}\bar{\text{b}}\tau^+\tau^-$ $\text{hA} \to \tau^+\tau^-\text{b}\bar{\text{b}}$				
Bkg. Exp.	2.7	7.1	1.3	7.7
Data	1	6	0	6
Limits (GeV)				
m_{h} Obs.	91.5	87.7	83.5	79.2
m_{A} Obs.	91.9	87.2	83.7	80.2
m_{h} Exp.	87.7	85.8	85.5	83.7
m_{A} Exp.	88.1	87.3	85.7	85.4

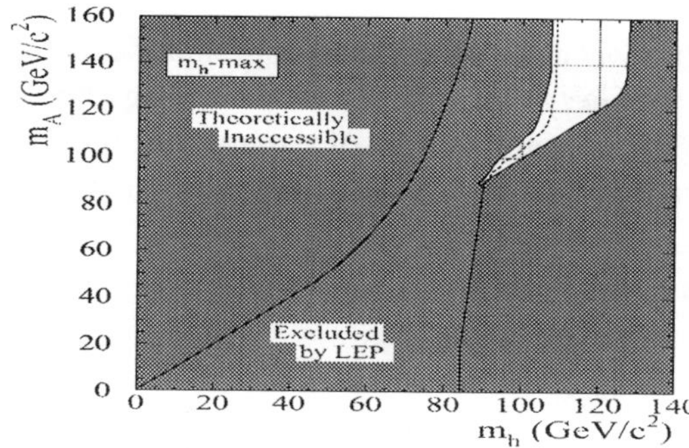

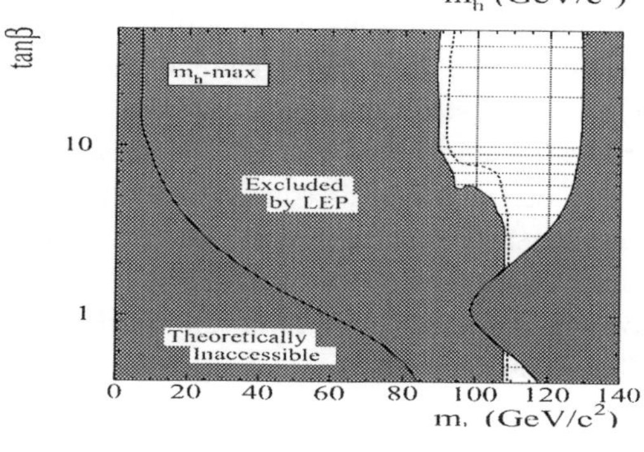

Figure 3. The exclusion plane for the maximal h mass benchmark for tan(β) verses m_{h} and m_{A} verses m_{h} for the combined results of all four LEP experiments.

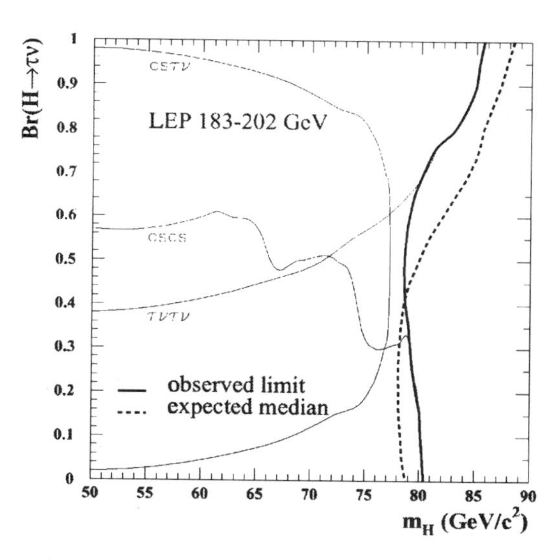

Figure 4. The exclusion plane for the charged Higgs boson as a function of branching fraction to $\tau; \nu_\tau$ for the four LEP experiments combined.

limit as a function of the Higgs branching to $\tau\nu_\tau$ is shown in Figure 4 [7]. The limit on the charged Higgs mass, independent of branching fraction, is $m_{\text{H}+-} > 78.6$ GeV. The expected limit is $m_{\text{H}+-} > 78.0$ GeV [7].

References

1. DELPHI Collaboration, ICHEP00 PA11-7A 619, DELPHI 2000-092 CONF 391.
2. ALEPH Collaboration, ICHEP00 PA11-7A 258, ALEPH 00-006 CONF 00-003.
3. OPAL Collaboration, ICHEP00 PA11-7A 220, OPAL Physics Note PN426.
4. L3 Collaboration, ICHEP00 PA11-7A 414, L3 Note 2588.
5. OPAL Collaboration, ICHEP00 PA11-7A 156, OPAL Physics Note PR285.
6. L3 Collaboration, ICHEP00 PA11-7A 413, L3 Note 2573.
7. Bock, P; et al; ALEPH Collaboration; L3 Collaboration; Opal Collaboration; LEP working group for Higgs boson searches. CERN-EP-2000-055; 2000.

EXOTIC OR 2HDM HIGGS AT LEP

ARI KIISKINEN

Siltavuorenpenger 20 C, 00014 Helsinki Institute of Physics, Finland
E-mail: ari.kiiskinen@cern.ch

A variety of searches for exotic Higgs bosons, i.e. in models beyond the Standard Model or its Minimal Supersymmetric extension, have been performed on the data collected by the four LEP experiments. No evidence of the Higgs boson has been found.

1 Introduction

This paper describes recent LEP results of searches for exotic Higgs bosons, i.e. Higgs bosons in models beyond the Standard Model (SM) or the Minimal Supersymmetric Standard Model (MSSM). Results of all four LEP experiments are given and they include data collected at LEP collision energies up to 202 GeV. All results given here are preliminary and all exclusion limits are given at 95% confidence level.

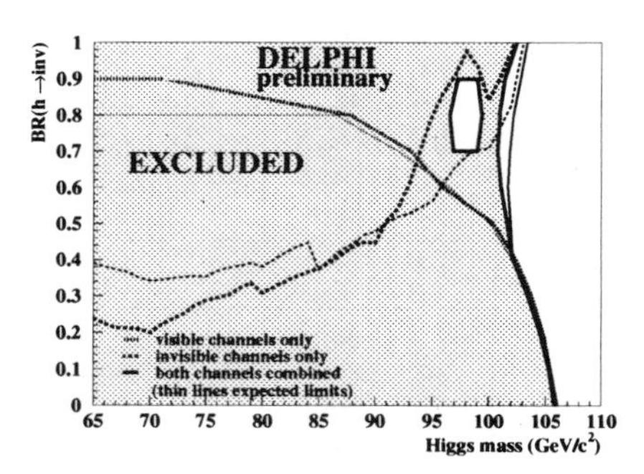

Figure 1. DELPHI exclusion for the Higgs boson mass as a function of the invisible branching fraction.

2 Invisible decays of the Higgs boson

If the Higgs boson decays to invisible particles, such as the lightest supersymmetric particles or Majorons, it will not be detected by standard Higgs boson searches. Specific analyses looking for such Higgs bosons produced in the Higgs Strahlung process have been developed by all four LEP experiments. These analyses look for a resonance in the recoil mass spectrum of the visible Z^0 part of the event. No significant excess of signal-like events has been observed and upper limits for the production cross-section of an invisibly decaying Higgs boson have been set. A lower mass limit for the invisibly decaying Higgs boson, assuming the SM production cross-section, has been set at 106.4 GeV/c^2 [1]. The result of the search for the invisible Higgs boson has also been combined with the results of analyses looking for the SM visible Higgs boson decays. Assuming the SM production cross-section, a lower mass limit of 96.6 GeV/c^2 has been set for a Higgs boson with any combination of the invisible branching fraction and the SM branching fraction [1].

3 Anomalous Higgs boson couplings

Anomalous couplings between the Higgs boson and vector bosons, i.e. couplings not included in the SM, could change the Higgs boson production cross-section from the SM prediction and could change the decay processes of the Higgs boson. For example, a coupling between the Higgs boson and photons would allow production of the Higgs boson in association with a photon. In this case Higgs bosons with mass of up to $\sqrt{s}$ could be produced at LEP, whereas the kinematical upper limit for the Higgs boson mass in the SM Higgs Strahlung process is $\sqrt{s} - m_{Z^0}$. Analyses looking for Higgs boson like objects produced in association with a photon have

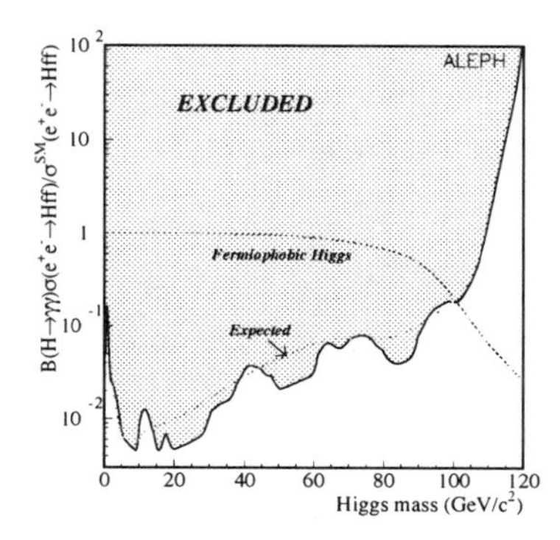

Figure 2. ALEPH exclusion for the cross-section of the fermiophobic Higgs boson.

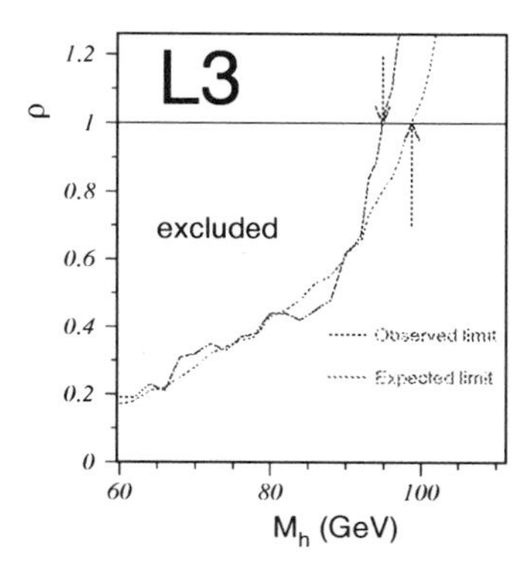

Figure 3. L3 exclusion for the cross-section of a Higgs boson decaying into quarks of any flavour. ρ is given in units of the SM Higgs production cross-section.

been performed in all four experiments, but no evidence of signal has been observed. Results have been given as upper limits for the photonic Higgs coupling as a function of the Higgs mass [2].

The enhanced photonic decay of Higgs bosons could be detected as a resonance in the di-photon mass spectrum. This has also been searched for, but no evidence of signal has been found. The results of these analyses have been interpreted in the context of the fermiophobic scenario, in which the coupling of the Higgs boson to the fermions is strongly suppressed and the dominant decay of a light Higgs boson would be to two photons. A lower mass limit of 103.3 GeV/c^2 has been set for the fermiophobic Higgs boson [2].

Limits for the strength of other possible anomalous couplings and different combinations of such couplings as a function of the Higgs mass have been also given [2].

4 2HDM model interpretations

The most common choice to extend the SM Higgs sector is to add a second Higgs doublet. Two Higgs Doublet Models (2HDM) must be consistent with the experimental exclusion of flavour changing neutral currents. This can be ensured in two ways: in models of Type I only the second Higgs doublet couples to fermions and in Type II the first Higgs doublet couples only to down-type fermions and the second one to up-type fermions. The MSSM Higgs sector is an example of model Type II.

The Higgs Strahlung process and the pair production of the lightest Higgs bosons are complementary to each other in 2HDM:

$$\sigma_{\mathrm{h^0Z^0}} = \sin^2(\beta-\alpha)\sigma_{\mathrm{HZ}}^{SM}$$
$$\sigma_{\mathrm{h^0A^0}} = \cos^2(\beta-\alpha)\bar{\lambda}\sigma_{\nu\nu}^{SM}$$

where $\bar{\lambda}$ is the phase space suppression factor near the kinematical threshold.

Because of this complementarity, the lightest Higgs boson should be detectable at LEP if the Higgs masses are low enough to make these processes kinematically accessible.

The results of the SM or the MSSM neutral Higgs boson searches have been reinterpreted in the context of more general 2HDM. Additional analyses looking for non-standard decay channels, such as Higgs boson decays to light quarks or gluons, have been included to obtain more general limits for model parameters. No evidence of Higgs bosons has been found and the results of these analyses exclude large fractions of the 2HDM parameter space [4].

5 Search for H^+H^- in 2HDM Type I

Previous searches for charged Higgs bosons have assumed a 2HDM Type II scenario in which the dominant decay modes of those H^+ which are accessible at LEP energies are to $c\bar{s}$ or $\tau\nu$. A new analysis searches for H^+H^- in 2HDM Type I in which a high value of $tan\beta$ results in the suppression of the fermionic decays leading in W^*A to be the dominant decay channel if A is light enough. In this analysis, the light A is assumed to decay into $b\bar{b}$ pairs leading to multijet final states. Results of this new analysis have been combined with the standard analysis of the fermionic $H^\pm$ decays and parts of previously unexplored parameter space with light A and light $H^\pm$ have been excluded [5].

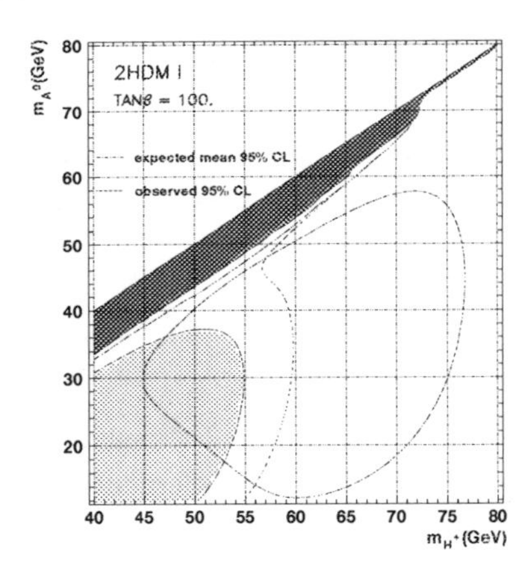

Figure 4. OPAL exclusion (shaded areas excluded) for the $H^\pm$ mass in 2HDM type I as a function of the A mass.

6 Search for $H^{++}H^{--}$

Left-right symmetric SUSY models with automatic R-parity conservation can lead to doubly charged Higgs bosons with a mass of the order of 100 GeV/c^2. At this mass range their dominant decay is to same sign lepton pairs. Stringent limits for $H^{++} \rightarrow e^+e^+$ and $H^{++} \rightarrow \mu^+\mu^+$ exist from high energy bhabbha exchange and from the absence of the muonium anti-muonium exchange. Therefore, an analysis looking for pairs of doubly charged Higgs bosons decaying into the four-tau final state has been performed. No sign of signal has been observed and a lower mass limit for the doubly charged Higgs bosons has been set at 92.8 GeV/c^2 [6].

7 Conclusions

The searches for the Higgs bosons of the SM or the MSSM have been complemented at LEP by a variety of searches for Higgs bosons in more general or alternative models. Many different scenarios have been explored but no evidence of Higgs bosons of any kind has been found so far. Negative search results have been used to set exclusion limits for the Higgs boson masses and other model parameters.

References

1. ALEPH Collab., ALEPH 2000-009 CONF 2000-006. DELPHI Collab., ICHEP2000 abst. #276. L3 Collab., ICHEP2000 abst. #434. OPAL Collab., ICHEP2000 abst. #226,#177.
2. ALEPH Collab., ICHEP2000 abst. #260. DELPHI Collab., ICHEP2000 abst. #367. L3 Collab., ICHEP2000 abst. #415,#432. OPAL Collab., ICHEP2000 abst. #157,#177.
3. J.F.Gunion et. al, *The Higgs Hunter's Guide*, (Addison-Wesley, 1990).
4. DELPHI Collab., ICHEP2000 abst. #102. L3 Collab., ICHEP2000 abst. #433. OPAL Collab., ICHEP2000 abst. #222,#227.
5. OPAL Collab., ICHEP2000 abst. #223.
6. OPAL Collab., ICHEP2000 abst. #244.

HIGGS SEARCHES AT THE TEVATRON RUN 1 RESULTS AND RUN 2 PROSPECTS

M. ROCO
Fermi National Accelerator Laboratory, Batavia, IL. USA
E-mail: roco@fnal.gov

This report summarizes the results of recent Higgs searches in $p\bar{p}$ collisions at $\sqrt{s} = 1.8$ TeV at the Tevatron using $\approx$100 pb^{-1} of integrated luminosity from Run 1. We also present estimates of the Higgs discovery and exclusion reach in Run 2 based on a study by the Fermilab Higgs working group [1].

1 Run 1 Results

1.1 SM Higgs Searches

The CDF and DØ experiments have searched for the Standard Model (SM) Higgs boson produced in association with a vector boson. The searches are restricted to a Higgs mass below 140 GeV/c^2 where $H \to b\bar{b}$ dominates. Results for $p\bar{p} \to WH \to \ell\nu b\bar{b}$ ($\ell = e, \mu$) and $p\bar{p} \to VH \to q\bar{q}b\bar{b}$ ($V = W, Z$) were published earlier by CDF [2]. Recently the channels $ZH \to \nu\bar{\nu}b\bar{b}$ and $ZH \to \ell^+\ell^-b\bar{b}$ ($\ell = e\mu$) have also been investigated.

The search for $\nu\bar{\nu}b\bar{b}$ events requires a large missing transverse energy $\not{E}_T \geq 40$ GeV, a lepton veto and 2 or 3 jets, at least one jet has to be b-tagged. The main backgrounds are QCD events with $\not{E}_T$ from jet energy mismeasurements, $W/Z+$ jets, $t\bar{t}$ and single top events and diboson production. A total of 40 (4) events is observed with 39 $\pm$ 4 (3.9 $\pm$ 0.6) expected from the single (double) tagged sample.

For the $\ell^+\ell^-b\bar{b}$ search, two leptons are required with a dilepton invariant mass consistent with the Z mass, and 2 or 3 jets with at least one jet b-tagged. There are 5 events observed in the data consistent with 3.2 $\pm$ 0.7 events expected from background consisting of $Z+$ heavy flavors, diboson production, and $t\bar{t}$ and single top events.

The 95% confidence level (CL) limit on the $p\bar{p} \to VH$ production cross section times the branching ratio BR($H \to b\bar{b}$) for the individual channels discussed above as well as the combined limit are shown in Figure 1 as a function of the SM Higgs mass. These limits are more than an order of magnitude above the expected SM Higgs associated production cross section in $p\bar{p}$ collisions therefore no mass limits can be set.

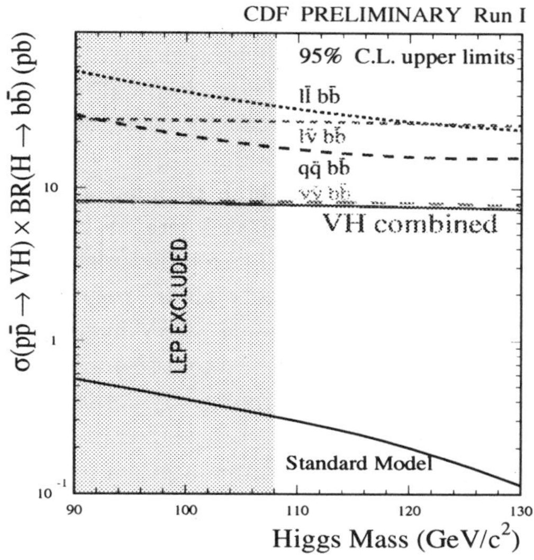

Figure 1. Run 1 individual and combined 95% CL limits on $\sigma(p\bar{p} \to VH)\times$BR($H \to b\bar{b}$) as a function of the SM Higgs mass.

1.2 SUSY Higgs Searches

The Higgs sector in supersymmetric (SUSY) extensions to the Standard Model includes at least five physical Higgs bosons: two CP-even scalars (h and H, with $m_h < m_H$), one CP-odd scalar A and a charged Higgs pair ($H^{\pm}$).

At tree-level all Higgs masses can be computed in terms of two parameters typically chosen to be m_A (or m_h) and $\tan\beta = v_u/v_d$, where v_u and v_d describe the Higgs coupling to *up*-type and *down*-type fermions, respectively. CDF has performed a search for the process $p\bar{p} \to b\bar{b}\phi \to b\bar{b}b\bar{b}$ ($\phi = h, H, A$) in Run 1. This channel becomes important at large $\tan\beta$ values due to the strongly enhanced Yukawa couplings between the Higgs scalars and the b quarks leading to production rates roughly a factor of $\tan^2\beta$ larger than the SM expectations.

The $b\bar{b}b\bar{b}$ final state is characterized by two clear signatures, the four-jet topology and a high b-quark content. The selection requires four or more high p_T jets, at least three should be tagged as b jets. Figures 2 and 3 show the 95% CL excluded regions in the $\tan\beta$ vs M_h and M_A plane, respectively. Results are shown for a SUSY mass scale of 1 TeV using two stop mixing scenarios, no mixing and maximal mixing. By convention, maximal mixing refers to a choice of SUSY parameters which gives the largest predicted value for the Higgs mass.

Recently DØ and CDF have published results on searches for charged Higgs bosons in decays of top quark pairs [3]. A charged Higgs boson lighter than the top mass could allow a large BR($t \to H^+b$) which competes with the SM prediction requiring the top quark to decay almost exclusively via $t \to W^+b$.

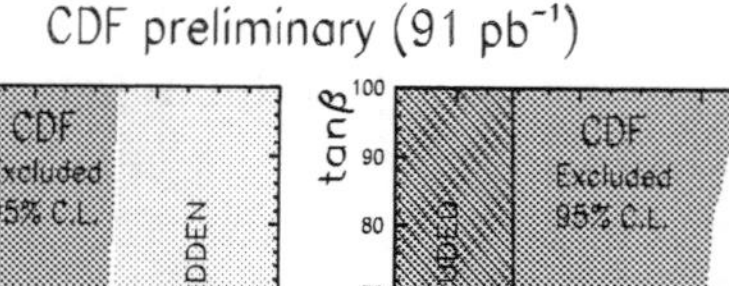

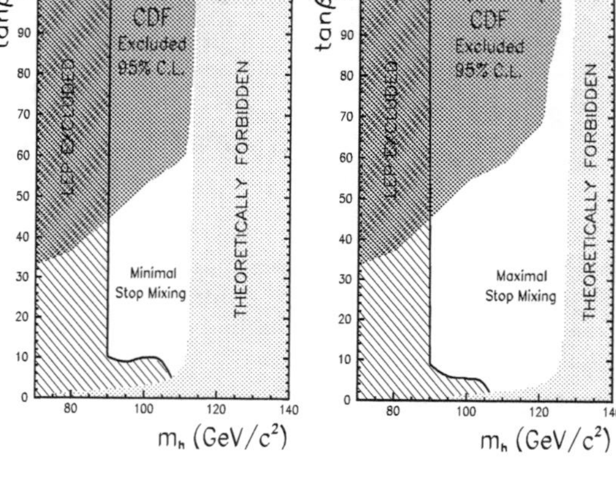

Figure 2. 95% CL exclusion regions in the $\tan\beta$ vs M_h plane. Also shown are the present LEP exclusion and theoretically forbidden regions.

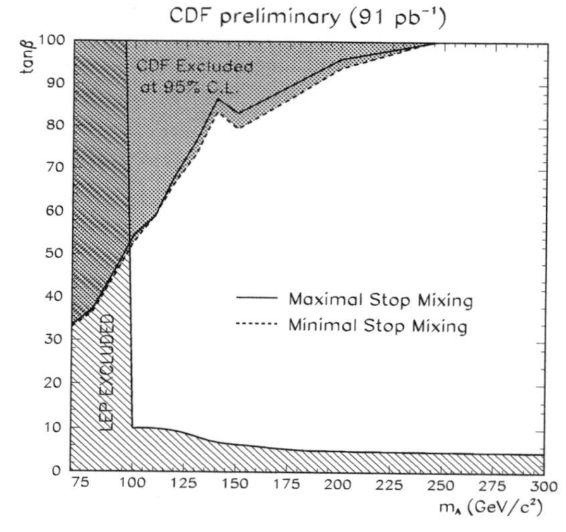

Figure 3. 95% CL exclusion regions in the $\tan\beta$ vs M_A plane. Also shown is the present LEP exclusion region.

2 Run 2 Prospects

With a center of mass energy $\sqrt{s} = 2$ TeV and instantaneous luminosities $\mathcal{L} = 2 \times 10^{32}\ cm^{-2}s^{-1}$ the next collider run, scheduled to begin in March 2001, promises an exciting physics program. To estimate the Higgs discovery and exclusion reach of the Tevatron in Run 2 an extensive study was performed by the Fermilab Higgs working group. This study extends previous Tevatron results by including additional SM Higgs decay modes in the previously explored Higgs mass region, considering the production of high mass Higgs bosons, systematically combining results from all possible search channels and considering additional decay modes arising from SUSY Higgs production. In addition, a detector simulation program based on parameterized calorimeter resolutions and particle identification efficiencies was developed to provide a realistic estimate of the geometric and kinematic acceptances of the upgraded detectors. Results are presented as the integrated luminosities required to exclude a Higgs boson at 95% CL, or to establish either 3σ or 5σ excesses over the predicted SM backgrounds.

2.1 SM Higgs Searches

Searches for Higgs masses below 135 GeV focused on associated Higgs production with final states determined by the decay mode of the accompanying W or Z: (1) $p\bar{p} \to WH \to \ell\nu b\bar{b}$, (2) $p\bar{p} \to ZH \to \ell^+\ell^- b\bar{b}$, (3) $p\bar{p} \to ZH \to \nu\bar{\nu}b\bar{b}$, or (4) $p\bar{p} \to WH/ZH \to q\bar{q}b\bar{b}$. Similar criteria as in the previous Run 1 analyses were used. A multivariate analysis using neural network techniques was performed for the leptonic modes resulting in significantly improved sensitivities [4].

For masses above 135 GeV the decay mode $H \to WW^*$ dominates. Searches were performed in the following production modes where V represents either a W or Z boson: (1) $p\bar{p} \to VH \to \ell^\pm\ell^\pm jj$ (2) $p\bar{p} \to H \to \ell^+\ell^-\nu\bar{\nu}$ (3) $p\bar{p} \to VH \to \ell^\pm\ell^\pm\ell^\mp$ leading to final states with like-sign lepton pairs with jets, opposite-sign leptons with a large $\not{E}_T$ and trileptons, respectively. The main SM backgrounds are from vector boson pair production WW, WZ, ZZ as well as $W/Z + j$, $t\bar{t}$ production and multijet events with jets misidentified as electrons. After some initial selection cuts on the p_T of the leptons and $\not{E}_T$, additional requirements on angular correlations and the cluster mass $M_C \equiv \sqrt{p_T^2(\ell\ell) + m^2(\ell\ell)} + |\not{E}_T|$ were applied. Sensitivity is maximized by fine tuning these cuts and using likelihood methods as described in detail in [5].

The extraction of the Higgs signal from the large background depends critically on the resolution we can attain for the Higgs mass, reconstructed from the measured b jet energies. The final results shown in Figure 4 include a 30% improvement in the mass resolution. This level of improvement is possible by combining calorimeter-based energies with information from charged particle momenta measurements and shower maximum detectors.

Figure 4 gives the results for the low mass and high mass Higgs analyses, combining all the SM search channels and the data from both CDF and DØ experiments. The contours show the required luminosities for 95% exclusion, 3σ evidence and 5σ discovery as a function of the SM Higgs mass.

The statistical method to combine the channels uses a Bayesian approach based on calculating the joint likelihood for a given experimental outcome as a function of the Higgs cross section. Systematic errors on the background estimate for each channel is taken into account by including into the likelihood a relative uncertainty on the background which is the smaller of the 10% of the expected background or $1/\sqrt{\mathcal{L}B}$, where B is the expected background in 1 fb^{-1} and $\mathcal{L}$ is the integrated luminosity. The luminosity thresholds are between 30-50% smaller if these systematic errors are not included.

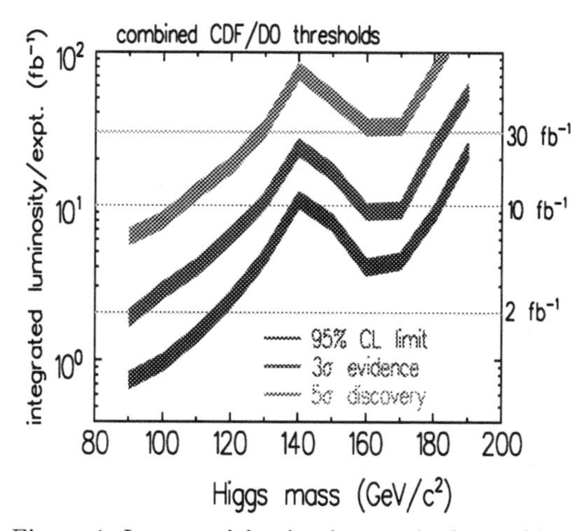

Figure 4. Integrated luminosity required to achieve 95% CL exclusion, 3σ evidence and 5σ discovery as a function of SM Higgs mass.

2.2 SUSY Higgs Searches

A study to evaluate the sensitivity reach for the neutral SUSY Higgs bosons via $p\bar{p} \to b\bar{b}\phi \to b\bar{b}b\bar{b}$ in Run 2 has been performed. Fig. 5 shows the 95% CL exclusion contours for one experiment in the $\tan\beta$ vs. m_A plane for several values of the integrated luminosities in the maximal mixing scenario. These results indicate that assuming $\tan\beta = 40$, the sensitivity reach is about 160 GeV/c^2 at 95% CL with an integrated luminosity of 2 fb^{-1}, extending up to 225 GeV/c^2 with 10 fb^{-1}.

The results of a study extending previous Run 1 searches [3] for charged Higgs bosons produced in top quark decays are shown in Figure 6 for a luminosity of 2 fb^{-1}.

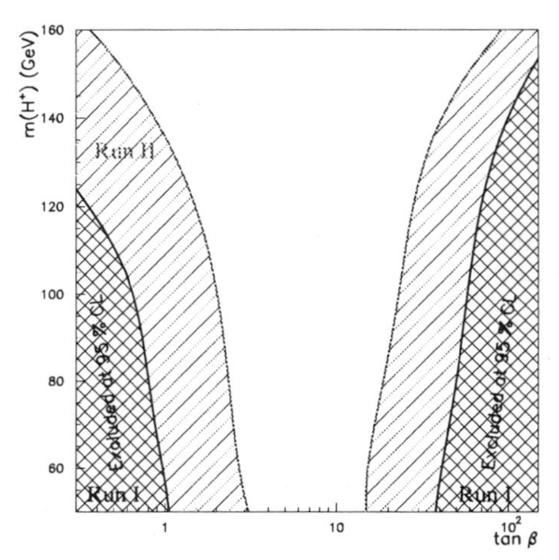

Figure 6. 95% CL exclusion regions in the M_H^+ vs $\tan\beta$ plane. for $m_t = 175$ GeV/c^2 and an integrated luminosity of 2 fb^{-1}.

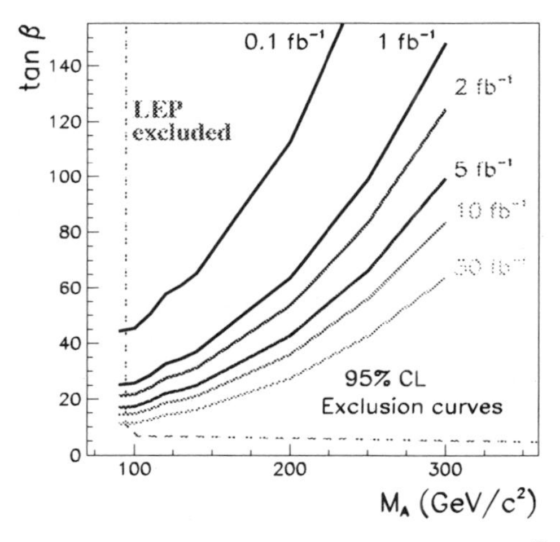

Figure 5. 95% CL exclusion curves for $p\bar{p} \to b\bar{b}\phi \to b\bar{b}b\bar{b}$ with $\phi = h, H, A$. The curves show the sensitivity reach for the MSSM neutral Higgs bosons in the $\tan\beta$ and m_A parameter space. The LEP excluded region is also shown for comparison. The results are shown for the maximal mixing scenario.

3 Summary

Recent results from searches for the SM and SUSY Higgs bosons in Run 1 at the Tevatron have been shown. The sensitivity of the present SM Higgs searches at the Tevatron is limited by statistics leading to a reach in $p\bar{p} \to VH$ production more than an order of magnitude higher than the SM prediction. The search for the neutral SUSY Higgs via $p\bar{p} \to b\bar{b}\phi \to b\bar{b}b\bar{b}$ excludes regions in the $\tan\beta$ vs M_A space which extend significantly those previously probed by LEP.

We have presented results from studies of the discovery and sensitivity reach for the Higgs bosons in Run 2. Combining all search channels and the data from both experiments a SM Higgs can be excluded at 95% CL over the full mass range $M_H < 190$ GeV with 20 fb^{-1}. The sensitivity to neutral SUSY Higgs production with $b\bar{b}$ has been shown. The $p\bar{p} \to b\bar{b}\phi \to b\bar{b}b\bar{b}$ channel serves as the most important mode for discovering or ruling out the MSSM Higgs at large $\tan\beta$.

References

1. http://fnth37.fnal.gov/higgs.html
   To be published in the Fermilab Higgs Working Group Final Report.
2. F. Abe et al., *Phys. Rev. Lett.* **79**, 3819.
   F. Abe et al., *Phys. Rev. Lett.* **81**, 5748.
3. B. Abbot et al., *Phys.Rev.Lett.* **82**, 4975.
   F. Abe et al., *Phys. Rev.* D **62**, 012004.
4. P. Bhat et al., *Fermilab-Pub-00-006.*
5. T. Han et al., *Phys. Rev.* D **59**, 093001.

CHARGINO, NEUTRALINO, HIGGSINO AT LEP

KOICHI NAGAI
Queen Mary and Westfield College, University of London,
London E1 4NS, UK
E-mail: Koichi.Nagai@cern.ch

Searches for charginos and neutralinos, predicted by supersymmetric theories, have been reviewed using data samples of 228pb^{-1} at centre-mass-of-energies of 192 − 202 GeV collected by each LEP experiment.

1 Introduction

In 1999 the LEP collider was operated at centre of mass energies, $\sqrt{s}$, of 192, 196, 200 and 202 GeV, where each of four LEP experiments (ALEPH, DELPHI, L3 and OPAL) collected data with integrated luminosities of 28, 79, 82 and 39pb^{-1} at respective centre of mass energies. These high statistics samples at high energies provide an opportunity to extend the region to search for charginos and neutralinos, predicted by supersymmetric theories (SUSY), such as minimal supersymmetric standard model (MSSM).

Charginos, $\tilde{\chi}_j^{\pm}$, are mass eigenstates formed by mixing of the fields of the fermionic partners of the W boson (winos) and those of the charged Higgs bosons (charged Higgsinos). Fermionic partners of the γ (photino), Z (zino) and neutral Higgs bosons mix to form mass eigenstates called neutralinos, $\tilde{\chi}_i^0$. In each case, the index j or i is ordered by increasing mass. The LSP is usually considered to be the lightest neutralino, $\tilde{\chi}_1^0$, although it could be the scalar neutrino (sneutrino), $\tilde{\nu}$, if it is sufficiently light.

If charginos exist and sufficiently light, they will be produced mainly via s-channel through a γ or Z exchange. For the wino component, there is a production process via t-channel $\tilde{\nu}$ exchange. The production cross-section for charginos is large unless the sneutrino ($\tilde{\nu}$) is light, in which case the cross-section is reduced by destructive interference between the s-channel and the t-channel. The details of chargino decay depend on the parameters of the mixing and the masses of the scalar partners of the ordinary fermions. The lightest chargino $\tilde{\chi}_1^+$ can decay into $\tilde{\chi}_1^0\ell^+\nu$, or $\tilde{\chi}_1^0 q\bar{q}'$, via a W boson, scalar lepton ($\tilde{\ell}$, $\tilde{\nu}$) or scalar quark (squark, $\tilde{q}$). In much of the MSSM parameter space, $\tilde{\chi}_1^+$ decays via a W boson are dominant. For small scalar lepton masses decays to leptons become important. Due to the energy and momentum carried away by the LSP (and possibly by neutrinos), the experimental signature for $\tilde{\chi}_1^+\tilde{\chi}_1^-$ events is large missing energy and large missing momentum transverse to the beam axis. If the sneutrino is lighter than the chargino, the two-body decay $\tilde{\chi}_1^+ \to \tilde{\nu}\ell^+$ dominates.

Neutralino pairs ($\tilde{\chi}_1^0\tilde{\chi}_2^0$) can be produced through an s-channel virtual Z, or by t-channel scalar electron (selectron, $\tilde{e}$) exchange. The MSSM prediction for the $\tilde{\chi}_1^0\tilde{\chi}_2^0$ production cross-section can vary significantly depending on the choice of MSSM parameters. It is typically a fraction of a picobarn and usually much lower than the cross-section for $\tilde{\chi}_1^+\tilde{\chi}_1^-$ production. The $\tilde{\chi}_2^0$ will decay into $\tilde{\chi}_1^0\nu\bar{\nu}$, $\tilde{\chi}_1^0\ell\ell$ or $\tilde{\chi}_1^0 q\bar{q}$, through a $Z^{(*)}$ boson, sneutrino, slepton, squark or a neutral SUSY Higgs boson (h^0 or A^0). The decay via $Z^{(*)}$ is the dominant mode in most of the parameter space. For small slepton masses decays to a lepton pair are important. For the cases of $\tilde{\chi}_2^0 \to \tilde{\chi}_1^0\ell\ell$ or $\tilde{\chi}_1^0 q\bar{q}$, this leads to an experimental signature consisting either of an acoplanar pair of particles or jets, or a mono-jet if the two jets in the final state have merged. The radiative decay process $\tilde{\chi}_2^0 \rightarrow \tilde{\chi}_1^0\gamma$ is also possible and can dominate for some regions of the parameter space.

Motivated by Grand Unification and to simplify the physics interpretation, the Constrained Minimal Supersymmetric Standard Model (CMSSM) is used to guide the analysis, but more general cases are also studied. In the CMSSM, introducing unification of masses and couplings at the GUT scale, the number of parameters is reduced to six, which are SU(2) gaugino mass parameter at EW-scale M_2, common scalar mass at GUT-scale m_0, ratio of VEV's of two Higgs doublets v_2/v_1, mixing parameter of two Higgs doublet field μ, common trilinear coupling A_0, and mass of the CP odd Higgs boson, m_A.

2 Analysis

The standard selection to search for charginos and neutralinos at LEP [1,2,3,4] are categorised by event topologies and mass differences ($\Delta M_+ = m_{\tilde{\chi}_1^+} - m_{\tilde{\chi}_1^0}$ and $\Delta M_0 = m_{\tilde{\chi}_2^0} - m_{\tilde{\chi}_1^0}$). For example, analysis for chargino search in OPAL is divided into three categories according to the charged multiplicity and number of isolated leptons and then each category has four ΔM_+ regions. For neutralino search, analysis is divided into two categories by the multiplicity and then each has four ΔM_0 regions. For each case topologies of signal tend to be low multiplicity in small ΔM region and to form hadron jets in large ΔM region. The contribution from each background process also depends on the ΔM region. In the small ΔM region, two photon processes are dominant which have cross-sections of $\mathcal{O}$(nb). In this region the standard analysis has low sensitivity because the topology of signal events is very similar to that of background. To obtain sensitivity in this region an analysis for charginos mass-degenerate with $\tilde{\chi}_1^0$, has been introduced [6], which reduce the contribution from two photon processes by tagging a photon from initial state radiation (ISR). Increasing ΔM, four fermion processes is becoming dominant with a cross-section of about 20 pb.

In the CMSSM analyses reported here possible cascade decay processes are taken into account. However, the number of considered cascade decays are different among experiments.

3 Results

No evidence of chargino and neutralino have been found in data taken until 1999. Limits on production cross-section for charginos and neutralinos were calculated. Figure 1 (a)

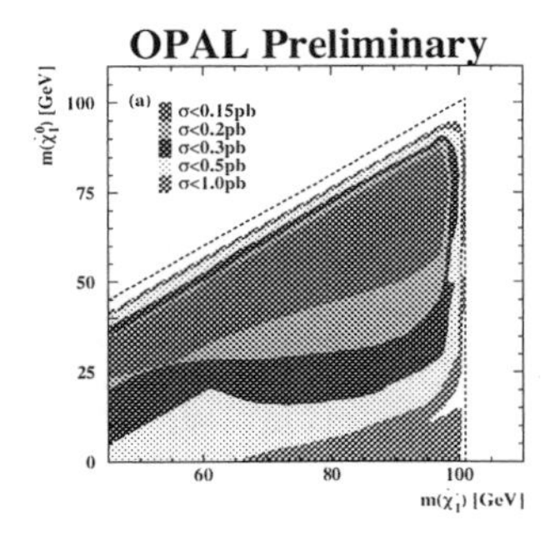

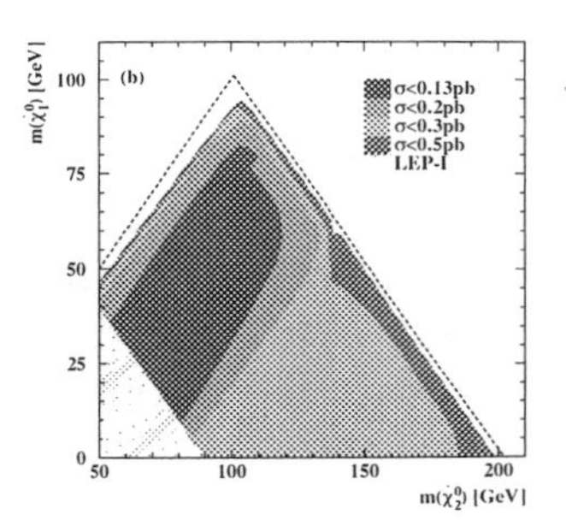

Figure 1. Upper limits on production cross-section at $\sqrt{s}$ = 202 GeV at 95% C.L. for the $e^+e^- \rightarrow \tilde{\chi}_1^+\tilde{\chi}_1^-$ process (a) and the $e^+e^- \rightarrow \tilde{\chi}_2^0\tilde{\chi}_1^0$ (b), assuming BR($\tilde{\chi}_1^+ \rightarrow \tilde{\chi}_1^0 W^{+(*)}$) = 100% and BR($\tilde{\chi}_2^0 \rightarrow \tilde{\chi}_1^0 Z^{(*)}$) = 100%, respectively.

shows a limit on the chargino production cross-section in a mass plane of $\tilde{\chi}_1^0$ and $\tilde{\chi}_1^+$.

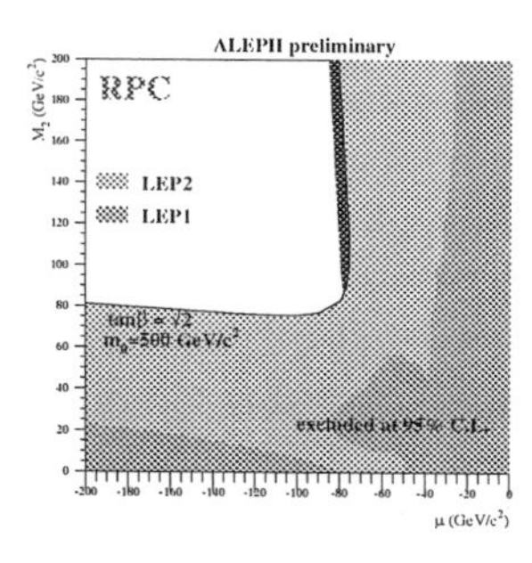

Figure 2. Exclusion region at 95% C.L. in the (M_2, μ) plane with $\tan\beta = \sqrt{2}$ and $m_0 = 500$ GeV. The lightest shaded area shows region excluded by chargino search, which reaches at the kinematical boundary of $\tilde{\chi}_1^+\tilde{\chi}_1^-$ production at $\sqrt{s} = 202$ GeV. The darkest shaded region extending beyond the kinematical boundary is obtained by results from the direct neutralino searches.

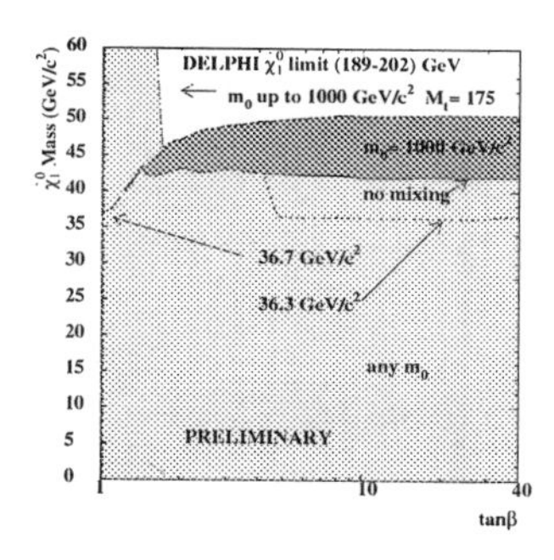

Figure 3. The lower limit at 95% C.L. on the mass of the lightest neutralino as a function of $\tan\beta$. The dashed and solid curves show the limit obtained for $m_0 = 1000$ GeV/c^2 and any m_0 assuming no mixing in the third family, respectively. The dash-dotted curve shows the limit obtained for any m_0 allowing for the mixing. The steep dotted curve shows the effect of the results from the maximal M_{h^0} scenario in Higgs boson searches.

Limits have been set close to the kinematic limit. However, there small spaces are left near the edge, because only a part of data have been used to calculate the limit (for instance, only 11pb^{-1} out of 38 pb^{-1} at $\sqrt{s} = 202$ GeV is used). The space left in small ΔM_+ region shows that the standard searches don't have good sensitivity due to large background from two photon processes. A cross-section limit on neutralino is shown in Figure 1 (b).

3.1 Interpretation in MSSM

The obtained cross-section limits are interpreted into CMSSM parameter space, using results from photonic event searches [5,7] and acoplanar lepton searches [1,5,8]. Figure 2 shows the excluded region at 95% C.L. in M_2-μ space for large m_0 case. The dark shaded area beyond the kinematic limit of chargino production are extended by direct neutralino searches. For small m_0, the exclusion region is reduced due to the destructive interference in chargino production and the two body decays.

The results are also translated into mass limits as a function of several parameters in MSSM. Figure 3 shows a mass limit for neutralino as a function of $\tan\beta$. In large m_0 case, almost limit curve is determined by chargino results except for near $\tan\beta = 1$, where results from direct neutralino search provide constraint on neutralino mass. In small m_0 case, the shape of curve is changed dramatically in high $\tan\beta$ region due to two body decays of chargino and neutralino with the slepton and sneutrino, where results from slepton search determine limits.

3.2 Results in small ΔM region

In the region of $\Delta M \lesssim 5$ GeV, the sensitivity of the standard searches is low due to the huge background from two photon processes. In the most extremely small ΔM region, chargino won't decay in a detector, which can be observed as a stable charged particle. In this case, we can use the results of search for heavy charged particles [5,11]. The analysis has been carried out for low multiplicity events using particle ID information. These results will be translated into MSSM parameter space.

In the region between standard search and extremely small ΔM, event topologies of chargino are very close to those of two photon

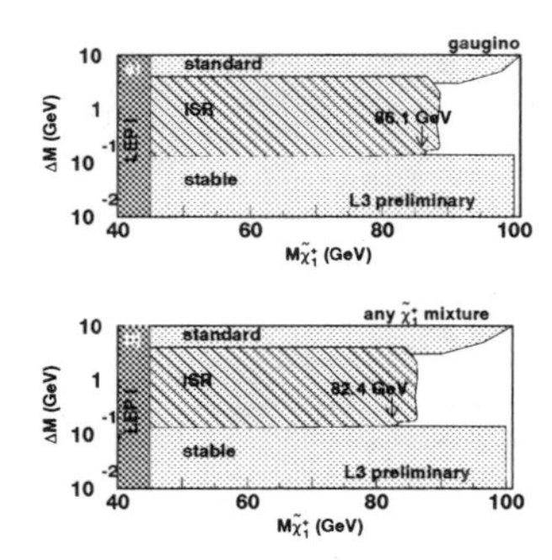

Figure 4. Exclusion regions in the plane (ΔM, $M_{\tilde{\chi}_1^+}$) at 95% C.L. for (a) gaugino-like charginos and (b) any mixture of charginos. The exclusion region by analyses with tagging ISR photon is shown as the grey hatched area on each plot. The top, left and bottom regions are excluded by the standard, LEP I and long-lived particle searches, respectively.

events. However the contribution from two photon processes can be reduced by tagging events accompanied by photons from initial state radiation. Analyses have shown no evidence of chargino production. The results are translated into MSSM parameter space as shown in Figure 4, where excluded area are shown for higgsino-like and gaugino-like charginos separately.

3.3 Impact of Higgs boson searches

LEP Higgs searches have excluded region of low $\tan\beta$ between 0.7 and 1.8 in a given MSSM parameters. These results have been interpreted with the results from direct SUSY particle searches [1,9]. Figure 3 also shows the effect from the MSSM Higgs searches on the lightest neutralino mass limit. The area excluded by the constraint from Higgs searches are shown by a dotted line. The mixing of $\tilde{\tau}$ was also taken into account in the figure, which causes that the mass limit becomes weak at high $\tan\beta$ region.

4 Conclusion

Each of the four LEP experiments collected data with an integrated luminosity of 228pb^{-1} at $\sqrt{s}$ = 192, 196, 200 and 202 in 1999. Searches have been performed, covering a variety of final states expected from charginos and neutralinos production. No evidence of these signals were found. The results are translated into upper limit of production cross-section for chargino and neutralino as well as limits in the MSSM parameter space. In small ΔM region analyses have been carried out and the results are interpreted in to MSSM parameter space. The constraint on $\tan\beta$ from the MSSM Higgs boson searches in a given parameter space has been taken into account in the MSSM interpretation.

References

1. The ALEPH collab., paper submitted to this conference, ICHEP **436** (2000).
2. The DELPHI collab., paper submitted to this conference, ICHEP **191** (2000).
3. The L3 collab., paper submitted to this conference, ICHEP **438** (2000).
4. The OPAL collaboration, G. Abbiendi et al., Eur. Phys. J. **C14** (2000) 187, the OPAL collab., G. Abbiendi et al., Eur. Phys. J. **C8** (1999) 255.
5. The OPAL collab., paper submitted to this conference, ICHEP **185** (2000).
6. The DELPHI collab., paper submitted to this conference, ICHEP **366** (2000), the L3 collab., paper submitted to this conference, ICHEP **439** (2000),
7. The OPAL collab., G. Abbiendi et al., Eur. Phys. J. **C8** (1999) 23.
8. The OPAL collab., G. Abbiendi et al., Eur. Phys. J. **C14** (2000) 51.
9. The DELPHI collab., paper submitted to this conference, ICHEP **372** (2000).
10. The DELPHI collab., paper submitted to this conference, ICHEP **628** (2000).
11. The DELPHI collab., paper submitted to this conference, ICHEP **370** (2000), the OPAL collab., paper submitted to this conference, ICHEP **200** (2000).

SQUARK AND SLEPTON SEARCHES AT LEP

M. ANTONELLI

Lab. di Frascati dell'INFN, Via E. Fermi 40,
I-00044 Frascati (Roma), Italy
E-mail: Mario.Antonelli@lnf.infn.it

The data collected up to a centre-of-mass energy of 202 GeV by the LEP experiments, corresponding to an integrated luminosity of about 0.7 fb^{-1}, are analysed in search for the scalar partners of quarks and leptons predicted by supersymmetric models. No evidence for any such particles was found. Lower mass limits have been obtained in the framework of the Minimal Supersymmetric Standard Model.

1 Introduction

In the Minimal Supersymmetric extension of the Standard Model (MSSM) [1], each Standard Model fermion chirality state has a scalar supersymmetric partner. The sfermions $\tilde{f}_R$ and $\tilde{f}_L$ are the supersymmetric partners of the right-handed and left-handed fermions, respectively. These are weak eigenstates which can mix to form the mass eigenstates. The mixing is a unitary transformation of the $\tilde{f}_R$ and $\tilde{f}_L$ states, parameterised by a mixing angle θ. Because the off-diagonal elements of the sfermion mass matrix are proportional to the SM partner masses, the mixing is expected to be relevant for the scalar top (stop, $\tilde{t}$), scalar bottom (sbottom, $\tilde{b}$) and scalar tau (stau, $\tilde{\tau}$).

Sfermions can be produced at LEP in pairs, $e^+e^- \rightarrow \tilde{f}\bar{\tilde{f}}$, via s-channel exchange of a virtual photon or a Z, whereas the t-channel exchange of neutralinos can contribute in the case of selectron ($\tilde{e}$) production, making possible, for this flavour, a mixed production $\tilde{e}_R\tilde{e}_L$ when kinematically allowed.

The searches for sfermions described here assume that all supersymmetric particles except the neutralino χ and possibly the sneutrino $\tilde{\nu}$ are heavier than the sfermions. The conservation of R-parity is also assumed; this implies that supersymmetric particles are produced in pairs and that the LSP is stable.

Under these assumptions, all the sfermions but the stop decay predominantly as $\tilde{f} \rightarrow f\chi$. The stop can decay as $\tilde{t} \rightarrow c\chi$ or $\tilde{t} \rightarrow b\ell\tilde{\nu}$ [2]. The first decay can proceed only via loop diagrams and thus has a very small width, of the order of 0.01–1 eV [2]. For low mass differences Δm between the stop and the neutralino, the stop lifetime becomes sizeble, and has to be taken into account.

The $\tilde{t} \rightarrow b\ell\tilde{\nu}$ channel proceeds via virtual chargino exchange and has a width of the order of 0.1–10 keV [2]. This second decay is dominat when it is kinematically allowed.

Direct searches for squarks and sleptons were performed by the LEP experiments with the data collected at $\sqrt{s}$ = 192-202 GeV and corresponding to an integrated luminosity of about 0.7 fb^{-1}. Since the production cross section for scalar particles is very small near threshold most of the single experiment results are combined in order to increase the sensitivity for the searches [3]. The analyses have been combined by means of a Likelihood Ratio method.

2 The Searches

Several selection algorithms have been developed to search for squarks and sleptons. Events with acoplanar jets and acoplanar jets plus two leptons are signatures for squark production. Events with acoplanar lepton pairs or with single electrons are expected from slepton production. All these channels are characterised by missing energy.

The event properties depend significantly on Δm, the mass difference between the decaying sfermion and the produced χ or $\tilde{\nu}$. When Δm is large, there is a substantial amount of energy available for the visible system and the signal events can look like WW, $We\nu$, $Z\gamma^*$, or $q\bar{q}(\gamma)$ events.

When Δm is small, the energy available for the visible system is small and the signal events are therefore similar to $\gamma\gamma$ events.

Acoplanar jet selections are used to search for the processes $e^+e^- \rightarrow \tilde{t}\bar{\tilde{t}}$ ($\tilde{t} \rightarrow c\chi$). For $\tilde{b} \rightarrow b\chi$ searches the analyses take the advantage of the lifetime content in b-jets.

The experimental signature for $e^+e^- \rightarrow \tilde{t}\bar{\tilde{t}}$ ($\tilde{t} \rightarrow b\ell\tilde{\nu}$) production is two acoplanar jets plus two leptons with missing momentum. Indirect constraints on sneutrino mass form the invisible Z width measurements set a lower limit on the stop mass of about 50 GeV/c^2.

The results shows a good agreement with the SM expectation. The non-observation of any excess is interpreted as lower limit on the squark masses [3]:

- $\tilde{t} \rightarrow c\chi$, $m_{\tilde{t}} > 97\,\mathrm{GeV}/c^2$ (*95* GeV/c^2) for $\theta_{\tilde{t}} = 0°$, ($\theta_{\tilde{t}} = 56°$) and $\Delta m > 5\,\mathrm{GeV}/c^2$.
- $\tilde{t} \rightarrow b\ell\tilde{\nu}$, $m_{\tilde{t}} > 93\,\mathrm{GeV}/c^2$ (*89* GeV/c^2) for $\theta_{\tilde{t}} = 0°$, ($\theta_{\tilde{t}} = 56°$) and $\Delta m > 7\,\mathrm{GeV}/c^2$
- $\tilde{b} \rightarrow b\chi$, $m_{\tilde{b}} > 97\,\mathrm{GeV}/c^2$ (*88* GeV/c^2) for $\theta_{\tilde{b}} = 0°$, ($\theta_{\tilde{b}} = 68°$) and $\Delta m > 7\,\mathrm{GeV}/c^2$

The mixing angle $\theta_{\tilde{q}}$ controls the squark production cross section trough the $\tilde{q}\bar{\tilde{q}}Z$ coupling.

In all these squark searches there is an improvement over the Tevatron exclusions [4] in the region of small mass differences ($\Delta m < 40\,\mathrm{GeV}/c^2$).

Slepton pair production leads to a final state characterized by two acoplanar leptons of the same flavour, missing mass and energy; No excesses have been found in the searches for acoplanar electrons and acoplanar muons. Lower mass limits are derived in the framework of the MSSM for $\mu = -200\,\mathrm{GeV}/c^2$, $\tan(\beta) = 1.5$ and $\Delta m > 15\,\mathrm{GeV}/c^{2}$[3]:

- $m_{\tilde{\mu}} > 92\,\mathrm{GeV}/c^2$ for $\tilde{\mu} \rightarrow \mu\chi$
- $m_{\tilde{e}} > 95\,\mathrm{GeV}/c^2$ for $\tilde{e} \rightarrow e\chi$

2.1 The case of stau

The preliminary data for the stau analyses provided for the combination of the summer 2000 conferences show an excess spreaded over all the experiments for LEP energies between 189 and 202 GeV [3]. The excess is found by the high-Δm analysis where 268 events have been observed with 221.8 events of expected background, 80% from WW, 12% from $\gamma\gamma$, and 5% from $\tau\tau$ processes. Systematic effects on the background evaluation have been studied. The value of the uncertainty found 2-4% is unlike to explain the observed excess.

Two possibilities are considered. Assuming a fluctuation of the background a lower limit on the stau mass has been set at 75 GeV/c^2 for $\Delta m > 4\,\mathrm{GeV}/c^2$.

The energy evolution of the excess has been fitted in order to study the signal hypothesis. The most likely value obtained by this multichannel fit results for the masses of $m_{\tilde{\tau}} = 88$ GeV/c^2 and $m_\chi = 45$ GeV/c^2. In this point, the C.L. for an upward fluctuation of the background is 1.6% for the data at 189-202 GeV.

Unfortunately, the analysis of the data collected in 2000 supports the hypothesis of a background fluctuation. For more details see [3,5].

2.2 Search for stable and metastable scalar top

Dedicated searches for stable and metastable hadrons have been performed by ALEPH [6] to cope with the peculiar phenomenology of the $\tilde{t} \to c/u\chi$ process in very small Δm regime where the $\tilde{t}$ can have a sizeble lifetime. Three different selections have been used, each designed to cope with a specific $\tilde{t}$ decay lenght range.

If the stop lifetime is negligible, the signature is the one covered by the "low-Δm" acoplanar jet.

The intermediate lifetime selection addresses the case of stop-hadrons decaying well within the tracking volume, but at a significant distance from the interaction vertex. It is based essentially on the tagging of tracks with a large impact parameter, originating from the stop decay.

When the stop decay lenght is larger than the detector size, a charged stop-hadron behaves like a heavy stable charged particle. It can be identified using the kinematic characteristics of stop pair-production and the high specific ionization that is expected to release in the TPC.

In all cases the number of candidates observed are found to be consistent with the background expected from SM processes. In the MSSM a 95% C.L. lower limit of 63 GeV/c^2 is obtained for the stop mass, valid for any range of Δm.

2.3 Absolute limit on selectron mass

Searches for mixed selectron production are used to extend the sensitivity of the selectron searches to situations where the acoplanar electron search is ineffective because of the small mass difference between the lightest selectron and the lightest neutralino. In this case the outgoing lepton from the decay of the heaviest selectron (typically the $\tilde{e}_L$) has enough energy to be detected, leading to a single electron topology.

Sfermions cascade decays such as $\tilde{\ell}_R \to \ell\chi'$ play an important role if the sfermion is not the Next to Lightest Supersimmetric Particle (NLSP). Similar analyses have been performed by the experiments[7,9,8] to cope with this scenario.

The results of the new searches are combined with the one from chargino and neutralino searches in the MSSM framework. A lower mass limit of 72 GeV/c^2 is obtained for the selectron mass [7] under the assumption of gaugino mass unification, scalar mass unification and no mixing in the stau sector ($A_{\tilde{\tau}} = 0$ GeV/c^2).

3 Conclusions

In the data sample of about 0.7 fb^{-1} collected in 1999 by the detectors at LEP at a centre-of-mass energy of 192-202 GeV, searches for signals from pair-production of the scalar partners of quarks and leptons have been performed. In all channels the number of candidates observed is consistent with the background expected from Standard Model processes. In the MSSM lower mass limits are set at 95% C.L..

References

1. For a collection of reviews see *Supersymmetry and Supergravity,* Ed. M. Jacob, North-Holland and World Scientific, 1986.
2. K. Hikasa and M. Kobayashi, Phys. Rev. **D 36** (1987) 724;
   M. Drees and K. Hikasa, Phys. Lett. **B 252** (1990) 127.
3. Lep Susy Working Group WWW page: http://lepsusy.web.cern.ch/lepsusy/
4. Thomas J. Lecompte, in this proceeding
5. M. Maggi, in this proceeding
6. ALEPH Collaboration, Phys. Lett. **B 488** (2000) 234.
7. ALEPH Collaboration, ALEPH 2000-061, CONF 2000-040.
8. L3 Collaboration, Phys. Lett. **B 471** (1999) 280.
9. DELPHI Collaboration, DELPHI 2000-087, CONF 286.

PHOTONIC AND GRAVITINO SEARCHES AT LEP

VINCENT HEDBERG

Fysiska institutionen, Lund University, Lund, Sweden

The four LEP experiments have updated searches within the framework of Gauge Mediated Supersymmetry Breaking models. No deviations from the expectations from Standard Model sources were observed in the data recorded during 1999 and new cross section limits and exclusion plots have been produced.

1 Introduction

In Gauge Mediated Supersymmetry Breaking (GMSB) models[1] the Gravitino ($\tilde{G}$) is predicted to be the Lightest SUSY Particle (LSP) and the Next-to-Lightest SUSY Particle (NLSP) is typically either a slepton, decaying to a lepton and a gravitino, or a neutralino decaying to a photon and a gravitino. The gravitinos are stable and neutral and can be observed only as missing energy by the experiments.

The mass of the gravitino ($m_{\tilde{G}}$) is proportional to the square of the SUSY breaking scale and a typical scale of something like 10-10,000 TeV corresponds to a gravitino mass within the range 0.02 eV-20 keV. The decay lengths of the NLSP (i.e. the slepton or the neutralino) in turn depends on the square of the gravitino mass and so the scale, the gravitino mass and the NLSP decay length are all related.

The experiments usually do different searches depending on the assumed decay length of the NLSP. If it is very short one is looking mainly for events with acoplanar lepton- or photon-pairs, if the decay length is of the same size as the experiment one can search for leptons and photons that do not point to the Interaction Point (IP) and finally if the decay length is larger than the size of the experiment one can (in the slepton-NLSP case) look for heavy stable particles.

None of the four LEP experiments has reported an excess in any of the GMSB search channels using the 1999 LEP data. The results from these searches is therefore presented as cross section limits and exclusion plots. Only results including data up to 202 GeV center-of-mass energy will be presented.

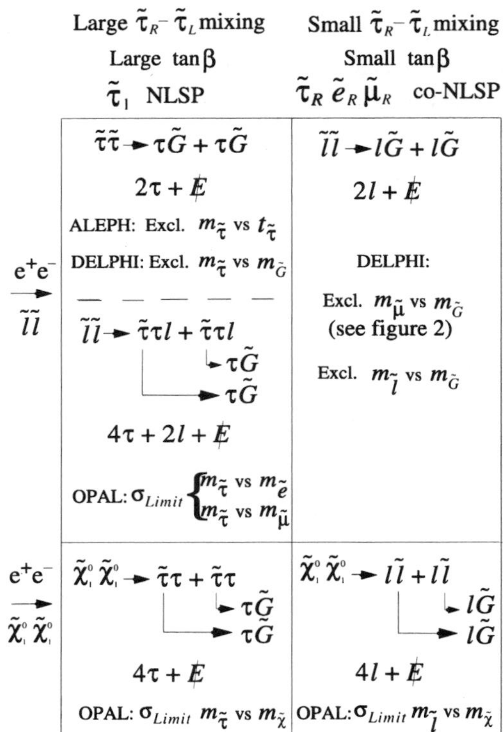

Figure 1. Slepton NLSP search scenarios and event topologies. Searches with data up to 202 GeV, where new exclusion plots or cross section limits have been produced, are indicated.

2 Slepton NLSP

In the slepton-NLSP scenario the experiments are usually doing the searches under the assumption that either the stau is the NLSP (i.e. large tanβ and large mixing in the stau sector) or that the NLSP is

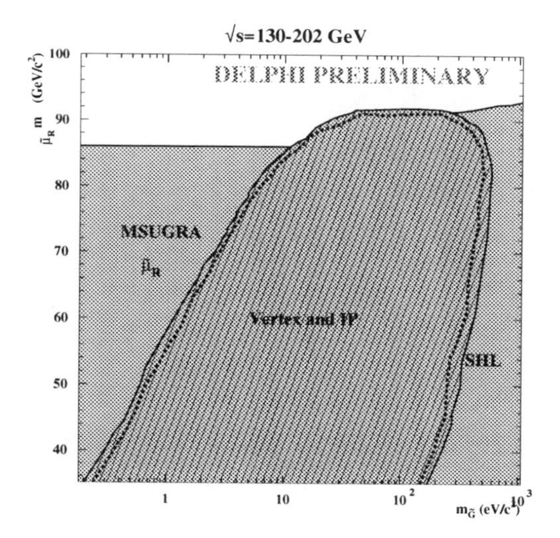

Figure 2. Exclusion plot in the $m_{\tilde{\mu}}$- $m_{\tilde{G}}$ plane by DELPHI using all data up to 202 GeV. The dashed line shows the expected limits.

a mass-degenerate state in which case also smuon and selectron searches become relevant. Searches have been made both for the case of direct slepton production and for the case of cascade decays in which first a pair of neutralinos are produced which then decays into sleptons and leptons.

Figure 1 summarizes the different search channels and the different event topologies studied by ALEPH[2], DELPHI[3] and OPAL[4]. The figure also indicates in which search channels the three collaborations have produced new exclusion plots or cross section limits using the 1999 data. One of these plots is shown in Figure 2 which illustrates a search for smuon pairs by DELPHI. Three different search analysis have been combined in this exclusion plot. For small gravitino masses ($m_{\tilde{G}}$<10 eV), i.e. a short slepton decay length, a MSUGRA search for $e^+e^- \to \tilde{\mu}\tilde{\mu} \to \tilde{\chi}^0_1\mu\tilde{\chi}^0_1\mu$ has been used to set a limit on $e^+e^- \to \tilde{\mu}\tilde{\mu} \to \tilde{G}\mu\tilde{G}\mu$ by using the MSUGRA-limit at $m_{\tilde{\chi}}$=0. This analysis gives a limit on the smuon mass of 86 GeV. In an intermediate gravitino mass region, for which the smuon is expected to decay inside the experiment, a search for muon-pair events where the muon tracks have kinks or do not point to the IP has been done. This gives a limit on the smuon mass which is between 86 and 91 GeV. For gravitino masses above a few hundred eV, searches for heavy stable particles (i.e. the smuon directly) by using dE/dx and the RICH gives an even higher limit which is between 91 and 93 GeV.

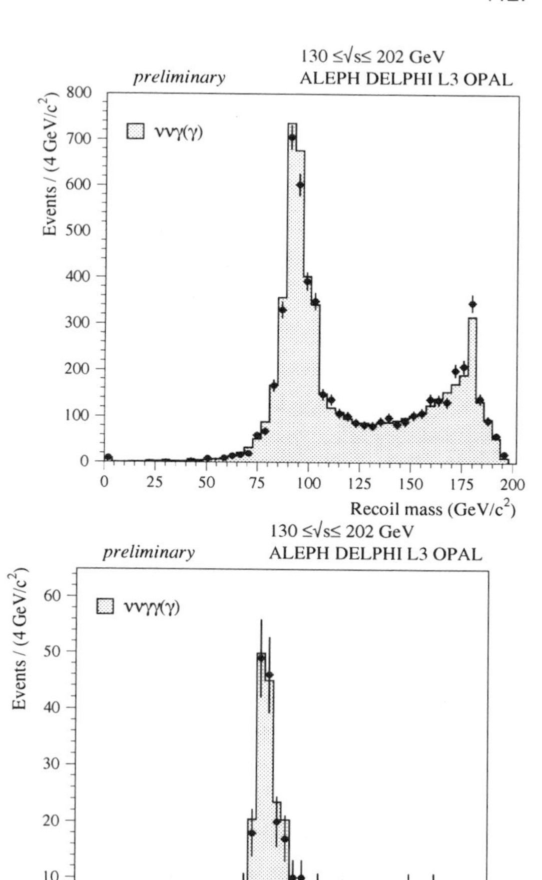

Figure 3. Recoil (or missing) mass distribution for single photon events (top) and two photon events (bottom) from all four LEP experiments[5].

3 Neutralino NLSP

GMSB models predict that if the neutralino is the NLSP, events with one or two photons + missing energy can be important discovery topologies. Figure 3 shows the recoil

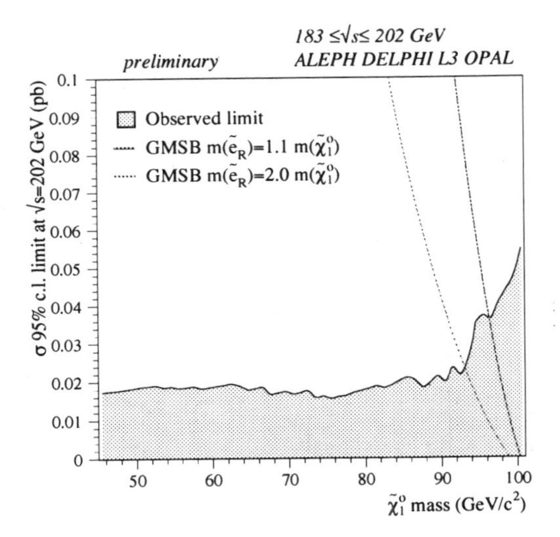

Figure 4. Cross section limit at $\sqrt{s}$ =202 GeV of the process $e^+e^- \to \tilde{\chi}_1^0\tilde{\chi}_1^0 \to \tilde{G}\gamma\tilde{G}\gamma$ as a function of $m_{\tilde{\chi}}$[5].

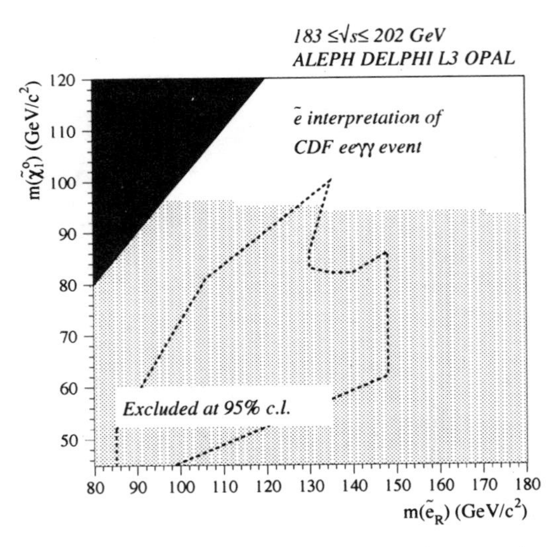

Figure 5. The shaded area shows the exclusion region in the $m_{\tilde{\chi}}$ versus $m_{\tilde{e}_R}$ plane[5].

mass (or missing mass) distribution for one and two photon events after combining the data from all four LEP experiments[5]. Neutrino pair production, $e^+e^- \to \nu\bar{\nu}\gamma(\gamma)$, is the only Standard Model process that can produce single and two photon events. None of the experiments claim to see a significant excess of events above SM expectations.

The $e^+e^- \to \tilde{\chi}_1^0\tilde{\chi}_1^0 \to \tilde{G}\gamma\tilde{G}\gamma$ process is the main GMSB search channel with the neutralino as the NLSP. Figure 4 shows the cross section limit versus neutralino mass for this process (after combining the data from all four LEP experiments[5]). The corresponding exclusion plot in the neutralino-selectron mass plane is shown in Figure 5. The limits in Figure 4 and 5 assumes a short decay length i.e. a low gravitino mass. Updated cross section limits have also been calculated by ALEPH[6] and DELPHI[7] under the assumption that the neutralinos decay inside the experiment and that the photons do not point towards the *IP*. In this case non-pointing single photon event are searched for since the probability is low that both photons decay inside the experiment.

The cross section for the process $e^+e^- \to \tilde{G}\tilde{\chi}_1^0 \to \tilde{G}\tilde{G}\gamma$, which produces true single photon events, is only large for very light gravitinos in the mass range $10^{-4} - 10^{-5}$ eV and the experiments are therefore in this case only looking for prompt photon production. DELPHI[7] and L3[8] have produced new cross section limits as a function of neutralino mass from the single photon events and DELPHI[7] and ALEPH[6] have calculated new limits for the gravitino mass under the assumption that all SUSY particles except the gravitino is too heavy to be produced. The results are $m_{\tilde{G}} > 1.1 \cdot 10^{-5}$ eV for ALEPH[6] and $m_{\tilde{G}} > 1.2 \cdot 10^{-5}$ eV for DELPHI[7].

References

1. S. Ambrosiano et al., *Phys. Rev.* D **56**, 1761 (1997);
   S. Dimopoulos et al., *Nucl. Phys.* B **488**, 39 (1997).
2. ALEPH 2000-10 CONF 2000-007.
3. DELPHI 2000-079 CONF 378.
4. OPAL PN418.
5. LEP SUSY working group note 00-04.1 and 00-05.1
6. ALEPH 2000-008 CONF 2000-005.
7. DELPHI 2000-94 CONF 393.
8. L3 Note 2584.

SUPERSYMMETRY SEARCHES AT THE TEVATRON

THOMAS J. LECOMPTE

High Energy Physics Division, Argonne National Laboratory, 9700 S. Cass Ave., Argonne IL 60439, USA

E-mail: lecompte@anl.gov

FOR THE CDF AND D0 COLLABORATIONS

The CDF and D0 experiments have each collected over 110 pb^{-1} of proton-antiproton collision data with $\sqrt{s} = 1800$ GeV during the period 1992-1995. Limits on the production of supersymmetric particles are presented here.

1 Introduction

The Tevatron Collider experiments, CDF and D0, bring unique capabilities to the search for supersymmetric partners to the known Standard Model particles. Until the LHC turns on, the Tevatron is the highest energy accelerator in the world, with a center of mass energy of 1.8 TeV. Because it collides protons and antiprotons, both of which are composite particles, the partonic center of mass energy is lower and variable, reaching approximately 600 GeV for quark-antiquark collisions and 400 GeV for gluon-gluon collisions. This allows for production of heavier superparticles than at LEP, as well as enhanced production of colored superparticles (squarks and gluinos) via QCD processes.

With these advantages comes a price, and that is the increased backgrounds from Standard Model processes, particularly QCD. These backgrounds can be many orders of magnitude larger than the expected signals, so the experiments have to reduce this by concentrating on signatures that are unusual or rare in Standard Model processes. Some of these are high transverse momentum (p_T) leptons, heavy flavor quark jets, missing transverse energy ($\not\!\!E_T$) and energetic photons. In this paper I will discuss two categories of these signatures.

2 Events with Leptons

One of the cleanest signatures for supersymmetry is associated production of the lightest chargino and second-lightest neutralino via $q + \overline{q}' \to W^{*\pm} \to \tilde{\chi}_1^{\pm}\tilde{\chi}_2^0$ followed by the decays $\tilde{\chi}_1^{\pm} \to l^{\pm}\nu\tilde{\chi}_1^0$ and $\tilde{\chi}_2^0 \to \tilde{\chi}_1^0 l^+ l^-$. In this model the $\tilde{\chi}_1^0$ is the lightest superparticle, and is stable, so manifests itself as missing energy. So the final signature is three isolated leptons.

Both experiments search for trilepton events, requiring minimum lepton transverse momenta between 5 and 11 GeV, not all with the same charge (CDF only) and with missing $\not\!\!E_T$ from the unseen $\tilde{\chi}_1^0$'s. Additionally, kinematic and topological requirements are imposed. Neither experiment sees any events[1], over an expected background of 1.2 ± 0.2 for CDF and 1.3 ± 0.4 for D0. This data allows the experiments to set limits on associated chargino-neutralino production (Figure 1) which are complimentary to limits set by the LEP experiments: the LEP limits exclude the low mass region, whereas the Tevatron limits exclude higher masses, provided the couplings are large enough for associated chargino-neutralino production to be significant.

A variation on this is trileptons from R-Parity violating (RPV) decays. R-Parity is defined to be even for particles and odd for superparticles, and in many models (such as

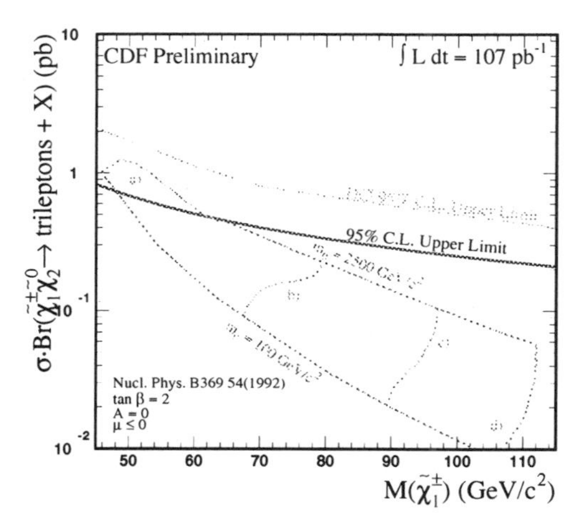

Figure 1. Tevatron limits on chargino and neutralino production from an analysis of trilepton events. The region to the left of the plot has been excluded by LEP.

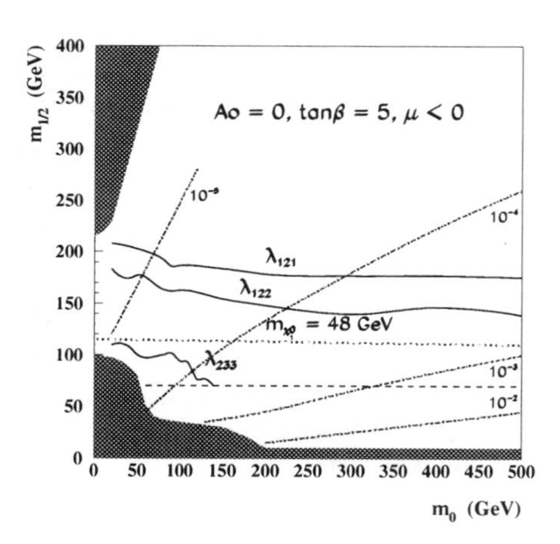

Figure 2. D0 limits on RPV 4-lepton events based on their trilepton limits.

the one used for the above analysis) is assumed to be conserved. However, it is not associated with a gauge symmetry, so it may not be conserved: there is no *a priori* reason to prefer R-parity conserving over R-parity violating models. In the RPV model investigated by D0, the lightest neutralinos are pair produced via an R-parity conserving process, and then decay via an R-parity violating process: $\tilde{\chi}_1^0 \to \nu l^+ l^-$. This therefore gives four leptons in the final state. D0 has reinterpreted their trilepton limits in the context of this model and the results are shown in figure 2.

CDF has taken a different approach to searching for the same process. They require identification of all four leptons, which lowers their efficiency, but allows them to recover it by relaxing their lepton selection requirements. For example only one lepton has to have p_T above 12 GeV; the other three can be as low as 5 GeV. They see one signal event over an expected background of 1.2 ± 0.2. This event, an $ee\mu\mu$ event has both muons near jets - a topology more consistent with heavy flavor production than supersymmetry. Limits are shown in figure 3.

Another possible neutralino decay would be semileptonic: $\tilde{\chi}_1^0 \to e^+ d\bar{u}$. D0 searches for this decay, requiring two electrons not from $Z^0 \to e^+e^-$ and at least four jets, and they observe two events over an expected background of 1.8 ± 0.4. Limits are shown in figure 4. CDF looks for a similar signature (with somewhat different motivation) from the decays $\tilde{g} \to \bar{c}e^+d$ and $\tilde{q} \to qe^+q\bar{q}$ [4] and their limits are shown in figure 5.

3 Stop and Sbottom Squarks

There are a number of possible R-parity conserving decays of stop and sbottom squarks, such as:

- $\tilde{t} \to b + \tilde{\chi}_1^+ \to Wb + \tilde{\chi}_1^0$
- $\tilde{t} \to b + \tilde{\chi}_1^+ \to bl + \tilde{\nu}$
- $\tilde{t} \to t + \tilde{\chi}_1^0$
- $\tilde{t} \to c + \tilde{\chi}_1^0$
- $\tilde{b} \to b + \tilde{\chi}_1^0$

The signatures vary from mode to mode, and can include decays that look very much like Standard Model top quark decays; such decays will be difficult to separate from the $\bar{t}t$ background without substantially more data. Note the similarity of the last two decays: pair produced squarks will leave a signature

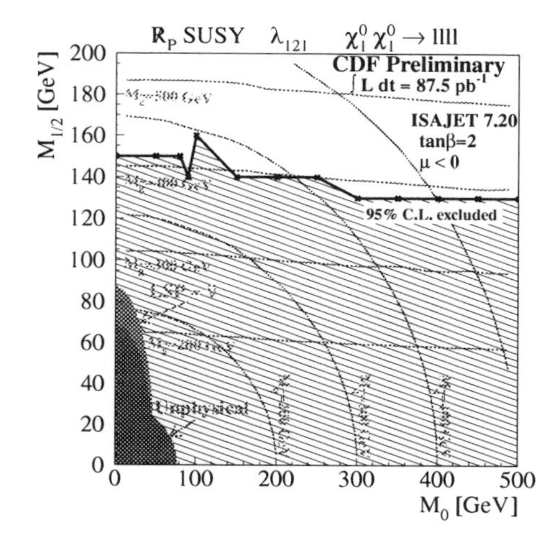

Figure 3. CDF limits on RPV 4-lepton events based on their direct search for such events.

of two heavy flavor jets (charm for $\bar{\tilde{t}}\tilde{t}$ production and bottom for $\bar{\tilde{b}}\tilde{b}$ production), plus missing E_T carried off by the neutralinos.

Both experiments search for these decays. The CDF search involves looking for 2 (or 3) acolinear jets with at least 40 GeV of missing E_T. Leptons are removed, and then the jets are searched for secondary vertices using the silicon detector. They observe 11 charm jets over a background of 15 ± 4, and 5 bottom jets over a background of 6 ± 2. From these events, limits (shown in Figures 6 and 7) can be extracted.[2] The CDF limits are more stringent than the D0 limits, because they used a larger dataset, but also because their secondary vertex detector improves their heavy flavor identification. The $\tilde{b}$ limits are better than the $\tilde{t}$ limits because identifying bottom is easier than identifying charm.

An interesting possibility for future stop searches is to consider R-Parity violating *production* of stop squarks.[3] This allows production of a single stop squark via a process like $d + s \to \tilde{t}$. Of course, you pay the price of the coupling constant (which could potentially still be large without being in disagreement with the data), but you gain in terms of phase space by having to produce only one squark instead of a pair. This could potentially increase the cross section by a factor of 100.

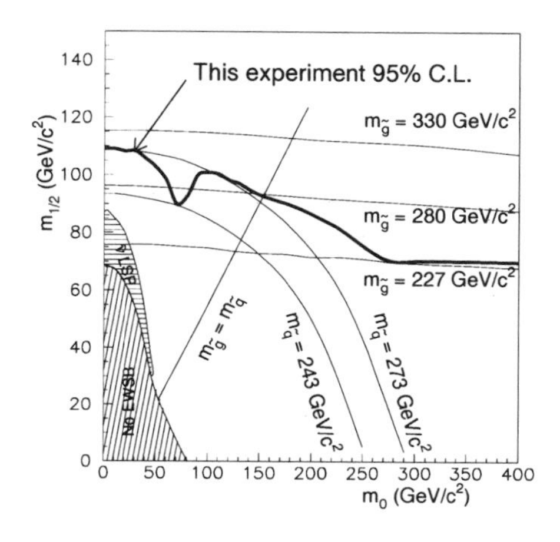

Figure 4. D0 limits on common scalar and fermion masses based on RPV dilepton events and the model described in the text.

4 The Future

The results of these searches have so far been negative, but inconclusive: there is still a sizable fraction of SUSY parameter space available. In March of 2001 the Tevatron will begin Run II, an ambitious program to deliver at least 20 times (and possibly as much as 150 times) the data to the two experiments as in Run I. Additionally, both experiments will be running with substantially improved detectors: D0 will have (for the first time) a central magnetic tracker and a silicon vertex detector, and CDF will have a completely new tracker, additional silicon, and improved lepton identification. It's not an accident that the two experiments are growing more similar - each is incorporating the best features of the other. Together, these two experiments provide the best discovery potential before the LHC.

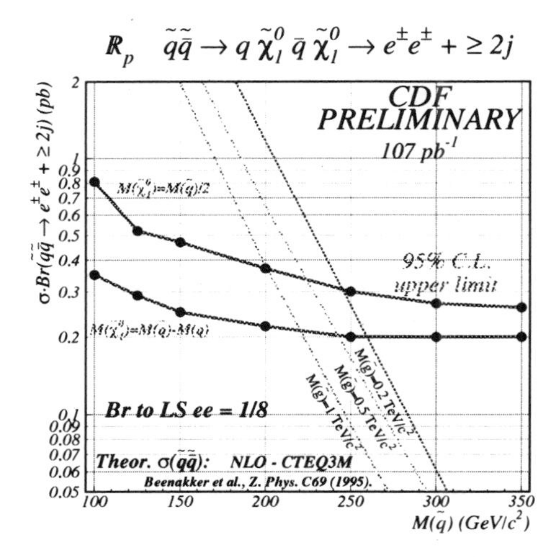

Figure 5. CDF limits on common squark masses based on the RPV model described in the text.

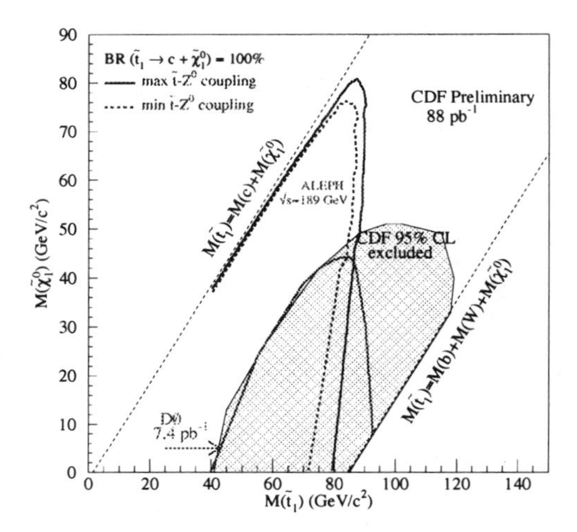

Figure 6. Tevatron limits on stop squark production, overlaid on LEP limits. The region in the top left is kinematically disallowed.

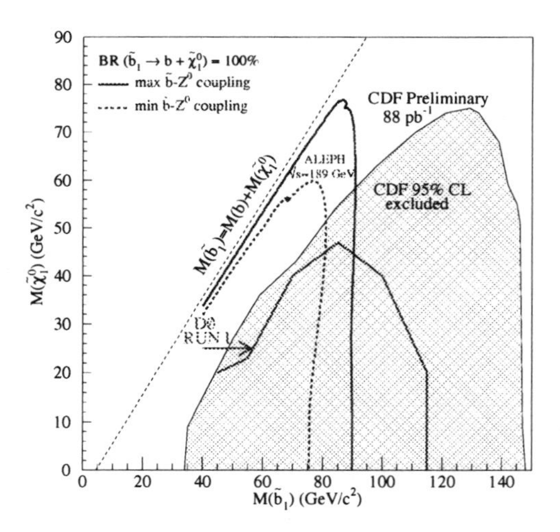

Figure 7. Tevatron limits on sbottom squark production, overlaid on LEP limits. The regions in the top left and bottom right are kinematically disallowed.

Acknowledgments

The author wishes to thank the members of the CDF and D0 Collaborations for the opportunity to present these results. The support and encouragement of the management of the Fermilab is also acknowledged. Finally, the design, construction and installation of the detectors has been made possible by the efforts of many people and their respective funding agencies who are not necessarily listed as authors here.

References

1. B. Abbott *et al.* Phys. Rev. Lett. **80**, 1591 (1998) and F. Abe, *et. al* Phys. Rev. Lett. **80**, 5275 (1998)
2. T. Affolder, *et al.* Phys. Rev. Lett. **84**, 5273 (2000)
3. E. Berger, B. Harris and Z. Sullivan, Phys. Rev. Lett. **83**, 4472 (1999) ANL-HEP-PR-00-062
4. B. Abbott *et al.* Phys. Rev. Lett. **83**, 4476 (1999)
5. F. Abe, *et al.* Phys. Rev. Lett. **83**, 2133 (1999)

SEARCH FOR SUPERSYMMETRY AT THE LHC

S. RAJAGOPALAN
Brookhaven National Laboratory, Upton, N.Y. 11973, USA
E-mail: srinir@bnl.gov

Supersymmetry is one of the best motivated extensions to the Standard Model that provides a solution to the hierarchy problem. The discovery of supersymmetry at the weak scale is straightforward at the LHC and the parameters of the underlying model can be determined. Extensive studies have been performed for minimal supergravity, gauge mediated and R-parity violating models. We review these studies in this paper.

1 Introduction

Supersymmetric models postulate the existence of superpartners for all presently observed particles. If SUSY exists at the weak scale, it will produce a rich spectrum of particles in the mass range that will be explored by the LHC. The SUSY cross-section is dominated by gluinos and squarks, which are strongly produced with cross-sections comparable to the Standard Model (SM) backgrounds. Their distinctive signatures render these SM backgrounds negligible. While the discovery of SUSY is straightforward, the main challenge for the LHC experiments, ATLAS and CMS, lie in extracting the parameters of the underlying SUSY model. Each experiment has performed detailed studies[1,2] for several particular choices of parameters in the minmal SUGRA, minimal GMSB and R-parity violating models.

SUSY must be broken in a hidden sector that communicates to the MSSM sector via some messenger interactions. In supergravity models, gravity is the sole messenger and the SUSY breaking scale is of the order of $(10^{11}\ \text{GeV})^2$. In the m-SUGRA model, the squarks, sleptons and the Higgs bosons have a common mass m_0 and the gauginos have a common mass $m_{1/2}$ at the GUT scale. The squarks and gluinos are strongly produced and decay in several steps into the lightest stable SUSY particle (LSP), the $\tilde{\chi}_1^0$. In gauge mediated models, the SUSY breaking scale in the hidden sector is much lower and the MSSM particles acquire their mass through $SU(3) \times SU(2) \times U(1)$ gauge interactions. The lightest SUSY particle is the gravitino ($\tilde{G}$). The next to lightest supersymmetric particle (NLSP) can either be $\tilde{\chi}_1^0$ or a $\tilde{\tau}_1$. Furthermore, there are no theoretical constraints on the lifetime of the NLSP : they may decay promptly or outside the detector volume. Finally, R-parity violating models have been considered where the LSP ($\tilde{\chi}_1^0$) is allowed to decay, violating either the baryon number (such as $\tilde{\chi}_1^0 \rightarrow qqq$) or the lepton number (such as $\tilde{\chi}_1^0 \rightarrow ll\nu$).

2 Minimal SUGRA models

A deviation from the SM is the first step in the search for supersymmetry. SUSY particles are produced in pairs at the LHC and decay to stable SM particles and the LSP ($\tilde{\chi}_1^0$) which escapes detection. SUSY events are harder than SM events and characterised by multiple jets, zero, one or more leptons and a large missing transverse energy ($\not{E}_T$). The effective mass (M_{eff}), defined as the sum of $\not{E}_T$ and the E_T of the four leading jets, provides a measure of this hardness. Figure 1 shows the M_{eff} distribution for SUSY signal events and the sum of all SM backgrounds ($t\bar{t}$, W+jets, Z+jets, QCD). The S/B exceeds 10 for large effective masses where SUSY cross-sections can be of the order of 1 pb. Furthermore, the effective mass can also provide

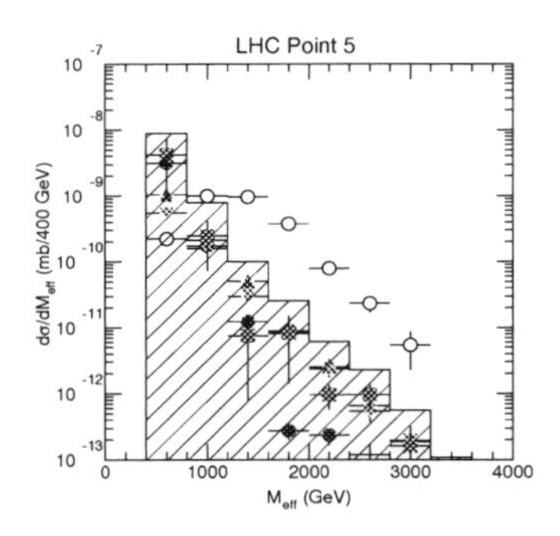

Figure 1. M_{eff} distribution for signal (open circles) and sum of SM backgrounds (hatched area).

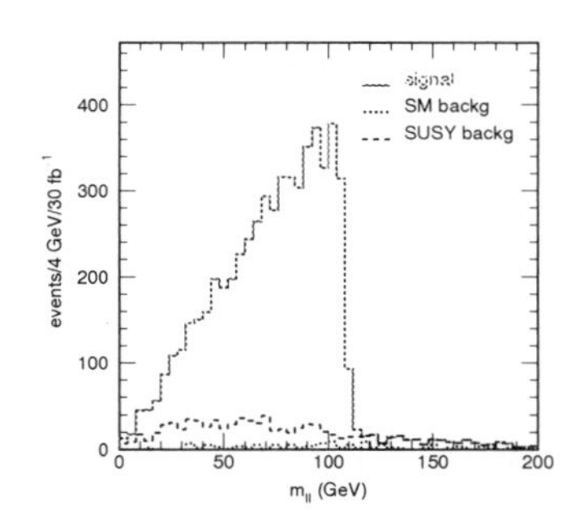

Figure 2. Dilepton invariant mass from $\tilde{\chi}_2^0 \rightarrow \tilde{l}_R^{\pm} l^{\mp} \rightarrow \tilde{\chi}_1^0 l^+ l^-$ (solid), other SUSY sources (dashed) and SM background (dotted).

a measure of the SUSY mass scale to within $\pm 10\%$ for $\int L = 10\ \text{fb}^{-1}$. The LHC reach, defined as the observation of at least 10 signal events with a $S/\sqrt{B} > 5$ for $\int L = 100\ \text{fb}^{-1}$, extends to squark and gluino masses greater than 2 TeV. Thus, evidence for SUSY and an estimate of the mass scales involved can be quickly inferred from the data coming out of LHC as a first step in the analysis process.

The second step in the analysis chain is to form the invariant masses of different combinations of the final states to obtain the necessary constraints to extract the sparticle masses. As an example, consider the decay $\tilde{q}_L \rightarrow \tilde{\chi}_2^0 q \rightarrow \tilde{l}_R^{\pm} l^{\mp} q \rightarrow \tilde{\chi}_1^0 l^+ l^- q$ studied by ATLAS for $m_0 = 100$ GeV, $m_{1/2} = 300$ GeV, $A_0 = 0$, $\tan\beta = 2.1$, $\text{sgn}(\mu) > 0$. The end point of the dilepton invariant mass, shown in Figure 2, can be measured in this case to an accuracy of 0.1% and is kinematically related to the sparticle masses:

$$M_{ll}^{max} = \frac{1}{M_{\tilde{l}}} \sqrt{(M_{\tilde{\chi}_2}^2 - M_{\tilde{l}}^2)(M_{\tilde{l}}^2 - M_{\tilde{\chi}_1}^2)}$$

SM backgrounds and combinatorics are suppressed by plotting the flavour-subtracted combination: $e^+e^- + \mu^+\mu^- + e^{\pm}\mu^{\mp}$.

Additional constraints can be obtained by measuring the endpoints of the llq and lq mass distributions. The resolution of the llq and lq edges are dominated by the jet energy scale, of the order of 1%. Using the measured constraints, a model-independent analysis has been performed to extract the masses of the SUSY particles involved in the decay: $\tilde{q}_L$, $\tilde{\chi}_2^0$, $\tilde{l}_R^{\pm}$ and $\tilde{\chi}_1^0$, to within an accuracy of 3%, 6%, 9% and 12% respectively for an $\int L = 100\ \text{fb}^{-1}$.

The SUSY particle masses thus measured can be used in a global fit to determine the parameters of the underlying m-SUGRA model. SUSY masses are generated for several points in the m-SUGRA parameter space. A χ^2 fit of the generated SUSY masses against the measurements for the above discussed example yields $m_0 = 100^{+4.1}_{-2.2}$ GeV, $m_{1/2} = 300 \pm 2.7$ GeV, $\tan\beta = 2.00 \pm 0.10$. Several other points in the parameter space have also been extensively studied using a similar strategy and produce comparable results.

3 Gauge Mediated Models

In the GMSB models, the gravitino ($\tilde{G}$) gets its mass through gravitational couplings at the Planck scale and hence is the LSP ($\ll 1$ GeV). The minimal model is characterised by Λ (ratio of the SUSY breaking scale to the messenger scale), M_m (scale of the messenger interactions), N_5 (number of messenger fields), $\tan\beta$ (ratio of the vev's), C_{grav} (dependent on the gravitino mass and hence the lifetime of the NLSP), and $\text{sgn}(\mu)$. The NLSP can either be a $\tilde{\chi}_1^0 \rightarrow \tilde{G}\gamma$ or $\tilde{\tau}_1 \rightarrow \tilde{G}\tau$ depending on the number of messenger fields. The lack of any theoretical constraints on the

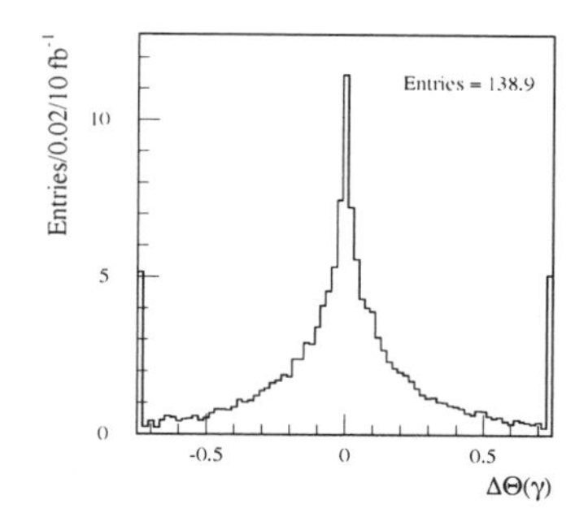

Figure 3. Non-pointing angles for photons in EM barrel calorimeter from $\tilde{\chi}_1^0 \to \tilde{G}\gamma$ for $c\tau$=1.1 km.

lifetime of the NLSP leads to different characteristic signatures depending on whether the NLSP decays promptly or outside the detector volume.

The prompt decay of $\tilde{\chi}_1^0$ provides a clear two-photon signature in addition to multiple jets and leptons. A $c\tau$ of the order the tracker radius ($\sim$ 1.5 m) will produce photon showers that do not point back to the vertex. For large $c\tau$, the NLSP will decay outside the detector resembling the signature of m-SUGRA models. However a fraction of these events will decay within the tracker (1% for $c\tau$ = 1 km) leading to non-pointing photon signatures. Since the cross-sections can be as high as tens of pb, this acceptance loss can be bearable. The EM calorimeter in ATLAS has been designed with narrow strips in rapidity in its first compartment that provides a pointing resolution of 60 mrad/$\sqrt{E}$. The non-pointing angle for photons in the barrel calorimeter from $\tilde{\chi}_1^0 \to \tilde{G}\gamma$ is shown in Figure 3. Though peaked at zero, such photons can be detected with over 80% efficiency with a 5σ cut on the reconstructed non-pointing angle. Studies have been conducted to determine that the LHC is sensitive up to a $c\tau$ of 100 km for $\int L = 30$ fb^{-1}.

For the case when $\tilde{\tau}$ is the NLSP, the mass splitting between the NLSP and the right handed sleptons ($\tilde{l}_R$) can be small, so $\tilde{e}_R$, $\tilde{\mu}_R$ and $\tilde{\tau}_1$ are effectively co-NLSP's decaying directly to gravitino. Decays of $\tilde{\chi}_1^0 \to \tilde{l}^{\pm}l^{\mp} \to \tilde{G}l^+l^-$ produce opposite-sign same-flavour dileptons with the characteristic sharp end-point in the invariant dilepton mass distribution. The strategy for the extraction of the underlying GMSB model parameters is similar to the m-SUGRA model as discussed in the previous section. For long lifetimes, where the $\tilde{\tau}$ decays outside the detector volume, the signature is a pair of quasi-stable heavy particles which resemble muons but have $\beta <$ 1. The slepton masses can be determined using the time of flight measurement in the muon system.

Fits to the underlying GMSB parameters have been performed for several cases. In each case, the model parameters can be extracted with a precision of 2% for Λ and N_5, and 10-40% for M_m and tanβ. The lifetime measurement of the slepton is based on identifying the kink from its decay in the tracker. This is a complex pattern recognition problem and should be feasible at low luminosity.

4 Conclusion

The discovery of SUSY is straightforward at the LHC if it exists at the weak scale. Since the main background for SUSY arise from other SUSY processes and not SM physics, the extraction of the underlying model is the real challenge. The studies performed thus far used minimal models with a small set of parameters using strategies based on kinematic endpoints of mass distributions. While it is impossible to explore the entire parameter space, such a strategy provides a good starting point which in turn can be used to mature the theoretical understanding.

References

1. ATLAS Detector and Physics Performance Technical Design Report, CERN-LHC-99-15, ATLAS TDR **15**, (1999).
2. S. Abdullin *et. al.*, CMS NOTE 1998-006, (1998).

R–PARITY VIOLATION SEARCHES AT LEP

IVOR FLECK

Universität Freiburg, Fakultät für Physik, Hermann-Herder-Str. 3, 79104 Freiburg, Germany
E-mail: Ivor.Fleck@cern.ch

At LEP many searches for physics beyond the Standard Model are performed. In this talk searches for SUSY particles decaying via R–parity violating couplings are presented for data taken at LEP2 for $\sqrt{s} \leq 202$ GeV. No evidence for any such particle has been observed.

1 R–Parity Violation

In SUSY theories a new discrete multiplicative symmetry, called R-parity is introduced, $R_p = (-1)^{3B+L+2S}$, which is 1 for SM particles and -1 for SUSY particles. If R-parity is conserved, the lightest supersymmetric particle (LSP) does not decay and the experimental signature is large missing momentum. If R-parity is violated, the LSP does decay and the experimental signature shows **no** missing momentum. Therefore exclusion limits for SUSY particles made under the assumption of R-parity conservation are not valid under R-parity violation.

The R-parity violating super–potential W is given by [1]:

$$W = \lambda_{ijk} \mathrm{L^i_L L^j_L \overline{E}^k_R} + \lambda'_{ijk} \mathrm{L^i_L Q^j_L \overline{D}^k_R} + \lambda''_{ijk} \mathrm{\overline{U}^i_R \overline{D}^j_R \overline{D}^k_R} + \epsilon_\mathrm{i} \mathrm{L^i_L H_u} \quad (1)$$

$\mathrm{L}^i(\mathrm{Q}^i)$ are the lepton (quark) $SU(2)_L$ doublet superfields, $\overline{\mathrm{E}}^i(\overline{\mathrm{U}}^i, \overline{\mathrm{D}}^i)$ are the singlet superfields, $\mathrm{H^u}$ is the Higgs field coupling to up-quarks; λ, λ', and λ'' are Yukawa couplings, ϵ an effective coupling and i, j, and k are generation indices. The first two terms break L and the third term breaks B.

From non LEP experiments limits on the size of the Yukawa couplings λ, λ', and λ'' exist and are, for a sparticle mass of 100 GeV, of $\mathcal{O}(10^{-2})$ for most of the couplings [1]. Much more stringent limits exist on the product of two of these couplings. Therefore it is assumed that only one of the couplings is significantly different from zero. At LEP these couplings can be probed down to values of 10^{-7} for sparticle masses up to the beam energy. Sneutrinos and squarks can also be produced singly and a direct search for these particles up to the centre-of-mass energy is possible. These sparticles contribute also to two fermion final state production via t-channel exchange.

Under R-parity violation any sparticle can decay directly via the R-parity violating potential. This kind of decay will be called direct decay. On the other hand decays like under R-parity conservation are allowed with only the LSP decaying via the R-parity violating potential. These decays are called indirect decays. The event topologies depend on which coupling is assumed to be dominant. Consequently the analyses are separated into topologies involving either a λ, λ', or λ'' coupling.

2 Experimental Results

The analyses from all four LEP experiments are designed to look for the decay products of supersymmetric particles. Depending on the sparticle and on the λ-coupling considered the final state visible inside the detector varies strongly. Therefore the analyses are designed to look for certain topologies, with topologies ranging from as little as two acoplanar leptons up to 10 jets. As no signal has been observed so far, limits on the masses of the SUSY particles are assigned, depending on the size of the λ-coupling only in the case of single production of the sparticle.

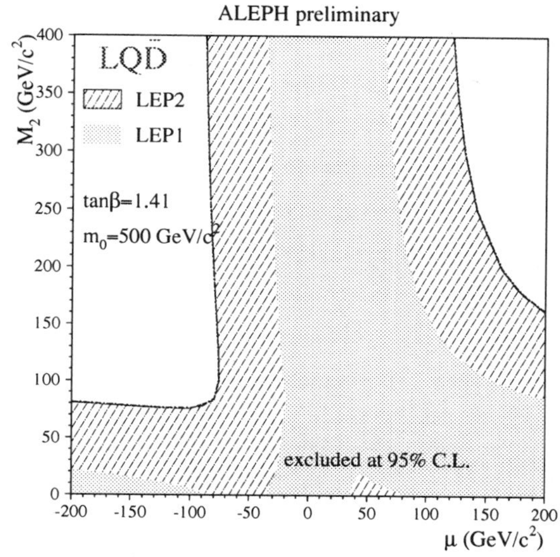

Figure 1. Area excluded in the $M_2 - \mu$ plane for $\tan\beta = 1.41$ and $m_0 = 500$ GeV for any coupling λ'. The outer solid line shows the kinematic limit for chargino pair production at $\sqrt{s} = 202$ GeV.

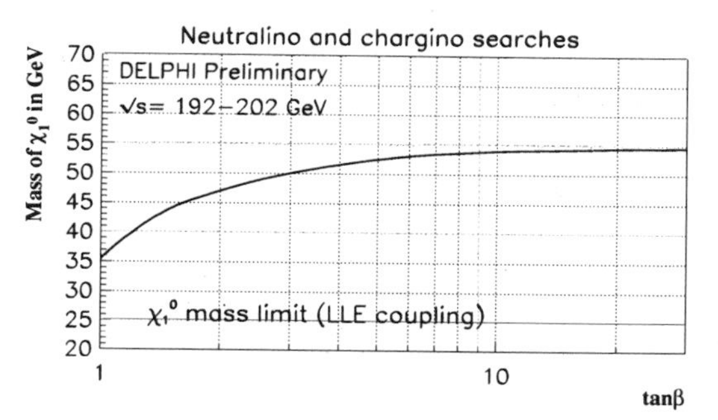

Figure 2. Limit on the mass of the lightest neutralino for any coupling λ.

The results presented in the following sections are valid under the assumption that only one of the λ-couplings is different from zero, unless otherwise stated. Results from bilinear terms are presented in the last chapter.

2.1 Gauginos

Final states resulting from the decays of gauginos consist of six primary fermions in the case of direct decays and at least ten primary fermions in the case of indirect decays. For cascade decays of the lightest chargino even more particles are expected.

From the searches for the production of chargino and neutralino pairs areas in the MSSM space are excluded. These exclusions go very close to the kinematic limit of the pair-production of the lightest chargino and an example is shown in Fig. 1 [2]. For small values of m_0 and large values of $\tan\beta$ areas above this kinematic limit are excluded from the pair-production of the lightest neutralino.

From the excluded areas lower mass limits for the gauginos are derived. The lower limit on the $\tilde{\chi}_1^0$ mass is shown in Fig. 2 [3] as a function of $\tan\beta$ for any value of the other MSSM parameters, M_0, μ and m_0. The lower mass limits are: $M(\tilde{\chi}_1^0) > 35$ GeV, $M(\tilde{\chi}_2^0) > 68$ GeV, $M(\tilde{\chi}_1^\pm) > 99$ GeV [2 3 4 5] for any λ-coupling.

2.2 Sleptons

The decay products from the pair-production of sfermions are 4 primary particles for the direct and eight for the indirect decay. In the case of SUSY partners of the charged leptons, denoted as sleptons, at least two charged leptons are in the final state, except for the direct decay via λ' which results in a four jet final state.

Again topological searches are performed, separately for each slepton flavour and for direct and indirect decays via the different λ-couplings. The upper cross-section limits achieved this way are translated into excluded masses, in the case of the indirect decays as a function of the lightest neutralino mass, as shown for example in Fig. 3 [5].

Another way of excluding sparticle masses comes from excluded areas in the MSSM parameter space. Combining results from searches for gauginos and sfermions larger areas in the MSSM parameter space can be excluded than by just searching for each sparticle separately. From this excluded

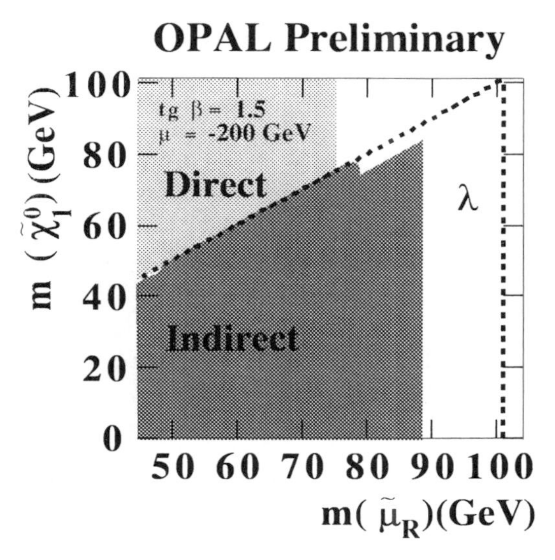

Figure 3. Excluded area in the m($\tilde{\mu}$)–m($\tilde{\chi}_1^0$) mass plane for any coupling λ.

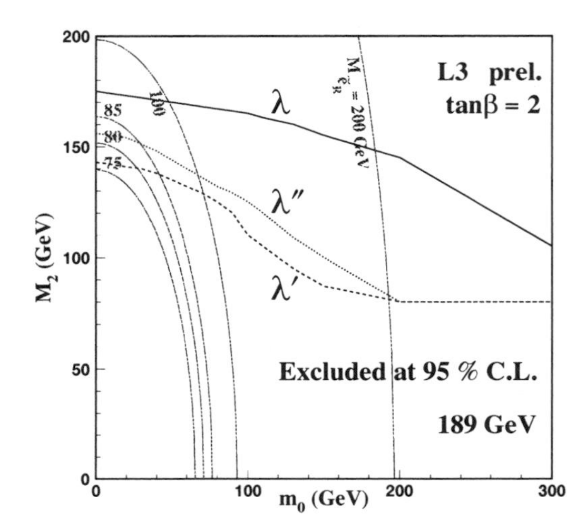

Figure 4. Excluded area in the $M_2 - m_0$ plane, shown separately for decays via λ, λ' or λ''.

area mass limits e.g. on selectron masses can be given. In Fig. 4 [4] the selectron isomass lines are shown together with the excluded areas, separately for decays via λ, λ' or λ''. The lower mass limits depend, as also for the topological analyses, on the λ-coupling considered and are about 90 GeV for selectrons and slightly lower for smuons and staus.

2.3 Sneutrinos

The pair-production of sneutrinos leads to 4 charged lepton or 4 jets in the final state for direct decays and to lepton and/or jets plus missing energy for indirect decays. Searches for these final states lead to lower mass limits of about 80 GeV for the second and third generation and due to the higher production cross-section in large parts of the parameter space to higher limits for the first generation.

In addition single production of sneutrinos is possible in the s-channel process $e^+e^- \rightarrow \tilde{\nu}_j$ via a coupling $\lambda_{1j1}, j = 2,3$. In this process sneutrinos masses up to the centre-of-mass energy are accessible, but only for large ($\mathcal{O}(10^{-2})$) values of the coupling λ. The direct decay leads to an electron-positron pair in the final state and the process is visible in a deviation in the e^+e^- cross-section measurement. If an additional coupling $\lambda_{iji}, i = 2,3, i \neq j$ is non zero then the process is also visible in the $\mu-$pair and the $\tau-$pair cross-section. Sneutrino masses between 130 and 202 GeV have been probed [6] [7] [8] [9], as shown in Fig. 5 [10] for λ_{121}. The spiked structure of the exclusion curve results from the different centre-of-mass energies of LEP.

For the single sneutrino production also the indirect decay into a neutralino and a neutrino is possible The final state then consists of missing energy and two lepton, at least one of them being an electron. The excluded sneutrino mass as a function of the coupling λ_{121} is shown in Fig. 5. The exclusions look similar for λ_{131}.

2.4 Squarks

Scalar quarks of the third generation can have large mass differences between the two mass eigenstates $\tilde{q}_1$ and $\tilde{q}_2$, with $\tilde{q}_1 = \cos\theta \cdot \tilde{q}_L + \sin\theta \cdot \tilde{q}_R$, where θ is the mixing angle. For $\theta = 0.98(1.17)$ the $\tilde{t}_1(\tilde{b}_1)$ decouples from the Z^0 and the production cross-section becomes minimal. Limits are therefore given as a function of the mixing angle.

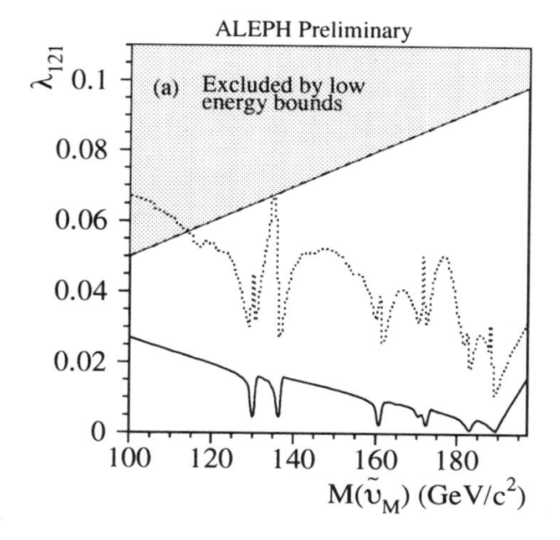

Figure 5. Excluded sneutrino masses for given values of λ_{121} for the direct decay (dashed line) and the indirect decay (solid line) compared wit existing limits on λ_{121}.

Direct and indirect decays from all λ-couplings have been studied and lower mass limits of about 80 GeV have been achieved for any mixing angle for stop and and slightly lower limits for sbottom production.

Single production of squarks proceeds via the interaction of a beam electron with a virtual quark inside a photon radiated from the other beam electron. This process is also sensitive to squark masses up to the centre-of-mass energy, but the production cross-section at masses close to the kinematic limit is small due to the momentum spectrum of the photon. Limits achieved at LEP are superseded by the results from Tevatron, except when assuming a branching ratio of the squark into a neutrino and a quark of more than 85 %. The lower mass limit is 165 GeV for values of the coupling of greater than the electromagnetic strength ($\sqrt{4\pi\alpha_{em}}$) [11].

Further an exchange of a squark in the t-channel is possible. Here the analyses are sensitive to masses up to several hundred GeV for couplings of $\mathcal{O}(10^{-1})$, with scalar down-quark masses excluded up to 400 GeV for a coupling of 0.3 [8].

2.5 Bilinear R-Parity Violation

The results presented so far all stem from the trilinear terms in the R-parity violating superpotential. An additional bilinear term is present in the superpotential. This term is always present due to the breaking of SUSY and has also been considered in the LEP analyses. Many topologies arising from the bilinear terms are identical to those from the trilinear terms. Those analyses can consequently be reused and are just interpreted differently.

A possible consequence of the spontaneous R-parity violation is the existence of a physical massless Nambu-Goldstone boson (Majoron J). Mixing of the gaugino and the lepton sector leads to new decays like $\tilde{\chi}^{\pm} \rightarrow \tau^{\pm} J$. Assuming this decay mode and a sneutrino mass above 300 GeV a lower chargino mass limit of 94.4 GeV is achieved [12].

References

1. B.C. Allanach, A. Dedes, H.K. Dreiner, hep-ph/9906209
2. ALEPH Collaboration, ALEPH 2000-013 CONF 2000-010
3. DELPHI Collaboration, DEPLHI 2000-015 CONF 336, 2000-016 CONF 337
4. L3 Collaboration, L3 Note 2585
5. OPAL Collaboration, PN 418
6. ALEPH Collaboration, ALEPH 2000-047 CONF 2000-030,
7. DELPHI Collaboration, DEPLHI 2000-036 CONF 355,
8. L3 Collaboration, L3 Note 2543
9. OPAL Collaboration, PN 368
10. ALEPH Collaboration, ALEPH 99-012 CONF 99-007
11. OPAL Collaboration, PN 451
12. DELPHI Collaboration, DEPLHI 2000-021 CONF 342

SEARCH FOR SQUARK PRODUCTION IN R–PARITY VIOLATING SUSY AT HERA

L. STANCO
Sezione I.N.F.N. di Padova
Via Marzolo,8 I-35131 Padova, Italy
stanco@pd.infn.it
On behalf of H1 and ZEUS Collaborations

Searches for squarks produced via R-parity violating interactions in e^+p collisions at a center–of–mass energy of 300 GeV have been performed at HERA using the two detectors, H1 and ZEUS, and an integrated luminosity of 37 and 48 pb^{-1}, respectively. Squarks produced in e^+–quark fusion could decay either to e^+–quark or via a supersymmetric gauge decay, resulting in many possible final states. The signal has been searched for in most of R–parity violating decays and gauge decays of the squarks. No evidence for squark production was found and limits were set on the R–parity violating coupling as a function of the squark mass and the SUSY parameters, extending to domains unexplored in other direct or indirect searches. For a fixed value of the coupling, HERA results are interpreted for the first time in terms of constraints on the parameters of the mSUGRA model.

1 Introduction

The study of high E_T phenomena at the HERA *ep* collider, at a center of mass energy of 300 GeV and with both baryonic and leptonic quantum numbers in the initial state, offers an excellent possibility to search for new particles which couple to electron–parton pairs. Such particles could be *squarks* in supersymmetric (SUSY) extensions of the Standard Model (SM) which violate R–parity. R–parity, defined as $R_p = (-1)^{3B+L+2S}$, is equal to 1 for standard particles and -1 for *sparticles*, where B,L and S denote baryon number, lepton number and spin, respectively. In the most general SUSY theory which preserves gauge invariance of the SM, R_p violation can arise through Yukawa couplings between one scalar squark or slepton and two Standard Model fermions. Such R_p violating ($\not{R}_p$) couplings allow the single production of sparticles and induce the decay of the lightest supersymmetric particle (LSP), thus leading to a phenomenology which widely differs from that of models where R_p conservation is imposed. Of special interest for HERA are those Yukawa vertices which couple a squark to a lepton–quark pair. These are described in the superpotential by the terms $\lambda'_{ijk}L_iQ_j\overline{D}_k$, with i,j,k being generation indices. The corresponding part of the Lagrangian, expanded in fields, is:

$$\mathcal{L} = \lambda'_{ijk}[-\tilde{e}^i_L u^j_L \overline{d}^k_R - e^i_L \tilde{u}^j_L \overline{d}^k_R - (\overline{e}^i_L)^c u^j_L \tilde{d}^{k*}_R + \tilde{\nu}^i_L d^j_L \overline{d}^k_R + \nu^i_L \tilde{d}^j_L \overline{d}^k_R + (\overline{\nu}^i_L)^c d^j_L \tilde{d}^{k*}_R] + \text{h.c.}$$

The second and third term describe the processes $e^+d^k \to \tilde{u}^j_L$ and $e^+\overline{u}^j \to \tilde{d}^{k*}$, respectively. With an e^+ incident beam, HERA is best sensitive to couplings λ'_{1j1} and the production of $\tilde{u}^j_L$ squarks via processes involving a valence d quark. On the contrary, HERA e^-p data taken in 1998-1999 will allow to better probe couplings λ'_{11k} and the $\tilde{d}^k_R$ squarks production.

The two contributed papers to the conference from H1[1] and ZEUS[2] show preliminary results on the search for squark production in the data collected in 1994–1997, for a total luminosity of 37 and 48 pb^{-1}, respectively. This represents for H1 an increase of statistics of a factor $\sim$ 13 compared to earlier published $\not{R}_p$ squark searches at HERA[3] and it supersedes preliminary results presented at ICHEP98[4] based on the same dataset. ZEUS results supersedes the preliminary results presented at EPS99[5].

2 Analysis and models assumptions

The $R\!\!\!/_p$ supersymmetric models provide a framework in which squarks can decay into $e + q$ depending on the unknown Yukawa coupling, while competing with gauge decay modes. Since there are four neutralinos $\tilde{\chi}_i^0$ (the mass eigenstates of mixtures of $\tilde{\gamma}$, $\tilde{Z}$ and the two neutral Higgsinos) and two charginos $\tilde{\chi}_i^{\pm}$ (mixtures of $\tilde{W}$ and the charged Higgsino), there can be as many as six gauge decays of the squark, followed by the subsequent decays of the χ's. H1 also considers decays of squarks into gluinos. The final states result into a multitude of various multi–jet topologies.

The most general Supersymmetric models have more than one hundred undetermined parameters. However, under certain assumptions concerning the Supersymmetric breaking, the low–energy effective theory which emerges is substantially simpler. In the minimal supersymmetric model (MSSM) the branching ratios for $\tilde{q}$ decay depend on the usual parameters μ, $\tan\beta$, M_1, M_2 and on the squark mass $M_{\tilde{q}}$ and the coupling λ'_{ijk}.

In both the H1 and ZEUS study additional simplifying assumptions have been made:

- at most one of the couplings λ'_{ijk} is non–zero;
- obtained limits will be valid only for 1st and 2nd generation squarks;

together with the assumptions on the models:

- the LSP is a χ or $\tilde{g}$ (H1) or imposed to be the lightest neutralino (Zeus);
- the gaugino mass parameters M_1 and M_2 are related by $M_1 = (5/3)\tan^2\theta_W\ M_2$;
- slepton mass is taken equal to squark mass (ZEUS,H1); or equal to 90 GeV; or a general common scalar mass for the sfermions is assumed at high scale (H1);
- the soft SUSY breaking trilinear couplings A_t, A_b and A_τ were assumed to vanish.

ZEUS also considers the additional assumption that the gluino be heavier than the squark, so that the decay $\tilde{q} \to q\tilde{g}$ is forbidden.

The resulting degrees of freedom allow the description of the SUSY parameter spaces in term of μ, M_2 and $\tan\beta$ other than the squark mass. Both experiments have undergone an exhaustive study of a large part of the SUSY space described by the latter parameters, for the first time.

Other constraints are singled out in the plots shown below.

Squark decay channels in $R\!\!\!/_p$ SUSY are classified per distinguishable event topologies. In addition to the Neutral Current (NC) and Charged Current (CC) –like topologies (e^+ or missing p_T and 1 jet) other final states correspond to $e^{\pm}$ or missing p_T and multi-jets. H1 also looks at gaugino cascade decays where two of the jets are replaced by a lepton (e or μ) and a neutrino.

The H1 and ZEUS detectors, together with several other aspects of the selection analysis, are reported in [1,2].

3 Results

In all final states analyzed, no discrepancy with the Standard Model expectations has been observed and limits on the λ'_{1j1} couplings as a function of the squark mass have been obtained.

As an example of invariant mass distribution, figure 1 shows the final result for the H1 e+*multijets* analysis: 160 events are present in the data, while 152.6 ± 13.4 are foreseen by the SM. Similar result is obtained e.g. for the missing p_T and jets from ZEUS (see figure 2).

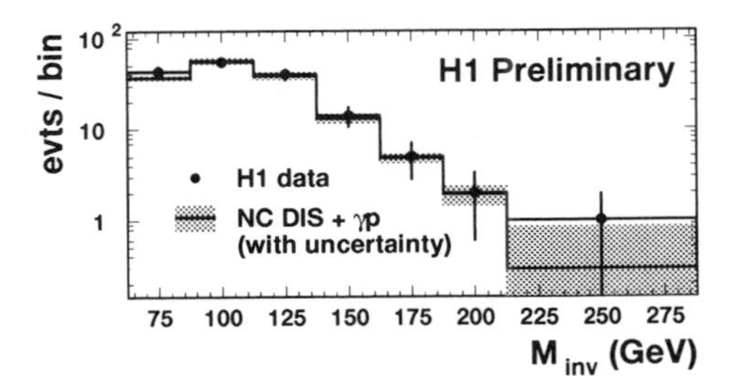

Figure 1. Mass spectrum for $e+multijets$ final state for data (symbols) and NC DIS expectation (histogram). The gray band indicates the uncertainty of the SM prediction.

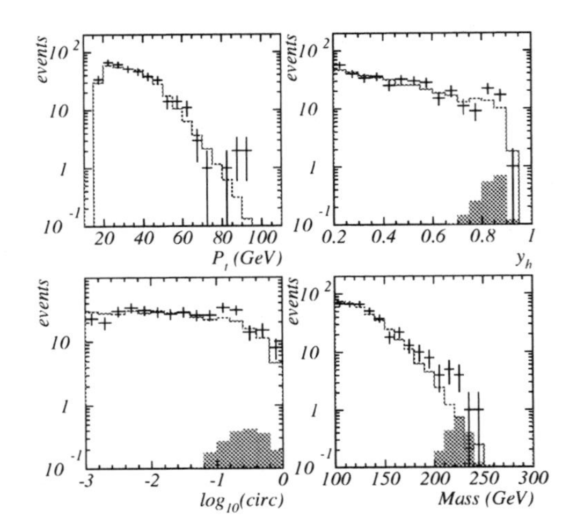

Figure 2. ZEUS preliminary: distributions of p_T, the DIS variable y_h, circularity in terms of $\log_{10}(c)$ and mass for the missing p_T sample with mass above 100 GeV. The points with error bars show the data (377 events) and the open histogram represents the expected background (352 events). The shaded histogram shows a hypothetical signal from a 220 GeV squark for $\mu = -180$ GeV, $M_2 = 100$ GeV and $\tan\beta = 2$. The signal is normalized to the expected sensitivity.

Limits are then derived in three different frameworks: unconstrained MSSM (H1 and ZEUS), constrained MSSM where a common mass m_0 is assumed at high energy for all the sfermions masses (H1), mSUGRA model which has a specific breaking mechanism (H1) where the number of free parameters is reduced assuming a radiative breaking of the electroweak symmetry (REWSB).

An example for the first kind of limit is shown in figure 3 from the ZEUS analysis and in figure 4 from H1.

Typical observations are as follows:

- upper limits on λ'_{1j1} go from 0.02 at $M_{\tilde{q}} = 140$ GeV to 0.8 at $M_{\tilde{q}} = 280$ GeV;
- a large part of the branching ratios has been covered;
- no strong variation of the limits within the SUSY parameter space;
- for a λ'_{1j1} value equal to the electromagnetic coupling (0.3) squark masses up to ~ 260 GeV are excluded.

In figure 4 it is also shown that HERA is able to obtain better limits than the indirect limits from Atomic Parity Violation experiments up to $M_{\tilde{q}} \sim 240$ GeV, for the case j=2.

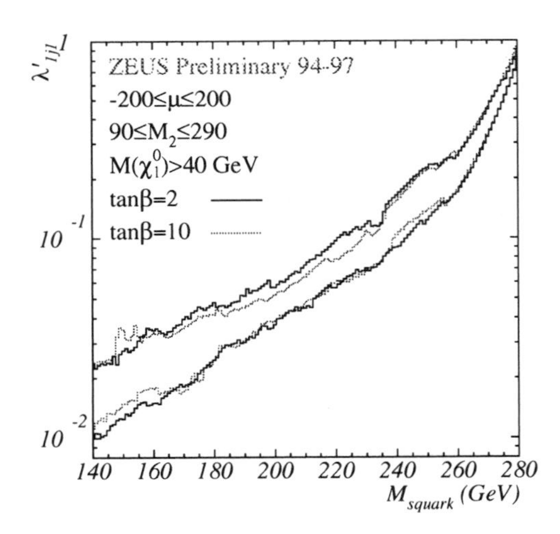

Figure 3. Shown for $\tan\beta = 2$ (solid) and 10(dotted) are the minimum and maximum 95% C.L. limits on λ'_{1j1} vs $M_{\tilde{q}}$ obtained in the region -200 GeV $\leq \mu \leq$ 200 GeV and 90 GeV $\leq M_2 \leq$ 290 GeV.

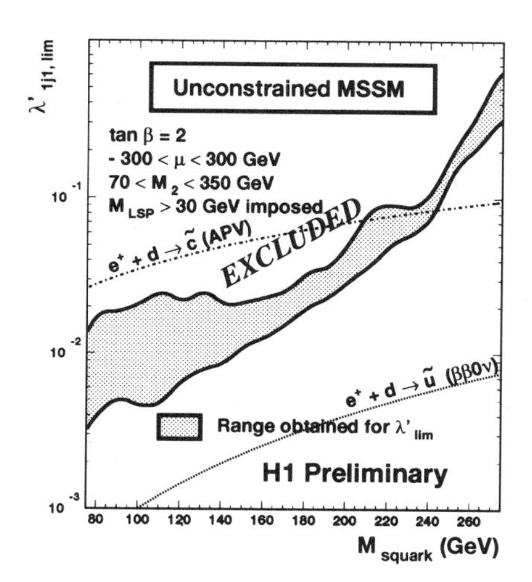

Figure 4. Exclusion upper limits at 95% C.L. on the coupling λ'_{1j1} as a function of the squark mass for $\tan\beta = 2$, in the unconstrained MSSM framework. For each squark mass, a scan on the MSSM parameters M_2 and μ has been performed and the largest (smallest) value for the coupling limit is shown by the upper (lower) curve.

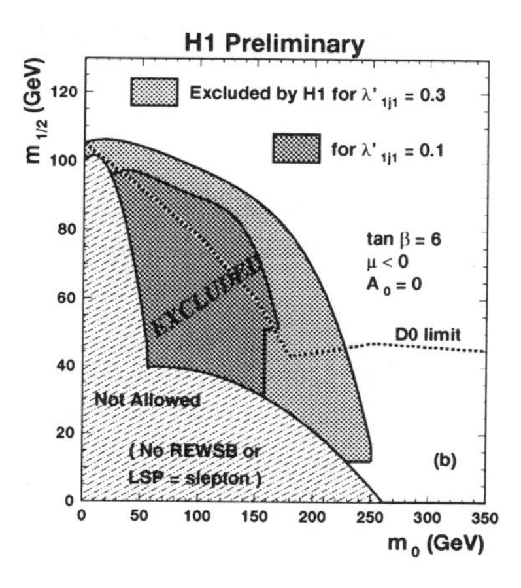

Figure 5. Domain of the plane $(m_0, m_{1/2})$ excluded by H1 with the negative sign for μ, $A_0 = 0$ and $\tan\beta = 6$ for a $R\!\!\!/_p$ coupling $\lambda'_{1j1} = 0.3$ (light shaded area) or $= 0.1$ (dark shaded area). The hatched domains correspond to values of the parameters where no REWSB is possible or where the lightest superparticle is a sfermion. The region below the dashed curve is at present excluded by the D0 experiment.

In the H1 study on the constrained MSSM no strong variation is observed within the SUSY parameter space.

Finally, H1 reported for the first time exclusion limits on the mass terms for scalars (m_0) and gauginos ($m_{1/2}$) in the framework of the mSUGRA model (see figure 5). Such limits complement and for large values of $\tan\beta$ extend the D0 limits, which however do not depend on the chosen λ'_{1j1} coupling.

Acknowledgments

It is a pleasure to thank the warmfulness of the organizers of the conference, despite the awful climate of Osaka city.

No result from those reported here would have been possible without the constant and heroic effort of many people from the H1 and the ZEUS collaborations which profited from the data offered by the HERA machine group.

References

1. H1 Collaboration, Contr. paper # 957 to ICHEP2000.
2. ZEUS Collaboration, Contr. paper # 1042 to ICHEP2000.
3. H1 Collaboration, S.Aid *et al.*, *Z.Phys.* **C71**, 211 (1996).
4. H1 Collaboration, Contr. paper # 580 to ICHEP98.
5. ZEUS Collaboration, Contr. paper # 548 to EPS98.

EXCITED LEPTONS AND LEPTOQUARKS AT LEP

M. A. FALAGAN
CIEMAT, Avda. Complutense 22, 28040 Madrid, Spain
E-mail: Miguel.Angel.Falagan@cern.ch

I summarise the search for excited leptons and leptoquarks in e^+e^- collisions at centre-of-mass energies up to 202 GeV, using the data taken by the Aleph, Delphi, L3 and Opal detectors. No evidence is found for their existence. Limits at the 95% confidence level are set on the mass and the couplings.

1 Excited leptons

The existence of excited leptons would provide evidence for fermion substructure. Composite models could explain the number of families and make the fermion masses and weak mixing angles calculable [2].

The interactions of excited leptons are studied with the model described in Reference [2]. This model assumes spin 1/2, and isospin doublets with left and right handed components.

$$L^* = \begin{pmatrix} \nu^* \\ \ell^* \end{pmatrix}_L + \begin{pmatrix} \nu^* \\ \ell^* \end{pmatrix}_R$$

Excited leptons can be produced in pairs or singly. Pair production, $e^+e^- \rightarrow L^*L^*$, proceeds mainly through s-channel Z/γ exchange. The coupling between a gauge boson and a pair of excited leptons is determined by their isospin and hypercharge. Therefore, the cross section for pair production depends only on the centre of mass energy, $\sqrt{s}$, and the mass of the excited lepton, m_*. In the case of single production of excited leptons, $e^+e^- \rightarrow L^*L$, the cross section also depends on the effective coupling parameters f/Λ and f'/Λ (associated with the SU(2) and U(1) gauge groups of the Standard Model) which determine the strength of the L^*LV interaction ($V = \gamma$, Z or W).

Once an excited lepton has been produced, it is expected to decay immediately into a standard lepton and a gauge boson, $L^* \rightarrow LV$. The decay fractions depend on the ratio f/f'.

The t-channel exchange of a virtual excited electron can contribute to photon pair production, $e^+e^- \rightarrow \gamma\gamma$. This process is sensitive to the existence of an excited electron with mass above the kinematical limit for single production.

Excited leptons are searched for at LEP in all the experimental topologies corresponding to the dominant decay modes [1]. In all selections, the number of observed events in data is consistent with the Standard Model expectation. From pair production searches, lower limits on the masses of excited leptons are obtained by Aleph, Delphi ($\sqrt{s} \leq 189$ GeV), L3 and Opal ($\sqrt{s} \leq 202$ GeV), as shown in Table 1. All values are close to the kinematical limit ($m_* = \sqrt{s}/2$).

Table 1. Opal and L3 95% confidence level lower mass limits for excited leptons.

L^*	95% CL Mass Limit (GeV)			
	$f = f'$		$f = -f'$	Any $\frac{f}{f'}$
	L3	OPAL	L3	L3
e^*	100.1	100.1	96.2	96.0
μ^*	100.3	100.1	96.2	96.2
τ^*	99.9	100.0	96.2	94.9
ν_e^*	99.3	99.5	99.5	98.5
ν_μ^*	99.4	99.5	99.5	98.5
ν_τ^*	93.9	91.9	99.4	92.7

Searches for single production of excited leptons are performed by Aleph ($\sqrt{s} \leq 189$ GeV), Delphi, L3 and Opal ($\sqrt{s} \leq 202$ GeV). Since no signal is found, upper limits on the coupling constant f/Λ are obtained in the

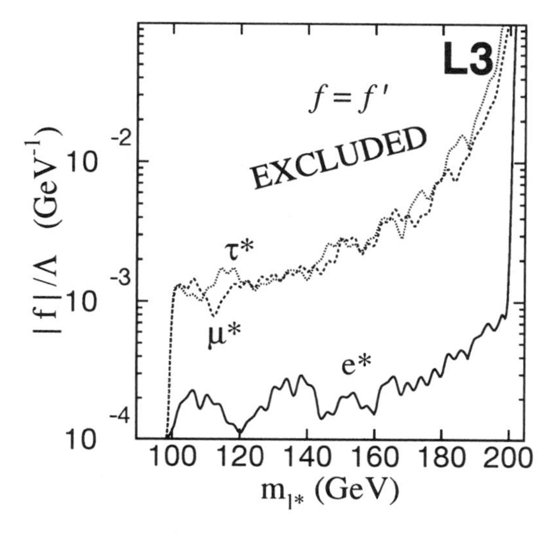

Figure 1. L3 95% confidence level upper limit on the coupling constant f/Λ, as a function of the excited lepton mass for e*, μ^* and τ^* with $f = f'$.

range from 10^{-4} GeV^{-1} to 1 GeV^{-1} as a function of the mass of the excited lepton, up to $m_* \simeq \sqrt{s}$. In Figure 1 the L3 limits on f/Λ for the case of charged excited leptons are shown. Similar limits are obtained for excited neutrinos. Upper limits to the coupling constant are obtained assuming different scenarios, $f = f'$ and $f = -f'$, and also independently from the coupling scenario.

Excited electrons are searched for indirectly, measuring the $e^+e^- \rightarrow \gamma\gamma$ differential cross section, by Aleph, L3 ($\sqrt{s} \leq 189$ GeV), Delphi and Opal ($\sqrt{s} \leq 202$ GeV), They find no significant deviation from the Standard Model. Upper limits are obtained on the coupling constant f/Λ, in the $f = f'$ scenario, in the range from 10^{-3} GeV^{-1} to 10^{-2} GeV^{-1} as a function of the mass of the excited electron, up to $m_* \simeq 400$ GeV.

2 Leptoquarks

Quarks and leptons show an apparent symmetry with respect to the family and multiplet structure of electroweak interactions. This symmetry is realised in some theories beyond the Standard Model, which predict the existence of new bosonic fields, called leptoquarks (LQs), mediating interactions between quarks and leptons [3].

These particles carry baryon and lepton number and therefore decay into a lepton and a quark. They have color, electric charge and weak isospin, and couple to the standard gauge bosons. Assuming dimensionless interactions with Standard Model fermions and gauge invariance, there could be 9 scalar states, S, and 9 vector states, V, grouped in two singlets, two doublets and a triplet of isospin. Only the mass, $m_{\rm LQ}$, and the coupling to fermions, λ, remain as free parameters. Leptoquarks are searched for at the LEP e^+e^- collider [1].

Pair production, $e^+e^- \rightarrow$ LQ LQ, proceeds mainly through s-channel Z/γ exchange, Figure 2. In the case of scalar leptoquarks the coupling to gauge bosons is determined by their quantum numbers, and the cross section depends only on the centre of mass energy, $\sqrt{s}$, and the mass of the leptoquark, $m_{\rm LQ}$. In the case of single production of leptoquarks, $e^+e^- \rightarrow$ e q LQ, Figure 2, the cross section also depends on the coupling between leptoquark, electron and quark. At LEP, first generation leptoquarks (coupling to fermions of the first generation) could be produced singly with mass close to the centre of mass energy, $m_{\rm LQ} \simeq \sqrt{s}$.

Once a leptoquark has been produced, it decays immediately into a lepton and a quark (provided that $\lambda > \mathcal{O}(10^{-6})$).

Pair produced leptoquarks are searched for by Delphi, L3 ($\sqrt{s} \simeq 91$ GeV) and Opal ($\sqrt{s} \leq 202$ GeV) in the three possible experimental signatures: *jet jet* ℓ ℓ, *jet jet* ℓ ν and *jet jet* ν ν. First generation singly produced leptoquarks are searched for by Aleph, L3 ($\sqrt{s} \simeq 91$ GeV), Opal ($\sqrt{s} \leq 189$ GeV) and Delphi ($\sqrt{s} \leq 202$ GeV), in the two possible experimental signatures: *jet* ℓ and *jet* (monojet). Note that in single production the leptoquark is produced in association with an electron and a quark which are lost in the

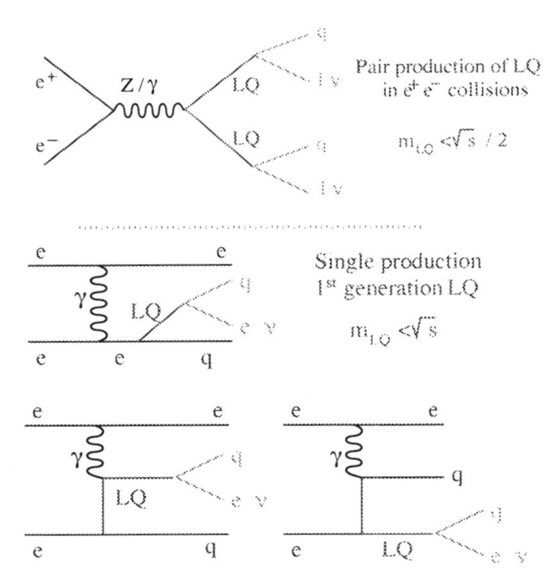

Figure 2. Feynman diagrams for leptoquark production in e^+e^- interactions.

beam pipe, and only the decay products of the leptoquark may be detected.

In all selections, the number of observed events in data is consistent with the Standard Model expectation. From pair production searches, lower mass limits on scalar leptoquarks are obtained by Opal, as shown in Table 2. From single production searches upper limits are obtained on the coupling constant λ in the range from 0.05 to 1.0 as a function of m_{LQ} up to $m_{\mathrm{LQ}} \simeq \sqrt{s}$.

Table 2. Opal 95% confidence level lower mass limits for scalar leptoquarks (GeV).

LQ	1st gen.	2nd gen.	3rd gen.
S_0	44.2	44.2	41.4
$\tilde{S}_0$	94.7	96.4	95.2
$S_{1/2}$	88.5	88.9	89.4
	95.6	97.0	96.1
$\tilde{S}_{1/2}$	83.3	83.3	83.3
	90.5	94.6	93.1
S_1	91.7	91.7	91.7
	44.2	44.2	43.2
	96.0	97.4	96.5

The t-channel exchange of a virtual leptoquark can contribute to hadron production, $e^+e^- \to q\bar{q}$. This process is sensitive to the existence of leptoquarks with masses above the kinematical limit for single production. Aleph, Delphi, Opal ($\sqrt{s} \leq 189$ GeV) and L3 ($\sqrt{s} \leq 202$ GeV), search for leptoquarks indirectly, measuring the $e^+e^- \to q\bar{q}$ differential cross section. They find no significant deviation from the Standard Model. Assuming $\lambda = \sqrt{4\pi\alpha}$, Aleph and L3 lower bounds on leptoquark masses are shown in Table 3.

Table 3. L3 and Aleph 95% confidence level lower mass limits for leptoquarks (GeV).

Leptoquark	L3	ALEPH
S_0(L/R)	413 / 322	380 / 56
$\tilde{S}_0$(R)	84	128
$S_{1/2}$(L/R)	64 / 117	120 / 99
S_1(L)	208	319
V_0(L/R)	584 / 136	618 / 137
$\tilde{V}_0$(R)	288	331
$V_{1/2}$(L/R)	202 / 183	144 / 169
$\tilde{V}_{1/2}$(L)	145	105
V_1(L)	394	515

References

1. Opal Collab. G. Abbiendi *et al.*, Abstracts 127, 128, 151, 185, 199, 215, 236, 237 and 238;
   L3 Collab. M. Acciarri *et al.*, Abstracts 498, 528, 530;
   Delphi Collab. P. Abreu *et al.*, Abstracts 192, 368, 454, 627;
   Aleph Collab. R. Barate *et al.*, Abstract 654;
   Contributed Papers to this Conference and references therein.
2. K. Hagiwara *et al.*, Z. Phys. **C29** (1985) 115.
3. M. Tanabashi, Eur. Phys. J. **C15** (2000) 294.

W PRODUCTION AND SEARCH FOR TOP AT HERA

A. MEHTA
Oliver Lodge Laboratory, University of Liverpool Liverpool L69 7ZE, England
E-mail: mehta@mail.desy.de

Investigations of W production and events with missing transverse momentum and an isolated lepton are performed in ep collisions at HERA. The observed events are compared with the expectations from the Standard Model and are used to set limits on the reaction $ep \to etX$ and on a possible flavour changing neutral current $tu\gamma$ coupling.

1 W production

The HERA collaborations H1 and ZEUS have investigated events which fulfil the topology of a W produced in ep collisons ($ep \to eWX$) at centre of mass energy ≈ 300 GeV[1,2,3]. In the analyses reported here only the electron and muon decay of the W are discussed but see[4] for a study on the τ and hadronic decay channels. W candidates are separated from the very large background by searching for events with missing transverse momentum and an isolated electron or muon.

ZEUS performs a search for events with isolated tracks in association with missing transverse momentum, followed by lepton identification. H1 performs a similar search but applies extra cuts against Standard Model processes other than W production. The advantage of the ZEUS selection is that a simpler set of cuts is employed. It does mean, however, that there is a higher expectation from non-W processes when compared to the H1 selection. The other major difference between the selections is that H1 searches in a larger range of polar angle[a] of the decay lepton, $5° < \theta_l < 145°$, compared to ZEUS, $17.2° < \theta_l < 114.6°$.

ZEUS observes 11 events in 82 pb^{-1} of data in good agreement with the Standard Model expectation of 9.8. Of these 11 events 3 were observed in the e^+p data from 1994-1997[5] and the remainder were recorded in more recent data. The breakdown of the events into each lepton channel and the different running periods is listed in table 1.

H1 observes 14 events in the 82 pb^{-1} of e^+p scattering data compared to a Standard Model expectation of 8.2. No events were seen in the 14 pb^{-1} of e^-p data. The observed excess above Standard Model expectations becomes more striking at large values of transverse momentum of the hadronic system P_T^X as can be seen in table 2. In the region $P_T^X > 25$ GeV 9 events are seen when only 2.26 are expected from the Standard Model. Of these 9 events 5 were first seen in the e^+p data from 1994-1997[6] and the remainder were recorded in more recent data.

Various distributions are compared for the selected events. Two examples are shown here. In Fig. 1 the distribution in P_T^X of ZEUS events from the muon channel are compared to the Standard Model expectation. As can be seen the 4 muons events lie at low values in good agreement with the Standard Model expectation, which is dominated by the two photon process ($ep \to e\mu^+\mu^-X$). All other distributions of the ZEUS events show good agreement with the Standard Model. In Fig. 2 the transverse mass of the neutrino-lepton system M_T is shown for the H1 events. The events generally have large values of M_T that are centred about the W mass. Although there is a difference in normalisation, the shape of the distribution is well described by the Monte Carlo prediction. The shapes of all other distributions of the H1 events, be-

[a] Polar angle is defined in the LAB frame with the proton beam direction at $\theta = 0$.

ZEUS	Electrons observed/expected (W)	Muons observed/expected (W)
94-97 e^+p 48 pb^{-1} published	3 / 3.5±0.7 (0.9)	0 / 2.0±0.4 (0.4)
98-99 e^-p 16 pb^{-1} preliminary	2 / 0.8±0.4 (0.5)	0 / 0.8±0.1 (0.2)
99 e^+p 18 pb^{-1} preliminary	2 / 1.8±0.4 (0.5)	4 / 0.9±0.1 (0.2)
Total 82 pb^{-1}	7 / 6.1±0.9 (1.9)	4 / 3.7±0.4 (0.8)

Table 1. The number of observed events from ZEUS compared to the Standard Model prediction and its W component for each data taking period.

Electron and Muon	H1 Prelim. Data	SM expectation	W	Other SM processes
$P_T^X > 0$ GeV	14	8.16 ± 1.97	6.36 ± 1.91	1.80 ± 0.46
$P_T^X > 12$ GeV	12	4.07 ± 1.03	3.30 ± 0.99	0.77 ± 0.27
$P_T^X > 25$ GeV	9	2.26 ± 0.57	1.83 ± 0.55	0.43 ± 0.15
$P_T^X > 40$ GeV	6	0.79 ± 0.22	0.74 ± 0.22	0.05 ± 0.03

Table 2. H1 observed and predicted event rates in the electron and muon decay channels combined for all e^+p data (82 pb^{-1}). Only the electron channel contributes for $P_T^X < 12$ GeV.

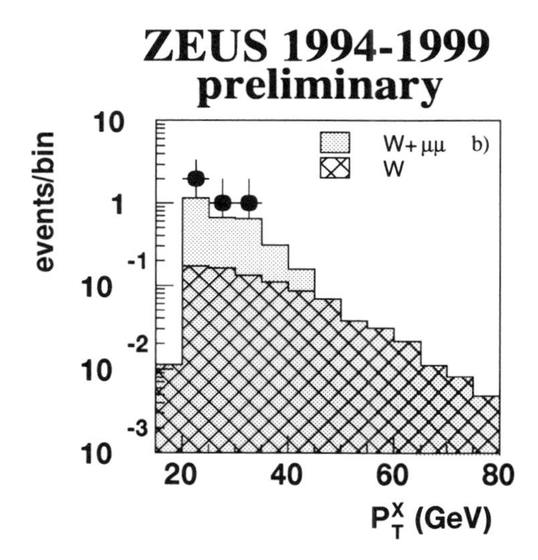

Figure 1. Distribution of P_T^X of the ZEUS muon events, compared with the Standard Model expectation.

sides that of P_T^X, show good agreement with the expectation.

Since the analysis cuts that are used by the two collaborations are rather different, the Standard Model expectations are also different. In order to facilitate a clearer comparison between the experiments ZEUS repeated their analysis with some extra cuts against non-W processes and for $P_T^X > 25$ GeV. H1 repeated their analysis but restricted the measurement to range of lepton polar angle of the ZEUS analysis ($17.2° < \theta_l < 114.6°$). ZEUS sees 1 event for a Standard Model expectation of 1.6. H1 sees 9 events compared to a Standard Model expectation 1.8. In both experiments the Standard Model is dominated by W production.

From the above results it is clear that the differences in event yield between ZEUS and H1 cannot be put down to differences in experimental technique. The most likely explanation is a statistical fluctuation in one experiment. We must wait until after the luminosity upgrade of HERA, when each experiment is expected to receive 1 fb^{-1} of data, to see whether the events are a product of new physics or merely a statistical fluctuation.

2 Top production

Since top decays dominantly via $t \to bW$, the W results may also be used to set lim-

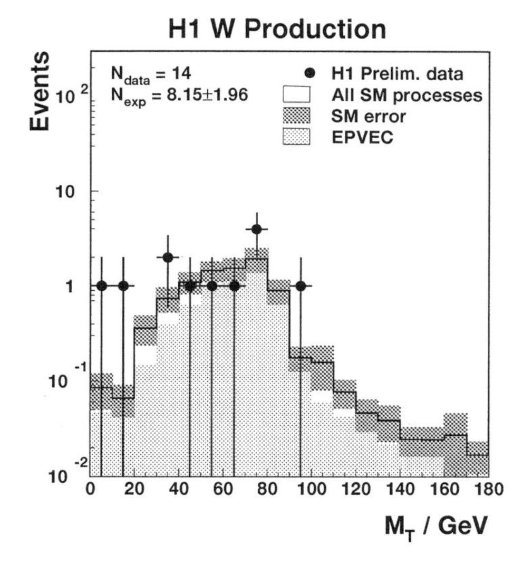

Figure 2. Distribution of M_T of the H1 events, compared with the Standard Model expectation.

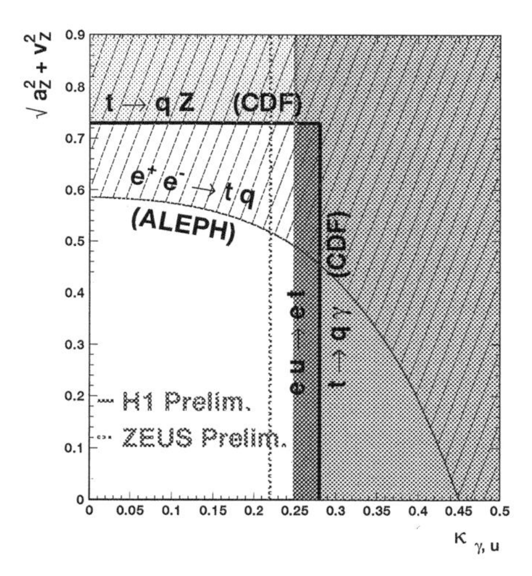

Figure 3. Limits on the anomolous $tu\gamma$ (magnetic) coupling $\kappa_{\gamma,u}$ and on the anomolous tuZ (vector) coupling $\sqrt{a_Z^2 + v_Z^2}$ obtained at the Tevatron (CDF), LEP (ALEPH) and HERA (H1 and ZEUS).

its on top production at HERA. In the Standard Model the cross section is extremely low, but certain extensions predict anomalous flavour changing neutral current coupling $\kappa_{\gamma,u}$ [7], which could result in an observable rate of single top production at HERA. The experiments select events which fulfil the W requirements and in addition have $P_T^X > 25(40)$ GeV for H1 (ZEUS). H1 applies some extra cuts to further reject Standard Model W production. H1 also searches in the hadronic decay of the W by selecting events with 3 or more jets any two of which are close in mass to M_W.

No clear evidence for top production was found, although H1 sees an excess of 5 data events in the leptonic decay channel compared with an expectation from the Standard Model of 1.2. ZEUS sees 0 events with an expectation of 0.65 from the Standard Model. In the hadronic decay channel H1 sees 10 events compared with 8.3 expected from the Standard Model. These results are used to place limits on $\kappa_{\gamma,u}$, which are shown in Fig.3. As can be seen the HERA results are very competitive with those from other colliders.

References

1. H1 Collab., "W production in ep collisions at HERA", abstract #974 submitted to ICHEP2000 Osaka, Japan.
2. ZEUS Collab., J. Breitweg et al., "Search for events with isolated high-energy leptons and missing transverse momentum at HERA", abstract #1041 submitted to ICHEP2000 Osaka, Japan.
3. H1 Collab., "Search for Single Top Production in $e^{\pm}$ collsions at HERA", abstract #961 submitted to ICHEP2000 Osaka, Japan.
4. H1 Collab., "W production in $e^{\pm}p$ collisions at HERA", prepared for Lepton Photon 99, Stanford, USA.
5. ZEUS Collab., J. Breitweg et al., *Phys. Lett.* **B 471** (2000) 411.
6. H1 Collab., C. Adloff et al., *Eur. Phys. J.* **C5** (1998) 575.
7. T. Han and J.L. Hewitt, *Phys. Rev.* **D 60** (1999) 074015;
   H. Fritzsch and D. Holtmannspötter, *Phys. Lett.* **B 457** (1999) 186.

MINI-REVIEW ON EXTRA DIMENSIONS

GREG LANDSBERG
Brown University, Department of Physics, 182 Hope St., Providence, RI 02912, USA
E-mail: landsberg@hep.brown.edu

One of the most stimulating recent ideas in particle physics involves a possibility that our universe has additional compactified spatial dimensions, perhaps as large as 1 mm. In this mini-review, we discuss the results of recent experimental searches for such large extra dimensions.

The possibility that the universe has more than three spatial dimensions has been a long-discussed issue.[1] Developments in string theory suggest that there could be up to $n = 7$ additional dimensions, compactified at very small distances on the order of 10^{-32} m. In a new model,[2] inspired by string theory, several of the compactified extra dimensions are suggested to be as large as 1 mm. These large extra dimensions (LED) are introduced to solve the hierarchy problem of the standard model (SM) by lowering the Planck scale, $M_{\rm Pl}$, to the TeV energy range. We refer to this effective Planck scale as M_S.

Since Newton's law of gravity is modified in the presence of compactified extra dimensions for interaction distances below the size of the LED, current gravitational observations rule out possibility of only a single LED. Recent preliminary results from gravity experiments at submillimeter distances,[3] as well as cosmological constraints,[4] indicate that the case of $n = 2$ is likely ruled out as well. However, for $n \geq 3$, the size of the LED becomes microscopic and therefore eludes the reach of direct gravitational measurements or cosmological constraints. However, high energy colliders, capable of probing very short distances, can provide crucial tests of the LED hypothesis, in which effects of gravity are enhanced at high energies due to accessibility of numerous excited states of graviton, corresponding to multiple winding modes of the graviton field around the compactified dimensions.

LED phenomenology at colliders has already been studied in detail.[5,6,7,8] One of the primary observable effects is an apparent non-conservation of momentum caused by the direct emission of gravitons, which leave the three flat spatial dimensions. A typical signature is the production of a single jet or a vector boson at large transverse momentum. The other observable effect is the anomalous production of fermion-antifermion or diboson pairs with large invariant mass stemming from the coupling to virtual gravitons. Direct graviton emission is expected to be suppressed by a factor $(1/M_S)^{n+2}$, while virtual graviton effects depend only weakly on the number of extra dimensions.[5,6,8] Virtual graviton production therefore offers a potentially more sensitive way to search for manifestations of LED.[a]

The effects of direct graviton emission, including production of single photons or Z's, were sought at LEP.[11,12,13] The following signatures were used: $\gamma \not\!\!E_T$ or $Z(jj)\not\!\!E_T$, where $\not\!\!E_T$ is the missing transverse energy in the detector, and j stands for jet. The former topology is typical of searches for GMSB supersymmetry, and the latter of searches for "invisible" Higgs. The negative results of these searches can be expressed in terms of limits on the effective Planck scale, as summarized in Table 1. Both the CDF and DØ

[a] Strictly speaking, virtual graviton effects are sensitive to the ultraviolet cutoff required to keep the divergent sum over the graviton modes finite.[5,6,8] This cutoff is expected to be of the order of the effective Planck scale. Dependence on the value of the cutoff is discussed, e.g., in Refs.[9,10]

Collaborations at the Fermilab Tevatron are also looking for direct graviton emission in the "monojet" ($j\not{E}_T$) channel, which is quite challenging due to large instrumental background from jet mismeasurement and cosmic rays. Although no results have been reported as yet, the sensitivity of these searches is expected to be similar to those at LEP.[17].

While the formalism for calculating direct graviton emission is well established, different formalisms have been used to describe virtual graviton effects.[5,6,8] Since difermion or diboson production via virtual graviton exchange can interfere with the SM production of the same final state particles, the cross section in the presence of LED is given by:[5,6,8] $\sigma = \sigma_{\rm SM} + \sigma_{\rm int}\eta_G + \sigma_G\ \eta_G^2$, where $\sigma_{\rm SM}$, $\sigma_{\rm int}$, and σ_G denote the SM, interference, and graviton terms, and the effects of ED are parametrized via a single variable $\eta_G = \mathcal{F}/M_S^4$, where $\mathcal{F}$ is a dimensionless parameter of order unity, reflecting the dependence of virtual graviton coupling on the number of extra dimensions. Several definitions exist for $\mathcal{F}$:

$$\mathcal{F} = 1,\ (\text{GRW}^5);$$

$$\mathcal{F} = \begin{cases} \log\left(\frac{M_S^2}{M}\right), & n = 2 \\ \frac{2}{n-2}, & n > 2 \end{cases},\ (\text{HLZ}^6);$$

$$\mathcal{F} = \frac{2\lambda}{\pi} = \pm\frac{2}{\pi},\ (\text{Hewett}^8).$$

Here, λ is a dimensionless parameter of order unity, conventionally set to be either $+1$ or -1 in cross section calculations within Hewett's formalism. Only the HLZ formalism has $\mathcal{F}$ depending explicitly on n.

Because different experiments have set limits on virtual graviton exchange using different formalisms, it is worthwhile to specify relationship between the three definitions of effective Planck scale, referred to as Λ_T, after the original[5] notation, M_S(Hewett), and M_S(HLZ):

$$M_S(\text{Hewett})\,|_{\lambda=+1} = \sqrt[4]{\frac{2}{\pi}}M_S(\text{HLZ})\,|_{n=4}$$

$$\Lambda_T = M_S(\text{HLZ})\,|_{n=4}\ .$$

Unless noted otherwise, we will express limits on the effective Planck scale in terms of M_S(Hewett), and they all will be given at 95% CL.

Among the many difermion and diboson final states tested for presence of virtual graviton effects at LEP,[11,13,14,15,16] the most sensitive channels involve the dielectron (both Drell-Yan and Bhabha scattering) and diphoton ppocesses.[b] None of the experiments see any significant deviation from the SM in the analyzed channels. This is translated into the limits on M_S(Hewett), listed in Table 2. They are of the order of 1 TeV for both signs of the interference term.

Virtual graviton effects have also been sought at HERA in the t-channel of $e^\pm p \to e^\pm p$ scattering, similar to Bhabha scattering at LEP.[5,8] A search carried out by the H1 Collaboration[19] with 82 pb^{-1} of e^+p and 15 pb^{-1} of e^-p data, have set limits on M_S between 0.5 and 0.8 TeV (see Table 3). Although these limits are somewhat inferior to those from LEP, the ultimate sensitivity of HERA at the end of the next run is expected to be similar to that at LEP.

Recently, the DØ Collaboration reported the first search for virtual graviton effects at a hadron collider,[20] based on the analysis of a two-dimensional distribution in the invariant mass and scattering angle of dielectron or diphoton systems, as suggested in Ref.[10] The results, corresponding to 127 pb^{-1} of data collected at $\sqrt{s} = 1.8$ TeV, agree well with the SM predictions, and provided the limits on the effective Planck scale, shown in Table 4 for all three formalisms.[5,6,8] These lim-

[b]Recent preliminary results from L3 at $\sqrt{s} > 200$ GeV indicate that the best sensitivity is found in the ZZ channel,[15] but details of the experimental analysis are not yet available. These results differ from those of an earlier L3 publication,[13] where the sensitivity in the ZZ channel at $\sqrt{s} = 189$ GeV was significantly lower than that in the $\gamma\gamma$ channel, as well as from recent OPAL results in the ZZ channel at the highest LEP energies,[16] consistent with Ref.[13] It may therefore be prudent to await final results from L3 on this issue.

Table 1. Lower limits at the 95% CL on the effective Planck scale, M_S(Hewett), in TeV, from searches for direct graviton production at LEP. Limits from $\sqrt{s} > 200$ GeV data are shown in normal font; limits from 189 GeV data are in *italics*; limits from 184 GeV data are in **bold** script.

Experiment	$e^+e^- \to \gamma G_{\rm KK}$					$e^+e^- \to Z G_{\rm KK}$				
	n=2	n=3	n=4	n=5	n=6	n=2	n=3	n=4	n=5	n=6
ALEPH	1.10	0.86	0.70	0.60	0.52	**0.35**	**0.22**	**0.17**	**0.14**	**0.12**
DELPHI	1.25	0.97	0.79	0.68	0.59	N/A	N/A	N/A	N/A	N/A
L3	*1.02*	*0.81*	*0.67*	*0.58*	*0.51*	*0.60*	*0.38*	*0.29*	*0.24*	*0.21*
OPAL	*1.09*	*0.86*	*0.71*	*0.61*	*0.53*	N/A	N/A	N/A	N/A	N/A

Table 2. Lower limits at the 95% CL on the effective Planck scale, M_S(Hewett), in TeV, from searches for virtual graviton effect at LEP. Upper (lower) rows correspond to $\lambda = +1$ ($\lambda = -1$). The ALEPH Collaboration used a different formalism for their analysis,[5] so their limits were translated into Hewett's formalism.[8] The L3 Collaboration used formalism[18] for diboson production,[13,15] in which the sign of λ is reversed, compared to Hewett.[8] To correct for that, we reverse the sign of λ when quoting the L3 limits in the $\gamma\gamma$, WW, and ZZ channels. Combined L3 limits are nevertheless affected by the mixture of two signs of λ in difermion and diboson channels. (See also footnote on the previous page for a discussion of the ZZ results.) Limits from $\sqrt{s} > 200$ GeV data are shown in normal font; limits from 189 GeV data are in *italics*; limits from 184 GeV data are in **bold** script. Some of the older limits obtained within the formalism[5] before an important revision was made, are not directly comparable with the results at the highest LEP energies.

Experiment	e^+e^-	$\mu^+\mu^-$	$\tau^+\tau^-$	$q\bar{q}$ ($b\bar{b}$)	$f\bar{f}$	$\gamma\gamma$	WW	ZZ	Combined
ALEPH	0.81	0.67	0.62	0.57 (0.44)	0.84	0.82	N/A	N/A	*1.00*
	1.05	0.65	0.60	0.53 (0.44)	1.05	0.81	N/A	N/A	*0.75*
DELPHI	N/A	0.73	0.65	N/A (N/A)	0.76	0.71	N/A	N/A	N/A
	N/A	0.59	0.56	N/A (N/A)	0.60	0.69	N/A	N/A	N/A
L3	0.99	*0.69*	*0.54*	**0.49** (N/A)	*1.00*	*0.80*	*0.68*	1.2	1.3
	0.91	*0.56*	*0.58*	**0.49** (N/A)	*0.84*	*0.79*	*0.79*	1.2	1.2
OPAL	N/A	*0.60*	*0.63*	N/A (N/A)	*0.68*	0.82	N/A	0.80	0.90
	N/A	*0.63*	*0.50*	N/A (N/A)	*0.61*	0.85	N/A	0.59	0.83

its are similar to and complementary to those from LEP, as different energy regimes are probed at the two colliders. A similar analysis in the dielectron channel is being pursued by the CDF Collaboration,[21] but no results have yet been reported. As the current Tevatron sensitivity is limited by statistics, rather than machine energy, we expect combined Tevatron limits to yield an improvement over the currently excluded range of M_S.

Although no evidence for LED has been found so far, we are looking forward to the next generation of collider experiments to shed more light on the mystery of large extra dimensions. The sensitivity of the upgraded Tevatron experiments in the next run is expected to double (2 fb^{-1}) or even triple (15 fb^{-1}), which offers a unique opportunity to see LED effects in the next 5 years. The ultimate test of the theory of large extra dimensions will become possible at the LHC, where effective Planck scales as high as 10 TeV will be able to be probed.

Table 3. Lower limits at the 95% CL on the effective Planck scale, M_SHewett, in TeV, from the H1 experiment.[19] The limits have been translated into Hewett's formalism[8] from the original formalism[5] used in the H1 analysis.

H1	e^+p	e^-p	Combined
$\lambda = +1$	0.45	0.61	0.56
$\lambda = -1$	0.79	0.43	0.83

Table 4. Lower limits at 95% CL on the effective Planck scale, M_S, in TeV, from the DØ experiment.[20]

GRW[5]	HLZ[6]						Hewett[8]	
	n=2	n=3	n=4	n=5	n=6	n=7	$\lambda = +1$	$\lambda = -1$
1.21	1.37	1.44	1.21	1.10	1.02	0.97	1.08	1.01

References

1. G.F.B. Riemann, *"Über die Hypothesen, welche der Geometrie zu Grunde liegen,"* Abh. Ges. Wiss. Gött. **13**, 1 (1868).
2. N. Arkani-Hamed, S. Dimopoulos, and G. Dvali, Phys. Lett. B**429**, 263 (1998); I. Antoniadis, N. Arkani-Hamed, S. Dimopoulos, and G. Dvali, Phys.Lett. B**436**, 257 (1998).
3. E. Adelberger, APS Meeting, Bulletin Am. Phys. Soc., Long Beach, April 2000.
4. S. Cullen and M. Perelstein, Phys. Rev. Lett. **83**, 268 (1999); L. Hall and D. Smith, Phys. Rev. D **60**, 085008 (1999); C. Hanhart *et al.*, nucl-th/0007016.
5. G. Giudice, R. Rattazzi, and J. Wells, Nucl. Phys. **B544**, 3 (1999) and revised version 2, e-print hep-ph/9811291.
6. T. Han, J.D. Lykken, and R.-J. Zhang, Phys. Rev. D **59**, 105006 (1999) and revised version 4, e-print hep-ph/9811350.
7. E.A. Mirabelli, M. Perelstein, and M.E. Peskin, Phys. Rev. Lett. **82**, 2236 (1999).
8. J.L. Hewett, Phys. Rev. Lett. **82**, 4765 (1999).
9. S. Nussinov and R. Shrock, Phye. Rev. D **59**, 105002 (1999).
10. K. Cheung and G. Landsberg, Phys. Rev. D **62**, 076003 (2000).
11. R. Barate *et al.* (ALEPH Collaboration), ALEPH Note CONF-99-027; *ibid.* 2000-008; M. Acciarri *et al.* (L3 Collaboration), Phys. Lett. B**470**, 268 (1999).
12. P. Abreu *et al.* (DELPHI Collaboration), CERN-EP-2000-021, to appear in Eur. Phys. J C; E. Anashkin *et al.* (DELPHI Collaboration), DELPHI Note 2000-094; G. Abbiendi *et al.* (OPAL Collaboration), hep-ex/0005002, submitted to Eur. Phys. J. C.
13. M. Acciarri *et al.* (L3 Collaboration), Phys. Lett. B**464**, 135; *ibid.* **470**, 281 (1999).
14. R. Barate *et al.* (ALEPH Collaboration), ALEPH Note 2000-047; P. Abreu *et al.* (DELPHI Collaboration), Phys. Lett. B**485**, 45 (2000); CERN-EP-2000-094, submitted to Phys. Lett. B; A. Behrmann *et al.* (DELPHI Collaboration), DELPHI Note 2000-036; *ibid.* 2000-128; M. Acciarri *et al.* (L3 Collaboration), L3 Note 2543 (2000); G. Abbiendi *et al.* (OPAL Collaboration), Phys. Lett. B**465**, 303 (1999); Eur. Phys. J. C**13**, 553 (2000).
15. M. Acciarri *et al.* (L3 Collaboration), L3 Note 2579 (2000); *ibid.* 2590 (2000).
16. G. Abbiendi *et al.* (OPAL Collaboration), OPAL Note PN 440 (2000).
17. M. Spiropulu, private communication.
18. K. Agashe and N.G. Deshpande, Phys. Lett. B**456**, 60 (1999).
19. C. Adloff *et al.* (H1 Collaboration), Phys. Lett. B**479**, 358 (2000); C. Niebuhr (H1 Collaboration), this proceedings.
20. B. Abbott *et al.* (DØ Collaboration), hep-ex/0008065, submitted to Phys. Rev. Lett.
21. D.W. Gerdes, to appear in Proc. of the XXXVth Recontres de Moriond, Les Arcs, March 18–25, 2000.

GENERAL PARTICLE SEARCHES AT LEP

WOLFANG LOHMANN
DESY Zeuthen and CERN,
E-mail: wlo@ifh.de

Electron-positron annihilations at highest LEP energies are used to search for contact interactions, additional heavy neutral gauge bosons Z′, neutral and charged heavy leptons and technicolour particles. No signals are found. Limits on new physics scales, masses and couplings are given.

1 Introduction

The LEP accelerator delivered e^+e^- annihilation data in a new energy domain exceeding $\sqrt{s}$= 200 GeV in 1999. Manifestations of contact interactions and additional neutral heavy gauge bosons are searched for by investigating the e^+e^- cross sections into leptons and hadrons. Heavy leptons can be produced directly pairwise or due to mixing with light neutrinos singly. Technicolor particles produced in pairs or associated with longitudinal polarised W bosons or photons are searched for. Also virtual effects on the e^+e^- annihilation cross section into $\mu^+\mu^-$ and W^+W^- are considered. Limits given throughout the paper are set at 95% C.L..

2 Contact Interactions

Contact interactions [1] occur if leptons and quarks are bound states of more fundamental constitutents. They alter the e^+e^- cross section well below the binding energy Λ. The most general ansatz for the Lagrangian is:

$$\mathcal{L} \sim \eta_{(L,R)} \frac{g^2}{\Lambda^2} \sum_{L,R} (\bar{e}\gamma^\nu e)(\bar{f}\gamma_\nu f),$$

where e and f are the chiral projections of the fermion spinors. The parameter $\eta_{(L,R)}$ is chosen to be $\pm$ 1 for a certain chirality combination of e and f. The cross section, $\sigma_{\mathrm{f}\bar{\mathrm{f}}}$, for $e^+e^- \to \mathrm{f}\bar{\mathrm{f}}(\gamma)$ results as a superposition of Standard Model (SM) and contact interactions contributions. Fixing $\frac{g^2}{4\pi} = 1$, Λ is the only free parameter.

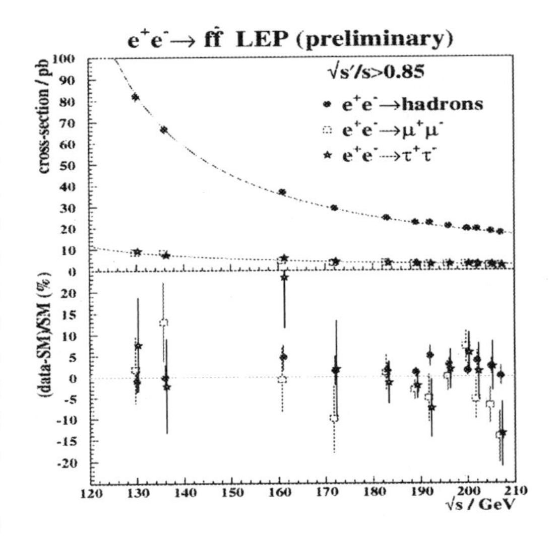

Figure 1. The dependence of $\sigma_{\mathrm{f}\bar{\mathrm{f}}}$ on $\sqrt{s}$ for several final states combined for the LEP experiments.

The search was done by all LEP experiments for $e^+e^- \to \mathrm{f}\bar{\mathrm{f}}(\gamma)$, where f is e, μ, τ or q up to $\sqrt{s}$= 202 GeV [2]. In addition, the processes $e^+e^- \to c\bar{c}(\gamma)$ and $e^+e^- \to b\bar{b}(\gamma)$, where the quarks are tagged by dedicated methods, are probed separately. Since there are no significant deviations from the SM, limits on Λ are derived. As an example, the lowest and highest limits obtained in the single experiments for leptonic and hadronic final states and the corresponding model are given in Table 1.

The dependence of $\sigma_{\mathrm{f}\bar{\mathrm{f}}}$ on $\sqrt{s}$ for several final states combined from all LEP experiments is shown in Figure 1. The solid lines are the predictions of the SM. The results for Λ obtained from the combined data for $\sigma_{\mathrm{f}\bar{\mathrm{f}}}$

Table 1. The values of the highest and lowest limits on Λ obtained in the single experiments for different models and channels. The $q\bar{q}$ final state was also investigated assuming contact interactions for u- or d-quarks only.

final state	highest		lowest	
	model	Λ (TeV)	model	Λ (TeV)
$\ell^+\ell^-$	VV	17.8	RR	5.9
$q\bar{q}$	AA	11.2	RL	2.7
$u\bar{u}$	VV	12.9	RR	1.5
$d\bar{d}$	AA	9.8	RR	1.8
$f\bar{f}$	VV	17.9	RR	5.5

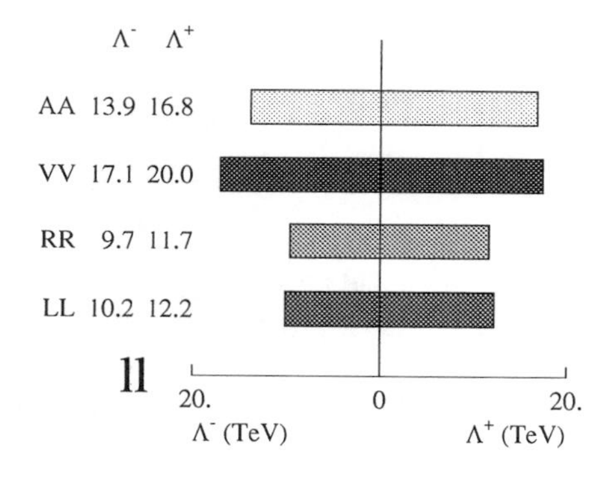

Figure 2. Limits on Λ from leptonic final states combined from all LEP experiments.

and the forward-backward asymmetry, $\mathcal{A}_{\rm FB}$, for leptons and b-quarks are shown in Figures 2 and 3. Slightly lower limits are obtained for c-quarks.

3 Heavy Neutral Gauge Bosons Z′

Additional heavy neutral gauge bosons are expected in many GUTs and most superstring models [3]. In e^+e^- annihilation the lightest can affect the Z lineshape via mixing or contribute to $\sigma_{\rm f\bar{f}}$ and $\mathcal{A}_{\rm FB}$ even if its mass is larger then $\sqrt{s}$. Measurements near the Z bound the mixing angle to very small values, hence it is set in most of the analyses of the high energy data to zero. From the ansatz:

$$\mathcal{L} \sim eA_\mu J^\mu_{(\gamma)} + gZ_\mu J^\mu_{(Z)} + g' Z'_\mu J^\mu_{(Z')}$$

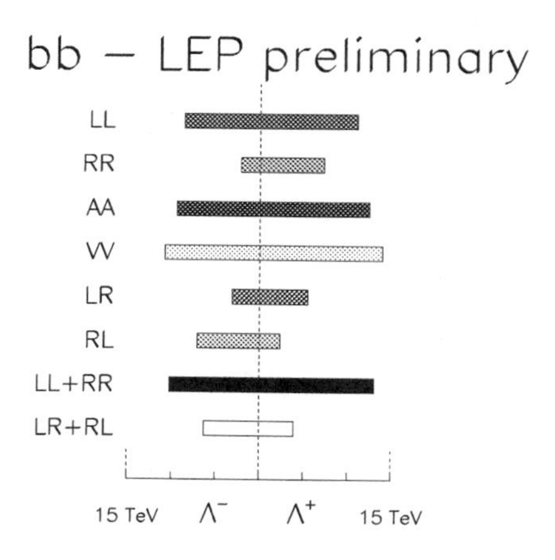

Figure 3. Limits on Λ from $e^+e^- \rightarrow b\bar{b}(\gamma)$ combined from all LEP experiments.

Table 2. Lower limits on $m_{Z'}$ in GeV from the single experiments and LEP combined in several models

model	single experiments	LEP combined
χ	460 - 753	679
ψ	275 - 410	600
η	315 - 486	425
LR	360 - 635	
SM	700 - 1170	

the prediction for $\sigma_{\rm f\bar{f}}$ is derived. It can be compared to the data after having specified the couplings to fermions. The mass of the Z′ is then the only free parameter. This is done using models obtained from E_6 GUTs or the left-right gauge group. In both cases a mixing angle, denoted as Θ_6 or α_{LR}, respectively, appears. Specific choices of Θ_6 are known as χ (Θ_6=0), ψ (Θ_6=$\pi/2$) or η (Θ_6=-arctan$\sqrt{5/3}$) models. A special choice in left-right models is the case $g_L = g_R$. For sake of completeness, also limits for a Z′ with the same couplings as in the SM are determined. The mass limits obtained in the single experiments and from LEP combined data for the specified models are summarised in Table 2.

An alternative approach is done by DELPHI and L3. Using leptonic final states and

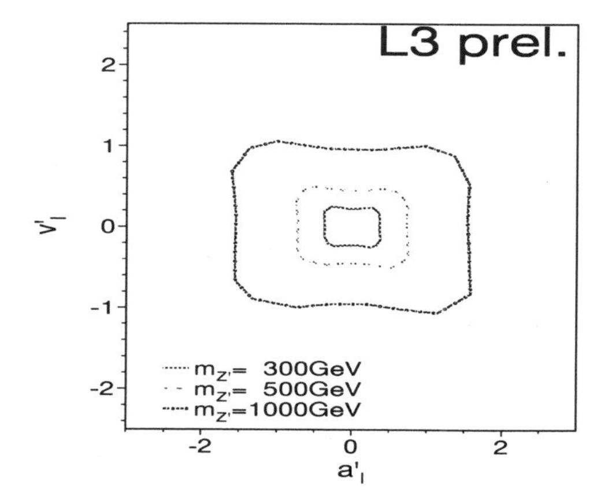

Figure 4. Limits on the couplings of a Z′ to leptons.

assuming lepton universality, limits for the couplings of a Z′ are determined for given $m_{Z'}$. This is illustrated in Figure 4 for L3. Similar results are obtained by DELPHI.

4 New Leptons

New leptons and quarks arise in theories with quark-lepton unification [4]. They are classified as sequential (same quantum numbers as SM fermions), mirror (chiral properties opposite to SM), vector ($g_L = g_R$) and singlet leptons.

Pairwise produced neutral and charged heavy leptons are assumed to decay into known leptons via charged or neutral current processes. Charged current decays are $L^0\bar{L}^0 \rightarrow \ell^+\ell^-W^+W^-$, $\ell = e, \mu, \tau$ and $L^+L^- \rightarrow \nu_e\bar{\nu}_eW^+W^-$. In the first case we search for two isolated leptons of the same family and hadronic jets from the W decays, in the second, depending on the decay of the W's, for one isolated lepton and two jets or four jets and missing momentum [5]. No signal is found and mass limits as summarised in Table 3 are obtained.

OPAL has also searched for the pair production of a new lepton decaying into its own heavy neutrino, $L^+L^- \rightarrow N_L\bar{N}_LW^+W^-$. No signal is seen, and the excluded mass domain is shown in Figure 5.

Table 3. The lower mass limits in GeV obtained for neutral and charged leptons. The ranges cover the results for the different lepton classes. Neutral leptons are treated as of Dirac and Majorana type.

decay	L3		OPAL	
	Dirac	Majorana	Dirac	Majorana
$L^0 \rightarrow e\ W$	96.9 - 97.1	85.7	88	76
$L^0 \rightarrow \mu\ W$	96.7 -99.4	88.0	88.1	76
$L^0 \rightarrow \tau\ W$	89 - 96	79.6	71.1	53.8
$L^\pm \rightarrow \nu_e\ W$	98.7 - 99.1		84.1	

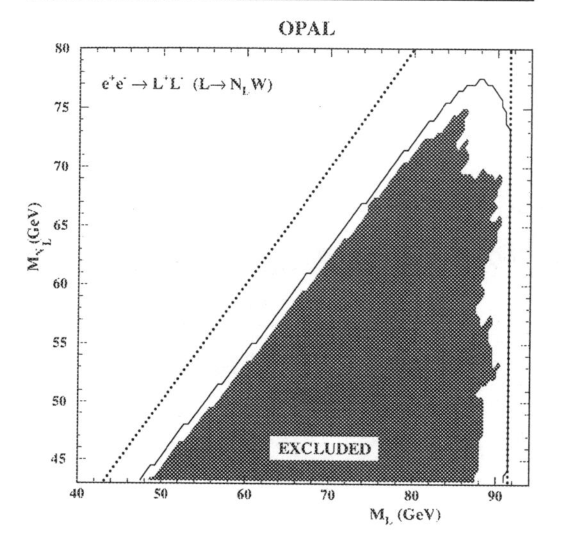

Figure 5. The mass ranges excluded for a new lepton generation.

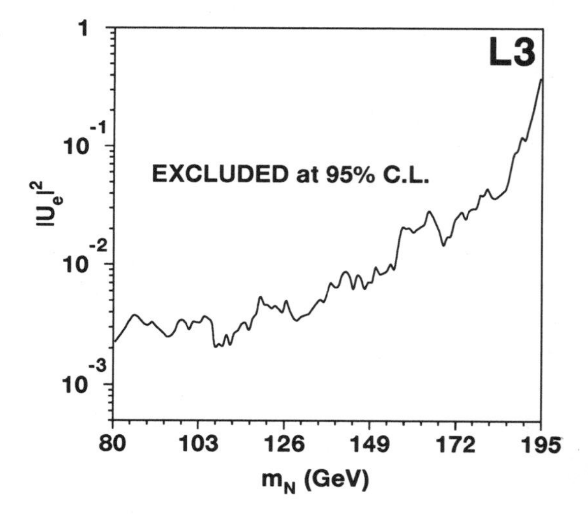

Figure 6. Upper limit on the mixing amplitude as a function of the neutrino mass.

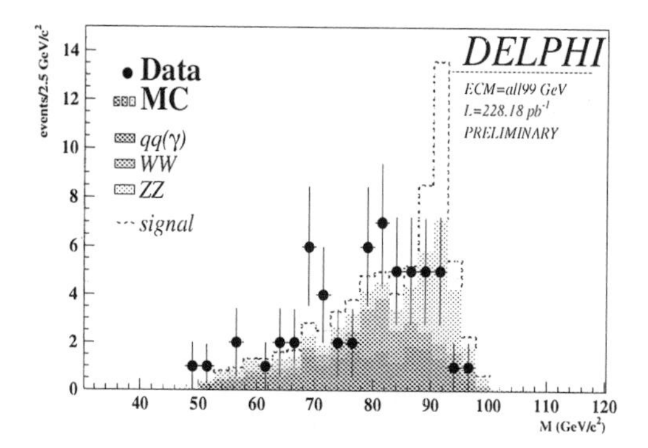

Figure 7. Dijet invariant mass of four jet final states containing b-quarks. A possible signal resulting from a π_T with a mass of 89 GeVis also shown (arbitrary normalisation).

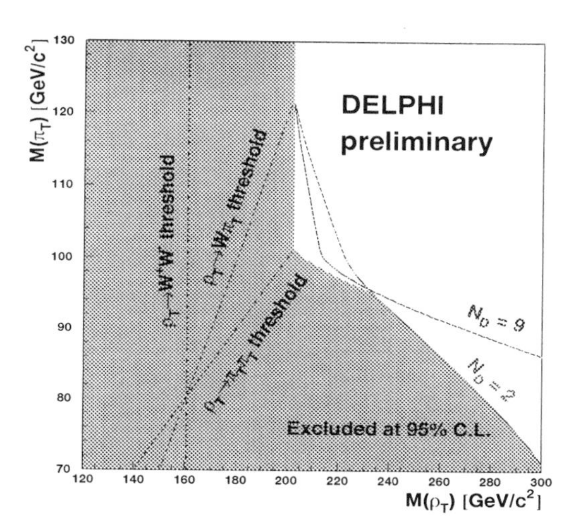

Figure 8. The excluded region in the m_{π_T}, m_{ρ_T} for any $W_L\pi_T$ mixing angle.

L3 made a search for a single heavy neutrino, which may be produced via mixing with a standard neutrino. The decay $N_e \to W^+e^-$ is considered. No signal is found, and limits on the mixing parameter U_e, as shown in Figure 6, are derived as a function of the mass of the heavy neutrino.

5 Technicolour

Technicolour provides electroweak symmetry breaking dynamically by strong interactions of gauge bosons. Recent developments of technicolour models avoid discrepancies with electroweak precision measurements and predict light scalar (π_T) and vector (ρ_T) techniparticles [6]. DELPHI [7] made a detailed study searching for techniparticles in the following channels:

$$\begin{aligned} e^+e^- \to \rho_T^{(*)} &\to \pi_T\pi_T \to \mathrm{b\bar{c}c\bar{b}} \\ &\to W_L\pi_T \to \mathrm{qq'c\bar{b}} \\ &\to W_LW_L \to \text{4jets, 2jets } (\ell\nu),\ 2\ (\ell\nu) \\ &\to \mathrm{f\bar{f}} \\ &\to \pi_T^0\gamma \to \mathrm{b\bar{b}}\gamma \end{aligned}$$

No signal is found. As an example, in Figure 7 the proper dijet mass spectrum for four jet final states containing b-quarks is shown, in which $\rho_T^{(*)} \to (W_L\pi_T), (\pi_T\pi_T)$ decays are expected. From the results of the search in all final states an exclusion plot in the (m_{π_T}, m_{ρ_T}) mass plane as shown in Figure 8 is obtained. N_D is the number of technidoublets, which also governs the mixing angle between W_L and π_T. The exclusion contour is valid for any mixing angle.

Acknowledgments

I thank all LEP collaborations and the 'two-fermion' working group for providing their results. Special thanks to Sabine Riemann for performing the LEP combined Z′ fits and reading carefully this text.

References

1. E.J. Eichten, K.D. Lane and M.E. Peskin, *Phys. Rev. Lett.* **50** 811 (1983).
2. ALEPH 2000-047, CONF 2000-030. DELPHI 2000-128 CONF 427, L3 Note 2543, OPAL Physics Note PN 424.
3. P. Langacker and M. Luo, *Phys. Rev.* D **45**, 278 (1992).
4. see e.g. A. Djouadi *at* al., SLAC-PUB-95-6772, in *Electroweak Symmetry Breaking at the TEV Scale*, Singapore, World Scientific (1997).
5. L3 Note 2550, OPAL Coll. CERN-EP/99-169.
6. see e.g. K. Lane, BUHEP-96-03, hep-ph/9610463 (1996).
7. DELPHI 2000-074 CONF 373.

EXCITED FERMIONS AT HERA

ALI SABETFAKHRI
Representing the ZEUS and H1 Collaborations
Department of Physics, University of Toronto, 60 St. George, Toronto, M5S 1A7, CANADA
E-mail: sabetf@physics.utoronto.ca

Heavy excited states of fermions have been searched for in e^+p collisions at a center-of-mass energy of 300 GeV with the ZEUS and H1 detectors at HERA using an integrated luminosity of 38 pb^{-1}. Excited neutrinos have been searched for in e^-p collisions at a center-of-mass energy of 318 GeV using an integrated luminosity of 16 pb^{-1}. No signal was observed and constraints on such states were set, which complement or extend beyond those obtained at other colliders.

1 Introduction

The existence of the excited fermions would be an evidence for fermion substructure. At HERA, excited fermions (f^*) up to the kinematic limit could be directly produced via the t-channel exchange of photon, Z or W bosons as shown in Fig. 1.

We use a specific phenomenological model which assumes that the excited leptons have spin and isospin 1/2, and both left-handed and right-handed components are in weak isodoublets[1]. The f^* production cross section and the decay branching ratios to a Standard Model fermion and a gauge boson depend on the parameters f, f', and f_s associated with the $SU(2)$, $U(1)$, and $SU(3)$ gauge groups and on the compositeness scale Λ.

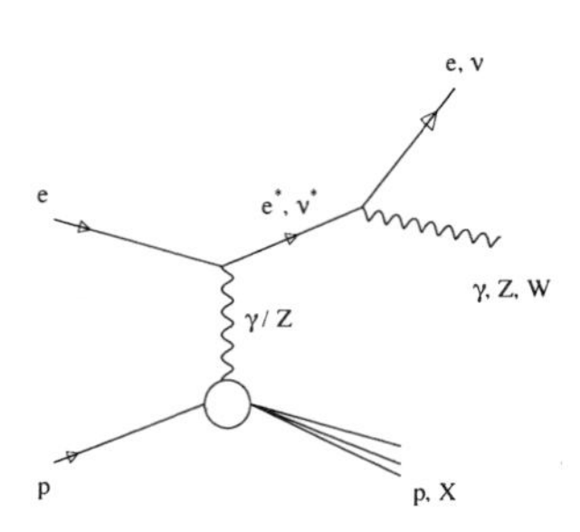

Figure 1. Diagram for the production of excited electrons or neutrinos in ep collisions with their subsequent decay. An excited quark could be produced at the boson-proton vertex.

2 Search for Excited Fermions

The ZEUS[2] and H1[3] collaborations have searched for excited electrons (e^*), neutrinos (ν^*) and quarks (q^*) as a narrow resonance[a] in the invariant-mass spectrum of standard fermions and gauge bosons, see Table 1. All kinds of f^* have been searched for in e^+p collisions at a center-of-mass energy of 300 GeV using an integrated luminosity of about 38 pb^{-1} per experiment. Since the production cross section in e^+p at high ν^* mass is strongly suppressed compared to that in e^-p,[b] a search for excited neutrinos using the recent e^-p collision data at 318 GeV center-of-mass energy corresponding to an integrated luminosity of about 16 pb^{-1} per experiment is also performed.

3 Results

No evidence for any such states was found and upper limits at 95% confidence level (C.L.) as a function of the f^* mass, M_{f^*}, were set on the characteristic coupling f/Λ. Figures 2-5 show the limits on f/Λ obtained from e*, v* and q* analysis, respectively. For

[a] The effect of a possibly large width for very massive excited neutrinos has been studied in [3].

[b] This is due to the smaller d-quark density compared to u-quark in the proton and chiral nature of W exchange.

the excited-electron search $f = f'$ were chosen. For the excited neutrino, two different scenarios were considered; one with $f = f'$ where the photonic decay of ν^* is forbidden, the other with $f = -f'$. The limits for the q^* shown in Fig. 5 assume $f = f'$ and only electroweak couplings (i.e. $f_s = 0$), complementing those from Tevatron, where q^* production requires non-zero f_s. The H1 limits[3] for q^* assuming $f_s \neq 0$ indicate that HERA is more competitive than Tevatron for $f_s < 0.1$ and q^* with masses less than 130 GeV as shown in Fig. 6.

Assuming $f/\Lambda = 1/M_{f^*}$, excited fermions with masses below 229, 163, and 203 GeV, for e^*, ν^* (for $f = -f'$), and q^* productions, respectively, are excluded.

Figures indicate that the domain probed by HERA extends beyond the reach of other colliders. 2001-2006 upgraded HERA data (~ 1 fb^{-1}) will bring much extended discovery potential, especially for ν^*.

Table 1. Decay modes and event signatures considered in the f^* searches by ZEUS and H1. The abbreviations e, γ, μ, and $\not{P}_t$ stand for electron, photon, muon, and missing transverse momentum, respectively. (Some ν^* decay modes were studied in the e^+p data set only.)

Channel	Topology	Studied by
$e^* \to e\gamma$	$e + \gamma$	ZEUS, H1
$e^* \to eZ \to eq\bar{q}$	e + 2jets	ZEUS, H1
$e^* \to eZ \to eee$	3 e	H1
$e^* \to eZ \to e\nu\bar{\nu}$	$e + \not{P}_t$	H1
$e^* \to eZ \to e\mu\bar{\mu}$	e+ 2μ	H1
$e^* \to \nu W \to \nu q'\bar{q}$	$\not{P}_t$+ 2jets	ZEUS, H1
$\nu^* \to \nu\gamma$	$\gamma + \not{P}_t$	ZEUS, H1
$\nu^* \to eW \to e\nu e$	$2e + \not{P}_t$	H1
$\nu^* \to eW \to eq'\bar{q}$	e + 2jets	ZEUS, H1
$\nu^* \to eW \to e\mu\nu$	$e + \mu + \not{P}_t$	H1
$\nu^* \to \nu Z \to \nu q\bar{q}$	$\not{P}_t$+ 2jets	ZEUS, H1
$\nu^* \to \nu Z \to \nu ee$	$2e + \not{P}_t$	H1
$q^* \to q\gamma$	γ + jet	ZEUS, H1
$q^* \to qZ \to qq'\bar{q}'$	3 jets	H1
$q^* \to qZ \to qee$	$2e$ + jet	H1
$q^* \to qZ \to q\mu\bar{\mu}$	2μ + jet	H1
$q^* \to qW \to q\nu e$	e + jet + $\not{P}_t$	ZEUS, H1
$q^* \to qW \to qq'\bar{q}$	3 jets	H1
$q^* \to qW \to q\mu\nu$	μ + jet + $\not{P}_t$	H1

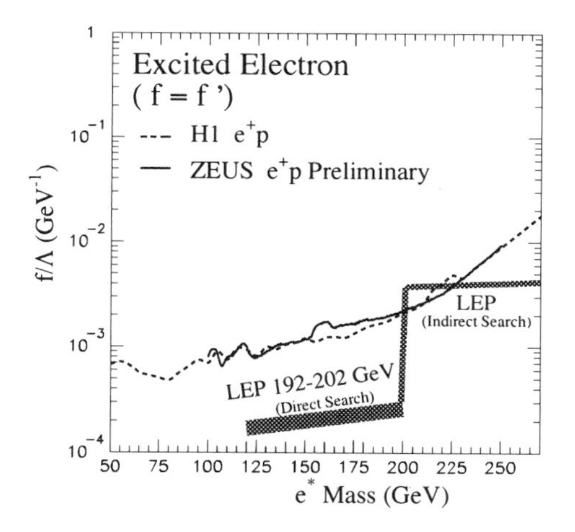

Figure 2. Upper limits at 95% C.L. on f/Λ as a function of the e^* mass assuming $f = f'$ from ZEUS and H1 searches for excited electrons (based on 38 pb^{-1} of e^+p data). The thick curve indicates the exclusion limits from LEP[4]. Above the LEP center-of-mass energy the LEP limits come from searching for the signs of virtual e^* exchange in the process $e^+e^- \to \gamma\gamma$. In all cases the areas above the curves are excluded.

References

1. K. Hagiwara, S. Komamiya and D. Zeppenfeld, Z. Phys. **C29** (1985) 115;
   U. Baur, M. Spira and P. M. Zerwas, Phys. Rev. **D42** (1990) 815.
   F. Boudjema, A. Djouadi and J. L. Kneur, Z. Phys. **C57** (1993) 425;
2. ZEUS Collaboration, Contributed paper **1040** to ICHEP2000, Osaka.
3. H1 Collaboration, DESY-00-102, accepted by Eur. Phys. J.
   H1 Collaboration, Contributed paper **956** to ICHEP2000. Osaka.
4. DELPHI Collaboration, Contributed paper to Moriond 2000, DELPHI 2000-035 CONF 354.
5. DELPHI Collaboration, Contributed paper **115** to EPS'99, Tampere.

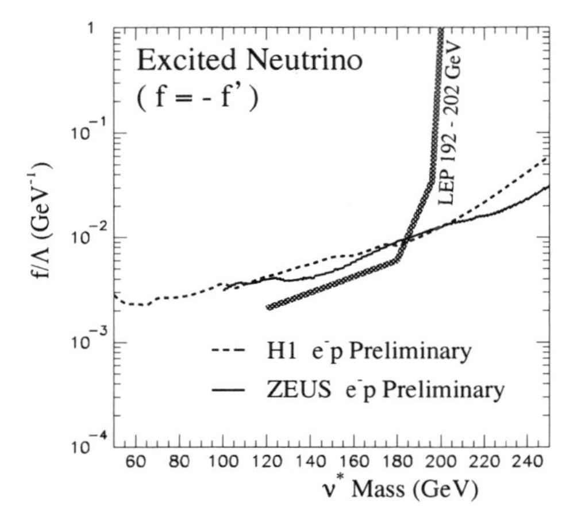

Figure 3. Upper limits at 95% C.L. on f/Λ as a function of the ν^* mass assuming $f = -f'$ from ZEUS and H1 searches for excited neutrinos (based on 16 pb^{-1} of e^-p data). The thick curve indicates the exclusion limits from LEP[4]. In all cases the areas above the curves are excluded.

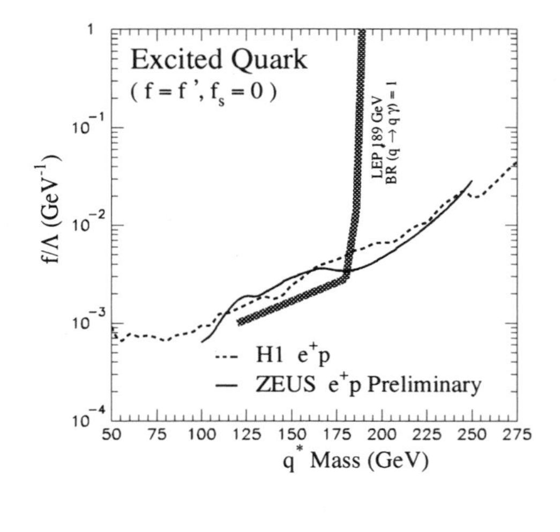

Figure 5. Upper limits at 95% C.L. on f/Λ as a function of the q^* mass assuming $f = f'$ and $f_s = 0$ from ZEUS and H1 searches for excited quarks assuming $f = f'$ and $f_s = 0$ (based on 38 pb^{-1} of e^+p data). The thick curve indicates the exclusion limits from LEP[5] derived assuming $BR(q^* \to q\gamma) = 1$. In all cases the areas above the curves are excluded.

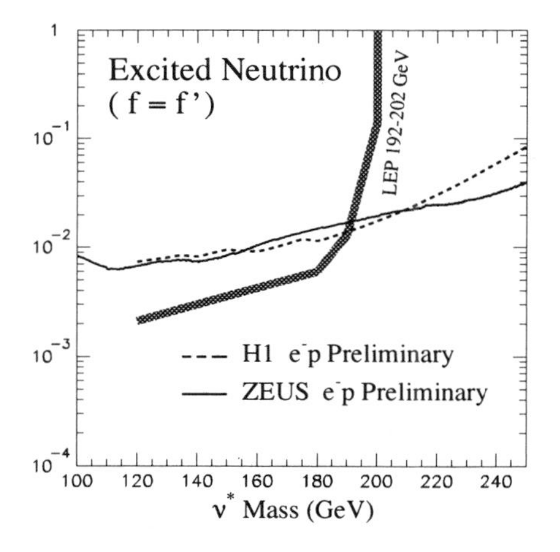

Figure 4. Upper limits at 95% C.L. on f/Λ as a function of the ν^* mass assuming $f = f'$ from ZEUS and H1 searches for excited neutrinos (based on 16 pb^{-1} of e^-p data). The thick curve indicates the exclusion limits from LEP[4]. In all cases the areas above the curves are excluded.

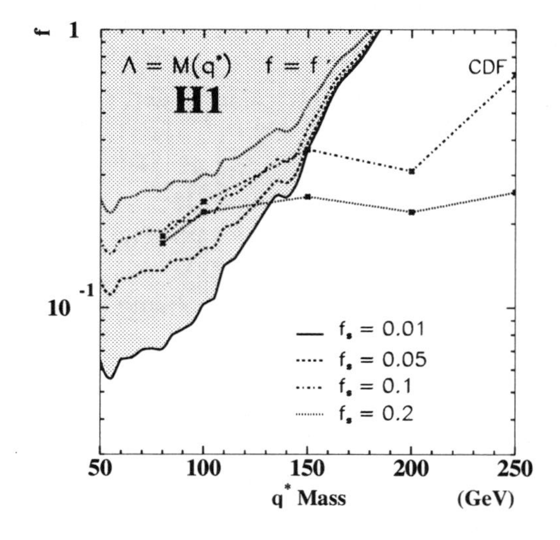

Figure 6. Exclusion limits on f values at 95% C.L. as a function of the mass of the excited quark, assuming $\Lambda = M(q^*)$, $f = f'$ and for different f_s values[3]. Also the exclusion limits from CDF (the two right curves) for two f_s values are shown. Values of the couplings above the curves are excluded.

SEARCH FOR CONTACT INTERACTIONS AND LEPTOQUARKS AT HERA

C. NIEBUHR
on behalf of the H1 and ZEUS collaborations
DESY, Notkestr. 85, D-22603 Hamburg
E-mail: niebuhr@mail.desy.de

Recent results on searches for contact interactions and for leptoquarks from the ep collider HERA are presented which are based on about 100 pb^{-1} of data collected in the years 1994-2000 by the H1 and ZEUS collaborations. No signals are observed which allows upper limits to be placed on parameters of theories extending beyond the Standard Model which are more restrictive than existing limits obtained at LEP and the TeVatron.

1 Introduction

The kinematic domain for deep-inelastic scattering (DIS) accessible at the ep collider HERA leads to sensitivity to new physics beyond the Standard Model (SM) competitive and in several areas beyond what has been obtained at LEP or at the TeVatron. At HERA electrons or positrons with an energy of 27.5 GeV have been brought into collision with protons of 820 or 920 GeV. Up to now both colliding beam experiments, H1 and ZEUS, each have collected ≈ 45 pb^{-1} of e^+p data at a center of mass energy of $\sqrt{s} = 300$ GeV. After raising the proton beam energy to 920 GeV in 1998 both experiments further collected ≈ 15 pb^{-1} of e^-p and another ≈ 45 pb^{-1} of e^+p data at $\sqrt{s} = 318$ GeV. As the appearance of new physics may be dependent on the lepton charge or on a given energy threshold the analysis of the recent data is of particular interest.

2 Search for Contact Interactions

In many scenarios for new physics anomalous eq interactions are indistinguishable from SM Neutral Current deep-inelastic scattering (NC DIS) on an event by event basis. However, at sufficiently large scale they can manifest themselves in distortions of the differential NC DIS cross section through their interference with the Standard Model. Pos-

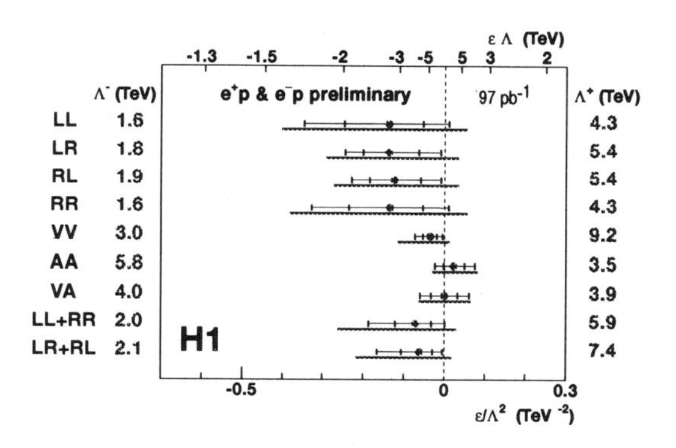

Figure 1. CI limits at 95% CL for ε/Λ^2 for a number of models resulting from the analysis of 97 pb^{-1} $e^{\pm}p$ data from H1.

sible deviations can be analysed in terms of four-fermion pointlike $eq \to eq$ contact interactions (CI) representing the low energy limit of a new interaction. The general chiral invariant (vectorlike) contact interaction Lagrange density can be written as:

$$\mathcal{L}_{CI} = \frac{g^2}{\Lambda^2} \sum_{a,b=L,R} \varepsilon_{ab}^q (\bar{e}_a \gamma^\mu e_a)(\bar{q}_b \gamma_\mu q_b) \quad (1)$$

where Λ is the effective mass scale, g the coupling strength and ε_{ab}^q determines the relative sign of the individual terms ($g^2 = 4\pi$ and $|\varepsilon_{ab}^q| = 0, \pm 1$). A large variety of different models have been tested by ZEUS[1] and H1[2]. No significant deviations from the SM have been observed by the two collabo-

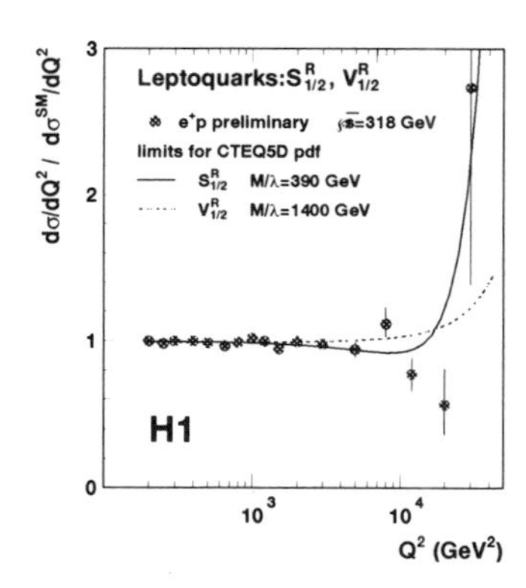

Figure 2. NC cross section at $\sqrt{s} = 318$ GeV normalised to the SM expectation using CTEQ5D parton distributions. The H1 data are compared with 95% CL exclusion limits of leptoquarks for the cases $S^R_{1/2}$ and $V^R_{1/2}$.

rations. By combining up to 97pb^{-1} of $e^{\pm}p$ data exclusion limits on ε/Λ^2 between 1.6 and 9.2 TeV depending on the chiral structure of the model are obtained as shown in fig. 1.

2.1 Searches for Leptoquarks in Contact Interactions

Leptoquarks (LQs) are scalar or vector bosons which carry baryon (B) and lepton (L) numbers and can couple to both quarks and leptons. In the Buchmüller-Rückl-Wyler model [3] the 14 possible LQs are classified according to the fermion number, $F = 3B + L$, and their branching ratios into eq or νq (0, 1/2 or 1). As an illustration of the sensitivity of the existing data to the effect of LQs fig. 2 shows the measured e^+p NC DIS cross section normalised to the SM expectation (using CTEQ5D) as a function of Q^2 together with the 95% exclusion limits for the two examples $S^R_{1/2}$ ($F = 0$) and $V^R_{1/2}$ ($F = 2$). Using this indirect method limits on the ratio of leptoquark mass and coupling, M_{LQ}/λ in the range $0.3 - 1.7$ TeV can be set, depending on the LQ type.

2.2 Limits on Low Scale Gravitational Effects

Recently an interesting idea on how to circumvent the hierarchy problem in $4 + n$ dimensional string theory was proposed [4]: the fundamental Planck scale, where gravity becomes comparable in strength to the other interactions is assumed to be near the weak scale $M_S = \mathcal{O}(\text{TeV})$. The weakness of gravity results from the fact that only gravitons can propagate into the bulk while SM particles are confined to the 3+1 dimensional subspace. The huge $1/M_{Planck}$ suppression of the graviton coupling to SM particles gets in this model compensated by the summation over a large multiplicity of Kaluza-Klein modes leading to an effective CI coupling $\eta_G = \lambda/M_S^4$. Using the combination of all data sets H1 derives [2] 95% CL limits of $M_S > 0.93(0.63)$ TeV assuming $\lambda = -1(+1)$.

3 Leptoquarks in resonance decays

Below the kinematical limit $\sqrt{s}$ at HERA LQs can be produced in the s-channel and decay with a distinct resonant structure in the Bjorken-x distribution at a value $x \simeq M^2_{LQ}/s$. Interference with the SM and the finite width of the LQ resonance lead to sensitivity even somewhat beyond $\sqrt{s}$. DIS background can be suppressed by employing the characteristics of the y (inelasticity) distribution of the different processes. In the 1994-1997 e^+p NC data a slight excess of events was observed around masses of 200 GeV which however is not confirmed by the new data taken in 1999-2000 (see fig. 3). ZEUS also observed a few outstanding events in the high-x, high-y region in the earlier e^+p data, but the new data do not contain such events [5], as shown in fig. 3. The ZEUS collaboration has combined the searches [6] in the $e^+p \to e^+X$ and $e^+p \to \bar{\nu}X$ channel and converted the resulting cross section limits to mass dependent limits on the coupling strength as shown for example in fig. 4.

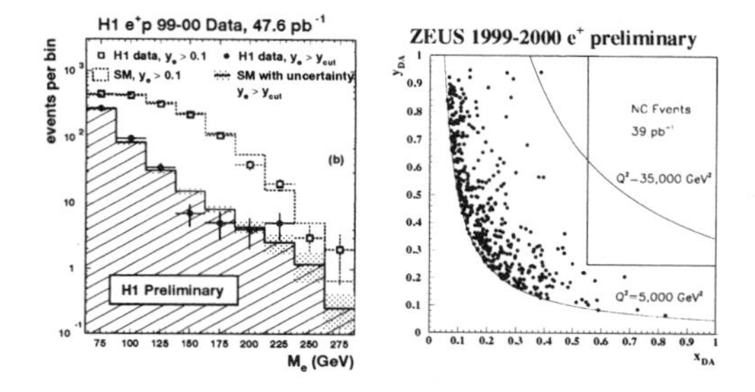

Figure 3. Mass spectra for NC DIS like final states e^+p H1 data and MC expectation before and after a cut on y (left). Event distribution in x and y for $Q^2 > 5000\ \mathrm{GeV}^2$ for 1999-2000 ZEUS data (right).

For a coupling of electromagnetic strength,

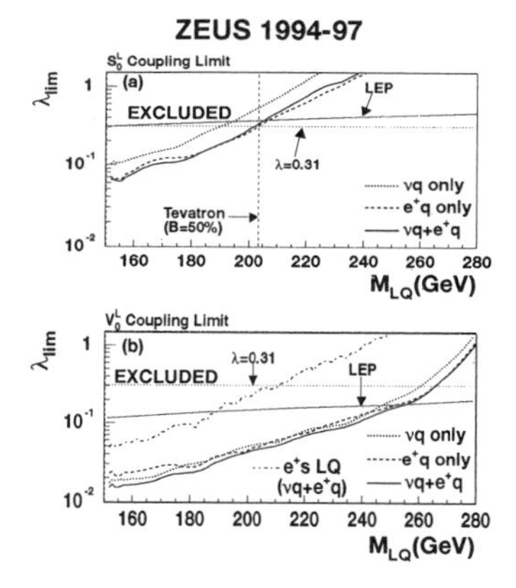

Figure 4. Limits on the coupling λ_{lim} for S_0^L (a) and V_0^L (b). The results are shown seperately for the $\bar{\nu}q$ and e^+q channel as well as for the combination of both.

$\lambda = 0.31$, LQ masses in the range $M_{LQ} \geq 150 - 290$ GeV can be excluded depending on the LQ type. Finally fig. 5 summarizes the present HERA, LEP and TeVatron mass limits [7] for two LQ types as derived from direct searches and indirect methods. A significant part of the area is presently only excluded by the HERA data.

Both experiments also searched for lepton-flavor-violating LQs which lead to a μ or a τ in the final state. No significant signal is observed which is translated [8] into limits on the ratio of coupling and LQ mass, $\lambda_{eq}\lambda_{\mu/\tau q}/M_{LQ}^2$.

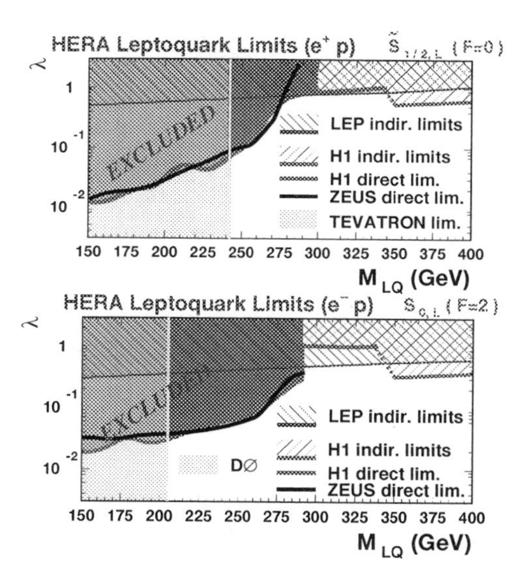

Figure 5. Summary of exclusion limits at 95% CL on the Yukawa coupling λ as a function of the LQ mass for a scalar LQ with $F = 0$ (top) and for a scalar LQ with $F = 2$ (bottom) in the framework of the BRW model.

Enhanced sensitivity to new physics can be expected after the year 2001 when the upgraded HERA will be operated at higher luminosity and $e^{\pm}$ polarization will be available.

References

1. *Eur. Phys. J.* **C14** (2000) 239.
2. H1 Coll., Contr. Paper **305**
3. W.Buchmüller, R.Rückl and D.Wyler, *Phys. Lett.* B **191**, 442 (1987). *Erratum Phys. Lett.* B **448**, 320 (1999)
4. N.Arkani-Hamed, S.Dimopolous and G.Dvali *Phys. Lett.* B **429**, 263 (1998) and *Phys. Rev.* D **59**, 086004 (1999).
5. ZEUS Coll., Contr. Paper **452**
6. ZEUS Coll., *Eur. Phys. J.* **C16** (2000) 253
   DESY 00-133, subm. to *Phys. Rev. D.*
7. H1 Coll., Contr. Paper **322**
8. H1 Coll., *Eur. Phys. J.* **C11** (1999) 447, *Erratum* **C14** (2000) 553
   ZEUS Coll., Contr. Paper **453**

SEARCHES FOR SINGLE TOP PRODUCTION AT LEP

V. F. OBRAZTSOV

Institute for High Energy Physics, 142284 Protvino, Moscow region, Russia
E-mail: obraztsov@mx.ihep.su

Single top production via flavour changing neutral currents in the reaction $e^+e^- \rightarrow t\bar{c}(\bar{u})$ is searched for in $\sim 1.4fb^{-1}$ of data collected by four LEP experiments at c.m.s energies in the range 189-202 GeV. In total, 148 candidate events are selected in the data to be compared with expected background of 151.8 events. Upper limits on the single top production cross sections are derived as well as a model dependent limit $Br(t \rightarrow Zc) + Br(t \rightarrow Zu) < 7\%$.

1 Introduction

Since summer of 1997, the LEP-2 e^+e^- collider has been operating at the cms energy of $\sqrt{s} > 183$ GeV. At this energy the production of single top quark in the reaction $e^+e^- \rightarrow Z^*(\gamma^*) \rightarrow t\bar{c}(\bar{u})$ [a] becomes kinematically possible. This is a flavour changing neutral current (FCNC) process, forbidden in the Standard Model (SM) at tree level. It can proceed via loops , but then the GIM [1] mechanism enormously suppresses the cross section down to $\sigma_{SM} < 10^{-9}$ fb [2]. Another possible SM process for the single top production is $e^+e^- \rightarrow e^-\bar{\nu}_e t\bar{b}$. Here a fine cancellation of several diagrams leads to the cross section of $\sigma \sim 10^{-4}$ fb for $\sqrt{s}$=200 GeV, as it was first shown in [3].

Extensions of SM , such as multiple higgs doublet models [4] or super-symmetric models [5]could lead to substantial enhancements of single top production. Some exotic models as compositeness [6] or dynamical breaking of EW symmetry [7] give rise to cross section potentially detectable at LEP2.

The first search for FCNC in the top sector was carried out by CDF. They looked for the decays $t \rightarrow \gamma(Z, g) + c(u)$ in the process $\bar{p}p \rightarrow \bar{t}tX$ at $\sqrt{s} = 1.8$ TeV and set limits [8]:

$$Br(t \rightarrow \gamma c) + Br(t \rightarrow \gamma u) < 3.2\%; \quad (1)$$
$$Br(t \rightarrow Zc) + Br(t \rightarrow Zu) < 33\%$$

Fenomenologically, the neutral current vertices for $t \rightarrow c(u)\gamma$ and $t \rightarrow c(u)Z$ can be expressed in a "minimal" way as:

$$\Gamma^{\gamma}_{\mu} = \kappa_{\gamma}\frac{ee_q}{\Lambda}\sigma_{\mu\nu}q^{\nu}; \Gamma^{z}_{\mu} = \kappa_z\frac{e}{sin2\theta_w}\gamma_{\mu}; \quad (2)$$

Here $\sigma^{\mu\nu} = \frac{1}{2}(\gamma^{\mu}\gamma^{\nu} - \gamma^{\nu}\gamma^{\mu})$; Λ is the scale parameter; $\kappa_{\gamma}, \kappa_z$ - anomalous coupling constants. See [9], [10] for more genral form. Using (2) it is straightforward to obtain for the top decay width:

$$\Gamma_{tc\gamma} = \kappa_{\gamma}^2\frac{\alpha e_q^2}{4}(\frac{m_t^2}{\Lambda^2})m_t; \quad (3)$$
$$\Gamma_{tcZ} = \frac{\kappa_z^2 \alpha m_t^3}{8sin^2 2\theta_w M_z^2}(1 - \frac{M_z^2}{m_t^2})(1 + 2\frac{M_z^2}{m_t^2})$$

From the CDF limit and assuming $\Lambda = m_t$ [9], one can obtain: $\kappa_{\gamma}^2 < 0.176$;$\kappa_z^2 < 0.533$. For the top production cross section at LEP:

$$\sigma_{t\bar{c}} = \frac{\pi\alpha^2}{s}(1 - \frac{m_t^2}{s})^2[\frac{\kappa_z^2(1 + \alpha_w^2)(2 + m_t^2/s)}{4sin^4 2\theta_w(1 - M_z^2/s)^2}$$
$$+\kappa_{\gamma}^2 e_q^2\frac{s}{m_t^2}(1 + \frac{2m_t^2}{s}) - \frac{3\kappa_{\gamma}\kappa_z\alpha_w e_q}{sin^2 2\theta_w(1 - M_z^2/s)}]$$

Here $\alpha_w = 1 - 4sin^2\theta_w$. The sign of the interference term is set to "-", which corresponds to the "pessimistic" scenario in a large class of models. Anyway, the effect of the interference is very small for the LEP-2 region. The cross section versus $\sqrt{s}$ is shown on Fig.1, for κ_z, κ_{γ} from the CDF limit. It reaches ~0.1 pb at $\sqrt{s}$ = 200 GeV, and should be multiplied by a factor of 4 to take into account $\bar{t}c; t\bar{u}; \bar{t}u$ channels. Given the integrated luminosity of $\sim 1.4fb^{-1}$ accumulated by the four LEP experiments, it is evi-

[a] The notation $t\bar{c}$ is used for both $t\bar{c}$ and $\bar{t}c$.

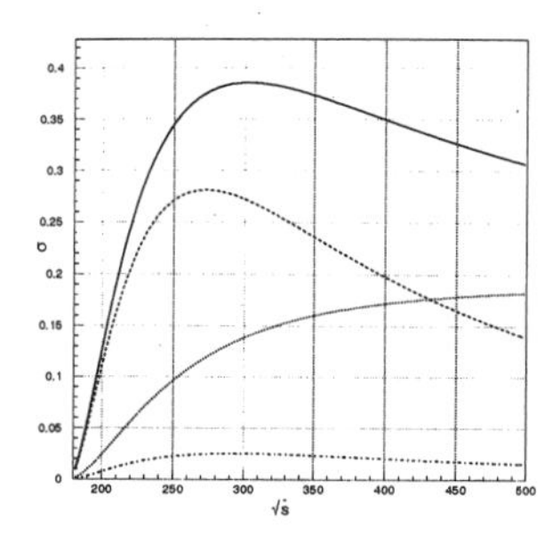

Figure 1. σ (pb) for $e^+e^- \rightarrow t\bar{c}$ versus $\sqrt{s}$ (solid curve). The dashed, dotted and dashed-dotted curves represent the contribution from Z, γ exchange and $Z\gamma$ interference.

Table 1. The statistics presented by the LEP experiments.

$\sqrt{s}$(GeV)	189	192	196	200	202
$\mathcal{L}(pb^{-1})$					
ALEPH	174	29	80	86	42
DELPHI	158	30	78	84	41
L3	177	30	84	83	37
OPAL	172				

dent that LEP-2 can improve the CDF limits significantly.

2 Data presented to the conference

The first results of the search for $e^+e^- \rightarrow t\bar{c}$ for $\sqrt{s} = 183$ GeV, $\mathcal{L}$ 50 pb^{-1} was published by DELPHI [11].

Four LEP collaborations presented the following data to the Osaka conference: ALEPH, DELPHI, L3 - a preliminary analysis of the statistics accumulated in 1998-1999 at $\sqrt{s} = 189 \div 202$ GeV(*189 GeV analysis was submitted before to ICHEP99 Tampere conference*); OPAL- the analysis of the 1998 year data at $\sqrt{s} = 189$ GeV, for both hadronic and leptonic channels. (*189 GeV hadronic channel analysis was submitted to Moriond-2000 conference.*) The detailes of the statistics are presented in Table.1

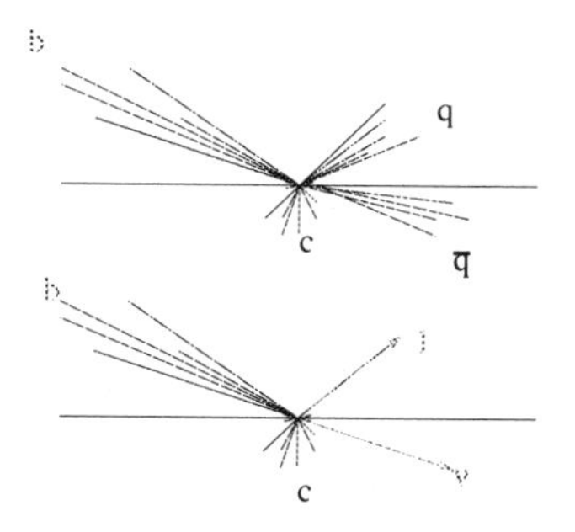

Figure 2.

3 Event selection

Fig.2 shows two topologies relevant for the process $e^+e^- \rightarrow t\bar{c}; t \rightarrow Wb$ corresponding to the hadronic/leptonic decay of W. At first look the topologies are similar to that of WW events. Indeed, the first steps of the selection are similar(the main background is $Z\gamma;\gamma\gamma$). For the **hadronic(4q)** channel it is: standard hadronic events selection with 3-5 jets; $E_{vis} > 0.7 \times \sqrt{s}$, Sph > 0.2 and/or Thrust < 0.82, then event is forced into 4 jets. For the **semileptonic(** $q\bar{q}l\nu$**)** channel: $E_{vis} >\sim 0.2\sqrt{s}$; isolated lepton(slim jet) requirement; ALEPH - e, μ identification, DELPHI- just one isolated track; L3,OPAL- e, μ, τ(1 or 3 tracks slim jet), then event is forced into 3 jets(one jet - lepton).

The difference with WW events is due to the almost at-rest production of top, which leads to the hierarchy of jet energies: $E_c \sim \sqrt{s} - m_t$- the lowest energy; $E_b \sim \frac{m_t^2 - m_W^2 + m_b^2}{2m_t}$- the highest energy. To improve the energy resolution, a 4C or 5C kinematic fit (4-jet topology) or 1C fit (semileptonic topology) is issential. The main cuts at this stage are for the **hadronic** channel: b- tag for the "b" jet ; cuts on $E_b, E_c, E_b/E_c, \beta_W$. **Semileptonic** channel: $m_{l\nu}; m_{qq}$ are calculated after 1C-fit; $m_{l\nu} \sim M_W$; $m_{qq} < 70$ GeV(anti WW cut !); $E_b > 50$GeV; b- tagging. ALEPH, DELPHI are using sequential cuts, L3, OPAL construct combined variable (binned likelihood function). To illustrate

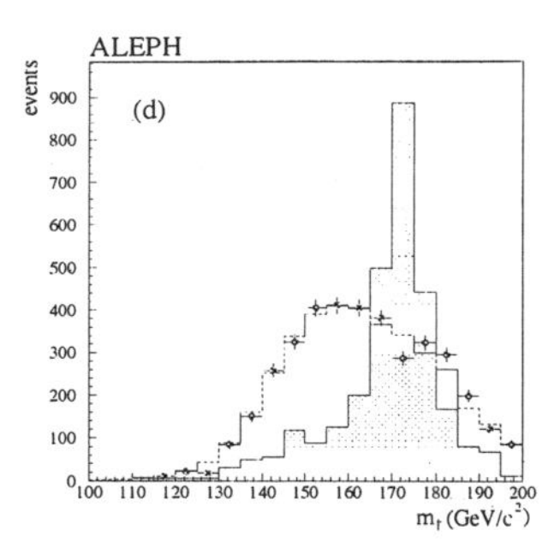

Figure 3. ALEPH $q\bar{q}l\nu$ analysis: $l\nu b$ mass for the data(points); background(open histogram) and top signal (shaded histogram with arbitrary normalization).

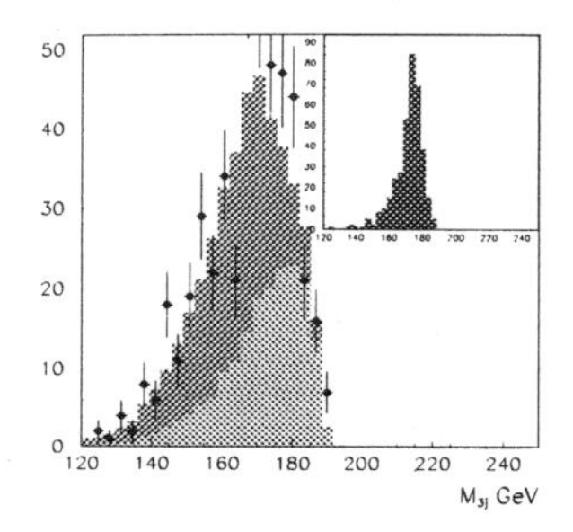

Figure 4. DELPHI 4q analysis: $q\bar{q}b$ mass for the data(points); background(histogram); top signal(histogram in the top-right corner).

the data-MC agreement and the top-signal resolution, Fig.3 and Fig.4 show the reconstructed top mass for the semileptonic and hadronic channels respectively. The final statistics of the selected candidates, calculated background and average selection efficiencies times branching are presented in Table.2 The total number of observed candidate events is 148, with expected background of 151.8, i.e no exess is found with respect to SM expectations.

4 Preliminary results

As it is obvious from the previous section, no evidence for single top production is observed at LEP-2. The results of the search are presented as a 95% upper limits on the top

Table 2. Summary of the events selection. The left|right number in a column corresponds to $q\bar{q}l\nu|4q$ channels; ϵ stands for the average $\epsilon \times Br$. A,D,L,O stand for ALEPH, DELPHI, L3, OPAL.

	A	D	L	O
ev	6|52	3|14	10|50	3|10
bkg	5.7|44.5	6.1|18.8	9.0|54.6	4.0|9.0
$\epsilon\%$	5.9|15.4	6.2|6.8	7.5|10.5	6.8|14.8
σ(pb)	0.23	0.12	0.136	0.35

quark production cross section(see Table.2). ALEPH and DELPHI present separate limits for each energy point, L3 derives a combined limit for the weighted average energy of $\bar{E}$=193.8 GeV; OPAL limit corresponds to 189 GeV. We have to follow L3 when calculating the preliminary combined LEP-2 limit, which is $\sigma < 90$ fb at $\bar{E} = 193.8$ GeV. It is interesting to compare the result with the prediction of [7]: $\sigma_{t\bar{c}+\bar{t}c} \sim 40$ fb at $\sqrt{s}$=192 GeV. The prediction looks within reach, when all LEP-2 statistics will be available. The results of the analysis can be also presented as 95% exclusion regions in κ_γ, κ_z plane. ALEPH, DELPHI do that by a procedure which combines results at different energies, taking into account the dependence of the x-section on $\sqrt{s}$. The QCD correction for the x-section is taken into account as well as the dependence of Br($t \to Wb$) on κ_γ, κ_z. L3 is using simplified procedure, when all the data are assigned to one average energy of $\bar{E}$=193.8 GeV. OPAL, L3 assume Br($t \to Wb$)=1. As an example, Fig.5 shows the ALEPH result together with the CDF limits. Preliminary combination of the ALEPH, DELPHI and L3 data is done using a Bayesian method, described in [12]. Because of the L3, all the data were assigned to the average energy of $\bar{E}$=193.8 GeV. The result is show on Fig.6.

The main systematics in the present search comes from the uncertainty in the top mass: $m_t = 174.3 \pm 5.1$ GeV [13]. Both

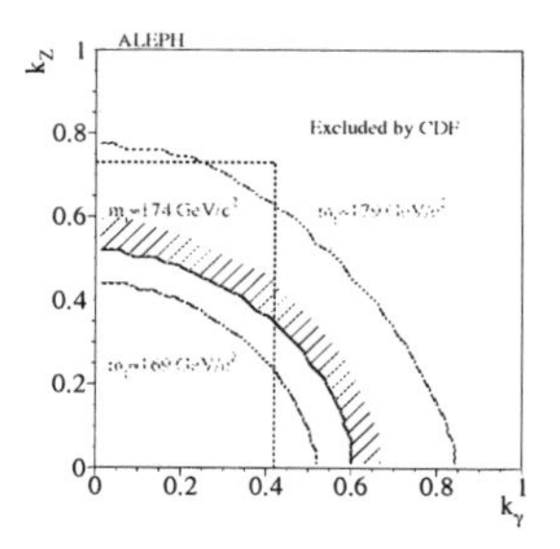

Figure 5. ALEPH κ_γ, κ_Z exclusion plot.

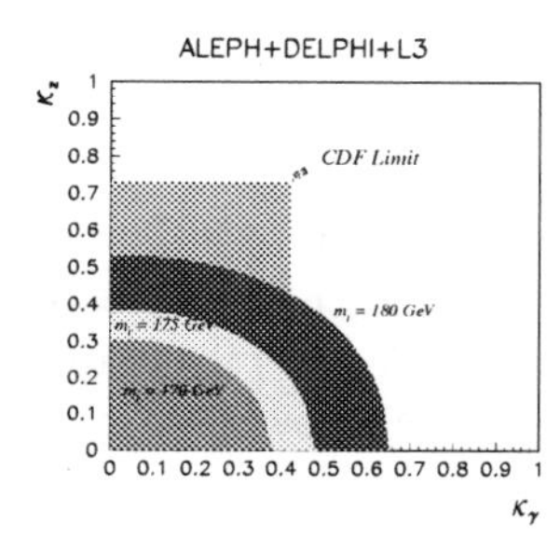

Figure 6. Preliminary combined ALEPH, DELPHI and L3 exclusion plot.

$e^+e^- \rightarrow t\bar{c}$ cross section and selection efficiencies strongly depend on it. That is why all the limits are shown for the most probable m_t mass, and $\pm\sigma$ values. As it is seen from the combined result, the limit on κ_z is significantly improved as compared with CDF, while κ_γ is not. Another way of the presentation is to fix κ_γ to zero and to plot the limit on κ_z or, what is equivalent, on the Br($t \rightarrow cZ + uZ$) versus the top mass. The result is shown on Fig.7.

5 Conclusions

An impressive progress has been achieved in the searches for single top production at LEP2 since the Tampere conference: if compared with CDF [8], the upper limit on $Br(t \rightarrow cZ)$ is improved by a factor of 3. Still a progress is expected in the nearest future: during the run of Y2K the LEP ex-

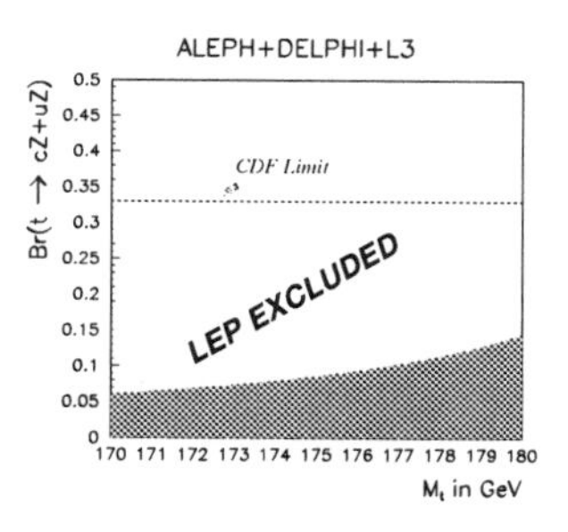

Figure 7. $Br_{t \rightarrow cZ+uZ}$ 95% upper limit versus m_{top}.

periments have already accumulated an integrated luminosity of more than $100pb^{-1}$. The expected total integrated luminosity is $\sim 2.5fb^{-1}$, which is a factor of 2 more than in the present studies. As a result, it will become possible to constraint several theoretical models [6], [7].

Acknowledgments

G.Alemanni, A.Bellerive, C.Geweninger, A.Onofre, F.Palla, O.Yuschenko, provided the essential material for this contribution. Author would like to thank O.Yuschenko and S.Slabospitsky for the discussions and invaluable help during preparation of the talk.

References

1. S.L.Glashow *et al.*,*Phys.Rev.*D**2**,1285(1970)
2. C.S.Huang *et al.*,*Phys.Lett.*B**452**,143(1999)
3. K.Hagiwara *et al*,*Phys.Lett.*B**325**,521(1994)
   E.Boos *et al.*,*Phys.Lett.*B**326**,190(1994)
4. M.Luke *et al.*,*Phys.Lett.*B**307**,387(1993)
5. J.L.Lopez *et al.*,*Phys.Rev.*D**56**,3100(1997)
6. H.Georgy *et al.*,*Phys.Rev.*D**51**,3888(1995)
7. B.A.Arbuzov *et al.*, *Yad.Fiz.***62**,528(1999)
8. CDF *Phys.Rev.Lett.***80**,2525(1998)
9. V.Obraztsov *et al*,*Phys.Let.*B**426**,393(1998)
10. Proceedings CERN 2000-004, 9 May 2000
11. DELPHI *Phys. Lett.* B**446**, 62(1999)
12. V.Obraztsov *NIM* A**316**,388(1992)
    Erratum *NIM* A**399**,500(1997)
13. PDG *Eur.Phys.J.*C**15**,389(2000)

NEW PHYSICS AT FUTURE LINEAR ELECTRON–POSITRON COLLIDERS

K.DESCH

University of Hamburg, Luruper Chaussee 149, 22761 Hamburg, Germany
E-mail: Klaus.Desch@desy.de

Future high energy linear electron positron colliders with centre–of–mass energies between 500 and 1000 GeV offer a unique opportunity to precisely study phenomena from new physics processes. Most important, the mechanism of electroweak symmetry breaking can be established in full depth if one or more Higgs bosons are observed. If supersymmetric particles are found within the kinematical reach of the machine, their properties can be studied with high precision allowing to obtain information about the breaking mechanism of supersymmetry and thus about energy scales far above the available centre–of–mass energy. Alternative scenarios including large extra space dimensions, leptoquarks and lepton flavour violation decays can also be studied.

1 Framework

The technical feasibility of a new generation of linear electron–positron colliders (LC) has been intensively studied in the past years in Japan (JLC), the US (NLC) and Europe (TESLA)[2]. The energy of such machines would be 500 GeV in a first phase with upgrade potential to approximately 1 TeV. A proposal for a multi-TeV machine (CLIC) exists as well but its physics potential is not subject of this contribution.

A LC has a very rich physics potential covering both the study of new phenomena and precision tests of the Standard Model (SM) (top quark physics, electroweak physics and QCD). Here only prospects for new phenomena will be described. The results presented are based on integrated luminosities of 500 to 1000 fb^{-1} and on realistic simulations of detector performance and background conditions.

2 Higgs Physics

In the SM and many of its extensions, the breaking of electroweak symmetry proceeds via the Higgs mechanism leading to one or more physical Higgs bosons. The mass of the lightest Higgs boson is bounded from above on theoretical grounds by perturbativity of the Higgs coupling up to a high energy scale and experimentally through electroweak precision measurements. The latter indicate a Higgs boson mass below 170 GeV at the 95% confidence level[3].

While a Higgs boson, if it exists, is very likely to be found at LEP2, Tevatron, or LHC, the precise measurement of its properties and couplings can only be performed at the LC.

The main production mechanisms for Higgs bosons at the LC are Higgsstrahlung ($\mathrm{e^+e^-\to Z^0H^0}$) and WW–fusion ($\mathrm{e^+e^-\to \nu_e\bar{\nu}_e H^0}$). Furthermore the Yukawa process $\mathrm{e^+e^-\to t\bar{t}H^0}$ and the double Higgsstrahlung process $\mathrm{e^+e^-\to Z^0H^0H^0}$ can be exploited. The Higgs boson itself can be detected independent of its decay through the observation of a mass peak in the invariant mass recoiling against a lepton pair from a $\mathrm{Z^0}$–decay in the Higgsstrahlung process. Furthermore all relevant Higgs boson decays ($\mathrm{b\bar{b}}$, $\mathrm{c\bar{c}}$, $\tau^+\tau^-$, gg, $\mathrm{W^+W^-}$, $\mathrm{Z^0Z^0}$, $\gamma\gamma$) can be observed exclusively in the detectors under study.

This large variety of accessible production mechanisms and decay modes in conjunction with the well defined initial state and clean environment of $\mathrm{e^+e^-}$–collisions allows the determination of a complete profile of the observed Higgs particle.

The coupling g_{HZZ} of the Higgs boson to

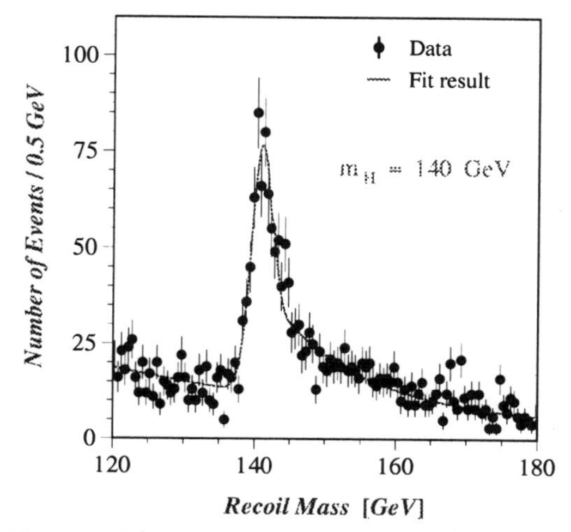

Figure 1. Higgs boson signal in the recoil mass distribution to a muon pair compatible with a Z^0 decay for a simulated Higgs boson mass $m_{\rm H} = 140$ GeV and 500 fb^{-1} at $\sqrt{s} = 350$ GeV.

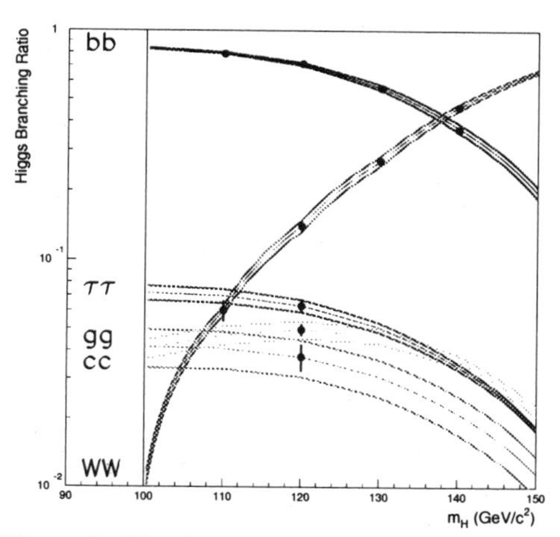

Figure 3. The SM prediction for the decay branching ratios of the Higgs boson. The points with error bars indicate the achievable experimental precision for 500^{-1}. The bands indicate the theoretical uncertainties of the prediction.

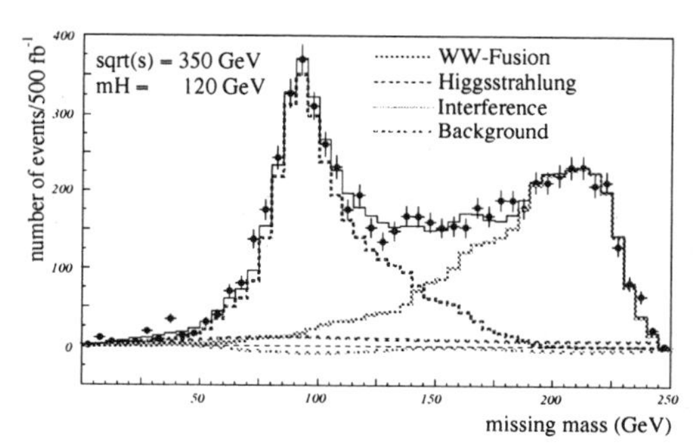

Figure 2. Missing mass distribution in events with two b-jets with an invariant mass compatible with the Higgs boson mass.

the Z^0 can be derived in a model independent way[4] from the observed cross section of the Higgsstrahlung process with leptonically decaying Z^0–bosons (Fig. 1). The coupling $g_{\rm HWW}$ can be derived from the cross section for WW–fusion using H$^0\to$b$\bar{\rm b}$ decays. This cross section can be disentangled from the Higgsstrahlung contribution with Z$^0\to\nu\bar{\nu}$ exploiting the different distribution of the missing invariant mass (Fig. 2).

An essential prediction of the Higgs mechanism is the Yukawa coupling of the Higgs boson to fermions being proportional to the fermion mass. This prediction can be accurately tested through the measurement of the Higgs boson decay branching ratios. The different hadronic Higgs boson decays b$\bar{\rm b}$, c$\bar{\rm c}$, and gg (being sensitive to the top–Higgs coupling) as well as $\tau^+\tau^-$ can be disentangled using the excellent flavour tagging capabilities of a LC detector[5]. The expected precision on the branching ratio measurement is ranging from 2% (for b$\bar{\rm b}$) to approximately 10% (for gg) for a light Higgs boson ($m_{\rm H} < 160$ GeV) (see Fig. 3). Furthermore the W$^+$W$^-$–decay can be observed already at $m_{\rm H} = 120$ GeV (10% precision) and precisely measured along with the Z^{0}Z^0–decay at higher Higgs boson masses (2% precision at 160 GeV)[6]. The rare loop-induced $\gamma\gamma$–decay can also be observed (15% precision). Finally, invisible Higgs boson decays can be studied through their clear signal of missing energy recoiling against two charged leptons. At 800 GeV centre–of–mass energy, the Yukawa process e$^+$e$^-\to$t$\bar{\rm t}$H^0 can be studied in order to extract the top-Yukawa coupling directly[7]. Studies indicate that a 6% measurement is possible with 1 ab^{-1}.

The mass of the Higgs boson, being the only free parameter of the Higgs sector in the SM, can be accurately determined in the Higgsstrahlung process using constrained kinematical fits to $q\bar{q}b\bar{b}$–events and $\ell^+\ell^-X$–events. The achievable precision is 50 MeV (or 0.04 %) for $m_H = 120$ GeV[4,8].

The total decay width of the Higgs boson is predicted to be less than 1 GeV for $m_H < 2m_Z$ being to small to be resolved experimentally. However indirect access to the total width is given through the combination of the measurement of a Higgs-gauge boson coupling with the corresponding decay branching ratio. The best determination is obtained from the combination of the branching ratio for $H^0 \to W^+W^-$ with the measurement of g_{HWW} from the WW–fusion process yielding a 5% uncertainty for 500 fb^{-1} for $m_H < 160$ GeV.

The spin and CP quantum numbers can be determined from the angular distributions in Higgsstrahlung events[9]. Here the fact that both the production and decay angles of the H^0 and the Z^0 can be nicely measured is advantageous.

The double Higgsstrahlung process $e^+e^- \to Z^0H^0H^0$ gives access to the triple Higgs coupling λ_{HHH} of the Higgs boson. This coupling determines the shape of the Higgs potential which is exactly predicted in terms of the Higgs mass in the SM and thus provides a rigorous test of the Higgs mechanism. Since the cross section for this process is small (0.35 fb for $m_H = 120$ GeV at $\sqrt{s} = 500$ GeV) a very high luminosity is essential. Preliminary studies indicate that a 20% measurement of λ_{HHH} seems possible for a light Higgs boson[10].

3 Supersymmetry

If supersymmetry is realised in nature and supersymmetric particles are observed at the LHC many open questions arise. The aim of the LC is to disentangle the kinematically accessible part of the particle spectrum and provide precise mass and cross section measurements. To achieve this, the following features of the LC are beneficial: 1. The large cross sections for s-particle production in e^+e^-–collisions allow a high statistical accuracy for mass and cross section measurements. 2. The tunable centre-of-mass energy and the high luminosity allow threshold scans for all accessible s-particles in a reasonable amount of running time yielding mass measurements with typical precision 50 MeV. 3. The possibility of polarised electron *and* positron beams allows to disentangle the different s-particle states[11].

Once these measurements are available, they can be used to determine the high energy behaviour of the theory and allow to conclude on the SUSY breaking mechanism. This can be done in a model-independent approach exploiting the supersymmetric renormalisation group equations to extrapolate the measured low–energy model parameters to a high energy scale. For example gauge coupling unification at the GUT scale (Fig. 4) can be tested.[12] Different SUSY breaking mechanism can be also distinguished by different dynamical signatures[1,13].

4 Alternative Physics

Supersymmetric theories solve the hierarchy problem through the intrinsic cancellation of quadratic divergences. A popular alternative are models with large extra space dimensions. These extra space dimensions are populated by gravity interactions but not by the electroweak interactions making the large Planck scale only an effective 3+1 dimension scale while the true gravity scale could be as low as the electroweak scale. This scenario has significant phenomenological consequences due to real and virtual effects from graviton interactions. One example is the process of real graviton production in association with a photon yielding a single photon plus miss-

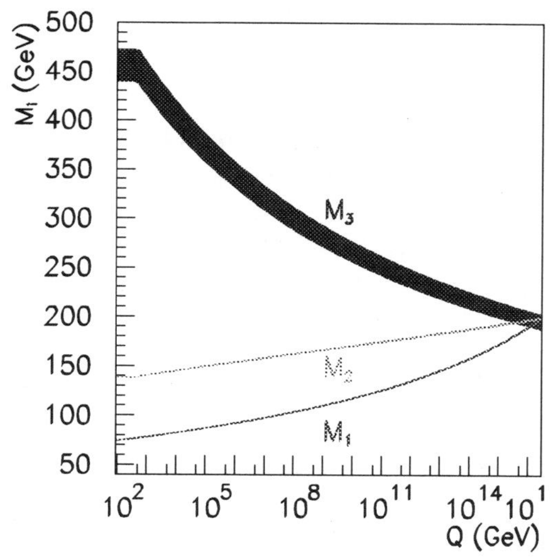

Figure 4. Extrapolation of the measured gaugino mass parameters M_1, M_2 and M_3 (taken from a gluino mass measurement at LHC) using the supersymmetric renormalisation group equations. Unification can be tested with high accuracy.

ing energy final state. The sensitivity to this process reaches several TeV for the scale M of the gravity interaction[14].

Prospects for leptoquark searches at the LC have also been carried out. Beam polarisation provides a powerful means to determine the leptoquark type and their observation in eγ collisions allows to determine their couplings[1].

The option of the TESLA design to operate at $\sqrt{s} = m_Z$ with a luminosity corresponding to 1 billion Z^0 per year offers the opportunity to search for very rare Z^0–decays. One example is the search for lepton flavour violation inspired by the possibility of heavy massive neutrinos[1].

5 Summary and Conclusion

An electron positron linear collider in the 500 - 1000 GeV regime allows very significant insights in the physics at the electroweak scale. The essential parts of the Higgs mechanism of electroweak symmetry breaking can be fully established through precision measurements of Higgs boson properties. Supersymmetric models can be precisely studied and precision is the key to obtain sensitivity to the high energy behaviour of the observed phenomena. Sensitivity to alternative models at the electroweak scale is also high.

References

1. D.K.Ghosh,P.Roy,S,Roy, Abstract 323, Signal of Anomaly Mediated Wino LSP in a Linear Collider;
J.I.Illana,M.Jack,T.Riemann, Abstract 469, Lepton Flavour Violation in Z decays, hep-ph/0001273;
A.F.Zarnecki, Abstract 451, Leptoquark Searches at Future Colliders, hep-ph/0003271; ECFA/DESY Higgs LC Working Group, M.Battaglia et al., Abstract 728, Studies in Higgs Physics with the TESLA e+e- Linear Collider.
2. for a summary see: N.Toge, these proceedings.
3. A.Gurtu, these proceedings; but see also A.D.Martin, J.Outhwaite, M.G.Ryskin, hep-ph/0008078.
4. P.Garcia-Abia,W.Lohmann, hep-ex/9908065.
5. M.Battaglia, hep-ph/9910271.
6. G.Borisov, F.Richard, hep-ph/9905413; E. Boos et al., hep-ph/9908487 .
7. H.Baer, S.Dawson, L.Reina, Phys.Rev **D61**, 13002 (2000);
S. Moretti, hep-ph/9911501.
8. A.Juste, hep-ex/9912041;
9. K.Hagiwara et al., Eur.Phys.J.**C14** 457 (2000).
10. A.Djouadi et al., Eur.Phys.J.**C10** 27 (1999).
11. G.A.Blair, H-U.Martyn, hep-ph/9910416.
12. G.A.Blair, W.Porod, P.M.Zerwas, hep-ph/0007107;
W. Porod, these proceedings.
13. S. Ambrosanio, G.A. Blair, Eur.Phys.J.**C12** 287 (2000).
14. M.Besancon, hep-ph/9909364.

OBSERVATION OF ANOMALOUS DIMUON EVENTS IN THE NUTEV DECAY DETECTOR

M. H. SHAEVITZ[2], T. ADAMS[4], A. ALTON[4], S. AVVAKUMOV[8], L. DE BARBARO[5], P. DE BARBARO[8], R. H. BERNSTEIN[3], A. BODEK[8], T. BOLTON[4], J. BRAU[6], D. BUCHHOLZ[5], H. BUDD[8], L. BUGEL[3], J. CONRAD[2], R. B. DRUCKER[6], B. T. FLEMING[2], R. FREY[6], J. FORMAGGIO[2], J. GOLDMAN[4], M. GONCHAROV[4], D. A. HARRIS[8], R. A. JOHNSON[1], J. H. KIM[2], S. KOUTSOLIOTAS[2], M. J. LAMM[3], W. MARSH[3], D. MASON[6], J. MCDONALD[7], C. MCNULTY[2], K. S. MCFARLAND[3], D. NAPLES[7], P. NIENABER[3], A. ROMOSAN[2], W. K. SAKUMOTO[8], H. SCHELLMAN[5], P. SPENTZOURIS[2], E. G. STERN[2], N. SUWONJANDEE[1], M. VAKILI[1], A. VAITAITIS[2], U. K. YANG[8], J. YU[3], G. P. ZELLER[5], AND E. D. ZIMMERMAN[2]

[1] *University of Cincinnati, Cincinnati, OH 45221*
[2] *Columbia University, New York, NY 10027*
[3] *Fermi National Accelerator Laboratory, Batavia, IL 60510*
[4] *Kansas State University, Manhattan, KS 66506*
[5] *Northwestern University, Evanston, IL 60208*
[6] *University of Oregon, Eugene, OR 97403*
[7] *University of Pittsburgh, Pittsburgh, PA 15260*
[8] *University of Rochester, Rochester, NY 14627*

A search for long-lived neutral particles (N^0) which decay into at least one muon has been performed using an instrumented decay channel at the E815 (NuTeV) experiment at Fermilab. The decay channel was composed of helium bags interspersed with drift chambers, and was used in conjunction with the NuTeV neutrino detector to search for N^0 decays. The data were examined for particles decaying into the muonic final states $\mu\mu$, μe, and $\mu\pi$. Three $\mu\mu$ events were observed over an expected background of 0.040 ± 0.009 events; no events were observed in the other modes. Although the observed events share some characteristics with neutrino interactions, the observed rate is a factor of 75 greater than expected. No Standard Model process appears to be consistent with this observation.

1 INTRODUCTION

In various extensions to the Standard Model, new particles exist which have reduced couplings to normal quarks and leptons. These new particles may have zero electric charge, long lifetimes, and small interaction rates with normal matter. We shall refer to these as N^0 particles in the following text. Examples of such N^0 particles include neutral heavy leptons (NHLs) or heavy sterile neutrinos[1,2,3] and neutral supersymmetric particles[4] such as neutralinos and sneutrinos. The N^0 particles can be produced either by pair production in hadronic interactions or via weak decays of mesons through mixing with standard neutrinos. The decays of the N^0 to normal hadrons and/or leptons can proceed through weak decays with mixing, or via R-parity violating supersymmetric processes.

High energy neutrino beamlines are ideal places to produce N^0 particles, since very large numbers of protons interact in these beamlines. N^0's may be produced via a number of mechanisms, including primary interactions of the protons either in the target or the beam dump, through prompt decays of charmed or bottom mesons, by decays of pions or kaons in the decay region, or in neutrino interactions in the shielding downstream of the decay region. A particle detector placed downstream of this sort of beamline (i.e., in the neutrino beam itself) can be used to search for N^0 decays.

We report here the results of a search using Fermilab's E815 (NuTeV) detector for N^0 particles in the mass region above 2.2 GeV/c^2 which decay into final states with at least one muon and one other charged particle. (A more complete writeup on this analysis can be found in Ref. 5.) For the search described here, the NuTeV neutrino beamline was used in conjunction with a low mass decay detector called the decay channel. NuTeV has previously reported results of searches for N^0's in the mass region between 0.3 to 3.0 GeV/c^2 with at least one final state muon[6], and in the mass region below 0.3 GeV/c^2 for decays to electrons[7].

2 THE BEAMLINE AND DETECTOR

During the 1997 fixed-target run at Fermilab, NuTeV received 2.54×10^{18} 800 GeV/c protons with the detector configured for this search. The proton beam was incident on a one-interaction-length beryllium oxide target at a targeting angle of 7.8 mr with respect to the detector. A sign-selected quadrupole train (SSQT)[8] focused either positive (for 1.13×10^{18} protons) or negative (for 1.41×10^{18} protons) secondary π and K mesons into a 440 m evacuated decay region pointed towards the NuTeV decay channel and neutrino detector hall. Surviving neutrinos (and possibly also N^0's) traversed ~850 meters of earth-berm shielding before reaching the NuTeV decay channel.

The decay channel region (Figure 1), located 1.4 km downstream of the production target, was designed to contain minimal material (in order to suppress neutrino interactions) and to have tracking sufficient to isolate two-track decays of neutral particles. A $4.6 \text{ m} \times 4.6 \text{ m}$ double array of plastic scintillation counters vetoed charged particles entering from upstream of the decay channel. The channel itself measured 34 m in length

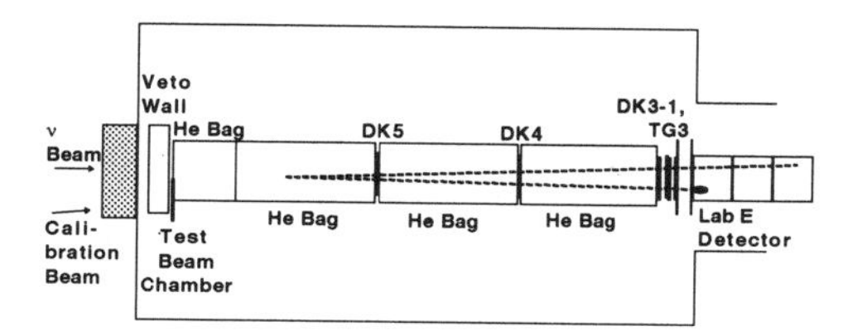

Figure 1. Schematic of the NuTeV decay channel with example $N^0 \to \mu\pi$ decay.

and was interspersed with $3 \text{ m} \times 3 \text{ m}$ argon-ethane drift chambers positioned at 14.5 m, 24 m, and 34 m downstream of the veto array in stations of 1, 1, and 4 chambers, respectively. The regions between the drift chamber stations were occupied by helium-filled cylindrical plastic bags 4.6 m in diameter in order to minimize neutrino interactions. The Lab E neutrino detector[9,10] provided final state particle energy measurement and identification.

3 EVENT SELECTION

Event selection criteria were developed to minimize known backgrounds while maintaining efficiency for a possible N^0 signal. These criteria were set using various studies of background and signal Monte Carlo samples before the data were analyzed in detail. However, one candidate decay channel event with two muon tracks and a mass greater than 2.2 GeV/c^2 was observed and studied in detail early in the analysis development, but the collaboration went to considerable effort to minimize bias caused by this observation.

The main sources of conventional events in the decay channel are: 1) deep-inelastic scattering (DIS) of neutrinos or anti-neutrinos in the drift chamber material; 2) DIS in the helium; and 3) DIS in the material surrounding the decay channel with a mis-reconstructed vertex in the fiducial volume. Other sources of background are small compared to DIS, including neutral kaon and diffractive π, K, ρ, and charm production from neutrino interactions.

The cuts used to isolate N^0 decays fell into two broad categories: reconstruction and "clean" cuts. Reconstruction cuts isolated events with exactly two tracks forming a vertex within the decay channel fiducial volume and having no charged particle identified in the upstream veto system. Chamber interactions were removed by demanding that the longitudinal (z) distance from the vertex position to any drift chamber was greater than ±101.6 cm. Special "clean" cuts were applied to reduce the deep-inelastic neutrino scattering (DIS) backgrounds. DIS events typically have large track multiplicities with many drift chamber hits and the "clean" cuts effectively reduced this background by requiring minimal extra tracks and hits in the event. In order to isolate high mass events, an additional cut was applied on the "transverse mass", $m_T > 2.2$ GeV, where $m_T = |p_T| + \sqrt{p_T^2 + m_V^2}$, with p_T the component of the total momentum perpendicular to the beam direction, and m_V the invariant mass for the two charged tracks.

4 BACKGROUND ESTIMATION USING MONTE CARLO

Detailed Monte Carlo simulations of both physics processes and detector effects were used to quantify the background from neutrino interactions after cuts. Input to the simulation was provided from several event generators including LEPTO/Jetset, resonance and continuum production models in the low-W region, and diffractive production using various Vector Meson Dominance (VMD) and Partially-Conserved Axial Current (PCAC) models. The event generators fed a GEANT-based detector simulation that produced hit-level simulations of raw data including noise and accidental activity in the detector. Monte Carlo events were processed using the same analysis routines used for the data.

Table 1. Estimated rates of background to the $N^0 \to \mu\mu(\nu)$ search

Source	$\mu\mu(\nu)$ events
DIS events	$(3.9 \pm 0.9) \times 10^{-2}$
Diffractive charm	$(1.1 \pm 0.1) \times 10^{-3}$
Diffractive π	$(1.7 \pm 0.1) \times 10^{-4}$
Diffractive K	$(3.3 \pm 0.3) \times 10^{-7}$
K_L^0 decays	$(3.9 \pm 3.9) \times 10^{-4}$
Other sources	$\ll 2.5 \times 10^{-4}$
Total $\mu\mu(\nu)$ Bkgnd	$(4.0 \pm 0.9) \times 10^{-2}$

After all cuts, the preliminary expected background is 0.040 ± 0.009 events in $\mu\mu$ mode, 0.14 ± 0.02 events in μe mode and 0.13 ± 0.02 events in $\mu\pi$ mode. As an example of the relative sizes of the contributions discussed above, the background sources for the $N^0 \to \mu\mu(\nu)$ mode are broken down in Table 1.

Monte Carlo events were compared to data outside of the signal region as a check on the background calculations. A prime example of one of the checks was the comparison of chamber region events with all the signal criteria cuts except for a z-vertex required to be within ±15.2 cm of a chamber. This is a very powerful data sample because if any observed events in the decay region were due to neutrino-He interactions, then there should be a factor of 28 more events in the chambers, after scaling for acceptance and mass. The numbers of observed events for the $\mu\mu$, μe, and $\mu\pi$ modes are listed in Table 2 along with the prediction for neutrino deep-inelastic scattering in the chambers. The observed events are consistent with the prediction, giving no indication of unexpected "clean", two-track neutrino interactions in the chambers.

5 RESULTS OF THE SEARCH

Using the signal event selection criteria given above, the Monte Carlo background predictions are given in Table 3. The number of

Table 2. Number of events in drift chambers which pass N^0 topology and "clean" cuts.

Decay Mode	Pred. Evts	Obs. Evts
$\mu\mu$ (chamber)	1.6	0
μe (chamber)	1.8	1
$\mu\pi$ (chamber)	2.7	2

Table 3. Predicted and observed events passing all signal cuts

Decay Mode	Pred. Events	Obs. Events
$\mu\mu(\nu)$	0.040 ± 0.009	3
$\mu e\,(\nu)$	0.14 ± 0.02	0
$\mu\pi$	0.13 ± 0.02	0

observed events is also shown. Three $\mu\mu(\nu)$ events were observed, which is considerably above the predicted background. No μe or $\mu\pi$ events are observed, consistent with expectation. The three $\mu\mu(\nu)$ events are shown in Figs. 2-4, and a summary of the event reconstruction characteristics is given in Table 4.

6 DISCUSSION AND CONCLUSIONS

In many ways, the three $\mu\mu$ events are consistent with a N^0 decay hypothesis. The events pass the analysis cuts, where the background is estimated to be 0.04 events. As expected for a decay relative to an interaction

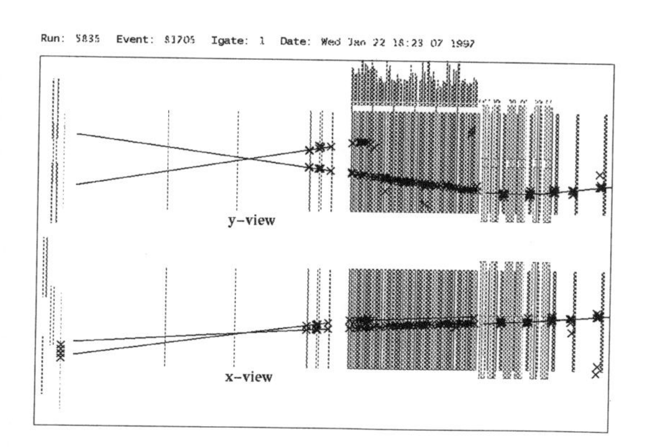

Figure 2. Run/Event 5835/81705: $\mu\mu(\nu)$ data event passing final cuts.

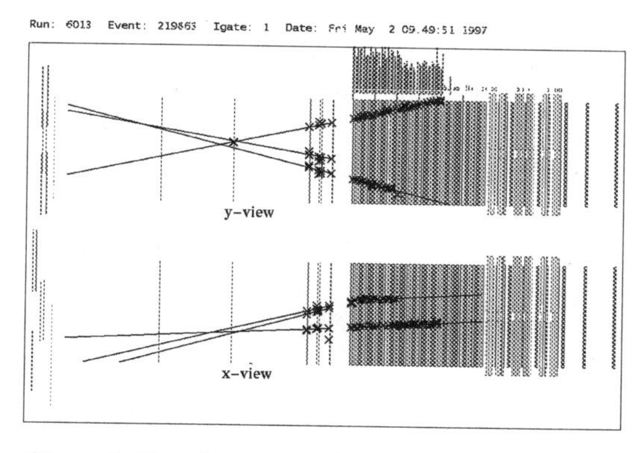

Figure 3. Run/Event 6013/219863: $\mu\mu(\nu)$ data event passing final cuts.

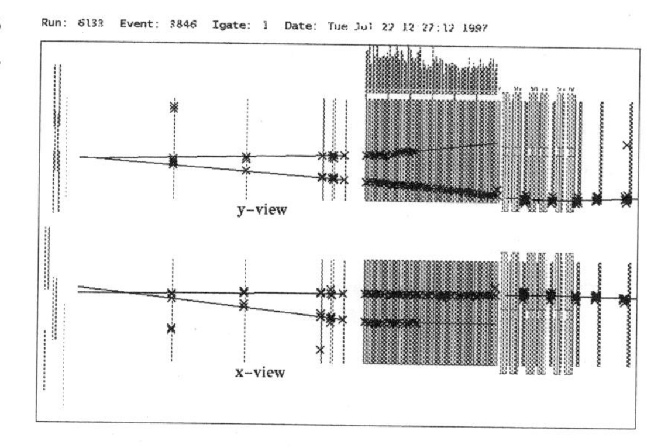

Figure 4. Run/Event 6133-03846: $\mu\mu(\nu)$ data event passing final cuts.

Table 4. Kinematic and reconstruction quantities for the three candidate $N^0 \to \mu\mu(\nu)$ events. The sign on the muon energy refers to the charge of the muon (if measured).

Run/Event	$E_{\mu 1}$ (GeV)	$E_{\mu 2}$ (GeV)	$P_{T\text{miss}}$ (GeV/c)	m_T (GeV/c^2)
5835/81705	−77.7	2.56	2.42	5.08
6133/3846	−92.0	5.85	1.41	3.08
6013/219863	48.0	4.34	2.07	4.66

hypothesis, all three events occur well within the fiducial volume away from the chambers and are evenly distributed throughout the decay channel. The transverse mass, invariant mass, and missing P_T are all consistent with a 5 GeV/c^2 N^0 decay.

Globally, the events have one feature which is improbable for an N^0 decay hypothesis. All three events have a muon energy asymmetry $(|E_1 - E_2|)/(E_1 + E_2)$ which is greater than 0.85. The probability that this occurs in a weak decay hypothesis[11] is less than 0.5% (including acceptance).

Interesting limits on the production of NHLs and neutralinos can be set from these results. The limit on NHLs reaches mixing parameter values below $|U|^2 = 10^{-5}$ for masses between 3 and 5 GeV/c^2. NuTeV is also the first experiment to set limits on the production of long-lived neutralinos in this mass range which decay by R-parity violation. The limit, although motivated by a neutralino hypothesis, is a generic limit applicable for any model of neutral particle production at the target[4] and reaches a differential cross section of $d\sigma/d\Omega \approx 4 \times 10^{-11}$ mb for decay lengths of ≈ 1.5 km.

On the other hand, several aspects of the candidate events are similar to those from neutrino interaction backgrounds, and might be indicative of unaccounted-for sources. For example, if the events were produced by neutrino interactions, then one would expect a high energy leading μ^- for ν-mode running. In the two cases where the charge of the leading muon can be measured, the event was in ν-mode with a leading μ^- and the observed energy asymmetry for the three events has a probability of 25 to 35%.

But it is difficult to explain the three events as neutrino interactions. If the events are due to charged-current neutrino interactions in the helium which somehow produce a second muon, then one would expect 28 times more chamber events in the data where none are observed. Various null hypotheses have been considered including unsimulated sources of prompt muon production, misreconstructions of chamber events, and unsimulated pion and kaon decay events. In all cases, the null hypotheses predict large numbers of other types of events that are not observed.

In conclusion, the rate corresponding to the observed three events is not consistent with Standard Model processes we have identified and the source of the events is not clear.

References

1. M. Gronau, C. N. Leung, and J. L. Rosner, Phys. Rev. D **29**, 2539 (1984).
2. R. E. Shrock, Phys. Rev. D **24**, 1232 (1981).
3. L. M. Johnson, D. W. McKay, and T. Bolton, Phys. Rev. D **56**, 2970 (1997).
4. L. Borissov *et al.*, hep-ph0007195.
5. T. Adams *et al.*, hep-ex/0009007.
6. A. Vaitaitis *et al.*, Phys. Rev. Lett. **83**, 4943 (1999).
7. J. A. Formaggio *et al.*, Phys. Rev. Lett. **84**, 4043 (2000).
8. J. Yu, *et al.*, "Technical Memorandum: NuTeV SSQT performance," Report No. FERMILAB-TM-2040, 1998.
9. W. Sakumoto *et al.*, Nucl. Instrum. Methods A **294**, 179 (1990); B. King *et al.*, Nucl. Instrum. Methods A **302**, 254 (1991).
10. D. Harris, J. Yu, *et al.*, Nucl. Instrum. Methods A **447**, 377 (2000).
11. J. A. Formaggio *et al.*, Phys. Rev. D **57**, 7037 (1998).

THE KARMEN TIME ANOMALY: SEARCH FOR A NEUTRAL PARTICLE OF MASS 33.9 MEV IN PION DECAY

P.-R. KETTLE[1], M. DAUM[1], M. JANOUSCH[2], J. KOGLIN[1,3], D. POČANIĆ[3], J. SCHOTTMÜLLER[1], C. WIGGER[1], Z.G. ZHAO[4]

[1] *PSI, Paul Scherrer Institute, Villigen, Switzerland.*

[2] *IPP, Institut für Teilchenphysik, Eidgenössische Technische Hochschule Zürich, Villigen-PSI, Switzerland.*

[3] *Physics Department, University of Virginia, Charlottesville, USA.*

[4] *IHEP, Institute of High Energy Physics, Chinese Academy of Science, Beijing, The People's Republic of China.*

E-mail: Peter-Raymond.Kettle@psi.ch

We have searched the muon momentum spectrum from pion decay-in-flight for evidence of a hitherto unknown neutral particle X, with a mass of 33.905 MeV. This process was suggested by the KARMEN collaboration as an explanation for a 'long-standing' anomaly in their data of neutrino induced reactions originating from stopped pion and muon decays. Using the advantages of a decay-in-flight experiment to kinematically separate muons associated with an X-particle from those originating from normal pion decay, we find no evidence for such a process and place an upper limit on the branching fraction $\eta \leq 6.0 \cdot 10^{-10}$of such decays at a confidence level of 95%.

1 Introduction

Based on the experimental findings of the KARMEN Collaboration and the hypothesis put forward by them to explain an anomaly seen in their time spectrum of neutrino induced reactions[1], we have searched for the pion decay

$$\pi^+ \to \mu^+ X, \qquad (1)$$

where X is a neutral particle of mass 33.905 MeV. Various theoretical explanations as to the nature of the X-particle have since been put forward. They range from a sterile neutrino scenario proposed by Barger et al.[2], through a supersymmetric solution involving the lightest neutralino, proposed by Choudhary et al.[3], to an exotic boson produced in muon decay, proposed by Gninenko and Krasnikov[4].

2 Experimental Principle and Method

Since the mass of X in decay (1) is very close to the mass difference between the charged pion and the muon, the resulting small Q-value makes it prohibitive to look for such a particle in *heavy neutrino searches* from pions decaying at rest. However, for a decay-in-flight experiment it has several advantages: the velocity and hence the energy-loss of the decay muon is very similar to that of the original pion, yielding a well-defined momentum, or so-called *magic momentum*, $p_\mu \approx p_\pi \cdot m_\mu / m_\pi \approx 0.757 p_\pi$. In addition, the flight direction of the muon differs only slightly from that of the pion. Thus, the pion beamline can itself be used as a spectrometer to separate muons from (1) from the dominant background of muons from the normal pion decay $\pi^+ \to \mu^+ \nu_\mu$.

In 1995, we measured the momentum spectrum of muons from π^+-decay in flight, using the high intensity pion channel at PSI, and placed an upper limit on the branching fraction η for this decay at $2.6 \cdot 10^{-8}$ (c.l. 95%) [5].

In order to increase our sensitivity on η, a new experiment, the detailed setup of which is described in [6], was started with substantial improvements made in suppressing the background.

A well-defined pion beam of 150 MeV/c decays in a 4.5m field-free region containing phase-space defining active collimators. Owing to the difference in the decay kinematics between reaction (1) and the normal pion decay, the associated muons from such decays are well separated by about 250 mrad in decay angle at the *magic momentum* of 113.5 MeV/c for muons from decay (1). This allows for a clean separation in the decay region which is further enhanced in the following muon spectrometer part of the beam-line.

The experimental method consists of tuning the pion transport system to a momentum of 150 MeV/c. The muon spectrometer part of the beam-line is then scanned over a momentum region of 103 MeV/c to 124 MeV/c, centered on the *magic momentum* of muons expected from decay (1). At each momentum setting, the pulse-heights and timing information of the three beam counters, as well as the twenty-two phase-space defining active veto-counters, are recorded on an event-by-event basis for a given number of protons on the pion production target. The number of events in a muon counter telescope, placed at an angle of 10° to the axis of the decay region, are also recorded as a monitor of the pion flux. The data so recorded for the forty-two momentum settings make-up a *decay muon scan.*

3 Analysis & Results

A momentum spectrum of candidate muon events is produced from each *decay muon scan* by applying appropriate timing and veto cuts to the raw events at each momentum setting and normalizing to the number of pion decays in the decay region. As the velocity of the muons from decay (1) would be the same as that of 150 MeV/c pions, by tuning the entire beam-line to 150 MeV/c, the pion timing at the three beam counters can be used to define strigent two-dimensional timing cuts, appropriate for selecting candidate muon events at the *magic momentum* of a *decay muon scan.* An example of such timing cuts, showing the particle identification power is shown in Figure 1.

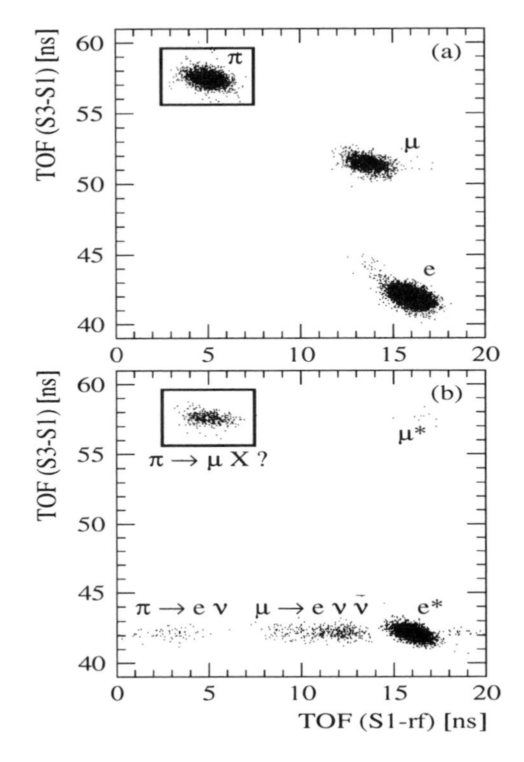

Figure 1. Experimental TOF-distributions with the relevant background event types (* refers to scattered particles). The upper plot (a) shows the situation when the whole beam-line is set to accept pions of 150 MeV/c. Plot (b) shows the situation for the central momentum of a 'decay muon scan' (113.5 MeV/c). The horizontal axes indicate the TOF of particles from the production target to the counter S1 relative to the accelerator time structure of 19.75 ns. The vertical axes indicate the TOF between the counters S1 and S3. The events in the box in (b) have the same timing as pions in (a), and thus may contain signal as well as background events. It should be noted that the number of protons for which the data were recorded is seven orders of magnitude higher in (b) than in (a).

Each momentum spectrum is then fitted with a hyperbolic function describing the smooth background due predominantly to scattered particles, and a Gaussian function describing the expected line-shape of muons from decay (1). An example of such a fit is shown in Figure 2 along with the effect one would expect to see for a branching fraction of $\eta = 5{\cdot}10^{-9}$. Clearly, no indication for the existence of the hypothetical decay (1) is evi-

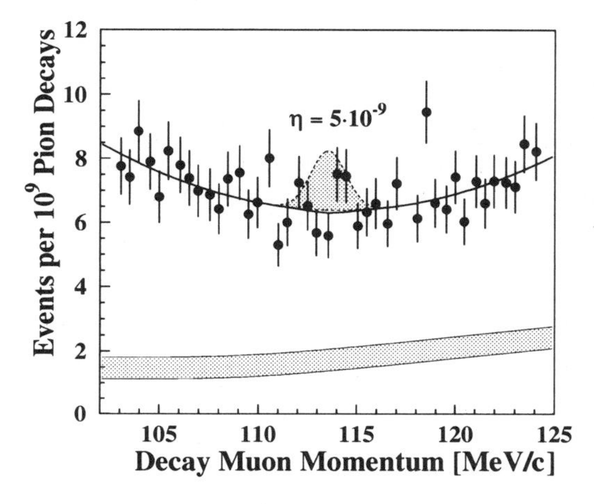

Figure 2. Experimental data of one 'decay muon scan'. The solid curve is the fit to the data (hyperbola plus Gaussian); the χ^2 is 33.6 for 37 degrees of freedom. The shaded peak indicates the expected effect of the hypothetical decay (1) for $\eta = 5.0{\cdot}10^{-9}$. The shaded band shows the expected level of background from the radiative pion decay, $\pi^+ \to \mu^+\nu_\mu\gamma$, within computational uncertainties ($\pm 1\sigma$). The minimum in the event distribution around 113.5 MeV/c arises from the fact that at this momentum, muons from the normal pion decay have a maximal emission angle and thus a minimal chance to enter the acceptance of the spectrometer.

dent in these data, nor is there any indication of its existence in the other momentum spectra.

The weighted mean of η from our twenty-eight separate momentum spectra give $\eta = (1.27 \pm 2.27) \cdot 10^{-10}$. Our systematic uncertainty was found to be 5% and accounts for the spectrometer acceptance and overall normalization uncertainty. In order to obtain a conservative result, we applied a factor of 1.05 multiplicatively to both η and its uncertainty which yields the branching fraction

$$\eta = (1.3 \pm 2.4) \cdot 10^{-10}. \qquad (2)$$

Following the Frequentist's approach we find an upper limit of

$$\eta \leq 6.0 \cdot 10^{-10} \text{ (c.l. } = 95\ \%). \qquad (3)$$

Figure 3 shows the KARMEN X-Particle branching ratio vs. lifetime plot together with the latest exclusion limits.

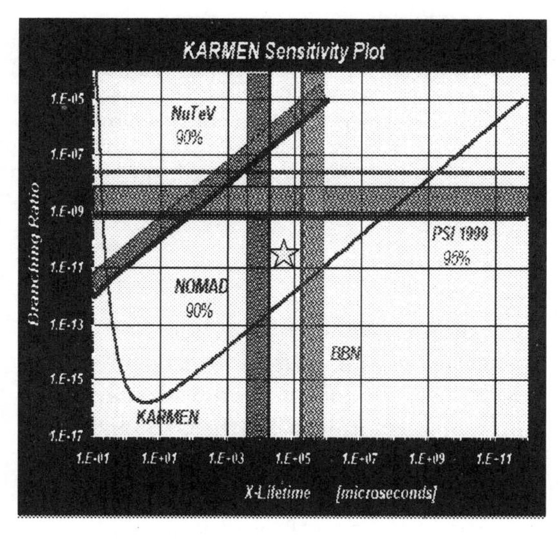

Figure 3. KARMEN X-Particle Sensitivity plot of Branching Ratio vs. Lifetime. The two limits, from NOMAD and Big-Bang Nucleosynthesis (BBN), apply to X being a sterile neurino. The star denotes the region still open

References

1. KARMEN Collaboration, B. Armbruster et al., *Phys. Lett.* B 348, 19 (1995).
2. V. Barger et al., *Phys. Lett.* B **352**, 365 (1995);
   V. Barger et al., *Phys. Lett.* B **356**, 617(E) (1995).
3. D. Choudhary and S. Sarkar, *Phys. Lett.* B **374**, 87 (1996);
   D. Choudhary et al., hep-ph/9911365 v2 (2000).
4. S.N. Gninenko and N.V. Krasnikov, *Phys. Lett.* B **434**, 163 (1998).
5. M. Daum et al., *Phys. Lett.* B **361**, 179 (1995).
6. M. Daum et al., *Phys. Rev. Lett.* **85**, 1815 (2000).

NEW LIMITS ON THE PRODUCTION OF MAGNETIC MONOPOLES AT FERMILAB

K. A. MILTON, G. R. KALBFLEISCH, M. G. STRAUSS, L. GAMBERG, W. LUO, AND E. H. SMITH

Department of Physics and Astronomy, University of Oklahoma, Norman, OK 73019-0225 USA

First results from an experiment (Fermilab E882) searching for magnetically charged particles bound to elements from the CDF and DØ detectors are reported. The experiment is described, and limits on magnetic monopole pair production cross sections for magnetic charges 1, 2, 3, and 6 times the Dirac pole strength are presented. These limits ($\sim$ 1 pb), hundreds of times smaller than those found in previous direct accelerator-based searches, use simple model assumptions for the photonic production of monopoles, as does the extraction of mass limits in the hundreds of GeV range.

1 Introduction

The most obvious reason for introducing magnetic charge into electrodynamic theory is the symmetry thereby imparted to Maxwell's equations. Further, the introduction of fictitious magnetic charge simplifies many calculations, as Bethe and Schwinger realized in their work on waveguides during World War II.[1]

Henri Poincaré first studied the classical dynamics of an electron moving in the field of a magnetic monopole,[2] while J. J. Thomson in lectures at Yale demonstrated that a classical static system consisting of electric (e) and magnetic (g) charges separated by a distance R had an intrinsic angular momentum pointing along the line separating the charges[3]: $\mathbf{J} = \frac{eg}{c}\hat{\mathbf{R}}$. Requiring that the radial component of this angular momentum be a multiple of $\hbar/2$ leads to Dirac's celebrated quantization condition, $eg = \frac{n}{2}\hbar c, \quad n = \pm 1, \pm 2, \pm 3, \ldots$ In fact, Dirac obtained this quantization condition by showing that quantum mechanics with magnetic monopoles was consistent only if this quantization condition held.[4] Thus, the existence of a single monopole in the universe would explain the empirical fact of the quantization of electric charge. Schwinger generalized this quantization condition to dyons, particles carrying both electric and magnetic charge.[5] He further argued that n had to be an even integer (sometimes even 4 times an integer).[5] Thus the smallest positive value of n could be 1 or 2, or 3 or 6 if it is the quark electric charge which quantizes magnetic charge.

2 Experiment Fermilab E882

The concept of the present experiment is that low-mass monopole–anti-monopole pairs could be produced by the proton–anti-proton collisions at the Tevatron. The monopoles produced would travel only a short distance through the elements of the detector surrounding the interaction vertex before they would lose their kinetic energy and become bound to the magnetic moments of the nuclei in the material making up the detector. We have obtained a large portion of the old detector elements (Be, Al, Pb) from the DØ and CDF experiments, and are in the process of searching for monopoles in these materials using an induction detector. A first paper describing our analysis of a large part of the DØ Al and Be samples has appeared.[6]

The model for the production process is that the monopole pairs are produced through a Drell-Yan process, which includes one factor of the velocity β to account for the phase space, and two additional factors of β to simulate the velocity suppression of the magnetic coupling. We use this rather simple model, the best available, because a proper field theoretical description of monopole interactions still does not exist.[7]

Any monopoles produced by the Tevatron are trapped in surrounding detector elements with 100% probability, and will be bound in that material permanently provided it is not melted down or dissolved.[8] Although the theory of binding is also in a crude state, monopole binding energies to nuclei are at least in the keV range, which is of the same order as the energy trapping the nucleus-monopole complex to the material lattice, more than adequate to insure permanent binding (and to preclude the extraction of monopoles from the sample by available magnetic fields).

We can set much better limits than those given by previous direct accelerator-based searches[9] because the integrated luminosity of Fermilab has increased by a factor of about 10^4 to 172 ± 8 pb^{-1} for DØ.

A schematic of the apparatus is available: *www.nhn.ou.edu/%7Egrk/apparatus.pdf.* The Fermilab samples are cut to a size of approximately $(7.5 \text{ cm})^3$ and are repeatedly moved up and down through a warm bore in a magnetically shielded cryogenic detector. The active elements are two superconducting loops connected to SQUIDs, which convert any current in the loops into a voltage signal. In empty space, the persistent current set up in the loop having inductance L by a monopole of charge g passing through it is $LI = 4\pi g/c$. A more exact expression was used in fitting data.

A pseudopole was constructed by making a long solenoid, which could either be physically moved through the detector loop, or turned on and off. It was also attached to an actual sample, so the background due to magnetic dipoles in the sample could be seen. Results of such tests are shown in Fig. 1. This demonstrates that we could easily detect a Dirac monopole.

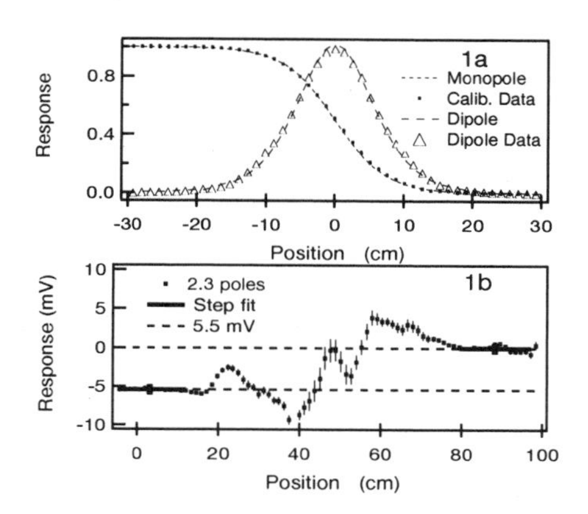

Figure 1. "Pseudopole" curves. a) Comparison of theoretical monopole response to an experimental calibration and of a simple point dipole of one sample with that calculated from the theoretical response curve. b) The observed "step" for a pseudopole current, corresponding to 2.3 minimum Dirac poles, embedded in an Al sample.

The monopole signal is a step in the output of the SQUID after that output has returned from its relatively large excursions resulting from dipoles in the sample. 222 Al and 6 Be samples were analyzed, and the distribution of steps had a mean of 0.16 mV and and rms spread of 0.73 mV, as shown in Fig. 2. We use the Feldman-Cousins analysis[10]: Because 8 samples were found within 1.28 σ of $n = \pm 1$, where 10.4 were expected, we can say at the 90% confidence level that the upper limit to the number of signal events with $n = \pm 1$ is 4.2. We also remeasured those outlying events, and found that all were within 2σ of $n = 0$, so we have no monopole candidates in this set. Similarly, the upper limit to the number of $|n| \geq 2$ events is 2.4.

We use the β^3 modified Drell-Yan production model together with the evolved CTEQ5m parton distribution functions[11] to estimate the acceptance of our experiment, as shown in Table 1. Using the total luminosity delivered to DØ, the number limit of monopoles, the mass acceptance so calculated, and the solid angle coverage of our

Table 1. Acceptances, upper cross section limits, and lower mass limits, as determined in this work (at 90% CL).

Magnetic Charge	$\|n\| = 1$	$\|n\| = 2$	$\|n\| = 3$	$\|n\| = 6$
Sample	Al	Al	Be	Be
$\Delta\Omega/4\pi$ acceptance	0.12	0.12	0.95	0.95
Mass Acceptance	0.23	0.28	0.0065	0.13
Number of Poles	< 4.2	< 2.4	< 2.4	< 2.4
Upper limit on cross section	0.88 pb	0.42 pb	2.3 pb	0.11 pb
Monopole Mass Limit	> 285 GeV	> 355 GeV	> 325 GeV	> 420 GeV

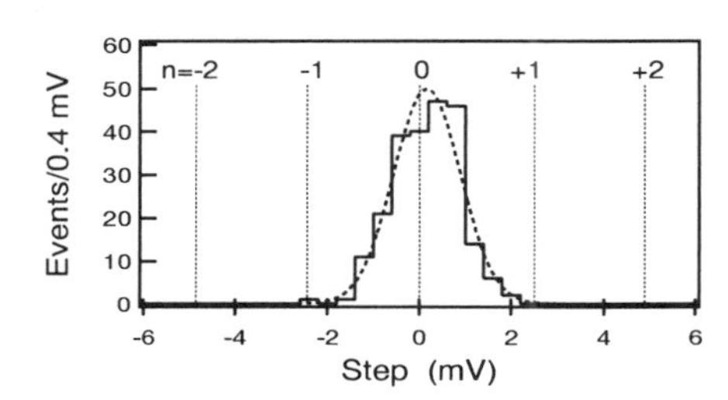

Figure 2. Histogram of steps. Vertical lines define the expected positions of signals for various n.

samples, we can obtain the $p\bar{p}$ cross section limits for the production of monopoles as given in Table 1. These are better by a factor of 200 than the earlier results of Bertani.[9] Using the production model again, we can convert these cross section limits into mass limits, simply by scaling the Drell-Yan cross section by the monopole enhancement factor of $n^2(137/2)^2$. These mass limits, also indicated in Table 1, are some 3 times larger than those of prior searches for accelerator-produced monopoles trapped in matter.[9]

3 Conclusions

This experiment to detect low-mass, accelerator produced magnetic monopoles is continuing. Over the next year, we will analyze the remaining Pb and Al samples from CDF, and extract better limits. In the future, such an experiment carried out using LHC exposures could reach monopole masses of a few TeV.

Acknowledgments

This work was supported in part by the US Department of Energy.

References

1. H. A. Bethe, J. Schwinger, J. F. Carlson, and L. J. Chu, 'Transmission of Irises in Waveguides,' 1942, in *Bethe Papers*, Archives, Olin Library, Cornell University, Ithaca, NY.
2. H. Poincaré, *Compt. Rend.* **123**, 530 (1896).
3. J. J. Thomson, *Electricity and Matter* (Scribners, New York, 1904), p. 26.
4. P. A. M. Dirac, *Proc. Roy. Soc. London* **A133**, 60 (1931).
5. J. Schwinger, *Phys. Rev.* **144**, 1087 (1966); **173**, 1536 (1969); D **12**, 3105 (1975); *Science* **165**, 757 (1969).
6. G. R. Kalbfleisch, *et al.*, submitted to *Phys. Rev. Lett.*, hep-ex/0005005.
7. L. Gamberg and K. A. Milton, *Phys. Rev.* D **61**, 075013 (2000).
8. L. Gamberg, G. R. Kalbfleisch, and K. A. Milton, *Found. Phys.* **30**, 543 (2000).
9. M. Bertani, *et al.*, *Eur. Phys. Lett.* **12**, 613 (1990).
10. G. J. Feldman and R. D. Cousins, *Phys. Rev. D* **57**, 3873 (1998), Table IV and V.
11. H. L. Lai et al., *Eur. Phys. J. C* **12**, 375 (2000).

MINI-REVIEW ON PROTON DECAY SEARCHES

M. SHIOZAWA

Kamioka Observatory, ICRR, Univ. of Tokyo
Higashi-mozumi, Kamioka-cho, Yoshiki-gun, Gifu 506-1205, JAPAN
E-mail: masato@icrkm4.icrr.u-tokyo.ac.jp

Among several proton decay (or nucleon decay in general) search experiments, most recent two experiments and their results are presented. First one in this article is Super–Kamiokande which utilizes a largest water Cherenkov detector with 50,000 tons of ultra pure water. Second one is Soudan 2 which is a iron tracking calorimeter with a total mass of 974 tons. In both experiments, various decay modes have been studied and no evidence for nucleon decays has yet been found. Experimental lower limit for nucleon partial lifetime was obtained for each decay mode and is 4.4×10^{33} years for $p \to e^+\pi^0$ mode and 1.9×10^{33} years for $p \to \bar{\nu}K^+$ mode at 90% confidence level from Super-Kamiokande.

1 Introduction

Most of Grand Unified Theories (GUTs) allow baryon number violated transitions between leptons and quarks and proton decay channels into lighter leptons and mesons become open. Therefore, the decay of the proton is one of the most dramatic predictions of various GUT models[1]. In the past two decades, several large mass underground detector experiments have looked for proton decay but no clear evidence has been reported[2].

It has been noted that there are several indirect evidence of GUT such as the observed family-structure of elementary particles and the meeting of the three gauge couplings. Moreover, recent discovery of finite, small neutrino mass[3] also suggests the physics at the energy scale far beyond the standard model[4]. Proton decays would provide the window for viewing the new physics and it is important to push up the experimental sensitivity for this processes. In this article, most recent two experiments, Super–Kamiokande and Soudan 2, are presented.

2 Super–Kamiokande

2.1 The Super–Kamiokande Detector

Super–Kamioka is a large water Cherenkov detector located in a mine 2700 meters-water-equivalent below the peak of mountain[5]. The detector holds 50 ktons of ultra-pure water contained in a cylindrical stainless steel tank and separated into two regions: a primary inner volume viewed by 11146 50 cm photomultiplier tubes (PMTs) and a veto region, surrounding the inner detector, viewed by 1885 20 cm PMTs. The fiducial volume is 22.5 kiloton and total detector livetime for physics analysis is now 1144 days corresponding to 70.4 kt·year exposure. Physical quantities of an event are automatically measured such as vertex position, the number of Cherenkov rings, momentum, particle type and the number of decay electrons by reconstruction algorithms[5].

2.2 $p \to e^+\pi^0$ search

This decay mode has a characteristic event signature, in which the electromagnetic shower caused by the positron is balanced against the two showers caused by the gamma rays from the decay of the π^0. This signature enables us to discriminate the signal events clearly from atmospheric neutrino background. To extract the $p \to e^+\pi^0$ signal from the event sample, these selection criteria are defined[5,6]: (A1) the number of rings is 2 or 3, (A2) all rings have a showering particle identification (PID), (A3) 85 MeV/c^2 < π^0 invariant mass < 185 MeV/c^2, (A4) no decay electron, (A5) 800 MeV/c^2 < total invariant mass < 1050 MeV/c^2 and total momentum < 250 MeV/c. Criterion (A2) selects $e^{\pm}$ and γ. Criterion (A3) only applies to 3-ring events. Criterion (A5) checks that the total invariant mass and total momentum correspond to the mass and momentum of the source proton, respectively. The total invariant mass and total momentum distributions after criteria (A1)–(A4) are shown in Figure 1.

From $p \rightarrow e^+\pi^0$ Monte Carlo sample, detection efficiency is estimated as 43%. Expected number of backgrounds from atmospheric neutrino interactions is estimated from atmospheric neutrino Monte Carlo sample as 0.1 events. Finally, there is no candidate events found in data sample. From these results, the lower limit on partial lifetime of proton is obtained as 4.4×10^{33} years at 90% confidence level (CL).

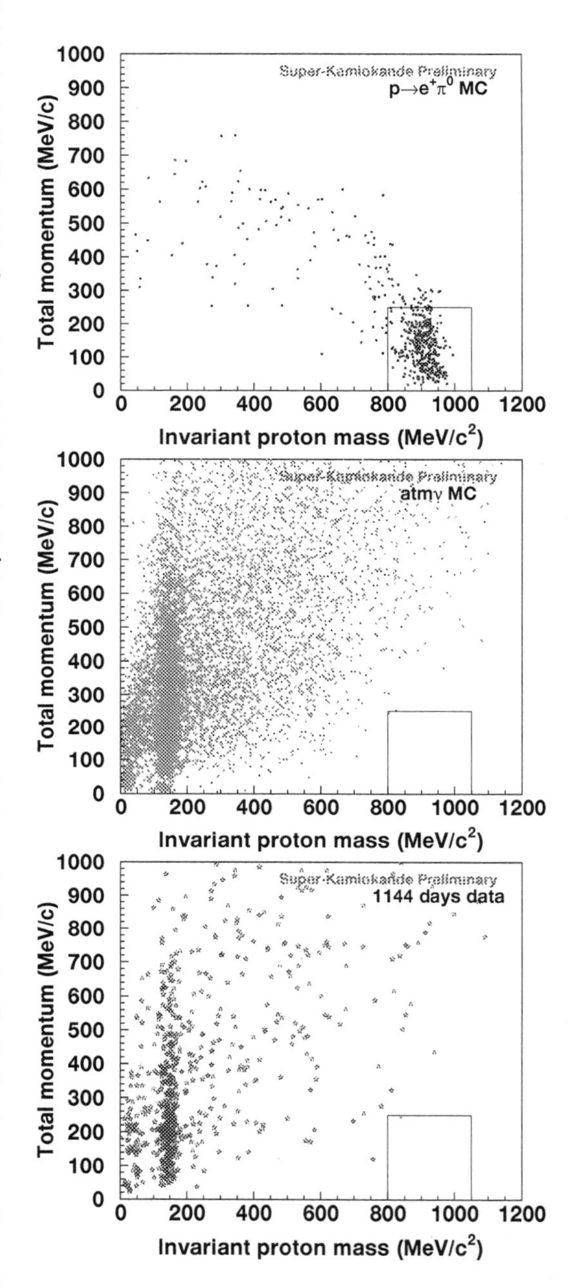

Figure 1. The total invariant mass and total momentum distributions after criteria (A1)–(A4) (see text) for 3 samples: (top) $p \rightarrow e^+\pi^0$ Monte Carlo, (middle) atmospheric neutrino Monte Carlo corresponding to 900 kton·year, (bottom) data corresponding to 70.4 kton·year. The boxed region in each figure shows the criterion (A5) for the $p \rightarrow e^+\pi^0$ signal.

2.3 $p \rightarrow \bar{\nu}K^+$ search

The $p \rightarrow \bar{\nu}K^+$ mode is generally favored by GUT models implemented with supersymmetry[7]. Because produced K^+ is expected to have momentum below Cherenkov threshold, the K^+ is generally invisible in a water Cherenkov detector. Therefore, experimental searches are performed by looking for decay products of the K^+. There are two prominent decay channels of K^+; $K^+ \rightarrow \mu^+\nu$ and $K^+ \rightarrow \pi^+\pi^0$ and three search methods for $p \rightarrow \bar{\nu}K^+$ have been developed[8].

In the first method, K^+ decays into μ^+ are looked for. The μ^+ is expected to have monochromatic momentum of 236 MeV/c. Selection criteria for this decay mode are defined as: (B1) the number of rings is one, (B2) the ring has a nonshowering PID, (B3) one decay electron, (B4) 215 MeV/c < muon momentum < 260 MeV/c, Figure 2 shows muon momentum distributions. Because there is no significant excess in the signal region, we applied spectrum fitting to ob-

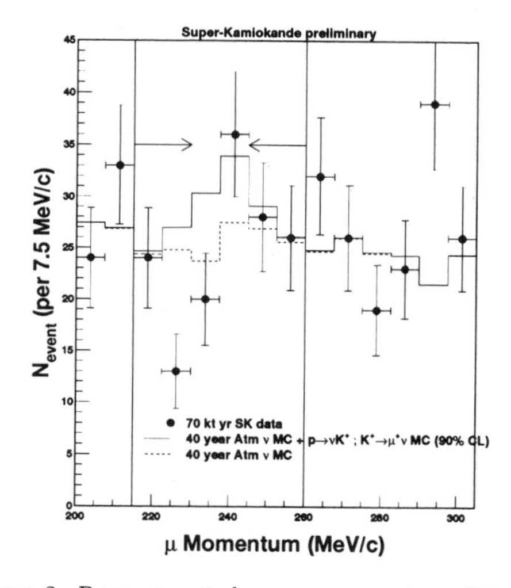

Figure 2. Reconstructed muon momentum distributions after criteria (B1–B3) (see text). Solid (dashed) line shows the estimated 90% CL number of proton decays + atmospheric neutrinos (atmospheric neutrinos). The black points with error bars show the data with the statistical errors.

tain upper limit of signal events. The dashed line shows the fitting result for backgrounds and solid line shows proton decay signal + backgrounds which gives the 90% CL upper limit of the number of signals. From this analysis, we obtained the partial lifetime limit for $p \to \bar{\nu}K^+$ decay mode as 4.3×10^{32} years at 90% CL.

In the second method, additional criterion is required to eliminate the remaining backgrounds. This criterion requires nuclear deexcitation γ from the residual ^{15}N nucleus. We expect the γ to be observed proceeding to the K^+ decay with the time difference corresponding to the K^+ lifetime ($\tau_{K^+} = 12$ nsec). By this criterion along with criteria (B1–B4), expected number of backgrounds is reduced to 1.1 events while detection efficiency including the kaon decay branching ratio is 9.3%. Candidate events are looked for in the data sample but no candidate is found. Obtained partial lifetime from this method is 9.5×10^{32} years at 90% CL.

In the third method, K^+ decays into two pions are used. Selection criteria for this method are: (C1) the number of rings is 2, (C2) all rings have a showering PID, (C3) 85 MeV/c^2 < π^0 invariant mass < 185 MeV/c^2, (C4) 175 MeV/c < π^0 momentum < 250 MeV/c, (C5) 40 p.e.s < photo electrons emitted by π^+ < 100 p.e.s, (C6) one decay electron. The criteria (C1–C4) select desired π^0 and the criteria (C5–C6) are defined for produced π^+. Detection efficiency including the kaon branching ratio is 6.8% and expected number of backgrounds is 1.9%. Again, there is no candidate remaining after these criteria and partial lifetime limit is 6.9×10^{32} years at 90% CL. In summary, we cannot find any candidate events for $p \to \bar{\nu}K^+$ decay mode in three methods. Combined lifetime limit from the three methods is obtained as 1.9×10^{33} years at 90% CL.

3 Soudan 2

3.1 The Soudan 2 Detector

The Soudan 2 detector is a time projection, modular iron tracking calorimeter with a total mass of 974 tons and fiducial mass of 770 tons[9]. 1 m long drift tubes fill the spaces in the stacked steel sheets with 1.6 mm thick to detect ionization electrons of charged particles.

3.2 $p \to \bar{\nu}K^+$ search

One of advantages of the Soudan 2 detector is that K^+ track from the proton decay can be imaged. Proton decay searches via $p \to \bar{\nu}K^+$ mode have been performed using two K^+ decay channels 3.56 kt·year exposure data are used for the $p \to \bar{\nu}K^+$ searches[10].

Selection criteria for $K^+ \to \mu^+\nu$ channel are: (D1) two charged tracks with common vertex (no proton), (D2) K^+ range < 50 cm, (D3) 28 cm < muon range < 58 cm, (D4) decay electron. Detection efficiency after all criteria is estimated as 9.0% and expected backgrounds is 0.4 events and one candidate event was found.

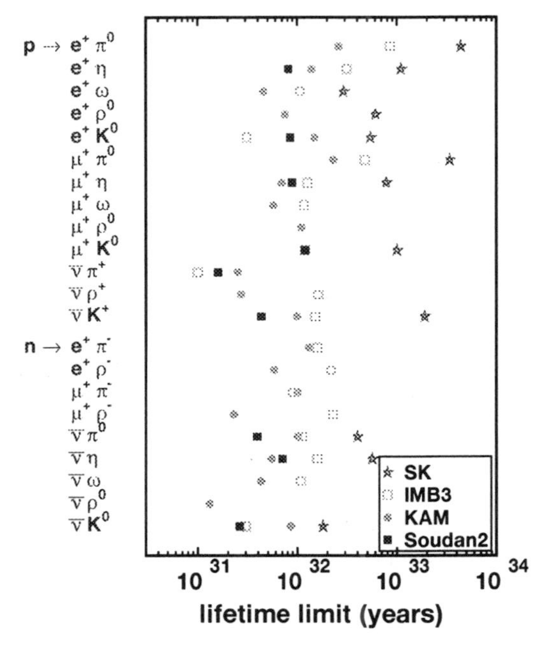

Figure 3. Obtained lifetime limits for various decay modes are summarized for Super–Kamiokande, IMB3, Kamiokande, and Soudan 2.

Moreover, selection criteria $K^+ \to \pi^+\pi^0$ channel are defined: (E1) two charged tracks and two showers, (E2) K^+ range < 50 cm, (E3) 100 MeV/c^2 $<$ invariant K^+ mass < 660 MeV/c^2, (E4) 80 MeV/c $< \pi^+$ momentum $<$ 400 MeV/c, (E5) 40 MeV/c $< \pi^0$ momentum $<$ 390 MeV/c, (F5) 10 MeV/c^2 $<$ invariant π^0 mass $<$ 290 MeV/c^2, Detection efficiency after all criteria is estimated as 5.5% and expected backgrounds is 1.1 events and no candidate event was found. From two methods, combined lower limit is obtained as 4.3×10^{31} years at 90% CL.

Summary

In this article, proton decay searches via decay modes of $p \to e^+\pi^0$ and $p \to \bar{\nu}K^+$ are presented. Figure 3 shows obtained lifetime limits for these decay modes and other various decay modes from Super–Kamiokande, IMB3, Kamiokande, and Soudan 2. In conclusion, there is no evidence for nucleon decays so far. We need to keep watching nucleons to open new physics beyond the standard model.

Acknowledgments

The author thanks the conference organizers and the participants for the comfortable conference in Osaka. He also appreciates the Super–Kamiokande collaborators for much help in preparing the latest results and his talk.

References

1. Jogesh C. Pati and Abdus Salam, Phys. Rev. Lett. **31**, 661 (1973); H. Georgi and S. L. Glashow, Phys. Rev. Lett. **32**, 438 (1974).
2. C. McGrew *et al.*, Phys. Rev. D **59**, 052004 (1999); K. S. Hirata *et al.*, Phys. Lett. **B220**, 308 (1989); C. Berger *et al.*, Z. Phys. **C50**, 385 (1991).
3. Y.Fukuda *et al.*, Phys. Rev. Lett. **81**, 1562 (1998).
4. For example, Jogesh C. Pati, hep-ph/0005095.
5. M. Shiozawa, PhD thesis, University of Tokyo (1999).
6. M. Shiozawa *et al.*, Phys. Rev. Lett. **81**, 3319 (1998).
7. N. Sakai and T. Yanagida, Nucl. Phys. **B197**, 533 (1982); S. Weinberg, Phys. Rev. **D26**, 287 (1982); J. Ellis *et al.*, Nucl. Phys. **B202**, 43 (1982).
8. Y. Hayato *et al.*, Phys. Rev. Lett. **83**, 1529 (1999).
9. W. W. M. Allison *et al.*, Nucl. Instr. Meth. **A376**, 36 (1996); W. W. M. Allison *et al.*, Nucl. Instr. Meth. **A381**, 385 (1996).
10. W. W. M. Allison *et al.*, Phys. Lett. **B427**, 217 (1998).

PRELIMINARY RESULTS OF THE L3 EXPERIMENT ON NEW PARTICLE SEARCHES AT $\sqrt{s} > 203$ GeV

SIMONETTA GENTILE

CERN, Geneva, Switzerland, Universitá "La Sapienza", Rome, Italy
E-mail: Simonetta.Gentile@cern.ch

A brief status of the searches for new particles based on approximately 100 pb^{-1} of data collected by the L3 experiment at $\sqrt{s} \simeq 203-208$ GeV is presented. The observed event rates in the studied search channels are consistent with the Standard Model expectations. All quoted results are preliminary.

1 Introduction

The increased center-of-mass energy during year 2000, at $\sqrt{s} \simeq 203-208$ GeV, in e$^+$e$^-$ collisions at LEP, opens a new window for the search of new physics. A preliminary analysis[1] searching for the Higgs Boson and also for supersymmetric particles in a data sample of approximately 100 pb^{-1}integrated luminosity collected by the L3 experiment is presented.

2 Higgs Searches

The Higgs particles searched are performed in the framework of different models.

In the Standard Model, the dominant Higgs production modes is e$^+$e$^-$ $\rightarrow Z^* \rightarrow HZ$. All significant signal decay modes are considered. The analysis procedure is optimized for $H \rightarrow b\bar{b}$, which accounts about 85% of the Higgs branching fraction in the mass range of interest and Z decaying in q$\bar{\text{q}}$, $\nu\bar{\nu}$, e$^+$e$^-$, $\mu^+\mu^-$. The number of observed events is consistent with the background expectation over the entire studied interval of Higgs masses. No evidence of a signal is seen in any channel.

For illustrative purposes, in Fig. 1(a) the reconstructed Higgs mass is shown for a sample of signal-like events selected, with mass-independent requirements, such as a large b-tag value. The exclusion confidence level as a function of the Higgs mass hypothesis is shown in Fig. 1(b). Combining these new re-

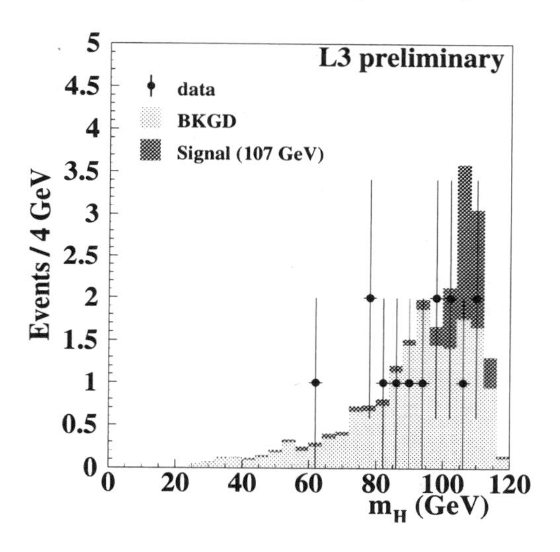

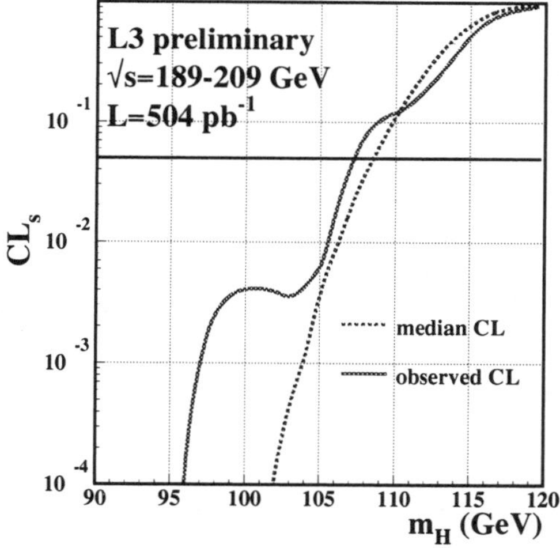

Figure 1. a) Reconstructed Higgs mass distributions in $\sqrt{s} = 203$-208 GeV data b) Observed and expected signal confidence level as a function of the Higgs mass.

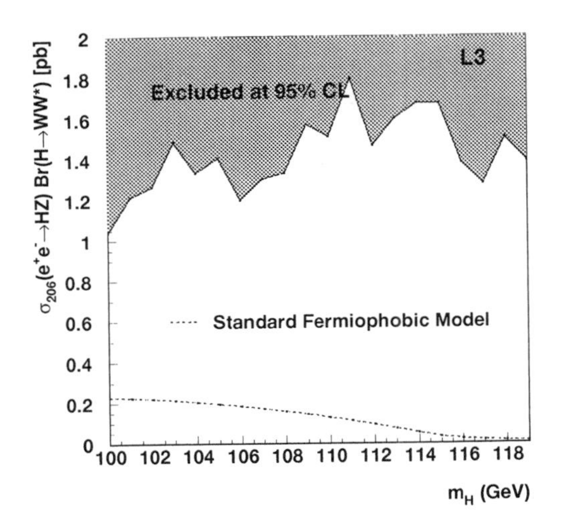

Figure 2. Observed upper limits on the Higgs cross section times the branching fraction into WWW as function of the Higgs boson mass.

sults with those of previous years a new limit on the Higgs boson mass is derived:

$m_{\rm H} > 107.6$ GeV at 95% CL.

In the Minimal Supersymmetric Standard Model framework the search for the lightest neutral scalar and neutral pseudoscalar Higgs bosons is also updated. Lower mass limits for large values of $\tan\beta$ are set at the 95% confidence level. The observed limits are $m_{\rm h}$, $m_{\rm A} > 86.4$ GeV(expected 87.3 GeV). The search for pair produced charged Higgs bosons and for Higgs decaying in invisible particles is also updated. I want conclude with few remarks on a search performed in a new Higgs decay channel for the first time presented: the fermiophobic Higgs. The search for a fermiophobic Higgs has been carried out primarily in the H→ $\gamma\gamma$ channel, where the Higgs couples to the photons through a W loop. For Higgs scalars heavier than 90 GeV, however, the branching fraction of $\gamma\gamma$ predicted relative to WW* becomes very small. This motivates a search for H→ WW^*. Fig. 2 shows the observed upper limits on the Higgs cross section times the Higgs branching ratio into WW^* as function of the Higgs boson mass.

3 SUSY searches

3.1 Scalar Leptons, Scalar Quarks, Charginos and Neutralinos

In this section, the search for new particles predicted by supersymmetric theories (SUSY) is reported and limits in the Minimal Supersymmetric Standard Model (MSSM) framework are derived.

Besides the main characteristic of missing energy, supersymmetric particle signals can be further specified according to the number of leptons or the multiplicity of hadronic jets in the final state. Charginos, $\tilde{\chi}_1^{\pm}$, the supersymmetric partners of $W^{\pm}, H^{\pm}$ decay mainly to final states similar to W W, decays. For neutralinos, $\tilde{\chi}_i^0$, the supersymmetric partner of Z, γ and neutral Higgs boson we distinguish two classes of processes: $e^+e^- \rightarrow \tilde{\chi}_1^0\tilde{\chi}_2^0$ and $e^+e^- \rightarrow \tilde{\chi}_2^0\tilde{\chi}_2^0$. Both for charginos and neutralinos, the event energy is directly related to ΔM, ($\Delta M = M_{\rm SUSY} - M_{\tilde{\chi}_1^0}$), where the $\tilde{\chi}_1^0$ is supposed to be the lightest stable supersymmetric particle (LSP).

Five types of selection criteria have been optimized for all possible decays of charginos and used also for the selection of $\tilde{\chi}_2^0\tilde{\chi}_2^0$ events, which give rise to final states very similar to those of chargino pair production, even if with very different branching ratios. The signal topologies and the associated background sources depend strongly on ΔM. Therefore all six selections are optimised separately for four different ΔM ranges.

The scalar leptons, the supersymmetric partners of the right- and left- handed leptons, decay into their partner lepton and neutralino, $\tilde{\ell} \rightarrow \tilde{\chi}_1^0\ell$. Their signatures are simpler since most of the time the final state is given by two acoplanar leptons of the same family. To account for the three lepton types three different selections are performed. The

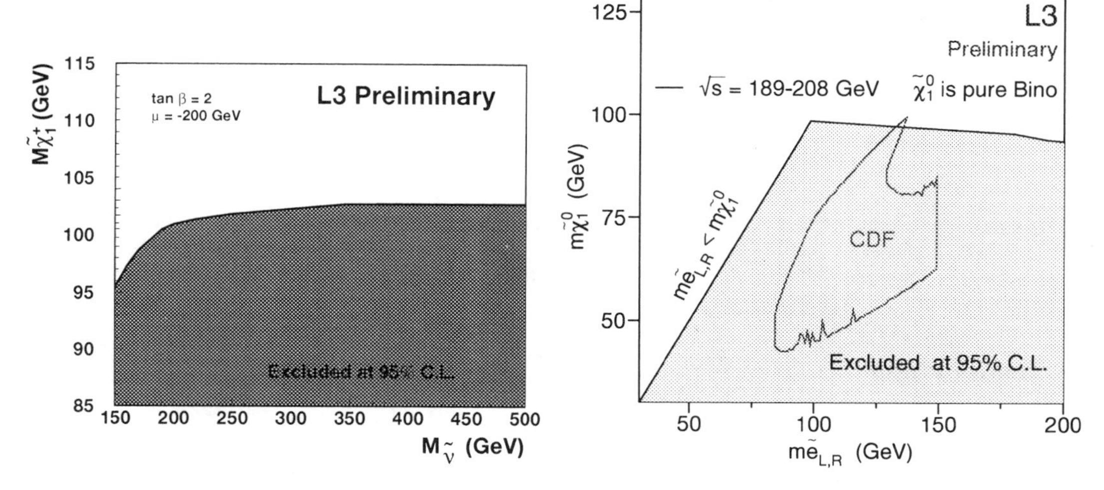

Figure 3. a) A lower limit on the chargino mass as a function of the s-neutrino mass. b) Excluded region for a pure bino neutralino model compared to the region consistent with supersymmetric interpretation of the CDF event in the scalar electron scenario.

dependence of the signals on ΔM is taken into account with three different optimisations for every selection. from Figure 3a) a limit value of $M_{\tilde{\chi}_1^1} > 102.6$ GeV is derived for high s-neutrino mass.

At LEP energies the most important process involving scalar quarks are the production of $\tilde{t}_1$ and $\tilde{b}_1$. The signal events of $\tilde{t}_1 \rightarrow c\tilde{\chi}_1^0$ and $\tilde{t}_1 \rightarrow b\tilde{\chi}_1^0$ are characterised by two high multiplicity acoplanar jets containing c- or b-quarks. In addition to the jets two leptons are present for $\tilde{t}_1 \rightarrow b\tilde{\ell}\tilde{\nu}$ and $\tilde{t}_1 \rightarrow b\tau\tilde{\nu}$ modes. The two $\tilde{\chi}_1^0$ in the final state escape the detection leading to missing energy in the event.

No excess of events relative to what is expected from Standard Model processes is observed in scalar leptons and scalar quarks searches. The collected luminosity does not allow to improve our previous limits.

3.2 Models with light gravitinos

Single or two-photon events with missing energy could provide evidence for various processes involving supersymmetric particles like pair-production of neutralinos ($\tilde{\chi}_1^0\tilde{\chi}_1^0$, $\tilde{\chi}_1^0\tilde{\chi}_2^0$,$\tilde{\chi}_2^0\tilde{\chi}_2^0$, etc.), gravitinos ($\tilde{G}\tilde{G}$), or associated production of a neutralino and a gravitino ($\tilde{\chi}_1^0\tilde{G}$). Neutralinos, according to different supersymmetric models, can either decay as $\tilde{\chi}_2^0 \rightarrow \tilde{\chi}_1^0\gamma$ or $\tilde{\chi}_2^0 \rightarrow \tilde{G}\gamma$. Pair-production of gravitinos can be accompanied by initial state radiation. The missing energy is carried away by the weakly interacting lightest supersymmetric particle (LSP). In these models the LSP is the gravitino. In the Standard Model single or two-photon events with missing energy are produced via the reaction $e^+e^- \rightarrow \nu\bar{\nu}\gamma(\gamma)$.

The supersymmetric interpretation, with gravitino as LSP, of the $ee\gamma\gamma$ event, observed by the CDF collaboration, is almost completly ruled out by our L3 data,as shown from figure 3b.

Searches for excited leptons and for new heavy sequential leptons have been updated. No excess has been found in any channel.

References

1. L3 Collab., L3 Note 2601 (2000), contrib. paper 410 to this Conference and reference therein.

UPDATE OF THE DELPHI SEARCHES RESULTS USING YEAR 2000 DATA

M.C. ESPIRITO SANTO
CERN, EP Division, 1211 Geneva 23, Switzerland
E-mail: maria.espirito.santo@cern.ch

Results of the DELPHI searches for new particles and phenomena are updated using the recent data collected in the year 2000. Higgs searches, Supersymmetry searches and searches for other exotic particles are covered. The luminosity used in the different analyses is in the range 90 to 100 pb^{-1}. All the results are preliminary and all limits are given at 95% confidence level (CL).

1 Introduction

Updates of DELPHI searches results using the data collected in the year 2000 are briefly presented. This report is bound to be a selection of topics, a more complete description can be found in [1] and [2] (submited to this conference) and in the references therein. The luminosity used in most analyses ranges from 90 to 100 pb^{-1}, collected at a centre-of-mass energy ($\sqrt{s}$) ranging from 205 GeV to 208.8 GeV ($< \sqrt{s} > \simeq 205.5$ GeV).

In this last year of the LEP running, the emphasis is put in exploiting as much as possible the discovery potential of the machine, by reaching the highest possible $\sqrt{s}$. In this spirit, two search paths have to be kept in mind: on one hand, the detailed exploration of the phenomenological consequences of proposed models; on the other hand, the coverage of all possibilities through topology-based (model independent) searches.

2 Supersymmetry searches

Many Supersymmetry (SUSY) searches in a variety of scenarios are conducted. The updated results are presented in [1]. In the framework of the minimal supersymmetric standard model (MSSM) assuming R parity conservation and gravity mediated SUSY breaking the pair production of charginos, neutralinos, sleptons and squarks is searched for. No evidence for a signal is found in any of the channels and cross-section limits can be derived, which can generally be translated into lower limits on the mass of the SUSY particle as a function of the mass difference (ΔM) with respect to the Lightest SUSY Particle (LSP). For the lightest chargino a mass limit of $m_{\tilde{\chi}_1^\pm} > 101.7$ GeV/c^2 is found for $\Delta M > 10$ GeV/c^2, assuming a relatively large sneutrino mass (>300 GeV/c^2). This value improves to 102.2 GeV/c^2 for mostly wino-like chargino. It degrades to $m_{\tilde{\chi}_1^\pm} > 89.3$ GeV/c^2 for $\Delta M > 3$ GeV/c^2. From the neutralino searches, limits in the range 0.1 pb to 0.8 pb for $\tilde{\chi}_1^0\tilde{\chi}_2^0$ production are obtained.

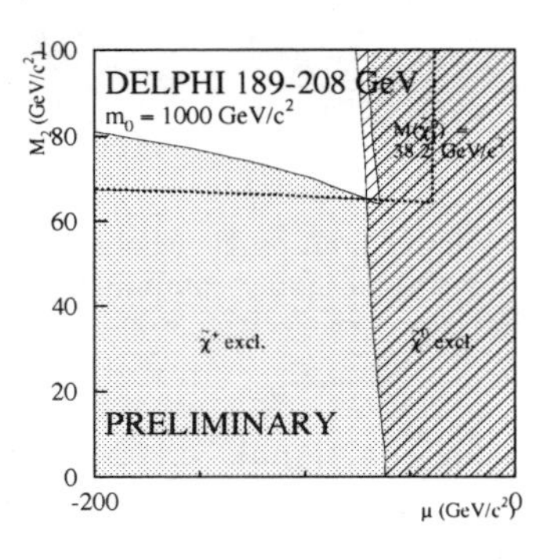

Figure 1. Excluded regions in the (μ,M_2) plane for $tan\beta = 1$ for $m_0 = 1000$ GeV/c^2. The shaded areas show regions excluded by searches for charginos and the hatched areas show regions excluded by searches for neutralinos. The thick dashed curve shows the isomass contour for $m_{\tilde{\chi}_1^0} = 38.2$ GeV/c^2, the lower limit on the LSP mass obtained at $tan\beta = 1$. The chargino exclusion is close to the isomass contour for $m_{\tilde{\chi}_1^\pm}$=102.2 GeV/c^2.

Putting together the results of the searches for charginos and neutralinos, and

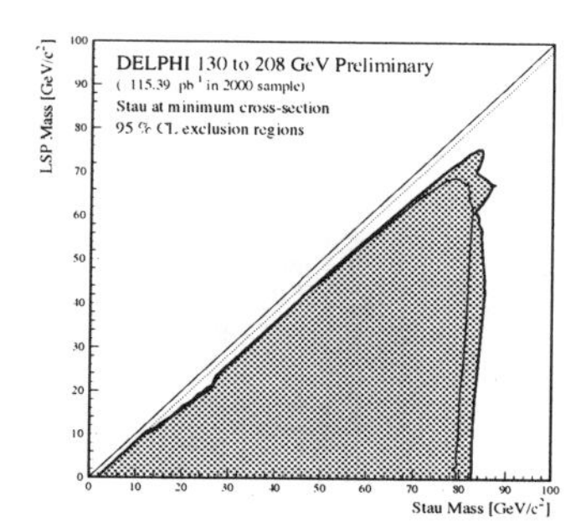

Figure 2. Exclusion limit at 95% CL for $\widetilde{\tau}_{min}$. The thick line represents the obtained exclusion limit and the thin line the mean limit expected from background-only experiments.

under the assumption of GUT scale unification, a limit on the LSP mass valid for any tanβ and for high m_0 (1 TeV) can be obtained: $m_{\tilde{\chi}_1^0} > 38.2$ GeV/c^2. This limit is reached for $tan\beta = 1$, and the regions in the (μ, M_2) plane excluded by each of the searches are shown in figure 1. For m_t= 175 GeV/c^2 and $m_0 < 1000$ GeV/c^2, the $0.8 < tan\beta < 2$ range is excluded by Higgs searches [2] and, if these constraints are used, the LSP limit for high m_0 is $m_{\tilde{\chi}_1^0} > 49$ GeV/c^2 for $tan\beta > 2.0$ and $m_{\tilde{\chi}_1^0} > 42$ GeV/c^2 for $tan\beta < 0.8$.

Sfermions are also searched for at LEP. In particular, the stau could be light in case of large third family mixing. In figure 2, the excluded region in the plane of the stau and LSP masses is shown for the mixing angle minimizing the cross-section.

Many other SUSY scenarios are considered. Also presented in [1] are the stau search in gauge mediated SUSY breaking scenarios and the chargino search in R parity violating models (coupling of UDD type).

3 Exotic searches

Examples of more topological approaches are the search for events with photons only and for monojet events. These final states are rather clean in the Standard Model (SM)

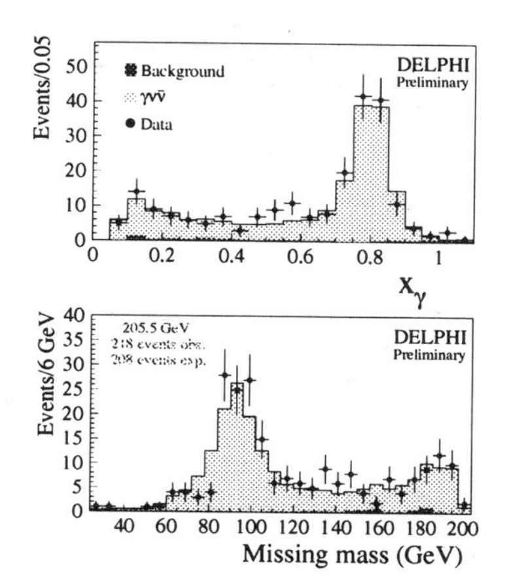

Figure 3. The distributions of $x_\gamma = E_\gamma/\sqrt{s}$ and the recoiling mass against the detected photon for the single photon events at 205.3 GeV. The light shaded area is the expected spectrum from $e^+e^- \rightarrow \nu\bar{\nu}\gamma$ and the dark shaded area is the total background from other sources.

and would constitute a rather clear signature of some SUSY processes. As an example, the single photon energy spectrum is shown in figure 3 showing a good agreement with the expectation. In the monojet analysis described in [3] no excess is found with respect to the expectations and the cross-section for the process $q\bar{q}\nu\bar{\nu}$ is measured to be:

$$\sigma(q\bar{q}\nu\bar{\nu}) = 0.190 \pm 0.058 \pm 0.024 \text{ pb},$$

where the first error is statistical and the second one is systematic.

The DELPHI searches for compositeness, technicolor and flavour changing neutral currents were also updated using the year 2000 data [1]. In particular excited electrons can be searched for directly, through the process $e^+e^- \rightarrow \ell\ell^* \rightarrow \ell\ell' V$, and indirectly, looking for deviation in the differential cross-section of the QED process $e^+e^- \rightarrow \gamma\gamma$ [4]. Figure 4 shows the upper limit on the ratio between the coupling and the mass of the excited particle as a function of its mass obtained by combining the two searches.

4 Higgs bosons searches

Higgs bosons were also searched for in different scenarios. In Two Higgs doublets models

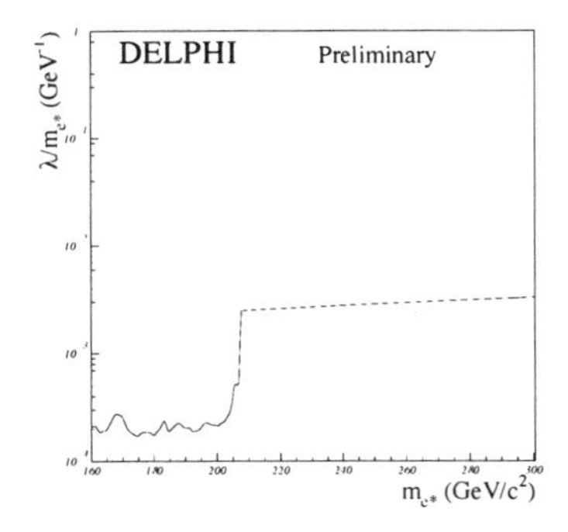

Figure 4. Combined excited electron limits for $f = f'$ from direct and indirect searches. The line shows the upper limits at 95% CL on the ratio λ/m_{e^*} as a function of the mass. Up to the kinematic limit the result is dominated by the single production direct search. Above this value the limit is the one coming from the indirect search using $e^+e^- \rightarrow \gamma\gamma$.

(2HDM), neutral Higgs bosons with important branching ratios into invisible particles (e.g. LSP in the MSSM) or photons (fermiophobic scenarios) can arise naturally. From the updates of such searches, and assuming branching-ratios of 1, the mass limits $m_h >$ 104.3 GeV/c^2 [2] and $m_h >$ 112.0 GeV/c^2 [1] are obtained for invisible and fermiophobic Higgs bosons respectively.

The data collected by the LEP experiments at $\sqrt{s} > 204$ GeV allows to extend the range of searches for the SM Higgs boson. The result from the DELPHI analysis are discussed in [2] for the different search channels. In general, agreement with the expectations is found, with a small excess seen in the 4 jets final state. This excess corresponds to events with a high probability of containing b quarks. The plot of the reconstructed mass combining all channels is shown in figure 5. In the figure the new data is combined with the data of the previous year and compared with the simulation expectations (including a Higgs signal of 110 GeV/c^2 mass). From these results a limit on the SM Higgs mass of $m_H >$ 109 GeV/c^2 (in agreement with the expected limit of 109.2 GeV/c^2) was derived.

In the run of this year, the emphasis is put in the exploration of the highest possible

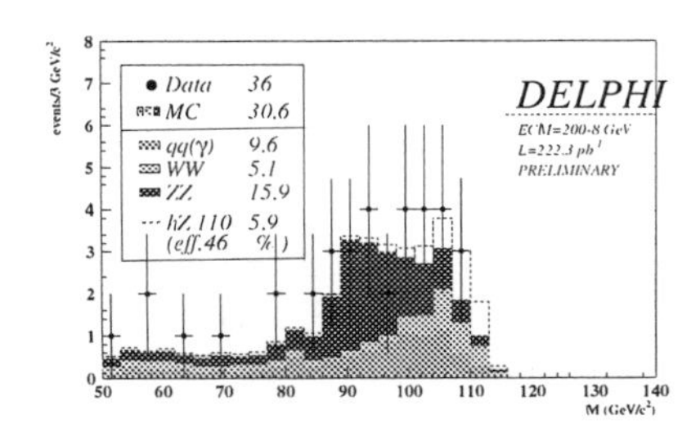

Figure 5. Distribution of the reconstructed Higgs boson mass when combining all HZ analyses at 200-208 GeV. Data (dots) are compared with SM background expectations (coloured histograms) and with the normalized signal spectrum added to the background contribution (white histogram). The mass hypothesis is 110 GeV/c^2.

mass region. With this purpose the analyses were checked to cover efficiently the mass range 110-115 GeV/c^2 and it is crucial to combine as early as possible the results from the four LEP experiments. The present LEP combined result was presented to this conference. However, the LEP running is ongoing at the collected luminosity is expected to double with respect to the present values.

References

1. DELPHI Collab., "Update of results on new particle searches with recent LEP data from the year 2000 run", contributed paper to ICHEP2000.
2. DELPHI Collab., "Search for Higgs bosons at LEP in the year 2000", contributed paper to ICHEP2000.
3. DELPHI Collab., "Measurements of the cross sections of WW, ZZ and other 4 fermion processes from data taken during the year 2000", contributed paper to ICHEP2000.
4. DELPHI Collab., "Determination of the $e^+e^- \rightarrow \gamma\gamma(\gamma)$ cross-section using data collected by DELPHI up to the year 2000", contributed paper to ICHEP2000.

LATEST RESULTS FROM ALEPH AND 2000 LEP SUSY WORKING GROUP RESULTS

MARCELLO MAGGI

INFN Sezione di Bari, I-70126 Bari, Italy and CERN, CH-1211 Geneve 23, Switzerland.
E-mail: Marcello.Maggi@cern.ch

First results based on the 2000 data sample collected with the ALEPH detector are presented. They are based on an integrated luminosity of about 62 pb$^-1$ collected at an average centre-of-mass energy of 205 GeV and of about 31 pb^{-1} at an average centre-of-mass energy of 207 GeV. Preliminary LEP combined results on searches for supersymmetric particles are also given.

1 Introduction

The LEP accelerator in summer 2000 was operated up to a centre-of-mass energy of about 209 GeV, where unexplored domains in high energy physics were probed.

The ALEPH detector collected about 62 and 31 pb$^-1$ at an average centre-of-mass energy of 205 and 207 GeV respectively. Very preliminary results on standard model (SM) processes, and on searches for Higgs bosons and supersymmetric particles are presented. More detailed descriptions of the results reported can be found in the contributed papers[1,2,3].

The LEP SUSY working group[4] has combined the preliminary results on searches for supersymmetric particles as obtained from the 2000 data collected, until 10 July, by the four LEP experiments: ALEPH, DELPHI, L3 and OPAL.

2 ALEPH results on 2000 data

2.1 Standard model processes

The comparison between the measurements of standard model processes performed with the latest data and the theoretical expectations gives the idea of the quality and the understanding of the detector performance.

The cross sections and asymmetries of fermion pairs are measured for both exclusive processes, for which radiative returns to the Z resonance are removed by requiring $\sqrt{s'/s} > 0.9$, and for inclusive processes, defined requiring $\sqrt{s'/s} > 0.1$ (for more details see[5]). Figure 1 shows all the cross sections measurements performed with the ALEPH detector, compared to the SM expectations.

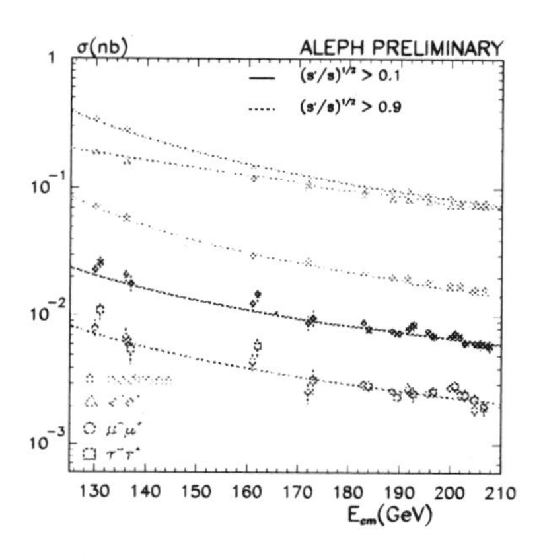

Figure 1. Cross section measurements for fermion pair production versus centre-of-mass energy. The curves are the predicted cross sections.

Studies of W and Z pair production were also performed at energies above 202 GeV. In Table 1 the cross section measurements are

Table 1. WW and ZZ cross sections measured at two different centre-of-mass energies. The errors are statistical only.

	$\sqrt{s} = 204.9$ GeV	$\sqrt{s} = 206.7$ GeV
σ^{WW}_{CC03}	16.70 ± 0.61 pb	17.01 ± 0.86 pb
σ^{ZZ}_{NC02}	$0.86^{+0.29}_{-0.26}$ pb	$0.51^{+0.35}_{-0.28}$ pb

given for the two processes at 205 and 207 GeV nominal centre-of-mass energies. While for the W pair production cross section a very good agreement with SM prediction is found, Z pair production cross section measurements are slightly below the theoretical prediction.

2.2 Searches for the Higgs bosons

Searches for various types of Higgs bosons have been performed. The analyses for the SM Higgs boson, the neutral Higgs boson of the minimal supersymmetric model (MSSM), the charged Higgs bosons and the invisible Higgs boson follow closely those designed for the data collected with the ALEPH detector in 1999 with few improvements[2]. Here I will detail only few results concerning the SM Higgs boson search since all ALEPH results were combined together with those obtained by the other experiments and this is elsewhere discusses[6].

In the search for the SM Higgs the overall number of observed candidates is in good agreement with the expected number of background events. The analyses, without taking into account systematic uncertainties, exclude masses below 111.1 GeV/c^2 with 95% confidence level (CL) while the expected exclusion is 112.3 GeV/c^2.

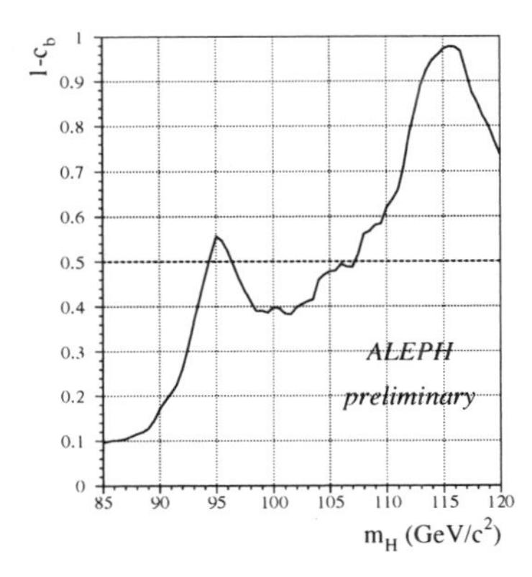

Figure 2. Observed (solid) and expected (dashed) CL curves for the background hypothesis as a function of the hypothesized standard model Higgs boson mass.

A promising candidate was selected, with a reconstructed Higgs boson mass of 114.3 GeV/c^2, in the four jet topology. Both Higgs boson jets are well b-tagged and show displaced vertices. The 14 GeV of missing energy in this event points in the same direction of one of the b-tagged jets where a muon is also indentified, compatible with a b semi-leptonic decay. The properties of this event cause almost entirely the peak in the observed background hypothesis confidence level distribution as a function of the Higgs boson mass at 115 GeV/c^2 shown in Figure 2. This corrisponds to a probability for this, or even less background-like, observation of 2%.

2.3 SUSY searches

Supersymmetric particles have been searched for in the framework of the MSSM, both with R-parity conservation and violation, and of gauge mediated symmetry breaking theories (GMSB). The various selections are similar to those optimised for the data collected at lower centre-of-mass energies[3]. Details of the analyses and the results are discussed by several speakers in this conference. Here the results with the latest data are only flashed.

Searches for squarks lead to the observation of 8 candidates in the data sample in agreement with the 6 events expected from background events. Sleptons decaying to

Table 2. Number of observed candidates and expectation from standard model background processes for different channels.

channel	Obs. events	Exp. events
$\tilde{t} \to c\chi$	7	5.2
$\tilde{t} \to b\ell\tilde{\nu}$	1	0.7
$\tilde{b} \to b\chi$	2	1.1
$\tilde{e} \to e\chi$	29	18.4
$\tilde{\mu} \to \mu\chi$	19	16.5
$\tilde{\tau} \to \tau\chi$	10	10.0

their standard model partner and the latest neutralino are also searched for. The observed events are in agrement with the background expected from SM processes as summarised in Table 2.

Searches for charginos and neutralinos have been updated only in topologies relevant in the case of heavy sleptons. Six chargino and one neutralino candidates were selected, to be compared with expectations of 4.8 and 1.2 events from background processes, respectively. Assuming heavy sleptons, chargino searches alone allow a mass lower limit of 37.2 GeV/c^2 to be set at 95% CL, which is extended to 38.5 GeV/c^2 when searches for neutralinos are taken into account.

Various searches adressing very different final states resulting from R-parity violation scenarios have been performed. R-parity is considered to be violated via a dominant LLE, LQ$\bar{\text{D}}$ or $\bar{\text{U}}\bar{\text{D}}\bar{\text{D}}$-type Yukawa coupling. A good agreement between expected standard model background and observed candidates has been found in all searched topologies.

In the GMSB scenario, the lightest neutralino is expected to decay mainly into a photon plus a gravitino. If the lifetime associated to this decay is small, the experimental topology consists of two acoplanar and energetic photons. This search has been updated and no candidates have been observed, to be compared with 0.7 background events expected from SM processes. Another characteristic topology of GMSB are long lived sleptons. This analysis has been updated on 192-196 GeV sample and no events survived the cuts, to be compared with about 0.9 expected background events.

3 LEP SUSY combined results

3.1 photonic final states

In the context of SUSY, single photon events are expected from several possible processes. The SM background is mainly originated by $e^+e^- \to \gamma\nu\bar{\nu}$. The total accepted single-photon rate, observed on data collected so far, agrees with SM expectation. To simulate the SM processes two different programs are in use, KORALZ and NUNUGPV, which agree on the predicted total rate. However, they differ in the predicted shape of the recoil mass distribution. Data seem to favour NUNUGPV. It is important to notice that the 4-5% systematic uncertainty, assigned to the simulation of SM processes, needs to be improved, since it can mask presence of new physics.

Acoplanar photon pairs have also been combined. The number of observed candidates is 38, in agreement with the SM expectations of 46.1 events. The measured recoil mass distribution is also well reproduced by SM simulated events.

3.2 Charginos and sfermions

The chargino pair production cross section is large for heavy sfermions ($m_{\tilde{\nu}} >$ 300 GeV/c^2). In this scenario, the chargino mass lower limit at 95% CL is 103.1 GeV/c^2. Collecting 10 pb−1 per experiment at the highest centre-of-mass energy, will extend the experimental sensitivity by 500 MeV/c^2 and close the mass window accessible at LEP.

Table 3. Sfermion mass lower limit at 95% CL for different channels.

channel	Mass limit for ΔM > 15 GeV/c^2
$\tilde{t} \to c\chi$	95(92) GeV/c^2 for $\theta_{\rm mix} = 0^o(56^o)$
$\tilde{t} \to b\ell\tilde{\nu}$	97(94) GeV/c^2 for $\theta_{\rm mix} = 0^o(56^o)$
$\tilde{b} \to b\chi$	96(85) GeV/c^2 for $\theta_{\rm mix} = 0^o(68^o)$
$\tilde{e} \to e\chi$	98 GeV/c^2
$\tilde{\mu} \to \mu\chi$	94 GeV/c^2
$\tilde{\tau} \to \tau\chi$	79 GeV/c^2

Results from the four LEP experiments for scalar quarks and scalar lepton searches have been combined. In Table 3 the obtained mass lower limits for the different channels are summarized. They refer to the hypothesis that the mass difference between the scalar and the LSP is greater than 15 GeV/*cr*.

3.3 More on search for $\tilde{\tau}$

Combining the data collected by the four LEP experiments at lower energies, an excess of acoplanar tau was observed [7] compatible with stau pair production of a mass of about 85 GeV/c^2, each stau decaying into tau χ this last being the LSP and having a mass of around 20 GeV/c^2.

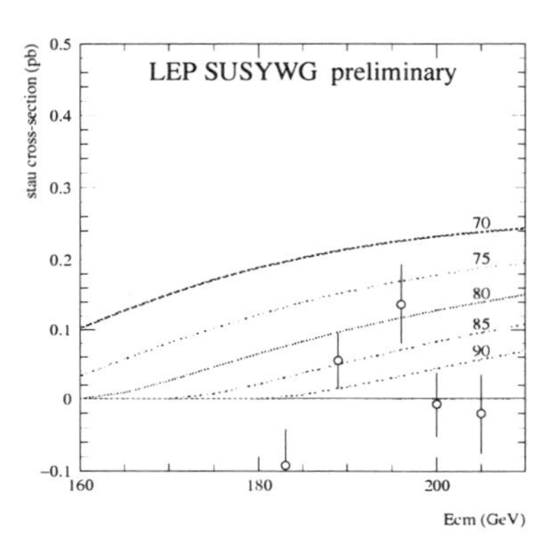

Figure 3. Measured cross sections of the acoplanar tau events in excess with respect to the expected SM processes as a function of the centre-of mass energy. The continuous curves are the prediction for stau pair production for various stau mass hypotheses.

In this year data, the excess is not confirmed. For the signal hypothesis of $m_{\tilde{\tau}}$ = 85 GeV/c^2 and m_χ = 22 GeV/c^2, taken as reference mass point, the four experiments show good agreement between data and SM background expectation. This is visible in Figure 3, which shows the measured stau cross-section as a function of the centre-of-mass energy. The last point is this year observation.

The considered signal mass point is excluded at 90% CL using the 2000 data sample only, indicating that the previous observation can be interpreted as a statistical fluctuation.

4 Conclusions

The outstanding LEP perfomance in this beginning of 2000 run allowed ALEPH to extend the physics reach significantly. Good agreement between data and standard model expectations is found in all the performed measurements and searches. However, an intriguing standard model Higgs boson candidate has been found with a reconstructed mass of 114.3 GeV/c^2.

The results of the searches for supersymmetric particles from the four experiments have been combined. The slight excess of acoplanar taus observed with previous years data is not confirmed. No evidence for new physics has been found. However, the discovery potential will still increase, expecially for chargino searches.

References

1. ALEPH Collaboration, ALEPH-CONF 2000-049, HEP-EPS'00 Abstract No. 188.
2. ALEPH Collaboration, ALEPH-CONF 2000-042, HEP-EPS'00 Abstract No. 262.
3. ALEPH Collaboration, ALEPH-CONF 2000-050, HEP-EPS'00 Abstract No. 397.
4. All informations and results are stored on the following internet adress: http://www.cern.ch/LEPSUSY
5. ALEPH Collaboration, European Physical Journal C12 (2000) 183.
   ALEPH 2000-028, CONF 2000-023.
6. Kara Hoffman, these proceeding.
7. G. Ganis "Standard supersymmetry at LEP" and A. Favara "Sleptons and, in particular, staus at LEP" to appear in the proceding of SUSY2K 8th International on Supersymmetries in Physics, CERN, Geneva, Switzerland 26 June- 1 July 2000.
   M. Antonelli, these proceeding.

NEWS FROM THE YEAR 2000: UPDATE ON OPAL SEARCHES FOR NEW PHENOMENA AND HIGGS AND COMBINED LEP HIGGS RESULTS

KARA HOFFMAN
CERN, EP Division, CH 1211 Geneva 23, Switzerland
E-mail: Kara.Hoffman@cern.ch

Here we update the results of searches for physics beyond the Standard Model and searches for Higgs bosons using the latest data collected with OPAL detector, including 80-100 pb^{-1} collected this year at the highest energies yet achieved at LEP, as well as the limits on Higgs bosons in a variety of models obtained by combining the latest results from the four LEP experiments.

1 Introduction

Thirty years after the first experimental confirmation of the Standard Model, the mechanism of electroweak symmetry breaking remains elusive, however, precision measurements of the electroweak sector still favor a light Higgs boson. In its final year of running, LEP, the Large Electron Positron Collider at CERN, is being driven to the highest attainable energies in an effort to maximize the combined Higgs discovery potential of the four LEP experiments. The absence thus far of a Standard Model Higgs signal has fueled interest in many disparate extensions of the Standard Model, motivating parallel searches for new phenomena and the non-minimal Higgs signatures they invoke. Here I update the results of such new particle and Higgs searches using the latest data collected by the OPAL collaboration, as well as the latest combined Higgs results from the four LEP experiments interpreted in a wide variety of models including, for the first time, fermiophobic and invisible Higgs decays.

2 Searches for New Phenomena Using the OPAL Detector

The most familiar extension of the Standard Model, the Minimal Supersymmetric Model (MSSM), results in a wide spectrum of new particles which have been sought by all four LEP experiments, and some combined re-

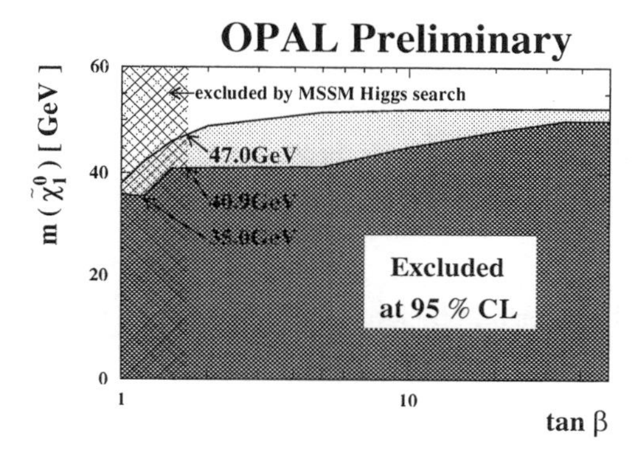

Figure 1. The excluded mass of the lightest neutralino $\tilde{\chi}_1^0$ as a function of $\tan\beta$ for $m_0 \geq 500$ GeV (lighter shading) and any m_0 (darker shading).

sults have been presented at this conference[1]. Not included in this combination, however, are model-independent searches for long lived ($\tau \geq 1\mu s$) heavy charged particles, which can be interpreted in the MSSM framework to obtain lower mass limits on scalar leptons of $\approx$ 95 GeV [2], assuming that the mass of the lightest supersymmetric particle (LSP) is heavy enough to suppress the scalar lepton decays. Also not included are neutralino searches, which OPAL interprets for two different values of the universal scalar mass, m_0, as shown in Figure 1. There has also been a recent surge of interest in semileptonic decays of a light (3.6 GeV) scalar bottom, $\tilde{b}$. OPAL has searched for such signatures using 540 pb^{-1} of data at energies of 161-206 GeV, and has observed 15 events with 20.5

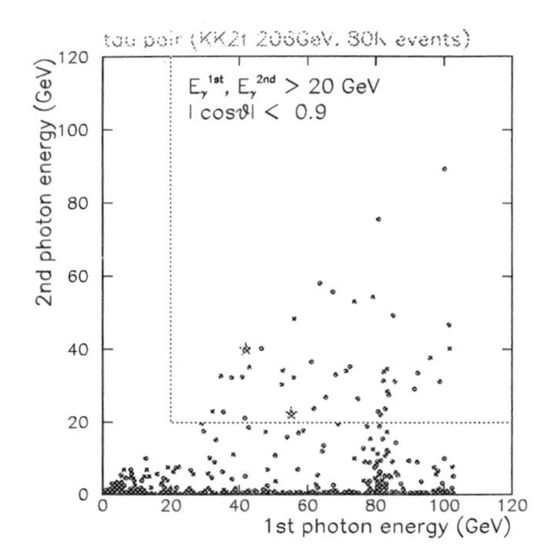

Figure 2. The distribution of photon energies in $\tau^+\tau^-\gamma\gamma$ events generated using the KK2f Monte Carlo (dots) compared to the two most energetic candidates selected by the $\tilde{\ell} \to \ell\tilde{\chi}_1^0 \to \ell\gamma\tilde{G}$ search (stars).

background events expected, where the signal detection efficiency for simulated $\tilde{b}$ events is 15%.

Other supersymmetric models which have been proposed, which include R-parity violating SUSY and gauge mediated SUSY breaking (GMSB), can lead to vastly different topologies. If R-parity is not conserved, supersymmetric particles can decay directly to Standard Model particles. Searches for R-parity violating decays of pair-produced sleptons, which proceed via the triple lepton vertices, $\tilde{\ell} \to \tilde{\chi}_1^0\ell$, $\tilde{\chi}_1^0 \to \ell^+\ell^-\nu$ (indirect decays), and, if $m_{\tilde{\chi}_1^0} > m_{\tilde{\ell}}$, $\tilde{\ell} \to \ell\nu$ (direct decays), have been updated using year 2000 data. As a result, direct decays are excluded below $\approx$ 76 GeV for both right-handed smuons, $\tilde{\mu}_R$, and staus, $\tilde{\tau}_R$. A larger mass range is excluded for indirect decays leaving only a small unexcluded region in the $(m_{\tilde{\ell}}, m_{\tilde{\chi}_1^0})$ mass plane along the kinematic limit $(m_{\tilde{\chi}_1^0} > m_{\tilde{\ell}})$ below $\tilde{\mu}_R \approx 92$ GeV, and $\tilde{\tau}_R \approx 88$ GeV. The signatures of GMSB models, which employ a light gravitino, $\tilde{G}$, as the LSP, are determined by the next-to-lightest SUSY particle, NLSP. Candidates for the NLSP include the neutralino, $\tilde{\chi}_1^0$, and the right-handed slepton, $\tilde{\ell}_R$, resulting in decay chains that terminate with $\tilde{\ell}_R \to \ell\tilde{G}$ or $\tilde{\chi}_1^0 \to \gamma\tilde{G}$. A search for pair-produced sleptons decaying via $\tilde{\ell} \to \ell\tilde{\chi}_1^0 \to \ell\gamma\tilde{G}$ in the mass region 10 GeV $< m_{\tilde{\ell}} - m_{\tilde{\chi}_1^0} < m_{\tilde{\ell}}$ uncovered an excess of events, selecting 3 events where 1.3 ± 0.3 were expected. Two of these events have two high energy photons rather than the single high energy photon expected from the background, which is thought to be nearly 100% $e^+e^- \to Z^0\gamma\gamma \to \gamma\gamma\tau^+\tau^-$. The distribution of the photon energies for a generated $\gamma\gamma\tau^+\tau^-$ sample are compared to the two anomalous data events in Figure 2.

This list of searches is not exhaustive, and further results may be found in Reference 2.

3 OPAL Searches for Higgs and Combined LEP Higgs Limits

The last missing piece of the Standard Model is the heavy neutral scalar prescribed to endow the observed spectrum of particles with mass. OPAL has searched for such scalar particles produced via the "Higgs-strahlung" process and a summary of the results obtained using the present year's data, classified according to the decay mode of the associated Z^0, is summarized in Table 1. OPAL's cumulative data alone excludes the existence of a Higgs boson with Standard Model couplings below a mass of 109.5 GeV. This sensitivity can be increased by combining OPAL's result with the results from the other three LEP experiments which have been presented by the previous three speakers[4]. The mass distribution of the Higgs candidates recorded at a number of center-of-mass energies from the four LEP experiments is shown superimposed on the expected background in Figure 3. The data matches the background distribution to within statistical errors and, when combined with the data collected previously, excludes a Standard Model Higgs boson with a mass below 113.3 GeV at the 95% confidence level (C.L.), where a limit of 113.4 is expected in

Table 1. Summary of OPAL searches for neutral Higgs bosons in year 2000 data.

Signature		Background	Data
$Z^0H^0 \rightarrow$	$q\bar{q}b\bar{b}$	8.9 ± 1.1	7
	$\nu\bar{\nu}b\bar{b}$	13.6 ± 1.4	6
	$e^+e^-b\bar{b}$	2.0 ± 0.3	2
	$\mu^+\mu^-b\bar{b}$	1.4 ± 0.3	6
	$\tau^+\tau^-b\bar{b},\ q\bar{q}\tau^+\tau^-$	1.3 ± 0.2	0
$A^0h^0 \rightarrow$	$\tau^+\tau^-b\bar{b}$	1.7 ± 0.2	0
	$b\bar{b}b\bar{b}$	3.7 ± 0.4	4

the absence of any signal[5].

Supersymmetric models require a minimum of two Higgs doublets, giving rise to at least five physical Higgs bosons, the neutral scalars H^0 and h^0 (where $m_{H^0} > m_{h^0}$, by definition), the CP-odd A^0, and the charged Higgs pair $H^{\pm}$. Dedicated searches for the CP-odd Higgs produced in association with a light neutral scalar, summarized for OPAL 2000 data in Table 1, can be combined with the above Higgs-strahlung channels, which can be interpreted as h^0 and H^0 production in the MSSM framework to scan the MSSM parameter space. Since the MSSM has many free parameters which strongly influence the limits, several "benchmarks" are considered. The limits in two of them, one which allows no scalar top mixing and one in which the parameters are chosen to maximize the mass of the h^0, are shown in Figure 4. In some SUSY models, it is also possible for the Higgs to couple to neutralinos allowing it to decay "invisibly". Searches for Higgs-strahlung with invisible decays performed by ALEPH, DELPHI, and OPAL, can be used to exclude a neutral Higgs below 107.6 GeV, assuming $\mathrm{Br}(h^0 \rightarrow \tilde{\chi}_1^0\tilde{\chi}_1^0) = 100\%$, where the mass of the Higgs is inferred from the decay products of the recoiling Z^0. This limit is weakened by a 2σ fluctuation near the expected limit of ≈ 110 GeV.

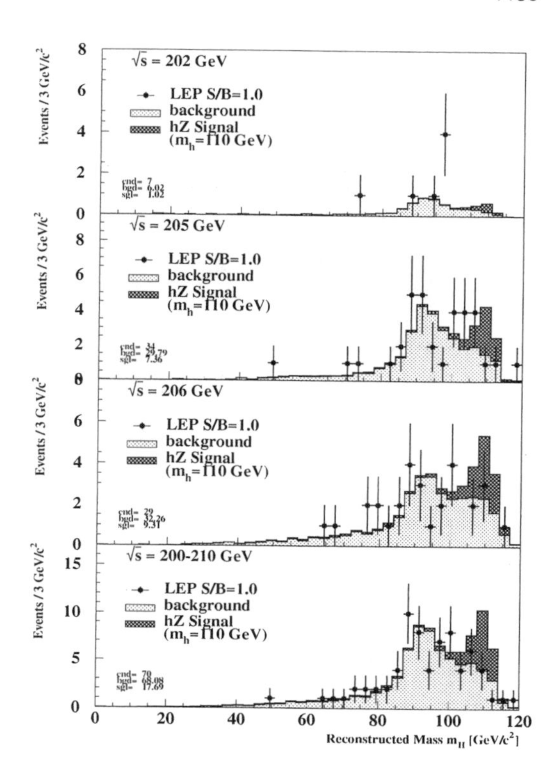

Figure 3. The mass distribution of neutral Higgs candidates from the four LEP experiments which satisfy the requirement $S/B > 1.0$ compared to the expected Standard Model Higgs boson signal (dark shaded region) superimposed on the expected background (light shaded region) at various center-of-mass energies.

Two Higgs doublet models (2HDM's) leave a pair of charged Higgs bosons after electroweak symmetry breaking, however, theoretical constraints on the SUSY parameters put the mass of such a boson out of the kinematic reach of LEP. There are, however, few general theoretical constraints on the form of the Higgs sector, thus it is important to search for nonminimal Higgs signatures without SUSY model assumptions. The four LEP experiments have searched for charged Higgs pair production, assuming only that $\mathrm{Br}(H^{\pm} \rightarrow q\bar{q}) + \mathrm{Br}(H^{\pm} \rightarrow \tau\nu) = 100\%$. The excluded region is shown as a function of $\mathrm{Br}(H^{\pm} \rightarrow \tau\nu)$ in Figure 5. In two Higgs doublet models, it is also possible for the neu-

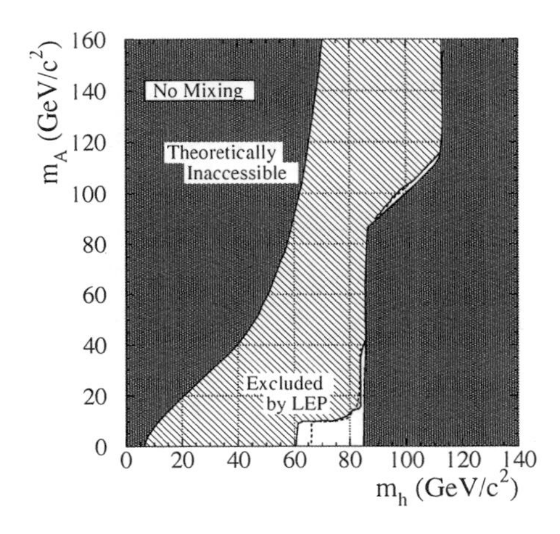

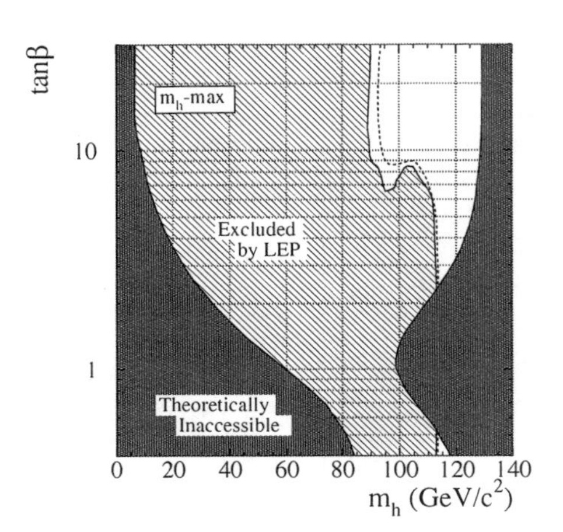

Figure 4. The region in the MSSM parameter space excluded at the 95% C.L. projected on the m_h *vs.* m_A plane for the no stop mixing benchmark (left), and projected on the m_h *vs.* $\tan\beta$ plane for the $max - m_h$ benchmark (right).

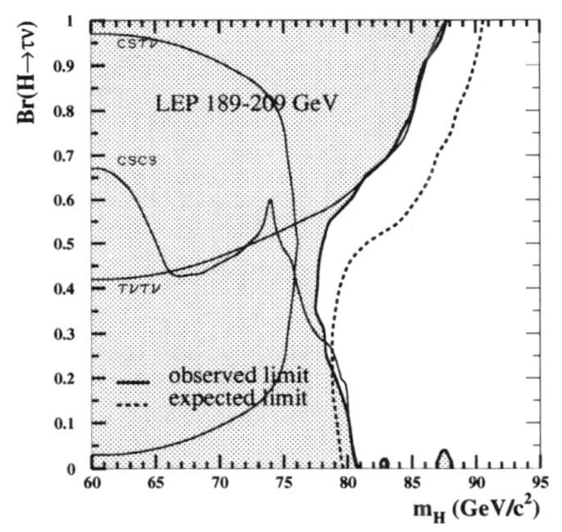

Figure 5. The excluded charged Higgs mass as a function of the branching fraction $\mathrm{Br}(H^+ \to \tau\nu)$ obtained by combining data from the four LEP experiments (shaded region) compared to the expected limit (dashed line). The thin solid contours outline the area excluded using each of the final state searches independently.

tral Higgs to decouple from the fermions for certain values of α and $\tan\beta$. Combining $h^0 \to \gamma\gamma$ searches from the four LEP experiments produced a limit of 106.4 GeV for such "fermiophobic" decays.

4 Conclusions

In the final year of LEP running, many signatures of the Higgs and physics beyond the Standard Model have been sought both by OPAL and the other LEP experiments. Compelling evidence of new physics has yet to be observed, however, LEP will continue to accumulate high energy data in the coming months, possibly leaving a window for discovery.

References

1. M. Maggi, these proceedings.
2. The OPAL Collaboration, *"New Particle Searches in e^+e^- Collisions at $\sqrt{s} = 200 - 209$ GeV"*, OPAL Physics Note PN435, July 18, 2000.
3. The OPAL Collaboration, *"Searches for Higgs Bosons in e^+e^- Collisions at the Highest LEP Energies"*, OPAL Physics Note PN450, July 21, 2000.
4. M.C. Espirito-Santo, S. Gentile, and M. Maggi, these proceedings.
5. The LEP working group for Higgs boson searches, ALEPH, DELPHI, L3 and OPAL Collaborations, *"Searches for Higgs bosons: Preliminary combined results using LEP data collected at energies up to 109 GeV"* ALEPH 2000-074 CONF 2000-051, DELPHI 2000-148 CONF 447, L3 Note 2600, OPAL Technical Note TN661, July 28, 2000.

SEARCH STRATEGIES FOR NON-STANDARD HIGGSES AT E^+E^- COLLIDERS

J. KALINOWSKI

Instytut Fizyki Teoretycznej UW, ul. Hoża 69, 00681 Warsaw, Poland
E-mail: kalino@fuw.edu.pl

The Higgs search strategies in minimal non-supersymmetric extensions of the SM are discussed.

1 Motivation

If no new physics is assumed up to the grand unification or M_{Pl} scales, the requirement of perturbativity and vacuum stability of the Standard Model constraints[1] the Higgs boson mass to lie within the range of 130 – 190 GeV. This is in perfect agreement with the electroweak precision fits[2] which strongly point to a light Higgs boson with $m_{H_{SM}} = 62^{+53}_{-30}$ GeV, and with the 95% CL upper limit 170 GeV. This mass range, well above the ultimate LEP2 reach (the current experimental LEP limit[3] is $m_{H_{SM}} > 113.2$ GeV) and rather difficult at Tevatron (particularly in its upper part), will be fully covered at the LHC by exploiting the $gg \to H \to \gamma\gamma$ or associate production $t\bar{t}H$, WH processes. For the future e^+e^- colliders this mass range is particularly easy. The "standard" Higgs hunting strategies at e^+e^- collisions rely on the Higgs-strahlung, $e^+e^- \to ZH$, and (for higher energies and heavier Higgs bosons) on the WW fusion, $e^+e^- \to \nu\bar{\nu}$, processes [4].

It should be stressed that the above implications for a light Higgs boson *with* substantial ZZh coupling can be altered if we admit new physics. By adding $\mathcal{O}_i^{NEW}$ to the electroweak observables $\mathcal{O}_i$, the SM contributions can be compensated resulting in a higher value of the Higgs mass.

In fact, the Higgs sector may turn out to be more complicated than just one doublet, as realised in the SM. Even in non-supersymmetric world, and adding additional SU(2) singlet or doublet Higgs fields only (to keep the tree-level $\rho = 1$), Higgs boson couplings may change considerably and thus complicate the Higgs boson searches. Particularly worrisome is the case of a light Higgs h with suppressed ZZh and WWh couplings; we will refer to it as a "bosophobic" Higgs. If such a Higgs boson with mass below 113 GeV exists, negative searches at LEP2 in $e^+e^- \to ZH$ translate into an upper limit on the g_{ZZh} coupling. Are we guaranteed to discover the bosophobic Higgs with other Higgs bosons too heavy to be produced? The answer turns out to be model dependent. The absence of ZZ coupling implies that the h will not be detectable at the Tevatron, and very difficult, if not impossible, at the LHC. Therefore we will consider a $\sqrt{s} = 500 - 800$ GeV e^+e^- linear collider (LC) assuming an integrated luminosity $L \gtrsim 500$ fb^{-1}.

2 Adding singlets

Adding singlet Higgs fields does not pose any particular theoretical problems nor benefits. However, if many singlet fields mix with the SM doublet in such a way that the physical Higgs bosons h_i share the SM WW/ZZ-Higgs coupling, the cross sections in $e^+e^- \to Zh_i$ $(i = 1, \ldots, N)$ for individual channels will be suppressed. The scenario considered in [5] assumes h_i spaced more closely than the experimental mass resolution and spread out over some substantial range around 200 GeV. The individual resonance peaks will overlap making a diffuse signal not much different from the background. If in addition Higgs bosons decay to a large number of different channels, identification of individual final states will not be possible nor useful due to large background. Another possibility, the so called stealthy Higgs, is considered in [6], where the usual Higgs doublet couples to many singlets (called Phions) which interact among themselves strongly. The net effect is that the SM-like Higgs boson is very broad and decays invisibly into Phions.

At hadron colliders such scenarios are real nightmare. On the other hand, it has been demonstrated[5] that by looking for an excess in the recoil mass m_X distribution due to a "continuum" of Higgses in $e^+e^- \to ZX$, the signal can be observed at an e^+e^- collider with $\sqrt{s} = 500$ GeV and integrated luminosity $> 100\ \text{fb}^{-1}$. Since the inclusive $e^+e^- \to ZX$ process can be used irrespectively of Higgs decay modes, the stealthy Higgs can cleanly be detected[6] by looking for a signal of leptons and missing energy.

3 Adding one Higgs doublet

Even the simplest two-Higgs-doublet model (2HDM) extension of the SM exhibits a rich Higgs sector structure. The CP-conserving (CPC) 2HDM predicts the existence of two neutral CP-even Higgs bosons (h^0 and H^0, with $m_{h^0} \leq m_{H^0}$ by convention), one neutral CP-odd Higgs (A^0) and a charged Higgs pair ($H^\pm$). The same spectrum of Higgs bosons is found in the minimal supersymmetric model (MSSM), where it has been demonstrated[7] that the detection of at least one of the Higgs bosons is possible either at LEP2 or LHC.

The situation is more complex in the non-supersymmetric 2HDM. Here we consider the type-II 2HDM, wherein one of the doublets couples to down-type quarks and leptons and the other to up-type quarks. The 2HDM allows for spontaneous and/or explicit CP violation (CPV) in the scalar sector[8] at the tree level. In the CPV case the physical mass eigenstates, h_i ($i = 1, 2, 3$), are mixtures

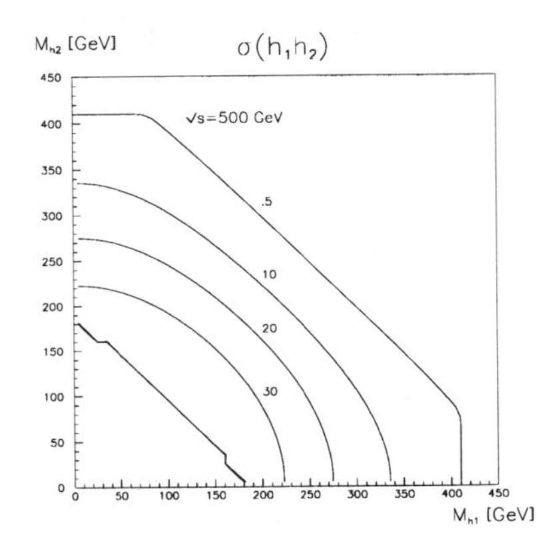

Figure 1. Contour lines for $\min[\sigma(e^+e^- \to h_1h_2)]$ in units of fb's. The contour lines are plotted for $\tan\beta = 0.5$; the plots are virtually unchanged for larger values of $\tan\beta$. The contour lines overlap in the inner corner as a result of excluding mass choices inconsistent with experimental constraints from LEP2 data. [From [10]]

(specified by three mixing angles α_i, in addition to the mixing angle β related to Higgs vev's) of the real and imaginary components of the original neutral Higgs doublet fields; as a result, the h_i have undefined CP properties.

If there are two light Higgs bosons h_1 and h_2, in the sense that Zh_1, Zh_2 and h_1h_2 channels are kinematically open, then at least one will be observable in Zh_1 or Zh_2 production or both in h_1h_2 pair production. This is because of the sum rule[9] for the couplings of any two of neutral Higgses to the Z boson

$$C_i^2 + C_j^2 + C_{ij}^2 = 1, \tag{1}$$

where $g_{ZZh_i} \equiv \frac{gm_Z}{c_W} C_i$ and $g_{Zh_ih_j} \equiv \frac{g}{2c_W} C_{ij}$, which says that all three couplings cannot be simultaneously suppressed. For example, if both C_1 and C_2 are dynamically suppressed, then from the above sum rule it follows that Higgs pair production is at full strength, $C_{12} \sim 1$. In Fig.1 contour lines are shown for the minimum value of the pair production cross section, $\sigma(e^+e^- \to h_1h_2)$ as a function of Higgs boson masses. The mimimum of $\sigma(h_1h_2)$ is found[10] by scanning over the mixing angles α_i consistent with present experimental constraints on C_i (which roughly exclude $m_{h_1} + m_{h_2} \lesssim 180$ GeV) and the assumption of less that 50 Zh_i events. With $L = 500$ fb^{-1} a large number of events is predicted for a broad range of Higgs boson masses. If 50 $h_1 h_2$ events before cuts and efficiencies prove adequate (i.e. $\sigma > 0.1$ fb), one can probe reasonably close to the kinematic boundary.

The main question, however, is whether a *single* neutral Higgs boson h_1 will be observed in e^+e^- collisions if it is sufficiently light, regardless of the masses and couplings of the other Higgs bosons. Such a scenario can easily be arranged by choosing model parameters so that the ZZ/WWh_1 couplings are too weak for its detection in Higgs-strahlung or WW fusion processes, and all other Higgs bosons are too heavy to be produced via Zh_i or h_ih_j processes at a given energy. In the CPC for example, one can simply choose $h_1 = A^0$ (the tree-level ZZ/WWA^0 coupling is zero), or in the general CPV model choose mixing angles α_i to zero the ZZ/WWh_1 coupling. Since the other Higgs bosons are assumed to be quite heavy to avoid production, implying no *light* Higgs with substantial ZZ/WW couplings, it would seem that the fit to precision electroweak constraints is likely to be poor. However, as shown in [11], a good global fit to EW data is possible even for very light h^0 or A^0 (with $m \sim 20$ GeV) in the CPC 2HDM.

If one of the two processes, Zh_2 and h_1h_2, is beyond the LC's kinematical reach, the sum rule in Eq. (1) is not sufficient to guarantee h_1 discovery if $C_1 \ll 1$. However, in this case we can exploit other sum rules[10] which constrain the Yukawa couplings of any Higgs boson h_i. For $C_i \ll 1$ they read (for obvious reasons we consider the third generation fermions)

$$(S_i^t)^2 + (P_i^t)^2 = \cot^2\beta$$
$$(S_i^b)^2 + (P_i^b)^2 = \tan^2\beta \qquad (2)$$

where the fermionic Higgs couplings are given by $\frac{gm_f}{2m_W}\bar{f}(S_i^f + i\gamma_5 P_i^f) f h_i$, i.e. S_i^f and P_i^f are

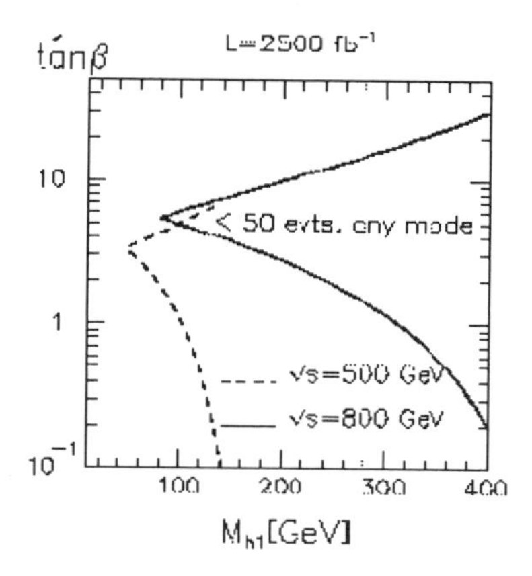

Figure 2. The maximum and minimum $\tan\beta$ values between which $t\bar{t}h_1$, $b\bar{b}h_1$ and Zh_1 final states all have fewer than 50 events assuming $L = 2500$ fb^{-1} at $\sqrt{s} = 500$ GeV (dashes) and $\sqrt{s} = 800$ GeV (solid). Masses of the remaining Higgs bosons are assumed to be 1000 GeV. [From [12]]

defined relative to the SM strength. Combining the two sum rules we find that the Yukawa couplings to top and bottom quarks cannot be simultaneously suppressed, *i.e.* at least one h_i Yukawa coupling must be large. Therefore the Higgs hunting strategies should include not only the Higgs-strahlung and Higgs-pair production but also Yukawa processes with Higgs radiation off top and bottom quarks in the final state. The current experimental limits in the m_{h_1}-$\tan\beta$ parameter space are rather weak, see [13].

It turns out that for large (small) $\tan\beta$, the $b\bar{b}h_1$ ($t\bar{t}h_1$) cross sections are comfortably large for h_1 discovery. However, scanning over mixing angles α_i we find[12] the difficult region of moderate $\tan\beta$, where even at very high integrated luminosity of 2500 fb^{-1} none of the Zh_1, $t\bar{t}h_1$ and $b\bar{b}h_1$ processes yields more than 50 events, see Fig.2.

The non-discovery wedge begins at $m_{h_1} \sim 50$ GeV at $\sqrt{s} = 500$ GeV (~ 80 GeV for $\sqrt{s} = 800$ GeV) and expands rapidly as m_{h_1} increases. Thus, it is apparent that, despite the sum rules guaranteeing significant fermionic couplings for a light 2HDM Higgs boson that is unobservable in Z+Higgs production, $\tan\beta$ and the α_i mixing angles can be chosen so that the cross section magnitudes of the two Yukawa processes are simultaneously so small that detection of such an h_1 cannot be guaranteed for integrated luminosities that are expected to be available.

Is the whole wedge consistent with electroweak constraints? This question, in the context of CPC 2HDM, is analysed in [14] with the general result that for LC $\sqrt{s} = 500$ (800) GeV, the $\tan\beta \sim 2$ portions of the 2HDM no-discovery wedges in m_{h_1}-$\tan\beta$ parameter space have $\Delta\chi^2 < 1$ (< 1.5) (relative to the best SM fit) and all of the no-discovery wedges' portions with $\tan\beta \gtrsim 1$ have $\Delta\chi^2 < 2$. Thus the discrimination from current EW data between the SM and the no-discovery scenarios in the 2HDM is rather weak at the LC with $\sqrt{s} = 500 - 800$ GeV.

4 Conclusions

In a general CPV 2HDM a light bosophobic Higgs boson, with all other Higgs bosons heavier than the kinematical reach of a 500-800 GeV e^+e^- collider, may escape detection. If $\sqrt{s}$ is pushed beyond 1 TeV, and the next lightest Higgs H is still not seen in ZH or $\nu\bar{\nu}H$, implying $m_H \gtrsim 1$ TeV, one would expect to see strong WW scattering behavior at both the LHC and the LC. As a result, only an LC with sufficiently large energy to probe a strongly interacting WW sector could be certain of seeing a Higgs signal, unless the electroweak fits really *do* indicate a relatively light Higgs boson.

Acknowledgments

I am grateful to P. Chankowski, B. Grzadkowski, J. Gunion, M. Krawczyk and P. Zerwas for many discussions. Work partially supported by the KBN Grant 2 P03B 052 16.

References

1. M. Lindner, *Z. Phys.* **C31** 295 (1986); T. Hambye and K. Riesselmann, *Phys. Rev.* **D55** 7255 (1997).
2. A. Gurtu, these Proceedings; see also B. Pietrzyk, these Proceedings.
3. K. Hoffmann, these Proceedings.
4. E. Accomando et al., *Phys. Rep.* **299** 1 (1998), and LC CDR Report DESY/ECFA 97-048/182.
5. J.R. Espinosa and J.F. Gunion, *Phys. Rev. Lett.* **82** 1084 (1999).
6. T. Binoth and J.J. van der Bij, *Z. Phys.* **C75** 17 (1997).
7. J.F. Gunion, A. Stange and S. Willenbrock, in *Electroweak Symmetry Breaking and New Physics at the TeV Scale*, ed. T.L. Barklow, S. Dawson, H.E. Haber and J.L. Siegrist, (World Scientific 1996).
8. T.D. Lee, *Phys. Rev.* **D8** 1226 (1973); S. Weinberg, *Phys. Rev.* **D42** 860 (1990).
9. A. Mendez and A. Pomarol, *Phys. Lett.* **B272** 313 (1991); J.F. Gunion, B. Grzadkowski, H.E. Haber and J. Kalinowski, *Phys. Rev. Lett.* **79** 982 (1997).
10. B. Grzadkowski, J.F. Gunion and J. Kalinowski, *Phys. Rev.* **D60** 075011 (1999).
11. P.H. Chankowski, M. Krawczyk and J. Zochowski, *Eur. Phys. J.* **C11** 661 (1999)
12. B. Grzadkowski, J.F. Gunion and J. Kalinowski, *Phys. Lett.* **B480** 287 (2000).
13. M. Krawczyk, J. Zochowski and P. Mättig, *Eur. Phys. J.* **C8** 495 (1999); for the effect of GigaZ, see preprint IFT/00-22.
14. P. Chankowski, T. Farris, B. Grzadkowski, J.F. Gunion, J. Kalinowski and M. Krawczyk, preprint UCD-2000-15, IFT/00-18.

BROKEN R PARITY, NEUTRINO ANOMALIES AND COLLIDER TESTS

M. HIRSCH, W. POROD, J. ROMÃO * & J. W. F. VALLE

Instituto de Física Corpuscular – C.S.I.C. – Universitat de València
Ed. de Institutos de Paterna – Apartado de Correos 22085 - 46071 València, Spain

* *Inst. Superior Tecnico, Depto. de Fisica, Av. Rovisco Pais, 1 1096 Lisboa Codex, Portugal*

The solar and atmospheric neutrino anomalies constitute the only solid and most remarkable evidence for physics beyond the Standard Model, indicating that the lepton mixing matrix is fundamentally distinct from that describing the quarks. Here I will report on how supersymmetry with spontaneously or bilinearly broken R Parity provides a predictive theory for neutrino mass and mixing which leads to a solution of neutrino anomalies which can be clearly tested at high energy accelerators.

1 Motivation

The simplest interpretation of the solar and atmospheric neutrino data [1,2,3] indicate that, in contrast to quark mixing, possibly two of the lepton mixing angles are large. Here I discuss how supersymmetry with broken R Parity provides a predictive theoretical model for neutrino mass and mixing which solves the solar and atmospheric neutrino anomalies in a way that allows the leptonic mixing angles to be probed at high energy accelerators.

R-parity conservation is an `ad hoc` assumption in the MSSM and $\not{R}_p$ may arise `explicitly` as unification remnant or `spontaneously` by $SU(2) \otimes U(1)$ doublet left sneutrino vacuum expectation values (VEVS) $\langle\tilde{\nu}_i\rangle$ as originally suggested [4,5] but with an `ad hoc` set of explicit breaking terms [6] to comply with LEP data on Z width. Preferably we break R-parity spontaneously through `singlet right sneutrino VEVS`, either by gauging L-number, in which case there is an additional Z [7] or within the $SU(2) \otimes U(1)$ scheme, in which case the `majoron` is an $SU(2) \otimes U(1)$ singlet, with suppressed Z coupling [8]. Spontaneous R-parity violation may lead to a successful electroweak baryogenesis [9].

If R-parity is broken spontaneously then `only bilinear` $\not{R}_p$ `terms arise in the effective theory` below the $\not{R}_p$ violation scale. Bilinear R–parity violation may also be assumed `ab initio` as the fundamental theory. For example, it may be the only violation permitted by higher Abelian flavour symmetries [10]. Moreover the bilinear model provides a theoretically self-consistent scheme in the sense that trilinear $\not{R}_p$ implies, by renormalization group effects, that also bilinear $\not{R}_p$ is present, but `not` conversely. The simplest $\not{R}_p$ model (we call it $\not{R}_p$ MSSM) is characterized by three independent parameters in addition to those specifying the minimal MSSM model. As shown in ref. [11] this leads to a predictive pattern for neutrino masses and mixing angles which provides a solution to the solar and atmospheric neutrino problems. It also predicts a well specified pattern of $\not{R}_p$ phenomena that can be tested at collider experiments, providing an independent determination of neutrino mixing angles at high energy accelerator experiments.

2 Bilinear $\not{R}_p$ MSSM

The minimal supergravity version of R-parity breaking MSSM [12] is specified by the superpotential,

$$W = W_{MSSM} + \epsilon_i L_i \widehat{H}_u \qquad (1)$$

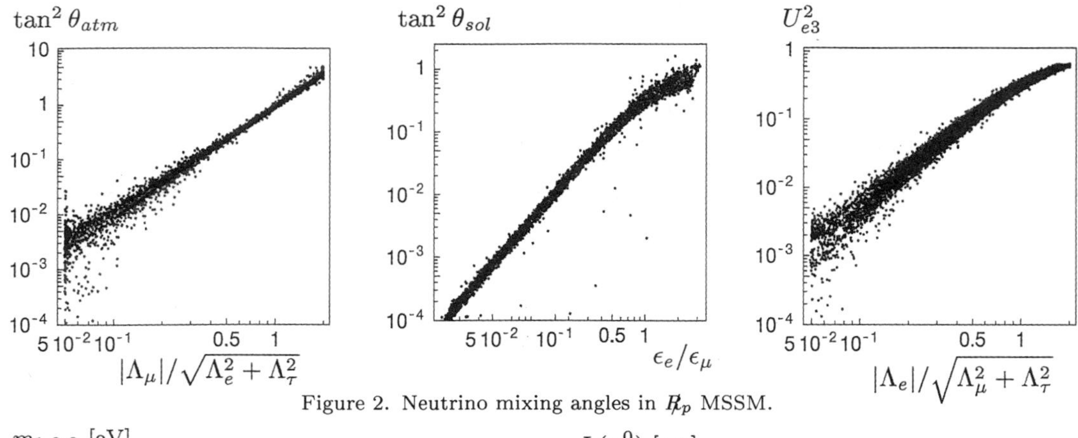

Figure 2. Neutrino mixing angles in $\not{R}_p$ MSSM.

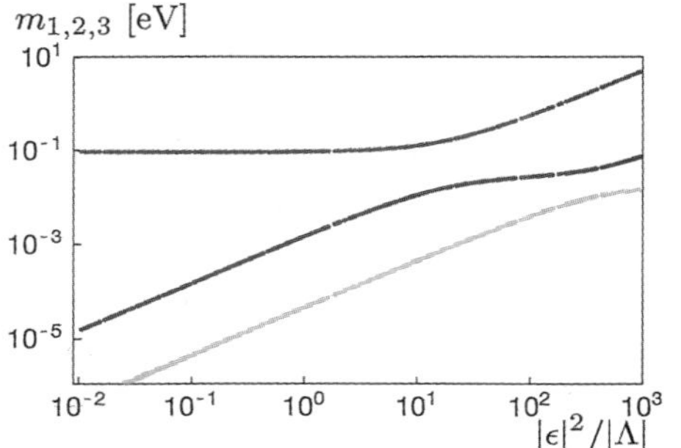

Figure 1. Typical neutrino masses in $\not{R}_p$ MSSM.

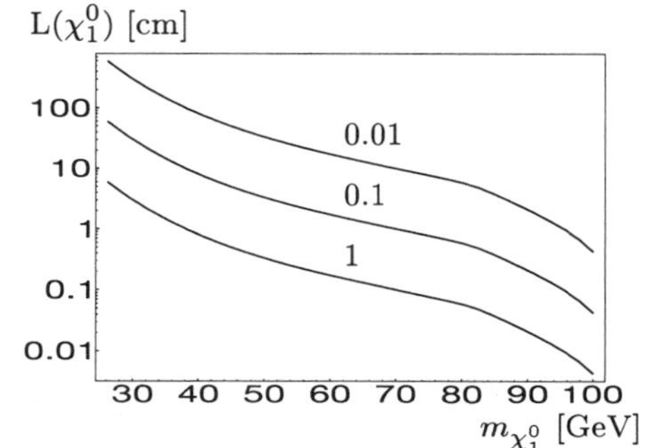

Figure 3. Typical neutralino decay length in $\not{R}_p$ MSSM.

Since lepton number is broken, neutrinos pick up a mass. The expected neutrino mass pattern is illustrated in Fig. (1), taken from [11]. It is typically hierarchical since only one neutrino acquires mass at the tree level, while the others get mass from calculable radiative corrections [11]. As a result neutrino masses can account for the solar and atmospheric neutrino problems [*]. Having only bilinear R-parity violating terms as the origin of the neutrino masses implies also that the three neutrino mixing angles (assuming CP conservation in the lepton sector) are determined as functions of the three bilinear $\not{R}_p$ terms, leading to a predictive scenario, independently of any particular form for the charged lepton mass matrix. This is illustrated in Fig. (2), taken from [11]. As can be seen, large angle solar solutions, LMA and LOW, now preferred by solar spectrum data and by the global fit of all solar neutrino data, as well as small angle solution (preferred by the rates) can be accounted for within the theory. However, as explained in [11], for the very particular case of strictly universal boundary conditions at the unification scale, consistency with the reactor experiments [14] implies the SMA solar solution.

[*] For such small masses neutrinoless double beta decay has been shown to be too small to observe [13]

3 Implications

There are a variety of implications of $\not{R}_p$ models [15]. The most obvious is that, unprotected by any symmetry, the lightest supersymmetric particle (LSP), produced with MSSM-like cross sections, will typically decay inside the detector, as shown in Fig. (3), taken from [16] Such decays are mainly into `visible modes`. Just as the neutrino mixing angles characterizing the neutrino anomalies, in our bilinear $\not{R}_p$ MSSM model also the neutralino de-

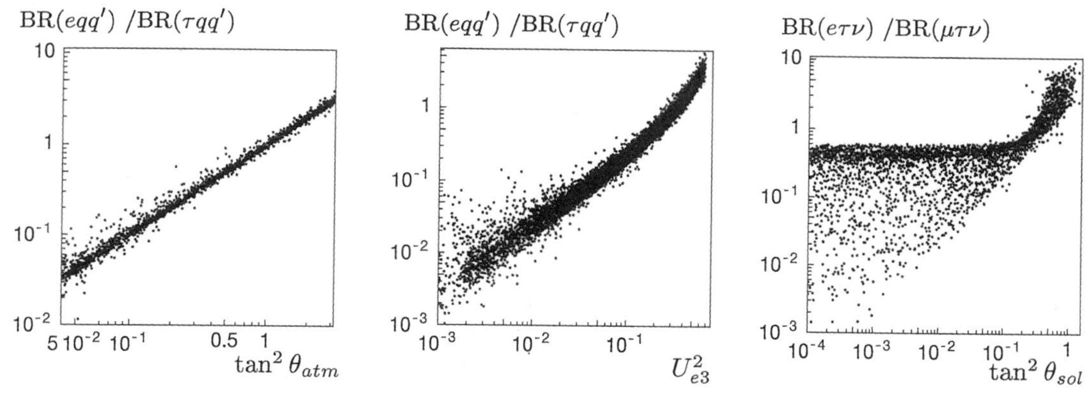

Figure 4. Neutralino BR in bilinear $R\!\!\!/_p$ MSSM.

cay branching ratios are determined by the same three fundamental $R\!\!\!/_p$ parameters in eq. (1). More exactly the neutrino mixing angles are correlated with ratios of $R\!\!\!/_p$ parameters. These may be taken as the Λ_μ/Λ_τ for the atmospheric angle, ϵ_e/ϵ_μ for the solar angle, and Λ_e/Λ_τ for the angle which is probed by the reactor experiments [14]. Here $\Lambda_i \equiv \epsilon_i \langle H_d \rangle + \mu \langle \widetilde{\nu}_i \rangle$, μ being the standard Higgsino mixing term. As shown in ref. [11] due to the minimization conditions the Λ ratios do not introduce independent parameters, hence the predictivity of the theory is manifest. As Fig. (4) indicates, the LSP decay branching ratios are strongly correlated with the leptonic mixing angles [†].

Neutralino decays can have remarkable consequences for gluino cascade decays at the LHC, enhancing high lepton multiplicity event rates and, correspondingly, thus decreasing the missing momentum signal expected in the R-parity conserving MSSM [18]. If R parity is broken particles other than the neutralino can be the LSP. One example is the stop [19]. In Fig. (5) we illustrate how two-body $R\!\!\!/_p$ decays of the lightest stop can be sizeable when compared with standard decays [20]. R parity violation can also affect gauge and Yukawa unification [21], texture predictions for V_{cb} [22] as well as $b \to s\gamma$ [23]. Turning to accelerators, $R\!\!\!/_p$ can affect the physics of the top quark [24] and it can lead to new signals for chargino production at LEP2 [25], and affect the phenomenology of supersymmetric scalars due to Higgs boson/slepton mixing [26].

[†] The possibility of probing leptonic mixing angles at accelerator experiments in $R\!\!\!/_p$ models has been previously considered in refs. [5,17].

Acknowledgments

This work was supported by DGICYT grant PB98-0693 and by the EEC under the TMR contract ERBFMRX-CT96-0090. M.H. was supported by the Marie-Curie program under grant No ERBFMBICT983000 and W.P. by a fellowship from the Spanish Ministry of Culture under the contract SB97-BU0475382.

References

1. For updated analyses of solar and atmospheric data see talk by M.C. Gonzalez-Garcia, plots available from http://neutrinos.uv.es. For details and references see [2] and [3].
2. N. Fornengo, M. Gonzalez-Garcia & J. W. F. Valle, Nucl. Phys. **B580** (2000) 58 [hep-ph/0002147]; M.C. Gonzalez-Garcia, et. al. Nucl. Phys. **B543**, 3 (1999) and Phys. Rev. **D58** (1998) 033004.
3. M.C. Gonzalez-Garcia, P.C. de Holanda, C. Peña-Garay and J. W. F. Valle, Nucl. Phys. **B573**, 3 (2000) [hep-ph/9906469].

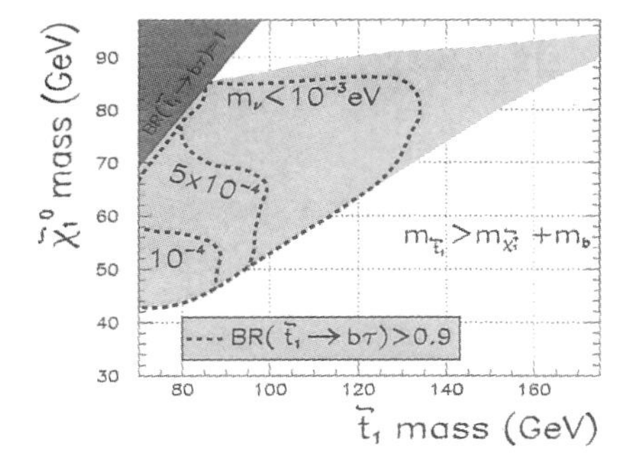

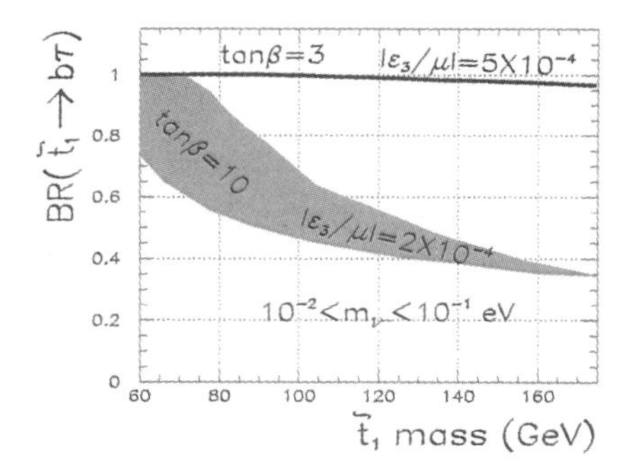

Figure 5. Stop decay BR in bilinear R_p MSSM.

4. C. S. Aulakh & R. N. Mohapatra, Phys. Lett. **B121**, 14 (1983).
5. A. Santamaria & J. W. F. Valle, Phys. Lett. **B195**, 423 (1987); Phys. Rev. **D39**, 1780 (1989) and Phys. Rev. Lett. **60**, 397 (1988).
6. G. G. Ross & J. W. F. Valle, Phys. Lett. **B151**, 375 (1985); J. Ellis, et. al. Phys. Lett. **B150**, 142 (1985).
7. M. C. Gonzalez-Garcia & J. W. F. Valle, Nucl. Phys. **B355** (1991) 330.
8. A. Masiero & J. W. F. Valle, *Phys. Lett.* **B 251** (1990) 273; J.C. Romão, C.A. Santos & J. W. F. Valle, *Phys. Lett.* **B 288** (1992) 311; J. C. Romao, A. Ioannisian & J. W. F. Valle, Phys. Rev. **D55**, 427 (1997) [hep-ph/9607401].
9. T. Multamaki & I. Vilja, Phys. Lett. **B433** (1998) 67.
10. J. M. Mira, E. Nardi, D. A. Restrepo & J. W. F. Valle, hep-ph/0007266.
11. M. Hirsch, these proceedings, J. C. Romao et. al. Phys. Rev. **D61** (2000) 071703 and hep-ph/0004115, Phys. Rev. **D** (2000), to appear.
12. M.A.Diaz, J. C. Romao & J.W.F.Valle, Nucl. Phys. **B524** (1998) 23 [hep-ph/9706315].
13. M. Hirsch, J. C. Romao & J. W. F. Valle, Phys. Lett. **B486** (2000) 255; M. Hirsch & J. W. F. Valle, Nucl. Phys. **B557** (1999) 60 [hep-ph/9812463].
14. M. Apollonio et al, Phys.Lett. **B466** (1999) 415; F. Boehm et al, hep-ex/9912050.
15. B. Allanach *et al.*, hep-ph/9906224; J. W. F. Valle, hep-ph/9808292 and hep-ph 9603307.
16. A. Bartl, et. al. hep-ph/0007157.
17. B. Mukhopadhyaya, S. Roy and F. Vissani, Phys. Lett. **B443** (1998) 19.
18. A. Bartl, et. al. Nucl. Phys. **B502** (1997) 19 [hep-ph/9612436].
19. A. Bartl, et. al. Phys. Lett. **B384** (1996) 151 [hep-ph/9606256].
20. M. Diaz, D. Restrepo & J. W. F. Valle, Nucl. Phys. **B583** (2000) 182.
21. M. A. Diaz, et. al. hep-ph/9906343, Nucl. Phys. **B** in press; Phys. Lett. **B453** (1999) 263.
22. M. Diaz, J. Ferrandis & J. W. F. Valle, Nucl. Phys. **B573** (2000) 75.
23. M. A. Diaz, E. Torrente & J. W. F. Valle, Nucl. Phys. **B551** (1999) 78.
24. H. Dreiner & R. J. Phillips, Nucl. Phys. **B367** (1991) 591; L. Navarro, W. Porod and J. W. F. Valle, Phys. Lett. **B459** (1999) 615; T. Han & M. B. Magro, Phys. Lett. **B476** (2000) 79; F. de Campos *et al.*, hep-ph/9903245.
25. F. de Campos, O. J. Eboli, M. A. Garcia-Jareno & J. W. F. Valle, Nucl. Phys. **B546** (1999) 33 [hep-ph/9710545].
26. A. Akeroyd et. al. Nucl. Phys. **B529** (1998) 3 [hep-ph/9707395]; F. Campos et. al. Nucl. Phys. **B451** (1995) 3.

PRECISION MEASUREMENTS, EXTRA GENERATIONS AND HEAVY NEUTRINO

V.A. ILYIN[1], M. MALTONI[2], V.A. NOVIKOV[3], L.B. OKUN[3], A.N. ROZANOV[3,4] AND M.I. VYSOTSKY[3]

[1] *SINP Moscow State Univ., Moscow, Russia*

[2] *Instituto de Física Corpuscular – C.S.I.C., Universitat de València, Valencia, Spain*

[3] *ITEP, Moscow, Russia*

[4] *CPPM, IN2P3-CNRS, Univ. Méditerranée, Marseilles, France*

Presented by A.Rozanov, E-mail: rozanov@cppm.in2p3.fr

The existence of extra chiral generations with all fermions heavier than M_Z is strongly disfavored by the precision electroweak data. The exclusion of one additional generation of heavy fermions in SUSY extension of Standard Model is less forbidden if chargino and neutralino have low degenerate masses with $\Delta m \simeq 1$ GeV. However the data are fitted nicely even by a few extra generations, if one allows neutral leptons to have masses close to 50 GeV. Such heavy neutrino can be searched in the reaction $e^+e^- \to N\bar{N}\gamma$ at LEP-200 with total final luminosity of $2600pb^{-1}$.

1 Introduction

The straightforward generalization of the Standard Model (SM) through inclusion of extra chiral generation(s) of heavy fermions, quarks ($q = U, D$) and leptons ($l = N, E$), is an example of New Physics at high energies which does not decouple at "low" ($\sim m_Z$) energies. New particles contribute to physical observables through self-energies of vector and axial currents. This gives corrections [1] δV_i to the functions $V_i (i = A, R, m)$ which determine [2] the values of physical observables (axial coupling g_A, the ratio $R = g_V/g_A$, and the ratio m_W/m_Z).

We consider the case of several lepton and quarks $SU(2)_L$ doublets and their right-handed singlet companions: $(UD)_L$, U_R, D_R, $(NE)_L$, N_R, E_R. In what follows we will assume that the mixing among new generations and the three existing ones is small, hence new fermions contribute only to oblique corrections (vector boson self energies).

2 LEPTOP fit to experimental data

We compare theoretical predictions for the case of the presence of extra generations with experimental data [3] with the help of the code LEPTOP [4]. These experimental data are the latest updates presented at this conference and they are well fitted by Standard Model. We perform the four parameter $(m_t, m_H, \alpha_s, \bar{\alpha})$ fit [a] to 18 experimental observables.

The fitted parameters [b] together with the values of the predicted observables and their pulls from the experimental data are given in the Table 1. Only the experimental value of the forward-backward assymetry in the Z decay into the pair of b-quarks A^b_{FB} shows a hint for disagreement with Standard Model. We take $m_D = 130$ GeV – the lowest value allowed for the new quark mass from Tevatron search [7] and take $m_U \gtrsim m_D$. As for the leptons from the extra generations, their masses are independent parameters. To simplify the analyses we start with $m_N = m_U$,

[a] The mass of Z-boson in the fit was fixed to the latest experimental value $M_Z = 91.1875(21)$ GeV

[b] During this conference the new results on the electron-positron annihilation into hadrons in the range $\sqrt{s} = 2 - 5$ GeV from BES [5] were released. With $\bar{\alpha}^{-1} = 128.945(60)$ [6] recalculated using these new BES results, we get from LEPTOP fit slighly higher prediction for the higgs mass $m_H = 78^{+53}_{-32}$ GeV, $m_t = 174.1(4.5)$ GeV, $\alpha_s = 0.1182(27)$, $\bar{\alpha}^{-1} = 128.927(58)$ and $\chi^2/ndf = 21.1/14$.

Table 1. LEPTOP fit of the precision observables.

Observ.	Exper. data	LEPTOP fit	Pull
Γ_Z [GeV]	2.4952(23)	2.4964(16)	-0.5
σ_h [nb]	41.541(37)	41.479(15)	1.7
R_l	20.767(25)	20.739(18)	1.1
A^l_{FB}	0.0171(10)	0.0164(3)	0.7
A_τ	0.1439(42)	0.1480(13)	-1.0
A_e	0.1498(48)	0.1480(13)	0.4
R_b	0.2165(7)	0.2157(1)	1.2
R_c	0.1709(34)	0.1723(1)	-0.4
A^b_{FB}	0.0990(20)	0.1038(9)	-2.4
A^c_{FB}	0.0689(35)	0.0742(7)	-1.5
s^2_l (Q_{FB})	0.2321(10)	0.2314(2)	0.7
s^2_l (A_{LR})	*0.2310(3)*	*0.2314(2)*	-1.5
A_b	0.911(25)	0.9349(1)	-1.0
A_c	0.630(26)	0.6683(6)	-1.5
m_W [GeV]	80.434(37)	80.397(23)	1.0
s^2_W (νN)	0.2255(21)	0.2231(2)	1.1
m_t [GeV]	174.3(5.1)	174.0(4.2)	0.1
m_H [GeV]		55^{+45}_{-26}	
$\hat{\alpha}_s$		0.1183(27)	
$\bar{\alpha}^{-1}$	128.88(9)	128.85(9)	0.3
χ^2/n_{dof}		21.4/14	

$m_E = m_D$. Any value of higgs mass above 113.3 GeV is allowed [8] in our fits, however χ^2 appears to be minimal for $m_H = 113$ GeV.

In Figure 1 the excluded domains in coordinates (N_g, Δm) are shown (here $\Delta m = (m_U^2 - m_D^2)^{1/2}$). Minimum of χ^2 corresponds to $N_g = 0.1$. We see that one extra generation corresponds to 2σ approximately.

We checked that similar bounds are valid for the general choice of heavy masses of leptons and quarks. In particular we found that for $m_N = m_D = 130$ GeV and $m_E = m_U$ one extra generation is excluded at 1.5 σ level, while for $m_E = m_U = 130$ GeV and $m_N = m_D$ the limits are even stronger than in Fig. 1. So the extra generations are excluded by the electroweak precision data, if all extra fermions are heavy: $m \gtrsim m_Z$.

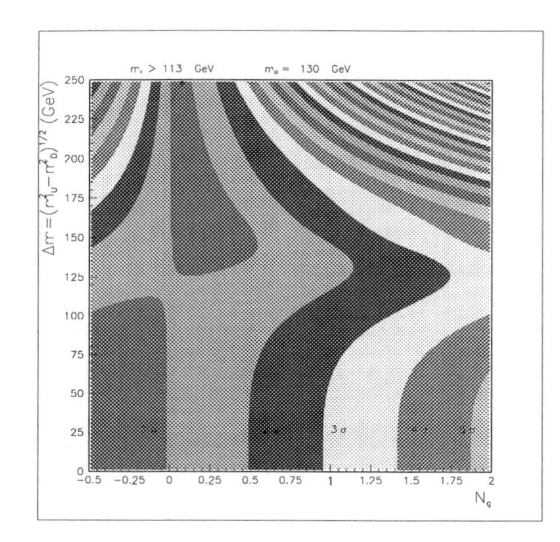

Figure 1. Constraints on the number of extra generations N_g and the mass difference in the extra generations Δm. The lowest allowed value $m_D = 130$ GeV from Tevatron search was used and $m_E = m_D$, $m_N = m_U$ was assumed.

3 Extra generations in SUSY

When SUSY particles are heavy they decouple [9,10] and the same standard model exclusion plots shown in Fig. 1 are valid.

One possible exception is the case of light chargino and neutralino. The latter are still not excluded - dedicated search at LEP II still allows the existence of such particles with masses as low as 68 GeV (gaugino region with light sneutrino) [11] or 77 GeV (higgsino case) [12] if their mass difference is ≈ 1 GeV. Analytical formulas for corrections to the functions V_i from quasi degenerate chargino and neutralino were derived and analyzed in [13]. Corrections are big and this allows one to get lower bounds on masses of chargino and neutralino: $m_\chi > 54$ GeV for the case of higgsino domination and $m_\chi > 61$ GeV for the case of wino domination at 95% CL.

The presence of chargino-neutralino pair (dominated by higgsino) with mass 80 GeV slightly relaxes the bounds on extra generations. We see that one extra generation of heavy fermions is allowed within 1.5σ domain

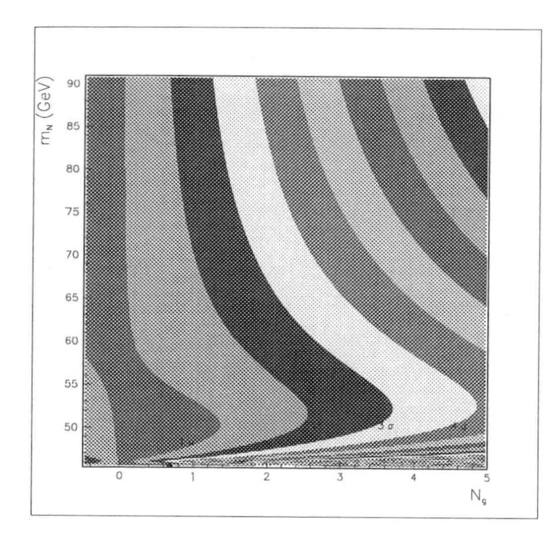

Figure 2. Constraints on the number of extra generations N_g and the mass of the neutral heavy lepton m_N. The values $m_U = 220$ GeV, $m_D = 200$ GeV, $M_E = 100$ GeV were used.

in case of the light chargino.

4 Heavy neutrino with $m_N < m_Z$

For particles with masses of the order of $m_Z/2$ oblique corrections drastically differ from what we have for masses $\gtrsim m_Z$. In particular, renormalization of Z-boson wave function produces large negative contribution to V_A. Quasi-stable neutral lepton N should have the mass slightly above $m_Z/2$ to avoid increasing the invisible Z-width and it should have the mixing angle with three known generations smaller than 10^{-6} to avoid desintegration in the detector. We consider new heavy neutrino with Dirac mass and we suppose that the Majorana mass of N_R is negligible. From the analysis of the initial set of precision data in papers [14,15] it was found that the existence of additional light fermions with masses ≈ 50 GeV is allowed. Now analyzing all precision data and using bounds from direct searches we conclude, that the only presently allowed light fermion is neutral lepton N.

As an example we take $m_U = 220$ GeV, $m_D = 200$ GeV, $m_E = 100$ GeV and draw exclusion plot in coordinates (m_N, N_g), see Fig. 2. From this plot it is clear that for the case of fourth generation with $m_N \approx 50$ GeV description of the data is not worse than for the Standard Model and that even two new generations with $m_{N_1} \approx m_{N_2} \approx 50$ GeV are allowed within 1.5σ.

5 Possibility for the direct search of the 50 GeV heavy neutrino

The direct search of the heavy neutrino is possible in e^+e^--annihilation into a pair of heavy neutrinos with the emission of initial state bremsstrahlung photon

$$e^+e^- \to \gamma + \mathrm{N\bar{N}} \tag{1}$$

The main background is the production of the pairs of conventional neutrinos with initial state bremsstrahlung photon

$$e^+e^- \to \gamma + \nu_i\bar{\nu}_i \tag{2}$$

where $i = e, \mu, \tau$. These background neutrinos are produced in decays of real and virtual Z. In case of $\nu_e\bar{\nu}_e$, two mechanisms contribute, through s-channel Z boson and from t-channel exchange of W boson. We calculated the signal and background distributions and rates [16] using CompHEP [17] computer code.

In Fig. 3 the distribution on "invisible" mass M_{inv} (invariant mass of the neutrino pair) is represented for SM background and the $N\bar{N}$ signal for $\sqrt{s} = 200$ GeV and different values of N masses, $M_N = 46 - 100$ GeV. Here we applied kinematical cuts on the photon polar angle and transverse momentum, $|\cos\vartheta_\gamma| < 0.95$ and $p_T^\gamma > 0.0375\sqrt{s}$, being the ALEPH selection criteria [18]. The photon detection efficiency 74% is assumed. For highest significance of the $N\bar{N}$ signal, evaluated as $N_S/\sqrt{N_B}$, one should include whole interval on M_{inv} allowed kinematically, so we applied $M_{inv} > 2m_N$ cut.

From the calculated signal significance [16] one can derive that only the analysis based

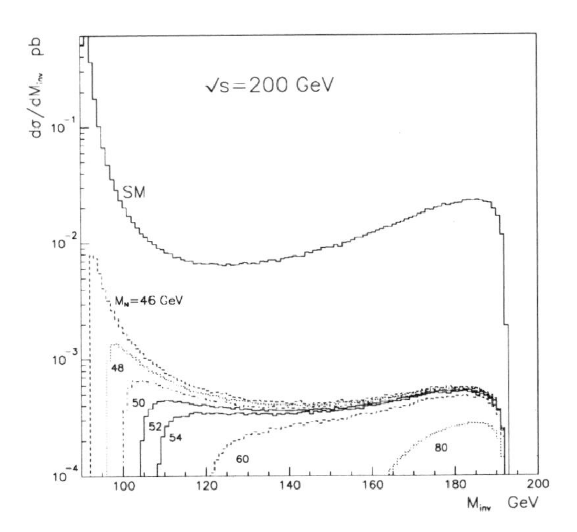

Figure 3. $d\sigma/dM_{inv}$ (in pb) for Standard Model and for the different values of m_N.

on combined data from all four experiments both from 1997-1999 runs ($\sqrt{s} = 182 - 202$ GeV) and from the current run, in total $\sim$ 2600 pb^{-1}, can exclude at 95% CL the interval of N mass up to $\sim$ 50 GeV.

Another possibility is to search for 50 GeV neutrino at the future TESLA $e^+ - e^-$ electron-positron linear collider. The increase in energy leads to the decrease both of the signal and the background, but it is compensated by the proposed increase of luminosity of 300 fb^{-1}/year [19]. Further advantage of the linear collider is the possibility to use polarized beams. This is important in suppressing the cross section of $e^+e^- \to \nu_e\bar{\nu}_e\gamma$ as this reaction goes mainly through the t-channel exchange of the W boson. However, even without exploiting the beam polarization the advantage of TESLA in the total number of events is extremely important. Thus, Standard Model is expected to give approximately 0.3 million single photon events for $M_{inv} > 100$ GeV while the number of 50 GeV neutrino pairs would be about 4000. Although the signal over background ratio is still small (2.3-0.5% for $m_N = 45 - 100$ GeV correspondingly) the significance of the signal is excellent, higher than 5 standard deviations for $m_N < 60$ GeV.

References

1. M. Maltoni, V. Novikov, L. Okun et al., *Phys. Lett.* B **476**, 107 (2000).
2. V.A.Novikov, L.B.Okun, A.N.Rozanov et al., *Rep. Prog. Phys.* **62**, 1275 (1999).
3. EWWG, lepewwg.web.cern.ch/LEPEWWG/plots/summer2000; A.Straessner, Moriond-2000.
4. V.Novikov et al., Preprint ITEP 19-95; Preprint CPPM -1-95; cppm.in2p3.fr./leptop/intro_leptop.html
5. Zhengguo Zhao, these proceedings
6. Bolek Pietrzyk, these proceedings.
7. From charged current decays $m_D > 128$ GeV, PDG, D.E. Groom et al *The European Physical Journal* C **15**, (2000).
8. Kara Hoffman,, these proceedings
9. I.V.Gaidaenko et al., *JETP Lett.* **67**, 761 (1998).
10. I.V.Gaidaenko et al., *Phys. Rep.* **320**, 119 (1999.
11. L3 note 2583, paper contributed to this conference.
12. DELPHI 2000-081, CONF 380, paper contributed to this conference.
13. M.Maltoni and M.I.Vysotsky, *Phys. Lett.* B **463**, 230 (1999).
14. N.Evans, *Phys. Lett.* B **340**, 81 (1994).
15. P.Bamert and C.P.Burgess, *Z. Phys.* C **66**, 495 (1995).
16. V. Ilyin, M. Maltoni, V. Novikov et al., hep-ph/0006324.
17. A.Pukhov et al, CompHEP user's manual v.3.3, Preprint INP MSU 98-41/542 (1998); hep-ph/9908288.
18. ALEPH Coll., Phys. Lett. **B429** (1998) 201.
19. www.desy.de/njwalker/ecfa-desy-wg4/parameter_list.html

A HIGGS OR NOT A HIGGS?

C.P. BURGESS,[1,2] J. MATIAS[3] AND M. POSPELOV[4]

[1] *Physics Department, McGill University*
3600 University St., Montréal, Québec, Canada, H3A 2T8.

[2] *Institute for Advanced Study, Princeton NJ, USA, 08540.*

[3] *Institut für Theoretische Physik E, RWTH Aachen, 52056 Aachen, Germany.*

[4] *Theoretical Physics Institute, University of Minnesota, Minneapolis MN, USA 55455.*

This talk summarizes a method for analyzing the properties of any new scalar particle, which is systematic in the sense that it minimizes *apriori* theoretical assumptions about the properties of the scalar particle, leading to very model-independent results. This kind of analysis lends itself to systematic survey through the terrain of candidate theories, which we find has vast unpopulated areas. It is also useful for quantifying the comparison of the goodness of fit of competing descriptions of data, should a new scalar be found.

1 Motivation

This talk[a] is a telegraphic summary of the much more detailed discussion of the physics of a new scalar presented in ref. [1]. We encourage interested readers to look to this reference, which fills in the fine pencil work behind the broad brush strokes presented here. (Lack of space also necessarily limits the number of papers we can cite, so please see [1] for more extensive referencing.)

Much has been written about the properties of the Higgs boson, both in its Standard Model (SM) guise, or within one of the more popular variant models, such as two doublet models (THDMs), left-right symmetric models (LRSMs) or supersymmetric generalizations of these.[2,3] Considerable experimental effort also has gone into Higgs searches, partly guided by the many detailed theoretical studies. The recent indications for a Higgs having a mass of order 115 GeV has led to an extension of LEP's running time, and may yet bring news of a final discovery.

But if a new scalar is indeed found, how can we know if it is our friend the Higgs rather than some other kind of scalar imposter? Ideally, this is answered by measuring all of the scalar's couplings and comparing the results to the well-known SM predictions. Unfortunately, the precision required to distinguish the SM Higgs from its popular close cousins is not likely to be available soon after discovery.

This talk addresses what we can do in the meantime. Instead of being glum due to the cup being half-empty – *i.e.* over our likely inability to distinguish scalars coming from well-motivated, but closely related models – we would like to rejoice at it being half-full: there will be numerous theories which predict scalars which are experimentally distinguishable from the SM very early on. It was the purpose of Ref. [1] to provide the first systematic roadmap to these dark and poorly explored corners of theory space.

2 The Framework

Of course any analysis must come with working assumptions, our goal is to minimize ours and to tie them closely to physical questions. We assume that at first only a new scalar is discovered, and all other new particles are reasonably heavy compared to it. *E.g.:* if the new scalar has mass 115 GeV, we imagine all other particles being much heavier (say > 200

[a] Presented by C. Burgess.

GeV). This assumption permits the analysis of the scalar's properties within the effective theory obtained by integrating out all other heavier particles. The lowest-dimension effective couplings of such a scalar are the most important at low energies. Up to dimension 4 the complete list of couplings is:

$$\frac{m_h^2}{2}\, h^2 + \frac{\nu}{3!}\, h^3 + \frac{a_Z}{2}\, Z_\mu Z^\mu\, h + a_W\, W_\mu^* W^\mu\, h,$$

and

$$\sum_{Q(f)=Q(f')} \overline{f}\Big(y_{ff'} + i\gamma_5 z_{ff'}\Big) f'\, h + \frac{\lambda}{4!}\, h^4 + \left(\frac{b_Z}{4}\, Z_\mu Z^\mu + \frac{b_W}{2}\, W_\mu^* W^\mu\right)\, h^2.$$

Some dimension-five interactions can also be important:

$$c_g\, G^\alpha_{\mu\nu} G^{\mu\nu}_\alpha\, h + \tilde{c}_g\, G^\alpha_{\mu\nu} \tilde{G}^{\mu\nu}_\alpha\, h + c_\gamma\, F_{\mu\nu} F^{\mu\nu}\, h$$
$$\tilde{c}_\gamma\, F_{\mu\nu} \tilde{F}^{\mu\nu}\, h + c_{Z\gamma} Z_{\mu\nu} F^{\mu\nu}\, h + \tilde{c}_{Z\gamma} Z_{\mu\nu} \tilde{F}^{\mu\nu}\, h.$$

Ref. [1] gives expressions for how observables depend on these couplings without making common theoretically-motivated assumptions (like $y_f \propto m_f/v \ll 1$). It also collects current experimental limits on their size.

3 Consequences

1. Map of Model Space The kinds of experimental distinctions likely to be possible soon after discovery can be summarized by the answers provided to four questions. (*i*) Q1: Are trilinear hWW and hZZ couplings of order electromagnetic in size ($O(e)$ or larger)? (*ii*) Q2: Is the same true for Yukawa couplings? (*iii*) Q3: Are electromagnetic $h\gamma\gamma$ couplings $O(e^2/16\pi^2)$ or larger? (*iv*) Are gluonic hgg couplings $O(g^2/16\pi^2)$ or larger?

There are 12 possible combinations of answers to these 4 yes/no questions because a 'yes' answer to Q1 generally implies a 'yes' answer to Q3. Table 1 enumerates the 12 options, and places the most popular models. Three features emerge: 1. The most popular models tend to cluster together, making them difficult to easily distinguish from one another. 2. Models are not *completely* clustered so experiments can immediately provide *some* information about the viability of *some* popular models. 3. Some categories are empty in Table 1, indicating a failure of theoretical imagination. Should experiments point us to the empty slots, theorists will fill them, so we must bear in mind they can exist.

Similarly general statements may be made concerning the finer distinctions amongst models sharing one of the entries of the Table. For instance loop corrections to Yukawa couplings are known to distinguish supersymmetric models from some 2HDMs. We show how these arguments rely on an underlying chiral symmetry, and so apply more generally than to these two alternatives. Alternatively, by comparing general expressions for hWW and hZZ couplings in multi-Higgs models, we find general inequalities which these couplings must satisfy, depending on the electroweak representation filled out by the various Higgses.

Acknowledgments

Support from N.S.E.R.C., F.C.A.R., DoE Grant No. DE-FG02-94ER40823, the Ambrose Monell Foundation and a Marie Curie EC grant (TMR-ERBFMBICT 972147) is gratefully acknowledged.

References

1. *A Higgs or Not a Higgs? What to Do if You Discover a New Scalar Particle*, C.P. Burgess, J. Matias and M. Pospelov, (hep-ph/9912459).
2. Theoretical discussions may be found in the contributions of Klaus Desch and Jan Kalinowski to this volume.
3. News about experimental Higgs searches in this volume are by Alessandra Caner, Ian Fisk, Peter Igo-Kemenes, Shan Jin, Ari Kiiskinen, Wolfgang Lohmann, Kaori Maeshima and Maria Roco.

Table 1

Class	Examples	Q1	Q2	Q3	Q4
I	SM, 2HDM (+), LRSM (+), SUSY (+)	Y	Y	Y	Y
II	Triplet (ν,+)	Y	Y	Y	N
III		Y	N	Y	Y
IV	TechniPGBs, LRSM (−), 2HDM (−), SUSY (−)	N	Y	Y	Y
V	Higher Representation (+)	Y	N	Y	N
VI	Triplet (ν, −)	N	Y	Y	N
VII		N	Y	N	Y
VIII		N	N	Y	Y
IX	Singlet w. RH ν (ν), Cons. Q.No. (ν)	N	Y	N	N
X		N	N	Y	N
XI		N	N	N	Y
XII	Higher Representation (−)	N	N	N	N

Table 1. The twelve categories of models, based on the size of their effective couplings. The positions of some representative models are indicated, where CP conserving scalar couplings are assumed for simplicity. ($\pm$) denotes the CP quantum number of the observed light scalar state. A ν in brackets indicates that the large Yukawa coupling may be restricted to neutrinos only. Categories XIII through XVI are not listed because models having $O(e)$ couplings to the W and Z generally also have $O(\alpha/2\pi)$ effective couplings to photons. Triplet indicates a doublet-triplet model for which the observed light scalar is dominantly from the triplet component.

LEPTOQUARK SIGNAL FROM GLOBAL ANALYSIS

A.F. ŻARNECKI

Institute of Experimental Physics, Warsaw University, Hoża 69, 00-681 Warszawa, Poland
E-mail: zarnecki@fuw.edu.pl

Data from HERA, LEP and the Tevatron, as well as from low energy experiments are used to constrain masses and Yukawa couplings for scalar and vector leptoquarks in the Buchmüller-Rückl-Wyler effective model. Some leptoquark models are found to describe the existing experimental data much better than the Standard Model. Increase in the global probability observed for models including S_1 or $\tilde{V}_\circ$ leptoquark production/exchange corresponds to more than a 3σ effect.

New result on the atomic parity violation (APV) in caesium[1] and the unitarity of the CKM matrix,[2] as well as recent LEP2 hadronic cross-section measurements[3] indicate possible deviations from the Standard Model (SM) predictions. Observed deviations can be described by the electron-quark contact interactions with an effective mass scale of the order of 13 TeV.[4] Exchange of the leptoquark type objects has been proposed as a possible explanation for these effects.

The global leptoquark analysis[5] combines relevant data from HERA, Tevatron and LEP2, results from low-energy eN, μN and νN scattering experiments, constraints on the CKM matrix unitarity and electron-muon universality, and the APV measurements. The Buchmüller-Rückl-Wyler model[6,7] is used for general classification of the first-generation leptoquarks. The analysis is based on the global probability function $\mathcal{P}(\lambda_{LQ}, M_{LQ})$, which describes the probability that the data come from the leptoquark model with coupling λ_{LQ} and mass M_{LQ}. It is defined in such a way that the SM probability $\mathcal{P}_{SM} \equiv 1$.

In the limit of very high leptoquark masses, constraints on the coupling to the mass ratio were studied using the contact interaction approximation.[8] The best description of the data is obtained for the S_1 and the $\tilde{V}_\circ$ leptoquark models with $\lambda_{LQ}/M_{LQ} \sim 0.3\ \text{TeV}^{-1}$. Observed increase in the global probability corresponds to more than 3σ deviation from SM. The effect is mostly resulting from the new data on APV in caesium, but is also supported by other measurements.

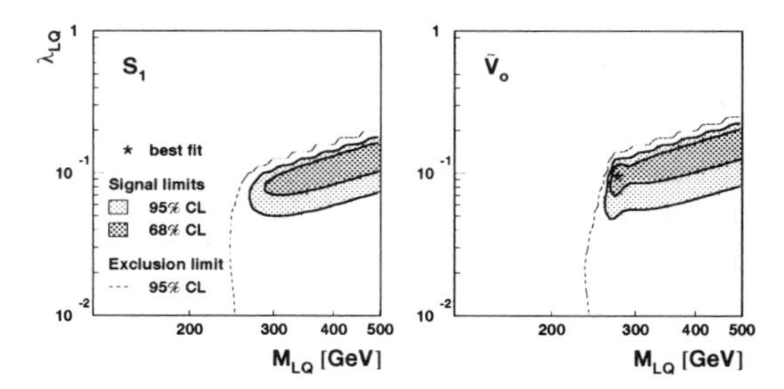

Figure 1. Signal limits on 68% and 95% CL for S_1 and $\tilde{V}_\circ$ leptoquarks. Dashed lines indicate the 95% CL exclusion limits. For the $\tilde{V}_\circ$ model a star indicates the best fit parameters. For the S_1 model the best fit is obtained in the contact interaction limit $M_{LQ} \to \infty$.

Constraints on the leptoquark couplings and masses were studied also for finite leptoquark masses, with mass effects correctly taken into account. Shown in Figure 1 are the 95% exclusion limits (corresponding to the global probability equal to 5% of the SM probability: $\mathcal{P}(\lambda_{LQ}, M_{LQ}) = 0.05$) as well as the 68% and 95% CL signal limits (corresponding to $\mathcal{P}(\lambda_{LQ}, M_{LQ}) = 0.32\ \mathcal{P}_{max}$ and $0.05\ \mathcal{P}_{max}$, respectively) for the S_1 and the $\tilde{V}_\circ$ leptoquark models. The best description of the data for the $\tilde{V}_\circ$ model is obtained for $M_{LQ} = 276 \pm 7$ GeV and $\lambda_{LQ} = 0.095 \pm 0.015$. Maximum at the low $\tilde{V}_\circ$ leptoquark mass results from the excess of high Q^2 NC DIS

Table 1. Results of the global leptoquark analysis: the 95% CL exclusion limits on the leptoquark coupling to the mass ratio λ_{LQ}/M_{LQ} (upper limit) and the leptoquark mass M_{LQ} (lower limit), the coupling to the mass ratio $(\lambda_{LQ}/M_{LQ})_{max}$ resulting in the best description of the experimental data and the corresponding model probability $\mathcal{P}_{max}$, and the 95% CL signal limits on λ_{LQ}/M_{LQ} and M_{LQ}, for models with $\mathcal{P}_{max} > 20$.

Model	95% CL excl. limits		best description		95% CL signal limits	
	$\frac{\lambda_{LQ}}{M_{LQ}}$ TeV^{-1}	M_{LQ} GeV	$\left(\frac{\lambda_{LQ}}{M_{LQ}}\right)_{max}$ TeV^{-1}	$\mathcal{P}_{max}$	$\frac{\lambda_{LQ}}{M_{LQ}}$ TeV^{-1}	M_{LQ} GeV
$S_\circ^L$	0.27	213				
$S_\circ^R$	0.25	242				
$\tilde{S}_\circ$	0.28	242				
$S_{1/2}^L$	0.29	229				
$S_{1/2}^R$	0.49	245	0.32 ± 0.06	35.8	0.09–0.44	258
$\tilde{S}_{1/2}$	0.26	233				
S_1	0.41	245	0.28 ± 0.04	367.	0.15–0.36	267
$V_\circ^L$	0.12	230				
$V_\circ^R$	0.44	231	0.28 ± 0.07	11.7		
$\tilde{V}_\circ$	0.52	235	0.34 ± 0.06	122.	0.16–0.46	259
$V_{1/2}^L$	0.47	235	0.30 ± 0.06	31.7	0.08–0.42	254
$V_{1/2}^R$	0.13	262				
$\tilde{V}_{1/2}$	0.47	244	0.30 ± 0.07	14.8		
V_1	0.14	254				

events observed in the 1994-97 HERA data.

Table 1 summarizes the results of the global leptoquark analysis.[5] For all models the 95% CL exclusion limits are given, both for λ_{LQ}/M_{LQ} (upper limit) and for M_{LQ} (lower limit). For leptoquark models which describe the existing experimental data better than SM the maximum value of the global probability $\mathcal{P}_{max}$ and the corresponding coupling to the mass ratio $(\lambda_{LQ}/M_{LQ})_{max}$ are included (in the contact interaction approximation). 95% CL signal limits for λ_{LQ}/M_{LQ} and M_{LQ} are given for models with $\mathcal{P}_{max} > 20$.

If the observed leptoquark signal is real it should become clearly visible in future colliders.[9]

This work has been partially supported by the Polish State Committee for Scientific Research (grant No. 2 P03B 035 17).

References

1. S.C. Bennett and C.E. Wieman, *Phys. Rev. Lett.* **82**, 2484 (1999).
2. D.E. Groom et al, *Euro. Phys. J.* C **15**, 1 (2000).
3. LEP Electroweak Working Group, C.Geweniger *et al*, LEP2FF/00-01.
4. A.F. Żarnecki hep-ph/0006196.
5. A.F. Żarnecki hep-ph/0003271, ICHEP 2000 paper #125.
6. W. Buchmüller, R. Rückl and D. Wyler, *Phys. Lett.* B **191**, 442 (1987); *Phys. Lett.* B **448**, 320 (1999)(E).
7. A. Djouadi, T. Köhler, M. Spira, J. Tutas, *Z. Phys.* C **46**, 679 (1990).
8. J. Kalinowski *et al*, *Z. Phys.* C **74**, 595 (1997).
9. A.F. Żarnecki hep-ph/0006335, ICHEP 2000 paper #193.

Parallel Session 12

New Detectors and Techniques

Convener: Ronaldo Bellazzini (INFN Pisa)

A PRECISE MEASUREMENT OF THE PION BETA DECAY RATE

S. RITT
for the PIBETA *Collaboration*
University of Virginia, Charlottesville, Virginia 22901-2458, U.S.A., and
Paul Scherrer Institute, CH-5232 Villigen PSI, Switzerland
E-mail: Stefan.Ritt@psi.ch

The PIBETA project at PSI, Switzerland, is a program of measurements with the aim of making a precise determination of the $\pi^+ \to \pi^0 e^+ \nu_e$ ($\pi\beta$) decay rate, which provides a new constraint on the Cabibbo-Kobayashi-Masakava matrix element V_{ud}. The PIBETA decay is measured with a stopped pion beam and normalized to $\pi^+ \to e^+\nu$ events. A new detector has been built which is described with details of the trigger and front-end electronics.

1 Introduction

The PIBETA collaboration has proposed a program[1] with the goal of making a precise determination of the $\pi^+ \to \pi^0 e^+ \nu_e$ decay rate with an accuracy of $\sim 0.5\%$, improving the present uncertainty by almost one order of magnitude.

The $\pi\beta$ decay is one of the most fundamental weak interaction processes. It is directly related to the Cabibbo-Kobayashi-Maskava (CKM) quark mixing matrix element V_{ud}. The most accurate extraction of V_{ud} is based on super-allowed Fermi transition in nuclei. Recent measurements of different nuclear transitions rates violate the three-generation CKM unitarity by more than two sigma. The analysis of nuclear β decay involves nuclear corrections which are uncertain at the level of a few tenths of a percent and do not appear in the $\pi\beta$ process, which therefore presents a more stringent test of the weak interaction theory. It could be used to constrain masses and couplings of additional neutral gauge bosons in grand unified theories.

2 Experimental Method

The experiment is performed using a stopped pion beam. The small branching ratio of $\sim 10^{-8}$ requires $\sim 10^6 s^{-1}$ pions being stopped in an active target surrounded with tracking detectors and a 240 element CsI (pure) calorimeter. The experiment uses the $\pi^+ \to e^+\nu_e$ decay which has been measured with a uncertainty of 0.4 % as a normalization.

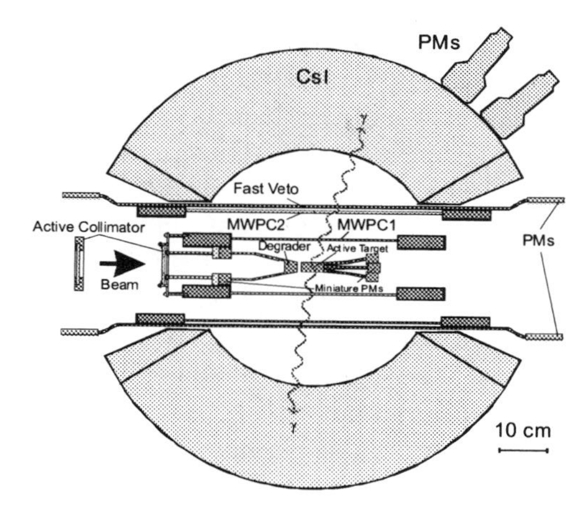

Figure 1. Cross section of the PIBETA detector.

The $\pi\beta$ decay is identified by detecting the two $\gamma's$ from the π^0 decay together with the positron in the active target. A cross section of the PIBETA detector is shown in Fig. 1.

Due to the high pion stopping rate, an efficient trigger scheme is needed to suppress the background from ordinary Michel decays and waveform digitizing is necessary to identify pile-up in the active target and the calorimeter.

3 Trigger and DAQ

A $\pi\beta$ event is detected by two simultaneous back-to-back particles each above 52 MeV. Since the back-to-back logic for the 240 CsI signals requires extensive logic, a new fast trigger module "LB500" has been developed at PSI[2]. It is a single width CAMAC module and provides 64 ECL lines which can be programmed as inputs or outputs. It features a Lattice FPGA[3] which runs unlike other modules in a "non-clocked" mode. Therefore it can be programmed via CAMAC to simulate traditional coincidence units which carry the timing of the input signals to the outputs with a typical propagation time of 10ns. This signal can then be used to generate ADC gates and TDC stops. Since the FPGA contains 6000 gates and 128 registers, it can replace various CAMAC and NIM electronics. It is used in the PIBETA experiment as a logic unit to provide 12 different triggers to the computer, as a 16-bit programmable synchronous prescaler and as an ECL shaper/fan-out.

The PIBETA experiment uses the MIDAS data acquisition system[4] which has become the standard DAQ at PSI and TRIUMF. It uses a PC to read out VME, FASTBUS and CAMAC. An integrated slow control and alarm system is connected to all HV, temperature and gas systems. This integration ensures a high flexibility and long term stability of the experiment. The MIDAS system contains a Web interface to control and monitor both the event based DAQ and the slow control system. It produces history plots dynamically and contains a sophisticated electronic logbook.

4 The Domino Sampling Chip

At a pion stopping rate of 1 MHz even a fast segmented calorimeter shows some significant pile-up. To reduce this problem, waveform digitizing is necessary. Instead of using expensive flash ADCs, a new analog sampling chip has been developed[5] at PSI. It uses 128 capacitors to sample a PM signal at high speed, which is then read out at much slower speed and digitized with a 5 MHz VME ADC. Since 48 channels are daisy-chained using a zero supression mechanism, an overall cost of less than 50 USD/channel is achieved. The sampling speed can be varied between 500 MHz and 1.2 GHz. A next generation of this chip is planned which has a higher sampling depth fo 1024 bins and a sampling speed of up to 10 GHz.

5 Conclusions

A new detector has been built to measure the $\pi\beta$ decay rate and various other pion and muon decays, using new trigger and frontend electronics. It has started taking data in summer 1999. In fall 2000, a new branching ratio with a $\sim 1\%$ error is expected, which will be further improved in the following year.

Acknowledgments

The author would like to thank Peter Dick, who engineered the LB500 unit and the PSI staff for their help to assemble the detector. This work is supported by the US National Science Foundation and PSI.

References

1. D. Počanić et al. *Proposal for an Experiment at PSI*, R 89-01.1, 1991.
2. LB500 module, PSI electronics pool, www1.psi.ch/~dick/lb500.html
3. Lattice Semicond. ispLSI 2128E-180, www.latticesemi.com/products/devices/isp2000e.html
4. S. Ritt *The MIDAS DAQ system*, midas.psi.ch and midas.triumf.ca
5. C. Bronnimann, R. Horisberger and R. Schnyder, Nucl. Instrum. Meth. **A420**, 264 (1999).

THE BTEV DETECTOR

SHELDON STONE
Physics Department, Syracuse Univeristy, Syracuse, NY 13244-1130, USA
E-mail: stone@phy.syr.edu

The BTeV Program has recently been approved by Fermilab. The physics goals and the detector are breifly described.

1 Introduction

BTeV is a program designed to challenge the Standard Model (SM) explanation of CP Violation, mixing and rare decays in the b and c quark systems. Exploiting the large number of b's and c's produced at the Fermilab Tevatron collider, we will make precise measurements of SM parameters and an exhaustive search for physics beyond the SM. A complete description of the physics goals, detector and simulation results can be found in the BTeV proposal at http://www-btev.fnal.gov. BTeV was recently approved, and comments by the director and the PAC can also be found here.

2 Physics Goals

A primary goal of BTeV is to make precise measurements that determine the fundamental parameters of nature with little or no theoretical uncertainty. The four CP violating angles α, β, γ and χ, present such an opportunity. Here α is measured using $B^o \to \rho\pi \to \pi^+\pi^-\pi^o$, β via $B^o \to J/\psi K_S$, γ from $B_s \to D_s K$ (four time dependent rates), or from a time independent analysis of $B^\mp \to D^o K^\mp$ and χ using $B_s \to J/\psi\eta$ (or η') (or ϕ). While we expect that β will have been measured, we can measure it much more precisely; it is unlikely that α, γ or χ will have been determined.

A complete program includes measuring B_s oscillations, searching for anomalous rates in "rare" decays and searching for mixing and CP violation in the charm sector, where SM rates are expected to be small and new physics could have large signals.

3 Detector Description

A sketch of the detector is shown in Fig. 1. The geometry is complementary to that used in current collider experiments. The detector looks similar to a fixed target experiment, but has two arms, one along the proton direction and the other along the antiproton direction.

The key design features of BTeV include: (1) A dipole located on the IR, which gives BTeV an effective "two arm" acceptance. The two-"foward" directions provide us with fast moving b's and correlated production of both b and $\bar{b}$. (2) A precision pixel vertex detector. This solution is radiation tolerant and vastly reduces combinatoric background and allows for fast tiggering.[1] (3) A detached vertex trigger at Level 1 that makes BTeV efficient for most final states, including purely hadronic modes. (4) Excellent particle identification using a Ring Imaging Cherenkov Detector (RICH), that has flourine gas and aerogel radiators and use hybrid-photdiodes for photon detection. (5) A high quality $PbWO_4$ crystal electromagnetic calorimeter capable of reconstructing final states with single photons, π^o's, η's or η''s, and of identifying electrons. (Based on the work of CMS.[2]) (6) Precision tracking using straw tubes and silicon microstrip detectors, which provide excellent momentum and mass resolution. (7) Excellent identification of muons using a dedicated detector with the ability to supply a dimuon trigger. (8) A very high speed

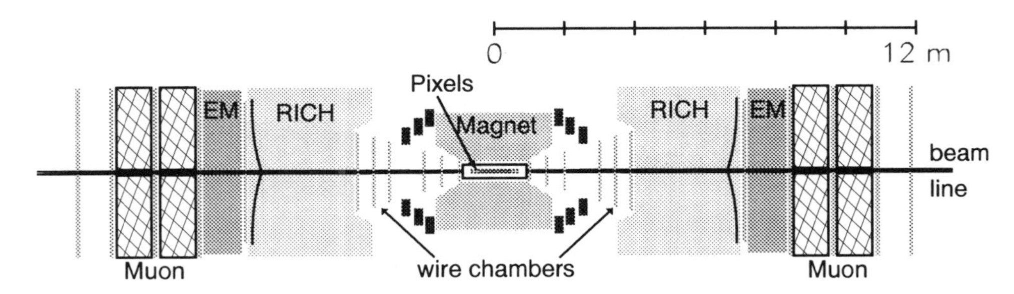

Figure 1. A sketch of the BTeV detector. The two arms are identical.

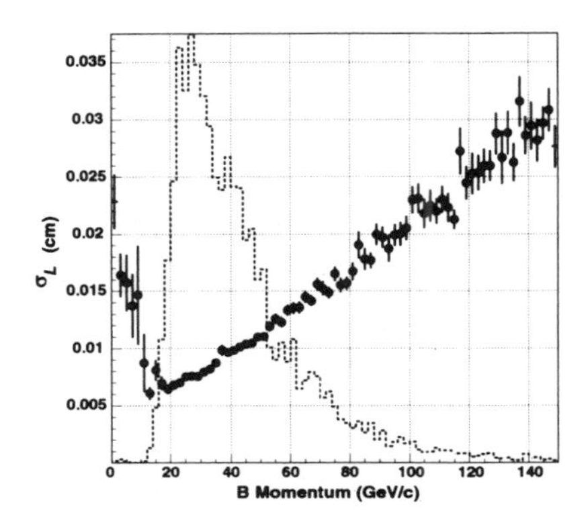

Figure 2. The B momentum distribution for $B^o \to \pi^+\pi^-$ events (dashed) and the error in decay length σ_L as a function of momentum.

and high throughput data acquisition system which eliminates the need to tune the experiment to specific final states.

The critical quantity for a b experiment is L/σ_L, where L is the distance between the primary (interaction) vertex and the secondary (decay) vertex, and σ_L is its error.

The efficacy of this geometry is illustrated by considering the σ_L distribution for the decay $B^o \to \pi^+\pi^-$ as a function of the B^o momentum in Fig. 2. We also show the momentum distribution of the B's accepted by BTeV. The following features are noteworthy: (1) The B's used by BTeV peak at $p =$ 30 GeV/c and average about 40 GeV/c. (2) The decay length is equal to 450 μm $\times p/M_B$. (3) The error on the decay length is smallest near the peak of our accepted momentum distribution. It increases at lower values of p, due to multiple scattering, and increases at larger values of p due to the smaller angles of the Lorentz-boosted decay products. The low momentum regime is characteristic of central detectors, while the higher momenta are the provence of LHCb. Since σ_L increases linearly with momentum as does L, it does no good to go to higher momentum.

4 Conclusions

Based on the importance of the physics and our reach,[3] the PAC recommended approval of BTeV and the director said:[4] *"It is important for Fermilab and the US HEP program to have such an excellent experimental program late in the decade."*

Please contact us if you want to participate in this exciting program of detector development and future physics results.

Acknowledgments

Support from the National Science Foundation is greatly appreciated.

References

1. Pixel tests are described in M. Artuso and J. Wang, [hep-ex/0007054].
2. CMS, CERN/LHCC 97-33 (1997).
3. J. Butler, in these proceedings.
4. The directors transparencies can be viewed at http://www-btev.fnal.gov/public_documents/Approval/index.html

DETECTION OF ANTIHYDROGEN WITH A SI-μ-STRIP AND CSI-CRYSTAL DETECTOR AT CRYOGENIC TEMPERATURE

C. REGENFUS
University of Zürich, Physics Institute, CH–8057 Zürich, Switzerland
E-mail: regenfus@cern.ch

ATHENA[1], one of 3 experiments at the new low energy antiproton facility at CERN (AD[2]), is designed for testing fundamental physic principles (CPT, Gravitation) to a high degree of precision by comparing cold antihydrogen to hydrogen. To monitor the production of the antihydrogen atoms and their spectroscopic response, a new detector dedicated for the endproducts of antihydrogen annihilations was developed. To meet the requirements of low temperature operation (77 K) in a high magnetic field, compact size, low power consumption and high granularity, a combination of two layers of each 16 double sided Si-μ-strip modules (16 cm long) was chosen, surrounded by 192 pure-CsI crystals (each $\approx 4\,\mathrm{cm}^3$), which are read by UV sensitive photo diodes. The frontend electronics (working point 77 K), realised in VLSI CMOS technique, features a self triggering capability of independent sub modules.

1 Introduction

The study of CPT invariance with the highest achievable precision is of fundamental importance for physics under all interactions. Equally important is the question of the gravitational acceleration of antimatter. Measurements of the energy levels of antihydrogen and its normal matter counterpart, the hydrogen atom, offer the possibility to compare directly matter and antimatter. A precision at a level of 1 part in 10^{18} could in principle be reached[3] by two photon spectroscopy of the 1S – 2S transition, due to the long lifetime of the excited state (122 ms) and thereby a natural line width of 5 parts in 10^{16}.

In the ATHENA[a] experiment $\overline{\mathrm{H}}$ atoms are supposed to be produced in high quantities ($\gg 1000$) at low kinetic energies (< 1 K), captured and hold in a magnetic trap for the scanning of their atomic levels. The first phase of the experiment is devoted to investigate $\overline{\mathrm{H}}$ formation in a nested penning trap, holding cold $\overline{\mathrm{p}}$ and e^+ plasmas in close vicinity. In a second phase a neutral trap will be added for confinement and spectroscopy of the $\overline{\mathrm{H}}$ atoms.

A compact cylindrical detector (ϕ 14 x 25 cm^2) for end products of $\overline{\mathrm{p}}$ and e^+ annihilations will be used as the general monitor for recombination studies as well as for $\overline{\mathrm{H}}$ spectroscopy. It will be placed in vacuum inside the main superconducting solenoid (6 T) covering the recombination area in the centre of the apparatus. It consists of 32 double sided Si-μ-strip modules[b] and 192 pure-CsI[c] scintillation crystals arranged in 16 rows (Fig. 1) around the tracker. One of the sides of each crystal is covered by a 4-segment Si-photodiode. To enhance the light yield a wavelength shifting paint is used on the other 5 sides. Working temperature for all detector

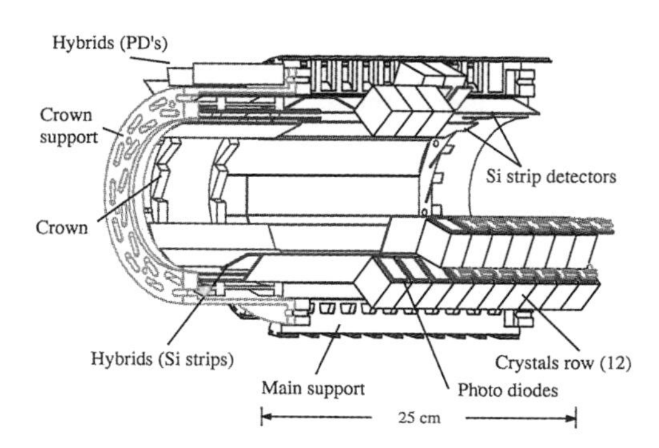

Figure 1. Explosion diagram of the $\overline{\mathrm{H}}$ detector.

[a] A complete and detailed description of the experiment is given in the proposal [1]

[b] Upper side, 382 strips, 16 cm long, 47 μm pitch (every 3rd. strip is read out); lower side, 128 pads, 18 mm wide, 1 mm pitch; in total 32 x 256 channels.
[c] CsI(Tl) is not suited for cryogenic temperatures

components including front end electronics is 77 K.

2 Detection principle

When no magnetic field gradient is present, neutral $\overline{\mathrm{H}}$ atoms, formed in the nested penning trap, will immediately escape the confinement region and annihilate on the trap walls (Fig. 2). On the average, three charged pions, three high energy γ's and two 511 keV

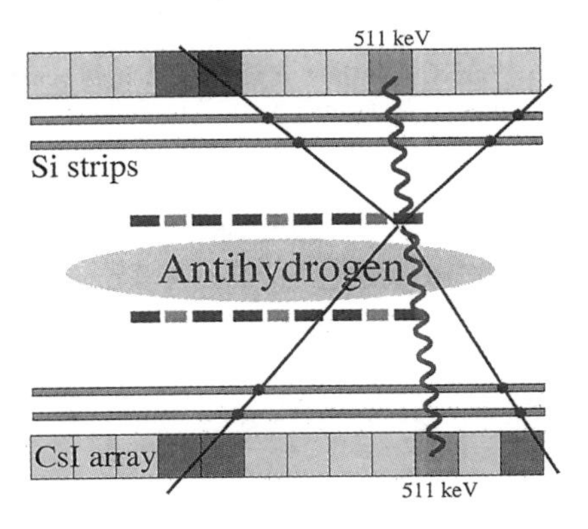

Figure 2. Simulated event of the annihilation of a $\overline{\mathrm{H}}$ atom on the walls of the nested penning trap. The solid lines represent charged pions, the wave lines the two photons from a positronium annihilation.

γ's will be produced. The detector is designed to be able to reconstruct these single $\overline{\mathrm{H}}$ annihilation events (see also [4]). Charged particles will give entries in the two Si-μ-strip layers allowing for a three dimensional reconstruction of the $\overline{\mathrm{p}}$ annihilation vertex[d]. The 511 keV gammas from a positronium annihilation will be detected in the crystal detector. Its high granularity takes in account the expected occupancy of roughly 5% of all crystals by the end products of one $\overline{\mathrm{p}}$ annihilation per μs (1 MeV is measured over the background of 2 GeV). Determined by the spatial resolution in the crystals (roughly 1 cm) the $\overline{\mathrm{p}}$ annihilation vertex can be matched along the directions of the two photons, which are supposed to be emitted back to back. Both detector components are self triggering, whereby coincidences between $\overline{\mathrm{p}}$ and e^+ annihilations can be resolved below the μs level.

The four dimensional reconstruction of events is necessary to distinguish $\overline{\mathrm{H}}$ annihilations from background reactions (e.g. residual gas collisions or $\overline{\mathrm{p}}$ annihilations). A number of about 1000 $\overline{\mathrm{H}}$ annihilation events is estimated to be needed as a unambiguous proof of the formation of antihydrogen in the ATHENA apparatus.

[d]Spatial resolution averages to about 1 mm

3 Outlook

Shortly after the startup of the AD (summer 2000) the ATHENA apparatus proved its ability of reliably capturing and cooling antiprotons. In parallel the full prototypes of strip detector modules and crystal units were successfully tested in the lab. Using cosmics and a beam telescope a S/N ratio of 50 (for MIPs) and a spatial resolution[e] of 28 μm was found for the Si-μ-strip modules. The time jitter of the fast trigger signal was less than 50 ns. The crystal readout yields a resolution of better 10% (FWHM) for the 511 keV line of a ^{22}Na test source. This corresponds to more than 30k detected photons per MeV. We look forward to exciting physical results to be expected from the research on antimatter at low kinetic energies, being however aware of the enormous technical challenge.

References

1. M. Holzscheiter et al., *Nucl. Phys.* B 56A, 336 (1997).
2. S. Maury at al., Cern Doc. *CERN/PS* 96-43 (AR), (1996).
3. M. Charlton et al., *Phys. Rep.* 241, 65 (1994).
4. C. Regenfus, *Hyp. Interactions* 119, 301 (1999).

[e]On the strip side

BEAM TEST RESULTS OF THE BTEV SILICON PIXEL DETECTOR

JEFFREY A. APPEL[1]
Fermilab, PO Box 500, Batavia, IL 60510, USA

We report the results of the BTeV silicon pixel detector tests carried out in the MTest beam at Fermilab in 1999-2000. The pixel detector spatial resolution has been studied as a function of track inclination, sensor bias, and readout threshold.

1 Introduction

BTeV has beam-tested single-chip silicon pixel detector prototypes and front-end readout chips in developing its vertex detector[2]. Of particular interest was a comparison of the resolution obtained, using 8 bit and 2 bit charge information, for a variety of incident beam angles (from 0 to 30 degrees). Spatial resolution was studied as a function of sensor bias and readout threshold. Here, only a very brief summary of the major results is presented. A more detailed discussion can be found in Ref. 3.

2 Experimental Setup

The tests were performed with a Fermilab 227 GeV/c pion beam incident on a 6 plane silicon microstrip telescope. The pixel sensors have 50 μm $\times$ 400 μm pixel size and are all from the "first ATLAS prototype submission" (both p-stop and p-spray types).[4] Up to four pixel detectors could be tested simultaneously.

Two types of readout chips were used, called FPIX0 and FPIX1.[5] Each FPIX0 readout pixel contains an amplifier, a comparator, and a peak sensing circuit. The analog output is digitized by an external 8-bit flash ADC. FPIX1 is the first implementation of a high speed readout architecture designed for BTeV. Each FPIX1 cell contains an amplifier, very similar to the FPIX0 amplifier, and four comparators, which form an internal 2-bit flash ADC. Each readout chip was indium bump-bonded to its sensor. The pixel detectors were calibrated using a pulser and two x-ray sources (Tb and Ag foils excited by an ^{241}Am α-emitter).

The readout threshold for FPIX0 was typically 2200-2500 equivalent e^- at the front end. The FPIX1 threshold was typically 3800 e^-. The FPIX0 and FPIX1 amplifier noise levels were typically 80 to 185 and 110 e^-, respectively. The relatively high FPIX1 readout threshold was due to noise and pickup problems in a circuit-board interface.

The extrapolation accuracy of the silicon microstrip telescope at the pixel detectors location was $\sim$ 2.1 μm. In order to select tracks incident on the active area of the pixel detectors, the FAST_OR output signal from one of the FPIX0-instrumented pixel detectors was required in the on-line trigger.

3 Results

The coordinate measured by a pixel detector is obtained by the position of the center of the cluster of hit pixels associated with a track, plus a linear "head-tail" correction which uses only the charge deposited on the edges of the cluster[8]. "Digital" positions are calculated without head-tail correction. By 10 degrees from normal incidence, there is always charge sharing across adjacent pixels. Resolution is somewhat degraded by charge loss near pixel boundaries as seen by ATLAS,[6] but these regions are included in all results.

The residual distribution widths, obtained for several track angles and various detectors, are shown in Fig. 1. The exper-

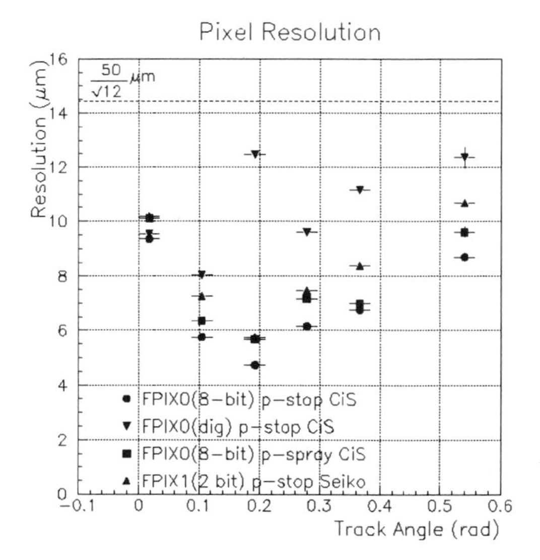

Figure 1. Position resolution along the short pixel dimension as a function of beam incidence angle for several detectors.

imental results are in good agreement with the simulation results described in Ref. 9. Due to diffusion, even the digital-position-calculation resolution is better than the pixel pitch divided by $\sqrt{12}$. The resolution of the FPIX1-instrumented p-stop detector is slightly worse than the results that we obtained by degrading by software the FPIX0-instrumented p-stop pulse height information to 2-bit equivalents. This is because the main effect degrading the resolution is the high threshold, with the 2-bit analog information only a minor effect. Comparing the p-spray and p-stop detectors in Fig. 1 (thresholds of $\sim$2200 e^-) show the charge losses in the p-spray sensor degrading the spatial resolution by a half to one micron.

For large track angles, there is not too much sensitivity to the bias voltage because the charge-sharing is dominated by the track inclination. At normal beam incidence, when the diffusion gives a substantial contribution to the charge-sharing, the sensor bias is important. The spatial resolution is still better than $10\mu m$ up to a threshold of $3800e^-$ (FPIX1 in Fig. 1). It does deteriorate rapidly for thresholds above $4000e^-$.

4 Summary

The spatial resolution achieved with FPIX0 and FPIX1 readout of ATLAS sensor prototypes is $< 10\mu m$ for a large range of incident track inclination, even using only 2-bit charge information. The resolution has a relatively small dependence on bias voltage, but does depend significantly on the discriminator threshold.

References

1. Representing authors G. Chiodini, J.N. Butler, C. Cardoso, H. Cheung, D.C. Christian, E.E. Gottschalk, B.K. Hall, J. Hoff, P.A. Kasper, R. Kutschke, S.W. Kwan, A. Mekkaoui, R. Yarema, and S. Zimmermann (Fermilab), C. Newsom (University of Iowa), A. Colautti, D. Menasce, and S. Sala (INFN - Milano), R. Coluccia and M. Di Corato (Universita di Milano), and J.C. Wang and M. Artuso (Syracuse University).
2. C. Newsom, *Overview of the BTeV Pixel Detector*, Proceedings of Pixel 2000, see www.ge.infn.it/Pix2000/Pixel2000.html.
3. G. Chiodini, et al., *Beam Test Results of the BTeV Silicon Pixel Detector*, Proc. Pixel 2000, *op. cit.*,hep-ex/0009023.
4. T. Rohe, et al., *Nucl. Instr. and Meth.* **A409** (1998) 224.
5. D.C. Christian, et al., *Nucl. Instr. and Meth.* **A 435** (1999) 144.
6. F. Ragusa, *Nucl. Instr. and Meth.* **A 447** (2000) 184.
7. S. Hancock, et al., *Nucl. Instr. and Meth.* **B1:16** (1984) 16.
8. R. Turchetta, *Nucl. Instr. and Meth.* **A 335** (1993) 44.
9. M. Artuso, *Spatial resolution predicted for the BTeV pixel sensor*, Proc. Pixel 2000, *op. cit.*

DIAMOND PIXEL DETECTORS

S. SCHNETZER
(FOR THE RD42 COLLABORATION)
Department of Physics, Rutgers University, Piscataway NJ 08854, USA
E-mail: steves@physics.rutgers.edu

Radiation hard diamond pixel detectors are an attractive alternative to silicon detectors for use at the LHC. Recent results on the pulse height and efficiency of a diamond pixel detector using CMS pixel readout electronics are presented.

The CMS and ATLAS pixel detectors will be subjected to particle fluences several orders of magnitude greater than that of any current particle detector. The detector components, both sensor material and electronics, will need to be radiation hard. The radiation hardness of chemical vapor deposited (CVD) diamond has been measured up to fluences greater than 10^{15} hadrons per cm^2 and has been found to be sufficient to allow diamond detectors to operate for several years at the highest design luminosity of the LHC.[1] In order for diamond pixel detectors to be viable for use at the LHC, hit efficiencies close to 100% must be achieved using electronics developed for LHC pixel detectors. In addition, the spatial resolution due to the charge sharing resulting from both inclined tracks and from Lorentz drift in a magnetic field should be comparable to that for silicon pixels.

We have initiated a series of studies to measure the efficiency and spatial resolution of diamond pixel detectors. We present here recent measurements made at a test beam at CERN using diamond pixel detectors with CMS pixel electronics.

The diamond sensor[2] used for these tests was 500 μm thick and was patterned with a 22 (column) $\times$ 32 (row) array of 125 μm $\times$ 125 μm sputtered titanium/tungsten electrodes. It was then bump-bonded with indium to the Honeywell version of the CMS PSI30 pixel readout chip.[3] The detector was tested in a high energy pion beam using a tracking telescope consisting of four horizontal and four vertical planes of silicon microstrip detectors. Tracks were extrapolated with a precision better than 2 μm in the pixel column direction and 7 μm in the pixel row direction onto the plane containing the pixel detector under test. The active area of the pixel detector used consisted of the 12 central columns $\times$ 30 rows. The diamond sensor was biased at a field of 1 V/μm.

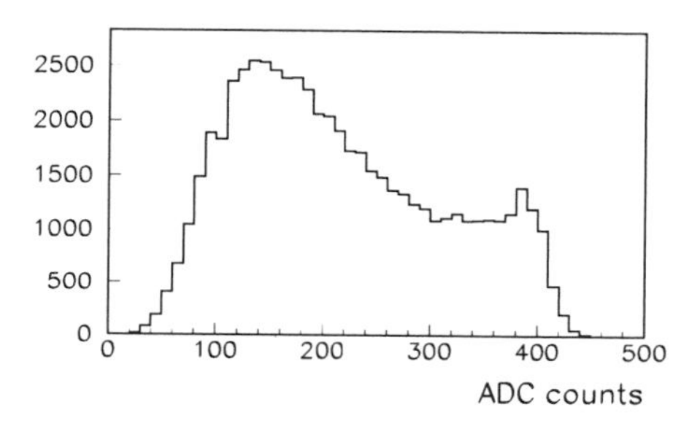

Figure 1. Pulse height distribution.

The pulse height of the diamond pixel detector is shown in Fig. 1 for tracks that passed through the active area. The measured calibration corresponds to 25 electrons of charge deposited per ADC count. The readout chip gain is nonlinear for input signals larger than 8,000 electrons and saturates at an input signal of about 10,000 electrons leading to the peak at around 400 ADC counts. The low-side tail of the pulse height distribution is partially due to threshold variations. The average threshold was approximately 2,000 electrons.

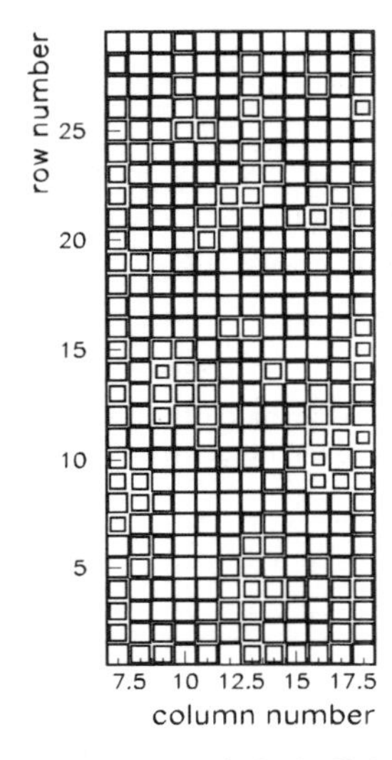

Figure 2. Map of pixel efficiencies.

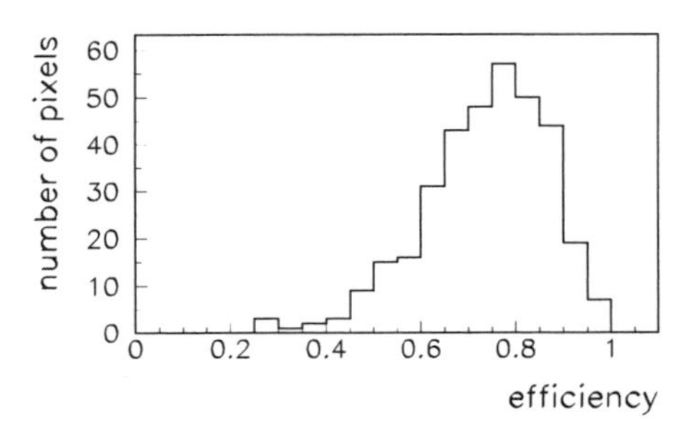

Figure 3. Distribution of efficiency.

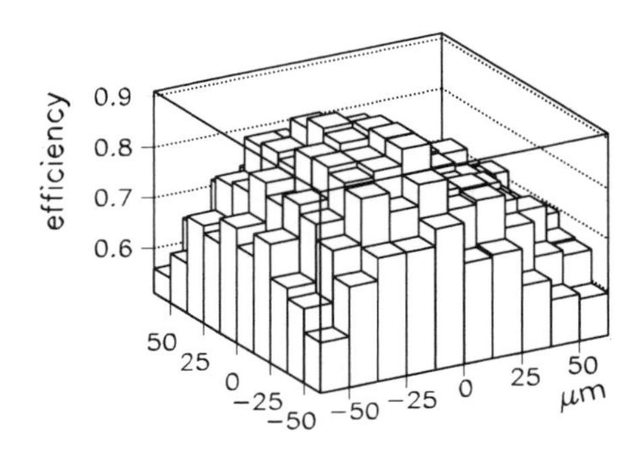

Figure 4. Efficiency as function of distance from pixel center averaged over all of the pixels in Fig. 3.

The efficiency of individual pixels is shown in Fig. 2 where the efficiency is given by the fraction of time that a given pixel or any of its nearest neighbors were above threshold when a track extrapolated to the pixel. The number of tracks per pixel was about 100. The distribution of pixel efficiencies is shown in Fig. 3 and the efficiency as a function of position within a pixel, averaged over all pixels, is shown in Fig. 4. The average efficiency is about 85% near the center of the pixel cell, about 70% near the pixel edges and about 60% in the pixel corners. The lower efficiency near the edges and corners is likely due to the charge spreading over two or more pixels, resulting in the charge on each pixel being below threshold.

Although the average efficiency is approximately 75%, the current diamond pixel detector has not yet met the criteria set above. In future tests, the efficiency will be increased by lowering the chip threshold. In addition, within the coming year, higher quality diamond is expected from the manufacturer. Measurements of the spatial resolution both for inclined tracks and for Lorentz drift in a 3 Tesla field are also planned.

This research was supported in part by a Major Research Instrumentation (MRI) award from the US National Science Foundation.

References

1. D. Meier *et al.* "Proton Irradiation of CVD Diamond Detectors for High Luminosity Experiments at the LHC", Nucl. Instr. Meth. A426 (1999) 173.
2. De Beers Industrial Diamond Division(UK) Ltd. Charters, Sunninghill, Ascot, Berkshire, SL5 9PX England
3. R. Bauer, Proceedings of 5th Workshop on Electronics for LHC Experiments, Snowmass, Colorado (1999).

DETECTION OF MINIMUM IONISING PARTICLES WITH CMOS SENSORS

M. WINTER
IReS, IN2P3/ULP, 23 rue du loess, BP 28, F-67037 Strasbourg, France
E-mail: marc.winter@cern.ch

A novel technique for detecting minimum ionising particles (i.e. m.i.p.) was designed and a first prototype fabricated in a standard CMOS technology, guided by the very high vertex detector performances demanded in future collider experiments. The device architecture resembles CMOS cameras, recently proposed as an alternative to CCD sensors for visible light imaging. The performances of the first prototype were evaluated with a 15 GeV/c π^- at CERN. Preliminary results of these tests show evidence for the detection of m.i.p.s with high detection efficiency and signal-to-noise ratio.

1 Introduction

Future High Energy Physics experiments require improved vertexing and tracking performances in order to pursue fundamental studies such as those of the top quark and of the Higgs boson [1]. Semi-conducting pixel detectors play a central role for this issue, as they allow for the high granularity needed to reconstruct accurately the impact parameters of charged tracks. However, the two existing semi-conductor techniques have hampering limitations: Charged Coupled Devices (i.e. CCD) reach the necessary granularity and introduce only little multiple scattering but are not fast and radiation hard enough; Hybrid Active Pixel Sensors (i.e. HAPS), on the other hand, are much faster and radiation harder but do not perform well enough in terms of granularity and multiple scattering.

Following the idea exposed in [2], a first prototype of Monolithic Active Pixel Sensor (MAPS) for m.i.p. detection was designed and fabricated, aiming to combine the specific advantages of CCDs and HAPSs in a single chip. MAPSs are produced in a fully standard VLSI technology, which translates into modest production costs; taking advantage of submicronic CMOS technology, they may become radiation hard; furthermore, several functionalities can be integrated on the sensor substrate, including random access.

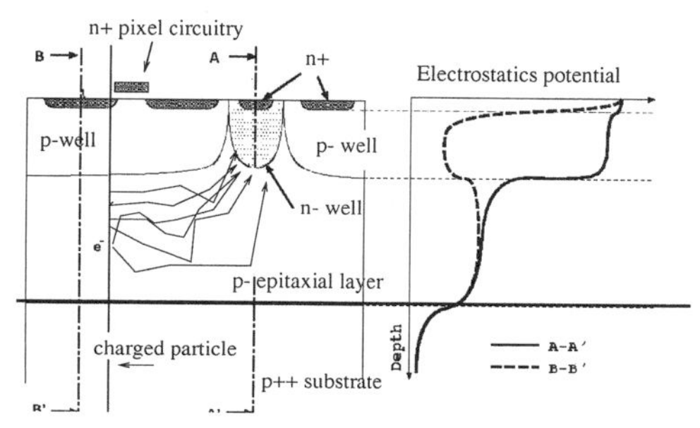

Figure 1. Internal structure of a pixel designed for charged particle tracking.

2 Principle of operation and main design features

The basic idea of MAPSs optimised for m.i.p. detection consisted in integrating a sensor in a twin-well technology with an n-well / p-substrate diode in order to achieve close to 100 % detection efficiency (see Fig.1).

The signal created by m.i.p.s in CMOS sensors originates in a low resistivity epitaxial silicon layer by producing excess carriers at the rate of about 80 electron-hole pairs per micron. The electrons liberated diffuse towards the n-well diode contact within a typical time of a few tens of nanoseconds. Because of the three orders of magnitude between the dopping levels of the p- epitaxial layer and of the p^{++} wells and substrate, potential barriers are created at the region boundaries, which act like mirros for the excess electrons (minority charge carrier).

The pixel array tested was fabricated in standard 0.6-μm technology by AMS. It is made of 64x64 square, 20 μm wide, pixels. The epitaxial layer of the sensor was measured to be about 14 μm thick. The charge collection time of the sensor was measured with laser shots and found to be less than 150 ns. The results of the device simulation, principle of operation and tests with an X-ray source can be found in [3].

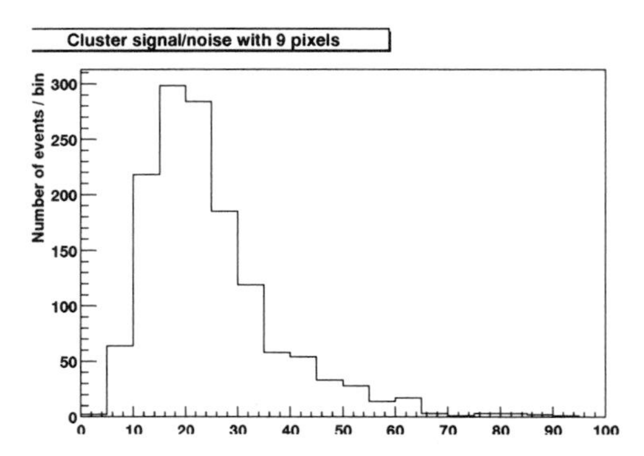

Figure 2. Signal-to-noise ratio.

3 Beam test conditions and results

The chip was mounted on a telescope made of 8 planes of silicon strip detectors, grouped in pairs of planes providing two orthogonal coordinates. The device was exposed to a 15 GeV/c π^- beam delivered by the CERN-PS.

Tracks were reconstructed in the telescope by requiring at least one hit per plane and by adjusting a straight line to the 8 telescope coordinates. Their intersections with the sensor plane were compared to the result of an independent cluster algorithm applied to the chip data, merging neighbour pixels with a $S/N > 5$.

The noise was estimated from the first 250 events of the run analysed. The kTC and fixed pattern components (FPN) were eliminated applying correlated double sampling between two consecutive frames.

1392 tracks were reconstructed in the telescope and interpolated within the sensor geometrical acceptance. For 1385 of them, a cluster was found within 50 μm of the interpolated point, translating into a preliminary value of the detection efficiency of 99.5 ± 0.2 %. It is noticeable that no substantial consequence of the FPN was observed.

As anticipated from simulations and from calibrations with a ^{55}Fe X-ray source, the charge produced by m.i.p.s was spread over several pixels. The number of pixels in a cluster was therefore allowed to go up to 9. The measured average S/N of a cluster was found to be about 25 - 30 (see Fig.2). This value, combined with the observed charge spread, indicates that the sensors provide excellent pattern recognition conditions for rejecting fake clusters due to noise fluctuations and for determining m.i.p. impact positions with an accuracy well beyond the single pixel digital precision (5.8 μm in the present case).

4 Conclusions and outlook

The first prototype of a CMOS sensor for m.i.p. detection was designed and fabricated. Its tests demonstrate that the detection technique works very efficiently and that it is likely to combine the specific performances of CCDs and HAPSs in a near future. The next steps of the developpement aim for macroscopic detector modules and for integrating various functionalities on the sensor substrate, in particular random access.

References

1. M.Caccia et al., *Proceedings of the International Workshop on Linear Colliders*, Sittges (Spain), 28 April - 5 May 1999;
2. R.Turchetta, *talk presented at the 2nd SOCLE Workshop*, LAL-Orsay (France), 23 February 1999;
3. G.Deptuch et al., *preprint LEPSI 99-15 (December 1999), accepted by Nucl.Inst.Meth..*, contributed paper Nr.716.

THE HERA-B TRACKING DETECTORS

REINHARD ECKMANN

For the HERA-B Inner and Outer Tracker Collaborations

Physics Department, University of Texas at Austin RLM5.208, Austin, TX, USA, 78712
E-mail: eckmann@desy.de

The HERA–B tracking system has to cover a large detection area with high granularity detectors. The new microstrip gas chamber technology with gas electron multiplication (MSGC–GEM) is applied for the first time in a large system and is used in the inner part of the acceptance. The outer part is equipped with large size, small diameter honeycomb drift chambers. The detectors are installed and preliminary results from the first year of operation are reported.

1 Introduction

HERA-B is an experiment at the HERA storage ring (DESY) designed to measure CP violation in the B system[1]. The B–mesons are produced in fixed target proton–nucleon interactions using an internal wire target in the 920 GeV HERA proton ring. To acquire about 2000 decays $B^0 \rightarrow J/\psi K_S^0$ per year HERA–B has to operate with an extremly high interaction rate of 40 MHz, leading to 4–5 superimposed interactions per bunch crossing (96 ns) with of order 100 charged tracks per event. Since such a B^0–decay takes place only once in $\approx 10^{12}$ interactions, a highly selective triggering scheme[2] is needed.

2 The Tracking Detectors

The HERA–B experiment is a magnetic forward spectrometer with a large acceptance ranging from 10 mrad up to 160 mrad (250 mrad in the bending plane). The granularity of the detectors has been adapted to the strong radial dependence ($\approx 1/R^2$) of the particle flux on the distance from the beam pipe. The Inner Tracker covers the region from 10 mrad up to 20(25) cm distance from the beam and is built in the MSGC–GEM [3] technology with a 300 μm pitch and 140,000 analog readout channels. The Outer Tracker uses honeycomb chambers to fill the remaining acceptance with in total 1000 m^2 of active area and around 115,000 electronic channels providing the digitized drift time. The inner diameter of the honeycomb drift cells is 5 mm for the part closest to the beam and 10 mm for the outermost region. In this way the occupancies of the Inner Tracker are kept below 5%, while for the "hottest" region of the Outer Tracker, 30% is reached.

The large size and the radiation environment of HERA–B is a challenge to detector technologies. Several design iterations were necessary to develop the present solutions. A detailed description of the detector technologies and how the final designs evolved can be found for the inner MSGC–GEM detector in [4,5,6] and for the outer honeycomb detector in [7,8].

The tracking is performed using three stereo orientations: vertical $0°$–layers and layers with a $\pm 5°$ angle. A hit resolution of $200 \mu m$ allows a momentum measurement with a resolution of $\delta p/p = 5.4 \cdot 10^{-3} \oplus 5.1 \cdot 10^{-5}\, p[GeV]$. The single strip (cell) efficiency is required to be >95% (>98%) for the Inner (Outer) Tracker. Both detectors are used in the First Level Trigger, which requires an efficient reconstruction of space points in four sets of $0°, \pm 5°$–layers. This is achieved by forming the "or" of two adjacent single layers and in the case of the Inner Tracker by an additional "or" of four neighboring strips.

3 The Commissioning Phase

The Outer Tracking system is completely installed, while for the Inner Tracker two trigger stations are missing. A feedback of the trigger output signal of the HELIX chip [9] to its input signal has prevented the operation of the trigger part. This problem has been solved by reducing the swing of the trigger output and improving the ground connection of the chamber. The installation of new trigger chambers is planned for the upcoming shutdown. All 136 Inner Tracker MSGC–GEM chambers are in stable operation, hit efficiencies of up to 90 % are reached after a period of careful training. The chambers are operated with a Ar:CO_2 (70:30) gas mixture at a drift voltage of -2.5 kV and a cathode voltage of -510 V. The GEM–voltages are in a range of 420 V – 460V, individually adjusted for each chamber to reduce the efficiency variations to about 10%.

The Outer Tracker system is operated with a Ar:CF_4:CO_2(65:30:5) gas mixture at a high voltage of 1700V (1800V) for the 5 mm (10 mm) cells and thresholds in the range of 3.5 – 4.5 fC. High voltage instabilities in a small amount of cells – $\approx$ 0.5% – made it necessary to reduce the voltage, so that the gas gain is lowered by a factor two. A feedback of the low level TTL trigger output to the ASD8 [10] input had to be suppressed by doubling the threshhold voltage. Therefore the hit efficiencies are around 90 % and the intrinsic hit resolution is measured to be 350 μm. The unstable cells introduce a considerable amount of dead channels due to a 1:16 grouping in the high voltage supply lines. It is expected, having access to the modules, that a large fraction of the dead channels can be recovered.

The data taken is being used so far for the commissioning of the experiment [11]. The alignment is being studied intensively and the reconstruction algorithm [12] is being improved to reduce the impact of dead channels

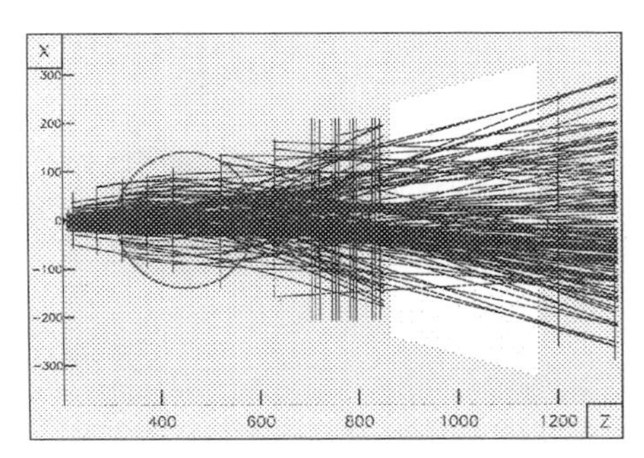

Figure 1. Reconstructed event

and the lower hit resolution and efficiency, so that a track reconstruction efficiency of 90% is reached.

Since the analysis of the data taken is ongoing all quoted numbers are preliminary.

References

1. T. Lohse et al. , DESY–PRC 94/02, 1994.
2. J. Flammer et al. , The HERA–B Level 1 Trigger, *Electronics for LHC experiments, Snowmass 1999.*
3. F. Sauli, *Nucl. Instr. and Meth. A* **386** *(1997) 531.*
4. B. Schmidt, MSGC Development for HERA–B, *Proceedings of Erice Workshop, 1997.*
5. T. Hott, *Nucl. Instr. and Meth. A* **408** *(1998) 258.*
6. T. Zeuner, *Nucl. Instr. and Meth. A* **446** *(2000) 324.*
7. M. Capeans, *Nucl. Instr. and Meth. A* **446** *(2000) 317.*
8. U. Uwer, The HERA-B Outer Tracker, *International Euroconference on High Energy Physics, Tampere 1999.*
9. C. Bauer et al. , Performance and radiation tolerance of the HELIX128-2.2 and 3.0 readout chips for the HERA-B microstrip detectors, *Electronics for LHC experiments, Snowmass 1999.*
10. K. Berkhan et al. , Large system experience with the ASD-8 chip in the HERA–B experiment, *Electronics for LHC experiments, Snowmass 1999.*
11. B. Schmidt, The status of the HERA–B experiment, *These proceedings.*
12. R. Mankel and A. Spiridonov, *Nucl. Instr. and Meth. A* **426** *(1999) 268.*

PHYSICS POTENTIAL AND THE STATUS OF DØ UPGRADE AT FERMILAB

JAEHOON YU

MS357, FNAL, P.O.Box 500, Batavia, IL60510

The DØ experiment is one of the two collider experiments at Fermilab. The DØ detector is a multi-purpose detector and took its data during Fermilab TeVatron collider run in 1992-1996. Both the DØ detector and the Tevatron accelerator at Fermilab are currently undergoing significant upgrade to extend the reach to new physics and to further probe Standard Model. In this paper, physics potential of the upgraded DØ detector and the upgrade status are discussed.

In the past few decades, Standard Model (SM) has undergone series of tests and has been extremely successful. The theory unifies three of the four known forces in nature and provides mechanism of which masses are generated from. Despite the successes, there are outstanding issues that cannot be explained within the context of current Standard Model framework. Three most outstanding issues are: neutrino masses whose evidence is becoming clearer from neutrino oscillation experiments[1], unobserved Higgs particle[2], the mediator of electroweak symmetry breaking which is the mechanism to generate masses, and the degree of CP violation which has been observed greater than SM prediction[3].

Therefore, the question becomes two fold: whether the SM is the theory of everything but we just did not discover the Higgs particle or the SM is flawed and there are other models that describe nature better and replace the SM. The upgraded TeVatron collider and its detectors could provide answers to two of the three above outstanding questions, Higgs particle and CP violation, in addition to information on physics beyond SM. In this paper, we present the physics potential of the DØ detector[4] and the status of its upgrade[5].

The most recent global Electroweak fits[6] performed by LEP Electroweak Working group presented at this conference put the limits on the SM Higgs mass to be above 113 GeV/c^2. All the evidences point to single neutral Higgs particle with low mass within the SM framework. However, none of the experiments has observed such particle yet.

In addition, there are general arguments for models beyond SM at Electroweak scale ($\sim$ 250 GeV). SM fits suggest the new physics is weakly coupled and might be indirectly pointing to supersymmetry (SUSY). On the contrary, all direct searches of SUSY particles have been negative. Searches for SUSY particles at LEP has put limits on super-partner masses; $m_{\tilde{b},\tilde{t}}$, $m_{\tilde{e},\tilde{\mu},\tilde{\tau}}$, $m_{\tilde{\chi}^{\pm}} > 70 \sim 90\text{GeV/c}^2$, and $m_{LSP} > 36\text{GeV/c}^2$. Similar results are obtained at TeVatron Run I from both DØ and CDF experiments. Despite the fact that the mass limits are getting more stringent, no evidence have been seen from the measurements.

In order to provide extended phase space and to enhance the possibility of inching closer to finding Standard Model Higgs particle, Tevatron accelerator has been upgraded to increase its luminosity by a factor of 10 to $>\sim 10^{32}\text{cm}^{-2}\text{sec}^{-1}$ and the center of mass energy by 10% to 2 TeV, with respect to the previous run. The increase in instantaneous luminosity is going to result in a total integrated luminosity of 2 fb^{-1}/experiment for the first two years of the run, Run IIa, and to ultimately result in over 15 fb^{-1}/experiment in Run IIb that follows Run IIa and that runs till the LHC experiments comes on-line. The increased accelerator capability and the physics goals naturally raise issues for detector upgrade to accommodate the increased event rates, decreased bunch spacing, and the emphasis in physics goals.

In order to provide adequate functionality for observing Higgs particles and CP violating processes, it is absolutely necessary to implement detectors to enhance b-tagging capability. In addition, the detector response must be fast to react to shorter bunch spacing of the Tevatron collider that starts out with 396 ns and will eventually become 132 ns.

The fundamental philosophy of the DØ upgrade is to retain as much of the excellent performance of the original detector, such as calorimetry, as possible. This philosophy increases cost effectiveness of the detector upgrade. The primary change of the detector is in the tracking system. While RunI detector has a drift chamber and a jet chamber vertex tracking system, without a central solenoid magnet, Run II detector has a 2 Tesla super conducting solenoid magnet[7] in the central tracking volume together with a new tracking detector systems.

In order to strengthen displaced vertex detection for b-tagging, a silicon micro-strip vertex detector (SMT)[8] surrounds the interaction region and is read out through the SVX-II chips[9]. The central cylinder of the SMT consists 6 four layer barrels of double and single sided detectors; two in 2 degree and the other two in 90 degree stereo angles. These barrels are interspersed with 12 disks (F-disks). Each F-disk consists of 12 double sided wedge shaped micro-strip detectors that each covers 15 degrees in ϕ. Large angle coverage is obtained by four large diameter disks (H-disks), two on either side of the central barrel-disk cylinder. Each H-disk consists of 24 single sided wedge shaped detectors. The angular coverage of SMT, including the H-disks, extends out to $\eta = \pm 3$. Total number of readout channel of the SMT system is approximately 800,000.

SMT is then surrounded by the central

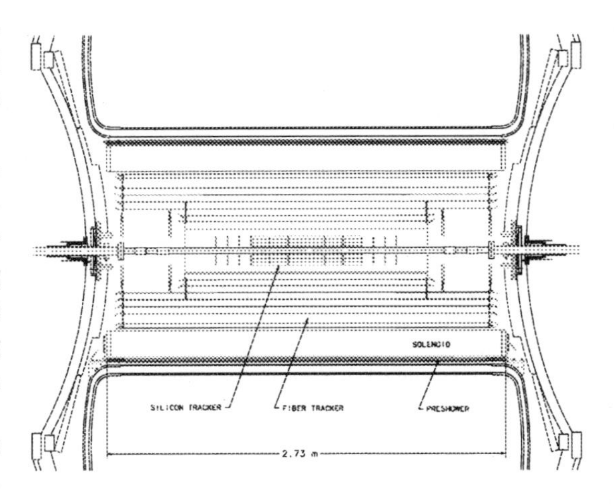

Figure 1. A schematic side view of the upgraded DØ tracking system.

scintillation fiber tracker (CFT)[10] which has a total of 74,000 readout channels. CFT consists of 8 cylindrical layers of scintillating fiber ribbon doublets that cover the directions along the beam and $\pm 5°$ alternate stereo angles with respect to the beam direction. The fibers are read out through low temperature VLPC (Visible Light Photon Counter) readout system whose operating temperature is $\sim 7°$k.

These two central tracking detectors, along with the 2 Tesla solenoid magnet, provide charged particle momentum resolution of $\Delta P_T/P_T \sim 5\%$ at $P_T = 10$GeV/c. Figure 1 shows a schematic side view of the upgraded DØ tracking system.

Since the solenoid has been added to the central tracking system, the total thickness of the material before the electromagnetic section of the calorimeter has been increased by about a factor of two to $\sim 2X_0$. In order to keep the electromagnetic calorimeter energy resolution as good as before, a central and a forward preshower detectors[11] before the first layer of the electromagnetic sections of calorimeters have been added. These preshower detectors consist of scintillation counter strips and $\sim 1X_0$ lead converters, and read out through the same readout chain as CFT. These detectors enable the DØ calorimeter system to retain its energy resolution to within 10% of the original resolution of $\sigma/E = 21\%/\sqrt{E}$. The calorimeter system is currently undergoing an electronics upgrade to sample signal fast to minimize pile-up effects that come from long intrinsic drift time ($\sim 460ns$) liquid argon (LAr).

In addition to the detector upgrade, the DØ experiment upgrades its trigger systems to maximally exploit the improved capability of the upgraded detector. The most important upgrade of the trigger system is the use of CFT and preshower detectors, exploiting quick response time of scintillation light detectors. The muon system also added three layers of scintillation counter layers to enhance muon trigger capability, in addition to upgraded forward muon system. Thick steel shielding blocks have been added on either sides of the muon system, surrounding the beam pipe, to reduce background from beam halo.

As of this conference, more than 90% of the detector construction has been completed, and installation and commissioning have begun since the end of last year. Just before this conference in June, 2000, the DØ central scintillating fiber tracker has been installed into its final position, and wave guide installation has begun. Commissioning effort, thus far, has been concentrated on preparation of DAQ system to adequately support detector debugging and commissioning effort. The DØ experiment is currently planning to begin a cosmic ray commissioning run with all available systems in December, 2000, till before the detector roll-in in January, 2001. This period should provide invaluable opportunity to integrate and to debug the detector system.

In conclusion, Tevatron RunII will provide an order of magnitude higher luminosity at the center of mass energy of 2 TeV and will significantly extend physics reach. This also enables extended search for physics beyond the Standard Model. Currently both accelerator and DØ detector upgrades are progressing well and will be ready for Run IIa which is scheduled to begin March 1, 2001. There is little doubt that the DØ experiment will make a significant impact in understanding of SM and search for new physics beyond SM. The DØ has already started preparing for Run-IIb upgrade, beyond 2fb^{-1} expected from Run-IIa.

References

1. Y. Fukuda *et al.*, Super-Kamiokande Collaboration, Phys. Rev. Lett. **81**, 2561 (1998); Toshiyuki Toshito, in these proceedings (2000)
2. http://lepewwg.web.cern.ch/LEPEWWG/
3. A. Alav-Harati *et al.*, KTeV Collaboration, Phys. Rev. Lett. **84**, 408 (2000); A. Alav-Harati *et al.*, KTeV Collaboration, Phys. Rev. Lett. **83**, 22 (1999)
4. S.Abachi, et al, DØ collaboration,"The DØ detector," Nucl. Instr. Meth. **A338**, 185 (1994).
5. "DØ Upgrade Technical Design Report," DØ Note 2962
6. Bolek Pietrzyk, in these proceedings (2000)
7. "Conceptual Design of a 2 Tesla Superconducting Solenoid for the Fermilab DØ Detector Upgrade," Fermilab TM-1886 (1994)
8. "DØ Silicon Tracker Technical Design Summary," DØ Note 2169 (1994)
9. T. Zimmerman *et al.*, "The SVX II Readout Chip," IEEE Tractions on Nuclear Science, Vol. 42, No. 4, August (1995)
10. M. Atac *et al.*, Nucl. Instrum. Meth. **A320**, 155 (1992)
11. M. Adams *et al.*, Nucl. Instrum. Meth. **A366**, 263 (1995)

THE DØ SILICON MICROSTRIP TRACKER: CONSTRUCTION AND TESTING

PETROS A. RAPIDIS
For the DØ Collaboration
Fermi National Accelerator Laboratory, Batavia, Illinois 60510, USA
E-mail: rapidis@fnal.gov

The DØ collaboration is nearing completion of the DØ Silicon Microvertex Detector, a 793,000 channel silicon strip tracking system for the DØ Upgrade. The production of this detector, which is one of the two largest detectors to date, included burn-in of electronics, testing with an IR laser, mechanical assembly under coordinate measuring machines, and a test of a significant portion of the readout system. We have observed various failure modes of the silicon sensors, and a particular failure mode involving microdischarges at the coupling capacitor edge will be described.

Fermilab's Tevatron is slated to begin a new run of high energy $p\overline{p}$ collisions, the so called Run II, in March 2001. During this run the center of mass energy will be 2 TeV and the expected luminosity for the first two years will be 2-4 fb^{-1}, i.e. 20 to 40 times more than Run I. Most of the processes of interest (top quark physics, Higgs boson searches, SUSY searches) have b quarks in the final state, and tagging of b quarks through a displaced vertex is a required feature of a detector optimized for such physics. The DØ Upgrade has as a major component a silicon strip vertex detector[1] , the SMT (Silicon Microvertex Tracker), to identify tracks with non-zero impact parameters for $|\eta| < 3$. A silicon track trigger that reconstructs online displaced vertices is also being built [2].

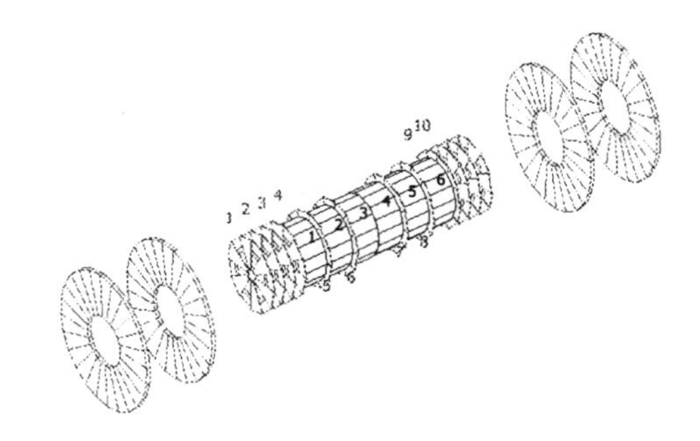

Figure 1. Overall layout of the DØ SMT. The separation between disk 1 and disk 12 is 1.1 m .

The SMT,shown in figure 1, consists of six cylindrical barrels of four layers of silicon sensors as shown in figure 2. The sensors for the four barrels near the z=0 position have double sided detectors with axial and 2° stereo strips for the second and fourth layers, and for the first and third layers they have sensors with axial and 90° large angle strips using double-metal readout. For the two outermost barrels single sided axial strip sensors are used instead of the 90° variety. The innermost(outermost) radius of the barrels is 2.7 cm(10.3 cm). Interspersed between the barrels, and also at the two ends, are twelve disks formed of double sided wedge sensors with 30° stereo strips which run at $\pm 15^o$ of the radial direction. Two larger disks at each end, built of single sided wedges glued back-to-back that provide a $\pm 7.5^o$ stereo readout, complete the SMT.

The 768 sensor assemblies that comprise the SMT have a silicon area of 3.0 m^2, of which half is double sided, and have $792,576$ AC-coupled readout channels. The size of this detector is of intermediate size between the largest silicon vertex detectors built to date (LEP, CDF Run I, CLEO) and the ones envisioned for LHC (CMS,ATLAS).

Extensive testing is required to assure adequate performance of such a large system. Each sensor module with its on-board

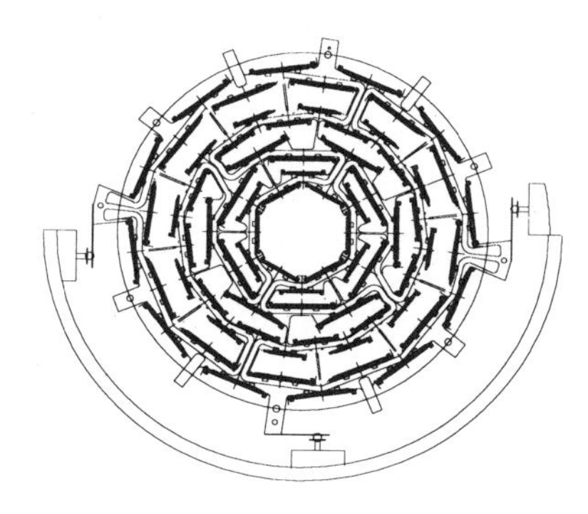

Figure 2. Cross section through the cylindrical portion of the SMT.

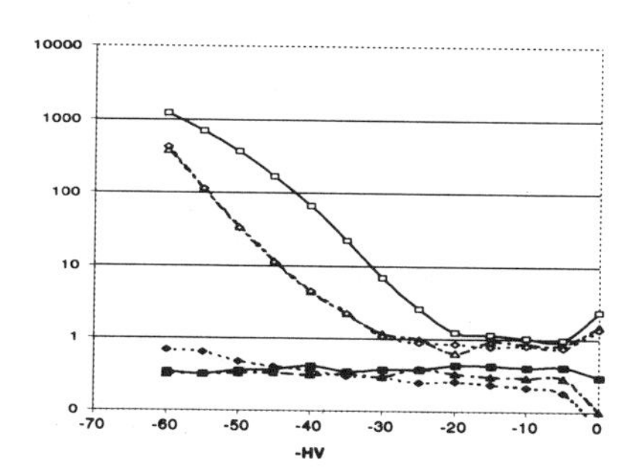

Figure 3. Strip current (nA) for p-side biased unirradiated detector, with p-side AC coupling capacitor floating (open points), or grounded (solid points).

mounted front electronics is tested for at least 72 hours and strips with excessive current/noise, or with shorted coupling capacitors are disconnected. Finally the light beam from an IR laser (λ=1064 nm) is used to scan across each sensor.

Early tests raised concerns about the noise performance of the detector and readout systems. Such issues have been studied with a rather large test setup of a barrel and disk assembly using the same electronics that will be used in the DØ detector. In this setup, which has proven invaluable in debugging the readout, 83,328 channels were run error free (error rate <1 per 3×10^{13} bits).

Final assembly of the detector is proceeding and half of it has been fully assembled. Alignment tolerances of less than 10μm, dictated by the requirements of the trigger processor, have been met.

We saw a variety of silicon sensor failures. In particular, for double sided double metal detectors, defects of the p-stop isolation used on the n-side were a concern. Another failure were microdischarges similar those observed by a KEK group [3]. Figure 3 illustrates the excessive currents seen with double sided detectors when a negative potential is applied on the p-side, if the AC coupling capacitor on the p-side is grounded. This behaviour is correlated with mask misalignments indicating a microdischarge due to high fields

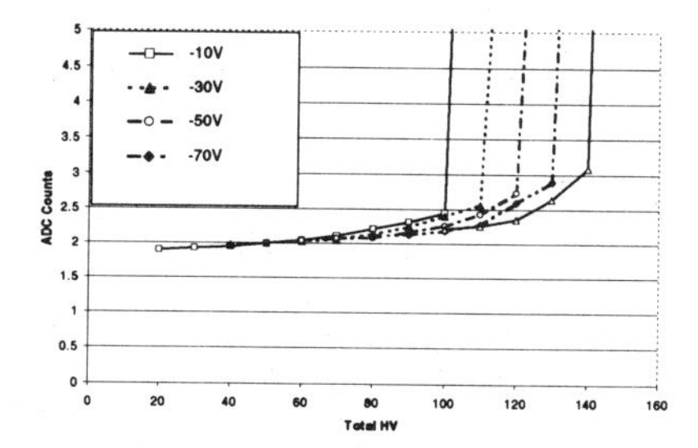

Figure 4. Noise of n-side strips for an irradiated double sided detector with bias on both the p- and n-sides; the four lines correspond to different p-side bias voltages.

in the isolating layer of SiO_2. This phenomenon moves to the n-side following irradiation and type inversion. This is clearly shown in figure 4 where large noise increases are observed for n-side strips with grounded AC coupling capacitors in a detector irradiated by 10^{14}neutrons/cm^2; one sees that the noise clearly depends on the n-side bias (total HV - neg HV), i.e the discharge happens at the n-side that is now the junction side.

References

1. For details see F. Lehner, *Nucl. Instrum. Meth.* A **447**, 9, (2000).
2. M. Narain, *Nucl. Instrum. Meth.* A **447**, 223, (2000).
3. T. Ohsugi *et al., Nucl. Instrum. Meth.* A **342**, 22, (1994), and **383**, 116, (1996).

THE ALL-SILICON TRACKER OF THE CMS EXPERIMENT

K. FREUDENREICH

ETH Zürich and CERN, EP Division, CH-1211 Geneva 23, Switzerland

E-mail: Klaus.Freudenreich@cern.ch

On behalf of the CMS Silicon Tracker Collaboration

Recently, the CMS collaboration has decided to equip the central tracker exclusively with silicon detectors. The new layout and its performance will be discussed.

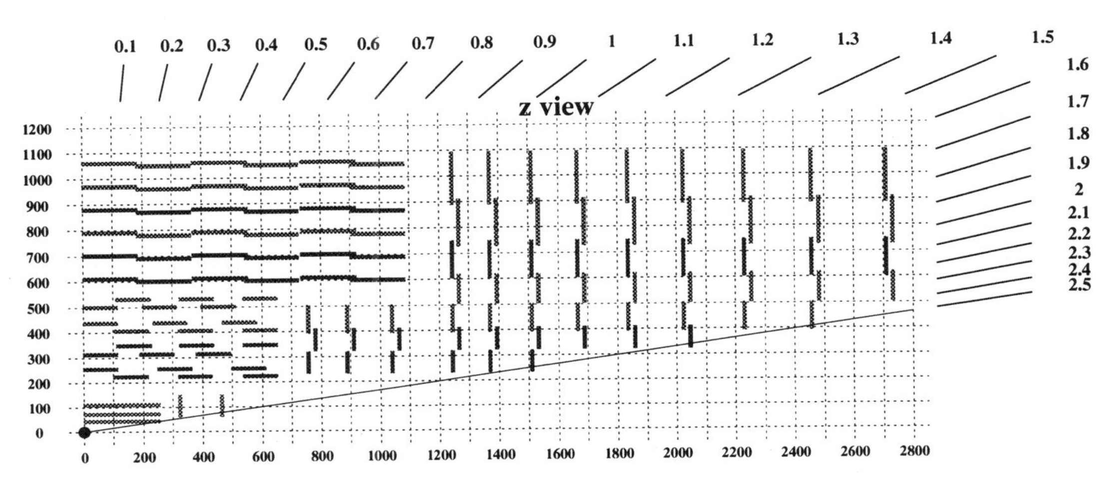

Figure 1. Cut view of one quarter of the CMS central tracker. Horizontal axis = distance to the IP along the beam line, vertical axis = radius (all dimensions in millimeters),

1 Design Considerations and Detector Layout

The main design goals for the CMS tracker are:

• Momentum resolution for isolated leptons in the central rapidity region of $\frac{\Delta p_T}{p_T} = 0.1 \times p_T$ with p_T in TeV.

• Ability both to tag and to reconstruct in detail b-jets and B-hadrons within jets.

• Reconstruction efficiency $>$ 95 % for isolated high p_T tracks and $>$ 90 % for high p_T tracks within jets.

In addition, the following detector requirements have to be fulfilled as well:

• Resistance to high radiation dose.

• Fast detector response ($<$ 25 ns) to reduce pile-up effects.

• Minimal amount of material in front of the calorimeters.

• Capability to cope with the CMS trigger requirements.

At the end of 1999, the CMS collaboration decided to equip the whole central tracker with silicon detectors exclusively [1]. Fig. 1 shows a cut view of this all-silicon layout. Starting from the IP there are up to 3 pixel detectors in the barrel having a pixel size of $150 \times 150\mu m$. In addition, there are two pixel end-cap disks on each side. The pixel detector has 50×10^6 channels in total.

The silicon strip detectors consist of 5 single-sided (layer nb. 6, 7, 10, 12 + 13) and 5 stereo layers (layer nb. 4, 5, 8, 9 + 11) in the barrel. On each side there are 3 mini-end-cap and 9 large end-cap disks. In order to keep the occupancy below the 1 % level, the silicon strip length is limited to 11.9 cm for radii (distance to the beam axis) of ≤ 55 cm. For the larger radii the strip length is increased to 18.9 cm. In order to compensate for the increased noise, these longer detectors will have thicker (500 μm instead of 320 μm) sensors whose signal scales with the detector thickness: 500 μm/320 μm $\approx$ 18.9 cm/11.9 cm, thus keeping the signal to noise ratio constant.

In total, there are 7,888 single-sided and 4032 double-sided detector modules which correspond to 15,952 single-sided equivalent modules. The read-out is done with APV25 chips in deep sub-micron technology with 128 channels/APV.

Table 1 compares the nb. of detectors, channels and silicon area of the CMS silicon strip tracker to silicon trackers of some past and future experiments. There is a thousand-fold increase in area from L3 to CMS!

experiment	nb. of detectors	nb. of channels	silicon area [m^2]
CMS	15.95 k	10 $\times 10^6$	223
ATLAS	16.0/2 k	6.15 $\times 10^6$	60
AMS 2	2.3 k	196 k	6.5
D0 2		793 k	4.7
CDF II	720	405 k	1.9
Babar		140 k	0.95
Aleph	144	95 k	0.49
L3	96	86 k	0.23

Table 1. Comparison of some silicon strip detectors.

2 Expected Detector Performance

The performance of the all-silicon detector remains about the same with respect to the original layout which used micro-gas-strip detectors in the outer layers. Fig. 2 and 3 show the momentum and impact parameter resolution as a function of η. The momentum resolution for $p_t = 1000$ GeV is close to 10 % at $\eta = 0$ and agrees with an analytic calculation using a pitch of 100 μm, a measured track length of 1.1 m and a magnetic field of 4 T. The transverse impact parameter resolution is < 30 μm for p $\geq$ 10 GeV.

3 Schedule and Conclusions

A preproduction of 200 silicon strip detectors will be done at the end of 2000. The module production will be highly automatized and is expected to take 3 years.

The availability of 6" technology allows to build a fast, homogeneous tracking detector. The 0.25 μm electronics yields better performance w.r.t. previous technologies at reduced price. The CMS all-silicon detector will have excellent performance.

I would like to thank my colleagues from the CMS Tracker Group for their support.

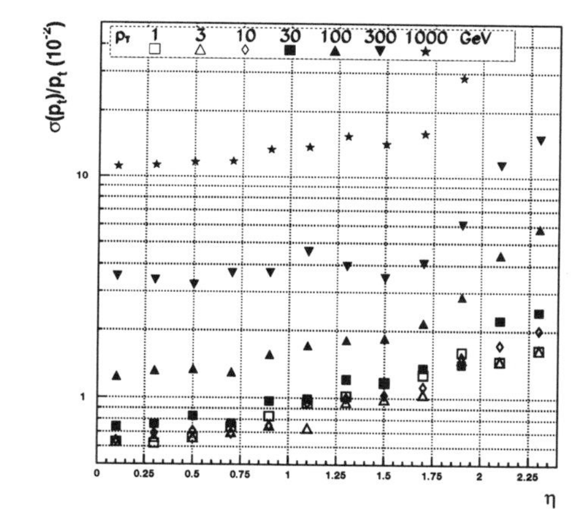

Figure 2. Momentum resolution as a function of η.

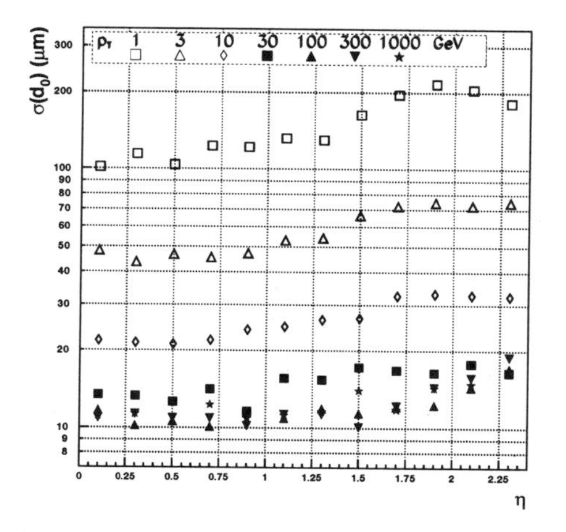

Figure 3. Impact parameter resolution versus η.

References

1. CMS Collaboration, Addendum to the CMS Tracker TDR, CERN/LHCC 2000-016.

THE AMS SILICON MICROSTRIPS TRACKER

DIVIC J. RAPIN
The AMS tracker collaboration
D.P.N.C., Université de Genève, 24 quai E.-Ansermet, CH-1211 Genève 4, Switzerland
E-mail: Divic.Rapin@physics.unige.ch

The Alpha Magnetic Spectrometer is a detector designed to search for anti-matter and dark matter by the analysis of cosmic rays in space. Its installation is scheduled on the International Space Station (ISS) for an operational period of 3 years. A precursor flight on a Space Shuttle occured in 1998 with a silicon surface of 2.1 m^2. The construction of the 7 m^2 second phase is starting. The AMS silicon tracker has a large dynamic range of up to 100 MIPS for dE/dx measurement. We discuss the design, construction and performance of the detector.

1 Introduction

Results from the STS91 precursor flight of AMS were presented at this conference [1] and in references [2,3,4].

2 The AMS-01 Tracker design

The design of the detector was guided by the very severe conditions of space such as mechanical stress at launch, large temperature range and relatively low electrical power available ($\sim$ 1 kW) for the whole experiment. Detailed descriptions are available in [5,6].

The tracker consists of 6 layers of double sided silicon microstrip detectors mounted on very light supporting disks made of carbon fiber and Al honeycomb (X_o = 0.65 % per layer). Four inner layers of 1 m diameter are in a magnetic field of 0.14 Tesla provided by a permanent magnet and two, with a diameter of 1.25 m, are placed outside of the field. The AMS-01 tracker was equipped with 38 % of the total number of sensors (2.1 m^2 and 58,368 channels) for the precursor flight.

The sensors ($40 \times 72 \times 0.3$ mm^3) have a strip pitch of 27.5 μm (110 for read-out) on the p-side (bending coordinate). The pitches of the n-side (non-bending) are 26 and 208 μm. Sensors are organized in 57 *ladders* (see fig 1) connecting up to 15 wafers to the front-end electronics (TFE hybrids) described in ref [7]. The routing of lines is made by kapton (upilex) microstrip cables directly glued on the silicon. The connections are made by microbonds. The mechanical rigidity is provided by a reinforcement made of Airex foam and carbon fiber. A thin EMI shielding is wrapped around the ladder.

The cooling of the front-end electronics is provided by Al bars filled with a highly thermo-conductive material (TPG).

3 Performance

Prototypes were tested with 50 GeV electrons at CERN [8] and with C ions at GSI [9,10] for dE/dx studies. We obtained position resolutions of $\sim$10 μm for p-side and $\sim$30 μm for n-side and a *signal/noise* $\simeq$ 8 for MIPs.

This performance was confirmed for the whole tracker during the STS91 shuttle flight of 1998 [5,6] and by subsequent tests with particle and ion beam at CERN and GSI. No failure occured during the 10 days of mission in space.

The momentum resolution for ^4He is shown in figure 2. Good agreement is seen between the expected resolution for flight data (histogram) and from ^4He beam of known momentum (points).

Figure 3 shows the quality of the dE/dx measurement for the identification of cosmics with Z>2.

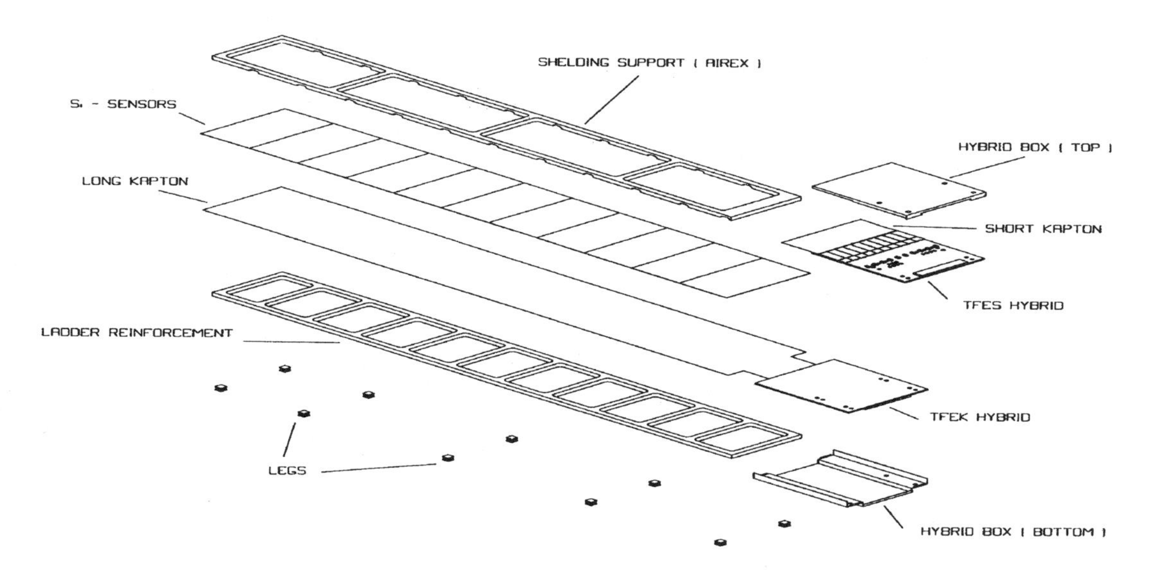

Figure 1. Exploded view of an AMS silicon ladder

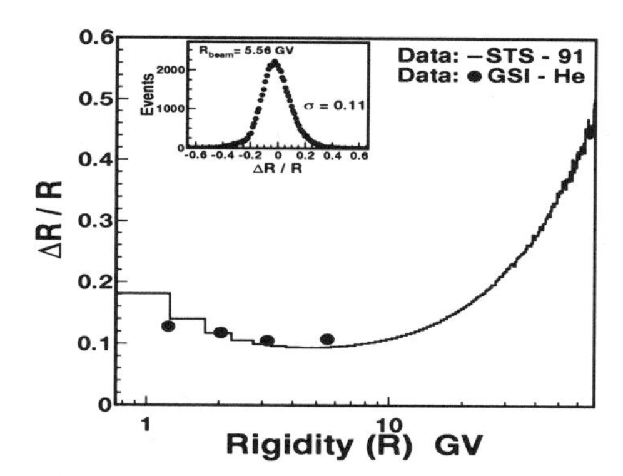

Figure 2. Rigidity resolution for ^{4}He

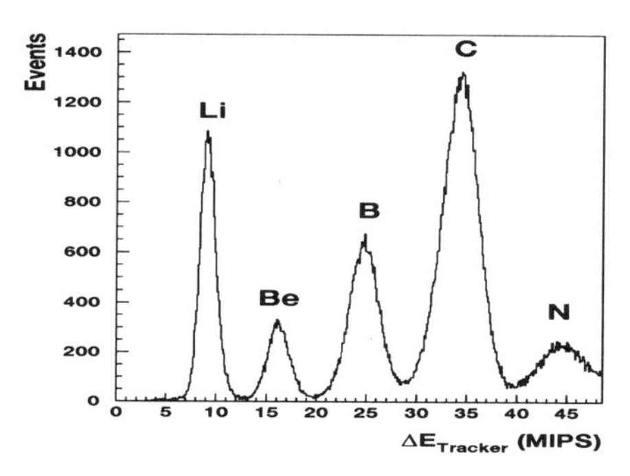

Figure 3. dE/dx spectrum from the AMS tracker

4 The tracker for AMS-02

AMS is being significantly upgraded (e.g. with a superconducting magnet) for its installation on ISS scheduled for fall 2003. The tracker is improved keeping the same concept, with a complete coverage reaching a total silicon surface of 7 m^2 (192 ladders) and organized in 8 layers for better redundancy. Electronics and cooling will be improved, read-out strips on n-side modified and the silicon wafers surface passivated.

References

1. V. Choutko in *Results from AMS*, Talk PA-09c-03 at this conference.
2. J. Alcaraz *et al.* PhL **B461**, 387 (1999).
3. J. Alcaraz *et al.* PhL **B472**, 21 (2000).
4. J. Alcaraz *et al.* PhL **B484**, 10 (2000).
5. J. Alcaraz *et al.* Nuovo Cim. **112A**, 1325 (1999).
6. W.J. Burger, NIM **A435**, 202 (1999).
7. G. Ambrosi, NIM **A435**, 215 (1999).
8. B. Alpat *et al.* NIM **A439**, 53 (1999).
9. B. Alpat *et al.* NIM **A446**, 522 (2000).
10. S.R. Hou *et al.* NIM **A435**, 169 (1999).

LORENTZ ANGLE MEASUREMENTS IN IRRADIATED SILICON DETECTORS BETWEEN 77 K AND 300 K

V. BARTSCH, W. DE BOER, J. BOL, A. DIERLAMM, E. GRIGORIEV, F. HAULER, S. HEISING, O. HERZ, L. JUNGERMANN, R. KERÄNEN, M. KOPPENHÖFER, F. RÖDERER

Institut für Experimentelle Kernphysik, Universität Karlsruhe, Germany

T. SCHNEIDER

Institut für Technische Physik, Forschungszentrum Karlsruhe, Germany

Future experiments are using silicon detectors in a high radiation environment and in high magnetic fields. The radiation tolerance of silicon improves by cooling it to temperatures below 180 K. However, at low temperatures the mobility increases, which leads to larger deflections of the charge carriers by the Lorentz force. We present measurements of the Lorentz angle between 77 K and 300 K before and after irradiation with a primary beam of 21 MeV protons to a flux of $10^{13}/cm^2$.

1 Experimental setup

The Lorentz angle Θ_L under which charge carriers are deflected in a magnetic field perpendicular to the electric field is defined by:

$$\tan(\Theta_L) = \frac{\Delta x}{d} = \mu_H B = r_H \mu B \qquad (1)$$

where the drift length corresponds to the detector thickness d and the shift of the center of charge is Δx. The Hall mobility is denoted by μ_H, the conduction mobility by μ. The Hall mobility differs from the conduction mobility by the Hall scattering factor r_H. This factor describes the influence of the magnetic field on the mean scattering time of carriers of different energy and velocity[1]. The Hall scattering factor has a value of 1.15 (0.7) for electrons (holes) at room temperature and decreases (increases) towards 1.0 with decreasing temperature.[2] Note that Θ_L represents an effective Lorentz angle, since the mobility is not constant in a depleted detector.

If the detector is irradiated with a red laser (660 nm, absorption length 3 μm), one type of carriers immediately recombines at the nearest electrode, whereas the other type drifts towards the opposite side. This allows to measure the Lorentz angle for electrons and holes separately by either injecting laser light on the n- or p-side. For our measurements the 10 T JUMBO magnet from the Forschungszentrum Karlsruhe was used. Details about the measurements and more extensive references can be found in Ref.[3].

2 Results

Instead of the Lorentz angle the shift in a 300μm thick detector is plotted for a 4 T magnetic field, which is the one of interest for future experiments. The dependence on temperature and bias voltage is shown in figures 1 and 2 together with the simulations from the Davinci software package by TMA[4].

For holes the temperature dependence is well described, but for electrons the Lorentz angle first falls below the simulation for decreasing temperature, as expected for a decreasing value of the Hall scattering factor r_H at lower temperatures[2]. However, below $T = 160$ K the Lorentz angle for electrons rapidlyincreases and is a factor two *above* the simulation at liquid nitrogen temperature. This is not understood, since it would imply a Hall scattering factor around two at low temperatures, where around one is expected[2].

The decrease of the Lorentz shift for the

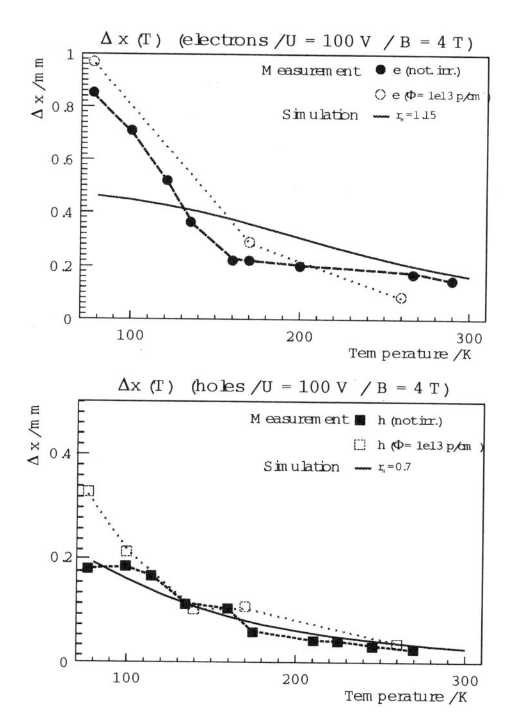

Figure 1. Lorentz shift for electrons (top) and holes (bottom) in a 4 T magnetic field for 300 μm detector as function of temperature. Data for an unirradiated detector and a detector irradiated with protons of 21 MeV to a fluence of $1.0 \cdot 10^{13}$ p/cm^2 are shown. The lines connecting the data points are drawn to guide the eye. For comparison, the temperature dependence from the Davinci simulation program with a constant Hall scattering factor of 1.15 (0.7) for electrons (holes) is shown by the solid lines.

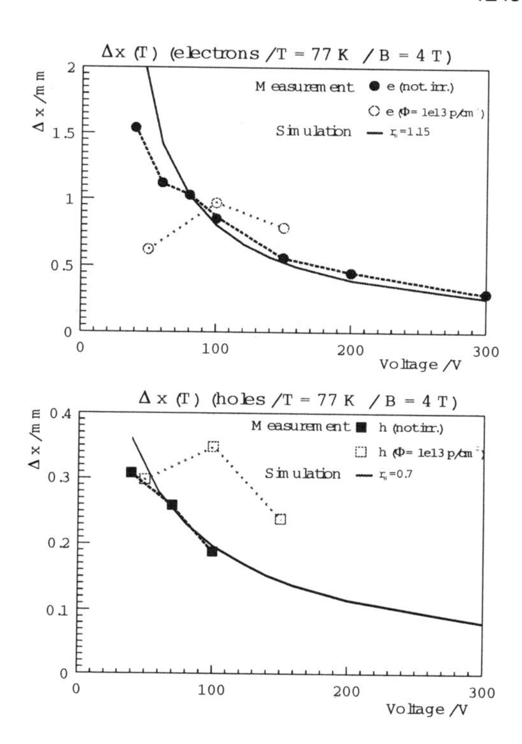

Figure 2. As Fig. 1, but now for the bias voltage dependence.

irradiated sample below 100 V in figure 2 is due to the reduced effective thickness of the partially depleted detector, since for the irradiated detector the depletion voltage increased from 40 to 100 V implying that the detector bulk underwent type inversion from n- to p-type.

3 Acknowledgements

This work was done within the framework of the RD39 Collaboration and supported by the BMBF under contract 7KA16P. We thank Dr. Iris Abt from the MPI, Munich, Germany for supplying us with double sided strip detectors from the HERA-B production by Sintef.

References

1. R.A. Smith, Semiconductors, Cambridge Univ. Press, 1968
2. Landolt-Börnstein, Numerical Data and Functional Relationships in Science and Technology, Group III, Band 17a, Springer Verlag, Berlin, 1982.
3. W. de Boer et al., http://xxx.lanl.gov/abs/physics/0007059
4. Technology Modelling Associates, Inc.: Davinci, Version 4.1, July 1998.

RADIATION HARD OPTICAL LINKS FOR THE ATLAS SCT AND PIXEL DETECTORS

J. D. DOWELL, D. G. CHARLTON, R. J. HOMER, P. JOVANOVIC, I. R. KENYON, G. MAHOUT, H. R. SHAYLOR, J. A. WILSON
School of Physics and Astronomy, University of Birmingham, Birmingham, B15 2TT, UK
E-mail: jdd@hep.ph.bham.ac.uk

I. M. GREGOR, R. WASTIE, A. R. WEIDBERG
Physics Department, Oxford University, Keble Road, Oxford, OX1 3RH, UK

S. GALAGEDERA, M. C. MORRISSEY, J. TROSKA, D. J. WHITE
CLRC Rutherford Appleton Laboratory, Chilton, Didcot, Oxon, OX11 OQX, UK

A. RUDGE
CERN, Geneva, Switzerland

M. L. CHU, S. C. LEE, P. K. TENG
Institute of Physics, Academia Sinica, Taipei, Taiwan 11529

A radiation hard optical readout system designed for the ATLAS Semiconductor Tracker (SCT) is described. Two independent versions of the front-end optical package housing two VCSEL emitters and an epitaxial Si PIN photodiode have been irradiated with neutron fluences over 10^{15} n cm^{-2}, the level encountered in the ATLAS pixel detector. Environmental tests have been performed down to -20 ^{o}C. Extensive radiation and lifetime tests have also been carried out on the opto-electronic components and the front-end VCSEL driver and timing/ control ASICs. Bit error rate, cross-talk and single event upset measurements using irradiated devices show that the system meets the performance specification.

1 Introduction

Optical links[1] will be used in the ATLAS Semiconductor Tracker (SCT) and pixel detector to transmit timing and control (TTC) information to the front-end electronics, and tracking data to the off-detector electronics some 90m away. The LHC beam crossing frequency is 40 MHz with approximately 20 collisions per crossing at the full luminosity of 10^{34} cm^{-2}s^{-1} producing a 1 MeV equivalent neutron fluence of 10^{15} n cm^{-2}(Si) and a total ionising dose of 500 kGy(Si) in the pixel detector over the nominal ten-year lifetime of the experiment. The corresponding figures for the SCT are 2×10^{14} n cm^{-2} and 100 kGy. For GaAs, the 1 MeV equivalent neutron fluences are six times higher. Radiation hard, low mass, non-magnetic packages have been developed to house the epitaxial Si PIN photodiode and two Vertical Cavity Surface Emitting Laser diodes (VCSELs) operating at 850 nm that are used to receive and transmit the data. The TTC link uses Bi-Phase Mark encoding to combine the 40 MHz clock with the control data on a single multimode fibre. SCT data links operate at 40 Mb/s and use NRZ coding of the binary readout data, with provision also to operate at 80 Mb/s for the pixel detector. Two data links are provided for each silicon strip detector module, one for each side. In the event of a link failure, the data can be rerouted through the other link. The control data may also be taken from a neighbouring module in the event of a TTC link failure. Several packaging schemes for the front end opto-electronic components have been studied and at present two alternative packages with similar characteristics have been successfully developed. Marconi (UK) use V-groove fibre mounting with a 45° mirror to reflect the light into or from the opto-electronic components which are precision mounted. The Taiwan group use instead 45° cleaving of the fibres which are glued into position after active alignment.

2 Radiation Hardness

Mitel VCSELs show good recovery after a short annealing period at 20 mA following irradiation with 2.9×10^{15} 1 Mev equivalent n cm^{-2}(GaAs), $\sim$ 1/2 pixel level. The total light output before irradiation is typically 1 mW at 10 mA. The main effect of the neutron irradiation is to shift the threshold current upwards by $\sim$2 mA. VCSELs from other manufacturers (Honeywell, Truelight) show similar behaviour. Enhanced ageing studies at 50^oC have given no failure, corresponding to a <0.2% failure rate (90% C.L.) over a 10 year running period in ATLAS. The epitaxial Si PIN photdiodes, made by Centronic, show a drop in responsivity of about 30% to 0.3 A/W after 10^{15} 1 MeV equivalent n cm^{-2}(Si). Ageing studies at 60^oC give similar results to the VCSELs. Pure silica core SIMM 50/60/125/250 fibre from Fujikura shows a total loss of <0.05 dB/m after 330 kGy ionising radiation.

The ASIC used to decode the TTC signals (DORIC) and the VCSEL driver chip use bipolar npn transistors in the AMS 0.8 micron BiCMOS technology. Samples have been exposed to 3×10^{14} n cm^{-2} and 500 kGy of gammas. All devices work correctly after irradiation. They are designed to be insensitive to changes in beta of the transistors, working with values as low as 10. Ageing studies at 100^oC show <0.3% failure (90% C.L.) over 10 years.

3 Environmental Tests

The optical packages will have to operate at temperatures as low as -15^oC. Irradiated VCSELs show an increase of 4 mA in laser threshold from room temperature to -18.3^o. Further work is needed to understand this. The irradiated PIN diode reponsivity falls at a rate of $0.25\%/^oC$ with decreasing temperature which is an expected trend.

4 Cross-talk and Single Event Upset

Bit error rate and cross talk measurements have been carried out using packaged devices after irradiation. The performance is well within specification (BER $< 10^{-9}$) for an optical power of 200 μW into the photodiode. However, single event upset tests using pion beams between 300 and 465 MeV/c at PSI show that, at a flux of 3×10^7 $cm^{-2}s^{-1}$ (pixel rate), up to 1 mW of optical power may be needed to meet the BER specification. While this is close to the limit, the lower fluxes expected in the SCT will allow more comfortable operation. The upsets are caused by particles interacting in the relatively large volume ($15\mu m\times550\mu m^2$) of the photodiode.

5 Conclusion

A successful optical readout scheme has been developed for the ATLAS SCT. The optical components and packaging are also expected to be sufficiently radiation hard for use in the pixel detector. Single event upsets due to particle interactions in the photodiode have been detected at rates which are acceptable for the SCT.

References

1. Further details may be found at http://atlas.web.cern.ch/Atlas/GROUPS/INNER_DETECTOR/sctnew/Electronics/links/

RESULTS OF A HIGH INTENSITY TEST BEAM WITH A LARGE NUMBER OF GEM+MSGC DETECTOR MODULES

DIRK MACKE

III. Physikalisches Institut B, RWTH Aachen
e-mail: macke@physik.rwth-aachen.de

ON BEHALF OF THE CMS MSGC FORWARD TRACKER COLLABORATION

A system of 18 detector modules, designed for the use in the forward-backward tracker of the CMS experiment, has been tested in a high intensity pion beam at the Paul-Scherrer-Institut (PSI, Villigen, CH). Each module was equipped with four Micro Strip Gas Counter (MSGC) substrates and a large area Gas Electron Multiplier (GEM) as second amplification stage.
The robustness of the detector modules when operated within an LHC-like environment was studied. Voltage settings yielding 98% detection efficiency for minimum ionising particles were set, and the number of strips lost over a period of two weeks under permanent irradiation with a particle flux of $\approx 4 \cdot 10^3 Hz/mm^2$ was recorded. The corresponding loss of strips was well below 10% of the total number of channels read out. A maximum signal-to-noise ratio of 100 has been achieved.

1 Introduction

The CMS MSGC forward tracker collaboration[1] chose Micro Strip Gas Counters (MSGC) with a Gas Electron Multiplier (GEM) as second amplification stage to fulfill the requirements of the original CMS design[3].
In the original design, the central tracker of the CMS detector was made of MSGCs for the outer regions. Four wedge-shaped MSGC substrates are mounted side by side to form a detector module. It has been shown that a constant detection efficiency over the whole detector surface can be achieved by this method[2]. Together, these modules form rings which are attached to 11 disks on each side of the CMS central tracker, forming one 'supermodule' on each side of the tracker[3].
The idea of GEM+MSGC detectors is to reduce the risk of losing strips due to discharges in the counting gas by sharing the total amplification between the MSGC and GEM part of the modules. A description of the working principle can be found elsewhere[1].
The so-called 'milestone two' (MF2) experiment was carried out to prove the robustness of the detectors in an LHC-like environment. 18 detector modules with 72 substrates, corresponding to 1% of the total forward tracker area, were put into the 350MeV/c pion beam at PSI. A total number of $\approx$ 17.000 channels was read out.

2 Results

To set realistic working conditions, the detectors were operated at settings corresponding to 98% detection efficiency for minimum ionising particles. This working point was reached for a signal-to-noise ratio (SNR) of 17 as can be seen in figure 1. A working point exceeding this by a factor of 2.2 was set (SNR=37) to take into account the different noise and ballistic deficit of the final electronics[3].

The accumulated loss of strips for all detector modules for a period of 376h at LHC-like conditions with a particle flux of $\approx 4 \cdot 10^3 Hz/mm^2$ is shown in figure 2. The dotted line corresponds to a strip loss of 10% which is the ceiling set by the track reconstruction efficiency requirements of the CMS detector. The strip loss stayed well below that limit.

Figure 3 shows the SNR vs. time for in-

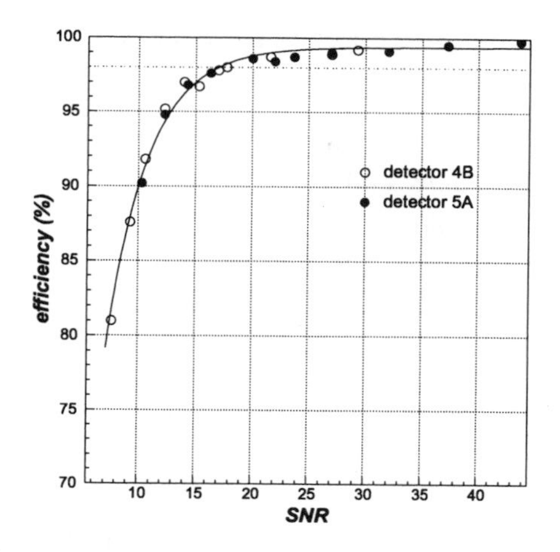

Figure 1. Efficiency vs. signal-to-noise ratio as measured in the PSI test beam.

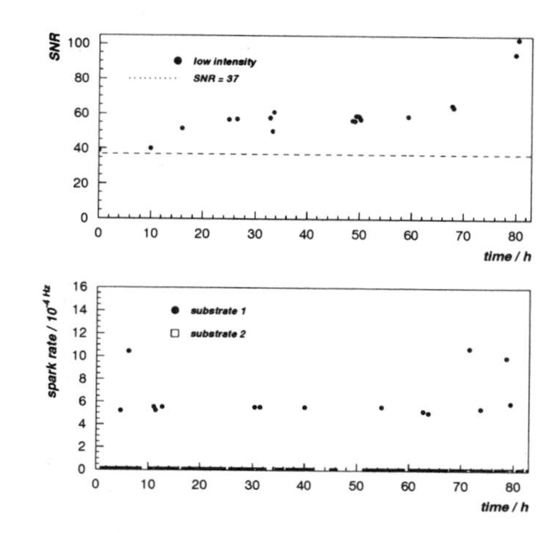

Figure 3. SNR vs. running time in the high intensity beam.

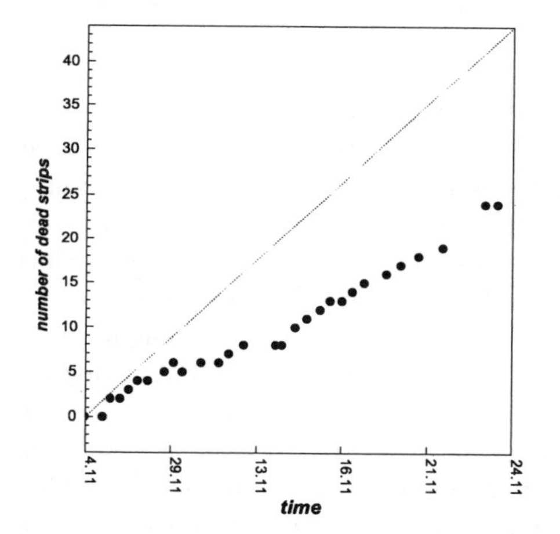

Figure 2. Lost strips during the MF2 test run.

creased detector amplifiaction. This study was carried out to prove that a comfortable safety margin can be achieved with GEM+MSGC detectors. Stable operation at a SNR ≈ 100, nearly three times higher than the required SNR=37, could be reached.

3 Conclusion

A large system of 18 detector modules with a total of 72 MSGC substrates and large area GEMs has been successfully operated for 376h in an LHC-like beam with particle fluxes of $\approx 4 \cdot 10^3 Hz/mm^2$. Stable operation with 98% particle detection efficiency has been shown. The required SNR could be exceeded by a factor of three to SNR ≈ 100. The total number of lost strips stayed well below the 10% strip loss allowed by the CMS detector's requirements on track reconstruction efficiency.

References

1. The CMS MSGC forward tracker collaboration, *Robustness test of a system of MSGC+GEM detectors at the cyclotron facility of the Paul Scherrer institute*, Preprint submitted to Elesevier Publishing (June 2000).
2. M. Ackermann et al., *Large scale test of wedge shaped Micro Strip Gas Counters*, NIM (436A) 313-325 (1999).
3. The CMS Collaboration, *Tracker Technical Design Report*, CERN/LHCC 98-6, CERN, Geneva (April 1998).
4. F. Sauli, *The Gas Electron Multiplier*, NIM (386A) 531ff (1997).

DIRC, THE PARTICLE IDENTIFICATION SYSTEM FOR BABAR

JOCHEN SCHWIENING
FOR THE BABAR-DIRC COLLABORATION
Stanford Linear Accelerator Center,
P.O. Box 20450, Stanford University, Stanford, CA 94309, USA
E-mail: jochen@slac.stanford.edu

The DIRC, a novel type of Cherenkov ring imaging device, is the primary hadronic particle identification system for the *BABAR* detector at the asymmetric B-factory, PEP-II at SLAC. *BABAR* began taking data with colliding beams mode in late spring 1999. This paper describes the performance of the DIRC during the first 16 months of operation.

The primary physics goal of the *BABAR* experiment[1] at the SLAC PEP-II asymmetric e^+e^- collider is to study CP violation in the B^0 meson system produced in $\Upsilon(4S)$ decays. At the $\Upsilon(4S)$, PEP-II collides 9 GeV electrons on 3.1 GeV positrons at $\beta\gamma(lab) = 0.56$. The study of CP-violation in hadronic final states of the B meson system requires the ability to tag the flavor of one of the B mesons via the cascade decay $b \to c \to s$, while fully reconstructing the final state of the other B. The momenta of the kaons used for flavor tagging extend up to about 2 GeV/c, with most below 1 GeV/c. On the other hand, pions from the rare two-body decays $B^0 \to \pi^-\pi^+(K^-\pi^+)$ must be well-separated from kaons, and have momenta between 1.5 and 4.5 GeV/c where high momentum tracks are srongly correlated with forward polar angles due to the c.m. system boost. Since the *BABAR* inner drift chamber tracker can provide π/K separation up to approximately 700 MeV/c, an additional dedicated particle identification system is required that must perform well over the range of 700 MeV/c to about 4 GeV/c.

The system being used in *BABAR* is a novel type of ring imaging Cherenkov detector, called the DIRC[2] (Detection of Internally Reflected Cherenkov light), which has been described in detail elsewhere[3]. Briefly, it uses 4.9 m long, rectangular bars made from synthetic fused silica as Cherenkov radiator and light guide. A charged particle with velocity v, traversing the fused silica bar with index of refraction n ($\sim$ 1.473), generates a cone of Cherenkov photons of half-angle θ_c with respect to the particle direction, where $\cos\theta_c = 1/\beta n$ ($\beta = v/c$, c = velocity of light). For particles with $\beta \approx 1$, some photons always lie within the total internal reflection limit, and are transported efficiently to either one or both ends of the bar, depending on the particle incident angle. Since the bar has a rectangular cross section and is made to optical precision, the magnitude of the Cherenkov angle is conserved during the reflection at the radiator bar surfaces. The photons are imaged via "pin-hole" focussing by expanding through a standoff region filled with 6000 litres of purified water onto an array of 10752 densely packed photomultiplier tubes placed at a distance of about 1.2 m from the bar end. Imaging in the *BABAR* DIRC occurs in three dimensions, by recording the location and the time at which a given PMT is hit. The expected single photon Cherenkov angle resolution is about 9 mrad, dominated by a geometric term that is due to the sizes of bars, PMTs and the expansion region, and a chromatic term from the photon production. The accuracy of the time measurement is limited by the intrinsic 1.5 ns transit time spread of the PMTs.

In the absence of correlated systematic errors, the resolution ($\sigma_{C,track}$) on

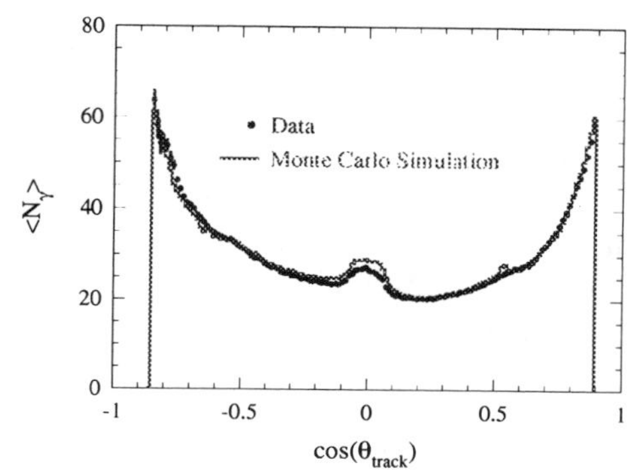

Figure 1. Number of detected photoelectrons *vs.* track dip angle θ_{track} for di-muon events.

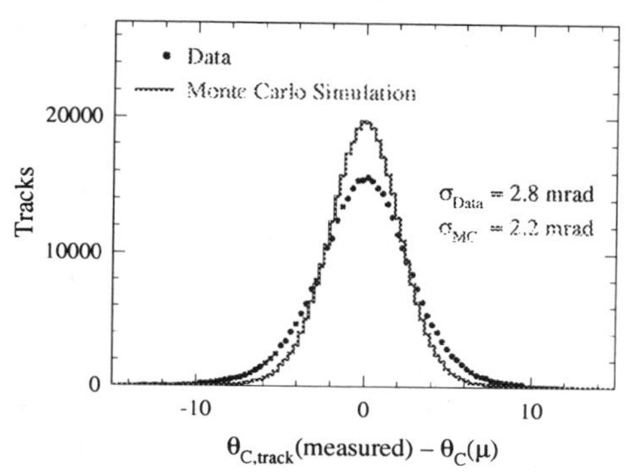

Figure 2. Resolution of the reconstructed Cherenkov polar angle per track for di-muon events.

the track Cherenkov angle should scale as $\sigma_{C,track} = \sigma_{C,\gamma}/\sqrt{N_{pe}}$, where $\sigma_{C,\gamma}$ is the single photon Cherenkov angle resolution, and N_{pe} is the number of photons detected. The average single photon resolution obtained for photoelectrons from di-muon events, $e^+e^- \to \mu^+\mu^-$, is 10.2 mrad, about 10% worse than the expected value. The time resolution obtained is 1.7 ns, close to the single-photon resolution of the PMTs.

The number of photoelectrons per track, shown in Fig. 1, varies from a minimum of about 20 for small dip angles at the center of the barrel to well over 50 at large dip angles. This is in good agreement with the value expected from the Monte Carlo simulation at all angles. This spectrum also demonstrates a very useful feature of the DIRC in the *BABAR* environment – the performance improves in the forward direction, as is needed to cope with the angle-momentum correlation of particles from the boost.

With the present alignment, the typical track Cherenkov angle resolution for di-muon events is shown in Fig. 2 to be 2.8 mrad. This is about 25% worse than the 2.2 mrad expected from simulation. From the measured single track resolution resolution *vs.* momentum in di-muon events and the difference between the expected Cherenkov angles of charged pions and kaons, the pion-kaon separation power of the DIRC can be inferred. The present separation between kaons and pions at 3 GeV/c is about 3.8 σ, approximately 10% worse than predicted by the Monte Carlo simulation, and is expected to improve with advances in tracking and detector alignment.

In summary, the DIRC was successfully commissioned, attained performance rather close to that expected from Monte Carlo, and has played a significant role in almost all *BABAR* analyses presented at this conference. The DIRC has been robust and stable and, almost 2 years after installation, about 99.7% of all PMTs and electronic channels are still operating with nominal performance.

Acknowledgments

Work supported by the Department of Energy under contracts DE-AC03-76SF00515 (SLAC), DE-AC03-76SF00098 (LBNL), DE-AM03-76SF0010 (UCSB), and DE-FG03-93ER40788 (CSU); the National Science Foundation grant PHY-95-11999 (Cincinnati).

References

1. The *BABAR* Collaboration, "'The *BABAR* Physics Book," SLAC-R-504 (1998).
2. B. N. Ratcliff, SLAC-PUB-6067 (1993); P. Coyle et al., *Nucl. Inst. Meth.* A **343** (1994) 292.
3. The *BABAR* Collaboration, "The *BABAR* Detector," to be published in *Nucl. Inst. Meth.* A.

MULTIANODE PHOTO MULTIPLIERS FOR RING IMAGING CHERENKOV DETECTORS

F. MUHEIM

The University of Edinburgh, Department of Physics and Astronomy
Mayfield Road, Edinburgh EH9 3JZ, Scotland/UK
E-mail: F.Muheim@ed.ac.uk

The 64-channel Multianode Photo Multiplier has been evaluated as a possible choice for the photo detectors of the LHCb Ring Imaging Cherenkov detector.

1 Introduction

The LHCb experiment will exploit the large rates of B hadrons that will be produced at the Large Hadron Collider and make precision measurements of CP violation. Excellent particle identification is needed for LHCb, e.g. three kaons in a large momentum range are produced by the decay $B_s^0 \to D_s^{\mp}K^{\pm}$, $D_s^+ \to \phi\pi^+$, $\phi \to K^+K^-$ which is sensitive to the CP violating phase γ. Charged particles will be identified by means of two Ring Imaging Cherenkov (RICH) detectors. The RICH photo detectors must be sensitive to single photons with a quantum efficiency $\int QEdE \sim \mathcal{O}(1eV)$ and provide spatial resolution with a granularity of about 2.5 x 2.5 mm^2 over a large area of $\sim$ 3m^2. The photo detectors must work in the magnetic fringe fields due to the LHCb dipole magnet and must cope with traversing charged particles.

2 Multianode Photo Multipliers

The multianode photo multiplier tube (MaPMT) consists of an array of 64 square anodes each with its own metal dynode chain incorporated into a single vacuum tube. The pixels have an area of 2.0×2.0 mm^2 and are separated by 0.3 mm gaps. The MaPMT, manufactured by Hamamatsu, has a 0.8 mm thick UV-glass window which transmits light down to a wavelength of 200 nm. The photons are converted in a Bialkali photo cathode with a quantum efficiency of maximum 22% at 380 nm. The mean gain of the 12-stage dynode chain is about 3×10^5 when operated at a voltage of 800 V.

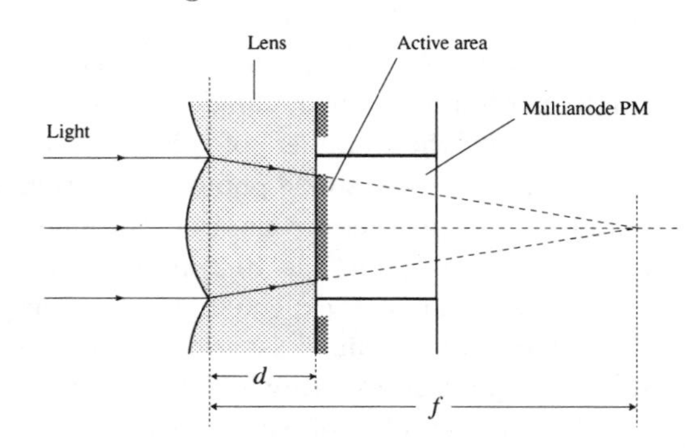

Figure 1. Schematic view of lens/MaPMT system. The focusing of normally incident light is illustrated.

The ratio of the sensitive photo cathode area to the total MaPMT area including the outer casing is only $\sim$ 48%. This geometrical coverage can be increased by placing a single lens with one refracting and one flat surface in front of each close-packed tube (Fig. 1). If the distance d of the refracting surface with radius-of-curvature R to the photo cathode is chosen to be equal to R the demagnification factor is $\approx 2/3$. Over the full aperture of the lens, light at normal incidence with respect to the photodetector plane is focused onto the photo cathode, thus restoring full geometrical acceptance.

3 R&D Results

The pulse height spectrum for the MaPMT is shown in Fig. 2, measured with a LED light source. The pedestal peak and the broad signal containing mostly one photo-electron are clearly visible. The signal to pedestal width ratio is 40:1.

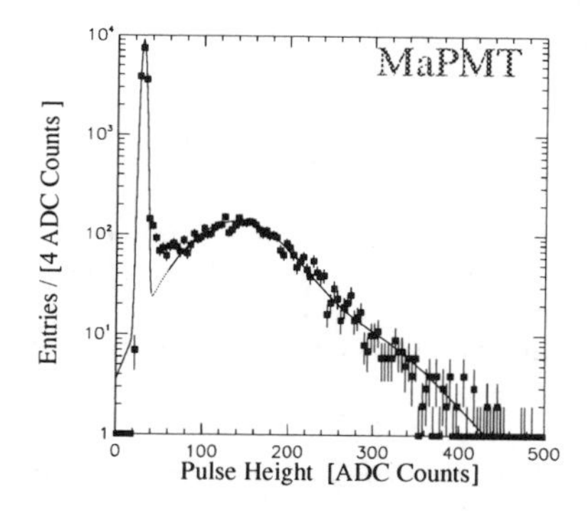

Figure 2. Pulse height spectrum of a pixel.

An array of 3x3 MaPMTs mounted onto the full-scale RICH 1 prototype[1] has been tested in a beam at the CERN SPS facility. The cathode voltage was set at −1000 V. Quartz lenses were mounted onto the front face of each MaPMT. The radiator was gaseous CF_4 at a pressure of 700 mbar. The data were recorded with a pipelined electronic read-out system based on the APVm chip[2] and running at LHC speed (40 MHz).

The data analysis included a common-mode baseline subtraction on a event-by-event basis. With the pipelined read-out electronics cross-talk was observed. This has been investigated using LED runs and several sources -all in the electronics - were identified. The cross-talk is removed by rejecting signals in a pixel if there is a larger signal in one of its cross-talk partner pixels. Genuine double hits are lost by this procedure and the photon yield is corrected for it.

The integrated signals of two runs of 6000 events each are shown in Fig. 3, one with and one without the lenses in front of the MaPMTs. The Cherenkov ring is clearly visible and the effect of the lenses is nicely demonstrated. The gain in in photo electrons by employing the lenses is 45%. The background is small. We measure 6.51 ± 0.34 photo electrons which is is in good agreement

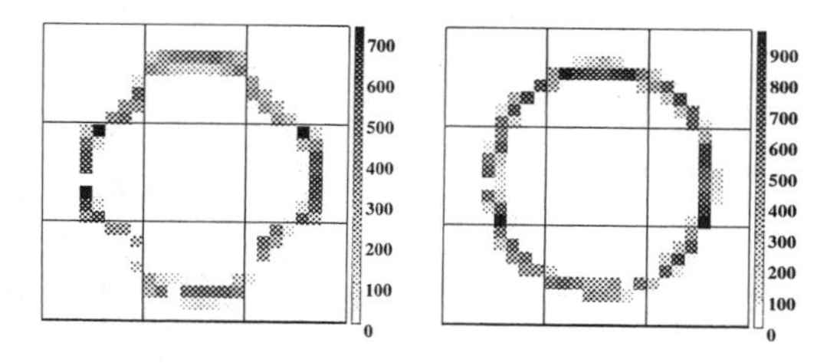

Figure 3. Cherenkov ring measured with the MaPMT array with (right plot) and without (left plot) quartz lenses mounted in front of the tubes.

with a full Monte Carlo simulation.

We have also exposed the MaPMT/lens to charged particles. The measured response is used to model the background in LHCb. The sensitivity of the MaPMT to magnetic fields has been studied. The MaPMT is affected by longitudinal magnetic fields but can be effectively shielded from the expected field strengths with a μ-metal structure.

4 Conclusions

We have successfully tested a 3x3 array of MaPMTs. Cherenkov light can be detected over the full area of closely packed tubes by means of quartz lenses focusing the light onto the sensitive area of the device. We have demonstrated that the MaPMT meets the performance requirements for charged particle identification in the LHCb experiment. The MaPMT has been chosen as the backup photo detector for LHCb.

Acknowledgments

I thank all my MaPMT collaborators for their excellent work presented here.

References

1. E. Albrecht *et al.*, Nucl. Instr. and Meth. **A 411** (1998) 249.
2. L.L. Jones *et al.*, "Electronics for LHC Experiments", Rome 1998, CERN/LHCC/98–36 (1998) 185.

THE CMS CRYSTAL CALORIMETER FOR THE LHC

F. ZACH
Institut de Physique Nucléaire de Lyon
43 Boulevard du 11 Novembre 1918
69622 Villeurbanne Cedex, France
E-mail: zach@in2p3.fr

The CMS crystal calorimeter, comprising about 80.000 scintillating lead tungstate crystals read out by avalanche photodiodes in the barrel and vacuum phototriodes in the endcap, is designed to give excellent energy resolution in the LHC environment. We are now entering in the construction phase. We present here a status report on the project, with recent results on tests beam, crystal production and photodetector choice.

1 Introduction

The Compact Muon Solenoid (CMS) experiment [1] will build a general-purpose detector designed to exploit the physics of proton-proton collisions at a centre-of-masse energy of 14 TeV over the full range of luminosities expected at LHC. The CMS detector is designed to measure the energy and momemtum of photons, electrons, muons and other charged particles with high precision.

1.1 Physics at CMS

CMS will be used for Higgs searches, on the context of both Standard model and non-minimal models. It will also be used for other new particle searches (for example supersymmetric particles), b physics and heavy ion physics.
The CMS electromagnetic calorimeter is well adapted for Higgs searches at low luminosity, in a mass region between 100 and 150 GeV/c^2. The particle is mainly produced via gluon-gluon fusion, and in this range of mass a clean signature of the Higgs will be its decay into two photons. The theoritical width of the Higgs is still relatively small ($< 20\ MeV$), so that measurement will be dominated by the experimental resolution. This need of excellent resolution has motivated the choice of an homogeneous rather than sampling calorimeter.

1.2 The choice of the electromagnetic calorimeter

The CMS ECAL comprises a barrel ($|\eta < 1.479$) and two endcaps. The active medium is made of lead tungstate ($PbWO_4$) scintillating crystals. The light produced by an incident particle is read by avalanche photodiodes (APD) in the barrel and vacuum phototriodes (VPT) in the endcaps. The photodetectors are followed by a preamplifier and a floating point sampling ADC. Then, the digitized samples are serialized and brought out of the detector via an optical link. All the electronic components are radiation hard.
The resolution of a calorimeter can be parametrized by the formula :

$$\frac{\sigma}{E} = \frac{a}{\sqrt{E}} \oplus b \oplus \frac{c}{E}$$

with the stochastic term a, the constant term b and the eletronic noise term c (2.7%,0.55%,200MeV for the barrel).
If these specifications are met, it will be possible to discover the Higgs in CMS during the low luminosity period in the 100-150 GeV region, using the 2-photon decay channel.

2 Status of ECAL

We have now started the production phase. The next items show the status of different ECAL components.

2.1 *The lead tungstate crystals*

Results from the batches of production crystals from Russia [2] (3.500 in April 2000) are very encouraging. The main crystals measurements, done at CERN by a machine named ACCOS, show that only few percent are rejected, due to bad light yield (< 8 pe/MeV at 8 X_0, see figure 1) or bad radiations hardness (Light Yield loss> 4 %).

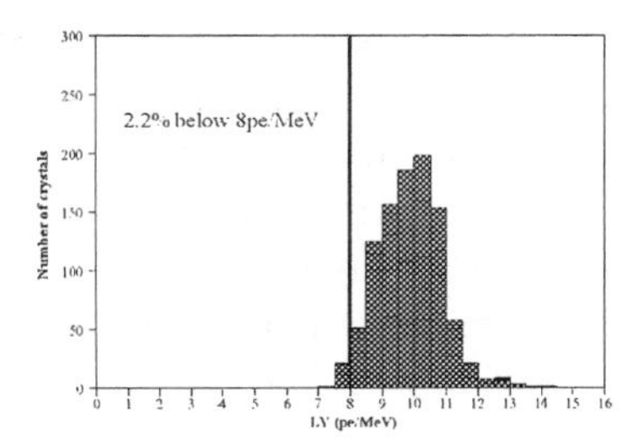

Figure 1. Distribution for 1000 crystals of light yield measurements done by ACCOS machine

2.2 *The photodetectors*

The photo-detectors have to operate in a 4T magnetic field. In the barrel, Avalanche Photodioes (APDs) delivered by Hamamastu are used. Two 25 mm^2 APDs are glued to the rear face on each crystal. The first production run (3.500 pieces) is arrived in June 2000. The main characteristics to control are gain versus bias, radiation hardness, breakdown bias and capacity.

2.3 *Readout of the photodetectors*

Scintillation light from the crystals is converted to a current by the photodetectors and shaped to a voltage pulse by the preamplifiers. The voltage pulse is digitized at 40 MHz by a floating point ADC. A 2^{17} dynamic range is achieved using a 12-bit ADC and four ranges. The digital values are transmitted to a counting room on high-speed optical links. The front-end electronics must all tolerate a harsh radiation environment - up to 100 kGy and 10^{14} n/cm^2 for ten years LHC running in the end cap. The complete front-end power consumption is a litlle over 1 W per channel. The design of the complete, radiation hard, electronics chain is now complete.

2.4 *ECAL Structure*

For the ECAL barrel, we have $360(\phi) \times 2 \times 85$ (θ) crystals. The smallest ensemble is a submodule of $2(\phi) \times 5$ (θ) crystals put in a reflective alveola. With 40 or 50 submodules, we obtain a module ; with 4 modules of 400 crystals and 1 module of 500 crystals, a supermodule. With 36 supermodules, we have the ECAL barrel.
A double cooling system enables the evacuation of the heat produced by the electronics. The temperature sensitivity of crystals and APDs requires a temperature stability of 0.1 ^oC at 18 ^oC.
The endcaps structure is more simple : 2 dees of 7810 crystals grouped into supercrystals of 5×5 crystals.

3 Beam Tests and Calibration

3.1 *Beam tests in 1999 and 2000*

In 1999, a prototype of 30 channels was build with pre-production crystals and APDs, dedicated to noise studies and energy scan. The noise was found of 36 Mev or 10.530 e^- by channel. The light yield measurement done in beam or in ACCOS are in good agreement. We obtain a mean stochastic term of 2.74 %, a constant term of 0.41 % and an electronic noise term of 142 MeV, for a sum of 9 channels, centering of deposited energy maximum. The figure 2 shows reconstructed energy distribution with 280 GeV electrons beam.
In May-June 2000, a similar prototype was done, but dedicated to monitoring and calibration studies. The figure 3 shows the laser and beam respons of a crystal during irradiations cycles. For all crystals, we obtain a very good correlation between laser and beam, with a laser stability better than 0.4 %. In the same way, in July 2000, a same pro-

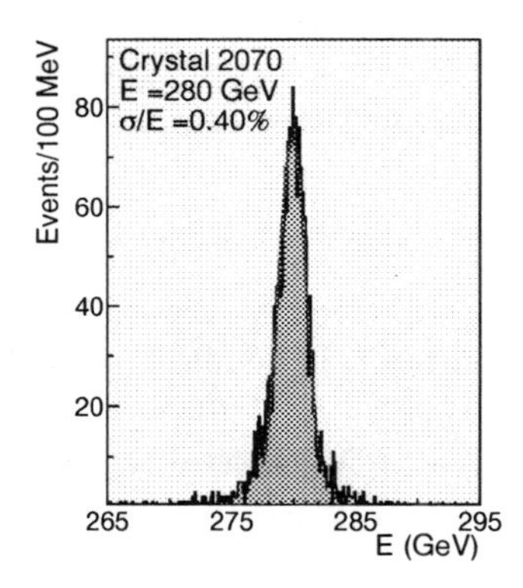

Figure 2. Reconstructed energy with 280 GeV electrons incident on barrel prototype matrix

totype with Chinese crystals will be tested, and in August 2000, a prototype with final electronics readout.

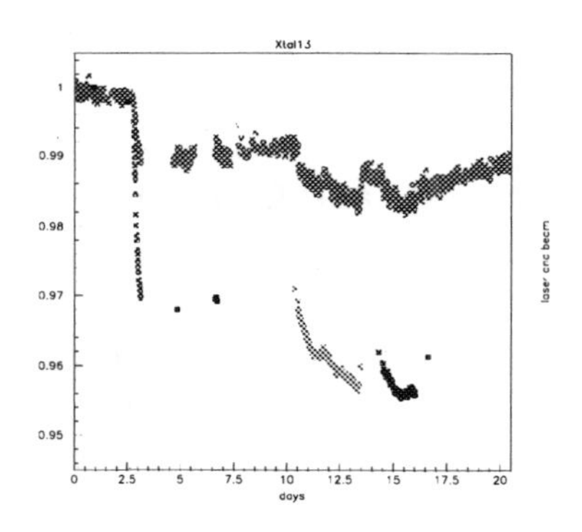

Figure 3. Laser and Beam Respons of a crystal during irradiations cycles

3.2 Monitoring and Calibration

To have the required resolution on the measured energy, it is very important to perform a precise calibration of the calorimeter. Before its installation on the LHC site, each module will be placed in electrons test beam for a precalibration of each crystal at 2 energy points.

During the run of the CMS experiment, a light monitoring system will track the behaviour of each channel. It is well establieshed that radiation only affects light transmission in the crystal, and not the scintillating process itself, so there is a relationship between the response to incident particles and the response to an injected light on the crystal front face.

A laser system will send two wavelengths (red and green) to the front face of each crystal and to reference PIN photodiodes. This sytem will provide a way to monitor with precision the calibration constants.

But the physics processes ($W \rightarrow e\nu$ and $Z \rightarrow e^- e^+$) will be used to calibrate the calorimeter during the running periods.

Combining these methods, it will take approximately 35 days at low luminosity to have a calibration at the level of 0.3 %.

4 Conclusion

Five years of intensive R&D on the CMS electromagnetic lead tungstate crystal calorimeter have resulted in a design that can meet the challenging LHC requirements. Mass productions of crystals, APDs and soon readout electronics is underway. This signals the start of the construction phase. In 2001, will be tested a Module of 400 crystals, with final mechanic, electronic and be constructed the first Super-Module of 1600 crystals.

Acknowledgments

I would like to acknowledge all CMS ECAL collaborators.

References

1. The CMS Electromagnetic Calorimeter project , Technical Design Report, CMS Collaboration *CERN/LHCC 97,33, CMS TDR 4*, 15 december 1997.
2. Status of barrel Crystal Preproduction in Bogorodisk,*private communication*, E. Auffray, June 2000.

PRELIMINARY RESULTS ON THE ATLAS LIQUID ARGON EM CALORIMETER

R. ZITOUN (FOR THE ATLAS LIQUID ARGON GROUP)
LAPP, CNRS-IN2P3-Université de Savoie, France
E-mail: zitoun@lapp.in2p3.fr

The ATLAS detector will operate at LHC in 2005. The construction of its electromagnetic calorimeter has started and modules 0 of the barrel and endcap parts have been tested at CERN in an electron beam in 1999 and 2000. Preliminary results of these tests and plans for the production are presented.

1 Introduction

The LHC collider will mainly be devoted to the search and the study of new particles: Higgs boson, supersymmetric particles, heavier bosons, etc. Such a physics program can only be done with detectors which are able to reconstruct an identify all types of particles (b quarks, e, μ and τ leptons, photons and missing transverse energy) in a very hostile environment: high luminosity, high collision rate, high radiation level background.

The liquid argon calorimeter of the ATLAS collaboration[1] has electrodes and absorbers with an accordion shape and will allow the study of the Higgs boson in the decay channels[2] H$\rightarrow \gamma\gamma$ and/or H $\rightarrow$ eeee.

2 Geometry and data readout

The calorimeter (Fig. 1) is made of 2 halves containing each a barrel and an endcap part:

i) The barrel covers the pseudo rapidity region below $\eta = 1.4$ ($|z| < 320$ cm) and extends between radii 140 cm and 200 cm. It is split in φ into 16 modules weighing about 3.4 tons each. In order to allow for e/γ/π^0/ jet separation, it is made of 3 compartments in depth (6+18+2 X_0 at $\eta = 0$). The first one is highly segmented in the η direction ($\Delta\eta \simeq 0.003$) while the second has $\Delta\eta\Delta\varphi = 0.025 \times 0.025$ towers. The segmentation is easily achieved by drawing strips and bands on the copper electrodes.

ii) an endcap is made of 2 coaxial wheels

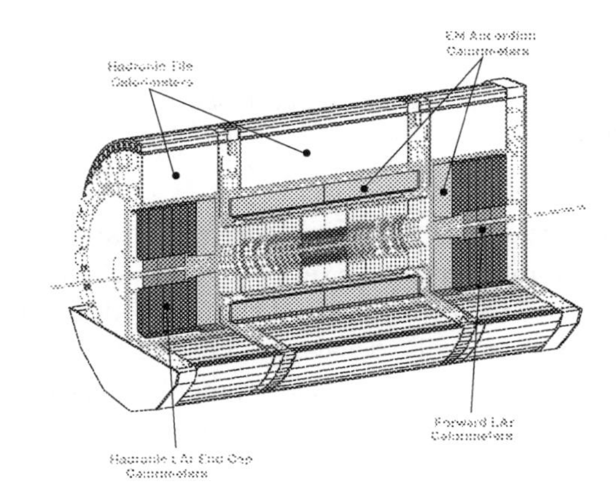

Figure 1. Schematic view of the ATLAS calorimeter system

covering the regions $1.4 < |\eta| < 2.5$ and $2.5 < |\eta| < 3.2$. The inner and outer radii are ~30 cm ~200 cm. The centre of the wheels lies at a distance of ~ 390 cm from ATLAS centre. Each wheel is splitted into 8 modules.

The calorimeter is supplemented by a presampler detector to achieve the goal energy resolution in the $|\eta| < 1.8$ region.

The primary signal is a triangular current of length ~ 450 ns (Fig. **??**). It is sent out of the cryostat containing the calorimeter to front end boards (FEB) located in the crack between the barrel and endcap parts. There, it is preamplified, shaped with CR-RC^2 filter giving a peaking time at about 50 ns and sampled at the LHC frequency (40 MHz). Samples are stored into an analog memory for at most ~ 1 μs. After level 1 trigger, 5 samples around the highest one are digitized on the

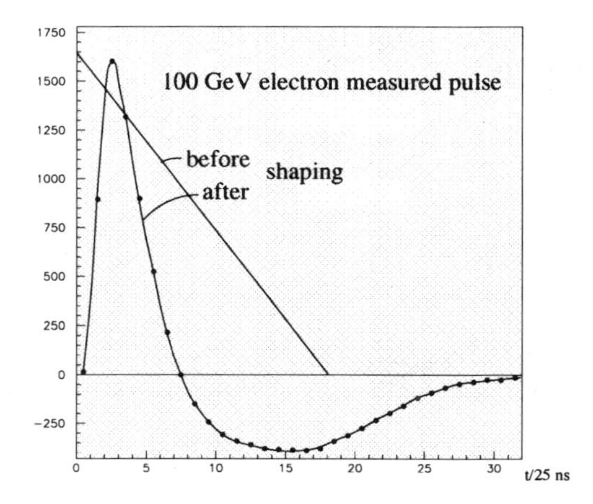

Figure 2. Signal shape before (triangle) and after shaping.

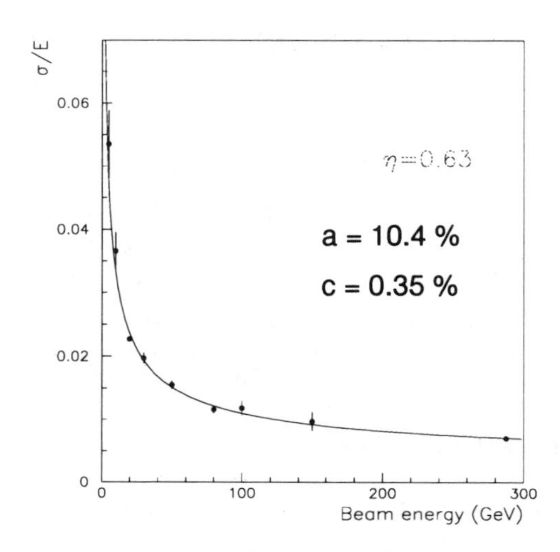

Figure 3. Energy resolution as a function of energy at $\eta = 0.63$.

FEB and sent to readout drivers located in the counting room. Every electronic channel is calibrated by injecting at the electrode level an exponentially decaying signal of varying amplitude simulating the triangular physics signal. All this system has been tested in the 1999-2000 test period.

3 Test beam results

The energy deposited by an electron is reconstructed from a cluster of size $\Delta\eta \times \Delta\varphi = 0.075 \times 0.075$ containing typically 30 cells. The noise level measured in such clusters is about 280 MeV, a value reduced by a factor 1.7 whith optimal filtering. Various corrections are applied to take into account the finite size of the cluster in η and in φ, the φ modulation due to the accordion shape, the energy lost in the dead material (cryostat, dead space) in front of the calorimeter. The resolution σ obtained for both barrel and endcap follows the law (Fig. 3)

$$\frac{\sigma}{E} = \frac{a}{\sqrt{E}} \oplus c$$

where a and the constant term c are about their expected values, 10% and 0.4% respectively. The electronic noise and the beam energy spread have been subtracted from the resolution. Position resolution at high energy ($\geq$ 100 GeV) has also been measured to be $\sigma_\eta = 570\mu$m a value which, in real ATLAS conditions is extrapolated to 400 μm.

4 Conclusion

The barrel and endcap modules 0 have been intensively tested in electron beams at CERN in 1999 and 2000. The performances agree with expectation. There are ongoing studies to assess calorimeter uniformity.

The lead absorbers and electrodes are under production; the module assembly has started. A stringent qualification series of tests is applied before and during stacking. HV tests are performed during stacking and, after cabling, there is an overall test of every module in liquid argon. The construction time is about 3 month per module so that completion is foreseen in 3 years from now.

References

1. The ATLAS Technical Proposal, CERN/LHCC/94-43 (1994).
2. The ATLAS Detector and Physics Performance, Technical Design Report, CERN/LHCC/99-14 and 99-15 (1999).

TEST BEAM RESULTS OF THE QUARTZ FIBRE CALORIMETER FOR THE H1 LUMINOSITY

B. ANDRIEU, V. BOUDRY, S. FERRON, F. MOREAU, A. SPECKA
LPNHE, Ecole Polytechnique, IN2P3-CNRS, F-91128 Palaiseau, FRANCE

E. BARRELET
LPNHE, Universités Paris VI et VII, F-75252 Paris, FRANCE

P. SMIRNOV, Y. SOLOVIEV
Lebedev Physical Institute, RU-117924 Moscow, RUSSIA

I. HERYNEK, J. HLADKY
Institute of Physics, Czech Academy of Sciences, CZ-18040 Praha, CZECH REPUBLIC

F. BONNIN, F. ZOMER
Laboratoire de l'Accélérateur Linéaire, Université de Paris Sud, F-91405 Orsay, FRANCE

To measure luminosity after HERA upgrade in 2000/2001, the H1 collaboration has chosen to build a tungsten/quartz-fibre Čerenkov calorimeter. The detector, completed in 1999, was tested at CERN-SPS shortly afterwards. The design of the detector is explained, and results of tests are shown.

1 A new luminometer for H1

In the H1 experiment at the HERA collider, luminosity is measured by detecting bremsstrahlung photons emitted by electrons. The upgrade of HERA will increase luminosity by a factor 5, and provide longitudinally polarized beams. A new photon detector has been built which will function in presence of GigaRad doses of synchrotron radiation. In addition, the detector will provide a bunch-by-bunch luminosity measurement with a precision of 1 % per minute to fully exploit the physics potential of polarization. The detector characteristics and performances in test-beam are briefly presented here. More details will be found in a paper[1] in preparation.

2 Description of the detector

Quartz-fibre Čerenkov calorimetry[2] was adopted to fulfill the above mentioned requirements. The Čerenkov threshold (0.7 MeV) suppresses the low-energy part of synchrotron radiation. Fibres with quartz core and polymer cladding were preferred because of their lower cost, higher flexibility, and sufficient resistance to radiation. To optimize the fiducial volume, tungsten was chosen as absorber for its low Molière radius.

The main parameters and a photograph of the detector are shown in figure 1 . Fibres are placed between tungsten plates in layers alternatively along $(x+z)$ and $(y+z)$ (z being the direction of the beam) and grouped in 12 bundles in each direction. This complicated geometry has two advantages:
- a 45° angle of all fibers w.r.t. beam (close to Čerenkov angle) to maximize light collection ; the detector is then also blind to background from the proton side (halo muons) ;
- the possibility of a precise measurement of the photon position in the transverse plane, to achieve required precision on luminosity.

number of layers:	69
number of fibres/layer:	224
fibre core diameter:	600 μm
fibre numerical aperture:	0.37
beam-fibre angle:	45°
beam incident angle:	54.7°
number of tungsten plates:	70
tungsten thickness:	0.7 mm
W/fibre volume ratio:	1.68
active volume:	$12 \times 12 \times 19$ cm^3
number of channels:	12+12
granularity:	10 mm
radiation length:	7.8 mm
sampling frequency:	0.36
total depth:	25 X_0
Molière radius:	17.2 mm

Figure 1. Main parameters and photograph of the quartz-fibre/tungsten calorimeter.

At the end of each bundle, a light mixer ensures spatial homogeneity of light collected in a photomultiplier (PMT). Finally LEDs can inject light in fibres or in PMTs, for optical monitoring of light yield and gains.

3 CERN tests results

The calorimeter was tested at CERN-SPS in 1999. The response to electrons from 8 to 100 GeV is linear and uniform within $\pm 1\%$. The energy resolution is 19 $\%/\sqrt{E/\mathrm{GeV}}$, with a constant term compatible with 0.

Using LED calibration events, the number of photoelectrons per deposited GeV is estimated to be 130$\pm$5. Subtracting quadratically the equivalent photostatistics term (8.8 $\%/\sqrt{E/\mathrm{GeV}}$) from the measured resolution, a sampling term of 16 $\%/\sqrt{E/\mathrm{GeV}}$ is deduced, in good agreement with expectations. This ensures that even a factor 2 loss of light yield will only degrade resolution from 19 to 20 $\%/\sqrt{E/\mathrm{GeV}}$.

The combination of tungsten absorber and of increased sensitivity to the shower core due to fibre orientation leads to very narrow showers. This translates into a position resolution in the transverse plane of 5 mm$/\sqrt{E/\mathrm{GeV}}$.

4 Conclusion

A tungsten/quartz-fibre Čerenkov calorimeter was built for H1 luminosity measurement. Its original design combines orientation of fibres at 45° w.r.t.beam and strip geometry in two orthogonal directions in the transverse plane. The results obtained at CERN-SPS in 1999 with electrons from 8 to 100 GeV are conform to specifications : linearity and spatial uniformity within 1 %, resolution of 19 $\%/\sqrt{E/\mathrm{GeV}}$ with a photostatistics term of 8.8 $\%/\sqrt{E/\mathrm{GeV}}$, position resolution of 5 mm$/\sqrt{E/\mathrm{GeV}}$.

Acknowledgements

I would like to thank all the people who contributed to the success of this project, in particular during the assembly phase.

References

1. B. Andrieu *et al.*, "Performances of the W/quartz-fibre calorimeter for the upgrade H1 luminosity detector", to appear in *Nucl. Instrum. Methods* A.
2. P. Gorodetzky *et al.*, *Nucl. Instrum. Methods* A361 (1995) 161-179.

RESISTIVE PLATE CHAMBERS IN HIGH ENERGY EXPERIMENTS

SERGIO P. RATTI*
Dip. di Fisica Nucleare and Sezione I.N.F.N. I27100 Pavia (Italy)
E-mail: ratti@pv.infn.it

The Resistive Plate Chamber (hereafter referred to as RPC) has been first developed by R. Santonico and his group[1] since the late seventies. They have been employed in several experiments at nuclear reactors[2], fixed target accelerators[3,4], colliders[5,6] as well as in cosmic rays[7,8].

In order to properly perform their functions, RPC's have to undergo a series of restricting working conditions, mostly connected to the severe radiaton backround environment in which they are designed to work.

The RPC is a very successful wireless detector which is able to ensure both a high efficiency and a fast response. The absence of wire technologies and the cheap material employed makes the cost production very low so that it can be used to build very large area detectors. A large RPC can have a sensitive area as large as $2m \times 2.5m$. In addition it can be operated in a low gas gain regime, providing a high rate capability. Finally it can provide time of flight measurements as well as position measurements.

1 RPC'c for LHC: performances and requirements

The RPC's play a significant role in the muon triggers of experiments designed at the future high energy colliders, particularly the CERN Large Hadron Collider (hereafter referred to as LHC)[9,10,11]. At this collider, the beam provide an average of 20 interaction per bunch crossing at a beam frequency of 40 MHz.

The environment will be heavily contaminated by γ ray and neutron backgrounds whose rate could vary -in different regions of the detectors- from ≈ 1 to $\approx 10^3$ Hz/cm^2. The overall doses during a 10 year exposure may vary in the range 1 Gy (barrel region) and 100 Gy (endcup region).

To meet this requirements the detector must provide a fast trigger with an overall time tolerance range $\sigma_{tot} \leq 3ns$; it has to have a low sensitivity to both γ rays and neutrons. Finally the radiation damage on *all* detector's components should be adequately affordable.

2 Rate Capability and Timing Properties

A prototipe of a double gap ($2.0m \times 2.5m$) RPC -with single gap readout- to be inserted into the CMS muon trigger system has been exposed to a $20Ci$ ^{137}Cs source at the CERN γ irradiation facility (GIF).

A $200GeV/c$ muon beam could be used to trigger the RPC in different γ background conditions. Different absorbers could be inserted between the source and the RPC inducing into the the RPC readout different overall hit counting rates with beam off. We selected two absorbers: one corresponding to a hit rate of 180 Hz/cm^2, the other to 550 Hz/cm^2 which is estimated to be safely larger than the one to the expected due to the background at the worst RPC position in the barrel region of CMS.

In fig.s 1 the results shown are twofold: fig. 1a shows the efficiency curves (ϵ vs. HV) in two different background conditions, compared to the no background (no osurce) situa-

*Co-authors are: S.Altieri, G.Belli, G.Bruno, R.Guida, M.Merlo, S.P.Ratti, C.Riccardi, P.Torre, P.Vitulo **(INFN and Univ. of Pavia)**, M.Abbrescia, A.Colaleo, G.Iaselli, F.Loddo, M. Maggi, B.Marangelli, S.Natali, S.Nuzzo, G.Pugliese, A.Ranieri, F.Romano **(INFN, Politechnique and Univ. of Bari)**

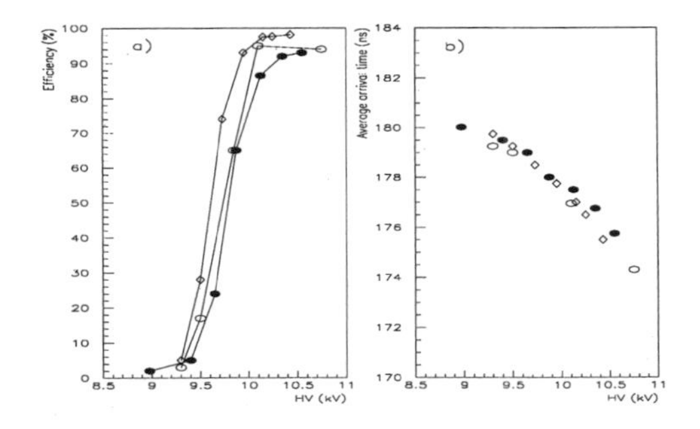

Figure 1. a- Efficiency vs. high voltage of a double gap RPC ($2.0m \times 2.5m$) in different background conditions; b- average arrival time; c- time resolution. ◇: no source; ○: background hit rate 180 Hz/cm^2; •: background hit rate 550 Hz/cm^2.

tion, while fig.s 1b,c show the time properties of the detector. From fig. 1b one sees that the average arrival time undergoes a time walk $\Delta\tau < 6ns$ over a high voltage range of $\approx 1.7KV$ from 9.0 KV to 10.7 KV. Fig. 1c shows on the other hand that the worse intrinsic time resolution σ_τ, over the same high voltage range in the same experimental conditions is $\sigma_\tau < 1.4ns$.

3 RPC Sensitivity to γ Rays and Neutrons

The RPC detector has to have low sensitivity to both γ rays and neutrons. A Monte Carlo algorithm has been exployted to simulate two different situations, i.e: exposure of a fully assembled RPC -including holding frames, electrodes, finishing foam- to:

1- isotropic flux of γ's or, separately, neutrons, evenly distributed on the chamber surface;

2- a parallel beam perpendicularly impinging the whole chamber surface.

The package MCNP-4b developed by the Los Alamos Nat'l Laboratory[12] was used for the γ's simulation, while GEANT was used for the neutron's simulation (MICAP for $E_n < 20MeV$ and FLUKA for higher energies were used as interface for cross sections).

The major assumption is that each produced charged particle reaching the RPC gas gap produces a signal into the readout output; if during any primary particle transport more than one charged particle reaches the gas gap, only the first is assumed to produce a signal.

The double gap sensitivity to γ rays is well below 8×10^{-2} for $E_\gamma < 100MeV$. The sensitivity is defined as: $s = N_I/N_o$, where N_I is the number of charged particles reaching any of the two gas gaps and N_o is the number of original primary particles impinging the chamber. If we use the simulated γ sensitivity and the expected γ spectrum in the region of the first RPC layer in the central barrel of the CMS setup, we get an expected hit rate in the RB1 region[9] of $\approx 5\ Hz/cm^2$.

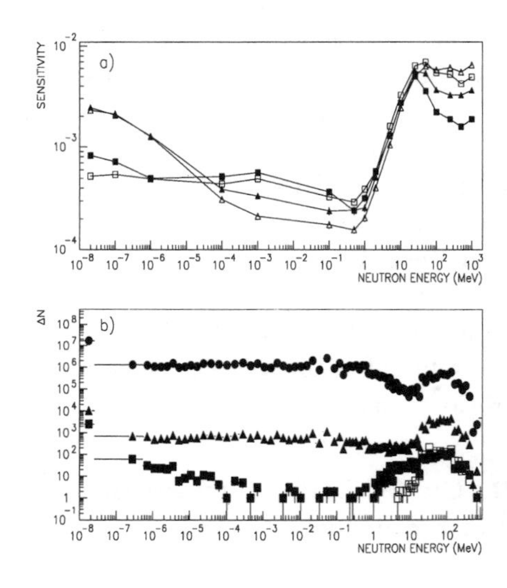

Figure 2. a- RPC neutron sensitivity vs. neutron energy. Full symbols for parallel beam; open symbols for isotropic beam. Squares: RPC without PVC foam; triangles: RPC with PVC foam. b- Charged particle spectra produced by primary neutrons into an RPC gas gap in the CMS barrel region, compared to the simulated neutron spectrum (full circles) in the RB1 region. Full squares: e^+/e^-; open squares: α particles; full triangles: protons.

The same definition of s holds for neutrons. The simulation program for a neutron source is more complex as it implies the use of neutron interaction cross sections for several nuclei and several energies.

The results obtained in this case are shown in fig.s 2. Fig. 2a shows the RPC sensitivity obtained in the conditions detailed in points 1- and 2- above, with or without PVC finishing foam. The sensitivity for $E_n < 10^{-5} MeV$ is mostly due to γ's coming from (γ, n) capture reactions whose cross section increases with decreasing energy ($\sigma \sim 1/\sqrt{E_n}$): thus the presence of the foam enhances s. The values, however remains below 2×10^{-3} for $E_n < 1\ MeV$. At higher energies the sensitivity reaches a maximum around $E_n \sim 50\ MeV$ around $s \sim 7 - 8 \times 10^{-3}$. At $\geq 15\ MeV$ incoming netrons the produced protons have a residual range of abuot 2 mm in bakelite so that thy can easily reach the gas gap.

Similar to the γ case, if we use the simulated neutron sensitivity and the expected neutron spectrum of fig. 2b we get an expected hit rate in the RB1 region[9] of ≈ 0.6 Hz/cm^2.

4 Front End Chip Radiation Damage

Intense radioactive environments may damage not only the material used to build the detector (namely, bakelite) but, most important, the components of the Front End (F.E.) electronic readout chps and boards which are located very close to the detector itself. The bakelite can easily stand high doses without changing its resistivity characteristics[13]. From this point of view a must is to keep the operating temperature under control. The damage to electronic components can be either severely disruptive (dislocations or threshold total dose effects) or reversible as tipically may be a *single event upset* produced by spurious transient hits. This phenomenon has been studied by exposing 4 F.E. final production 8 channel chips to the intense fluence of the 250 kW Triga Mark II reactor of the University of Pavia and by measuring reactor on/off hit rates. The measured neutron flux was about $6 \times 10^5\ n/cm^2 s$ and the neutro energy in the range $0.4\ eV - 10\ MeV$. The average chip rate remains stable for fluences up to 10 LHC year equivalent in the range $1.9 - 2.5\ 10^{-3}$ $Hz/channel$.

5 Mass Production Quality Control

Finally we tested the electrical resistivity of the first 10% mass production of bakelite-electrodes. The allowed values may range from $1to610^{10}\ \Omega cm$. Of the total 400 slabs produced about 77% have ρ in the range $1to4\ 10^{10}\ \Omega cm$ and about 8% are to be discarded. We will however use slabs of higher resistivity in the construction of the double gap RPC placed in regions of lower background level.

References

1. R.Santonico, R.Cardarelli, N.I.M. **A187**, 337 (1981) ;
2. E.Calligarich et al., N.I.M. in Phys. Res. **A337**, 350 (1994);
3. G.Cataldi et al., N.I.M. **A344**, 350 (1994);
4. P. Sheldon et al. Sci. Acta **11**, 437 (1996);
5. Babar Coll. TDR, (march 1995);
6. R.deAsmundis et al., Sci. Acta **11**, 139 (1996);
7. C.Bacci et al., N.I.M. **A443**, 342 (2000);
8. P.O.Mazur, Sci. Acta **11**, 331 (1996); ARGO Coll., Nucl.Phys.Proc.Suppl. **78**, 38 (1999);
9. CMS: Muon Project TDR, CERN/LHCC/ 97-32 (1997);
10. ATLAS: Muon Spectrometer TDR CERN/ LHCC/97-22 (1997)
11. LHCb: Technical Proposal CERN/LHCC/ 98-04 (1998);
12. MICAP Los Alamos report;
13. M.Abbrescia et al., RPC Barrel E.D.R., CMS doc. 1999/43 (1999).

THE CMS MUON TRIGGER

N. NEUMEISTER
CERN, CH-1211 Geneva 23, Switzerland
E-mail: Norbert.Neumeister@cern.ch

The LHC at CERN will provide proton-proton collisions at a centre-of-mass energy of 14 TeV and a bunch crossing interval of 25 ns. At design luminosity this will result in an inelastic pp interaction rate of 1 GHz. The CMS experiment will have a multi-level trigger system in order to reduce this rate to about 100 Hz for events written to a mass storage. This online data selection is an unprecedented experimental challenge. The architecture of the CMS muon trigger is described.

1 Introduction

CMS (Compact Muon Solenoid)[1] is a general-purpose experiment designed to study proton-proton and heavy-ion collisions at the Large Hadron Collider (LHC) of CERN. Its main feature is a strong solenoidal magnetic field ensuring high momentum resolution for charged particles. The detector consists of a silicon tracker with an embedded pixel detector, a crystal electromagnetic calorimeter, a copper-scintillator sandwich hadron calorimeter and a sophisticated four station muon system made up of tracking chambers and dedicated trigger chambers.

The LHC bunch crossing frequency for proton-proton interactions is 40 MHz and its design luminosity is 10^{34} cm^{-2}s^{-1}. At this luminosity approximately 25 inelastic collisions occur every 25 ns corresponding to an interaction rate of the order of 1 GHz. This input rate of 10^9 interactions every second must be reduced by a factor of at least 10^7 to 100 Hz, in order to match the capabilities of the mass storage and off-line computing systems.

CMS will have a multi-level trigger system to perform this rate reduction. The first stage of this rate reduction is performed by the Level-1 (L1) trigger which has to take a decision to accept or reject an event within a few microseconds after a collision, during which full detector information is stored in the detector front-end pipelines. The L1 trigger is a custom-built electronics system. The Higher Level Trigger (HLT) stages have much longer processing times and are therefore based solely on commercial computer farms.

2 Level-1 Trigger

The L1 trigger electronics has to provide a decision to retain or to reject an event for each bunch crossing, every 25 ns. It is pipelined and runs dead time free. The maximum output rate of the L1 trigger is 100 kHz, which is determined by the speed of the detector electronics readout and the input of the data acquisition system. To account for the limited reliability of rate predictions a safety factor of three is taken into account, therefore a maximum L1 rate of 30 kHz is foreseen in order to guarantee a large safety margin.[a] This rate is shared equally between muon and calorimeter triggers.

The L1 trigger system is organized into three subsystems: the L1 calorimeter trigger, the L1 muon trigger, and the L1 global trigger. A diagram of the L1 trigger system is shown in Fig. 1.

The L1 trigger decision will be based upon the presence of trigger objects such as muons, photons, electrons, and jets as well as global sums of E_T and missing E_T. Trigger objects are determined in three logical steps.

[a]These two rates will be reduced to 75 kHz and 25 kHz respectively at the start-up of LHC due to the deliberate incompleteness of the on-line processor farm which will be upgraded at a later stage.

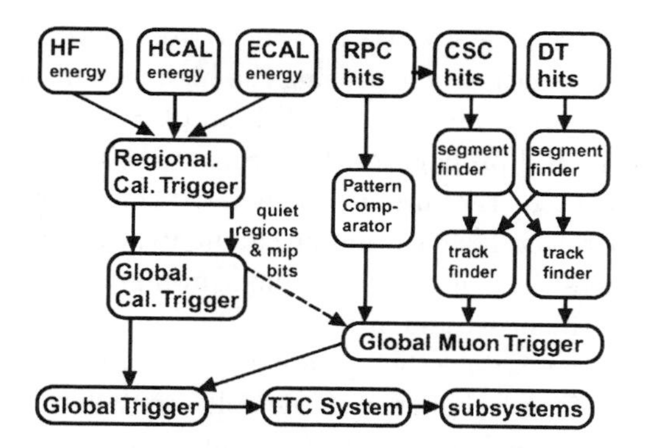

Figure 1. Overview of the L1 trigger system.

The first one is the production of local trigger information (Trigger Primitives) by the calorimeters and the muon system. The following two steps are calculations of regional and global sub-detector specific quantities. Only the global quantities are forwarded to the Global Trigger[2]. The trigger objects are based on coarsely segmented trigger data from the sub-detectors. The full-resolution data, which can be used by the Higher Level Triggers, are stored in pipeline memories in the front-end pipelines. All trigger objects are accompanied by their (η, ϕ) coordinates to allow the Global Trigger to vary thresholds based on the location of the trigger objects. In addition, the presence of the trigger object coordinate data in the trigger data, permits a quick determination of the regions of interest where the HLT analyses should focus.

2.1 Level-1 Muon Trigger

The L1 muon trigger is organized into subsystems representing the 3 different muon detector systems, the Drift Tube (DT) trigger in the barrel, the Cathode Strip Chamber (CSC) trigger in the endcaps up to $|\eta| < 2.4$ and the Resistive Plate Chamber (RPC) trigger covering both the barrel and the endcap regions up to $|\eta| < 2.1$. The L1 muon trigger also has a Global Muon Trigger (GMT) that combines the trigger information from the DT, CSC and RPC trigger systems and sends this to the L1 Global Trigger.

The RPC strips are connected to a Pattern Comparator Trigger (PACT)[3]. The PACT is based on spatial and time coincidence of hits in RPC chambers. Hit patterns are compared to predefined patterns corresponding to various p_T values in order to combine the hits into tracks and calculate their p_T.

The barrel DT chambers consist of three superlayers, each of which is composed of four staggered layers of drift cells. The wires of the inner and outer superlayers are parallel to the beam direction and measure the azimuthal coordinate ϕ in the magnetic bending plane. The central superlayer measures the track position along the beam. The drift tubes are equipped with Bunch and Track Identifier (BTI)[4] electronics that finds track segments from coincidences of aligned hits in four layers of one drift tube superlayer (SL). The track segments positions and angles are sent to the Track Correlator (TRACO)[4], which attempts to combine the segments from the two SL measuring the ϕ coordinate. The best combinations from all TRACOs of a single chamber together with the SL η segments are collected by the Trigger Server. The Trigger Server then sends the best two segments to the DT Track Finder[5].

The CSCs form Local Charged Tracks (LCT) from the cathode strips, which are combined with the anode wire information for bunch crossing identification[6]. The LCT pattern logic assigns a p_T and quality, which is used to sort the LCTs on the motherboard and the Muon Port Card (MPC) that collects LCTs from up to 9 CSC chambers. The best 3 LCTs from all the MPCs in a sector are transmitted to the CSC Track Finder[6].

The DT and CSC Track Finders are running in parallel. There is some information exchange between them in the overlap region joining the barrel and the endcaps. The Track Finders combine the segments found in the four muon stations into full muon tracks and assign p_T values to them. They are then sorted according to criteria taking into account transverse momentum and quality. The four best DT and the four best CSC muons are sent to the Global Muon Trigger.

The Global Muon Trigger receives the eight highest p_T muons found by the RPC trigger and four muons from each of the two Track Finders. It performs a DT/CSC-RPC matching based on distance in (η, ϕ)-space and determines the four best muon candidates in the entire CMS detector, which are finally sent to the Global Trigger. For a muon, the sign of the electric charge, isolation information and an indication whether there is compatibility with a signal delivered by a minimum-ionizing particle (MIP) are also available. The GMT greatly improves the overall muon trigger efficiency and rate capability compared to the individual muon subdetectors.

3 Higher Level Triggers

To perform further event filtering after a L1 accept decision, the data corresponding to each selected event must be moved from about 512 front-end buffers to a single location. This function will be performed by a system employing a high performance readout network to connect the sub-detector readout units via a switch fabric to the event filter units, which are implemented by a computer farm. The flow of event data will be controlled by an event manager system.

In order to optimize the data flow, the filter farm performs event selection in progressive stages by applying a series of Higher Level Trigger filters. The initial filtering decision (Level-2 trigger) is made using only data from the calorimeters and the muon systems. This avoids saturating the system bandwidth by reading out the large volume of tracking data at the full L1 output rate.

The Level-2 muon trigger is based on a local muon reconstruction algorithm using the L1 trigger data as seed. It redefines the L1 p_T measurement using the high resolution measurements from the DT and CSC chambers and propagates the reconstructed muon tracks to the interaction point. This allows to reject fake L1 muon candidates and reduces the contribution of non-prompt muons. For inclusive muon triggers the Level-2 trigger can reduce the L1 rate by a factor of $\sim$10.

The remainder of the full event data is only transferred to the farm after passing the Level-2 trigger and the final Higher Level Trigger algorithms use the full event data for the decision to keep an event. For the muon trigger this means matching muon and tracker data.

The Higher Level Trigger system is implemented as a processing farm that is designed to achieve a total rejection factor of 10^3 and writes up to 100 events/second to mass storage. The last stage of HLT processing does full event reconstruction and event filtering with the primary goal of making data sets of different signatures on easily accessed media.

References

1. CMS Collaboration, *CMS Technical Proposal*, CERN/LHCC 94-38 (1994).
2. N. Neumeister et al., CMS Note 1997/009 (1997); C.-E. Wulz, CMS Note 2000/052 (2000).
3. M. Andlinger et al., *Nucl. Instr. Meth.* **A370** (1996) 389.
4. F. Gasparini et al., *Nucl. Instr. Meth.* **A336** (1993) 91; R. Martinelli et al., CMS Note 1999/007 (1999); I. D'Antone et al., CMS Note 1996/078 (1996).
5. A. Kluge, T. Wildschek, CERN CMS Notes 1997/091, 092 and 093 (1997); G.M. Dallavalle et al., CMS Note 1998/042 (1998).
6. J. Hauser, CERN/LHCC/99-33 (1999) 304; D. Acosta et al., CERN/LHCC/99-33 (1999).

NOVEL DAQ AND TRIGGER METHODS FOR THE KLOE EXPERIMENT

The KLOE collaboration[1] *– presented by Paolo Branchini*
Via della Vasca Navale 84 Roma 00146, Italy
E-mail: Paolo.Branchini@roma3.infn.it

KLOE, a new state of the art detector recently commissioned for physics operation at DAΦNE, has many innovative interesting features, especially in the DAQ and Trigger areas. Custom electronics assert a trigger in a 2 microseconds decision time and distributes it to the FEE with a 50 ps time resolution. Data are read out using 10 front-end data acquisition chains and sent to a farm of online servers through a FDDI Gigaswitch. The design of the KLOE DAQ system allows to manage an input data rate as high as 50 Mbytes/s, and is completely scalable by extending the number of computers connected to the switch.

1 Introduction

The main goal of the KLOE[2] experiment at the DAΦNE e^+e^- collider in Frascati is to measure CP violation parameters in the Kaon system with a sensitivity of one part in ten thousand. The KLOE detector consists of three main parts; a large cylindrical tracking drift chamber DC; a hermetic lead-scintillator fibers calorimeter EmC; a large magnet surrounding the whole detector, consisting of a superconducting coil and an iron yoke. DAΦNE peak luminosity is expected to be $10^{33} cm^{-2} s^{-1}$, which corresponds to 10 kHz of estimated data rate with an average event size of 5 kbytes. About half of it is due to the ϕ decays while 5 kHz are expected to be due to prescaled Bhabha and cosmic events. The trigger system, first of all, must have a good rejection power for background, and be as close as possible to 100% efficiency on both the charged and the neutral decay of $K_l K_s$. Secondly it has to produce a valid signal within 350 ns from the bunch crossing, in order to start EmC ADCs and TDCs conversion. Moreover a fixed dead time of about 2 μs must be implemented in the trigger system to allow the drift mechanism in the chamber cells to take place. Finally all the relevant information provided by the Trigger system are acquired. DAQ requirements are as challenging as the trigger ones. In fact the DAQ system must be able to sustain 50 Mbytes/s, be scalable and flexible, asynchronous with respect to the trigger signal, have a fixed dead time, and be optimised for low channel occupancies. To keep up with these demands, the Trigger and DAQ systems have been designed as a mixture of custom specialised and high speed commercial electronics. Moreover an efficient software architecture has been designed and developed to deal with this complex environment.

2 Overview of the Trigger and DAQ system

The KLOE Trigger [3] system, is based on local energy deposits in the EmC and hit multiplicity information in the DC. A two level scheme has been adopted in order to produce a first early trigger signal (T1) to start EmC FEE digitisation while the second level trigger signal (T2), delayed by 2μs with respect to T1 is designed to fully exploit the information coming from the DC, which is used, together with the calorimetric information, to confirm the first level, to start digitisation of the drift chamber FEE and to start the DAQ read-out. The minimum time between bunch crossing in DAΦNE is 2.7 ns, therefore the trigger must operate in continuous mode. Signals from EmC and the DC are treated to generate two separate trigger decisions, that are OR-ed to form the final T1 and T2 signals. To avoid the intrinsic jiitter of the trigger signal formation the first level trigger is synchronised within 50 ps with the machine radiofrequency before being delivered to the FEE. The data taking process is event-driven and built around custom protocols and read-out controllers. Each controller hosts a trigger FIFO to store the information of the incoming trigger and a memory to store the event fragment coming from the FEE. Therefore the data taking process is asynchronous with respect to the T2.

3 Trigger hardware architecture

The trigger architecture is divided in two chains one for the EmC, one for the DC.

3.1 Calorimeter trigger chain

The signals coming from EmC's PM's are splitted in three different paths, through which the signals are delivered to the ADCs, TDCs and trigger boards. Signals delivered to the trigger chain are summed up according to a sector logic, the information corresponding to the outermost layer of the calorimeter is also formed to assert the cosmic veto. Once the analog signals are formed they are compared to a set of different thresholds and processed in a logic state machine (DISH). This stage provides output signals indicating local energy deposits above a certain energy threshold in the calorimeter. The number of hits in the calorimeter is determined in the stage which delivers the information to the "Trigger Organiser and Timing Analyser (TORTA)" for the final decision. Furthermore all the information provided by DISHes and the ones provided by the TORTA are acquired and allow us for continuous monitoring of the trigger system.

3.2 Drift Chamber trigger chain

DC signals are first shaped to a width of 250 ns, then summed in groups of 12 contiguous wires, finally these signals are sent to 20 sum units. The output signals from the SUPPLI are sent to three boards, which join the half-plane multiplicity information and pack the DC layers in groups of 4 or 6. These modules produce as an output signal the multiplicity from 10 DC superlayers. At the end of the chain the "Chamber Activity Fast Fetch (CAFFE)" board sums the signals of the superlayers to produce an output in current which is proportional to the number of fired wires in the DC. When this signal exceeds a given programmable threshold a T1D is issued. The same signal is also integrated during the following 1 μs and a T2D is issued if a different programmable threshold is exceeded. The informations coming from the CAFFE are acquired.

3.3 The Trigger logic box and the trigger DAQ interface

The TORTA receives all the trigger signals delivered from the EmC and the DC, it also merges all the informations and on the basis of programmable logic tables, takes the final decision and delivers the trigger signal. The TORTA distributes T1 to the Trigger Supervisor (TS) and receives back a signal which disables further T1 generation for a fixed dead time (2.6 μs). The T1 is also delivered to the Distributor module (TD), which performs the synchronisation with the machine RF and distributes the T1 to the EmC FEE. Whenever the TORTA generates a T2yes the TS asserts a T2. This signal is delivered to the FEE and DAQ process can start.

4 DAQ hardware architecture

The detector is divided in ten DAQ chains. Four chains are devoted to the acquisition of the EmC, four to the DC and two for the trigger. These chains are made up to six crates, each with sixteen slave boards and a local read-out controller -the ROCK. All ROCKs [4] in the chain are interconnected to a controller manager, ROCKM which resides in a VME crate with a VME processor board. A commercial bus (VIC bus) is used to perform initialisation tasks. The ROCK performs crate level read-out and gathers data from the AUX-bus, a custom protol developed to enhance the event-driven behaviour of the KLOE DAQ. A trigger signal (T2) starts transactions on the AUX-bus. Data transfer is carried out using high-speed random length block transfers. The ROCK then builds data frames consisting of an event number and slave data. In the same fashion the ROCKM performs chain level read-out through the custom C-bus. The C-bus tags data transaction with event numbers, making the entire chain an event-number-driven machine. After a system initialisation phase the data taking process is fully handled by the ROCKs and the ROCKM and CPU activity is required only to acquire data from ROCKM. The VME processor is then in charge of moving the frames from the VME-bus to the farm. Each chain is connected to the farm via an FDDI link. Finally data are archived by an IBM tape library.

5 DAQ software architecture

Two software processes - the Collector and the Sender - which run asynchronously with respect to each other on the VME processor manage the read-out activity. The Collector reads the sub-events framed in the ROCKM memory and pushes data frames containing a given set of sub-events in a shared memory with FIFO structure. The Sender retrieves from the FIFO packets of sub-events and dispatches them to the Receiver processes in the online computer farm using TCP/IP protocol over FDDI links. The assignment between packets of sub-events and Receivers is handled by a process according to the farm load [6]. Since the detector is divided in ten chain final event building must occur in the farm. This happens after receiving the ten different data streams by means of a process the Builder, which is decoupled from the one which receives the data, -the Receiver, by a shared memory. After rebuilding data are written on disk and archived. All DAQ processes, are co-ordinated by a central run-control and can communicate between them using the KLOE message passing system.

6 Event Filtering and Monitoring tasks

Using the shared memory filled by the farm Builder, the raw event structure is available to many other processes. The Trgmon, for instance, uses the pattern of the acquired information from the trigger chains to provide a fast monitor of relevant quantities such as instant luminosity, background levels, data rates. Histogram-server and Event-display also exploit the shared memory mechanism to fetch data. Dedicated histogram browsers check for "dead" detector channels and monitor the general behaviour comparing with reference histograms.

References

1. Refer to the e-Print hep-ex/0006039 for the complete author list.
2. The KLOE collaboration, LNF-94/028.
3. V. Bocci et al., "Trigger implementation in the KLOE experiment", LEB99, Snow mass, Colorado, and references threin.
4. A.Aloisio et al., "ROCK: The Readout Controller for the KLOE experiment", IEEE trans. On Nucl. Science, 43, 1, 167(1996).
5. P. Branchini et al., "Front end data acquisition for the KLOE experiment", Proc. of Int. Conf. on Computing in High Energy Phys. 1998, Chicago USA.
6. E. Pasqualucci et al., "Monitoring of the KLOE DAQ System", CHEP 98, FNAL, USA.

Parallel Session 13

Future Accelerators

Conveners: Stephen D. Holmes (Fermilab) and
Ian Wilson (CERN)

FUTURE HADRON COLLIDERS

T. M. TAYLOR

CERN, European Organization for Nuclear Research, Geneva, Switzerland
E-mail: Tom.Taylor@cern.ch

The status and projected performance of future hadron colliders are reviewed. While the major emphasis is placed on the Large Hadron Collider now under construction at CERN, prospects for machines in the closer and more distant future, that is the upgraded Tevatron and a very large collider, are also addressed.

1 Introduction

Hadron colliders provide the High Energy Physics community with excellent discovery potential. This is mainly due to the fact that it is at present easier to achieve a high centre-of-mass energy with these machines than with any other. The simultaneous advent of powerful computers also made it possible to unravel the complicated combination of "events" coming inevitably from the interaction of hadrons (as opposed to the much cleaner interaction of leptons). It is therefore of interest to keep up to date with the characteristics of machines which will be, or may be, commissioned in the near future. In this report the status of three such accelerators will be discussed, namely the upgrade of the Tevatron at Fermilab, the Large Hadron Collider (LHC) at CERN, and a possible future much larger facility usually referred to as a Very Large Hadron Collider (VLHC).

2 Tevatron Upgrade

The Fermilab Tevatron is the highest energy collider operating in the world today. The last Tevatron collider run was over the period 1993-96. An integrated luminosity of about 150 pb^{-1} was delivered to each detector at E_{cm} = 1800 GeV. At the end of this run the typical initial luminosity was 1.6 $10^{31} cm^{-2}s^{-1}$, the record being 2.5 $10^{31} cm^{-2}s^{-1}$. The machine is presently in the final stages of being upgraded with the aim of exploiting its capabilities to the fullest extent while it retains its supremacy[1].

2.1 Performance Goals and Schedule for Run II

The initial Run II goal is to deliver an integrated luminosity of at least 2 fb^{-1} by the end of 2002. The Run II luminosity goal is to exceed 10 $10^{31} cm^{-2}s^{-1}$. In order to achieve this performance the Fermilab complex will be required to support more protons in collision, and many more antiprotons in collision. This will be achieved by increasing the stacking rate of the antiproton, and recovering them for re-use at the end of the stores. The old main ring has been replaced by a 150 GeV Main Injector, which is capable of tripling the antiproton production targetting rate, and an 8.9 GeV Recycler, using permanent magnets, has been introduced to recover the unspent antiprotons. Commissioning of the complex is now under way and after a detector commissioning run (without silicon) this fall, the start-up of Run II with both detectors is expected in early spring 2001.

2.2 Longer Term Prospects

It is confidently expected that luminosity will be pushed to 2 $10^{32} cm^{-2}s^{-1}$ by 2003 and with further enhancements could reach 5 $10^{32} cm^{-2}s^{-1}$ by 2006, to give a total integrated luminosity of 15 fb^{-1} during the pre-LHC era. Such enhancements will be based on (a) improved antiproton availability, thanks to electron cooling, a liquid lithium lens and increased aperture, and (b) controlling the beam-beam interaction by use of electron beam compensation. R&D projects aimed at these improvements are already underway.

3 LHC

The Large Hadron Collider[2] (LHC) is a new facility under construction at CERN which will provide interactions between beams of protons at a centre of mass energy of 7 + 7 TeV. The accelerator will be installed in the circular tunnel of 27 km in circumference, presently occupied by the LEP electron-positron collider, and which will cease operation this fall. The collider was first discussed in the early 1980s, and the first design study appeared in 1987. Approval of the project was obtained in 1996, and the LHC is scheduled to come into operation in 2005. It is expected that it will take about 3 years to achieve the design luminosity of $10^{34} cm^{-2} s^{-1}$.

3.1 Accelerator Physics

The LHC will consist of two synchrotrons installed in the 27 km LEP tunnel. They will be filled with protons delivered from the SPS and its pre-accelerators at 0.45 TeV. The beams will be accelerated to 7 TeV in two superconducting magnetic channels, after which they will counter-rotate for several hours, colliding at the experiments.

Luminosity The design luminosity $10^{34} cm^{-2} s^{-1}$ will be achieved by filling each of the two rings with 2835 bunches, with 25 ns spacing, of 1.1 10^{11} particles each. The resulting beam current is 0.56 A.

Collective Instabilities Each proton bunch produces an electromagnetic wakefield which perturbs the succeeding bunches, and under certain phase conditions the resulting oscillations can be amplified and lead to beam loss. These collective instabilities can be severe in the LHC because of the large beam current and their effect is minimized by a careful control of the electromagnetic properties of the elements surrounding the beam. The bellows are shielded from the beam by thin fingers with sliding contacts, and the inner side of the stainless steel beam pipe is coated with pure copper to reduce its resistance to beam induced wall currents[3]. Feedback systems are also being designed.

Long term Stability The beams will be stored for about 10 hours. During these 4 10^8 revolutions the natural oscillations of particles around the central orbit should not increase significantly. Non-linear components of the guiding and focusing magnetic fields of the machine can perturb the motion of single beams. The destabilizing effects of magnetic imperfections is more pronounced at injection energy, because the errors are larger owing to persistent current effects in the superconducting cables, and also because the beams occupy a larger fraction of the coil cross section. During injection and ramping the beams will be separated. When they are brought into collision at high energy it is the beam-beam interaction which will dominate the perturbative forces, and which is expected to provide the ultimate limitation to useful stored beam current.

Control of Beam Loss The fraction of the particles which diffuses towards the beam pipe wall is lost, and the particle energy is converted into heat in the surrounding material, which can induce a quench of the superconducting magnets and interrupt operation for some hours. Two of the eight insertions are devoted to respectively betatron and momentum cleaning, where particles at greater than 7 sigma from the central trajectory are intercepted.

Flexibility of the Lattice As most of the components of the LHC are closely packed and embedded in a continuous cryostat, the adaptability of the machine to allow for upgrades and cope with unpredictable demands depends on the flexibility of the lattice. It has been found possible to include such flexibility by introducing individual powering of the quadrupoles in the dispersion suppressors and matching section of the insertions[6].

Synchrotron Radiation While synchrotron radiation at LHC is insufficient to provide damping of particle oscillations, the power emitted, about 3.7 kW per beam, has to be absorbed at cryogenic temperature. In addition it releases adsorbed gas molecules, which increase the residual gas pressure and liberate photo-electrons. These are accelerated across the beam pipe by the strong positive electric field of the proton bunches, adding to the cryogenic load, and may induce beam instabilities via the build up of an electron cloud.

3.2 Major Hardware

Magnets The magnetic channels are 194 mm apart and are housed in the same yoke and cryostat, a configuration that not only saves space but also gives a 25% reduction in cost[7]. It does however mean that the two beams must have the same energy. To bend 7 TeV protons around the ring, the 14.2 m long LHC dipoles must be able to produce a field of 8.36 T, and this can only be done with the commercially available NbTi superconductor by cooling it to below 2 K using superfluid helium. At this temperature the specific heat of the conductor is very low, putting new demands on cable quality and coil assembly. The magnets have to be "trained" to reach the highest quench fields, and individual training periods must be short for the efficient testing of the 1,232 dipoles. Should a magnet quench, the current in the corresponding circuit of one of the eight arcs is by-passed via a cold diode, heaters are fired to distribute the energy uniformly in the coil, and switches are opened across a resistor into which the 1.3 GJ of energy is dumped.

Following a long period of R&D and full-size prototyping, thirty pre-series magnets have been ordered from each of the three vendors which participated in the development program, with delivery to start this year.

In addition to the dipoles, more than 5000 magnets are needed to focus, adjust and bring the LHC beams into collision. All of these magnets will have been ordered by the end of this year.

Cryogenics for the LHC The cryogenic technology chosen for the LHC uses superfluid helium, which has highly efficient heat transfer properties, allowing kilowatts of refrigeration to be transported over more than three kilometers with a temperature difference of less than 0.1 K. LHC superconducting magnets incorporate a long 1.9 K bath of superfluid helium at just above atmospheric pressure. This is cooled by low pressure liquid helium flowing in heat exchanger tubes threaded along the string of magnets[8]. Refrigeration power equivalent to over 160 kW at 4.5 K is distributed around the 27 km ring. CERN will reuse the four existing LEP 18 kW, 4.5 K cryoplants: they will be adapted and 1.9 K stages will be added. Four new 18 kW plants have been ordered. Multi-stage cold centrifugal compressors are also needed, and following extensive prototype testing suitable equipment has been ordered from two suppliers. The design of the cryogenic distribution line is presently being validated on industrially supplied prototypes.

Thirteen million liters of liquid nitrogen will be vaporized during the initial cool-down of 36,000 tons of material and the total inventory of liquid helium will be 800,000 liters. The various helium enclosures will incorporate 40,000 leak-tight pipe junctions. All the major cryogenic equipment has been defined, and most has already been ordered.

Vacuum Achieving adequate beam lifetime for physics in the LHC requires ultrahigh vacuum. The vacuum chamber over most of the circumference of the LHC will be the inner wall of the magnet cold bore at

1.9 K, and therefore a very good cryo-pump. Due to the effect of the high-energy, high intensity beams, cryo-trapped molecules may be desorbed from the cold surfaces via several mechanisms - synchrotron radiation bombardment, photoelectrons resonantly accelerated by the beam potential, protons scattered by the residual gas - on a scale not experienced in previous accelerators. The cryogenic load at 1.9 K will be minimized by intercepting the heat due to synchrotron radiation and to beam image current on a beam screen, inserted into the magnet cold bore and cooled at 5 to 20 K. The beam/residual gas lifetime is dominated by nuclear scattering and, for a lifetime of 100 hours, requires an H_2 gas density of $< 10^{15}$ m^{-3}.

Recent work has revealed the importance of the electron cloud effect at the LHC, but this is now both theoretically and experimentally understood, and measures have been taken to ensure that machine performance should not be affected[9]. This will however remain a domain for study both before and during the commissioning of the machine.

Interface with the Experiments The beams intersect at four of the eight potential crossing points, namely at points designated 1, 2, 5 and 8, the other four points being used for accelerator equipment. The luminosity for each experiment will be controlled by independent low beta insertions, the closest elements of which being the inner quadrupole triplets at 23 m from the interaction point (IP). The radiation emanating from the interaction is such that for the high luminosity experiments, ATLAS and CMS, at points 1 and 5, it is necessary to protect the triplets with an absorber block installed at 19 m from the IP. A second neutral particle absorber is situated at 140 m from the IP, just in front of the beam separation dipole. These absorbers are ideally situated for luminosity monitors, which are also being studied. The absorbers are not required at points 2 and 8, which house the ALICE and LHCb experiments, but both of these experiments require an additional dipole magnet in front of the closest quadrupole as part of the compensating scheme for their dipole spectrometer magnets.

In order to shield the forward muon chambers from the background coming from the first absorber, the two big experiments will provide thick bunkers, which are being studied in close collaboration with the machine.

Another major interface in all four experiments is that of the vacuum pipe. It is planned to reduce the requirement for external pumps by taking advantage of the newly developed technology of non-evaporable getter (NEG) coating of the inner surface of the pipe[10]. The fragility of the thin chambers is such that before any local intervention it will be necessary to inject temporarily ultra-pure inert gas to minimize the risk of buckling.

3.3 Schedule

The design, specification and ordering of components for the collider is progressing according to plan, and costs are to be globally within the budget envelope. A delay of a few months has occurred with the civil engineering work at point 5, due to unexpectedly difficult conditions, but the way forward is more predictable and it may be possible to recover part of this. It is planned to have equipment and injection tests in a first octant in 2004, and to have the installation completed in 2005.

3.4 Upgrade Potential

All equipment which is being designed to last for the expected duration of operation of the LHC should be capable of working at up to 7.5 TeV, or at 7 TeV with the design beam current increased by up to a factor of two. At this level there is no safety margin. The energy of the LHC is essentially limited by its installation in the existing LEP tunnel.

Even if magnets with a higher field could be built on a large scale, the synchrotron radiation load, which, for a machine of constant radius increases with the fourth power of the energy, would call for a corresponding increase in cryogenic power. On a more modest scale, it may be possible to increase the luminosity to beyond the design luminosity of $10^{34} cm^{-2} s^{-1}$ by reducing the β-function at the interaction points. In order to exploit this route fully, new quadrupoles will be required for the low-β insertions. These will have to be based on the use of multi-filamentary Nb_3Sn or Nb_3Al, having the same gradient and length as in the present design of the inner triplets, but providing a larger aperture.

4 The Next Generation Hadron Collider

The lead time for providing a new collider is of the order of 20 years, and work is already in progress on the magnets which could populate a machine to follow the LHC. It is recognized that, while the next collider should ideally be capable of achieving an energy of five to ten times that of the LHC, its cost is an extremely sensitive issue. Thus present studies address first and foremost the issue of finding ways to get significant reductions in unit costs. A major cost driver being the price of the dipole magnets, this is where the thrust is being applied. Three main lines of approach are being followed: low, medium and high field. In the low field version, a single superconducting turn excites a field of up to 1.9 T across the gaps in a simple iron circuit[11]. Regular NbTi superconductor can be used, but the design could also take advantage of the work being done on 15 materials and on HTS being developed for utilities, if this drives the cost down sufficiently. It is presently assumed that the high field versions will rely on A15 conductors (Nb_3Sn, Nb_3Al) or eventually HTS. The regular cosine theta coil geometry, used in the Tevatron, RHIC and the LHC, is proposed for the medium field (5-10 T) version, and the virtues of a new design concept, called the common coil magnet[12], are being investigated for the high field (10-13 T) version, in competition with a cosine theta design. While the cosine theta geometry minimizes the quantity of expensive superconducting material required for a given field and aperture and capitalizes on past studies, the forces within the coils are such that rectangular blocks of winding are more appropriate than sectors. It should be noted that force retention is a major issue for magnets expected to run at above 10 T. The management of this stress is also the subject of ongoing R&D[13]. It should be possible to contain the problems associated with magnet field quality by limiting the dynamic range to 1:10, for example, instead of 1:15 of the LHC. The common coil magnet design provides a feature which makes this feasible.

The upgrade potential of the LHC is thus severely limited by the diameter of the existing tunnel. It was however the existence of the tunnel, built for LEP, which sealed the approval of the LHC project. A new, much larger tunnel will be required for a VLHC. Tunnels can also be expensive, so besides the magnets it is important to address the tunnel issue. More and more tunnels are being constructed, especially for flood control, and accelerator builders are monitoring the downward evolution of unit costs in the sector. The choice of a site is also important in this respect - although this must be offset by the value of existing laboratory infrastructure.

If one could rely on a reasonable level of regular funding over the period of about 20 years an obvious route to the next collider would be to put a low field ring into as large a tunnel as possible, to achieve perhaps double the energy of the LHC, and then, while this is being exploited for physics, construct high field magnets to necessary to get another factor of about five.

Acknowledgments

I wish to thank Steve Holmes and Bill Barletta for up-to-date information on the Tevatron upgrade and VLHC studies, without which this report would not have been possible.

References

1. S. Holmes, *1999 IEEE Particle Accelerator Conf.*, 43-47 (1999).
2. L. Evans, *IEEE Trans. on Applied Superconductivity* Vol. **10**, 44-48 (2000).
3. R. Veness et al., *1999 IEEE Particle Accelerator Conf.*, 1339-1341 (1999).
4. J.-P. Koutchouk, *1999 IEEE Particle Accelerator Conf.*, 372-376 (1999).
5. D. Kalchev et al., *1999 IEEE Particle Accelerator Conf.*, 2620-2622 (1999).
6. A. Faus-Golfe et al., *1997 IEEE Particle Accelerator Conf.*, 1442-1444 (1997).
7. C. Wyss, *1999 IEEE Particle Accelerator Conf.*, 149-153 (1999).
8. P. Lebrun, *IEEE Trans. on Appl. Superconduct.*, Vol. **10**, 1500-1505 (2000).
9. O. Bruning et al., *1999 IEEE Particle Accelerator Conf.*, 2629-2631 (1999).
10. C. Benvenuti and P. Chiggiato, *J. Vac. Science*, A , **14**, 3278-3282 (1996).
11. G.W. Foster, *1999 IEEE Particle Accelerator Conf.*, 182-184 (1999)
12. R. Gupta, *1999 IEEE Particle Accelerator Conf.*, 3239-3241 (1999)
13. P. McIntyre, *1999 IEEE Particle Accelerator Conf.*, 2936-2938 (1999)

FUTURE PROSPECTS FOR MUON FACILITIES

STEVE GEER
Fermi National Accelerator Laboratory, PO Box 500, Batavia, IL60510, USA
E-mail: sgeer@fnal.gov

The motivation, prospects, and R&D plans for future high–intensity muon facilities are described, with an emphasis on neutrino factories. The additional R&D needed for muon colliders is also considered.

1 Introduction

Recently there has been much interest in developing a very intense muon source capable of producing a millimole of muons per year. This interest is well motivated. A very bright muon beam that can be rapidly accelerated to high energies would provide a new tool for particle physics. The beam toolkit presently available to physicists interested in particle interactions at the highest energies is limited to beams of charged stable particles: electrons, positrons, protons, and antiprotons. The development of intense μ^+ and μ^- beams would significantly extend this toolkit, opening the way for multi-TeV muon colliders [1], lower energy muon colliders (Higgs factories [2]), muon–proton colliders, etc. In addition, all of the muons decay to produce neutrinos. A new breed of high energy high intensity neutrino beams would become possible [3]. Finally, there is the prospect of using the low energy (or stopped) muons to study rare processes with orders of magnitude more muons than currently available.

In response to the seductive vision of a millimole muon source an R&D collaboration was formed in the US in 1995, initially motivated by the desire to design a multi-TeV muon collider, and more recently by the desire to design a "neutrino factory" [3,4]. The motivation for neutrino factories is two-fold. First, the neutrino physics that could be pursued at a neutrino factory is compelling [5]. Second, a neutrino factory would provide a physics-driven project that would facilitate the development of millimole muon sources: the enabling technology for so many other goodies, including muon colliders.

2 The Neutrino Factory Concept

Conventional neutrino beams are produced from a beam of charged pions decaying in a long (typically several hundred meters) decay channel. If positive (negative) pions are selected, the result is an almost pure ν_μ ($\overline{\nu}_\mu$) beam from $\pi^+ \rightarrow \mu^+\nu_\mu$ ($\pi^- \rightarrow \mu^-\overline{\nu}_\mu$) decays. The neutrino oscillation physics community would like ν_e and $\overline{\nu}_e$ beams as well as ν_μ and $\overline{\nu}_\mu$ beams. For this we will need a different sort of neutrino source.

An obvious way to try to get ν_e and $\overline{\nu}_e$ beams is to exploit the decays $\mu^+ \rightarrow e^+\nu_e\overline{\nu}_\mu$ and $\mu^- \rightarrow e^-\nu_\mu\overline{\nu}_e$. To create a neutrino beam with sufficient intensity for a new generation of oscillation experiments will require a very intense muon source. With a millimole of muons per year we can imagine producing high energy beams containing $O(10^{20})$ neutrinos and antineutrinos per year. However, to achieve this a large fraction f of the muons must decay in a channel that points in the desired direction. Muons live 100 times longer than charged pions. Since the decay fraction f must be large we cannot use a linear muon decay channel unless we are prepared to build one that is tens of kilometers long. A more practical solution is to inject the muons into a storage ring with long straight sections. The useful decay fraction f is just the length of the straight section divided by the circumference of the ring. It has been shown that $f \sim 0.3$ is achievable [6]. The resulting muon

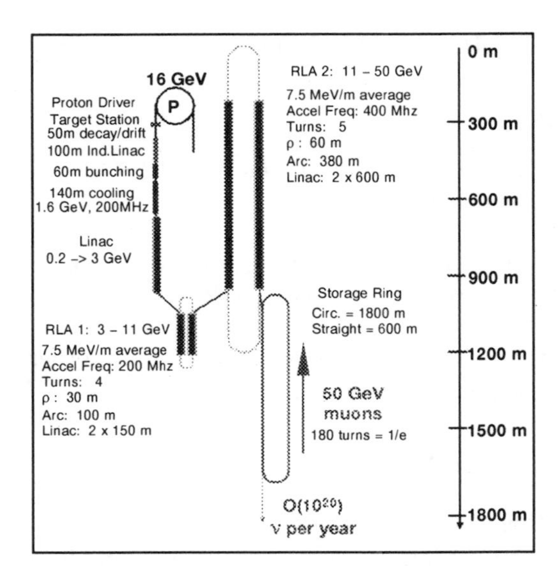

Figure 1. Schematic of the neutrino factory design from the 6 months feasibility study at Fermilab.

storage ring is sufficiently compact that it can be tilted downwards at a large angle so that the neutrino beam can pass through the Earth [3], and very long baseline experiments ($L \sim O(10^4)$ km) can be imagined.

Thus the "neutrino factory" concept [3,4], shown in Fig. 1, is to create a millimole/year muon source, rapidly accelerate the muons to the desired storage ring energy, and inject them into a storage ring with a long straight section that points in the desired direction.

3 Making Muons

Muons are produced in charged pion decays. Figure 2 shows a conceptual layout of the required pion source. The pion production target receives beam from a high–intensity multi–GeV proton driver. Most of the pions are produced with both longitudinal and transverse momenta of a few hundred MeV/c or less. These pions can be radially confined within a high–field solenoid. Therefore the target is within a 20 T solenoid, and is tilted at a small angle to reduce pion reabsorption.

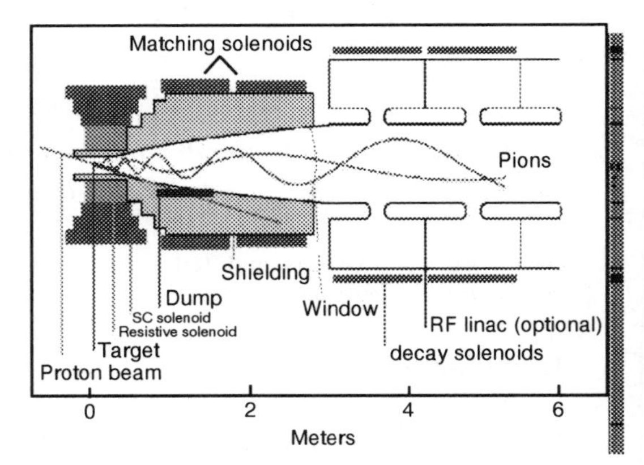

Figure 2. Conceptual layout of an intense pion source. The pion decay channel starts at 3 meters and continues for 50 meters to the right.

Downstream of the target solenoid the captured pions are transfered to a lower field larger radius solenoidal decay channel.

As an example, in the recent 6 months study [6] the design was based on a 16 GeV accelerator cycling at 15 Hz, and producing a 1.5 MW beam composed of 4×10^{13} protons/pulse, with each pulse consisting of 4 bunches separated by 500 ns. Detailed simulations predict 0.18 π^+/p and 0.15 π^-/p within the channel 9 m downstream of an 80 cm long carbon target tilted at 50 mrad, where the pions are required to have kinetic energies in the range 30–230 MeV. Hence, there are $\sim 10^{21}$ captured π^+ per 10^7 seconds of operation. If the pion source was upgraded to use a 4 MW proton beam incident on a liquid Hg jet target the pion yields increase by about a factor of 5.

The pions are produced with a large energy spread and small time spread (population A in Fig. 3). At the end of a 50 m long 1.25 T decay channel 95% of the pions have decayed. The daughter muons have a large energy spread, but their energies are correlated with their arrival time, with the faster muons arriving before the lower energy par-

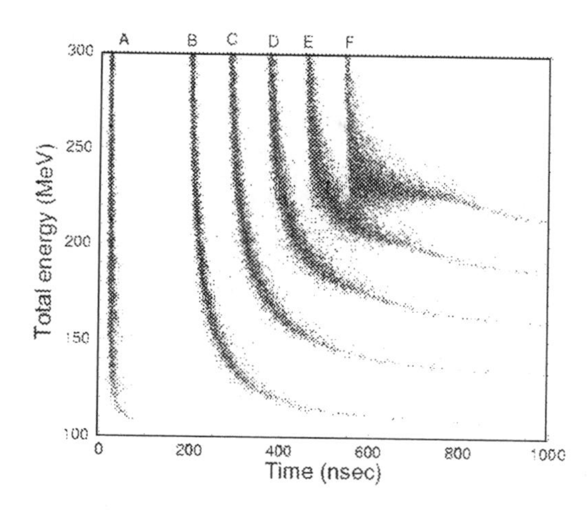

Figure 3. Simulated particle longitudinal phase–space populations: (A) pions at the production target, (B) Muons at the end of the decay channel, (C)–(F) Muons as they propagate down the induction linac.

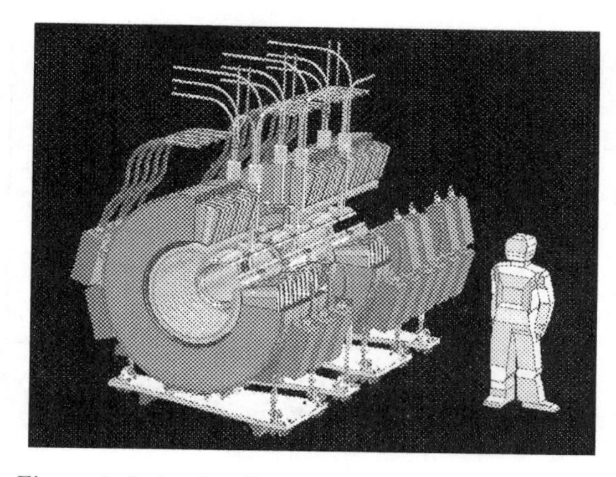

Figure 4. Induction linac design from the 6 months feasibility study at Fermilab.

ticles (population B in Fig. 3). Before we can capture the muons into RF bunches we must reduce their energy spread, which can be done either using RF cavities or using an induction linac to accelerate the late particles and deaccelerate the early particles.

In the induction linac design shown in Fig. 4 a 3 T superconducting solenoid channel (r = 20 cm) at the center of the linac keeps the muons radially confined. The linac consists of 100 modules, each 1 m long with a 10 cm gap providing a potential difference from -0.5 to +1.5 MV. A current pulse excites the induction cores outside of the solenoids. The changing toroidal magnetic field in the cores produces the accelerating gradient in the gap. The particle populations C–F in Fig. 3 show the evolution of the longitudinal phase–space occupied by the muons as they propagate down the linac.

After the energy spread has been reduced in the induction linac, the next task is to capture the muons with $E \sim 230$ MeV into a string of RF buckets. It is convenient to pass the muons through a 2.5 m long liquid hydrogen absorber, reducing their energy by $\sim$ 80 MeV. The muons can then be captured into $\sim$ 50 bunches with a 200 MHz buncher. Detailed simulations predict that at this point there will be 0.12μ captured per proton on the pion production target. However, with the design used for the 6 months study the RF buckets are full, dooming us to substantial particle losses as the beam enters the cooling channel, which is the last component of the muon source. A better RF capture design is currently being studied.

4 Cooling Muons

Before the muons can be accelerated to high energies we must reduce their transverse phase-space so that they fit within the acceptance of the first acceleration stage. This means we must "cool" the transverse phase–space by at least a factor of a few in each transverse plane. This must be done fast, before the muons decay. Stochastic– and electron–cooling are too slow, so we will need to use a new cooling technique. The technique proposed is "ionization cooling". In an ionization cooling channel the muons pass through an absorber in which they lose transverse– and longitudinal–momentum by dE/dx losses. The longitudinal momentum is then replaced using an RF cavity, and the

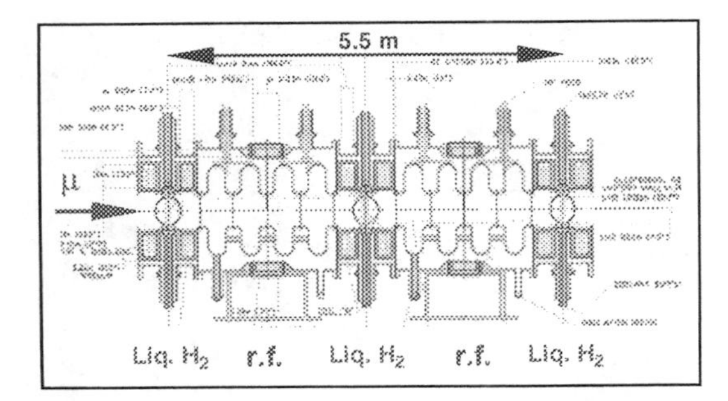

Figure 5. Cooling channel design.

process is repeated many times, removing the transverse muon momenta. This cooling process will compete with transverse heating due to Coulomb scattering. To minimize the effects of scattering we chose low–Z absorbers placed in the cooling channel lattice at positions of low–$\beta_\perp$ so that the typical radial focusing angle is large. If the focusing angle is much larger than the average scattering angle then scattering will not have much impact on the cooling process.

Figure 5 shows one design for a short section of an $\sim$ 100 m long cooling channel. To minimize scattering, liquid hydrogen absorbers are used with thin low–Z windows. The absorbers are located at low–$\beta_\perp$ locations within a high–field solenoid channel. The absorber sizes and the exact arrangement of solenoids vary with different cooling lattice designs. The channel shown in Fig. 5 has 30 cm long absorbers with a radius of 15 cm, within a 3.5 T axial field. The reaccelerating cavities operate at 200 MHz and provide a peak gradient of 15 MV/m. This deep RF bucket is needed to keep the large muon energy spread captured longitudinally, as well as providing the reacceleration.

The performance of the channel shown in Fig. 5 has been simulated. Results are shown in Fig. 6. Whilst the total number of muons decreases as the bunch travels down the channel, the number within the acceptance of the first acceleration stage increases by a factor

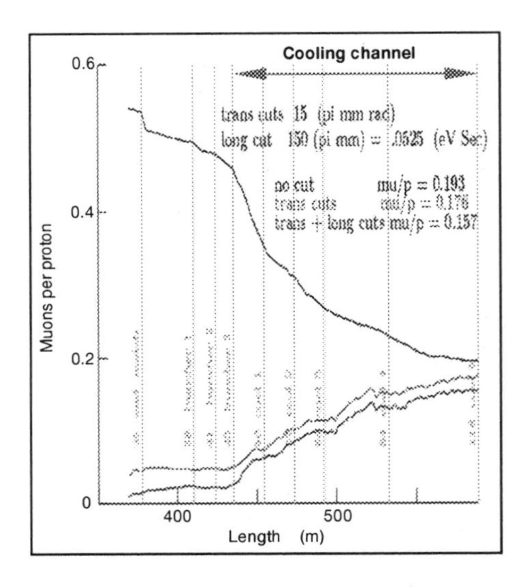

Figure 6. Simulated cooling channel performance. The total number of surviving muons (top curve) is shown versus position along the channel. The lower two curves show the number of muons within the acceptance of a large– and smaller–acceptance acceleration system. Lattice designed by R. Palmer.

of $\sim$ 5. Further optimization is in progress – there are lots of lattice variants to study !

5 Acceleration and Storage

The acceleration system is shown in Fig. 1. We need high gradients to accelerate the muons to high energy before they decay. An average gradient exceeding 5 MV/m is desirable (Fig. 7). After the first accelerating stage large solenoidal fields are no longer needed to confine the muons, and the desire for high gradients and an acceptable peak power leads us to choose superconducting RF. Once the muons have been accelerated to the desired energy, they are injected into a large acceptance storage ring with long straight sections ... and we are in business.

6 R&D

The goal for the 6 month feasibility study was to design a neutrino factory providing

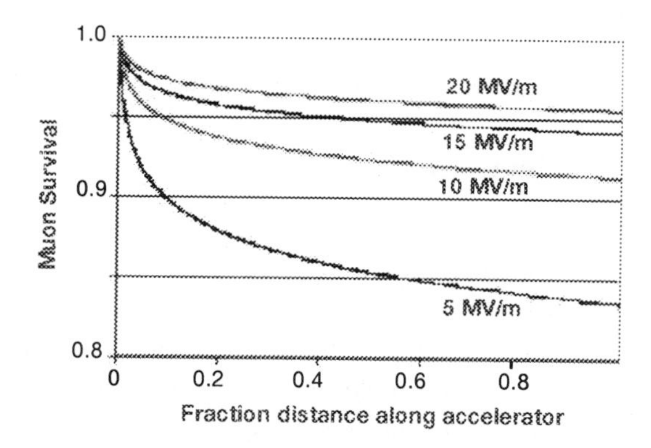

Figure 7. Muon survival as a function of distance along a 50 GeV acceleration system, shown for four different average accelerating gradients.

2×10^{20} muon decays per year in the beam–forming straight section. The study design yielded 6×10^{19} decays per year, not quite achieving the goal. With a more optimal RF buncher and cooling channel design and a slighter larger acceptance accelerator the design goal is probably achievable. With an upgraded 4 MW proton driver and a liquid Hg target an additional factor of 5 in intensity might be achieved. However, there are many R&D issues that must be addressed before a neutrino factory could be built. Below is a summary of the main issues and the R&D that is being pursued to address them.

6.1 Target Issues

A target experiment [7] is being prepared at BNL. The main goals of this endeavor are to (i) demonstrate a MW–level target in a 20T solenoid, (ii) measure pion and neutron yields to benchmark the simulation codes, and (iii) demonstrate the target lifetime in the high radiation environment. This activity is expected to proceed over the next 3 years. In addition there are particle production experiments (E910 at BNL, HARP at CERN, and a proposal P–907 at Fermilab) that are expected to benchmark the pion yields within the next couple of years.

6.2 Induction Linac Issues

The main issues are (i) can a suitable pulser system be built (4 pulses separated by 500 ns), and (ii) can the stray fields from the internal SC solenoid be kept low enough so that the inductive coils do not saturate. Within the neutrino factory/muon collider collaboration it is proposed to demonstrate the linac and pulser in the next few years.

6.3 Cooling Issues

The mission of the MUCOOL collaboration [8] is to design, prototype, and bench–test all cooling channel components, and eventually beam–test a cooling section. The component issues are (i) can sufficiently high gradient RF cavities be built and operated in the appropriate magnetic field and radiation environment, (ii) can liquid hydrogen absorbers with thin enough windows be built so that the dE/dx heating can be safely removed, and (iii) can the lattice solenoids be built to tolerance and be affordable? The MUCOOL collaboration has embarked on a design, prototyping, and testing program for all these components. This is expected to proceed over the next 3 years. Once the components have been developed the next step will be to build and bench test a cooling section to ensure all engineering issues have been resolved.

Detailed planning for the beam test phase must wait until the cooling channel design work is further along. In particular we must have a good quantitative understanding of how changes to the channel parameters effect the performance of the channel, and what the real issues are that need to be addressed by a muon beam experiment. We would hope that we will know enough to propose an experiment in about a year. The experiment is likely to be a large scale endeavor, and may be fertile ground for international collaboration. Whilst this is being worked out, the more modest MUSCAT experiment [9] is under way at TRIUMF to measure the scattering of low energy muons in various materials.

6.4 Acceleration Issues

The main R&D issue is whether superconducting RF cavities can be operated at 200 Mz (and perhaps 400 MHz) with sufficiently high gradients, using a practical power source. The neutrino factory/muon collider collaboration is planning to develop and test appropriate cavities. This activity will probably take more than 3 years.

6.5 Alternative Designs

There are two ongoing initiatives that could change the design details, and adjust the R&D described so far. First, a second generation design study has been launched at BNL with the participation of the neutrino factory/muon collider collaboration. Second, there is a parallel study being pursued at CERN [10] using similar design ideas but different technology choices. Finally, there is a Japanese initiative [11] which could radically change the design, in which $\sim$ 1 GeV/c pions are collected directly into a large acceptance accelerator (FFAG), and FFAGs do everything (no cooling etc).

7 A Final Word on Muon Colliders

Further R&D issues must be resolved to realize a muon collider, which needs brighter beams to obtain reasonable luminosity, and a cost effective high energy acceleration scheme. Perhaps the most challenging issue is that of cooling. To continue the cooling process beyond that needed for neutrino factories will require a further technology that enables the longitudinal phase–space occupied by the bunches to be reduced. It has been proposed to use "emittance exchange" in which some of the gain in the transverse phase–space is traded for a gain in longitudinal phase–space. Although concepts for emittance exchange exist, no practical realization has yet been worked out on paper. This is currently under active study, and is a make–or–break issue for muon colliders.

Acknowledgments

I am indebted to R. Palmer, A. Sessler, A. Tollestrup, and members of the Neutrino Factory/Muon Collider Collaboration, without which there would be little prospect of developing millimole per year muon sources. This work was supported at Fermilab under grant US DOE DE-AC02-76CH03000.

References

1. "$\mu^+\mu^-$ Collider: A Feasibility Study", Muon Collider Collab., FERMILAB-Conf.-96/092.
2. Muon Collider Collab., *Phys. Rev. ST Accel. Beams* **2**, 081001 (1999).
3. S. Geer, *Phys. Rev.* **D57**, 6989 (1998).
4. An older idea, ascribed to Kushkarev, Wojcicki, and Collins, was to inject pions into a storage ring with straight–sections, the pions decay in situ to create a captured muon beam. The resulting neutrino beam intensity is several orders of magnitude less than obtained with the modern neutrino factory scheme.
5. S. Geer, hep-ph/0008155, *Comments on Nuclear and Particle Physics.*
6. N. Holtkamp and D. Finley (editors) "A feasibility study of a neutrino source based on a muon storage ring", Report to the Fermilab Directorate, FERMILAB-PUB-00-108-E, http://www.fnal.gov/projects/muon_collider/nu-factory/nu-factory.html
7. BNL E951, spokesperson K. McDonald.
8. MUCOOL Collab., spokesperson S. Geer
9. MUSCAT Collab., Spokespeople: R. Edgecock and K. Nagamine.
10. http://muonstoragerings.cern.ch/
11. Y. Kuno and Y. Mori, talks at NUFACT00, May 2000.

FUTURE PROSPECTS FOR ELECTRON COLLIDERS

NOBU TOGE
KEK, High Energy Accelerator Research Organization, Tsukuba, Ibaraki 305-0801, Japan
E-mail: nobu.toge@kek.jp

An overview on the future prospects for electron colliders is presented. In the first part of this paper we will walk through the status of current development of next-generation electron linear colliders of sub-TeV to TeV energy range. Then we will visit recent results from technological developments which aim at longer term future for higher energy accelerators.

1 Introduction

Development of e^+e^- linear colliders (LCs) is motivated by the notion that their total linac lengths would scale as: $L_{\rm LC} \sim E_{CM}$. This is to be compared with storage rings whose cost-optimized circumference $L_{\rm ring}$ has to grow as: $L_{\rm ring} \sim E_{CM}^2$ for compensating the synchrotron radiation energy loss. This advantage of high-energy LCs was first discussed by Amaldi[1] and its implementation pioneered by SLC[2] at SLAC. Indeed, the typical site length required for LCs with $E_{CM} \sim 0.5$ TeV is currently considered to be $O(30)$ km or less with the required AC power of $O(100)$ MW for RF generation, while the tunnel length for the existing LEP is 30 km.

This paper presents a review on (1) the current status of development of next-generation (LCs), namely, TESLA[3], NLC[4], JLC[5] and CLIC[6], and (2) some description of technological developments relevant to the longer term future such as the efforts on W-band RF technologies, plasma- and laser-based particle acceleration.

2 Common Issues with Next Generation LCs

The luminosity at an LC is given by

$$L = f_{\rm rep} \frac{N_B N^+ N^-}{4\pi\sigma_x\sigma_y} H_D, \qquad (1)$$

where $f_{\rm rep}$ is the pulse repetition rate which is typically limited at $\sim O(100)$Hz, N_B the number of bunches to be accelerated in a single machine pulse, $N^\pm$ the bunch population, $\sigma_{x,y}$ the transverse beam size at the interaction point (IP), and H_D the luminosity enhancement factor due to beam disruption at the IP. This can be rewritten in terms of RF power efficiency as:

$$L \propto \frac{P_{AC}\eta_{AC\to RF}\eta_{RF\to beam}}{E_{CM}} \sqrt{\frac{\delta_B}{\epsilon_y}} H_D, \qquad (2)$$

where P_{AC} is the wall plug power, $\eta_{AC\to RF}$ and $\eta_{RF\to beam}$ are the efficiencies for the AC-to-RF power conversion and its transfer to beam energies, δ_B is average beam energy loss due to beamstrahlung, ϵ_y is the vertical beam emittance at the IP. To keep the environment for the detector facility clean, the δ_B is limited to a few %. Typical luminosity enhancement of 1.3 $\sim$ 1.5 is considered for H_D with "flat beams".

Realization of LCs thus critically depends on the availability of highly efficient and reliable RF power sources and high-gradient accelerator structures for use at linacs, together with robust sources of multi-bunch beams with ultra-low emittance. Table 1 summarizes the important parameters for the next-generation LCs that are currently under active development. More detailed parameters are compiled and periodically updated at the TRC site[7].

Fig. 1 shows the normalized horizontal and vertical beam emittance values that have been maintained at SLC (SLAC), and the ones recently achieved at the ATF (KEK) damping ring[5]. It is seen that beam emit-

Table 1. Some of the pertinent parameters for the next-generation LCs currently under development.

	TESLA	JLC (C)	NLC / JLC (X)	CLIC
Accelerator	S.C.	N.C.	N.C.	N.C./2-beam
RF freq [GHz]	1.3	5.7	11.4	30
E_{acc} [MV/m]	22	34	55	150
$N^{\pm}$ / Bunch [10^{10}]	2	1.11	0.95	0.4
N_B / Beam	2820	72	95	154
Bunch spacing [ns]	337	2.8	2.8	0.67
Bunch train length	950 μs	202 ns	270 ns	103 ns
x/y Beam Emittance [10^{-6} m]	10 / 0.03	3.3 / 0.05	4.5 / 0.1	2 / 0.02
x/y beam size at IP [nm]	553 / 5	318 / 4.33	30 / 4.9	202 / 2.5
z beam size [mm]	0.4	0.2	0.12	0.03
Two-linac length [km]	23	16	10.5	4.6
AC power to make RF [MW]	95	130	100	100

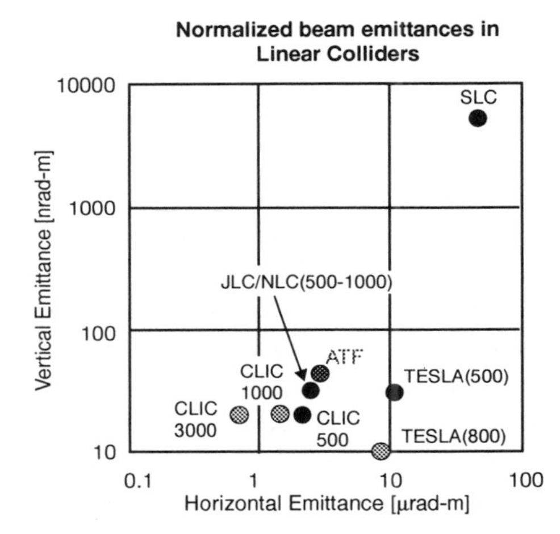

Figure 1. Beam emittance (x and y required at next-generation LCs, compared to those achieved at SLAC SLC and KEK ATF.

tance that is required for typical next-generation LCs is nearly achieved. Much work is still needed, however, to firmly establish the low beam emittance in multi-bunch beam operation at ATF. In addition, all the LCs require bunch lengths much shorter than 1 mm at the IP, because of the very small β_y^*. Thus, some kind of bunch length compression beam lines or a bunch source with short bunch length is also required.

The scale of the infrastructure for supporting LCs in Table 1 is roughly comparable to that of LEP or only slightly larger. However, the precision that is required in construction and operation of the complex RF and beam control systems is far more substantial than any existing accelerator facilities. The R&D topics for LCs constitute one of the major frontiers that the present day accelerator physicists and engineers are facing.

3 Specific Projects on Next Generation LCs

3.1 TESLA

The TESLA collaboration centered around DESY has been developing an LC concept based on superconducting (SC) niobium accelerating structures (Fig. 2), operated at 2 K and 1.3 GHz for an accelerating gradient of 22 MV/m or higher.

The low RF losses on the SC cavity walls are expected to lead to good power conversion efficiency of AC→ beam power (23 %). The long RF pulses associated with the SC RF system allow many bunches, spaced wide apart, to be accelerated in each machine pulse. This makes it possible to realize head-on beam collision at the IP, eliminating

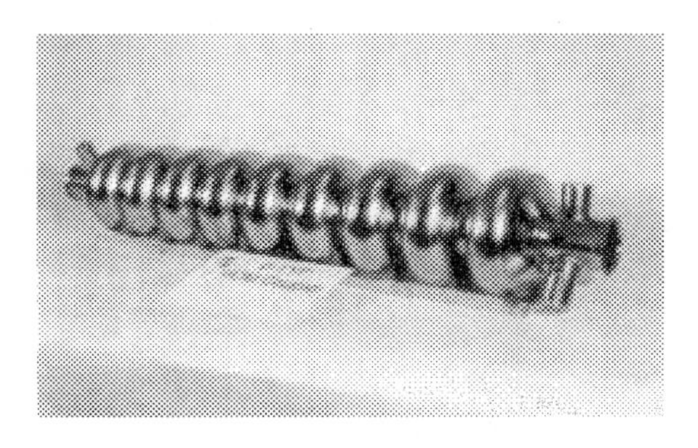

Figure 2. 9-cell TESLA Niobium cavity.

the need for Crab-crossing and relatively easy implementation of fast bunch-to-bunch orbit feedback.

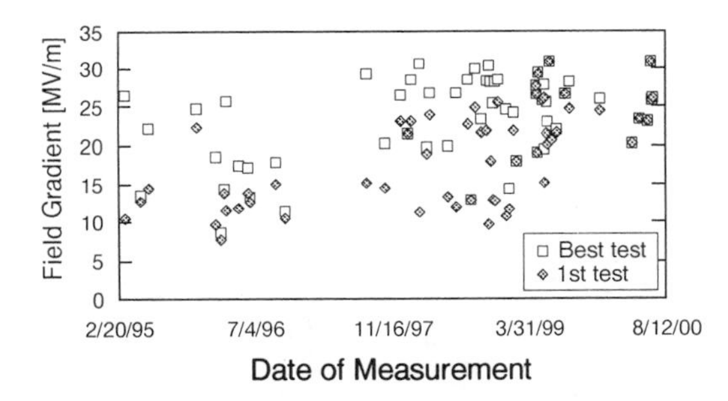

Figure 3. Cavity performance on the vertical test stand.

A major challenge exists in reducing the cost per unit accelerating voltage, compared to existing examples of LEP and CEBAF for the SC RF technology. Fig. 3 shows a summary of cavity performance on the vertical test stand, performed at the TESLA Test Facility (TTF) since 1995. Good performance improvement is visible. It has been observed by now whenever a newly built cavity reaches 20 MV/m in the first cw-test, it also satisfies the TESLA500 goal of $Q_0 = 10^{20}$ on average. While demonstrating up to 22 MV/m accelerating gradient, the TTF linac is also operated for studies of SASE FEL applications.

Ongoing research programs at TESLA include: a new "super-structure" scheme which allows to reduce the number of coupler components, fabrication technologies which do not rely on welding, development of compensation schemes for detuning from Lorentz-force-induced resonance of SC cavities during the RF pulse. Studies are also done on a wide range of beam dynamics issues with the damping ring, final focus and positron production. The TESLA technical design report is currently in preparation and to be published in spring of 2001.

3.2 C-band JLC

Development of main linac scheme based on the C-band (5.71 GHz) technology has been pursued for JLC in Japan. While positioned as a back-up scheme for the X-band JLC scenario, a fast-track R&D on the C-band scheme has been possible, due to relatively straightforward, if not trivial, extrapolation from the conventional S-band (2.856 GHz) technologies and hard work.

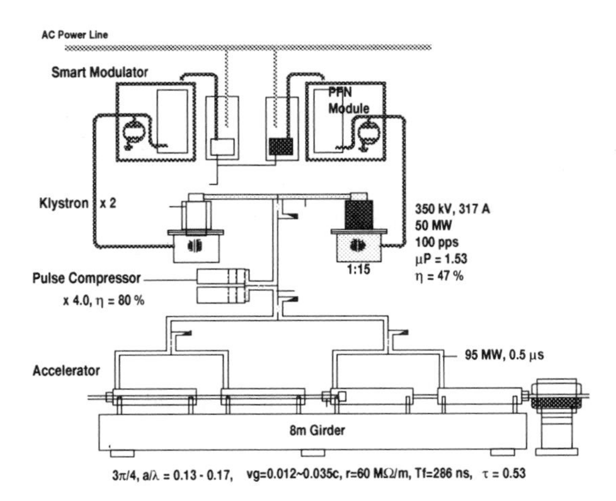

Figure 4. One unit of the C-band RF system.

One unit of the C-band RF system (Fig. 4) consists of a pair of 50 MW, 2.5 μs klystrons which are driven by a 350 kV modulator. RF pulse compression is achieved with an energy storage cavity with irises in it. The accelerating structure is made of room-temperature copper with heavy damping. Prototype testing of the RF components have been substantially complete, including 3 solenoid-focusing klystrons, which all satisfied the design specifications, and a modulator whose efficiency exceeded 50 %. Testing of a klystron with periodic permanent magnet (PPM) focusing is currently under way. High-power operation of the pulse compression cavity is being planned. Testing of high-power operation of the C-band accelerating structure (design gradient 40 MV/m) is needed.

3.3 NLC and X-band JLC

Both the US NLC collaboration centered around SLAC and the JLC group in Japan around KEK are pursuing an LC main linac scheme based on the X-band (11.424 GHz) technology. An R&D collaboration exists between KEK and SLAC, since 1998[4,8], to accelerate the efforts using the common LC design parameters, and is serving as a matching point of the work by the two groups.

While initially some regarded the X-band scheme as a rather ambitious extrapolation from the S-band technology, beam acceleration (up to 55 MV/m) and beam-loading compensation have been already experimentally demonstrated in NLCTA experiments at SLAC.

Development of new, highly efficient klystron modulators is under way while taking advantage of high-power semiconductor switching devices that are recently available on market. With successful operation of X-band klystrons with solenoid beam focusing, vigorous R&D work is being made to establish the design of PPM-focused X-band klystrons (Fig. 5). The latest prototype of PPM klystron at SLAC has shown > 75 MW 3 μs output, well exceeding the initial design goal.

Design, prototype construction and testing of highly optimized accelerating structures (Fig. 6) and efficient RF power distribution systems using delay lines are being made as joint efforts between NLC and JLC groups, yielding encouraging results[8].

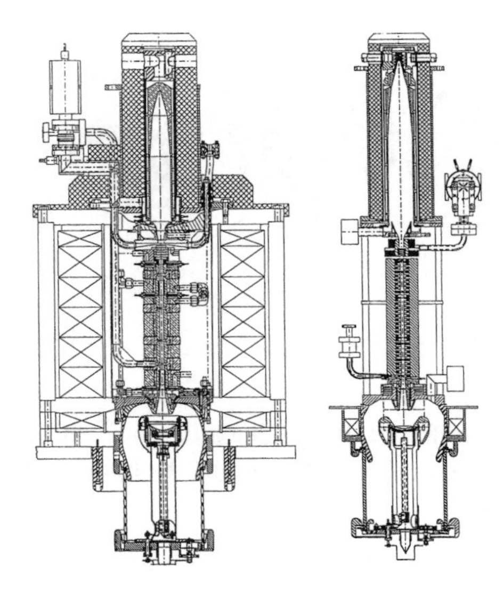

Figure 5. X-band klystrons. Solenoid-focusing type (left) and PPM (periodic-permanent magnet) focusing type (right).

Figure 6. X-band accelerator structure.

3.4 CLIC

The R&D on CLIC, centered around CERN, aims to achieve a high-energy (0.5 - 5 TeV), high-luminosity LC by using a two-beam acceleration (TBA) scheme, which, in a way, can be interpreted as an elongated kind of a klystron. As shown in Fig. 7, high current "drive" beam is passed through a drive beam accelerator (937 MHz). From there the RF power is extracted by transfer structures and is fed to the main beam accelerator (30 GHz) running parallel at a very high gradient up to 150 MV/m.

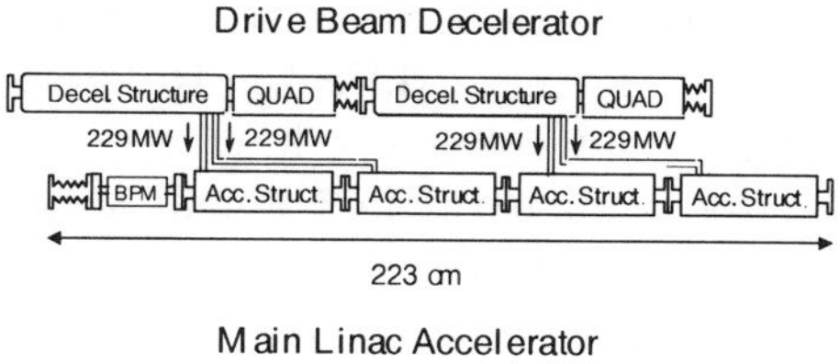

Drive Beam Decelerator

Decel. Structure | QUAD | Decel. Structure | QUAD

229MW | 229MW | 229MW | 229MW

BPM | Acc. Struct. | Acc. Struct. | Acc. Struct. | Acc. Struct.

223 cm

Main Linac Accelerator

Figure 7. Two-beam acceleration concept shown in diagram for one main-beam and drive-beam module at CLIC.

Because of the expected high gradient, the CLIC scheme has been considered the most suitable for a multi-TeV LC in the post-LHC era. However, its proponents also point out that it can be built in stages without major modifications for initial lower-energy operation.

The natural focus by the CLIC group has been on demonstration of the feasibility of the novel TBA concept. An early proof-of-principle experiment was conducted at the single-bunch-mode operation of CLIC Test Facility 1 (CTF1) in 1995, where a field gradient of 125 MV/m was created with a 76 MW, 30 GHz RF power. CTF2 is currently in operation for testing the multi-bunch drive beam operation with total charge 373 nC.

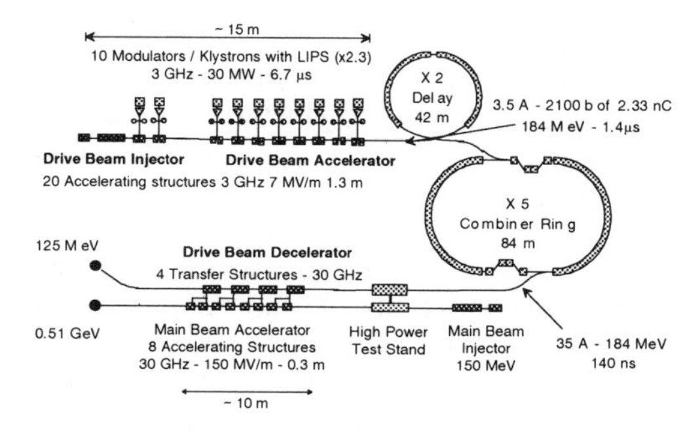

Figure 8. Schematic layout of CTF3 test facility planned at CERN.

Substantial design optimization and conceptual development have been recently done for CLIC on the bunch manipulation associated with creation of the multi-bunch drive beams. CTF3 (Fig. 8 is under study for testing all major parts of the CLIC RF power scheme.

4 Beyond Next-Generation LCs

The quest would not stop for energies higher than the regime that can be afforded with the CLIC scheme.

One possibility is to continue using solid-state structures for acceleration but at higher RF frequencies, such as W-band (90 MHz). This is for achieving higher accelerating gradient with improved power efficiency, which is inversely proportional to the square of the RF wavelength ($\sim \lambda^{-2}$). While the RF breakdowns are relatively reduced at this high frequency, the pulsed heating of the material due to ohmic heating is likely to be a limiting factor in operation[10]. Application of new material such as strengthened copper alloys may have to be considered. Development of fabrication technologies for structures, whose inner dimensions being a few mm, is critical[11]. Electro-discharge machining (EDM), UV-lithography and deep X-ray lithography (LIGA)[12] are being investigated.

Plasma- or laser-based beam acceleration is another set of concepts that are being investigated. Several variant schemes exist where the accelerating fields within plasma are produced by the precursor beam or a pre-fired laser or combination of both. Early work by Nakajima et al[13] demonstrated a proof-of-principle of laser wakefield acceleration where the particle acceleration corresponding to 1.7 GV/m field was observed over a length of $\sim$ 1 cm. A new experiment with a much larger scale is currently under way for a 1.4 m-long plasma wakefield acceleration at E157[14]. Other efforts include the activities at STELA[15] and LEAP[16].

5 Conclusions

We went through the status of TESLA, JLC, NLC and CLIC projects and longer-term R&D towards future LCs[17]. Steady progress is being made. However, it has to be noted that none of the next-generation LC schemes have yet demonstrated the beam operation of their latest versions of the main linac hardware for an extended period ($\sim$ one year) with a non-negligible length ($O(100)$ m) to their full specifications. Likewise, none of the groups have firmly cleared substantial portion of industrialization issues associated with mass production of high-precision RF (and other) hardware components at this point.

The break-through can occur, hopefully, though extension or expansion of existing or new test facilities, some of which have been mentioned in this review. To do so requires a major amount of funding, resources and manpower. While admittedly being arduous, it should be also realized that they are quite intellectually stimulating tasks.

It is the author's opinion that it would be much worthwhile creating a more active, world-wide network in our community which strongly encourages the cross-migration of scientists and engineers (particularly, the young and talented ones) across the three related fields for promoting our future, namely:

1. Traditional research in high energy physics,
2. Research of next-generation particle accelerators, including LCs and large hadron or muon machines,
3. Basic and applied research of accelerators with longer term prospects, such as W-band, plasma- and laser-acceleration.

This would require conscious efforts on the part of both the leaders and members of the high energy physics and accelerator communities. "Participation" is the keyword.

Acknowledgments

The author would like to thank the session conveners S.Holmes and I.Wilson for offering him the opportunity to attempt this review. Special thanks must go to R.Brinkmann, R.Siemann, R.Ruth, H.Henke, D.Edwards, M.Syphers and colleagues from KEK and SLAC, too numerous to individually identify, for kindly sharing the material that is used for preparing this review.

References

1. U.Amaldi, *Phys. Lett.* **61B**, 313 (1976). The very first concept of an LC appeared in M. Tigner, *Nuovo Cimento* **37**, 1228 (1965).
2. P.Raimondi, et al., invited talk (THZF101) at EPAC 2000, Vienna (2000).
3. http://tesla.desy.de.
4. http://www-project.slac.stanford.edu /lc/NLC-tech.html.
5. http://lcdev.kek.jp, http://c-band.kek.jp.
6. http://cern.web.cern.ch/CERN /Divisions/PS/CLIC/Welcome.html.
7. http://www.slac.stanford.edu/xorg/ilc-trc /ilc-trchome.html.
8. http://lcdev.kek.jp/ISG/index.html.
9. R.H.Siemann, invited talk (FRYE11) at EPAC 2000, Vienna (2000). D.Whittum, SLAC-PUB-7932 (1998).
10. D.P.Pritzkau and R.H.Siemann, SLAC-PUB-8554 (2000).
11. P.J.Chou, et al., SLAC-PUB-7339 (1996).
12. H.Henke, invited talk (TUXF202) at EPAC 2000, Vienna (2000).
13. K.Nakajima, et al., Physica Scripta **T52**, 61 (1994).
14. M.J.Hogan et al., SLAC-PUB-8352, (2000).
15. Presentations by W.D.Kimura et al., at Advanced Accelerator Concepts Workshop, Baltimore (1998).
16. T. Plettner et al., at Advanced Accelerator Concepts Workshop, Baltimore (1998). Y.C.Huang, et al., Nucl. Instr. Meth. **A407**, 316 (1998).
17. We did not cover the work on SBLC (DESY) and VLEPP (BINP, Russia), which are currently no longer pursued. However, their former members continue playing active roles in the ongoing LC R&D.

Parallel Session 14

Lattice Field Theory

Conveners: Karl Jansen (CERN) and
Tetsuya Onogi (Hiroshima)

APPLICATIONS OF NON-PERTURBATIVE RENORMALIZATION

JOCHEN HEITGER

Deutsches Elektronen-Synchrotron DESY, Platanenallee 6, D-15738 Zeuthen, Germany

(ALPHA Collaboration)

A short survey of the renormalization problem in QCD and its non-perturbative solution by means of numerical simulations on the lattice is given. Most emphasis is on scale dependent renormalizations, which can be reliably addressed via a recursive finite-size scaling procedure employing a suitable intermediate renormalization scheme. To illustrate these concepts we discuss some — partly recent — computations of phenomenologically relevant quantities: the running QCD gauge coupling, renormalization group invariant quark masses and the renormalization of the static-light axial current.

1 Introduction

Apart from its well established rôle as a non-perturbative framework to calculate relations between Standard Model parameters and experimental quantities from first principles [1], Lattice Field Theory is particularly designed to solve various renormalization problems in QCD [2,3]. Since renormalized perturbation theory as analytical tool is limited to high energy processes, where the QCD coupling is sufficiently small, but inadequate for bound states and momentum transfers of the order of typical hadronic scales, $\mu \simeq 1\,\mathrm{GeV}/c$, a genuinely non-perturbative solution of the theory is generally required. This is achieved by numerical Monte Carlo simulations of the Euclidean QCD path integral on a space-time lattice. Though renormalization is an ultraviolet phenomenon (relevant scales $\mu^{-1} \sim a$) and QCD asymptotically free, tolerable simulation costs prevent the lattice spacing a from becoming much smaller than the extent of physical observables so that a truncation of the lattice perturbative series is often not justified. Therefore, it is far more safe to perform renormalizations non-perturbatively.

In addition, Lattice QCD has a large potential to address the computation of fundamental parameters of the theory, which escape a direct determination by experiments. The most prominent ones among them are the QCD coupling constant itself and the quark masses, whose running with the energy scale is desirable to be understood on a quantitative level beyond perturbation theory — the central subject of the next sections. The knowledge of these quantities (e.g. at some common reference point, $\alpha_s(M_Z)$ or $\overline{m}(2\,\mathrm{GeV})$) might then also provide essential input to theoretical analyses of observables of phenomenological interest. For instance, the mixing ratio ϵ'/ϵ in the neutral kaon system incorporates the strange quark mass value.

The renormalization properties of many other quantities have been investigated with lattice methods, e.g. (bilinear) quark composite operators, $\Delta S = 2$ matrix elements and, as presented at this conference too, structure functions [4]. Later we will briefly discuss the non-perturbative renormalization of the static axial current as a further example.

During the last few years the lattice community has seen much theoretical and numerical advances [1,5]. Here it is worth to mention at least the issue of $\mathrm{O}(a)$ discretization effects inherent in the Wilson fermion action. In case of the quenched approximation to QCD, where all dynamics due to virtual quark loops is ignored, they have been systematically eliminated through a non-perturbative realization of Symanzik's improvement programme [2,3,6]. Hence, lattice artifacts can be extrapolated away linearly in a^2, which allows to precisely extract many physical quantities in the continuum limit, $a \to 0$.

$$L_{\rm max} = C/F_\pi\text{: O}(\tfrac{1}{2}\text{fm}) \quad \text{hadronic scheme} \hookrightarrow \text{SF} \quad \longrightarrow \quad \alpha_{\rm SF}\,(\mu = 1/L_{\rm max})$$

$$\downarrow$$

$$\alpha_{\rm SF}\,(\mu = 2/L_{\rm max})$$

$$\downarrow$$

$$\bullet\bullet\bullet$$

$$\downarrow$$

$$\alpha_{\rm SF}\,(\mu = 2^n/L_{\rm max})$$

$$\downarrow$$

$$\text{PT}$$

$$\downarrow$$

$$\text{jet physics } (e^+e^- \to q\,\overline{q}\,g) \hookleftarrow \quad \text{value for } \Lambda_{\rm QCD}/F_\pi \quad \overset{\text{PT}}{\longleftarrow} \quad \Lambda_{\rm SF} L_{\rm max}$$

2 Intermediate schemes

As a representative example for a non-perturbative renormalization problem we may consider the calculation of quark masses through the PCAC relation,

$$F_{\rm K} m_{\rm K}^2 = (\overline{m}_{\rm u} + \overline{m}_{\rm s})\langle 0|\overline{u}\gamma_5 s|{\rm K}\rangle \qquad (1)$$

$$(\overline{u}\gamma_5 s)_{\overline{\rm MS}} = Z_{\rm P}(g_0, a\mu)(\overline{u}\gamma_5 s)_{\rm lattice}\,, \qquad (2)$$

in which the scale and scheme dependent renormalization constant $Z_{\rm P}$ relates the lattice results to the $\overline{\rm MS}$ scheme and is computable in lattice perturbation theory. But since this expansion introduces errors which are difficult to control, a non-perturbative determination of the renormalization factor is needed. A non-perturbative renormalization condition between the two schemes can, however, not be formulated, because $\overline{\rm MS}$ is only defined perturbatively.

The idea to overcome this problem is the introduction of an intermediate renormalization scheme: the lattice observable is first matched at some fixed scale μ_0 to the corresponding one in the intermediate scheme, and afterwards it is evolved from μ_0 up to high energies, where perturbation theory (PT) is expected to work well. Nonetheless, as in a simulation one then has to cover many scales (the box size L, $\mu \simeq 0.2\,\text{GeV} - 10\,\text{GeV}$ and the lattice cutoff a^{-1}) simultaneously, the task to reliably match the low energy regime with the high energy one, i.e. the applicability domain of perturbation theory, gets quite complicated. In the present context two implementations of such schemes are available, the regularization independent approach [7] and the QCD Schrödinger functional (SF) [8,9]. Whereas the former may suffer from the scale hierarchy problem in practice, the basic strategy of the SF approach is to recourse to an intermediate finite-volume renormalization scheme, where one identifies two of the before-mentioned scales, $\mu = 1/L$, and takes low energy data as input in order to use the non-perturbative renormalization group to scale up to high energies [2,3].

A schematic view of a non-perturbative computation of short distance parameters on the lattice along these lines, here in case of the running QCD coupling $\alpha(\mu)$, is given in the diagram above; the same can also be set up for the running quark masses. It is important to note that all relations '$\rightarrow$' are accessible in the continuum limit and in this sense universal by construction.

3 $\Lambda_{\rm QCD}$ and $M_{\rm quark,\,RGI}$ via the SF

The Schrödinger functional is the QCD partition function with certain Dirichlet boundary conditions in time imposed on the quark and gluon fields, for which a renormalized coupling constant can be defined as the response to an infinitesimal variation of the boundary conditions [8]. By help of the so-called step scaling function, being a measure for the change in the coupling when changing the box size L (and thus having the meaning of a discrete β–function), one is now able in the SF scheme to make contact with the high-energy regime of perturbative scaling:

$$\Lambda \equiv \lim_{\mu\to\infty} \left\{ \mu (b_0 \bar{g}^2(\mu))^{-b_1/2b_0^2} \, \mathrm{e}^{-1/2b_0\bar{g}^2} \right\}$$
$$b_0 = 11/(4\pi)^2 \,, \; b_1 = 102/(4\pi)^4 \,. \qquad (3)$$

Every step during the non-perturbative evolution towards the perturbative regime has been extrapolated to the continuum limit in the quenched approximation [10], and upon conversion to the $\overline{\mathrm{MS}}$ scheme this results in a value for the Λ–parameter:

$$\Lambda^{(0)}_{\overline{\mathrm{MS}}} = 238(19)\,\mathrm{MeV}\,. \qquad (4)$$

An extension of this investigation to the situation with two dynamical quarks is already in progress by the ALPHA Collaboration.

In a very similar way, in terms of the current quark mass renormalization factor Z_{P} of eq. (2) replacing the SF coupling to build up another step scaling function, the scale and scheme independent renormalization group invariant (RGI) quark masses

$$M \equiv \lim_{\mu\to\infty} \left\{ (2b_0 \bar{g}^2(\mu))^{-d_0/2b_0} \, \overline{m}(\mu) \right\}$$
$$b_0 = 11/(4\pi)^2 \,, \; d_0 = 8/(4\pi)^2 \qquad (5)$$

were obtained in the same reference. Both evolutions are displayed in Fig. 1, and at the scale μ_0 (leftmost point in Fig. 1b) the matching between the lattice regularization and $\overline{\mathrm{MS}}$ via the SF is completed:

$$\frac{M}{\overline{m}^{\mathrm{SF}}(\mu_0)} = 1.157(12)\,, \; \mu_0 \simeq 275\,\mathrm{MeV}\,. \quad (6)$$

For the O(a) improved theory and a massless renormalization scheme as utilized here, these results can also be summarized as

$$M = Z_{\mathrm{M}}(g_0) \times m(g_0) + \mathrm{O}(a^2)$$
$$Z_{\mathrm{M}}(g_0) = \frac{M}{\overline{m}(\mu)} \times \frac{\overline{m}(\mu)}{m(g_0)}\,, \qquad (7)$$

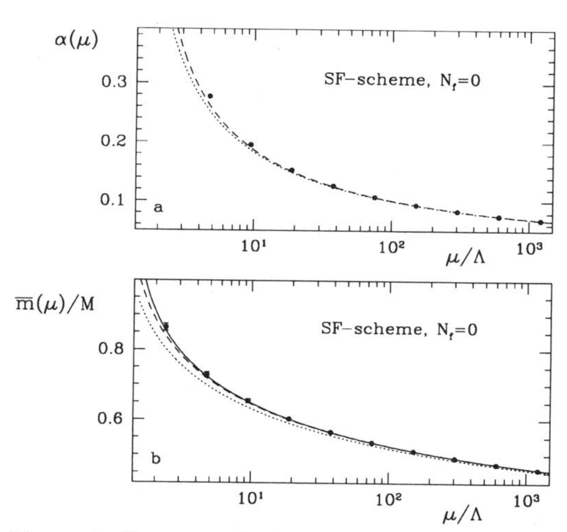

Figure 1. Non-perturbative scale evolution of α_{SF} and $\overline{m}^{\mathrm{SF}}/M$ computed from simulations of the SF in the quenched approximation. The lines represent perturbative predictions involving the 2– and 3–loop β–function (a) and 1/2–, 2/2– and 2/3–loop expressions for the τ– and β–functions, respectively (b).

where m is the bare current quark mass, and the flavour independent total renormalization factor Z_{M}, non-perturbatively known for a range of bare couplings g_0 in the quenched approximation [10], is composed of an universal part, $M/\overline{m}$, and of $\overline{m}/m = Z_{\mathrm{A}}/Z_{\mathrm{P}}$ depending on the lattice regularization.

4 The strange quark's mass

In order to illustrate the non-perturbative quark mass renormalization just explained in a concrete numerical application, we first sketch our strategy for the computation of light quark masses [14]. Their ratios are known from chiral perturbation theory (χPT) [11] as

$$\frac{M_{\mathrm{u}}}{M_{\mathrm{d}}} = 0.55(4)\,, \quad \frac{M_{\mathrm{s}}}{\hat{M}} = 24.4(1.5) \qquad (8)$$

with $\hat{M} = \frac{1}{2}(M_{\mathrm{u}} + M_{\mathrm{d}})$ [12]. Nevertheless there are still questions, which might be answered decisively only using Lattice QCD. They concern the applicability of χPT in general, i.e. in how far the lowest orders dominate the full result, and the problem that the parameters in the chiral Lagrangian (at a given order in the expansion) can not be inferred with great precision from experimental data alone. This statement holds in particular for the overall scale of the quark masses, which is only defined once the connection with the fundamental theory, QCD, is made. Since the parameters in the chiral Lagrangian (the so-called low energy constants) are independent of the quark masses, it is important to realize that these problems can be dealt with by working with unphysical — of course not too large — quark masses, where it is essential or at least of significant advantage to explore a certain range of quark masses. While a determination of some low energy constants based on these ideas has been recently tested in [15], we focus in the following on the computation of the renormalization-group invariant mass of the strange quark by combining χPT with lattice techniques.

To this end, and in the spirit of the considerations before, we define a reference quark mass $M_{\rm ref}$ implicitly through

$$m_{\rm PS}^2(M_{\rm ref})r_0^2 = (m_{\rm K}r_0)^2 = 1.5736\,. \quad (9)$$

Here $m_{\rm PS}^2(M)$ is the pseudoscalar meson mass as a function of the quark mass for mass-degenerate quarks, and $r_0 = 0.5\,\rm fm$ and $\frac{1}{2}(m_{\rm K^+}^2 + m_{\rm K^0}^2)\big|_{\rm pure\ QCD} = (495\,\rm MeV)^2$ enter the r.h.s. of eq. (9). χPT in full QCD relates $M_{\rm ref}$ to the other light quark masses viz.

$$2M_{\rm ref} \simeq M_{\rm s} + \hat{M}\,, \quad (10)$$

which has been substantiated also numerically in the case of quenched QCD [14]. The remaining task is now to calculate $M_{\rm ref}$ from Lattice QCD.

As the foregoing discussion holds true in mass independent renormalization schemes too, one arrives by virtue of the PCAC relation applied to the vacuum-to-pseudoscalar matrix elements at the central relation

$$2r_0M_{\rm ref} = Z_{\rm M}\frac{R\,|_{m_{\rm PS}^2r_0^2=1.5736}}{r_0}\,1.5736$$
$$R \equiv \frac{F_{\rm PS}}{G_{\rm PS}}\,, \quad (11)$$

where $Z_{\rm M}$ is the flavour independent renormalization factor of the previous section, which directly leads to the RGI quark masses, being pure numbers and not depending on the scheme. By means of numerical simulations of the SF in large volumes of size $(1.5\,\rm fm)^3 \times 3\,\rm fm$, the ratio R/a and the meson mass $m_{\rm PS}a$ can be computed accurately as a function of the bare quark mass and the bare coupling by evaluating suitable correlation functions [13,14]. With the values for the scale r_0/a from [16], a mild extrapolation yields R/a at the point $m_{\rm PS}^2r_0^2 = 1.5736$. Then the quantity $2r_0M_{\rm ref}$ is extrapolated to the continuum limit. Both fits are shown in Fig. 2. In view of the still significant slope in the latter, we discard the point furthest away from the continuum in this extrapolation as a safeguard against higher order lattice spacing effects. Moreover, the analysis was repeated for $M_{\rm ref}$ in units of the kaon decay constant, which amounts to substitute eq. (11) by

$$\frac{2M_{\rm ref}}{(F_{\rm K})_{\rm R}} = \frac{M}{\overline{m}}\,\frac{1}{Z_{\rm P}\,r_0^2\,G_{\rm PS}}\,1.5736\,. \quad (12)$$

Here we observe a weaker lattice spacing dependence. The final results of these analyses

$$2r_0M_{\rm ref} = 0.36(1)$$
$$\stackrel{r_0=0.5\,\rm fm}{\longrightarrow}\; 2M_{\rm ref} = 143(5)\,\rm MeV$$
$$\frac{2M_{\rm ref}}{(F_{\rm K})_{\rm R}} = 0.87(3)$$
$$\stackrel{(F_{\rm K})_{\rm R}=160\,\rm MeV}{\longrightarrow}\; 2M_{\rm ref} = 140(5)\,\rm MeV$$

are completely consistent with each other. But, as also pointed out in that reference, the assignment of physical units is intrinsically ambiguous in the quenched approximation. Consulting e.g. the recent results of the CP-PACS Collaboration [17], roughly 10 % larger numbers would be obtained, if the scale r_0 were replaced by one of the masses of the stable light hadrons. $\overline{\rm MS}$ masses for finite renormalization scales μ can be obtained through perturbative conversion factors known up to 4–loop precision. A typical result is

$$\overline{m}_{\rm s}^{\overline{\rm MS}}(2\,\rm GeV) = 97(4)\,\rm MeV\,, \quad (13)$$

where the uncertainty in $\Lambda^{(0)}_{\overline{\rm MS}}$, eq. (4), entering the relation of the running quark masses in the $\overline{\rm MS}$ scheme to the RGI masses, eq. (5), and the quark mass ratios from full QCD chiral perturbation theory, eq. (8), were taken into account [14].

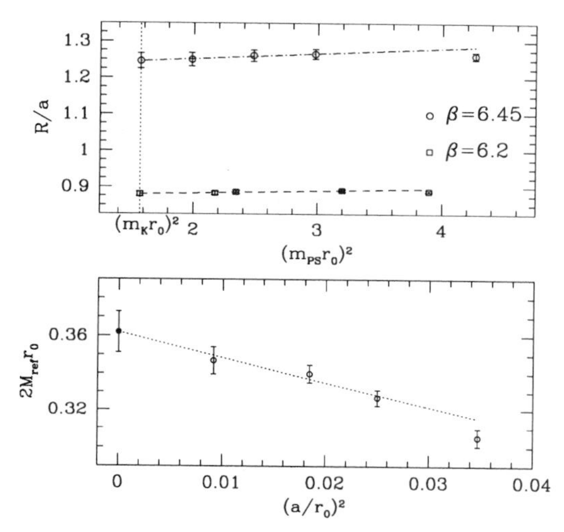

Figure 2. Extrapolations of the ratio R to the kaon mass scale and, in units of r_0, to the continuum limit.

A compilation of lattice results on the strange quark mass in the quenched approximation can be found e.g. in [18]. Most of these differ in the Ward identity used and in whether non-perturbative renormalization and a continuum extrapolation has been performed or not; also systematic errors often are not estimated uniformly either. Our result (13) includes all errors except quenching. Finally it is interesting to note that, as reported by the CP-PACS Collaboration in their comprehensive study about simulations with two dynamical flavours [19], dynamical quark effect appear to decrease the estimates for the strange quark mass by $\sim 20\,\%$ or less.

5 The static-light axial current

Let us turn to another example, where a scale and scheme dependent renormalization is encountered, i.e. the matrix element $\langle 0|(A_{\rm R})_\mu|{\rm B}(p)\rangle = ip_\mu F_{\rm B}$ describing leptonic B–decays in the theory with heavy quarks. It involves the renormalized axial current, $(A_{\rm R})_\mu = Z_{\rm A}\bar{b}\gamma_\mu\gamma_5 d$, and the decay constant $F_{\rm B}$, which is by its own an interesting quantity for a first principles computation on the lattice. Since $m_{\rm b} \simeq 4\,{\rm GeV} \gg \Lambda_{\rm QCD}$ implies large discretization errors of ${\rm O}\left((am_{\rm b})^2\right)$, a direct treatment assuming a relativistic b quark is difficult on the lattice. Therefore, in the first place one may restrict to an effective theory, one possibility being the static approximation, where the b quark is taken to be infinitely heavy.

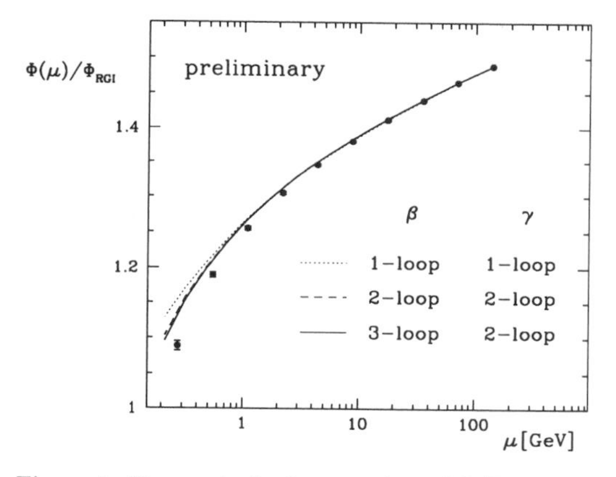

Figure 3. Non-perturbative running of $\Phi/\Phi_{\rm RGI}$ with the energy scale in the static approximation, computed in the SF scheme, compared to perturbation theory based on several combinations of orders, to which the β– and γ–functions have been evaluated.

As at the end we want to relate the physical matrix element Φ at a scale $\mu = m_{\rm b}$,

$$F_{\rm B}\sqrt{m_{\rm B}} \equiv \Phi(\mu) + {\rm O}\left(\frac{\Lambda_{\rm QCD}}{m_{\rm b}}\right), \qquad (14)$$

to the one determined on the lattice at some matching scale μ_0, a crucial ingredient is its (scale and scheme independent) renormalization group invariant counterpart

$$\Phi_{\rm RGI} \equiv \lim_{\mu\to\infty}\left\{(2b_0\bar{g}^2(\mu))^{-\gamma_0/2b_0}\,\Phi(\mu)\right\}$$
$$b_0 = 11/(4\pi)^2\,,\ \gamma_0 = -1/4\pi^2 \qquad (15)$$

to be passed into the factorization

$$\Phi(\mu) = \frac{\Phi(\mu)}{\Phi_{\rm RGI}}\,\frac{\Phi_{\rm RGI}}{\Phi_{\rm SF}(\mu_0)}\,\Phi_{\rm SF}(\mu_0)\,. \qquad (16)$$

As already anticipated in the notation, the further strategy is basically analogous to that explained when considering the coupling and the quark masses: we again adopt the SF framework and invoke an appropriate step scaling function, while everything is meant in the static approximation now.

The definition of the renormalized static axial current and the step scaling function, together with the 2–loop anomalous dimension, has recently been worked out perturbatively [20]. The preliminary status of the outcome of the corresponding non-perturbative investigation by numerical simulations of the SF in the quenched approximation is depicted in Fig. 3. The leftmost factor in eq. (16) is then supposed to be inferable in that region, where perturbation theory is feasible again.

6 Conclusions

Numerical simulations on the lattice can be applied to renormalization problems in QCD. In particular, the Schrödinger functional scheme offers a clean and flexible approach to deal with the accompanying scale differences. As a consequence of good control over statistical, discretization and systematic errors, non-perturbative coupling and quark mass renormalization can be performed with confidence, and solid results for $\Lambda^{(0)}_{\overline{\rm MS}}$ and $\overline{m}_{\rm s}^{\overline{\rm MS}}$ with high precision of the order of a few % were reached in the quenched approximation. Similar ideas are now carried over to the heavy quark sector of QCD, where first steps towards a computation of renormalization group invariant matrix elements in the static approximation are under way.

The presented concepts will be valuable also for full QCD. Despite more powerful (super-)computers continuously being developed, a quantitative understanding of dynamical sea quark effects is a great challenge which, albeit in sight, still demands for much effort on the theoretical as well as on the technical/implementational side of Lattice QCD.

References

1. R. Kenway, these proceedings.
2. R. Sommer, *Schladming Lectures 1997*, hep-ph/9711243, and references therein.
3. M. Lüscher, *Les Houches Lectures 1997*, hep-ph/9802029, and references therein.
4. K. Jansen, these proceedings.
5. M. Campostrini et al. (ed), *Proc. of Int. Symp. on Lattice Field Theory 1999, Nucl. Phys. (Proc. Suppl.)* B **83-84**.
6. R. Sommer, *Nucl. Phys. (Proc. Suppl.)* A **60**, 279 (1998), hep-lat/9705026.
7. G. Martinelli et al., *Nucl. Phys.* B **445**, 81 (1995), hep-lat/9411010.
8. M. Lüscher et al., *Nucl. Phys.* B **384**, 168 (1992), hep-lat/9207009.
9. S. Sint, *Nucl. Phys.* B **421**, 135 (1994), hep-lat/9312079.
10. S. Capitani, M. Lüscher, R. Sommer and H. Wittig, *Nucl. Phys.* B **544**, 669 (1999), hep-lat/9810063.
11. J. Gasser and H. Leutwyler, *Phys. Rept.* **87**, 77 (1982).
12. H. Leutwyler, *Phys. Lett.* B **378**, 313 (1996), hep-ph/9602366.
13. M. Guagnelli, J. Heitger, R. Sommer and H. Wittig, *Nucl. Phys.* B **560**, 465 (1999), hep-lat/9903040.
14. J. Garden, J. Heitger, R. Sommer and H. Wittig, *Nucl. Phys.* B **571**, 237 (2000), hep-lat/9906013.
15. J. Heitger, R. Sommer and H. Wittig, hep-lat/0006026.
16. M. Guagnelli, R. Sommer and H. Wittig, *Nucl. Phys.* B **535**, 389 (1998), hep-lat/9806005.
17. S. Aoki et al., *Phys. Rev. Lett.* **84**, 238 (2000), hep-lat/9904012.
18. H. Wittig, *at Int. Euro. Conf. on High-Energy Physics 1999*, hep-ph/9911400.
19. A. Ali Khan et al., hep-lat/0004010; K. Kanaya, these proceedings.
20. M. Kurth and R. Sommer, hep-lat/0007002.
21. J. Heitger, M. Kurth and R. Sommer, in preparation.

DYNAMICAL QUARK EFFECTS IN QCD ON THE LATTICE — RESULTS FROM THE CP-PACS —

KAZUYUKI KANAYA FOR THE CP-PACS COLLABORATION
Institute of Physics, University of Tsukuba, Tsukuba, Ibaraki 305-8571, Japan
E-mail: kanaya@rccp.tsukuba.ac.jp

Results of a systematic lattice QCD simulation with two degenerate flavors of sea quarks, identified as dynamical u and d quarks, are presented. The simulation was performed on a dedicated parallel computer, called CP-PACS, developed at the University of Tsukuba. Clear dynamical quark effects are observed in the light hadron mass spectrum and in the light quark masses: In the light hadron mass spectrum, major parts of the discrepancy between quenched QCD and experiment are shown to be removed by introducing two flavors of dynamical quarks. For the averaged mass of u and d quarks, we find $m_{ud}^{\overline{\rm MS}}(2{\rm GeV}) = 3.44^{+0.14}_{-0.22}$ MeV using the π and ρ meson masses as physical input, and for the s quark mass, we obtain $m_s^{\overline{\rm MS}}(2{\rm GeV}) = 88^{+4}_{-6}$ MeV or 90^{+5}_{-11} MeV with the K or ϕ meson mass as additional input. These values are about 20–30% smaller than the previous estimates in the quenched approximation. We also discuss the U(1) problem and B meson decay constants.

1 Introduction

CP-PACS is a dedicated parallel computer designed and developed at the University of Tsukuba for simulations in the physics of fields[1]. With 2048 node processors interconnected with a three-dimensional hypercrossbar network, the CP-PACS achieves a peak performance of 614.4 GFLOPS. Since 1996, intensive calculations of lattice QCD have been performed on the CP-PACS. Among others, the first systematic study including both chiral and continuum extrapolations was attempted for lattice QCD with two flavors of dynamical quarks. In this paper, we report on the results of these studies, focusing on the topics of dynamical quark effects in QCD.

We study lattice QCD[2] formulated on a 4-dimensional hyper-cubic lattice with a finite lattice spacing a. Continuum physics is defined in the limit of large lattice volume and vanishing lattice spacing. Therefore, in order to extract predictions for the real world from the simulations on finite lattices, we have to extrapolate data obtained on a sufficiently large lattice to vanishing lattice spacing (*the continuum extrapolation*). Furthermore, because the contribution of quarks in the calculation is quite computer-time intensive as we decrease the quark mass, with the current computers and current algorithms, we also have to extrapolate to the physical point of light u and d quarks using data at around the s quark mass region (*the chiral extrapolation*). It is important to have good control of the systematic errors due to both these extrapolations.

Because of the huge computational power required, majority of calculations have been made in the quenched approximation, in which the effects of dynamical quark loops are ignored. As the first project on the CP-PACS, we made an extensive simulation of quenched QCD[3]. The quality of extrapolations and therefore the precision of the final hadron spectrum were significantly improved over previous studies. From this study, the existence of systematic errors due to the quenched approximation was clearly demonstrated in the continuum limit.

Therefore, as the next logical step, we then performed a series of "full QCD" simulations, in which the effects of dynamical quarks are taken into account, on the CP-PACS[4,5]. After chiral and continuum extrapolations, clear dynamical quark effects are observed in the light hadron mass spectrum and

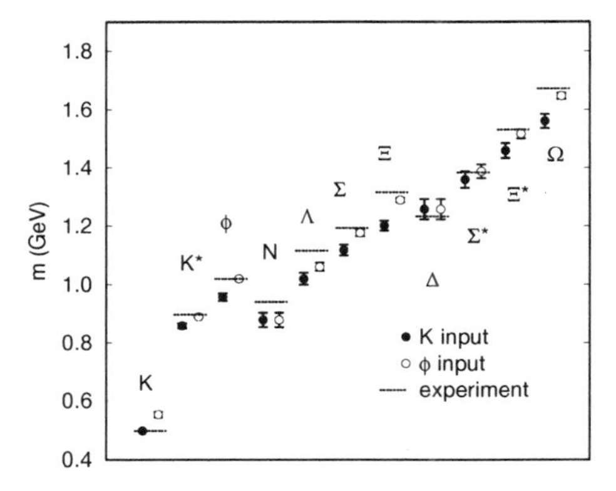

Figure 1. Quenched light hadron spectrum for ground state mesons and baryons in octet and decuplet representations of flavor SU(3).

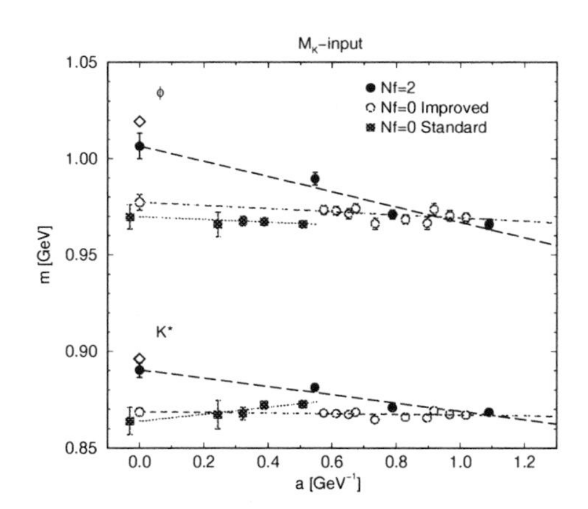

Figure 2. Continuum extrapolation of vector meson masses m_ϕ and m_{K^*} in $N_f = 2$ and $N_f = 0$ (quenched) QCD, using the K meson mass as input.

in the light quark masses. We also found noticeable effects in B meson decay constants.

In Sec. 2, we summarize the results for the light hadron spectrum from these studies. In Sec. 3, light quark masses are discussed. Sections 4 and 5 are devoted to the U(1) problem and B meson decay constants, respectively. Conclusions are given in Sec. 6.

2 Light hadron spectrum

The precise computation of the hadronic mass spectrum, directly from the first principles of QCD, is one of the main goals of lattice QCD. This provide us with a direct and non-perturbative test of the validity of QCD as the fundamental theory for strong interactions.

In Fig. 1, the latest results for the light hadron spectrum in the quenched approximation of QCD are summarized[3]. Simulations are made on four lattices $32^3 \times 56$ to $64^3 \times 112$ with lattice spacings in the range $a \approx$ 0.1–0.05 fm. The spacial lattice size was fixed to be about 3 fm, with which the finite size effects are estimated to be maximally 0.5% in the spectrum. The u and d quarks were treated as degenerate. On each lattice, five quark masses, corresponding to the pseudoscalar-to-vector mass ratio $m_{\rm PS}/m_{\rm V} \approx$ 0.75–0.4, were studied. The u, d quark mass m_{ud} and the lattice spacing a were fixed using the experimental values for m_π and m_ρ as inputs, while the s quark mass was fixed either by m_K (K-input) or m_ϕ (ϕ-input). Errors in Fig. 1 include statistical as well as systematic errors from chiral and continuum extrapolations, but do not include the errors from the quenched approximation.

From Fig. 1, we see that, although the global pattern of the experimental spectrum is correctly reproduced, there remain systematic discrepancies of up to about 10% (7 standard deviations). The resulting spectrum is different depending on the choice of input for s quark mass; the K-input or the ϕ-input. These discrepancies and ambiguities are due to the quenched approximation.

Because this limitation of the quenched approximation was made clear, the next logical step is to perform a "full QCD" calculation removing the quenched approximation. As the first step towards the realistic QCD, we performed a series of QCD simulations with two degenerate flavors of sea quarks, identified as dynamical u and d quarks, while the s quark is treated in the quenched approximation ($N_f = 2$ QCD)[4,5].

A key ingredient in avoiding a rapid increase of the computer time is the improvement of the lattice theory, with which lattice artifacts are reduced on computationally less intensive coarse lattices. We adopted the combination of an RG-improved gauge action and a "clover"-type improved Wilson quark action, and carried out the first systematic investigation of full QCD to perform both continuum and chiral extrapolations. Our preparatory full QCD study[6] shows that this action leads already to small lattice artifacts at $a \sim 0.2$ fm. Therefore, we have chosen the simulation parameters as summarized in Table 1. The spacial lattice size was fixed to be about 2.5 fm for all lattices.

Recently, we have doubled the statistics on the finest lattice at $a \approx 0.1$fm. All results, except for the B meson decay constatnts, presented in this paper are based on this full statistics.

Figure 2 shows the lattice spacing dependence of K^* and ϕ meson masses from $N_f = 2$ QCD, compared with the results of quenched calculations. For the quenched masses, two different data sets are shown: Those denoted as "$N_f = 0$ Standard" are the results of the quenched simulation, mentioned before, using the standard lattice action[3]. Because the action used in the full QCD calculation is different from the original quenched calculation, we carried out an additional quenched simulation using the same improved action as for the full QCD runs. The results are denoted as "$N_f = 0$ Improved" in the figure.

Our data for hadron spectrum confirms the expectation that both quenched calcu-

Table 1. Simulation parameters for $N_f = 2$ QCD on the CP-PACS. L_s is the spacial size of the lattice. On each lattice, four sea quark masses in the range $m_{\rm PS}/m_{\rm V} \approx 0.8$–0.6 were simulated. For each sea quark mass, we studied hadrons using five valence quarks in the range $m_{\rm PS}/m_{\rm V} \approx 0.8$–0.5.

lattice	a (fm)	L_s (fm)	$N_{\rm trajectory}$
$12^3 \times 24$	0.215(2)	2.58(3)	5000–7000
$16^3 \times 32$	0.153(2)	2.48(3)	5000–7000
$24^3 \times 48$	0.108(2)	2.58(3)	4000

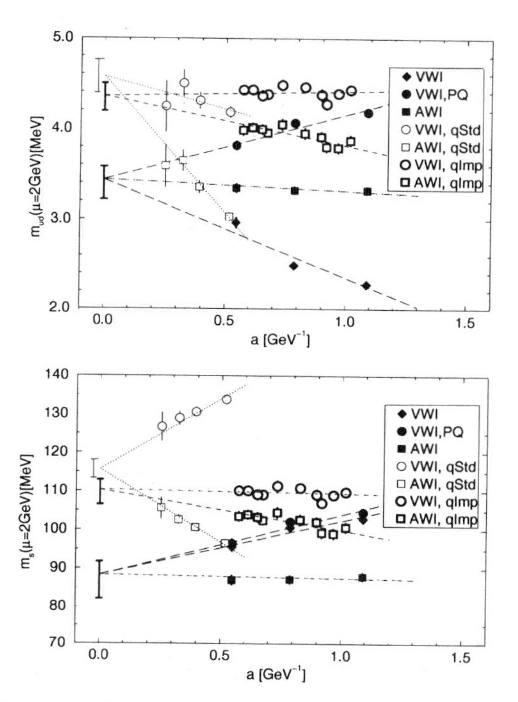

Figure 3. Continuum extrapolation of the average u and d quark mass m_{ud} and the s quark mass m_s in the $\overline{\rm MS}$ scheme at 2 GeV. m_s is from the K-input. Filled symbols are for $N_f = 2$ QCD. Quenched results with the standard action (qStd) and the improved action (qImp) are shown with thin and thick open symbols, respectively.

lations must lead to universal values in the continuum limit. The quenched results, however, show discrepancies from the experimental values, as discussed before. On the other hand, when we introduce two flavors of dynamical quarks, the discrepancies are much reduced. This means also that the ambiguities from the choice of input for s quark mass are much reduced in $N_f = 2$ QCD. The remaining small difference might be caused by the quenching of the s quark.

3 Light quark masses

Although quark massses are the most fundamental parameters of QCD, due to the confinement, it is impossible to measure them directly by an experiment. They have to be

Table 2. Light quark masses in the $\overline{\rm MS}$ scheme at 2 GeV.

	m_{ud} (MeV)	m_s (MeV) (K-input)	m_s (MeV) (ϕ-input)
$N_f = 0$ standard	4.57±0.18	116±3	144±6
$N_f = 0$ improved	$4.36^{+0.14}_{-0.17}$	110^{+3}_{-4}	132^{+4}_{-6}
$N_f = 2$	$3.44^{+0.14}_{-0.22}$	88^{+4}_{-6}	90^{+5}_{-11}

indirectly inferred from hadronic observables using a non-perturbative theoretical relation between these hadronic quantities and QCD parameters. A lattice QCD determination of the hadron spectrum provides us with such a theoretical relation directly from the first principles of QCD.

Fig. 3 summarizes the lattice spacing dependence of the average u and d quark mass m_{ud} and the s quark mass m_s, in $N_f = 2$ full QCD and in quenched QCD[5]. On the lattice, there exist several alternative definitions for the quark mass. In the figures, they are denoted as VWI (vector Ward identity quark masses), AWI (axial-vector Ward identity quark masses), etc. See [5] and [3] for details. While different definitions of quark masses lead to results that differ at finite lattice spacing, they should converge to a universal value in the continuum limit. Results in Fig. 3 clearly demonstrate that this is actually the case.

Values for the light quark masses in the continuum limit are summarized in Table 2. Errors include our estimates for systematic errors from chiral and continuum extrapolations and renormalization factors. First, we note that the two quenched calculations lead to universal values, as in the case of the light hadron spectrum. However, the quenched value for m_s differs by about 20% between K-input and ϕ-input. We find that this discrepancy between the inputs disappears within an error of 10% by the inclusion of two flavors of sea quarks.

The most interesting point is that the values predicted through $N_f = 2$ QCD are 20–30% smaller than those in the quenched QCD. In particular, our s quark mass in

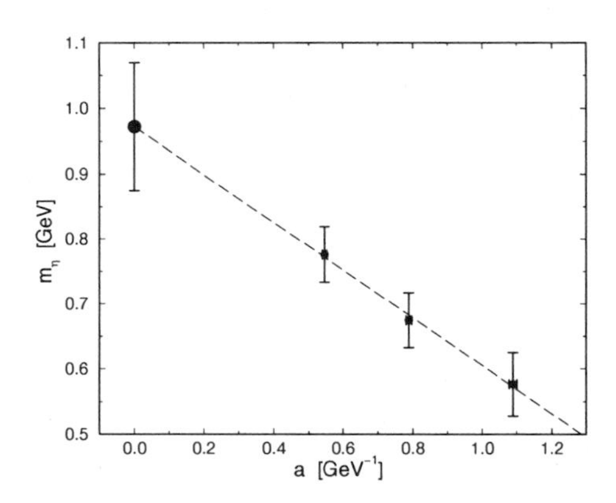

Figure 4. Continuum extrapolation of the flavor-singlet $u\bar{u} + d\bar{d}$ meson mass.

$N_f = 2$ QCD is about 90 MeV, which is significantly smaller than the value ≈ 150 MeV often used in hadron phenomenology, and almost saturating an estimate of the lower bound from QCD sum rules using the positivity of spectral functions[7]. On the other hand, our result for the u, d to s quark mass ratio, $m_s^{\overline{\rm MS}}/m_{ud}^{\overline{\rm MS}} = 26 \pm 2$, is consistent with 24.4 ± 1.5 from one loop chiral perturbation theory[8].

4 U(1) problem

The clarification of the mechanism for a large η' meson mass is an important issue in QCD. Propagators of a flavor non-singlet meson consist of a loop of a valence quark propagator, while propagators of the flavor singlet η' meson have an additional contribution with two disconnected valence quark loops. The fact that the η' is much heavier than the corresponding non-singlet meson π means that the two-loop contribution should exactly cancel the π pole of the one-loop contribution, leaving the heavy η' pole. This phenomenon is considered to be related with the anomalous violation of the flavor singlet axial U(1) symmetry and with the topological structure of gauge field configurations.

The calculation of the two-loop contribution requires a large amount of computations on the lattice. For this reason only limited results are available. In an approximation ignoring the mixing with the $s\bar{s}$ state, we studied one and two-loop contributions, and performed, for the first time, both chiral and continuum extrapolations[4]. We obtain $m_{u\bar{u}+d\bar{d}} = 972 \pm 97$ MeV for the flavor-singlet $u\bar{u}+d\bar{d}$ meson. See Fig. 4. In the real world, the $u\bar{u}+d\bar{d}$ state mixes with the $s\bar{s}$ state to lead to $\eta(547)$ and $\eta'(958)$ mesons. We are extending the study to inspect the mixing with the $s\bar{s}$ state and the relation to the topological structures.

5 B mesons on the lattice

The decay constant for the B_q meson is defined by $\langle 0|\bar{b}\gamma_\mu\gamma_5 q|B_q(p)\rangle = if_{B_q}p_\mu$ where q denotes either d or s quark. The non-perturbative determination of f_{B_q}, and also the bag parameters B_{B_q}, is quite important for a precise determination of CKM matrix elements. Therefore, intensive lattice calculations have been made[9].

On the lattice, however, the simulation of the heavy b quark is not a trivial extension of light quark simulations, because $m_b \sim 4$ GeV is larger than the lattice cutoff $\sim$ 1–4 GeV to date. Two methods have been developed to simulate heavy quarks on the lattice. One is based on a non-relativistic effective theory of QCD (NRQCD) defined through an expansion in the inverse heavy quark mass[10]. Another employs a relativistic action and reinterprets it in terms of a non-relativistic Hamiltonian (Fermilab method)[11]. Because the both methods include an effective treatment of heavy quarks, the consistency of the results among them should be checked.

Majority of the lattice studies are done in the quenched approximation. On the other hand, a chiral perturbation theory[12] suggests sizable corrections from dynamical quarks in the values of f_{B_q}. The first full QCD calculations of f_{B_q} were made by the MILC Collaboration[13] using the Fermilab method, and by Collins et al.[14] using the NRQCD method. In these studies, configurations were generated using the staggered sea quarks, which is different from the valence light quark (the Wilson quark[13] or the clover quark[14]).

Using the CP-PACS computer, we studied heavy meson decay constants applying a consistent formulation for sea and valence light quarks, the clover quark, and applied both the NRQCD method and the Fermilab method[15]. Our best estimates of heavy meson decay constants for $N_f = 2$ and $N_f = 0$ (with improved action) are summarized in Table 3. Because the Fermilab method is applicable also for the c quark, we also computed f_{D_d} and f_{D_s} with this method. The fact that our f_{D_s} for $N_f = 2$ is consistent with the recent experimental results, 285±20±40 MeV (ALEPH[16]) and 280±19±44 MeV (CLEO[17]), is quite encouraging.

From the table, we see that Fermilab and NRQCD methods are consistent with each other. We also note that $N_f = 2$ results for B mesons are about 10–15% larger than

Table 3. Heavy meson decay constants in MeV. The lattice scale was fixed by the ρ meson mass. Two errors are statistical and systematic. For B_s and D_s, the s quark mass is fixed from the K-input; the difference between K and ϕ-input is found to be smaller than the systematic error.

		$N_f=0$	$N_f=2$
f_{B_d}	Fermilab	188±3±9	208±10±11
	NRQCD	191±5±11	205±8±15
f_{B_s}	Fermilab	220±2±15	250±10±13
	NRQCD	220±5±13	242±8±17
f_{D_d}	Fermilab	218±2±15	225±14±14
f_{D_s}	Fermilab	250±1±18	267±13±17

the quenched values, while f_{D_d} and f_{D_s} are less sensitive to N_f. Increase of the B meson decay constants affects the determination of several CKM matrix elements through the $B_q - \overline{B}_q$ mass difference ΔM_q. Our results for f_{B_d} and f_{B_s} are consistent with the hypothesis that the Wolfenstein parameter ρ is positive.

6 Conclusions

We performed the first systematic study of lattice QCD with two flavors of dynamical quarks. We found that dynamical quark effects are quite important in the hadron physics. The effect is as large as 20–30% in the values of light quark mass and about 10–15% in B meson decay constants. Both of the shifts has significant implications to phenomenological studies of the standard model. It is urgent to evaluate dynamical quark effects in other hadronic quantities, such as the bag parameters B_{B_q}. It is also important to study the effects of dynamical s quark. Further intensive studies on the lattice are under way to clarify the precise structure of the standard model.

Acknowledgments

The studies presented here were performed by the CP-PACS Collaboration. I thank other members of the Collaboration; A. Ali Khan, S. Aoki, Y. Aoki, G. Boyd, R. Burkhalter, S. Ejiri, M. Fukugita, S. Hashimoto, N. Ishizuka, Y. Iwasaki, T. Izubuchi, T. Kaneko, Y. Kuramashi, T. Manke, K. Nagai, J. Noaki, M. Okamoto, M. Okawa, H.P. Shanahan, Y. Taniguchi, A. Ukawa, and T. Yoshié, for discussions. This paper is in part supported by the Grants-in-Aid of Ministry of Education, Science and Culture (No. 10640248) and JSPS Research for Future Program.

References

1. Y. Iwasaki, Nucl. Phys. B (Proc. Suppl.) 60A (1998) 246; T. Boku *et al.*, in Proc. Supercomputing '97 (1997) 108. For further details, see `http://www.rccp.tsukuba.ac.jp`.
2. R.D. Kenway, these proceedings.
3. S. Aoki *et al.* (CP-PACS Collaboration), Phys. Rev. Lett. 84 (2000) 238.
4. R. Burkhalter for the CP-PACS Collaboration, Nucl. Phys. B (Proc. Suppl.) 73 (1999) 3; S. Aoki *et al.* (CP-PACS Collaboration), *ibid.* 192; 216.
5. A. Ali Khan *et al.* (CP-PACS Collaboration), hep-lat/0004010, to appear in Phys. Rev. Lett.
6. S. Aoki *et al.* (CP-PACS Collaboration), Phys. Rev. D60 (1999) 114508.
7. S. Narison, Nucl. Phys. B (Proc. Suppl.) 86 (2000) 242.
8. H. Leutwyler, Phys. Lett. B378 (1996) 313.
9. A. Kronfeld, these proceedings; S. Hashimoto, Nucl. Phys. B (Proc. Suppl.) 83-84 (2000) 3.
10. B.A. Thacker and G.P. Lepage, Phys. Rev. D43 (1991) 196.
11. A.X. El-Khadra, A.S. Kronfeld and P.B. Mackenzie, Phys. Rev. D55 (1997) 3933.
12. M.J. Booth, Phys. Rev. D51 (1995) 2338; S.R. Sharpe and Y. Zhang, *ibid.* D53 (1996) 5125.
13. C. Bernard *et al.* (MILC Collaboration), Nucl. Phys. B (Proc. Suppl.) 83-84 (2000) 289.
14. S. Collins *et al.*, Phys. Rev. D60 (1999) 074504.
15. A. Ali Khan *et al.* (CP-PACS Collaboration), in preparation. Preliminary results are given in Nucl. Phys. B (Proc. Suppl.) 83-84 (2000) 265; 331.
16. ALEPH Collaboration, ALEPH 2000-062, contributed paper for ICHEP 2000.
17. CLEO Collaboration (M. Chadha *et al.*), Phys. Rev. D58 (1988) 32002.

WEAK MATRIX ELEMENTS AND K-MESON PHYSICS

MASSIMO TESTA

Dipartimento di Fisica, Universita' di Roma "La Sapienza",
P.le A.Moro 2, 00185 Roma,Italy
INFN-Sezione di Roma, Italy
E-mail: massimo.testa@roma1.infn.it

An overview is presented about old and recent methods to compute the $K \to \pi\pi$ decay amplitude.

1 Introduction

Kaon Physics is a very complicated blend of Ultraviolet and Infrared effects which still defies complete physical understanding.

The problem consists in the large enhancement (≈ 20) of the $\Delta I = \frac{1}{2}$ amplitude with respect to the $\Delta I = \frac{3}{2}$ one.

Being a process involving hadrons, K-decay must be treated non-perturbatively, so that lattice discretization is the ideal tool to deal with this problem. In fact lattice regularization is the *only* convergent (as $a \to 0$) approximation scheme to QCD.

Due to the difficulties of putting the Standard Model on the lattice, one can use weak interaction perturbation theory, which, together with Asymptotic Freedom of Strong Interactions, allows the definition of an effective low energy actions for non-leptonic decays:

$$\mathcal{H}^{eff}_{\Delta S=1} = \qquad (1)$$
$$= \lambda_u \frac{G_F}{\sqrt{2}} \left[C_+(\mu) O^{(+)}(\mu) + C_-(\mu) O^{(-)}(\mu) \right]$$

where $\lambda_u = V_{ud} V^*_{us}$ and:

$$O^{(\pm)} = \qquad (2)$$
$$= \left[(\bar{s}\gamma^L_\mu d)(\bar{u}\gamma^L_\mu u) \pm (\bar{s}\gamma^L_\mu u)(\bar{u}\gamma^L_\mu d) \right] - [u \leftrightarrow c]$$

$O^{(-)}$ is a pure $I = \frac{1}{2}$, while $O^{(+)}$ is a mixture of $I = \frac{1}{2}$ and $I = \frac{3}{2}$.

The $O^{(\pm)}$'s transform as $(8,1) \oplus (1,8)$ and $(27,1) \oplus (1,27)$ under the $SU(3) \otimes SU(3)$ chiral group and some discrete symmetries. The coefficients $C_\pm(\mu)$ reliably computed in Perturbation Theory, show a slight octet enhancement[1]:

$$\left| \frac{C_-(\mu \approx 2\ Gev)}{C_+(\mu \approx 2\ Gev)} \right| \approx 2 \qquad (3)$$

The rest of the enhancement (≈ 10) should, then, be provided by the matrix elements of $O^{(\pm)}$ and is a non perturbative, infrared effect.

The difficulty with Lattice regularization lies in the fact that naive discretization of Dirac fermions entails a multiplication of low energy degrees of freedom (Doublers) whose elimination complicates the scheme.

There are, essentially, two possibilities:

- Wilson Fermions[2]

  A term is added to the Lagrangian, breaking explicitly the chiral symmetry, which can be restored, as $a \to 0$, by the inclusion of appropriate counterterms. This formulation is ultra-local (at the lagrangian level only near neighbors interactions are involved) and it is very convenient for numerical purposes.

- Ginsparg-Wilson Fermions[3,4,5,6,7]

  This discretization is much more respectful of the chiral properties of the (continuum) QCD lagrangian, at the expense of being non local at the lattice level, which makes it, at the moment, numerically very demanding.

My remarks on renormalization will, therefore, be addressed to Wilson fermions formulation.

The difficulty of the problem consists, first of all, in giving the correct definition of the operators $O^{(\pm)}$.

In order to construct finite composite operator of dimension 6, $\tilde{O}_6(\mu)$, we must mix the original bare operator, $O_6(a)$, with bare operators of equal ($O_6^{(i)}(a)$) or smaller ($O_3(a)$) dimension, in general with different naive chiralities[8].

A general non perturbative technique to to construct composite operators is based on the systematic exploitation of Chiral Ward Identities [9].

It turns out that, in order to minimize the renormalization procedure, the best strategy is to compute $K \to \pi\pi$ in the world in which $m_K = 2m_\pi$ or $m_K = m_\pi$, with pions at rest (see section(2.1)) and then extrapolate to the real world through chiral perturbation theory[10,11,12]. In these cases the ultraviolet subtractions are limited to an overall renormalization which could be determined non perturbatively[13].

2 Infrared Problems

Approaches requiring the construction of an asymptotic two pion state face the problems due to the fact that the theory is defined, through the functional integral, in the euclidean region. In the next two subsections we will briefly discuss the nature of the problem and possible proposals to solve it.

2.1 *Infinite volume*

In order to compute the $K \to \pi\pi$ width we have to evaluate the matrix element ${}_{(out)}\langle \pi(\underline{p})\pi(-\underline{p})|\, \mathcal{H}_W\, |K\rangle$ with two interacting hadrons in the final state. This is not easy to do in the euclidean region[14]. It can be shown that:

$$\left\langle \varphi_{\underline{p}}(t_1)\varphi_{-\underline{p}}(t_2)\mathcal{H}_W(0)K(t_K)\right\rangle \underset{\substack{t_K\to-\infty\\ t_1>>t_2>>0}}{\approx}$$

$$\approx e^{m_K t_K - E_{\underline{p}}t_1 - E_p t_2}\sqrt{\frac{Z_K}{2m_K}\frac{Z_\pi}{2m_\pi}} \times$$

$$\left\{\left[\frac{{}_{(out)}\langle \underline{p}, -\underline{p}|\mathcal{H}_W|K\rangle + {}_{(in)}\langle \underline{p}, -\underline{p}|\mathcal{H}_W|K\rangle}{2}\right]\right.$$

$$+ P_{\underline{q}}(t_2) \qquad (4)$$

where

$$P_{\underline{q}}(t_2) = -\mathcal{P}\sum_n \exp[-(E_n - 2E_{\underline{q}})t_2](2\pi)^3\delta^3(\underline{P}_n)$$

$$\times N_n \frac{[\mathcal{M}(\underline{q}, -\underline{q}; n)]^* \langle n, out|\mathcal{H}_W(0)|K\rangle}{E_n(E_n - 2E_{\underline{q}})} \qquad (5)$$

In eq.(5), $\mathcal{M}^{(o.s.)}_{\underset{2m_\pi}{\pi\pi} \to \underset{2E_\pi}{\pi\pi}}$ denotes an off-shell extrapolation of the $\underset{2m_\pi}{\pi\pi} \to \underset{2E_\pi}{\pi\pi}$ scattering amplitude, which represents the euclidean version of the final state interaction.

The problem with eqs.(4),(5) is that, for large, positive t_2:

$$P_{\underline{q}}(t_2) \approx e^{2(E_{\underline{q}} - m_\pi)t_2} \qquad (6)$$

so that, asymptotically in t_2, $P_{\underline{q}}(t_2)$ dominates over the physically relevant matrix elements in eq.(4).

If we choose $\underline{p} = 0$ in eq.(4), we have:

$$\langle \pi_{\underline{0}}(t_1)\pi_{\underline{0}}(t_2)\mathcal{H}_W(0)K(t_K)\rangle \underset{\substack{t_K\to-\infty\\ t_1>>t_2>>0}}{\approx} \qquad (7)$$

$$\approx e^{m_K t_K - m_\pi t_1 - m_\pi t_2}\sqrt{\frac{Z_K}{2m_K}\frac{Z_\pi}{2m_\pi}}$$

$$\langle \pi(\underline{0})\pi(\underline{0})|\, \mathcal{H}_W\, |K\rangle\,(1 + \frac{c}{\sqrt{t_2}}\mathcal{M}^{(o.s.)}_{\underset{2m_\pi}{\pi\pi} \to \underset{2m_\pi}{\pi\pi}})$$

One sees from eq.(7) that for π's at rest and weakly interacting it is possible to extract a meaningful matrix element.

2.2 *Finite volume*

Lellouch and Lüscher have recently formulated a strategy based on the exploitation of the finiteness of volume in lattice simulations[15]. Their proposal is based on the following relation between finite and infinite volume matrix elements:

$$|\langle \pi\pi, E = m_K|\, \mathcal{H}_W(0)\, |K\rangle|^2 = \qquad (8)$$

$$= V^2\, |{}_V\langle \pi\pi, E|\, \mathcal{H}_W(0)\, |K\rangle_V|^2 \left(\frac{m_K}{k_\pi}\right)^3 \times$$

$$\times 8\pi[q\phi'(q) + k\delta_0'(k)]$$

In eq.(8) $|\pi\pi, E\rangle_V$ denotes a finite volume two pion state with zero total momentum and "angular momentum" and energy E, while $|K\rangle_V$ denotes a single finite volume kaon state with zero momentum. Both states are normalized to 1. $|\pi\pi, E\rangle$ and $|K\rangle$ denote the corresponding infinite volume states covariantly normalized according to the usual convention which, for single particle states reads as:

$$\langle \underline{p} \mid \underline{q}\rangle = (2\pi)^3 2\omega_{\underline{p}} \delta^{(3)}(\underline{p} - \underline{q}) \qquad (9)$$

In a finite volume the allowed values, k, of the 'radial' relative momentum of a zero total momentum s-wave two particle state obey the relation[16]:

$$n\pi - \delta_0(k) = \phi(q) \qquad (10)$$

where $\delta_0(k)$ is the s-wave phase-shift, $q \equiv \frac{kL}{2\pi}$, k is related to the center of mass energy, E as:

$$E = 2\sqrt{m_\pi^2 + k^2} \qquad (11)$$

and:

$$\tan\phi(q) = -\frac{\pi^{3/2} q}{Z_{00}(1; q^2)} \qquad (12)$$

$$Z_{00}(s; q^2) = \frac{1}{\sqrt{4\pi}} \sum_{\underline{n}\in Z^3} (\underline{n}^2 - q^2)^{-s} \qquad (13)$$

Eqs.(10)-(13) completely define the quantities appearing in eq.(8).

I will now present a different approach[17] to the relation between finite and infinite volume matrix elements, which may lead to a better understanding of the nature of eq.(8). The argument goes as follows.

In order to relate the states at finite volume with those at infinite volume we take the two-point Green function of a scalar operator $\sigma(x)$, $\int_V d^3x \, \langle \sigma(\underline{x}, t)\sigma(0)\rangle$, and consider its behavior as the space volume V becomes large. We have:

$$\int_V d^3x \, \langle \sigma(\underline{x}, t)\sigma(0)\rangle_V \underset{V\to\infty}{\longrightarrow} \qquad (14)$$

$$\frac{(2\pi)^3}{2(2\pi)^6} \int \frac{d\underline{p}_1}{2\omega_1} \frac{d\underline{p}_2}{2\omega_2} \delta(\underline{p}_1 + \underline{p}_2) e^{-(\omega_1+\omega_2)t} \times$$
$$\times \left| \langle 0 | \sigma(0) \left| \underline{p}_1, \underline{p}_2 \right\rangle \right|^2 =$$
$$= \frac{1}{2(2\pi)^3} \int dE e^{-Et} \left| \langle 0 | \sigma(0) \left| \pi\pi, E \right\rangle \right|^2 \times$$
$$\times \int \frac{d\underline{p}_1}{2\omega_1} \frac{d\underline{p}_2}{2\omega_2} \delta(\underline{p}_1 + \underline{p}_2) \delta(E - \omega_1 - \omega_2) =$$
$$= \frac{\pi}{2(2\pi)^3} \int \frac{dE}{E} e^{-Et} \left| \langle 0 | \sigma(0) \left| E \right\rangle \right|^2 k(E)$$

where:

$$k(E) = \sqrt{\frac{E^2}{4} - m_\pi^2} \qquad (15)$$

On the other hand:

$$\int_V d^3x \, \langle \sigma(\underline{x}, t)\sigma(0)\rangle = \qquad (16)$$
$$V \sum_n \left| \langle 0 | \sigma(0) \left| \pi\pi, n \right\rangle_V \right|^2 e^{-E_n t} \underset{V\to\infty}{\longrightarrow}$$
$$\underset{V\to\infty}{\longrightarrow} V \int_0^\infty dE \rho(E) \left| \langle 0 | \sigma(0) \left| \pi\pi, E \right\rangle_V \right|^2 e^{-Et}$$

where $|\pi\pi, n\rangle_V$ denote the finite volume two pion states classified according to the quantum number n defined in eq.(10) and:

$$\rho(E) \equiv \frac{\Delta n}{\Delta E} = \frac{q\phi'(q) + k\delta_0'(k)}{4\pi k^2} E \qquad (17)$$

denotes the density of states of energy E.

Comparing eqs.(14) and (16), we get the correspondence:

$$|\pi\pi, E\rangle \Leftrightarrow 4\pi \sqrt{\frac{VE\rho(E)}{k(E)}} \, |\pi\pi, E\rangle_V \qquad (18)$$

In a similar way it is easy to show:

$$|\underline{p} = 0\rangle \Leftrightarrow \sqrt{2mV} \, |\underline{p} = 0\rangle_V \qquad (19)$$

From eqs.(18) and (19) we get:

$$\left| \langle \pi\pi, E = m_K | \mathcal{H}_W(0) | K \rangle \right|^2 = \qquad (20)$$
$$= 32\pi^2 V^2 \frac{\rho(m_K) m_K^2}{k_\pi} \left| {}_V\langle \pi\pi, E | \mathcal{H}_W(0) | K \rangle_V \right|^2$$

where:

$$k_\pi \equiv \sqrt{\frac{m_K^2}{4} - m_\pi^2} \qquad (21)$$

Using the expression of $\rho(E)$ given by eq.(17), eq.(20) looks the same as eq.(8). There is, however an important difference. In fact the derivation[15] of eq.(8) requires to work at a fixed volume V and at a fixed value of n, defined in eq.(10)[a]. Eq.(20), on the contrary, is valid at fixed energy E, asymptotically in V, so that, while we let $V \to \infty$, we must allow simultaneously $n \to \infty$. A possible relation between the two approaches will be discussed in a forthcoming paper[17].

The strategy proposed by Lellouch and Lüscher[15] consists in tuning the volume V so that the first excited two-pion state ($n = 1$) is degenerate in energy with the kaon state ($L \approx 5 \div 6\ Fm$) and compute the finite volume Green´s function:

$$\int_V d^3x d^3y \left\langle \sigma(\underline{x},t)\mathcal{H}_W(0)K(\underline{y},t')\right\rangle_V \underset{t'\to-\infty}{\approx}$$
$$= e^{m_K t'}{}_V \langle K|\, K(0)\, |0\rangle\, V^2 \times$$
$$\times \sum_n \langle 0|\, \sigma(0)\, |\pi\pi, n\rangle_V \times$$
$$\times_V \langle \pi\pi, n|\, \mathcal{H}_W(0)\, |K\rangle_V\, e^{-E_n t} =$$
$$= e^{m_K t'}{}_V \langle K|\, K(0)\, |0\rangle\, V^2 \times$$
$$\times \sum_n |\langle 0|\, \sigma(0)\, |\pi\pi, n\rangle_V| \times$$
$$\times\, |_V \langle \pi\pi, n|\, \mathcal{H}_W(0)\, |K\rangle_V|\, e^{-E_n t} \qquad (22)$$

The last equality in eq.(22) is justified by the cancellation of the final state interactions phases in $\langle 0|\, \sigma(0)\, |\pi\pi, n\rangle_V$ and ${}_V \langle \pi\pi, n|\, \mathcal{H}_W(0)\, |K\rangle_V$.

Then, from

$$\int_V d^3x \left\langle \sigma(\underline{x},t)\sigma(0)\right\rangle = \qquad (23)$$
$$V \sum_n |\langle 0|\, \sigma(0)\, |\pi\pi, n\rangle_V|^2\, e^{-E_n t}$$

we compute $|\langle 0|\, \sigma(0)\, |\pi\pi, 1\rangle_V|$ and, finally, $|_V \langle \pi\pi, 1|\, \mathcal{H}_W(0)\, |K\rangle_V|$.

In the case of a $\Delta I = \frac{1}{2}$ transition we face a further complication[11,12]: independently of the chosen procedure, a subtraction has to be performed, due to the fact that the relevant correlator $\langle \sigma(t)\mathcal{H}_W(0)K(t')\rangle$ is dominated, for large t, by the vacuum insertion between $\sigma(t)$ and $\mathcal{H}_W(0)K(t')$. As a consequence, the relevant physical information about the K decay is contained in the connected correlator:

$$\langle \sigma(t)\mathcal{H}_W(0)K(t')\rangle_{conn} \equiv \qquad (24)$$
$$\equiv \langle \sigma(t)\mathcal{H}_W(0)K(t')\rangle - \langle \sigma(0)\rangle\, \langle \mathcal{H}_W(0)K(t')\rangle$$

[a] in fact $n < 8$

References

1. M.K. Gaillard, B. Lee, Phys. Rev. Lett. 33 (1974) 108;
   G. Altarelli, L. Maiani, Phys. Lett. 52B (1974) 351.
2. K. Wilson, *Phys. Rev.* D **14**, 2455 (1974); in *New phenomena in subnuclear physics*, ed. A. Zichichi (Plenum, New York, 1977)
3. P. H. Ginsparg, K. G. Wilson, *Phys. Rev.* D **25**, 2649 (1982).
4. R. Narayanan, H. Neuberger, *Phys. Rev. Lett.* **71**, 3251 (1993).
5. H. Neuberger, *Phys. Rev.* D **59**, 085006 (1999).
6. M. Lüscher, *Nucl. Phys.* B **549**, 295 (1999).
7. P. Hasenfratz, *Nucl. Phys.* B **525**, 409 (1998).
8. M. Bochicchio, L. Maiani, G. Martinelli, G.C. Rossi, M. Testa, Nucl. Phys. B262 (1985) 331.
9. L. Maiani, G. Martinelli, G.C. Rossi, M. Testa, Phys. Lett. 176B (1986) 445;
   L. Maiani, G. Martinelli, G.C. Rossi, M. Testa, Nucl. Phys. B289 (1987) 505.
10. C. Bernard, T. Draper, G. Hockney, A. Soni, Nucl. Phys. (Proc. Suppl.) 4 (1988) 483.
11. C. Dawson, G. Martinelli, G.C. Rossi, C.T. Sachrajda, S. Sharpe, M. Talevi, M. Testa, *Nucl. Phys.* B **514**, 313 (1998)

12. M. Testa, Nucl. Phys. B (Proc. Suppl.) 63 (1998) 877.
13. G. Martinelli, C. Pittori, C.T. Sachrajda, M. Testa, A. Vladikas, Nucl. Phys. B445 (1995) 81.
14. L. Maiani, M. Testa, Phys. Lett. B245 (1990) 585.
15. L. Lellouch, M. Lüscher, hep-lat/0003023.
16. M. Lüscher, *Nucl. Phys.* B **354**, 531 (1991)
17. D. Lin, G. Martinelli, C. Sachrajda, M. Testa, in preparation.

STRUCTURE FUNCTIONS ON THE LATTICE

K. JANSEN

CERN, TH Division, 1211 Geneva 23, Switzerland
E-mail: Karl.Jansen@cern.ch

We report on a lattice computation of the second moment of the pion matrix element of the twist-2 non-singlet operator corresponding to the average momentum of parton densities. We apply a fully non-perturbatively evaluated running renormalization constant as well as a careful extrapolation of our results to the continuum limit. Thus the only limitation of our final result is the quenched approximation.

1 Introduction

A reliable computation of parton distribution functions from first principles would be very important for future experiments planned e.g. at the LHC. The results of such calculations would offer a unique way to test whether QCD is indeed the correct theory for the strong interactions.

In principle, the lattice regularization offers such a calculational scheme, where the only starting point is the QCD lagrangian, and indeed is able to give results for moments of parton density distributions *in the continuum* and *fully non-perturbatively*, as we will show in this contribution. The work presented here is a summary of a series of publications [1,2,3,4,5] that provide the essential ingredients to reach the above ambitious aim.

The moments of parton density distributions are related to expectation values of certain local operators, which are renormalized multiplicatively by applying appropriate renormalization factors $Z(1/\mu)$ that depend on the energy scale μ. This leads to consider renormalized matrix elements $O_{\rm SF}^{\rm ren}(\mu)$ to be computed in a certain, at this stage not specified, renormalization scheme SF. If the energy scale μ is chosen large enough, it is to be expected that the scale evolution is very well described by perturbation theory, giving rise to the following definition of a *renormalization group invariant matrix element:*

$$O_{\rm INV}^{\rm ren} = O_{\rm SF}^{\rm ren}(\mu) \cdot f^{\rm SF}(\bar{g}^2(\mu)) \qquad (1)$$

with $\bar{g}(\mu)$ the running coupling and

$$f^{\rm SF}(\bar{g}^2) = (\bar{g}^2(\mu))^{-\gamma_0/2b_0} \cdot \exp\left\{ - \int_0^{\bar{g}(\mu)} dg \left[\frac{\gamma(g)}{\beta(g)} - \frac{\gamma_0}{b_0 g} \right] \right\} \qquad (2)$$

where $\beta(g)$ and $\gamma(g)$ are the β and anomalous-dimension functions computed to a given order of perturbation theory in the specified scheme, i.e. here the SF scheme. Once we know the value of $O_{\rm INV}^{\rm ren}$ evaluated non-perturbatively, the running matrix element in a preferred scheme can be computed, for example in the $\overline{MS}$ scheme:

$$O_{\overline{\rm MS}}^{\rm ren}(\mu) = O_{\rm INV}^{\rm ren} / f^{\overline{\rm MS}}(\bar{g}^2(\mu)) \qquad (3)$$

with now, of course, the β and γ functions computed in the $\overline{MS}$ scheme.

A non-perturbatively obtained value of the renormalization group invariant matrix element is hence of central importance. Its calculation has to be performed in several steps. The reason is that we have to cover a broad range of energy scales – from the deep perturbative to the non-perturbative region. With the scale-dependent renormalization factor $Z(1/\mu)$ we can write the renormalized matrix element of eq. (1) as

$$O_{\rm INV}^{\rm ren} = \frac{\langle \pi | \mathcal{O}_{\rm NS} | \pi \rangle}{Z^{\rm SF}(1/\mu)} \cdot f^{\rm SF}(\bar{g}^2(\mu)) \,, \qquad (4)$$

with $\langle \pi | \mathcal{O}_{\rm NS} | \pi \rangle$ the expectation value of the (non-singlet) operator under consideration in given states, here the pion states.

So far, all our discussions have been in the continuum. However, if we think of the lattice regularisation and eventual numerical simulations to obtain non-perturbative results, it would be convenient to compute the renormalized matrix element at only one convenient (i.e. small hadronic) scale μ_0. We therefore rewrite the r.h.s. of eq. (4) as:

$$\frac{\langle\pi|\mathcal{O}_{\rm NS}|\pi\rangle}{Z^{\rm SF}(1/\mu_0)}\cdot\underbrace{\frac{Z^{\rm SF}(1/\mu_0)}{Z^{\rm SF}(1/\mu)}}_{\equiv\sigma(\mu/\mu_0,\bar{g}(\mu))}\cdot f^{\rm SF}(\bar{g}^2(\mu)) \quad (5)$$

where we introduce the step scaling function $\sigma(\mu/\mu_0,\bar{g}(\mu))$, which describes the evolution of the renormalization factor from a scale μ_0 to a scale μ. The advantage of concentrating on the step scaling function instead of the renormalization factor itself, is that the step scaling function is well defined in the continuum and hence suitable for eventual continuum extrapolations of lattice results.

We finally write the r.h.s. of eq. (4) as

$$O_{\rm SF}^{\rm ren}(\mu_0)\underbrace{\sigma(\mu/\mu_0,\bar{g}(\mu))\cdot f^{\rm SF}(\bar{g}^2(\mu))}_{\equiv\mathfrak{S}_{\rm INV}^{\rm UV}(\mu_0)} \quad (6)$$

with $O_{\rm SF}^{\rm ren}(\mu_0)$ the renormalized matrix element, which is to be computed only once at a scale μ_0 and the (ultraviolet) invariant step scaling function $\mathfrak{S}_{\rm INV}^{\rm UV}(\mu_0)$, which still depends on the infrared scale μ_0. The following sections are devoted to a description of how these two basic ingredients can be reliably computed on the lattice using non-perturbative methods, i.e. numerical simulations.

2 The renormalization group invariant step scaling function

Let us start this section by disclosing what is hidden behind the fictitious SF scheme mentioned in the introduction. SF stands for Schrödinger functional and denotes a finite physical volume, $V = L^3\cdot T$, renormalization scheme where the energy scale μ is identified with the inverse spatial length of the box itself, e.g. $\mu = 1/L$. The peculiarity of the Schrödinger functional set-up is that fixed boundary conditions in time x_0 are imposed with classical fields at the time boundaries at time $x_0 = 0$ and $x_0 = T$. For a more detailed discussion we refer the reader to [7].

To discuss the renormalization of operators related to moments of parton distribution functions, we first have to provide a renormalization condition. Denoting by $|SF\rangle$ a classical SF state, i.e. a classical quark field at a time boundary with external momentum $\mathbf{p}$, the renormalization condition that we will use reads

$$\langle SF|\mathcal{O}^{\rm ren}\left(\mu=\frac{1}{L}\right)|SF\rangle = \langle SF|\mathcal{O}^{\rm tree}|SF\rangle\ . \quad (7)$$

The relation between the expectation value of the bare operator and the renormalized one is established through a scale-dependent renormalization constant:

$$O^R(\mu) = Z^{-1}(1/\mu)O^{\rm bare}(1/L)\ . \quad (8)$$

In perturbation theory, on the 1-loop level, we have $Z(1/\mu) = 1-\bar{g}^2(\mu)\left[\gamma^{(0)}\ln(\mu)+B_0\right]$ with $\gamma^{(0)}$ the anomalous dimension and B_0 the constant part. Up to this point, the discussion is given solely in the continuum where the SF renormalization scheme is a perfectly acceptable one. Different schemes such as the $\overline{\rm MS}$ scheme can be related to the SF scheme as usual in perturbation theory.

If we are interested, however, in a non-perturbative calculation, we have to detour for a short time (which means, however, a substantial computer time) to a finite lattice with non-zero lattice spacing a that allows for numerical simulations. A lattice representation of the twist-2 non-singlet operator, which is the only case we are considering here, is given by

$$\mathcal{O}_{12}(x) = \frac{1}{4}\bar{\psi}(x)\gamma_{\{1}\overleftrightarrow{D}_{2\}}\frac{\tau^3}{2}\psi(x)\ , \quad (9)$$

where $\overleftrightarrow{D}_\mu$ is the covariant derivative and the bracket around indices means symmetrization. The operator is probed by boundary quark fields ζ and $\bar{\zeta}$, which reside at $x_0 = 0$ and a correlation function is constructed

$$f_{O_{12}}(\frac{x_0}{a}) = \sum_{\mathbf{x},\mathbf{y},\mathbf{z}} \langle e^{i\mathbf{p}(\mathbf{y}-\mathbf{z})} \mathcal{O}_{12}(\mathbf{x}) \bar{\zeta}(\mathbf{y}) \gamma_2 \tau^3 \zeta(\mathbf{z}) \rangle \tag{10}$$

with $\mathbf{p}$ the spatial 3-momentum.

To take into account the effects of the boundary fields, we also consider the boundary operators defined at the time boundaries $x_0 = 0$ and $x_0 = T$:

$$\mathcal{O}_0 = \frac{a^6}{L^3} \sum_{\mathbf{y},\mathbf{z}} \bar{\zeta}(\mathbf{y}) \gamma_5 \frac{\tau^3}{2} \zeta(\mathbf{z}),$$
$$\mathcal{O}_T = \frac{a^6}{L^3} \sum_{\mathbf{y},\mathbf{z}} \bar{\zeta}'(\mathbf{y}) \gamma_5 \frac{\tau^3}{2} \zeta'(\mathbf{z}) \tag{11}$$

from which we construct the correlation function:

$$f_1 = -\langle \mathcal{O}_0 \mathcal{O}_T \rangle \ . \tag{12}$$

The boundary wave-function contribution can then be taken out by considering the ratio $f_{O_{12}}(x_0/a)/\sqrt{f_1}$.

There are several physical scales in our problem which all have to be given in units of L, which is the only scale that is to be changed. Therefore, in order to arrive at a definition of the renormalization constant, we have to specify the spatial momentum $\mathbf{p}$, the quark mass m_q and the time x_0 where we read off the expectation value of the operator between SF states from the correlation function. The final physical result does, of course, not depend on our choice of these quantities, but we choose them solely for convenience. In particular we select

$$m_q = 0 \ , \quad x_0 = T/4 \ , \tag{13}$$
$$\mathbf{p} \equiv (p_1, p_2, p_3) = (p_1 = 2\pi/L, 0, 0) \ .$$

The choice of a zero quark mass results in using a massless renormalization scheme and the choice of the smallest available momentum on the lattice minimizes lattice artefacts. With the above choice, it is indeed only the physical box length L (assuming $T = L$) that we identify with the inverse scale, which is varied in the problem.

We are now in a position to give the precise definition of the renormalization constant

$$Z(L) = \bar{Z}(L)/Z_1(L) \ , \tag{14}$$

with

$$\bar{Z}(L) = f_{O_{12}}(T/4)/f_{O_{12}}^{\mathrm{tree}}(T/4) \ ,$$
$$Z_1(L) = \sqrt{f_1(L)}/\sqrt{f_1^{\mathrm{tree}}(L)} \ , \tag{15}$$

where we divide by the corresponding tree level expression as required by the renormalization condition, eq. (7). Instead of computing the Z-factors, we concentrate on the step scaling functions

$$\sigma_{\bar{Z}} = \frac{\bar{Z}(2L)}{\bar{Z}(L)}, \ \sigma_{f_1} = \frac{Z_1(2L)}{Z_1(L)}, \ \sigma_Z = \frac{Z(2L)}{Z(L)} \ , \tag{16}$$

because, in contrast to the Z-factors, the step scaling functions have a well-defined continuum limit. The strategy is now to compute the step scaling functions at various values of the lattice spacing while keeping fixed the conditions in eqs. (13) and the physical scale $\mu = L^{-1}$ (determined by the running coupling $\bar{g}(\mu)$) and to extrapolate the results thus obtained to $a = 0$.

It is one of the basic ingredients and characteristics of our work that almost all simulation results at non-zero lattice spacings have been obtained by employing the standard Wilson action and the non-perturbatively improved clover action. Since these two formulations lead to different lattice artefacts, it is a very crucial test of our results that their continuum extrapolations give consistent results. That this is indeed the case is demonstrated in fig. 1. It shows that for the two step scaling functions σ_{f_1} and $\sigma_{\bar{Z}}$ the continuum limit of both discretizations agree within the error bars. We note that in the case of $\sigma_{\bar{Z}}$ a quadratic extrapolation in the lattice spacing a is necessary while for σ_{f_1} a linear extrapolation is sufficient. After checking that a similar behaviour is found at all values of the coupling we have simulated, we performed

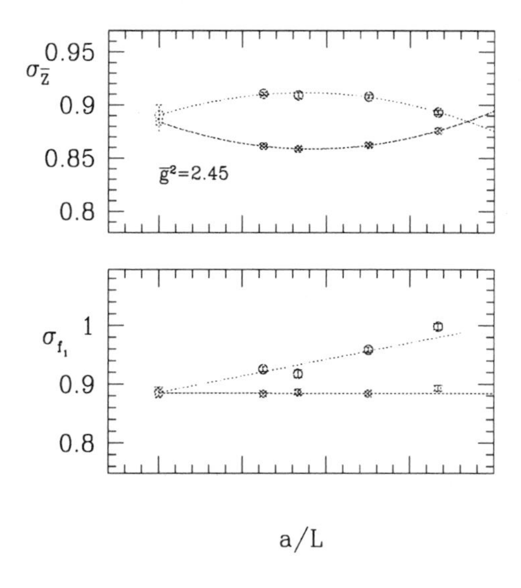

Figure 1. Continuum extrapolation of the step scaling functions $\sigma_{\bar{Z}}$ and σ_{f_1} performed separately for the Wilson action (circles and dotted lines) and the non-perturbatively improved action (squares and full lines) at a fixed value of the running coupling $\bar{g}^2 = 2.45$.

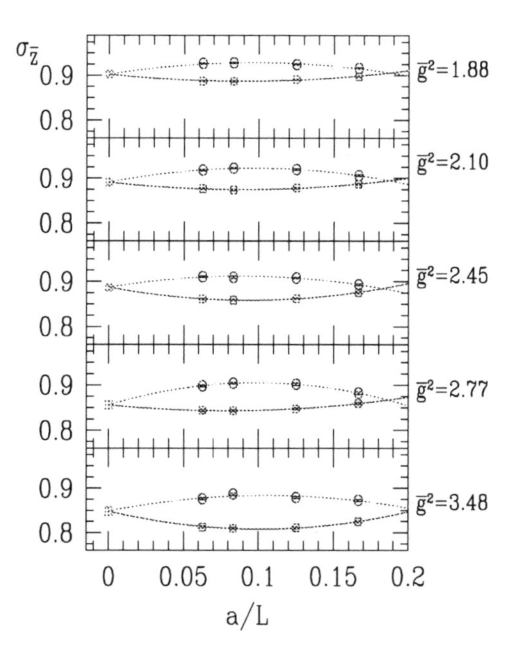

Figure 2. Constraint continuum extrapolation of $\sigma_{\bar{Z}}$ (notation as in fig. 1).

constraint fits, demanding that the continuum value of the step scaling functions be the same from both actions. A summary of our results for $\sigma_{\bar{Z}}$ is shown in fig. 2.

At this point, having performed all necessary continuum extrapolations, we end our detour on a lattice with non-zero lattice spacing and come back to the discussion in the continuum. With the results on the step scaling function, extrapolated to the continuum limit, which were obtained at 9 values of the running coupling constant, we can now compute the (ultraviolet) invariant step scaling function

$$\mathfrak{S}_{\mathrm{INV}}^{\mathrm{UV}}(\mu_0) = \sigma(\mu/\mu_0, \bar{g}^2(\mu_0)) \cdot f(\bar{g}^2(\mu)) \,, \tag{17}$$

which still depends on the infrared scale μ_0. This scale dependence will only be cancelled when multiplying with the matrix element, renormalized at the scale μ_0. The function $f(\bar{g}^2(\mu))$ is the same as in eq. (3) and the β and γ functions are taken up to 3 loops in the SF scheme[a].

In fig. 3 we show $\mathfrak{S}_{\mathrm{INV}}^{\mathrm{UV}}(\mu_0)$ as a function of μ/μ_0. For large enough energy scales, $\mu/\mu_0 > 100$, $\mathfrak{S}_{\mathrm{INV}}^{\mathrm{UV}}(\mu_0)$ does not change within the errors and we can determine a value for it by fitting the last, say, 4 points to a constant. Although there still is a scheme dependence in the invariant step scaling function through the remaining dependence on μ_0, the value should be independent of the choice of coupling used in the analysis. This is nicely illustrated in fig. 3, where the choices of $\bar{g}(L/4)$ and $\bar{g}(L)$ give consistent values for the invariant step scaling function.

We can therefore now give the first piece of information for the invariant matrix ele-

[a] For the γ function we have taken an effective 3-loop parametrization as obtained by fitting our data to an effective 3-loop form [2].

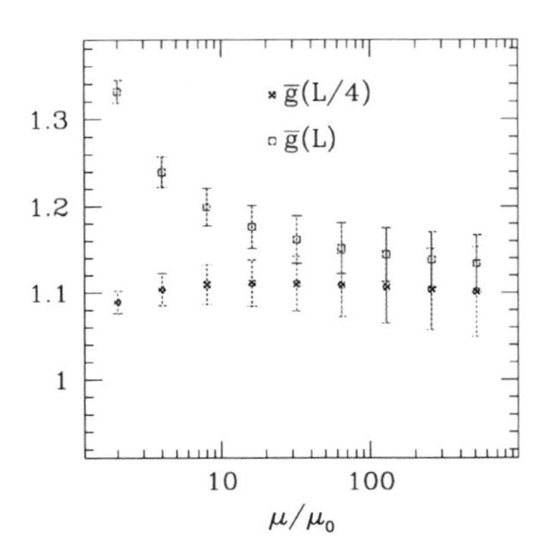

Figure 3. The values of $\mathfrak{S}^{\mathrm{UV}}_{\mathrm{INV}}(\mu_0)$ for two choices of the running coupling.

ment itself, as needed in eq. (6), and quote

$$\mathfrak{S}^{\mathrm{UV}}_{\mathrm{INV}}(\mu_0) = 1.11(4) \ . \tag{18}$$

3 Renormalized matrix element

As a next step we have to compute the renormalized matrix element itself. Again, we will always use the SF scheme and we remark that this is the first time that a calculation of a 2-quark matrix element is attempted in this set-up. We first tried to compute the matrix element (within pion states) of the operator of eq. (9). However, since this operator needs a non-vanishing momentum, we found, similar as in [6], that the signal is very noisy. We therefore decided to switch to the operator

$$\mathcal{O}_{00}(x) = \bar{\psi}(x)\left[\gamma_0 \overleftrightarrow{D}_0 - \frac{1}{3}\sum_{k=1}^{3}\gamma_k \overleftrightarrow{D}_k\right]\frac{\tau^3}{8}\psi(x) \tag{19}$$

which has the advantage that it can be computed at zero momentum. Taking the boundary operators of eq. (11) we construct a correlation function

$$f_{\mathrm{M}}(x_0) = a^3 \sum_{\mathbf{x}} \langle \mathcal{O}_0 \mathcal{O}_{00}(x) \mathcal{O}_T \rangle \ , \tag{20}$$

which again is to be normalized by f_1, eq. (12), to take out the boundary wave-function contributions. Performing a transfer matrix decomposition, we find that for large enough values of x_0 and staying far enough from both boundaries

$$\begin{aligned} f_1 &\simeq \rho^2 e^{-m_\pi T} , \\ f_{\mathrm{M}}(x_0) &\simeq \rho^2 e^{-m_\pi T} \langle \pi | \mathcal{O}_{00} | \pi \rangle \ . \end{aligned} \tag{21}$$

Assuming that there is a plateau region where $f_M(x_0)/f_1 = \mathrm{const} \equiv \langle \pi|\mathcal{O}_{00}|\pi\rangle$, and in which the first excited state gives essentially no contributions, we obtain the physical matrix element $\langle x \rangle$ after a suitable normalization (see [6]):

$$\langle x \rangle \equiv \frac{2\kappa}{m_\pi} \langle \pi | \mathcal{O}_{00} | \pi \rangle \ . \tag{22}$$

In order to extract the matrix element of eq. (22) we have chosen lattices with $T = 3$ fm and followed the correlation function $f_{\mathrm{M}}(x_0)$ up to a distance of 1 fm in time direction. At this distance we are sure that we project on the pion states as an inspection of the pseudoscalar and axial-vector correlation functions (from which we also extracted the pion masses) showed. Indeed, for 1 fm $< x_0 <$ 2 fm the correlation function exhibits a plateau behaviour as can be seen in fig. 2 of ref. [5].

Once we have the bare matrix element we need to renormalize it. To this end we computed $Z(1/\mu_0)$ with $\mu_0^{-1} = 1.436 r_0$, $r_0 \approx 0.5$ fm. We repeated such a calculation for various lattice sizes, choosing the values of β such that $L_{\mathrm{max}} = \mu_0^{-1}$ is kept fixed. Interpolating the numerical simulation data we obtain in this way $Z(1/\mu_0)$ in a range of lattice spacing $0.05 \leq a \leq 0.1$, i.e. the range of a where the bare matrix element itself has been computed.

This allowed us to apply the renormalization factor at exactly the same values of β where the matrix element has been computed. We show the matrix element renormalized with $Z_{12}(1/\mu_0)$ in fig. 4. Again all calculations have been performed with two

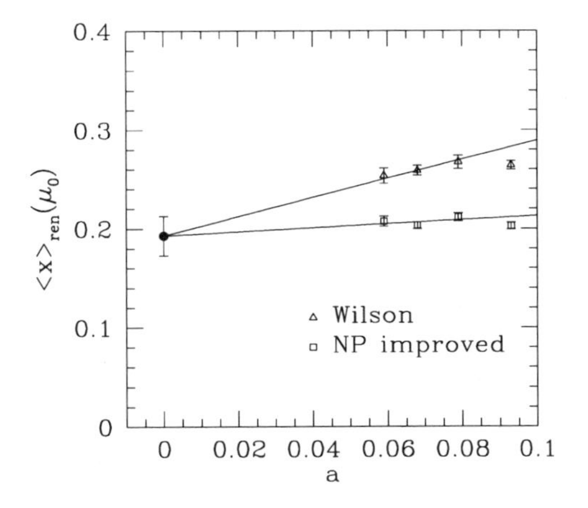

Figure 4. Constrained continuum extrapolation of the renormalized matrix element

different discretizations, and we find quite convincingly the same continuum limit.

It might, however, not have escaped the reader's attention that we have used the wrong renormalization factor, namely Z_{12}, for renormalizing the operator $\mathcal{O}_{00}(x)$ of eq. (19). The continuum extrapolation of the such renormalized operator needs a correction factor. In ref. [5] we have demonstrated that this correction factor can be taken from perturbation theory and amounts to a shift of the continuum renormalized matrix element by a few per cent. Taking the correction factor to be the same as found in [5], we can finally give our main result:

$$O_{\mathrm{INV}}^{\mathrm{ren}} = 0.222(24) \qquad (23)$$

4 Conclusion

In this contribution we have demonstrated how we can compute on the lattice, in a fully non-perturbative fashion, a renormalization group invariant matrix element $O_{\mathrm{INV}}^{\mathrm{ren}}$ for the second moment of parton distributions of the twist-2 non-singlet operator in the pion. A preliminary value for $O_{\mathrm{INV}}^{\mathrm{ren}}$ is given in eq. (23) and can be used now as integration constant to obtain the renormalized matrix element at any scale in the preferred renormalization scheme.

We want to emphasize that the value of the renormalized matrix element is shifted substantially from a value of about $\langle x\rangle(a = 0.093) = 0.30$ at $\beta = 6$ to $\langle x\rangle(a = 0) = 0.2$ in the continuum limit. Hence we experience strong lattice artefacts in the renormalized matrix element. Still, when the matrix element is run in the $\overline{\mathrm{MS}}$ scheme to a scale of $\mu = 2.4$ GeV, we find $\langle x\rangle(\mu = 2.4\mathrm{GeV}) \approx 0.3$. Thus we find that, by a conspiration of two effects, our result agrees with the number quoted in the pioneering work of [6]. Therefore, the fact that within the quenched approximation, used here exclusively, the number from the lattice simulations is higher than the experimental value persists.

Acknowledgements

The work presented in this talk is part of a most enjoyable and lively collaboration with M. Guagnelli and R. Petronzio. I also want to thank A. Bucarelli, F. Palombi, and A. Shindler for useful discussions and essential contributions.

References

1. A. Bucarelli, F. Palombi, R. Petronzio, A. Shindler, Nucl. Phys. **B552** (1999) 379
2. M. Guagnelli, K. Jansen, R. Petronzio, Nucl. Phys. **B542** (1999) 395
3. M. Guagnelli, K. Jansen, R. Petronzio, Phys. Lett. **B457** (1999) 153
4. M. Guagnelli, K. Jansen, R. Petronzio, Phys. Lett. **B459** (1999) 594
5. M. Guagnelli, K. Jansen, R. Petronzio, preprint, CERN-TH/2000-259
6. C. Best *et al.*, Phys. Rev. **D56** (1997) 2743
7. M. Lüscher, R. Narayanan, P. Weisz and U. Wolff, Nucl. Phys. **B384** (1992) 168 S. Sint, Nucl. Phys. **B421** (1994) 135

EXACT CHIRAL SYMMETRY / THE GINSPARG-WILSON RELATION AND DOMAIN WALL FERMION

Y. KIKUKAWA

Department of Physics, Nagoya University, Nagoya 464-8602, Japan
E-mail: kikukawa@eken.phys.nagoya-u.ac.jp

We discuss the relation between the exact chiral symmetry based on the Ginsparg-Wilson relation and the chiral properties of domain-wall fermion. We derive low energy effective action for the light fermion mode of the domain-wall fermion. For vector-like theories, the inverse of the effective Dirac operator turns out to be identical to the inverse of the truncated Neuberger's lattice Dirac operator, except a local contact term which would give the chiral symmetry breaking in the Ginsparg-Wilson relation. For chiral theories, when the domain wall fermion couples to an interpolating five-dimensional gauge field, it reproduces the effective action for the Weyl fermion defined based on the Ginsparg-Wilson relation. The complex phase of the determinant of the domain-wall fermion can be regarded as a lattice implementation of the η-invariant. A lattice expression for the five-dimensional Chern-Simons term is obtained.

1 Introduction

It has become clear recently that chiral symmetry can be implemented exactly on the lattice and that the gauge interaction of the Weyl fermions can be described in the framework of lattice gauge theory. The clue to this development is the construction of gauge-covariant and local Dirac operators [1,2,3] which solve the Ginsparg-Wilson relation [4].

In this talk, I would like to discuss the chiral property of the low energy effective action of domain wall fermion and its relation to the exact chiral symmetry based on the Ginsparg-Wilson relation. This talk is based on the works done in collaborations with T. Noguchi and T. Aoyama of Kyoto University [5,6].

2 Chiral Symmetry on the Lattice

2.1 Nielsen-Ninomiya Theorem

It has been known that it is not straightforward to maintain chiral symmetry in lattice gauge theory due to the species doubling. This problem of the species doubling is generic on the lattice, as stated in the Nielsen-Ninomiya theorem[7]. Any lattice action for free Dirac fermion may be expressed by a lattice Dirac operator D,

$$S = a^4 \sum_{x,y} \bar{\psi}(x)\, D(x-y)\, \psi(y) = \int \frac{d^4k}{(2\pi)^4}\, \bar{\psi}(-k)\, \tilde{D}(k)\, \psi(k). \quad (1)$$

Then it can be shown that the following four conditions on the properties of the lattice Dirac operator D cannot be satisfied simultaneously.

1. $\tilde{D}(k)$ is a periodic and analytic function in terms of momentum k_μ. (locality)
2. For small momentum $|k_\mu| a \ll \pi$, $\tilde{D}(k) \simeq i\gamma_\mu k_\mu$.
3. $\tilde{D}(k)$ is invertible for all k_μ except $k_\mu = 0$. (no species doublers)
4. $\gamma_5 D + D\gamma_5 = 0$. (chiral invariance)

Therefore, in order to eliminate the species doublers, chiral symmetry must be broken explicitly in lattice fermion actions.

2.2 The Ginsparg-Wilson relation

This problem of the breakdown of chiral symmetry on the lattice has been examined from the point of view of the renormalization group by Ginsparg and Wilson [4]. Although chiral symmetry is broken explicitly, the Wilson-Dirac fermion [8] contains a single mode of a massless Dirac fermion for sufficiently small momentum,

$$|k_\mu| \ll \pi/a. \tag{2}$$

Then one may consider the low energy effective action for this massless mode, which describes the small momentum region of the Wilson-Dirac fermion.

In order to obtain the effective action, one may invoke the block-spin transformation. By averaging the field variables over the minimal hyper-cubic box, new field variables can be defined, which lives on the lattice with the lattice spacing $a' = 2a$ twice as large as the original lattice spacing a:

$$\psi'(x') = Z\frac{1}{2^4}\sum_{x\in b(x')}\psi(x) = B(x';\psi). \tag{3}$$

The effective action for the blocked variables is evaluated through the functional integral of the original variables, by introducing the kernel of the block-spin transformation:

$$e^{-S'[\psi',\bar\psi']} = \int \prod_x d\psi(x)d\bar\psi(x)\, e^{-S_W[\psi,\bar\psi]} \times e^{-\alpha\sum_{x'}\left(\bar\psi'(x')-B(x';\bar\psi)\right)\left(\psi'(x')-B(x';\psi)\right)}. \tag{4}$$

There is a fixed point of this block-spin transformation. The effective action at the fixed point can be written in terms of an effective Dirac operator, which we denote by D^*,

$$S^* = a^4\sum_x \bar\psi(x)D^*\psi(x). \tag{5}$$

It turns out that this effective Dirac operator satisfies the following relation,

$$\gamma_5 D^* + D^*\gamma_5 = aD^*\gamma_5 R D^*, \quad R = \frac{4}{\alpha}. \tag{6}$$

This is the relation which was first obtained by Ginsparg and Wilson [4] and is referred as the Ginsparg-Wilson relation.

In terms of the fermion propagator $S_F = D^{*-1}$, this relation can be written as

$$\gamma_5 S_F(x,y) + S_F(x,y)\gamma_5 = \gamma_5 R\delta(x,y). \tag{7}$$

It is clear from this expression that chiral symmetry is broken only in local contact terms. Such local terms would not contribute to the physical amplitudes evaluated at long-distance, $x - y \neq 0$.

2.3 Exact chiral symmetry on the lattice

The Ginsparg-Wilson relation in fact implies an exact symmetry of the fermion action.[9] Under the following transformation,

$$\delta\psi(x) = \gamma_5\left(1 - aRD^*\right)\psi(x), \tag{8}$$
$$\delta\bar\psi(x) = \bar\psi(x)\gamma_5, \tag{9}$$

the action is invariant. As long as D^* is local, this transformation is a local transformation. Then this symmetry can be regarded as the lattice counterpart of chiral symmetry in the continuum theory. Thus chiral symmetry can be maintained exactly if a lattice Dirac operator satisfies the Ginsparg-Wilson relation.

2.4 Neuberger's lattice Dirac operator

In order to incorporate the gauge interaction, it is necessary to make a lattice Dirac operator gauge-covariant. Then the question is how to construct a lattice Dirac operator which is local, is gauge-covariant and satisfies the Ginsparg-Wilson relation.

The explicit gauge-covariant solution of the Ginsparg-Wilson relation has been derived recently by Neuberger [2] from the overlap formulation of chiral determinant.[10,11] It is defined through the Wilson-Dirac operator with a negative mass in a certain range as follows:

$$D = \frac{1}{2a}\left(1 + X\frac{1}{\sqrt{X^\dagger X}}\right) \tag{10}$$
$$= \frac{1}{2a}\left(1 + \gamma_5\frac{H}{\sqrt{H^2}}\right), \tag{11}$$

where

$$X = \left(D_{\rm w} - \frac{m_0}{a}\right), \; H = \gamma_5 X, \; 0 < m_0 < 2. \tag{12}$$

$D_{\rm w}$ is the gauge-covariant form of the Wilson-Dirac operator, which is obtained by replacing the difference operator with the covariant one:

$$D_{\rm w} = \sum_{\mu} \left\{ \gamma_\mu \frac{1}{2} \left(\nabla_\mu - \nabla_\mu^\dagger\right) + \frac{a}{2} \left(\nabla_\mu \nabla_\mu^\dagger\right) \right\}. \tag{13}$$

It is straightforward to check that this lattice Dirac operator satisfies the Ginsparg-Wilson relation Eq. (6) with $R = 2$, using the second expression.

Since Neuberger's lattice Dirac operator contains the inverse square root of the hermitian Wilson-Dirac operator H, locality properties of the Dirac operator is not quite trivial. This question has been examined by Hernándes, Jansen and Lüscher.[3]

One of the striking result of the Ginsparg-Wilson relation is that the chiral anomaly associated with the chiral transformation [9,12,13] satisfies the index theorem at a finite lattice spacing.[14] The topological properties of the gauge anomaly, which also follows from the Ginsparg-Wilson relation, plays a crucial role to establish the exact gauge invariance in the construction of chiral gauge theories.[22,23]

3 Domain wall fermion and the Ginsparg-Wilson relation

The domain-wall fermion[15] is the basis of Neuberger's Dirac operator. In the continuum theory the domain wall fermion is defined by the five-dimensional Dirac fermion coupled to a scalar field condensate with a kink-like topological defect, which field equation is given by

$$\left\{\gamma_\mu D_\mu + \gamma_5 D_5 + \langle\phi(x_5)\rangle\right\} \psi(x, x_5) = 0, \tag{14}$$

where $\langle\phi(x_5)\rangle = m_0 \epsilon(x_5)$. This kink-like singularity leads to the delta-function potential which is attractive only for the right-handed component of the fermion and produces a chiral zero mode which is bounded to the domain wall at $x_5 = 0$,

$$-\gamma_5 m_0 \delta(x_5) \in D^\dagger D \Longrightarrow \psi_R(x)\, e^{-m_0|x_5|}. \tag{15}$$

Kaplan pointed out that this system can be implemented on the lattice using the Wilson mass term.[15] When $0 < m_0 < 2$, the mass term

$$\sum_\mu \frac{2}{a} \sin^2\left(\frac{p_\mu a}{2}\right) - \frac{m_0}{a} \mathrm{sgn}(\mathrm{x}_5 - \frac{1}{2}) \tag{16}$$

shows the kink-like singularity only for the physical mode with $|p_\mu| << \frac{\pi}{a}$.

In a simplified formulation[16,17], domain wall fermion can be defined by the five-dimensional Wilson-Dirac fermion with a negative mass ($0 < m_0 < 2$) in a finite extent of the fifth-dimension, $x_5 = t a_5$, $t \in [1, N]$. (a_5 denotes the lattice spacing of the fifth dimension.) Gauge fields are assumed to be four-dimensional.

$$S_{\rm DW} = a_5 \sum_{t=1}^{N} a^4 \sum_x \bar\psi(x,t) \Big(D_{\rm w} - P_L \partial_t - P_R \partial_t^\dagger - \frac{m_0}{a}\Big) \psi(x,t). \tag{17}$$

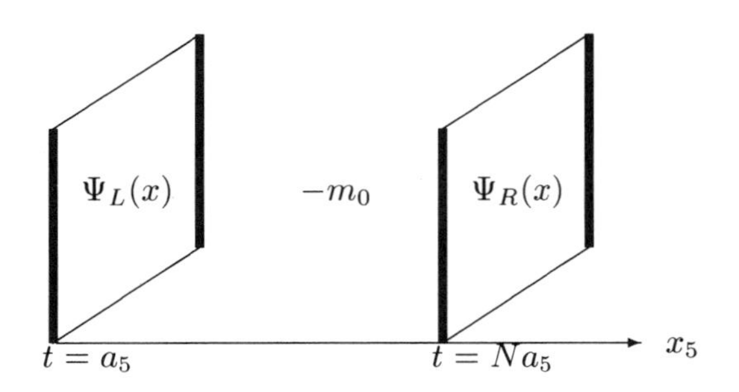

At a finite N, a single light Dirac fermion appears on the boundary walls and it can be probed by the field variables at the boundaries, which are referred as $q(x)$ and $\bar q(x)$ by Furman and Shamir:

$$q(x) = \psi_L(x, 1) + \psi_R(x, N), \tag{18}$$

$$\bar q(x) = \bar\psi_L(x, 1) + \bar\psi_R(x, N). \tag{19}$$

It has been argued that chiral symmetry of the light fermion is preserved up to corrections suppressed exponentially in the lattice size N of the fifth dimension[17,18,19].

3.1 *Domain-wall fermion and Neuberger's lattice Dirac operator*

In order to see the relation of the light fermion mode of the domain wall fermion to Neuberger's Dirac operator, we evaluate the low energy effective action for the light fermion mode, $q(x)$ and $\bar{q}(x)$ by integrating out heavy modes [20,5]. The inverse of the effective Dirac operator, that is, the propagator of $q(x)$ and $\bar{q}(x)$ is given as follows [5]:

$$\langle q(x)\bar{q}(y)\rangle = \frac{1}{a^4}\left(\frac{1}{a}D_{\text{eff}}^{(N)^{-1}} - \delta(x,y)\right), \tag{20}$$

where

$$D_{\text{eff}}^{(N)} = \frac{1}{2a}\left(1+\gamma_5 \tanh\frac{N}{2}a_5\widetilde{H}\right). \tag{21}$$

$\widetilde{H}$ is defined through the transfer matrix of the five-dimensional Wilson fermion,

$$T = e^{-a_5\tilde{H}} = \begin{pmatrix} \frac{1}{B} & -\frac{1}{B}C \\ -C^\dagger\frac{1}{B} & B+C^\dagger\frac{1}{B}C \end{pmatrix}, \tag{22}$$

where

$$C = a_5\,\sigma_\mu\,\frac{1}{2}\left(\nabla_\mu - \nabla_\mu^\dagger\right), \tag{23}$$

$$B = 1 + a_5\left(\frac{a}{2}\nabla_\mu\nabla_\mu^\dagger - \frac{m_0}{a}\right). \tag{24}$$

In the limit $N\to\infty$, the effective Dirac operator Eq. (21) then reduces to Neuberger's lattice Dirac operator using $\widetilde{H}$ and turns out to satisfy the Ginsparg-Wilson relation. The contact term in the propagator is subtracting the chiral symmetry breaking in the Ginsparg-Wilson relation. Thus the light fermion mode of the domain-wall fermion is related directly to Neuberger's lattice Dirac operator.

In order to obtain the low energy effective action, we need to subtract the contribution of the massive modes properly. For this purpose, we note that the five-dimensional Wilson fermion subject to the anti-periodic boundary condition in the finite fifth-dimension[21] is obtained from the domain wall fermion by adding the mass term to the boundary field variables, $q(x)$ and $\bar{q}(x)$, as

$$+\frac{1}{a_5}\bar{q}(x)q(x) \tag{25}$$

and, as clearly seen, that all the fermion modes become massive. By virtue of this simple relation, the partition function of the domain wall fermion can be shown to be factorized in the following form:

$$\det D_{5\text{w}} = \det D_{\text{eff}}^{(N)}\cdot\det D_{5\text{w}}{}^{[AP]}. \tag{26}$$

The first factor in the r.h.s. is a four-dimensional determinant and it gives the effective action for the light fermion mode of the domain wall fermion, which is indeed given by Neuberger's local Dirac operator.

4 Domain wall fermion and Chiral gauge theory based on the Ginsparg-Wilson relatin

Next, we discuss how to formulate chiral gauge theories using domain wall fermion and examine the relation to the Weyl fermion defined based on the Ginsparg-Wilson relation.

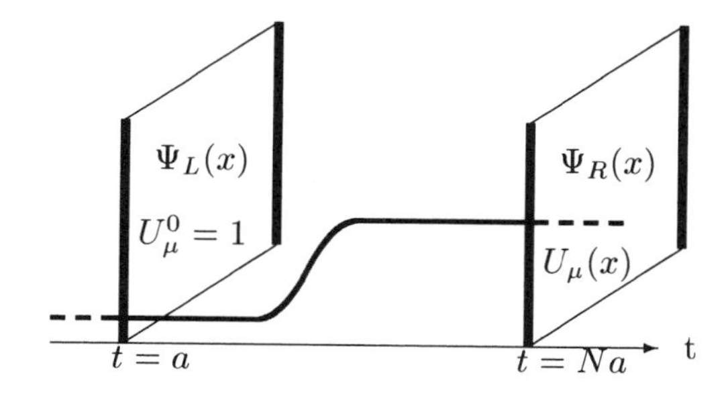

In order to introduce the chiral gauge coupling to the light modes on the domain walls, one may introduce the gauge field which is varying in the fifth direction and is interpolating smoothly between the four-dimensional gauge fields at the boundaries. To examine whether this "chiral" domain wall fermion could describe the Weyl fermions interacting with the four-dimensional gauge fields on the walls, we consider the variation of the determinant of the domain wall fermion with respect to the gauge field. For this purpose, it is convenient to introduce another parameter u which interpolates the five-dimensional gauge fields as follows.

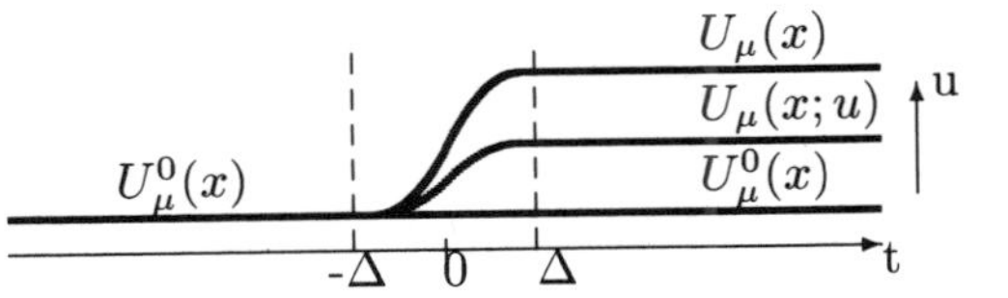

Then by taking the variation with respect to u, we can derive the following identity:

$$\lim_{T\to\infty} \frac{d}{du} \ln\det\left(D_{5\mathrm{w(T)}} - \frac{m_0}{a}\right)$$
$$= \lim_{T\to\infty} \sum_{t=-T+1}^{t=T} \sum_x \mathrm{tr}\left(\frac{d}{du} D_{5\mathrm{w}} \frac{1}{D_{5\mathrm{w}} - \frac{m_0}{a}}\right)$$
$$+\mathrm{Tr}_x\, P_L \frac{d}{du} D \frac{1}{D} \tag{27}$$

The second term in the r.h.s. of the identity is the contribution from the boundary walls. It is indeed the chiral gauge current (in the gauge-covariant form) written in terms of Neuberger's lattice Dirac operator. This term can be related to the effective action for the Weyl fermions defined based on the Ginsparg-Wilson relation. In fact, the variation of the effective action for the (right-handed) Weyl fermion is given by

$$\frac{d}{du} \Gamma_{\mathrm{eff}}[U^u_\mu] = \mathrm{Tr}_x\, P_L \frac{d}{du} D \frac{1}{D}$$
$$+\sum_i \left(v_i, \frac{d}{du} v_i\right), \tag{28}$$

where the second term in the r.h.s. is the so-called measure term. [22,23]

In order to obtain a consistent four-dimensional description of the chiral fermins from the domain wall fermion, we should require the path-independence of the five-dimensional interpolation, that is, the integrability. In this respect, the "chiral" domain wall fermion produces, in the bulk five-dimensional space, the Chern-Simons current which carries the information about the path-dependence of the interpolation and the gauge anomaly. It is given by the first term of the r.h.s. of the identity, Eq.(27). From this current, one may define the lattice counterpart of the Chern-Simons term by

$$2\pi Q_5' \left[U_\mu(x,t)\right]$$
$$\equiv \int_0^1 du \sum_{x,t}{}' \mathrm{tr}\left(\frac{d}{du} D_{5\mathrm{w}} \frac{1}{D_{5\mathrm{w}} - \frac{m_0}{a}}\right)$$
$$- \int_0^1 du \sum_k \left(v_k, \frac{d}{du} v_k\right), \tag{29}$$

where $\sum_{x,t}{}' = \sum_{t=-T+1}^{t=T} \sum_x$. Then we obtain the five-dimensional formula of the effective action for the Weyl fermions (defined based on the Ginsparg-Wilson relation) using the domain wall fermion as

$$\Gamma_{\mathrm{eff}}\left[U_\mu\right] - \Gamma_{\mathrm{eff}}\left[U^0_\mu\right]$$
$$= \lim_{T\to\infty} \left\{ \ln\det\left(D_{5\mathrm{w(T)}} - \frac{m_0}{a}\right)_{u=1} - \ln\det\left(D_{5\mathrm{w(T)}} - \frac{m_0}{a}\right)_{u=0} - 2\pi Q_5'\left[U_\mu(x,t)\right]\right\}, \tag{30}$$

Note that in the r.h.s. the path-dependence of the five-dimensional interpolation cancels out. This formula can be regarded as the lattice counterpart of the formula using the η-invariant.[24,25] It realizes the argument given by Kaplan and Shmaltz in the continuum theory [26] at a finite lattice spacing.

In order to estabish the exact gauge invariance of the effective action for anomaly-free chiral gauge theories[22,23], we need to examine the topological structure of the Chern-Simons term.

5 Conclusion

In conclusion,

- Domain wall fermion can produce chiral fermions bounded to the 4D walls, under certain conditions for lattice gauge fields

- Low energy effective theories of these chiral fermions can be constructed so that both locality and chiral symmetry are maintained by virtue of the Ginsparg-Wilson relation

- Chiral gauge-coupling can be introduced for these chiral fermions, through the interpolating five-dimensional gauge fields

- The lattice counterpart of the Cherns-Simons term can be defined and it should take account of the integrability and the gauge invariance

Acknowledgments

The author would like to thank H. Neuberger, M. Lüscher and H. Suzuki, for valuable discussions. This work is supported in part by Grant-in-Aid for Scientific Research of Ministry of Education (#10740116).

References

1. P. Hasenfratz, Nucl. Phys. B (Proc. Suppl.) 63 (1998) 53.
2. H. Neuberger, Nucl. Phys. B417 (1998) 141; Phys. Lett. B427 (1998) 353.
3. P. Hernándes, K. Jansen, M. Lüscher, Nucl. Phys. B552 (1999) 363.
4. P. H. Ginsparg and K. G. Wilson, Phys. Rev. D25 (1982) 2649.
5. Y. Kikukawa, T. Noguchi, `hep-lat/9902022`.
6. T. Aoyama, Y. Kikukawa, `hep-lat/9905003`.
7. N.B. Neilsen and M. Ninomiya, PL B105 (1981) 219; Nucl. Phys. B185 (1981) 20 [E: B195 (1982) 541]; *ibid* B193 (1981) 173. Friedan, Commun. Math. Phys. 85 (1982) 481.
8. K.G. Wilson, Phys. Rev. D10 (1974) 2445; *in* New phenomena in subnuclear physics, ed. A. Zichichi (Plenum, New York, 1977) (Erice, 1975)
9. M. Lüscher, Phys. Lett. B428 (1998) 342.
10. R. Narayanan and H. Neuberger, Nucl. Phys. B412 (1994) 574; Phys. Rev. Lett. 71 (1993) 3251; Nucl. Phys. B443 (1995) 305.
11. Y. Kikukawa and H. Neuberger, Nucl. Phys. B513 (1998) 735.
12. Y. Kikukawa and A. Yamada, Phys. Lett. B448 (1999) 265.
13. K. Fujikawa, Nucl. Phys. B546 (1999) 480. D.H. Adams, `hep-lat/9812003`. H. Suzuki, Prog. Theor. Phys. 102 (1999) 141.
14. P. Hasenfrats, V. Laliena and F. Niedermayer, Phys. Lett. B427 (1998) 125.
15. D. B. Kaplan, Phys. Lett. B288 (1992) 342.
16. Y. Shamir, Nucl. Phys. B406 (1993) 90.
17. V. Furman, Y. Shamir, Nucl. Phys. B439 (1995) 54.
18. S. Aoki, Y. Taniguchi, Phys. Rev. D59 (1999) 054510.
19. Y. Kikukawa, H. Neuberger, A. Yamada, Nucl. Phys. B526 (1998) 572.
20. H. Neuberger, Phys. Rev. **D57** (1998) 5417.
21. P. Vranas, Phys. Rev. D57 (1998) 1415.
22. M. Lüscher, Nucl. Phys. B549 (1999) 295.
23. M. Lüscher, Nucl. Phys. B568 (2000) 162.
24. L. Alvarez-Gaumé, S. Della Pietra and V. Della Pietra, Phys. Lett. 166B (1986) 177; Commun. Math. Phys. 109, (1987) 691-700.
25. R.D. Ball and H. Osborn, Phys. Lett. B165 (1985) 410; Nucl. Phys. B263 (1986) 243. R.D. Ball, Phys. Lett. B171 (1986) 435; Phys. Rept. 182 (1989) 1.
26. D. B. Kaplan and M. Schmaltz, Phys. Lett. B368 (1996) 44.

B AND D MESONS IN LATTICE QCD

ANDREAS S. KRONFELD

Theoretical Physics Department, Fermi National Accelerator Laboratory, Batavia, IL, USA

Computational and theoretical developments in lattice QCD calculations of B and D mesons are surveyed. Several topical examples are given: new ideas for calculating the HQET parameters $\bar{\Lambda}$ and λ_1; form factors needed to determine $|V_{cb}|$ and $|V_{ub}|$; bag parameters for the mass differences of the B mesons; and decay constants. Prospects for removing the quenched approximation are discussed.

1 Introduction

In the standard model, interactions involving the Cabibbo-Kobayashi-Maskawa (CKM) matrix violate CP, with strength proportional to the area of the "unitarity triangle." Fig. 1 shows a recent summary[1] of the triangle. The dominant uncertainties are theoretical, coming from non-perturbative QCD. Each blob shows experimental uncertainties for fixed theoretical inputs, so the range of blobs illustrates the theoretical uncertainties. If measuring the apex $(\bar{\rho}, \bar{\eta})$ were the only goal, one might conclude from Fig. 1 that the most pressing issue is to reduce the theoretical uncertainties, which would require greater investment in computing for lattice QCD.

Measuring $(\bar{\rho}, \bar{\eta})$ is not the most exciting goal, however. "High-energy physics is exciting and will remain exciting, precisely because it exists in a state of permanent revolution."[2] That means we would prefer to discover additional, non-KM sources of CP violation. Indeed, "it is possible, likely, unavoidable, that the standard model's picture of CP violation is incomplete."[3]

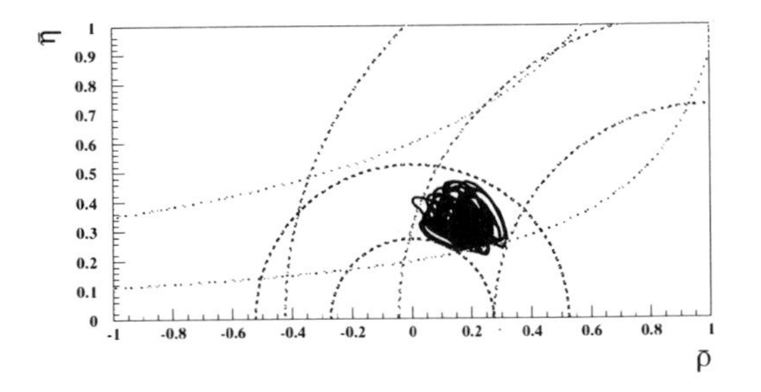

Figure 1. Constraints on the unitarity triangle from CP conserving B decays and indirect CP violation in the kaon. From Plaszczynski and Schune.[1]

Lattice QCD can aid the discovery of new sources of CP violation and may be essential. A lot of information will be necessary to figure out what is going on at short distances. One way to think about this is sketched in Fig. 2. The triangle $B\gamma A$ is determined from (quark-level) tree processes. The side B requires $|V_{ud}|$ from $n \to pe^-\bar{\nu}$, and $|V_{ub}|$ from $B^- \to \rho^0 l^- \bar{\nu}$ or $\bar{B}^0 \to \pi^+ l^- \bar{\nu}$; the angle γ requires the CP asymmetry of $B^\pm \to D^0_{CP} K^\pm$ (or $B_s \to D_s^\pm K^\mp$); the side A requires $|V_{cd}|$ from $D^0 \to \pi^- l^+ \nu$, and $|V_{cb}|$ from $B^- \to D^{0(*)} l^- \bar{\nu}$. One could call this the "tree triangle". The triangle $\alpha C \beta$ is determined from mixing processes (including interference of decays with and without mixing). The angle α requires the asymmetry of $B \to \rho\pi$; the side C requires $|V_{td}|$ from $\Delta m_{B^0_d}$, and $|V_{tb}|$ from $t \to W^+ b$; the angle β requires the asymmetry of $B^0 \to J/\psi K_S$. One could call this the "mixing triangle".

Checking whether the mixing triangle agrees with the tree triangle tests for new physics in the amplitude of B^0_d-$\bar{B}^0_d$ mixing. New physics in the magnitude muddles the

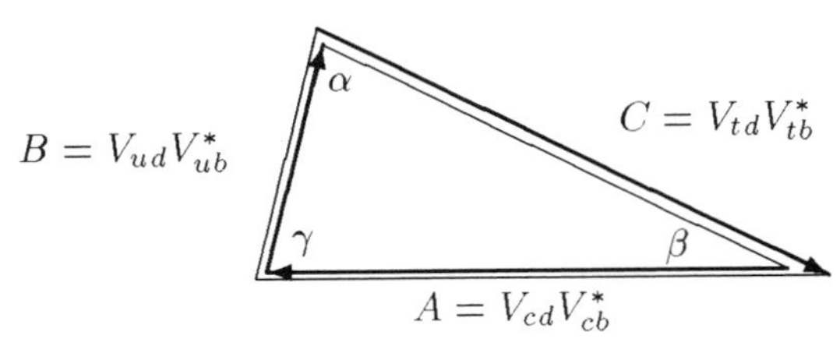

Figure 2. Two different unitarity triangles: the tree triangle $B\gamma A$, and the mixing triangle $\alpha C \beta$.

extraction of $|V_{td}|$, and new physics in the phase muddles the extraction of angles α and β. Similarly, taking $\Delta m_{B_s^0}$, $B_s \to D_s^{\pm} K^{\mp}$, and $B_s \to J/\psi\, \eta^{(\prime)}$ (or $J/\psi\, \phi$) sorts out new physics in B_s^0-$\bar{B}_s^0$ mixing.

These tests are impossible without knowing the sides accurately, so hadronic matrix elements are needed. In a few cases a symmetry provides it, e.g., isospin cleanly yields the matrix element for $n \to pe^- \bar{\nu}$. For the others, we must "solve" non-perturbative QCD and, therefore, *you* need lattice calculations.

You are probably tired of waiting and may ask why results should come any time soon. The fastest of today's computers are now powerful enough to eliminate the sorest point: the quenched approximation. Also, (lattice) theorists have slowly developed a culture of estimating systematic uncertainties, which is now not bad and would improve if more non-practitioners became sufficiently informed about the methods to make constructive suggestions.

The rest of this talk starts with some theoretical aspects that might make it easier for the outsider to judge the systematic errors of heavy quarks on the lattice. Then I show recent results needed for the sides A, B, and C,.

2 Lattice Spacing Effects (Theory)

Lattice QCD calculates matrix elements by computing the functional integral, using a Monte Carlo with importance sampling. Hence, there are statistical errors. This part of the method is well understood and, these days, rarely leads to controversy. When conflicts do arise, they usually originate in the treatment of systematics. The non-expert does not need to know how the Monte Carlo works, but can develop some intuition of how the systematics work. Don't be put off by lattice jargon: the main tool is familiar to all: it is effective field theory.

Lattice spacing effects can be cataloged with Symanzik's local effective Lagrangian (LE$\mathcal{L}$).[4] Finite-volume effects can be controled and exploited with a general, massive quantum field theory.[5] The computer algorithms work better for the strange quark than for down or up, but the dependence on m_q can be understood and controlled via the chiral Lagrangian.[6] Finally, discretization effects of the heavy-quark mass m_Q are treated with HQET[7] or NRQCD.[8] In each case one can control the extrapolation of artificial, numerical data, if one generates numerical data close enough to the real world.

Volume effects are unimportant in what follows, and chiral perturbation theory is a relatively well-known subject. Therefore, here I will focus on the effective field theories that help us control discretization effects.

2.1 Symanzik's LE$\mathcal{L}$

Symanzik's formalism[4] describes the lattice theory with continuum QCD:

$$\mathcal{L}_{\text{lat}} \doteq \mathcal{L}_{\text{cont}} + \sum_i a^{s_i} C_i(a;\mu)\mathcal{O}_i(\mu), \qquad (1)$$

where the symbol $\doteq$ means "has the same physics as". The LE$\mathcal{L}$ on the right-hand side is defined in, say, the $\overline{\text{MS}}$ scheme at scale μ. The coefficients C_i describe short-distance physics, so they depend on the lattice spacing a. The operators do not depend on a.

If $\Lambda_{\text{QCD}}a$ is small enough the higher terms can be treated as perturbations. So, the a dependence of the proton mass is

$$m_p(a) = m_p + aC_{\sigma F}\langle p|\bar{\psi}\sigma \cdot F\psi|p\rangle, \qquad (2)$$

taking the leading operator for Wilson fermions as an example. To reduce the second term one might try to reduce a greatly, but CPU time goes as $a^{-(5 \text{ or } 6)}$. It is more effective to combine several data sets and extrapolate, with Eq. (2) as a guide. It is even better to adjust things so $C_{\sigma F}$ is $O(\alpha_s^\ell)$ or $O(a)$, which is called Symanzik improvement of the action. For light hadrons, a combination of improvement and extrapolation is best, and you should look for both.

2.2 HQET for large m_Q

The Symanzik theory, as usually applied, assumes $m_q a \ll 1$. The bottom and charm quarks' masses in lattice units are at present large: $m_b a \sim$ 1–2 and $m_c a$ about a third of that. It will not be possible to reduce a enough to make $m_b a \ll 1$ for many, many years. So, other methods are needed to control the lattice spacing effects of heavy quarks. There are several alternatives:

1. static approximation[9]
2. lattice NRQCD[10]
3. extrapolation from $m_Q \approx m_c$ up to m_b

3′. combine 3 with 1

4. normalize systematically to HQET[11]

All use HQET in some way. The first two discretize continuum HQET; method 1 stops at the leading term, and method 2 carries the heavy-quark expansion out to the desired order. Methods 3 and 3′ keep the heavy quark mass artificially small and appeal to the $1/m_Q$ expansion to extrapolate back up to m_b. Method 4 uses the same lattice action as method 3, but uses the heavy-quark expansion to normalize and improve it. Methods 2 and 4 are able to calculate matrix elements directly at the b-quark mass.

The methods can be compared and contrasted by *describing* the lattice theories with HQET.[12] This is, in a sense, the opposite of *discretizing* HQET. One writes down a (continuum) effective Lagrangian

$$\mathcal{L}_{\text{lat}} \doteq \sum_n \mathcal{C}_{\text{lat}}^{(n)}(m_Q a; \mu)\mathcal{O}_{\text{HQET}}^{(n)}(\mu), \qquad (3)$$

with the operators defined exactly as in continuum HQET, so they do not depend on m_Q or a. As long as $m_Q \gg \Lambda_{\text{QCD}}$ this description makes sense. There are two short distances, $1/m_Q$ and the lattice spacing a, so the short-distance coefficients $\mathcal{C}_{\text{lat}}^{(n)}$ depend on $m_Q a$. Since all dependence on $m_Q a$ is isolated into the coefficients, this description shows that heavy-quark lattice artifacts arise only from the mismatch of the $\mathcal{C}_{\text{lat}}^{(n)}$ and their continuum analogs $\mathcal{C}_{\text{cont}}^{(n)}$.

For methods 1 and 2, Eq. (3) is just a Symanzik LE$\mathcal{L}$. For lattice NRQCD we recover the well-known result that some of the coefficients have power-law divergences.[10] So, to take the continuum limit one must add more and more terms to the action. This leaves a systematic error, which, in practice, is usually accounted for conservatively.

Eq. (3) is more illuminating for methods 3 and 4, which use Wilson fermions (with an improved action). Wilson fermions have the same degrees of freedom and heavy-quark symmetries as continuum QCD, so the HQET description is admissible for all $m_Q a$. Method 4 matches the coefficients of Eq. (3) term by term, by adjusting the lattice action. In practice, this is possible only to finite order, so there are errors $(\mathcal{C}_{\text{lat}}^{(n)} - \mathcal{C}_{\text{cont}}^{(n)})\langle\mathcal{O}_{\text{HQET}}^{(n)}\rangle$, starting with some n. Method 3 reduces $m_Q a$ until the mismatch is of order $(m_Q a)^2 \ll 1$ (or $\lesssim 1$). This runs the risk of reducing m_Q until the heavy-quark expansion falls apart.

The non-expert can get a feel for which methods are most appropriate by asking himself what order in Λ_{QCD}/m_b is needed. For zeroth order, method 1 will do. For the first few orders, the others are needed, although with method 3 one should check that the calculation's Λ_{QCD}/m_Q is small enough too.

3 New Results

3.1 $\bar{\Lambda}$ and λ_1

The matching of lattice gauge theory to HQET provides a new way to calculate matrix elements of the heavy-quark expansion.[13] The spin-averaged B^*-B mass is given by[14]

$$\bar{M} = m + \bar{\Lambda} - \lambda_1/2m, \qquad (4)$$

where m is the heavy quark mass, and $\bar{M} = \frac{1}{4}(3M_{B^*} + M_B)$. The lattice changes the short-distance definition of the quark mass:[12]

$$\bar{M}_1 - m_1 = \bar{\Lambda}_{\text{lat}} - \lambda_{1\,\text{lat}}/2m_2. \qquad (5)$$

Because the lattice breaks Lorentz symmetry, $m_1 \neq m_2$, but they are still calculable in perturbation theory.[15] The lambdas in Eq. (5) are labeled "lat" because they suffer lattice artifacts from the gluons and light quark.

After fitting a wide range of lattice data to Eq. (5) *and* taking the continuum limit, we find[13] $\bar{\Lambda} = 0.68^{+0.02}_{-0.12}$ GeV and $\lambda_1 = -(0.45 \pm 0.12)$ GeV^2 in the quenched approximation. The lambdas appear also in the heavy-quark expansion of inclusive decays. Although the current analysis is thorough, there are several ways to improve it.[13] For example, $\bar{\Lambda}_{\text{lat}}$ has an unexpectedly large a dependence, so the analysis should be repeated with an action for which $C_{\sigma F}$ in Eq. (2) is $O(a)$.

3.2 $B \to \pi l \nu$ form factors and V_{ub}

It is timely to discuss $\bar{B}^0 \to \pi^+ l^- \bar{\nu}$, because there are three calculations to compare, using lattice NRQCD (method 2),[16] the extrapolation method (method 3),[17] and the HQET matching method (method 4).[18] UKQCD's work is final,[17] and the other two are preliminary.[16,18] The decay rate requires a form factor, called $f_+(E)$, which depends on the pion's energy in the B's rest frame, $E = v \cdot p_\pi$. It is related to the matrix element $\langle \pi | V^\mu | B \rangle$, which can be computed in lattice QCD. The systematics are smallest when the pion's three-momentum is small.

The three recent results are compared in Fig. 3. The error bars shown are statistical only. For NRQCD these are larger than expected.[16] For the other two, the comparison gives a fair idea of systematics that are not common to both, because the extrapolation of heavy quark mass needed with method 3 amplifies the statistical error.[17] The other two works[16,18] compute directly at the b quark mass and, thus, circumvent this problem. The heavy quark masses of UKQCD[17] are all below 1.3 GeV, and as low as 500 MeV, so one might worry whether the heavy-quark expansion applies.

3.3 $B^- \to D^{0(*)} l^- \bar{\nu}$ and V_{cb}

The form factors $\mathcal{F}_{B \to D^{(*)}}$ of the decays $B \to D^{(*)} l \nu$ are normalized to unity for infinite quark masses. What is needed from lattice QCD, therefore, is the deviation from the unity for physical quark masses. Hashimoto *et al.*[19,20] have devised methods based on double ratios, in which all the uncertainties cancel in the symmetry limit. Consequently, all errors scale as $\mathcal{F} - 1$, not as $\mathcal{F}$.

For $B \to D l \nu$ they find[19] (published)

$$\mathcal{F}_{B \to D}(1) = 1.058 \pm 0.016 \pm 0.003^{+0.014}_{-0.005}, \quad (6)$$

where error bars are from statistics, adjusting the quark masses, and higher-order radiative corrections. For $B \to D^* l \nu$ they find[20] (still preliminary)

$$\begin{aligned} \mathcal{F}_{B \to D^*}(1) = 0.935 \pm 0.022^{+0.008}_{-0.011} \\ \pm 0.008 \pm 0.020, \end{aligned} \quad (7)$$

where now the last uncertainty is from $1/m_Q^3$. In both results, an ongoing test of the lattice spacing dependence is not included, but that will probably not be noticeable. More seriously, these results are, once again, in the quenched approximation, but the associated uncertainty are still only a fraction of $\mathcal{F} - 1$. Both results will be updated soon, with calculations at a second lattice spacing and refinements in the radiative corrections.

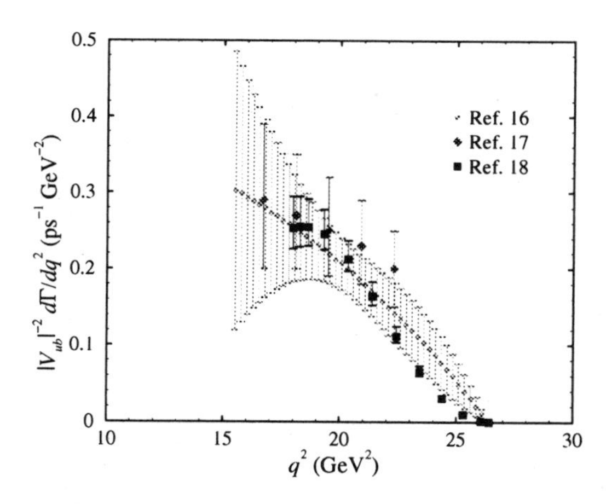

Figure 3. Recent results for the decay $B \to \pi l \nu$.

group	method	$B_{B_d}(4.8\text{ GeV})$
JLQCD[21]	2	$0.85 \pm 0.03 \pm 0.11$
APE[23]	3	$0.93 \pm 0.08^{+00}_{-06}$
UKQCD[24]	3	$0.92 \pm 0.04^{+03}_{-00}$

Table 1. Recent quenched results for B_B.

3.4 B^0_q-$\bar{B}^0_q$ mixing: B_B, f_B, and V_{tq}

The mass difference of CP eigenstates is

$$\Delta m_{B^0_q} = \frac{G_F^2 m_W^2 S_0}{16\pi^2} |V_{tq}^* V_{tb}|^2 \eta_B \langle Q_q^{\Delta B=2} \rangle, \tag{8}$$

where the light quark q is d or s, S_0 is an Inami-Lim function, and $\langle Q_q^{\Delta B=2} \rangle$ is

$$\langle \bar{B}^0_q | Q_q^{\Delta B=2} | B^0_q \rangle = \tfrac{8}{3} m_{B_q}^2 f_{B_q}^2 B_{B_q}. \tag{9}$$

The μ dependence in η_B and $Q_q^{\Delta B=2}$ cancels. New physics could compete with the W and t box diagrams and change Eq. (8).

Lattice QCD gives matrix elements, so the basic results are $\langle Q_q^{\Delta B=2} \rangle$ and f_{B_q}. It is often stated that uncertainties in B_{B_q} and $\xi^2 = f_{B_s}^2 B_{B_s} / f_{B_s}^2 B_{B_s}$ should be small, because they are ratios. Some cancelation should occur, but only if one can show that the errors are under control. At present there are unresolved issues in method 3, so one should be cautious.

With that warning, results for B_B from three groups are in Table 1. Note that JLQCD now includes the short-distance part of the $1/m_Q$ contribution.[22] APE[23] (final) and UKQCD[24] (preliminary) both extrapolate linearly in $1/m_Q$ from charm (e.g., $1.75\text{ GeV} < m_{\text{“}B\text{”}} < 2.26\text{ GeV}$ for APE). It is not clear whether the first term of the heavy-quark expansion is adequate here; everyone working in B physics can and should form his or her own opinion. It is also not clear how the $(m_Q a)^2$ lattice artifacts of method 3 fare through the $1/m_Q$ extrapolation. It is likely that the systematic error is not well controlled and, thus, possibly underestimated in for the last two rows Table 1. At present one should prefer the JLQCD results.

$f_P(n_f)$	MILC[25]	CP-PACS[26]
$f_B(0)$	$171 \pm 6 \pm 17^{+21}_{-\ 4}$	$190 \pm 3 \pm 9$
(2)	$190 \pm 6^{+20+9}_{-15-0}$	$215 \pm 11 \pm 11$
$f_{B_s}(0)$	$197 \pm 5 \pm 23^{+25}_{-\ 6}$	$224 \pm 2 \pm 15$
(2)	$218 \pm 5^{+26+11}_{-23-0}$	$250 \pm 10 \pm 13$
$f_D(0)$	$199 \pm 6 \pm 12^{+14}_{-\ 0}$	$224 \pm 2 \pm 15$
(2)	$213 \pm 4^{+14+7}_{-13-0}$	$236 \pm 14 \pm 14$
$f_{D_s}(0)$	$222 \pm 5^{+19+15}_{-17-\ 0}$	$252 \pm 1 \pm 18$
(2)	$240 \pm 4^{+25+9}_{-23-0}$	$275 \pm 10 \pm 17$

Table 2. Preliminary unquenched results for f_B, etc., in the continuum limit. All values in MeV.

The MILC[25] and CP-PACS[26] groups have new, *preliminary* unquenched calculations of the heavy-light decay constants f_B, f_{B_s}, f_D, and f_{D_s}. Both use method 4. Both have results at several lattice spacings, so they can study the continuum limit. The status for Osaka is tabulated in Table 2. The first error is statistical, the second systematic. MILC also provides an estimate of the error from quenching. (With $n_f = 2$ the strange quark is still quenched.) CP-PACS[26] also has results with method 2, which agree very well with method 4. One should not, at this time, take the differences between the two groups' central values very seriously. It is more important to understand the different systematics of methods 2, 3, and 4.

4 Prospects

For B physics it is important to remove the quenched approximation, more so than to reduce the lattice spacing much further. To do so, we need more computing. Fermilab, MILC, and Cornell are building a cluster of PCs to tackle the problem.[27] Our pilot cluster has 8 nodes with a Myrinet switch. We plan to go up to 48–64 nodes, and then hope to assemble a cluster of thousands of nodes. The large cluster would evolve, by upgrading a third or so of the nodes every year. This is an ambitious plan, but not more ambitious than the experimental effort to understand flavor mixing and CP violation.

The last few years have seen significant strides in understanding heavy quarks in lattice QCD. The progress has been both computational and theoretical, with one guiding the other. Calculations shown here, for $\bar{B}^0 \to \pi^+ l^- \bar{\nu}$, $D^0 \to \pi^- l^+ \nu$, $B^- \to D^{0(*)} l^- \bar{\nu}$, and B_d^0-$\bar{B}_d^0$ mixing are a subset, but in Fig. 2 they are as basic as A, B, C. With the right amount of support from the rest of the community, we hope to obtain the tools needed to resolve the few outstanding problems and to produce excellent unquenched results. Indeed, the example of f_B shows that this is already beginning.

Acknowledgments

I would like to thank Arifa Ali Khan, Claude Bernard, Rüdi Burkhalter, Tetsuya Onogi, Hugh Shanahan and Akira Ukawa for correspondence. I have benefited greatly from collaboration with Shoji Hashimoto, Aida El-Khadra, Paul Mackenzie, Sinéad Ryan, and Jim Simone. Fermilab is operated by Universities Research Association Inc., under contract with the U.S. Department of Energy.

References

1. S. Plaszczynski and M. Schune, "Overall determination of the CKM matrix," hep-ph/9911280.
2. J. D. Lykken, "Physics needs for future accelerators," hep-ph/0001319.
3. Y. Nir, "Future lessons from CP violation," `http://www-theory.fnal.gov/ /people/ligeti/Brun2/` (Feb., 2000).
4. K. Symanzik, in *Recent Developments in Gauge Theories*, ed. G. 't Hooft *et al.* (Plenum, New York, 1980).
5. M. Lüscher, Commun. Math. Phys. **104**, 177 (1986); **105**, 153 (1986).
6. J. Gasser and H. Leutwyler, Annals Phys. **158**, 142 (1984).
7. For a popular review, see M. Neubert, Phys. Rept. **245**, 259 (1994).
8. W.E. Caswell and G.P. Lepage, Phys. Lett. **167B**, 437 (1986).
9. E. Eichten, Nucl. Phys. B Proc. Suppl. **4**, 170 (1987); E. Eichten and B. Hill, Phys. Lett. **B234**, 511 (1990).
10. G. P. Lepage and B. A. Thacker, Nucl. Phys. B Proc. Suppl. **4**, 199 (1987); B. A. Thacker and G. P. Lepage, Phys. Rev. **D43**, 196 (1991).
11. A. X. El-Khadra, A. S. Kronfeld, and P. B. Mackenzie, Phys. Rev. **D55**, 3933 (1997).
12. A. S. Kronfeld, Phys. Rev. **D62**, 014505 (2000).
13. A. S. Kronfeld and J. N. Simone, hep-ph/0006345.
14. A. F. Falk and M. Neubert, Phys. Rev. **D47**, 2965 (1993).
15. B. P. Mertens, A. S. Kronfeld and A. X. El-Khadra, Phys. Rev. **D58**, 034505 (1998).
16. S. Aoki *et al.* [JLQCD Collaboration], Nucl. Phys. Proc. Suppl. **83**, 325 (2000); T. Onogi, private communication.
17. K. C. Bowler *et al.* [UKQCD Collaboration], Phys. Lett. **B486**, 111 (2000).
18. S. M. Ryan *et al.*, Nucl. Phys. Proc. Suppl. **83**, 328 (2000).
19. S. Hashimoto *et al.*, Phys. Rev. **D61**, 014502 (2000).
20. J. N. Simone *et al.*, Nucl. Phys. Proc. Suppl. **83**, 334 (2000).
21. S. Hashimoto *et al.*, Phys. Rev. **D60**, 094503 (1999).
22. S. Hashimoto *et al.*, hep-lat/0004022.
23. D. Becirevic *et al.*, hep-lat/0002025.
24. L. Lellouch and C. J. Lin [UKQCD Collaboration], hep-ph/9912322.
25. C. Bernard *et al.* [MILC Collaboration], Nucl. Phys. Proc. Suppl. **83**, 289 (2000); C. Bernard, private communication.
26. A. Ali Khan, R. Burkhalter, and H. Shanahan, private communication.
27. For more details, visit our web-site at `http://www-theory.fnal.gov/pcqcd/`.

Parallel Session 15

Field Theory

Conveners: Thomas Appelquist (Yale) and
Kenichi Konishi (Pisa)

QUANTUM AND CLASSICAL GAUGE SYMMETRIES

KAZUO FUJIKAWA AND HIROAKI TERASHIMA
Department of Physics, University of Tokyo, Bunkyo-ku, Tokyo 113, Japan

The use of the mass term of the gauge field as a gauge fixing term, which was discussed by Zwanziger, Parrinello and Jona-Lasinio in a large mass limit, is related to the non-linear gauge by Dirac and Nambu. We have recently shown that this use of the mass term as a gauge fixing term is in fact identical to the conventional local Faddeev-Popov formula without taking a large mass limit, if one takes into account the variation of the gauge field along the entire gauge orbit. This suggests that the classical massive vector theory, for example, could be re-interpreted as a gauge invariant theory with a gauge fixing term added in suitably quantized theory. As for massive gauge particles, the Higgs mechanics, where the mass term is gauge invariant, has a more intrinsic meaning. We comment on several implications of this observation.

1 Introduction

The Faddeev-Popov formula[1] and the resulting BRST symmetry[2] provide a basis for the modern quantization of gauge theory. On the other hand, a modified quantization scheme[34]

$$\int \mathcal{D}A_\mu\{\exp[-S_{YM}(A_\mu) - \int f(A_\mu)dx] \Big/ \int \mathcal{D}g\exp[-\int f(A^g_\mu)dx]\} \quad (1)$$

with, for example, $f(A_\mu) = (m^2/2)(A_\mu)^2$ has been recently analyzed in a large mass limit in connection with the analysis of Gribov-type complications[5]. This gauge fixing in the large mass limit is related to the limit $\lambda = 0$ of the non-linear gauge

$$A^2_\mu = \lambda = \quad const. \quad (2)$$

discussed by Dirac and Nambu[6] many years ago. Nambu used the above gauge to analyze the possible spontaneous breakdown of Lorentz symmetry. In his treatment, the limit $\lambda = 0$ is singular, and thus the present formulation is not quite convenient for an analysis of the possible breakdown of Lorentz symmetry. Some aspects of this non-linear gauge have been discussed in Ref.[7]. The above gauge fixing (1) has also been used in lattice simulation[8].

We have recently shown[9] that the above scheme is in fact identical to the conventional *local* Faddeev-Popov formula

$$\int \mathcal{D}A_\mu\delta(D^\mu\frac{\delta f(A_\nu)}{\delta A_\mu})\det\{\delta[D^\mu\frac{\delta f(A^g_\nu)}{\delta A^g_\mu}]/\delta g\} \times\exp[-S_{YM}(A_\mu)] \quad (3)$$

without taking the large mass limit, if one takes into account the variation of the gauge field along the entire gauge orbit parametrized by the gauge parameter g. The above equivalence is valid only if the Gribov-type complications are ignored.

We here comment on the possible implications[10] of the above equivalence in a more general context of quantum gauge symmetry, namely, BRST symmetry.

2 Abelian example

We first briefly illustrate the proof[9] of the above equivalence of (1) and (3) by using an example of Abelian gauge theory, $S_0 = -(1/4)\int dx(\partial_\mu A_\nu - \partial_\nu A_\mu)^2$, for which we can work out everything explicitly. In this note we exclusively work on Euclidean theory with metric convention $g_{\mu\nu} = (1,1,1,1)$. As a simple and useful example, we choose the gauge fixing function $f(A) \equiv (1/2)A_\mu A_\mu$ and

$$D_\mu(\frac{\delta f}{\delta A_\mu}) = \partial_\mu A_\mu. \quad (4)$$

Our claim above suggests the relation

$$Z = \int \mathcal{D}A^\omega_\mu\{e^{-S_0(A^\omega_\mu)-\int dx\frac{1}{2}(A^\omega_\mu)^2}$$

$$/\int \mathcal{D}h\, e^{-\int dx \frac{1}{2}(A_\mu^{h\omega})^2}\}$$

$$= \int \mathcal{D}A_\mu^\omega \mathcal{D}B\mathcal{D}\bar{c}\mathcal{D}c \times e^{-S_0(A_\mu^\omega)+\int[-iB\partial_\mu A_\mu^\omega+\bar{c}(-\partial_\mu\partial_\mu)c]dx} \quad (5)$$

where the variable A_μ^ω stands for the field variable obtained from A_μ by a gauge transformation parametrized by the gauge orbit parameter ω. To establish this result, we first evaluate

$$\int \mathcal{D}h\, e^{-\int dx \frac{1}{2}(A_\mu^{h\omega})^2}$$
$$= \int \mathcal{D}h\, e^{-\int dx \frac{1}{2}(A_\mu^\omega+\partial_\mu h)^2}$$
$$= \int \mathcal{D}h\, e^{-\int dx \frac{1}{2}[(A_\mu^\omega)^2-2(\partial_\mu A_\mu^\omega)h+h(-\partial_\mu\partial_\mu)h]}$$
$$= \int \mathcal{D}B \frac{1}{\det\sqrt{-\partial_\mu\partial_\mu}} \times e^{-\int dx \frac{1}{2}[(A_\mu^\omega)^2-2(\partial_\mu A_\mu^\omega)\frac{1}{\sqrt{-\partial_\mu\partial_\mu}}B+B^2]}$$
$$= \frac{1}{\det\sqrt{-\partial_\mu\partial_\mu}} \times e^{-\int dx\frac{1}{2}(A_\mu^\omega)^2+\frac{1}{2}\int\partial_\mu A_\mu^\omega\frac{1}{-\partial_\mu\partial_\mu}\partial_\nu A_\nu^\omega dx} \quad (6)$$

where we defined $\sqrt{-\partial_\mu\partial_\mu}h = B$. Thus

$$Z = \int \mathcal{D}A_\mu^\omega\{\det\sqrt{-\partial_\mu\partial_\mu}\} \times e^{-S_0(A_\mu^\omega)-\frac{1}{2}\int\partial_\mu A_\mu^\omega\frac{1}{-\partial_\mu\partial_\mu}\partial_\nu A_\nu^\omega dx}$$
$$= \int \mathcal{D}A_\mu^\omega\mathcal{D}B\mathcal{D}\bar{c}\mathcal{D}c \times \exp\{-S_0(A_\mu^\omega) - \frac{1}{2}\int B^2 dx + \int[-iB\frac{1}{\sqrt{-\partial_\mu\partial_\mu}}\partial_\mu A_\mu^\omega + \bar{c}\sqrt{-\partial_\mu\partial_\mu}c]dx\} \quad (7)$$

which was derived in Refs.[3] and [4] and is invariant under the BRST transformation

$$\delta A_\mu^\omega = i\lambda\partial_\mu c, \quad \delta c = 0$$
$$\delta\bar{c} = \lambda B, \quad \delta B = 0 \quad (8)$$

with a Grassmann parameter λ. Note the appearance of the imaginary factor i in the term $iB\frac{1}{\sqrt{-\partial_\mu\partial_\mu}}\partial_\mu A_\mu^\omega$ in (7).

We next rewrite the expression (7) as

$$\int \mathcal{D}A_\mu^\omega\mathcal{D}B\mathcal{D}\Lambda\mathcal{D}\bar{c}\mathcal{D}c\,\delta(\frac{1}{\sqrt{-\partial_\mu\partial_\mu}}\partial_\mu A_\mu^\omega - \Lambda) \times e^{-S_0(A_\mu^\omega)-\frac{1}{2}\int(B^2+2i\Lambda B)dx+\int\bar{c}\sqrt{-\partial_\mu\partial_\mu}c dx}$$
$$= \int \mathcal{D}A_\mu^\omega\mathcal{D}\Lambda\mathcal{D}\bar{c}\mathcal{D}c\,\delta(\frac{1}{\sqrt{-\partial_\mu\partial_\mu}}\partial_\mu A_\mu^\omega - \Lambda) \times e^{-S_0(A_\mu^\omega)-\frac{1}{2}\int\Lambda^2 dx+\int\bar{c}\sqrt{-\partial_\mu\partial_\mu}c dx}. \quad (9)$$

We note that we can compensate any variation of $\delta\Lambda$ by a suitable change of gauge parameter $\delta\omega$ inside the δ-function as

$$\frac{1}{\sqrt{-\partial_\mu\partial_\mu}}\partial_\mu\partial_\mu\delta\omega = \delta\Lambda. \quad (10)$$

By a repeated application of infinitesimal gauge transformations combined with the invariance of the path integral measure under these gauge transformations, we can re-write the formula (9) as

$$\int \mathcal{D}A_\mu^\omega\mathcal{D}\Lambda\mathcal{D}\bar{c}\mathcal{D}c\,\delta(\frac{1}{\sqrt{-\partial_\mu\partial_\mu}}\partial_\mu A_\mu^\omega) \times e^{-S_0(A_\mu^\omega)-\frac{1}{2}\int\Lambda^2 dx+\int\bar{c}\sqrt{-\partial_\mu\partial_\mu}c dx}$$
$$= \int \mathcal{D}A_\mu^\omega\mathcal{D}\bar{c}\mathcal{D}c\,\delta(\frac{1}{\sqrt{-\partial_\mu\partial_\mu}}\partial_\mu A_\mu^\omega) \times e^{-S_0(A_\mu^\omega)+\int\bar{c}\sqrt{-\partial_\mu\partial_\mu}c dx}$$
$$= \int \mathcal{D}A_\mu^\omega\mathcal{D}B\mathcal{D}\bar{c}\mathcal{D}c \times e^{-S_0(A_\mu^\omega)+\int[-iB\frac{1}{\sqrt{-\partial_\mu\partial_\mu}}\partial_\mu A_\mu^\omega+\bar{c}\sqrt{-\partial_\mu\partial_\mu}c]dx}$$
$$= \int \mathcal{D}A_\mu^\omega\mathcal{D}B\mathcal{D}\bar{c}\mathcal{D}c \times e^{-S_0(A_\mu^\omega)+\int[-iB\partial_\mu A_\mu^\omega+\bar{c}(-\partial_\mu\partial_\mu)c]dx}. \quad (11)$$

In the last stage of this equation, we redefined the *auxiliary* variables B and $\bar{c}$ as

$$B \to B\sqrt{-\partial_\mu\partial_\mu}, \quad \bar{c} \to \bar{c}\sqrt{-\partial_\mu\partial_\mu} \quad (12)$$

which is consistent with BRST symmetry and leaves the path integral measure invariant. We have thus established the desired result (5). We emphasize that the integral over the entire gauge orbit, as is indicated in (10), is essential to derive a local theory (11) without taking a large mass limit[9].

It is shown that this procedure works for the non-Abelian case also[9], though the actual procedure is much more involved, if the (ill-understood) Gribov-type complications can be ignored such as in perturbative calculations.

3 Possible Implications

In the classical level, we traditionally consider

$$\mathcal{L} = -\frac{1}{4}(\partial_\mu A_\nu - \partial_\nu A_\mu)^2 - \frac{1}{2}m^2 A_\mu A^\mu \quad (13)$$

as a Lagrangian for a massive vector theory, and

$$\mathcal{L}_{eff} = -\frac{1}{4}(\partial_\mu A_\nu - \partial_\nu A_\mu)^2 - \frac{1}{2}(\partial_\mu A^\mu)^2 \quad (14)$$

as an effective Lagrangian for Maxwell theory with a Feynman-type gauge fixing term added. The physical meanings of these two Lagrangians are thus completely different.

However, the analysis in Section 2 shows that the Lagrangian (13) could in fact be interpreted as a gauge fixed Lagrangian of *massless* Maxwell field in quantized theory. To be explicit, by using (5), the Lagrangian (13) may be regarded as an effective Lagrangian in

$$Z = \int \mathcal{D}A^\omega_\mu \{ e^{\int dx[-\frac{1}{4}(\partial_\mu A_\nu - \partial_\nu A_\mu)^2 - \frac{1}{2}m^2 A^\omega_\mu A^{\omega\mu}]} \Big/ \int \mathcal{D}h\, e^{-\int dx \frac{m^2}{2}(A^{h\omega}_\mu)^2} \}$$
$$= \int \mathcal{D}A^\omega_\mu \mathcal{D}B\mathcal{D}\bar{c}\mathcal{D}c \times \exp\{ \int dx[-\frac{1}{4}(\partial_\mu A_\nu - \partial_\nu A_\mu)^2 - iB\partial_\mu A^\omega_\mu + \bar{c}(-\partial_\mu\partial_\mu)c]\}. \quad (15)$$

where we absorbed the factor m^2 into the definition of B and $\bar{c}$.

One can also analyze (14) by defining $f(A_\mu) \equiv \frac{1}{2}(\partial_\mu A^\mu)^2$ in the modified quantization scheme (1). The equality of (1) and (3) then gives

$$\int \mathcal{D}A_\mu \delta(D^\mu \frac{\delta f(A_\nu)}{\delta A_\mu}) \times \det\{\delta[D^\mu \frac{\delta f(A^g_\nu)}{\delta A^g_\mu}]/\delta g\} \exp[-S_0(A_\mu)]$$
$$= \int \mathcal{D}A_\mu \delta(\partial_\nu\partial^\nu(\partial^\mu A_\mu)) \times \det[\partial_\nu\partial^\nu\partial_\mu\partial^\mu] \exp[-S_0(A_\mu)]$$
$$= \int \mathcal{D}A_\mu \mathcal{D}B\mathcal{D}\bar{c}\mathcal{D}c \times \exp\{-S_0(A_\mu) + \int dx[-iB\partial_\nu\partial^\nu(\partial^\mu A_\mu) - \bar{c}(\partial_\nu\partial^\nu\partial_\mu\partial^\mu)c]\} \quad (16)$$

After the re-definition of *auxiliary* variables, $B\partial_\nu\partial^\nu \to B$, $\bar{c}\partial_\nu\partial^\nu \to \bar{c}$, which preserves BRST symmetry, (16) becomes

$$\int \mathcal{D}A_\mu \mathcal{D}B\mathcal{D}\bar{c}\mathcal{D}c \times \exp\{-S_0(A_\mu) + \int dx[-iB(\partial^\mu A_\mu) + \bar{c}(-\partial_\mu\partial^\mu)c]\} \quad (17)$$

which agrees with (11) and (15). We can thus assign an identical physical meaning to two classical Lagrangians (13) and (14) in suitably *quantized* theory.

Similarly, the two classical Lagrangians related to Yang-Mills fields

$$\mathcal{L} = -\frac{1}{4}(\partial_\mu A^a_\nu - \partial_\nu A^a_\mu + gf^{abc}A^b_\mu A^c_\nu)^2 - \frac{m^2}{2}A^a_\mu A^{a\mu} \quad (18)$$

and

$$\mathcal{L}_{eff} = -\frac{1}{4}(\partial_\mu A^a_\nu - \partial_\nu A^a_\mu + gf^{abc}A^b_\mu A^c_\nu)^2 - \frac{1}{2}(\partial_\mu A^{a\mu})^2 \quad (19)$$

could be assigned an identical physical meaning as an effective gauge fixed Lagrangian associated with the quantum theory defined by[9]

$$\int \mathcal{D}A^a_\mu \mathcal{D}B^a \mathcal{D}\bar{c}^a \mathcal{D}c^a \times \exp\{-S_{YM}(A^a_\mu) + \int dx[-iB^a(\partial^\mu A^a_\mu) + \bar{c}^a(-\partial_\mu(D^\mu c)^a]\} \quad (20)$$

which is invariant under BRST symmetry.

We have illustrated that the apparent "massive gauge field" in the classical level has no intrinsic physical meaning. It can be interpreted either as a massive (non-gauge) vector theory, or as a gauge-fixed effective Lagrangian for a massless gauge field in quantized theory. In the framework of path integral, these different *interpretations* may also be understood as a more flexible choice of the path integral measure than the classical Poisson bracket analysis suggests[10]: One choice of the measure

$$\begin{aligned}\int d\mu\ \exp\{\int dx[-\frac{1}{4}(\partial_\mu A^a_\nu - \partial_\nu A^a_\mu \\ +gf^{abc}A^b_\mu A^c_\nu)^2 - \frac{m^2}{2}A^a_\mu A^{a\mu}]\} \\ \equiv \int \mathcal{D}A_\mu \frac{1}{\int \mathcal{D}g \exp[-\int \frac{m^2}{2}(A^{ag}_\mu)^2 dx]\}} \\ \times \exp\{\int dx[-\frac{1}{4}(\partial_\mu A^a_\nu - \partial_\nu A^a_\mu \\ +gf^{abc}A^b_\mu A^c_\nu)^2 - \frac{m^2}{2}A^a_\mu A^{a\mu}]\} \quad (21)\end{aligned}$$

gives rise to a renormalizable massless gauge theory, and the other naive choice

$$\begin{aligned}\int d\mu\ \exp\{\int dx[-\frac{1}{4}(\partial_\mu A^a_\nu - \partial_\nu A^a_\mu \\ +gf^{abc}A^b_\mu A^c_\nu)^2 - \frac{m^2}{2}A^a_\mu A^{a\mu}]\} \\ \equiv \int \mathcal{D}A_\mu \exp\{\int dx[-\frac{1}{4}(\partial_\mu A^a_\nu - \partial_\nu A^a_\mu \\ +gf^{abc}A^b_\mu A^c_\nu)^2 - \frac{m^2}{2}A^a_\mu A^{a\mu}]\} \quad (22)\end{aligned}$$

gives rise to a non-renormalizable massive *non-gauge* theory. A somewhat analogous situation arises when one attempts to quantize the so-called anomalous gauge theory: A suitable choice of the measure with a Wess-Zumino term gives rise to a consistent quantum theory in 2-dimensions, for example[11]. ¿From a view point of classical-quantum correspondence, one can define a classical theory uniquely starting from quantum theory by considering the limit $\hbar \to 0$, but not the other way around in general.

In the context of the present general interpretation of apparently massive classical gauge fields, the massive gauge fields generated by the Higgs mechanism are exceptional and quite different. Since all the terms including the mass term are gauge invariant, one can assign an intrinsic meaning to the massive gauge field in Higgs mechanism. In view of the well known fact that the massive non-Abelian gauge theory is inconsistent as a quantum theory (22), it may be sensible to treat all the classical massive non-Abelian Lagrangians as a gauge fixed version of pure non-Abelian gauge theory and to restrict the massive non-Abelian gauge fields to those generated by the Higgs mechanism.

It is a long standing question if one can generate gauge fields from some *more* fundamental mechanism. To our knowledge, however, there exists no definite convincing scheme so far. On the contrary, there is a no-go theorem or several arguments against such an attempt[12]. Apart from technical details, the basic argument against the "dynamical" generation of gauge fields is that the Lorentz invariant positive definite theory cannot simply generate the negative metric states associated with the time components of massless gauge fields. In contrast, the dynamical generation of the Lagrangian of the structure

$$\begin{aligned}\mathcal{L} = -\frac{1}{4}(\partial_\mu A^a_\nu - \partial_\nu A^a_\mu + gf^{abc}A^b_\mu A^c_\nu)^2 \\ -\frac{m^2}{2}(A^a_\mu)^2 \quad (23)\end{aligned}$$

does not appear to be prohibited by general arguments so far. If one considers that the induced Lagrangian such as (23) is a *classical* object which should be quantized anew, one could regard $\frac{m^2}{2}(A^a_\mu)^2$, which breaks classical gauge symmetry, as a gauge fixing term in the modified quantization scheme[34]. In this interpretation, one might be allowed to say that massless gauge fields are generated dynamically. Although a dynamical generation of pure gauge fields is prohibited, a *gauge fixed* Lagrangian might be allowed to be generated. (In this respect, one may recall that much of the arguments for the no-go theorem[12] would be refuted if one could generate a gauge fixed Lagrangian with the Faddeev-Popov term added.) The mass for the gauge field which has an intrinsic unambiguous physical meaning is then further induced by the spontaneous symmetry breaking of the gauge symmetry thus defined (the Higgs mechanism).

We next comment on a mechanism for generating gauge fields by the violent random fluctuation of gauge degrees of freedom at the beginning of the universe[13]; this scheme is based on the renormalization group flow starting from an initial chaotic theory. In such a scheme, it is natural to think that one is always dealing with quantum theory, and thus no room for our way of re-interpretation of the induced theory. Nevertheless, we find a possible connection in the following sense: To be precise, an example of massive Abelian gauge field in *compact* lattice gauge theory

$$\int \mathcal{D}U \frac{\mathcal{D}\Omega}{vol(\Omega)} \exp[-S_{inv}(U) - S_{mass}(U^{\Omega})] \tag{24}$$

is analyzed in Ref.[13]. Here $S_{inv}(U)$ stands for the gauge invariant part of the lattice Abelian gauge field U, and $S_{mass}(U^{\Omega})$ stands for the gauge non-invariant mass term with the gauge freedom Ω. In compact theory, one need not fix the gauge and instead one may take an average over the entire gauge volume of Ω. They argued that the mass term, which breaks gauge symmetry softly, disappears in the long distance limit when one integrates over the entire gauge freedom Ω. Their scheme is apparently dynamical one, in contrast to the kinematical nature of our re-interpretation. Nevertheless, the massive Abelian theory is a free theory in continuum formulation, and the disappearance of the mass term by a mere smearing over the gauge volume may suggest that the mass term in their scheme is also treated as a kind of gauge artifact, just as in our kinematical re-interpretation.

In conclusion, the equivalence of (1) and (3) allows a more flexible *quantum interpretation* of various classical Lagrangians such as massive gauge theory.

As for a recent BRST analysis of the observation in Ref.[10], see Ref.[14].

References

1. L.D. Faddeev and V.N. Popov, Phys. Lett. **B25**(1967)29.
2. C. Becchi, A. Rouet and R. Stora, Comm. Math. Phys. **42** (1975)127.
   J. Zinn-Justin, Lecture Notes in Physics, **37** (Springer-Verlag, Berlin, 1975)2.
3. D. Zwanziger, Nucl.Phys. **B345** (1990) 461 ; **B192** (1981) 259.
4. G. Parrinello and G. Jona-Lasinio, Phys.Lett.**B251**(1990)175.
5. V.N. Gribov, Nucl. Phys. **B139**(1978)1.
6. Y. Nambu, Suppl. Prog. Theor. Phys. Extra Number (1968)190.
   P.A.M. Dirac, Proc. Roy. Soc.(London) **A209**(1951)291.
7. K. Fujikawa, Phys. Rev. **D7**(1973)393.
8. W. Bock, M. Golterman, M. Ogilvie and Y. Shamir, hep-lat/0004017.
9. K. Fujikawa and H. Terashima, Nucl. Phys.**B577**(2000)405.
10. K. Fujikawa and H. Terashima, hep-th/0004190.
11. R. Jackiw and R. Rajaraman, Phys. Rev. Lett. **54** (1985) 1219; ibid., **55** (1985) 224..
    K. Harada and I. Tsutsui, Phys. Lett. **183B** (1987) 311.
    O. Babelon, F. A. Schaposnik, and C.M. Viallet, Phys. lett. **177B** (1986) 385.
12. K.M. Case and S. Gasiorowicz, Phys. Rev. **125**(1962)1055.
    S. Coleman and E. Witten, Phys. Rev. Lett. **45**(1980)100.
    S. Weinberg and E. Witten, Phys. Lett. **B96**(1980)59.
13. D. Foerster, H.B. Nielsen and M. Ninomiya, Phys. Lett. **B94**(1980)135, and references therein.
14. R. Banerjee and B.P. Mandal, Phys. Lett. **B488**(2000)27.

ABELIAN DOMINANCE IN LOW-ENERGY GLUODYNAMICS DUE TO DYNAMICAL MASS GENERATION

K.-I. KONDO AND T. SHINOHARA

Department of Physics, Faculty of Science,
Chiba University, Chiba 263-8522, Japan
E-mail: kondo, sinohara@cuphd.nd.chiba-u.ac.jp

We propose a modified version of the maximal Abelian (MA) gauge. By adopting the modified MA gauge in QCD, we show that the off-diagonal gluons and Faddeev-Popov ghosts acquire their masses through the ghost–anti-ghost condensation due to four ghost interaction coming from the gauge-fixing term of the modified MA gauge. The asymptotic freedom of the original non-Abelian gauge theory is preserved in this derivation.

1 Introduction

The main subject of this talk is to understand the Abelian dominance in low-enegy QCD which was found by Suzuki and Yotsuyanagi[4] based on Monte Carlo simulations of lattice gauge theory under the maximal Abelian (MA) gauge proposed by Kronfeld et al.[3]. The Abelian dominance was predicted by Ezawa and Iwazaki[2] immediately after the proposal of the Abelian projection by 't Hooft[1]. The Abelian dominance and the subsequent magnetic monopole dominance is quite important to understand quark confinement from the viewpoint of the dual superconductor picture[5] of the QCD vauum, since the condensation of magnetic monopole can lead to the dual superconductivity based on the electro-magnetic duality argument.

For the Abelian dominance to hold in low-energy region of QCD, it is sufficient to show that the off-diagonal gluons become massive and they can be in a sense neglected in the low-energy region (although the latter statement is not necessarily true as shown in this talk). In order to really understand the Abelian dominance, we need to know the mechanism of the mass generation for the off-diagonal gluons. We propose a modified version of the MA gauge. Then we show that mass generation can be understood as a consequence of taking the (modified) MA gauge. We will give two kinds of explanations. One is given by a rather formal argument based on a novel reformulation of the gauge theory (as a perturbative deformation of a topological quantum field theory (TQFT)) proposed by the author[7]. Another is to examine the explicit form of the gauge fixing term in the MA gauge[13]. We give a prediction on the off-diagonal gluon mass.

2 The modified MA gauge

When we calculate the expectation value $\langle W_C[\mathcal{A}]\rangle_{YM}$ of the Wilson loop, we must specify the procedure of the gauge fixing. In order to incorporate the magnetic monopole in the non-Abelian gauge theory without the elementary scalar (Higgs) field, we adopt the modified MA gauge to define the gauge-fixed QCD. The gauge fixing (GF) and the Faddeev-Popov (FP) term of the modified MA gauge is given by

$$S_{GF+FP} = \int d^4x \; i\delta_B\bar{\delta}_B\left[\frac{1}{2}A^a_\mu(x)A^{\mu a}(x) - \frac{\alpha}{2}iC^a(x)\bar{C}^a(x)\right], \qquad (1)$$

where δ_B ($\bar{\delta}_B$) is the BRST (anti-BRST) transformation. The special case $\alpha = -2$ was discussed by several papers[7,9,11]. The modified MA gauge fixing term which is the BRST and anti-BRST exact and

FP conjugation[18] invariant has a hidden $OSp(4|2)$ supersymmetry[7]. Due to this supersymmetry, the dimensional reduction of Parisi-Sourlas[23] type takes place.

For simplicity, we discuss only the SU(2) case. For SU(3) case, see ref.[13].

$$\begin{aligned}S'_{GF+FP} = \int d^4x \Big\{ & B^a D_\mu[a]^{ab} A^{\mu b} + \frac{\alpha}{2} B^a B^a \\ & + i\bar{C}^a D_\mu[a]^{ac} D^\mu[a]^{cb} C^b \\ & - ig^2 \epsilon^{ad}\epsilon^{cb} \bar{C}^a C^b A^{\mu c} A^d_\mu \\ & + i\bar{C}^a g \epsilon^{ab} (D_\mu[a]^{bc} A^c_\mu) C^3 \\ & - \alpha g \epsilon^{ab} i B^a \bar{C}^b C^3 \\ & + \frac{\alpha}{4} g^2 \epsilon^{ab}\epsilon^{cd} \bar{C}^a \bar{C}^b C^c C^d \Big\}. \end{aligned} \quad (2)$$

Integrating out the B^a field leads to

$$\begin{aligned}S'_{GF+FP} = \int d^4x \Big\{ & -\frac{1}{2\alpha} (D_\mu[a]^{ab} A^{\mu b})^2 \\ & + i\bar{C}^a D_\mu[a]^{ac} D^\mu[a]^{cb} C^b \\ & - ig^2 \epsilon^{ad}\epsilon^{cb} \bar{C}^a C^b A^{\mu c} A^d_\mu \\ & + \frac{\alpha}{4} g^2 \epsilon^{ab}\epsilon^{cd} \bar{C}^a \bar{C}^b C^c C^d \Big\}. \end{aligned} \quad (3)$$

A crucial difference of the modified MA gauge from the conventional Lorentz gauge is the necessity of four ghost interactions for renormalizability. Even for $\alpha = 0$, the four ghost interaction term is induced through radiative corrections due to the existence of the $c\bar{c}AA$ vertex.

3 Ghost self-interaction and dynamical mass generation

Integrating out off-diagonal field components $(A^a_\mu, B^a, C^a, \bar{C}^a)$ in Yang-Mills theory in the MA gauge, we can obtain an effective Abelian gauge theory written in terms of only the diagonal components $(a^i_\mu, B^i, C^i, \bar{C}^i, B^i_{\mu\nu})$. This theory called the Abelian projected effective gauge theory (APEGT) which is regarded as a low-energy effective theory (LEET) of QCD. The coupling constant of APEGT has the μ(renormalization-scale) dependence governed by the β function which is the same as the original Yang-Mills theory.[6,14] This reflects the asymptotic freedom of the original non-Abelian gauge theory. The other RG functions and the anomalous dimensions have been calculated recently[14].

In the MA gauge, the renormalizability requires the existence of four ghost interactions. The modified MA gauge determines the strength of four ghost interaction where the modified MA gauge is obtained from the viewpoint of pursuing the maximal symmetry, namely, BRST, anti-BRST, FP conjugation and $OSp(4|2)$ supersymmetry. The attractive four ghost interaction leads to two types of ghost–anti-ghost condensations,[13,21]

$$\epsilon^{ab} \langle i\bar{C}^a(x) C^b(x) \rangle \neq 0, \quad (4)$$

$$\delta^{ab} \langle i\bar{C}^a(x) C^b(x) \rangle \neq 0. \quad (5)$$

In the condensed vacuum, the ghost-gluon 4-body interaction,

$$-ig^2 \epsilon^{ad}\epsilon^{cb} \bar{C}^a C^b A^{\mu c} A^d_\mu, \quad (6)$$

leads to a mass term of the off-diagonal gluons,

$$-ig^2 \epsilon^{ad}\epsilon^{cb} \langle \bar{C}^a C^b \rangle A^{\mu c} A^d_\mu = \frac{1}{2} g^2 \langle i\bar{C}^c C^c \rangle A^{\mu a} A^a_\mu, \quad (7)$$

Thus this condensation leads to the mass for the off-diagonal gluons

$$m_A^2 = g^2 \langle i\bar{C}^a(x) C^a(x) \rangle > 0. \quad (8)$$

On the other hand, the off-diagonal ghost (and anti-ghost) acquires the mass,[13]

$$m_C^2 = \alpha g^2 \langle i\bar{C}^a C^a \rangle, \quad (9)$$

through the four ghost interaction,

$$\begin{aligned} & \frac{\alpha}{4} g^2 \epsilon^{ab}\epsilon^{cd} \bar{C}^a \bar{C}^b C^c C^d \\ = & \frac{\alpha}{2} g^2 (i\epsilon^{ab} \bar{C}^a C^b)^2 = \frac{\alpha}{2} g^2 (i\bar{C}^a C^a)^2 \\ \to & \alpha g^2 \langle i\bar{C}^a C^a \rangle i\bar{C}^b C^b. \end{aligned} \quad (10)$$

Note that the introduction of the explicit mass term spoils the renormalizability. It can be shown that the diagonal gluons remain massless. The mass obtained in this way provides the scale which is comparable to the QCD scale Λ_{QCD}.[13] This result is consistent with the lattice simulations performed by Amemiya and Suganuma[20]. The dynamical mass generation for the off-diagonal components strongly supports the Abelian dominance in low-energy (or long-distance) QCD.

At least for $G = SU(2)$, the Lagrangian in the modified MA gauge has a novel (continuous) global symmetry, $SL(2,R)$ as found by Schaden[21]. Then the mass generation can be considered as a spontaneous breaking of this symmetry from $SL(2,R)$ to the non-compact Abelian subgroup corresponding to the ghost number charge Q_c. This mechanism of mass generation can be called the dynamical Higgs mechanism, since QCD has no elementary scalar field. The associated massless Nambu-Goldstone particles can not be observed, since they have zero norms due to the extended quartet mechanism.[22] In these analyses, we have assumed that the vacuum satisfies the physical condition,

$$Q_B|0\rangle = 0, \quad Q_c|0\rangle = 0, \quad \bar{Q}_B|0\rangle = 0. \quad (11)$$

For $G = SU(3)$, the ghost condensation scenario for mass generation of the off-diagonal components can be applied and leads to two different masses for off-diagonal gluons; two of them are heavier than the remaining four off-diagonal gluons, e.g.,

$$m_{A^1} = m_{A^2} = \sqrt{2} m_{A^4} = \sqrt{2} m_{A^5} = \sqrt{2} m_{A^6} = \sqrt{2} m_{A^7}. \quad (12)$$

4 Discussion

Finally, we raise the problems to be solved in the future investigations.

1. All the results obtained above are invariant under the residual $U(1)^{N-1}$ Abelian gauge symmetry. However, they may depend on the gauge parameter α of the MA gauge. In the recent work, α was determined by requiring the μ independence of the effective potential of the order parameter of the ghost condensation as

$$\alpha = b_0/N = 11/3. \quad (13)$$

2. The proof of renormalizability of QCD in the (modified) MA gauge has not yet been given when the ghost condensation takes place. In the absence of ghost condensation, the proof of renormalizability was given 15 years ago[19].

3. For $SU(3)$, no one has shown the existence of a global symmetry whose spontaneous breaking leads to the mass generation of off-diagonal fields (through the ghost condensation). Hence the relationship between the mass generation and the spontaneous symmetry breaking is not yet understood in a satisfactory level.

4. It is important to show how the dynamical mass of the off-diagonal gluons is related to the mass of the dual gauge field in the dual Abelian gauge theory (e.g., dual Ginzburg-Landau theory[26]) which is expected to be a LEET of QCD.

Acknowledgments

This work is supported in part by the Grant-in-Aid for Scientific Research from the Ministry of Education, Science and Culture (10640249).

References

1. G. 't Hooft, Nucl.Phys. B 190 [FS3], 455-478 (1981).
2. Z.F. Ezawa and A. Iwazaki, Phys. Rev. D 25, 2681-2689 (1982).
3. A. Kronfeld, M. Laursen, G. Schierholz and U.-J. Wiese, Phys. Lett. B 198, 516-520 (1987).
4. T. Suzuki and I. Yotsuyanagi, Phys. Rev. D 42, 4257-4260 (1990).
5. Y. Nambu, Phys. Rev. D 10, 4262-4268 (1974).
   G. 't Hooft, in: High Energy Physics,

edited by A. Zichichi (Editorice Compositori, Bologna, 1975).
S. Mandelstam, Phys. Report 23, 245-249 (1976).
6. K.-I. Kondo, hep-th/9709109, Phys. Rev. D 57, 7467-7487 (1998).
K.-I. Kondo, hep-th/9803063, Prog. Theor. Phys. Supplement, No. 131, 243-255.
7. K.-I. Kondo, hep-th/9801024, Phys. Rev. D 58, 105019 (1998).
8. K.-I. Kondo, hep-th/9803133, Phys. Rev. D 58, 085013 (1998).
9. K.-I. Kondo, hep-th/9805153, Phys. Rev. D 58, 105016 (1998).
10. K.-I. Kondo, hep-th/9810167, Phys. Lett. B 455, 251 (1999).
11. K.-I. Kondo, hep-th/9904045, Intern. J. Mod. Phys. A, to be published.
12. K.-I. Kondo and Y. Taira, hep-th/9906129, Mod. Phys. Lett. A 15, 367-377 (2000).
K.-I. Kondo and Y. Taira, hep-th/9911242.
13. K.-I. Kondo and T. Shinohara, hep-th/0004158, Phys. Lett. B, to be published.
14. K.-I. Kondo and T. Shinohara, hep-th/0005125.
15. D.I. Diakonov and V.Yu. Petrov, Phys. Lett. B 224, 131 (1989); hep-th/9606104.
16. M. Hirayama and M. Ueno, hep-th/9907063.
17. M. Quandt and H. Reinhardt, hep-th/9707185, Int. J. Mod. Phys. A13, 4049-4076 (1998).
18. G. Curci and R. Ferrari, Phys. Lett. B 63, 91-94 (1976).
I. Ojima, Prog. Theor. Phys. 64, 625-638 (1980).
19. H. Min, T. Lee and P.Y. Pac, Phys. Rev. D 32, 440-449 (1985).
20. K. Amemiya and H. Suganuma, Phys. Rev. D 60, 114509 (1999).
21. M. Schaden, hep-th/9909011, version 3.
22. T. Kugo and I. Ojima, Prog. Theor. Phys. Suppl. 66, 1-130 (1979).
23. G. Parisi and N. Sourlas, Phys. Rev. Lett. 43, 744-745 (1979).
24. L. Bonora and M. Tonin, Phys. Lett. B 98, 48-50 (1981).
25. G. Prosperi, private communications.
26. T. Suzuki, Prog. Theor. Phys. 80, 929 (1988); 81, 752 (1989).
S. Maedan and T. Suzuki, Prog. Theor. Phys. 81, 229 (1989).
H. Suganuma, S. Sasaki and H. Toki, Nucl. Phys. B 435, 207 (1995).

EXACT SOLUTIONS OF THE SU(2) YANG-MILLS-HIGGS THEORY

ROSY TEH
Universiti Teknologi MARA, Permatang Pasir, 13500 Permatang Pauh, Penang, Malaysia
E-mail: rosyteh@penang.itm.edu.my

Some exact static solutions of the SU(2) Yang-Mills-Higgs theory are presented. These solutions satisfy the first order Bogomol'nyi equations, and possess infinite energies. They are axially symmetric and could possibly represent monopoles and an antimonopole sitting on the z-axis.

1 Introduction

The theory of the SU(2) Yang-Mills-Higgs (YMH) field became of interest when 't Hooft and Polykov [1] discovered the monopole solution in the mid-seventies. This field theory has been shown to possess both the magnetic monopole [2] and multimonopole solutions [3]. In the limit of vanishing Higgs potential, these solutions are exact, satisfy the first order Bogomol'nyi equations [4] and have minimal energies. However, only numerical solutions are known when the Higgs potential is finite [5]. Numerical monopole-antimonopole solutions which do not satisfy the Bogomol'nyi condition are recently shown to exist [6].

Here we examined the SU(2) YMH theory when the scalar field is taken to have no mass or self interaction. We found that this theory possesses static solutions which are both exact as well as partially exact. These solutions satisfy the first order Bogomol'nyi equations. They are axially symmetric and could possibly represent monopoles and an antimonopole sitting on the z-axis.

The SU(2) Yang-Mills-Higgs Lagrangian in 3+1 dimensions is

$$\begin{aligned}\mathcal{L} &= -\frac{1}{4}F^a_{\mu\nu}F^{a\mu\nu} + \frac{1}{2}D^\mu\Phi^a D_\mu\Phi^a \\ &\quad -\frac{1}{4}\beta(\Phi^a\Phi^a - \frac{m^2}{\beta})^2; \\ D_\mu\Phi^a &= \partial\mu\Phi^a + \epsilon^{abc}A^b_\mu\Phi^c, \\ F^a_{\mu\nu} &= \partial_\mu A^a_\nu - \partial_\nu A^a_\mu + \epsilon^{abc}A^b_\mu A^c_\nu, \end{aligned} \quad (1)$$

where m is the Higgs field mass, and β is the strength of the Higgs potential. The gauge field coupling constant g is set to one and the metric used is $g_{\mu\nu} = (-+++)$. The SU(2) group indices a, b, c run from 1 to 3 and the spatial indices $\mu, \nu, \alpha = 0, 1, 2,$ and 3 in Minkowski space. The equations of motion that follow from the Lagrangian (1) are

$$\begin{aligned} D^\mu F^a_{\mu\nu} &= \epsilon^{abc}\Phi^b D_\nu\Phi^c, \\ D^\mu D_\mu\Phi^a &= -\beta\Phi^a(\Phi^b\Phi^b - \frac{m^2}{\beta}). \end{aligned} \quad (2)$$

't Hooft proposed that the tensor,

$$F_{\mu\nu} = \partial_\mu A_\nu - \partial_\nu A_\mu - \epsilon^{abc}\hat{\Phi}^a\partial_\mu\hat{\Phi}^b\partial_\nu \hat{Phi}^c (3)$$

where $A_\mu = \hat{\Phi}^a A^a_\mu$, the unit vector $\hat{\Phi}^a = \frac{\Phi^a}{|\Phi|}$ and the Higgs field magnitude $|\Phi| = \sqrt{\Phi^a\Phi^a}$, be identified with the electromagnetic field tensor. The abelian electric field is $E_i = F_{0i}$, and the abelian magnetic field is $B_i = -\frac{1}{2}\epsilon_{ijk}F_{ij}$. The topological magnetic current is $k_\mu = \frac{1}{8\pi}\ \epsilon_{\mu\nu\rho\sigma}\ \epsilon_{abc}\ \partial^\nu\Phi^a\ \partial^\rho\Phi^b\ \partial^\sigma\Phi c$, and the corresponding conserved magnetic charge is $M = \frac{1}{4\pi}\oint d^2\sigma_i\ B_i$.

2 The Exact Solutions

We use the static axially symmetric purely magnetic ansatz [6] which is given by

$$\begin{aligned} A^a_\mu &= \frac{1}{r}R_1\hat{\phi}^a\hat{r}_\mu - \frac{1}{r}R_2\hat{r}^a\hat{\phi}_\mu + \\ &\frac{1}{r}(1-\tau_1)\hat{\phi}^a\hat{\theta}^\mu - \frac{1}{r}(1-\tau_2)\hat{\theta}^a\hat{\phi}_\mu, \\ \Phi^a &= \Phi_1\ \hat{r}^a + \Phi_2\ \hat{\theta}^a. \end{aligned} \quad (4)$$

Simplifying the above ansatz (4) to $R_1 = R_2 = R(\theta)$ and $\tau_1 = \tau_2 = \tau(r)$, leads to the gauge field strength and covariant derivative of the Higgs field, given respectively by

$$\begin{aligned}
&F^a_{\mu\nu} = \\
&(\hat{\theta}^a(R\tau) + \hat{r}^a(\dot{R} + R\cot\theta + (\tau^2 - 1))) \\
&.(\hat{\phi}_\mu\hat{\theta}_\nu - \hat{\phi}_\nu\hat{\theta}_\mu)\frac{1}{r^2} + \\
&(\hat{\theta}^a(r\tau' + R\cot\theta + R^2) + \hat{r}^a(R\tau)) \\
&.(\hat{r}_\mu\hat{\phi}_\nu - \hat{r}_\nu\hat{\phi}_\mu)\frac{1}{r^2} - \\
&\hat{\phi}^a(r\tau' + \dot{R})(\hat{r}_\mu\hat{\theta}_\nu - \hat{r}_\nu\hat{\theta}_\mu)\frac{1}{r^2}, \\
&D_\mu\Phi^a = \hat{\phi}^a\hat{\phi}_\mu(\Phi_2\cot\theta + R\Phi_2 + \tau\Phi_1)\frac{1}{r} + \\
&\hat{\theta}^a\hat{\theta}_\mu(\dot{\Phi}_2 + \tau\Phi_1)\frac{1}{r} + \hat{r}^a\hat{r}_\mu(r\Phi_1' + R\Phi_2)\frac{1}{r} + \\
&\hat{\theta}^a\hat{r}_\mu(r\Phi_2' - R\Phi_1)\frac{1}{r} + \hat{r}^a\hat{\theta}_\mu(\dot{\Phi}_1 - \tau\Phi_2)\frac{1}{r}. \quad (5)
\end{aligned}$$

Prime means $\frac{\partial}{\partial r}$ and dot means $\frac{\partial}{\partial\theta}$. By allowing the Higgs field to be $\Phi_1 = \frac{1}{r}(\tau - 1) = \frac{1}{r}\psi(r)$ and $\Phi_2 = \frac{1}{r}R(\theta)$, the equations of motion (2) can be simplified to just two coupled ordinary differential equations,

$$\begin{aligned}
&-r^2\psi'' + 2\psi(\psi+1)^2 + \\
&2(\dot{R} + R\cot\theta + R^2)(1+\psi) = 0, \\
&\ddot{R} + \dot{R}\cot\theta - R(1+\cot^2\theta) - 2R^2\cot\theta - 2R^3 \\
&+ 2(r\psi' + \psi(1+\psi))R = 0, \quad (6)
\end{aligned}$$

which can further be reduced to two ordinary differential equations of first order,

$$\begin{aligned}
r\psi' + \psi + \psi^2 &= -p, \\
\dot{R} + R\cot\theta + R^2 &= p, \quad (7)
\end{aligned}$$

where p is an arbitary constant. Eq. (7) for ψ is exactly solvable for all values of p. However Eq. (7) for R is only exactly solvable when p takes the values 0 and -2 and for other values of p, it can only be numerically solved. Eq. (7) are first order differential equations and they are found to satisfy the Bogomol'nyi first order equations , $B^a_i = D_i\Phi^a$.

In this paper, we would like to focus only on the exact solution when $p = -2$. In this case, the exact solutions to Eq. (7) are

$$\begin{aligned}
\psi &= \left(\frac{c_1r^3 - 2}{c_1r^3 + 1}\right), \\
R &= R_{(1)} = -\tan\theta. \quad (8)
\end{aligned}$$

With solution (8), Eq. (7) for R can be reduced to the Bernoulli equation to obtained the second exact solution,

$$\begin{aligned}
&R = R_{(2)} = -\tan\theta + \\
&(\sin\theta\cos^2\theta(c_2 + \frac{1}{\cos\theta} + \ln\tan\frac{\theta}{2}))^{-1}. \quad (9)
\end{aligned}$$

In solutions (8) and (9), c_1 and c_2 are arbitrary constants, and solution $R_{(2)}$ is more singular than solution $R_{(1)}$.

The solutions (8) lead to the exact gauge fields,

$$\begin{aligned}
A_\mu &= A^a_{(1)\mu} + A^a_{(2)\mu}; \\
A^a_{(1)\mu} &= \frac{1}{r}\tan\theta(\hat{r}^a\hat{\phi}_\mu - \hat{\phi}^a\hat{r}_\mu), \\
A^a_{(2)\mu} &= \frac{1}{r}\left(\frac{r^3-2}{r^3+1}\right)(\hat{\theta}^a\hat{\phi}_\mu - \hat{\phi}^a\hat{\theta}_\mu), \quad (10)
\end{aligned}$$

where the integration constant c_1 is set to one. When r tends to infinity, the gauge potentials (10) do not tend to a pure gauge, and when r approaches zero, only $A^a_{(2)\mu}$ tends to a pure gauge but not $A^a_{(1)\mu}$. The energy of the system is not finite at the point $r = 0$ and along the plane $z = 0$.

It is noted that with the ansatz (4), $A_\mu = \hat{\Phi}^a A^a_\mu = 0$. Hence the abelian electric field is zero and the abelian magnetic field is independent of the gauge fields (10),

$$\begin{aligned}
B_i &= B_r\hat{r}_i + B_\theta\hat{\theta}_i; \\
B_r &= \frac{9\left((r^3-2) - (r^3+1)\tan^2\theta\right)}{r^2\left((r^3-2)^2 + (r^3+1)^2\tan^2\theta\right)^{3/2}}, \\
B_\theta &= \frac{27r\tan\theta}{\left((r^3-2)^2 + (r^3+1)^2\tan^2\theta\right)^{3/2}}. \quad (11)
\end{aligned}$$

3 The Magnetic Flux

We would like to define the abelian field magnetic flux as

$$\Omega = 4\pi M = \oint d^2\sigma_i B_i$$
$$= 2\pi \int B_i (r^2 \sin\theta d\theta)\hat{r}_i. \quad (12)$$

where M is the magnetic charge. We would also like to rewrite the Higgs field of Eq. (4) from the spherical coordinate system to the cylindrical coordinate system [6],

$$\Phi^a = \Phi_1\ \hat{r}^a + \Phi_2\ \hat{\theta}^a,$$
$$= \tilde{\Phi}_1\ \hat{\rho}^a + \tilde{\Phi}_2\ \delta^{a3};$$
$$\tilde{\Phi}_1 = \Phi_1 \sin\theta + \Phi_2 \cos\theta = |\Phi|\ \cos\alpha,$$
$$\tilde{\Phi}_2 = \Phi_1 \cos\theta - \Phi_2 \sin\theta = |\Phi|\ \sin\alpha. \quad (13)$$

Hence $\sin\alpha$ and the unit vector of the Higgs field can be calculated and shown respectively to be

$$\sin\alpha = \frac{(r^3-2)\cos\theta + (r^3+1)\sin\theta\tan\theta}{\sqrt{(r^3-2)^2 + (r^3+1)^2\tan^2\theta}},$$
$$\hat{\Phi}^a = \cos\alpha\ \hat{\rho}^a + \sin\alpha\ \delta^{a3}. \quad (14)$$

The abelian magnetic field and the magnetic charge can be written respectively as

$$B_i = -\frac{1}{\rho r}\frac{\partial}{\partial\theta}(\sin\alpha)\,\hat{r}_i + \frac{1}{\rho}\frac{\partial}{\partial r}(\sin\alpha)\,\hat{\theta}_i,$$
$$M = -\frac{1}{2}\sin\alpha|_{0,r}^{\pi}. \quad (15)$$

Hence it can be shown that the magnetic charge enclosed in the upper hemisphere and the lower hemisphere is one each and the magnetic charge at the origin is negative one. Therefore the system carries a net magnetic charge of one.

4 Comments

1. The exact magnetic solutions (8) have been shown to represent two monopoles and an antimonopole sitting on the z-axis, with the antimonopole at the origin and the two monopoles at $z = \pm\sqrt[3]{\frac{2}{c_1}}$. The positions of the monopoles can be varied by changing the value of the parameter c_1 but the antimonopole's position is fixed at the origin.
2. In these exact solutions, the magnitude of the Higgs field is zero at the positions of the two monopoles. The singularity at $r = 0$ of the Higgs field corresponds to the antimonopole.
3. The energy density of the abelian magnetic field (11), are concentrated at the points where the antimonopole and the two monopoles are located.
4. The next exact solutions, when $p = 0$, is $\psi_0 = \frac{1}{(c_3 r - 1)}$, $R_0 = \frac{1}{(c_4 \sin\theta + \sin\theta \ln\tan\frac{\theta}{2})}$, where c_3 and c_4 are arbitrary constants. In this case we notice that $(\psi_0, R = 0)$ and $(\psi = 0, R_0)$ are also solutions of the equations of motion (2). Hence we can linear superposed these two sets of solutions to get the solutions (ψ_0, R_0). Therefore linear superposition of nonlinear solutions is possible, when $p = 0$.
5. Only numerical solutions can be obtained for $R(\theta)$, when p takes value other than 0 and -2. The solution $\psi(r)$ for Eq. (7) is exact for all values of p.

Acknowledgments

The authors would like to thank Bahagian Latihan dan Pembangunan Staf, Universiti Teknologi MARA, Shah Alam, Malaysia for the travel grant and leave of absence to attend the conference, and the organizing committee of the ICHEP 2000 Conference for the financial assistance and the local hospitality thus making this work possible.

References

1. G. 't Hoof, Nucl. Phy B **79**, 276 (1974); A.M. Polyakov, Sov. Phys. - JETP, **41**, 988 (1975); Phys. Lett. B **59**, 82 (1975).

2. M.K. Prasad and C.M. Sommerfield, Phys. Rev. Lett. **35**, 760 (1975).
3. C. Rebbi and P. Rossi, Phys. Rev. D **22**, 2010 (1980); R.S. Ward, Commun. Math. Phys. **79**, 317 (1981); P. Forgacs, Z. Horvarth and L. Palla, Phys. Lett. B **99**, 232 (1981); Nucl. Phys. B **192**, 141 (1981); M.K. Prasad, Commun. Math. Phys. **80**, 137 (1981); M.K. Prasad and P. Rossi, Phys. Rev. D **24**, 2182 (1981).
4. E.B. Bogomol'nyi, Sov. J. Nucl. Phys. **24**, 449 (1976).
5. B. Kleihaus, J. Kunz and D. H. Tchrakian, Mod. Phys. Lett. A **13**, 2523 (1998).
6. B. Kleihaus and J. Kunz, Phys. Rev. D **61**, 025003 (2000).

TOWARD CONVERGENCE OF THE VARIATIONAL MASS EXPANSION IN ASYMPTOTICALLY FREE THEORIES

J.-L. KNEUR AND D. REYNAUD

Physique Mathématique et Théorique, UMR CNRS, F–34095 Montpellier Cedex 5, France

We re-examine a modification of perturbative expansions, valid for asymptotically free theories, producing "variationally improved" expansions of physical quantities relevant to dynamical (chiral) symmetry breaking. The large order behaviour of this expansion is shown to be drastically improved, for reasons analogous to the convergence properties of the delta-expansion of the anharmonic oscillator.

1 Introduction

In many field theories, the convergence of ordinary perturbation can be systematically improved by a variational-like procedure, in which the perturbative expansion is reorganized so as to depend on arbitrary adjustable parameters, to be fixed by some optimization prescription. This so-called "delta-expansion" (DE)[1], closely related to the "order-dependent mapping" (ODM) method[2], was even proved rigorously to converge[3], at least for the energy levels of the anharmonic oscillator and related D=1 models. However, the latter convergence properties crucially rely on the known analyticity properties of the oscillator[4], and it is a priori difficult to extend directly such methods to theories in higher dimensions. Moreover, in QCD, or similar $D \geq 2$ asymptotically free theories (AFT), the problem of determining non-perturbative quantities that characterise interesting aspects like e.g. dynamical symmetry breaking (DSB), is not solely a technical problem of summing a certain class of graphs. As is well known, there are intrinsic obstacles to a first principle determination of such quantities, in particular inherent ambiguities appear when attempting to sum perturbative expansions, as indicated by the non Borel summability[5].

We shall discuss here a variant of DE proposed some time ago[6], but re-investigated recently with new results[7] that we summarize here. The basic idea is to perform a modification in two stages, first rewriting the ordinary perturbation in the coupling g of physical quantities (depending also on a mass m) in the alternative form of "mass power" expansions (MPE) in $(\hat{m}/\Lambda)^\alpha$, where $\hat{m}$ is the scale invariant mass, Λ the basic scale and the power α is given by known renormalization group (RG) coefficients. This leads to a non-trivial massless (chiral) limit for the relevant DSB (χSB) order parameters (or for analogous quantities like e.g the "mass gap" in D=2 theories). In a second stage, a specific form of the (scaled) delta-expansion is performed on the MPE series.

In [7] we investigate in details the behaviour of the large orders of such alternative mass expansions. We find that the ODM (scaled) DE summation performed in a second stage produces a renormalization scheme (RS) dependent factorial damping of the original perturbative coefficients at large orders. Our variational MPE series can be convergent, however in the approximation where only the leading $m \to 0$ contributions are taken into account, or Borel summable, otherwise. Those results exhibit many properties similar to what happens with the scaled delta-expansion[3] in the anharmonic oscillator.

2 Mass gap from mass expansion

To illustrate the basic mechanism, consider in a "generic" AFT the first RG order evolution for the (renormalized) "current" mass:

$$M_1 = m(\mu)\,[1 + 2b_0 g^2(\mu) \ln(\frac{M_1}{\mu})]^{-\frac{\gamma_0}{2b_0}}\ , \quad (1)$$

where the "self-consistent" condition $M \equiv m(\mu' \equiv M)$ was used. Equivalently

$$M_1 = \hat{m}\,[\ln(M_1/\Lambda)]^{-A} \equiv \hat{m}\,F^{-A} \qquad (2)$$

with $\Lambda = \mu \exp[-1/2b_0 g^2(\mu)]$ the RG invariant scale and $\hat{m} \equiv m(\mu)[2b_0 g^2(\mu)]^{-A}$ the scale invariant mass ($A \equiv \frac{\gamma_0}{2b_0}$); and where b_0, γ_0 are the one-loop RG-coefficients, with $b_0 > 0$ for an AFT [a]. Eq. (2) defines

$$F(\hat{m}/\Lambda) = \ln(\hat{m}/\Lambda) - A \ln F \qquad (3)$$

with the remarkable property that $F \simeq \hat{m}^{1/A}$ for $\hat{m} \simeq 0$, in contrast to the ordinary Log function (asymptotic to $F(\hat{m})$ for $\hat{m} \to \infty$). On its principal branch, $F(x)$ has an alternative series expansion[7]

$$F(x) = \sum_{p=0}^{\infty} (\frac{-1}{A})^p \frac{(p+1)^p}{(p+1)!}\, x^{(p+1)/A} \qquad (4)$$

with finite convergence radius $R_c = e^{-A} A^A$. Accordingly, $M_1(\hat{m})$ in Eq. (2) exhibits dif-

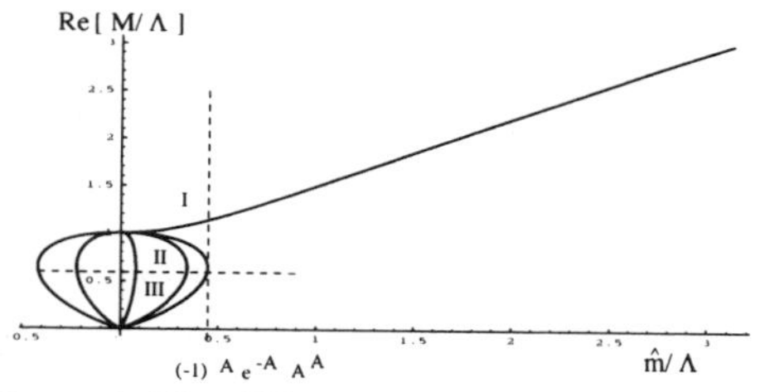

Figure 1. The different branches of the real part of $M(\hat{m})/\Lambda = (\hat{m}/\Lambda)\,F^{-A}$, for the choice $A = 4/9$ (corresponding to QCD with 3 light quark flavours).

ferent branches according to the values of the RG parameter A (see Fig. 1). Now, for most values of A, there is only one such branch which, for *real* $\hat{m}$ values, is real and continuously matching the asymptotic perturbative behaviour of F for large $\hat{m}$ (region I in Fig. 1). Algebraically, the non-zero mass gap $M_1 = \Lambda$ for $\hat{m} \to 0$ is also obtained directly from Eq. (4) in Eq. (2). In that sense $F(\hat{m})$ provides a well-defined bridge between the "non-perturbative" massless regime, where F has power expansion (4), and the ordinary perturbative regime.

[a] Our normalization is $\beta(g) = -b_0 g^3 - b_1 g^5 - \cdots$, $\gamma_m(g) = \gamma_0 g^2 + \gamma_1 g^4 + \cdots$.

3 Higher orders

The true pole mass is evidently not given simply by Eq. (1), but includes RG and non-log perturbative contributions of higher orders:

$$M^P(\hat{m}) = \frac{2^{-C}\,\hat{m}}{F^A [C+F]^B} \sum_{n=0}^{\infty} \frac{d_n}{(2b_0 F)^n}, \qquad (5)$$

$$F = \ln[\hat{m}/\Lambda] - A\ln F - (B-C)\ln[C+F], \qquad (6)$$

$$A = \frac{\gamma_1}{2b_1},\ B = \frac{\gamma_0}{2b_0} - A,\ C = \frac{b_1}{2b_0^2} \qquad (7)$$

which generalizes Eqs. (2)-(3) [b]. The coefficients d_n include the (dominant) non-log perturbative contributions from the n-loop graphs and (subdominant) contributions from higher (≥ 3 loop) RG orders. Similarly to Eq. (4), $F(\hat{m})$ in (6) has a (A, B, C dependent) power expansion in $(\hat{m}/\Lambda)^{1/A}$ for sufficiently small $\hat{m}$, with A given in Eq. (7). It is also possible to obtain from a similar construction[6] other quantities than the pole mass: typically in QCD the (gauge and RG invariant) quantities F_π/Λ (the pion decay constant) and $m_q\langle\bar{q}q\rangle/\Lambda^4$, related to χSB. However, the dominant contributions d_n in (5) are expected generically to behave as[5]

$$d_n \sim (2b_0)^{(n-1)}(n-1)! \quad (n \to \infty) \qquad (8)$$

so that the series in Eq. (5) is not Borel summable, leading to ambiguities of $\mathcal{O}(\Lambda)$[5,7]

4 Variational mass expansion

We shall see now that it is in fact possible to cure the latter potential ambiguities of our DSB estimates, by restarting from the basic Lagrangian, and combining in a natural manner the previous MPE construction with a specific form of DE. As mentioned above,

[b] Strictly speaking Eq. (5) applies directly only if $C = b_1/(2b_0^2) \geq 0$. If $C < 0$ (as in the $O(N)$ GN model[8]) a slightly different approach is necessary[6].

delta expansion is essentially a reorganization of the interaction part of a Lagrangian, by first adding and subtracting a "trial" mass term, to be used at each order as an adjustable parameter. In our case we define DE as the (gauge invariant) substitution

$$m(\mu) \to (1-x)\, m(\mu);\; g^2(\mu) \to x\, g^2(\mu) \quad (9)$$

at arbitrary order, where x is the new expansion parameter and m the *arbitrary* adjustable mass. The x series can be resummed, e.g. for (5), by contour integration[6]:

$$M^P = \sum_{p=0}^{n} \oint dv \frac{e^{v/m''}}{F^A [C+F]^B} \frac{n^{\gamma-1} d_p}{(2b_0 F)^p} \quad (10)$$

where $(\hat{m}/\Lambda) \equiv m''$ and the (massless) limit of interest, $x \to 1$, is performed via

$$x \equiv 1 - v/n\;;\quad \hat{m} \to n^\gamma\, \hat{m} \quad (11)$$

$(n \to \infty)$ with $0 < \gamma \le 1$[7]. Eq. (10) with (4) and (8) gives after some algebra

$$M^P/\Lambda \sim \sum_n \frac{n!\, n^{[(1-\gamma)n/A]}}{\Gamma[1+n/A]} (m'')^{-n/A} \quad (12)$$

where in (10) and (12) we omit unessential overall factors and $\sim$ indicates that (12) is only valid for $n \to \infty$ and for the leading $m'' \to 0$ term of (4). One obtains that (12) converges for any $m'' > 0$ iff

$$0 < A < \gamma < 1\,. \quad (13)$$

Note moreover that if $A < \gamma$ any dominant (and a forciori any subdominant) renormalon behaviour, of the form[5] $\sim r^n n!$ (with r a finite number) is damped similarly.

Now, A is a RS *dependent* quantity. The idea thus is to perform appropriate scheme changes $A \to A'$ in (5) etc, such that $0 < A' \le \gamma$, for a given rescaling of $\hat{m}$ in (11).

But F involves also subdominant terms, Eq.(4), giving additional contributions, for finite m'', to the $(m'')^{-n/A}$ terms for arbitrary n. It is however possible to show, similarly to the damping operating in (12), that the full series remains convergent provided that expansion (4) is truncated at *finite* order $p \ll n$, while it diverges from terms $p \sim n$, but remains Borel-summable[7].

5 Conclusions

It is quite remarkable that the above damping properties may be the appropriate generalization for D>1 renormalizable theories of the established convergence properties of the standard delta-expansion for the oscillator[3]. In D≥2 theories, the damping of the original renormalon behaviour exhibited here is not totally surprising: intuitively, what happens is that our variational reorganization of perturbative expansions is such that $\hat{m}/\Lambda$ has a certain arbitrariness, *and* dependence on $\hat{m}/\Lambda$ is power-like (rather than log-like), for $\hat{m} \to 0$. We conclude by arguing that such a method provides a well-defined basis to estimate the DSB order parameters in AFT.

References

1. W.E. Caswell, Ann. Phys. (N.Y) 123 (1979) 153; I.G. Halliday and P. Suranyi, Phys. Lett. B85 (1979) 421; P. M. Stevenson, Phys. Rev. D23 (1981) 2916; Nucl. Phys. B203 (1982) 472; J. Killinbeck, J. Phys. A 14 (1981) 1005; A. Neveu, Nucl. Phys. B18(1990) 242.
2. R. Seznec and J. Zinn-Justin, J. Math. Phys. 20(1979) 1398.
3. A. Duncan and H.F. Jones, Phys. Rev. D47 (1993) 2560; C.M. Bender, A. Duncan and H.F. Jones, Phys. Rev. D 49 (1994) 4219; C. Arvanitis, H.F. Jones and C. Parker, Phys. Rev. D 52 (1995) 3704; R. Guida, K. Konishi and H. Suzuki, Ann. Phys. 241 (1995) 152; Annals Phys. 249 (1996) 109.
4. C. Bender and T.T. Wu, Phys. Rev. D7 (1973) 1620.
5. for a recent review see M. Beneke, Phys. Rept. 317(1999) 1.
6. C. Arvanitis et al, Int.J.Mod.Phys. A12 (1997)3307; Phys.Lett. B390(1997) 385; J.-L. Kneur, Phys.Rev. D57(1998) 2785.
7. J.-L. Kneur and D. Reynaud, hep-th/0009
8. D. Gross and A. Neveu, Phys. Rev. D10(1974) 3235.

SYMMETRY BREAKING/RESTORATION IN A NON-SIMPLY CONNECTED SPACE-TIME

HISAKI HATANAKA

Institute for Cosmic Ray Research, University of Tokyo, Kashiwanoha 5-1-5, Kashiwa 277-8582, Japan

SEIHO MATSUMOTO AND KATSUHIKO OHNISHI

Graduate School of Science and Technology, Kobe University, Rokkodai, Nada, Kobe 657-8501, Japan

MAKOTO SAKAMOTO

Department of Physics, Kobe University, Rokkodai, Nada, Kobe 657-8501, Japan

Field theories compactified on non-simply connected spaces, which in general allow to impose twisted boundary conditions, are found to unexpectedly have a rich phase structure. One of characteristic features of such theories is the appearance of critical radii, at which some of symmetries are broken/restored. A phase transition can occur at the classical level, or can be caused by quantum effects. The spontaneous breakdown of the translational invariance of compactified spaces is another characteristic feature. As an illustrative example, the $O(N)$ ϕ^4 model on $M^3 \otimes S^1$ is studied and the novel phase structure is revealed.

The parameter space of field theories compactified on non-simply connected spaces is, in general, wider than that of ordinary field theories on the Minkowski space-time, and is spanned by twist parameters specifying boundary conditions,[1,2] in addition to parameters appearing in the actions. Physical consequences caused by twisted boundary conditions turn out to be unexpectedly rich and many of them have not been uncovered yet. The purpose of this talk is to report some of interesting properties of such theories overlooked so far.

One of characteristic features of such theories is the appearance of critical radii of compactified spaces, at which some of symmetries are broken/restored.[3] Symmetry breaking patterns are found to be unconventional. A phase transition can occur at the classical level, or can be caused by quantum effects. Radiative corrections would become important when a compactified scale is less than the inverse of a typical mass scale, and then some of broken symmetries could be restored, or conversely some of symmetries could be broken. Another characteristic and probably surprising feature is the spontaneous breakdown of the translational invariance of compactified spaces.[4] Twisted boundary conditions do not allow vacuum expectation values of twisted bosons to be non-vanishing constants. In other words, vacuum expectation values of twisted bosons have to vanish or to be coordinate-*dependent* if they are non-vanishing. If the minimum of a potential does not lie at the origin, twisted bosons could acquire non-vanishing vacuum expectation values, which should be coordinate-dependent. Then, we have to minimize the total energy, which consists of both the kinetic term and the potential term, to find the vacuum configuration. When non-vanishing vacuum expectation values of twisted bosons are energetically preferable, they should be coordinate-dependent and hence the translational invariance is broken spontaneously. Among other characteristic features, a phenomenologically important observation is that twisted boundary conditions can break supersymmetry spontaneously.[5a] This will give a new type of spontaneous supersymmetry breaking mechanisms and it would be of great interest to investigate a possibility to construct realistic supersymmetric models with this supersymmetry breaking mechanism, though this subject will not be treated in this talk.

As an illustrative example, we here concentrate on the $O(N)$ ϕ^4 model on $M^3 \otimes S^1$.[b] The action which consists of N real scalar fields ϕ_i $(i = 1, \cdots, N)$ is given by

$$S = \int d^3x \int_0^{2\pi R} dy \Big[\frac{1}{2} \partial_A \phi_i \partial^A \phi_i - \frac{m^2}{2} \phi_i^2 - \frac{\lambda}{8} \left(\phi_i^2\right)^2 \Big], \qquad (1)$$

where y and R denote the coordinate and the radius of S^1, respectively. Since S^1 is multiply-connected, we can impose a twisted boundary condition on ϕ_i such as

$$\phi_i(y + 2\pi R) = U_{ij} \phi_j(y). \qquad (2)$$

The matrix U must belong to $O(N)$, otherwise the action would not be single-valued. We shall below consider various boundary conditions and discuss physical consequences.

(1) $U = \mathbf{1}$

In this case, the fields $\phi_i(y)$ obey the periodic boundary condition. For $m^2 > 0$, the phase structure is trivial: The $O(N)$ symmetry is unbroken in a whole range of R. For $m^2 < 0$, $O(N)$ would be broken to $O(N-1)$. It is well known that the leading correction to the squared mass is proportional to $1/R^2$ for small radius R [6,7] and that the broken symmetry $O(N-1)$ can be restored for $R \le R^* = \mathcal{O}(\sqrt{\lambda}/\mu)$ $(\mu^2 \equiv -m^2)$, just like the symmetry restoration at high temperature.[8] Thus, we have found no new interesting phenomena with $U = \mathbf{1}$.

[a]This subject was discussed in the talk given by M. Tachibana at this Conference.

[b]The classical analysis of this model has been done in Ref.3.

(2) $U = -\mathbf{1}$

In this case, $\phi_i(y)$ obey the antiperiodic boundary condition. For $m^2 > 0$, nothing happens and the $O(N)$ symmetry remains unbroken in a whole range of R, while for $m^2 < 0$, several interesting phenomena occur.[3] For $R > R^* \sim 1/(2\mu)$, the $O(N)$ symmetry is spontaneously broken to $O(N-2)$ but *not* $O(N-1)$! The translational invariance of S^1 is also broken spontaneously because of the y-dependent vacuum expectation values of $\phi_i(y)$. For $R \le R^*$, all the broken symmetries are restored. It should be emphasized that the mechanism of this symmetry restoration is different from the previous case of $U = \mathbf{1}$ and that the present symmetry restoration has a classical origin. This may be seen from the fact that R^* is of order $1/\mu$ but not $\sqrt{\lambda}/\mu$. Radiative corrections in this case are less important.

The nontrivial phase structure for $m^2 < 0$ may be understood as follows: We first note that since $\phi_i(y)$ obey the twisted (antiperiodic) boundary condition, a non-vanishing vacuum expectation value of $\langle \phi_i(y) \rangle$ immediately implies that it is y-dependent, otherwise it would not satisfy the boundary condition. The y-dependent configuration of $\langle \phi_i(y) \rangle$ will induce the kinetic energy proportional to $1/R^2$. It follows that for large radius R, non-vanishing $\langle \phi_i(y) \rangle^2$ is preferable because the origin is not the minimum of the potential for $m^2 < 0$ and because the contribution from the kinetic energy is expected to be small. Therefore, for large radius R, the $O(N)$ symmetry and also the translational invariance of S^1 are spontaneously broken. Since $\langle \phi_i(y) \rangle$ must obey the antiperiodic boundary condition, non-vanishing $\langle \phi_i(y) \rangle$ cannot be constants and turn out to be of the form $\langle \phi_i(y) \rangle = (v \cos(y/2R), v \sin(y/2R), 0, \cdots, 0)$, where $v = \sqrt{(2\mu^2 - 1/(2R^2))/\lambda}$ at the tree level. Since two of $\langle \phi_i(y) \rangle$'s have non-vanishing expectation values, the $O(N)$ symmetry should be broken to $O(N-2)$, but not $O(N-1)$.[c] On the other hand, for small radius, the contribution from the kinetic energy becomes large, so that the y-independent configuration of $\langle\phi_i(y)\rangle$ is preferable and this implies that $\langle\phi_i(y)\rangle$ should vanish.

(3) $U = \begin{pmatrix} \mathbf{1}_L & 0 \\ 0 & -\mathbf{1}_{N-L} \end{pmatrix}$

Since the twist matrix U is not proportional to the identity matrix, the boundary condition (2) explicitly breaks $O(N)$ down to $O(L) \times O(N-L)$, which is the subgroup of $O(N)$ commuting with U. For $m^2 > 0$, the $O(L) \times O(N-L)$ symmetry is unbroken in a whole range of R if $N > L > (N-4)/3$, but is broken to $O(L-1) \times O(N-L)$ for $R < R^* = \mathcal{O}(\sqrt{\lambda}/m)$ if $0 < L < (N-4)/3$, in spite of positive m^2. This symmetry breaking for $R < R^*$ comes from the fact that a one-loop self-energy diagram in which a boson obeying the antiperiodic boundary condition propagates gives a negative contribution to the squared mass.[7]

For $m^2 < 0$, the $O(L) \times O(N-L)$ symmetry is broken to $O(L-1) \times O(N-L)$ in a whole range of R if $0 < L < (N-4)/3$, but is restored for $R \leq R^* = \mathcal{O}(\sqrt{\lambda}/\mu)$ if $N > L > (N-4)/3$. It should be noticed that the translational invariance is not broken in this model because the vacuum expectation values of the untwisted bosons are always y-independent and because no twisted bosons acquire non-vanishing vacuum expectation values.

(4) General $U \in O(N)$

We can show that the twisted boundary condition (2) generally breaks $O(N)$ to $O(L_0) \times U(L_1/2) \times \cdots \times U(L_{M-1}/2) \times O(L_M)$ with $L_0 + L_1 + \cdots + L_M = N$.[3] A new phenomenon is that in some class of models phase transitions could occur several times when the radius R varies from 0 to ∞. The full details of the phase structure will be reported elsewhere.[9]

[c]The exception is the model with $N = 1$. In this case, there is no continuous symmetry and the $O(1)$ model has only a discrete symmetry Z_2. The vacuum expectation value of $\phi(y)$ is found to be a kink-like configuration for $R > R^*$.[4]

Acknowledgments

We would like to thank to J. Arafune, C.S. Lim, M. Tachibana and K. Takenaga for valuable discussions and useful comments. This work was supported in part by JSPS Research Fellowship for Young Scientists (H.H) and by Grant-In-Aid for Scientific Research (No.12640275) from the Ministry of Education, Science, and Culture, Japan (M.S).

References

1. C.J. Isham, *Proc. R. Soc. London* A**362** (1978) 383; A**364** (1978) 591.
2. Y. Hosotani, *Phys. Lett.* B**126** (1983) 309; *Ann. Phys.* **190** (1989) 233.
3. K. Ohnishi and M. Sakamoto, *Phys. Lett.* B**486** (2000) 179.
4. M. Sakamoto, M. Tachibana and K. Takenaga, *Phys. Lett.* B**457** (1999) 33.
5. M. Sakamoto, M. Tachibana and K. Takenaga, *Phys. Lett.* B**458** (1999) 231; *Prog. Theor. Phys.* **104** (2000) 633.
6. L.H. Ford and T. Yoshimura, *Phys. Lett.* A**70** (1979) 89.
7. D.J. Toms, *Phys. Rev.* D**21** (1980) 928; D**21** (1980) 2805.
8. L. Dolan and R. Jackiw, *Phys. Rev.* D**9** (1974) 3320; S. Weinberg, *Phys. Rev.* D**9** (1974) 3357.
9. H. Hatanaka, S. Matsumoto, K. Ohnishi and M. Sakamoto, in preparation.

LIGHT-FRONT-QUANTIZED QCD IN LIGHT-CONE GAUGE

PREM P. SRIVASTAVA

Instituto de Física, Universidade do Estado de Rio de Janeiro, Rio de Janeiro, RJ 20550-013, Brasil
E-mail: prem@uerj.br, prem@lafex.cbpf.br

STANLEY J. BRODSKY

Stanford Linear Accelerator Center, Stanford University, Stanford, CA 94309, USA
E-mail: sjbth@slac.stanford.edu

The light-front (LF) quantization[1] of QCD in light-cone (l.c.) gauge is discussed. The Dirac method is employed to construct the LF Hamiltonian and canonical quantization of QCD. The Dyson-Wick perturbation theory expansion based on LF-time ordering is constructed. The framework automatically incorporates the Lorentz condition as an operator equation. The propagator of the dynamical ψ_+ part of the free fermionic propagator is shown to be causal, while the gauge field propagator is found to be transverse. The interaction Hamiltonian is re-expressed in a form closely resembling the one in conventional theory, except for additional instantaneous interactions. The fact that gluons have only physical degrees of freedom in l.c. gauge may provide an analysis of coupling renormalization similar to that of the pinch technique, which is currently being discussed as a physical and analytic renormalization scheme for QCD.

1 Introduction

The quantization of relativistic field theory at fixed light-front time $\tau = (t - z/c)/\sqrt{2}$, was proposed by Dirac[2] half a century ago. It has found important applications[3,4,5,6] in gauge theory and string theory. The light-front (LF) quantization of QCD in its Hamiltonian form provides an alternative approach to lattice gauge theory for the computation of nonperturbative quantities. We discuss here[7] the LF quantization of QCD gauge field theory in l.c. gauge employing the Dyson-Wick S-matrix expansion based on LF-time-ordered products. The case of covariant gauge has been discussed in our earlier work[8].

2 QCD action in light-cone gauge

The LF coordinates are defined as $x^\mu = (x^+ = x_- = (x^0 + x^3)/\sqrt{2}, x^- = x_+ = (x^0 - x^3)/\sqrt{2}, x^\perp)$, where $x^\perp = (x^1, x^2) = (-x_1, -x_2)$ are the transverse coordinates and $\mu = -, +, 1, 2$. The coordinate $x^+ \equiv \tau$ will be taken as the LF time, while x^- is the longitudinal spatial coordinate.

The quantum action of QCD in l.c. gauge is described in the standard notation by

$$\mathcal{L}_{QCD} = -\frac{1}{4} F^{a\mu\nu} F^a{}_{\mu\nu} + B^a A^a{}_- + \bar{c}^a \mathcal{D}_-^{ab} c^b + \bar{\psi}^i (i\gamma^\mu D_\mu^{ij} - m\delta^{ij})\psi^j .$$

Here $\bar{c}^a, c^a$ are anticommuting ghost fields and auxiliary fields $B^a(x)$ are introduced in the *linear* gauge-fixing term. The action is invariant under BRS symmetry transformations. Since B^a carries canonical dimension three, no quadratic terms in them are permitted.

3 Spinor field propagator

The quark field term in LF coordinates reads

$$i\sqrt{2}\bar{\psi}_+^i \gamma^0 D_+^{ij} \psi_+^j + \bar{\psi}_+^i (i\gamma^\perp D_\perp^{ij} - m\delta^{ij})\psi_-^j + \bar{\psi}_-^i \left[i\sqrt{2}\gamma^0 D_-^{ij}\psi_-^j + (i\gamma^\perp D_\perp^{ij} - m\delta^{ij})\psi_+^j \right]$$

where[8] $\psi_\pm = \Lambda^\pm \psi$.

This shows that the minus components ψ^j_- are in fact nondynamical fields without kinetic terms. Their equations of motion in l.c. gauge lead to the constraint equation

$$i\sqrt{2}\psi^j_-(x) = -\frac{1}{\partial_-}(i\gamma^0\gamma^\perp D^{kl}_\perp - m\gamma^0\delta^{kl})\psi^l_+(x)$$

The free field propagator of ψ_+ is determined from the quadratic terms (suppressing the color index) $i\sqrt{2}{\psi_+}^\dagger\partial_+\psi_+ + {\psi_+}^\dagger(i\gamma^0\gamma^\perp\partial_\perp - m\gamma^0)\psi_-$ where $2i\partial_-\psi_- = (i\gamma^\perp\partial_\perp + m)\gamma^+\psi_+$. The equation of motion for the independent component ψ_+ is nonlocal in the longitudinal direction. In the quantized theory we find the following nonvanishing *local* anticommutator $\{\psi_+(\tau, x^-, x^\perp), {\psi_+}^\dagger(\tau, y^-, y^\perp)\} = \frac{1}{\sqrt{2}}\Lambda^+\delta(x^- - y^-)\delta^2(x^\perp - y^\perp)$. They may be realized in momentum space through the following Fourier transform

$$\psi(x) = \frac{1}{\sqrt{(2\pi)^3}}\sum_{r=\pm}\int d^2p^\perp dp^+\theta(p^+)\sqrt{\frac{m}{p^+}}$$

$$\left[b^{(r)}(p)u^{(r)}(p)e^{-ip\cdot x} + d^{\dagger(r)}(p)v^{(r)}(p)e^{ip\cdot x}\right]$$

where[8]

$$u^{(r)}(p) = \frac{\left[\sqrt{2}p^+\Lambda^+ + (m + \gamma^\perp p_\perp)\Lambda^-\right]}{(\sqrt{2}p^+m)^{\frac{1}{2}}}\tilde{u}^{(r)}$$

and the nonvanishing anticommutation relations are given by: $\{b^{(r)}(p), {b^\dagger}^{(s)}(p')\} = \{d^{(r)}(p), {d^\dagger}^{(s)}(p')\} = \delta_{rs}\delta^3(p - p')$. The free propagator $< T\psi^i_+(x)\psi^{\dagger j}_+(0) >_0$ is then shown[8] to be causal

$$\frac{i\delta^{ij}}{(2\pi)^4}\int d^4q\frac{\sqrt{2}q^+\Lambda^+}{(q^2 - m^2 + i\epsilon)}e^{-iq\cdot x}$$

and it contains no instantaneous term.

4 Gauge field propagator in l.c. gauge

In the l.c. gauge the ghost fields decouple. We can obtain the free propagator using the Lagrangian density of abelian gauge theory

$$\frac{1}{2}\left[(F_{+-})^2 - (F_{12})^2 + 2F_{+\perp}F_{-\perp}\right] + BA_-$$

where $F_{\mu\nu} = (\partial_\mu A_\nu - \partial_\nu A_\mu)$. Following the Dirac procedure we show that the phase space constraints remove all the canonical momenta from the theory. The surviving variables are $A_\perp$ and A_+. The latter, however, is a dependent variable satisfying $\partial_-(\partial_- A_+ - \partial_\perp A_\perp) = 0$. The construction of the Dirac bracket shows that in the l.c. gauge on the LF we simultaneously obtain the Lorentz condition $\partial \cdot A = 0$ as an operator equation as well. The *reduced* Hamiltonian is found to be ${H_0}^{LF} = \frac{1}{2}\int d^2x^\perp dx^- \left[(\partial_- A_+)^2 + \frac{1}{2}F_{\perp\perp'}F^{\perp\perp'}\right]$.

The equal-τ commutators are found to be $[A_\perp(x), A_\perp(y)] = i\delta_{\perp\perp'}K(x, y)$ where $K(x, y) = -(1/4)\epsilon(x^- - y^-)\delta^2(x^\perp - y^\perp)$. They are *nonlocal* in the longitudinal coordinate, like the ones for scalar field, but there is no violation of the *microcausality* principle on the LF. Their momentum space realization is obtained by the Fourier transform[7]

$$A^\mu(x) = \frac{1}{\sqrt{(2\pi)^3}}\int d^2k^\perp dk^+\frac{\theta(k^+)}{\sqrt{2k^+}}\sum_{(\perp)}$$

$$E^\mu_{(\perp)}(k)\left[b_{(\perp)}(k)e^{-ik\cdot x} + b^\dagger_{(\perp)}(k)e^{ik\cdot x}\right]$$

where k^- is shown[7] to be defined through the dispersion relation, $2k^-k^+ = k^\perp k^\perp$ corresponding to a *massless* photon. Here the nonvanishing commutators are given by $[b_{(\perp)}(k), {b^\dagger}_{(\perp')}(k')] = \delta_{(\perp)(\perp')}\,\delta^3(k - k')$. The free gluon propagator is hence found to be[7]

$$\frac{i\delta^{ab}}{(2\pi)^4}\int d^4k\, e^{-ik\cdot x}\,\frac{D_{\mu\nu}(k)}{k^2 + i\epsilon}$$

where

$$D_{\mu\nu}(k) = -g_{\mu\nu} + \frac{n_\mu k_\nu + n_\nu k_\mu}{(n\cdot k)} - \frac{k^2}{(n\cdot k)^2}n_\mu n_\nu$$

with $n_\mu = \delta^+_\mu$, $E^\mu_{(\perp)}(k) = E^{(\perp)\mu}(k) = -D^\mu_\perp(k)$, $n^\mu D_{\mu\nu}(k) = 0$, and $k^\mu D_{\mu\nu}(k) = 0$.

5 QCD Hamiltonian in l.c. gauge

The interaction Hamiltonian in the l.c. gauge, $A^a_- = 0$, may be rewritten[7] as

$$-g\bar{\psi}^i\gamma^\mu A^{ij}_\mu\psi^j + \frac{g}{2}f^{abc}(\partial_\mu A^a_\nu - \partial_\nu A^a_\mu)A^{b\mu}A^{c\nu}$$

$$+\frac{g^2}{4}f^{abc}f^{ade}A_{b\mu}A^{d\mu}A_{c\nu}A^{e\nu}-\frac{g^2}{2}j_a^+\frac{1}{(\partial_-)^2}j_a^+$$
$$-\frac{g^2}{2}\bar{\psi}^i\gamma^+(\gamma^{\perp'}A_{\perp'})^{ij}\frac{1}{i\partial_-}(\gamma^{\perp}A_{\perp})^{jk}\psi^k$$

where $j_a^+=\bar{\psi}^i\gamma^+(t_a)^{ij}\psi^j+f_{abc}(\partial_-A_{b\mu})A^{c\mu}$ and a sum over distinct flavours, not written explicitly, is to be understood.

The Dyson-Wick perturbation expansion based on the time ordering with respect to the LF time τ, is built[7] straightforwardly. There are no ghost interaction terms in l.c. gauge. The instantaneous interactions, which can be treated systematically, are seen to be required, say, from the computation of the classical Thomson scattering limit or of electron-muon scattering. While using LF coordinates together with the dimensional regularization the causal prescription for the $1/k^+$ singularity, as given by Mandelstam and Leibbrandt, is mathematically consistent with the causal form of, say, the fermionic propagator. Computations of the divergent parts of the 1-loop gluon and quark self-energy and a three-gluon vertex corrections have been discussed in reference[7].

The fact that gluons have only physical degrees of freedom in l.c. gauge may provide an analysis of coupling renormalization similar to that of the pinch technique, which is currently being discussed[9] as means to define a physical and analytic renormalization scheme for QCD. In addition, the couplings of gluons in the l.c. gauge provides a simple procedure for the factorization of soft and hard gluonic corrections in high momentum transfer inclusive and exclusive reactions.

Acknowledgments

The financial aids from CNPq and FAPERJ for the participation at ICHEP2000 and the hospitality offered at the Theory Division of SLAC are gratefully acknowledged by PPS.

References

1. Research partially supported by the Department of Energy under contract DE-AC03-76SF00515.
   Presented at *ICHEP2000*, Osaka, Japan, Parallel Session PA-15b, July 27-02 August 2000. Scanned *transparencies*, [TALK 15b-03], are available at: http://ichep2000.hep.sci.oaka-u.ac.jp
2. P.A.M. Dirac, Rev. Mod. Phys. **21**, 392 (1949).
3. S.J. Brodsky, *Light-Cone Quantized QCD and Novel Hadron Phenomenology*, SLAC-PUB-7645, 1997; S.J. Brodsky and H.C. Pauli, *Light-Cone Quantization and QCD*, Lecture Notes in Physics, vol. 396, eds., H. Mitter et. al., Springer-Verlag, Berlin, 1991.
4. S.J. Brodsky, H. Pauli and S.S. Pinsky, Phys. Rept. **301**, 299 (1998).
5. K.G. Wilson et. al., Phys. Rev. **D49**, 6720 (1994); K.G. Wilson, Nucl. Phys. B (proc. Suppl.) **17**, (1990). R.J. Perry, A. Harindranath, and K.G. Wilson, Phys. Rev. Lett. **65**, 2959 (1990).
6. P.P. Srivastava, *Perspectives of Light-Front Quantum Field Theory: Some New Results*, in *Quantum Field Theory: A 20th Century Profile*, pgs. 437-478, Ed. A.N. Mitra, Indian National Science Academy, New Delhi, India, 2000; Slac preprint, SLAC-PUB-8219, August 1999; hep-ph/9908492, 9911450.
7. P.P. Srivastava and S.J. Brodsky, *Light-front-quantized QCD in light-cone gauge revisited*, SLAC-PUB-8543, July 2000.
8. P.P. Srivastava and S.J. Brodsky, Phys. Rev. **D61**, 25013 (2000); Slac preprint SLAC-PUB-8168, hep-ph/9906423.
9. J.M. Cornwall, Phys. Rev. **D 26**, 1453 (1982); N.J. Watson, Phys. Lett. **B349**, 155 (1995); J. Papavassiliou, Phys. Rev. **D 62**, 045006 (2000); S.J. Brodsky, E. Gardi, G. Grunberg, and J. Rathsman, hep-ph/0002065.

THERMAL FORWARD SCATTERING AMPLITUDES IN TEMPORAL GAUGES

F. T. BRANDT, J. FRENKEL AND F. R. MACHADO
Instituto de Física, Universidade de São Paulo
São Paulo, SP 05315-970, BRAZIL

We employ the thermal forward scattering amplitudes technique in order to compute the gluon self-energy in a class of temporal gauges. The leading T^2 and the sub-leading $\ln(T)$ contributions are obtained for temperatures high compared with the external momentum. The logarithmic contributions have the same structure as the ultraviolet pole terms which occur at zero temperature (we have recently extended this result to the Coulomb gauge). We also show that the prescription poles, characteristic of temporal gauges, do not modify the leading and sub-leading high-temperature behavior. The one-loop calculation shows that the thermal self-energy *is transverse*. This result has also been extended to higher orders, using the BRS identities

There have been many investigations of thermal gauge field theories in the temporal gauge, both in the imaginary and in the real time formalisms [1,2,3,4]. One of the main advantages of the *non-covariant* temporal gauge is that it is physical and effectively ghost-free. At finite temperature, it may be considered a more natural choice, since the Lorentz invariance is already broken by the presence of the heat bath. It is also convenient for calculating the response of the QCD plasma to a chromo-electric field [1,5]. Despite these advantages, explicit calculations are known to be more involved than in covariant gauges, mainly because of the extra poles at $q \cdot n = 0$ in the propagator, where q is the loop momentum and $n = (n_0, \vec{0})$ $(n_0^2 > 0)$ is the temporal axial four-vector.

The standard method of calculation in the imaginary time formalism, employs the *contour integral formula* [5]

$$T \sum_{n=-\infty}^{\infty} f(q_0 = i\,\omega_n^\sigma) = \frac{1}{2\pi i}\oint_C dq_0 f(q_0)\frac{1}{2}\left[\coth\left(\frac{1}{2}\beta q_0\right)\right]^{i^{(2\sigma)}}, \quad (1)$$

where

$$\omega_n^\sigma = \pi T\,(2n+\sigma); \begin{cases} \sigma = 0 \ (Bosons) \\ \sigma = 1 \ (Fermions) \end{cases}$$

and the contour C is formed by two anti-parallel straight lines equidistant from the imaginary axis. Of course, this formula cannot be employed when the function $f(q_0)$ has poles along the imaginary axis. This is the reason why one should use some prescription for the poles in the temporal gauge [4] gluon propagator

$$\frac{1}{q^2}\left\{-i\delta^{ab}\left[g_{\mu\nu} - \frac{1}{q\cdot u}(q_\mu u_\nu + q_\nu u_\mu) + \frac{q_\mu q_\nu}{(q\cdot u)^2}\left(\frac{\alpha}{n_0^2}q^2+1\right)\right]\right\}, \quad (2)$$

where $u = n/n_0$ is the heat bath four-velocity.

For general covariant gauges, it is possible to show that all the thermal Green functions can be expressed (after the Cauchy integration in the complex plane) in terms of forward scattering amplitudes of on-shell thermal particles [6,7]. The main purpose of this work is to extend the forward scattering method to a class of temporal gauges. As an illustration of this technique, we will compute the *full tensor structure* of the one-loop gluon self-energy. (Previous calculations have considered only the static limit of the component Π_{00} of the self-energy [1,4]). In this way, we will be able to investigate the properties of transversality and gauge invariance of the leading contributions proportional to T^2. We also verify that *the gauge dependent* sub-leading logarithmic part shares with the previous calculations in general covariant gauges [7] the interesting property of having the same structure as the *ultraviolet pole contributions* which occur at zero temperature.

The details involved in the forward scattering technique are explained in the appendix A of reference [8]. An important condition in order to be able to apply this technique is that the gluon propagator should have only the mass-shell poles at $q^2 = 0$. Therefore, our first task when using the axial gauge gluon propagator is to separate the contributions which can potentially have poles at $n \cdot q = 0$ from the normal mass-shell poles. The simplest contribution having poles only at $n \cdot q = 0$ is the ghost loop diagram shown in Fig. 1(c). At finite temperature it is proportional to

$$\int d^3\vec{q} \sum_{q0} \left[\frac{t_{\mu\nu}}{n \cdot q\, n \cdot (q+k)} + q \leftrightarrow -q \right], \quad (3)$$

where $t_{\mu\nu}$ is a momentum independent quantity and $q_0 = 2\pi i n T$ ($n = 0, \pm 1, \pm 2, \cdots$). Using partial fractions the integrand in (3) can be wirtten as

$$\frac{1}{n \cdot k} \left[\frac{1}{n \cdot q} - \frac{1}{n \cdot (q+k)} \right]. \quad (4)$$

Performing a shift $q \to q - k$ in the second term, we can easily see that the ghosts *effectively decouple.* This shows that there is a simple mechanism for *some* cancellations of the temporal gauge poles, which involves only simple algebraic manipulations, *before* the computation of the sum over the Matsubara frequencies q_0. As we will see next, there are other contributions from the diagrams in Figs. 1(a) and 1(b) sharing same property.

The separation of the temporal gauge poles can be accomplished in a more systematic and physical way, using the following well known tensor decomposition of the gluon self-energy

$$\Pi^{ab}_{\mu\nu} = \delta^{ab} \left(\Pi_T P^T_{\mu\nu} + \Pi_L P^L_{\mu\nu} + \Pi_C P^C_{\mu\nu} + \Pi_D P^D_{\mu\nu} \right), \quad (5)$$

where

$$P^T_{\mu\nu} = g_{\mu\nu} - P^L_{\mu\nu} - P^D_{\mu\nu},$$
$$P^L_{\mu\nu} = \frac{(u \cdot k k_\mu - k^2 u_\mu)(u \cdot k k_\nu - k^2 u_\nu)}{-k^2 |\vec{k}|^2},$$
$$P^C_{\mu\nu} = \frac{2k \cdot u k_\mu k_\nu - k^2(k_\mu u_\nu + k_\nu u_\mu)}{k^2 |\vec{k}|},$$
$$P^D_{\mu\nu} = \frac{k_\mu k_\nu}{k^2}, \quad (6)$$

where $k^\mu P^{T,L}_{\mu\nu} = 0$, $k^i P^T_{i\nu} = 0$, $k^i P^L_{i\nu} \neq 0$ ($i = 1,2,3$) and $k^\mu P^{C,D}_{\mu\nu} \neq 0$.

The calculation of structures Π_A, Π_B, Π_C and Π_D and the following algebraic manipulations were performed with the help of computer algebra. After partial fraction decompositions [as in Eq. (4)] and shifts $q \to q - k$, the rather involved expressions for Π_C and Π_D simplify considerably and all the temporal gauge poles at $n \cdot q = 0$ cancel. We then proceed using the contour integral formula [Eq. (1)] and show that $\Pi_C = \Pi_D = 0$. It is interesting to note that, from the general relation [9]

$$\Pi_D = \frac{\Pi_C^2}{k^2 - \Pi_L} \quad (7)$$

the vanishing of Π_C to one loop order implies, in fact, that Π_D should vanish up to the three-loops order. Using the Becchi-Rouet-Stora identities [10] we have extended this result to all orders [8].

For the structures Π_A and Π_B the temporal gauge poles do not cancel at the integrand level and we have to employ a prescription in order to be able to use the Eq. (1). Using the procedure described in the appendix of reference [4] we were able to show, by explicit calculation, that the *prescription poles* do not contribute to the leading and the subleading high temperature limit of Π_A and Π_B. Therefore, the thermal gluon self-energy can be represented, in the limit of high temperatures, in terms of *forward scattering amplitudes* of on-shell thermal gluons, as given by Eq. (6) of reference [8].

In conclusion, our results show that the full tensor structure of the thermal gluon self-

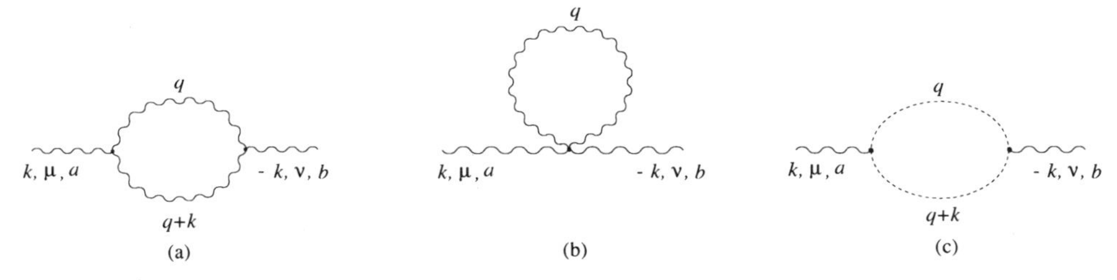

Figure 1. One-loop diagrams which contribute to the gluon-self energy. Wavy and dashed lines denotes respectively gluons and ghosts. All external momenta are inwards.

energy can be consistently computed in a class of temporal gauges, and expressed in terms of forward scattering amplitudes of on-shell thermal gluons. Using this approach, we have obtained the known gauge invariant result for the leading high temperature T^2 contribution. We also have shown, by explicit calculation, that the one-loop thermal self-energy *is exactly transverse* for any temperature regime. Motivated by this result, we were able to prove the transversality to all orders. This property seems to be very peculiar to the temporal gauges. It is not valid, for instance, in general covariant gauges, except for the Feynman gauge, where the transversality has been verified only to one-loop order [7].

Our approach also gives sub-leading contributions which are in agreement with the conjecture proposed in [11], according to which the ultraviolet divergent contributions which arises at $T = 0$ are identical to the thermal contributions proportional to $\ln(1/T)$. We have recently verified this conjecture also in the *Coulomb gauge* so that the divergent part of the Coulomb gauge gluon self-energy [12,13] can be alternatively obtained from our $\ln(1/T)$ contribution. The details of this analysis will be reported elsewhere.

Acknowledgments

F.T.B and F.R.M. acknowledge the financial support from FAPESP (Brazil). F.T.B and J.F. would like to thank CNPq for a grant.

References

1. K. Kajantie and J. Kapusta, Ann. Phys. **160**, 477 (1985).
2. R. Kobes, G. Kunstatter, and K. W. Mak, Z. Phys. **C45**, 129 (1989).
3. K. A. James and P. V. Landshoff, Phys. Lett. **B251**, 167 (1990).
4. G. Leibbrandt and M. Staley, Nucl. Phys. **B428**, 469 (1994).
5. M. L. Bellac, *Thermal Field Theory* (Cambridge University Press, Cambridge, England, 1996).
6. J. Frenkel and J. C. Taylor, Nucl. Phys. **B334**, 199 (1990); **B374**, 156 (1992).
7. F. T. Brandt and J. Frenkel, Phys. Rev. D **56**, 2453 (1997).
8. F. T. Brandt, J. Frenkel and F. R. Machado, Phys. Rev. D **61**, 125014 (2000).
9. H. A. Weldon, Annals Phys. **271**, 141 (1999).
10. C. Becchi, A. Rouet, and R. Stora, Commun. Math. Phys. **42**, 127 (1975).
11. F. T. Brandt and J. Frenkel, Phys. Rev. Lett. **74**, 1705 (1995).
12. J. Frenkel, J.C. Taylor Nucl. Phys. **B109**, 439 (1976).
13. G. Leibbrandt and J. Williams, Nucl. Phys. **B475**, 469 (1996).

MACROCONSTRAINTS FROM MICROSYMMETRIES

LUDWIK TURKO
Institute of Theoretical Physics, University of Wroclaw,
pl. Maksa Borna 9, 50-204 Wrocław, Poland
E-mail: turko@ift.uni.wroc.pl

The dynamics governing the evolution of a many body system is constrained by a nonabelian local symmetry. We obtain explicit forms of the global macroscopic condition assuring that at the microscopic level the evolution respects the overall symmetry constraint.

Let us consider a multiparticle quantum system with the local interaction invariant with respect to the internal symmetry group G. We shall call this underlying symmetry of microscopic interactions a microsymmetry of the system. The system transforms under a given representation of the symmetry group. We shall call this property a macrosymmetry of the system. Let us try to answer a question: is macrosymmetry preserved during a time evolution of the system? This problem is treated with more details by J. Rafelski and the author[1].

The system consists of particles belonging to multiplets of the symmetry group. One denotes $f_{(\zeta)}^{(\alpha_i,\nu_i)}(\Gamma,\vec{r},t)$ a distribution function of the particle, transforming under α_i representation of the symmetry group, with quantum numbers of ν_i member of the multiplet. The variables $(\Gamma,\vec{r})$ denotes a set of the phase - space variables such as $(\vec{p},\vec{r})$ and t is time. A subscript ζ denotes other quantum numbers characterizing different multiplets of the same representation α.

The number of particles of the specie $\{\alpha,\nu_\alpha\}$ is:

$$N_{\nu_\alpha;(\zeta)}^{(\alpha)}(t) = \int dV d\Gamma f_{(\zeta)}^{(\alpha,\nu_\alpha)}(\Gamma,\vec{r},t). \quad (1)$$

Let us consider the corresponding state vector in particle number representation: $\left|N_{\nu_{\alpha_1}}^{(\alpha_1)},\ldots,N_{\nu_{\alpha_n}}^{(\alpha_n)}\right\rangle$. All other variables, related to phase-space properties of the system are suppressed here. This vector describes symmetry properties of our systems and transforms as a direct product representation of the symmetry group G. This representation is of the form:

$$\alpha_1^{N^{(\alpha_1)}} \otimes \alpha_2^{N^{(\alpha_2)}} \otimes \cdots \otimes \alpha_n^{N^{(\alpha_n)}}. \quad (2)$$

A multiplicity $N^{(\alpha_j)}$ of the representation α_j in this product is equal to a number of particles which transform under this representation:

$$N^{(\alpha_j)} = \sum_j \left(\sum_{\zeta_j} N_{\nu_{\alpha_j};(\zeta_j)}^{(\alpha_j)} \right) = \sum_j N_{\nu_{\alpha_j}}^{(\alpha_j)}. \quad (3)$$

The representation given by Eq. (2) can be decomposed into direct sum of irreducible representations Λ_k. Corresponding states are denoted as $|\Lambda_k,\lambda_{\Lambda_k};\mathcal{N}\rangle$ where λ_{Λ_k} is an index numbering members of the representation Λ and $\mathcal{N}$ is a total number of particles

$$\mathcal{N} = \sum_k N_{\nu_{\alpha_k}}^{(\alpha_k)}. \quad (4)$$

Each physical state can be decomposed into irreducible representation base states $|\Lambda_k,\lambda_{\Lambda_k};\mathcal{N};\xi_{\Lambda_k}\rangle$. Variables ξ_Λ are degeneracy parameters required for the full description of a state in the "symmetry space". Let us consider a projection operator $\mathcal{P}^\Lambda$ on the subspace spanned by all states transforming under representation Λ.

$$\mathcal{P}^\Lambda \left|N_{\nu_{\alpha_1}}^{(\alpha_1)},\ldots,N_{\nu_{\alpha_n}}^{(\alpha_n)}\right\rangle = \sum_{\xi_\Lambda}^{\oplus} |\Lambda,\lambda_\Lambda;\xi_\Lambda\rangle\, C_{\{N_{\nu_{\alpha_1}}^{(\alpha_1)},\ldots,N_{\nu_{\alpha_n}}^{(\alpha_n)}\}}^{\Lambda,\lambda_\Lambda}(\xi_\Lambda). \quad (5)$$

This operator has the generic form (see e.g. [2]):

$$\mathcal{P}^{\Lambda} = d(\Lambda) \int_G d\mu(g) \bar{\chi}^{(\Lambda)}(g) U(g). \quad (6)$$

Here $d(\Lambda)$ is the dimension of the representation Λ, $\chi^{(\Lambda)}$ is the character of the representation Λ, $d\mu(g)$ is the invariant Haar measure on the group, and $U(g)$ is an operator transforming a state under consideration. We will use the matrix representation:

$$U(g) \left| N^{(\alpha_1)}_{\nu_{\alpha_1}}, \ldots, N^{(\alpha_n)}_{\nu_{\alpha_n}} \right\rangle$$
$$= \sum_{\nu_1^{(1)}, \ldots, \nu_n^{(N_{\nu_n})}} D^{(\alpha_1)}_{\nu_1^{(1)} \nu_1} \cdots D^{(\alpha_1)}_{\nu_1^{(N_{\nu_1})} \nu_1} \cdots D^{(\alpha_n)}_{\nu_n^{(1)} \nu_n}$$
$$\cdots D^{(\alpha_n)}_{\nu_n^{(N_{\nu_n})} \nu_n} \left| N^{(\alpha_1)}_{\nu_{\alpha_1}}, \ldots, N^{(\alpha_n)}_{\nu_{\alpha_n}} \right\rangle. \quad (7)$$

$D^{(\alpha_n)}_{\nu,\nu'}$ is a matrix elements of the group element g corresponding to the representation α. Notation convention in Eq. (7) arises since there are $N^{(\alpha_j)}_{\nu_{\alpha_j}}$ states transforming under representation α_j and having quantum numbers of the ν_{α_j}-th member of a given multiplet.

The probability $\overline{P^{\Lambda,\lambda_\Lambda}_{\{N^{(\alpha_1)}_{\nu_{\alpha_1}}, \ldots, N^{(\alpha_n)}_{\nu_{\alpha_n}}\}}}$ that $N^{(\alpha_1)}_{\nu_{\alpha_1}}, \ldots, N^{(\alpha_n)}_{\nu_{\alpha_n}}$ particles transforming under the symmetry group representations $\alpha_1, \ldots, \alpha_n$ combine into $\mathcal{N}$ particle state transforming under representation Λ of the symmetry group is given by

$$\left\langle N^{(\alpha_1)}_{\nu_{\alpha_1}}, \cdots, N^{(\alpha_n)}_{\nu_{\alpha_n}} \right| \mathcal{P}^{\Lambda} \left| N^{(\alpha_1)}_{\nu_{\alpha_1}}, \ldots, N^{(\alpha_n)}_{\nu_{\alpha_n}} \right\rangle$$
$$= \sum_{\xi_\Lambda} |\mathcal{C}^{\Lambda,\lambda_\Lambda}_{\{N^{(\alpha_1)}_{\nu_{\alpha_1}}, \ldots, N^{(\alpha_n)}_{\nu_{\alpha_n}}\}}(\xi_\Lambda)|^2. \quad (8)$$

Left hand side of this equation can be calculated directly from Eqs.(6) and (7). One gets finally

$$\overline{P^{\Lambda,\lambda_\Lambda}_{\{N^{(\alpha_1)}_{\nu_{\alpha_1}}, \ldots, N^{(\alpha_n)}_{\nu_{\alpha_n}}\}}}$$
$$= \mathcal{A}^{\{\mathcal{N}\}} d(\Lambda) \int_G d\mu(g) \bar{\chi}^{(\Lambda)}(g) [D^{(\alpha_1)}_{\nu_1 \nu_1}]^{N^{(\alpha_1)}_{\nu_{\alpha_1}}}$$
$$\cdots [D^{(\alpha_n)}_{\nu_n \nu_n}]^{N^{(\alpha_n)}_{\nu_{\alpha_n}}}. \quad (9)$$

where $\mathcal{A}^{\{\mathcal{N}\}}$ is a permutation normalization factor. For particles of the kind $\{\alpha, \zeta\}$ we included in Eq. (9) the permutation factor:

$$\mathcal{A}^{\alpha}_{(\zeta)} = \frac{\mathcal{N}^{(\alpha)}_{(\zeta)}!}{\prod_{\nu_\alpha} \mathcal{N}^{(\alpha)}_{\nu_\alpha;(\zeta)}!}. \quad (10)$$

The permutation factor $\mathcal{A}^{\{\mathcal{N}\}}$ is a product of all "partial" factors

$$\mathcal{A}^{\{\mathcal{N}\}} = \prod_j \prod_{\zeta_j} \mathcal{A}^{\alpha_j}_{(\zeta_j)}. \quad (11)$$

Because of macrosymmetry all weights in Eq. (9) are constant in time. It provides subsidiary constraints on distribution functions $f^{(\alpha_i,\nu_i)}$. These conditions assure that in a dynamical evolution the symmetry of the system is preserved.

We now convert the global constraint into a time evolution condition and consider:

$$\frac{d}{dt} \overline{P^{\Lambda,\lambda_\Lambda}_{\{N^{(\alpha_1)}_{\nu_{\alpha_1}}, \ldots, N^{(\alpha_n)}_{\nu_{\alpha_n}}\}}} = 0. \quad (12)$$

Introducing here the result of Eq. (9) one obtains:

$$0 = \frac{d \mathcal{A}^{\{\mathcal{N}\}}}{dt} d(\Lambda) \times$$
$$\int_G d\mu(g) \bar{\chi}^{(\Lambda)}(g) [D^{(\alpha_1)}_{\nu_1 \nu_1}]^{N^{(\alpha_1)}_{\nu_{\alpha_1}}} \cdots [D^{(\alpha_n)}_{\nu_n \nu_n}]^{N^{(\alpha_n)}_{\nu_{\alpha_n}}}$$
$$+ \sum_{j=1}^{n} \sum_{\nu_{\alpha_j}} \frac{d N^{(\alpha_j)}_{\nu_{\alpha_j}}}{dt} \mathcal{A}^{\{\mathcal{N}\}} d(\Lambda)$$
$$\times \int_G d\mu(g) \bar{\chi}^{(\Lambda)}(g) [D^{(\alpha_1)}_{\nu_1 \nu_1}]^{N^{(\alpha_1)}_{\nu_{\alpha_1}}}$$
$$\cdots [D^{(\alpha_n)}_{\nu_n \nu_n}]^{N^{(\alpha_n)}_{\nu_{\alpha_n}}} \log[D^{(\alpha_j)}_{\nu_j \nu_j}]. \quad (13)$$

All integrals which appear in Eq. (9) and Eq. (13) can be expressed explicitly in an analytic form for any compact symmetry group.

To write an expression for the time derivative of the normalization factor $\mathcal{A}^{\{\mathcal{N}\}}$ we perform analytic continuation from integer to continuous values of variables $N^{(\alpha_n)}_{\nu_{\alpha_n}}$.

Thus we replace all factorials by the Γ–function of corresponding arguments. We encounter here also the digamma function ψ [3]:

$$\psi(x) = \frac{d \log \Gamma(x)}{d x}. \tag{14}$$

This allows to write:

$$\frac{d \mathcal{A}^{\{\mathcal{N}\}}}{dt}$$

$$= \mathcal{A}^{\{\mathcal{N}\}} \sum_j \sum_{\zeta_j} \left[\frac{d \mathcal{N}^{(\alpha_j)}_{(\zeta_j)}}{dt} \psi(\mathcal{N}^{(\alpha_j)}_{(\zeta_j)} + 1) \right.$$

$$\left. - \sum_{\nu_{\alpha_j}} \frac{d \mathcal{N}^{(\alpha_j)}_{\nu_{\alpha_j};(\zeta_j)}}{dt} \psi(\mathcal{N}^{(\alpha_j)}_{\nu_\alpha;(\zeta_j)} + 1) \right]. \tag{15}$$

The time derivatives $d\mathcal{N}^{(\alpha)}_{\nu_\alpha;(\zeta)}/dt$ are obtained from the integrated kinetic equation fulfilled by a set of distribution functions $f^{(\alpha_i,\nu_i)}_{(\zeta)}(\Gamma, \vec{r}, t)$. The case of the generalized Vlasov - Boltzmann kinetic equations was considered in [1].

These subsidiary conditions fulfilled by the microscopic kinetic equations are the necessary conditions for preserving the internal symmetry on the macroscopic level. Rates of change $d\mathcal{N}^{(\alpha)}_{\nu_\alpha;(\zeta)}/dt$ are related to "macrocurrents", which are counterparts of "microcurrents" related directly to a symmetry on a microscopic level via the Noether theorem. This can be considered as a set of conditions on macrocurrents to provide consistency with the overall symmetry of the system. These conditions leads to nontrivial results only for nonabelian symmetry groups. In the abelian case all charges are additive ones. The charge conservation on the microscopic level is equivalent to the global charge conservation of the multiparticle system. This is not the case for nonabelian symmetries, where nonabelian charges can combine to different representations of the symmetry group.

New constraints on kinetic equations lead to decreasing number of available states for the system during its time evolution. One can expect that such a system when approaching its equilibrium would produce less entropy. This should give some observable effects, e.g. for particle production processes in heavy-ion collision.

It should be noted that these new constraints on evolution equations are purely quantum effect. In the case of classical systems a concept of representations of the symmetry group is not applicable.

Acknowledgments

Work supported in part by the Polish Committee for Scientific Research under contract KBN - 2P03B 030 18.

References

1. L. Turko and J. Rafelski, *Dynamics of Multiparticle Systems with non – Abelian Symmetry* **hep-th/0003079**; to be published in **EPJ**
2. E. P. Wigner, *Group Theory and Its Application to the Quantum Mechanics of Atomic Spectra*, (Academic Press, New York and London, 1959).
3. M. Abramowitz and I. A. Stegun (eds.), *Handbook of Mathematical Functions*, (National Bureau of Standards, Applied Mathematics Series · 55, 1964).

CONFINEMENT AND FLAVOR SYMMETRY BREAKING VIA MONOPOLE CONDENSATION

HITOSHI MURAYAMA

Department of Physics, University of California, Berkeley CA 94720, USA

Lawrence Berkeley National Laboratory, Cyclotron Road, Berkeley CA94720 USA
E-mail: murayama@lbl.gov

We discuss dynamics of $N = 2$ supersymmetric $SU(n_c)$ gauge theories with n_f quark hypermultiplets. Upon $N = 1$ perturbation of introducing a finite mass for the adjoint chiral multiplet, we show that the flavor $U(n_f)$ symmetry is dynamically broken to $U(r) \times U(n_f - r)$, where $r \leq [n_f/2]$ is an integer. This flavor symmetry breaking occurs due to the condensates of magnetic degrees of freedom which acquire flavor quantum numbers due to the quark zero modes. We briefly comment on the $USp(2n_c)$ gauge theories. This talk is based on works with Giuseppe Carlino and Ken Konishi.[1,2]

1 Introduction

There have been at least two main dynamical issues in gauge theories: confinement and flavor symmetry breaking. The former is an obvious requirement in understanding the real-world strong interaction dynamics, namely the lack of observation of isolated quarks. The spectrum of light hadrons demands a linear potential with respect to the distance beween the quark and the antiquark in meson boundstates. The latter is a more subtle issue. Nambu pointed out that the lightness of the pions can be understood if they are what we now call Nambu–Godstone bosons of spontaneously broken symmetries. We need the $SU(3)_L \times SU(3)_R$ flavor symmetry of the QCD to be dynamically broken down to $SU(3)_V$ by quark bilinear condensates

$$\langle \bar{u}u \rangle = \langle \bar{d}d \rangle = \langle \bar{s}s \rangle \neq 0. \tag{1}$$

An important question is what microscopic mechanism is behind the confinement and dynamical flavor symmetry breaking. The seminal work by Seiberg and Witten[3] showed that, using $N = 2$ supersymmetric gauge theories, confinement can be understood as a consequence of the magnetic monopole condensation as conjectured a long time ago by 't Hooft and Mandelstam.[4] Our aim is to bring the understanding of the dynamical flavor symmetry breaking to the same level, done in collaboration with Giuseppe Carlino and Ken Konishi[1,2]. Surprisingly, this question had not been addressed systematically so far. Seiberg and Witten themselves studied the case with flavor,[5] but there were only two examples which exhibited dynamical flavor symmetry breaking ($SU(2)$ with $n_f = 2, 3$) and it was not possible to draw a general lesson. Later works on general gauge groups[6,7] focused on the appearance of the dual gauge group, and did not discuss the issue of flavor symmetry breaking.

We start with $N = 2$ supersymmetric $SU(n_c)$ QCD with n_f hypermultiplet quarks in the fundamental representation. We later add a perturbation which leaves only $N = 1$ supersymmetry, $W = \mu \text{tr} \Phi^2$, a mass term for the adjoint chiral superfield in the $N = 2$ vector multiplet. This theory has $U(n_f)$ flavor symmetry. We found that the flavor symmetry is in general dynamically broken as

$$U(n_f) \to U(r) \times U(n_f - r). \tag{2}$$

There are isolated vacua for $0 \leq r \leq [n_f/2]$. We have shown that this dynamical flavor symmetry breaking is caused by condensation of magnetic degrees of freedom. For the vacuum $r = 0$, there is no breaking of the flavor $U(n_f)$ symmetry. For the vacuum $r = 1$, what condenses is nothing but the magnetic monopoles, which belong to the fundamental representatin of the $U(n_f)$ flavor group. For the vacua $r > 1$, magnetic monopoles "break up" into "dual quarks" before reaching the singularities where they become massless; it is the "dual quark" which condenses and breaks the flavor symmetry. In any case, the flavor symmetry breaking and the confinement[a] are both caused by the condensation of magnetic degerees of freedom.

Thanks to holomorphy, there is no phase transition by varying μ from small ($\mu \ll \Lambda$) to large ($\mu \gg \Lambda$). Therefore one can study the theory in both limits and compare the results; this would not only provide us non-trivial cross checks but also insight into the dynamics of the theory. We can also resort to completely different techniques to analyze the theory in the different limits.

In the limit (1) ($\mu \gg \Lambda$), we can integrate the adjoint chiral multiplet Φ out from the theory, and study the resulting $N = 1$ low-energy theory. The low-energy theory has a superpotential term

$$W = -\frac{1}{\mu}(\tilde{Q}T^a Q)(\tilde{Q}T^a Q), \qquad (3)$$

where Q ($\tilde{Q}$) are the quark chiral superfields in the fundamental (anti-fundamental) representation of the gauge group. Then we can use analysis by Seiberg[9] on $N = 1$ supersymmetric QCD together with the above effective superpotential (3). It is then easy to identify the vacua of the theory by solving for the extrema of the superpotential with respect to the gauge-invariant composites such as $M^{ij} = \tilde{Q}^i Q^j$ or $B^{i_1 \cdots i_{n_c}} = Q^{i_1} Q^{i_2} \cdots Q^{i_{n_c}}$, $\tilde{B}^{i_1 \cdots i_{n_c}} = \tilde{Q}^{i_1} \tilde{Q}^{i_2} \cdots \tilde{Q}^{i_{n_c}}$. This makes it easy to identify the flavor symmetry breaking patterns.

In the other limit (2) ($\mu \ll \Lambda$), we start with $N = 2$ limit ($\mu = 0$) where the low-energy effective theory is known exactly. In this limit, we can identify monopole degrees of freedom etc which become massless at singularities. We then turn on $\mu \neq 0$. This way we obtain information on the microscopic dynamics of magnetic degrees of freedom.

When considering the theory in various limits, a very powerful check is provided by counting the number of vacua, by further perturbing the theory by finite masses of hypermultiplet quarks. Quark masses make the vacua discrete and countable, and we must obtain the same number of vacua in different limits. In fact, we considered four such limits in total. Two of them have large μ. In the limit (1A), we regard both μ and m_i large and solve for vacua semi-classically (*i.e.*, including the effects of gaugino condensates in unbroken pure Yang-Mills factors). In the limit (1B), we integrate out Φ and use known $N = 1$ dynamics together with the effective superpotential Eq. (3), further combined with the mass terms for the quarks. The other two have small μ, namely setting $\mu = 0$ first, and then reintroduce $\mu \neq 0$ later on. In the limit (2A), we approach singularities from large Φ on the Coulomb branch. In the limit (2B), we approach singularities from large $Q, \tilde{Q}$ on the Higgs branch. All these approaches should give the identical number of vacua, and the consistency among them tell us, for example, which singularity on the Coulomb branch corresponds to which symmetry breaking pattern.

I will not discuss the limit (1A) in this talk and simply refer interested parties to our paper.[2] I first discuss the limit (1B) and identify the flavor symmetry breaking patterns. Then I will briefly review how the monopoles acquire flavor quantum numbers. I will move on to the analyses with small μ next, first on the Coulomb branch, and next on the Higgs branch. The consistency among different approaches gives us full understanding of the dynamics.

2 Large μ Analysis

$N = 2$ supersymmetric QCD can be viewed as a special version of $N = 1$ supersymmetric gauge theories with the following superpotential

$$W = \sqrt{2}\tilde{Q}_i \Phi Q_i + m_i \tilde{Q}_i Q_i + \mu \mathrm{tr}\Phi^2, \qquad (4)$$

where the last term breaks $N = 2$ to $N = 1$. When μ is large, we can integrate out Φ field, and obtain

$$W = -\frac{1}{\mu}(\tilde{Q}_i T^a Q_i)(\tilde{Q}_j T^a Q_j) + m_i \tilde{Q}_i Q_i. \qquad (5)$$

---

[a] We use the term "confinement" somewhat loosely, as in "s-confinement" in [8].

Doing Fierz transformation on the first term, we obtain

$$W = \frac{1}{2\mu}\left[\mathrm{tr}M^2 - \frac{1}{n_c}(\mathrm{tr}M)^2\right] + \mathrm{tr}mM, \quad (6)$$

where $M_{ij} = \tilde{Q}_i Q_j$ is the meson chiral superfield, and the mass $m = \mathrm{diag}(m_1, \cdots, m_{n_f})$ and the meson field are in the matrix notation.

Due to lack of time, I concentrate on the case $n_f < n_c$. I refer to our papers[1,2] for larger number of flavors. In this case, the non-perturbative superpotential[10] is added to Eq. (6):

$$\Delta W = (n_c - n_f)\frac{\Lambda_1^{(3n_c-n_f)/(n_c-n_f)}}{(\mathrm{det}M)^{1/(n_c-n_f)}}, \quad (7)$$

where $\Lambda_1^{3n_c-n_f} = \mu^{n_c}\Lambda^{2n_c-n_f}$ is the scale of the low-energy $N=1$ theory.

By solving for the meson matrix $M = \mathrm{diag}(\lambda_1, \cdots, \lambda_{n_f})$, we find that λ_i satisfy quadratic equations and hence there are two solutions for each of them. We obtain

$$\lambda_i = \frac{1}{2}(Y \pm \sqrt{Y^2 + 4\mu X}) + O(m), \quad (8)$$

where the signs $\pm$ indicate two solutions for each $i = 1, \cdots, n_f$ and hence there are 2^{n_f} possibilities. For the choice of r plus signs and $n_f - r$ minus signs, we can further determine X and Y, which can take $(2n_c - n_f)$ possible phases. Avoiding double counting for $r \leftrightarrow n_f - r$, we find $(2n_c - n_f)2^{n_f-1}$ vacua in total. The most important outcome from this analysis is that, in $m \to 0$ limit, r eigenvalues with plus sign are degenerate, and $n_f - r$ eigenvalues with minus sign are also degenerate. Such a vacuum for the meson field exhibits dynamical flavor symmetry breaking, $U(n_f) \to U(r) \times U(n_f - r)$.

3 Semi-classical Monopoles

As was shown by 't Hooft and Polyakov, there are solitonic solutions to the gauge-Higgs system which appear as magnetic monopoles under the low-energy gauge group. The canonical example is the $SU(2)$ gauge theory with the adjoint Higgs Φ, where the expectation value of $\Phi = a\sigma_3$ breaks $SU(2) \to U(1)$. The mass of the magnetic monopole is given roughly as $M \sim 4\pi a/g$, while the mass of the W-boson is $m_W \sim ga$. Therefore for the weakly-coupled case, the magnetic monopole is heavy and W (electric monopole) is light, while for the strongly-coupled case, the magnetic monopole is light and the W-boson is heavy.

The case with flavor is quite interesting.[11] The cancellation of the $SU(2)$ Witten anomaly requires an even number of flavors: $2n_f$ doublet quarks. They can couple to the adjoint Higgs as $\mathcal{L}_{\mathrm{Yukawa}} = q_i \Phi q_i$, which produces Majorana-type mass terms. The largest possible flavor symmetry is $SO(2n_f)$. Solving Dirac equation for the quarks in the monopole background, there is one zero-energy mode for each flavor. The important question is what statistics the fermion zero modes follow. Surprisingly, they are bosons. The reasoning is simple. The way to judge if an excitation is bosonic or fermionic is by studying the 2π rotation of space and asking if the excitation produces a minus sign (fermion) or not (boson). In the presence of a 't Hooft–Polyakov monopole, a naive 2π spatial rotation is not a symmetry because the isospin space is tied to the real space ("hedgehog"). Therefore one needs to make a 2π rotation both for the real space and the isospin (gauge) space to determine statistics. For fermion zero modes in the $SU(2)$ doublet representation, spatial rotation produces a minus sign, while the isospin rotation produces another minus sign. The fermion zero mode does not produce a minus sign under the true 2π rotation and hence is a boson. Therefore monopole states with or without the fermion zero mode have the same statistics and the same energy. In other words, the monopole states form a multiplet.

For the $SU(2)$ gauge theory, or in general $USp(2n_c)$ gauge theories, we have $SO(2n_f)$ flavor symmetry. Fermion zero mode operators q^i follow the anti-commutation relation $\{q^i, q^j\} = \delta^{ij}$ upon canonical quantization, and they are represented as gamma matrices $q^i = \gamma^i/\sqrt{2}$. The monopole Hilbert space is the representation space of the anti-commutation relation, and is hence a spinor representation of $SO(2n_f)$, with 2^{n_f} states. For the $SU(n_c)$ gauge theories ($n_c > 2$), however, the flavor symmetry is only as large as $U(n_f) \subset SO(2n_f)$, and hence the monopole multiplet (spinor under $SO(2n_f)$) is decomposed into irreducible multiplets under $U(n_f)$: totally anti-symmetric tensor representations. One can easily check that the the dimensions match: $\sum_r {}_{n_f}C_r = 2^{n_f}$ using the binomial theorem.

We have learned that the monopoles acquire flavor quantum numbers of the rank-r totally anti-symmetric representations, while the theory breaks the $U(n_f)$ flavor symmetry dynamically to $U(r) \times U(n_f - r)$ in the previous section. We are naturally led to a conjecture that the flavor symmetry breaking is caused by the condensation of the magnetic monopoles which causes confinement at the same time.

4 Moduli Space of $N = 2$ Theories

The classical moduli space of the theory is determined by solving the vacuum equations

$$\Phi Q_i = 0, \tag{9}$$

$$\tilde{Q}_i \Phi = 0, \tag{10}$$

$$\sum_i \left\{ Q_i \tilde{Q}_i - \frac{1}{n_c} \mathrm{tr} Q_i \tilde{Q}_i \right\} = 0, \tag{11}$$

$$\left[\Phi, \Phi^\dagger\right] = 0, \tag{12}$$

$$Q^\dagger T^a Q - \tilde{Q}^T T^a \tilde{Q}^* = 0. \tag{13}$$

There are three types of "branches" to the vacuum solutions[6]: (1) Coulomb branch, (2) Non-baryonic (or mixed) branch, and (3) Baryonic branch. The baryonic branch appears only for $n_f \geq n_c$ and we will not discuss it.

The solution to Eq. (12) is given by $\Phi = \mathrm{diag}(\phi_1, \cdots, \phi_{n_c})$ with the constraint $\mathrm{tr}\Phi = \sum_k \phi_k = 0$. This defines the complex $(n_c - 1)$-dimensional Coulomb branch. At a generic point on the Coulomb branch, the theory is a free $U(1)^{n_c - 1}$ gauge theory while there appear massless particles on singular submanifolds. The singularities can be found where the auxiliary curve[12]

$$y^2 = \prod_{k=1}^{n_c} (x - \phi_k)^2 + 4\Lambda^{2n_c - n_f} \prod_{i=1}^{n_f} (x + m_i) \tag{14}$$

is maximally degenerate.

The non-baryonic branch is given by the following field configurations:

$$Q = \left(\begin{array}{ccc|ccc|c} \kappa_1 & & & 0 & & & 0 \\ & \ddots & & & \ddots & & \vdots \\ & & \kappa_r & & & 0 & 0 \\ \hline 0 & & & 0 & & & 0 \\ & \ddots & & & \ddots & & \vdots \\ & & 0 & & & 0 & 0 \end{array}\right) \tag{15}$$

$$\tilde{Q} = \left(\begin{array}{ccc|ccc|c} 0 & & & \kappa_1 & & & 0 \\ & \ddots & & & \ddots & & \vdots \\ & & 0 & & & \kappa_r & 0 \\ \hline 0 & & & 0 & & & 0 \\ & \ddots & & & \ddots & & \vdots \\ & & 0 & & & 0 & 0 \end{array}\right) \tag{16}$$

$$\Phi = \left(\begin{array}{ccc|ccc} 0 & & & & & \\ & \ddots & & & 0 & \\ & & 0 & & & \\ \hline & & & \phi_{r+1} & & \\ & 0 & & & \ddots & \\ & & & & & \phi_{n_c} \end{array}\right) \tag{17}$$

Because both hypermultiplets $Q, \tilde{Q}$ and the vector multiplet Φ have expectation values, it is also called the mixed branch. There are separate r-branches for each choice of the integer r. The integer r can range from 1 to $\min\{[\frac{n_f}{2}], n_c - 2\}$, and hence this branch exists only for $n_f \geq 2$ and $n_c \geq 3$. It is important to note that the limit $\kappa_k \to 0$ recovers $U(r)$ gauge symmetry and the branch touches the Coulomb branch ("root" of the no-baronic branch). The theory at the root is a $U(r) \times U(1)^{n_c - r - 1}$ gauge theory which is asymptotically non-free, and hence the gauge fields survive as dynamical degrees of freedom in the low-energy limit. Along the root, there are special isolated points where we can find $n_c - r - 1$ massless monopole multiplets so that the curve becomes maximally degenerate.

5 Coulomb Branch Description

One can locate points on the Coulomb branch where the curve is maximally degenerate, so that the points survive after $\mu \neq 0$ perturbation. They do lie on the roots of r-branches. After mass perturbation for the hypermultiplets, one can count the number of vacua by identifing the points on the Coulomb brach which coalese to the same point when the masses are turned off. This is a technically involved analysis which required us many pages of the paper [2]. Nonetheless the result is simple. Starting from the maximally degenerate point on the r-branch root, the mass perturbation splits the point into ${}_{n_f}C_r$ vacua. Comparing this counting to the large μ analysis, we can say that the vacuum at the r-branch root breaks the flavor symmetry as $U(n_f) \to U(r) \times U(n_f - r)$.

Therefore, the following picture appears true. The semi-classical monopoles far away from the singularities on the Coulomb branch acquire the flavor quantum number of the rank-r totally anti-symmetric tensor prepresentation under the $U(n_f)$ flavor group. They become massless at the maximally degenerate point along the r-branch root and condense upon $\mu \neq 0$ perturbation. This picture, however, leaves a paradoxical situation. The low-energy effective Lagrangian of the monopoles would have a large accidental symmetry $U({}_{n_f}C_r)$, and upon condensation of one of the components, all the others remain massless. Even though it is logically not impossible, it casts some doubts about this naive picture. Indeed, something non-trivial happens between the semi-classical regime and the r-branch root as I will describe in the next section.

6 Low-energy Effective Lagrangians

Now we aproach the singularities from the Higgs branch. As remarked earlier, by turning off $\kappa_k \to 0$, we can approach the roots where we recover infrared-free $U(r) \times U(1)^{n_c - r - 1}$ gauge theory. At the maximally degenerate point along the roots, we have also $n_c - r - 1$ additional magnetic monopole hypermultiplets e_k, $\tilde{e}_k$ coupled to the $U(1)$ factors. The fundamental quarks still couple to the $SU(r)$ gauge factor as the fundamental representation q_i and $\tilde{q}_i$ because the non-renormalization theorem guarantees the flat hyper-Kähler metric for the quarks. The unique effective Lagrangian obtained this way is[6]

$$W = \sqrt{2}\tilde{q}_i \phi q_i + \sqrt{2}\psi_0 \tilde{q}_i q_i + \sqrt{2} \sum_{k=1}^{n_c - r - 1} \psi_k \tilde{e}_k e_k, \tag{18}$$

where ϕ, ψ_0 belong to the $U(r)$ vector multiplet and ψ_k to each of the $U(1)$ vector multiplets. The $N = 1$ perturbation then is given by

$$\Delta W = \mu\Lambda \sum_{j=0}^{n_c - r - 1} x_j \psi_j + \mu \mathrm{tr}\phi^2, \tag{19}$$

where x_j are $O(1)$ constants, and we find vacua

$$q = \tilde{q} = \left(\begin{array}{ccc|c} 1 & & & 0 \\ & \ddots & & \vdots \\ & & 1 & 0 \end{array} \right) \sqrt{-\frac{\mu\Lambda}{\sqrt{2}r}}, \tag{20}$$

$$e_k = \tilde{e}_k = \sqrt{-\mu\Lambda}, \quad \psi_0 = \psi_k = 0. \tag{21}$$

Note that the expectation values of $q, \tilde{q}$ break the flavor symmetry $U(n_f)$ to $U(r) \times U(n_f - r)$, where the unbroken $U(r)$ is the diagonal subgroup of the flavor group $U(n_f)$ and the $U(r)$ gauge group. And there are ${}_{n_f}C_r$ choices to pick r flavors out of n_f quark flavors for vacuum expectation values.

The semi-classical monopoles in the rank-r anti-symmetric tensor representation therefore must have "broken up" into "dual quarks" in the way that the monopole $M_{i_1 \cdots i_r} = q_{i_1} \cdots q_{i_r}$ is matched to the baryonic composite before one reaches the singularities on the Coulomb branch. For the special case of $r = 1$, the "dual quarks" themselves are the magnetic monopoles. The evidence for this identification is the following. The singularities, after $m_i \neq 0$ perturbation, have $U(1)^{n_c - 1}$ gauge group and one can study the monodromy around the singularities. It can be seen that there are one massless magnetic monopole for each $U(1)$ factors. By sending quark mass to zero, ${}_{n_f}C_r$ singularities coalesce into a point where the massless monopoles belong to the $U(r)$ quark multiplet. Therefore, the quarks, which are continuously connected to the "electric quarks" at the large VEVs along the Higgs branches, are indeed magnetic degrees of freedom at the non-baryonic branch roots.

7 Conclusion

We have studied the issues of confinement and the dynamical flavor symmetry breaking in gauge theories, by starting with $N = 2$ $SU(n_c)$ QCD with n_f flavors and perturbing it by the adjoint mass term. We have shown that both confinement and flavor symmetry breaking are caused by a single mechanism: condensation of magnetic degrees of freedom which carry flavor quantum numbers.

We have also studied $USp(2n_c)$ theories. There the magnetic monopoles are spinors under the $SO(2n_f)$ flavor group, and cannot "break up" into quarks. Therefore quarks and monopoles coexist at the singularity on the moduli space and the theory becomes superconformal. No local effective Lagrangian can be written and one cannot discuss it along the same line as in the $SU(n_f)$ gauge theories. However, the flavor symmetry is broken to $U(n_f)$, and this is consistent with the condensation of the spinor monopoles. This strongly suggests that the overall picture of flavor symmetry breaking via monopole condensation is correct in this case as well.

Acknowledgments

I thank Beppe Carlino and Ken Konishi for collaboration. This work was supported in part by the Department of Energy under contract DE–AC03–76SF00098, and in part by the National Science Foundation under grant PHY-95-14797.

References

1. G. Carlino, K. Konishi and H. Murayama, JHEP **0002**, 004 (2000) [hep-th/0001036].
2. G. Carlino, K. Konishi and H. Murayama, hep-th/0005076.
3. N. Seiberg and E. Witten, Nucl. Phys. **B426**, 19 (1994) [hep-th/9407087].
4. G. 't Hooft, Nucl. Phys. **B190**, 455 (1981); S. Mandelstam, Phys. Lett. **B53**, 476 (1975); Phys. Rept. **23**, 245 (1976).
5. N. Seiberg and E. Witten, Nucl. Phys. **B431**, 484 (1994) [hep-th/9408099].
6. P. C. Argyres, M. Ronen Plesser and N. Seiberg, Nucl. Phys. **B471**, 159 (1996) [hep-th/9603042].
7. P. C. Argyres, M. Ronen Plesser and A. D. Shapere, Nucl. Phys. **B483**, 172 (1997) [hep-th/9608129].
8. C. Csaki, M. Schmaltz and W. Skiba, Phys. Rev. Lett. **78**, 799 (1997) [hep-th/9610139]; Phys. Rev. **D55**, 7840 (1997) [hep-th/9612207].
9. N. Seiberg, Nucl. Phys. **B435**, 129 (1995) [hep-th/9411149].
10. I. Affleck, M. Dine and N. Seiberg, Nucl. Phys. **B241**, 493 (1984).
11. P. Hasenfratz and G. 't Hooft, Phys. Rev. Lett. **36**, 1119 (1976); R. Jackiw and C. Rebbi, Phys. Rev. **D13**, 3398 (1976).
12. A. Hanany and Y. Oz, Nucl. Phys. **B452**, 283 (1995) [hep-th/9505075]; P. C. Argyres, M. R. Plesser and A. D. Shapere, Phys. Rev. Lett. **75**, 1699 (1995) [hep-th/9505100].

FATE OF CHIRAL SYMMETRIES IN SUPERSYMMETRIC QUANTUM CHROMODYNAMICS*)

YASUHARU HONDA

Department of Physics, Tokai University, Hiratsuka, Kanagawa 259-1292, JAPAN
E-mail: 8jspd005@keyaki.cc.u-tokai.ac.jp

MASAKI YASUÈ

Department of Natural Science School of Marine Science and Technology, Tokai University, Shimizu, Shizuoka 424-8610, JAPAN
and
Department of Physics, Tokai University, Hiratsuka, Kanagawa 259-1292, JAPAN
E-mail: yasue@keyaki.cc.u-tokai.ac.jp

In supersymmetric quantum chromodynamics with N_c-colors and N_f-flavors of quarks, our effective superpotential provides the alternative description to the Seiberg's $N = 1$ duality at least for $N_f \geq N_c+2$, where spontaneous breakdown of chiral symmetries leads to $SU(N_c)_{L+R} \times SU(N_f - N_c)_L \times SU(N_f - N_c)_R$ as a nonabelian chiral symmetry. The anomaly-matching is ensured by the presence of Nambu-Goldstone superfields associated with this breaking and the instanton contributions are properly equipped in the effective superpotential.

*) Talk given by M.Y.

1 Prologue

In supersymmetric quantum chromodynamics (SQCD) with N_c-colors and N_f-flavors of quarks, we have chiral $SU(N_f)$ symmetry. At low energies, we have its dynamical breakdown to vectorial $SU(N_f)$ symmetry, for $N_f \leq N_c$. For remaining cases, we have restoration of chiral $SU(N_f)$ symmetry including the case with $N_f = N_c$. For $N_f \geq N_c+2$, we need "magnetic" degrees of freedom, namely, "magnetic" quarks.[1] And, especially, for $3N_c/2 < N_f < 3N_c$, the well-defined $N = 2$ duality supports this description based on the $N = 1$ duality.[2,3]

In this talk, I will add the "electric" description expressed in terms of mesons and baryons instead of "magnetic" quarks to SQCD with $N_c+2 \leq N_f \leq 3N_c/2$.[4] In the "electric" phase, however, since anomaly-matching[5] is not satisfied, we expect spontaneous breakdown of chiral symmetries.[6] The residual symmetries will include vectorial $SU(N_c)$ symmetry and chiral $SU(N_f - N_c)$ symmetry, which are found by inspecting vacuum structure of our effective superpotential to be discussed.

2 Anomalous $U(1)$ Symmetry and Superpotential

We follow the classic procedure to construct our effective superpotential, which explicitly uses S composed of two chiral gauge superfields.[7] We impose on the superpotential the relation: $\delta\mathcal{L} \sim F^{\mu\nu}\tilde{F}_{\mu\nu}$ under the anomalous $U(1)$ transformation, where $\mathcal{L}$ represents the lagrangian of SQCD and $F^{\mu\nu}$ ($\tilde{F}_{\mu\nu} \sim \epsilon_{\mu\nu\rho\sigma}F^{\rho\sigma}$) is a gluon's field strength. As a result, the anomalous $U(1)$-term is reproduced by the F-term of S.

We find a superpotential, where mesons and baryons are denoted by T, B and $\bar{B}$:[8]

$$W_{\rm eff} = S\left\{\ln\left[\frac{S^{N_c-N_f}\det(T)\,f(Z)}{\Lambda^{3N_c-N_f}}\right] + N_f - N_c\right\} \qquad (1)$$

with an arbitrary function, $f(Z)$, to be determined, where Λ is the scale of SQCD and Z is defined by (with abbreviated notations)

$BT^{N_f-N_c}\bar{B}/\det(T)$. This is the superpotential to be examined. It looks familiar to you except for the function $f(Z)$ here. In the classical limit, Z is equal to one and the function $f(Z)$ can be parametrized by $f(Z) = (1-Z)^\rho$, where ρ is a positive parameter. The parameter ρ is probably equal to 1. If $\rho=1$, we recover the superpotential for $N_f = N_c+1$ given by $W_{\rm eff} = S\left\{\ln\left[(\det(T) - BT\bar{B})/S\Lambda^{3N_c-N_f}\right] + 1\right\}$.[1]

3 Strategy

To examine dynamical properties of our superpotential, we 1) first go to slightly broken SUSY vacuum, where symmetry behavior of the superpotential is more visible, 2) use universal scalar masses of μ_L and μ_R, which respect global symmetry and only break SUSY, and 3) check the consistency with SQCD in its SUSY limit, after determining the SUSY-broken vacuum. The SQCD defined in the SUSY limit of the so-obtained SUSY-broken SQCD should exhibit the consistent anomaly-matching property and yield the compatible result with instanton calculus.

Let π_i be $\langle 0|T_i^i|0\rangle$, π_λ be $\langle 0|S|0\rangle$ and z be $\langle 0|Z|0\rangle$. Since the dynamics requires that some of the π acquire non-vanishing VEV's, suppose that one of the π_i (i=1 $\sim$ N_f) develops a VEV, and let this be labeled by i = 1: $|\pi_1| \sim \Lambda^2$. This VEV is determined by solving $\partial V_{\rm eff}/\partial\pi_i = 0$, yielding

$$G_T W^*_{{\rm eff};a}\frac{\pi_\lambda}{\pi_a}(1-\alpha) = G_S W^*_{{\rm eff};\lambda}(1-\alpha) + \beta X + M^2\left|\frac{\pi_a}{\Lambda}\right|^2, \qquad (a=1{\sim}N_c)\,, \tag{2}$$

where $\alpha = zf'(z)/f(z)$; $\beta = z\alpha'$; $M^2 = \mu_L^2 + \mu_R^2 + G'_T\Lambda^2\sum_{i=1}^{N_f}\left|W_{{\rm eff};i}\right|^2$; $X = G_T\sum_{a=1}^{N_c}W^*_{{\rm eff};a}(\pi_\lambda/\pi_a) - G_B\sum_{x=B,\bar{B}}W^*_{{\rm eff};x}(\pi_\lambda/\pi_x)$; $W_{{\rm eff};i(\lambda)} = \partial W_{\rm eff}/\partial\pi_{i(\lambda)}$; G's come from field-dependent Kähler potentials. The SUSY breaking effect is specified by $(\mu_L^2 + \mu_R^2)|\pi_1|^2$ through M^2 because of $\pi_1 \neq 0$.

Without knowing the details of solutions to these equations, we can find that

$$\left|\frac{\pi_a}{\pi_1}\right|^2 = 1 + \frac{\frac{M^2}{\Lambda^2}\left(\left|\pi_1\right|^2 - \left|\pi_a\right|^2\right)}{G_S W^*_{{\rm eff};\lambda}(1-\alpha) + \frac{M^2}{\Lambda^2}\left|\pi_a\right|^2 + \beta X}, \tag{3}$$

which cannot be satisfied by $\pi_{a\neq 1}$=0. In fact, $\pi_{a\neq 1}$=π_1 is a solution to this problem, leading to $|\pi_a|$=$|\pi_1|$. Then, you can see the emergence of the vectorial $SU(N_c)$ symmetry.

4 Symmetry Breaking

Using the input of $|\pi_{i=1\sim N_c}| \equiv \Lambda_T^2 \sim \Lambda^2$ just obtained, we reach the solutions given by $|\pi_B| = |\pi_{\bar{B}}| \equiv \Lambda_B^{N_c} \sim \Lambda^{N_c}$, $|\pi_{i=N_c+1\sim N_f}| = \epsilon|\pi_{1\sim N_c}|$ and $|\pi_\lambda| \sim \epsilon^{1+\frac{\rho}{N_f-N_c}}\Lambda^3$. Notice that π_i (i = N_c+1 $\sim$ N_f) and π_λ accompany the factor ϵ. This parameter ϵ, defined to be $|1-z|$, measures the SUSY breaking effect. So, taking the SUSY limit with $\epsilon \to 0$, we reach the SUSY vacuum specified by these VEV's. The solutions clearly show the presence of vectorial $SU(N_c)$ symmetry and chiral $SU(N_f-N_c)$ symmetry. The resulting breaking pattern is described by $G = SU(N_f)_L \times SU(N_f)_R \times U(1)_V \times U(1)_A$ down to $H = SU(N_c)_{L+R} \times SU(N_f-N_c)_L \times SU(N_f-N_c)_R \times U(1)'_V \times U(1)'_A$.

We find consistent anomaly-matching property due to the emergence of the Nambu-Goldstone superfields associated with $G \to H$, where massless bosons responsible for the anomalies of the broken part, G/H, and massless fermions for those of the unbroken part, H. Therefore, the anomaly-matching is a purely dynamical consequence. We have further checked that our superpotential is consistent with holomorphic decoupling and instanton calculus for SQCD with $N_f = N_c$ reproduced by massive quarks with flavors of $SU(N_f-N_c)$. The detailed description can be found in the literature.[4]

5 Summary

Dynamical breakdown of chiral symmetries are shown to be determined by the effective superpotential: $W_{\rm eff} = S\ \{\ln[\ S^{N_c-N_f}\ \det(T)\ f(Z)\ \Lambda^{N_f-3N_c}] + N_f - N_c\}$ with $f(Z)$ dynamically determined to be $(1-Z)^\rho$ ($\rho > 0$), It will be realized at least in SQCD with $N_c + 2 \leq N_f \leq 3N_c/2$. This superpotential exhibits 1) holomorphic decoupling property, 2) spontaneously breakdown of chiral $SU(N_c)$ symmetry and restoration of chiral $SU(N_f - N_c)$ symmetry described by $SU(N_f)_L \times SU(N_f)_R \times U(1)_V \times U(1)_A \rightarrow SU(N_c)_{L+R} \times SU(N_f - N_c)_L \times SU(N_f - N_c)_R \times U(1)'_V \times U(1)'_A$, 3) consistent anomaly-matching property due to the emergence of the Nambu-Goldstone superfields, and and 4) correct vacuum structure for $N_f = N_c$ reproduced by instanton contributions when all quarks with flavors of $SU(N_f - N_c)$ become massive.

In this end, we have two phases in SQCD: one with chiral $SU(N_f)$ symmetry for "magnetic" quarks and the other with spontaneously broken chiral $SU(N_f)$ symmetry for the Nambu-Goldstone superfields. This situation can be compared with the case in the ordinary QCD with two flavors: one with proton and neutron and the other with pions.

Finally, I mention related three works here, which are characterized by 1) Dynamical evaluations of condensates,[9] 2) Instable SUSY vacuum in the "magnetic" phase,[10] and 3) Slightly different effective superpotential in the "electric" phase.[11] All these works indicate spontaneous chiral symmetry breaking in the "electric" phase.

References

1. N. Seiberg, Phys. Rev. D **49**, 6857 (1994); Nucl. Phys. B **435**, 129 (1995).
2. K. Intriligator and N. Seiberg, Nucl. Phys. **B431** (1994) 551; P.C. Argyres, M.R. Plesser and A.D. Shapere, Phys. Rev. Lett. **75** (1995) 1699; A. Hanany and Y. Oz, Nucl. Phys. **B452** (1995) 283.
3. R.G. Leigh and M.J. Strassler, Nucl. Phys. **B447** (1995) 95; P.C. Argyres, M.R. Plesser and N. Seiberg, Nucl. Phys. **B471** (1996) 159; M.J. Strassler, Prog. Theor. Phys. Suppl. **No.123** (1996) 373; N.Evans, S.D.H. Hsu, M. Schwetz and S.B. Selipsky, Nucl. Phys. Proc. Suppl **52A** (1997) 223; P.C. Argyres, Nucl. Phys. Proc. Suppl **61A** (1998) 149; T. Hirayama, N. Maekawa and S. Sugimoto, Prog. Theor. Phys. **99** (1998) 843.
4. Y. Honda and M. Yasuè, Prog Theor. Phys. **101**, 971 (1999); Phys. Lett. B **466**, 244 (1999).
5. G. 't Hooft, in *Recent Development in Gauge Theories*, Proceedings of the Cargese Summer Institute, Cargese, France, 1979, edited by G. 't Hooft *et al.*, NATO Advanced Study Institute Series B: Physics Vol. 59 (Plenum Press, New York, 1980).
6. T. Banks, I. Frishman, A. Shwimmer and S. Yankielowicz, Nucl. Phys. **B177**, 157 (1981).
7. G. Veneziano and S. Yankielowicz, Phys. Lett. **113B**, 321 (1983); T. Taylor, G. Veneziano and S. Yankielowicz, Nucl. Phys. **B218**, 493 (1983).
8. A. Masiero, R. Pettorino, M. Roncadelli and G. Veneziano, Nucl. Phys. **B261** (1985) 633; M. Yasuè, Phys. Rev. D **35** (1987) 355 and D **36** (1987) 932; Prog Theor. Phys. **78** (1987) 1437.
9. T. Appelquist, A. Nyffeler and S.B. Selipsky, Phys. Lett. B **425** (1998) 300.
10. N. Arkani-Hamed and R. Rattazzi, Phys. Lett. B **454** (1999) 290.
11. P.I. Pronin and K.V. Stepanyantz, hep-th/9902163 (Feb., 1999).

AUXILIARY FIELD FORMULATION OF SUPERSYMMETRIC NONLINEAR SIGMA MODELS

KIYOSHI HIGASHIJIMA

Department of Physics, Graduate School of Science, Osaka University, Toyonaka, Osaka 560-0043, Japan
E-mail: higashij@phys.sci.osaka-u.ac.jp

MUNETO NITTA

Department of Physics, Tokyo Institute of Technology, Oh-okayama, Meguro, Tokyo 152-8551, Japan
E-mail: nitta@th.phys.titech.ac.jp

Two dimensional $\mathcal{N}=2$ supersymmetric nonlinear sigma models on hermitian symmetric spaces are formulated in terms of the auxiliary superfields. If we eliminate auxiliary vector and chiral superfields, they give D- and F-term constraints to define the target manifolds. The integration over auxiliary vector superfields, which can be performed exactly, is equivalent to the elimination of the auxiliary fields by the use of the classical equations of motion.

1 Introduction

Two dimensional nonlinear sigma models (NLσM) have been interested in, since they have many similarities to four dimensional QCD such as asymptotic freedom, the mass gap, instantons and so on. Non-perturbative analyses of NLσM can be easily done by the large-N method. In the $O(N)$ model, the mass gap appears as non-perturbative effect. In the $\mathbf{C}P^{N-1}$ model, a gauge boson is dynamically generated. $\mathcal{N}=1$ SUSY NLσM (SNLσM) have been also investigated. The $\mathcal{N}=1$ $O(N)$ model is simply a combination of the bosonic $O(N)$ NLσM and the Gross-Neveu model which shows dynamical chiral symmetry breaking.[1]

Along this line it is interesting to discuss non-perturbative analyses of $\mathcal{N}=2$ SNLσM, since they may have similarities to four dimensional $\mathcal{N}=1$ QCD. To discuss non-perturbative effects of NLσM by the large-N method, it is necessary to reformulate them by the auxiliary field method. However there was no auxiliary field formulation of $\mathcal{N}=2$ SNLσM except for the $\mathbf{C}P^{N-1}$ and the Grassmann models.[2,3] In this talk, we formulate $\mathcal{N}=2$ SNLσM on hermitian symmetric spaces (HSS) G/H (see Table 1) by the auxiliary field method.[4,5,6] Since $\mathcal{N}=2$ SUSY in two dimensions is equivalent to $\mathcal{N}=1$ SUSY in four dimensions, we use the notation of four dimensions.

2 Auxiliary Field Formulation

2.1 SNLσM without F-term constraints

It was recognized in Ref. [2] that the $\mathbf{C}P^{N-1}$ model can be constructed by an auxiliary vector superfield V: Let ϕ be dynamical chiral superfields belonging to $\mathbf{N}$ of $SU(N)$. Then its Kähler potential can be written as

$$K(\phi,\phi^\dagger,V)=e^V\phi^\dagger\phi-cV, \tag{1}$$

where cV is an Fayet-Iliopoulos (FI) D-term. (c is a positive constant called an FI-parameter.)

This model can be immediately generalized to the Grassmann model, $G_{N,M}(\mathbf{C})$ ($N>M$) by replacing ϕ by an $N\times M$ matrix chiral superfield Φ and V by an $M\times M$ matrix vector superfield $V=V^AT_A$, where T_A are generators of $U(M)$ gauge group:[3]

$$K(\Phi,\Phi^\dagger,V)=\operatorname{tr}(\Phi^\dagger\Phi e^V)-c\operatorname{tr}V. \tag{2}$$

Table 1. Hermitian symmetric spaces (HSS).

Type	G/H	$\dim_{\mathbf{C}}(G/H)$
AIII$_1$	$\mathbf{C}P^{N-1} = SU(N)/SU(N-1) \times U(1)$	$N-1$
AIII$_2$	$G_{N,M}(\mathbf{C}) = U(N)/U(N-M) \times U(M)$	$M(N-M)$
BDI	$Q^{N-2}(\mathbf{C}) = SO(N)/SO(N-2) \times U(1)$	$N-2$
CI	$Sp(N)/U(N)$	$\frac{1}{2}N(N+1)$
DIII	$SO(2N)/U(N)$	$\frac{1}{2}N(N-1)$
EIII	$E_6/SO(10) \times U(1)$	16
EVII	$E_7/E_6 \times U(1)$	27

After we integrate out V and fix a gauge in these models, we obtain Kähler potentials of the Fubini-Study metric of $\mathbf{C}P^{N-1}$ and its generalization to the Grassmann manifold.

2.2 SNLσM with F-term constraints

In this section, to obtain the rest of HSS, we introduce auxiliary chiral superfields ϕ_0 or Φ_0 besides auxiliary vector superfields and construct G-invariant superpotentials as summarized in Table 2.[4] Integration over the auxiliary chiral superfields gives F-term constraints, which are holomorphic.

The simplest example is $Q^{N-2}(\mathbf{C})$, where dynamical fields constitute an $SO(N)$ vector ϕ. It can be embedded into $\mathbf{C}P^{N-1}$ by an F-term constraint $\phi^2 = 0$. Hence an auxiliary chiral superfield is taken to be a singlet ϕ_0 and a superpotential to be $W = \phi_0 \phi^2$.

Both $SO(2N)/U(N)$ and $Sp(N)/U(N)$ can be embedded into $G_{2N,N}$ by F-term constraints $\Phi^T J \Phi = 0$, where Φ is a $2N \times N$ matrix chiral superfield. Here $J = \begin{pmatrix} \mathbf{0} & \mathbf{1}_N \\ \epsilon \mathbf{1}_N & \mathbf{0} \end{pmatrix}$, where $\epsilon = +1$ (or -1) for $SO(2N)$ (or $Sp(N)$). Then auxiliary chiral superfields constitute an $N \times N$ matrix Φ_0 belonging to the symmetric (anti-symmetric) tensor representation of $SO(2N)$ ($Sp(N)$), and the superpotential is $W = \mathrm{tr}\,(\Phi_0 \Phi^T J \Phi)$.

It is less trivial to find F-term constraints for E_6 and E_7 models. The F-term constraints are G-invariants (I_2 or $I_2{}'$ in Table 2) for the classical groups; on the other hand they are not G-invariants but the differentiation of G-invariants (∂I_3 or ∂I_4 in Table 2) for the exceptional groups. We must introduce auxiliary chiral superfields ${\phi_0}^i$ and ${\phi_0}^\alpha$ ($i = 1, \cdots, 27$; $\alpha = 1, \cdots, 56$) belonging to the fundamental representations of E_6 and E_7. Superpotentials can be written as $W = \Gamma_{ijk} {\phi_0}^i \phi^j \phi^k$ and $W = d_{\alpha\beta\gamma\delta} {\phi_0}^\alpha \phi^\beta \phi^\gamma \phi^\delta$ for E_6 and E_7 models, where Γ and d are rank-3 and rank-4 symmetric tensors of E_6 and E_7, respectively. At first sight one might consider the number of the F-term constraints are too large. However some of them are not independent due to the identities of invariant tensors Γ and d. (Only 10 of 27 equations and 28 of 56 equations are independent for E_6 and E_7 cases, respectively.)

3 Integration over Auxiliary Fields

The path integration over auxiliary chiral superfields ϕ_0 or Φ_0 is easy since they are linear in the lagrangian; on the other hand, the path integration over auxiliary vector superfields V is nontrivial. However we can perform integration over V *exactly* although they are not quadratic.[5] The result for Abelian V is

$$\int [dV] \exp\left[i \int d^D x d^4\theta \left(\phi^\dagger \phi\, e^V - cV \right) \right] = \exp\left[i \int d^D x d^4 \theta\, c \log(\phi^\dagger \phi) \right]. \qquad (3)$$

This coincides with the result obtained by the equation of motion of V. Its coincidence is highly nontrivial since there are infinite num-

Table 2. F-term constraints and embedding.

G/H	G-invariants	superpotentials	constraints	embedding
$\frac{SO(N)}{SO(N-2)\times U(1)}$	$I_2=\phi^2$	$\phi_0 I_2$	$I_2=0$	$\mathbf{C}P^{N-1}$
$\frac{SO(2N)}{U(N)}, \frac{Sp(N)}{U(N)}$	$I_2{}'=\Phi^T J\Phi$	$\mathrm{tr}\,(\Phi_0 I_2{}')$	$I_2{}'=0$	$G_{2N,N}$
$\frac{E_6}{SO(10)\times U(1)}$	$I_3=\Gamma_{ijk}\phi^i\phi^j\phi^k$	$\Gamma_{ijk}\phi_0{}^i\phi^j\phi^k$	$\partial I_3=0$	$\mathbf{C}P^{26}$
$\frac{E_7}{E_6\times U(1)}$	$I_4=d_{\alpha\beta\gamma\delta}\phi^\alpha\phi^\beta\phi^\gamma\phi^\delta$	$d_{\alpha\beta\gamma\delta}\phi_0{}^\alpha\phi^\beta\phi^\gamma\phi^\delta$	$\partial I_4=0$	$\mathbf{C}P^{55}$

ber of corrections in bosonic cases. Eq. (3) can be proved by the following theorem.

Theorem. Let $\sigma(x,\theta,\bar\theta)$ and $\Phi(x,\theta,\bar\theta)$ be vector superfields and W be a function of σ. Then,

$$\int [d\sigma]\exp\left[i\int d^Dxd^4\theta\,(\sigma\Phi - W(\sigma))\right] = \exp\left[i\int d^Dxd^4\theta\; U(\Phi)\right]. \quad (4)$$

Here $U(\Phi)$ is defined as $U(\Phi)=\hat\sigma(\Phi)\Phi - W(\hat\sigma(\Phi))$, where $\hat\sigma$ is a solution of the stationary equation $\frac{\partial}{\partial\sigma}(\sigma\Phi - W(\sigma))|_{\sigma=\hat\sigma} = \Phi - W'(\hat\sigma)=0$.

To prove Eq. (3), put $\Phi=\phi^\dagger\phi$, $\sigma=e^V$ and $W=c\log\sigma$, and note that $[d\sigma]=[dV]$. This theorem can be generalized to many variables and to matrix variables, which can be applied to integration over non-Abelian vector superfields. As an application of the theorem, we can show that

$$\int [dV]\exp\left[i\int d^Dxd^4\theta\,\left(f(\phi^\dagger\phi\, e^V) - c\,V\right)\right] = \int [dV]\exp\left[i\int d^Dxd^4\theta\,\left(\phi^\dagger\phi\, e^V - c\,V\right)\right],$$

where f is an *arbitrary* function.

4 Discussion

Non-perturbative analyses of $\mathcal{N}=2$ SNLσM on HSS are possible, which are in progress.[8]

Let us discuss a generalization to an arbitrary Kähler G/H. It is known that H must be of the form $H=H_{\rm s.s}\times U(1)^n$, where $H_{\rm s.s}$ is the semi-simple subgroup of H and $n=\mathrm{rank}\,G-\mathrm{rank}\,H_{\rm s.s}$, and a Kähler potential has n free parameters.[7] As seen in Eq. (3), the FI-parameter c represents a size of G/H after integration over V. Hence to obtain a Kähler G/H, it is needed to introduce n FI-terms by considering a gauge group including n Abelian factors.

Acknowledgements

The work of M. N. is supported in part by JSPS Research Fellowships.

References

1. E. Witten, Phys. Rev. **D16** (1977) 2991; O. Alvarez, Phys. Rev. **D17** (1978) 1123.
2. E. Witten, Nucl. Phys. **B149** (1979) 285; A. D'adda, P. Di Vecchia and M. Lüscher, Nucl. Phys. **B152** (1979) 125.
3. S. Aoyama, Nuovo Cim. **57A** (1980) 176.
4. K. Higashijima and M. Nitta, Prog. Theor. Phys. **103** (2000) 635, hep-th/9911139.
5. K. Higashijima and M. Nitta, Prog. Theor. Phys. **103** (2000) 833, hep-th/9911225.
6. K. Higashijima and M. Nitta, hep-th/0006025, to appear in Proceedings of Confinement 2000.
7. K. Itoh, T. Kugo and H. Kunitomo, Nucl. Phys. **B263** (1986) 295.
8. K. Higashijima, T. Kimura, M. Nitta and M. Tsuzuki, in preparation.

MASSIVE GHOSTS IN SOFTLY BROKEN SUSY GAUGE THEORIES

D.I.KAZAKOV AND V.N.VELIZHANIN

Bogoliubov Laboratory of Theoretical Physics, Joint Institute for Nuclear Research, Dubna, Moscow Region, 141980, Russia. E-mail: kazakovd@thsun1.jinr.ru

It is shown that, due to soft supersymmetry breaking in gauge theories within the superfield formalism, there appears the mass for auxiliary gauge fields. It enters into the RG equations for soft masses of physical scalar particles and can be eliminated by solving its own RG equation. Explicit solutions up to the three-loop order in the general case are given.

1 Introduction

Recently, it has been realized[1,2,3] that renormalizations in a softly broken SUSY theory are not independent but follow from those of an unbroken SUSY theory. According to the approach advocated in[3,4], one can perform the renormalization of a softly broken SUSY theory in the following straightforward way:

One takes renormalization constants of a rigid theory, calculated in some massless scheme, substitutes for the rigid couplings (gauge and Yukawa) their modified expressions, that depend on a Grassmannian variable, and expands over this variable.

This gives renormalization constants for the soft terms. Differentiating them with respect to a scale, one can find corresponding renormalization-group equations.

Thus, the soft-term renormalizations are not independent but can be calculated from the known renormalizations of a rigid theory with the help of the differential operators. Explicit form of these operators has been found in a general case and in some particular models like SUSY GUTs or the MSSM[3]. The same expressions have been obtained also in a somewhat different approach in[2,5].

There is, however, some minor difference. The authors of[2,5] have used the component approach, while in[1,3,4], the superfield formalism is exploited. This creates the usual difference in gauge-fixing and ghost field terms and in the renormalization scheme. The latter is related to the choice of regularization. In[2,5] the dimensional reduction (DRED) regularization is used. In this case, one is bounded to introduce the so-called ϵ-scalars to compensate the lack of bosonic degrees of freedom in 4-2ϵ dimensions. These ϵ-scalars in due course of renormalization acquire a soft mass that enters into the RG equations for soft masses of physical scalar particles. If one gets rid of the ϵ-scalar mass by changing the renormalization scheme, DRED $\to$ DRED$'$, there appears an additional term in RG equations for the soft scalar masses, which is absent in RG equations in[1,3,4].

We have to admit that, indeed in our approach, though the ϵ-scalars in the superfield formalism are absent, that term appears in higher orders and is related to the soft masses of other unphysical particles, the auxiliary gauge fields.

2 Massive Auxiliary Fields

Consider an arbitrary $N = 1$ SUSY gauge theory with unbroken SUSY within the superfield formalism. The Lagrangian of a rigid theory is given by

$$\mathcal{L}_{rigid} = \int d^2\theta \, \frac{1}{4g^2} \mathrm{Tr} W^\alpha W_\alpha + h.c. \quad (1)$$
$$+ \int d^2\theta d^2\bar\theta \;\; \bar\Phi^i (e^V)^j_i \Phi_j + \int d^2\theta \;\; \mathcal{W} + h.c.,$$

where

$$W_\alpha = -\frac{1}{4}\bar D^2 e^{-V} D_\alpha e^V,$$

is the gauge field strength tensor and the superpotential $\mathcal{W}$ has the form

$$\mathcal{W}=\frac{1}{6}y^{ijk}\Phi_i\Phi_j\Phi_k+\frac{1}{2}M^{ij}\Phi_i\Phi_j. \quad (2)$$

To fix the gauge, the usual gauge-fixing term can be introduced. It is useful to choose it in the form

$$\mathcal{L}_{g.fix.}=-\frac{1}{16}\int d^2\theta d^2\bar{\theta}\mathrm{Tr}\left(\bar{f}f+f\bar{f}\right) \quad (3)$$

where the gauge fixing condition is taken as

$$f=\bar{D}^2\frac{V}{\sqrt{\xi g^2}}\,,\quad \bar{f}=D^2\frac{V}{\sqrt{\xi g^2}}\,. \quad (4)$$

Then, the corresponding ghost term for our choice of the gauge-fixing condition, takes the form

$$\mathcal{L}_{ghost}=\int d^2\theta d^2\bar{\theta}\ \mathrm{Tr}\left(\frac{b+\bar{b}}{\sqrt{\xi g^2}}\right)\left((c-\bar{c})\right.$$
$$\left.+\frac{1}{2}\left[V,c+\bar{c}\right]+\frac{1}{12}\left[V,\left[V,c-\bar{c}\right]\right]+\ldots\right) \quad (5)$$

To perform the SUSY breaking, that satisfies the requirement of "softness", one can introduce a gaugino mass term as well as cubic and quadratic interactions of scalar superpartners of the matter fields[7]

$$-\mathcal{L}_{soft.br.}=\left[\frac{M}{2}\lambda\lambda+\frac{1}{6}A^{ijk}\phi_i\phi_j\phi_k\right. \quad (6)$$
$$\left.+\ \frac{1}{2}B^{ij}\phi_i\phi_j+h.c.\right]+(m^2)^i_j\phi^*_i\phi^j,$$

where λ is the gaugino field, and ϕ_i is the lowest component of the chiral matter superfield.

One can rewrite the Lagrangian (6) in terms of N=1 superfields introducing the external spurion superfields[7] $\eta=\theta^2$ and $\bar{\eta}=\bar{\theta}^2$, where θ and $\bar{\theta}$ are Grassmannian parameters, as[1]

$$\mathcal{L}_{soft}=\int d^2\theta\ \frac{1}{4g^2}(1-2M\theta^2)\mathrm{Tr}W^\alpha W_\alpha$$
$$+\int d^2\theta d^2\bar{\theta}\ \ \bar{\Phi}^i(\delta^k_i-(m^2)^k_i\eta\bar{\eta})(e^V)^j_k\Phi_j$$
$$+\int d^2\theta\left[\frac{1}{6}(y^{ijk}-A^{ijk}\eta)\Phi_i\Phi_j\Phi_k\right.$$
$$\left.+\ \frac{1}{2}(M^{ij}-B^{ij}\eta)\Phi_i\Phi_j\right]+h.c.$$

Thus, one can interpret the soft terms as the modification of the couplings of a rigid theory. The couplings become external superfields depending on Grassmannian parameters θ and $\bar{\theta}$. We suggest the following modification of the gauge coupling

$$\frac{1}{g^2}\rightarrow\frac{1}{\tilde{g}^2}=\frac{1-M\theta^2-\bar{M}\bar{\theta}^2-\Delta\theta^2\bar{\theta}^2}{g^2}.$$

In papers[3,4], the Δ term was absent. Indeed, it is self-consistent to put $\Delta=0$ in the lowest order of perturbation theory, but it appears in higher orders due to renormalizations.

One has to take into account, however, that, since the gauge-fixing parameter ξ may be considered as an additional coupling, it also becomes an external superfield and has to be modified. The soft expression can be written as

$$\tilde{\xi}=\xi\left(1+x\theta^2+\bar{x}\bar{\theta}^2+(x\bar{x}+z)\theta^2\bar{\theta}^2\right),$$

where the parameters x and z can be obtained by solving the corresponding RG equation.

Now, if one replaces all the couplings in the Lagrangian of the rigid theory by appropriate external superfields and rewrites the Lagrangian in components, one finds that Δ plays a role of a soft mass of the scalar components of supersymmetric ghosts and scalar auxiliary fields from vector supermultiplet. Thus, this extra Δ term is associated with unphysical, ghost, degrees of freedom, just like in the component approach, one has the mass of unphysical ϵ-scalars. When going down with energy, all massive fields decouple, and one gets the usual nonsupersymmetric Yang-Mills theory.

The Δ-term is renormalized and obeys its own RG equation which can be obtained from the corresponding expression for the gauge coupling via Grassmannian expansion. In due course of renormalization, this term is mixed with the soft masses of scalar superpartners and gives an additional term in RG equations for the latter (the X term of Jack *et al* [5] mentioned above).

3 RG Equations for the Soft Parameters.

Thus, following the procedure described in papers[3,4], to get the RG equations for the soft terms, one has to modify the gauge (g_i^2) and Yukawa (y_{ijk}) couplings replacing them by external superfields :

$$\tilde{g}_i^2 = g_i^2(1 + M_i\eta + \bar{M}_i\bar{\eta} + (2M_i\bar{M}_i + \Delta_i)\eta\bar{\eta}),$$

$$\tilde{y}^{ijk} = y^{ijk} - A^{ijk}\eta + \frac{1}{2}(y^{njk}(m^2)^i_n + y^{ink}(m^2)^j_n + y^{ijn}(m^2)^k_n)\eta\bar{\eta}.$$

Then, the β functions of RG equations for the soft masses of scalar superpartners of the matter fields and for the mass of the auxiliary gauge field are given by [3]

$$[\beta_{m^2}]^i_j = D_2\gamma^i_j, \qquad \beta_{\Sigma_{\alpha_i}} = D_2\gamma_{\alpha_i}, \qquad (7)$$

where γ^i_j and $\gamma_{\alpha_i} = \beta_{\alpha_i}/\alpha_i$ are the anomalous dimensions of the matter fields and of the gauge coupling, respectively, and we have introduced the notation

$$\Sigma_{\alpha_i} = M_i\bar{M}_i + \Delta_i.$$

The modified expression for the operator D_2 is

$$D_2 = \bar{D}_1 D_1 + \Sigma_{\alpha_i}\alpha_i\frac{\partial}{\partial\alpha_i} + \frac{1}{2}(m^2)^a_n\left(y^{nbc}\frac{\partial}{\partial y^{abc}} + permutations\right).$$

It coincides now with that of Jack *et al* [5] with $X_i = \Delta_i$.

To find Σ_{α_i}, one can use equation (7). In particular, using the expression for the anomalous dimension γ_α and the anomalous dimension of the matter fields γ^j_i one can get the solution

$$\Sigma^{(1)}_\alpha = M^2$$

$$\Sigma^{(2)}_\alpha = -2\alpha[\frac{1}{r}(m^2)^i_j C(R)^j_i - M^2C(G)],$$

$$\Sigma^{(3)}_\alpha = \frac{\alpha}{2r}[\frac{1}{2}(m^2)^i_n y^{nkl}y_{jkl} + \frac{1}{2}(m^2)^n_j y^{ikl}y_{nkl} + 2(m^2)^m_n y^{ikn}y_{jkm} + A^{ikl}A_{jkl} - 8\alpha M^2C(R)^i_j]C(R)^j_i - 2\alpha^2QC(G)M^2 - 4\alpha^2C(G)[\frac{1}{r}(m^2)^i_j C(R)^j_i - M^2C(G)].$$

These expressions for Δ_α coincide with those obtained Jack *et al* [6] for the mass of the ϵ-scalars.

The nonzero Δ-term modifies the expression for the β function of the soft scalar mass starting from the second loop. Substituting $\Sigma^{(2)}_\alpha$ into the expression for the differential operator D_2 gives in two loops RGE for soft scalar mass, which coincides with component calculations.

4 Conclusion

Summarizing, we would like to stress once again that soft breaking of supersymmetry can be realized via interaction with an external superfield that develops nonzero v.e.v.'s for its F and D components. In the superfield notation, it can be reformulated as a modification of the rigid couplings that become external superfields. The same is true for the gauge-fixing parameter that can also be considered as a rigid coupling. The soft masses of scalar particles obtain their contribution from the D-components of external superfields. The latter also lead to nonzero masses for unphysical degrees of freedom, ghost and gauge auxiliary fields. These unphysical masses enter into the RG equations for the physical scalars and have to be eliminated. This creates an ambiguity in the running scalar masses; it can be resolved by passing to the pole masses.

References

1. Y.Yamada, *Phys.Rev.*, **D50** (1994) 3537.
2. I.Jack and D.R.T.Jones, *Phys. Lett.* **B415** (1997) 383.
3. L.A.Avdeev, D.I.Kazakov and I.N.Kondrashuk, *Nucl.Phys.*, **B510** (1998) 289.
4. D.I.Kazakov,*Phys.Lett.*,**B448**(1998)201.
5. I.Jack, D.R.T.Jones and A.Pickering, *Phys.Lett.*, **B426** (1998) 73.
6. I. Jack, D. R. T. Jones and A. Pickering, *Phys.Lett.*, **B432** (1998) 114.
7. L. Girardello and M. T. Grisaru, *Nucl. Phys.*, **B194** (1982) 65.

SUSY BREAKING THROUGH COMPACTIFICATION

M. SAKAMOTO

Department of Physics, Kobe University, Rokkodai, Nada, Kobe 657-8501, Japan
E-mail: sakamoto@phys.sci.kobe-u.ac.jp

M. TACHIBANA

Yukawa Institute for Theoretical Physics, Kyoto University, Kyoto 606-8502, Japan
E-mail: motoi@yukawa.kyoto-u.ac.jp

K. TAKENAGA

The Niels Bohr Institute, Blegdamsvej 17, DK-2100, Copenhagen ϕ, Denmark
E-mail: takenaga@alf.nbi.dk

We propose a new mechanism of spontaneous supersymmetry breaking. The existence of extra dimensions with nontrivial topology plays an important role. We investigate new features resulting from this mechanism. One noteworthy feature is that there exists a phase in which the translational invariance for the compactified directions is broken spontaneously. The mechanism we propose also yields a variety of non-trivial vacuum structures depending on topology of compactified spaces and also on symmetries of Lagrangians.

It is thought that (super)string theories (and/or more fundamental theories, such as M-theory) are plausible candidates to describe physics at the Planck scale. In general these theories are defined in more than 4 space-time dimensions because of the consistency of the theories.

In the region of sufficiently low energy, however, it is known that our space-time is 4 dimensions, so that extra dimensions must be compactified by some mechanism,[1] and supersymmetry (SUSY) that is usually possessed by these theories has to be broken, because it is not observed in the low energy region. At this stage, mechanisms of compactification and SUSY breaking are not fully understood.

It may be interesting to consider the quantum field theory in space-time with some of the space dimensions being multiply-connected. This is because such a study may shed new light on unanswered questions concerning the (supersymmetric) standard model and/or elucidate some new dynamics which are useful to seek and understand new physics beyond it. Actually, it has been reported that flavour-blind SUSY breaking terms can be induced through compactification by taking account of possible topological effects of a multiply-connected space.[2,3,4] Therefore, it is important to investigate the physics possessed by the quantum field theory considered in such a space-time.

We consider supersymmetric field theories in space-time with one of the space dimensions being compactified. One has to specify the boundary conditions of fields for the compactified direction when the compactified space is multiply-connected. In contrast to finite temperature field theory, we do not know *a priori* what they should be. We shall relax the conventional periodic boundary condition to allow nontrivial twisted boundary conditions for the fields.[5,6] Due to the topology of the space, the configuration space also has a nontrivial topology. This causes an ambiguity in the quantization of the theory on the space to yield an undetermined parameter in the theory.[7] The boundary conditions of the fields are twisted with this parameter.[a]

As a noteworthy consequence of our mechanism, there appears a nontrivial phase structure with respect to the size of the compactified space. Namely, the translational invariance for the compactified direction is broken spontaneously,[9] when the size of the compactified space exceeds a certain critical value. The curious vacuum structures resulting from the mechanism also have an influence on the mass spectrum of the theory. The mass spectrum includes Nambu-Goldstone bosons (fermions) corresponding to the breakdown of the global symmetries of the theory.

Let us discuss a basic idea of our SUSY-breaking mechanism briefly. Let $W(\Phi)$ be a superpotential consisting of the chiral superfields Φ_j. The scalar potential is then given by $V(A) = \sum_j |F_j|^2 = \sum_j \left|\frac{\partial W(A)}{\partial A_j}\right|^2$, where $A_j(F_j)$ denotes the lowest (highest) component of Φ_j. Supersymmetry would be unbroken if there exist solutions to the F-term conditions

$$-F_j^* = \left.\frac{\partial W(A)}{\partial A_j}\right|_{A_k=\bar{A}_k} = 0 \qquad \text{for all} \quad j, \tag{1}$$

since such solutions would lead to $V(\bar{A}) = 0$. Our idea for supersymmetry breaking is simple: We impose nontrivial boundary conditions on the superfields for the compactified direction. They must be consistent with the single-valued nature of the Lagrangian but inconsistent with the F-term conditions (1). Then, no solution to the F-term conditions will be realized as a vacuum configuration of the model. Thus, we expect that $V(\langle A\rangle) > 0$ (because $\langle A_j\rangle \neq \bar{A}_j$) and that supersymmetry is broken spontaneously.

In order to realize such a mechanism, let us consider a theory in which one of the space coordinates, say y, is compactified on a circle S^1 whose radius is R. Since S^1 is multiply-connected, we must specify boundary conditions for the S^1 direction. Let us impose nontrivial boundary conditions on superfields defined by

$$\Phi_j(x^\mu, y + 2\pi R) = e^{2\pi i\alpha_j}\Phi_j(x^\mu, y). \tag{2}$$

The phase α_j should be chosen such that the Lagrangian density is single-valued, *i.e.* $\mathcal{L}(x^\mu, y + 2\pi R) = \mathcal{L}(x^\mu, y)$. In other words, the phase has to be one of the degrees of freedom of the symmetries of the theory. Suppose that $\bar{A}_j$, which is a solution to the F-term conditions, is a nonzero constant for some j. It is easy to see that if $e^{2\pi i\alpha_j} \neq 1$, then the vacuum expectation value $\langle A_j\rangle$ is strictly forbidden to take the nonzero constant value $\bar{A}_j$, because this is inconsistent with the boundary condition (2). In this way, our idea is realized by the mechanism that the nontrivial boundary conditions imposed on the fields play the role of preventing the vacuum expectation values of the fields from being solutions to the F-term conditions.

Note that the above result does not always lead us to the conclusion that $\langle A_j\rangle = 0$, though it is always consistent with (2). Certainly, if the translational invariance for the S^1 direction is not broken, $\langle A_j\rangle$ has to vanish because of (2). If the translational invariance for the S^1 direction is broken, however, vacuum expectation values will no longer be constants, and some of the $\langle A_j\rangle$ can depend on the coordinate of the compactified space as an energetically favourable configuration. One should include the contributions from the kinetic terms of the scalar fields in addition to the scalar potential in order to find the true vacuum configuration.

Let us show how to realize the above idea below. For simplicity, here we consider $j = 1$ case. First we expand the complex scalar field $A(x^\mu, y)$ in Fourier modes

$$A(x^\mu, y) = \frac{1}{\sqrt{2\pi R}}\sum_{n=-\infty}^{\infty} A^{(n+\alpha)}(x^\mu)e^{i\frac{n+\alpha}{R}y}. \tag{3}$$

[a] In a SUSY quantum mechanical system on a multiply-connected space, it has been shown that the ground state wave function is actually twisted by the parameter.[8]

Note here that due to the twisted boundary condition (2) there are no zero modes. Moreover, we introduce a dimensionally reduced "effective" potential defined by

$$\begin{aligned}\mathcal{V}_{eff}[A] &\equiv \int_0^{2\pi R} dy\Big[|\partial_y A|^2 + V(A)\Big] \\ &= \sum_{n=-\infty}^{\infty}\Big[\Big(\frac{n+\alpha}{R}\Big)^2 - \mu^2\Big]|A^{(n+\alpha)}|^2 + \cdots \\ &= \Big[\Big(\frac{\alpha}{R}\Big)^2 - \mu^2\Big]|A^{(\alpha)}|^2 + \cdots, \end{aligned} \tag{4}$$

where $A^{(\alpha)}$ is the lowest mode and the dots denote higher order terms including the quartic term of $A^{(\alpha)}$. The μ^2 term comes from the potential $V(A)$ which is assumed to include a negative mass squared term. From this equation it is easy to find that if $R < R^* \equiv \frac{\alpha}{\mu}$, the curvature at the origin of the potential is positive so that $\langle A(x^\mu, y)\rangle$ vanishes. Thus in this case, the translational invariance for the S^1 direction is not broken. On the other hand, if $R > R^*$, the curvature becomes negative and the potential is given as a double-well type. Then the vacuum expectation value of $A(x^\mu, y)$ does not vanish but rather has y-dependence. In this case, the translational invariance for the S^1 direction is spontaneously broken.

It may be interesting to ask how our mechanism works on more complex manifolds, such as a torus. We expect that there will appear complicated phase structures, depending on the size of their compactified spaces and on how we impose nontrivial boundary conditions on superfields. In the previous papers,[10,11] we studied the Z_2 and $U(1)$ models and found that the vacuum structures of the models are quite different form each other. We think this is the general feature of our mechanism. We can also study models with non-abelian global symmetries, such as $SU(N)$. In addition, it is important to study gauge theories and to see how our mechanism works and what new dynamics are hidden in them. It may also be interesting to investigate how partial SUSY breaking occurs in gauge theories in connection with the well-known BPS objects.

Acknowledgments

This work was supported in part by Grant-In-Aid for Scientific Research (No.12640275) from the Ministry of Education, Science, and Culture, Japan (M.S) and by a Grant-in-Aid for Scientific Research, Grant No. 3666 (M.T). K. T. would like to thank The Niels Bohr Institute and INFN, Sezione di Pisa for warm hospitality.

References

1. For a review, see for example, D. Bailin and A. Love, Rep. Prog. Phys. **50** (1987) 1087.
2. J. Scherk and J.H. Schwarz, Phys. Lett. **B82** (1979) 60.
3. P. Fayet, Phys. Lett. **B159** (1985) 121, Nucl. Phys. **B263** (1986) 87.
4. K. Takenaga, Phys. Lett. **B425** (1998) 114, Phys. Rev. **D58** (1998) 026004-1; **D61** (2000) 129902(E).
5. Y. Hosotani, Phys. Lett. **B126** (1983) 309, Ann. Phys. **190** (1989) 233.
6. C. J. Isham, Proc. R. Soc. London. **A346** (1978) 591.
7. K. D. Rothe and J. A. Swieca, Nucl. Phys. **B149** (1979) 237.
8. K. Takenaga, Phys. Rev. **D62** (2000) 065001.
9. M. Sakamoto, M. Tachibana and K. Takenaga, Phys. Lett. **B457** (1999) 231 and references therein.
10. M. Sakamoto, M. Tachibana and K. Takenaga, Phys. Lett. **B458** (1999) 33.
11. M. Sakamoto, M. Tachibana and K. Takenaga, Prog. Theor. Phys. **104** (2000) 633.

MULTI-INSTANTONS AND GAUGE/STRING THEORY DYNAMICS

N. DOREY[1], T.J. HOLLOWOOD[1] AND V.V. KHOZE[2]

[1] *Department of Physics, University of Wales Swansea, Swansea, SA2 8PP, UK*

[2] *Department of Physics and IPPP, University of Durham, Durham, DH1 3LE, UK*

The ADHM construction of Yang-Mills instantons can be very naturally understood in the framework of D-brane dynamics in string theory. In this point-of-view, the mysterious auxiliary symmetry of the ADHM construction arises as a gauge symmetry and the instantons are modified at short distances where string effects become important. By decoupling the stringy effects, one can recover all the instanton formalism, including the all-important volume form on the instanton moduli space. We describe applications of the instanton calculus to the AdS/CFT correspondence and higher derivative terms in the D3-brane effective action.

1 Introduction

Instantons are solutions of the classical equations of gauge theory with finite action. A single instanton in $SU(N)$ is constructed by taking an $SU(2)$ instanton, which has a scale size and position in $\mathbb{R}^4$, and then embedding it inside $SU(N)$, which involves $4N-5$ additional "coset" parameters. Rather perversely, we want to describe these moduli in the following way: firstly a'_n, which is (minus) the position of the instanton. To this we add 2 complex N vectors $w_{u\dot{\alpha}}$ subject to the 3 constraints

$$(\tau^c)^{\dot{\alpha}}{}_{\dot{\beta}}\bar{w}^{\dot{\beta}}w_{\dot{\alpha}} = 0 \ . \qquad (1)$$

The instanton solution is actually independent of an auxiliary $U(1)$ which rotates $w_{\dot{\alpha}}$ by a phase. The physical meaning of the parameters is

$$\begin{array}{lcc} a'_n & \longrightarrow & -\text{position in } \mathbb{R}^4 \\ \rho^2 = \bar{w}^{\dot{\alpha}}w_{\dot{\alpha}} & \longrightarrow & \text{size}^2 \\ \rho^{-2}w_{u\dot{\alpha}}(\tau^c)^{\dot{\alpha}}{}_{\dot{\beta}}\bar{w}^{\dot{\beta}}_v & \longrightarrow & SU(2) \subset SU(N) \end{array} \qquad (2)$$

Multi-instantons are described by a non-abelian generalization of this construction. The instanton position a'_n becomes a 4-vector of $k \times k$ hermitian matrices and there are $2k$ N-vectors $w_{ui\dot{\alpha}}$, $i = 1, \ldots, k$. The generalization of (1) is the famous set of ADHM [1] constraints:

$$\mathcal{B}^c \equiv (\tau^c)^{\dot{\alpha}}{}_{\dot{\beta}}\big(\bar{w}^{\dot{\beta}}w_{\dot{\alpha}} + \bar{a}'^{\dot{\beta}\alpha}a'_{\alpha\dot{\alpha}}\big) = 0 \ , \qquad (3)$$

where $a'_{\alpha\dot{\alpha}} = a'_n\sigma^n_{\alpha\dot{\alpha}}$. The moduli space of k instantons, $\mathfrak{M}_{k,N}$, is then given by $\{a'_n, w_{\dot{\alpha}}\}$ modulo the ADHM constraints and modulo an auxiliary $U(k)$ symmetry which acts as $w_{\dot{\alpha}} \to w_{\dot{\alpha}}U$, $a'_n \to U^\dagger a'_n U$.

Three important coments are in order:

(i) In a supersymmetric theory, instantons also have Grassmann moduli which arise from the fermion fields (see [2–4]). In a SUSY gauge theory with $\mathcal{N}$ supersymmetries ($\mathcal{N} = 1, 2, 4$), there are $\mathcal{N}$ gluino fields and the corresponding Grassmann collective coordinates are $k \times k$ matrices $\mathcal{M}'^A_\alpha$, $k \times N$ matrices μ^A and $N \times k$ matrices $\bar{\mu}^A$, where $A = 1, \ldots, \mathcal{N}$. These are subject to fermionic analogues of the ADHM constraints:

$$\mathcal{F}^A_{\dot{\alpha}} \equiv \bar{\mu}^A w_{\dot{\alpha}} + \bar{w}_{\dot{\alpha}}\mu^A + [\mathcal{M}'^\alpha, a'_{\alpha\dot{\alpha}}] = 0 \ . \qquad (4)$$

(ii) In order to do instanton calculations, we need to known how to change variables in the path integral from the fields to the collective coordinates. A direct approach to this problem has only been achieved in the cases $k = 1, 2$ [5, 6]. An alternative and tractable approach [7] is to use the symmetries of the theory, and in this respect SUSY is a very powerful symmetry, along with cluster decomposition to deduce the measure on the SUSY instanton moduli space at arbitrary k. The resulting expression for the measure

$\mathcal{Z}_{k,N}$ is fortunately rather simple:

$$\int \frac{d^4a'\, d^2w\, d^{2\mathcal{N}}\mathcal{M}'\, d^{\mathcal{N}}\mu\, d^{\mathcal{N}}\bar{\mu}\, \delta(\mathcal{B}^c)\, \delta(\mathcal{F}^A_{\dot\alpha})}{\mathrm{Vol}\, U(k)\big(\det \boldsymbol{L}\big)^{\mathcal{N}-1}} \ . \tag{5}$$

Here $\boldsymbol{L}$ is an operator on $k\times k$ matrices:

$$\boldsymbol{L}\cdot\Omega = \{\bar{w}^{\dot\alpha}w_{\dot\alpha},\Omega\} + [a'_n,[a'_n,\Omega]] \ . \tag{6}$$

In an $\mathcal{N}=4$ SUSY gauge theory the measure (5) is not the complete story because in these theories the action evaluated on the instanton solution is not just the constant and depends on instanton moduli. In these theories all but the 8 SUSY and 8 superconformal fermion zero modes, which are protected by the corresponding symmetries, are lifted beyond linear order at the classical level by the Yukawa interactions of the theory [4]. This leads to a 4-fermion term in the instanton action:

$$\frac{\pi^2}{2g^2}\epsilon_{ABCD}\mathrm{tr}_k(\bar{\mu}^A\mu^B + \mathcal{M}'^{\alpha A}\mathcal{M}'^B_{\alpha}) \times \boldsymbol{L}^{-1}(\bar{\mu}^C\mu^D + \mathcal{M}'^{\beta C}\mathcal{M}'^D_{\beta}) \ . \tag{7}$$

(iii) The moduli space $\mathfrak{M}_{k,N}$ is not a smooth manifold: it has orbifold-type singularities that occur when $U(k)$ does not act freely. Physically these are points where an instanton shrinks to zero size, *i.e.* $w_{iu\dot\alpha}=0$ for a given i. We can illustrate this for the case of a single instanton in $SU(2)$. In this case, $\mathfrak{M}_{1,2}=\mathbb{R}^4\times\mathbb{R}^4/Z_2$, where $\mathbb{R}^4$ corresponds to position of the instanton while the angular coordinates of the second $\mathbb{R}^4$ parameterize the $SU(2)$ gauge orientation and finally the radial coordinate of this factor is the scale size. It is important to emphasize that these singularities are *not* evidence of any sickness in the instanton calculus. In fact when calculating the instanton contribution to any physical quantity in field theory these short-distance singularities are prefectly harmless. There is a natural way to smooth, or blow up, the singularities of $\mathfrak{M}_{k,N}\to\mathfrak{M}^{(\zeta)}_{k,N}$: simply modify the ADHM constraints by adding a term proportional to the identity matrix to the right-hand side:

$$\mathcal{B}^c \equiv (\tau^c)^{\dot\alpha}{}_{\dot\beta}\big(\bar{w}^{\dot\beta}w_{\dot\alpha} + \bar{a}'^{\dot\beta\alpha}a'_{\alpha\dot\alpha}\big) = \zeta^c 1_{[k]\times[k]} \ . \tag{8}$$

The new term prevents any component $w_{ui\dot\alpha}\to 0$ and so instantons cannot shrink to zero size. For example, in the case $k=1$ and $N=2$ described above, it is possible to show that the orbifold factor $\mathbb{R}^4/Z_2$ becomes the Eguchi-Hanson manifold. Remarkably, the smoothed moduli space $\mathfrak{M}^{(\zeta)}_{k,N}$ describes instantons in non-commutative gauge theory on a spacetime with non-commuting coordinates [8]:

$$[x_n,x_m] = -i\bar{\eta}^c_{nm}\zeta^c \ , \tag{9}$$

where $\bar{\eta}^c_{nm}$ is a 't Hooft eta symbol.

2 Meaning of ADHM

The curious set of ADHM data and an auxiliary $U(k)$ symmetry naturally arises in the dynamics of D-branes in string theory [4,9]. The low energy collective dynamics of N coincident D$(p{+}4)$-branes in Type II string theory is described by a $U(N)$ SUSY gauge theory in $p{+}5$-dimensions with 16 supercharges. An instanton in the world-volume theory of the D$(p+4)$-branes is a soliton which has 4 transverse directions in the higher dimensional brane, *i.e.* is some kind of p-brane. The remarkable thing is that it is precisely a Dp-brane bound to the D$(p{+}4)$-brane. In general k Dp-branes bound to the N higher dimensional D$(p{+}4)$-branes correspond to a charge k instanton in a $U(N)$ SUSY gauge theory. In order to see how this works out, we have to consider the low energy collective dynamics of the Dp-branes. This is described by a SUSY $U(k)$ gauge theory with 16 supercharges, but with additional matter fields arising from the higher dimensional branes which break half of these supersymmetries. To be more specific, let us suppose that $p=3$. In this case a theory with 16 supercharges is

$\mathcal{N} = 4$ SUSY gauge theory. Let us analyse the spectrum of fields in terms of $\mathcal{N} = 1$ supermultiplets. Along with the $\mathcal{N} = 1$ vector multiplet containing the gauge field, there are 3 adjoint-valued chiral superfields Φ, X and $\tilde{X}$. The 6 real scalars of these chiral multiplets describe the transverse positions of the D3-branes and in particular X and $\tilde{X}$ describe the positions of the D3-branes within the D7-branes, while Φ describes the separation between the D3- and D7-branes. Open string going between the D3-branes and D7-branes give rise a N chiral multiplets Q and $\tilde{Q}$ in, respectively, the $\boldsymbol{k}$ and $\bar{\boldsymbol{k}}$ representations of the gauge group. The resulting theory has $\mathcal{N} = 2$ supersymmetry and X and $\tilde{X}$ form an adjoint hypermultiplet while Q and $\tilde{Q}$ form N fundamental hypermultiplets.

This gauge theory then describes the low energy dynamics of the D3-branes (in the presence of D7-branes). Let us consider the space of vacua of this theory. The theory has a Higgs branch where the gauge group is completely broken (the scalar components of) $\Phi = 0$ and Q, $\tilde{Q}$, X and $\tilde{X}$ are non zero. The equations describing the Higgs branch follow from the D and F-flatness conditions and these precisely the ADHM constraints (3) with the identifications

$$w_{\dot{\alpha}} = \begin{pmatrix} Q^\dagger \\ \tilde{Q} \end{pmatrix} , \qquad a'_{\alpha\dot{\alpha}} = \begin{pmatrix} X^\dagger & \tilde{X} \\ -\tilde{X}^\dagger & X \end{pmatrix} . \tag{10}$$

Hence there is a natural identification of $\mathfrak{M}_{k,N}$ and the Higgs branch of our $\mathcal{N} = 2$ gauge theory. Notice that this gauge theory, with gauge group $U(k)$, is *not* the original $\mathcal{N} = 4$ gauge theory that lives on the D7-branes, which has gauge group $U(N)$. The Higgs branch describes a situation in which the D3-branes lie inside the D7-branes ($\Phi = 0$). On the contrary the Coulomb branch, on which $Q = \tilde{Q} = 0$, while Φ, X and $\tilde{X}$ are non zero, describes a situation in which the D3-branes have moved off the D7-branes. There are mixed branches which describe situations in which some of the D3-branes are on the D7-branes while some have moved off into the bulk. The points where Q_i and $\tilde{Q}_i$ go to zero connect the different phases and correspond to points where the D3-branes can move off into the D7-branes. These are precisely the points where an instanton shrinks to zero size. So in a certain respect, that we will make explicit shortly, this stringy context leads to a certain resolution of the orbifold singularities of $\mathfrak{M}_{k,N}$.

However, there is more to this than an identification between $\mathfrak{M}_{k,N}$ and the Higgs branch of the gauge theory. If we dimensionally reduce the system to $p = -1$, so that we are describing a system of D-instantons and D3-branes, then the the partition function of the $U(k)$ gauge theory (which is now a 0-dimensional field—or matrix—theory) can be identified with the measure on the ADHM moduli space in the limit where bulk effects decouple from the branes, $\alpha' \to 0$. The 4D gauge field and Φ can be amalgamated into χ_a, $a = 1, \ldots, 6$, an adjoint-valued 6-vector. The bosonic part of the partition function is

$$\mathcal{Z}_{k,N} = \int \frac{d^4a'\, d^2w\, d^6\chi\, d^3D}{\mathrm{Vol}\, U(k)} \exp\big(- \mathrm{tr}_k \chi_a \boldsymbol{L} \chi_a + \alpha'^4 \mathrm{tr}_k [\chi_a, \chi_b]^2 - 2\alpha'^4 \mathrm{tr}_k D^2 + i \mathrm{tr}_k D^c \mathcal{B}^c \big) \times \cdots \tag{11}$$

Here, D^c is an adjoint-valued 3-vector that arises as an auxiliary field of the 4D theory. Now if we take $\alpha' = 0$, then the integral over χ_a is Gaussian and gives rise to a factor $(\det \boldsymbol{L})^{-3}$, while the D^c are nothing but Lagrange multipliers for the ADHM constraints! Notice that the resulting partition function in this limit gives precisely the the bosonic parts of the measure on the ADHM moduli space in an $\mathcal{N} = 4$ SUSY theory (5).

If we don't take the $\alpha' = 0$ limit, then in a sense we resolve the singularities of $\mathfrak{M}_{k,N}$ since the D^c no longer act as Lagrange multipliers for the ADHM constraints, rather, the constraints are smeared over a scale $\sqrt{\alpha'}$. How does this kind of resolution relate to the blow-up $\mathfrak{M}^{(\zeta)}_{k,N}$? The modifications of the ADHM constraints by the parameters ζ^c can naturally be incorporated into the stringy construction since they correspond to Fayet-Illiopolos (FI) couplings in the $U(k)$ gauge theory, *i.e.* add $-i\zeta^c \mathrm{tr}_k D^c$ to the exponent in (11). There are consequently two different ways to smooth $\mathfrak{M}_{k,N}$, via stringy corrections or via FI couplings; however, we shall argue later that they lead to the same effect.

Before we leave this section there are three further issues that we mention.

(i) It is important that χ_a also couples to a fermion bilinear:

$$\Sigma^a_{AB} \mathrm{tr}_k \chi_a (\bar{\mu}^A \mu^B + \mathcal{M}'^{\alpha A} \mathcal{M}'^{B}_{\alpha}) , \qquad (12)$$

for, when $\alpha' = 0$ and χ_a is integrated-out, a the 4-fermion interaction (7) is generated.

(ii) Hitherto, we have been considering the situation where the N D($p+4$)-branes are coincident; however, what happens when they separate? Consider the case with $p = -1$. From the point-of-view of the D3-branes the answer is straightforward: the scalars which correspond to the positions of the branes gain a VEV $\langle \varphi_a \rangle$, a 6-vector of $N \times N$ matrices, and one moves out onto the Coulomb branch of the $U(N)$ gauge theory. This effect is then easily incorporated into the D-instanton $U(k)$ theory, by modifying the following couplings:

$$\begin{aligned} w_{\dot\alpha}\chi_a &\to w_{\dot\alpha}\chi_a + \langle\varphi_a\rangle w_{\dot\alpha} , \\ \mu^A\chi_a &\to \mu^A\chi_a + \langle\varphi_a\rangle \mu^A . \end{aligned} \qquad (13)$$

These couplings have the form of mass terms for $w_{\dot\alpha}$. It turns out that the new couplings precisely reproduce the *constrained instanton formalism* [10] that describes instantons in theories with VEVs. The effect of the extra coupling to the VEVs is to suppress instantons of large size in the instanton measure and superconformal invariance is explicitly broken.

(iii) It is worth commenting on the case when $N = 1$ and the original gauge theory has gauge group $U(1)$. It is well known that abelian theories do not have instantons; however, we can still define the ADHM construction. In this case, the ADHM constraints are explicitly solved by taking $w_{\dot\alpha} = 0$ and $a'_n = -\mathrm{diag}(X^1_n, \dots, X^k_n)$. In other words, these "abelian instantons" are point like and moreover

$$\mathfrak{M}_{k,1} = \mathrm{Sym}^k(\mathbb{R}^4) . \qquad (14)$$

This space has singularities whenever 2 instantons coincide. However, one finds that the gauge potential that arises from the ADHM data is pure gauge. Nevertheless, when we modify the ADHM construction as in (8), in other words consider instantons in a non-commutative theory, then the abelian instanton solutions become non-trivial. The deformed space $\mathfrak{M}^{(\zeta)}_{k,1}$ is smooth; for example, $\mathfrak{M}_{2,1} = \mathbb{R}^4 \times \mathbb{R}^4/Z_2$, while the deformation replaces the orbifold factor with the Eguchi-Hanson manifold. So we see $\mathfrak{M}_{1,2} = \mathfrak{M}_{2,1}$ and $\mathfrak{M}^{(\zeta)}_{1,2} = \mathfrak{M}^{(\zeta)}_{2,1}$, a property that does *not* generalize to $N > 1$ and $k > 2$.

3 Calculations with Instantons

In this section we shall summarize a number of interesting applications of the instanton calculus. We will primarily be interested in the $\mathcal{N} = 4$ theory with possible stringy corrections, FI couplings and VEVs.

3.1 The AdS/CFT correspondence

The AdS/CFT correspondence realizes the old idea that string theory describes large-N of gauge theory [11]. In fact it is much stronger: $\mathcal{N} = 4$ SUSY gauge theory is equivalent to Type IIB string theory compactified on $AdS_5 \times S^5$. As usual with a duality it is hard to prove, since calculations can only be done at weak coupling in the gauge theory, $g^2N \ll 1$, while—presently at least—calculations on the string theory side can only be done in the classical supergravity limit where the radius of curvature $R \gg \sqrt{\alpha'}$; which means $g^2 \ll 1$ while $g^2N \gg 1$. Some quantities, however, are protected against renormalization in g^2N, and the value calculated in the gauge theory can be compared directly to the value extracted from the supergravity approximation to string theory.

For us the relevant correlation functions involve 16 dilatinos Λ on the supergravity side that correspond to a certain composite operator in the gauge theory. These correlation functions receive contributions from D-instantons in the string theory whose coupling dependence singles them out as instanton contributions in the gauge theory. The correspondence requires that the k instanton contribution to the correlator, in the infrared and at leading order in $\frac{1}{N}$, should be, schematically, [12, 13]

$$\langle\Lambda(x_1)\cdots\Lambda(x_{16})\rangle \sim \sqrt{N}\,g^{-24}\,q^k\,k^{25/2}\sum_{d|k} d^{-2} \times\int\frac{d^4X\,d\rho}{\rho^5}\prod_{i=1}^{16}\mathcal{F}(x_i - X,\rho)\ . \tag{15}$$

Here, $\{X_n,\rho\}$ parameterizes a point in AdS_5 and the details of the expression for the integrand may be found in [4]. What is remarkable about (15) is that the k dependence is only through the numerical prefactor $q^k k^{25/2}\sum_{d|k} d^{-2}$. This looks like a disaster because there seems little chance that the integral over k instantons, with its intrinsic complexity, would reduce to something that is simply a number times a one instanton contribution.

The k-instanton contribution to the correlators involves inserting into the measure, (5) with $\mathcal{N} = 4$ along with the 4-fermion coupling (7), the 16 composite operators. This contribution turns out to be calculable at leading order in $1/N$ in a way that we summarize below [4]:

(i) For $N \geq 2k$, and so certainly at large N, the ADHM constraints can be solved by a simple change of variables: the biggest impediment to progress with the ADHM construction proves to be entirely benign. The idea involves changing variables from the $w_{\dot\alpha}$ to quadratic gauge invariant variables

$$W^{\dot\alpha}_{\ \dot\beta} = \bar w^{\dot\alpha} w_{\dot\beta}\ . \tag{16}$$

The ADHM constraints are then linear in $W^{\dot\alpha}_{\ \dot\beta}$, as is apparent from (3), and the the δ-function constraints in (5) may trivially be solved.

(ii) The 4-fermion term in the instanton action (7) can be bilinearized by introducing a 6-vector of $U(k)$-adjoint variables χ_a. The Grassmann collective coordinates can then integrated-out. The χ_a variables are precisely those that arise naturally in the D-instanton/D3-brane system described previously.

(iii) The remaining expression is then amenable to a saddle-point approximation at large N. The saddle-point solution has a very simple interpretation. Each of the k instantons are embedded in mutually commuting $SU(2)$ subgroups of the gauge group, as one might have expected on statistical grounds alone. Furthermore, and less intuitive, is that they have the same size ρ and sit at the same point X_n in spacetime; so at the saddle point

$$\bar w^{\dot\alpha} w_{\dot\beta} = \rho^2\delta^{\dot\alpha}_{\ \dot\beta}1_{[k]\times[k]}\ ,\qquad a'_n = -X_n 1_{[k]\times[k]}\ . \tag{17}$$

Furthermore, the auxiliary variables χ_a have the saddle-point value

$$\chi_a = \rho^{-1}\hat\Omega_a 1_{[k]\times[k]}\ , \tag{18}$$

where $\hat\Omega_a$ is a unit 6-vector. So the saddle point is parameterized by a point in $AdS_5 \times S^5$! Amazingly, instantons in the gauge theory act as a probe that feel the ten-dimensional geometry of the dual theory. Notice that the S^5 part of the geometry arises from the auxiliary variables χ_a.

(iii) The integral of the fluctuations around the saddle-point solution assembles into something that is known: precisely the partition function of $\mathcal{N} = 1$ 10D $SU(k)$ Yang-Mills dimensionally reduced to 0 dimensions, where the 10D gauge field is formed from the

traceless parts of a'_n and χ_a. This is known to be proportional to $\sum_{d|k} d^{-2}$ [14, 15].
Putting all of this together immediately solves the puzzle alluded to above: any correlation function will look one instanton-like up to an overall k dependent factor. In addition, one can show that the k-dependence and overall factor of $\sqrt{N}$ are exactly reproduced. Instantons consequently provide one of the most convincing pieces of evidence in favour of the AdS/CFT correspondence.
We can also couch our result in terms of the partition function of the D-instanton/D3-brane system:

$$\mathcal{Z}_{k,N} \underset{N\to\infty}{=} 2^{3-2k}\pi^{6k-25/2} \sqrt{N}k^{3/2}\sum_{d|k} d^{-2}\int \frac{d^4X\,d\rho\,d^5\hat{\Omega}}{\rho^5}\cdot d^8\xi\,d^8\bar{\eta}\ , \tag{19}$$

where X and ρ are the overall position and scale size, respectively, while ξ and $\bar{\eta}$ are the 8 SUSY and superconformal fermion zero modes, respectively. It is possible to generalize these kinds of calculations to other AdS/CFT duals.

3.2 *Instanton effects in D3-branes*

The collective excitations of N D3-branes are described at low energies by an $\mathcal{N}=4$ SUSY gauge theory with gauge group $U(N)$. However, the minimal $\mathcal{N}=4$ Lagrangian is only valid at low energy and there is an infinite tower of the higher derivative interactions that come with powers of α', the string length scale. Some of these, *but not all* are encoded in the Born-Infeld Lagrangian. In [16], it was argued that in the case of a single D3-brane, instantons contribute to certain terms of order α'^4, including one of the form $(\partial F)^4$, where F is the abelian field strength. Furthermore, the $SL(2,Z)$ modular symmetry of the Type IIB string theory, which is realized as electro-magnetic duality in the D3-brane theory, fixes the instanton contributions exactly. In fact the coupling to this term in the effective action involves the logarithm of the Dedekind eta function:

$$\ln|\eta(\tau)^4| = -\tfrac{\pi}{3}\mathrm{Im}\tau - 2\sum_{k=1}^{\infty}(q^k+\bar{q}^k)\Big[\sum_{d|k} d^{-1}\Big]\ . \tag{20}$$

Here, the first term is a tree-level contribution, while the other terms come from k instantons and k anti-instantons, respectively.
We can relate the k-instanton terms in the effective action of the D3-brane predicted by Green and Gutperle [16] for the case $N=1$ (without FI and VEV terms) to the k-instanton partition function modded out by the integral over the overall k-instanton position in R^4 and its superpartners (the 8 supersymmetric fermion zero modes)

$$\widehat{\mathcal{Z}}_{k,1}(\zeta,\alpha') = \sum_{d|k} d^{-1}\ . \tag{21}$$

Here, the FI coupling ζ, absent at the start, arises as a source.
What is interesting about the string result (21) is that in order to have a non-trivial contribution when $\zeta=0$, it is absolutely essential to have the α' corrections in the D-instanton/D3-brane system. Another way of seeing this is that superconformal invariance must be broken. It turns out that when $\zeta\neq 0$ we can legitimately set $\alpha'=0$ to yield:

$$\widehat{\mathcal{Z}}_{k,1}(\zeta,0) = \sum_{d|k} d^{-1}\ , \tag{22}$$

which is then a statement about the integral over the resolved *centered* moduli space $\widehat{\mathfrak{M}}^{(\zeta)}_{k,1}$, where $\mathfrak{M}^{(\zeta)}_{k,N} = R^4\times\mathfrak{M}^{(\zeta)}_{k,N}$. Note that the integral (21) does not actually depend on the α' coupling.
The whole story of D-instanton corrections to the D3-brane effective action generalizes to the non-abelian case of N D3-branes [17]. In this case, it is necessary that the D3-branes are separated by adding VEVs $\langle\varphi\rangle$ so that theory is in a Coulomb phase. In this case, there is generalization of (21):

$$\widehat{\mathcal{Z}}_{k,N}(\zeta,\alpha',\langle\varphi\rangle) = N\sum_{d|k} d^{-1}\ . \tag{23}$$

It is possible to prove this using Morse theory arguments. First of all, one can set α' to zero in (23). If the VEVs vanished, the latter quantity is Gauss-Bonnet-Chern integral on $\widehat{\mathfrak{M}}^{(\zeta)}_{k,N}$. Turning on the VEV has the effect of introducing a Morse function on $\widehat{\mathfrak{M}}^{(\zeta)}_{k,N}$ and using standard arguments the integral $\widehat{\mathcal{Z}}_{k,N}(\zeta,0,\langle\varphi\rangle)$ localizes onto the critical point set. These are the submanifold of $\mathfrak{M}^{(\zeta)}_{k,N}$ where

$$w_{\dot{\alpha}}\chi_a + \langle\varphi_a\rangle w_{\dot{\alpha}} = 0 \ . \qquad (24)$$

The VEV $\langle\varphi_a\rangle$ is a 6-vector of diagonal $N\times N$ matrices. The critical points correspond to associating each instanton with a particular D3-brane; in other words a partition $k = \{k_1,\dots,k_n\}$, where $k_u \geq 0$. For each partition the critical point set is a product of abelian, $N=1$, instanton moduli spaces:

$$\mathfrak{M}_{k_1,1}\times\cdots\times\mathfrak{M}_{k_n,1} \ . \qquad (25)$$

Localizing the integral on the critical point sets gives us a relation of the form

$$\mathcal{Z}_{k,N}(\zeta,0,\langle\varphi\rangle) = \sum_{\{k_j\}} \mathcal{Z}_{k,1}(\zeta,0)\times\cdots\times\mathcal{Z}_{k,1}(\zeta,0) \ . \qquad (26)$$

Notice that we have not separated out the center-of-mass integrals yet. Each factor $\mathcal{Z}_{k,1}(\zeta,0)$ leaves 8 unsaturated Grassmann integrals; hence, most of the partitions in the sum have more than 8 unsaturated Grassmann integrals and will not contribute to $\widehat{\mathcal{Z}}_{k,N}(\zeta,0,\langle\varphi\rangle)$. Only the partitions where all k of the instantons live on the same D3-brane will survive, and there are N of these; so

$$\widehat{\mathcal{Z}}_{k,N}(\zeta,0,\langle\varphi\rangle) = N\widehat{\mathcal{Z}}_{k,1}(\zeta,0) = N\sum_{d|k} d^{-1} \ . \qquad (27)$$

What is striking about this result is that for $N>1$ it holds only for nonvanishing VEVs, *i.e.* in the Coulomb phase, but nevertheless the right-hand side of (23) is independent of $\langle\varphi\rangle$.

References

1. M.F. Atiyah, N.J. Hitchin, V.G. Drinfeld and Y.I. Manin, Phys. Lett. **65A** (1978) 185.
2. E. Corrigan, D. Fairlie, P. Goddard and S. Templeton, Nucl. Phys. B140 (1978) 31.
3. V.V. Khoze, M.P. Mattis and M.J. Slater, Nucl. Phys. **B536** (1998) 69 [hep-th/9804009].
4. N. Dorey, T.J. Hollowood, V.V. Khoze, M.P. Mattis and S. Vandoren, Nucl. Phys. **B552** (1999) 88 [hep-th/9901128].
5. H. Osborn, Ann. of Phys. **135** (1981) 373.
6. N. Dorey, V.V. Khoze and M.P. Mattis, Phys. Rev. **D54** (1996) 2921 [hep-th/9603136]; Phys. Rev. **D54** (1996) 7832 [hep-th/9607202].
7. N. Dorey, V.V. Khoze and M.P. Mattis, Nucl. Phys. **B513** (1998) 681 [hep-th/9708036]
   N. Dorey, T.J. Hollowood, V.V. Khoze and M.P. Mattis, Nucl. Phys. **B519** (1998) 470 [hep-th/9709072].
8. N. Nekrasov and A. Schwarz, Commun. Math. Phys. **198** (1998) 689 [hep-th/9802068].
9. M. Douglas, hep-th/9512077, hep-th/9604198
   E. Witten, Nucl. Phys. **B460** (1996) 541, hep-th/9511030.
10. I. Affleck, M. Dine and N. Seiberg, Nucl. Phys. **B241** (1984) 493.
11. O. Aharony, S. S. Gubser, J. Maldacena, H. Ooguri and Y. Oz, Phys. Rept. **323** (2000) 183 [hep-th/9905111].
12. T. Banks and M.B. Green, JHEP **05:002** (1998) [hep-th/9804170]
13. M. Bianchi, M. B. Green, S. Kovacs and G. Rossi, JHEP **9808** (1998) 013 [hep-th/9807033].
14. G. Moore, N. Nekrasov and S. Shatashvili, Commun. Math. Phys. **209** (2000) 77 [hep-th/9803265].
15. W. Krauth, H. Nicolai and M. Staudacher, Phys. Lett. **B431** (1998) 31, hep-th/9803117
16. M. B. Green and M. Gutperle, hep-th/0002011.
17. N. Dorey, T.J. Hollowood and V.V. Khoze, *to appear*.

MULTI-INSTANTON CALCULUS IN SUPERSYMMETRIC THEORIES

F. FUCITO

INFN, sez. di Roma 2, Via della Ricerca Scientifica, 00133 Roma, Italy
E-mail: fucito@roma2.infn.it

In this talk I review some recent results concerning multi-instanton calculus in supersymmetric field theories. More in detail, I will show how these computations can be efficiently performed using the formalism of topological field theories.

1 Introduction

Our understanding of the non–perturbative sector of field and string theories has greatly progressed in recent times. Recently, for the first time, the entire non–perturbative contribution to the holomorphic part of the Wilsonian effective action was computed for $N = 2$ globally supersymmetric (SUSY) theories with gauge group $SU(2)$, using ansätze dictated by physical intuitions [1]. A few years later, a better understanding of non–perturbative configurations in string theory led to the conjecture that certain IIB string theory correlators on an $AdS_5 \times S^5$ background are related to Green's functions of composite operators of an $N = 4$ $SU(N_c)$ Super Yang–Mills (SYM) theory in four dimensions in the large N_c limit [2]. Although supported by many arguments, these remarkable results remain conjectures and a clear mathematical proof seems to be out of reach at the moment. In our opinion this state of affairs is mainly due to the lack of adequate computational tools in the non–perturbative region. To the extent of our knowledge, the only way to perform computations in this regime in SUSY theories and from first principles is via multi–instanton calculus. Using this tool, many partial checks have been performed on these conjectures, both in $N = 2$ and $N = 4$ SUSY gauge theories [3], [4], [5], [6], [7]. The limits on these computations come from the exploding amount of algebraic manipulations to be performed and from the lack of an explicit parametrization of instantons of winding number greater than two [8]. In order to develop new computational tools that might allow an extension to arbitrary winding number, I revisit instanton computations for $N = 2$ in the light of the topological theory built out of $N = 2$ SYM [9], *i.e.* the so–called Topological Yang–Mills theory (TYM) [10].

2 Topological Yang–Mills Theory

Here I collect some results which will be relevant to our discussion. I use the same notation of [9] to which I refer the reader for a detailed exposition of this material. As it is well known [11], after the twisting procedure, the Lagrangian of $N = 2$ SYM is invariant under

$$\begin{aligned} sA &= \psi - Dc \quad , \\ s\psi &= -[c, \psi] - D\phi \quad , \\ s\phi &= -[c, \phi] \quad , \\ sc &= -\frac{1}{2}[c, c] + \phi. \end{aligned} \tag{1}$$

The BRST operator, s, defined in (1) is such that $s^2 = 0$ and when the set of equations in (1) is restricted to the solutions of the Euler–Lagrange classical equation (*the zero modes*) it gives the derivative on the space M^+. In terms of the parameters of the ADHM construction, these solutions look like

$$A = U^\dagger dU, \tag{2}$$

$$\psi = U^\dagger \mathcal{M} f (d\Delta^\dagger) U + U^\dagger (d\Delta) f \mathcal{M}^\dagger U, \tag{3}$$

$$c = U^\dagger (s + C) U, \tag{4}$$

$$\phi = U^\dagger \mathcal{M} f \mathcal{M}^\dagger U + U^\dagger \mathcal{A} U. \tag{5}$$

Plugging (5) into (1) leads to the action of the operator s on the elements of the ADHM construction

$$\mathcal{M} = s\Delta + C\Delta = \mathcal{S}\Delta \ , \quad (6)$$

$$\mathcal{A} = s\mathcal{M}\Delta + C\mathcal{M} = \mathcal{S}\mathcal{M} \ , \quad (7)$$

$$s\mathcal{A} = -[C, \mathcal{A}] \ , \quad (8)$$

$$sC = \mathcal{A} - CC \ , \quad (9)$$

i.e. this is the realization of the BRST algebra on the instanton moduli space. C is a connection I must introduce to have a nilpotent s.

Let us see now how, at the semi–classical level, any correlator which is expressed as a polynomial in the fields, becomes after projection onto the zero–mode subspace, a well–defined differential form on $M^{+\ 10}$. Symbolically

$$\langle fields \rangle = \int_{M^+} \left[(fields)\ e^{-S_{\mathrm{TYM}}} \right]_{zero-mode} \ . \quad (10)$$

A generic function on the zero–mode subspace will then have the expansion

$$g(\widehat{\Delta}, \widehat{\mathcal{M}}) = g_0(\widehat{\Delta}) + g_{i_1}(\widehat{\Delta})\widehat{\mathcal{M}}_{i_1} + \ldots + \frac{1}{p!} g_{i_1 i_2 \ldots i_p}(\widehat{\Delta})\widehat{\mathcal{M}}_{i_1}\widehat{\mathcal{M}}_{i_2}\cdots\widehat{\mathcal{M}}_{i_p} \quad (11)$$

the coefficients of the expansion being totally antisymmetric in their indices. Now $\widehat{\mathcal{M}}_i$'s and the $s\widehat{\Delta}_i$'s are related by a (moduli–dependent) linear transformation K_{ij}, which is completely known once the explicit expression for C is plugged into the $\widehat{\mathcal{M}}_i$'s:

$$\widehat{\mathcal{M}}_i = K_{ij}(\widehat{\Delta}) s\widehat{\Delta}_j \ . \quad (12)$$

It then follows that

$$\widehat{\mathcal{M}}_{i_1}\widehat{\mathcal{M}}_{i_2}\cdots\widehat{\mathcal{M}}_{i_p} = = \epsilon_{j_1 \ldots j_p} K_{i_1 j_1} K_{i_2 j_2} \cdots K_{i_p j_p}\ s^p\widehat{\Delta} = = \epsilon_{i_1 \ldots i_p} (\mathrm{det} K)\ s^p \widehat{\Delta} \ , \quad (13)$$

where $s^p\widehat{\Delta} \equiv s\widehat{\Delta}_1 \cdots s\widehat{\Delta}_p$. I then conclude that

$$\int_{M^+} g(\widehat{\Delta}, \widehat{\mathcal{M}}) = \frac{1}{p!} \int_{M^+} g_{i_1 \ldots i_p}(\widehat{\Delta})\widehat{\mathcal{M}}_{i_1}\cdots\widehat{\mathcal{M}}_{i_p} = \int_{M^+} s^p\widehat{\Delta}\ |\mathrm{det} K| g_{12\ldots p}(\widehat{\Delta}) \ . \quad (14)$$

The determinant of K naturally stands out as *the instanton integration measure for* $N = 2$ *SYM theories.*

3 Conclusions

In this talk i have argued that the results of multi-instanton calculus at the semiclassical level can be easily recovered in the formalism of topological field theories. More benefits come from this reformulation than what I have presented until now. It is for example possible to show that correlators of the type of (10) can be written as a total derivative on the moduli space. The only contributions to these quantities, come from zero-size instantons. Given the peculiar properties of the ADHM construction at the boundary of the moduli space, this might lead to recursion relations among correlators computed at different winding number. To get to this conclusion, we probably have to better understand what is the geometrical role of the action, projected on the subspace of the zero-modes. The connection between moduli spaces of instantons and their construction in terms of D-branes of string theory can be very helpful in this respect.

References

1. N. Seiberg and E. Witten, Nucl. Phys. **B426** (1994) 19; *ibid.* **B431** (1994) 484.
2. J. Maldacena, Adv. Theor. Math. Phys. **2** (1998) 231.
3. D. Finnell and P. Pouliot, Nucl. Phys. **B453** (1995) 225.
4. N. Dorey, V. V. Khoze and M. P. Mattis, Phys. Rev. **D54** (1996) 2921; *ibid.* **D54** (1996) 7832.
5. F. Fucito and G. Travaglini, Phys. Rev. **D55** (1997) 1099.
6. M. Bianchi, S. Kovacs, M. Green and G. C. Rossi, JHEP **9808** (1998) 013.
7. N. Dorey, T. J. Hollowood, V. V. Khoze, M. P. Mattis and S. Vandoren, Nucl.

Phys. **B552** (1999) 88.
8. M. Atiyah, V. Drinfeld, N. Hitchin and Yu. Manin, Phys. Lett. **65A** (1978) 185.
9. D. Bellisai, F. Fucito, A. Tanzini and G. Travaglini, Phys.Lett. **B480** (2000) 365; *"Instanton Calculus, Topological Field Theories and $N = 2$ Super Yang–Mills Theories"*, `hep-th/0003272` to appear in JHEP.
10. E. Witten, Commun. Math. Phys. **117** (1988) 353.
11. L. Baulieu and I. M. Singer, Nucl. Phys. Proc. Suppl. **B5** (1988) 12.

SPONTANEOUS SUSY BREAKING IN N=2 SUPER YANG-MILLS THEORIES

PETER MINKOWSKI IN COLLABORATION WITH LUZI BERGAMIN

E-MAIL: MINK@ITP.UNIBE.CH

Institute for Theoretical Physics, University of Bern, Sidlerstrasse 5, CH-3012 Bern, Switzerland

It is shown that the same essentially non-semiclassical mechanism, which generates in the nonsupersymmetric pure Yang-Mills theory the binary condensate of gauge field strengths, is responsible for the spontaneous breaking of the two supersymmetries in N=2 super Yang-Mills systems. A detailed discussion is presented in ref. [1]

1 Introduction

Following the work of Seiberg and Witten [2] the semiclassical modifications due to multiinstanton configurations within N=2 theories were shown not to induce any spontaneous breaking of supersymmetry [3], [4]. General Ward identities for N=1 Yabg-Mills systems with matter fields were derived in ref. [5] *under the assumption* of exactly unbroken susy.

The above situation is widely interpreted as indication, not to say proof, that nonperturbative effects - as represented by instantons - do not break supersymmetries. However the above configurations are semiclassical, nonperturbative, while binary condensate formation is nonsemiclassical, nonperturbative. The latter can in conjunction with spontaneous symmetry breaking also generate spontaneous symmetry restoration as is the case for CP in nonsupersymmetric QCD [a]

We propose to discuss the extension of the effective potential, representing Greens functions of composite local operators to the theories under study.

2 Short discussion

A universal feature of susy in connection with spontaneous effects is revealed through the 'once local' form of N=2 susy algebra

$$\left\{ j^{\ i}_{\mu\alpha} (x) \ , \ \overline{Q}_{k \ \dot\beta} \right\} \ = \ \delta^{\ i}_{\ k} \ \vartheta_{\ \mu \ \alpha\dot\beta} \ (\ x \ )$$
$$\vartheta_{\ \mu \ \alpha\dot\beta} \ = \ \vartheta_{\ \mu \ \nu} \ \sigma^{\ \nu}_{\alpha\dot\beta} \tag{1}$$

In eq. 1 $j^{\ i}_{\mu\alpha}$, $i = 1,2$, $\vartheta_{\ \mu \ \nu}$ denote supercurrents and energy momentum tensor respectively. Now if we consider the spontaneous parameters to imply a nonvanishing expectation value for the energy momentum (-density) operator - at this stage just a logical possibility - we obtain

$$\begin{aligned} &\langle \ \Omega \ | \ \vartheta_{\ \mu\nu} \ | \ \Omega \ \rangle \ = \ \varepsilon \ g_{\ \mu\nu} \quad \rightarrow \\ &\langle \ \Omega \ | \left\{ \ j^{\ i}_{\ \mu\alpha} \ (x) \ , \ \overline{Q}_{k \ \dot\beta} \ \right\} \ | \ \Omega \ \rangle \ = \\ &= \ \varepsilon \ \delta^{\ i}_{\ k} \ \sigma_{\ \mu\alpha\dot\beta} \quad ; \quad \varepsilon \ > \ 0 \end{aligned} \tag{2}$$

From eq. 2 the universal spontaneous breakdown of both (N=2) supersymmetries follows. In addition the spectral function of the two supercurrents exhibits the equally universal contribution from two massless goldstinos of the form

---

[a] For a detailed discussion we refer to the appendix in ref. [1].

$$\langle \Omega | \left\{ j^{\,i}_{\mu\alpha}(x) \, , \, j^{\,*k}_{\nu\dot\beta}(y) \right\} | \Omega \rangle =$$
$$\delta^{ik} (2\pi)^{-3} \int d^4 q$$
$$\exp(-iqz)\, \varepsilon(q^0)\, \Gamma_{\mu\nu\varrho}(q)\, \sigma^{\varrho}_{\alpha\dot\beta}$$
$$\Gamma_{\mu\nu\varrho} = \delta(q^2)\, \gamma_{\mu\nu\varrho} + \cdots$$
$$\gamma_{\mu\nu\varrho} = \varepsilon\, (g_{\mu\varrho}\, q_\nu + g_{\nu\varrho}\, q_\mu - g_{\mu\nu}\, q_\varrho)$$
$$z = x - y \tag{3}$$

The Christoffel-symbol like structure of the quantity $\gamma_{\mu\nu\varrho}$ in eq. 3 is not accidental.
The spontaneous energy density ε in eqs. 2 and 3 has to be positive, as implied by susy and thus opposite to the same quantity in QCD.
For further details we are forced here to refer to ref. [1]. This in order to focus on the essential features and to remain within the space requirements.

Conclusions

A universal connection between the ground state expected value of the energy momentum tensor and spontaneous breaking of supersymmetries in N=2 super Yang-Mills theories is demonstrated. Contrary to QCD the vacuum energy density ε is necessarily positive (nonnegative) as implied by susy. As a consequence the coupling of both goldstino modes to the supercurrents is completely determined by the vacuum energy density.

References

1. L. Bergamin and P. Minkowski, hep-th/hep-th/0003097.
2. N. Seiberg and E. Witten, Nucl. Phys. B426 (1994) 19, Nucl. Phys. B431 (1994) 484.
3. V. V. Khoze, Multi-instanton contributions to gauge/string theory dynamics, see these proceedings.
4. F. Fucito, Multi instanton calculus in supersymmetric theories, see these proceedings.
5. D. Amati, K. Konishi, Y. Meurice, G.C. Rossi and G. Veneziano, Phys.Rept.162 (1988) 169.

AN EXACT SOLUTION OF THE RANDALL-SUNDRUM MODEL AND THE MASS HIERARCHY PROBLEM

S. ICHINOSE

Laboratory of Physics, School of Food and Nutritional Sciences, University of Shizuoka, Yada 52-1, Shizuoka 422-8526, Japan
E-mail: ichinose@u-shizuoka-ken.ac.jp

An exact solution of the Randall-Sundrum model for a simplified case (one wall) is obtained. It is given by the $1/k^2$-expansion (thin wall expansion) where $1/k$ is the thickness of the domain wall. The vacuum setting is done by the 5D Higgs potential and the solution is for a family of the Higgs parameters. The mass hierarchy problem is examined. Some physical quantities in 4D world such as the Planck mass, the cosmological constant, and fermion masses are focussed. Similarity to the domain wall regularization used in the chiral fermion problem is explained. We examine the possibility that the 4D massless chiral fermion bound to the domain wall in the 5D world can be regarded as the real 4D fermions such as neutrinos, quarks and other leptons.

In nature there exists the mass hierarchy such as the Planck mass (10^{19}GeV), the GUT scale (10^{15}Gev), the electro-weak scale (10^2GeV), the neutrino mass ($10^{-11}-10^{-9}$GeV) and the cosmological size(10^{-41}GeV). How to naturally explain these different scales ranging over 10^{60} (so huge !) order has been the long-standing problem (the mass hierarchy problem). Triggered by the development of the string and D-brane theories, some interesting new approaches to the compactification mechanism have recently been proposed and are applied to the hierarchy problem. Here we examine the Randall-Sundrum (RS) model[1] which has some attractive features compared with the Kaluza-Klein compactification. The model is becoming a strong candidate that could solve the mass hierarchy problem.

We take the 5D gravitational theory with the 5D Higgs potential: $S[G_{AB},\Phi]=\int d^5X\sqrt{-G}(-\frac{1}{2}M^3\hat{R}-\frac{1}{2}G^{AB}\partial_A\Phi\partial_B\Phi-V(\Phi))$, $V(\Phi)=\frac{\lambda}{4}(\Phi^2-{v_0}^2)^2+\Lambda$, $\cdots$(A), where $X^A(A=0,1,2,3,4)$ is the 5D coordinates and we also use the notation $(X^A)\equiv(x^\mu,y),\mu=0,1,2,3.$ The three parameters λ, v_0 and Λ in $V(\Phi)$ are called here *vacuum parameters.* Following [1], we take the line element: $ds^2=\mathrm{e}^{-2\sigma(y)}\eta_{\mu\nu}dx^\mu dx^\nu+dy^2$, where $\eta_{\mu\nu}=\mathrm{diag}(-1,1,1,1)$.

We consider the simple case : $\Phi=\Phi(y)$. The 5D Einstein equation reduces to $-6M^3(\sigma')^2=-\frac{1}{2}(\Phi')^2+V, 3M^3\sigma''=(\Phi')^2$, $\cdots$(B), where $'=\frac{d}{dy}$. As the extra space, we take the real number space $\mathbf{R}=(-\infty,+\infty)$. This is a simplified version of the original RS-model[1] where $S^1/\mathbf{Z}_2$ is taken. We impose the following asymptotic behaviour for the (classical) vacuum of $\Phi(y)$: $\Phi(y)\rightarrow\pm v_0$, $y\rightarrow\pm\infty$. This means $\Phi'\rightarrow 0$, and from (B), $\sigma''\rightarrow 0$. From eq.(B), we are led to $\sigma'\rightarrow\pm\omega,\sigma\rightarrow\omega|y|$ as $y\rightarrow\pm\infty$. $\omega(>0)$ is determined, by considering $y\rightarrow\pm\infty$ in (B), as $\omega=\sqrt{\frac{-\Lambda}{6}}M^{-\frac{3}{2}}$, where we see the sign of Λ must be *negative*, that is, the 5D geometry must be *anti de Sitter* in the asymptotic regions. We may set $M=1$ without ambiguity.

Let us take the following form for $\sigma'(y)$ and $\Phi(y)$ as an *exact* solution.

$$\sigma'(y)=k\sum_{n=0}^{\infty}\frac{c_{2n+1}}{(2n+1)!}\{\tanh(ky+l)\}^{2n+1},$$

$$\Phi(y)=v_0\sum_{n=0}^{\infty}\frac{d_{2n+1}}{(2n+1)!}\{\tanh(ky+l)\}^{2n+1}\quad(1)$$

where c's and d's are coefficient-constants to be determined. The free parameter l comes from the *translation invariance* of (B). A *new mass scale* $k(>0)$ is introduced here to make the quantity ky dimensionless. The physical meaning of $1/k$ is the "thickness" of the domain wall. We call M, k and r_c(defined later) *fundamental parameters.* The distortion of 5D space-time by the existence of the domain wall should be small so that the quantum effect of 5D gravity can be ignored and the present *classical* analysis is valid. This requires the condition[1] $k \ll M$. The coefficient-constants c's and d's have the following constraints

$$\sqrt{\frac{-\Lambda}{6}} = k\sum_{n=0}^{\infty}\frac{c_{2n+1}}{(2n+1)!}, 1 = \sum_{n=0}^{\infty}\frac{d_{2n+1}}{(2n+1)!}, (2)$$

which are obtained by considering the asymptotic behaviours $y \to \pm\infty$ in (1).

We first obtain the recursion relations between the expansion coefficients, from the field equations (B). For $n \geq 2$, they are given by

$$\begin{aligned} & c_{2n+1}/(2n)! - c_{2n-1}/(2n-2)! \\ & = {v_0}^2/3(D'_n - 2D'_{n-1} + D'_{n-2}) , \\ -6C_{n-1} & = -({v_0}^2/2)(D'_n - 2D'_{n-1} + D'_{n-2}) \\ + \frac{\lambda}{4}\frac{{v_0}^4}{k^2}&(E_{n-2} - 2D_{n-1}) , E_n = \sum_{m=0}^{n} D_{n-m}D_m, \\ D_n & = \sum_{m=0}^{n}\frac{d_{2n-2m+1}d_{2m+1}}{(2n-2m+1)!\,(2m+1)!} , \\ D'_n & = \sum_{m=0}^{n}\frac{d_{2n-2m+1}d_{2m+1}}{(2n-2m)!\,(2m)!}, \\ C_n & = \sum_{m=0}^{n}\frac{c_{2n-2m+1}c_{2m+1}}{(2n-2m+1)!\,(2m+1)!} . (3) \end{aligned}$$

The first few terms, $(c_1, d_1), (c_3, d_3)$, are explicitly given as

$$\begin{aligned} d_1 & = \frac{\pm\sqrt{2}}{v_0 k}\sqrt{\Lambda + \frac{\lambda {v_0}^4}{4}} , c_1 = \frac{2}{3k^2}(\Lambda + \frac{\lambda {v_0}^4}{4}), \\ \frac{d_3}{d_1} & = 2 + \frac{1}{k^2}\{\frac{8}{3}(\Lambda + \frac{\lambda {v_0}^4}{4}) - \lambda {v_0}^2\}, \\ \frac{c_3}{c_1} & = 2 + \frac{1}{k^2}\{\frac{16}{3}(\Lambda + \frac{\lambda {v_0}^4}{4}) - 2\lambda {v_0}^2\}, (4) \end{aligned}$$

where $\pm$ sign in d_1 reflects $\Phi \leftrightarrow -\Phi$ symmetry in (B). We take the positive one in the following. It is confirmed that the above relations determine all c's and d's recursively in the order of increasing n. They are described by the three dimensionless vacuum parameters: $(\Lambda + \frac{\lambda {v_0}^4}{4})/k^2M^3$, $\lambda {v_0}^2/k^2$, ${v_0}^2/M^3$. In order for this solution to make sense, as seen from the expression for d_1, Λ should be bounded also from below, in addition to from above: $-\frac{\lambda {v_0}^4}{4} < \Lambda < 0$.

Assuming the convergence of the infinite series (1), we can obtain *leading* behaviour of the vacuum parameters in the dimensional reduction: As $k \to \infty$,

$$-\Lambda \sim M^3k^2 \ , \ v_0 \sim M^{3/2} \ , \ \lambda \sim M^{-3}k^2 . (5)$$

We can evaluate the asymptotic forms of the Planck mass M_{pl} and the cosmological term as

$$\begin{aligned} {M_{pl}}^2 & \sim M^3 \int_{-r_c}^{r_c} dy\, e^{-2\omega|y|} \sim \frac{M^3}{k} , \\ \Lambda_{4d} & \sim \Lambda \int_{-r_c}^{r_c} dy\, e^{-4\sigma(y)} \sim -M^3k < 0 , (6) \end{aligned}$$

where the *infrared regularization* parameter r_c is introduced and we have used the *4D reduction condition*: $kr_c \gg 1$.

Using the value $M_{pl} \sim 10^{19}$GeV , the "rescaled" cosmological parameter $\tilde{\Lambda}_{4d} \equiv \Lambda_{4d}/{M_{pl}}^2$ has the relation: $\sqrt{-\tilde{\Lambda}_{4d}} \sim k \sim M^3 \times 10^{-38}$ GeV, where the relations (6) are used. The mass unit of M is GeV here andin the following. Some typical cases are 1) ($k = 10^{-41}, M = 0.1$), 2) ($k = 10^{-13}, M = 10^8$) 3) ($k = 10, M = 10^{13}$) 4) ($k = 10^4, M = 10^{14}$) and 5) ($k = 10^{19}, M = 10^{19}$). Cases 1) and 5) are typical extreme ones and have serious defects. Cases 3) and 4) are some intermediate cases which are acceptable except for the cosmological constant.

We point out the mechanism presented here has a strong similarity to that in the chiral fermion determinant. The condition on k in the RS model, from previous relations , is given as $\frac{1}{r_c} \ll k \ll M$. The corresponding one of the chiral fermion is given by[5,6] $|k^\mu| \ll M_F \ll \frac{1}{t}$. Both conditions guarantee the mechanism effectively works. As

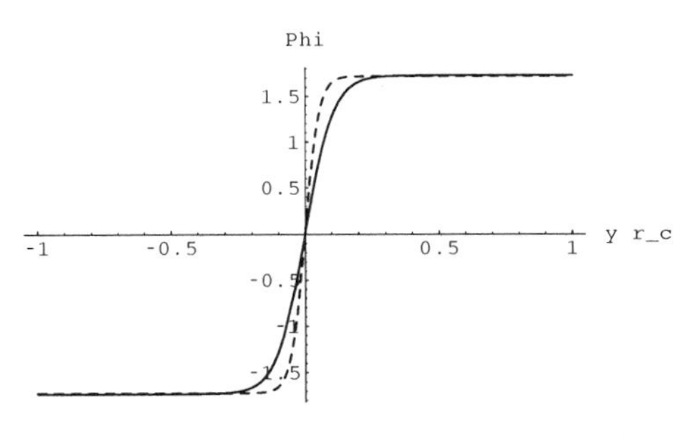

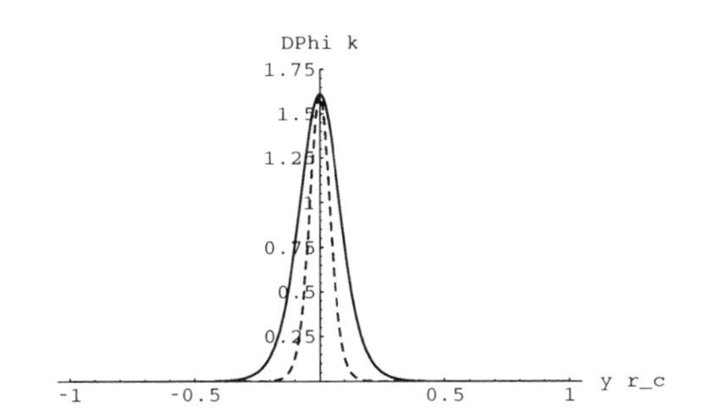

Figure 1. Vertical axes: 5D Higgs field ($\Phi(y)$, Left) and its derivative ($\Phi'(y)/k$, Right); Horizontal axes: y/r_c; Configuration of (Vac.3); Solid line: $kr_c = 10$, Dotted line: $kr_c = 20$.

in the Callan and Harvey's paper[2], we can have the 4D *massless chiral* fermion bound to the wall by introducing 5D *Dirac* fermion ψ into (A): $+\int d^5X\sqrt{-G}(\bar{\psi}\nabla\psi + g\Phi\bar{\psi}\psi)$. If we *regulate* the extra axis by the finite range $-r_c \leq y \leq r_c$, the 4D fermion is expected to have a small mass $m_f \sim k\mathrm{e}^{-kr_c}$ (This is known for the two-walls case in [3,4]). If we take the case 3) and regard the 4D fermion as a neutrino ($m_\nu \sim 10^{-11} - 10^{-9}$GeV), we obtain $r_c = 2.76 - 2.30\text{GeV}^{-1}$. If we take case 4), we obtain $r_c = (3.45 - 2.99) \times 10^{-3}\text{GeV}^{-1}$. When the quarks or other leptons ($m_q, m_l \sim 10^{-3} - 10^2$GeV) are taken as the 4D fermion, and take the case 4), we obtain $r_c = (1.61 - 0.461) \times 10^{-3}\text{GeV}^{-1}$. It is a quite fascinating idea to *identify the chiral fermion zero mode bound to the wall with the neutrinos, quarks or other leptons.*

In terms of new parameters $\Omega \equiv \Lambda + \frac{\lambda}{4}{v_0}^4 (0 < \Omega < \frac{\tau}{4}{v_0}^2), \tau \equiv \lambda {v_0}^2$, instead of Λ and λ, the precise forms are obtained by the $\frac{1}{k^2}$-expansion for the case $kr_c \gg 1$ as $\Omega = M^3k^2\sum_{n=0}^{\infty}\frac{\alpha_n}{(kr_c)^{2n}}$, $\tau = k^2\sum_{n=0}^{\infty}\frac{\gamma_n}{(kr_c)^{2n}}$, $v_0 = M^{3/2}\sum_{n=0}^{\infty}\frac{\beta_n}{(kr_c)^{2n}}$. Here α's,γ's and β's are some numerical (real) numbers to be consistently chosen using (2). For general M, k, r_c except the condition $kr_c \gg 1$, the coefficients are determined as, by safe truncation up to $n = 2$: $(\alpha_0, \alpha_1, \alpha_2) \equiv (1, 1, 1)$input; $(\beta_0, \beta_1, \beta_2; \gamma_0, \gamma_1, \gamma_2) = (1.6, 1.1, 0.45; 4.2, 1.8, 3.5) \cdots$(Vac.3). We notice our solution has one *free parameter* for each n-th set $(\alpha_n, \beta_n, \gamma_n)$. Using this freedom we can adjust one of the three vacuum parameters in the way the observed physical values are explained. In (Vac.3), we take α's as the input. Taking the value $kr_c = 10$, (Vac.3) has the vacuum expectation value $v_0M^{-3/2} = 1.6$, the cosmological constant $\Lambda k^{-2}M^{-3} = -1.7$ and the coupling $\lambda k^{-2}M^3 = 1.6$.

For the Vac.3, we plot $\Phi(y)$ and $\Phi'(y)$ for two cases $kr_c = 10$ and $kr_c = 20$ in Fig.1.

For the case of the S^1/Z_2 extra space, see ref.[7]. The RS model has given us richer possibilities for the mass hierarchy problem than before.

Acknowledgments

The author thanks G.W.Gibbons for much discussions about the RS-model. He also thanks M.Yamaguchi for some comment.

References

1. L.Randall and R.Sundrum,Phys.Rev. Lett.**83**(1999)3370;ibid.**83**(1999)4690.
2. C.G.Callan and J.A.Harvey,Nucl.Phys.**B250**(1985)427.
3. Y.Shamir,Nucl.Phys.**B406**(1993)90.
4. P.M.Vranas, Phys.Rev.**D57**(1998)1415.
5. S.Ichinose, Phys.Rev.**D61**(2000)055001.
6. S.Ichinose, Nucl.Phys.**B574**(2000)719.
7. S.Ichinose, US-00-07, hep-th/0008245.

GRAVITATIONAL BENDING OF LIGHT AND EXTRA DIMENSIONS

B. H. J. MCKELLAR

School of Physics, University of Melbourne, Parkville, Vic Australia 3052
E-mail: b.mckellar@physics.unimelb.edu.au

XIAO-GANG HE

Department of Physics, National Taiwan University, Taipei, 10617, Taiwan.
E-mail: hexg@phys.ntu.edu.tw

GIRISH C. JOSHI

School of Physics, University of Melbourne, Parkville, Vic Australia 3052
E-mail: g.joshi@physics.unimelb.edu.au

We study gravitational lensing and the bending of light in low energy scale (M_S) gravity theories with extra space-time dimensions n. We find a correction to the deflection angle with a strong quadratic dependence on the photon energy. The observation of the bending of visible light for photons grazing the sun confirms Einstein's predictions to 15% at the 90% confidence level. Restricting the KK correction to the deflection angle for visible photons grazing the Sun to 15% demands that the scale M_S has to be larger than $1.4(2/(n-2))^{1/4}$ TeV for $n \neq 2$, and approximately 4 TeV for n=2.

The gravitational bending of light, is one of the most important experimental results which support the Einstein General Relativity (EGR) theory. Photons are deflected when passing by a massive object. In EGR theory at grazing incidence the deflection angle is predicted to be $\theta_0 = 4G_N m/R$, where m is the mass and R the radius of the massive object. For the Sun the deflection angle is $1.75''$. This prediction provides an important test for different theories of gravity[1]. It is usual to measure deviation from the EGR theory in terms of the post-Newtonian parameter γ defined by $\theta = (4G_N m/R)(1+\gamma)/2$. In EGR theory $\gamma = 1$. The EGR theory is in agreement within a level better than one percent for observations in the radio band to visible band, for sources which are at a large angular distance from the sun[2].

In this paper we study gravitational lensing in theories with extra space-time dimensions. Additional details are available in ref [3]

It has recently been proposed that gravitational effects can become large at a scale M_S near the weak scale due to effects from extra dimensions[4], which is quite different from the traditional concept that gravitational effects only become large at the Planck scale $M_{Pl} = \sqrt{1/G_N} \sim 10^{19}$ GeV. In this proposal the total space-time has $D = 4 + n$ dimensions. The relation between the scale M_S and the Planck scale M_{Pl}, assuming all extra dimensions are compactified with the same size R, is given by $M_{Pl}^2 \sim R^n M_S^{2+n}$. For $n \geq 2$, M_S can be of order one TeV and R can be in the sub-millimeter region[4]. When the extra dimensions are compactified there are towers of states, the Kaluza-Klein (KK) states with spin-2, spin-1 and spin-0, which interact with ordinary matter fields. These theories have many interesting consequences[5] from which we can obtain information about the allowed value for M_S. The lower bound for M_S is constrained, typically, to be of order one TeV from collider experimental data[5]. There are also constraints on the lower bound for M_S of up to a few hundred TeV depending on the models from cosmological and astrophyiscal considerations[5].

At a quantum level, gravitational lensing is due to exchange of a massless graviton between photons and massive objects in EGR theory. A calculation on this basis gives the standard results. This approach has been exploited to estimate loop corrections to the classical result[7], and in 5 dimensional Kaluza-Klein theories[8] and we will use it to investigate the corrections from extra dimensions.

In theories with extra dimensions, gravitational lensing also receives contributions from the exchange of massive KK states in addition to that from the usual graviton exchange. The massive KK states couple to matter fields in a way similar to the massless graviton. This makes gravitational lensing a sensitive test of theories with extra dimensions.

After compactifying the extra n dimensions, for a given KK level $\vec{l}$ there are one spin-2, n-1 spin-1 and n(n-1)/2 spin-0 states[6]. Assuming that all standard fields are confined to a four dimensional world-volume and gravitation is minimally coupled to standard fields, it was found that the spin-1 KK states decouple while the spin-2 and spin-0 KK states couple to all standard fields[6]. We, however, found that only spin-2 KK states can interact with both the photon and the Sun.

The graviton and the spin-2 KK states couple to the energy momentum tensor of the Sun which is similar to the coupling of a spin-2 particle to a scalar. We will therefore treat the Sun as a scalar S and obtain the deflection angle by matching the scattering cross section and the impact parameter. Using the Feynman rules given in Ref.[6], we can obtain the scattering amplitude for, $\gamma(\epsilon_1(p_1)) + S(k_1) \to \gamma(\epsilon_2(p_2)) + S(k_2)$.

Gauge invariance permits us to obtain the scattering amplitude from that of a pure massless graviton[9] by replacing $1/q^2$ by $1/q^2 + \sum_l 1/(q^2 - m_l^2)$.

For small deflection angles the initial and final photon energies are approximately the same, and we denote their common value by ω, and use the small angle approximation

All possible KK states have to be summed over. The masses for the KK states are given by $m_l^2 = 4\pi^2 \vec{l}^2/R^2$, where $\vec{l}$ represents the hyper-cubic lattice sites in n-dimensions. For M_S in the multi-TeV range the KK states are nearly degenerate and the sum can be approximated by integral in n-dimensions. With the further approximation that $|q^2|/M_S^2 << 1$, restricting the present consideration to $n > 2$, and using $G_N = (4\pi)^{n/2}\Gamma(n/2)R^{-n}M_S^{-(n+2)}$, the differential cross section for small scattering angle $\tilde{\theta}$ can be written as

$$\frac{d\sigma}{d\Omega} = 16 G_N^2 m^2 \left(\frac{1}{\tilde{\theta}^2} + \omega^2 \Delta\right)^2, \quad (1)$$

with

$$\begin{aligned} \Delta &= \sum_l 1/(q^2 - m_l^2) \\ &= (1/(M_S^4 G_N))(2/(n-2)) \end{aligned} \quad (2)$$

Keeping the leading correction to the deflection angle θ, we obtain

$$\theta = \theta_0 \left(1 - 2\omega^2 \Delta\, \theta_0^2 \ln \theta_0\right). \quad (3)$$

We note that effect of extra dimensions is always to increase the deflection angle, and also to introduce a ω dependence in the deflection angle. One may be surprised at the appearence of an ω dependence. We therefore emphasise that in the EGR theory an ω dependence can be generated at one loop order[7], and in the 5-dimensional KK theory, Delbourgo and Weber[8] found that the effective potential (in 5 dimensions) from exchange of KK towers is quadratically dependent on the energy. In these cases the frequency dependent contribution is extremely small, but the contribution from extra dimensions obtained here can be very large–close to the present experimental reach. For easy comparison with data we work with the post-Newtonian parameter γ. The expression for θ gives the correction $\Delta\gamma = \gamma - 1$ as

$$\Delta\gamma = -4\omega^2 \Delta\, \theta_0^2 \ln(\theta_0). \quad (4)$$

Experimental observations have found no deviations from the EGR theory prediction for γ from radio waves to visible light. For $M_S = 1$ TeV, the typical limit set by most of the collider experiments, there is no conflict for photons with frequencies below the visible. Experimental observations of gravitational lensing by the Sun in visible light from whole sky survey of Hipparcos have found[2] $\gamma = 0.997 \pm 0.003$. This is a very impressive result. Unfortunately this value can not be used directly in our case because in the Hipparcos analysis, γ was assumed to be constant in the whole range of frequency ω and impact parameter b and most of the data was at large $b \geq r_E/2$. In our analysis we find that the largest deviation from EGR is reached for light grazing the Sun. In this region the accuracy of the observations is not as good as the whole sky result. The result for visible light near the solar limb is $\gamma = 0.95 \pm 0.11$[10], which is only at the 15% level. However even with such accuracy, we find that the mass M_S is constrained to be larger than about $1.4(2/(n-2))^{1/4}$ TeV at 2σ level for $n > 2$ and a factor of approximately 3 larger for $n = 2$. This bound is comparable with the limit obtained from collider data.

We suggest that future studies of the parameter γ should vigorously investigate its frequency dependence and its impact parameter dependence. Theories of the type considered here, with mass M_S about 3 TeV scale, suggest $\gamma - 1$ is negligible for radio frequencies, is positive of order 3×10^{-3} in the visible and is so large at γ-ray frequencies that our approximation are no longer valid for light grazing the Sun. For larger impact parameters, the KK effect can become much smaller.

Acknowledgments

This work was supported in part by the National Science Council of R.O.C under Grant NSC 89-2112-M-002-016 and by the Australian Research Council. We thank Dr. R. Webster, Professor K Wali and Professor R Delbourgo for helpful discussions.

References

1. For reviews see for example, R. Narayan and M. Bartelmann, e-print: `ast-ph/9606001` ; C. M. Will, *Int. J. Mod. Phys.* D **1**, 13 (1992); I. Shapiro, in *General Relativity and Gravitation*, ed N. Ashby, D. Bartlett and W. Wyss, (Cambridge, 1990) p313.
2. M. Froeschle, F. Mignard and F. Arenon, in Proceedings from the Hipparcos Venice '97 symposium, Session 1; `http://astro.estec.esa.nl/Hipparcos/venice.html`.
3. X-G. He, G. C. Joshi, and B. H. J. Mckellar, *Gravitational Lensing and Extra Dimensions* `hep-ph/9908469`.
4. N. Arkani-Hamed, S. Dimopoulos, and G. Dvali, *Phys. Lett.* B **429**, 263(1998); I. Atoniadis et al., *Phys. Lett.* B **436**, 257(1998).
5. K. R. Dienes, E. Dudas and T. Ghẹrghetta, *Phys. Lett.* B **436**, 55 (1998); E. Mirabelli, M. Perlestein and M. Peskin, *Phys. Rev. Lett.* **82**, 2236 (1999); J. Hewett, *Phys. Rev. Lett.* **82**, 4765 (1999); K.R. Dienes et al., *Nucl. Phys.* B **543**, 387 (1999); S. Cullen and M. Perelstein, *Phys. Rev. Lett.* **83**, 268 (1999); L. Hall and D. Smith, *Phys. Rev.* D **60**, 085008 (1999).
6. T. Han, J. Lykken and R-J. Zhang, *Phys. Rev.* D **59**, 105006 (1999)
7. R. Delbourgo and P. Phaocas-cosmetatos, *Phys. Lett.* B **41**, 533 (1972).
8. R. Delboourgo and R. O. Weber, *Il Nuovo Cimento* **92A**, 347 (1986).
9. T. Rizzo, *Phys. Rev.* D **60**, 075001 (1999); X.-G. He, *Phys. Rev.* D **60**, 115017 (1999).
10. B. Jones, *Astronomical J.* **81**, 455 (1976);

LEPTON TRANSMUTATIONS FROM A ROTATING MASS MATRIX

TSOU SHEUNG TSUN
Mathematical Institute, University of Oxford,
24-29 St. Giles', Oxford OX1 3LB, United Kingdom
E-mail: tsou@maths.ox.ac.uk

Fermion mass matrices generally rotate in generation space under scale changes, which can lead to fermions of different generations transmuting into one another. The effect is examined in detail and its cross-section calculated for $\gamma + \ell_\alpha \longrightarrow \gamma + \ell_\beta$ with $\ell_\alpha \neq \ell_\beta$ the charged leptons e, μ, or τ. For the (conventional) Standard Model, this is weak and probably undetectable, though with some notable exceptions. But for the Dualized Standard Model, which we advocate and have already used quite successfully to explain quark mixing and neutrino oscillation, the effect is larger and could be observable. Estimates of transmutational decays are also given.

1 Introduction

By a rotating mass matrix we mean one which undergoes unitary transformations through scale changes, as a result of the renormalization group equation, as can happen in many gauge theories. This means that even if the mass matrix m is diagonal (in generation space) at a certain scale, it will not remain so as the scale changes. Hence in general we can expect nonzero transition between fermions of different generations:

$$\ell_\alpha \longrightarrow \ell_\beta, \quad \alpha \neq \beta, \tag{1}$$

such as $e \to \mu$, $e \to \tau$, $\mu \to \tau$. We shall use the term 'transmutation' for this direct transition to distinguish it from e.g. $e \to \mu$ conversion via FCNC.

Here I shall concentrate on transmutational processes[1,2] in the standard model (SM) and, in greater detail, in the dualized standard model (DSM).

2 Mass matrix rotation

In the SM, because the leptonic MNS mixing matrix[3] U is nontrivial[4], the mass matrix L for the charged leptons will rotate as a result of the following term in the linearized RGE[5]:

$$\frac{dL}{d\mu} = \frac{3}{128\pi^2}\frac{1}{246^2}(ULU^\dagger)(ULU^\dagger)^\dagger L + \cdots, \tag{2}$$

where $ULU^\dagger = N$ the neutrino (Dirac) mass matrix. Therefore L cannot be diagonal at all scales. The magnitude of the off-diagonal elements will depend on poorly known or unknown quantities such as the mixing U and the Dirac mass m_3 of the heaviest neutrino. If we take the present popular theoretical biases, namely that U is bimaximal[6] and that m_3 is around the t quark mass, then (2) gives

$$\langle\mu|\tau\rangle \text{ changes by} \sim 5.5 \times 10^{-3}\ \text{GeV}$$
$$\langle e|\tau\rangle \text{ changes by} \sim 1.8 \times 10^{-7}\ \text{GeV}$$
$$\langle e|\mu\rangle \text{ changes by} \sim 1.1 \times 10^{-8}\ \text{GeV}$$

per decade change in energy.

In the DSM[7], the fermion mass matrix is of the following factorized form:

$$m = m_T \begin{pmatrix} x \\ y \\ z \end{pmatrix} (x, y, z), \tag{3}$$

where m_T is essentially the mass of the heaviest generation. Under renormalization m remains factorized, but the vector (x, y, z) changes as

$$\frac{d}{d\mu}\begin{pmatrix} x \\ y \\ z \end{pmatrix} = \frac{5}{32\pi^2}\rho^2 \begin{pmatrix} x_1 \\ y_1 \\ z_1 \end{pmatrix}, \tag{4}$$

where ρ is a (fitted) constant and

$$x_1 = \frac{x(x^2 - y^2)}{x^2 + y^2} + \frac{x(x^2 - z^2)}{x^2 + z^2}, \quad \text{cyclic.} \tag{5}$$

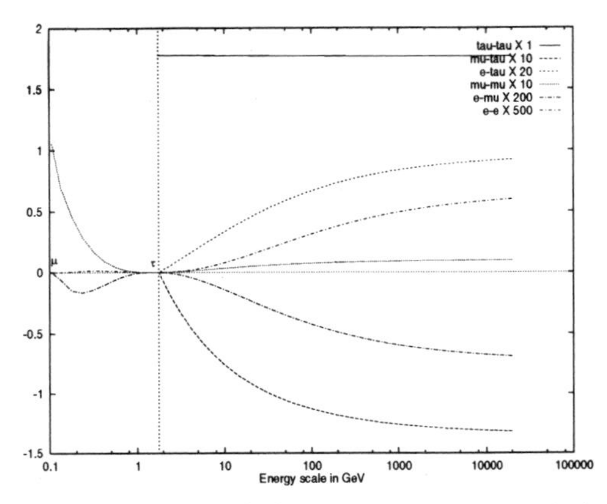

Figure 1. Elements of the rotating mass matrix in GeV for charged leptons in the DSM scheme.

The off-diagonal elements have been calculated explicitly, using 3 free parameters determined by fitting experimental mass and mixing parameters (giving sensible predictions for the remaining paramenters)[8]. These are shown in Figure 1. Hence the results we report below are entirely parameter-free.

3 Lepton states

To define lepton states one must diagonalize the mass matrix, but since the eigenvectors depend on scale, there is no canonical recipe.

We suggest two quite different schemes for exploration: fixed scale diagonalization (FSD) which is applicable to SM, and step-by-step diagonalization (SSD) which is an inherent part of DSM[7,1,2].

4 Transmutational decays

Using the results of §2 we give estimates for the branching ratios of transmutational decays[1]. With some exceptions, SM (with FSD) estimates are all far below present experimental bounds and are hence not so interesting. One exception is the process $\mu^- \rightarrow e^-e^+e^-$, where one could get a branching ratio of 10^{-3} (experimental limit 10^{-12}), if one applied FSD naively. This shows how sensitive these calculations are to transmutational model and/or diagonalization scheme.

Table 1. Branching ratios of transmuational decays.

Decays	DSM est.	Expt limit
$Z^0 \rightarrow \tau^-\mu^+$	4×10^{-8}	1.2×10^{-5}
$\pi^0 \rightarrow \mu^-e^+$	3×10^{-9}	1.7×10^{-8}
$\psi \rightarrow \mu^+\tau^-$	6×10^{-6}	not given
$\Upsilon \rightarrow \mu^+\tau^-$	2×10^{-6}	not given
$\mu^- \rightarrow e^-\gamma$	0	4.9×10^{-11}
$\mu^- \rightarrow e^-e^+e^-$	0	1.0×10^{-12}

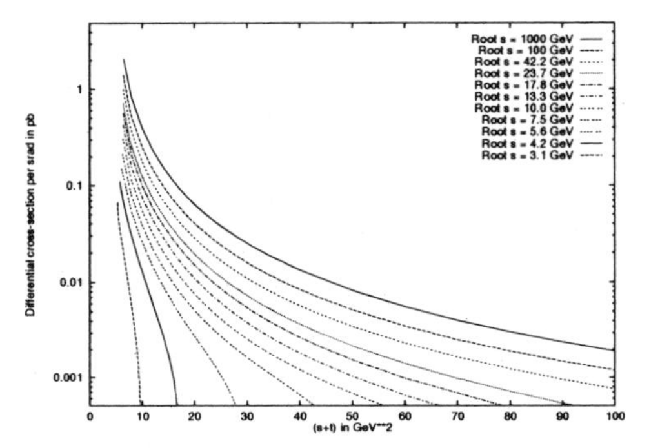

Figure 2. Differential cross sections for $\gamma e \rightarrow \gamma\tau$.

The parameter-free estimates in DSM give branching ratios which are in general larger but still below present experimental limits, as indicated in Table 1. It is important to note that because of SSD the branching ratios of transmutational leptonic decays are automatically zero to first order (Figure 1). The π^0 decay is of particular interest as being less than one order from the experimental bound.

5 Photo-transmutation

We studied[2] (mainly for DSM) the following

$$\gamma + \ell_\alpha \longrightarrow \gamma + \ell_\beta, \quad \alpha \neq \beta. \qquad (6)$$

We calculated the cross sections for: $\gamma e \rightarrow \gamma\mu$, $\gamma e \rightarrow \gamma\tau$, $\gamma\mu \rightarrow \gamma\tau$, leaving out τ-initiated reactions as being experimentally unrealistic at present. A sample of the DSM

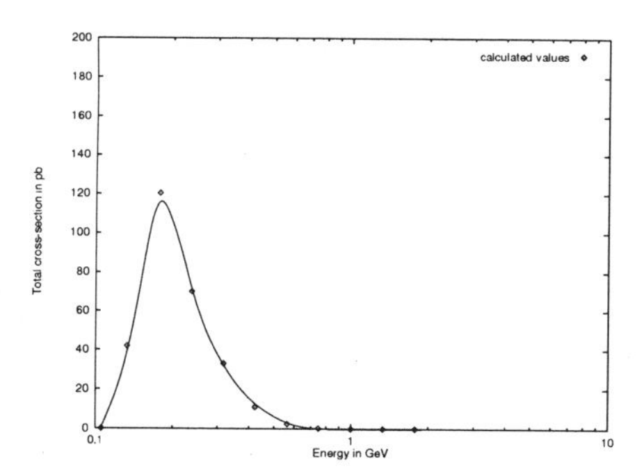

Figure 3. Total cross section for $\gamma e \to \gamma\mu$.

results is presented in Figure 2, for a range of c.m. energies $\sqrt{s}$. Because of the calculated form of the rotation matrix, we get in general: $\gamma\mu \to \gamma\tau > \gamma e \to \gamma\tau > \gamma e \to \gamma\mu$. However, at low energies $\gamma e \to \gamma\mu$ becomes quite sizeable, as seen in Figure 3, where the total cross section has a peak of $\sim$ 100 pb at c.m. energy $\sim$ 200 MeV.

SM calculations depend on further assumptions, under which e.g. $\mu \to \tau$ at $\sqrt{s} =$ 17.8 GeV is $\sim$ 3 orders smaller than DSM.

6 Possible experimental tests

The DSM predictions for the transmutational decay modes: $\pi^0 \to \mu^- e^+$, $\psi \to \tau^-\mu^+$, $\Upsilon \to \tau^-\mu^+$ could be near experimental limits and sensitivites, for LEP, BEPC and B-factories.

For photo-transmutations, one may consider virtual γ from e^+e^- colliders. Above τ, it is more profitable to look for[9] $e \to \tau$. Below τ, it is more profitable to look for $e \to \mu$, at around 200 MeV c.m. energy. Again, LEP and/or BEPC may provide tests.

7 Conclusions

- Lepton transmutation is a necessary consequence in both SM and DSM.
- The SM results are in general smaller than DSM results, but there are uncertainties and further assumptions.
- The DSM calculations are entirely parameter-free. There are no violations of data in all the cases we were able to consider.
- Experimental tests of mainly DSM predictions seem feasible in the near future, for the following:
  - decays of π^0, ψ, Υ,
  - photo-transmutation of $e \to \tau$ at high and $e \to \mu$ at low energy,
  - other processes e.g. $e^+e^- \to e^+\mu^-$.
- Whether our results will pass these tests or not, there is exciting new physics that has to be explored.

Acknowledgments

I thank the Royal Society for a travel grant.

References

1. José Bordes, Chan Hong-Mo and Tsou Sheung Tsun, hep-ph/0006338.
2. José Bordes, Chan Hong-Mo, Jacqueline Faridani, and Tsou Sheung Tsun, hep-ph/0007004.
3. Z. Maki, M. Nakagawa and S. Sakata, Prog. Theor. Phys. 28 (1962) 247.
4. Soudan, Chooz and Superkamiokande data: full references in our papers[1,2].
5. E.g. B. Grzadkowski, M. Lindner and S. Theisen, Phys. Lett. B198, 64, (1987).
6. See e.g. references in our papers[1,2].
7. The model was proposed in Chan Hong-Mo and Tsou Sheung Tsun, Phys. Rev. 57D, (1998) 2507, hep-th/9701120; an up-to-date summary can be found in Chan Hong-Mo, hep-th/0007016, invited lecture at the Intern. Conf. on Fund. Sciences, March 2000, Singapore.
8. José Bordes, Chan Hong-Mo and Tsou Sheung Tsun, hep-ph/9901440, Eur. Phys. J. C10 (1999) 63.
9. under study.

CAN CPT BE VIOLATED THROUGH EXTENDED TIME REVERSAL?

RECAI ERDEM AND ÜNAL UFUKTEPE
Izmir Institute of Technology, Izmir, Turkey
E-mail:erdem@likya.iyte.edu.tr ufuktepe@likya.iyte.edu.tr

We consider the implications of the extension of time reversal through Wigner types and group extensions. We clarify its physical content and apply the results in a toy model. Finally we point out the possibility of violation of CPT in this framework

Detailed studies on direct experimental observation of time reversal violation has begun only in the last few years[1]. So it is not clear yet if its magnitude is in perfect agreement with what predicted from CPT theorem or if there may be other sources of time reversal violation other than the ones already anticipated. In this study first we point out that there are two ways to extend the standard time reversal transformation. One of them is the method of group extensions which first studied in detail by L. Michel[2]. The simplest analog of group extension of Poincare group by an internal group is the situation of parity in the standard model of electroweak interactions. The other way is the doubling of the Hilbert space in projective representation of time reversal due to the antiunitarity of time reversal, which is first studied by Wigner[3] and extended later S. Weinberg[4]. We also consider a combination of these two alternative ways and we concentrate our attention to spin $\frac{1}{2}$ particles. Next we further clarify the physical content of the model by embeding the scheme into higher dimensional spaces and we give an explicit realization of these results through a toy model. Finally we show that CPT theorem does not apply in general in the case of extended time reversal.

The fact that symmetry transformations in quantum theory, $U(g)$ are defined up to a phase factor allows introduction of representations of symmetry transformations, other than the usual one, known as projective representations whose multiplication law is $U(g_1)U(g_2) = e^{i\phi(g_1,g_2)}U(g_1g_2)$ with $g_1, g_2 \in G$. Wigner Noticed that if one uses projective representation of time reversal there are two types of time reversal operators: i)$\mathcal{T}^2 = -1$, ii)$\mathcal{T}^2 = 1$ since the above rule implies, $\mathcal{T}^2 = \omega I$ (where ω is some phase) and the associativity property of $\mathcal{T}$) results in $\mathcal{T}^2\mathcal{T}=\omega\mathcal{T}= \mathcal{T}\mathcal{T}^2 = \omega^*\mathcal{T}$. Later S. Weinberg has shown that one can mix these two cases in one general time reversal operator, that is,

$$\mathcal{T} == \begin{pmatrix} 0 & e^{i\frac{\phi}{2}}U(T)K \\ e^{-i\frac{\phi}{2}}U(T)K & 0 \end{pmatrix} \quad (1)$$

where $U(T)$ is some unitary matrix and K stands for the antilinear part of $\mathcal{T}$, which takes complex conjugate of c-numbers. For spin 1/2 fields $U(T)$ is a 4×4 matrix. So the minumum dimension where the unusual Wigner types can arise is an 8-component spinor representation.

Once we have adopted 8-component spinors there is another way, to extend time reversal, known as the method of group extension[2]. One can assign the upper and lower 4-components of the 8-component spinor to different representations of internal group. Furthermore one can take the upper and lower 4-components to be related by time reversal as the upper and lower 2-components of 4-component spinor are related by parity in Weyl representation.

8-component spinors naturally arise in space-times with dimensions higher than 5. For example consider a 6 dimensional Minkowski vector, with the metric $(g_{AB}) = diag(1, -1, -1, -1, -1, -1)$, embeded in the corresponding Clifford algebra

$$X = \Gamma^A x_A = \begin{pmatrix} V & ix_6\gamma^0 \\ ix_6\gamma^0 & -\gamma^0 V \gamma^0 \end{pmatrix} \quad (2)$$

here $V = i\gamma_5\, x_4 + \gamma^\mu\, x_\mu$ where $\mu = 0, 1, 2, 3$. The time reversal of X, $X^{(T)} = \Gamma^0\, X \Gamma^{0\dagger}$ can be generated by many different operators, for example, the time reversal operator $\Gamma^0 = (\Gamma^0_{ij})$ corresponds to the usual time reversal operator for $\Gamma^0_{11} = G^0_{22} = i\gamma^1\gamma^3\, K$, $\Gamma^0_{12} = G^0_{21} = 0$ while the case $\Gamma^0_{12} = G^0_{21} = \gamma^2\, K$, $\Gamma^0_{11} = G^0_{22} = 0$ corresponds to the unusual Wigner types with $\mathcal{T}^2 = 1$ accomponied with a reversal of the upper and lower components of the 8-component spinor. If the upper and lower components are assigned to different representations of the internal group then the last two time reversal operators may include a transformation in the internal space as in the case of parity. We assume that particles are localized in the extra space-time directions and their coordinates determine the internal quantum numbers of the particles. Then the general wave expansion of a solution of the free Dirac equation, Ψ for space-time dimensions higher than five is the same as the usual four dimensional case except the spinor parts of the field, $u(v)$ have 8-components, that is they consist of two 4-component spinors $u_1(v_1)$, $u_2(v_2)$. After acting $\mathcal{J}$ on Ψ and using the similar identities as in 4-component case[4] one finds that the time reversal operator for one particle theory is applicable in field theory as well.

In the light of the above discussion we shall discuss some of the implications of extended time reversal. The extended time reversal given by S. Weinberg (Eq. (1) corresponds to the mixing of the Wigner types with $\mathcal{T}^2 = -1$ and $\mathcal{T}^2 = 1$, respectively, that is for spin 1/2 particles

$$\Psi = \cos\eta \begin{pmatrix} e^{-i\frac{\eta}{2}}\psi_1^{(1)} \\ e^{i\frac{\eta}{2}}\psi_2^{(1)} \end{pmatrix} + \sin\eta \begin{pmatrix} e^{-i(\alpha)}\psi_1^{(-1)} \\ e^{i(\alpha)}\psi_2^{(-1)} \end{pmatrix}$$

where $\alpha = \frac{\eta}{2} + \frac{\pi}{4}$ and the superscripts (1) and (−1) denote the $\mathcal{T}^2 = -1$ and $\mathcal{T}^2 = 1$ subspaces, respectively. The fact that the $\mathcal{T}^2 = 1$ fermions are not observed yet means that one should take $\eta \simeq 0$. Meanwhile if one identifies the upper 4-components of the 8-component spinors with the fermions of the standard model the absence of additional fermions enforces to take the lower 4-components (at least effectively) as singlets under standard model gauge group. This means that either we should assign the lower components to a different gauge group or take it very heavy to be observed at low energies. However in this case it may have effect on physical observables through internal propagators. So taking it be singlet under the gauge group seems more reasonable. Of course if the upper and lower 4-components do not mix at all and its observablity (at least low energies) is impossible. So we must mix these two components unless they have a common standard model subgroup. The simplest way to do so is to couple ψ_1, ψ_2 through the following interaction[5]

$$m\bar{\psi}_1\psi_1 + m'\bar{\psi}_2\psi_2 + M\psi_1^\dagger\psi_2 + M\psi_2^\dagger\psi_1$$
$$= \bar{\Psi}\tilde{M}\Psi\,, \quad \tilde{M} = \begin{pmatrix} m & \gamma^0 M \\ \gamma^0 M & m' \end{pmatrix} \quad (3)$$

where Ψ is an 8-component spinor consisting of two 4-component spinors ψ_1 and ψ_2. To make mixing small one must take $M \simeq 0$ or $m' >> m$. There will be an additional source of time reversal (other than that predicted from CPT theorem) for $m' \neq m$. One can suppress this violation by taking $|m - m'| << m$. The physical content of the model can be seen better by the diagonalization of the corresponding $\tilde{M}$ which results in two 4-component spinors ψ, ψ^c,

$$\psi \propto m''\psi_1 - M\gamma^0\psi_2\,, \; \psi^c \propto M\gamma^0\psi_1 + m''\psi_2$$
$$m'' = \frac{1}{2}(m - m' + \sqrt{(m-m')^2 + 4M^2})$$

If the Lagrangian is invariant under the interchange of the subscripts 1 and 2 of the upper and lower components then T transformation will be equivalent to time reversal so that CPT theorem necessarily holds.

However if one takes $m \neq m'$ or introduces another Lagragian term with such an asymmetry under this interchange one can introduce additional source of time reversal violation due to the extension of time reversal or one can take the general time reversal operator introduced by S. Weinberg (which corresponds to the mixture of two Wigner types) to make the applicablity of CPT theorem questionable. CPT theorem states that the condition of weak local commutivity for fields (which is satisfied by all reasonable fields) is enough for the following equation to be satisfied[6]

$$(\Psi_0, \phi_\mu(x_1)\phi_\nu(x_2)....\phi_\rho(x_n)\Psi_0) = i^F(-1)^J(\Psi_0, \phi_\mu(-x_1)....\phi_\rho(-x_n)\Psi_0) \quad (4)$$

(where F stands for the number of fermion fields, J is the total number of undotted indices; $\phi_\eta(x_k)$ is the field operator (or a fermion bilinear in the case of fermion) at the point x_k; the subscripts $\mu, \nu, ..$ stand for possible additional degrees of freedom for the fields (or bilinears) as the components of vectors for vector fields.) provided the fields satisfy some axioms including

$$U(\Lambda, a)\phi_j(f)U(\Lambda, a)^{-1} = \sum S_{jk}(a^{-1})\phi_k(\{\Lambda, a\}f) \quad (5)$$

where $\{\Lambda, a\}f = f(\Lambda^{-1}(x-a))$ and $U(\Lambda, a)$ is the unitary representation of Poincare Group with Λ corresponding to the homogeneous Lorentz group and a to the translations. In the case of 4-component spinors CPT invariance reduces to the validity of Eq. (4). For example $\bar{\psi}^A\Gamma\psi^A$, under CPT, transforms as $CPT : \bar{\psi}^A\Gamma\psi^B \rightarrow \bar{\psi}^B\Gamma'\psi^A$ where $\Gamma' = \Gamma$ for $\Gamma = 1$, $i\gamma_5$, $[\gamma_\mu, \Gamma_\nu]$ while $\Gamma' = -\Gamma$ for γ_μ, $\gamma_5\gamma_\mu$ and the superscripts A, B denote different species of fermions. After contraction of X^{BA} with itself or with other X^{AB}'s and comparing it with Hermitian conjugate of the original terms result in Eq. (4). One can perform a similiar and simpler procedure for the other type of fields (i.e. for scalars, vector fields, etc.) to get the same conclusion. So in the usual 4-component spinor case the validity of Eq. (4) is equivalent to CPT invariance. On the other hand in the case of extended-T this is not the case. For example take the operator Γ^0 with $\Gamma^0_{12}=G^0_{21} = \gamma^2 K$ $\Gamma^0_{11}=G^0_{22}=0$ as the relavant extended time reversal operator while all other discrete space-time transformation remain the same. Then time reversal and CPT become

$$T : \psi_{1(2)}(\vec{x}, t) \rightarrow -(+)i\gamma^2\psi_{2(1)}(\vec{x}, -t)$$
$$CPT : \Psi(\vec{x}, t) \rightarrow iL^0\gamma^0\bar{\Psi}^T(-\vec{x}, -t)$$
$$L^0 = \begin{pmatrix} 0 & e^{i\eta}I_4 \\ e^{-i\eta}I_4 & 0 \end{pmatrix} \quad (6)$$

up to arbitrary phases. One can see from the above derivation that CPT invariance is not equivalent to the validity of Eq. (4) in this case. For example

$$CPT : \bar{\Psi}^A\tilde{M}\Psi^B \rightarrow -\bar{\Psi}^B L^0\tilde{M}L^0\Psi^A \quad (7)$$

So the requirement of the equality of the Hermitian conjugate of Eq. (7) to the original term does not imply Eq. (4). If one takes T as the usual time reversal one obtains Eq. (4) but then Eq. (5) is not satisfied because ψ_1 mixes with ψ_2 so ψ_1 does not span the whole physical Hilbert space so that one can not guarantte a Fourier series in terms of ψ_1 is enough to experess the left hand side of Eq. (5). Therefore it seems that CPT theorem does not hold in this case.

Our results can be understood in more physical terms as follows: i) In the case of of extension by Wigner types the mixture of different types causes mixture of $\mathcal{T}^2$ odd and even terms in fermion bilinears. ii) In the case of extension by group extensions $\mathcal{T}$ causes a rotation in the internal space while there is no additional transformation similar to C to compansate this rotation. (After one introduces an intrinsic parity a PT transformation (on the spinor part of the fermion field) in the spinor representation of the Lorentz group is not equivalent to identity because P is not the simple space inversion anymore. However one can introduce the charge conjugation, C so that CP is effectiveley equivalent to space inversion so that CPT on the spinor part of the fermion field is equivalent to identity. Therefore, in anology with the 4-component case, after introduction of an intrinsic time reversal degree of freedom one expects to have a $\tilde{C}$ transformation so that $CPT\tilde{C}$ is the exact symmetry of nature.

References

1. CPLEAR Collaboration, CERN-EP/98-153, *Phys.Lett.* **B 444**, 43 (1998); Particle Data Group, Reviews of Particle Properties, *Phys.Rev.* **D 54**, 1 (1996)
2. L. Michel, in *Group Theoretical Concepts and Methods in Elementary Particle Physics, Lectures of the Istanbul Summer School of Theoretical Physics*, edited by F. Gürsey (Gordon and Breach Science Publishers, New York, 1964)
3. E.P. Wigner, in *Group Theoretical Concepts and Methods in Elementary Particle Physics, Lectures of the Istanbul Summer School of Theoretical Physics*, edited by F. Gürsey (Gordon and Breach Science Publishers, New York, 1964); R.M.F. Houtappel, H. Van Dam and E.P. Wigner, Rev.Mod.Phys. **37**, 595 (1965)
4. S. Weinberg, *The Quantum Theory of Fields Vol.I Foundations* (Cambridge University Press, 1995), Appendix C of Section 2.
5. R. Erdem, in *Photon and Poincare Group*, Editor V.V. Dvoeglazov (Nova Science Publishers, New York, 1999)
6. R.F. Streater, A.S. Wightman *PCT, Spin and Statistics, and All That* (W.A. Benjamin, New York, 1964)

ILLUSORY ARE THE CONVENTIONAL ANOMALIES IN THE CONFORMAL-GAUGE TWO-DIMENSIONAL QUANTUM GRAVITY

N. NAKANISHI*

12-20 Asahigaoka-cho, Hirakata 573-0026, Japan
E-mail: nbr-nakanishi@msn.com

The exact solution in terms of Wightman functions is given to the BRS-formulated conformal-gauge two-dimensional quantum gravity coupled with D massless scalar fields. The solution is seen to be *free of various anomalies.* Its anomalous feature appears only in the B-field equation. The nilpotency violation of BRS charge for $D \neq 26$, found by Kato and Ogawa, is shown to have been caused by their elimination of B-field. Feynman-diagrammatic method and path-integral formalism are shown to yield misleading results (e.g., FP-ghost number current anomaly), owing to the fact that these approaches are based on T*-product, which does not generally respect field equations.

My talk is based on the work done in collaboration with M. Abe.[1–3]

1 I discuss the BRS-formulated conformal-gauge two-dimensional quantum gravity coupled with $D(\neq 26$ in general) massless scalar fields. The purpose of my talk is to point out that the conventional belief of various anomalies must be reconsidered. To avoid possible misunderstanding, I stress here that I do not intend to criticize non-field-theoretical approaches to string theory such as conformal field theory, path-integral approach without ghosts, etc. I restrict the path-integral approach to the one which reproduces all the Green's functions obtainable by Feynman-diagrammatic method.

As is well known, the conformal-gauge two-dimensional quantum gravity reduces to a free field theory if the B-field (= BRS daughter of FP antighost) is eliminated (or path-integrated) at the starting point. This fact is usually discarded because it contradicts the expectation from string theory. Indeed, Fujikawa's path-integral analysis[4] and Kato-Ogawa's operator-theoretical analysis,[5] in which also the B-field is eliminated at a certain step, are believed to give the right answer. It has never been discussed seriously the reason why the reduction to the free-field theory must be regarded as inadmissible. This point should be clarified *mathematically.* The truth can be found by constructing explicitly the exact solution to the model without eliminating the B-field.

2 Although what I discuss explicitly is a very particular model, the problem itself is the one *at very deep level.* Hence I first discuss the basic aspect of two fundamental approaches to the Lagrangian quantum field theory — operator-formalism approach and path-integral approach — in the general framework.

In the operator formalism, given field equations and equal-time commutation relations, one sets up a *q-number* Cauchy problem for all *full-dimensional commutators* between fundamental fields. Solving it, one obtains an infinite-dimensional (nonlinear) Lie algebra of fields. Then one constructs its matrix representation in terms of Wightman functions (= vacuum expectation values of simple products). It is important to note here that this representation is of the (nonlinear) Lie algebra but *not of its universal enveloping algebra.* Hence, the original field equations are not necessarily respected at the representation level. If such a situation happens, the violation is called *"field-equation anomaly"*.

In the path-integral or Feynman-dia-

*Professor Emeritus of Kyoto University

grammatic approach, operator level and representation level are not discriminated. The results obtained by this approach coincide with those of the operator-formalism approach *if every fundamental field has its own canonical conjugate and if there is no field-equation anomaly.* The solution is given directly in terms of the vacuum expectation values of T*-products (= "covariantized" T-products). I emphasize that *T*-product is different from T-product* (in T*-product, all time differentiations are made *after* taking time ordering of fields). Hence *T*-product does not generally respect field equations*, so that *Noether theorem does not necessarily hold inside T*-product.*

3 Now, I discuss the BRS-formulated conformal-gauge two-dimensional quantum gravity coupled with $D(\neq 26$ in general) massless scalar fields.[1] The gravitational field $g_{\mu\nu}$ is gauge-fixed to $\rho\eta_{\mu\nu}$, where $\rho(x) > 0$. The matter fields are D massless scalar fields ϕ_M $(M = 0, 1, \ldots, D-1)$. Parametrizing $g_{\mu\nu}$ by a traceless symmetric tensor $h^{\mu\nu}$, one has an *action independent of* ρ.

In terms of light-cone variables, fundamental fields are $h_\pm$, $\tilde{b}^\pm$, $c^\pm$, $\bar{c}^\pm$, and ϕ_M. The Lagrangian density is given by

$$\mathcal{L} = \Big[-\frac{1}{2}\tilde{b}^+ h_+ - i\bar{c}^+\partial_- c^+ + (+ \leftrightarrow -)\Big] + \partial_+\phi_M \cdot \partial_-\phi^M + \mathcal{L}_\mathrm{I}, \qquad (1)$$

$$\mathcal{L}_\mathrm{I} = \frac{1}{2}h_+\Big[-2i\bar{c}^+\partial_+ c^+ - i(\partial_+\bar{c}^+ \cdot c^+ + \partial_-\bar{c}^+ \cdot c^-) + \partial_+\phi_M \cdot \partial_+\phi^M\Big] + (+ \leftrightarrow -) + O(h^2), \qquad (2)$$

where $O(h^2)$ denotes higher-order terms of $h_\pm$. Note that *the B-field has no canonical conjugate.*

Field equations are

$$h_\pm = 0, \qquad (3)$$

$$\tilde{b}^\pm = 2\frac{\delta\mathcal{L}_\mathrm{I}}{\delta h_\pm}, \qquad (4)$$

$$\partial_\mp c^\pm = 0, \qquad \partial_\mp \bar{c}^\pm = 0, \qquad (5)$$

and $\partial_+\partial_-\phi_M = 0$. Only Eq. (4) is nonlinear. From it, one has $\partial_\mp\tilde{b}^\pm = 0$; therefore any of $\tilde{b}^\pm$, $c^\pm$, $\bar{c}^\pm$ and $\partial_\pm\phi_M$ is a function of $x^\pm$ *alone.* I emphasize that *this remarkable simplicity is valid only in operator formalism, but not in path-integral formalism* because of the use of T*-product.

The exact solution in terms of Wightman functions is constructed explicitly. Nonvanishing truncated Wightman functions are n-point functions consisting of *either* ϕ_M, ϕ_N and $n-2$ B-fields *or* $c^\pm$, $\bar{c}^\pm$ and $n-2$ B-fields. Thus if B-field is not considered at all, the solution is a free-field one.

The solution is completely consistent with BRS invariance and FP-ghost number conservation. But *field-equation anomaly for Eq. (4) appears* for $D \neq 26$.

4 In the path-integral or Feynman-diagrammatic approach, one has

$$\langle \mathrm{T}^*\tilde{b}^\pm(x_1)h_\pm(x_2)\rangle = -2i\delta^2(x_1 - x_2) \qquad (6)$$

in spite of Eq. (3). Owing to Eq. (6), *one-loop diagrams are nonvanishing* for the Green's functions consisting of B-fields only,[3] though the corresponding Wightman functions vanish. Since the B-field is the BRS daughter of FP-antighosts, this fact implies *the appearance of BRS anomaly.* Furthermore, since T*-product violates field equations, many connected Green's functions consisting of *both* + components *and* − components remain nonvanishing in contrast to truncated Wightman functions. Hence *the decoupling of right mover and left mover is no longer valid.* Finally, owing to Eq. (6), an infinite number of higher-order terms in $\mathcal{L}_\mathrm{I}$ do contribute. Thus *the Feynman-diagrammatic results are very crazy.*

5 The BRS Noether current is given by

$$j_b{}^\mp = j_b'{}^\mp + \left(\tilde{b}^\pm - 2\frac{\delta\mathcal{L}_\mathrm{I}}{\delta h_\pm}\right)c^\pm \qquad (7)$$

with

$$j_b'{}^\mp \equiv -\tilde{b}^\pm c^\pm + i\bar{c}^\pm c^\pm \partial_\pm c^\pm. \qquad (8)$$

Kato and Ogawa[5] defined the BRS charge by using Eq. (7) (Note that Eq. (7) does not involve the B-field explicitly) and found its nilpotency violation for $D \neq 26$. However, the second term of Eq. (7) suffers from the field-equation anomaly. If one defines the BRS charge by using Eq. (8), *the nilpotency is not violated for any D*. Thus *the validity of Kato-Ogawa's claim depends on the way of defining the BRS charge.* This conclusion remains valid even if the finiteness effect of string length is taken into account.[2]

6 The FP-ghost number current is given by

$$j_c{}^{\mp} = -i\bar{c}^{\pm}c^{\pm}. \tag{9}$$

Its conservation is a simple consequence of Eq. (5). Nevertheless, its first-order perturbation term is *not* conserved. Explicitly, one-loop calculation yields

$$\left\langle \mathrm{T}^* j_c{}^{\mp}(x_1)\, 2\frac{\delta\mathcal{L}_\mathrm{I}}{\delta h_\pm(x_2)} \right\rangle = -12\partial_\pm D_\mathrm{F}(x_1-x_2)\cdot\partial_\pm{}^2 D_\mathrm{F}(x_1-x_2). \tag{10}$$

This result has been regarded as the appearance of FP-ghost number current anomaly.[6] The current "non-conservation" is, however, merely due to the use of T*-pruduct. Indeed, without T*, D_F is replaced by $D^{(+)}$ in the right-hand side of Eq. (10), and since $\partial_\pm D^{(+)}$ is a function of $x^\pm$ *alone*, the conservation law of $j_c{}^{\mp}$ is seen to be perfectly all right. Thus *the FP-ghost number current anomaly is nothing but an illusion caused by T*-product.*

7 In conclusion, my claims are as follows:

(a) Owing to the use of T*-product, Feynman-diagrammatic approach yields *very crazy results* in the BRS-formulated conformal-gauge two-dimensional quantum gravity.

(b) The nilpotency violation of BRS charge for $D \neq 26$, claimed by Kato and Ogawa, is *not an intrinsic result.* One can construct another BRS charge *nilpotent for any D*. What is intrinsic is the existence of the field-equation anomaly for the B-field equation.

(c) The FP-ghost number current is conserved perfectly; the "anomaly", whose existence is unreasonably believed to be required by the Riemann-Roch theorem, is an *illusion caused by T*-product.* Similar misleading discussions are found concerning Virasoro anomaly and gravitational anomaly in the literature.[7]

References

1. M. Abe and N. Nakanishi, *Int. J. Mod. Phys.* **A 14**, 521 (1999).
2. M. Abe and N. Nakanishi, *Int. J. Mod. Phys.* **A 14**, 1357 (1999).
3. M. Abe and N. Nakanishi, *Prog. Theor. Phys.* **102**, 1187 (1999). Further references are contained therein.
4. K. Fujikawa, *Phys. Rev.* **D 25**, 2584 (1982).
5. M. Kato and K. Ogawa, *Nucl. Phys.* **B 212**, 443 (1983).
6. U. Kraemmer and A. Rebhan, *Nucl. Phys.* **B 315**, 717 (1989).
7. M. B. Green, J. H. Schwarz and E. Witten, *Superstring theory. 1*, pp.141–142 and pp.145–146 (Cambridge University Press, 1987).

CHAMSEDDINE, FRÖHLICH, GRANDJEAN METRIC AND LOCALIZED HIGGS COUPLING ON NONCOMMUTATIVE GEOMETRY

R.KURIKI

Laboratory for Particle and Nuclear Physics,
High Energy Accelerator Research Organization (KEK),
Tsukuba, Ibaraki 305-0801, Japan

A.SUGAMOTO

Department of Physics,
Ochanomizu University, Tokyo 112-0012, Japan

1 Historical introduction

A.Connes and J.Lott write the standard model in terms of the noncommutative geometry [2]. They reconstruct the model on a discrete space, $M_4 \times Z_2$. As the result, they find that

1. The Higgs boson is a gauge field corresponding to a finite difference between two M_4,
2. The distance between two M_4 is proportional to the inverse of the fermion mass.

From these results the following question naturally comes. *Can we define a gravitational theory in the language of the noncommutative geometry ?*

Chamseddine, Fröhlich, Grandjean introduce the gravitational sector into the Connes-Lott formulation of the standard model [3]. In $M_4 \times Z_2$, they introduce a Dirac operator in which the non-diagonal part contains local functions of x which is the local coordinate of M_4 and the differential operator of the diagonal part becomes the covariant derivative of M_4. Then they write vielbein in six dimensional space (M_4 plus $SU(2)$ doublet isospace). The vielbein is represented in the Hilbert space of the leptonic sector. The set of vielbein is equal to the set of generators of one forms, $\Omega(\mathcal{A})^1{}_D$. In their formulation the $SU(2)$ doublet isospace is on the same stage with M_4 and they develop the noncommutative differential geometry on the six dimensional space. Then they impose the unitarity condition and the torsion-less condition on the geometry and finally find a gravitational action. The form of the gravitational action is similar to the Liouville action of non-critical superstring theory.

2 Our motivation and the idea

In the future we would like to express the huge mass difference among families of quarks (leptons) based on the noncommutative geometry by following [3]. As the first step, here we only treat one-family of quarks and consider the mass difference between u and d. We want to propose a formulation in which the mass difference can be represented as a geometrical information of an extended space, M_4 plus the flavor space. Remember that the $SU(2)$ isospace is on the same stage with M_4 in [3].

Our idea comes from the above Connes-Lott's results, namely the geometrical interpretation of the Higgs field and the relation between the distance and the fermion mass. We elevate Higgs (Yukawa) couplings to some components of 'vielbein' in the representation space of our algebra. Our Higgs (Yukawa) couplings depend on the local coordinate of M_4.

3 Our results and discussion

We adopt the algebra $\mathcal{A}$ as

$$\mathcal{A} = (\mathcal{A}_1 \oplus \mathcal{A}_2 \oplus \mathcal{A}_3) \otimes C^\infty(M_4), \qquad (1)$$

where M_4 is a smooth, compact, four-dimensional Riemannian spin manifold; $\mathcal{A}_3$ is the algebra of complex 2×2 matrices. $\mathcal{A}_1$ and $\mathcal{A}_2$ are C. The representation of $\mathcal{A}$ is selected as $a = \begin{pmatrix} a_1 & 0 & 0 \\ 0 & a_2 & 0 \\ 0 & 0 & a_3 \end{pmatrix}$, where a_i is a C^∞-function on M_4 with values in $\mathcal{A}_i$, $i=1,2,3$.

We assume our Hilbert space as $L = \begin{pmatrix} u_R \\ d_R \\ u_L \\ d_L \end{pmatrix}$. We want to regard u_R, d_R as an $SU(2)$ singlet and u_L, d_L as an $SU(2)$ doublet.

In the Hilber space the set of our vielbein is represented as

$$\mathcal{E}^a = \gamma^a \begin{pmatrix} 1&0&0&0 \\ 0&1&0&0 \\ 0&0&1&0 \\ 0&0&0&1 \end{pmatrix}, \qquad (2)$$

$$\mathcal{E}^{\bar r} = \gamma^5 \begin{pmatrix} \mathbf{0} & \bar{\mathcal{A}} e_{\bar r} \\ -e_{\bar r}^{\top} \bar{\mathcal{A}} & \mathbf{0} \end{pmatrix}, \qquad (3)$$

where

$$\bar{\mathcal{A}} = \begin{pmatrix} 0 & \frac{v}{\sqrt{2}} \\ \frac{v}{\sqrt{2}} & 0 \end{pmatrix}, \qquad (4)$$

$$e_u = e_{\bar u} = \begin{pmatrix} 0&0 \\ 1&0 \end{pmatrix}, \quad e_d = e_{\bar d} = \begin{pmatrix} 0&1 \\ 0&0 \end{pmatrix} \qquad (5)$$

$a = 1, \ldots, 4$ and $\bar r = \bar u, \bar d$. (3.25) and (3.26) in [3] have inspired us (2) and (3). v is a constant. We treat (2) and (3) as generators of one-form for our algebra though these may contain auxiliary parts [a].

[a] As the authors of [3] have described, in order to consistently define the noncommutative differential geometry one must exclude auxiliary parts from general functions in Dirac operator. We now ignore it since we want to only present our formulation.

Let us introduce a local basis $e^a_\mu(x)$, $\mu = 1, \ldots, 4$ of orthonormal tangent vectors to M_4 for each a. To depict a 'curved' discrete space, we introduce gamma matrices by $\gamma^\mu \equiv e^\mu_a \gamma^a$ and define the set of generators for the 'curved discrete space' as

$$\mathcal{E}^\mu = \gamma^\mu \begin{pmatrix} 1&0&0&0 \\ 0&1&0&0 \\ 0&0&1&0 \\ 0&0&0&1 \end{pmatrix}, \qquad (6)$$

$$\mathcal{E}^r = \gamma^5 \begin{pmatrix} \mathbf{0} & \mathcal{A}^* e_r \\ -e_r^* \mathcal{A} & \mathbf{0} \end{pmatrix}, \qquad (7)$$

where

$$\mathcal{A} = \begin{pmatrix} 0 & f(x)\frac{v}{\sqrt{2}} \\ \tilde f(x)\frac{v}{\sqrt{2}} & 0 \end{pmatrix}, \qquad (8)$$

where $\mathcal{A}^*$ is the complex conjugation of $\mathcal{A}$. $f(x)$,$\tilde f(x)$ are localized Higgs (Yukawa) couplings. We express the dual basis, following (3.29) in [3],

$$\omega = \begin{pmatrix} \gamma^\mu \omega_{1\mu,ij}(x) & & \omega'_2(x) & 0 \\ & & 0 & \omega_2(x) \\ \omega'_1(x) & 0 & \gamma^\mu\omega_{2\mu,ij}(x), & \\ 0 & \omega_1(x) & & \end{pmatrix},$$

where $\omega_{1\mu,ij}(x)$ and $\omega_{2\mu,ij}(x)$ are two-times-two matrices, $i,j = 1,2$.

Following (3.28) and (3.29) in [3], we write down the connection coefficients of (6) and (7) as

$$\Omega^A{}_B = \begin{pmatrix} \gamma^\mu \omega_{1\mu}{}^A{}_{B,ij} & \mathcal{A}\gamma^5 e^{-\sigma}\Omega_2 \\ \Omega_1 \gamma^5 e^{-\sigma} \mathcal{A}^* & \gamma^\mu \omega_{2\mu}{}^A{}_{B,ij} \end{pmatrix}, \qquad (9)$$

where $A, B, C = 1,2,3,4,5,6$ and $5,6$ respectively, refers to u, d. Ω_1,Ω_2 are expressed by

$$\Omega_1 = \begin{pmatrix} \omega'_1{}^A{}_B & 0 \\ 0 & \omega_1{}^A{}_B \end{pmatrix},$$

$$\Omega_2 = \begin{pmatrix} \omega'_2{}^A{}_B & 0 \\ 0 & \omega_2{}^A{}_B \end{pmatrix}.$$

We define the Dirac operator on the above curved space as

$$\hat{D} = \begin{pmatrix} i\nabla_R & 0 & \gamma^5 \mathbf{M}^* \\ 0 & i\nabla_R & \\ & & \\ \gamma^5 \mathbf{M} & & i\nabla_L \end{pmatrix}, \quad (10)$$

$$\mathbf{M}^* = \begin{pmatrix} f^*(x)(\ \phi^{0*}\ ,\ \phi^+\ ) \\ \tilde{f}^*(x)(\ -\phi^-\ ,\ \phi^0\ ) \end{pmatrix},$$

$$\mathbf{M} = \left(f(x) \begin{pmatrix} \phi^0 \\ \phi^- \end{pmatrix} \tilde{f}(x) \begin{pmatrix} -\phi^+ \\ \phi^{0*} \end{pmatrix} \right),$$

where we regard ϕ^0 as the neutral Higgs-like field and $\phi^{+*} = \phi^-$.

$$i\nabla_R = \gamma^\mu (i\partial_\mu - \frac{g'}{2} Y B_\mu),$$

$$i\nabla_L = \gamma^\mu (i\partial_\mu - \frac{g}{2}\tau_a A_{a\mu} - \frac{g'}{2} Y B_\mu),$$

where $A_{a\mu}(x)$ is the $SU(2)_W$ gauge field and $B_\mu(x)$ is the $U(1)_Y$ gauge field. τ_a are the Pauli matrices.

We impose the unitarity condition and the torsion-less condition on the above gravitational components by following [3]. Finally solving some components of the above two conditions, we find that

$$|f(x)|^2 \propto \exp \int^x dy^\mu (2\omega_{1\nu}{}^u{}_{u,\,11}$$
$$-\text{combination of gauge fields})(y)\ . \quad (11)$$

Here the localized Higgs (Yukawa) coupling $f(x)$ is equal to the integration of the spin connection $\omega_{1\nu}{}^u{}_{u,\,11}$ and gauge fields. What we want to stress is that the spin connection has been introduced in the flavor isospace of u [b]. This is that we want to obtain in our formulation. We find that *localized Higgs couplings (Yukawa) are represented by the geometrical object of the flavor isospace.* By the virtue of noncommutative geometry we can logically combine geometries of additional space, the flavor isospace, with things which are in four dimensional manifold.

Moreover in the virtue of Eq.(11) we can *geometrically* present the conditions that the Higgs couplings become zero. The choices of the topological properties for the four dimensional continuum M_4 are not free, but are restricted by the other conditions,

$$|\tilde{f}|^2 = -\frac{2}{v^2}\frac{\omega_{1\mu}{}^{ua}{}_{,11}}{\omega_{1\mu}{}^a{}_{u,11}} = -\frac{2}{v^2}\frac{\omega_{1\mu}{}^{ua}{}_{,\,12}}{\omega_{1\mu}{}^a{}_{u,21}} = \dots$$

Notice that these restrictions contain the connections which spread over the whole space of $M_4 \times Z_2$.

To solve the spin connection from the Einstein equation of our gravitational action, we have calculated the candidate terms. We have checked that

- There are no differential terms of the localized Higgs couplings in $\sqrt{g}R$,
- There are $\partial f(x)$ and $\partial^2 f(x)$ in $\sqrt{g}R^2$.

There are many future works to be expected. To increase the number of families is not difficult in our formulation. However to control the mixing among families in term of noncommutative geometry may not be in the direct extension of our work.

Acknowledgments

This talk is given by R.K. based on hep-th/0004127. She is supported by the Research Fellowships of the Japan Society for the Promotion of Science.

References

1. R.Kuriki and A.Sugamoto,preprint hep-th/0004127 v3, *Chamseddine, Fröhlich, Grandjean Metric and Localized Higgs Coupling on Noncommutative Geometry*
2. A.Connes,J.Lott, *Nucl.Phys.***B(Proc.Suppl.)**,29-47(1990), *PARTICLE MODELS AND NONCOMMUTATIVE GEOMETRY*
3. A.H.Chamseddine, J.Fröhlich, O.Grandjean, *J.Math.Phys.***36**, 6255-6275(1995), *The Gravitational Sector in the Connes-Lott Formulation of the Standard Model*

[b] We have other equations like Eq.(11). These depend on $\omega_{2\nu}{}^u{}_{u,\,11}$, $\omega_{1\nu}{}^d{}_{d,\,22}$, $\omega_{2\mu}{}^d{}_{d,\,22}$.

MEASURE FIELDS, THE COSMOLOGICAL CONSTANT AND SCALE INVARIANCE

E.I.GUENDELMAN

Physics Department, Ben Gurion University of the Negev, Beer Sheva 84105, Israel
E-mail: guendel@bgumail.bgu.ac.il

The consequences of considering the measure of integration in the action to be defined by degrees of freedom independent of the metric are studied. Models without the cosmological constant problem, new ways of spontaneously breaking scale symmetry which have an interesting cosmology and theories of extended objects (string, branes) without a fundamental scale appear possible.

When formulating generally covariant theories, the form for the action $S = \int d^D x\sqrt{-g}L$ is considered. Here L is a scalar and $g = det(g_{\mu\nu})$, $g_{\mu\nu}$ being the metric defined on space-time manifold of interest. In this case the metric determines the measure of integration to be $d^D x\sqrt{-g}$. The fact is however that one can still maintain general coordinate invariance and replace $\sqrt{-g}$ by another density Φ, which depend on degrees of freedom independent of the metric, obtaining then the modified measure $d^D x\Phi$. If Φ is also a total divergence, then the action $S = \int d^D x\Phi L$ is invariant up to the integral of a total divergence under the shift $L \rightarrow L + constant$.

A particular realization for Φ is obtained by considering D scalar fields φ_j and defining $\Phi = \varepsilon^{\mu_1 \ldots \mu_D}\varepsilon_{j_1 \ldots j_D}\partial_{\mu_1}\varphi_{j_1} \ldots \partial_{\mu_D}\varphi_{j_D}$.

Realistic gravitational theories without the cosmological constant problem can be constructed using actions of the form $S = \int d^D x\Phi L_1 + \int d^D x\sqrt{-g}L_2$, where L_1 and L_2 are both independent of the measure fields [1] φ_j. There is a good reason not to consider mixing of Φ and $\sqrt{-g}$, like for example using $\frac{\Phi^2}{\sqrt{-g}}$. This is because S is invariant (up to the integral of a divergence) under the infinite dimensional symmetry $\varphi_a \rightarrow \varphi_a + f_a(L_1)$ where $f_a(L_1)$ is an arbitrary function of L_1 if L_1 and L_2 are φ_a independent. Such symmetry is absent if mixed terms are present. Models with two measures could be related to brane scenarios (one measure associated to the brane, the other to the bulk).

We will study now the dynamics of a scalar field ϕ interacting with gravity as given by the choice of lagrangians (for $D = 4$) [2]

$L_1 = \frac{-1}{\kappa}R(\Gamma, g) + \frac{1}{2}g^{\mu\nu}\partial_\mu\phi\partial_\nu\phi - V(\phi), L_2 = U(\phi)$ where $R(\Gamma, g) = g^{\mu\nu}R_{\mu\nu}(\Gamma), R_{\mu\nu}(\Gamma) = R^\lambda_{\mu\nu\lambda}, R^\lambda_{\mu\nu\sigma}(\Gamma) = \Gamma^\lambda_{\mu\nu,\sigma} - \Gamma^\lambda_{\mu\sigma,\nu} + \Gamma^\lambda_{\alpha\sigma}\Gamma^\alpha_{\mu\nu} - \Gamma^\lambda_{\alpha\nu}\Gamma^\alpha_{\mu\sigma}$. In the variational principle $\Gamma^\lambda_{\mu\nu}, g_{\mu\nu}$, the measure fields scalars φ_a and the scalar field ϕ are all to be treated as independent variables.

If we perform the global scale transformation ($\theta =$ constant) $g_{\mu\nu} \rightarrow e^\theta g_{\mu\nu}$ then S is invariant provided $V(\phi)$ and $U(\phi)$ are of the form $V(\phi) = f_1 e^{\alpha\phi}, U(\phi) = f_2 e^{2\alpha\phi}$ and φ_a is transformed according to $\varphi_a \rightarrow \lambda_a\varphi_a$ (no sum on a) which means $\Phi \rightarrow \left(\prod_a \lambda_a\right)\Phi \equiv \lambda\Phi$ such that $\lambda = e^\theta$ and $\phi \rightarrow \phi - \frac{\theta}{\alpha}$.

Let us consider the equations which are obtained from the variation of the φ_a fields. We obtain then $A^\mu_a\partial_\mu L_1 = 0$ where $A^\mu_a = \varepsilon^{\mu\nu\alpha\beta}\varepsilon_{abcd}\partial_\nu\varphi_b\partial_\alpha\varphi_c\partial_\beta\varphi_d$. Since det $(A^\mu_a) = \frac{4^{-4}}{4!}\Phi^3$, if $\Phi \neq 0$ we obtain that $\partial_\mu L_1 = 0$, or that $L_1 = M$, where M is constant. This constant M appears in a self-consistency condition of the equations of motion [2] that allows us to solve for $\chi \equiv \frac{\Phi}{\sqrt{-g}} = \frac{2U(\phi)}{M+V(\phi)}$. M produces ssb of scale invariance.

To get the physical content of the theory, it is convenient to go to the Einstein conformal frame where $\overline{g}_{\mu\nu} = \chi g_{\mu\nu}$. In terms of $\overline{g}_{\mu\nu}$ the non Riemannian contribution (defined as $\Sigma^\lambda_{\mu\nu} = \Gamma^\lambda_{\mu\nu} - \{^{\lambda}_{\mu\nu}\}$ where $\{^{\lambda}_{\mu\nu}\}$ is the Christof-

fel symbol), disappears from the equations, which can be written then in the Einstein form ($R_{\mu\nu}(\overline{g}_{\alpha\beta})$ = usual Ricci tensor)
$R_{\mu\nu}(\overline{g}_{\alpha\beta}) - \frac{1}{2}\overline{g}_{\mu\nu}R(\overline{g}_{\alpha\beta}) = \frac{\kappa}{2}T^{eff}_{\mu\nu}(\phi)$,
$T^{eff}_{\mu\nu}(\phi) = \phi_{,\mu}\phi_{,\nu} - \frac{1}{2}\overline{g}_{\mu\nu}\phi_{,\alpha}\phi_{,\beta}\overline{g}^{\alpha\beta} + \overline{g}_{\mu\nu}V_{eff}(\phi)$ and $V_{eff}(\phi) = \frac{1}{4U(\phi)}(V+M)^2$. Notice that for generic smooth functions V and U, if $V+M=0$ at some point, then generically $V_{eff} = V'_{eff} = 0$ at such point, that is, the ground state has zero vacuum energy without fine tuning!. No assumption of scale invariance in involved in this result.

If $V(\phi) = f_1 e^{\alpha\phi}$ and $U(\phi) = f_2 e^{2\alpha\phi}$ as required by scale invariance, we obtain

$$V_{eff} = \frac{1}{4f_2}(f_1 + Me^{-\alpha\phi})^2 \qquad (1)$$

We see that as $\alpha\phi \to \infty$, $V_{eff} \to \frac{f_1^2}{4f_2} = const.$ providing an infinite flat region, so we expect a slow rolling inflationary scenario to be viable. Also a minimum is achieved at zero cosmological constant for the case $\frac{f_1}{M} < 0$ at the point $\phi_{min} = \frac{-1}{\alpha} ln \mid \frac{f_1}{M} \mid$. Finally, the second derivative of the potential V_{eff} at the minimum is $V''_{eff} = \frac{\alpha^2}{2f_2} \mid f_1 \mid^2 > 0$ if $f_2 > 0$. That is, a realistic scalar field potential, with massive excitations when considering the true vacuum state, is achieved in a way consistent with scale invariance.

Furthermore, one can consider this model as suitable for the present day universe rather than for the early universe, after we suitably reinterpret the meaning of the scalar field ϕ. This can provide a long lived almost constant vacuum energy for a long period of time, which can be small if $f_1^2/4f_2$ is small. Such small energy density will eventually disappear when the universe achieves its true vacuum state. Notice that a small value of $\frac{f_1^2}{f_2}$ can be achieved if we let $f_2 >> f_1$. In this case $\frac{f_1^2}{f_2} << f_1$, i.e. a very small scale for the energy density of the universe is obtained by the existence of a very high scale (that of f_2) the same way as a small neutrino mass is obtained in the see-saw mechanism from the existence also of a large mass scale. In what follows, we will take $f_2 >> f_1$.

We can also include a fermion ψ, where the kinetic term of the fermion is chosen to be part of L_1, $S_{fk} = \int L_{fk}\Phi d^4x$ and $L_{fk} = \frac{i}{2}\overline{\psi}[\gamma^a V^\mu_a(\overrightarrow{\partial}_\mu + \frac{1}{2}\omega^{cd}_\mu\sigma_{cd}) - (\overleftarrow{\partial}_\mu + \frac{1}{2}\omega^{cd}_\mu\sigma_{cd})\gamma^a V^\mu_a]\psi$ there V^μ_a is the vierbein, $\sigma_{cd} = \frac{1}{2}[\gamma_c, \gamma_d]$. For self-consistency, the curvature scalar is taken to be (if we want to deal with the spin connection ω^{ab}_μ instead of $\Gamma^\lambda_{\mu\nu}$ everywhere) $R = V^{a\mu}V^{b\nu}R_{\mu\nu ab}(\omega)$, $R_{\mu\nu ab}(\omega) = \partial_\mu\omega_{\nu ab} - \partial_\nu\omega_{\mu ab} + (\omega^c_{\mu a}\omega_{\nu cb} - \omega^c_{\nu a}\omega_{\mu cb})$.

Global scale invariance is obtained provided ψ also transforms, as in $\psi \to \lambda^{-\frac{1}{4}}\psi$. Mass terms consistent with scale invariance exist: $m_1 \int \overline{\psi}\psi e^{\alpha\phi/2}\Phi d^4x + m_2 \int \overline{\psi}\psi e^{3\alpha\phi/2}\sqrt{-g}d^4x$.

If we consider the situation where $m_1 e^{\alpha\phi/2}\overline{\psi}\psi$ or $m_2 e^{3\alpha\phi/2}\overline{\psi}\psi$ are much bigger than $V(\phi) + M$, i.e. a high density approximation, we obtain that the consistency condition is $\chi = -\frac{3m_2}{m_1}e^{\alpha\phi}$. Using this, we obtain, after going to the Conformal Einstein Frame (CEF), which involves the transformations to the scale invariant fields $\overline{V}^a_\mu = \chi^{\frac{1}{2}}V^a_\mu$ and $\psi' = \chi^{-\frac{1}{4}}\psi$ and they lead to a mass term,

$$-2m_2(\frac{|m_1|}{3|m_2|})^{3/2}\int \sqrt{-\overline{g}}\overline{\psi}'\psi' d^4x \qquad (2)$$

The ϕ dependence of the mass term has disappeared, i.e. masses are constants. Low density of matter can also give results which are similar to those obtained in the high density approximation, in that the coupling of the ϕ field disappears and that the mass term becomes of a conventional form in the CEF. In this is the case, we study the limit $\alpha\phi \to \infty$. Then $U(\phi) \to \infty$ and $V(\phi) \to \infty$. In this case, taking $m_1 e^{\alpha\phi/2}\overline{\psi}\psi$ and $m_2 e^{3\alpha\phi/2}\overline{\psi}\psi$ much smaller than $V(\phi)$ or $U(\phi)$ respectively, we get $\chi = \frac{2f_2}{f_1}e^{\alpha\phi}$. If this is used, we obtain the mass term $m \int \sqrt{-\overline{g}}\overline{\psi}'\psi' d^4x$, where

$$m = m_1(\frac{f_1}{2f_2})^{\frac{1}{2}} + m_2(\frac{f_1}{2f_2})^{\frac{3}{2}} \qquad (3)$$

The mass term is independent of ϕ again.

So the interaction of ϕ with matter dissapears in the CEF due to scale invariance. This is an explicit realization of the symmetry which Carroll was looking for avoiding interactions of the 'quintessential scalar' and the rest of the world [3] and which would apply if we consider the universe being totally in the the region $\alpha\phi \to \infty$. The effects of terms which explicitly (rather than spontaneously) break scale invariance have also been studied [4].

Other scenarios can be explored: taking m_1 and m_2 of the same order of magnitude, we see that the mass of the Dirac particle is much smaller in the region $\alpha\phi \to \infty$, for which (3) is valid, than it is in the region of high density of the Dirac particle relative to $V(\phi)+M$, as displayed in eq. (2), if the "see-saw" assumption $\frac{f_1}{f_2} << 1$ is made. Therefore if space is populated by diluted Dirac particles of this type, the mass of these particles will grow substantially if we go to the true vacuum valid in the absence of matter, i.e. $V + M = 0$. In the region $\alpha\phi \to \infty$, we minimize the matter energy, but maximize the potential energy V_{eff} and at $V + M = 0$, we minimize V_{eff}, and particle masses are big. The true vacuum state must be in a balanced intermediate stage. Clearly how much above $V + M = 0$ such true vacuum is located must be correlated to how much particle density is there in the Universe. A non zero vacuum energy, which must be of the same order of the particle energy density, has to appear and this could explain the "accelerated universe" that appears to be implied by the most recent observations, together with the "cosmic coincidence", that requires the vacuum energy be comparable to the matter energy.

One can also develop a modified measure theory for the case of extended objects [5] including super symmetric strings and branes [6]. In the case of strings, we can replace in the Polyakov action, the measure $\sqrt{-\gamma}d^2x$ (where γ_{ab} is the metric defined on the world sheet, $\gamma = det(\gamma_{ab})$ and a, b indices for the world sheet coordinates) by Φd^2x, where $\Phi = \varepsilon^{ab}\varepsilon_{ij}\partial_a\varphi_i\partial_b\varphi_j$. Then for the bosonic string , we consider the action

$$S = -\int d\tau d\sigma \Phi[\gamma^{ab}\partial_a X^\mu \partial_b X^\nu g_{\mu\nu} - \frac{\varepsilon^{ab}}{\sqrt{-\gamma}}F_{ab}] \tag{4}$$

where $F_{ab} = \partial_a A_b - \partial_b A_a$ and A_a is a gauge field defined in the world sheet of the string. The term with the gauge fields is irrelevant if the ordinary measure of integration is used, since in that case it would be a divergence, but is needed for a consistent dynamics in the modified measure reformulation of string theory. This is due to the fact that if we avoid such a contribution to the action, the variation of the action with respect to γ^{ab} leads to the vanishing of the induced metric on the string. The equation of motion obtained from the variation of the gauge field A_a is $\varepsilon^{ab}\partial_a(\frac{\Phi}{\sqrt{-\gamma}}) = 0$. From which we obtain that $\Phi = c\sqrt{-\gamma}$ where c is a constant which can be seen is the string tension. The string tension appears then as an integration constant and does not have to be introduced from the beginning. The string theory Lagrangian in the modified measure formalism does not have any fundamental scale associated with it. Extensions to both the super symmetric case [6] and to higher branes [5, 6] are possible.

References

1. For a review and further refs. see E.I.Guendelman and A.B.Kaganovich, Phys. Rev. D60, 065004 (1999).
2. E.I.Guendelman, Mod. Phys. Lett. A14, 1043; Mod. Phys. Lett. A14, 1397; Class.Quantum Grav. 17, 361 (2000); gr-qc/9901067; gr-qc/0004011.
3. S.M.Carroll, Phys. Rev. Lett. 81, 3067 (1998).
4. A.B.Kaganovich, hep-th/0007144.
5. E.I.Guendelman, hep-th/0005041 (to appear in Class. Quantum Grav.).
6. E.I.Guendelman, hep-th/0006079.

POWER CORRECTIONS IN EIKONAL CROSS SECTIONS

ERIC LAENEN
NIKHEF Theory Group, Kruislaan 409
1098 SJ Amsterdam, The Netherlands

GEORGE STERMAN
C.N. Yang Institute for Theoretical Physics, SUNY Stony Brook
Stony Brook, New York 11794 - 3840, U.S.A.

WERNER VOGELSANG
RIKEN-BNL Research Center, Brookhaven National Laboratory,
Upton, NY 11973, U.S.A.

We explore power corrections in eikonal approximations to hadronic cross sections.

1 Introduction

Power corrections [1] are important phenomenologically in a large class of QCD hard-scattering cross sections for which the operator product expansion is not directly available. Examples that have received considerable attention include event shapes in electron-positron annihilation and transverse momentum distributions in Drell-Yan cross sections. In each of these cases, a purely perturbative description of the cross section leads to integrals of the form $I_p \equiv Q^{-p} \int_0^Q d\mu\, \mu^{p-1}\, \alpha_s(\mu)$ with Q the hard scale and $p \geq 1$. In perturbation theory with a fixed coupling, I_p is just a number, but when the coupling runs, the integral becomes ill-defined at its lower limit. This observation requires us to introduce a "minimal" set of power corrections of the form λ_p/Q^p, one for each ambiguous I_p that we encounter. The perturbative expression is cut off, or otherwise regularized to make it finite without changing the set of exponents p. The values of the coefficients λ_p are then to be found by comparison with experiment; they will depend on the nature of perturbative regularization that is employed. In any case, it is only the sum of regularized perturbation theory and power corrections that has physical meaning.

The first step in this process is to show that in some self-consistent approximation the cross section at hand may be written in terms of integrals like the I_p above. In many cases, this step involves the resummation of logarithms associated with soft gluon emission, for which the eikonal approximation is useful. In this talk, we discuss an expression for the eikonal approximation in hadronic collisions where the analysis of power corrections through the running coupling is particularly transparent.

2 The Eikonal Cross Section

To be specific, we discuss the eikonal approximation as it appears when partons a and b combine through an electroweak current, such as the annihilation of quark with antiquark or gluon fusion to an electroweak vector or Higgs boson,

$$\sigma_{ab}^{(\mathrm{eik})}(q) = \int d^4x\; \mathrm{e}^{-iq\cdot x} \langle 0|W_{ab}^{\dagger}(0)W_{ab}(x)|0\rangle \tag{1}$$

where the operators W_{ab} are defined by

$$W_{ab}(0) \equiv \Phi^{\dagger}_{\beta'}(0)\Phi_{\beta}(0) \qquad (2)$$

in terms of nonabelian phase operators for a and b, $\Phi_\beta(0) = P\ \mathrm{e}^{-ig\int_0^\infty d\lambda\beta\cdot A(\lambda\beta)}$, with lightlike velocities β and β', $\beta\cdot\beta'=0$.

The eikonal cross sections reproduce the logarithms, as singular as $(Q/q_0)\alpha_s^n \ln^{2n-1}(q_0/Q)$ and $(Q/q_T)\alpha_s^n \ln^{2n-1}(q_T/Q)$, that characterize the edges of partonic phase space at which the energy of radiation, q_0, or its total transverse momentum, q_T, vanish. The resummation of these logarithms is most convenient in terms of transforms,

$$\tilde{\sigma}_{ab}^{(\mathrm{eik})}(N,\mathbf{b}) = \int d^4q\ \mathrm{e}^{-Nq_0 - i\mathbf{b}\cdot\mathbf{q}}\,\sigma_{ab}^{(\mathrm{eik})}(q)\,. \qquad (3)$$

In the transformed functions we find logarithms at each order up to $\alpha_s^n \ln^{2n} N$ and $\alpha_s^n \ln^{2n}(bQ)$, which exponentiate. The exponentiation of energy logarithms is known as threshold resummation [2], of transverse momentum logarithms as k_T resummation [3].

3 Exponentiation

Transforms of the eikonal cross section may be written in exponential form on the basis of algebraic considerations that have been known for a long time,

$$\tilde{\sigma}_{ab}^{(\mathrm{eik})}(N,\mathbf{b}) = \exp\left[E_{ab}^{(\mathrm{eik})}(N,\mathbf{b}Q,\epsilon)\right]\,, \qquad (4)$$

where the exponent is an integral over functions w_{ab}, sometimes called "webs" [4], which may be defined by a modified set of diagrammatic rules,

$$\begin{aligned} E_{ab}^{(\mathrm{eik})} &= 2\int^Q \frac{d^{4-2\epsilon}k}{\Omega_{1-2\epsilon}} \\ &\times w_{ab}\left(k^2, \frac{k\cdot\beta k\cdot\beta'}{\beta\cdot\beta'}, \mu^2, \alpha_s(\mu^2), \epsilon\right) \\ &\times \left(\mathrm{e}^{-N(k_0/Q)+i\mathbf{k}\cdot\mathbf{b}} - 1\right)\,. \qquad (5) \end{aligned}$$

The variable k in this expression may be thought of as the momentum contributed by the web to the final state. The webs factor from each other under the transforms, and indeed along any symmetric integral over phase space.

Webs have a number of restrictive properties. At fixed k, they are invariant under rescalings of the velocities in the eikonal phases, which corresponds to boost invariance under the axis defined by the two. In addition, at *any* fixed order, the web function has *only one* overall collinear and IR divergence, from $k_T \to 0$ and $k_0 \to 0$, respectively. Finally, the web functions have no overall renormalization:

$$\mu\frac{d}{d\mu}\,w_{ab}\left(k^2, \frac{k\cdot\beta k\cdot\beta'}{\beta\cdot\beta'}, \mu^2, \alpha_s(\mu^2), \epsilon\right) = 0\,. \qquad (6)$$

Using boost invariance in the large-N limit, we find that the exponent takes the form

$$\begin{aligned} E_{ab}^{(\mathrm{eik})} &= 2\int \frac{d^{2-2\epsilon}k_T}{\Omega_{1-2\epsilon}} \qquad (7) \\ &\times \int_0^{Q^2-k_T^2} dk^2\ w_{ab}\left(k^2, k_T^2 + k^2\right) \\ &\times \left[\mathrm{e}^{-i\mathbf{b}\cdot\mathbf{k}_T}\, K_0\left(2N\sqrt{\frac{k_T^2+k^2}{Q^2}}\right)\right. \\ &\left. - \ln\sqrt{\frac{Q^2}{k_T^2+k^2}}\right] + \mathcal{O}\left(\mathrm{e}^{-N}\right)\,. \end{aligned}$$

This form, which requires dimensional regularization for its collinear divergences, is completely general for the eikonal cross section.

4 Factorization

The factorized eikonal cross section $(\hat{\sigma}_{ab}^{(\mathrm{eik})})$ may be constructed to $\mathcal{O}(1/N)$ by dividing by moments of eikonal distributions, which may be defined as

$$\begin{aligned} &\tilde{\phi}_f^{(\mathrm{eik})}(N,\mu,\epsilon) \qquad (8) \\ &= \exp\left[-\ln\left(N\mathrm{e}^{\gamma_E}\right)\int_0^{\mu^2}\frac{d\mu'^2}{\mu'^2}\,A_f\left(\alpha_s(\mu'^2)\right)\right]\,, \end{aligned}$$

where A_a, with $A_a^{(1)} = C_a$, is the coefficient of $\ln N$ in the Nth moment of the $a \to a$ splitting function. Factorization theorems ensure the cancellation of collinear divergences in the resulting hard-scattering functions. Invoking this requirement, we find an explicit relation between the webs and the anomalous dimensions,

$$\int_0^{Q^2-k_T^2} dk^2\, w_{ab}\left(k^2, k_T^2+k^2\right) \tag{9}$$
$$= \frac{A_a\left(\alpha_s(k_T^2)\right) + A_b\left(\alpha_s(k_T^2)\right)}{(k_T^2)^{1-2\epsilon}} + \ldots .$$

In this fashion, we derive a general form for the eikonal approximation to the hard-scattering functions $\hat{\sigma}_{ab}$ of electroweak annihilation,

$$\hat{\sigma}_{ab}^{(\mathrm{eik})}(N,\mathbf{b},Q,\mu) = \frac{\sigma_{ab}^{(\mathrm{eik})}(N,\mathbf{b},Q)}{\tilde{\phi}_a(N,\mu)\,\phi_b(N,\mu)}$$
$$= \exp\left[\hat{E}_{ab}^{(\mathrm{eik})}(N,b,Q)\right], \tag{10}$$

where the collinear-finite exponent (here shown to NLL) is

$$\hat{E}_{ab}^{(\mathrm{eik})} = \int_0^{Q^2} \frac{dk_T^2}{k_T^2} \sum_{i=a,b} A_i\left(\alpha_s(k_T^2)\right) \times$$
$$\left[J_0\left(bk_T\right) K_0\left(\frac{2Nk_T}{Q}\right) + \ln\left(\frac{Ne^{\gamma_E}k_T}{Q}\right)\right]. \tag{11}$$

This result is the basis of the joint threshold-k_T resummation [5] described in Ref. [6].

As described above, the resummation of logarithms as in (11) requires the inclusion of power corrections in both Q^{-1} as well as b, to compensate for the ill-defined behavior of the strong coupling at low scales. In [7] it was shown that only integer powers of Q^{-1} are necessary; in [8] models of the running coupling were invoked to suggest that power corrections begin at order Q^{-2}. Eq. (7) implies that only even powers of Q are present in all generality for the eikonal approximation. This is because, up to a single log, the expansion of the Bessel function $K_0(z)$ at small z involves only even powers of z. Other consequences of this approach have been discussed in Ref. [6].

Acknowledgements

The work of G.S. was supported in part by the National Science Foundation, grant PHY9722101. The work of E.L. is part of the research program of the Foundation for Fundamental Research of Matter (FOM) and the National Organization for Scientific Research (NWO). W.V. is grateful to RIKEN, Brookhaven National Laboratory and the U.S. Department of Energy (contract number DE-AC02-98CH10886) for providing the facilities essential for the completion of this work.

References

1. Yu.L. Dokshitzer, talk at 11th Rencontres de Blois, hep-ph/9911299.
2. G. Sterman, *Nucl. Phys.* **B281**, 310 (1987); S. Catani and L. Trentadue, *Nucl. Phys.* **B327**, 323 (1989), **B353**, 183 (1991).
3. J.C. Collins, D.E. Soper and G. Sterman, *Nucl. Phys.* **B223**, 381 (1983).
4. J.G.M. Gatheral, *Phys. Lett.* **B133**, 9 (1983).
5. H.-n. Li, *Phys. Lett.* **B454**, 328 (1999), hep-ph/9812363.
6. E. Laenen, G. Sterman and W. Vogelsang, *Phys. Rev. Lett.* **84**, 4296 (2000), hep-ph/0002078, and in preparation.
7. H. Contopanagos and G. Sterman, *Nucl. Phys.* **B419**, 77 (1994), hep-ph/9310313; G.P. Korchemsky and G. Sterman, *Nucl. Phys.* **B437**, 415 (1995), hep-ph/9411211.
8. M. Beneke and V. Braun, *Nucl. Phys.* **B454**, 253 (1995), hep-ph/9506452; Yu.L. Dokshitzer, G. Marchesini and B.R. Webber, *Nucl. Phys.* **B469**, 93 (1996), hep-ph/9512336.

Parallel Session 16

Superstring Theory

Conveners: Nobuyuki Ishibashi (KEK) and
Eliezer Z. Rabinovici (Hebrew)

CONFORMAL ANOMALY AND C-FUNCTION FROM ADS/CFT CORRESPONDENCE

SACHIKO OGUSHI
Department of Physics, Ochanomizu University
Tokyo 112-8610, Japan
E-mail: g9970503@edu.cc.ocha.ac.jp

The bosonic part of supergravity with single scalar (dilaton) and arbitrary scalar potential is considered to describe special renormalization group flows in dual quantum field theory.

The dilaton-dependent conformal anomaly from five-dimensional dilatonic gravity with arbitrary potential is calculated by using AdS/CFT correspondence. Such anomaly should correspond to four-dimensional dual quantum field theory. We suggested the candidate c-functions from such dilatonic gravity. These c-functions which have fixed points in asymptotically AdS region are expressed in terms of dilatonic potential and they are positively defined and monotonic for some examples of scalar potential.

1 Introduction

In the last few years, there were a lot of attention related to AdS/CFT duality[1]. It is a duality between the large N limit of conformal field theories in d-dimensions and supergravity in $d+1$-dimensional Anti-de-Sitter space with a compact manifold. Due to this conjecture we can study some properties of large N CFT by simple calculations in supergravity side[2]. The main idea of this proposal is that $d+1$-dimensional supergravity on AdS should be supplemented by the fields on the d-dimensional boundary of AdS, which is a conformal field theory in d-dimensions. As the most interesting example, we can see the duality between $\mathcal{N}=4, SU(N)$ SYM theory in four-dimensions and the type IIB string theory compactified on $AdS_5 \times S^5$.

Recently there are much attention for studying of Renormalization Group (RG) flow from supergravity side[3,4,5](and refs.therein). To describe c-function from AdS/CFT correspondence, we review briefly a general discussion of deformations in field theory and in the dual description based on the report[6]. The deformation of CFT are made by adding the terms which break conformal invariance but keep lorentz invariance. $S_{\rm CFT} = S_{\rm CFT} + \int d^dx\ \phi\Phi(\phi)$. $\Phi(\phi)$ is the local operator which has conformal dimension Δ and the coefficient ϕ has $d-\Delta$ dimension. This ϕ is the coupling constant of the operator in CFT. But it is the field in supergravity in AdS background. The running of the coupling constant represents RG flow in CFT side, and this corresponds to the radial coordinate dependence of the field in AdS. Near the boundary of AdS, the field ϕ behaves as

$$\phi(x,U) \stackrel{U\to\infty}{\longrightarrow} U^{\Delta-d}\phi_{(0)}(x),$$

where $\phi_{(0)}$ is the boundary value of AdS background. If there are scalar mass terms in AdS side, the classical equation of motion leads to the relation between conformal dimention Δ and scalar mass m, the radius l of AdS as

$$\Delta = \frac{d}{2} + \sqrt{\frac{d^2}{4} + l^2m^2}.$$

If supergravity theory has only massless scalar, the deformation of CFT is marginal which does not break conformal invariance, but other cases including mass terms in AdS side correspond to the relevant or irrelevant deformations in CFT. So the mass term is important here. Where do the mass terms come from? The terms come from the scalar potential terms in supergravity.

In general, the scalar potential in supergravity side has a very complicated form (the construction of five-dimensional gauged supergravity is given in the theses[7]). So then we need to expand the potential around the stationary points which is given by the variation of the potential with respect to scalar ϕ. Thus we can obtain mass terms of the scalar, and this scalar has the dependence of radial coordinate U. On the CFT side, ϕ is the coupling constant which has fixed points given by the variation of ϕ with respect to energy scale U. This U corresponds to the radial coordinate on the AdS side. In this fixed point, the theory is conformal field theory having central charge c. The RG flow represents the changing of the central charges. Understanding the relationship between different conformal field theories in two dimension, the Zamolodchikov's c-function is a great tool[8]. The properties of c-function is known as c-theorem given by Zamlolodchikov as follows. First, this function is positive. Second, c-function is monotonically increasing function of the energy scale. This represents that RG transformation leads to a loss of information about the short distance degrees of freedom in the theory. Third, the function has fixed points which agree to the central charges. The extensions of above c-theorem (c-function) for four-dimensional field theory have been proposed by Cardy[9].

On the AdS side, renomalization group flow corresponds to the geometrical changes. For examples, the compact manifold S^5 has $SO(6)$ symmetry which represents R-symmetry of $\mathcal{N} = 4$ in four-dimensional CFT. The changing of the compact manifold corresponds to the changing of SUSY number of CFT.

Now let us move on to the problem how to define c-function from AdS/CFT. It is well known in two-dimensions that the central charge can be determined by the anomaly calculation as

$$\langle T^{\mu}_{\mu} \rangle = \frac{c}{24\pi} R \, .$$

The coefficient of the scalar curvature R corresponds to the central charge.

What is c-function in AdS side? Getting the idea from the relation between the central charge and the anomaly, we calculated the conformal anomaly by using the method which based on the work[10]. This method is focused on the conformal invariance of d + 1-dimensional AdS gravity action and the breaking of this invariance corresponds to the conformal anomaly in d-dimensional CFT. Then we tried to define c-function by the anomaly calculation with arbitrary scalar potential. Having examined some examples of scalar potentials, we checked the c-theorem and compared our c-function with the other proposals for it. The detailed discussions of conformal anomaly and c-function were summarized in the previous works[11].

2 Conformal anomaly and c-function from AdS/CFT

We start from the bulk action of d + 1-dimensional gravity with a scalar;ϕ and the arbitrary scalar potential $\Phi(\phi)$

$$S = \frac{1}{16\pi G} \int_{M_{d+1}} d^{d+1}x \sqrt{-\hat{G}} \times \Big\{ \hat{R} + X(\phi)(\hat{\nabla}\phi)^2 + Y(\phi)\hat{\Delta}\phi + \Phi(\phi) + 4\lambda^2 \Big\} \, . \tag{1}$$

Here M_{d+1} is d + 1-dimensional manifold whose boundary is d-dimensional manifold M_d and we choose $\Phi(0) = 0$. Such action corresponds to (bosonic sector) of gauged supergravity with single scalar. In other words, one considers RG flow in extended supergravity when scalars lie in one-dimensional submanifold of complete scalars space. Note also that classical vacuum stability restricts the form of dilaton potential [12]. As well-known, we also need to add the surface terms [13] to the bulk action in order to have well-defined variational principle. At the moment, for the purpose of calculation of Weyl anomaly (via AdS/CFT correspondence) the surface terms are irrelevant. We choose the AdS like metric $\hat{G}_{\mu\nu}$ on M_{d+1} and the metric $\hat{g}_{ij}$ on M_d in the following form

$$ds^2 \equiv \hat{G}_{\mu\nu}dx^\mu dx^\nu$$
$$= \frac{l^2}{4}\rho^{-2}d\rho d\rho + \sum_{i=1}^{d}\hat{g}_{ij}dx^i dx^j. \quad (2)$$

Here l is related with λ^2 by $4\lambda^2 = d(d-1)/l^2$ and $\hat{g}_{ij} = \rho^{-1}g_{ij}$. ρ is the radial coordinate in AdS. If $g_{ij} = \eta_{ij}$, the boundary of AdS lies at $\rho = 0$. The action (1) diverges in general since it contains the infinite volume integration on M_{d+1}. The action is regularized by introducing the infrared cutoff ϵ and replacing

$$\int d^{d+1}x \to \int d^dx \int_\epsilon d\rho. \quad (3)$$

We also expand g_{ij} and ϕ with respect to ρ:

$$g_{ij} = g_{(0)ij} + \rho g_{(1)ij} + \rho^2 g_{(2)ij} + \cdots,$$
$$\phi = \phi_{(0)} + \rho\phi_{(1)} + \rho^2\phi_{(2)} + \cdots . \quad (4)$$

Then the action is also expanded as a power series on ρ. The subtraction of the terms proportional to the inverse power of ϵ does not break the invariance under the scale transformation $\delta g_{\mu\nu} = 2\delta\sigma g_{\mu\nu}$ and $\delta\epsilon = 2\delta\sigma\epsilon$. When d is even, however, the term proportional to $\ln\epsilon$ appears. This term is not invariant under the scale transformation and the subtraction of the $\ln\epsilon$ term breaks the invariance. The variation of the $\ln\epsilon$ term under the scale transformation is finite when $\epsilon \to 0$ and should be canceled by the variation of the finite term (which does not depend on ϵ) in the action since the original action (1) is invariant under the scale transformation. Therefore the $\ln\epsilon$ term $S_{\ln}$ gives the Weyl anomaly T of the action renormalized by the subtraction of the terms which diverge when $\epsilon \to 0$ ($d = 4$)

$$S_{\ln} = -\frac{1}{2}\int d^4x\sqrt{-g}T . \quad (5)$$

The equations of motion given by the variation of (1) with respect to ϕ and $G^{\mu\nu}$ lead to $g_{(1)ij}$ and $\phi_{(1)}$ written by the boundary values;$g_{(0)ij}$,$\phi_{(0)}$. We focused on four-dimensional case here which is considered to the most interesting case. The anomaly terms which proportional to $\ln\epsilon$ are obtained by the following form.

$$T = -\frac{1}{8\pi G}\Big[h_1R^2 + h_2R_{ij}R^{ij} + h_3R^{ij}\partial_i\phi\partial_j\phi + h_4Rg^{ij}\partial_i\phi\partial_j\phi + h_5R\frac{1}{\sqrt{-g}}\partial_i(\sqrt{-g}g^{ij}\partial_j\phi) + h_6(g^{ij}\partial_i\phi\partial_j\phi)^2 + h_7\left(\frac{1}{\sqrt{-g}}\partial_i(\sqrt{-g}g^{ij}\partial_j\phi)\right)^2 + h_8g^{kl}\partial_k\phi\partial_l\phi\frac{1}{\sqrt{-g}}\partial_i(\sqrt{-g}g^{ij}\partial_j\phi)\Big] \quad (6)$$

The explicit forms of h_1 and h_2 are

$$h_1 = \Big[3\ \{(24 - 10\ \Phi)\ \Phi'^6 + (62208 + 22464\ \Phi + 2196\ \Phi^2 + 72\ \Phi^3 + \Phi^4)\ \Phi''\ (\Phi'' + 8\ V)^2 + 2\ \Phi'^4\ \{(108 + 162\ \Phi + 7\ \Phi^2)\ \Phi'' + 72\ (-8 + 14\ \Phi + \Phi^2)\ V\} - 2\ \Phi'^2 \times \{(6912 + 2736\ \Phi + 192\ \Phi^2 + \Phi^3)\ \Phi''^2 + 4\ (11232 + 6156\ \Phi + 552\ \Phi^2 + 13\ \Phi^3)\ \Phi''\ V + 32\ (-2592 + 468\ \Phi + 96\ \Phi^2 + 5\ \Phi^3)\ V^2\} - 3\ (-24 + \Phi), \times\ (6 + \Phi)^2\ \Phi'^3\ (\Phi''' + 8\ V')\}\Big] / \Big[16\ (6 + \Phi)^2\ \{-2\ \Phi'^2 + (24 + \Phi)\ \Phi''\} \times \{-2\ \Phi'^2 + (18 + \Phi)\ (\Phi'' + 8\ V)\}^2\Big]$$

$$h_2 = -\big[3\ \{(12 - 5\ \Phi)\ \Phi'^2 + (288 + 72\ \Phi + \Phi^2)\ \Phi''\}\big] / \big[8\ (6 + \Phi)^2\ \{-2\ \Phi'^2 + (24 + \Phi)\ \Phi''\}\big] . \quad (7)$$

We also give the explicit forms of $h_3, \cdots, h_8$ in the paper[11]. Here $V(\phi) \equiv X(\phi) - Y'(\phi)$ and "$'$" denotes the derivative with respect to ϕ. For the simplicity, we choose $l = 1$, denote $\Phi(\phi_{(0)})$ by Φ and abbreviate the index (0). Thus, we found the complete conformal anomaly from bulk side. Taking the limit of $\Phi \to 0$, we obtain

$$h_1 \to \frac{3 \cdot 62208\Phi''(8V)^2}{16 \cdot 6^2 \cdot 24 \cdot 18\Phi''(8V)^2} = \frac{1}{24},$$

$$h_2 \to -\frac{3\cdot 288\Phi''}{8\cdot 6^2\cdot 24\Phi''} = -\frac{1}{8}, \tag{8}$$

and we find that the standard result (conformal anomaly of $\mathcal{N}=4$ super YM theory covariantly coupled with $\mathcal{N}=4$ conformal supergravity [14,15]) is reproduced [16]. In order that the region near the boundary at $\rho=0$ is asymptotically AdS, we need to require $\Phi\to 0$ and $\Phi'\to 0$ when $\rho\to 0$. One can also confirm that $h_1\to\frac{1}{24}$ and $h_2\to -\frac{1}{8}$ in the limit of $\Phi\to 0$ and $\Phi'\to 0$ even if $\Phi''\neq 0$ and $\Phi'''\neq 0$. In the AdS/CFT correspondence, h_1 and h_2 are considered to relate with the central charge c of the conformal field theory (or its analog for non-conformal theory). Since we have two functions h_1 and h_2, there are two ways to define the candidate c-function when the conformal field theory is deformed:

$$c_1 = \frac{24\pi h_1}{G}\ , \quad c_2 = -\frac{8\pi h_2}{G}\ . \tag{9}$$

To include cosmological constant $4\lambda^2$ when $\Phi=0$, we consider the generalization of potential as $V(\phi)=4\lambda^2+\Phi(\phi)$. Using the relation $4\lambda^2 = d(d-1)/l^2$, then we obtain the relation $V(0)$ and l as $l=\left(\frac{12}{V(0)}\right)^{\frac{1}{2}}$. One should note that it is chosen $l=1$ in (7). We can restore l by changing $h_1\to l^3h_1$ and $h_2\to l^3h_2$ and $\Phi'\to l\Phi'$, $\Phi''\to l^2\Phi''$ and $\Phi'''\to l^3\Phi'''$ in (7). Then in the limit of $\Phi\to 0$, one gets

$$c_1\ ,\quad c_2 \to \frac{\pi}{G}\left(\frac{12}{V(0)}\right)^{\frac{3}{2}}\ , \tag{10}$$

which agrees with the proposal of the previous work [17] in the limit. The c-function c_1 or c_2 in (9) is, of course, more general definition. It is interesting to study the behaviour of candidate c-function for explicit values of dilatonic potential at different limits.

3 The properties of c-function

The definitions of the c-functions in (9) are not always good ones since our results are too wide. That is, we have obtained the conformal anomaly for arbitrary dilatonic background which may not be the solution of original $d=5$ gauged supergravity. As only solutions of five-dimensional gauged supergravity describe RG flows of dual QFT, it is not strange that above candidate c-functions are not acceptable. They quickly become non-monotonic and even singular in explicit examples. They actually measure the deviations from supergravity description and should not be taken seriously. As pointed in the work[18], it might be necessary to impose the condition $\Phi'=0$ on the conformal boundary. Such condition follows from the equations of motion of five-dimensional gauged supergravity. Anyway as $\Phi'=0$ on the boundary in the solution which has the asymptotic AdS region, we can add any function which proportional to the power of $\Phi'=0$ to the previous expressions of the c-functions in Ref.(9). As a trial, if we put $\Phi'=0$, we obtain instead of (9)

$$\begin{aligned} c_1 &= \frac{2\pi}{3G} \\ &\quad\times \frac{62208+22464\Phi+2196\Phi^2+72\Phi^3+\Phi^4}{(6+\Phi)^2(24+\Phi)(18+\Phi)}, \\ c_2 &= \frac{3\pi}{G}\frac{288+72\Phi+\Phi^2}{(6+\Phi)^2(24+\Phi)}. \end{aligned} \tag{11}$$

We should note that there disappear the higher derivative terms like Φ'' or Φ'''. That will be our final proposal for acceptable c-function in terms of dilatonic potential. The given c-functions in (11) also have the property (10) and reproduce the known result for the central charge on the boundary. Since $\frac{d\Phi}{d\rho}\to 0$ in the asymptotically AdS region even if the region is UV or IR, the given c-functions in (11) have fixed points in the asymptotic AdS region $\frac{dc}{dU}=\frac{dc}{d\Phi}\frac{d\Phi}{d\phi}\frac{d\phi}{dU}\to 0$, where $U=\rho^{-\frac{1}{2}}$ is the radial coordinate in AdS or the energy scale of the boundary field theory.

We can now check the monotonity in the c-functions. For this purpose, we consider some examples. In previous works[3,4], the potentials which have the following properties appeared. First, those potential are the monotonic increasing function of the absolute value $|\phi|$. Second, scalar ϕ is the monotonically decreasing function of the energy scale U and $\phi = 0$ at the UV limit corresponding to the boundary. Third, potential has minimum zero at the scalar is zero. The last two properties come from the classical equation of motions.

Then in order to know the energy scale dependences of c_1 and c_2, we only need to investigate the Φ dependences of c_1 and c_2. From (11), we find

$$\frac{d\,(\ln c_1)}{d\Phi} < 0, \quad \frac{d\,(\ln c_2)}{d\Phi} < 0 . \tag{12}$$

Using the above properties, we can understand that c-function are monotonically increasing functions of the energy scale U as

$$\frac{dc}{dU} = \underbrace{\frac{dc}{d\Phi}}_{<0}\underbrace{\frac{d\Phi}{d\phi}}_{>0}\underbrace{\frac{d\phi}{dU}}_{<0} > 0, \tag{13}$$

which agree with Zamolodchikov c-theorem.

In the thesis[17], another c-function has been proposed in terms of the metric as follows:

$$c_{\rm GPPZ} = \left(\frac{dA}{dz}\right)^{-3} , \tag{14}$$

where the metric is given by

$$ds^2 = dz^2 + e^{2A} dx_\mu dx^\mu . \tag{15}$$

Here z is the radial coordinate in AdS like space and A is the function of z. The c-function (14) is positive and has a fixed point in the asymptotically AdS region again and the c-function is also monotonically decreasing function of the energy scale. The c-functions (11) proposed in our papers are given in terms of the dilaton potential, not in terms of metric, but it might be interesting that the c-functions in (11) have the similar properties (positivity, monotonity and fixed point in the asymptotically AdS region). These properties could be understood from the equations of motion. When the metric has the form (15), the equations of motion are:

$$\phi'' + dA'\phi' = \frac{\partial\Phi}{\partial\phi}, \tag{16}$$

$$dA'' + d(A')^2 + \frac{1}{2}(\phi')^2 = -\frac{4\lambda^2 + \Phi}{d-1}, \tag{17}$$

$$A'' + d(A')^2 = -\frac{4\lambda^2 + \Phi}{d-1}. \tag{18}$$

Here $' \equiv \frac{d}{dz}$. From (16) and (17), we obtain

$$0 = 2(d-1)A'' + \phi'^2 \tag{19}$$

If $A'' = 0$, then $\phi' = 0$, which tells that if we take $\frac{dc_{\rm GPPZ}}{dz} = 0$, then $\frac{dc_1}{dz} = \frac{dc_2}{dz} = 0$. Thus $c_{\rm GPPZ}$ has a fixed point, c_1 and c_2 have a fixed point. From (16) and (17), we also obtain

$$0 = d(d-1)A'^2 + 4\lambda^2 + \Phi - \frac{1}{2}\phi'^2 . \tag{20}$$

Then at the fixed point where $\phi' = 0$, we obtain

$$0 = d(d-1)A'^2 + 4\lambda^2 + \Phi. \tag{21}$$

Taking $c_{\rm GPPZ}$ and A' is the monotonic function of z, potential V and also c_1 and c_2 are also monotonic function at least at the fixed point.

4 Conclusion

In summary, we obtained the conformal anomaly from five-dimensional AdS bulk side with arbitrary scalar potential. ¿From this holographic conformal anomaly, we suggested the candidate c-function for four-dimensional boundary QFT. It is shown that such proposal gives monotonic and positive c-function for few examples of dilatonic potential. The works described here can be generalized also to the cases that include the large number of scalar, that is, 42 scalars in $\mathcal{N} = 8$ five-dimensional supergravity. We dicussed this case and summarized in the paper[19]. And the extension of our works to more higher dimension are also possible.

Acknowledgments

The author would like to thank S. Nojiri and S.D. Odintsov for fruitful collaborations which this talk based upon. This work is supported in part by Japan Society for the Promotion of Science.

References

1. J.M. Maldacena, *Adv.Theor.Math.Phys.* **2** (1998) 231.
2. E. Witten, *Adv. Theor. Math. Phys.* **2** (1998) 253; S. Gubser, I.R. Klebanov and A.M. Polyakov, *Phys. Lett.* **B428** (1998) 105.
3. D. Freedman, S. Gubser, K. Pilch and N.P. Warner, hep-th/9904017; *JHEP* 0007 (2000) 038.
4. L. Girardello, M. Petrini, M. Porrati and A. Zaffaroni, *JHEP* **9812** (1998) 022; *JHEP* **9905** (1999) 026; *Nucl. Phys.* **B569** (2000) 451.
5. J. Distler and F. Zamora, *JHEP* **0005** (2000) 005; S. Nojiri and S.D. Odintsov, *Phys. Lett.* **449** (1999) 39; *Phys. Rev.* **D61**(2000) 044014; K. Behrndt and D. Lüst, *JHEP* **9907**(1999)019; N. Constable and R.C. Myers, *JHEP* **9911** (1999) 020; K. Behrndt and M. Cvetič, hep-th/9909058; A. Chamblin and G. Gibbons, *Phys. Rev. Lett.* **84**(2000)1090; N. Evans, C. Johnson and M. Petrini, hep-th/0008081.
6. O. Aharony, S.S. Gubser, J. Maldacena, H. Ooguri and Y. Oz, *Phys. Rept.* **323** (2000) 183.
7. M. Gunaydin, L. Romans and N. Warner, *Phys. Lett.* **B154** (1985) 268; M. Pernici, K. Pilch and P. van Nieuwenhuizen, *Nucl.Phys.* **B259** (1985) 460.
8. A.B. Zamolodchikov, *JETP Lett.* **43**, 730(1986).
9. J.L. Cardy, *Phys. Lett.* **B215**,749(1988).
10. M. Henningson and K. Skenderis, *JHEP* **07**(1998) 023.
11. S. Nojiri, S.D. Odintsov and S.Ogushi, hep-th/9912191; hep-th/0001122, (to appear in *Phys. Rev.* **D**).
12. P.K. Townsend, *Phys.Lett.* **B148** (1984) 55.
13. G.W. Gibbons and S.W. Hawking, *Phys. Rev.* **D15** (1977) 2752.
14. M. Kaku, P.K. Townsend and P. van Nieuwenhuizen, *Phys. Rev.* **D17**(1978) 3179; E. Bergshoeff, M. de Roo and B. de Wit, *Nucl.Phys.* **B182** (1981) 173; E.S. Fradkin and A. Tseytlin, *Phys.Rept.* **119**(1985) 233.
15. H. Liu and A. Tseytlin, *Nucl.Phys.* **B533**(1998) 88.
16. S. Nojiri and S.D. Odintsov, *Phys. Lett.* **B444** (1998) 92.
17. L. Girardello, M. Petrini, M. Porrati and A. Zaffaroni, *JHEP* **9812** (1998) 022.
18. M. Taylor-Robinson, hep-th/0002125.
19. S. Nojiri, S.D. Odintsov and S.Ogushi, hep-th/0005197, (to appear in *Prog. Teor.Phys.*).

ON THE CONSISTENCY OF NON-CRITICAL NON-ORIENTABLE OPEN-CLOSED STRING FIELD THEORIES

H. KAWABE

Yonago National College of Technology, Yonago 683-8502, Japan
E-mail: kawabe@yonago-k.ac.jp

N. NAKAZAWA

Department of Physics, Faculty of Science and Engineering, Shimane University
Matsue 690-8504, Japan
E-mail: nakazawa@riko.shimane-u.ac.jp

We propose a real symmetric matrix-vector model which describes the non-orientable open-closed strings for $0 < c \leq 1$. The string field theory is constructed by the stochastic quantization method. The algebraic structure of the field theory hamiltonian ensures the integrability of the stochastic time evolution. The Schwinger-Dyson equations are consistent to the orientable string theory.

1 Introduction

Recently, non-perturbative definition of string field theory has been proposed by the large N reduction of the super Yang-Mills theories. It gives simple systems of matrix variables as the constructive definition of superstrings.[1,2,3] Type I superstring is a theory of non-orientable open-closed strings, in which the finiteness of amplitudes and anomaly cancellation specify the gauge group $SO(32)$. It is important to study the consistency of non-orientable string theories from matrix model viewpoint. In matrix model approach, continuum string field theories are realized at the double scaling limit.[4] Application of the stochastic quantization method systematically deduces string interactions.[5] To describe non-orientable 2D surfaces, one introduces real symmetric matrices instead of hermitian matrices.[6] Orientable open-closed string theories have been studied by loop-gas models.[7,8] Furthermore Chan-Paton factor has been introduced for the $c = 0$ non-orientable string field theory.[9] In this case, "coloured" boundaries are described by vector variables. The partition function corresponds to the sum over all dynamically triangulated 2D non-orientable surfaces with boundaries.

In this note we propose a loop-gas model on non-orientable 2D surfaces. With this model, a string field theory is derived based on stochastic quantization method. We investigate the algebraic structure of string field theory hamiltonian at the continuum limit. The Schwinger-Dyson equations imply the consistency of the non-orientable string field theory.

We firstly propose a matrix-vector model which describes non-orientable open-closed strings. It is based on the loop gas model with a modification for the non-orientability. Matrices and vectors are localized on the discretized 1-dimensional target space, $x, x' \in \mathbf{Z}$. $N \times N$ matrices $A_{xx'}$ are link variables. While the N-vectors V_x^a, site variables, correspond to the open string end-points with Chan-Paton factor "a", which runs from 1 to r. We start with the $O(N)$ gauge invariant action,

$$S = \frac{1}{2}\mathrm{tr}\Big(\frac{1}{2}\sum_{x,x'} A_{xx'}A_{x'x} - \frac{1}{3}\frac{g}{\sqrt{N}}\sum_{x,x',x''} A_{xx'}A_{x'x''}A_{x''x}\Big)$$

$$+\frac{1}{2}\sum_{x,x'}\sum_{a=1}^{r} V_x^a\big(\delta_{xx'} - \frac{g_B^a}{\sqrt{N}}A_{xx'}\big)V_{x'}^a \ . \tag{1}$$

where vectors V_x^a are real and matrices $A_{xx'}$ are non-zero only for $x' = x, x \pm 1$. The matrices A_{xx} are real symmetric and the ones connecting neighbouring sites satisfy the condition $A_{xx+1} = A_{x+1x}^t$.

By integrating out the link matrices $A_{xx\pm 1}$, we obtain the effective action of non-orientable string interactions. By the redefinition, $M_{x\ ij} \equiv A_{xx\ ij} - \frac{\sqrt{N}}{2g}\delta_{ij}$, we regard collective fields,

$$\phi_x(L) \equiv \frac{1}{N}\mathrm{tr}(\mathrm{e}^{L\frac{M_x}{\sqrt{N}}})$$

and

$$\psi_x^{ab}(L) \equiv \frac{1}{N}V_x^a \mathrm{e}^{L\frac{M_x}{\sqrt{N}}}V_x^b \ ,$$

as non-orientable closed strings and open strings with Chan-Paton factors a, b, respectively. With these string variables, the effective action is expressed as

$$\begin{aligned} S_{\rm eff} = & \frac{Ng}{3}\mathrm{tr}\sum_x\big(\frac{1}{2g} + \frac{M_x}{\sqrt{N}}\big)^3 \\ & -\frac{N}{2}\mathrm{tr}\sum_x\big(\frac{1}{2g} + \frac{M_x}{\sqrt{N}}\big)^2 \\ & + \sum_x\sum_a\Big\{V_x^a\big(\frac{g_B^a}{2g} - 1 + g_B^a\frac{M_x}{\sqrt{N}}\big)V_x^a\Big\} \\ & + \frac{1}{2}\sum_{x,x'}\int_0^\infty dL\Big\{\frac{1}{2L}N^2 C_{xx'}^{(p_0)}\phi_x(L)\phi_{x'}(L) \\ & + N\sum_{a,b}\frac{g_B^a g_B^b}{g}C_{xx'}^{(p_0/2)}\psi_x^{ab}(L)\psi_{x'}^{ba}(L)\Big\}, \end{aligned} \tag{2}$$

where the coefficients of string bilinear terms, $C_{xx'}^{(p_0)} = C_{xx'}^{(p_0/2)} = \delta_{x'x+1} + \delta_{x'x-1}$ are called as adjacency matrices. In order to realize the model of the non-critical strings with the central charge $0 < c \leq 1$, the adjacency matrices should be replaced by

$$\begin{aligned} C_{xx'}^{(p_0)} &= \cos(\pi p_0)\big(\delta_{x'x+1} + \delta_{x'x-1}\big) \ , \\ C_{xx'}^{(p_0/2)} &= \cos(\frac{\pi}{2}p_0)\big(\delta_{x'x+1} + \delta_{x'x-1}\big) \ . \end{aligned} \tag{3}$$

The background momentum $p_0 = 1/m$ corresponds to $c = 1 - 6/m(m+1)$.

2 Stochastic Quantization of the Model

Now we apply the stochastic quantization method. In Ito calculus, Langevin equations define the one step deformation of matrices and vectors with respect to the discretized stochastic time $\Delta\tau$,

$$\begin{aligned} \Delta M_{xij}(\tau) &= -\frac{\partial S_{\rm eff}}{\partial M_{xji}}\Delta\tau + \Delta\xi_{xij}(\tau) \ , \\ \Delta V_{xi}^a(\tau) &= -\lambda_x^a\frac{\partial S_{\rm eff}}{\partial V_{xi}^a}\Delta\tau + \Delta\eta_{xi}^a(\tau) \ , \end{aligned} \tag{4}$$

where λ_x^a is the scaling parameter of the stochastic time evolution on the boundary, *i.e.*, the open string end-point with the "colour" index a. $\Delta\xi_{xij}$ and $\Delta\eta_{xi}^a$ are white noises defined by the correlations,

$$\begin{aligned} & < \Delta\xi_{xij}(\tau)\Delta\xi_{x'kl}(\tau) >_\xi \\ & \qquad = \Delta\tau\delta_{xx'}(\delta_{il}\delta_{jk} + \delta_{ik}\delta_{jl}) \ , \\ & < \Delta\eta_{xi}^a(\tau)\Delta\eta_{x'j}^b(\tau) >_\eta = 2\lambda_x^a\Delta\tau\delta^{ab}\delta_{xx'}\delta_{ij} \ . \end{aligned} \tag{5}$$

The Langevin equations (4) describe the "microscopic" processes of the following one-step stochastic time evolution of string fields,

$$\begin{aligned} \phi_x(L)|_\tau + \Delta\phi_x(L) &= \phi_x(L)|_{\tau+\Delta\tau} \ , \\ \psi_x^{ab}(L)|_\tau + \Delta\psi_x^{ab}(L) &= \psi_x^{ab}(L)|_{\tau+\Delta\tau} \ . \end{aligned}$$

Up to the order of $\Delta\tau$, we determine $\Delta\phi_x(L)$ and $\Delta\psi_x^{ab}(L)$ in Ito calculus. For closed string fields,

$$\begin{aligned} & \Delta\phi_x(L) = \\ & \quad \Delta\tau L\Big\{\Big(g\frac{\partial^2}{\partial L^2} - \frac{1}{4g}\Big)\phi_x(L) \\ & \quad + \frac{1}{2}\int_0^L dL'\phi_x(L')\phi_x(L-L') \\ & \quad + \frac{1}{2}\sum_{x'} C_{xx'}^{(p_0)}\int_0^\infty dL'\phi_x(L+L')\phi_{x'}(L') \\ & \quad + \frac{1}{Ng}\sum_{a,b}\sum_{x'} C_{xx'}^{(p_0/2)}\int_0^\infty dL' L'\psi_x^{ab}(L+L')\psi_{x'}^{ab}(L') \end{aligned}$$

$$+\frac{1}{2N}L\phi_x(L)+\frac{1}{N}\sum_a\psi_x^{aa}(L)\Big\}$$
$$+\Delta\zeta_x(L)\ . \qquad (6)$$

Here, $(1/2N)L\phi_x(L)$ is the characteristic term for non-orientable interactions, which remains in the continuum limit. For open string fields, we have

$$\Delta\psi_x^{ab}(L)=$$
$$2\lambda_x^a\Delta\tau\Big\{\Big(\frac{\partial}{\partial L}+\frac{1}{2g}-\frac{1}{g_B^a}\Big)\psi_x^{ab}(L)$$
$$+\frac{1}{g}\sum_c\sum_{x'}C_{xx'}^{(p_0/2)}\int_0^\infty dL'\psi_x^{bc}(L+L')\psi_{x'}^{ac}(L')\Big\}$$
$$+(a\leftrightarrow b)$$
$$+2\lambda^a\Delta\tau\delta^{ab}\phi_x(L)$$
$$+\Delta\tau\Big\{L\Big(g\frac{\partial^2}{\partial L^2}-\frac{1}{4g}\Big)\psi_x^{ab}(L)+\frac{L^2}{2N}\psi_x^{ab}(L)$$
$$+\sum_c\int_0^L dL'\psi_x^{ac}(L')\psi_x^{cb}(L-L')$$
$$+\frac{1}{2}L\sum_{x'}C_{xx'}^{(p_0)}\int_0^\infty dL'\psi_x^{ab}(L+L')\phi_{x'}(L')$$
$$+\frac{1}{g}\sum_{c,d}\sum_{x'}C_{xx'}^{(p_0/2)}\int_0^L dL'\int_0^\infty dL''\int_0^\infty dL'''$$
$$\times\psi_x^{ad}(L'+L'')\psi_x^{cb}(L-L'+L''')\psi_{x'}^{cd}(L''+L''')$$
$$+\int_0^L dL'L'\psi_x^{ab}(L')\phi_x(L-L')\Big\}$$
$$+\Delta\zeta_x^{ab}(L)+\Delta\eta_x^{ab}(L)\ . \qquad (7)$$

Here we find the characteristic term, $(1/2N)L^2\psi_x^{ab}(L)$. In eqs. (6), (7), $\Delta\zeta_x(L)$, $\Delta\zeta_x^{ab}(L)$ and $\Delta\eta_x^{ab}(L)$ are collective fields which contain a white noise,

$$\Delta\zeta_x(L)\equiv\frac{1}{N}L\mathrm{tr}(\mathrm{e}^{L\frac{M_x}{\sqrt{N}}}\frac{\Delta\xi_x}{\sqrt{N}})\ ,$$
$$\Delta\zeta_x^{ab}(L)\equiv\frac{1}{N}\int_0^L dL'V_x^a\mathrm{e}^{L'\frac{M_x}{\sqrt{N}}}\frac{\Delta\xi_x}{\sqrt{N}}\mathrm{e}^{(L-L')\frac{M_x}{\sqrt{N}}}V_x^b,$$
$$\Delta\eta_x^{ab}(L)\equiv\frac{1}{N}(\Delta\eta_x^a\mathrm{e}^{L\frac{M_x}{\sqrt{N}}}V_x^b+V_x^a\mathrm{e}^{L\frac{M_x}{\sqrt{N}}}\Delta\eta_x^b).$$
$$(8)$$

The correlations of these new collective fields are given by

$$<\Delta\zeta_x(L)\Delta\zeta_x(L')>=\Delta\tau\frac{2LL'}{N^2}\phi_x(L+L'),$$

$$<\Delta\eta_x^{ab}(L)\Delta\eta_x^{cd}(L')>$$
$$=\lambda_x^a\Delta\tau\frac{2}{N}\{\delta^{ac}\psi_x^{bd}(L+L')+\delta^{ad}\psi_x^{bc}(L+L')\}$$
$$+\lambda_x^b\Delta\tau\frac{2}{N}\{\delta^{bd}\psi_x^{ac}(L+L')+\delta^{bc}\psi_x^{ad}(L+L')\},$$
$$<\Delta\zeta_x^{ab}(L)\Delta\zeta_x^{cd}(L')>=\Delta\tau\frac{1}{N}\int_0^L ds\int_0^{L'}ds'$$
$$\times\{\psi_x^{ad}(s+s')\psi_x^{cb}(L+L'-s-s')$$
$$+\psi_x^{ac}(s+s')\psi_x^{bd}(L+L'-s-s')\},$$
$$<\Delta\zeta_x(L)\Delta\zeta_x^{ab}(L')>=\Delta\tau\frac{2LL'}{N^2}\psi_x^{ab}(L+L')\ .$$
$$(9)$$

The correlation functions of these collective fields must be taken in Ito's sense. Namely, $<\Delta\zeta_x(L)>=<\Delta\zeta_x^{ab}(L)>=<\Delta\eta_x^{ab}(L)>=0$. With this condition, the Langevin equations deduce the Schwinger-Dyson(S-D) equations at the equilibrium limit, $\lim_{\tau\to\infty}\Delta\phi_x(L)=\lim_{\tau\to\infty}\Delta\psi_x^{ab}(L)=0$. We will consider the S-D equations later in the Laplace transformed form.

The string field theory hamiltonian is given by the corresponding Fokker-Planck(F-P) hamiltonian. In the stochastic time evolution of an observable $O(\phi,\psi)$, the F-P hamiltonian operator $\hat{H}_{\mathrm{FP}}$ is defined by following identity:

$$<\phi(0),\psi(0)|\mathrm{e}^{-\tau\hat{H}_{\mathrm{FP}}}O(\hat{\phi},\hat{\psi})|0>$$
$$\equiv<O(\phi_{\xi\eta}(\tau),\psi_{\xi\eta}(\tau))>_{\xi\eta}\ . \qquad (10)$$

In r.h.s., $\phi_{\xi\eta}(\tau)$ and $\psi_{\xi\eta}(\tau)$ are the solutions of the Langevin equations and the expectation value is taken by the noise correlation. In l.h.s., $\phi_x(0)$ and $\psi_x(0)$ are the initial condition of the string fields. The F-P hamiltonian is equivalent to the differential operator appearing in the well known F-P equation for the probability distribution functional except for the operator ordering. In (10), $\hat{H}_{\mathrm{FP}}$ is given by replacing string variables to the creation operators and the differential of the string variables to the annihilation operators.

In order to take the continuum limit, we introduce a length scale "ε", then consider the limit $\varepsilon\to 0$. We define the physical length of strings as $\ell \equiv L\varepsilon$. With the definition of infinitesimal stochastic time $d\tau \equiv \varepsilon^{-2+D}\Delta\tau$ and $\lambda^a \equiv \varepsilon^{3/2-D/2}\lambda_x^a$, the renormalization of the string field operators are expressed as

$$\begin{aligned}
\Phi_x(\ell) &\equiv \varepsilon^{-D}\phi_x(L) , \\
\Psi_x^{ab}(\ell) &\equiv \varepsilon^{-1/2-D/2}\psi_x^{ab}(L) , \\
\Pi_x(\ell) &\equiv \varepsilon^{-1+D}\frac{\partial}{\partial\phi_x(L)} , \\
\Pi_x^{ab}(\ell) &\equiv \varepsilon^{-1/2+D/2}\frac{\partial}{\partial\psi_x^{ab}(L)} .
\end{aligned} \tag{11}$$

where $D = 2 + p_0 = 2 + 1/m$ is the scaling dimension of closed string creation operators. Under the double scaling limit, the string coupling constant $G_{\rm st} \equiv N^{-2}\varepsilon^{-2D}$ is kept finite. The scaling of g and g_B^a are related with the cosmological constant Λ and the mass μ at the string end-points, respectively. Namely, $g^* - g \sim \varepsilon^2\Lambda$ and $g_B^{a*} - g_B^a \sim \varepsilon^{p_0+1}\mu$. At the continuum limit, the relevant commutation relations of the renormalized field operators are

$$\begin{aligned}
\left[\Pi_x(\ell), \Phi_{x'}(\ell')\right] &= \delta_{xx'}\delta(\ell-\ell') , \\
\left[\Pi_x^{ab}(\ell), \Psi_{x'}^{cd}(\ell')\right] &= \frac{1}{2}(\delta^{ac}\delta^{bd} + \delta^{ad}\delta^{bc})\delta_{xx'}\delta(\ell-\ell').
\end{aligned} \tag{12}$$

3 Algebraic Structure of the String Field Theory Hamiltonian

The continuum F-P hamiltonian $\mathcal{H}_{\rm FP}$ is expressed as a linear combination of three types of generators $\mathcal{L}_x(\ell)$, $\mathcal{J}_x^{ab}(\ell)$, and $\mathcal{K}_x^{ab}(\ell)$ as

$$\begin{aligned}
\mathcal{H}_{\rm FP} = \sum_x \int_0^\infty d\ell\, \Big\{ &-G_{\rm st}\mathcal{L}_x(\ell)\; \ell\; \Pi_x(\ell) \\
&- \sqrt{G_{\rm st}}\sum_{a,b}^r \Big(\lambda^a \mathcal{J}_x^{ab}(\ell) + \lambda^b \mathcal{J}_x^{ba}(\ell)\Big)\Pi_x^{ab}(\ell) \\
&+ \sum_{a,b}^r \mathcal{K}_x^{ab}(\ell)\; \ell\; \Pi_x^{ab}(\ell) \\
&+ \sqrt{G_{\rm st}}\sum_{a,b,c}^r \int_0^\ell d\ell'\ell'\,\mathcal{J}_x^{cb}(\ell')\Psi_x^{ac}(\ell-\ell')\Pi_x^{ab}(\ell)\Big\}.
\end{aligned} \tag{13}$$

The explicit forms of the generators are given by

$$\begin{aligned}
-G_{\rm st}\mathcal{L}_x(\ell) &= \frac{1}{2}\int_0^\ell d\ell'\Phi_x(\ell')\Phi_x(\ell-\ell') \\
&\quad + \frac{1}{2}\sqrt{G_{\rm st}}\;\ell\;\Phi_x(\ell) \\
&\quad + \int_0^\infty d\ell'\Phi_x(\ell+\ell')F_x(\ell') \\
&\quad + \sqrt{G_{\rm st}}\sum_{a,b}^r \int_0^\infty d\ell'\ell'\,\Psi_x^{ab}(\ell+\ell')F_x^{ab}(\ell'),
\end{aligned}$$

$$\begin{aligned}
-\sqrt{G_{\rm st}}\mathcal{J}_x^{ab}(\ell) &= \frac{1}{2}\Phi_x(\ell)\delta^{ab} \\
&\quad + \sum_c^r \int_0^\infty d\ell'\Psi_x^{cb}(\ell+\ell')F_x^{ca}(\ell'),
\end{aligned}$$

$$\begin{aligned}
\mathcal{K}_x^{ab}(\ell) &= \int_0^\ell d\ell'\Psi_x^{ab}(\ell')\Phi_x(\ell-\ell') \\
&\quad + \frac{1}{2}\sqrt{G_{\rm st}}(r+2)\ell\Psi_x^{ab}(\ell) \\
&\quad + \int_0^\infty d\ell'\Psi_x^{ab}(\ell+\ell')F_x(\ell') \\
&\quad + \sum_{c,d}^r \int_0^\infty d\ell' \int_0^{\ell+\ell'} d\ell''\Psi_x^{ac}(\ell'')\Psi_x^{db}(\ell+\ell'-\ell'')F_x^{cd}(\ell'),
\end{aligned} \tag{14}$$

where $F_x(\ell')$ and $F_x^{ab}(\ell')$ are composed of a pair of an annihilation operator and a creation operator,

$$\begin{aligned}
F_x(\ell') &= G_{\rm st}\ell'\Pi_x(\ell') + \frac{1}{2}\sum_{x'} C_{xx'}^{(p_0)}\Phi_{x'}(\ell'), \\
F_x^{ab}(\ell') &= \sqrt{G_{\rm st}}\Pi_x^{ab}(\ell') + \frac{1}{g}\sum_{x'} C_{xx'}^{(p_0/2)}\Psi_{x'}^{ab}(\ell').
\end{aligned}$$

The physical meaning of these generators are as follows. $\mathcal{L}_x(\ell)$, Virasoro generator, extends the string with the length ℓ' to one with the length $\ell+\ell'$. $\mathcal{J}_x^{ab}(\ell)$ extends only the open string by changing its end-points with the index "b" to "a". $\mathcal{K}_x^{ab}(\ell)$ extends both closed and open strings by cutting them and assigns the indices "a" and "b" to the open string new end-points. These generators satisfy the following commutation relations:

$$
\begin{aligned}
&[\mathcal{L}_x(\ell),\mathcal{L}_{x'}(\ell')] = (\ell-\ell')\delta_{xx'}\mathcal{L}_x(\ell+\ell')\ ,\\
&[\mathcal{L}_x(\ell),\mathcal{J}^{ab}_{x'}(\ell')] = -\ell'\delta_{xx'}\mathcal{J}^{ab}_x(\ell+\ell')\ ,\\
&[\mathcal{J}^{ab}_x(\ell),\mathcal{J}^{cd}_{x'}(\ell')] = \delta^{cb}\delta_{xx'}\mathcal{J}^{ad}_x(\ell+\ell') \\
&\qquad -\delta^{ad}\delta_{xx'}\mathcal{J}^{cb}_x(\ell+\ell')\ ,\\
&[\mathcal{L}_x(\ell),\mathcal{K}^{ab}_{x'}(\ell')] = (\ell-\ell')\delta_{xx'}\mathcal{K}^{ab}_x(\ell+\ell') \\
&\quad +\frac{\sqrt{G_{\rm st}}}{2}\delta_{xx'}\sum_c\int_0^\ell du(\ell-u) \\
&\quad \times\Big\{\mathcal{J}^{cb}_x(\ell+\ell'-u)\Psi^{ac}_x(u)+(a\leftrightarrow b)\Big\},\\
&[\mathcal{J}^{ab}_x(\ell),\mathcal{K}^{cd}_{x'}(\ell')] \\
&\quad = -\delta^{ad}\delta_{xx'}\mathcal{K}^{bc}_x(\ell+\ell')-\delta^{ac}\delta_{xx'}\mathcal{K}^{bd}_x(\ell+\ell') \\
&\quad -\frac{\sqrt{G_{\rm st}}}{2}\delta_{xx'}\int_0^\ell du\Big\{\{\mathcal{J}^{ad}_x(\ell+\ell'-u)\Psi^{cb}_x(u) \\
&\quad +\sum_e\delta^{ad}\mathcal{J}^{ec}_x(\ell+\ell'-u)\Psi^{be}_x(u)\}+\{c\leftrightarrow d\}\Big\},\\
&[\mathcal{K}^{ab}_x(\ell),\mathcal{K}^{cd}_{x'}(\ell')] \\
&\quad = \frac{G_{\rm st}}{4}\delta_{xx'}\sum_e\int_0^{\ell'} du\int_0^{\ell'-u} dv \\
&\quad \times\Big\{\{\mathcal{J}^{ae}_x(\ell+\ell'-u-v)\Psi^{cb}_x(u)\Psi^{ed}_x(v) \\
&\qquad +(c\leftrightarrow d)\}+\{a\leftrightarrow b\}\Big\} \\
&\quad -\frac{G_{\rm st}}{4}\delta_{xx'}\sum_e\int_0^{\ell} du\int_0^{\ell-u} dv \\
&\quad \times\Big\{\{\mathcal{J}^{ce}_x(\ell+\ell'-u-v)\Psi^{ad}_x(u)\Psi^{eb}_x(v) \\
&\qquad +(a\leftrightarrow b)\}+\{c\leftrightarrow d\}\Big\}.
\end{aligned}
\tag{15}
$$

The similar algebraic structure has been observed in the orientable string case.[10] In the precise sense, three generators do not construct an algebra. Properly, the algebraic structure is understood as the consistency condition on the equilibrium expectation value or "integrable condition" of the stochastic time evolution. We notice the antisymmetrized linear combination $\mathcal{J}^{ab}_x(\ell)-\mathcal{J}^{ba}_x(\ell)$ satisfies the $SO(r)$ current algebra. Therefore this model contains the Virasoro and $SO(r)$ current algebras.

4 Schwinger-Dyson Equations

The S-D equations are derived from the Langevin equations. For the analyticity, we consider the Laplace transformed string variables,

$$
\begin{aligned}
\tilde\phi_x(z) &= \frac{1}{N}{\rm tr}\frac{1}{z-\frac{M_x}{\sqrt N}}\ ,\\
\tilde\psi^{ab}_x(z) &= \frac{1}{N}V^a_x\frac{1}{z-\frac{M_x}{\sqrt N}}V^b_x\ .
\end{aligned}
\tag{16}
$$

Then we derive the Langevin equations for these new variables. By assuming the existence of the equilibrium limit, we obtain the S-D equations as follows:

$$
\begin{aligned}
&\frac{1}{2}\tilde\phi_x(z)^2+\frac{1}{2N}\partial_z\tilde\phi_x(z) \\
&+\frac{1}{2}\sum_{x'}C^{(p_0)}_{xx'}\int\frac{dz'}{2\pi i}\frac{1}{z-z'}\tilde\phi_x(z')\tilde\phi_{x'}(-z') \\
&+\frac{1}{Ng}\sum_{x',ab}C^{(\frac{p_0}{2})}_{xx'}\int\frac{dz'}{2\pi i}\frac{1}{(z-z')^2}\tilde\psi^{ab}_x(z')\tilde\psi^{ba}_{x'}(-z') \\
&+\ \text{(potential terms)}\ =0\ ,\\
&\delta^{ab}\tilde\phi_x(z) \\
&+\frac{1}{g}\sum_{x',c}C^{(\frac{p_0}{2})}_{xx'}\int\frac{dz'}{2\pi i}\frac{1}{(z-z')}\tilde\psi^{bc}_x(z')\tilde\psi^{ca}_{x'}(-z') \\
&+\ \text{(potential terms)}\ =0,\\
&\frac{1}{2N}\partial_z^2\tilde\psi^{ab}_x(z)-\partial_z\tilde\psi^{ab}_x(z)\tilde\phi_x(z) \\
&+\frac{1}{2}\sum_{x'}C^{(p_0)}_{xx'}\int\frac{dz'}{2\pi i}\frac{1}{(z-z')^2}\tilde\psi^{ab}_x(z')\tilde\phi_{x'}(-z') \\
&+\frac{1}{g}\sum_{x',cd}C^{(\frac{p_0}{2})}_{xx'}\int\frac{dz'}{2\pi i}\frac{1}{(z-z')^2} \\
&\qquad\times\tilde\psi^{ac}_x(z')\tilde\psi^{db}_x(z')\tilde\psi^{cd}_{x'}(-z') \\
&+\ \text{(potential terms)}\ =0\ .
\end{aligned}
\tag{17}
$$

The contour of integration encloses the singularities of $\tilde\phi_x(z')$ and leaves outside the point $z'=z$ as well as the singularities of $\tilde\phi_x(-z')$. We notice that the large N limit of the S-D equations coinside with those in the orientable string case. Thus the solutions of the large N limit of the S-D equations correctly give the same disc amplitudes as in the orientable string theory which have been obtained by Kazakov and Kostov.

5 Conclusion

We have proposed the real symmetric matrix-vector model defined on the discretized 1D target space which describes the conformal matter with the central charge $0 < c \leq 1$ on the non-orientable 2D random surfaces with boundaries. The non-critical non-orientable open-closed string field theory with Chan-Paton factor is derived from the underlying stochastic process defined by the matrix-vector model. F-P hamiltonian is a linear combination of three constraints, $\mathcal{L}_x(\ell)$, $\mathcal{J}_x^{ab}(\ell)$ and $\mathcal{K}_x^{ab}(\ell)$. They satisfy the algebraic relation including the Virasoro and $SO(r)$ current algebras. The closure of the constraints implies the integrability of the time evolution of the underlying stochastic process. The large N limit of the S-D equations are consistent to the orientable case. The scaling behaviour belongs to the same universality class as for the orientable case. We conjecture that at the central charge $c \to 1$ limit, the non-critical model may be equivalent to the 2D string theory. The non-orientable open-closed string theory with $SO(r)$ gauge symmetry by Chan-Paton method is consistent only if the gauge group is $SO(2)$. To prove the conjecture, we have to investigate more carefully the cancellation of the logarithmic singularities in the disc and mobius amplitudes.

References

1. T. Banks, W. Fischler, S. Shenker, L. Susskind, Phys. Rev. **D55**(1997) 5112.
2. N. Ishibashi, H. Kawai, Y. Kitazawa, A. Tsuchiya, Nucl. Phys. **B498**(1997) 469.
3. H. Itoyama, A. Tsuchiya, Prog. Theor. Phys. **101**(1999) 1371.
4. N. Ishibashi, H. Kawai, Phys. Lett. **B314**(1993) 190;
   Phys. Lett. **B322**(1994) 67.
5. N. Nakazawa, Mod. Phys. Lett. **A10**(1995) 2175;
   A. Jevicki, J. Rodrigues, Nucl. Phys. **B421**(1994) 278.
6. E. Brézin, H. Neuberger, Phys. Rev. Lett. **65**(1990) 2098; Nucl. Phys. **B350**(1991) 513.
7. V. A. Kazakov Phys. Lett. **B237**(1990) 212;
   I. Kostov, Phys. Lett. **B238**(1990)181;
   V. A. Kazakov, I. Kostov, Nucl. Phys. **B386**(1992)520.
8. I. Kostov, Phys. Lett. **B344**(1995) 135; Phys. Lett. **B349**(1995) 284.
9. N. Nakazawa, D. Ennyu, Phys. Lett. **B417**(1998)247.
10. D. Ennyu, H. Kawabe, N. Nakazawa, Phys. Lett. **B454**(1999) 43.

NONCOMPACT GEPNER MODELS FOR TYPE II STRINGS ON A CONIFOLD AND AN ALE INSTANTON

SHUN'YA MIZOGUCHI

Institute of Particle and Nuclear Studies, KEK
1-1 Oho, Tsukuba, Ibaraki 305-0801, JAPAN
E-mail: shunya.mizoguchi@kek.jp

We construct modular invariant partition functions for type II strings on a conifold and a singular Eguchi-Hanson instanton by means of the $SL(2,\mathbb{R})/U(1)$ version of Gepner models. In the conifold case, we find an extra massless hypermultiplet in the IIB spectrum and argue that it may be identified as a soliton. In the Eguchi-Hanson case, our formula is new and different from the earlier result, in particular does not contain graviton. The lightest IIB fields are combined into a six-dimensional $(2,0)$ tensor multiplet with a negative mass square. We give an interpretation to it as a doubleton-like mode.

1 Introduction

The recognition of the importance of physics near the singularity in the moduli space is one of the highlights in the recent developments of string theory. Well-known examples are the conifold singularity of Calabi-Yau threefolds and the ADE singularities of K3 surfaces. In both cases, type IIA or IIB strings acquire extra massless solitons due to various D-branes wrapped around the vanishing cycles of the singularity[1,2].

In this contribution we construct modular invariant partition functions for the CFTs which describe those theories on such singularities.[a] At first sight, such an attempt might look like nonsense because one would expect non-perturbative effects near the singularity, where the CFT description of the strings on a *compact* space breaks down. However, we may still expect some dual perturbative CFT description for those theories[3], where the non-perturbative effects in one theory are studied perturbatively in the other theory. The idea is to "pinpoint" those theories by using an abstract CFT approach, just as we can use ordinary Gepner models to describe some special points of moduli space of ordinary Calabi-Yau compactifications.

[a] We do not consider any D-brane probe.

This paper is organized as follows. In section 2, we construct a partition function for type II strings on a conifold[4] and argue that the extra massless hypermultiplet that appears in the spectrum may be identified as a soliton. In section 3, we address an issue in the known partition function formulas for the singular ALE spaces. In section 4, we present a new modular invariant partition function for type II strings on a singular Eguchi-Hanson instanton, the simplest (A_1) four-dimensional ALE space in the ADE classification. Our formula is new and different from the earlier result[5,6]. The last section is devoted to some concluding remarks.

2 Conifold

Let us begin with four-dimensional type II theories "compactified" on a conifold. The CFT for the four-dimensional part is a free $N=2$ SCFT, while the conifold part is the $c=9$, $SL(2,\mathbb{R})/U(1)$ Kazama-Suzuki model[7]. The clue that leads to the relation between a conifold and the $SL(2,\mathbb{R})/U(1)$ SCFT may be found in the equation of the deformed conifold (See ref.[5] for further explanations and references.):

$$z_1^2+z_2^2+z_3^2+z_4^2=\mu, \tag{1}$$

where z_i $(i=1,\ldots,4)$ are the coordinates of $\mathbb{C}^4$. $\mu\to 0$ is the conifold limit. To apply the abstract CFT approach we replace μ with μz_5^{-1}, where z_i ($i = 1, \ldots, 5$) are now thought of as the coordinates of $\mathbb{P}^4$. The negative power of -1 has been determined by the Calabi-Yau condition, and interpreted as the $SL(2,\mathbb{R})/U(1)$ $N = 2$ model with level $k = -(-1 - 2) = 3$. Its central charge is $c = \frac{3k}{k-2} = 9$. Thus, unlike ordinary Gepner models, the necessary central charge for the internal CFT is supplied by a single $SL(2,\mathbb{R})/U(1)$ CFT.

In fact, if $c > 3$, *any* unitary representation for the $N = 2$ superconformal algebra can be constructed from the $SL(2,\mathbb{R})/U(1)$ coset[8]. It means that we are allowed to use any $c = 9$ representations. What's the criteria for the representations to be chosen? Our guideline is modular invariance and spacetime supersymmetry.

The generic (nondegenerate) $N = 2$, $c = 9$ superconformal characters are given by

$$\mathrm{Tr} q^{L_0} y^{J_0} = q^{h+1/8} y^Q \frac{\vartheta_3(z|\tau)}{\eta^3(\tau)} \tag{2}$$

(NS sector), where $q = \exp(2\pi i\tau)$ and $y = \exp(2\pi i z)$. To improve the modular behavior of the monomial factor $q^{h+1/8}$, we consider a countable set of infinitely many representations so that the sum of the monomial factors forms a certain theta function. (If the $c = 9$ CFT is realized as an $N = 2$ Liouville$\times S^1$ system[9,10], this operation amounts to the momentum summation along S^1.) It is still not enough to construct a modular invariant combination because the numbers of theta and η functions are not the same. (The $\sqrt{\tau}$ factors do not cancel in the modular S transformation.) To cure this problem, we consider *continuously many* representations for each $U(1)$ charge in the above theta function and integrate over them with the Gaussian weight (Liouville momentum integration).

Which theta functions should we use? To determine them we require spacetime supersymmetry. We now have three theta functions: one from the free complex fermion for the two transverse spacetime dimensions, one from the Jacobi theta function in the $N = 2$ characters, and one from the character summation above. They must be GSO projected in an appropriate way to give a vanishing partition function for the theory to be supersymmetric. Therefore, we need some identities among the products of three theta functions. One solution has been known for a long time[11]:

$$\begin{aligned}\Lambda_1(\tau) \equiv \Theta_{1,1}(\tau,0)\left(\vartheta_3^2(0|\tau) + \vartheta_4^2(0|\tau)\right) \\ -\Theta_{0,1}(\tau,0)\,\vartheta_2^2(0|\tau) = 0,\end{aligned} \tag{3}$$

where

$$\Theta_{m,1}(\tau,z) = \sum_{n\in\mathbb{Z}} q^{(n+m/2)^2} y^{(n+m/2)}, \tag{4}$$

($m = 0, 1$) are the level-1 $SU(2)$ theta functions. There is another solution[4]:

$$\begin{aligned}\Lambda_2(\tau) \equiv \Theta_{0,1}(\tau,0)\left(\vartheta_3^2(0|\tau) - \vartheta_4^2(0|\tau)\right) \\ -\Theta_{1,1}(\tau,0)\,\vartheta_2^2(0|\tau) = 0,\end{aligned} \tag{5}$$

which is nothing but the modular S transform of (3). Their modular transformations are

$$\Lambda_1(\tau+1) = i\Lambda_1(\tau), \quad \Lambda_2(\tau+1) = -\Lambda_2(\tau), \tag{6}$$

and

$$\begin{aligned}\Lambda_1\left(-\frac{1}{\tau}\right) &= \frac{e^{-\frac{3\pi i}{4}}\tau^{\frac{3}{2}}}{\sqrt{2}}\left(-\Lambda_1(\tau) + \Lambda_2(\tau)\right), \\ \Lambda_2\left(-\frac{1}{\tau}\right) &= \frac{e^{-\frac{3\pi i}{4}}\tau^{\frac{3}{2}}}{\sqrt{2}}\left(\Lambda_1(\tau) + \Lambda_2(\tau)\right).\end{aligned} \tag{7}$$

Thus $\left|\Lambda_1(\tau)/\eta^3(\tau)\right|^2 + \left|\Lambda_2(\tau)/\eta^3(\tau)\right|^2$ is modular invariant. Including the other contributions as well, we obtain the following modular invariant partition function:

$$\begin{aligned}Z_{\text{conifold}} = \int \frac{d^2\tau}{(\mathrm{Im}\tau)^2} \frac{1}{(\mathrm{Im}\tau)^{\frac{3}{2}} |\eta(\tau)|^6} \\ \cdot \left[\left|\Lambda_1(\tau)/\eta^3(\tau)\right|^2 + \left|\Lambda_2(\tau)/\eta^3(\tau)\right|^2\right].\end{aligned} \tag{8}$$

One may easily read off from Z_{conifold} which $N = 2$, $c = 9$ representations are used (Figure 1; see ref.[4] for the R-sector.).

The following observations support that Z_{conifold} is really the partition function for type II strings on a conifold:

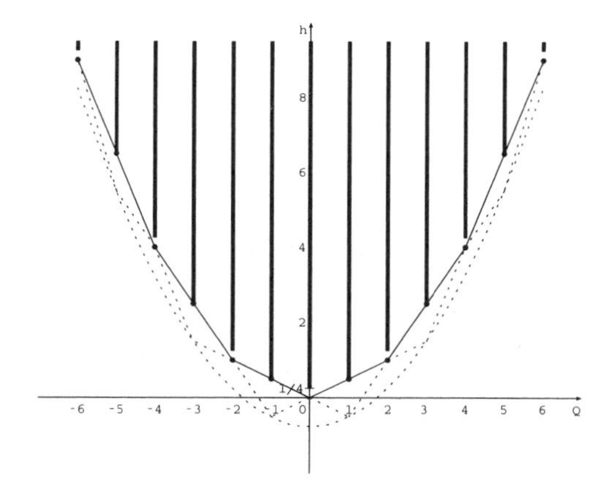

Figure 1. $N = 2$, $c = 9$ representations used as the internal CFT (NS-sector).

• First of all, the spectrum exhibits a continuum. This is due to the integration over the $N = 2$ representations (Liouville momentum integration), which is a consequence of the requirement from modular invariance. Therefore, spacetime, which was initially supposed to be four-dimensional, becomes five-dimensional effectively. Another related observation is that the graviton, the dilaton and the B-field correspond to the $(h, Q) = (1/4, 0)$ point, and hence are massive. This is a generic $N = 2$ representation and corresponds to a non-normalizable ("principal unitary series") $SL(2, \mathbb{R})$ representation. This is certainly different from the ordinary CFT description of the Calabi-Yau "*compact*"-ifications.

• Consider type IIB theory. We have one chiral- and one anti-chiral primary fields of $h = 1/2$ in the NS-sector (Figure 1). They (together with their spectral flows) give rise to a massless $N = 2$, $U(1)$ vector multiplet and a hypermultiplet. The vector multiplet agrees with the Hodge number $h_{2,1} = 1$ of the deformed conifold, while the hypermultiplet is an extra one.[b] We will argue that this hypermultiplet may be identified as the famous massless soliton[1] coming from the wrapped D3 brane. The extra massless fields are due to the second chiral primary field $(h, Q) = (1, 2)$, which exhibit a gap above. Technically, the representations on the boundary of the unitarity region are degenerate representations, and hence are smaller than the generic ones. The irreducible character for $(h, Q) = (1/2, 1)$ is given by

$$\mathrm{Tr} q^{L_0} y^{J_0}|_{(h,Q)=(\frac{1}{2},1)} = \frac{q^{\frac{1}{2}+\frac{1}{8}} y}{1 + q^{\frac{1}{2}} y} \frac{\vartheta_3(z|\tau)}{\eta^3(\tau)} \qquad (9)$$

and not in the generic form like (2). In fact, their difference is $((h, Q) = (\frac{1}{2}, 1))$

$$(2) - (9) = \frac{q^{1+\frac{1}{8}} y^2}{1 + q^{\frac{1}{2}} y} \frac{\vartheta_3(z|\tau)}{\eta^3(\tau)}, \qquad (10)$$

which is precisely the irreducible character for $(h, Q) = (1, 2)$[12].[c] (And similarly for the anti-chiral primary fields; they are shown by the dots in Figure 1.) Therefore, a generic representation with $Q = 1$ splits into two chiral primaries on reaching the boundary of the unitarity region. This implies that the massless fields made out of the $(h, Q) = (1/2, \pm 1)$ (anti-)chiral primary fields cannot acquire masses (= Liouville momenta) alone, without "eating" the fields from the $(h, Q) = (1, \pm 2)$ representations,[d] and both are trapped near the singularity. This will support the identification of the extra hypermultiplet as the massless soliton.

• An $N = 2$, $U(1)$ vector multiplet and a hypermultiplet are also the massless excitations of the intersecting M5-branes[13]. This will agree with the duality of a conifold to the intersecting NS5-brane system proposed in refs.[3,14].

---

[b]This is neither the universal hypermultiplet, nor the one coming from the Kähler form on the conifold; the dilaton is paired with the $(h, Q) = (1/4, 0)$ representation, while the Kähler form does not have a compact support and hence will correspond to a continuous representation of $SL(2, \mathbb{R})$.

[c]Degenerate representations on the boundary of the unitarity region correspond to discrete series of $SL(2, \mathbb{R})$[8].

[d]This is a phenomenon reminiscent of the BPS saturation or the Higgs mechanism. We thank T. Eguchi and S.-K. Yang for comments.

3 ALE Instantons: a Puzzle

Let us next consider the "blow-down" limit of the ALE instantons. A similar guess from the defining equations suggests that the corresponding CFT will be a certain Landau-Ginzburg orbifold of a tensor product of $N=2$ minimal models and a $SL(2,\mathbb{R})/U(1)$ Kazama-Suzuki model with total central charge 6[5]. For example, the equations for the A_{n-1}-series are given by

$$z_1^n + z_2^2 + z_3^2 = \mu z_4^{-n}, \qquad (11)$$

where z_i $(i = 1,\ldots,4)$ are now the coordinates of $\mathbb{P}^3$. Thus the corresponding CFT is the product of the minimal model of level $n-2$ and the $SL(2,\mathbb{R})/U(1)$ CFT of level $n+2$, *orbifolded by* $\mathbb{Z}_n$.

We will now clarify what is the puzzle in the known formulas for the partition functions obtained in the previous works[6]. (See also ref.[15].) The standard argument goes as follows: To construct a modular invariant, three different $N=2$ SCFTs must be taken into account: the free $N=2$ SCFT for the four transverse spacetime dimensions, the level-$(n-2)$ minimal model, and the level-$(n+2)$ $SL(2,\mathbb{R})/U(1)$ SCFT. Since the $SL(2,\mathbb{R})/U(1)$ construction scans the whole $N=2$ unitarity region[8], we may consider any $U(1)$ charge lattice and sum over the representations in the third $c = 3+\frac{6}{n}$ SCFT. Thus the building blocks must consist of three Jacobi theta functions (two from the transverse dimensions and one from $SL(2,\mathbb{R})/U(1)$), an $N=2$ minimal character, and a certain theta function.

The crucial observation is that the following theta identity holds[5,6] (See also ref.[16], eq.(2.24).):

$$\sum_{m\in\mathbb{Z}_{2(k+2)}} \Theta_{m,k+2}(\tau,\frac{kz}{k+2}) \Big[\vartheta_3^3(0)ch_{l,m}^{(\mathrm{NS})}(\tau,z) - \vartheta_4^3(0)ch_{l,m}^{\widetilde{(\mathrm{NS})}}(\tau,z) - \vartheta_2^3(0)ch_{l,m}^{(\mathrm{R})}(\tau,z)\Big] = \chi_l^{(k)}(\tau,z)\left(\vartheta_3^4(0)-\vartheta_4^4(0)-\vartheta_2^4(0)\right), \qquad (12)$$

where $ch_{l,m}(\tau,z)$ and $\chi_l^{(k)}(\tau,z)$ are the $N=2$ minimal and the affine $SU(2)$ characters, respectively. This means that if we take $\Theta_{m,k+2}(\tau,\frac{kz}{k+2})$ as the theta function for the $c = 3+\frac{6}{n}$, $SL(2,\mathbb{R})/U(1)$ SCFT and consider the combinations of operators specified in the LHS of (12), we get a product of a level-k $SU(2)$ WZW model and a free fermion theory. Modular invariants are easily constructed from those of the $SU(2)$ WZW models. This fact has been used to rationalize the duality between the singular K3 and the system of NS5-branes at the partition function level[5,6].

Now here comes the puzzle: Since the lightest fields always come from the identity operator of $SU(2)$ while Jacobi's identity is independent of the type of the singularity, the ground-state degeneracy does not change whatever the singularity is! To look into the situation more closely, let us consider the simplest A_1 $(n=2)$ case, the blow-down limit of the Eguchi-Hanson instanton. In this case the partition function is simply given by

$$Z_{\mathrm{EH}}^{(\mathrm{graviton})} = \int \frac{d^2\tau}{(\mathrm{Im}\tau)^2}\frac{1}{(\mathrm{Im}\tau)^{\frac{5}{2}}|\eta(\tau)|^{18}} \cdot \left|\frac{1}{2}(\vartheta_3^4-\vartheta_4^4-\vartheta_2^4+\vartheta_1^4)\right|^2 . (13)$$

This is almost identical to the partition function for $D=10$ type II(B) strings, but the essential difference is the number of η functions; since the free $c=6$, $N=2$ SCFT for the four-dimensional flat space is replaced by the $SL(2,\mathbb{R})/U(1)$ SCFT, the partition function (13) has only *nine* (rather than twelve) η functions in the denominator. Therefore, not the whole ghost ground-state energy $-1+\frac{1}{2} = -\frac{1}{2}$ is absorbed in the η functions, but the $-\frac{1}{8}$ due to the shortage of three ηs is canceled by the "Liouville ground-state energy" of $+\frac{1}{8}$ from the $SL(2,\mathbb{R})/U(1)$ internal SCFT. (Otherwise $q^{1/8}$ has no where to go and the modular invariance is lost. In the conifold case this was $q^{1/4}$ and can be seen as the gap above the origin (Figure 1).)

Thus the graviton is again paired with a non-normalizable state of the internal SCFT, and hence acquires a $(\text{mass})^2$ of $\frac{1}{8}$. This is always the case for other ADE singularities, being a common feature of non-critical strings formulated as such Gepner-like models.[e] No other field is lighter than the graviton in this partition function. This implies that the modular invariant constructed from the identity (12) does *not* respect the geometry of the ALE spaces, *nor* can it see the zeromodes of the dual NS5-branes. This is the puzzle.

Is there any other orbit than (12) that closes under modular transformations? In the next section we will construct a new modular invariant partition function[f] for the Eguchi-Hanson instanton by choosing a different initial condition in the β-method.

4 New Partition Function for the Eguchi-Hanson Instanton

A general method to obtain a modular invariant combination from a set of theta functions has been given in Gepner's original paper[16] and was called "β-method". We first note that the quartic polynomial of the theta functions in (13) can be written as follows:

$$\begin{aligned}&(\vartheta_3+\vartheta_4)^2(\vartheta_3^2-\vartheta_4^2)-(\vartheta_2+\vartheta_1)^2(\vartheta_2^2-\vartheta_1^2)\\&+(\vartheta_3-\vartheta_4)^2(\vartheta_3^2-\vartheta_4^2)-(\vartheta_2-\vartheta_1)^2(\vartheta_2^2-\vartheta_1^2)\\&=2(\vartheta_3^4-\vartheta_4^4-\vartheta_2^4+\vartheta_1^4).\end{aligned}\tag{14}$$

The LHS of the equation clearly shows how the GSO projection should be done in each sector; for each term, the first factor is the state sum of the transverse spacetime dimensions, while the second is the one coming from the internal $N=2$ SCFT. The first and the third terms are in the NS sector, and the second and the fourth ones are in the R sector, for both spacetime and internal SCFTs. We now consider a variant of this equation:

$$\begin{aligned}&(\vartheta_3+\vartheta_4)^2(\vartheta_2+\vartheta_1)^2-(\vartheta_2+\vartheta_1)^2(\vartheta_3-\vartheta_4)^2\\&+(\vartheta_3-\vartheta_4)^2(\vartheta_2-\vartheta_1)^2-(\vartheta_2-\vartheta_1)^2(\vartheta_3+\vartheta_4)^2\\&=16\vartheta_1\vartheta_2\vartheta_3\vartheta_4.\end{aligned}\tag{15}$$

Each term gives the same phase in the modular T transformation, and their sum returns to itself up to a phase by the S transformation. Using this, we write a new modular invariant:

$$\begin{aligned}Z_{\rm EH}^{\rm (doubleton)}&=\int\frac{d^2\tau}{({\rm Im}\tau)^2}\frac{1}{({\rm Im}\tau)^{\frac{5}{2}}|\eta(\tau)|^{18}}\cdot\frac{1}{16^2}\\&\cdot\Big|(\vartheta_3+\vartheta_4)^2\left((\vartheta_2+\vartheta_1)^2+(\vartheta_2-\vartheta_1)^2\right)\\&-(\vartheta_2+\vartheta_1)^2\left((\vartheta_3-\vartheta_4)^2+(\vartheta_3+\vartheta_4)^2\right)\\&+(\vartheta_3-\vartheta_4)^2\left((\vartheta_2-\vartheta_1)^2+(\vartheta_2+\vartheta_1)^2\right)\\&-(\vartheta_2-\vartheta_1)^2\left((\vartheta_3+\vartheta_4)^2+(\vartheta_3-\vartheta_4)^2\right)\Big|^2.\end{aligned}\tag{16}$$

In fact, the alternating sum inside $|\cdots|$ vanishes *trivially*; however, it can be interpreted as a consequence of the cancellation between the NS- and the R-sectors. Again, the first and the third terms are in the NS-sector for the transverse spacetime SCFT, but they are now paired with the R-sector of the internal SCFT (and similarly for the transverse R-sector)![g] Note that $Z_{\rm EH}^{\rm (graviton)}$ and $Z_{\rm EH}^{\rm (doubleton)}$ are separately modular invariant.

Remarkably, $Z_{\rm EH}^{\rm (doubleton)}$ does not contain any graviton because of its peculiar GSO projection. The lightest IIB fields are combined into a $(2,0)$ tensor multiplet in six dimensions, which coincides with the zeromode excitations on the IIA NS5-brane[18]!; this is a manifestation of T-duality. In fact, its $(\text{mass})^2$ is *negative* $(=-\frac{1}{8})$! It does not necessarily mean the instability of our vacuum because there is no reason to believe that the six-dimensional spacetime is flat any more. Perhaps it might be understood as a doubleton-like mode. The doubleton[19] is known to be the lowest Kaluza-Klein mode with a negative $(\text{mass})^2$ in (say) the $AdS_7\times S^4$ compactification of $D=11$ supergravity.

---

[e] We thank M. Bando, H. Kawai and T. Kugo for discussions on this point.

[f] The work done in collaboration with M. Naka.

[g] The change of the fermion boundary condition was discussed in ref.[17], but they were led to a different result.

It is a pure gauge mode in the bulk but has a holographic dual on the six-dimensional *AdS* boundary, on which an M5-brane sits. If the T-duality relation between a singular ALE and NS5-branes persists even in strong coupling, and if the abstract CFT approach can consistently describe physics near the singularity (as it seemed to be in the conifold case), the CFT must "see" some alternative dual to the strongly coupled type IIA theory; the one-loop CFT calculation for the latter itself is certainly inconsistent. Thus if the NS5-branes could be replaced by M5-branes, the mysterious tachyonic fields would then have a explanation as a doubleton-like mode. For the full understanding of this, we will need to generalize our formula to other ADE cases.

5 Concluding Remarks

We have constructed partition functions for type II strings on a conifold and a singular Eguchi-Hanson instanton by using a Gepner-like abstract CFT approach.

In the conifold case, we found an extra massless hypermultiplet and argued that this might be identified as the massless soliton. The appearance of the massless soliton in the spectrum comes as a surprise; to confirm the identification further, the nature of the extra states must be clarified in the boundary-state formulation of D-branes.

The doubleton-like mode in the Eguchi-Hanson case is also quite unexpected. The relation to the M-theory dual is still mysterious. It would be interesting to compare (reconcile?) our result with the works of refs.[10].

Acknowledgments

We thank M. Naka and T. Tani for valuable discussions.

References

1. A. Strominger, *Nucl. Phys.* **B451**, 96 (1995).
2. E. Witten, *Nucl. Phys.* **B443**, 85 (1995).
3. M. Bershadsky, C. Vafa and V. Sadov, *Nucl. Phys.* **B463**, 398 (1996).
4. S. Mizoguchi, *JHEP* **0004**, 014 (2000).
5. H. Ooguri and C. Vafa, *Nucl. Phys.* **B463**, 55 (1996).
6. T. Eguchi and Y. Sugawara, *Nucl. Phys.* **B577**, 3 (2000).
7. D. Ghoshal and C. Vafa, *Nucl. Phys.* **B453**, 121 (1995).
8. L.J. Dixon, M.E. Peskin and J. Lykken, *Nucl. Phys.* **B325**, 329 (1989).
9. O. Aharony, M. Berkooz, D. Kutasov and N. Seiberg, *JHEP* **9810**, 004 (1998). A. Giveon, D. Kutasov and O. Pelc, *JHEP* **9910**, 035 (1999).
10. A. Giveon and D. Kutasov, *JHEP* **9910**, 034 (1999); *JHEP* **0001**, 023 (2000).
11. A. Bilal and J.L. Gervais, *Nucl. Phys.* **B284**, 397 (1987).
12. T. Eguchi and A. Taormina, *Phys. Lett.* **B210**, 125 (1988).
13. A. Hanany and I.R. Klebanov, *Nucl. Phys.* **B482**, 105 (1996).
14. K. Dasgupta and S. Mukhi, *Nucl. Phys.* **B551**, 204 (1999).
15. S. Yamaguchi, hep-th/0007069.
16. D.Gepner, *Nucl.Phys.***B296**, 757 (1988).
17. D. Anselmi, M. Billo, P. Fre, L. Girardello and A. Zaffaroni, *Int. J. Mod. Phys.* **A9**, 3007 (1994).
18. C.G. Callan, J.A. Harvey and A. Strominger, *Nucl. Phys.* **B367**, 60 (1991).
19. M. Gunaydin, P. van Nieuwenhuizen and N.P. Warner, *Nucl. Phys.* **B255**, 63 (1985).

WORLD-SHEET OF THE DISCRETE LIGHT FRONT STRING

GORDON W. SEMENOFF
Niels Bohr Institute, Blegdamsvej 17, D-2100 Copenhagen Ø, Denmark
E-mail: semenoff@nbi.dk

Some aspects of light-like compactifications of superstring theory and their implications for the matrix model of M-theory are discussed.

1 Light-Like compactification

T-duality is one of the most profound of stringy phenomena. Closed string theory on a space-time which has a compact dimension with radius R has the same spectrum as a string theory on a space with radius α'/R. This leads to many interesting properties and is an essential part of the web of dualities which relate the different superstring theories and M theory. One might ask whether it is important that the dimension which is compactified is space-like. In this talk, I will review some recent work[1] which asks what happens when the dimension that is compactified is light-like, rather than space-like.

Of course, we could get a light-like circle by boosting a compactified spatial circle by an infinite amount[2]. Consider a closed string with compactified spatial direction, $X^{D-1} \sim X^{D-1} + 2\pi R$. In terms of light-cone coordinates, $X^{\pm} = \frac{1}{\sqrt{2}}\left(X^0 \pm X^{D-1}\right)$

$$(X^+, X^-) \sim \left(X^+ + \sqrt{2}\pi R, X^- - \sqrt{2}\pi R\right)$$

We consider a boosted reference frame, where $\tilde{X}^+ = \Lambda X^+$ and $\tilde{X}^- = \Lambda^{-1} X^-$. We fix $\Lambda = \sqrt{2}R^+/R$ and take the limit $\Lambda \to \infty$ with R^+ fixed, we finally get the light-like direction compactified,

$$\left(\tilde{X}^+, \tilde{X}^-\right) \sim \left(\tilde{X}^+ + 2\pi R^+, X^-\right) \qquad (1)$$

The original spatial circle is vanishingly small, $R = \sqrt{2}R^+/\Lambda$. The momenta transform to

$$P^+ = \Lambda \frac{\sqrt{(P^{D-1})^2 + \vec{P}^2 + M^2} + P^{D-1}}{\sqrt{2}}$$

$$P^- = \frac{\sqrt{(P^{D-1})^2 + \vec{P}^2 + M^2} - P^{D-1}}{\sqrt{2}\Lambda}$$

with $P^{D-1} = -N/R$ and $\vec{p}^2 = \sum_1^{D-2} p_i^2$. In the limit of infinite boost this becomes

$$(P^+, P^-) = \left(\frac{R^+}{2N}\left(\vec{P}^2 + M^2\right), N/R^+\right)$$

Here, P^+ is the infinite momentum frame Hamiltonian which generates translations of X^-. Also, since $X^+ \sim X^+ + 2\pi R^+$ the conjugate momentum is quantized, $P^+ = N/R^+$. Of the states with $N = 0$, all but the massless, low momentum ones get infinite energy.

Closed string theory on a space with one dimension compactified to a vanishingly small circle is T-dual to a closed string theory on the un-compactified space. As a consequence, the spatial compactification to a vanishingly small circle has no effect on the spectrum of the theory. For any light-like compactification radius R^+, the energy spectrum of the rest-frame states is just that of the decompactified theory.

Of course the states which have finite energy and momentum in the frame with compact light-cone must have infinite momentum in the compact spatial direction and infinite energy in the original rest frame. Under T-duality, states with monzero momentum in the vanishingly small compact direction are exchanged with states with fundamental strings wrapping the very large dual circle. Since they are very long, they have large energy. *Thus, light-like compactification does not alter the spectrum of the string theory. What it does is explores the theory which is*

T-dual to it in a kinematical regime where there are long fundamental strings wrapping an almost infinite compact direction. This a high energy state in the rest frame string theory and one could in principle study it there. In the infinite momentum frame it is a generic state with finite energy and momentum. This will be the reason why the zero temperature limit of the partition functions that we shall compute in the following are independent of R^+.

The thermodynamic partition function is obtained from the trace over physical states of the Boltzmann factor, $\exp(-\beta_\mu P^\mu)$ [a]

$$Z = \sum_{N=0}^{\infty} e^{-\beta_- N/R^+} \int \frac{d^{D-2}P}{(2\pi)^{D-2}} \cdot e^{-\vec{\beta}\cdot\vec{P}} \cdot e^{-\beta_+ R^+ \vec{P}^2/2N} \sum_{M^2} \rho(M^2) e^{-\beta_+ R^+ M^2/2N} \quad (2)$$

where we sum over states in the mass spectrum. These are conveniently found in the light cone gauge by imposing the constraints $L_0 + \tilde{L}_0 = 0$ and the level matching condition $L_0 - \tilde{L}_0 = N \cdot \text{integer}$. (Details are explained in [3].) The first constraint gives the mass shell condition. The second is the level matching condition and is imposed using an integer-valued Lagrange multiplier. The result for the NSR superstring is elegantly summarized in terms of the Hecke operator [4] acting on the partition function of the superconformal field theory with target space R^8,

$$-\frac{2\pi\beta_\mu R^\mu F}{V} = \mathcal{H}[e^{-\beta_\mu\beta^\mu/2\beta_\mu R^\mu}] * \mathcal{F}[\tau,\bar{\tau}] \quad (3)$$

where

$$\mathcal{F} = \left[\left(\frac{1}{4\pi^2\alpha'\tau_2} \right)^4 \frac{1}{|\eta(\tau)|^{24}} |\theta_2(0,\tau)|^8 \right]_{\tau=i\nu} \quad (4)$$

and $\nu = 2\pi\alpha'/\beta_\mu R^\mu$ is a fixed constant (β_μ is spacelike and R^μ is light-like). The factor in front contains the ratio of volumes of R^8 and $R^9 \times S^1$ with compactified light cone. The action of $\mathcal{H}[p]$ on a modular function $\phi(\tau,\bar{\tau})$ is defined by

$$\mathcal{H}[p] * \phi(\tau,\bar{\tau}) = \sum_{N=1}^{\infty} p^N \sum_{\substack{kr=N,\ r\ \text{odd} \\ s \bmod k}} \frac{1}{N} \phi\left(\frac{s+\tau r}{k}, \frac{s+\bar{\tau} r}{k} \right) \quad (5)$$

This is similar to other partition functions for conformal field theories on symmetric orbifolds. For a recent discussion see ref.[5].

This result is a discretization of the usual Teichmuller space which occurs in the genus 1 string amplitude. Recall that the genus zero contribution is insensitive to compactifications and for the superstring it vanishes. At genus 1, because of the modification of the GSO projection by the finite temperature boundary conditions, the superstring torus amplitude is non-zero. Here, we see that the usual integration over the Teichmuller space of tori is replaced by a summation over discrete parameters

$$\tau = \frac{s + i\nu r}{k}$$

2 A theorem about path integrals

It is interesting to see what happens when we compactify a null direction in the path integral representation of the string free energy. We can do this for the contribution from arbitrary genus. We begin with

$$F = -\sum_{g=0}^{\infty} g_s^{2g-2} \int [dh_g dX] e^{-S[h_g,X]} \quad (6)$$

where

$$S[h_g, X] = \frac{1}{4\pi\alpha'} \int \sqrt{h} h^{ab} \partial_a X^\mu \partial_b X^\mu \quad (7)$$

Here we will use the Bosonic string. Most of our considerations apply to the bosonic sector of any string theory. For the relationship with matrix theory which we shall discuss later, supersymmetry is important and

[a] Here, we have introduced a covariant temperature. β_μ should be understood as the inverse temperature $\beta = 1/k_B T = \sqrt{2\beta_+\beta_- - \vec{\beta}^2}$ times the D-velocity of the heat bath $v^\mu = \beta^\mu/\beta$. k_B is Boltzmann's constant.

that case is more closely related to the Green-Schwarz superstring. The only modification of our arguments for the superstring would be that either the world-sheet fermion boundary conditions in the Green-Schwarz case, or the GSO projection in the Neveu-Schwarz-Ramond case would be modified to depend on the winding numbers of the world-sheet in the compact time direction.

The string coupling constant is g_s and its powers weight the genus, of the string's world-sheet. For each value of the genus, g, $[dh_g]$ is an integration measure over all metrics of that genus and is normalized by dividing out the volume of the world-sheet re-parameterization and Weyl groups. We will assume that the metrics of both the world-sheet and the target spacetime have Euclidean signatures.

We wish to study the situation where the target space has particular compact dimensions. Two compactifications will be needed. The first compactifies the light-cone in Minkowski space by making the identification (1). In our Euclidean coordinates,

$$(X^0, X^9) \sim \left(X^0 + \sqrt{2}\pi i R, X^9 - \sqrt{2}\pi R\right) \tag{8}$$

In order to introduce temperature, we shall have to compactify Euclidean time,

$$\left(X^0, \vec{X}, X^9\right) \sim \left(X^0 + \beta, \vec{X}, X^9\right) \tag{9}$$

This compactification (with the appropriate modification of the GSO projection in the case of superstrings) introduces the temperature, $T = 1/k_B\beta$, so that (6) computes the thermodynamic free energy.

In order to implement this compactification in the path integral, we assume that the world-sheet is a Riemann surface Σ_g of genus g whose homology group $H_1(\Sigma_g)$ is generated by the closed curves,

$$a_1, a_2, \ldots, a_g \ , \ b_1, b_2, \ldots, b_g$$
$$a_i \cap a_j = \emptyset, \ b_i \cap b_j = \emptyset, \ a_i \cap b_j = \delta_{ij} \tag{10}$$

Furthermore, one may pick a basis of holomorphic differentials $\omega_i \in H^1(\Sigma_g)$ with the properties

$$\oint_{a_i} \omega_j = \delta_{ij} \quad , \quad \oint_{b_i} \omega_j = \Omega_{ij} \tag{11}$$

where Ω is the period matrix. It is complex, symmetric, $\Omega_{ij} = \Omega_{ji}$, and has positive definite imaginary part.

Compactification is implemented by including the possible windings of the string world-sheet on the compact dimensions. These form distinct topological sectors in the path integration in (6). In the winding sectors, the bosonic coordinates of the string should have a multi-valued part which changes by β·integer or $(i)\sqrt{2}\pi R$·integer as it is moved along a homology cycle. The derivatives of these coordinates should be single-valued functions. It is convenient to consider their exterior derivatives which can be expressed as linear combinations of the holomorphic and anti-holomorphic 1-forms and exact parts,

$$\begin{aligned} dX^0 &= \sum_{i=1}^{g} \left(\lambda_i \omega_i + \bar{\lambda}_i \bar{\omega}_i\right) + \text{exact} \\ dX^9 &= \sum_{i=1}^{g} \left(\gamma_i \omega_i + \bar{\gamma}_i \bar{\omega}_i\right) + \text{exact} \end{aligned} \tag{12}$$

Then, we require

$$\begin{aligned} \oint_{a_i} dX^0 &= \beta n_i + \sqrt{2}\pi R^+ i p_i \\ \oint_{b_i} dX^0 &= \beta m_i + \sqrt{2}\pi R^+ i q_i \\ \oint_{a_i} dX^9 &= \sqrt{2}\pi R^+ p_i \\ \oint_{b_i} dX^9 &= \sqrt{2}\pi R^+ q_i \end{aligned} \tag{13}$$

with p_i, q_i, m_i, n_i integers. With (11), we use these equations to solve for the constants in (12). With the formula $\int \omega_i \bar{\omega}_j = -2i\left(\Omega_2\right)_{ij}$, we compute the part of the string action which contains the winding integers,

$$S = \frac{\beta^2}{4\pi\alpha'}\left(n\Omega^\dagger - m\right)\Omega_2^{-1}\left(\Omega n - m\right) +$$
$$2\pi i \frac{\sqrt{2}\beta R^+}{4\pi\alpha'}\frac{1}{2}\left[\left(p\Omega^\dagger - q\right)\Omega_2^{-1}\left(\Omega n - m\right)\right.$$

$$+ \left(n\Omega^\dagger - m\right)\Omega_2^{-1}\left(\Omega p - q\right)\big] + \ldots (14)$$

Note that the integers p_i and q_i appear linearly in a purely imaginary term in the action. This is the only place that they appear in the string path integral (unlike m_i and n_i which could appear in the GSO projection). When the action is exponentiated and summed over p_i and q_i, the result will be periodic Dirac delta functions. These delta functions impose a linear constraint on the period matrix of the world-sheet. Thus, with the appropriate Jacobian factor, the net effect is to insert into the path integral measure the following expression,

$$\sum_{\substack{mn\\rs}} e^{-\frac{\beta^2}{4\pi\alpha'}(n\Omega^\dagger - m)\Omega_2^{-1}(\Omega n - m)} \left|\det \Omega_2\right|$$

$$\nu^{2g} \prod_{j=1}^{g} \delta\left(\left(n_i + i\nu r_i\right)\Omega_{ij} - \left(m_j + i\nu s_j\right)\right) (15)$$

where $\nu = 4\pi\alpha'/\sqrt{2}\beta R^+$, the same constant as in (4) if we specialize to the temperature D-vector $\beta_0 \equiv \beta$, all other components vanishing. Consequently, the integration over metrics in the string path integral is restricted to those for which the period matrix obeys the constraint

$$\sum_{i=1}^{g} \left(n_i + i\nu r_i\right)\Omega_{ij} - \left(m_j + i\nu s_j\right) = 0 \quad (16)$$

for all combinations of the $4g$ integers m_i, n_i, r_i, s_i such that Ω is in a fundamental domain of period matrices for surfaces of genus g.

Since the columns of the period matrix are linearly independent vectors, these are g independent complex constraints on the moduli space of Σ_g. Thus its complex dimension $3g-3$ is reduced to $2g-3$ and there is further discrete data contained in the integers. One would expect that, when the compactifications are removed, either $\beta \to \infty$ or $R^+ \to \infty$, the discrete data assembles itself to a "continuum limit" which restores the complex dimension of moduli space.

It is interesting to ask whether the Riemann surfaces with the constraint (16) can be classified in a sensible way. The answer to this question is yes, a Riemann surface obeys the constraint (16) if and only if it is a branched cover of the torus, T^2, with Teichmuller parameter $i\nu$. This is established through the

Theorem: Σ_g is a branched cover of T^2 if and only if the period matrix obeys (16), for some choice of integers m_i, n_i, r_i and s_i.

The *proof* can be found in ref. [1].

As a concrete example, the constraint (16) can be solved explicitly for genus one. The torus amplitude for the finite temperature type II superstring was given in the NSR formulation by Attick and Witten [6].

The modification of their formula by the null compactification can be found using (15),

$$\frac{F}{V} = -\sum_{\tau\in\mathcal{F}} \frac{\nu^2 e^{-\frac{\beta^2|n\tau-m|^2}{4\pi\alpha'\tau_2}}}{m^2+\nu^2 n^2}\left(\frac{1}{4\pi^2\alpha'\tau_2}\right)^5$$
$$\cdot\frac{1}{4\left|\eta(\tau)\right|^{24}}\left[\left(\theta_2^4\bar\theta_2^4 + \theta_3^4\bar\theta_3^4 + \theta_4^4\bar\theta_4^4\right)(0,\tau) +\right.$$
$$+ e^{i\pi(m+n)}\left(\theta_2^4\bar\theta_4^4 + \theta_4^4\bar\theta_2^4\right)(0,\tau) - e^{i\pi n}\left(\theta_2^4\bar\theta_3^4\right.$$
$$\left.\left. + \theta_3^4\bar\theta_2^4\right)(0,\tau) - e^{i\pi m}\left(\theta_3^4\bar\theta_4^4 + \theta_4^4\bar\theta_3^4\right)(0,\tau)\right] \quad (17)$$

where the solution of (16) yields the discrete Teichmuller parameter,

$$\tau = \frac{m + i\nu s}{n + i\nu r}$$

and one should sum over the integers so that τ is in the fundamental domain, $\mathcal{F}$,

$$\mathcal{F} \equiv \left\{\tau_1 + i\tau_2 \left| -\frac{1}{2} < \tau_1 \le \frac{1}{2}; |\tau| \ge 1; \tau_2 > 0\right.\right\} \quad (18)$$

Modular transformations and identities for theta functions can be used to rewrite (17) as the Hecke operator acting on the partition function of a superconformal field theory, with torus world-sheet and target space R^8 seen in the formulae (3), (4) and (5).

3 Implications for the matrix model

M-theory is a parameter free quantum mechanical system which has 11-dimensional super-Poincare symmetry, 11-dimensional supergravity as its low energy limit and produces the five known consistent superstring theories at various limits of its moduli space. Details of its dynamics are thus far unknown. The matrix model [7], [8] model is conjectured to describe the full dynamics of M-theory in a particular kinematical context, the infinite-momentum frame.

An important check of the matrix model conjecture would be to use it to reproduce perturbative string theory. Most straightforward is the IIA superstring which is gotten by compactification of a spatial direction of M-theory. With this compactification, the matrix model itself becomes 1+1-dimensional, maximally-supersymmetric Yang-Mills theory. According to Dijkgraaf, Verlinde and Verlinde [9], the string degrees of freedom which emerge in the perturbative string limit are simultaneous eigenvalues of the matrices. At finite temperature, the matrices are defined on a torus [3] and their eigenvalues, since they solve polynomial equations, are functions on branched covers of the torus. If the matrix model is to agree with perturbative string theory, these branched covers must be the full set of Riemann surfaces that contribute to the string path integral. In the work that we have reviewed here, we have indeed seen that this is the case, the moduli spaces of branched covers which occur in the matrix model and the world-sheets that occur in the measure in the string path integral are identical.

Acknowledgments

This work is supported in part by MaPhySto, Center for Mathematical Physics and Stochastics, funded by the Danish National Research Foundation. It is also supported in part by NSERC of Canada. I have benefited from conversations with Jan Ambjørn, Gianluca Grignani, Mark Goresky, Janosh Kollar, Yutaka Matsuo, Peter Orland, Lori Paniak, Savdeep Sethi, Graham Smith, Richard Szabo, Erik Verlinde and Kostya Zarembo.

References

1. G. Grignani, P. Orland, L. D. Paniak and G. W. Semenoff, "Matrix theory interpretation of DLCQ string world-sheets," hep-th/0004194, Phys Rev. Lett. in press.
2. N. Seiberg, "Why is the matrix model correct?," Phys. Rev. Lett. **79**, 3577 (1997) [hep-th/9710009].
3. G. Grignani and G. W. Semenoff, "Thermodynamic partition function of matrix superstrings," Nucl. Phys. **B561**, 243 (1999) [hep-th/9903246].
4. J.-P. Serre, *A Course in Arithmetic*, Springer-Verlag, New York, 1973.
5. H. Fuji and Y. Matsuo, "Open string on symmetric product," hep-th/0005111.
6. J. Atick and E. Witten, "The Hagedorn transition and the number of degrees of freedom of string theory," Nucl. Phys. **B310**, 291 (1988).
7. T. Banks, W. Fischler, S. H. Shenker and L. Susskind, "M theory as a matrix model: A conjecture," Phys. Rev. **D55**, 5112 (1997) [hep-th/9610043].
8. L. Susskind, "Another conjecture about M(atrix) theory," hep-th/9704080.
9. R. Dijkgraaf, E. Verlinde and H. Verlinde, "Matrix string theory," Nucl. Phys. **B500**, 43 (1997) [hep-th/9703030].

D-Branes in the Background of NS Fivebranes

Shmuel Elitzur[1], *Amit Giveon*[1], *David Kutasov*[3], *Eliezer Rabinovici*[1] *and Gor Sarkissian*[1]

[1]Racah Institute of Physics, The Hebrew University
Jerusalem 91904, Israel

[3]Department of Physics, University of Chicago
5640 S. Ellis Av., Chicago, IL 60637, USA

We study the dynamics of D-branes in the near-horizon geometry of NS fivebranes. This leads to a holographically dual description of the physics of D-branes ending on and/or intersecting $NS5$-branes. We use it to verify some properties of such D-branes which were deduced indirectly in the past, and discuss some instabilities of non-supersymmetric brane configurations. Our construction also describes vacua of Little String Theory which are dual to open plus closed string theory in asymptotically linear dilaton spacetimes.

1. Introduction

The work discussed is based on the results appearing in hep-th/0005052. Due to the restricted space all references should be searched for in the original paper. In the last few years it was found that embedding various supersymmetric gauge theories in string theory, as the low energy worldvolume dynamics on branes, provides an efficient tool for studying many aspects of the vacuum structure and properties of BPS states in these theories. One class of constructions involves systems of D-branes ending on and/or intersecting $NS5$-branes. For applications to gauge theory one is typically interested in taking the weak coupling limit $g_s \to 0$ (as well as the low energy limit). In this limit one might expect the system to be amenable to a perturbative worldsheet treatment, but the presence of the $NS5$-branes and branes ending on branes complicate the analysis.

In the absence of a derivation of the properties of D-branes interacting with $NS5$-branes from first principles, some of their low energy properties were postulated in the past based on symmetry considerations and consistency conditions. One of the main purposes of this talk is to derive some of these properties by a direct worldsheet study of D-branes in the vicinity of $NS5$-branes.

In the analysis, we will use the improved understanding of the dynamics of $NS5$-branes achieved in the last few years. It is now believed that in the weak coupling limit (but not necessarily at low energies) fivebranes decouple from gravity and other bulk string modes and give rise to a rich non-gravitational theory, Little String Theory (LST) . It was proposed to study LST using holography. The theory on a stack of $NS5$-branes was argued to be holographically dual to string theory in the near-horizon geometry of the fivebranes. Many properties of LST can be understood by performing computations in string theory in this geometry. In particular, we will find below that it is an efficient way to study properties of D-branes ending on or intersecting $NS5$-branes.

From the point of view of LST, D-branes in the vicinity of $NS5$-branes give rise to new particle and extended object states in the theory, as well as new vacua, typically with reduced supersymmetry (when the branes are space-filling). Another motivation of this work is to understand the non-perturbative spectrum and dynamics of extended objects in LST, and more generally study the interplay between the non-trivial dynamics on fivebranes and the worldvolume physics of D-branes in their vicinity.

The plan of the talk is the following. In section 2 we review some facts regarding the relevant brane configurations and the physical issues involved. In particular, we describe the conjectured properties of these configurations that we will try to verify. Section 3 is a review of the near-horizon geometry of $NS5$-branes (the CHS geometry) and its holographic relation to LST. We also describe a modification of this geometry corresponding to fivebranes positioned at equal distances around a circle, which plays a role in the analysis. In section 4 we review some facts about D-branes in flat space and on a three-sphere (or $SU(2)$ WZW model). In section 5 we study D-branes in the CHS geometry as well as its regularized version. We verify some of the properties described in section 2, and describe additional features which follow from our analysis.

2. Some properties of brane configurations

One class of brane constructions, that gave rise to many insights into gauge dynamics, involves D-branes suspended between $NS5$-branes. The construction realizes four dimensional $N = 1$ supersymmetric gauge theory with gauge group $G = U(N_c)$ and N_f chiral superfields in the fundamental representation (more precisely, N_F fundamentals Q^i, $i = 1, \cdots, N_F$, and N_F anti-fundamentals $\tilde{Q}_i$). This will help introduce the issues that will be discussed later.

Consider a configuration of N_c $D4$-branes stretched between an $NS5$-brane and an $NS5'$-brane separated by a distance L along the x^6 direction (fig. 1). At distances greater than L, the five dimensional theory on the $D4$-branes reduces to a four dimensional theory with $N = 1$ supersymmetry. The boundary conditions provided by the fivebranes imply that out of all the massless degrees of freedom on the fourbranes (described by $4-4$ strings), only the $4d$ $N = 1$ vector multiplets for $G = U(N_c)$ survive. To decouple the gauge dynamics from the complications of string theory one takes the limit $g_s \to 0$, $L/l_s \to 0$, with the four dimensional gauge coupling $g^2 = g_s l_s / L$ held fixed.

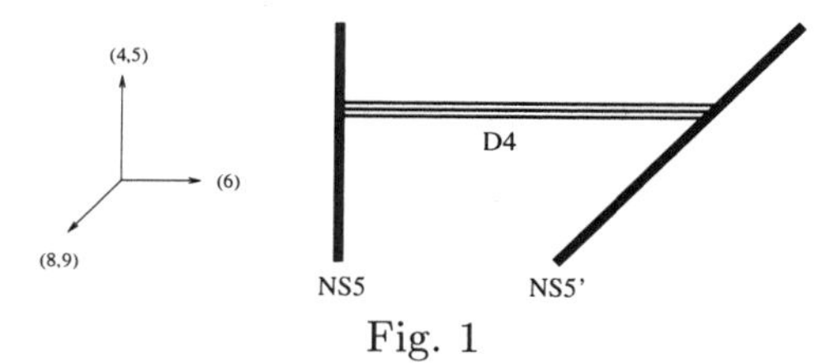

Fig. 1

One can add matter in the fundamental representation of G in one of the two ways illustrated in fig. 2 (which are related). One is to place N_f $D6$-branes between the fivebranes (fig. 2a). The $4-6$ strings give N_f fundamentals of $U(N_c)$, Q^i, $\tilde{Q}_i$, whose mass is proportional to the separation between the sixbranes and the fourbranes in (x^4, x^5). Alternatively, one can add to the configuration N_f $D4$-branes stretching from the $NS5$-brane to infinity (fig. 2b). In this case, the N_f fundamentals of $U(N_c)$ should arise from $4-4$ strings stretched between the two kinds of fourbranes. Their mass is the separation between the fourbranes in (x^4, x^5).

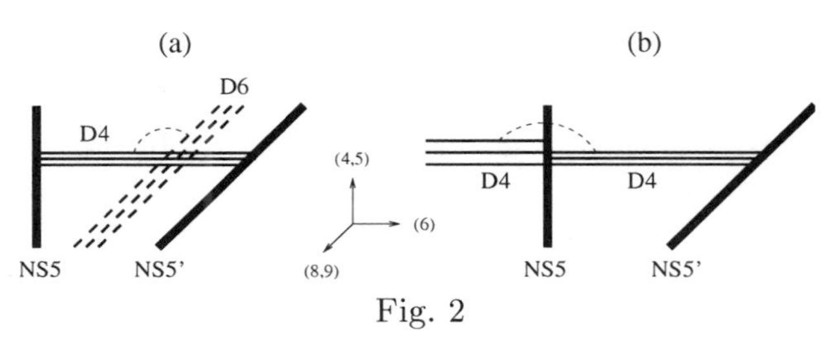

Fig. 2

In the configuration of fig. 2a, the low energy physics should be independent of the locations of the $D6$-branes along the interval between the fivebranes. It was pointed out that a particularly natural location for the sixbranes is at the same value of (x^4, x^5, x^6) as the $NS5'$-brane. In this case the $NS5'$-brane is embedded in the $D6$-branes; it divides them into two disconnected parts (fig. 3).

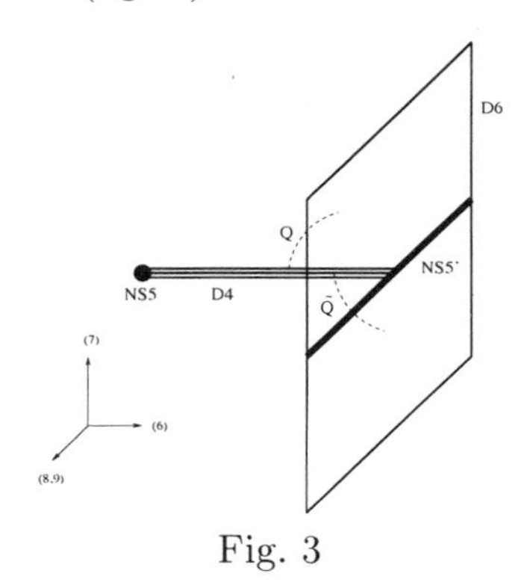

Fig. 3

Consequently, the configuration has two separate $U(N_f)$ symmetries, acting on the two semi-infinite sixbranes, and one may attempt to interpret them as the $U(N_f) \times U(N_f)$ global symmetry of $N = 1$ SQCD, under which Q and $\tilde{Q}$ transform as $(N_F, 1)$ and $(1, \overline{N}_F)$, respectively. This would imply that $4-6$ strings connecting the $D4$-branes to the upper part of the $D6$-branes give rise at low energies to the N_f chiral multiplets in the fundamental of $U(N_c)$, Q^i, while those that connect the fourbranes to the lower part of the sixbranes give the N_f chiral superfields in the anti-fundamental of $U(N_c)$, $\tilde{Q}_i$, as indicated in fig. 3.

The low-lying excitations of the brane configurations discussed above can be divided into two classes: those that are bound to one of the fivebranes, and those that are not. In this talk we will analyze the properties of the first class of excitations. It includes the following:

(a) $4-6$ strings connecting N_c $D4$-branes to N_F $D6'$-branes, all of which end on an

$NS5$-brane[1] (fig. 4a). The prediction is that they give rise to a chiral spectrum: a chiral superfield Q in the (N_c, N_F) of $U(N_c) \times U(N_F)$.

(b) $4-4$ strings connecting fourbranes ending on an $NS5$-brane from opposite sides (fig. 4b). They should give rise to chiral superfields, Q in the (N_c, N_F) and $\tilde{Q}$ in the $(\overline{N}_c, \overline{N}_F)$, or hypermultiplets $(Q, \tilde{Q})$.

(c) $4-6$ strings connecting $D4$-branes ending on $NS5$-branes to $D6$-branes intersecting the fivebranes (fig. 4c). They should also give rise to hypermultiplets $(Q, \tilde{Q})$.

We verify the predictions (a), (b), (c). In this talk we will describe the ingredients leading to the verification of the chiral symmetry.

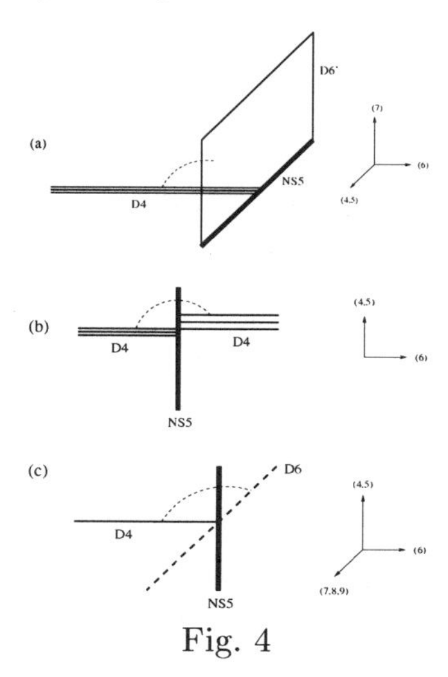

Fig. 4

3. The near-horizon geometry of $NS5$-branes and holography

The background fields around a stack of k parallel $NS5$-branes are :

$$
\begin{aligned}
e^{2(\Phi-\Phi_0)} &= 1 + \sum_{j=1}^{k} \frac{l_s^2}{|\vec{x} - \vec{x}_j|^2} \\
G_{IJ} &= e^{2(\Phi-\Phi_0)} \delta_{IJ} \\
G_{\mu\nu} &= \eta_{\mu\nu} \\
H_{IJK} &= -\epsilon_{IJKL} \partial^L \Phi
\end{aligned}
\tag{3.1}
$$

$I, J, K, L = 6, 7, 8, 9$ label the directions transverse to the fivebranes. $\mu, \nu = 0, 1, \cdots, 5$ are the directions along the brane. $\{\vec{x}_j\}$ are the locations of the fivebranes in $\vec{x} = (x^6, \cdots, x^9)$. H is the field strength of the NS-NS B-field; G, Φ are the metric and dilaton, respectively.

[1] The configuration of fig. 4a can be obtained from that of fig. 3 by exchanging $(x^4, x^5) \leftrightarrow (x^8, x^9)$ and removing some branes.

The background (3.1) interpolates between flat ten dimensional spacetime far from the fivebranes, and a near-horizon region, in which the 1 on the right hand side of the first line of (3.1) can be neglected (fig. 5).

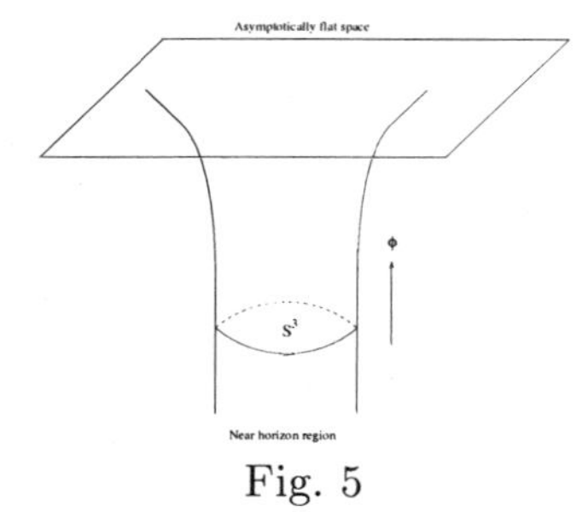

Fig. 5

String propagation in the near-horizon geometry can be described by an exact world-sheet Conformal Field Theory (CFT). The target space is

$$\mathbb{R}^{5,1} \times \mathbb{R}_\phi \times SU(2) \tag{3.2}$$

$\mathbb{R}_\phi$ corresponds to the radial direction $r = |\vec{x}|$:

$$\begin{aligned} \phi &= \frac{1}{Q} \log \frac{|\vec{x}|^2}{kl_s^2} \\ \Phi &= -\frac{Q}{2}\phi \end{aligned} \tag{3.3}$$

where we set $\Phi_0 = 0$ by rescaling $\vec{x}$. Q is related to the number of fivebranes via

$$Q = \sqrt{\frac{2}{k}}. \tag{3.4}$$

For some reasons it is more convenient to use the coset (cigar) description in this case. The full background now is

$$\mathbb{R}^{5,1} \times \frac{SL(2,\mathbb{R})_k}{U(1)} \times \frac{SU(2)_k}{U(1)} \tag{3.5}$$

Since the background (3.5) can be made arbitrarily weakly coupled. Consider the (NS,NS) sector of the theory. The observables are primaries of the $N = 1$ superconformal algebra with scaling dimension $(h, \bar{h}) = (\frac{1}{2}, \frac{1}{2})$. By analyzing correlation functions of $V_{j;m,\bar{m}}$, one finds that poles in correlators correspond to discrete representations of $SL(2)$

$$|m| = j + n, \quad |\bar{m}| = j + \bar{n}; \quad n, \bar{n} = 1, 2, 3, \cdots \tag{3.6}$$

This leads to a discrete spectrum of states in LST, which exhibits Hagedorn growth at high energy.

4. Some properties of D-branes

In this section we review some properties of D-branes in flat space and on S^3, in preparation for our discussion of D-branes in the CHS geometry (3.2), and its regularized version (3.5).

4.1. 4-6 strings in flat space

Later, we analyze $4-6$ strings connecting $D4$-branes and $D6'$-branes, both of which end on a stack of $NS5$-branes (fig. 4a).

Consider an open string, one of whose ends is on a $D4$-brane. The other end is on a $D6'$ brane. The vertex operator describing the (bosonic) ground state of a $4-6'$ string stretched between a $D4$-brane at $(x^8, x^9) = (0,0)$ and a $D6'$-brane at $(x^8, x^9) = (a,b)$,

$$V = e^{-\varphi}\sigma_{4567}S_{4567}e^{\frac{i}{\pi}(a(x_L^8 - x_R^8)+b(x_L^9 - x_R^9))}e^{ik_\mu x^\mu} \tag{4.1}$$

where k_μ ($\mu = 0,1,2,3$) is the $4d$ spacetime momentum. φ is the bosonized superconformal ghost. The vertex operator (4.1) is written in the -1 picture; one can check that the coefficient of $e^{-\varphi}$ in (4.1) is an $N=1$ superconformal primary, which is a necessary condition for its BRST invariance. The requirement that it has worldsheet dimension $1/2$, which is also necessary for BRST invariance, implies that the mass squared of the ground state of the $4-6'$ string is $-k_\mu^2 = \frac{1}{\pi^2}(a^2+b^2)$, as one would expect. (we work in a convention $\alpha' = 1/2$, in which the scalars x are canonically normalized on the boundary and the tension of the fundamental string is $T = 1/\pi$). In particular, when the $D6'$-brane intersects the $D4$-brane, *i.e.* when $a=b=0$, this mass vanishes. The vertex operator (4.1) describes a particle which transforms as a scalar under $3+1$ dimensional Lorentz rotations. The spin field S_{4567} has 4 components, half of which are projected out by the GSO projection, so (4.1) actually describes two real scalar particles. σ_{4567} is a collection of twist fields.

4.2. D-branes on the $SU(2)$ group manifold

We next turn to some facts regarding D-branes on a group manifold G, focusing on the case $G = SU(2)$. In the absence of D-branes, the WZW model has an affine $G_L \times G_R$ symmetry. If $g(z,\overline{z})$ is a map from the worldsheet to the group G, the symmetry acts on it as:

$$g \to h_L(z) g h_R(\overline{z}) \tag{4.2}$$

If the worldsheet has a boundary, there is a relation between left-moving and right-moving modes, and the $G_L \times G_R$ symmetry is broken. One can still preserve some diagonal symmetry G, say the symmetry

$$g \to hgh^{-1} \tag{4.3}$$

corresponding to $h_L = h_R^{-1} = h$ in (4.2). The presence of this symmetry constrains the boundary conditions that can be placed on g. Allowing $g(\text{boundary}) = f$ for some $f \in G$ we must also allow $g(\text{boundary}) = hfh^{-1}$ for every $h \in G$. This means that g on the boundary takes value in the conjugacy class containing f . For $G = SU(2)$, conjugacy classes are parametrized by a single angle θ, $0 \leq \theta \leq \pi$, corresponding to the choice $f = \exp(i\theta\sigma_3)$. Thinking of $SU(2)$ as the group of three dimensional rotations, the conjugacy class C_θ is the set of all rotations of angle 2θ about any axis. In addition not any value of θ gives rise to a consistent model. One finds that the possible conjugacy classes on which a boundary state can live are quantized, the corresponding θ must satisfy

$$\theta = 2\pi \frac{j}{k} \tag{4.4}$$

with j integer or half integer satisfying $0 \leq j \leq \frac{k}{2}$.

5. D-branes in the near-horizon geometry of $NS5$-branes

After assembling the necessary tools, we are now ready to study the physics of the configurations of fig. 4. consider the configuration of fig. 4a. A stack of N_c $D4$-branes ends from the left, *i.e.* from negative x^6, on k coincident $NS5$-branes. N_F $D6'$-branes end on the $NS5$-branes from above (positive x^7). From the point of view of the geometry (3.1) (fig. 5), the D-branes extend into the CHS throat, as indicated in fig. 6a.

The $D4$-branes intersect the three-sphere at the point $x^7 = x^8 = x^9 = 0$; the $D6'$-branes at $x^6 = x^8 = x^9 = 0$ (see fig. 6b). Thus, they correspond to the boundary states $g|_{\text{boundary}} = 1$ and $i\sigma_3$, respectively. The $D4$-branes are described by the boundary state with $\theta = 0$ and $f = 1$, while the $D6'$-branes correspond to a transformed state, with $\theta = 0$ and $f = \exp(i\pi\sigma_3/2) = i\sigma_3$. The $SU(2)$ currents J^a satisfy the boundary conditions for strings ending on the $D4$ and $D6'$-branes, respectively. In order to preserve worldsheet supersymmetry one has to impose analogous boundary conditions on the fermions. For example, for a string ending on the $D4$-branes one has

$$\chi^a = \overline{\chi}^a \tag{5.1}$$

while for a boundary on a $D6'$-brane the χ^a satisfy an analog of the sixbranes extend into the throat, the boundary conditions on ϕ are Neumann.

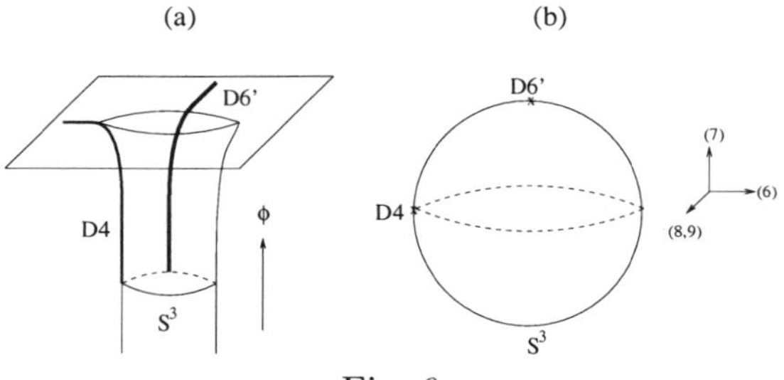

Fig. 6

We would like to construct the vertex operator for emitting the lowest lying $4-6'$ string in the geometry of fig. 6, *i.e.* generalize (4.1) to the fivebrane near-horizon geometry. Some parts of the discussion leading to (4.1) are unchanged. In particular, the geometry is the same as there in $(x^0, x^1, x^2, x^3, x^4, x^5)$. The presence of ϕ allows also a contribution $\exp(\beta\phi)$ to the vertex operator. Thus, the analog of (4.1) for this case has the form

$$V = e^{-\varphi}\sigma_{45}S_{45}e^{ik_\mu x^\mu}e^{\beta\phi}V_2 \tag{5.2}$$

where V_2 is the contribution of the $SU(2)$ group manifold to the vertex operator, to which we turn next.

The $4-6'$ vertex operator V_2 changes the worldsheet boundary conditions from $g=1$ to $g=f=\exp(i\alpha\sigma_3/2)$. The $D6'$-brane corresponds to $\alpha=\pi$, but it is instructive to discuss the general case, in which the angle between the $D4$ and $D6'$-branes is $\alpha/2$ (see fig. 7).

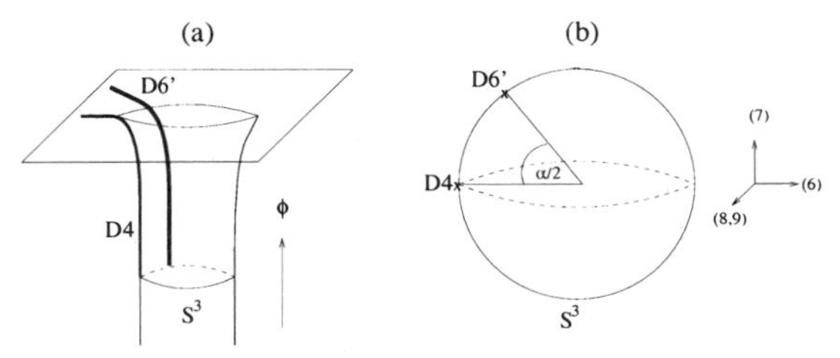

Fig. 7

We find that the lowest lying open string connecting the $D4$ and $D6'$-branes in the configuration of fig. 7 is described by the vertex operator

$$V^+_{4-6'} = e^{-\varphi}\sigma_{45}S_{45}e^{ik_\mu x^\mu}V_{jm} \tag{5.3}$$

with

$$\frac{k_\mu^2}{2} + \frac{m^2 - j(j+1)}{k} = \frac{1}{4} \tag{5.4}$$

To compute the spectrum of low-lying normalizable excitations of the $4-6'_+$ string, one computes the two point function of the operators (5.3) on the disk. This amplitude exhibits first order poles in j. Using the mass-shell condition (5.4), these can be interpreted as poles in k_μ^2; they correspond to on-shell particles in four dimensions, created from the vacuum by the operator (5.3).

The calculation of this two point function is very similar to its closed string analog, and is described in appendix A of the original paper. The result is, as in the closed string case, a series of poles corresponding to the discrete representations of $SL(2)$, (3.6). The lowest lying state corresponds to $n=1$, *i.e.* $j=|m|-1$. Plugging this together with into (5.4) we find that the mass of the lowest lying normalizable state of the $4-6'_+$ string is

$$M^2(\alpha) = \frac{1}{2}(\frac{\alpha}{\pi} - 1) \tag{5.5}$$

In particular, we find that as expected, for $\alpha=\pi$ the lowest lying state is massless. The vertex operator which creates this massless particle from the vacuum is

$$V^+_{4-6'}(k_\mu^2 = 0) = e^{-\varphi}\sigma_{45}S_{45}e^{ik_\mu x^\mu}V_{\frac{k}{4}-1,-\frac{k}{4}} \tag{5.6}$$

The degeneracy works out as well. The chiral structure is recaptured. This validates both the approximate picture and the LST methods used.
We note that for $\alpha < \pi$ the lowest lying state (5.5) is tachyonic; the stable vacuum is obtained by its condensation. This process has a very natural interpretation from the point of view of brane theory, which also makes it clear what is the endpoint of the condensation. For $\alpha \neq \pm\pi$, the configuration of fig. 7 is not supersymmetric, hence stability needs to be checked. For $\alpha < \pi$ the configuration of fig. 7 can reduce its energy by having the $D4$-brane slide away from the $NS5$-brane so that it ends on the $D6'$-brane instead (fig. 8). The resulting vacuum is stable. For $\alpha > \pi$, the configuration of figs. 7, 8 is stable under small deformations. Indeed, the lowest lying open string state is massive in this case (5.5).

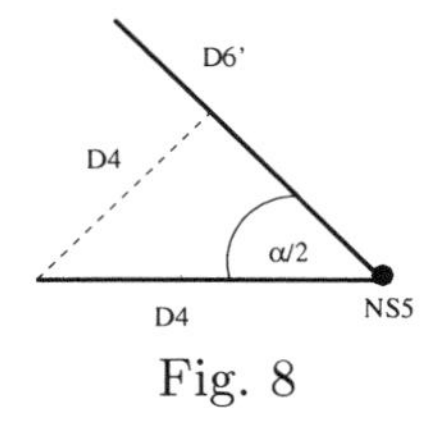

Fig. 8

From the point of view of the brane configurations describing four dimensional gauge theories like that of fig. 3, one can change α from π in a number of ways. One is to change the relative position of the $NS5$ and $NS5'$-branes in x^7 by the amount Δx^7. In the gauge theory on the $D4$-branes this corresponds to turning on a Fayet-Iliopoulos D-term. Depending on the sign of the FI term, either Q or $\tilde{Q}$ should condense to minimize the D-term potential

$$V_D \sim (Q^\dagger Q - \tilde{Q}^\dagger \tilde{Q} - r)^2 \tag{5.7}$$

where r is proportional to Δx^7. The gauge theory analysis is nicely reproduced by our string theory considerations[2] $\Delta x^7 > 0$ corresponds to $0 < \alpha < \pi$. In this case the ground state of the $4-6_+$ string is tachyonic, Q condenses and the $D4$-branes detach from the $NS5'$-branes and attach to the $D6$-branes. $\Delta x^7 < 0$ corresponds to $\alpha > \pi$ for the $4-6_+$ string which is hence massive, but since $0 < -\alpha < \pi$ for the $4-6_-$ strings ($\tilde{Q}$), a similar process of condensation to the above occurs for them.

Another way of changing α is to tilt the $D6$-branes by some angle in the (x^6, x^7) plane, which must lead to a potential similar to (5.7). The calculations in the cases (b) and (c) mentioned in the introduction can be found in the original paper.

Acknowledgements: This research was supported in part by NSF grant #PHY94-07194 and by the Israel Academy of Sciences and Humanities – Centers of Excellence Program. The work of A.G. and E.R. is also supported in part by BSF – American-Israel Bi-National Science Foundation. D.K. is supported in part by DOE grant #DE-FG02-90ER40560.

[2] Both in gauge theory and in string theory, in the context of the full configuration of fig. 3 the foregoing discussion is valid for $N_f \geq N_c$.

NONCOMMUTATIVE AND ORDINARY SUPER YANG-MILLS ON (D($P-2$),DP) BOUND STATES

RONG-GEN CAI AND NOBUYOSHI OHTA

Department of Physics, Osaka University, Toyonaka, Osaka 560-0043, Japan
E-mail: cai@het.phys.sci.osaka-u.ac.jp, ohta@phys.sci.osaka-u.ac.jp

We study properties of (D($p-2$), Dp) nonthreshold bound states ($2 \leq p \leq 6$) in the dual gravity description. These bound states can be viewed as Dp-branes with a nonzero NS B field of rank two. We find that in the decoupling limit, the thermodynamics of the N_p coincident Dp-branes with B field is the same not only as that of N_p coincident Dp-branes without B field, but also as that of the N_{p-2} coincident D($p-2$)-branes with two smeared coordinates and no B field, for $N_{p-2}/N_p = \tilde{V}_2/[(2\pi)^2\tilde{b}]$ with $\tilde{V}_2$ being the area of the two smeared directions and $\tilde{b}$ a noncommutativity parameter. We also obtain the same relation from the thermodynamics and dynamics by probe methods. This suggests that the noncommutative super Yang-Mills with gauge group $U(N_p)$ in ($p+1$) dimensions is equivalent to an ordinary one with gauge group $U(\infty)$ in ($p-1$) dimensions in the limit $\tilde{V}_2 \to \infty$. We also find that the free energy of a Dp-brane probe with B field in the background of Dp-branes with B field coincides with that of a Dp-brane probe in the background of Dp-branes without B field.

1 Introduction

Nowadays it is well-known that the worldvolume coordinates will become noncommutative if a Dp-brane is put in a nonvanishing NS B background [1], and the field theories on the worldvolume of such D-branes are called noncommutative field theories in order to distinguish the ordinary field theories on the worldvolume of D-branes without NS B field. The gauge theories on the noncommutative spacetimes can naturally be realized in string theories. According to the Maldacena conjecture [2], the string theories can be used to study the large N noncommutative field theories in the strong 't Hooft coupling limit.

More recently, Lu and Roy [3] have found that in the system of (D($p-2$), Dp) bound states ($2 \leq p \leq 6$), the noncommutative effects of gauge fields are actually due to the presence of *infinitely many* D($p-2$)-branes which play the dominant role over the Dp-branes in the large B field limit. The Dp-branes with a constant B field represent dynamically the system of infinitely many D($p-2$)-branes with two smeared transverse coordinates (additional isometrics) and no B field in the decoupling limit. With this observation, Lu and Roy further argued that there is an equivalence between the noncommutative super Yang-Mills theory in ($p+1$) dimensions and an ordinary one with gauge group $U(\infty)$ in ($p-1$) dimensions.

In this talk, we would like to discuss this equivalence from the viewpoint of the thermodynamics of the (D($p-2$), Dp) bound states in the dual gravity description with two dimensions compactified on a torus. The case discussed by Lu and Roy [3] corresponds to the infinite volume limit of the torus, and we would like to clarify some subtle questions in this analysis.

In the next section, we study the black (D($p-2$), Dp) configuration and some of its basic thermodynamic properties. As mentioned above, it has been noticed that the thermodynamics of the black Dp-branes with B field is the same as that of Dp-branes without B field. We show here that the thermodynamics of the Dp-branes with nonzero B field is also completely the same as that of D($p-2$)-branes with two smeared coordinates and zero B field. We obtain a relation [eq. (15) below] between the numbers of Dp-branes with B field and the D($p-2$)-branes without B field when this equivalence is valid. Since the worldvolume theory of Dp-branes with B field is the noncommutative super

Yang-Mills with gauge group $U(N_p)$ in $(p+1)$ dimensions with two dimensions compactified on a torus and that of D$(p-2)$-branes with two smeared coordinates and zero B field is an ordinary super Yang-Mills with gauge group $U(N_{p-2})$ in $(p+1)$ dimensions on the dual torus, as we will see, this implies that these theories are equivalent in the large N limit. When the volume of the torus is sent to infinity, the latter theory reduces to $U(\infty)$ gauge theory in $(p-1)$ dimensions for fixed N_p, in agreement with [3]. This equivalence is shown to be consistent with the Morita equivalence. The details and dynamics of probes in the background of this bound state can be found in [4,5].

2 The (D$(p-2)$, Dp) bound states

We start with the general solution

$$\begin{aligned} ds^2 &= H^{-1/2}[-fdt^2 + dx_1^2 + \cdots + dx_{p-2}^2 \\ &\quad + h(dx_{p-1}^2 + dx_p^2)] \\ &\quad + H^{1/2}(f^{-1}dr^2 + r^2 d\Omega_{8-p}^2), \\ e^{2\phi} &= g^2 H^{\frac{3-p}{2}} h, \quad B_{p-1,p} = \tan\theta H^{-1} h, \\ A^p_{012\cdots p} &= g^{-1}(H^{-1}-1)h\cos\theta\coth\alpha, \\ A^{p-2}_{012\cdots(p-2)} &= g^{-1}(H^{-1}-1)\sin\theta\coth\alpha \end{aligned} \tag{1}$$

and

$$\begin{aligned} H &= 1 + \frac{r_0^{7-p}\sinh^2\alpha}{r^{7-p}}, \; f = 1 - \left(\frac{r_0}{r}\right)^{7-p}, \\ h^{-1} &= \cos^2\theta + H^{-1}\sin^2\theta. \end{aligned} \tag{2}$$

Here g is the string coupling constant, r_0 is the non-extremal Schwarzschild mass parameter and α is the boost parameter. The solution (1) interpolates between the black D$(p-2)$-brane solution with two smeared coordinates x_{p-1} and x_p $(\theta = \pi/2)$, and the black Dp-brane with zero B field $(\theta = 0)$.

Denote the area of the 2-torus spanned by x_{p-1} and x_p by V_2 and the spatial volume of the D$(p-2)$-brane with worldvolume coordinates $(t, x_1, \cdots, x_{p-2})$ by V_{p-2}. The spatial volume of the Dp-brane with worldvolume coordinates $(t, x_1, \cdots, x_p)$ is then $V_p = V_{p-2}V_2$.

The charge density of the Dp-brane in the bound state system is given by

$$Q_p = \frac{(7-p)\Omega_{8-p}\cos\theta}{2\kappa^2 g} r_0^{7-p}\sinh\alpha\cosh\alpha, \tag{3}$$

where $2\kappa^2 = (2\pi)^7\alpha'^4$ is the gravity constant in ten dimensions and Ω_{8-p} is the volume of a unit $(8-p)$-sphere. The D$(p-2)$-brane charge density on its worldvolume is

$$Q_{p-2} = \frac{(7-p)\Omega_{8-p}V_2\sin\theta}{2\kappa^2 g} r_0^{7-p}\sinh\alpha\cosh\alpha. \tag{4}$$

In fact the D$(p-2)$-brane charge density on the worldvolume of Dp-brane is $\tilde{Q}_{p-2} = Q_{p-2}/V_2$. According to the charge quantization rule, and defining $\tilde{R}^{7-p} = r_0^{7-p}\sinh\alpha\cosh\alpha$, we have the relation between the number N_p of Dp-branes and N_{p-2} of D$(p-2)$-branes:

$$\begin{aligned} \tilde{R}^{7-p} &= N_p\frac{2\kappa^2 g T_p}{(7-p)\Omega_{8-p}\cos\theta} \\ &= N_{p-2}\frac{2\kappa^2 g T_{p-2}}{(7-p)\Omega_{8-p}V_2\sin\theta}, \end{aligned} \tag{5}$$

where the tensions T_p and T_{p-2} have a unified expression as $T_p = (2\pi)^{-p}(\alpha')^{-(p+1)/2}$. From (3) and (4), we can see that the asymptotic value $\tan\theta$ of the B field has the following relation:

$$\tan\theta = \frac{\tilde{Q}_{p-2}}{Q_p} = \frac{1}{V_2}\frac{T_{p-2}}{T_p}\frac{N_{p-2}}{N_p}. \tag{6}$$

The thermodynamics associated with the solution (1) is

$$\begin{aligned} M &= \frac{(8-p)\Omega_{8-p}V_p r_0^{7-p}}{2\kappa^2 g^2}\left(1 + \frac{7-p}{8-p}\sinh^2\alpha\right), \\ T &= \frac{7-p}{4\pi r_0\cosh\alpha}, \\ S &= \frac{4\pi\Omega_{8-p}V_p}{2\kappa^2 g^2} r_0^{8-p}\cosh\alpha. \end{aligned} \tag{7}$$

It is somewhat surprising that these thermodynamic quantities are completely the same as those for the Dp-branes without B field. Here we focus on another aspect of the thermodynamics of this black configuration:

These thermodynamic quantities are all independent of the parameter θ. Thus these thermodynamic quantities are also those of the black D$(p-2)$-branes with two smeared coordinates and no B field. Therefore there is the thermodynamic equivalence not only between black Dp-branes with B field and those without B field, but also between the black Dp-branes with B field and black D$(p-2)$-branes with two smeared coordinates and no B field. This fact is important in the following discussions.

Now we turn to the field theory limit (decoupling limit) of the bound state solution (1), in which the gravity decouples from the field theory on the worldvolume of Dp-branes. In the decoupling limit ($\alpha' \to 0$),

$$\begin{aligned} &\tan\theta = \frac{\tilde{b}}{\alpha'}, \quad g = \tilde{g}\alpha'^{(5-p)/2}, \\ &r = \alpha' u, \quad x_{0,1,\cdots,p-2} = \tilde{x}_{0,1,\cdots,p-2}, \\ &r_0 = \alpha' u_0, \quad x_{p-1,p} = \frac{\alpha'}{\tilde{b}}\tilde{x}_{p-1,p}, \end{aligned} \tag{8}$$

one has the following decoupling limit solution:

$$\begin{aligned} ds^2 &= \alpha' \left[\left(\frac{u}{R}\right)^{(7-p)/2} \left(-\tilde{f}dt^2 + d\tilde{x}_1^2 + \cdots \right.\right. \\ &\quad \left. + d\tilde{x}_{p-2}^2 + \tilde{h}(d\tilde{x}_{p-1}^2 + d\tilde{x}_p^2) \right) \\ &\quad \left. + \left(\frac{R}{u}\right)^{(7-p)/2} \left(\tilde{f}^{-1}du^2 + u^2 d\Omega_{8-p}^2 \right) \right], \\ e^{2\phi} &= \tilde{g}^2\tilde{b}^2\tilde{h}\left(\frac{R}{u}\right)^{(7-p)(3-p)/2}, \\ B_{p-1,p} &= \frac{\alpha'}{\tilde{b}}\frac{(au)^{7-p}}{1+(au)^{7-p}}, \end{aligned} \tag{9}$$

where the corresponding RR fields are not exposed explicitly,

$$\tilde{f} = 1 - \left(\frac{u_0}{u}\right)^{7-p}, \quad \tilde{h} = \frac{1}{1+(au)^{7-p}}, \tag{10}$$

and $a^{7-p} = \tilde{b}^2/R^{7-p}$,

$$R^{7-p} = \frac{1}{2}(2\pi)^{6-p}\pi^{-(7-p)/2}\Gamma[(7-p)/2]\tilde{g}\tilde{b}N_p. \tag{11}$$

The solution (9) is the dual gravity description of the noncommutative gauge field theory with gauge group $U(N_p)$ in $(p+1)$ dimensions. When $a=0$, the solution (9) reduces to the usual decoupling limit solution of Dp-branes without B field. This implies that the noncommutativity effect is weak in field theories for $au << 1$ or at long distance.

In the decoupling limit, these thermodynamic quantities become

$$\begin{aligned} E &= \frac{(9-p)\Omega_{8-p}\tilde{V}_p}{2(2\pi)^7(\tilde{g}\tilde{b})^2}u_0^{7-p}, \\ T &= \frac{7-p}{4\pi}R^{-\frac{7-p}{2}}u_0^{\frac{5-p}{2}}, \\ S &= \frac{2\Omega_{8-p}\tilde{V}_p}{(2\pi)^6(\tilde{g}\tilde{b})^2}R^{\frac{7-p}{2}}u_0^{(9-p)/2}. \end{aligned} \tag{12}$$

Here $\tilde{V}_p = V_{p-2}\tilde{V}_2$ is the spatial volume of the Dp-brane after taking the decoupling limit (8), and $\tilde{V}_2 = V_2\tilde{b}^2/\alpha'^2$ is the area of the torus. Using (12), one finds the free energy, defined as $F = E - TS$, of the thermal excitations:

$$\begin{aligned} F = &-\frac{\Omega_{8-p}V_{p-2}\tilde{V}_2}{(2\pi)^7\tilde{g}^2\tilde{b}^2}\frac{5-p}{2}\left(\frac{4\pi}{7-p}\right)^{\frac{2(7-p)}{5-p}} \\ &\times R^{\frac{(7-p)^2}{5-p}}T^{\frac{2(7-p)}{5-p}}, \end{aligned} \tag{13}$$

in terms of the temperature. We notice that these thermodynamic quantities, after rescaling the string coupling constant as $\tilde{g}\tilde{b} = \hat{g}$, are exactly the same as those of the black Dp-branes without B field in the decoupling limit. This means that in this supergravity approximation, the thermodynamics of the large N noncommutative and ordinary gauge field theories both in $(p+1)$ dimensions are equivalent to each other. This also implies that in the planar limit, the number of the degrees of freedom in the noncommutative gauge theories coincides with that in the ordinary field theories not only in the weak coupling limit [7], but also in the strong coupling limit.

The solution (9) is described by the quantities of Dp-branes. In fact the "radius"

R can also be expressed by quantities of D$(p-2)$-branes. Using (6), we obtain

$$R^{7-p} = \frac{1}{2}(2\pi)^{6-p}\pi^{-(7-p)/2}\Gamma[(7-p)/2] \times \tilde{g}\tilde{b}N_{p-2} \times \frac{(2\pi)^2\tilde{b}}{\tilde{V}_2}. \quad (14)$$

In the decoupling limit, we are thus led to the relation between the numbers of Dp- and D$(p-2)$-branes:

$$\tan\theta = \frac{\tilde{b}}{\alpha'} \quad \Longrightarrow \quad \frac{N_{p-2}}{N_p} = \frac{\tilde{V}_2}{(2\pi)^2\tilde{b}}. \quad (15)$$

Because $\tilde{V}_2$ and $\tilde{b}$ can be kept finite, we can thus conclude that in the decoupling limit of the Dp-branes with NS B field, the number of the D$(p-2)$-branes can be kept finite. This looks different from the claim by Lu and Roy [3] where they conclude that the number of the D$(p-2)$-branes becomes infinity in the decoupling limit. This is so because they take a little different decoupling limit and there $\tilde{x}_{p-1}$ and $\tilde{x}_p$ are infinitely extended. Mathematically our decoupling limit becomes the same as theirs by taking $\tilde{V}_2 \to \infty$ but keeping $N_{p-2}/\tilde{V}_2 = N_p/(2\pi)^2\tilde{b}$ finite.

In the decoupling limit (8), $\tan\theta \to \infty$ as $\alpha' \to 0$. The decoupling limit solution (9) for the black Dp-brane with NS B field is thus expected to be related with the solution of black D$(p-2)$-brane with two smeared coordinates and no B field in the same decoupling limit. For our convenience, we rewrite the black D$(p-2)$-brane with two smeared coordinates:

$$\begin{aligned} ds^2 &= H^{-1/2}[-fdt^2 + dx_1^2 + \cdots + dx_{p-2}^2 \\ &\quad + H(dx_{p-1}^2 + dx_p^2)] \\ &\quad + H^{1/2}(f^{-1}dr^2 + r^2 d\Omega_{8-p}), \\ e^{2\phi} &= g^2 H^{(5-p)/2}, \qquad B_{p-1,p} = 0, \\ A^{p-2}_{01\cdots(p-2)} &= g^{-1}(H^{-1}-1)\coth\alpha, \end{aligned} \quad (16)$$

where H and f are the same as those in (2). Note that here x_{p-1} and x_p are two smeared transverse coordinates for the D$(p-2)$-branes. In the decoupling limit (8), we reach

$$\begin{aligned} ds^2 &= \alpha' \left[\left(\frac{u}{R}\right)^{(7-p)/2} \left(-\tilde{f}dt^2 + d\tilde{x}_1^2 + \cdots \right.\right. \\ &\quad \left. + d\tilde{x}_{p-2}^2 + \frac{1}{(au)^{7-p}}(d\tilde{x}_{p-1}^2 + d\tilde{x}_p^2)\right) \\ &\quad \left. + \left(\frac{R}{u}\right)^{(7-p)/2} \left(\tilde{f}^{-1}du^2 + u^2 d\Omega_{8-p}^2\right)\right], \\ e^{2\phi} &= \tilde{g}^2\tilde{b}^{5-p}(au)^{(7-p)(p-5)/2}, \end{aligned} \quad (17)$$

where $B_{p-1,p} = 0$, $\tilde{f}$ and R^{7-p} are given in (10) and (14), respectively. Obviously for $au >> 1$, the decoupling solution (9) of the Dp-brane with NS B field is indeed equivalent to the decoupling limit solution (17) of the black D$(p-2)$-branes with two smeared coordinates and no NS B field, as noticed in [3]. Note that the coordinate u corresponds to an energy scale of worldvolume gauge field theories, and in (9) au reflects the noncommutative effect of gauge fields. It has been found that in order for the dual gravity description (9) of noncommutative gauge fields to be valid, $au >> 1$ should be satisfied [3], in which case N_p can be small and the noncommutativity effect is strong in the corresponding field theories.

We know that the solution (9) is a dual gravity description of a noncommutative super Yang-Mills theory with gauge group $U(N_p)$ in $(p+1)$ dimensions. What is the field theory corresponding to the supergravity solution (17) for large au? To see this, let us note that the supergravity description (17) breaks down for large au since the effective size of the torus shrinks. Nevertheless, we can make a T-duality along the directions $\tilde{x}_{p-1}$ and $\tilde{x}_p$. We then obtain a usual decoupling limit solution of N_{p-2} coincident Dp-branes without B field:

$$\begin{aligned} ds^2 &= \alpha' \left[\left(\frac{u}{R}\right)^{(7-p)/2} \left(-\tilde{f}dt^2 + d\tilde{x}_1^2 + \cdots \right.\right. \\ &\quad \left. + d\tilde{x}_{p-2}^2 + dx_{p-1}^2 + dx_p^2\right) \\ &\quad \left. + \left(\frac{R}{u}\right)^{(7-p)/2} \left(\tilde{f}^{-1}du^2 + u^2 d\Omega_{8-p}^2\right)\right], \end{aligned}$$

$$e^{2\phi} = \frac{(2\pi)^4 \tilde{g}^2 \tilde{b}^4}{\tilde{V}_2^2} \left(\frac{u}{R}\right)^{(7-p)(p-3)/2}. \quad (18)$$

This solution describes a $(p+1)$-dimensional ordinary super Yang-Mills theory with gauge group $U(N_{p-2})$ on the dual torus with area $\hat{V}_2 = (2\pi)^4 \tilde{b}^2/\tilde{V}_2$. Since the dual torus is characterized by the periodicity $x_{p-1,p} \sim x_{p-1,p} + \sqrt{\hat{V}_2}$, the radii of the dual torus go to zero for $\tilde{V}_2 \to \infty$ and the $(p+1)$-dimensional ordinary super Yang-Mills theory then reduces to a $(p-1)$-dimensional one. This means that if the torus in (17) is very large ($\tilde{V}_2 \to \infty$), the solution is a dual gravity description of a $(p-1)$-dimensional ordinary super Yang-Mills theory with gauge group $U(\infty)$. The $(p-1)$-dimensional theory has the Yang-Mills coupling constant

$$g_{\rm YM}^2 = (2\pi)^{p-4}\tilde{g}, \quad (19)$$

while the coupling constant of the $(p+1)$-dimensional noncommutative gauge field is $g_{\rm YM}^2 = (2\pi)^{p-2}\tilde{g}\tilde{b}$. Thus we reach the equivalence argued by Lu and Roy [3] between the noncommutative super Yang-Mills theory in $(p+1)$ dimensions and the ordinary one with gauge group $U(\infty)$ in $(p-1)$ dimensions.

This equivalence can also be understood from a T-duality of the decoupling limit solution (9) for the Dp-branes with B field. To see this, we use the general T-duality transformation $SL(2, \mathbf{Z})$ as described in [6]. Defining

$$\rho \equiv \frac{\tilde{V}_2}{(2\pi)^2\alpha'}\left(B_{p-1,p} + i\sqrt{G_{(p-1)(p-1)}G_{pp}}\right),$$

the duality transformation

$$\rho \to \frac{a\rho + b}{c\rho + d}, \quad (20)$$

on $\tilde{x}_{p-1}$ and $\tilde{x}_p$ gives a dual solution

$$ds^2 = \alpha'\left[\left(\frac{u}{R}\right)^{(7-p)/2}\left(-\tilde{f}dt^2 + d\tilde{x}_1^2 + \cdots + d\tilde{x}_{p-2}^2 + dx_{p-1}^2 + dx_p^2\right) + \left(\frac{R}{u}\right)^{(7-p)/2}\left(\tilde{f}^{-1}du^2 + u^2 d\Omega_{8-p}^2\right)\right],$$

$$e^{2\phi} = \frac{(2\pi)^4 \tilde{g}^2 \tilde{b}^4}{\tilde{V}_2^2} \left(\frac{u}{R}\right)^{(7-p)(p-3)/2}, \quad (21)$$

$\tilde{B}_{p-1,p} = \alpha'/\tilde{b}$, by choosing $c = -1$ and $d = \tilde{V}_2/(2\pi)^2\tilde{b}$ when the latter is an integer. Note that $d = N_{p-2}/N_p$ must be a rational number. If this is not an integer, after some steps of Morita equivalence transformation following [6], one can reach a solution like (21). For $p = 3$ and $\tilde{f} = 1$, the solution (21) reduces to the case discussed in [6]. Note that the solution (21) is the same as (18) except that the former has a nonvanishing constant B field while the latter has zero B field. The solution (21) describes a twisted ordinary super Yang-Mills theory with gauge group $U(N_{p-2})$ in $(p+1)$ dimensions, living on the dual torus with area $\hat{V}_2 = (2\pi)^4\tilde{b}^2/\tilde{V}_2$. For $\tilde{V}_2 \to \infty$, however, the theory reduces to a $(p-1)$-dimensional ordinary super Yang-Mills theory. Therefore we again arrive at the conclusion that the $(p+1)$-dimensional noncommutative gauge field is equivalent to an ordinary gauge field with gauge group $U(\infty)$ in $(p-1)$ dimensions for $\tilde{V}_2 \to \infty$, from the point of view of dual gravity description.

Next let us address another evidence to render support of the above equivalence from the viewpoint of thermodynamics. We find that the thermodynamics of decoupling limit solution (17) of the D$(p-2)$-branes is completely the same as those in (12). The worldvolume theory of the (D$(p-2)$, Dp) bound states (1) (or N_p coinciding Dp-branes with B field) is a noncommutative gauge field theory with gauge group $U(N_p)$ in $(p+1)$ dimensions with two dimensions compactified on a torus, while the worldvolume theory is an ordinary gauge field theory with the same gauge group $U(N_p)$ if the NS B field is absent. The above equivalence of the descriptions of the (D$(p-2)$, Dp) bound states implies that the bound states can also be described by ordinary gauge field theories in $(p+1)$ dimensions. Moreover, the equivalence is valid also between the $(p+1)$-dimensional noncommutative $U(N_p)$ theory and the $(p-1)$-dimensional $U(\infty)$ ordinary theory in the limit $\tilde{V}_2 \to \infty$ for the reason described above. For finite volume, the equivalence is between the $(p+1)$-dimensional noncommutative $U(N_p)$ gauge field and a (twisted) ordinary $U(N_{p-2})$ gauge field with the relation (15), the latter living on a dual torus. In this case, the Yang-Mills coupling constant is

$$g_{\rm YM}^2 = \frac{(2\pi)^p \tilde{g}\tilde{b}^2}{\tilde{V}_2}, \tag{22}$$

for the $(p+1)$-dimensional ordinary gauge field theory. As a self-consistency check, one may find that the coupling constant (19) can also be obtained from (22) after a trivial dimensional reduction. The equivalence of the descriptions and the relation (15) can be approached from the point of view of probe branes [5].

3 Conclusions

we have investigated two equivalent descriptions of the nonthreshold (D$(p-2)$, Dp) bound states in the dual gravity description. In the decoupling limit, the bound states can be described as Dp-branes with nonvanishing NS B field, and then the worldvolume theory is a noncommutative gauge field with gauge group $U(N_p)$ in $(p+1)$ dimensions (with two dimensions compactified on a torus in our case) if the number of the coincident Dp-branes is N_p. On the other hand, the nonthreshold (D$(p-2)$, Dp) bound state will reduce to the solution of D$(p-2)$-branes with two smeared coordinates and zero B field in the decoupling limit and the large $au >> 1$ limit. The latter condition is necessary for the validity of the dual gravity description. The worldvolume theory of the D$(p-2)$-branes should be an ordinary gauge field theory with gauge group $U(N_{p-2})$ in $(p+1)$ dimensions if the number of the coincident D$(p-2)$-branes is N_{p-2}. From the viewpoint of the thermodynamics of dual gravity solutions for the bound states (D$(p-2)$, Dp), we have found that N_p coincident Dp-branes with NS B field is equivalent to N_{p-2} coincident D$(p-2)$-branes with two smeared coordinates and no B field. In the equivalence, N_p and N_{p-2} must obey the relation (15), where $\tilde{V}_2$ is the area of the two additional dimensions and $\tilde{b}$ is a noncommutativity parameter. When the volume of the torus is sent to infinity keeping this relation, the ordinary super Yang-Mills theory reduces to the one with gauge group $U(\infty)$ in $(p-1)$ dimensions. This equivalence is consistent with the Morita equivalence of noncommutative SYM on the torus. We have identified the Yang-Mills coupling constant for the $(p-1)$-dimensional ordinary Yang-Mills theory.

Acknowledgments

This work was supported in part by the Japan Society for the Promotion of Science and by grant-in-aid from the Ministry of Education, Science, Sports and Culture No. 99020.

References

1. N. Seiberg and E. Witten, *JHEP* **09**, 032 (1999).
2. J. Maldacena, *Adv. Theor. Math. Phys.* **2**, 231 (1998).
3. J.X. Lu and S. Roy, *Nucl. Phys.* **B579**, 229 (2000).
4. R.G. Cai and N. Ohta, *Phys. Rev.* D **61**, 124012 (2000).
5. R.G. Cai and N. Ohta, *JHEP* **03**, 009 (2000).
6. A. Hashimoto and N. Itzhaki, *JHEP* **12**, 007 (1999).
7. D. Bigatti and L. Susskind, *Phys. Rev.* D **62**, 066004 (2000).

PERTURBATIVE ULTRAVIOLET AND INFRARED DYNAMICS OF NONCOMMUTATIVE QUANTUM FIELD THEORY

MASASHI HAYAKAWA
Theory Division, KEK, Tsukuba, Ibaraki 305-0801, Japan
E-mail: haya@post.kek.jp

Perturbative aspects of ultraviolet and infrared dynamics of noncommutative quantum field theory is examined in detail. It is observed that high loop momentum contribution to the nonplanar diagram develops a new infrared singularity with respect to the external momentum. This singular behavior is closely related to that of ultraviolet divergence of planar diagram. It is also shown that such a relation is precise in noncommutative Yang-Mills theory, but the same feature does not persist in noncommutative generalization of QED.

1 Introduction

The primary purpose here is to observe the perturbative aspects of noncommutative quantum field theory. After viewing in Sec. 2 one motivation why noncommutative quantum field theory becomes interesting, we argue ultraviolet (UV) property derived from perturbative consideration [1,2] with an introduction to perturbative framework of noncommutative quantum field theory in Sec. 3. In Sec. 4, we examine the infrared (IR) aspects, and show how it closely related to UV side, especially in noncommutative Yang-Mills (NCYM) theory. Final section is devoted to the discussion and conclusion.

2 Motivation

Noncommutative field theory appears in the matrix model [3,4]. The matrix model conjecture [5,6,7] is intended to provide a constructive definition of the superstring theory and to extract nonperturbative consequences of the interacting superstring dynamics, which will enable us to ask whether string theory is real or not. However, for instance, IIB matrix model [6] does not have any dimensionless coupling constant which can be decreased at will. Therefore, the direct perturbative analysis is not available in that model.

One way to see appearance of noncommutative Yang-Mills theory from matrix model is to expand IIB matrix model action around BPS solution [4]

$$[X^\mu, X^\nu] = -iC^{\mu\nu} 1_N \,. \qquad (1)$$

where the size of bosonic matrices X^M ($M = 1, \cdots, 10$) is taken to be infinite. $C^{\mu\nu} = -C^{\nu\mu}$ denotes the abelian part of the field strength $F_{\mu\nu}$, where μ is restricted to $1, \cdots, 4$. Reminding that IIB matrix proposes that the eigenvalues of X^M constitutes the points of the universe, at least semiclassically, the above relation (1) implies that the location of each point x^μ is uncertain in those four directions:

$$|x^\mu|\,|x^\nu| \geq 2\pi\,|C^{\mu\nu}| \quad \text{for } \mu \neq \nu \,, \qquad (2)$$

and that $C^{\mu\nu}$ characterizes the minimal area of accuracy in each two-dimensional plane.

IIB matrix model action gives an action with respect to the fluctuation a_μ (and the other six bosonic coordinates and fermionic variables) around the previous BPS solution (1), where $X^\mu = X^\mu_{(0)} + C^{\mu\nu} a_\nu$ and $X^\mu_{(0)}$ is the classical part satisfying eq. (1). Indeed, through the map called as "Weyl correspondence", the system can be described in terms of a four-dimensional field theory (See Ref. [4] on its detail.). The resulting theory is $\mathcal{N} = 4$ supersymmetric noncommutative Yang-Mills (NCYM) theory, where the fields in the action are multiplied by the star-product defined by

$$(f * g)(x) =$$

$$\exp\left(\frac{1}{2i}\partial_\mu C^{\mu\nu}\partial'_\nu\right) f(x)\, g(x')\Big|_{x'\to x}, \tag{3}$$

where $C^{\mu\nu}$ is the parameter appearing before. For the coordinates, for instance, we obtain

$$x^\mu * x^\nu - x^\nu * x^\mu = -iC^{\mu\nu}. \tag{4}$$

This algebraic relation is isomorphic to the original algebra (1) satisfied by the background represented by matrices.

Now the coupling constant $g_{\rm NCYM}$ in the resulting Yang-Mills theory is given by

$$g^2_{\rm NCYM} = \frac{4\pi^2 g^2_{\rm IIB}}{C^2}. \tag{5}$$

Here $g_{\rm IIB}$ is the coupling constant of the original matrix model and also has dimension of the length squared. In the canonical basis of $C^{\mu\nu}$, it has been assumed that $C^{\mu\nu} = C\,(\, i\sigma_2 \otimes 1 + 1 \otimes i\sigma_2\,)$ for simplicity. Thus, by taking $C^{\mu\nu}$ sufficiently large compared to $g_{\rm IIB}$, we get weak coupling NCYM theory. Hence, we can investigate the dynamics of matrix model by analyzing the quantum mechanical aspect of NCYM theory. The challenge is to show the existence of gravity and string in the quantized NCYM system. If this is shown, it will give a strong evidence that supports the entire matrix model as constructive definition of superstring. The structures of deformation of the open string algebra due to closed string background and the background independence [8] might also be further demonstrated.

However, since we do not know noncommutative quantum field theory itself so well, we are inclined to begin with examination of simpler systems, and capture the generic aspects possessed by noncommutative quantum field theory.

In the succeeding sections, we would like to see a few remarkable features of the perturbative noncommutative field theory.

3 Perturbative analysis of noncommutative field theory

In order to figure out the basic facets of the perturbative framework of noncommutative quantum field theory, we pay our attention to the noncommutative extension of a real scalar ϕ^4 theory in Euclidean four-dimensional space (see Ref. [2,9] on its detail)

$$S_{\phi^4_4} = \int d^4x \left[\frac{1}{2}\partial_\mu\phi * \partial_\mu\phi + \frac{1}{2}m^2\phi * \phi + \frac{\lambda}{4}\phi * \phi * \phi * \phi\right]. \tag{6}$$

The procedure for perturbation theory is the same as that in the ordinary field theory. The first task is to derive Feynman rule from the action (6). Then we apply Feynman rule to write down the diagrams relevant to the process and to the order of the coupling constant, and evaluate the associated contributions.

To derive Feynman rule, it is convenient to work in momentum space:

$$\phi(x) = \int \frac{d^4p}{(2\pi)^4} e^{ip\cdot x}\widetilde{\phi}(p). \tag{7}$$

Then the star product works on the basis elements $e^{ip\cdot x}$ in such a way that

$$\begin{aligned} e^{ip\cdot x} * e^{iq\cdot x} &= e^{\frac{1}{2i}\partial\wedge\partial'} e^{ip\cdot x} e^{iq\cdot x'} \\ &= e^{\frac{i}{2}p\wedge q}\, e^{i(p+q)\cdot x}, \end{aligned} \tag{8}$$

where $p \wedge q \equiv p_\mu C^{\mu\nu} q_\nu = -q \wedge p$. This extra phase factor is reminiscent of noncommutativity of the star product.

First, we consider the propagator. Due to the total momentum conservation of the system, only one momentum is linearly independent. Thus, there is no room for phase factors to enter; the propagator is the same as in the ordinary field theory

$$\left\langle \widetilde{\phi}(p)\widetilde{\phi}(q)\right\rangle = (2\pi)^4\delta^4(p+q)\,\frac{1}{p^2+m^2}. \tag{9}$$

However, the interaction vertex picks up nontrivial phase factor

$$\int \prod_{j=1}^{4} \frac{d^4p}{(2\pi)^4}\,(2\pi)^4\delta^4(p_1 + \cdots + p_4)$$

$$\times\frac{\lambda}{4}\exp\left(\frac{i}{2}\sum_{i<j}p_i\wedge p_j\right)\widetilde{\phi}(p_1)\cdots\widetilde{\phi}(p_4)\,, \tag{10}$$

from the star-product. Due to this phase factor, the interaction has only cyclic symmetry, in contrast to the point vertex in the ordinary real scalar ϕ^4 field theory which is invariant under the whole permutation group. Such a loss of symmetry of the vertex is better described if the vertex and legs get some width. The width does not permit us to exchange the two neighboring external legs, for instance. Alternatively, if multiple lines, rather than one, are assigned to each leg, they also retain only cyclic symmetry. The ordinary Yang-Mills theory is such an example [11]. It gives natural description of the propagator based on the double line representation, each line carrying the color degrees of freedom. In fact, also to the noncommutative field theory, the double line representation will turn out to be suitable. This aspect is most crucial to observe the important fact that noncommutative field theory gives the same UV structure as that of the corresponding ordinary large N field theory.

To pursue the best picture, we attempt to write each momentum p_j as a combination of outgoing and incoming momenta

$$p_j = k_j - k_{j-1}\,. \tag{11}$$

($k_0 = k_4$), and examine the consequences. By drawing the flow of each new momentum k_j, we get the double line representation for the vertex as shown in Fig. 1.

To see that this parametrization is natural, we rewrite the vertex (10) in terms of k_i

$$\int 4^4\prod_{j=1}^{4}\frac{d^4k_j}{(2\pi)^4}(2\pi)^4\delta^4(k_1+\cdots+k_4)$$
$$\times\frac{\lambda}{4}\left[e^{\frac{i}{2}k_4\wedge k_1}\widetilde{\phi}(k_1-k_4)\right]\times\cdots$$
$$\times\left[e^{\frac{i}{2}k_3\wedge k_4}\widetilde{\phi}(k_4-k_3)\right]\,. \tag{12}$$

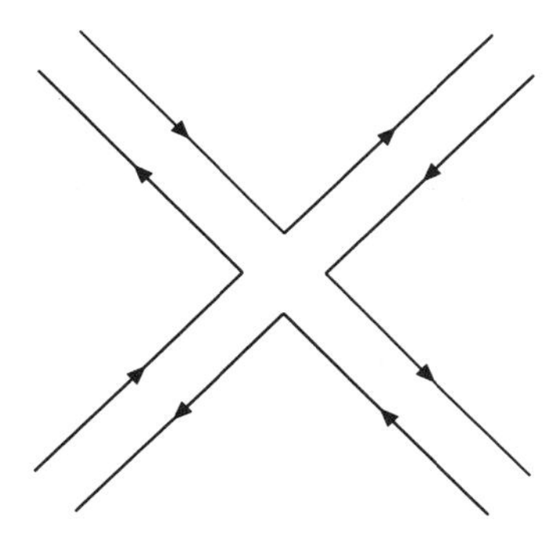

Figure 1. Double line representation for ϕ^4-vertex.

(4^4 is the Jacobian factor due to the change of the variables from p_i's to k_j's.) There, the expression of the phase factor has simplified: $\sum_{i<j}p_i\wedge p_j = \sum_{i=1}^4 k_i\wedge k_{i+1}$, and each piece has been placed in front of the field which depends on the same pair of momenta in eq. (12). Regarding the quantity in each bracket as a matrix element

$$\phi[k_1,k_2] = e^{\frac{i}{2}k_1\wedge k_2}\widetilde{\phi}(k_2-k_1) \tag{13}$$

labeled by two momenta, it is easy to see that it constitutes a "hermitian" matrix

$$(\phi[k_1,k_2])^* = \phi[k_2,k_1]\,, \tag{14}$$

from reality condition $\widetilde{\phi}(p)^* = \widetilde{\phi}(-p)$ in momentum space.

What we learn here is that, such a hermitian quantity is a building block of the interaction vertex in the noncommutative real scalar theory, and expresses Feynman rule compactly by the "double-line" representation.

Taking into account of these facts, we are inclined to recall the ordinary hermitian matrix field theory with quartic interaction

$$S_{[\Phi^4_4]_N} = \int d^4x\,\mathrm{tr}\left[\frac{1}{2}\partial_\mu\Phi\partial_\mu\Phi+\frac{1}{2}m^2\Phi^2 +\frac{\lambda_H}{4N}\Phi^4\right]\,, \tag{15}$$

where $\Phi(x)$ is an $N\times N$ hermitian matrix-valued field. The factor $1/N$ in front of quartic interaction is prepared for the future purpose to take large N limit. In terms of the momentum space variable $\widetilde{\Phi}(p)$, the interaction vertex of large N hermitian matrix field theory takes the form

$$\int 4^4 \prod_{j=1}^{4} \frac{d^4 k_j}{(2\pi)^4} (2\pi)^4 \delta^4(k_1 + \cdots + k_4) \times \frac{\lambda_H}{4N} \widetilde{\Phi}_{i_4}^{\ i_1}(k_1 - k_4) \times \cdots \times \widetilde{\Phi}_{i_3}^{\ i_4}(k_4 - k_3) \,. \quad (16)$$

Comparison of eq. (12) and eq. (16) shows that Feynman diagrams drawn in noncommutative ϕ^4 theory and large N hermitian matrix field theory coincide with each other, including their combinatoric factors.

Explicit evaluation of the diagrams show that the phase factor in noncommutative field theory plays the role of the color indices carried by the matrix field in the large N field theory; the phase factor distinguishes planar and nonplanar diagrams. To illustrate this aspect in more detail, we consider one loop contribution to the two point function in both theories. There are two types of diagrams as shown in Fig. 2.

As noted before, in both theories, we can draw the same diagrams. Thus, also in the side of noncommutative field theory, we can use the same terminology to distinguish these two types of the diagrams as in the ordinary field theory . That is, Fig. 2(a) is called as planar while Fig. 2(b) as nonplanar.

First we recall the situation in the side of large N field theory

$$\Pi_{\text{planar}}^{\text{large N}}(p, \lambda_H) = \frac{\lambda_H}{2} \int \frac{d^4 q}{(2\pi)^4} \frac{1}{q^2 + m^2} \,,$$
$$\Pi_{\text{nonplanar}}^{\text{large N}}(p, \lambda_H) = \frac{1}{N} \frac{\lambda_H}{4} \int \frac{d^4 q}{(2\pi)^4} \frac{1}{q^2 + m^2} \,. \quad (17)$$

Both diagrams diverge quadratically, but the large N limit extracts planar one (planar limit).

We return to the side of noncommutative field theory. There, the direct computation

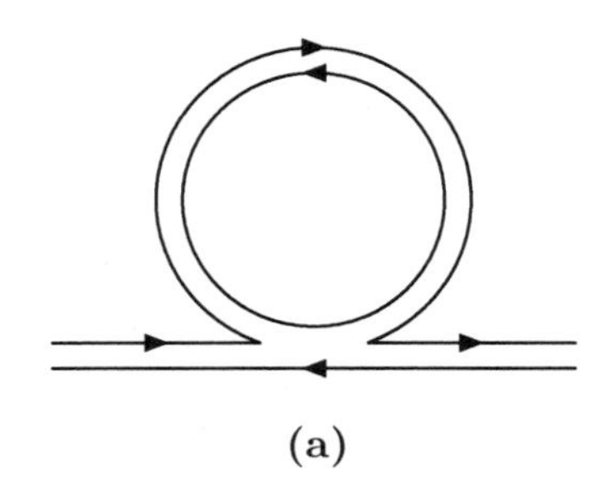

(a)

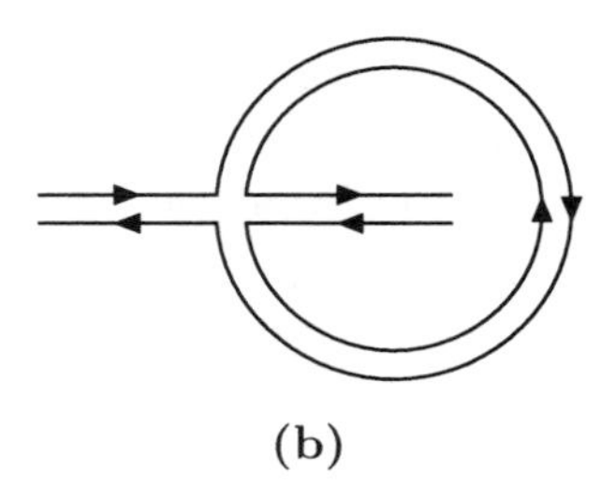

(b)

Figure 2. One-loop correction to two-point functions in ϕ^4 theory.

shows that

$$\Pi_{\text{planar}}^{\text{NC}}(p, \lambda) = \frac{\lambda}{2} \int \frac{d^4 q}{(2\pi)^4} \frac{1}{q^2 + m^2} \,,$$
$$\Pi_{\text{nonplanar}}^{\text{NC}}(p, \lambda) = \frac{\lambda}{4} \int \frac{d^4 q}{(2\pi)^4} e^{ip\wedge q} \frac{1}{q^2 + m^2} \,, \quad (18)$$

We see that the planar diagram in Fig. 2(a) gets no phase factors. Its contribution coincides with that of large N field theory

$$\Pi_{\text{planar}}^{\text{NC}}(p, \lambda) = \Pi_{\text{planar}}^{\text{large N}}(p, \lambda) \,, \quad (19)$$

and diverges quadratically. However, the nonplanar diagram in Fig. 2(b) gets nonzero phase factor. We would like to see what is the effect of such a phase factor.

The Schwinger parametrization of the propagator enables us to perform the momentum integration

$$\Pi_{\text{nonplanar}}^{\text{NC}}(p, \lambda) = \frac{\lambda}{4} \frac{1}{16\pi^2} \int_0^\infty \frac{d\alpha}{\alpha^2} \times \exp\left(-\alpha m^2 - \frac{\widetilde{p}^2}{4} \frac{1}{\alpha}\right) \,, \quad (20)$$

where $\widetilde{p}^{\mu} = C^{\mu\nu}p_{\nu}$. Then the UV limit is translated to the vanishing limit of α. Nonzero noncommutative parameter ensures that the integral converges since the exponentially suppression factor works when $\frac{1}{\alpha} \to \infty$. The conclusion is that nonplanar diagram is UV-finite in noncommutative field theory.

Recalling that planar diagram contributions are the same in both theories, the UV limit of noncommutative field theory is equivalent to the UV and planar limit of the corresponding large N field theory [1,2]. This is the aspects of UV limit of noncommutative field theory. It is determined by the planar diagrams.

4 IR aspect of noncommutative field theory

Next we would like to examine IR limit of noncommutative field theory. For that purpose, we return to the nonplanar contribution (20) to the two-point function. The simple rescaling $\alpha = \widetilde{p}^2 t$ shows that this diverges in the IR side quadratically

$$\Pi^{\rm NC}_{\rm nonplanar}(p,\lambda) \propto \frac{1}{\widetilde{p}^2} \quad \text{for } p_\mu \to 0 \,. \qquad (21)$$

To pursue its origin, we set $C^{\mu\nu}$ to zero. Then, the integral is that of the planar diagram, which diverges quadratically. It can be regularized by introducing the ultraviolet cut-off Λ [9]

$$\Pi^{\rm NC}_{\rm planar}(p,\lambda) \propto \frac{1}{16\pi^2}\int_0^\infty \frac{d\alpha}{\alpha^2} \times \exp\left(\alpha m^2 - \frac{1}{4}\frac{1}{\Lambda^2}\frac{1}{\alpha}\right) . \qquad (22)$$

The cutoff-dependence found above reflects the quadratic divergence of the planar diagram. Comparison of the planar contribution (22) to the nonplanar contribution (20) shows that nonzero $C^{\mu\nu}$ and p_μ replaces the cutoff dependence of the planar diagrams by $1/\widetilde{p}^2$ in nonplanar diagrams. The IR-divergent behavior generated by the nonplanar diagrams reflects the UV-divergent behavior of the planar diagrams [9,10].

We examine more interesting system, i.e, U(1) noncommutative Yang-Mills theory, in detail. We consider the transverse part of the renormalized vacuum polarization for the photon [a]. Its ultraviolet behavior is dominated by the planar diagrams

$$\Pi_{\mu\nu}(q)|_{\rm transverse} \to -\frac{g^2}{16\pi^2}\frac{10}{3}\ln(q^2) \times \left(\delta_{\mu\nu}q^2 - q_\mu q_\nu\right) \quad \text{for } |q| \gg 1/\sqrt{|C|}\,. \qquad (23)$$

The infrared limit, which is now dominated by nonplanar diagrams. can be computed [10]

$$\Pi_{\mu\nu}(q)|_{\rm transverse} \to -\frac{g^2}{16\pi^2}\frac{10}{3}\left(-\ln(\tilde{q}^2)\right) \times \left(\delta_{\mu\nu}q^2 - q_\mu q_\nu\right) \quad \text{for } |q| \ll 1/\sqrt{|C|}\,. \qquad (24)$$

Note that $\left(-\ln(\tilde{q}^2)\right)$ is positive for $|q| \ll 1/\sqrt{|C|}$. From those results, we see that the logarithmic nature of singularities coincide with each other. Furthermore, both limiting behaviors coincide with each other, including a numerical coefficient.

There is an example which does not give such a precise correspondence between the infrared and ultraviolet sides as found in NCYM theory. It is noncommutative QED, which is a noncommutative generalization of QED [10]. Another aspect of that theory is that we cannot find its counterpart of large N field theory associated with noncommutative QED.

5 Conclusion and discussion

Here we have observed a few fundamental properties of noncommutative quantum field

[a] There arises another hard singular term proportional to the Lorentz structure, which is intrinsic to the nonvanishing noncommutativity $C^{\mu\nu}$. Its implication is discussed in Ref. [12].

theory. Its UV limit is governed by planar diagrams, and usually also described by the corresponding large N field theory.

We have also seen that a new type of singularity in the infrared side is generated by the nonplanar diagrams, and it has a close relationship to the behavior at UV limit.

It is interesting to ask the practical issue whether noncommutative quantum field theory accommodates the quantum theory of gravity and string (See the recent attempts in Ref. [13]), especially whether the connection of IR and UV sides observed in NCYM theory is manifestation of some duality nature (e.g. closed-open channel duality) of string theory.

Acknowledgements

The author thanks to financial support from theory group at KEK.

References

1. D. Bigatti and L. Susskind, *Magnetic fields, branes and noncommutative geometry*, hep-th/9908056.
2. N. Ishibashi, S. Iso, H. Kawai and Y. Kitazawa, *Nucl. Phys.* **B573**, 573 (2000), hep-th/9910004.
3. M. Li, *Nucl. Phys.* **B499**, 149 (1997), hep-th/9612222; A. Connes, M. R. Douglas and A. Schwarz, *JHEP* **02**, 003 (1998), hep-th/9711162.
4. H. Aoki, N. Ishibashi, S. Iso, H. Kawai, Y. Kitazawa and T. Tada, *Nucl. Phys.* **B565**, 176 (2000), hep-th/9908141.
5. T. Banks, W. Fischler, S. H. Shenker and L. Susskind, *Phys. Rev.* **D55**, 5112 (1997), hep-th/9610043.
6. V. Periwal, *Phys. Rev.* **D55**, 1711 (1997), hep-th/9611103;
N. Ishibashi, H. Kawai, Y. Kitazawa and A. Tsuchiya, *Nucl. Phys.* **B498**, 467 (1997), hep-th/9612115.
7. L. Motl, *Proposals on nonperturbative superstring interactions*, hep-th/9701025; R. Dijkgraaf, E. Verlinde and H. Verlinde, *Nucl. Phys.* **B500**, 43 (1997), hep-th/9703030.
8. N. Seiberg, *JHEP* **0009**, 003 (2000), hep-th/0008013.
9. S. Minwalla, M. V. Raamsdonk and N. Seiberg, *Noncommutative perturbative dynamics*, hep-th/9912072.
10. M. Hayakawa, *Phys. Lett.* **B478**, 394 (2000), hep-th/9912094;
Perturbative analysis on infrared and ultraviolet aspects of noncommutative QED on R^4, hep-th/9912167.
11. G. 't Hooft, *Nucl. Phys.* **B72**, 461 (1974).
12. A. Matusis, L. Susskind and N. Toumbas, *The IR/UV connection in the non-commutative gauge theories*, hep-th/0002075.
13. N. Ishibashi, S. Iso, H. Kawai and Y. Kitazawa, *Nucl. Phys.* **B583**, 159 (2000), hep-th/0004038; D. J. Gross, A. Hashimoto and N. Itzhaki, *Observables of non-commutative gauge theories*, hep-th/0008075.

Conference Participants

Kazuo	Abe	KEK
Koya	Abe	Tohoku University
Toshinori	Abe	Stanford Linear Accelerator Center
Petar R.	Adzic	VINCA Institute of Nuclear Sciences
Hiroaki	Aihara	University of Tokyo
Ziad	Ajaltouni	Universite Blaise Pascal IN2P3/CNRS
Keiichi	Akama	Saitama Medical College
Andrew G.	Akeroyd	KEK
Thomas	Alderweireld	Universite de Mons-Hainaut
Reyes	Alemany	CERN
Jim	Alexander	Cornell University
James V.	Allaby	CERN
Claude	Amsler	University of Zurich
Bernard	Andrieu	Ecole Polytechnique
Antonella	Antonelli	Laboratori Nazionali di Frascati
Mario	Antonelli	INFN Laboratori Nazionali di Frascati
Masaharu	Aoki	KEK-IPNS
Mayumi	Aoki	KEK
Shigeki	Aoki	Kobe University
Sinya	Aoki	University of Tsukuba
Jeffrey A.	Appel	Fermilab
Jiro	Arafune	University of Tokyo
Firooz	Arash	Institute for Studies In Theoretical Physics and Mathematics
Silvia	Arcelli	University of Maryland
Richard	Arnowitt	Texas A&M University
Samuel H.	Aronson	BNL
Yuzo	Asano	University of Tsukuba
Alan	Astbury	TRIUMF
Jean-Jacques	Aubert	IN2P3
Patrick	Aurenche	Laboratoire d'Annecy le Vieux de Physique Theorique
Zekeriya Z.	Aydin	Ankara University
Kaladi S.	Babu	Oklahoma State University
Paul H.	Baillon	CERN
Andreas	Bamberger	University of Freiburg
Myron	Bander	University of California, Irvine
Swagato	Banerjee	LAPP, Annecy.
Elisabetta	Barberio	CERN
Artur J.	Barczyk	PSI
Anthony R.	Barker	University of Colorado, Boulder
Gary J.	Barker	Universitaet Karlsruhe
Michael	Barnett	Lawrence Berkeley National Laboratory
Michel	Baubillier	LPNHE Paris VI-VII

Laurent	Baulieu	CNRS/LPTHE
Guenter G.	Baum	Universitaet Bielefeld
Aurelio	Bay	University of Lausanne
Jules J.C.	Beckers	University of LIEGE
Eugene W.	Beier	University of Pennsylvania
Ronaldo	Bellazzini	INFN Pisa
Martin	Beneke	RWTH Aachen
Edmond L.	Berger	Argonne National Laboratory
Zvi	Bern	UCLA
Jose	Bernabeu	Universidad de Valencia
Christopher C.	Bernido	Central Visayan Institute
Werner	Bernreuther	RWTH Aachen
Alessandro	Bertolin	Padova University
Stefano	Bertolini	INFN and SISSA
Iain A.	Bertram	Lancaster University
Daniel C.	Bertrand	Brussels Free University
Siegfried	Bethke	MPI Munich
Alessandro	Bettini	Laboratori Nazionali del Gran Sasso
Sampa	Bhadra	York University
Anwar Ahmad	Bhatti	Rockfeller University
Otmar	Biebel	MPI for Physics, Munich
Ikaros I.	Bigi	University of Notre Dame
Gianmario	Bilei	INFN Perugia
Grahame A.	Blair	Royal Holloway and Bedford New College
Edward C.	Blucher	University of Chicago
Barry J.	Blumenfeld	Johns Hopkins University
Steve R.	Blusk	University of Rochester
Simon	Blyth	Carnegie Mellon University
Armin	Boehrer	Universitaet Siegen
Alexander E.	Bondar	Budker Institute of Nuclear Physics
Gerard R.	Bonneaud	Ecole Polytechnique Paris
Daniela	Bortoletto	Purdue University
Francesca M.	Borzumati	Sissa
Fawzi	Boudjema	Laboratoire d'Annecy le Vieux de Physique Theorique
Joseph F.	Boudreau	University of Pittsburgh
Andrzej Andrzej	Bozek	Institute of Nuclear Physics, Krakow
Ivanka C.	Bozovic-Jelisavcic	VINCA Institute of Nuclear Sciences
Concezio	Bozzi	INFN Sezione di Ferrara
Sylvie	Braibant	CERN
Paolo	Branchini	Sezione INFN Roma III
Fernando T.	Brandt	Universidade de Sao Paulo
James G.	Branson	UC San Diego
James E.	Brau	University of Oregon
Wolfgang B.	Braunschweig	RWTH Aachen
Roy A.	Briere	Carnegie Mellon University
Chiara	Brofferio	Sezione INFN Milano e Universita' di Milano-Bicocca

Nick	Brook	Bristol University
Thomas E.	Browder	University of Hawaii
David N.	Brown	Lawrence Berkeley National Laboratory
Pawel	Bruckman	CERN
Philippe	Bruel	LPNHE-Ecole Polytechnique
Graziano	Bruni	Sezione INFN Bologna
Patricia R.	Burchat	Stanford University
Cliff P.	Burgess	McGill University
Philip N.	Burrows	Oxford University
Joel N.	Butler	Fermilab
Jonathan M.	Butterworth	University College London
Rong-Gen	Cai	Osaka University
Alessandra	Caner	CERN
Robert M.	Carey	Boston University
Carl E.	Carlson	College of William and Mary
Per	Carlson	KTH Stockholm
Robert K.	Carnegie	Carleton University
Roger J.	Cashmore	CERN
David G.	Cassel	Cornell University
Carla	Cattadori	Sezione INFN Milano
Francesca R.	Cavallo	Sezione INFN Bologna
Fabio	Cerutti	CERN
Dhiman	Chakraborty	State University of New York at Stony Brook
Theresa J.	Champion	Birmingham University
Hong-Mo	Chan	Rutherford Appleton Laboratory
Paoti	Chang	National Taiwan University
Sanghyeon	Chang	Tohoku University
Wen-Chen	Chang	Academia Sinica
Hesheng	Chen	Institute of High Energy Physics, Chinese Academy of Sciences
Hai-Yang	Cheng	Academia Sinica
Chih-Yung	Chien	Johns Hopkins University
Yuichi	Chikashige	Seikei University
Gi-Chol	Cho	Ochanomizu Women's University
Suyong	Choi	University of California, Riverside
Young-Il	Choi	Sungkyunkwan University
Vitaly	Choutko	MIT
David	Cinabro	Wayne State University
Vitaliano	Ciulli	Scuola Normale Superiore and INFN Pisa
Carlo	Civinini	INFN Sezione di Firenze
Robert B.	Clare	MIT
Allan G.	Clark	University of Geneva
Barbara M.M.	Clerbaux	CERN
Thomas E.	Coan	Southern Methodist University
Cornelia	Coca	IFIN-HH, Bucharest
Jim	Cochran	Iowa State University
Jacques	Colas	Laboratoire d'Annecy le Vieux de Physique des Particules

Paul	Colas	CEA Saclay
John S.	Conway	Rutgers University
Francois	Corriveau	IPP/McGill University
Jose L.	Cortes	University of Zaragoza
Douglas F.	Cowen	University of Pennsylvania
Paschal A.	Coyle	IN2P3-CNRS
James A.	Crittenden	University of Bonn
Giovanni	Crosetti	Sezione INFN Genova
Donald	Cundy	CERN
Saverio	D'auria	Glasgow University
Samuel	Dagan	Tel-Aviv University
Mogens	Dam	Niels Bohr Institute
Mikhail V.	Danilov	Institute of Theoretical and Experimental Physics
Michel	Davier	LAL-Orsay
Alessandro	de Angelis	University of Udine
Wim	de Boer	Karlsruhe University
Catherine	de Clercq	Vrije Universiteit Brussel
Paul J.	de Jong	NIKHEF
Albert	de Roeck	CERN
Eddi A.	de Wolf	CERN
Pascal	Debu	CEA Saclay
Bernard J.	Degrange	IN2P3/CNRS France
Robert	Delbourgo	University of Tasmania
Giuseppe	Della Ricca	University and INFN - Trieste
Klaus	Desch	Universitaet Hamburg
Adriano	Di Giacomo	Sezione INFN Pisa
Michael	Dine	UC Santa Cruz
Guenther	Dissertori	CERN
Antonio	Dobado	University Complutense de Madrid
Masaru	Doi	Osaka Univ. of Pharmaceutical Sciences
Jiri	Dolejsi	Faculty of Mathematics and Physics of the Charles University
Jonathan	Dorfan	SLAC
Peter J.	Dornan	Imperial College
John D.	Dowell	Birmingham University
Manuel	Drees	TUM
Dongsheng	Du	Institute of High Energy Physics
Ehud	Duchovni	Weizmann Institute of Science
Guenter	Duckeck	Ludwig-Maximilans-Universitaet
Friedrich	Dydak	CERN
Anne	Ealet	C.P.P.M.
Oscar JP	Eboli	Instituto de Fisica Teorica-UNESP
Karl M.	Ecklund	Cornell University
Reinhard	Eckmann	University of Texas at Austin / DESY
Tohru	Eguchi	University of Tokyo
Richard D.	Ehrlich	Cornell University
Gerald	Eigen	University of Bergen

Gad Eilam	TECHNION - Israel Institute of Technology
Tord J.C. Ekelof	Uppsala University
Nicolas N. Ellis	CERN
Eckhard E. Elsen	DESY
Ryusuke Endo	Yamagata University
Recai Erdem	IYTE, Izmir
Pavel F. Ermolov	Institute of Nuclear Physics Moscow State University
Maria Catarina Espirito-Santo	CERN
Domenec Espriu	University of Barcelona
Erez Etzion	Tel Aviv University
David Evans	The University of Birmingham
Franco L. Fabbri	Laboratori Nazionali di Frascati
Marco Fabbrichesi	Sezione INFN Trieste
Miguel A. Falagan	CIEMAT
Laurent Favart	Universite Libre de Bruxelles
Lutz Feld	University of Freiburg
Tom Ferbel	University of Rochester
Erasmo Ferreira	Federal University of Rio de Janeiro
Ferruccio Feruglio	Padova University and INFN, Padova
Peter Filip	MPI
Ian M. Fisk	University of California, San Diego
Ivor Fleck	University of Freiburg
Robert Fleischer	DESY
Jeff R. Forshaw	Manchester University
Francesco Forti	Sezione INFN Pisa
Brian Foster	Bristol University
Paolo Franzini	Universita' di Roma,"La Sapienza"
Klaus Freudenreich	ETH-Zurich
Ariane Frey	CERN
Jerome I. Friedman	MIT
Yitzhak Frishman	Weizmann Institute of Science
Francesco Fucito	INFN Sezione di Roma 2
Keisuke Fujii	KEK-IPNS
Tadao Fujii	University of Tokyo
Kazuo Fujikawa	University of Tokyo
Mineo Fukawa	Naruto University of Education
Masataka Fukugita	University of Tokyo
Takeshi Fukuyama	Ritsumeikan University
Andreas Furtjes	CERN
Hideo Fusaoka	Aichi medical University
Pauline Gagnon	Indiana University
Lina Galtieri	Lawrence Berkeley National Laboratory
Diego Gamba	University of Turin & Sezione INFN Torino
Gerardo Ganis	MPI-Physik
John Garvey	The University of Birmingham
J.William Gary	University of California, Riverside

Paolo	Gauzzi	Rome University "La Sapienza"
M.Beatriz	Gay Ducati	IF-UFRGS
Joerg	Gayler	DESY
Neil I.	Geddes	Rutherford Appleton Laboratory
Stephen H.	Geer	Fermilab
Simonetta	Gentile	CERN
Ambar	Ghosal	University of Shizuoka
Paolo	Giacomelli	CERN-INFN Bologna
Jean-Francois	Glicenstein	CEA-Saclay
Nelly	Gogitidze	LPI/Moscow, DESY/Hamburg
Keith E.	Gollwitzer	Fermilab
Bostjan	Golob	University of Ljubljana
Gene	Golowich	University of Massachusetts
Andrey	Golutvin	ITEP/Moscow
Michel D.	Gonin	Ecole Polytechnique
Concha	Gonzalez Garcia	IFIC/Univ. de Valencia-CSIC
Benedetto	Gorini	CERN
Alfred T.	Goshaw	Duke University
Toru	Goto	KEK
Ricardo	Graciani	DESY
Per	Grafstrom	CERN
Francesco	Grancagnolo	Sezione INFN Lecce
Paul D.	Grannis	State University of New York at Stony Brook
Alan L.	Grant	CERN
Dirk	Graudenz	Paul Scherrer Institute
Massimiliano	Grazzini	ETHZ
Tony	Grifols	University Autonoma de Barcelona
Jean-Francois	Grivaz	Laboratoire de l'Accelerateur Lineaire
Eduardo I.	Guendelman	Ben Gurion University
Christophe	Guicheney	Universite Blaise Pascal IN2P3/CNRS
Atul	Gurtu	Tata Institute of Fundamental Research
Carlo	Gustavino	Laboratori Nazionali di Frascati
Laszlo J.	Gutay	Purdue University, West Lafayette
Andreas	Gute	Universitaet Erlangen-Nuernberg
Naoyuki	Haba	Mie University
Lawrence J.	Hall	University of California, Berkeley
Hideki	Hamagaki	University of Tokyo
Koichi	Hamaguchi	University of Tokyo
Shinji	Hamamoto	Toyama University
Kazunori	Hanagaki	Princeton University
John Renner	Hansen	Niels Bohr Institute
Gail G.	Hanson	Indiana University
Takanori	Hara	Osaka University
Toshio	Hara	Kobe University
Koji	Harada	Kyushu University
Masayasu	Harada	Nagoya University

Neville	Harnew	Oxford University
Paul F.	Harrison	Queen Mary & Westfield College
Masahiro	Haruyama	Hokkai-Gakuen Univesity
Takuya	Hasegawa	Tohoku University
Shoji	Hashimoto	KEK
Hisaki	Hatanaka	ICRR, University of Tokyo
Masashi	Hayakawa	KEK
Masahito	Hayashi	Osaka Institute of Technology
Mitsuo J.	Hayashi	Tokai University
Takemi	Hayashi	Kogakkan University
Hisaki	Hayashii	Nara Women's University
Avetik	Hayrapetyan	DESY
Masashi	Hazumi	Osaka University
Thomas	Hebbeker	Humboldt University Berlin, Institute of Physics
Vincent	Hedberg	University of Lund
Jochen	Heitger	DESY Zeuthen
Richard J.	Hemingway	Carleton University
Juan Jose	Hernandez	Universidad de Valencia
Stanley S.	Hertzbach	University of Massachusetts
Rolf-Dieter	Heuer	University of Hamburg
Joanne L.	Hewett	SLAC
Keisho	Hidaka	Tokyo Gakugei University
Kiyoshi	Higashijima	Osaka University
Ken-ichi	Hikasa	Tohoku University
Michael D.	Hildreth	University of Notre Dame
Thomas M.	Himel	SLAC
Zenro	Hioki	University of Tokushima
Shiro	Hirai	Osaka Elctro-Commuicatin Junior College
Martin	Hirsch	Universidad de Valencia
David G.	Hitlin	Caltech
Andreas	Hocker	Laboratoire de l'Accelerateur Lineaire
Kara	Hoffman	CERN
Wolfgang	Hollik	Institut fuer Theoretische Physik, Universitaet Karlsruhe
Stephen D.	Holmes	Fermilab
Sven-Olof H.	Holmgren	University of Stockholm
Walter	Hoogland	University of Amsterdam
Jiri	Horejsi	Faculty of Mathematics and Physics of the Charles university
Dezso	Horvath	CERN
Jiri	Hosek	Nuclear Physics Institute, Academy of Sciences of the Czech Republic
Yoshimoto	Hoshi	Tohoku Gakuin University
George W.S.	Hou	National Taiwan University
Jingliang	Hu	Rensselaer Polytechnic Institute
Tao	Huang	Chinese Academy of Sciences
Enzo	Iarocci	INFN
Shoichi	Ichinose	University of Shizuoka
Keiji	Igi	Kanagawa University

Peter	Igo-Kemenes	Heidelberg Univ. and CERN
Toru	Iijima	KEK
Jose I.	Illana	DESY Zeuthen
Masahiro	Imachi	Yamagata University
Jun	Imazato	KEK-IPNS
Richard L	Imlay	Louisiana State University
Takao	Inagaki	KEK
Takeo	Inami	Chuo University
Dharmavaram	Indumathi	Institute of Mathematical Sciences
Kenzo	Inoue	Kyushu University
Nobuyuki	Ishibashi	KEK
Nobuhiro	Ishihara	KEK-IPNS
Takanobu	Ishii	KEK
Kenzo	Ishikawa	Hokkaido University
Gino	Isidori	INFN Laboratori Nazionali di Frascati
Hitoshi	Ito	Kinki University
Hiroyuki	Iwasaki	KEK-IPNS
Masako	Iwasaki	University of Oregon
Yoichi	Iwasaki	University of Tsukuba
Takahiro	Iwata	Nagoya University
Joseph M.	Izen	University of Texas at Dallas
Barbara V.	Jacak	SUNY Stony Brook
David J.	Jackson	Osaka University
Andrew	Jaffe	Univeristy of California Berkeley
Michel	Jaffre	Laboratoire de l'Accelerateur Lineaire
Karl	Jansen	CERN
Goran L. B.	Jarlskog	Lund University
Colin P.	Jessop	SLAC
Stephane	Jezequel	Laboratoire d'Annecy le Vieux de Physique des Particules
Shan	Jin	University of Wisconsin
Erik K.	Johansson	Stockholm University
Roger WL	Jones	Lancaster University
Sonja	Kabana	University of Bern
Seiji	Kabe	KEK
Vladimir	Kadyshevsky	JINR
Harris	Kagan	Ohio State University
Alexei B.	Kaidalov	Institute of Theoretical and Experimental Physics
Ryoichi	Kajikawa	Nagoya University
Takaaki	Kajita	University of Tokyo
Jan	Kalinowski	Warsaw University
Peter IP	Kalmus	Queen Mary, University of London
Tuneyoshi	Kamae	Stanford Linear Accelerator Center and Hiroshima Univ.
Koichi	Kamata	Nishina Memorial Foundation
Yuri A.	Kamyshkov	University of Tennessee
Kazuyuki	Kanaya	University of Tsukuba
Shinya	Kanemura	Universitaet Karlsruhe

Kenji	Kaneyuki	University of Tokyo
J. S.	Kang	Korea University
Kyungsik	Kang	Brown Univesity
Dean A.	Karlen	Carleton University
Nobuhiko	Katayama	KEK
Kiyoshi	Kato	Kogakuin University
Yukihiro	Kato	Kinki University
Hiroshi	Kawabe	Yonago National College of Technology
Kiyotomo	Kawagoe	Kobe University
Tatsuo	Kawamoto	University of Tokyo
Hiroyuki	Kawamura	Hiroshima University
Yoshiharu	Kawamura	Shinshu university
Ken	Kawarabayashi	Tokyo International University
Masahiro	Kawasaki	University of Tokyo
Boris	Kayser	National Science Foundation
Dmitrii Igorevich	Kazakov	JINR
Yoichi	Kazama	University of Tokyo
Edward T.	Kearns	Boston University
Vladimir	Kekelidze	JINR
Richard D.	Kenway	Edinburgh University
Peter-Raymond	Kettle	Paul Scherrer Institute
Valentin V.	Khoze	University of Durham
Hiromichi	Kichimi	KEK-IPNS
Paul	Kienle	Technical University Munich
Maria Novella	Kienzle	Geneva University
Christian M.	Kiesling	Max-Planck-Institute for Physics
Ari P P	Kiiskinen	Helsinki Institute of Physics
Hisashi	Kikuchi	Ohu University
Yoshio	Kikukawa	Nagoya University
William B.	Kilgore	BNL
Choong Sun	Kim	Yonsei University
Jae Yool	Kim	Chonnam National University
Jihn E.	Kim	Seoul National University
Shinhong	Kim	University of Tsukuba
Yoshitaka	Kimura	KEK
Kay	Kinoshita	University of Cincinnati
Toichiro	Kinoshita	Cornell University
Andrew	Kirk	The University of Birmingham
David	Kirkby	Stanford University
Shoichi	Kitamura	Tokyo Metropolitan Univerity of Health Sciences
Noriaki	Kitazawa	Tokyo Metropolitan University
Niels Jorgen	Kjaer	CERN
Matthias	Kleifges	Forschungszentrum Karlsruhe
Joshua R.	Klein	University of Pennsylvania
Jean-Loic	Kneur	Universite de Montpellier
Pyungwon	Ko	KAIST

Dy-Holm	Koang	IN2P3/CNRS-Grenoble Univ
Keizo	Kobayakawa	Fukui University of Technology
Akizo	Kobayashi	Niigata University
Makoto	Kobayashi	KEK
Tetsuro	Kobayashi	Fukui Univesity of Engineering
Tomio	Kobayashi	University of Tokyo
Yoji	Kohara	Nihon University
Antje	Kohnle	MPI fuer Kernphysik
Yoshio	Koide	University of Shizuoka
Sachio	Komamiya	University of Tokyo
Masahiro	Komatsu	Nagoya University
Takeshi K.	Komatsubara	KEK-IPNS
Akira	Konaka	TRIUMF
Kei-Ichi	Kondo	Chiba University
Takahiko	Kondo	KEK
Otto C.W.	Kong	Academia Sinica
Kenichi	Konishi	University of Pisa
Susumu	Koretune	Shimane Medical University
Masatoshi	Koshiba	The University of Tokyo
Tsuneyuki	Kotani	Osaka University
Ashutosh V.	Kotwal	Duke University
Emi	Kou	Ochanimizu Univ. & Nagoya Univ.
Alexandre	Kourilin	JINR
Wolfgang D.	Kretschmer	Universitaet Erlangen-Nuernberg
Andreas S.	Kronfeld	Fermilab
Dirk	Kruecker	Humboldt University Berlin, Institute of Physics
Takahiro	Kubota	Osaka University
Yuichi	Kubota	University of Minnesota
Taichiro	Kugo	Kyoto University
Yoshitaka	Kuno	KEK-IPNS
Shuichi	Kunori	University of Maryland
Tzee-Ke	Kuo	Purdue University
Eduard Alekseevich	Kuraev	JINR
Hisaya	Kurashige	Kobe University
Rie	Kuriki	KEK
Takeshi	Kurimoto	Toyama University
Shin-Ichi	Kurokawa	KEK
Kiichi	Kurosawa	University of Tokyo
Alexander	Kusenko	UCLA
Masahiro	Kuze	KEK-IPNS
Eric	Laenen	NIKHEF
Bertrand	Laforge	LPNHE CNRS/IN2P3, Universite Paris VI-VII
Dominique	Lalanne	Accelerateur Lineaire Orsay
Armando	Lanaro	CERN - INFN Laboratori Nazionali di Frascati
Livio	Lanceri	Universita` di Trieste and INFN-Trieste
Greg	Landsberg	Brown University

Karol	Lang	The University of Texas at Austin
Andrew J.	Lankford	University of California, Irvine
Thomas J.	Lecompte	Argonne National Laboratory
Gerhard	Leder	Institute for High Energy Physics
Francois R.	Lediberder	LPNHE Paris VI-VII
Hong Seok	Lee	KAIST
K.B.	Lee	Louisiana State University
Juliet	Lee Franzini	Laboratori Nazionali di Frascati
George	Leibbrandt	University of Guelph
David W.G.S.	Leith	SLAC , Stanford University
Rupert	Leitner	Faculty of Mathematics and Physics of the Charles university
Vincent	Lemaitre	CERN
Jacques	Lemonne	Vrije Universiteit Brussel
Sergey V.	Levonian	DESY
Hsiang-Nan	Li	National Cheng-Kung University
Weiguo	Li	Institute for High Energy Physics
C.S.	Lim	Kobe University
Manfred B.	Lindner	Technische Universitaet Muenchen
Anna	Lipniacka	University of Stockholm
Keh-Fei	Liu	University of Kentucky
Bernd	Loehr	DESY
Wolfgang	Lohmann	CERN
Cai-Dian	Lu	Hiroshima University
Kam-Biu	Luk	UC Berkeley/LBL
Vera G.	Luth	SLAC
Pierre	Lutz	CEA Saclay
Louis	Lyons	Oxford University
Jukka J.	Maalampi	University of Helsinki
Anna	Macchiolo	Universite' de Montreal
David B.	Macfarlane	UC San Diego
Dirk	Macke	RWTH Aachen
Jimmy N.	MacNaughton	Institute for High Energy Physics
Kaori	Maeshima	Fermilab
Marcello	Maggi	CERN
Stephen R.	Magill	Argonne National Laboratory
K.T.	Mahanthappa	University of Colorado
Luciano	Maiani	CERN
Akihiro	Maki	KEK
Ziro	Maki	Kyoto University YITP
Alexandre	Malakhov	JINR
Italo	Mannelli	Sezione INFN Pisa and Scuola Normale Superiore Pisa
Marcello	Mannelli	CERN
Uri	Maor	Tel-Aviv University
Jesus	Marco	Instituto de Fisica de Cantabria
Nancy	Marinelli	Imperial College
John F.	Martin	University of Toronto

Mario	Martinez	DESY
Fernando	Martinez-Vidal	IN2P3-Paris 6 & 7
Hans-Ulrich	Martyn	RWTH Aachen, I. Physics Institute
Francesco	Marzano	I.N.F.N - Sezione di Roma I
Akira	Masaike	Fukui University of Technology
Toshihide	Maskawa	Kyoto University
Koichi	Matsuda	Ritsumeikan University
Masahisa	Matsuda	Aichi University of Education
Satoshi	Matsuda	Kyoto University
Tatsuro	Matsuda	Miyazaki University
Takayuki	Matsui	KEK
Takayuki	Matsuki	Tokyo Kasei University
Shigeo	Matsumoto	Chuo Univesity
Clara	Matteuzzi	Milano University and INFN
Giorgio	Matthiae	University of Roma II
Peter	Mattig	CERN
Chris J.	Maxwell	Durham University
John	McDonald	Glasgow University
Bruce H J	Mckellar	University of Melbourne
Andrew	Mehta	Liverpool University
Salvatore	Mele	CERN
Blazenka	Melic	Rudjer Boskovic Institute, Zagreb
Bruce	Mellado	Columbia University
Dario	Menasce	Sezione INFN Milano
Marcel M.H.M.	Merk	NIKHEF
Marco	Meschini	Sezione INFN Firenze
Wesley J.	Metzger	University of Nijmegen
Yasuo	Miake	University of Tsukuba
Bernard	Michel	Universite Blaise Pascal
Shoichi	Midorikawa	Aomori University
Ernesto	Migliore	CERN
Satoshi	Mihara	Univ. of Tokyo
Giora	Mikenberg	Weizmann Institute of Science
David H.	Miller	Purdue University
David J.	Miller	University College London
Kimball A.	Milton	University of Oklahoma
Yukihiro	Mimura	KEK
Peter C.	Minkowski	University of Bern
Shekhar C.	Mishra	Fermilab
Mikolaj K.	Misiak	CERN/Warsaw Univesity
Guenakh	Mitselmakher	University of Florida
Kenkichi	Miyabayashi	Nara Women's University
Akiya	Miyamoto	KEK
Shun'ya	Mizoguchi	KEK
Kohei	Mizutani	Saitama University
Klaus	Moenig	DESY Zeuthen

Glenn R. Moloney	The University of Melbourne
Toshinori Mori	University of Tokyo
Kazushige Morii	Hiroshima University
Masahiro Morii	Harvard University
Toshiyuki Morii	Kobe University
Takuya Morozumi	Hiroshima University
Guy Steve Muanza	Institut de Physique Nucleaire de Lyon
Franz Muheim	Edinburgh University
David John Munday	Cavendish Laboratory, Cambridge University
Akira Murakami	Saga University
Yasushi Muraki	Nagoya University
Akihiro Murayama	Shizuoka University
Hitoshi Murayama	University of California Berkeley
Trond Myklebust	University of Oslo
Koichi Nagai	QMW London
Kunihiro Nagano	DESY / KEK-IPNS
Yorikiyo Nagashima	Osaka University
Elemer Nagy	C.P.P.M.
Seichi Naito	Osaka City University
Hideo Nakajima	Utsunomiya University
Kenzo Nakamura	KEK
Mitsuhiro Nakamura	Nagoya University
Noboru Nakanishi	Kyoto University
Itsuo Nakano	Okayama University
Mikihiko Nakao	KEK
Tsuyoshi Nakaya	Kyoto University
Naohito Nakazawa	Shimane Univ.
Won Namkung	Pohang University of Science and Technology
Satyanarayan Nandi	Oklahoma State University
Rosario Nania	Sezione INFN Bologna
Vemuri S. Narasimham	T.I.F.R.
Beate Naroska	Universitaet Hamburg
Adriano A. Natale	Instituto de Fisica Teorica-UNESP
Sergio Natali	Sezione INFN Bari
Uriel Nauenberg	University of Colorado
Charles A. Nelson	SUNY AT BINGHAMTON
Harry N. Nelson	UC Santa Barbara
Norbert Neumeister	CERN
Harvey B. Newman	Caltech
Paul R. Newman	Birmingham University
Carsten Niebuhr	DESY
Takeshi Nihei	Lancaster University
Masao Ninomiya	Kyoto University
Kazuhiko Nishijima	Nishina Memorial Foundation
Koichiro Nishikawa	Kyoto University
Hiroyuki Nishiura	Junior College of Osaka Institute of Technology

Osamu	Nitoh	Tokyo University of Agriculture and Technology
Muneto	Nitta	Tokyo Institute of Technology
Seishi	Noguchi	Nara Women's University
Mihoko M.	Nojiri	YITP, Kyoto University
Tadashi	Nomura	Kyoto University
Yasunori	Nomura	University of Tokyo
Mitsuaki	Nozaki	Kobe University
Tadao	Nozaki	KEK
Gerald	Oakham	Carleton University
Vladimir F.	Obraztsov	Institute for High Energy Physics
Satoru	Odake	Shinshu University
Pier	Oddone	Lawrence Berkeley National Laboratory
Christian	Oehler	Forschungszentrum Karlsruhe
Satoru	Ogawa	Toho University
Sachiko	Ogushi	Ochanomizu Women's University
Benedict Y.	Oh	PENN STATE UNIVERSITY
Choo-Hiap	Oh	National University of Singapore
Sun Kun	Oh	Konkuk University
Ichiro	Ohba	Waseda University
Tommy	Ohlsson	Royal Institute of Technology
Fairouz	Ohlsson-Malek	CNRS/IN2P3/U.Joseph Fourier-Grenoble
Takayoshi	Ohshima	Nagoya University
Tokio K.	Ohska	KEK
Takashi	Ohsugi	Hiroshima University
Nobuyoshi	Ohta	Osaka University
Hisataka	Okabe	Osaka Prefectural Education Center
Yasuhiro	Okada	KEK
Toru	Okusawa	Osaka City University
Rolf G.C.	Oldeman	NIKHEF
Keith A.	Olive	University of Minnesota
Stephen L.	Olsen	Hawaii
Tetsuya	Onogi	Hiroshima University
Yona	Oren	Tel-Aviv University
Lynne H.	Orr	University of Rochester
Robert S.	Orr	University of Toronto
Noriyuki	Oshimo	Ochanomizu Women's University
Per	Osland	University of Bergen
Ilmar	Ots	University of Tartu
Farid	Ould-Saada	University of Oslo
Kazuhiko	Ozaki	Osaka Institute of Technology
Sandip	Pakvasa	University of Hawaii
Henryk A.	Palka	Institute of Nuclear Physics, Cracow
Elisabetta	Pallante	University of Barcelona
Marco	Pallavicini	Sezione INFN Genova
Carmen	Palomares	CIEMAT
Fernando	Palombo	Sezione INFN Milano

Luc	Pape	CERN
Adam	Para	Fermilab
Hwanbae	Park	Korea University
Giampiero	Passarino	Universita di Torino
Antonio	Passeri	Sezione INFN Roma III
Fernanda	Pastore	Sezione INFN Roma III e Dipartimento di Fisica
James E.	Paterson	SLAC
Sergio	Patricelli	Sezione INFN Napoli
Christoph M.E.	Paus	MIT
Ken	Peach	Rutherford Appleton Laboratory
Geoffrey F.	Pearce	Rutherford Appleton Laboratory
Antonio R.	Pellegrino	Argonne National Laboratory
Jose A.	Penarrocha	IFIC. Universitat de Valencia
Emmanuelle	Perez	CEA Saclay
Roberto	Petti	CERN
Julian P.	Phillips	Liverpool University
Bolek	Pietrzyk	LAPP, Annecy-le-Vieux
Herbert V.R.	Pietschmann	University of Vienna
Leo E.	Piilonen	Virginia Tech
David E.	Plane	CERN
Tomaz	Podobnik	University of Ljubljana and Institute "J.Stefan", Ljbuljana
Lee G.	Pondrom	University of Wisconsin
Bernard G.	Pope	Michigan State University
Werner	Porod	Universidad de Valencia
Frank C.	Porter	Caltech
Nelli	Poukhaeva	University of Antwerpen
Gilbert	Poulard	CERN
Joaquim	Prades	Universidad de Granada
Lawrence E.	Price	Argonne National Laboratory
Philipp A.	Puls	University of Vienna
Milind V.	Purohit	University of S. Carolina
Arnulf	Quadt	CERN
Eliezer Z.	Rabinovici	The Hebrew University
Ghita	Rahal	ETH-Zurich
Srini	Rajagopalan	BNL
Jiri	Rames	Academy of Sciences of the Czech RepublicInstitute of Physics
Pierre	Ramond	University of Florida
Patricia	Rankin	University of Colorado
Petros A.	Rapidis	Fermilab
Divic J.	Rapin	University of Geneva
Federico	Rapuano	Sezione INFN Roma I
Sergio P.	Ratti	Sezione INFN and UNIV. of Pavia
Gerhard	Raven	UC San Diego
Neville W.	Reay	Kansas State University
Pierpaolo	Rebecchi	CERN
Stefan	Recksiegel	University of Karlsruhe

Don D. Reeder	University of Wisconsin
Christian Regenfus	University of Zurich
James J. Reidy	University of Mississippi
Peter B. Renton	Oxford University
Wayne W. Repko	Michigan State University
Francois Richard	Laboratoire de l'Accelerateur Lineaire
Burton Richter	SLAC & IUPAP
Jan Ridky	Academy of Sciences of the Czech Republic
Michael M. Rijssenbeek	State University of New York at Stony Brook
Isabelle Ripp	IReS Strasbourg
Stefan Ritt	Paul Scherrer Institute
B. Lee Roberts	Boston University
Maria Teresa P. Roco	Fermilab
Chiara Roda	University of Pisa & I.N.F.N. Pisa
Francois G. J. Rohrbach	CERN
Michael Roney	University of Victoria
Francesco Ronga	INFN Laboratori Nazionali di Frascati
Jerome L. Rosen	Northwestern University
S. Peter Rosen	U. S. Dept. of Energy
Lawrence Rosenson	MIT
Sylvie J.M. Rosier	Laboratoire d'Annecy le Vieux de Physique des Particules
Giancarlo Rossi	University of Rome Tor Vergata
Probir Roy	T.I.F.R.
Alexandre N. Rozanov	C.P.P.M.
Roy Rubinstein	Fermilab
Kay Runge	University of Freiburg
James S. Russ	Carnegie Mellon University
Ali Sabetfakhri	University of Toronto
Ahren J. Sadoff	Ithaca College/Cornell University
Gerard Sajot	Universite Joseph Fourier
Norisuke Sakai	Tokyo Institute of Technology
Hiroshi Sakamoto	Kyoto University
Jiro Sakamoto	Shimane University
Makoto Sakamoto	Kobe University
Makoto Sakuda	KEK
Dorothea Samtleben	Universitat Hamburg
Miguel Angel Sanchis	Universidad de Valencia
A. I. Sanda	Nagoya University
Noboru Sasao	Kyoto University
Rodolfo Sassot	Universidad de Buenos Aires
Mamata Satapathy	Utkal University
Hikaru Sato	Hyogo University of Education
Tetsuo Sawada	Nihon University
Christoph Schaefer	CERN L3
Terry L. Schalk	UC Santa Cruz
Peter Schleper	Heidelberg / DESY

Michael	Schmelling	MPI for Nuclear Physics
Bernhard	Schmidt	DESY
Alexander Sascha	Schmidt-Kaerst	RWTH Aachen
Stephen	Schnetzer	Rutgers State University of New Jersey
Stephan	Schoenert	MPI
Fridger	Schrempp	DESY
Stephen E.	Schrenk	Virginia Tech
Henning	Schroeder	Rostock University
Alan J.	Schwartz	University of Cincinnati
Jochen	Schwiening	Stanford Linear Accelerator Center
Felix	Sefkow	University of Zurich
Michiko	Sekimoto	KEK-IPNS
Gordon W.	Semenoff	University of British Columbia
Michael H.	Shaevitz	Fermilab and Columbia University
Benjamin C.	Shen	University of California
Sergey	Shevchenko	Caltech
Hiroshi	Shibuya	Toho University
Kazunari	Shima	Saitama Institute of Technology
Tokuzo	Shimada	Meiji University
Takao	Shinkawa	National Defense Academy in Japan
Masaomi	Shioden	Ibaraki College of Technology
Masato	Shiozawa	ICRR, University of Tokyo
Junpei	Shirai	Tohoku University
Ken-ichi	Shizuya	Kyoto University
Ferenc	Sikler	CERN
Dennis	Silverman	University of California, Irvine
Vladislav J.	Simak	Institute of Physics Prague
Andris	Skuja	University of Maryland
Krzysztof J.	Sliwa	Tufts University
Gerard	Smadja	Institut de Physique Nucleaire de Lyon
Alasdair M.	Smith	CERN
Nigel JT	Smith	Rutherford Appleton Laboratory
Stefan	Soldner-Rembold	CERN
Dongchul	Son	Kyungpook National University
Hee Sung	Song	Seoul National University
Amarjit S.	Soni	BNL
Davison E.	Soper	University of Oregon
Stefania	Spagnolo	Rutherford Appleton Laboratory
Michel	Spiro	Saclay
Neil JC	Spooner	Sheffield University
Prem P.	Srivastava	UERJ- Universidade do Estado de Rio de Janeiro
Achim	Stahl	CERN
Luca	Stanco	INFN Padua
George F.	Sterman	SUNY Stony Brook
Ewan D.	Stewart	KAIST
Achille	Stocchi	Laboratoire de l'Accelerateur Lineaire

James L. Stone	Boston Unversity/U.S. Department of Energy
Sheldon Stone	Syracuse University
Raimund M. Strohmer	Ludwig-Maximilans-Universitaet
Paolo Strolin	CERN/University of Naples
Mark Strovink	University of California, Berkeley
Richard A. Stroynowski	SMU
Dong Su	SLAC
Daijiro Suematsu	Kanazawa University
Akio Sugamoto	Ochanomizu Women's University
Hirotaka Sugawara	KEK
Shojiro Sugimoto	KEK
Akira Sugiyama	Nagoya University
Lawrence Sulak	Boston University
Takayuki Sumiyoshi	KEK
Atsumu Suzuki	Kobe University
Jun-Ichi Suzuki	KEK
Shiro Suzuki	Nagoya University
Andre Sznajder	Universidade Estadual do Rio de Janeiro
Motoi Barner Tachibana	Yukawa Institute for Theoretical Physics
Hiroyasu Tajima	University of Tokyo
Tamotsu Takahashi	Osaka City University
Kunio Takamatsu	KEK/CROSS
Fumihiko Takasaki	KEK-IPNS
Eiichi Takasugi	Osaka University
Hiroshi Takeda	Kobe University
Tatsu Takeuchi	Virginia Tech
Yasuo Takeuchi	Kamioka Observatory, ICRR, Univ. of Tokyo
Yoshihiro Takeuchi	Nihon University
Masato Takita	Osaka University
Norio Tamura	Niigata University
Hidekazu Tanaka	Rikkyo University
Katsumi Tanaka	Ohio State University
Kazuhiro Tanaka	Juntendo University
Minoru Tanaka	Osaka University
Reisaburo Tanaka	Okayama University
Mieko Tanaka-Yamawaki	Miyazaki University
Yoshiaki Tanii	Saitama University
Morimitsu Tanimoto	Niigata University
Naho Tanimoto	Okayama University
Shogo Tanimura	Kyoto University
Enrico Tassi	NIKHEF
Xerxes R. Tata	University of Hawaii
Toshiaki Tauchi	KEK
Ludwig F.G. Tauscher	Basel University
Jean-Paul Tavernet	LPNHE Paris VI-VII
Thomas M. Taylor	CERN

Rosy Chooi Gim	Teh	University Technology MARA
Roberto	Tenchini	CERN / INFN - Pisa
Yoshiki	Teramoto	Osaka City University
Haruhiko	Terao	Kanazawa University
Kunihiko	Terasaki	Kyoto University
Tadayuki	Teshima	Chubu University
Massimo	Testa	University of Roma I
Jenny A.	Thomas	University College London
Mark A.	Thomson	CERN
Maury	Tigner	Cornell University
Charles C.W.J.P.	Timmermans	NIKHEF
Jan J.	Timmermans	NIKHEF
Vincent	Tisserand	Laboratoire d'Annecy le Vieux de Physique des Particules
Nobu	Toge	KEK, Accelerator Lab.
Perihan	Tolun	METU, Ankara
Toshiyuki	Toshito	Kamioka Observatory, ICRR, Univ. of Tokyo
Paul K.	Townsend	University of Cambridge
Minh-Tam	Tran	University of Lausanne
Stephen	Trentalange	University of California at Los Angeles
Gerard	Tristram	PCC College de France
Zoltan	Trocsanyi	University of Debrecen
Sheung Tsun	Tsou	Oxford University
Motomu	Tsuda	Saitama Institute of Technology
Wu-Ki	Tung	Michigan State University
M.	Turala	Institute of Nuclear Physics
Ludwik	Turko	Wroclaw University
Michael	Tyndel	Rutherford Appleton Laboratory
Ekaterini	Tzamariudaki	Werner Heisenberg Institut
Sadaharu	Uehara	KEK-IPNS
Tsuneo	Uematsu	Kyoto University
Thomas S.	Ullrich	Yale University
Yoshinobu	Unno	KEK
Tracy	Usher	SLAC
Noriyuki	Ushida	Aichi University of Education
Andrea	Valassi	CERN
German E.	Valencia	Iowa State University
Bertrand	Vallage	CEA Saclay
Jose W F	Valle	Universidad de Valencia - IFIC/CSIC
Karl A.	van Bibber	LLNL
Jochum J.	van Der Bij	University of Freiburg
Anja C.M.	van Dulmen	NIKHEF
Ann	van Lysebetten	Vrije Universiteit Brussel
Ger G.	van Middelkoop	NIKHEF
Nikos	Varelas	University of Illinois at Chicago
Georges	Vasseur	CEA Saclay
Giovanni	Venturi	Sezione INFN Bologna

Marco	Verzocchi	CERN
Pratibha	Vikas	University of Minnesota
Didier	Vilanova	CEA Saclay
Sotirios	Vlachos	University of Basel
Vaclav	Vrba	Academy of Sciences of the Czech Republic
Gloria	Vuagnin	Universita' di Trieste
Anders	Waananen	The Niels Bohr Institute
Klaus	Wacker	Universitaet Dortmund
Maneesh	Wadhwa	Basel University
Albrecht	Wagner	DESY
Yau W.	Wah	The University of Chicago
Helmut	Wahlen	Wuppertal University
Seiichi	Wakaizumi	University of Tokushima
Kameshwar C.	Wali	Syracuse University
Rainer	Wanke	Johannes Gutenberg-Universitaet Mainz
Andreas	Warburton	Cornell University
Bennie F.L.	Ward	University of Tennessee
Jason	Ward	CERN
Tsunetoshi	Watanabe	Asia University
Yasushi	Watanabe	Tokyo Institute of Technology
Nigel K.	Watson	University of Birmingham
Sylvain	Weisz	CERN
Thorsten	Wengler	CERN
Torsten	Wildschek	University of Wisconsin
Ian	Wilson	CERN
Isabelle	Wingerter-Seez	Laboratoire de l'Accelerateur Lineaire
Roland	Winston	The University of Chicago
Marc	Winter	Institut de Recherches Subatomiques
Michael	Witherell	Fermilab
Guy	Wormser	IN2P3
Jorg	Wotschack	CERN
Douglas M.	Wright	LLNL
Yue-liang	Wu	Institute of Theoretical Physics
Bruce D.	Yabsley	KEK
Haruichi	Yabuki	Hyogo University of Education
Mitsuru	Yamada	Ibaraki University
Norikazu	Yamada	KEK
Ryuji	Yamada	Fermilab
Sakue	Yamada	KEK-IPNS
Youichi	Yamada	Tohoku University
Taketora	Yamagata	Tokyo Metropolitan University
Akira	Yamaguchi	Tohoku University
Masahiro	Yamaguchi	Tohoku University
Tetsuji	Yamaki	Sugiyama Jogakuen Univ.
Katsuji	Yamamoto	Kyoto University
Kazuhiro	Yamamoto	Osaka City University

Taku Yamanaka	Osaka University
Satoru Yamashita	The University of Tokyo
Masanori Yamauchi	KEK-IPNS
Yuji Yamazaki	KEK-IPNS
Chiaki Yanagisawa	State University of New York at Stony Brook
Un-Ki Yang	University of Chicago
York-Peng E. Yao	University of Michigan
Osamu Yasuda	Tokyo Metropolitan University
Masaki Yasue	Tokai University
Shinjiro Yasumi	KEK
Minghan Ye	China Center of Advanced Science and Technology
Christophe Yeche	CEA Saclay DAPNIA/SPP
Yoshimitsu Yokoo	Fukui Medical University
Tamiaki Yoneya	University of Tokyo
Takuo Yoshida	Osaka City University
Tomoteru Yoshie	University of Tsukuba
Tadashi Yoshikawa	KEK
Motohiko Yoshimura	Tohoku University
Jaehoon Yu	Fermilab
Kazuya Yuasa	Waseda University
Tetsuyuki Yukawa	Graduate University for Advanced Studies (Sokendai)
Haruo Yuta	Aomori University
Francois Zach	Institut de Physique Nucleaire de Lyon
Alexandre M. Zaitsev	Institute for High Energy Physics
Aleksander F. Zarnecki	Warsaw University
Michael E. Zeller	Yale University
Zhengguo Zhao	Chinese Academy of Sciences
Zhipeng Zheng	Institute of High Energy Physics, CAS
Yongsheng Zhu	Chinese Academy of Sciences
Robert Zitoun	Laboratoire d'Annecy le Vieux de Physique des Particules
Fabian Zomer	LAL-ORSAY
Marek Zralek	University of Silesia
Gianni Zumerle	Sezione INFN Padova
Armand Zylberstejn	CEA Saclay

REGISTRATION DESK

記念講演会　ジェローム・フリードマン
べてのものはクォークから
主催：ICHEP 2000組織委員会、仁科記念財団　共催：(財)大阪国際交流センター

'00年 7月 9日

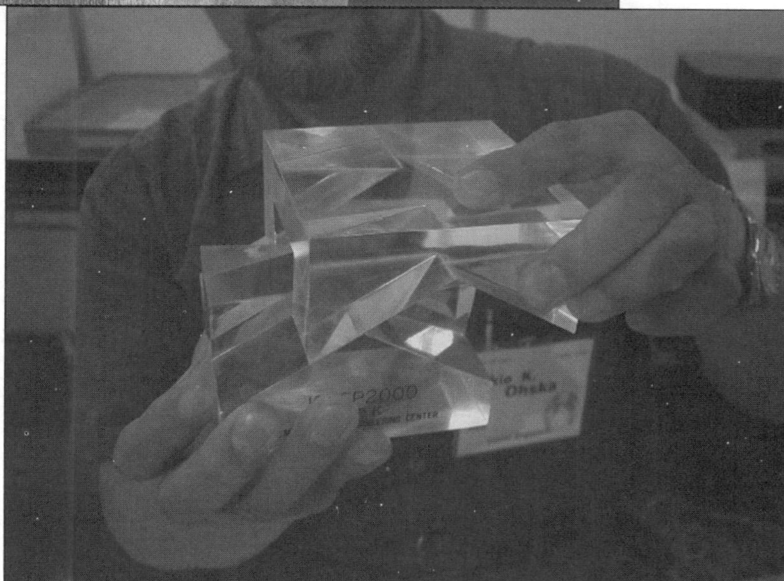

Bamberger

FOUND
FOUND
FOUND
FOUND